SHOCK COMPRESSION OF CONDENSED MATTER—2001

Previous Proceedings in the Series of Conferences of the American Physical Society Topical Group on Shock Compression of Condensed Matter

Year	Held in	Publisher	ISBN
1999	Snowbird, Utah, USA	AIP Conference Proceedings 505	1-56396-923-8
1997	Amherst, Massachusetts, USA	AIP Conference Proceedings 429	1-56396-738-3
1995	Seattle, Washington, USA	AIP Conference Proceedings 370	1-56396-566-6
1993	Colorado Springs, Colorado, USA	AIP Conference Proceedings 309	1-56396-219-5
1991	Williamsburg, Virginia, USA	North-Holland	0-444-89732-1
1989	Albuquerque, New Mexico, USA	North-Holland	0-444-88271-5
1987	Monterey, California, USA	North-Holland	0-444-87097-0
1985	Spokane, Washington, USA	Plenum Press	0-306-42276-X
1983	Santa Fe, New Mexico, USA	North-Holland	0-444-86904-2
1981	Menlo Park, California, USA	AIP Conference Proceedings 78	0-88318-177-0

Other Related Titles from AIP Conference Proceedings

615 Review of Progress in Quantitative Nondestructive Evaluation: Volume 21
Edited by Donald O. Thompson and Dale E. Chimenti, May 2002,
2 vol. hard cover set, CD-ROM included, 0-7354-0061-X

524 Nonlinear Acoustics at the Turn of the Millennium: ISNA 15;
15th International Symposium on Nonlinear Acoustics
Edited by Werner Lauterborn and Thomas Kurz, July 2000, 1-56396-945-9

497 Nondestructive Characterization of Materials IX
Edited by Robert E. Green, Jr., December 1999, 1-56396-911-4

489 Physics of Glasses: Structure and Dynamics
Edited by Philippe Jund and Rémi Jullien, October 1999, 1-56396-903-3

To learn more about these titles, or the AIP Conference Proceedings Series, please visit the webpage
http://proceedings.aip.org/proceedings

SHOCK COMPRESSION OF CONDENSED MATTER—2001

Proceedings of the Conference of the American Physical Society
Topical Group on Shock Compression of Condensed Matter
held in Atlanta, Georgia, June 24–29, 2001

PART TWO

Edited by:

MICHAEL D. FURNISH
Sandia National Laboratories
Albuquerque, New Mexico, USA

NARESH N. THADHANI
Georgia Institute of Technology
Atlanta, Georgia, USA

YASUYUKI HORIE
Los Alamos National Laboratory
Los Alamos, New Mexico, USA

SPONSORING ORGANIZATIONS
APS, SCCM Topical Group
APS, Education Division
Georgia Institute of Technology, CEISMC Office

CD-ROM INCLUDED

Melville, New York, 2002
AIP CONFERENCE PROCEEDINGS ■ 620

EDITORS

Michael D. Furnish
Sandia National Laboratories
MS 1168, P.O. Box 5800
Albuquerque, NM 87185-1168
USA
E-mail: mdfurni@sandia.gov

Yasuyuki Horie
Los Alamos National Laboratory
MS D413, P.O. Box 1663
Los Alamos, NM 87545
USA
E-mail: horie@lanl.gov

Naresh N. Thadhani
Georgia Institute of Technology
Materials Science and Engineering
771 Ferst Drive, N. W.
Atlanta, GA 30332-0245
USA
E-mail: naresh.thadhani@mse.gatech.edu

The articles on pp. 103–106, 142–145, 149–152, 165–168, 181–184, 279–282, 385–390, 519–522, 553–556, 563–566, 630–633, 657–660, 689–692, 701–704, 705–708, 725–728, 735–738, 779–782, 783–786, 817–820, 829–832, 837–840, 853–855, 860–863, 864–867, 894–897, 926–929, 942–945, 950–953, 973–978, 999–1002, 1003–1006, 1019–1022, 1051–1054, 1055–1058, 1059–1064, 1073–1076, 1077–1080, 1153–1156, 1196–1199, 1239–1242, 1247–1250, 1318–1322, and 1351–1354 were authored by U. S. Government employees and are not covered by the below mentioned copyright.

The articles on pp. 79–82, 303–306, 419–422, 507–510, 841–844, 991–994, 1027–1030, 1035–1038, 1302–1305, and 1359–1362 are © British Crown Copyright 2001/MOD and are published with the permission of the Controller of Her Britannic Majesty's Stationery Office.

The articles on pp. 407–410, 934–937, 946–949, and 1023–1026 were prepared on behalf of the Defence Research Establishment Suffield, Alberta, Canada, and therefore the copyright in these papers belongs to the Crown, i.e., to the Canadian Government.

Authorization to photocopy items for internal or personal use, beyond the free copying permitted under the 1978 U.S. Copyright Law (see statement below), is granted by the American Institute of Physics for users registered with the Copyright Clearance Center (CCC) Transactional Reporting Service, provided that the base fee of $19.00 per copy is paid directly to CCC, 222 Rosewood Drive, Danvers, MA 01923. For those organizations that have been granted a photocopy license by CCC, a separate system of payment has been arranged. The fee code for users of the Transactional Reporting Service is: 0-7354-0068-7/02/$19.00.

© 2002 American Institute of Physics

Individual readers of this volume and nonprofit libraries, acting for them, are permitted to make fair use of the material in it, such as copying an article for use in teaching or research. Permission is granted to quote from this volume in scientific work with the customary acknowledgment of the source. To reprint a figure, table, or other excerpt requires the consent of one of the original authors and notification to AIP. Republication or systematic or multiple reproduction of any material in this volume is permitted only under license from AIP. Address inquiries to Office of Rights and Permissions, Suite 1NO1, 2 Huntington Quadrangle, Melville, N.Y. 11747-4502; phone: 516-576-2268; fax: 516-576-2450; e-mail: rights@aip.org.

L.C. Catalog Card No. 2002105355
ISBN 0-7354-0068-7 Set
ISSN 0094-243X
Printed in the United States of America

Contents

Preface xxv
Foreword xxix
Topical Group on Shock Compression of Condensed Matter: 1999-2001 APS Fellows xxxi
Photograph: Recipient of the APS Shock Compression Science Award, 2001 xxxii
Conferences of the APS Topical Group on Shock Compression of Condensed Matter xxxiv

PART ONE

CHAPTER I

PLENARY

The Coupling between Shock Waves and Condensed Matter: Continuum Mechanics
to Quantum Mechanics 3
    Y. M. Gupta
The History of the APS Topical Group on Shock Compression of Condensed Matter 11
    J. W. Forbes
Traditional Analysis of Nonlinear Wave Propagation in Solids 20
    L. Davison
Shock Wave Paradigms and New Challenges 26
    J. R. Asay
Mechanical States of Solids 36
    J. J. Gilman
What Is a Shock Wave to an Explosive Molecule? 42
    C. M. Tarver

CHAPTER II

EQUATION OF STATE: NONENERGETIC MATERIALS

Shock Waves and Plasma Physics 53
    A. Ng
Nickel Critical Point Parameters from Shock Experiments with Porous Samples 59
    D. N. Nikolaev, V. Y. Ternovoi, and A. A. Pyalling
High-Pressure Vaporization and Boiling of Condensed Material: A Generalized
Clausius-Clapeyron Equation 63
    A. L. Gonor
Calculated Hugoniot Curves of Porous Metal: Copper, Nickel, and Molybdenum 67
    Y. Wang, R. Ahuja, and B. Johansson
Analysis of Isobaric Expansion Data Based on Soft-Sphere Equation of State for
Liquid Metals 71
    P. R. Levashov, V. E. Fortov, K. V. Khishchenko, and I. V. Lomonosov
EOS Data of Ti-6Al-4V to Impact Velocities of 10.4 km/s on a Three-Stage Gun 75
    N. A. Winfree, L. C. Chhabildas, W. D. Reinhart, D. E. Carroll, and G. I. Kerley

Measurements of the Equation of State of Lead under Varying Conditions by Multiple Methods 79
    S. D. Rothman, A. M. Evans, P. Graham, K. W. Parker, J. Palmer, T. Jalinaud, J.-P. Davis,
    J. Asay, M. Knudson, and C. Hall

Experimental Study of Highly Compressed Iron Using Laser Driven Shocks 83
    A. Benuzzi-Mounaix, G. Huser, M. Koenig, B. Faral, N. Grandjouan, D. Batani,
    E. Henry, M. Tomasini, B. Marchet, T. Hall, M. Boustie, T. De Rességuier, M. Hallouin,
    and F. Guyot

Equation of State and Phase Diagram of Iron 87
    V. V. Dremov, A. L. Kutepov, A. V. Petrovtsev, and A. T. Sapozhnikov

Density-Functional Molecular Dynamics Simulations of Shocked Molecular Liquids 91
    J. D. Kress, S. Mazevet, and L. A. Collins

Temperature Measurements of Single and Double Shock Compressed Liquid Nitrogen in Overtaking Shock Wave Configuration 95
    A. A. Pyalling, V. Y. Ternovoi, and A. S. Filimonov

Density Functional Calculation of the Hugoniot of Shocked Liquid Nitrogen 99
    S. Mazevet, J. D. Kress, L. A. Collins, W. W. Wood, J. D. Johnson, and P. Blottiau

Theoretical Equation of State for Water at High Pressures 103
    H. D. Jones

Thermophysical Properties of Helium under Multiple Shock Compression 107
    V. Y. Ternovoi, A. S. Filimonov, A. A. Pyalling, V. B. Mintsev, and V. E. Fortov

Phase Diagrams and Thermodynamic Properties of Metals at High Pressures, High Temperatures 111
    I. V. Lomonosov, V. E. Fortov, K. V. Khishchenko, and P. R. Levashov

Physical Interpretation of Mathematically Invariant $K(\rho,P)$ Type Equations of State for Hydrodynamically Driven Flow 115
    G. M. Hrbek

Thermodynamic Properties of Nonideal Strongly Degenerate Hydrogen Plasma 119
    P. R. Levashov, V. S. Filinov, V. E. Fortov, and M. Bonitz

Construction of Wide-Range Equations of State through "Merging" Local Equations Using Mixture Model 127
    L. F. Gudarenko and V. G. Kudelkin

On the Shock Response of Polychloroprene 131
    J. C. F. Millett, N. K. Bourne, G. T. Gray III, and G. Cooper

The Shock Hugoniot of an Epoxy Resin 135
    N. Barnes, N. K. Bourne, and J. C. F. Millett

Invariant Functional Forms for $K(\rho,P)$ Type Equations of State for Hydrodynamically Driven Flow 139
    G. M. Hrbek

Simulated, Theoretical, and Experimental Shock Trajectories in Cylindrical Geometry 143
    R. Kanzleiter, W. Atchison, R. Bowers, and J. Guzik

CHAPTER III

EQUATION OF STATE: ENERGETIC MATERIALS

Development of the LANL Sandwich Test 149
    L. G. Hill

Re-shock Experiments in LX-17 to Investigate Reacted Equation of State 153
    K. S. Vandersall, J. W. Forbes, C. M. Tarver, P. A. Urtiew, and F. Garcia

A Hybrid Monte Carlo Method for Equilibrium Equation of State of Detonation Products 157
    M. S. Shaw

Calculation of Chemical Detonation Waves with Hydrodynamics and a Thermochemical
Equation of State.. 161
    W. M. Howard, L. E. Fried, P. C. Souers, and P. A. Vitello

ANFO Cylinder Tests 165
    L. L. Davis and L. G. Hill

A New Temperature-Dependent Equation of State for Inert, Reactive, and
Composite Materials.. 169
    O. Heuzé, J. C. Goutelle, and G. Baudin

Shock Polar Calculation of Inert Nitromethane by Molecular Dynamics Simulation............ 173
    L. Soulard

Detonation Product EOS Studies: Using ISLS to Refine Cheetah........................... 177
    J. M. Zaug, W. M. Howard, L. E. Fried, and D. W. Hansen

Structural Studies and EOS of Diaminodinitroethylene (DADNE, FOX-7) under
Static Compression 181
    S. M. Peiris, G. I. Pangilinan, F. J. Zerilli, and T. P. Russell

Thermodynamic Representations for Solid/Melt Systems at High Pressure
and Temperature.. 185
    M. Braithwaite, C. E. Sims, and N. L. Allan

CHAPTER IV

PHASE TRANSITIONS

Recent Progress in Understanding the Shock Response of Ferroelectric Ceramics 191
    R. E. Setchell

Macro- and Meso-Scale Modeling of PZT Ferroelectric Ceramics 197
    R. M. Brannon, S. T. Montgomery, J. B. Aidun, and A. C. Robinson

Simulation of the Effects of Shock Stress and Electrical Field Strength on
Shock-Induced Depoling of Normally Poled PZT 95/5 201
    S. T. Montgomery, R. M. Brannon, J. Robbins, R. E. Setchell, and D. H. Zeuch

Multidimensional Validation Impact Tests on PZT 95/5 and ALOX 205
    M. D. Furnish, J. Robbins, W. M. Trott, L. C. Chhabildas, R. J. Lawrence,
    and S. T. Montgomery

Effects of Initial Porosity on the Shock Response of Normally Poled PZT 95/5 209
    R. E. Setchell, B. A. Tuttle, J. A. Voigt, and E. L. Venturini

The Shear Strength of Potassium Chloride above the B1-B2 Phase Transition
during Shock Loading 213
    J. C. F. Millett and N. K. Bourne

Spatial Evolution of Three-Wave Structure in Shocked Potassium Chloride 217
    E. Zaretsky

Investigation of Liquid-Solid Phase Transition Using Isentropic
Compression Experiments (ICE) 221
    J.-P. Davis, D. B. Hayes, J. R. Asay, P. W. Watts, P. A. Flores, and D. B. Reisman

Alpha-Omega Transition in Ti: Equation of State and Kinetics 225
    C. W. Greeff, D. R. Trinkle, and R. C. Albers

Shock Induced Melting of Lead (Experimental Study)..................................... 229
    C. Mabire and P.-L. Héreil

Electrical Conductivity Investigation of Graphite-Diamond Transition under
Multiple Shock-Wave Compression 233
    V. I. Postnov, V. E. Fortov, V. V. Yakushev, and T. I. Yakusheva

Abnormal Electric Conductivity of Lithium at High Dynamic Pressure 237
    V. E. Fortov, V. V. Yakushev, K. L. Kagan, I. V. Lomonosov, V. I. Postnov, T. I. Yakusheva, and A. N. Kuryanchik

A Kinetic Model of Multiple Phase Transitions in Ice 241
    G. Cruz León, S. Rodríguez Romo, and V. Tchijov

The Ab-Initio Study of Structural Stability of Uranium 245
    A. Kutepov and S. Kutepova

CHAPTER V

MODELING, SIMULATION, AND THEORY: NONREACTIVE MATERIALS

Superseismic Loading and Shock Polars: An Example of Fluid-Solid Coupling 251
    M. Arienti and J. E. Shepherd

Comparing Lagrangian Godunov and Pseudo-viscosity Schemes for Multidimensional
Impact Simulations 255
    G. Luttwak

Discrete Element Method Modeling of Gas 259
    W. Wang, Z. Tang, P. Gong, and Y. Horie

Nonlocal Theory of Macro-Meso-Level Energy Exchange in the
Shock Compressed Matter 263
    T. A. Khantouleva

Macro-Meso Energy Exchange in Dynamically Deformed Steels 267
    Y. I. Mescheryakov

Analysis of the Slowing of a High Energy Proton Shot through a Target in the
Frame of the Fokker-Plank Equation 271
    V. Molinari and F. Teodori

Anisotropic Failure Model Development and Implementation 275
    J. D. Walker, K. A. Dannemann, and C. E. Anderson, Jr.

Modeling Anisotropic Plasticity: 3D Eulerian Hydrocode Simulations of High Strain
Rate Deformation Processes 279
    M. W. Burkett, S. P. Clancy, P. J. Maudlin, and K. S. Holian

Interface Tracking in Eulerian and MMALE Calculations 283
    G. Luttwak

Numerical Investigation into the Performance of a Rarefaction Shock Wave Cutter
for Offshore Oil-Gas Platform Removal 287
    J. P. Morris, L. A. Glenn, T. H. Antoun, and I. N. Lomov

Analysis of Radiation-Driven Jetting Experiments on NOVA and Z 291
    R. J. Lawrence, T. A. Mehlhorn, T. A. Haill, K. G. Budge, T. G. Trucano, K. R. Cochrane, and J. J. MacFarlane

Non-Newtonian Viscosity Effects at Shocked Fluid Interfaces 295
    S. M. Valone

Sensitivities for Taylor-Test Model Parameters 299
    R. J. Henninger

Transmission of Shocks along Thin-Walled Tubes 303
    D. A. Salisbury, A. R. Giles, and R. E. Winter

Computational Characterization of Three-Stage Gun Flier Plate Launch 307
    D. E. Carroll, L. C. Chhabildas, W. D. Reinhart, N. A. Winfree, and G. I. Kerley
Modeling and Simulation of Explosively Driven Electromechanical Devices 311
    P. N. Demmie
Numerical Simulations of the Influence of Loading Pulse Shape on SHPB Measurements 315
    A. D. Resnyansky and G. T. Gray III
Shock Wave Effects in Copper: Design of an Experimental Device for Post Recovery Mechanical Testing 319
    F. Buy and F. Llorca
The Contribution of the Expanding Shell Test to the Modeling of Elastoplasticity at High Strain Rates 323
    F. Llorca and F. Buy
The Expanding Shell Test: Numerical Simulation of the Experiment 327
    F. Buy and F. Llorca

CHAPTER VI

MOLECULAR DYNAMICS MODELING: NONREACTIVE MATERIALS

Large-Scale Molecular Dynamics Simulations of Shock-Induced Plasticity, Phase Transformations, and Detonation 333
    T. C. Germann
Atomistic Simulations of the Motion of an Edge Dislocation in Aluminum Using the Embedded Atom Method 339
    N. Bhate, R. J. Clifton, and R. Phillips
Hugoniot Constraint Molecular Dynamics Study of a Transformation to a Metastable Phase in Shocked Silicon 343
    E. J. Reed, J. D. Joannopoulos, and L. E. Fried
Molecular Dynamics and Experimental Study of Shock Polarization of Nitromethane 347
    L. Soulard
Shock-Induced Structural Phase Transformations Studied by Large-Scale Molecular-Dynamics Simulations 351
    K. Kai, T. C. Germann, P. S. Lomdahl, and B. L. Holian
Atomistic Modeling of Orientation Dependence of Shock Wave Properties in Diamond 355
    S. V. Zybin, M. L. Elert, J. A. Harrison, and C. T. White
Shock Waves in Dusty Plasmas 359
    J. E. Hammerberg, T. C. Germann, and B. L. Holian
Continuum Properties from Molecular Simulations 363
    R. J. Hardy, S. Root, and D. R. Swanson
Uniaxial Hugoniostat: Method and Applications 367
    J.-B. Maillet and S. Bernard
Discrete Element Method Simulation of Nonlinear Viscoelastic Stress Wave Problems 371
    W. Wang, Z. Tang, and Y. Horie
Molecular Dynamics Simulation of Shock Wave Compression of Metals 374
    A. A. Selezenev, V. K. Golubev, A. Y. Aleinikov, O. I. Butnev, R. A. Barabanov, and B. L. Voronin
Large-Scale Molecular Dynamics Simulations of Shock Waves in Laves Crystals and Icosahedral Quasicrystals 378
    J. Roth

CHAPTER VII

MODELING AND SIMULATION: REACTIVE MATERIALS

Electronic Excitations Vibrational Spectra, and Chemistry in Nitromethane and HMX 385
    E. J. Reed, M. Riad Manaa, J. D. Joannopoulos, and L. E. Fried

A Study of Deflagration to Detonation Transition in a Model A-B System Using Molecular Dynamics 391
    J. Fellows, P. J. Haskins, and M. D. Cook

Steady Flow Detonations from Molecular Dynamics Simulations 395
    D. R. Swanson and C. T. White

Elastic Properties of HMX 399
    T. D. Sewell, D. Bedrov, R. Menikoff, and G. D. Smith

Molecular Dynamics Simulations of HMX Crystal Polymorphs Using a Flexible Molecule Force Field 403
    D. Bedrov, G. D. Smith, and T. D. Sewell

Ab Initio Molecular Dynamics Simulations of Molecular Collisions of Nitromethane 407
    D. Wei, F. Zhang, and T. K. Woo

Impact Response of PBX 9501 below 2 GPa 411
    K. Kline, Y. Horie, J. J. Dick, and W. Wang

Mesoscale Modelling of Shock Initiation Behavior in HMX-Based Explosives 415
    R. N. Mulford and D. C. Swift

Characterization of the Saturn Air Lens and Its Use in Foam Studies 419
    E. J. Harris, D. A. Salisbury, P. Taylor, and R. E. Winter

Steady-State Model of Heterogeneous Detonation with Inert Particles 423
    A. Gonor, I. Hooton, and S. Narayan

Modeling High Explosives with the Method of Cells and Mori-Tanaka Effective Medium Theories 427
    B. E. Clements and E. M. Mas

Numerical Simulations of Anti-tank Mine Detonations 431
    L. Laine, Ø. Ranestad, A. Sandvik, and A. Snekkevik

Simulation of Shaped-Charge with SPH Rezone Method 435
    J. Yao, M. E. Gunger, and D. A. Matuska

Modeling and Prediction of Sensitivity in Energetic Materials 439
    N. V. Garmasheva, V. P. Filin, B. G. Loboiko, A. N. Averin, D. Mathieu,
    P. Simonetti, and R. Belmas

Approximate Blast Theory: Application to Solids 442
    G. J. Hutchens

Effect of Reaction Rate Periodicity on Detonation Propagation 446
    E. O. Morano and J. E. Shepherd

A Complete Equation of State for Detonation Products in Hydrocodes 450
    O. Heuzé

How Point and Line Defects Affect Detonation Properties of Energetic Solids 454
    M. M. Kuklja

Hydro-Reactive Computations with a Temperature Dependent Reaction Rate 460
    Y. Partom

A Mechanistic Study of Delayed Detonation in Impact Damaged Solid Rocket Propellant 464
    E. R. Matheson and J. T. Rosenberg

Numerical Simulation on Laser Initiation of Thin Explosive 468
    S. Kubota, K. Nagayama, H. Shimada, and K. Matsui

CHAPTER VIII

SPALL, FRACTURE, AND FRAGMENTATION OF METALS

The Effect of Material Cleanliness on Dynamic Damage Evolution in 10100 Cu 475
    W. R. Thissell, A. K. Zurek, D. A. S. Macdougall, D. Miller, R. Everett,
    A. Geltmacher, R. Brooks, and D. Tonks

Influence of Microstructural Anisotropy on the Spallation of 1080 Eutectoid Steel 479
    G. T. Gray III, N. K. Bourne, J. C. F. Millett, M. F. Lopez, and K. S. Vecchio

Incipient Spall Studies in Tantalum—Microstructural Effects 483
    L. C. Chhabildas, W. M. Trott, W. D. Reinhart, J. R. Cogar, and G. A. Mann

The Spall Strength Measurement and Modelling of AQ80 Iron and Copper Systems 487
    P. D. Church, W. G. Proud, T. D. Andrews, and B. Goldthorpe

Grain Size and Pressure Effects on Spall Strength in Copper 491
    A. J. Schwartz, J. U. Cazamias, P. S. Fiske, and R. W. Minich

Cavitation in Compressible Visco-Plastic Materials 495
    C. Denoual and J. M. Diani

Dynamic Properties of Shock Loaded Thin Uranium Foils 499
    D. L. Robbins, A. M. Kelly, D. J. Alexander, R. J. Hanrahan, R. C. Snow, R. J. Gehr,
    T. D. Rupp, S. A. Sheffield, and D. B. Stahl

**Hugoniot Elastic Limit and Spall Strength of Aluminum and Copper Single Crystals
over a Wide Range of Strain Rates and Temperatures** 503
    S. V. Razorenov, G. I. Kanel, K. Baumung, and H. J. Bluhm

**A Flash X-Ray Technique to Measure Strain Distribution at Interfaces Sliding
at High Pressure and Velocity** 507
    R. E. Winter, P. Taylor, D. J. Carley, A. J. Barlow, H. Pragnell, and L. Markland

Spallation in the Alloy Ti-6Al-4V 511
    P. D. Church, T. Andrews, N. K. Bourne, and J. C. F. Millett

Cylinder Fragmentation Using Gas Gun Techniques 515
    T. F. Thornhill, W. D. Reinhart, L. C. Chhabildas, D. E. Grady, and L. T. Wilson

Dynamic Fracture Studies Using Sleeved Taylor Specimens 519
    M. R. Gilmore, J. C. Foster Jr., and L. L. Wilson

The Effect of Orientation on the Spall Strength of the Aluminum Alloy 7010-T6 523
    M. R. Edwards, N. K. Bourne, and J. C. F. Millett

Controlled Fragmentation 527
    W. Arnold

Ejecta Particle Size Distributions for Shock-Loaded Sn and Al Targets 531
    D. S. Sorenson, R. W. Minich, J. L. Romero, T. W. Tunnell, and R. M. Malone

Investigation of the Observed Anisotropic Fracture in Steels 535
    B. E. Clements, E. M. Mas, and G. T. Gray III

**Applying Micro-mechanics to Finite Element Simulations of Split Hopkinson
Pressure Bar Experiments on High Explosives** 539
    E. M. Mas, B. E. Clements, W. R. Blumenthal, C. M. Cady, and G. T. Gray III

**Effect of Oriented Elastic and Strength Characteristics on the Impact Fracture
of Anisotropic Materials** 543
    A. V. Radchenko, S. V. Kobenko, and M. N. Krivosheina

Experimental Study of Explosive Fragmentation of Metals Melts 547
    A. K. Zhiembetov, A. L. Mikhaylov, and G. S. Smirnov

CHAPTER IX

CONSTITUTIVE AND MICROSTRUCTURAL PROPERTIES OF METALS

Nonequilibrium Fluctuations in Shock Compression of Polycrystalline Copper α-Iron 553
    Y. Horie and K. Yano

On the Conversion of Plastic Work into Heat during High-Strain-Rate Deformation 557
    G. Ravichandran, A. J. Rosakis, J. Hodowany, and P. Rosakis

Crystal Failure and Crack Formation during Plastic Flow 563
    C. S. Coffey and J. Sharma

Evolution in the Patterning of Adiabatic Shear Bands 567
    M. A. Meyers, Q. Xue, and V. F. Nesterenko

Microstructural Evolution in Adiabatic Shear Localization in Stainless Steel 571
    M. A. Meyers, M. T. Perez-Prado, Q. Xue, Y. Xu, and T. R. McNelley

On the Measurement of Shear-Strength in Quasi-isentropic Loading 575
    Z. Rosenberg, N. K. Bourne, G. T. Gray III, and J. C. F. Millett

On the Shock Response of the Shape Memory Alloy NiTi 579
    J. C. F. Millett, N. K. Bourne, G. T. Gray III, and G. S. Stevens

Al and Cu Dynamic Strength at a Strain Rate of $5 \cdot 10^8$ s^{-1} 583
    M. Werdiger, S. Eliezer, E. Moshe, Z. Henis, E. Dekel, Y. Horovitz, and B. Arad

The Effects of Shear Banding in 6-4 Titanium on Round and Square Taylor Impacts 587
    J. U. Cazamias

Numerical Simulation of Elastic-Viscous-Plastic Properties, Polymorphous Transformations and Spall Fracture in Iron 591
    A. V. Petrovtsev, V. A. Bychenkov, and G. V. Kovalenko

Growth of Perturbations on Metals Interface at Oblique Collision with Supersonic Velocity of Contact Point Motion 595
    O. B. Drennov, A. L. Mikhaylov, P. N. Nizovtsev, and V. A. Raevskii

Numerical Simulation of the Vacancy Diffusion in Shocked Crystals 599
    Y. Skryl and M. M. Kuklja

Anomalous Behavior of Aluminum Near the Melting Temperature: Transition in the Rate Controlling Mechanism of Yielding and Realization of Superheated Solid States under Tension 603
    G. I. Kanel, S. V. Razorenov, K. Baumung, and H. Bluhm

Inertia and Temperature Effects in Void Growth 607
    L. Seaman and D. R. Curran

Void Coalescence Model for Ductile Damage 611
    D. L. Tonks, A. K. Zurek, and W. R. Thissell

Laser Driven High Pressure, High Strain Rate Materials Experiments 615
    D. H. Kalantar, A. M. Allen, F. Gregori, B. Kad, M. Kumar, K. T. Lorenz,
    A. Loveridge, M. A. Meyers, S. Pollaine, B. A. Remington, and J. S. Wark

Plastic Deformation in Laser-Induced Shock Compression of Monocrystalline Copper 619
    M. A. Meyers, F. Gregori, B. K. Kad, M. S. Schneider, D. H. Kalantar, B. A. Remington,
    J. S. Wark, T. Boehly, and G. Ravichandran

Formation and Morphology of Twinning in Titanium under High Strain Rate Deformation 623
    B. Herrmann, A. Venkert, G. Kimmel, A. Landau, D. Shvarts, and E. Zaretsky

Influence of the Structural Levels on the Elastic-Plastic Hardening of Metals under Submicrosecond Shock Loading 627
    Y. Sud'enkov

Laser-Driven Planar Impact of Miniature Specimens of HY-100 Steel............................ 630
   D. J. Alexander and D. L. Robbins
The Effect of Microstructure on the Shock Behaviour of γ-Titanium Aluminides............ 634
   J. C. F. Millett, I. P. Jones, N. K. Bourne, and G. T. Gray III
Experimental Analysis of Shock Wave Effects in Copper................................... 638
   F. Llorca, F. Buy, and J. Farre
On the Dependence of the Yield Strength of Metals on Temperature and Strain Rate:
The Mechanical Equation of the Solid State.. 642
   P. P. Milella

CHAPTER X

MECHANICAL PROPERTIES: POLYMERS

The Deviatoric Response of an Epoxy Resin to One-Dimensional Shock Loading.............. 649
   N. K. Bourne, J. C. F. Millett, N. Barnes, and I. Belcher
On the Strength Behaviour of Kel-F-800™ and Estane Polymers............................. 653
   N. K. Bourne, J. C. F. Millett, G. T. Gray III, and P. Mort
Thermal Activation Constitutive Model for Polymers Applied to Polytetrafluoroethylene...... 657
   F. J. Zerilli and R. W. Armstrong
A Viscoelastic Model for PBX Binders.. 661
   E. M. Mas, B. E. Clements, W. R. Blumenthal, C. M. Cady, G. T. Gray III, and C. Liu
Influence of Temperature and Strain Rate on the Compressive Behavior of PMMA
and Polycarbonate Polymers.. 665
   W. R. Blumenthal, C. M. Cady, M. F. Lopez, G. T. Gray III, and D. J. Idar
Effects of Initial Temperature on the Shock and Release Behavior of Filled and
Unfilled Epoxies... 669
   M. U. Anderson, R. E. Setchell, and D. E. Cox
Evolution of Stress Relaxation Structures for Several Polymers Subjected to
Plane Shock Compression around 0.5 GPa Shock Stress Measured by PVDF Gauge.......... 673
   Y. Mori and K. Nagayama

CHAPTER XI

MECHANICAL PROPERTIES: COMPOSITES

Discrete Element Modeling for Shock Processes of Heterogeneous Materials................. 679
   Z. P. Tang and W. W. Wang
Validation of an Advanced Material Model for Simulating the Impact and Shock
Response of Composite Materials.. 685
   R. A. Clegg, C. J. Hayhurst, and H. Nahme
Analytical and Computational Study of One-Dimensional Impact of Graded
Elastic Solids... 689
   M. Scheidler and G. Gazonas
Strain Rate Sensitivity of Graphite/Polymer Laminate Composites........................ 693
   I. H. Syed and N. S. Brar
Dynamic Tensile Response of Alumina-Al Composites................................... 697
   R. Atisivan, A. Bandyopadhyay, and Y. M. Gupta

Resolving Mechanical Response of Plastic Bonded Explosives at High Strain-Rate Using Split Hopkinson Pressure Bar 701
    V. S. Joshi and R. J. Lee

A Combined Experimental/Computational Approach for Assessing the High Strain Rate Response of High Explosive Simulants and Other Viscoelastic Particulate Composite Materials 705
    J. Corley, W. Riedel, S. Hiermaier, P. Weidemaier, and K. Thoma

Influence of Interface Scattering on Shock Waves in Heterogeneous Solids 709
    S. Zhuang, G. Ravichandran, and D. E. Grady

Mesoscale Descriptions of Shock-Loaded Heterogeneous Porous Materials 713
    M. R. Baer and W. M. Trott

Experiment and Theory for the Characterization of Porous Materials 717
    A. D. Resnyansky, N. K. Bourne, and J. C. F. Millett

Shock Wave Propagation Process in Epoxy Syntactic Foams 721
    J. Ribeiro, J. Campos, I. Plaksin, and R. Mendes

Compressive Properties of a Closed-Cell Aluminum Foam as a Function of Strain Rate and Temperature 725
    C. M. Cady, G. T. Gray III, C. Liu, C. P. Trujillo, B. L. Jacquez, and T. Mukai

The Mechanism of Strain Rate Strengthening during Dynamic Compression of Closed-Cell Aluminum Foam 729
    K. A. Dannemann, J. Lankford Jr., and A. E. Nicholls

PART TWO

CHAPTER XII

MECHANICAL PROPERTIES: CERAMICS AND GLASSES

The HEL Upper Limit 735
    J. P. Billingsley

On the HEL and the "Ramping" above HEL 739
    E. Bar-On, Y. Partom, M. B. Rubin, and D. Z. Yankelevsky

Factors Influencing the Shape of the Fracture Wave Induced by the Rod Impact of a Brittle Material 743
    A. D. Resnyansky and N. K. Bourne

Spall Strength of Ceramic in a Multilayer System 747
    B. A. M. Vaughan, N. H. Murray, W. G. Proud, and J. E. Field

Computer Simulation of the Propagation of Short Shock Pulses in Ceramic Materials 751
    V. A. Skripnyak, E. G. Skripnyak, and T. V. Zhukova

Influence of Microstructural Bias on the Hugoniot Elastic Limit and Spall Strength of Two-Phase $TiB_2+Al_2O_3$ Ceramics 755
    G. Kennedy, L. Ferranti, R. Russell, M. Zhou, and N. Thadhani

Shock Compression, Adiabatic Expansion, and Multi-phase Equation of State of Carbon 759
    K. V. Khishchenko, V. E. Fortov, I. V. Lomonosov, M. N. Pavlovskii, G. V. Simakov, and M. V. Zhernokletov

Thermodynamic Parameters and Equation of State of Low-Density SiO_2 Aerogel ... 763
    M. V. Zhernokletov, T. S. Lebedeva, A. B. Medvedev, M. A. Mochalov, A. N. Shuykin, and V. E. Fortov

The Hugoniot Elastic Limit of AlON 767
    J. U. Cazamias, P. S. Fiske, and S. J. Bless
The Failure of Aluminium Nitride under Shock 771
    I. M. Pickup and N. K. Bourne
On the Failure of Boron Carbide under Shock 775
    N. K. Bourne and G. T. Gray III
Spallation of Hot Pressed Boron Carbide Ceramic 779
    P. T. Bartkowski, D. P. Dandekar, and D. J. Grove
Shock Equation of State and Dynamic Strength of Tungsten Carbide 783
    D. P. Dandekar and D. E. Grady
Bar Impact Tests on Alumina (AD995) 787
    J. U. Cazamias, W. D. Reinhart, C. H. Konrad, L. C. Chhabildas, and S. J. Bless
Investigating Multi-dimensional Effects in Single-Crystal Sapphire 791
    W. D. Reinhart, L. C. Chhabildas, W. M. Trott, and D. P. Dandekar
Experimental Characterization of the Dynamic Failure Resistance of
TiB_2/Al_2O_3 Composites 795
    A. R. Keller and M. Zhou
Fragmentation of Expanding Cylinders and the Statistical Theory of N. F. Mott 799
    D. Grady
Digital Speckle Flash X-Ray Photography 803
    S. G. Grantham and W. G. Proud
The Deviatoric Response of Three Dense Glasses under Shock Loading Conditions ... 807
    D. D. Radford, W. G. Proud, and J. E. Field
Impact Induced Failure Zones in Homalite Bars 811
    R. Russell, S. J. Bless, and T. Beno

CHAPTER XIII

MECHANICAL PROPERTIES: REACTIVE MATERIALS

Elastic Precursor Decay in HMX Explosive Crystals 817
    J. J. Dick and A. R. Martinez
Influence of Polymer Molecular Weight, Temperature, and Strain Rate on the
Mechanical Properties of PBX 9501 821
    D. J. Idar, D. G. Thompson, G. T. Gray III, W. R. Blumenthal, C. M. Cady,
    P. D. Peterson, E. L. Roemer, W. J. Wright, and B. L. Jacquez
Moiré Interferometry Studies of PBX 9501 825
    P. J. Rae, H. T. Goldrein, S. J. P. Palmer, and W. Proud
Experimental Simulations of Dynamic Stress Bridging in Plastic Bonded Explosives ... 829
    K. M. Roessig and J. C. Foster Jr.
An Optical Microscopy and Small-Angle Scattering Study of Porosity in Thermally
Treated PBX 9501 833
    J. T. Mang, C. B. Skidmore, S. F. Son, R. P. Hjelm, and T. P. Rieker
Sub-molecular Fracture Steps in Shock-Shattered RDX Crystals and Follow-On
Nano-Indentation Evaluation of Early Stage Plasticity 837
    J. Sharma, C. S. Coffey, R. W. Armstrong, W. L. Elban, and S. M. Hoover
Reaction of Shocked but Undetonated HMX-Based Explosive 841
    P. Taylor, D. A. Salisbury, L. S. Markland, R. E. Winter, and M. I. Andrew

Investigation of Dispersive Waves in Low-Density Sugar and HMX Using Line-Imaging Velocity Interferometry 845
    W. M. Trott, L. C. Chhabildas, M. R. Baer, and J. N. Castañeda

Isentropic Compression of LX-04 on the Z Accelerator 849
    D. B. Reisman, J. W. Forbes, C. M. Tarver, F. Garcia, R. C. Cauble, C. A. Hall, J. R. Asay, K. Struve, and M. D. Furnish

Mechanical Behavior of Energetic Materials during High Acceleration 853
    Y. Lanzerotti and J. Sharma

Using Simultaneous Time-Resolved SHG and XRD Diagnostics to Examine Phase Transitions of HMX and TATB 856
    C. K. Saw, J. M. Zaug, and D. L. Farber

Use of High-Speed Photography to Augment Split Hopkinson Pressure Bar Measurements of Energetic Materials 860
    R. J. Lee and V. S. Joshi

Mechanical Behavior of Explosives at High Pressures 864
    J. M. Kelley, V. S. Joshi, and R. H. Guirguis

Investigation of Shock Wave Impulse Influence on Solid Propellant Combustion 868
    A. Y. Dolgoborodov and V. N. Marshakov

CHAPTER XIV

DETONATION PHENOMENA

Investigation of Isentrope for Detonation Products of TATB-Based Composition 875
    Y. A. Aminov, M. M. Gorshkov, V. T. Zaikin, G. V. Kovalenko, Y. R. Nikitenko, and G. N. Rykovanov

Observations on Type II Deflagration-to-Detonation Transitions 878
    M. J. Gifford, W. G. Proud, and J. E. Field

Pressure Wave Measurements from Thermal Cook-Off of an HMX Based High Explosive PBX 9501 882
    F. G. Garcia, J. W. Forbes, C. M. Tarver, P. A. Urtiew, D. W. Greenwood, and K. S. Vandersall

Measurement of Low Level Explosives Reaction in Gauged Multi-dimensional Steven Impact Tests 886
    A. M. Niles, F. Garcia, D. W. Greenwood, J. W. Forbes, C. M. Tarver, S. K. Chidester, R. G. Garza, and L. L. Swizter

The Effect of Additives on the Detonation Characteristics of a Liquid Explosive 890
    P. J. Haskins, M. D. Cook, and R. I. Briggs

Electromagnetic Properties of Pre-detonating Explosives 894
    G. P. Chambers, R. J. Lee, T. J. Oxby, and W. F. Perger

Effect of GMB on Failure and Reaction Regime of NM/PMMA-GMB Mixtures 898
    J. Góis, J. Campos, and I. Plaksin

Pressure Wave Measurements in Cylinders of Detonating LX-17 902
    J. W. Forbes, P. C. Souers, P. A. Urtiew, K. S. Vandersall, F. Garcia, D. W. Greenwood, and L. Green

Diameter Effect Curve and Detonation Front Curvature Measurements for ANFO 906
    R. A. Catanach and L. G. Hill

Experimental Investigation of Heterogeneous HE Decomposition Mechanism in Detonation Wave Front 910
    A. V. Fedorov

Experimental and Numerical Study of Temperatures in Cavity Collapse 914
    A. M. Milne and N. K. Bourne

Detonation Phenomena of PBX Microsamples 918
    I. Plaksin, J. Campos, J. Ribeiro, and R. Mendes

Detonation Meso-Scale Tests for Energetic Materials 922
    I. Plaksin, J. Campos, J. Ribeiro, R. Mendes, J. Góis, A. Portugal, P. Simões, and L. Pedroso

Convective Detonations 926
    R. H. Guirguis and A. M. Landsberg

Effect of Void Size on the Detonation Pressure of Emulsion Explosives 930
    Y. Hirosaki, K. Murata, Y. Kato, and S. Itoh

Momentum Transfer during Shock Interaction with Metal Particles in Condensed Explosives 934
    F. Zhang, P. A. Thibault, R. Link, and A. L. Gonor

Reaction Zone Transformation for Steady-State Detonation of High Explosives under Initial Density Increase 938
    A. V. Utkin, S. A. Kolesnikov, S. V. Pershin, and V. E. Fortov

The Effect of Variation of Aluminized Particle Size and Polymer on the Performance of Explosives 942
    D. Woody and J. J. Davis

Near-Field Impulse Effects from Detonation of Heterogeneous Explosives 946
    D. L. Frost, F. Zhang, S. McCahan, S. B. Murray, A. J. Higgins, M. Slanik, M. Casas-Cordero, and C. Ornthanalai

Effect of Metal Particle Size on Blast Performance of RDX Based Explosives 950
    J. J. Davis and P. J. Miller

Effect of an Inert Material's Thickness and Properties on the Ratio of Energies Imparted by a Detonation's 1st and 2nd Propulsion Stages 954
    J. E. Backofen and C. A. Weickert

Obtaining the Gurney Energy Constant for a Two-Step Propulsion Model 958
    J. E. Backofen and C. A. Weickert

Aluminised Explosive Compositions Based on NQ and BTNEN 962
    M. F. Gogulya, A. Y. Dolgoborodov, M. A. Brazhnikov, M. N. Makhov, and V. I. Arkhipov

Proton Radiography Examination of Unburned Regions in PBX 9502 Corner Turning Experiments 966
    E. N. Ferm, C. L. Morris, J. P. Quintana, P. Pazuchanic, H. Stacy, J. D. Zumbro, G. Hogan, and N. King

CHAPTER XV

EXPLOSIVE AND INITIATION STUDIES

Mesoscale Mechanics of Plastic Bonded Explosives 973
    K. M. Roessig

Compaction Wave Profiles in Granular HMX 979
    R. Menikoff

Mechanistic Model of Hot Spot: A Unifying Framework 983
    K. Yano, Y. Horie, and D. Greening

Microstructural Model of Ignition for Time Varying Loading Conditions 987
    R. V. Browning and R. J. Scammon

Development of a Simple Model of "Hot-Spot" Initiation in Heterogeneous
Solid Explosives.. 991
    N. J. Whitworth

Initiation of PETN Powder by Pulse Laser Ablation....................................... 995
    K. Nagayama, K. Inou, and M. Nakahara

Double Shock Initiation of the HMX Based Explosive EDC-37........................ 999
    R. L. Gustavsen, S. A. Sheffield, R. R. Alcon, R. E. Winter, P. Taylor, and D. A. Salisbury

Plastic Deformation Rate and Initiation of Crystalline Explosives 1003
    J. Namkung and C. S. Coffey

Factors Affecting Shock Sensitivity of Energetic Materials 1007
    A. Chakravarty, M. J. Gifford, M. W. Greenaway, W. G. Proud, and J. E. Field

The Burning Rate of Aluminium Particles in Cylinder Tests........................... 1011
    D. J. Evans, A. M. Milne, and I. Softley

First Results of Reaction Propagation Rates in HMX at High Pressure 1015
    D. L. Farber, A. P. Esposito, J. M. Zaug, J. E. Reaugh, and C. M. Aracne

Embedded Electromagnetic Gauge Measurements and Modeling of Shock Initiation
in the TATB Based Explosives LX-17 and PBX 9502 1019
    R. L. Gustavsen, S. A. Sheffield, R. R. Alcon, J. W. Forbes, C. M. Tarver, and F. Garcia

Detonation Initiation in Preshocked Liquid Explosives.................................. 1023
    A. J. Higgins, F. X. Jetté, A. C. Yoshinaka, J. H. S. Lee, and F. Zhang

Lagrangian Analysis of EDC37 Shock Initiation Data 1027
    J. R. Maw

Transient Detonation Processes in a Plastic Bonded Explosive 1031
    K. A. Thomas, E. S. Martin, J. E. Kennedy, I. A. Garcia, and J. C. Foster Jr.

An Investigation into the Initiation of Hexanitrostilbene by Laser-Driven Flyer Plates......... 1035
    M. W. Greenaway, M. J. Gifford, W. G. Proud, J. E. Field, and S. G. Goveas

Shock Initiation of UF-TATB at 250° C... 1039
    P. A. Urtiew, J. W. Forbes, F. Garcia, and C. M. Tarver

Manganin Gauge and Reactive Flow Modeling Study of the Shock Initiation of PBX 9501..... 1043
    C. M. Tarver, J. W. Forbes, F. Garcia, and P. A. Urtiew

Fragment Impact Characterization of Melt-Cast and PBX Explosives............. 1047
    M. D. Cook, P. J. Haskins, R. I. Briggs, C. Stennett, J. Fellows, and P. J. Cheese

Hugoniot and Shock Initiation Studies of Isopropyl Nitrate............................ 1051
    S. A. Sheffield, L. L. Davis, M. R. Baer, R. Engelke, R. R. Alcon, and A. M. Renlund

Reactive Stress Growth Measurements for the Explosive IRX-4..................... 1055
    G. T. Sutherland

The Combustion of Explosives 1059
    S. F. Son

Effect of Temperature Profile on Reaction Violence in Heated and Self-Ignited PBX 9501 1065
    B. Asay, P. Dickson, B. Henson, L. Smilowitz, and L. Tellier

Ignition Chemistry in HMX from Thermal Explosion to Detonation 1069
    B. F. Henson, B. W. Asay, L. B. Smilowitz, and P. Dickson

Instrumentation of Slow Cook-Off Events 1073
    H. W. Sandusky and G. P. Chambers

Kinetics of the β–δ Phase Transition in PBX 9501... 1077
    L. B. Smilowitz, B. F. Henson, B. W. Asay, P. M. Dickson, and J. M. Robinson

The Measurement of Hot-Spots in Granulated Ammonium Nitrate 1081
    W. G. Proud

CHAPTER XVI

SHOCK-INDUCED MODIFICATIONS AND MATERIAL SYNTHESIS

Computational Modeling of the Shock Compression of Powders 1087
    D. J. Benson, I. Do, and M. A. Meyers

Three-Scale Model for Numerical Simulation of Mechano-Chemical Processes in Shock-Compressed Powder Bodies 1093
    V. N. Leitsin, V. A. Skripnyak, and M. A. Dmitireva

Effect of Shock-Activation on Post-shock Reaction Synthesis of Ternary Ceramics 1097
    J. L. Jordan and N. N. Thadhani

Synthesis of Functional Ceramics Layers Using Novel Method Based on Impact of Ultra-fine Particles 1101
    J. Akedo and M. Lebedev

The Study of Internal Deformation Fields in Granular Materials Using 3D Digital Speckle X-Ray Flash Photography 1105
    H. T. Goldrein, S. G. Grantham, W. G. Proud, and J. E. Field

Investigation of Shock-Induced Chemical Reactions in Mo-Si Powder Mixtures Using Instrumented Experiments with PVDF Stress Gauges 1109
    K. S. Vandersall and N. N. Thadhani

Shock-Induced Cubic Silicon Nitride and Its Properties 1113
    T. Sekine

Dynamic Response of Titanium Carbide-Steel, Ceramic-Metal Composites 1119
    B. Klein, N. Frage, E. Zaretsky, and M. P. Dariel

Investigation of Shock-Induced Chemical Reactions in Ni-Ti Powder Mixtures Using Instrumented Experiments 1123
    X. Xu and N. N. Thadhani

TiC by SHS and Dynamic Compaction 1127
    E. P. Carton, M. Stuivinga, and A. Boluijt

Cooling Rate Threshold in Transformation of C_{60} Fullerene to Amorphous Diamond and Highly Disordered Carbon in SCARQ Experiments 1131
    T. Homae, A. Okamoto, K. G. Nakamura, K-I. Kondo, M. Yoshida, K. Hirabayashi, and K. Niwase

CHAPTER XVII

INSTRUMENTATION

Carbon Resistor Pressure Gauge Calibration at Low Stresses 1137
    B. Cunningham, K. S. Vandersall, A. M. Niles, D. W. Greenwood, F. Garcia, and J. W. Forbes, and W. H. Wilson

Advanced Cryogenic System Capabilities for Precision Shock Physics Measurements on Z 1141
    D. L. Hanson, R. R. Johnston, M. D. Knudson, J. R. Asay, C. A. Hall, J. E. Bailey, and R. J. Hickman

Temperature Controlled Vessel for Equation of State Measurements 1145
    T. D. Rupp, R. J. Gehr, D. B. Stahl, S. A. Sheffield, and D. L. Robbins

PVDF Gauge Piezoelectric Response under Two-Stage Light Gas Gun Impact Loading 1149
    F. Bauer

Outputs of Shock-Loaded Small Piezoceramic Disks 1153
    J. A. Charest and J. L. Mace

Improvements in the Signal Fidelity of the Manganin Stress Gauge 1157
D. Greenwood, J. Forbes, F. Garcia, K. Vandersall, P. Urtiew, L. Green, and L. Erickson

CHAPTER XVIII

EXPERIMENTAL TECHNIQUES

Recent Advances in Quasi-isentropic Compression Experiments (ICE) on the Sandia Z Accelerator 1163
C. A. Hall, J. R. Asay, M. D. Knudson, D. B. Hayes, R. L. Lemke, J.-P. Davis, and C. Deeney

Temperature Measurement of Isentropically Accelerated Flyer Plates 1169
T. Bergstresser and S. Becker

SYRINX Project: HPP Generators Devoted to Isentropic Compression Experiments 1173
C. Mangeant, F. Lassalle, P. L'Eplattenier, P.-L. Héreil, D. Bergues, and G. Avrillaud

Correcting Free Surface Effects by Integrating the Equations of Motion Backward in Space ... 1177
D. Hayes and C. Hall

Picosecond Time-Resolved X-Ray Diffraction: Estimation of Local Pressure 1181
Y. Hironaka, F. Saito, A. Yazaki, K. G. Nakamura, and K-I. Kondo

Laser Triggered Synchronizable X-Ray System for Real Time Study of Shock Waves in Condensed Materials 1185
J. P. Farrell, K. Batchelor, V. Dudnikov, T. Srinivasan-Rao, J. Smedley, and J. McDonald

0D Modelisation of the Magnetic Flux Compression Scheme for Isentropic Compression Experiments 1188
P. L'Eplattenier, G. Avrillaud, and J. Vanpoperynghe

Simultaneous VISAR and TXD Measurements on Shocks in Beryllium Crystals 1192
D. C. Swift, D. L. Paisley, G. A. Kyrala, and A. Hauer

Experiment to Capture Gaseous Products from Shock-Decomposed Materials 1196
W. H. Holt, W. Mock Jr., F. Santiago, and R. M. Gamache

Sound Velocity Doppler Laser Interferometry Measurements on Tin 1200
E. Martinez and J.-M. Servas

Projectile Acceleration Aiming at Velocities above 9 km/s by a Compact Gas Gun 1204
T. Moritoh, N. Kawai, K. G. Nakamura, and K.-I. Kondo

Characterization of Impact in Composite Laminates 1208
K. Minnaar and M. Zhou

Erratum: Friction in High-Speed Impact Experiments 1212
R. A. Pelak, P. Rightley, and J. E. Hammerberg

CHAPTER XIX

OPTICAL AND ELECTRICAL MEASUREMENTS

Shock Temperature of NaCl Measured with Wide-Band Optical Radiometry 1215
T. Ogura, K. G. Nakamura, H. Takenaka, and K.-I. Kondo

Ultrafast Spectroscopic Investigation of Shock Compressed Glycidyl Azide Polymer Films and Nitrocellulose Films 1219
J. H. Reho, D. S. Moore, D. J. Funk, G. L. Fisher, and R. L. Rabie

Emission Spectroscopy Applied to Shock to Detonation Transition in Nitromethane 1223
V. Bouyer, G. Baudin, C. Le Gallic, and P. Hervé

Ultrafast Measurement of the Optical Properties of Shocked Nickel and Laser Heated Gold ... 1227
  D. J. Funk, D. S. Moore, J. H. Reho, K. T. Gahagan, S. D. McGrane, and R. L. Rabie
Optical Extinction of Sapphire Shock-Loaded to 250–260 GPa ... 1231
  D. E. Hare, D. J. Webb, S-H. Lee, and N. C. Holmes
Temperature Measurement of Tin under Shock Compression ... 1235
  P.-L. Héreil and C. Mabire
Gated IR Images of Shocked Surfaces ... 1239
  S. S. Lutz, W. D. Turley, P. M. Rightley, and L. E. Primas
Optical Probing of the Electron Temperature Gradient ... 1243
  T. Ao, I. Vollrath, and A. Ng
Ellipsometry in the Study of Dynamic Material Properties ... 1247
  A. W. Obst, K. R. Alrick, W. W. Anderson, K. Boboridis, W. T. Buttler,
  S. K. Lamoreaux, B. R. Marshall, S. L. Montgomery, J. R. Payton,
  and M. D. Wilke
Shock-Induced Birefringence in Lithium Fluoride ... 1251
  J. H. Nguyen and N. C. Holmes
Vibrational Spectra of Nitro Compounds under Shock Compression ... 1255
  T. Kobayashi, T. Sekine, and H. He
Transient Bond Scission of Polytetrafluoroethylene under Laser-Induced Shock Compression Studied by Nanosecond Time-Resolved Raman Spectroscopy ... 1259
  K. G. Nakamura, K. Wakabayashi, K.-I. Kondo
Shock-Induced Orientation of Benzene Molecules Studied by Nanosecond Time-Resolved Raman Spectroscopy ... 1263
  K. Wakabayashi, K. G. Nakamura, and K.-I. Kondo
Measurements of the Conductivity of Shocked Polymethylmethacrylate ... 1267
  D. Townsend and N. K. Bourne

CHAPTER XX

IMPACT PHENOMENA, BALLISTICS, HYPERVELOCITY STUDIES, AND EXOTIC SHOCK CONFIGURATIONS

New Directions and New Challenges in Analytical Modeling of Penetration Mechanics ... 1273
  J. D. Walker
Ballistic Response of Fabrics: Model and Experiments ... 1279
  D. L. Orphal, J. D. Walker, and C. E. Anderson Jr.
Long-Rod Moving-Plate Interaction ... 1283
  Y. Partom
Conversion of Finite Elements into Meshless Particles for Penetration Computations Involving Ceramic Targets ... 1287
  G. R. Johnson, R. A. Stryk, S. R. Beissel, and T. J. Holmquist
Using the Penetration-Velocity Relationship to Correct for Variations in Target Hardness ... 1291
  S. J. Bless and J. Cazamias
Ballistic Testing and High-Strain-Rate Properties of Hot Isostatically Pressed Ti-6Al-4V ... 1294
  Y. Gu, V. F. Nesterenko, and S. S. Indrakanti
Deformation and Damage of Two Aluminum Alloys from Ballistic Impact ... 1298
  C. E. Anderson Jr. and K. A. Dannemann
Recovery of Uranium Fragments ... 1302
  H. R. James, D. H. McElrue, and R. E. Winter

Modeling of Uranium Alloy Response in Plane Impact and Reverse Ballistic Experiments 1306
    B. Herrmann, A. Landau, D. Shvarts, V. Favorsky, and E. Zaretsky

On the Entrance Phase in Long Rod Penetration 1310
    Z. Rosenberg and E. Dekel

Impact Interaction of Projectile with Conducting Wall at the Presence of Electric Current 1314
    V. T. Chemerys, A. I. Raychenko, and B. S. Karpinos

The Use of the Taylor Test in Exploring and Validating the Large-Strain,
High-Strain-Rate Constitutive Response of Materials 1318
    J. C. Foster Jr., M. Gilmore, and L. L. Wilson

Dynamic Characterization of Compliant/Brittle Materials Using Split Hopkinson Bar 1323
    N. S. Brar and V. S. Joshi

Yield and Strength Properties of the Ti-6-22-22S Alloy over a Wide Strain Rate
and Temperature Range... 1327
    L. Krüger, G. I. Kanel, S. V. Razorenov, L. Meyer, and G. S. Bezrouchko

CHAPTER XXI

LASER-DRIVEN SHOCKS

Sub-picosecond Laser-driven Shocks in Metals and Energetic Materials...................... 1333
    D. S. Moore, D. J. Funk, K. T. Gahagan, J. H. Reho, G. L. Fisher, S. D. McGrane,
    and R. L. Rabie

Time-Resolved Measurement of the Launch of Laser-Driven Foil Plate 1339
    H. He, T. Kobayashi, and T. Sekine

Laser-Launched Flyer Plates and Direct Laser Shocks for Dynamic Material
Property Measurements... 1343
    D. L. Paisley, D. C. Swift, R. P. Johnson, R. A. Kopp, and G. A. Kyrala

Development of Laser-Driven Flyer Techniques for Equation-of-State Studies of
Microscale Materials... 1347
    W. M. Trott, and R. E. Setchell, and A. V. Farnsworth Jr.

Ultrafast Time-resolved 2D Spatial Interferometry for Shock Wave Characterization in
Metal Films 1351
    K. T. Gahagan, J. H. Reho, D. S. Moore, D. J. Funk, and R. L. Rabie

A Computational Study of Laser Driven Flyer Plates 1355
    A. V. Farnsworth Jr., W. M. Trott, and R. E. Setchell

Modelling of Laser Spall Experiments on Aluminium.. 1359
    C. M. Robinson

Taking Thin Diamonds to Their Limit: Coupling Static-compression and Laser-shock
Techniques to Generate Dense Water... 1363
    K. K. M. Lee, L. R. Benedetti, A. Mackinnon, D. Hicks, S. J. Moon, P. Loubeyre,
    F. Occelli, A. Dewaele, G. W. Collins, and R. Jeanloz

Radiative Shock Experiment Using High Power Laser 1367
    M. Koenig, A. Benuzzi-Mounaix, N. Grandjouan, V. Malka, S. Bouquet, X. Fleury,
    B. Marchet, C. Stehlé, S. Leygnac, C. Michaut, J. Chièze, D. Batani, E. Henry,
    and T. Hall

1–10 Mbar Laser-driven Shocks Using the Janus Laser Facility 1371
    J. Dunn, D. F. Price, S. J. Moon, R. C. Cauble, P. T. Springer, and A. Ng

Transition from Expansion to Shock Compression in Laser Irradiated Si by Multiple Shots.... 1375
    A. Yazaki, H. Kishimura, Y. Hironaka, F. Saito, K. G. Nakamura, and K.-I. Kondo

CHAPTER XXII

EQUATION OF STATE AND GEOPHYSICS

Evidence for Kinetic Effects on Shock Wave Propagation in Tectosilicates 1381
  P. S. DeCarli, E. Bowden, T. G. Sharp, A. P. Jones, and G. D. Price
The Principal Hugoniot and Dynamic Strength of Dolerite under Shock Compression 1385
  K. Tsembelis, W. G. Proud, and J. E. Field
Explosion in the Granite Field: Hardening and Softening Behavior in Rocks 1389
  I. N. Lomov, T. H. Antoun, and L. A. Glenn
Depth of Cracking beneath Impact Craters: New Constraint for Impact Velocity 1393
  T. J. Ahrens, K. Xia, and D. Coker
Shock Flattening of Spheres in Porous Media: Implications for Flattened Chondrules 1397
  T. Sekine, N. Hirata, A. Yamaguchi, T. Kobayashi, H. He, and Z.-P. Tang
The Possible Composition and Thermal Structure of the Earth's Lower Mantle and Core 1401
  Z. Gong, X. Li, and F. Jing
Molecular Dynamics Modeling of Impact-Induced Shock Waves in Hydrocarbons 1406
  M. L. Elert, S. Zybin, and C. T. White
High Intensity X-Ray Coupling to Meteorite Targets 1410
  J. L. Remo and M. D. Furnish
The Dynamic Strength of Cement Paste under Shock Compression 1414
  K. Tsembelis, W. G. Proud, and J. E. Field

Participant List 1419
Author Index A1
Subject Index S1

CHAPTER XII

MECHANICAL PROPERTIES: CERAMICS AND GLASSES

THE HEL UPPER LIMIT

J.P. Billingsley

*U.S. Army AMCOM, AMSAM-RD-SS-AA,
Redstone Arsenal, Alabama 35898, U.S.A*

ABSTRACT. A threshold particle velocity, V_f, derived by Professor E.R. Fitzgerald for the onset of atomic lattice Disintegration Phenomena (LDP) is shown to exceed and/or compare rather well with the maximum experimental Hugoniot Elastic Limit (HEL) particle (mass) velocities (Up_{HEL}) for selected hard strong mineral/ceramic materials.

INTRODUCTION

This work can be considered as an extension to the results reported in References [1] and [2] which delineated the relevance of the Debroglie Monument-Wavelength Equation to the HEL decay phenomenon. Earlier, Reference [3] in 1989, Reference [4] in 1990, Reference [5] in 1994, and Reference [6] in 1995 had all suggested a tie in between the Debroglie relation and HEL phenomena. In particular, Reference [5] contains comparative information that provided the genesis and some of the data for Reference [7] in 2001 and the present document. In [5], it was suggested that V_f could be an upper limit for HEL particle velocities, Up_{HEL}.

ANALYSIS

There are two important particle (mass) velocities (V_1 and V_f) connected with the Debroglie-Fitzgerald Particle Momentum Wave (PMW) theory of stress-strain wave motion. They are associated with atomic lattice instability and final breakdown (LDP) caused by a conflict between TWO physical laws that are: (1) the Debroglie momentum (mV)-wavelength (λ) relation that is $\lambda = h/(mV)$, where h = Planck's constant, m = particle mass, and V = particle velocity. (2) The wavelength (λ) of a propagating disturbance along a row of particles CANNOT be less then $2d_1$ where d_1 is the repetitive spacing between particles, because there is nothing to vibrate in the space between the particles [8, 9, 10].

When $\lambda = 2d_1$, then $V_1 = h/(2md_1)$ = the largest possible particle velocity without violating Rule (2). The importance and influence of this conflict cannot be over emphasized. It led Dr. E.R. Fitzgerald [8, 9] to derive a relation for the maximum particle velocity, V_f, that the lattice could sustain before chaotic conditions (LDP) occur. This relation is:

$$V_f = \left(V_1 + \sqrt{V_1^2 + 4V_1C_s}\right)/2 \approx \sqrt{V_1C_s}$$

C_s is an elastic wave related velocity that is a measure of the lattice cohesive energy. It is given by

$$C_s = \sqrt{(2*C_t^2 + C_L^2)/6} \approx C_t$$

where C_L = the longitudinal elastic wave velocity and C_t = transverse elastic shear wave velocity.

APPLICATION

So V_f, for materials known to have very large experimental Up_{HEL} magnitudes, was computed and compared to these Up_{HEL} data. These hard strong mineral/ceramic materials were: diamond, single crystal and fused quartz, aluminum oxide (sapphire and alumina), silicon carbide, boron carbide, and three different types of Partially Stabilized Zirconia (PSZ). These comparisons are graphically summarized in Figures 1 and 2. Reference [7] contains more detailed comparative information and lists the sources of the experimental data.

DISCUSSION

Figure 1 shows the comparisons when V_f is greater than Up_{HEL} for eight examples. Figure 2 depicts the comparisons for three examples where V_f is exceeded by the experimental Up_{HEL} magnitudes.

When these cases, in Figure 2, are carefully considered, along with those where V_f is greater than Up_{HEL}, it is apparent that V_f is a good measure of these maximum Up_{HEL} values [7]. Thus even in these most extreme cases, judiciously computed V_f values provide some rationale for the anomalously large Up_{HEL} magnitudes. Consequently, V_f could, at least, be employed to estimate an upper bound on Up_{HEL} for untested materials. Finally, it is worth noting that "V_f effects," or LDP, may be responsible for "failure waves" experimentally observed [11] in glasses and ceramics.

REFERENCES

1. Billingsley, J.P., "The Hugoniot Elastic Decay Limit," Proceedings of the *Shock Compression of Condensed Matter – 1997* Conference, AIP Conference Proceedings 429, New York, 1998, Pages 199-202, Ed. by Schmidt, Dandekar, and Forbes.

2. Billingsley, J.P., "The Decay Limit of the Hugoniot Elastic Limit," *International Journal of Impact Engineering*, Vol. 21, No. 4, April 1998, Pages 267-281.

3. Balankin, A.S., "Synergetics and the Mechanics of a Deformable Body," *Soviet Technical Physics Letter*, Vol. 15, No. 11, November 1989, Pages 878-880.

4. Billingsley, J.P. and Oliver, J.M., "The Relevance of the Debroglie Relation to the Hugoniot Elastic Limit (HEL) of Shock Loaded Solid Material," U.S. Army AMCOM, Technical Report RD-SS-90-4, March 1990.

5. Billingsley, J.P., "Possible V_1 and V_f Effects in Shock Loaded Materials from Polymers to Diamond," U.S. Army AMCOM, Technical Report RD-SS-94-4, February 1994.

6. Cherepanov, G.P., Balankin, A.S., and Ivanova, V.S., "Fracture Mechanics – A Review," *Engineering Fracture Mechanics*, Vol. 51, No. 6, August 1995, Pages 997-1033.

7. Billingsley, J.P., "The Upper limit of the Hugoniot Elastic Limit," U.S. Army AMCOM, Special Report RD-SS-01-01, May 2001.

8. Fitzgerald, E.R., Particle Waves and Deformation of Crystalline Solids, Interscience Publishers, a Division of John Wiley and Sons, Inc., New York, 1966.

9. Fitzgerald, E.R., "Particle Waves and Phonon Fission in Crystals," *Physics Letters*, Vol. 10, No. 1, May 15, 1964, Pages 42-43.

10. Girifalco, L. A., Statistical Mechanics of Solids, Oxford University Press, New York, 2000, Pages 98-99.

11. Brar, N. S., "Failure Waves in Glass and Ceramics Under Shock Compression," Proceedings of the *Shock Compression Condensed Matter – 1999* Conference, AIP Conference Proceedings 505, 2000, Pages 601-606, Ed. by Furnish, Chabildas, and Hixon.

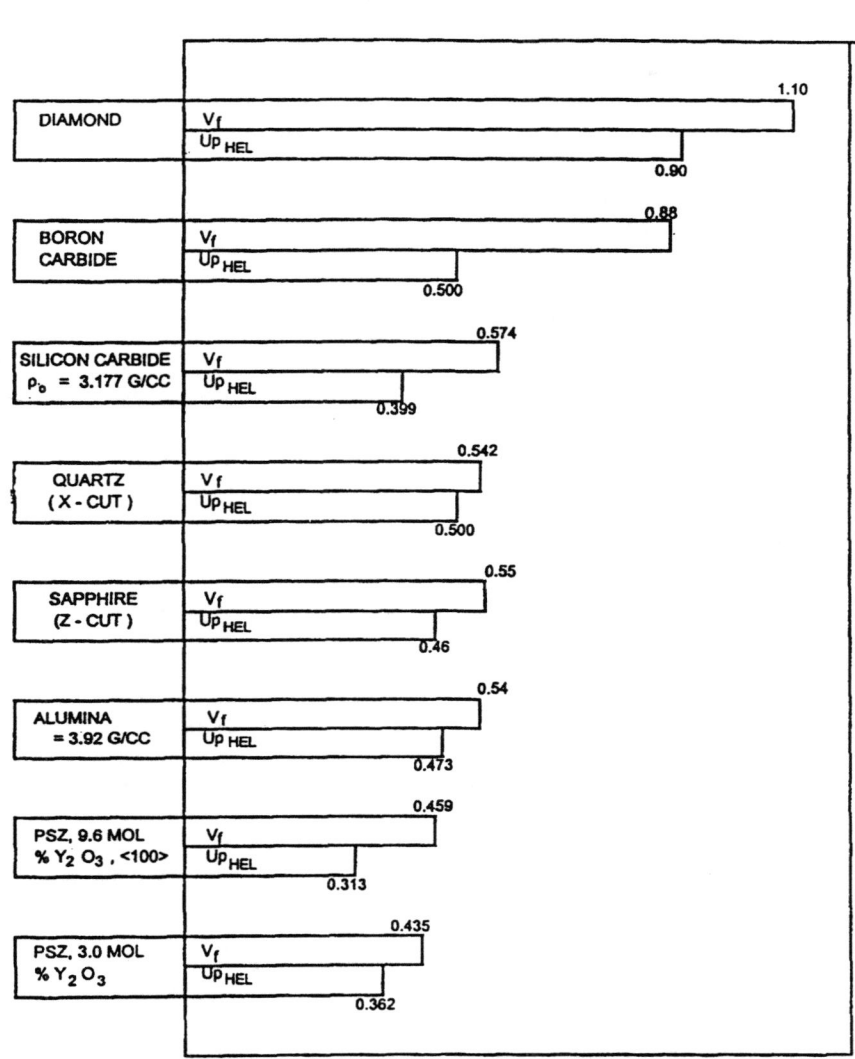

FIGURE 1. V_f and Up_{HEL} Comparison Where $V_f > Up_{HEL}$

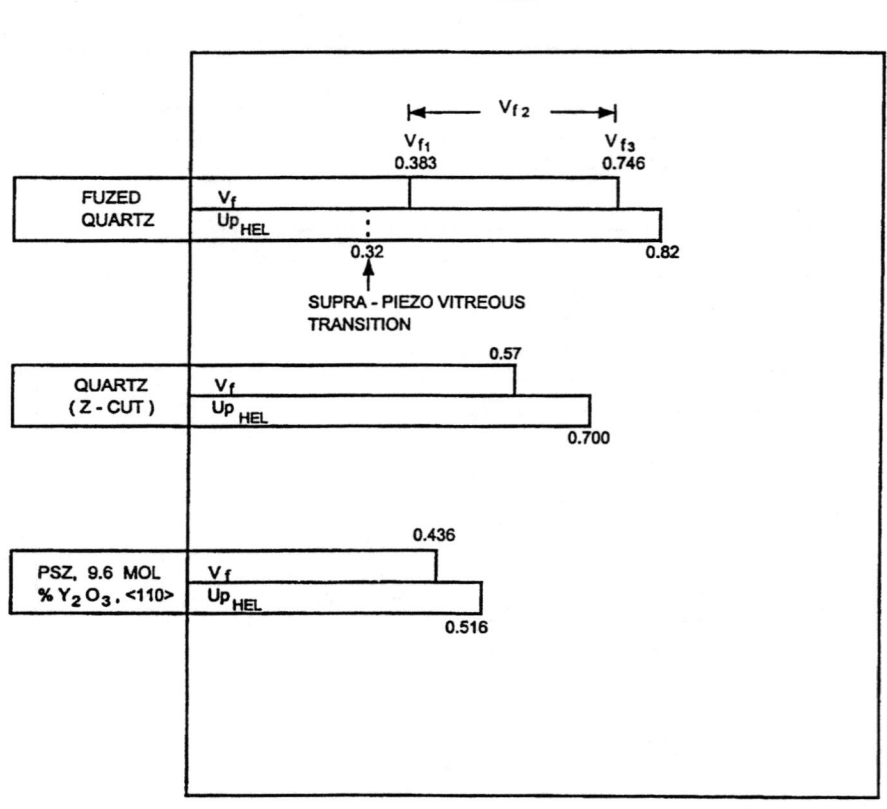

FIGURE 2. V_f and Up_{HEL} Comparison Where $V_f < Up_{HEL}$

ON THE HEL AND THE "RAMPING" ABOVE HEL

E. Bar-on[1], Y. Partom[1], M. B. Rubin[2] and D. Z. Yankelevsky[3]

[1]*Rafael Ballistics Center, P.O. Box 2250, Haifa 31021, Israel.*
[2]*Faculty of Mechanical Engineering, Technion-Israel Institute of Technology, Haifa 32000, Israel.*
[3]*Faculty of Civil Engineering, Technion-Israel Institute of Technology, Haifa 32000, Israel.*

Abstract. Unlike in metals, stress and particle velocity histories of shock waves in ceramic materials show a typical "ramping" above the Hugoniot Elastic Limit (HEL). Under the assumption that the HEL signifies the beginning of failure of the material, this "ramping" has been described by viscoplasticity or by moduli degradation. However, there is ample experimental evidence to rule out plastic flow at the HEL level of stress, and it seems improbable that the moduli would degrade significantly during the compressive phase of a plate-impact. The proposed micro-mechanical mechanism for the HEL, and for the ramping beyond the HEL, is based on the process of porous compaction due to pressure above a threshold pressure p_{crush}. Although the pore volume of high-grade ceramics is quite small, it is sufficient to cause the ramping observed in Hugoniot measurements from the seventies. Additional evidence for the effect of porosity on the HEL stress has been given in recent years. The experimental evidence is supported by simulations, in which a simple model for porous compaction was used. These simulations suggest that porous compaction is the main micro-mechanical mechanism causing the "ramping" and other features related to the HEL.

INTRODUCTION

The concept of the **HEL** (Hugoniot Elastic Limit) is used extensively in high velocity impact dynamics. For metals under impact the material undergoes a compression phase, which at first causes elastic deformations and at a specified value of stress causes inelastic deformations to evolve. This point on the Hugoniot curve is denoted as the HEL and for the case of metals it is related to the yield stress and to the onset of plastic deformations. The HEL in the compressive phase curve of metals is recognized by the sharp break of the stress-strain curve, which continues to increase to higher values of stress as a concave function (Fig. 1a).

In brittle materials behavior, the concept of the HEL is not so clear. For brittle materials the transition from the elastic phase to the inelastic phase occurs gradually, and the compressive phase curve describes a convex function (Fig. 1b).

FIGURE 1: Typical compressive phase:
(a) In metals; (b) In ceramics

Many researchers (e.g. works by Longy and Cagnoux [1], Beachump et al [2], Lankford [3], [4] and Lankford et al. [5]) suggest that micro-plasticity plays a major role as a limiting factor in the compressive failure of high strength ceramics. It is possible, that under high hydrostatic pressures due to confinement, or near crack tips due to high-pressure singularity points, one can find dislocations, which are associated with plasticity. However, the difference in the behavior at the transition between the elastic phase and the inelastic phase in ceramics, as compared with metals, suggests that there is a different physical mechanism acting in ceramics. Specifically, in the present work, the HEL characterizes a transition phase between elastic phase and the phase of inelastic strains due to pore crushing.

This conclusion is based on the following experimental observations. However, it is also consistent with the fact that the proposed model, which assumes pore collapse but no plasticity in the regular sense, produces results that are in good agreement with experimental data.

EXPERIMENTAL PROOF FOR THE POROSITY EFFECT

Hugoniot data for several ceramics were measured by Gust and Royce [6] who applied shock waves using explosives. The tested aluminas were distinguished according to their initial porosity, starting with 6.6% porosity AD-85 and ending with 0.2% porosity Lucalox. In Fig. 2, the measured steady-state Hugoniot stress and the best-fit curves for four of these ceramics (from [6]) are plotted. Studying this figure, one can conclude that as the value of porosity decreases (going from right to the left in the graph) - (1) the HEL value increases, and (2) the HEL cusp becomes indistinguishable and the curve converges to the Hugoniot of a single crystal $Al_2 O_3$ (sapphire). Another conclusion from these results is that the measured Hugoniots do not follow one curve in the stress-volume plane (like in a $p - \alpha$ model), but there is a family of curves, where each curve is associated with the initial porosity of a specific ceramic.

When they analyzed their data, Gust and Royce noted that (in [6], bottom of page 289) - "For the more porous materials,...it was noted that in the compaction regime, i.e., from the yield point to about 300 kbar, the value obtained for U_{p2} from impedance matching was consistently greater than $U_{fs2}/2$. This is the opposite of results obtained for metals and is consistent with a model in which the porosity is permanently crushed out of the material by the shock compaction".

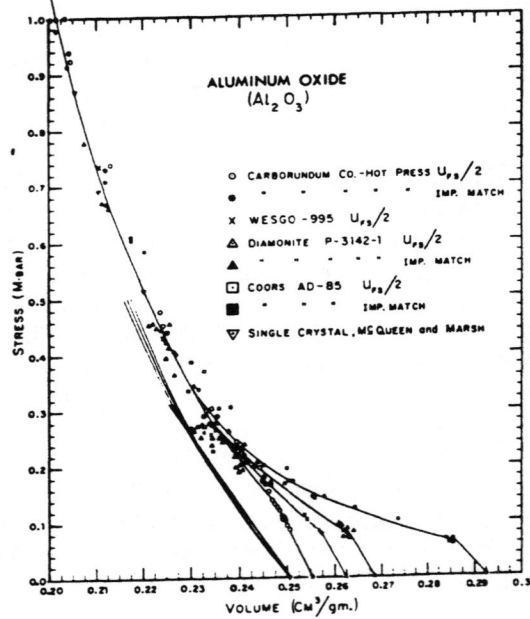

FIGURE 2: Experimental Hugoniots for $Al_2 O_3$ ceramics [6]

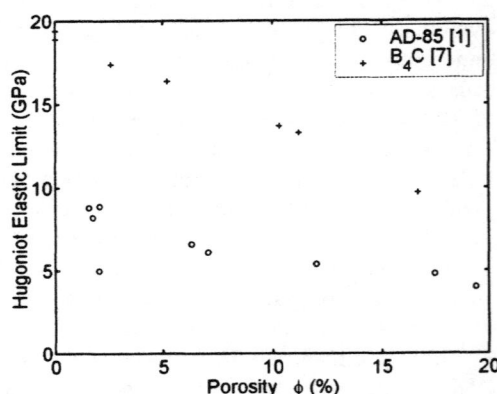

FIGURE 3: The effect of porosity on the HEL stress magnitude of AD-85 alumina [1] and Boron Carbide [7]

The effect of porosity on the magnitude of the HEL stress was studied over the years by different researchers. Rosenberg reported the effect of porosity on the HEL stress in Boron Carbide (B_4C) ceramics [7], and Longy and Cagnoux showed similar behavior in AD-85 alumina [1]. Both works have shown that increasing porosity decreases the magnitude of the HEL stress (Fig. 3).

POROSITY IN THE MODEL

In many constitutive models, a Representative Volume Element dv of brittle material in the present configuration (at time t) is decomposed into a solid part whose volume is dv_s and a pore volume dv_p, such that

$$dv = dv_s + dv_p, \quad dV = dV_S + dV_P \quad (1a,b)$$

where dV, dV_S, dV_P are the values of dv, dv_s, dv_p in a fixed reference configuration, and the indexes s, p denote the solid and pores, respectively. The porosity ϕ and its reference value Φ are then defined by

$$\phi = \frac{dv_p}{dv}, \quad \Phi = \frac{dV_P}{dV} \quad (2a,b)$$

In the compressive phase, the porosity ϕ is determined through integration of the following evolution equations:

$$\dot{\phi} = \begin{cases} f_3(p)\frac{\dot{v}}{v} & \text{for} \quad p > p_{crush} \\ 0 & \text{for} \quad p_{crush} \geq p \end{cases} \quad (3a,b)$$

Here p_{crush} is a threshold pressure to begin pore collapse in compaction, v is the current element volume and $f_3(p)$ is given by

$$f_3(p) = C_1\left(\frac{p - p_{crush}}{p_{crush}}\right) + C_2\left(\frac{p - p_{crush}}{p_{crush}}\right)^2 \quad (4)$$

p_{crush}, C_1, C_2 are material parameters to be determined, and are unique for each ceramic. Here they are taken to be constants. However, they are probably functions of the initial porosity and/or other variables. Figs. 4, 5 and 6 show the way that the calculated stress curve is affected according to the varied values of these material parameters.

Figure 7 demonstrates the good agreement, which can be achieved in simulating the stress curve in a plate-impact experiment [8], using this model.

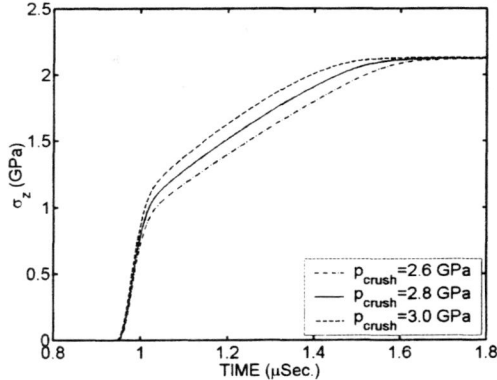

FIGURE 4: The effect of varying p_{crush} in Eq. (4)

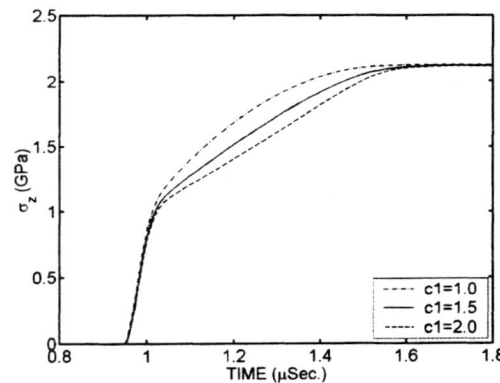

FIGURE 5: The effect of varying C_1 in Eq. (4)

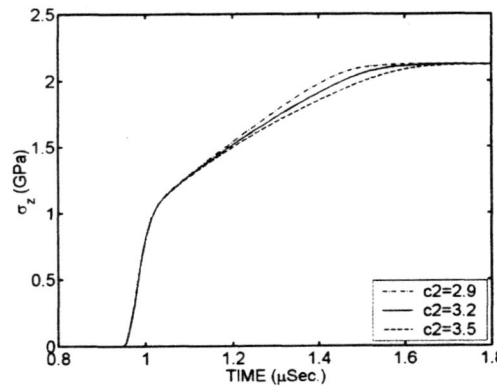

FIGURE 6: The effect of varying C_2 in Eq. (4)

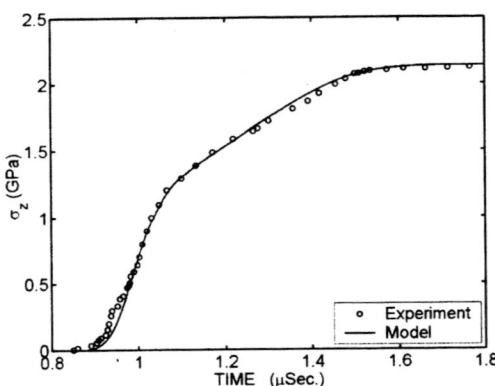

FIGURE 7: Comparison of the measured compressive phase in a plate-impact experiment [8] with the model calculation

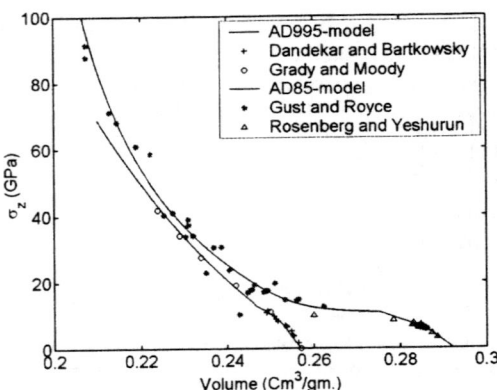

FIGURE 8: Experimental data and model simulations, proofs of the porosity effect on the HEL value and the "ramping" in the Hugoniots of two ceramics

DISCUSSION

This work shows that porous compaction is the main mechanism active during the compressive phase of shock loading of ceramics. This conclusion has been based on the analysis of experimental data that has been collected over the last 30 years. The proposed model of porous compaction correctly predicted the ramping observed in the compression wave, without causing precursor decay or wave separation, as observed in pseudo-plasticity models.

It is important to emphasize that here attention has been focused on the compression phase, and the unloading or tension phases have been omitted. Consequently, it has been possible to consider only a simple model for porous compaction, without the necessity to go into detail of a model to simulate the response of brittle materials to impact loads (compressive and tensile phases). For general material response it is necessary to consider a much more complicated phenomena associated with micro-cracking, and possibly a pseudo "plastic-flow" (related to the amount of crushed pores).

REFERENCES

[1] Longy F., Cagnoux J., "Plasticity and Microcracking in Shock-Loaded Alumina". J. Am. Ceram. Soc., Vol. 72(6), PP. 971-979. (1989).

[2] Beachump E. K., Carr M. J. and Graham R. A., "Plastic Deformation in Alumina by Explosive Shock Loading". J. of Am. Ceramics Soc., Vol. 68, No. 12. PP. 696-699. (1985).

[3] Lankford J., "Compressive Strength and Microplasticity in Polycrystalline Alumina". J. of Materials Science, Vol. 12. PP. 791-796. (1990).

[4] Lankford J., "High Strain Rate Compression and Plastic Flow of Ceramics". J. of Materials Science, Vol. 15. PP. 745-750. (1996).

[5] Lankford J., Predebon W. W., Staehler J. M., Subhash, Pletka B. J. and Anderson C. E., "The Role of Plasticity as a Limiting Factor in the Compressive Failure of High Strength Ceramics". Mechanics of Materials, Vol. 29. PP. 205-218. (1998).

[6] Gust W.H., and Royce E.B., "Dynamic Yield Strength of B_4C, B_eO and Al_2O_3 Ceramics". J. of Appl. Phys., Vol. 42, No. 1, PP. 276-295. (January 1971).

[7] Rosenberg Z., "The Response of Ceramic Materials to Shock Loading". Proceedings of the APS Shock Waves in Condensed Matter Conference, PP. 439-446. Williamsburg, VA. (1991).

[8] Rajendran A. M. and Grove D. J., "Modeling the Impact Behavior of AD-85 Ceramic". 24th Int. SAMPE Technical Conf. T925-T934. (1992).

FACTORS INFLUENCING THE SHAPE OF THE FRACTURE WAVE INDUCED BY THE ROD IMPACT OF A BRITTLE MATERIAL

A.D. Resnyansky[1] and N.K. Bourne[2]

[1] Weapons Systems Division, Aeronautical and Maritime Research Laboratory,
DSTO, PO Box 1500, Salisbury SA 5108, Australia
[2] Royal Military College of Science, Cranfield University, Shrivenham, Swindon, SN6 8LA, UK

Abstract. A fracture wave in a brittle material is a continuous fracture zone which may be associated with the damage accumulation process during the propagation of shock waves. In multidimensional structures the fracture wave may behave in an unusual way. The high-speed photography of penetration of a borosilicate (pyrex) glass block by a hemispherical-nosed rod (1) shows a visible flat wave forming as the fracture front. The role of the complex stress state and kinetic description of the damage accumulation are analysed to describe the process of the impact. The DYNA2D hydrocode and a kinetic strain-rate sensitive model (2) are employed.

INTRODUCTION

Studies of the fracture waves in brittle materials become increasingly popular due to the application of ceramics, high strength glass, concrete *etc*.

In considering the failure phenomenon, special attention is paid to the formation and behaviour of a continuous fracture zone within or behind shock wave which Kanel named a fracture wave (3). Feature of the process is the material failure in a compression zone, because fracture wave follows the shock wave.

The fracture wave phenomenon is associated with accumulation of damage that requires a kinetic constitutive description. In the multi-dimensional case of ballistic impact, fracture waves may behave in an unusual manner. Investigations have used high-speed imaging of impacted borosilicate glass (pyrex) in various papers (1, 4). In (4) a periodic structure of fracture waves in colliding glass rods has been observed. In (1) the ballistic impact of a pyrex glass block by a rod with a hemi-spherical nose resulted in development of a fracture wave that appeared to be headed by a flat region.

This behaviour is apparently related to the specific role of a complex stress state in the area of the impact superimposed by the kinetic accumulation of damage during the fracture wave formation.

The role of the complex stress state is analysed with a kinetic approach to the damage accumulation in order to describe the fracture wave formation. The damage kinetics is verified with available one-dimensional tests (5). The choice of an equivalent stress for the damage kinetics treating the complex stress state is discussed in detail. The present paper analyses the test results observed in (1). The work employs a damage accumulation model of the phase transition type developed earlier (2). The model was incorporated within the DYNA2D-hydrocode. Results of the calculations demonstrate that good agreement with the experiment may be taken into account with the treatment of complex stress state within the constitutive approach in the manner of a cup model.

MODEL

The model (2) uses the representation of a brittle material as a two-phase 'mixture' of two constituents. Each of the components is managed by its own strain-rate sensitive model of the relaxation type (6). A damage parameter c is responsible for the non-equilibrium transition of one phase to another. The 'undamaged' phase corresponds to the parameter value $c=0$. $c=1$ associates the material with a state of continuous 'totally damaged' material, which is a hypothetical material saturated with 'damage' that means that no further degradation can occur in the material. The 'damaged' state differs from the original one only in its mechanical properties. For example, in the given case the second phase of the pyrex glass has the same density and bulk modulus, but a significantly reduced shear modulus. Yield limits of the second constituent should be also essentially reduced for the whole range of strain rates. The combined material is governed by a model with transition kinetics for the damage parameter which allow smooth connection of the two phases during the fracture process.

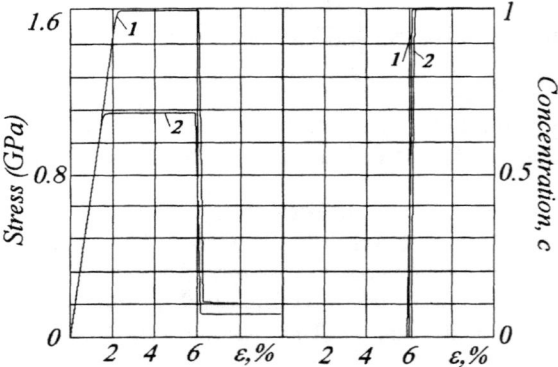

FIGURE 1. Calculated stress-strain curves (left) and associated concentration parameter (right) used for the modelling of the pyrex glass at the strain rates 10^3 s^{-1} (*1*) and 10^{-2} s^{-1} (*2*).

The model is an example of a two-phase phenomenological approach which can consider many situations where two phases described by their own properties can be connected through the transition/concentration parameter (7, 8).

The model for each of the phases is a strain-rate sensitive model described in (6). The model includes:
(i) conservation laws;
(ii) constitutive equation for relaxation of shear stress;
(iii) the two phase model is completed with a kinetics for the damage parameter c:

$$\frac{dc}{dt} = -\psi(c, \sigma, T),$$

here ψ is a kinetics dependence fit to experimental stress-strain curves, σ for stress, and T for temperature.

The equations of state for the damaged material are a combination of equations of state (EoS) for the phases according to a phenomenological approach (a substitute of the mixture rule).

The model is employed by an in-house one-dimensional code and is incorporated within LS-DYNA2D (9).

ONE-DIMENSIONAL ANALYSIS

The chosen damage kinetics has been verified with available experiments (the stress measurements by manganin gauges) of the plane impact of a block of pyrex glass with a copper flyer plate (5). Calculation of the stress states are shown in Fig. 2 for two velocities of impact. The calculations clearly demonstrate appearance of the transition related to damage and which can be associated with the fracture wave. The calculation results (curves *C1* and *C2*) are in a good agreement with the test data (curves *E1* and *E2* from (5)).

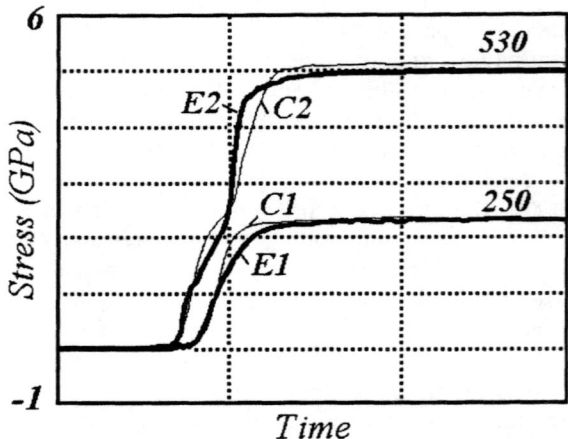

FIGURE 2. Stress profiles in plates of pyrex glass for impact by copper flyer plate with velocities 250 m s^{-1} (curves *C1* and *E1*) and 530 m s^{-1} curves *C2* and *E2*).

ANALYSIS OF BALLISTIC IMPACT

Treating stresses in a multi-dimensional geometry is not simple. Traditional fracture criteria for complex stress states elaborate an equivalent stress as a criterion of failure. For brittle materials two stress modes, which are the shear stress σ and pressure p, are considered essential. Therefore for engineering materials like concrete, soil, *etc.* a closed curve (a cap fracture model (10)) is usually drawn in the shear stress-pressure space which is a limiting surface for stresses the material can sustain at quasi-static loadings. In this case the equivalent stress $\sigma_{eqv}(\sigma, p)$ is a functional combination of the shear stress and pressure which restricts the set of accessible states within the zone $\sigma_{eqv} < \sigma_{crit}$. Here $\sigma_{eqv} = \sigma_{crit}$ can be considered as a fracture criterion.

For the case of shock wave loading there is not a single level curve $\sigma_{eqv} = \sigma_{crit}$. Conventionally for the processes of different duration a separate criterion $\sigma_{eqv} = \sigma_{crit}$ can be drawn in the (σ, p)-space. We consider a cap model presentation suitable for our current purpose. In this case a chosen quasi-static criterion could be drawn as curve 1 in Fig. 3. Then a theoretical strength curve would be a curve 2 in Fig. 3. However for impact processes at a finite duration corresponding curve would be between the curves 1 and 2.

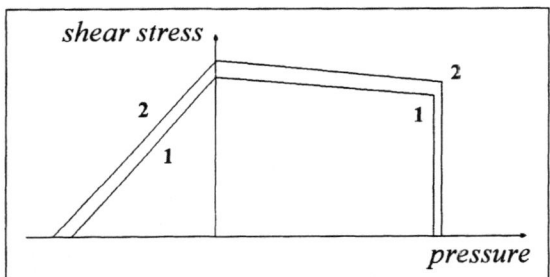

FIGURE 3. Schematic of contours of equivalent stress σ_{eqv} = const for the quasi-static and theoretical strength

Application of these intermediate criteria to a real problem is almost impossible but this idea may be easily realised within the kinetic approach we employ. We consider that the functional form of equivalent stress in the above form can be used instead of a simple state stress σ in the constitutive equations of the model. With this approach the stress states at which the material fail is between the curves 1 and 2 in Fig. 3. A simple loading confirmation of this can be found in Fig. 1 where two stress limits correspond to a high strain rate and a quasi-static case.

Now we analyse the case of impact of a block of pyrex glass by a steel rod with the impact velocity of 536 m/s (1). The rod is of 10 mm diameter and has a hemi-spherical nose. The failure wave photographs taken in (1) are shown in Fig. 4. The 5 mm cross-mark is used for location of the fracture wave position.

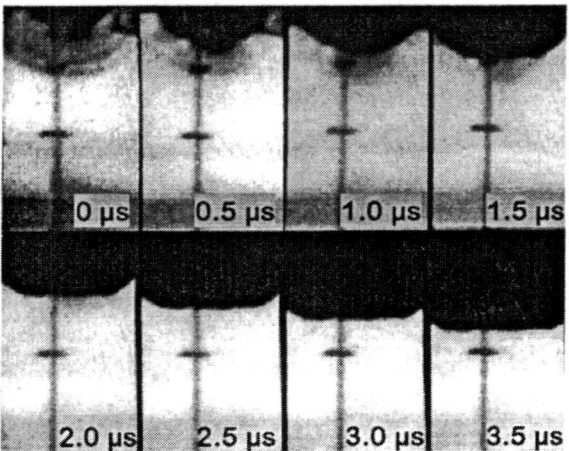

FIGURE 4. The high-speed photographs (1) of the high-velocity impact of a pyrex glass block by a semi-spherical-nose projectile.

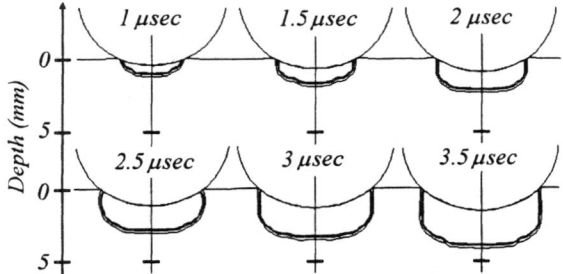

FIGURE 5. Development of the damage contours in a block of Pyrex glass after impact by a semi-spherical-nose rod with the velocity of 536 m/s.

Result of the corresponding calculation with DYNA2D-hydrocode employing the model implemented is shown in Fig. 5. The failure wave front is visible as three successive contours of the damage concentration levels ($c=3/6$; $c=4/6$, and $c=5/6$). The contours are very close to each other because of the specific kinetics chosen for this case; sharp stress drop and concentration rise in Fig. 1 for the case of uniform deformation illustrates this. The fracture wave in the calculation is being developed with approximately the same velocity of 1.8 km/s as observed in the test (1).

In order to understand the nature of the fracture wave a calculation was conducted for similar problem with a flat-nose projectile. Results of this calculation are shown in Fig. 6.

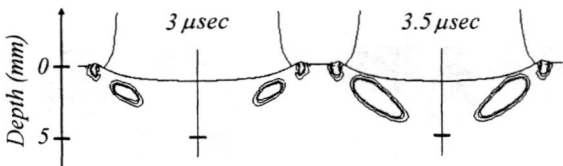

FIGURE 6. Development of the damage contours in a block of Pyrex glass after impact by a flat-nose rod with the velocity of 536 m/s.

The calculation demonstrates that two actual fracture zones are being formed during the impact. The first one attached to the free surface of the glass target is associated with tensile stress (negative pressure) and is typical for conventional materials. The second one, visible as bigger inner fracture zones, is in the tail of propagation of significant shear stresses and is not typical for conventional material because it is in the compression zone. However, this fracture zone in the 45° direction to the impact line is typical for ceramics and other brittle materials studied in detail. Apparently superposition of these effects at the impact by the hemi-spherical-nose projectile gives this unusual effect.

CONCLUSION

A phase-transition model developed for description of damaging brittle material can be used for calculation of fracture waves in glasses.

Formation of the flat fracture front in pyrex glass at the high-velocity impact by a hemi-spherically nosed rod can be a result of superposition of the fracture waves determined by two mechanisms. The first one is conventional and associated with tensile and shear deformation in the tensile zone. The second one is associated with shear in the compression zone. Taking both mechanisms with the kinetic approach can be done with suitable choice of equivalent stress in the constitutive equations.

REFERENCES

1. Bourne, N. K., Forde, L., and Field, J. E., "High-speed photography and stress-gauge studies of the impact and penetration of plates by rods," in *Proc. 22nd International Congress on High-Speed Photography and Photonics-1996*, edited by D. L. Paisley, Proc. SPIE Vol. 2869, 1997, pp. 626-635.
2. Imomnazarov, Kh. Kh., Resnyansky, A.D., and Romensky, E.I., "Dynamic Damage Model for Viscoelastic Material", in ACAM99, the Second Australasian Congress on Applied Mechanics, 10-12 February, 1999, Canberra, Australia, U-078, pp. 1-6.
3. Kanel, G.I., Molodets, A.M. and Dremin, A.N., *Combust. Explos. Shock Waves* **13**, 772-777 (1977).
4. Murray, N.H., Bourne, N.K., Field, J.E., and Rosenberg, Z. "Symmetrical Taylor Impact of Glass Bars", in *Shock Compression in Condensed Matter-1997*, edited by S.C. Schmidt *et al.*, AIP Conference Proceedings 429, New York, 1998, pp. 533-536.
5. Bourne, N. K., Rosenberg, Z., Mebar, Y., Obara, T., and Field, J. E., *J de Physique IV Colloq.* C8, **4**, 635-640 (1994).
6. Godunov, S.K. and Romensky, E.I., *Elements of Continuum Mechanics and Conservation Laws* (in Russian), Novosibirsk, Nauchnaya Kniga Publ., 1998.
7. Resnyansky, A.D., Milton, B.E., and Romensky, E.I., "A Two-Phase Shock-Wave Model of Hypervelocity Liquid Jet Injection into Air", in *Proc. JSME Centennial Grand Congress (Int. Conf. on Fluid Engnrg), JSME ICFE-97-228,* Tokyo, Japan, 13-16 July, 1997, pp 943-947.
8. Resnyansky, A.D. and Romensky, E.I., "Using a homogenization procedure for prediction of material properties and the impact response of unidirectional composite", in *Proceedings 11th Int. Conf. on Composite Materials, Vol. II: Fatigue, Fracture and Ceramic Matrix Composites,* (Ed. Murray L. Scott), Gold Coast, Queensland, Australia, 14th-18th July, 1997, Woodhead Publishing Ltd, 1997, pp. 552-561.
9. Hallquist, J.O., *User's manual for DYNA2D An explicit two-dimensional hydrodynamic finite-element code with interactive rezoning,* Lawrence Livermore National Laboratory, UCID-18756, Rev. 2, 1984.
10. Chen, W.F. and Baladi, G.Y. *Soil Plasticity,* Developments in Geotechnical Engineering, 38, Elsevier Sci. Publ., NY, 1985.

SPALL STRENGTH OF CERAMIC IN A MULTILAYER SYSTEM

B.A.M. Vaughan, N.H. Murray[1], W.G. Proud and J.E. Field

PCS, Cavendish Laboratory, Madingley Road, Cambridge, CB3 0HE. UK.
[1]*Now at Corus Group PLC. E-mail: nataliehmurray@hotmail.com*

Abstract. Investigations into the dynamic properties of alumina ceramic have been carried out for several years at the Cavendish Laboratory [1,2,3]. Previous work has demonstrated a reduction in spall strength with an increased width of the compression pulse using either thicker fliers, or a flier with longer double transit time for the shock wave. Here the variation in spall strength of an alumina ceramic as part of a multilayer system is investigated. Results indicate that the spall strength decreases with increasing time for which the target is under compression. There is some indication that spall strength may decrease faster under shock ring-up than a single shock taking the sample to the same ultimate pressure.

INTRODUCTION

Ceramics have been used in armour applications for several decades, first seeing use in the 1960's to protect American aircrews from small arms fire [4,5]. Their success is due to their inherent high compressive strength and low density, and are used whenever weight is a constraint [6]. Modern armour configurations are typically multi-layered structures consisting of ceramic and other materials such as metals and composites.

Plate impact experiments have been performed on several types of alumina ceramic [2] to determine Hugoniots and other properties, including spall strength [1] and lateral stress. In the case of lateral stress, studies of ceramics with and without a cover plate of metal or ceramic between the flier and the target were performed [3]. The results show the importance of geometry on measured stress in the target, and prompted the current study, which is to investigate the spall strength of 880 alumina ceramic in several multilayer systems.

The spall strength of a material can be determined experimentally by impact of plates of material, which are carried into tension through the interaction of rarefaction fans. These release fans are dispersive, which gives rise to a geometry dependence.

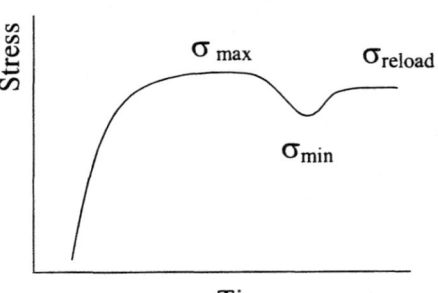

FIGURE 1. Schematic representation of a gauge trace showing a target undergoing spall.

The spall strength can be shown [7] to be given by

$$\sigma_{spall} = \left(\frac{Z_T + Z_W}{2Z_W}\right)(\sigma_{reload} - \sigma_{min}) \quad (1)$$

where Z_T and Z_W refer to the impedance of the target material and to the PMMA window respectively, σ_{reload} and σ_{min} are the stresses recorded by the gauge in the PMMA indicated in Fig. 1. For the target ceramic, the impedance is taken to be the elastic impedance, $Z_{880} = \rho_o c_L$, whereas for the PMMA window, the impedance at each stress state is determined from $Z_w = \sigma/u_p$, using data taken from Marsh [8]. Figure 1 shows a schematic of a stress trace. The solution given in Eqn. (1) assumes elastic behaviour and is more appropriate in the case when $\sigma_{reload} - \sigma_{min} \ll \sigma_{max} - \sigma_{reload}$.

EXPERIMENTAL

Experiments were carried out using the single stage 50-mm gas gun facility at the University of Cambridge [9]. Longitudinal stresses were measured by means of commercial manganin gauges (Micro Measurements type LM-SS-210FD-050). These were placed in the 'back-surface' configuration as shown in Fig. 2. The calibration data of Rosenberg et al. [10] were used to convert the voltage data into stress. In addition, the pressure dependence of the impedance of the PMMA backing has been taken into account when calculating the stress in the target ceramic. It should be noted that if the impedance matched stress profile is used to measure σ_{reload} and σ_{min}, then Eqn. (1) reduces to

$$\sigma_{spall} = \sigma_{reload} - \sigma_{min} \quad (2)$$

In some experiments, VISAR was used to record the PMMA-ceramic interface velocity, giving an independent measurement of the spall process. Three different target configurations were investigated and compared with previous studies [1]. The impact velocity in each case was selected to give rise to a *ca.* 4 GPa peak stress in the target. Table 1 shows the target configurations investigated.

TABLE 1. Target configuration

Configuration	Flier (mm)		Cover plate (mm)		Target thickness (mm)
1	1.8	Cu	1.2	Al	4.0
2	1.8	Cu	1.2	Al	11.0
3	1.8	Cu	2.4	Al	4.0
4	3.0	Al	None		12.0
5	3.0	880	None		6.0

RESULTS

With the generic experimental arrangement shown in Fig. 2 the traces shown in Fig. 3 were obtained for the configurations 1-3.

The back surface stress is converted to stress in 880 by an impedance matching technique that accounts for the change in impedance of PMMA with increasing stress.

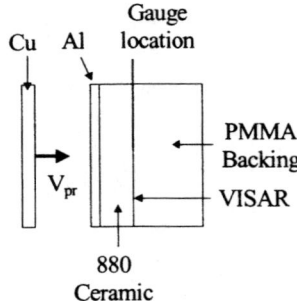

FIGURE 2. Experimental arrangement for multiple step shock. The fliers were 1.8 mm OFHC copper, the target was 880 alumina ceramic and the buffer material was Dural (dimensions given in Table 1). The backing was 12 mm thick PMMA.

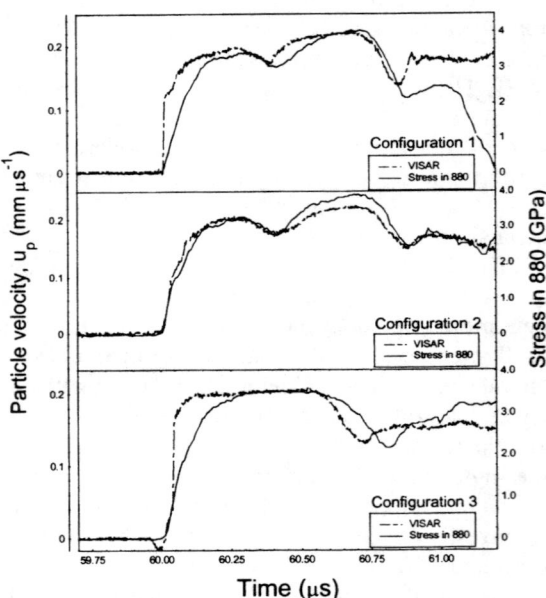

FIGURE 3. Traces obtained for the calculated stress in 880 alumina ceramic. The corresponding VISAR traces are shown alongside.

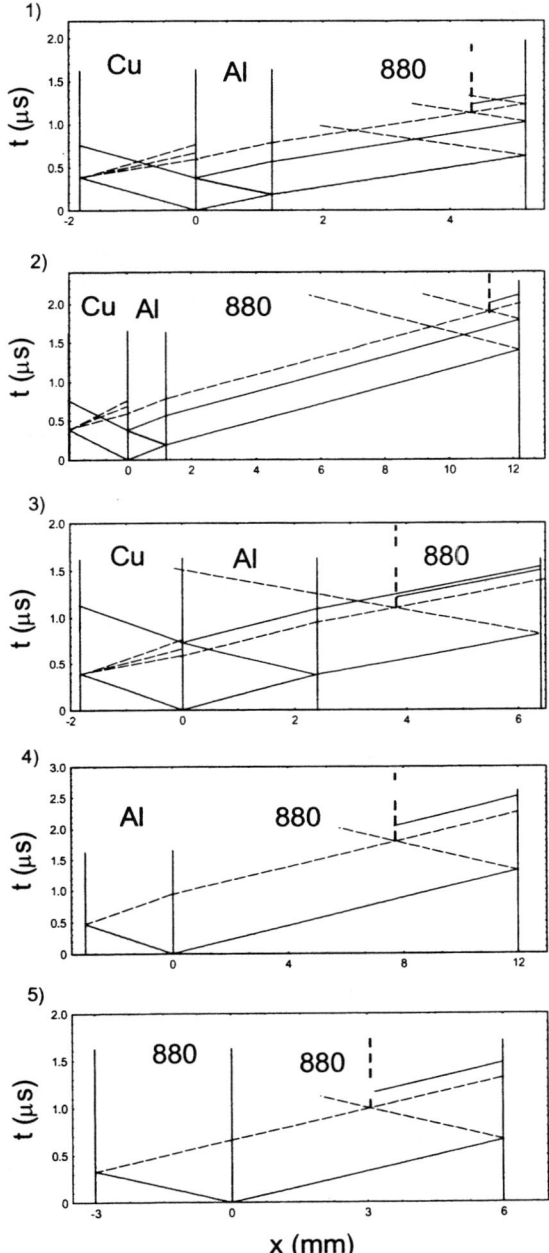

FIGURE 4. t-x diagrams for each of the configurations 1-5.

The t-x diagrams corresponding to each of the configurations in Table 1 are shown schematically in Fig. 4. The release processes will manifest as *release fans* and will include waves both faster and slower than the initial shock wave. The lines shown are approximate, but show the relative positions of wave speeds in the different targets. The time for which the sample is under compression until the release waves from the rear surfaces of the flier and target interact is denoted t_{spall}. Table 2 contains the values of t_{spall} for each of the arrangements and Fig. 5 shows a fit for this data.

TABLE 2. Results

Configuration	t_{spall} (µs)	Spall Strength (GPa)
1	0.600 ± 0.2	0.214 ± 0.08
2	0.604 ± 0.3	0.127 ± 0.156
3	0.573 ± 0.2	0.535 ± 0.07
4	0.942 ± 0.2	0.356 ± 0.08
5	0.659 ± 0.2	0.462 ± 0.09

It can be seen from Fig. 4 (configurations 1 and 2) that the thin aluminium buffer allows two compression pulses to reach the gauge plane before the release from the rear of the target arrives. Interactions deeper in the target and flier do not register on the gauge after the spall signal, and may be ignored. The situation at the gauge plane is similar to that of a single shock, as in [7] and cases 4,5. For the thick aluminium buffer (configuration 3), the spall signal reaches the PMMA-ceramic interface at approximately the same time as the second compression pulse, and may be thought of as a single shock of reduced strength.

Those cases, 3-5, being single or approximately single shocks, have a higher spall strength than the two-step compression pulses with similar t_{spall}. It is unclear at this stage whether this is a physical consequence of the loading process, or the result of experimental scatter in the data. A two step loading pulse may weaken the material more effectively before the arrival of the release fan, which may explain these findings.

CONCLUSION

Spall strength in 880 alumina ceramic is dependent on the geometry of the system under investigation. We have found a correlation between the time the material is under compression until the release waves interact to form a spall plane, and the measured spall strength. This is consistent with previous work [1], which attributed the effect to the accumulation of damage in the form of microcracks. The longer the compressive stage be-

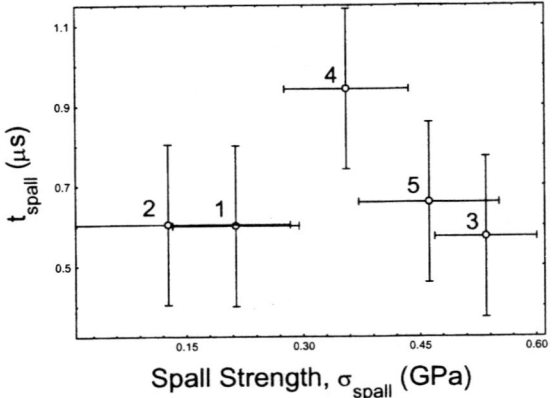

FIGURE 5. Measured spall strength using Eqn. (2), plotted against the time until release waves interact for each geometry. 1, 2 are 'double shock' results.

fore spallation occurs, the more the microcracks can grow, generating larger flaws so that the tensile strength of the material is degraded. Additionally, there may be an effect due to the loading history, so that for a given compression time, t_{spall}, a single shock produced a higher spall strength than a step-loading pulse caused by ringing up of stress in an intervening low impedance buffer. Detailed modelling of these situations using an appropriately sophisticated code and further experiments will be required to fully understand the data presented here.

ACKNOWLEDGEMENTS

The research was supported by DERA (Chertsey). We thank Drs I.M. Pickup and B.J. James for providing the materials and their continued interest in this work. D. L. A. Cross provided technical assistance.

REFERENCES

1. Murray, N.H., Bourne, N.K., Rosenberg, Z. and Field, J.E., *J. Appl. Phys.* **84**, No. 2, pp. 734-738 (1998).
2. Murray, N.H., "The Response of Alumina Ceramics to Plate Impact Loading", PhD Thesis, University of Cambridge (1997).
3. Murray, N.H., Millett, J.C.F., Proud, W.G. and Rosenberg, Z., "Issues Surrounding Lateral Stress Measurements in Alumina Ceramics", in *Shock Compression of Condensed Matter - 1999*, edited by Furnish, M.D., Chhabildas, L.C., and Hixson, R.S., American Institute of Physics, 2000 pp. 581-584.
4. Hannon, F. S. and Abbott, K. H., *Materials Engineering*, **68**, (Sept. 1968) pp. 42-43
5. Rolston, R.F., Bodine, E., Dunleavy, J., *Space/Astronautics*, (July 1968), pp. 55-63
6. den Reijer, P.C., 'On the Penetration of Rods into Ceramic Faced Armours', *Proc. 12th Int. Symp. Ballistics*, **1**, (1990) pp. 389-400
7. Grady, D.E. and Kipp, M.E., "Dynamic Fracture and Fragmentation" in *High-Pressure Shock Compression of Solids I*, edited by Asay, J.R. and Shahinpoor, M., Springer-Verlag, New York, 1993, pp. 265-322.
8. Marsh, S.P. "LASL Shock Hugoniot Data", University of California Press. 1980, pp. 446-451
9. Bourne, N.K., Rosenberg, Z., Johnson, D.J., Field, J.E., Timbs, A.E. and Flaxman, R.P., *Meas. Sci. Technol.* **6**, 1462-1470 (1995).
10. Rosenberg, Z., Yaviz, D. and Partom, Y. *J. Appl. Phys.* **63**, 3702 (1980).

COMPUTER SIMULATION OF THE PROPAGATION OF SHORT SHOCK PULSES IN CERAMIC MATERIALS

Vladimir A. Skripnyak, Evgeniya G. Skripnyak, and Tat'yana V. Zhukova

Department of Mechanics of Solids, Tomsk State University, 36, Lenin Ave., Tomsk 634050, Russia

Abstract. The propagation of shock pulses of submicrosecond duration in Al_2O_3, ZrO_2, and SiC ceramics is investigated by the numerical simulation method. Contributions of various mechanisms of structure evolution to stress relaxation in brittle ceramics are discussed. A theoretical formalism is suggested for the prediction of dynamic strength of polycrystalline and nanocrystalline ceramics with different porosity and grain size.

INTRODUCTION

Use of ceramic nanopowders allow materials to be synthesized with the required grain size, porosity, and structural homogeneity. It is well known that at static loads the fracture toughness, compressive strength, tensile strength, and flexural strength of modern nanocrystalline ceramic materials come close to those of metal alloys and steels. However, experimentally the mechanical behavior of nanocrystalline ceramics at dynamic loads has not yet been determined. Are there simultaneous increase of the dynamic strength and the dynamic fracture toughness of nanocrystal ceramics? Some researchers suggest that when the grain size is smaller than 100 nm, the dislocations can not support the inelastic deformation of structural elements in polycrystalline ceramics. At the dynamic loads, microcracks are nucleated in ceramics and cause the strength decrease. Therefore, a study of the influence of the structural factors on mechanical behavior of ceramic materials under high strain rates is of practical interest.

METHOD OF COMPUTER SIMULATION

The micromechanical approach was used for computer simulation of mechanical behavior of polycrystalline ceramics under shock loading. Ceramics are considered as a continuous medium at the macroscopic level, and as a structural medium at the mesoscopic level. A mechanical state at the macroscopic level is described by a system of conservation equations in continuum mechanics. The stress and strain fields are significantly nonuniform at the mesoscopic level. The stress and strain parameters at the macroscopic level are different from those at the mesoscopic level. Ceramic materials are considered as brittle when the temperature during deformation does not exceed 20% of the melting temperature. The model [1] relates the kinetics of inelastic deformation of ceramics to structural changes. The numerical method [2] is used for calculation in the present work.

MODEL

The main structural elements of polycrystalline ceramics are grains and crystalline blocks. Voids and cracks are considered as structural defects at the mesoscopic level. These structural defects significantly influence the mechanical behavior of ceramics, because they cause stress concentrations at the mesoscopic level. Two different stresses are used in the model. The average stress in the representative volume of the material is the stress at the macrolevel. The stress in the solid phase of ceramics is considered as stress at the mesoscopic level. It is denoted by the superscript m. The connection between the stresses at the macroscopic and the mesoscopic levels is given by the formula

$$\sigma_{ij} = \sigma_{ij}^m \exp(-\alpha/\alpha^*), \quad (1)$$

where α is the specific volume of pores and plane cracks and α^* is an empirical constant.

For Al_2O_3 and ZrO_2, $\alpha^* = 0.1$ describes well the dependence of the static strength of ceramics with porosity. The pressure P^m is calculated by the Mie-Gruneisen equation of state [3], and the deviator stress tensor is calculated by the relaxation equation:

$$DS_{ij}/Dt = 2\mu(\dot{e}_{ij} - \dot{e}_{ij}^n), \quad (2)$$

where D/Dt is Jaumann's derivative, μ is the effective shear modulus of the damaged medium, $\dot{e}_{ij}$ is the deviator for the effective strain rate tensor, and $\dot{e}_{ij}^n$ is the deviator for the effective inelastic strain rate tensor.

The damage parameter α during inelastic deformation can be written in the form

$$\dot{\alpha}/(1-\alpha) = \dot{\varepsilon}_{kk}^n, \quad (3)$$

where $\dot{\varepsilon}_{kk}^n$ is the inelastic volumetric strain rate.

The evolution of the inelastic shear strain at the mesoscopic level under compression causes the reduction of the initial volume of pores upon shock. The volumetric inelastic strain rate caused by the crack healing under compression (4) and opening of cracks under tension (5) is given by expression

$$\dot{\varepsilon}_{kk}^n = \{S_{ij}^{(m)} \dot{e}_{ij}^{n(m)} / [P^{(m)}(1-\alpha/\alpha^*)]\} H(\alpha)H(P^{(m)}), \quad (4)$$

$$\dot{\varepsilon}_{kk}^m = 8\pi R^2 \eta_1 N \dot{R} H(P^{(m)} - P^*), \quad (5)$$

where $H(\cdot)$ is the Heaviside function, R is the average crack size, N is the density of cracks, $\dot{R}$ is the velocity of the cracks size growth, and η_1 and P^* are the model parameters.

It is assumed that the shear inelastic strain caused by microcracks nucleation and growth is given by the expression [1]

$$\dot{e}_{ij}^{n(m)} = (2/3)R_0^3 \dot{N} + 2R^2 N \dot{R}, \quad (6)$$

where R_0 is the nucleated cracks size, and $\dot{N}$ is the rate of cracks nucleation.

Therefore, the kinetics of inelastic strain in brittle ceramics is defined by the kinetics of cracks nucleation and growth. The nucleation of microcracks is described by the formula

$$\dot{N} = [(S_u - S^{**})/\eta_2] H(N), \quad (7)$$

where $S_u = [(2/3) S_{ij}^m S_{ij}^m]^{1/2}$, S^{**} is the critical shear stress of cracks nucleation, N is the density of cracks, and η_2 is the model parameter. For a plane shock, the components of the deviator for the stress tensor are $S_2 = S_3 = -1/2 S_1$, therefore, $S_u = (4/3)\tau$.

In the model, the dynamic strength is related to the critical shear stress by the expression $S^{**} = (3/2)Y_0 \exp(-\alpha/0.1)$.

It is assumed, that the parameter Y_0 is connected with the grain size (d) and the Burgers vector (b) by the relation $Y_0 = 2\tau_{teor}(d/b)^{-0.1}$. Theoretical shear strength of the crystal lattice τ_{teor} is estimated in the context of Frenkel's theory [4].

The growth of the average microcracks size is described by the equation

$$\frac{\dot{R}}{R} = \frac{1}{\eta_3}(S_u - S^*)H(S_u - S^*) + \frac{1}{\eta_4}(P^{(m)} - P^*)H(P^{(m)} - P^*), \quad (8)$$

where η_3, η_4, P^*, and S^* are the model parameters.

We assume that the relaxation of shear stress is caused by two mechanisms, namely, by crack growth (curve 1 in Fig. 1) and by nucleation of new cracks (curve 2), which have different relaxation times. Critical stresses S^* and S^{**} of activation of this mechanisms are also assumed different. The critical shear stress S^* is given by the relation

$$S^* = S_0^* + \eta_5 P^{(m)} H(P^{(m)}), \quad (9)$$

where η_5 and S_0^* are parameters of the model.

The values of η_5 and S_0^* are estimated from experimental data on static strengths at tension and compression. The initial microcrack density N_0, R, and α_0 characterize the initial damage of polycrystalline ceramics. The parameters η_1, η_2, η_3, and η_4 control the kinetics of inelastic deformation.

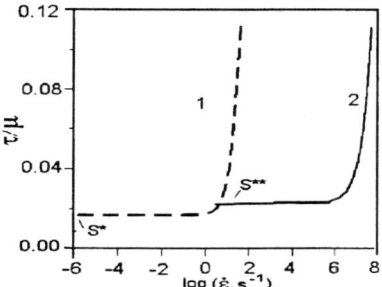

FIGURE 1. Calculated dependence of shear stress on the strain rate in $ZrO_2+3mol\%$ Y_2O_3 ceramics with $1-\mu m$ grain size.

Their values are determined by numerical simulation of the propagation of shock pulses. The criterion $\alpha \geq 0.3$ is used to detect macroscopic failure.

NUMERICAL RESULTS AND DISCUSSION

The model described by Eqs. (1)-(9) is used for computer modeling of the dynamic phenomena in ceramics with different initial porosity and grain size.

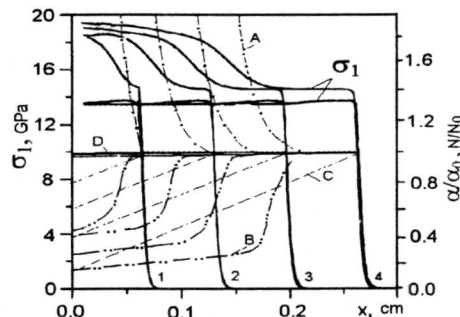

FIGURE 2. Calculated damage parameters in SiC ceramics under shock loading.

Figure 2 shows the calculated profiles of stress waves with amplitudes of 18 and 13 GPa in SiC ceramics. Ceramics had the initial damage $\alpha_0 = 0.015$.

Profiles (1-4) correspond to times of 0.05, 0.1, 0.15, and 0.2 μs after the impact. When the amplitude of the shock (13 GPa) was less than the Hugoniot elastic limit (HEL), the crack density (curve C) and damage parameter α (curve D) changed only slightly. A decrease in the initial crack density resulted in the increase of the calculated spall strength. The density of cracks (curve A) rose sharply and the damage parameter α (curve B) decreased when the amplitude of the shock (18 GPa) was higher than the HEL. A decrease in specific volume of voids α caused the growth of the shear strength in shock. A decrease in grain size and increase in void and crack size lowers the stress concentration at the mesoscopic level. In this case, the HEL (σ_{HEL}) and the dynamic strength $Y = -\sigma_{HEL}(1-2\nu)/(1-\nu)$ increased. Figure 3 shows the calculated dynamic strengths for ceramics with different initial porosity. The experimental data in Fig. 3 were borrowed from [5].

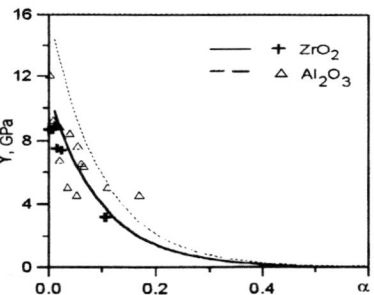

FIGURE 3. Solid and dashed curves are the results of calculation.

Equation (10) can be used to estimate the dynamic strength for polycrystalline and nanocrystalline Al_2O_3 and ZrO_2 ceramics

$$Y(\alpha) = Y_{teor}(d/b)^{-0.1}\exp(-\alpha/0.1) . \quad (10)$$

The Burgers vector b equal to $4.76 \cdot 10^{-8}$ and $5.14 \cdot 10^{-8}$ cm was used for Al_2O_3 and ZrO_2 respectively. The grain size d is in the same units as b. The estimate $Y_{teor} = 2\tau_{teor}$ gives $Y_{teor} = 20.67$ GPa for Al_2O_3 and $Y_{teor} = 10.96$ GPa for ZrO_2. The maximum theoretical estimates of the HEL are 20.6 and 29.9 GPa for ZrO_2 and Al_2O_3 ceramics, respectively. True values of the HEL will be much lower.

The model predicts high values of the HEL when crystalline blocks are considered as structural elements of the single crystals. The calculated HELs are in good agreement with experimental data for polycrystalline alumna and sapphire (12-21 GPa [5]). Figure 4 shows predicted σ_{HEL} for ceramics with different grain size. The experimental data were borrowed from [5].

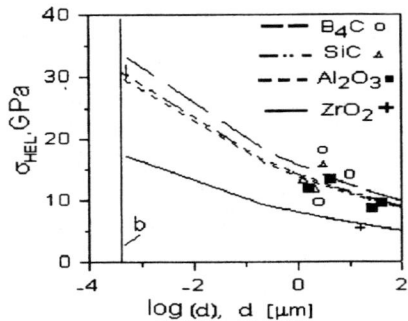

FIGURE 4. Calculated dependence of the Hugoniot elastic limit of ceramics on the grain size.

Simulations suggest that inelastic deformation is negligible in polycrystalline ceramics at high shock amplitude when the pulse duration is comparable to the relaxation time. Under these conditions, the actual spall strength of polycrystalline ceramics is comparable to the theoretical strength of the crystalline lattice at tension.

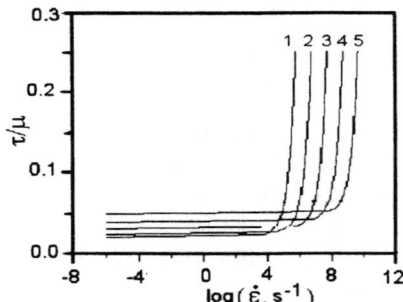

FIGURE 5. Calculated dependence of the shear stress on the strain rate in Al_2O_3 ceramics with grain sizes 100 (1), 10 (2), and 1 μm (3), 100 (4), and 10 nm (5).

Figure 5 shows the calculated shear stress for ceramics with different grain size as a function of strain rate. The calculated results predict not only the growth of shear strength but also the decrease of the relaxation time in nanocrystalline ceramics.

In single crystal sapphire and ruby and nanocrystalline ceramics based on Al_2O_3 and ZrO_2, the fast relaxation time of shear stress is about 10-20 ns. The shear stresses approach the theoretical shear strength of the crystal lattice when the duration of the strong shock is less than 50 ns. Under deformation in brittle ceramics, the dissipation of mechanical energy results in the work on inelastic strain and increase in the surface energy of cracks. Inelastic deformation of some ceramic materials may be caused by the martensitic phase transition.

CONCLUSIONS

1) The computer simulation predicts that the decrease of the grain size in polycrystalline alumna from 100 mm to 10 nm causes the 4-fold increase of the HEL.
2) When the amplitude of shock does not exceed the HEL, the density of cracks in ceramics slowly decreases in shock front. In this case, the shear strength and the spall strength of polycrystalline ceramics may increase.
3) Inelastic deformation of ceramic materials can be caused by the nucleation of microcracks at shock compression above the HEL.

REFERENCES

1. Skripnyak, V. A. and Skripnyak, E. G., "Computer Modeling of Mechanical Behavior of Constructional Ceramics under Shock Loading," in *New Models and Numerical Codes for Shock Waves Processes in Condensed Matter-1997*, edited by I.G. Cameron, AWE Hunting – BRAE, Oxford, UK, 1997, pp. 26-36.
2. Zhukova, T. V., Makarov, P. V., Platova, T. M., and Skripnyak, V. A., Fiz. Goreniya i vzryva **23**, N.1, 29-34, (1987). [in Russian].
3. Ahrens, T. J., "Equation of State," in *High-Pressure Shock Compression of Solids*, ed. J.R. Asay, M. Shahinpoor, Springer-Verlag, New-York, 1993, pp.75-113.
4. Macmillan, N. H., "The Ideal Strength of Solids," in *Atomistic of Fracture*, edited by R. Latanision , J. R. Pickens, Plenum Press, New.-York., 1983, pp. 95-164.
5. Kanel, G. I., Razorenov, S. V., Utkin, A. V., and Fortov, V. E., Shock-Waves Phenomena in Condensed Medium, [in Russian], Yanus, Moskow, 1996.

INFLUENCE OF MICROSTRUCTURAL BIAS ON THE HUGONIOT ELASTIC LIMIT AND SPALL STRENGTH OF TWO-PHASE $TiB_2+Al_2O_3$ CERAMICS

Greg Kennedy[1], Louis Ferranti[1], Rodney Russell[2], Min Zhou[3], Naresh Thadhani[1]

[1]Materials Science and Engineering, Georgia Institute of Technology, Atlanta, GA 30332-0245
[2]IAT, University of Texas at Austin, Austin, TX 78759
[3]Mechanical Engineering, Georgia Institute of Technology, Atlanta, GA 30332-0245

Abstract. The influence of microstructural bias on the Hugoniot Elastic Limit and spall strength of two-phase $TiB_2+Al_2O_3$ ceramics was investigated in this study. The microstructural bias includes differences in phase (grain) size and phase distribution such that in one case a continuous (interconnected) TiB_2 network surrounds the Al_2O_3 phase, and in another case the TiB_2 and Al_2O_3 phases are interdispersed and uniformly inter-twined with each other. Dynamic compression and tension (spall) properties were measured using plate-impact gas-gun experiments. The measurements used piezoelectric PVDF stress gauges to obtain the loading profile and determine the Hugoniot Elastic Limit, and VISAR interferometry to obtain the spall signal and determine tensile properties. The experimental results reveal that while the strength under dynamic compression (HEL) is more dominantly dependent on the phase size, the tensile spall strength scales with the connectivity of the TiB_2 phase.

INTRODUCTION

The objective of the present work is to characterize the high-strain-rate deformation and damage response of four types of microstructurally-biased, two-phase $TiB_2+Al_2O_3$ ceramics. The microstructures involve either a continuous (interconnected) TiB_2 network that surrounds Al_2O_3 (qualitatively termed 'T@A'), or where the TiB_2 and Al_2O_3 phases are interdispersed and uniformly inter-twined with each other (qualitatively termed 'TinA') [1]. Past work on these two-phase ceramics has revealed an 80% increase in compressive strength with increasing strain rate (3.5 GPa at 10^{-4} s^{-1} to 5.8 GPa at 10^3 s^{-1}) [1]. In addition, the two-phase $Al_2O_3+TiB_2$ ceramics have shown superior static and dynamic mechanical properties than their monolithic constituents [1-3]. The $Al_2O_3+TiB_2$ ceramics have also shown better penetration resistance than monolithic Al_2O_3, and the system in which TiB_2 is an interconnected phase surrounding Al_2O_3 has been shown to exhibit a superior ballistic performance compared with the system in which the two phases are simply uniformly interdispersed [1,4]. Micromechanical simulations have also demonstrated the effect of microstructural bias on failure resistance [5,6]. However, the influence of microstructural bias on the fundamental dynamic properties of these ceramics has not been fully established. In the present work, normal plate impact experiments were used to measure the HEL and the spall strength of the materials. The measured responses were then correlated with the microstructure characteristics.

EXPERIMENTAL PROCEDURE

Details of the processing approaches used for fabricating these two-phase $TiB_2+Al_2O_3$ ceramics are described elsewhere [1,7]. The impact experiments used an 80-mm diameter, single-stage gas gun. Measurements of the HEL and the shock wave speeds under dynamic compression were obtained from stress profiles recorded using PVDF stress gauges. A TiB_2 flyer plate backed by an air gap and mounted at the head of an aluminum projectile, was used to impact the target assembly consisting of ~3 mm thick ceramic sample plates backed by TiB_2 backer plates. Experiments were conducted at a nominal velocity of ~0.750 km/s (~15 GPa nominal impact stress), measured using arrival time pins.

One PVDF gauge package was placed at the impact surface and another was placed between the sample and TiB$_2$ backer plate. The experimental setup was designed such that planar-parallel shock waves propagate the target, and the input and propagated stress-profiles are measured with little interference from radial reflected waves.

Fig. 1 (b) shows the experimental setup, illustrating the target sample (without any backer) being impacted by a projectile consisting of a 4140 steel (or silicon carbide) flyer plate and an aluminum sabot. An air gap exists between the flyer plate and the sabot to allow full unloading required for the spall strength measurements. All sample surfaces were lapped for flatness and parallelism. The ceramic targets were polished with 5-μm diamond paste to ensure reflectivity required by the VISAR beam. The time-resolved longitudinal motion of the sample free surface was measured with a VALYN VISAR and recorded on a digital oscilloscope.

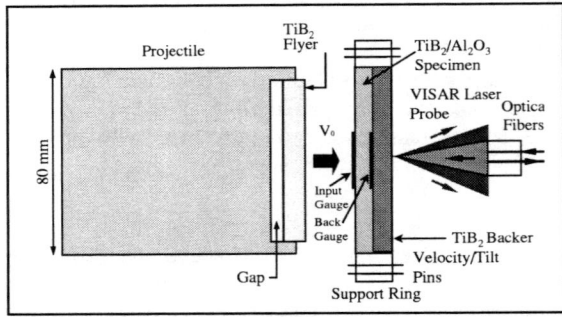

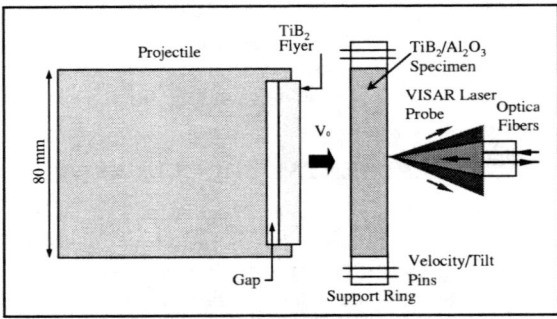

FIGURE 1. Schematics of setups used for measurements of (a) stress profiles with PVDF gauges and (b) free-surface velocity traces with VISAR interferometry.

RESULTS AND DISCUSSIONS
Microstructural Characteristics

Optical micrographs of the four types of microstructurally-biased samples studied, are shown in Figure 2 (a-d). It should be noted that the samples do not reveal 100% microstructural bias, i.e., while micrographs of Samples A and C generally illustrate a nearly continuous (interconnected) TiB$_2$ phase surrounding

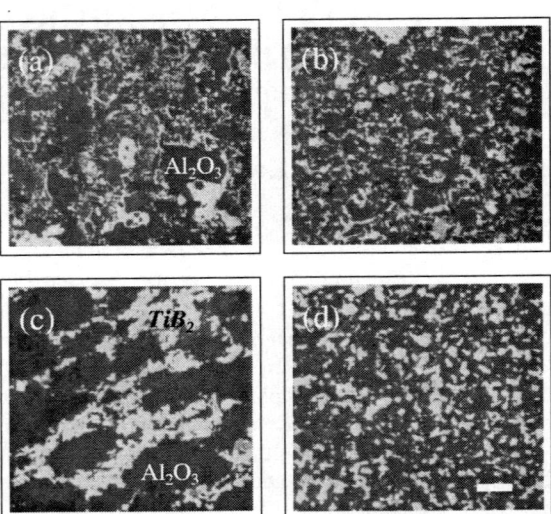

FIGURE 2. Optical micrographs of two-phase ceramics (TiB$_2$ white and Al$_2$O$_3$ dark phase) corresponding to Sample-A SHS-made with TiB$_2$ surrounding Al$_2$O$_3$ (T@A); Sample-B SHS-made with intermixed TiB$_2$ and Al$_2$O$_3$ (TinA) phases; Sample-C made by MM with TiB$_2$ surrounding Al$_2$O$_3$ (T@A); and Sample-D made by MM with intermixed TiB$_2$ + Al$_2$O$_3$ (TinA).

Al$_2$O$_3$, regions where the converse is true are also present. Likewise, Samples B and D show not only an intermixed structure but also the continuous phase. The TiB$_2$ phase connectivity was quantified based on average integral curvature measurements (described elsewhere [8]) and the grain size of the respective phases was obtained using the mean liner intercept method. The average grain (phase) sizes of TiB$_2$ and Al$_2$O$_3$ in both SHS samples (A and B) are smaller than those in the manually mixed samples (C and D), with Sample C showing the largest size for both constituents. Sample-C also represents the microstructure with the highest connectivity for TiB$_2$.

Elastic Properties

The elastic properties of the ceramics were characterized using an ULTRAN Laboratories ultrasonic test apparatus, which allows determination of longitudinal and shear wave velocities. Densities were measured on individual samples by the Archimedean method. The modulii (E = 474-505 GPa and μ = 172-175 GPa) of the various ceramic samples are relatively similar except in the case of Sample B (E = 421 GPa and μ = 149 GPa), which has ~4.4% porosity.

Dynamic Compression Behavior

Measurements of the Hugoniot Elastic Limit (σ_{HEL}) were obtained from normal planar impact experiments conducted at an impact velocity of 750 m/s. Table I lists the sample thickness, density, impact velocity measured using shorting pins, wave speed measured by considering the times of travel through the powder thickness as recorded by input and backer gauges, the Hugoniot Elastic Limit (σ_{HEL}), and the yield stress in simple tension (σ_{YS}) calculated from the σ_{HEL}.

The wave-speed in the material (a function of sample density and microstructure in addition to loading conditions) is measured to be similar for all four samples (within range of experimental scatter). The Hugoniot Elastic Limit (σ_{HEL}), the axial stress at which a solid loaded under conditions of uniaxial strain begins to exhibit plastic deformation, is observed to be a strong function of the microstructure including phase size. It can be seen that the σ_{HEL} is lowest for Sample-B (4.4±1.2 GPa), due to its high level of porosity (~4%) and highest for Sample-D (8.5±4.5 GPa) (with a large standard deviation). The σ_{HEL} values reported previously by Grady [9] is 9-18 GPa for TiB$_2$, and ~6.7 for Al$_2$O$_3$. The yield stress in simple tension was also calculated through σ_{YS} = 2 $(C_S^2 / C_L^2) \times (\sigma_{HEL})$ [9], and found to show the same trend for the four microstructurally-biased ceramics, with Sample D showing the highest and Sample B the lowest yield strength.

Tensile (spall) Strength

Spall experiments were performed (with VISAR interferometry) on samples of microstructure 'A' at impact velocities corresponding to input stresses above and below the σ_{HEL}, and on samples of microstructures 'C' and 'D' at stresses below the σ_{HEL}, to ensure that the compression-induced damage does not influence the tensile response. The velocity traces revealed a spall signal (velocity decrease, ΔU_{fs}) which was used to compute the spall strength ($\sigma_{spall}=1/2\rho_o C_o \Delta U_{fs}$). As shown in Table II, the spall strength for sample-'A' is 0.320 MPa at input stresses of 3.7 and 7.9 GPa. However, with input stress increasing to 11.8 GPa, the spall strength decreases to 0.160 GPa. It can be seen that while the spall strength decreases with increasing input stress, the two-phase TiB$_2$+Al$_2$O$_3$ ceramic maintains non-negligible spall strength even at input stresses exceeding the Hugoniot Elastic Limit (6.2±3.4 GPa).

Table I. Summary of Results of HSR Experiments under Dynamic Compression

SAMPLE TYPE AND EXPT. #	TARGET THICKNESS (MM)	DENSITY (TMD)	WAVE SPEED (KM/S)	HUGONIOT ELASTIC LIMIT σ_{HEL} (GPa)	TENSILE YIELD STR. σ_{YS} (GPa)
A-9914	3.03	99.4%	8.24±.83	6.2±3.1	4.2±2.1
B-9916	3.42	95.6%	9.67±1.0	4.4±1.2	3.11±0.84
C-9920	3.36	99.7%	9.08±.74	5.5±2.3	4.02±1.7
D-9921	3.06	99.1%	8.31±.78	8.5±4.5	6.23±3.3

Table II - Loading Conditions and Summary of Results of Spall Experiments

SAMPLE TYPE AND EXPT. #	FLYER PLATE MATERIAL	FLYER & TARGET TH. (MM)	TARGET DENSITY (%TMD)	IMPACT VELOCITY (M/S)	INPUT STRESS (GPA)	SPALL STRENGTH (GPA)
A-0005	4140 steel	2.66 / 16.22	98.8%	495	7.9	0.320
A-0007	4140 steel	2.66 / 16.20	98.8%	237	3.7	0.320
A-0008	4140 steel	2.66 / 7.99	98.8%	758	11.8	0.160
C-9925	SiC	4.30 / 7.54	99.7%	244	3.8	0.311
D-9926	SiC	4.77 / 7.10	99.1%	239	3.8	0.222

Tensile spall experiments were also performed on samples of microstructures 'C' and 'D', at an input stress of ~3.8 GPa. A spall strength of 0.311 GPa for Sample-C and 0.222 GPa for Sample-D, was observed. A comparison of spall test results between Sample-C and Sample-D shows that the latter sample with dispersed microstructure has a lower spall strength, while the microstructure with interconnected phases has a higher spall strength, even though Sample 'D' has a smaller phase size. The measured high value of spall strength of the two-phase ceramic is similar to the published spall strength of TiB_2 (0.33 GPa) but lower than that of Al_2O_3 (0.45 GPa).

DISCUSSION AND SUMMARY OF RESULTS

The two-phase $TiB_2+Al_2O_3$ ceramics made either by the SHS or mechanical milling methods, reveal differences in microstructure which qualitatively show TiB_2 as a continuous (interconnected) phase surrounding Al_2O_3 (T@A), or TiB_2 and Al_2O_3 intermixed with each other (TinA). Quantitative microscopy analysis based on the measurement of the integral curvature showed that the samples investigated do not exhibit 100% microstructural bias. However, the overall trend of the influence of microstructural bias emerging from the results of experiments performed to-date illustrates that the dynamic yield strength and the σ_{HEL} are more dominantly dependent on the phase size. Sample-C prepared by manual mixing and having the largest phase (grain) size shows the lowest values in contrast to the other samples of similar (~99%) density. In contrast, the tensile spall strength appears to scale with the continuity of the TiB_2 phase. Sample-C, which has the most interconnected TiB_2 phase, has the highest value of the tensile spall strength. The results therefore, illustrate that while the Hugoniot Elastic Limit and the dynamic compressive yield strength of $Al_2O_3+TiB_2$ are dependent on the average grain (phase) size, the tensile spall strength scales with the TiB_2-phase connectivity, and less so with the average phase size. It is possible that the interconnected phase morphology is more effective in impeding initiation and progression of fracture under tensile conditions. If so, TiB_2 as an interconnected phase has the ability to suppress and relax the cracks formed in Al_2O_3. Alternatively, it is also possible that in the two-phase ceramic with the microstructure containing dispersed Al_2O_3 and TiB_2 phases, the mechanical energy trapping due to scattering of waves from incoherent boundaries and interfaces, results in the lowering of the tensile (spall) strength. Further work is currently in progress to more clearly delineate the effects of interconnected versus dispersed TiB_2 phase, and to eventually fabricate two-phase $Al_2O_3+TiB_2$ ceramics with the microstructural bias that yields the most optimal dynamic properties.

ACKNOWLEDGEMENTS

Funding provided by U.S. Army Research Office, under Grant DAAG55-98-1-0454 (Dr. David Stepp program monitor). The authors thank Dr. K.V. Logan for providing the samples and valuable comments.

REFERENCES

1. K.V. Logan, Ph.D. dissertation, Georgia Institute of Technology, 1993.
2. A. Keller, "An Experimental Analysis of the Dynamic Failure Resistance of TiB_2/Al_2O_3 Composites," Georgia Tech M.S. Thesis, 2000.
3. Andrew Keller, Greg Kennedy, Louis Ferranti, Min Zhou, and Naresh Thadhani, "Correlation of Dynamic Behavior With Microstructural-Bias In Two-Phase $TiB_2+Al_2O_3$ Ceramics," to be published in Proc. of Fourth Int. Symp. on Impact Engineering, Japan, July 2001.
4. G. Gilde, J.W. Adams, M. Burkins, M. Motyka, P.J. Patel, E. Chin M. Sutaria, M. Rigali, and L. Prokurat Franks, "Processing of Aluminum oxide and Titanium Diboride Composites for Penetration Resistance," in Proc. of 25th Acers Cocoa Beach Conference, January 23-28, 2001.
5. J. Zhai and M. Zhou, "Finite Element Analysis of Micromechanical Failure Modes in Heterogeneous Brittle Solids", Int. J. of Fracture, special issue on *Failure Mode Transition in Solids*, R. C. Batra, Y. D. S. Rajapakse, and A. J. Rosakis, eds., **101**, pp. 161-180, 2000.
6. M. Zhou and J. Zhai, Micromechanical Characterization of the Fracture Resistance of A TiB_2/Al_2O_3 Ceramic Composite System, Acers Ann. Mtg. & Expo, St. Louis, MI, April 30, '01.
7. K.V. Logan, "Shaped Refractory Products and the Method of Making the Same," U.S. Patent, # 5,141,900, August 25, 1992.
8. L. Ferranti, C. Davis, G. Kennedy, and N.N. Thadhani, unpublished results.
9. D.E. Grady, Sandia National Laboratories Reports, SAND94-3266 (February 1995).

SHOCK COMPRESSION, ADIABATIC EXPANSION AND MULTI-PHASE EQUATION OF STATE OF CARBON

Konstantin V. Khishchenko[1], Vladimir E. Fortov[1], Igor' V. Lomonosov[2], Mikhail N. Pavlovskii[3], Gennadii V. Simakov[3], and Mikhail V. Zhernokletov[3]

[1]*Institute for High Energy Densities, RAS, Izhorskaya 13/19, Moscow 127412, Russia*
[2]*Institute for Chemical Physics Research, RAS, Chernogolovka 142432, Russia*
[3]*Russian Federal Nuclear Center – VNIIEF, Sarov 607190, Russia*

Abstract. The compressibility of graphite and diamond samples of various initial densities has been studied experimentally in shock waves at pressures in the range 5–74.5 GPa. The states of graphite samples have been investigated also under reflected shock loading up to 182 GPa and in adiabatic release waves at pressures down to 4.5 GPa. A semiempirical equation-of-state model, which takes into account the effects of polymorphs transformation and melting, is proposed. An equation of state is developed for the graphite, diamond, and liquid phases of carbon, and the critical analysis of calculated results in comparison with the newly acquired and available high-energy-density experimental data is made.

INTRODUCTION

The multi-phase equation of state (EOS) of carbon over wide range of pressures and temperatures is required for numerical simulation of hydrodynamic processes under extreme conditions of high energy densities [1–3].

In the present study we have obtained data on the compressibility of graphite and diamond samples of various initial densities in shock waves at pressures in the range 5–74.5 GPa. We also have measured the states of graphite samples under reflected shock loading up to 182 GPa and in adiabatic release waves at pressures down to 4.5 GPa. We propose the semiempirical EOS model, which takes into account the effects of polymorphs transformation and melting. EOS is constructed for the graphite, diamond, and liquid phases of carbon on the basis of the model developed. The critical analysis of calculated results in comparison with the newly acquired and available high-energy-density experimental data is made.

EXPERIMENTAL METHODS AND RESULTS

The samples we investigated experimentally consisted of synthetic diamond with initial densities ρ_{00} = 2.02, 1.789, 0.607, and 0.56 g/cm^3, and graphite with ρ_{00} = 1.87 g/cm^3. Shock waves were generated in the samples through copper, aluminum, or iron plates by steel projectiles accelerated by the detonation products of condensed explosives up to velocities of 5–6 km/s. By recording the velocity of the shock front U_s (with an error of ~ 1.5 %) in the samples, we were able to define the particle velocity U_p and pressure P [4] based on the known Hugoniot equation of the plate material and the preset parameters of the shock waves generated in them. Measurements were taken by an electric contact basis method; the signals from the detectors were displayed on a fast oscilloscope. The experimental data averaged over six to eight recordings are presented in Fig. 1.

The data obtained from the samples of graphite and diamond with densities ρ_{00} = 2.02–1.789 g/cm^3

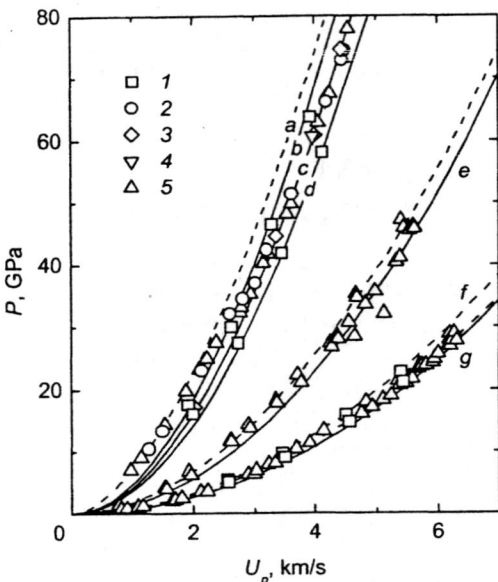

FIGURE 1. Shock Hugoniots of diamond and graphite samples with initial densities ρ_{00} = 1.87 (*a*, *c*), 2.02 (*b*), 1.789 (*d*), 1.011 (*e*), 0.607 (*f*), and 0.56 (*g*) g/cm^3. Solid lines correspond to results of calculation for diamond, dashed lines — for graphite. Experimental data: *1* and *2* — this work, *3* — [5], *4* — [7, 8], *5* — [9].

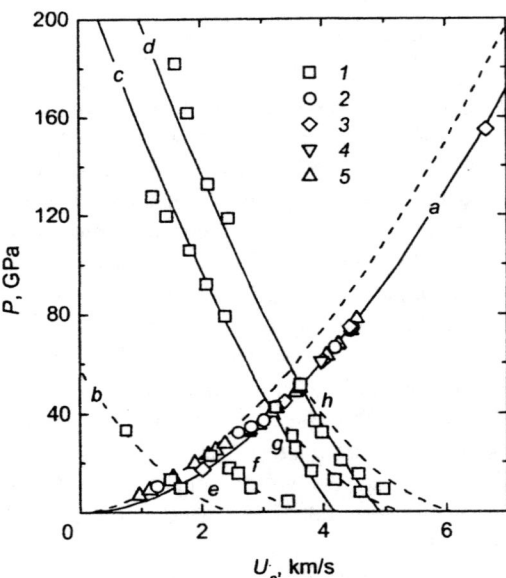

FIGURE 2. Shock Hugoniot (*a*), curves of second shock compression (*b–d*), and release isentropes (*e–h*) of graphite samples with initial density ρ_{00} = 1.87 g/cm^3. The notations are analogous to Fig. 1.

agree satisfactorily with the results of previous studies of the shock compression of porous natural diamond samples of ρ_{00} = 1.9 g/cm^3 [5] and graphite samples of ρ_{00} = 1.768–1.878 g/cm^3 (above the region of shock-induced graphite-to-diamond transformation) [6–9]. An analogous conclusion follows for the data obtained from the diamond samples of ρ_{00} = 0.56 g/cm^3 in comparison with the data for foamed carbon with the same initial density from Ref. [9].

In experiments where we studied reflected shock and adiabatic release waves, the measured quantity was shock velocity U_s in barriers with various dynamic impedances placed after the samples, allowing us to determine the parameters of the final states. A number of points on the second shock adiabats were obtained by reflection from barriers that were more rigid than graphite and with known Hugoniots (tungsten, tantalum, copper, zinc, and titanium). The region of decreased pressure was studied using soft dynamic barriers (polymethylmethacrylate, polyethylene, and foamed polystyrene with various initial densities), the shock Hugoniots of which are also known [10].

In order to monitor the parameters of the original states, in the majority of experiments on graphite samples with diameter 40 mm and thickness 2.5 mm we used both cylindrical tablets made of barrier materials with dimensions 12 and 3 mm respectively and tablets made from graphite itself. The determined values of pressure behind first shock front are $P \simeq$ 13.4, 23.1, 42.5, and 51.4 GPa. A detailed description of the methods and components used to perform the experiments is contained in Ref. [10]. The experimental data for second shock compression and adiabatic expansion of graphite averaged over five to seven independent recordings are presented in Fig. 2.

EOS MODEL

According to the EOS model, the Helmholtz free energy for matter is preassigned as a sum of three components

$$F(V,T) = F_c(V) + F_a(V,T) + F_e(V,T),$$

describing the elastic part of interaction at $T = 0$ K (F_c) and the thermal contribution by atoms (F_a) and electrons (F_e).

The selected form of F_c for the solid phases [11] provides for the correct values of the cohesive energy E_{coh} in the limit of $V \to \infty$, and the bulk compression modulus and its pressure derivative under normal conditions [12, 13], as well as agreement with results of calculation [14] using the corrected Thomas–Fermi model in the range $V_{0c}/V = 20$–500 (where V_{0c} is specific volume at $P = 0$, $T = 0$ K).

The lattice contribution to the free energy of the solid phases is defined by excitation of acoustic and optical modes of thermal vibrations of atoms

$$F_a(V,T) = F_a^{acst}(V,T) + \sum_{\alpha=1}^{3(v-1)} F_{a\alpha}^{opt}(V,T),$$

$$F_a^{acst}(V,T) = \frac{RT}{v}\left[3\ln\left(1 - e^{-\theta^{acst}/T}\right) - D(\theta^{acst}/T)\right] - \beta_{acst}\frac{T^2/\theta^{acst}}{e^{\theta^{acst}/T} - 1},$$

$$F_{a\alpha}^{opt}(V,T) = \frac{RT}{v}\ln\left(1 - e^{-\theta_\alpha^{opt}/T}\right) - \beta_{opt\alpha}\frac{T^2/\theta_\alpha^{opt}}{e^{\theta_\alpha^{opt}/T} - 1},$$

where R is the gas constant, v is the number of atoms in an elementary cell of the lattice, D is the Debye function [15], θ^{acst} and θ_α^{opt} are the characteristic temperatures of acoustic and optical modes of phonon spectrum [16]. The last terms in expressions for F_a^{acst} and $F_{a\alpha}^{opt}$ take into account the effects of anharmonicity of thermal lattice vibrations. These terms are exponentially small at low temperatures and provide for behavior of heat capacity $C_V - 3R \sim T^2$ at $T \to \infty$ [17]. The coefficients β_{acst} and $\beta_{opt\alpha}$ are found from high-temperature data for enthalpy of graphite under the normal pressure [18].

The electronic component of the free energy of the graphite phase is given by the expression

$$F_e(V,T) = -\frac{1}{2}\beta_0 T^2 \sigma^{-\gamma_0},$$

where β_0 is the coefficient of electronic heat capacity at $T = 0$ K [17], γ_0 is Gruneisen coefficient under normal conditions.

The electron contribution to the free energy of diamond, which is a dielectric with the energy gap between the valence and conduction bands $\Delta_0 \simeq 5.5$ eV [19], is negligible in comparison with F_a at temperatures $T \ll \Delta_0/2k$ (where k is Boltzman constant). Therefore $F_e = 0$ for the diamond phase.

The free energy F of liquid carbon has the form of thermodynamic potential of the liquid phase in the EOS model [1].

RESULTS OF CALCULATIONS

As follows from Fig. 1 and 2, the multi-phase EOS constructed for carbon adequately describe the experimental data obtained in this work and previously [5, 7–9] over the entire range of dynamic characteristics generated in shock-loading and adiabatic release waves.

The adequacy of the proposed form of contribution of thermal lattice vibrations to the thermodynamic potential is illustrated by Fig. 3, in which the calculated values of isobaric heat capacity of graphite and diamond under the atmospheric pressure are compared with experimental data [17, 20].

An analysis of the results of measurements of sound velocity in shock compressed graphite with $\rho_{00} = 2.2$ g/cm^3 [21] indicates that carbon is in the solid diamond phase at pressures in the range 80–143 GPa. On the calculated shock Hugoniot of the

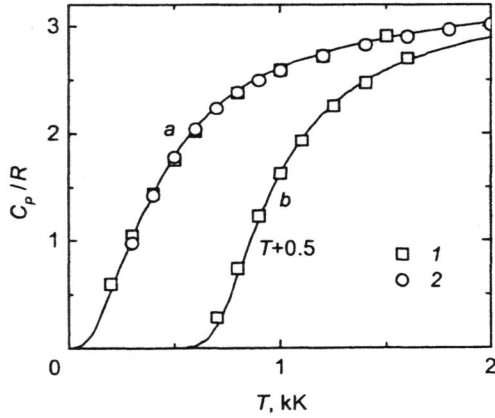

FIGURE 3. Isobaric heat capacity of graphite (*a*) and diamond (*b*) under the atmospheric pressure. Experimental data: *1* — [17], *2* — [20].

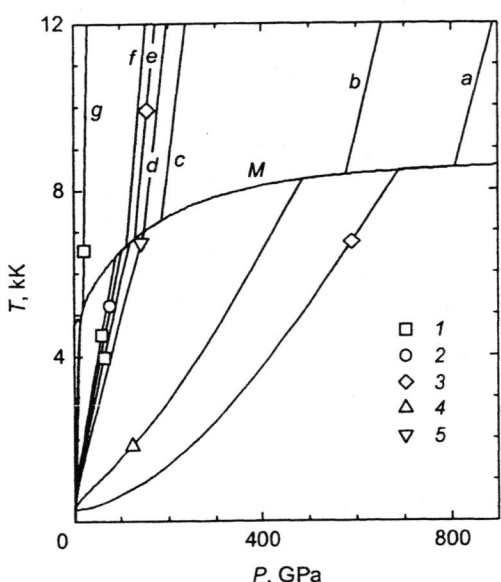

FIGURE 4. Phase diagram of carbon. M — diamond melting curve, a–g — shock Hugoniots of carbon samples with initial densities ρ_{00} = 3.51 (a), 3.191 (b), 2.2 (c), 2.02 (d), 1.87 (e), 1.789 (f), and 0.56 (g) g/cm^3. The level of pressures realized in experiments: 1 and 2 — this work, 3 — [5], 4 — [9], 5 — [21].

graphite with that initial density melting of diamond begins at $P \simeq 150$ GPa, and this value is consistent with experimental data [21].

The phase diagram of carbon calculated on the basis of the developed EOS is shown in Fig. 4. It reveals a region of states realized in the shock-wave experiments with traditional explosives systems [5, 6, 9] and a two-stage light-gas gun [21]. As can be seen, the data obtained in this work correspond to both solid and liquid phases of carbon.

Thus, the multi-phase EOS we have constructed for carbon describes consistently the collection of the newly acquired and available experimental data, and it can be employed effectively in numerical simulation of hydrodynamic processes at high energy densities.

ACKNOWLEDGMENTS

The work has been done due to financial support of the Russian Foundation for Basic Research, Grant 00-02-16324.

REFERENCES

1. Bushman, A. V., Fortov, V. E., Kanel', G. I., and Ni, A. L., *Intense Dynamic Loading of Condensed Matter*, Taylor & Francis, Washington, 1993.
2. Bushman, A. V., Vorob'ev, V. S., Korobenko, V. N., Rakhel, A. D., Savvatimskii, A. I., and Fortov, V. E., *Int. J. Thermophys.* **14**, 565–572 (1993).
3. Lomonosov, I. V., Fortov, V. E., Frolova, A. A., Khishchenko, K. V., Charakhch'yan, A. A., and Shurshalov, L. V., *Phys. Dokl.* **43**, 306–308 (1998).
4. Al'tshuler, L. V., *Sov. Phys. Usp.* **8**, 52 (1965).
5. Pavlovskii, M. N., *Sov. Phys. Solid State* **13**, 741–742 (1971).
6. Pavlovskii, M. N., and Drakin, V. P., *JETP Lett.* **4**, 116–118 (1966).
7. Trunin, R. F., Simakov, G. V., Moiseev, B. N., Popov, L. V., and Podurets, M. A., *Sov. Phys. JETP* **29**, 628–629 (1969).
8. Trunin, R. F., *Phys. Usp.* **37**, 1123–1146 (1994).
9. Marsh, S. P. (Ed.), *LASL Shock Hugoniot Data*, University of California Press, Berkeley, CA, 1980.
10. Al'tshuler, L. V., Bushman, A. V., Zhernokletov, M. V., Zubarev, V. N., Leont'yev, A. A., and Fortov, V. E., *Sov. Phys. JETP* **51**, 373–383 (1980).
11. Bushman, A. V., Lomonosov, I. V., and Fortov, V. E., *Sov. Tech. Rev. B. Therm. Phys.* **5**, 1–44 (1993).
12. Aleksandrov, I. V., Goncharov, A. F., Zisman, A. N., and Stishov, S. M., *Sov. Phys. JETP* **66**, 384–390 (1987).
13. Hanfland, M., Beister, H., and Syassen, K., *Phys. Rev. B* **39**, 12598–12603 (1989).
14. Kalitkin, N. N., and Kuz'mina, L. V., *Preprint IPM Akad. Nauk SSSR* No. 35, Moscow, 1975.
15. Landau, L. D., and Lifshits, E. M., *Statistical Physics*, Pergamon Press, Oxford, 1980.
16. Khishchenko, K. V., Lomonosov, I. V., and Fortov, V. E., *High Temp.–High Press.* **30**, 373–378 (1998).
17. Wunderlich, B., and Baur, H., *Fortschritte der Hochpolymeren Forshung* **7**, 151–368 (1970).
18. Buchnev, L. M., Smyslov, A. I., Dmitriev, I. A., Kuteinikov, A. F., and Kostikov, V. I., *Dokl. Akad. Nauk SSSR* **278**, 1109–1111 (1984).
19. Fahy, S., and Louie, S. G., *Phys. Rev. B* **36**, 3373–3385 (1987).
20. Bergman, G. A., Buchnev, L. M., Petrova, I. I., Senchenko, V. N., Fokin, L. R., Chekhovskoi, V. Ya., and Sheindlin, M. A., *Tables of Standard Reference Data. GSSSD 25–90*, Izdatel'stvo Standartov, Moscow, 1991.
21. Shaner, J. W., Brown, J. M., Swenson, C. A., and McQueen, R. G., *J. de Phys.* **45**, C8-235–C8-237 (1984).

THERMODYNAMIC PARAMETERS AND EQUATION OF STATE OF LOW-DENSITY SiO$_2$ AEROGEL

M.V.Zhernokletov[1], T.S.Lebedeva[1], A.B.Medvedev[1], M.A.Mochalov[1], A.N.Shuykin[1], V.E.Fortov[2]

1) *Russian Federal Nuclear Center – VNIIEF, Sarov, Russia, 607190*
2) *Institute of Extreme State Thermodynamics, Moscow, 127412*

Abstract. This paper studies properties of low-density SiO$_2$ aerogel of initial density ρ_o = 0.08 g/cm^3, 0.15 g/cm^3, and 0.19 g/cm^3 in shock compression up to ~ 13 GPa pressures in plane- and semispherical-geometry devices. Shock-wave velocities up to ~14 km/s, luminance temperatures up to ~20000 K, light absorptivity α ~ (1-4)10^3 cm^{-1} in a shock-compressed aerogel, and sound speed are measured with the optical method in visible spectrum (λ = 406, 498, 550, and 600 nm). Thermodynamic parameters of liquid-state aerogel are calculated by the equation of state using the modified van der Waals model for reactive mixtures.

INTRODUCTION

Silicon aerogels of a low initial density (0.008-0.36 g/cm^3) have been extensively used recently for producing and studying nonideal plasma at high local energy and temperature concentrations. Using porous samples extends the material phase diagram range accessible to dynamic experiments. Principal experimental data for highly porous samples has been obtained with dynamic methods through material compression and irreversible heating at the front of powerful shock waves generated by condensed and nuclear explosive detonation [1-7]. Refs. [5-6] measure luminance temperatures in addition to aerogel compressibility.

This paper measures thermodynamic parameters of shock-compressed aerogel using a pyrometer of visible spectrum (400...600 nm). A wide experimental data set is obtained with the optical method and used to test the equation of state of low-density SiO$_2$ aerogel.

MEASUREMENT METHOD AND EXPERIMENTAL RESULTS.

Samples of initial density 0.08, 0.15, and 0.19 g/cm^3 fabricated by Novosibirsk Catalysis Institute were used. The experimentally measured spectral transmission of aerogel in the 400...600nm range is 60...80%, which allowed using the optical method to measure cinematic and thermodynamic parameters of S$_i$O$_2$ aerogel during its shock compression.

The experimental scheme for simultaneous measurement of shock wave velocity and luminance temperatures at the front in the plane-wave experiments is presented in Fig. 1.

The ~3-mm-thick SiO$_2$ aerogel sample to be studied was fastened in a cell evacuated down to residual pressure no worse than 10^{-1} mm Hg and covered with a sapphire substrate 2 mm thick. All the measurements were performed with 4-channel photoelectronic pyrometer [8], whose design appears in Fig.1. The shock front emission in the aerogel was formed to a parallel light beam through diaphragm $\varnothing$ 10 mm in the chamber casing and external reflector (1) by objectives (2) and (4), then the beam was directed to pyrometer (5).

Glass plates (11-14) distribute emission among four photoelectric multipliers. To separate spectral ranges, interference filters of ~50% transmission at wavelengths λ = 406 nm, 498 nm, 550 nm, and 600 nm at 10 nm half-height bandwidth were used.

Typical oscillograms of radiation luminance buildup at the shock front in aerogel are presented in Fig. 2.

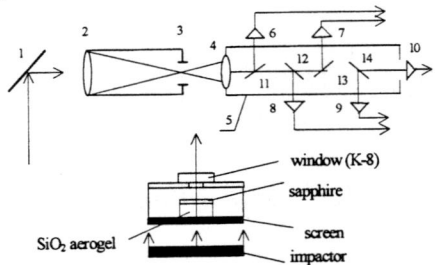

1- reflector; 2, 4 - objectives; 3 - diaphragm; 5 – pyrometer casing; 6, 7, 8, 9, 10 photoelectric multiplyer ; 11,12,13,14 deflection plates.
FIGURE 1. Experimental scheme and measuring cell design

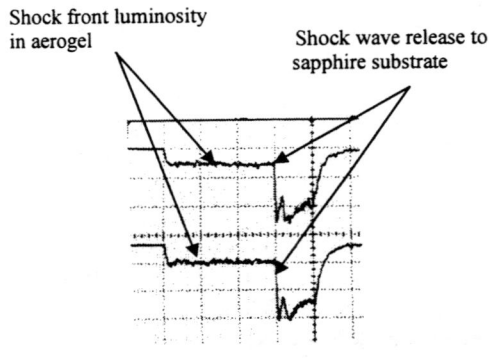

time 200ns/div
upper beam - $\lambda = 406$nm, lower beam - $\lambda = 498$nm

FIGURE 2. Oscillograms of shock front emission in SiO$_2$ aerogel and sapphire substrate

The oscillogram provides the strict time of the shock wave arrival at aerogel and an abrupt signal burst at the time of the shock wave arrival at the aerogel-sapphire interface, which allows the shock wave velocity to be measured. Material mass velocity behind the shock front was calculated through intersection of the wave beam with unloading isentropes of screen material (aluminum) that had been computed by the equation of state with taking into account melting [9,10].

Compressibility and luminance temperatures up to 13GPa were measured with semi-spherical geometry device MZ-4 [11].

The results of this paper along with the data on shock compressibility and luminance temperatures of aerogel with initial density 0.008, 0.27, and 0.36 g/cm^3 from refs. [5-6] are given in Fig. 3 and Fig. 4.

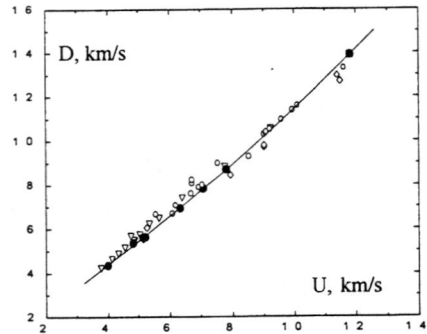

◊ - 0.008, ■ - 0.08, ● - 0.15, ▼ - 0.19, ○ - 0.27, ∇ - 0.36 g/cm^3
— fitted data from this paper:
$D(U) = 0.556 + 0.868 U + 0.022 U^2$
FIGURE 3. Low-density aerogel Hugoniot

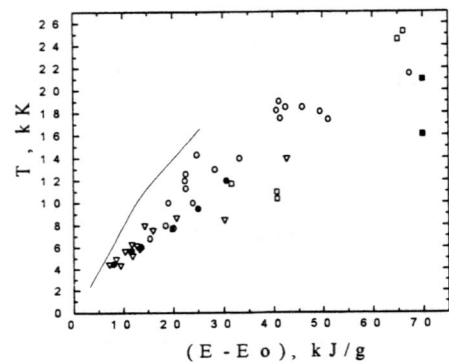

□ – 0.008, ■ – 0.08, ● - 0.15, ▼ - 0.19, ○ - 0.27, ∇ - 0.36 g/cm^3
—— - calculation by equation of state for $\rho_{oo}= 0.15$ g/cm^3;
FIGURE 4. Aerogel luminance temperatures vs. shock compression energy

The measurement accuracy is 0.5% for shock-wave velocity, ~ 1% for mass velocity, ~ 5% for luminance temperatures.

The measured luminance temperatures in the $2 \leq P \leq 13$ GPa pressure range were obtained by comparison between the front emission amplitude and the reference source emission.

Sound speed in the shock-compressed aerogel was measured with the "overtaking" method [8]. The experiment used a sample of $\rho_{oo} = 0.15$ g/cm^3 and thickness increased up to ~ 7 mm, so that the rarefaction wave on the side of the impactor definitely overtake the shock wave in the sample. A typical oscillogram for the emission in aerogel at shock compression pressure P = 6.59 GPa at wavelength λ = 406 nm appears in Fig. 5.

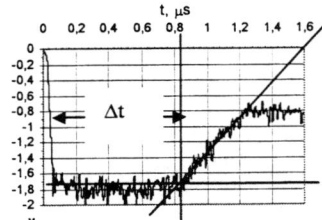

FIGURE 5. Oscillogram of shock front luminosity

The calculated data for motion of the shock wave (SW) and rarefaction waves (RW) in this device and experimentally measured time Δt = (799 ± 4) ns were used to estimate sound speed in shock-compressed aerogel as C_s = 3.8 km/s.

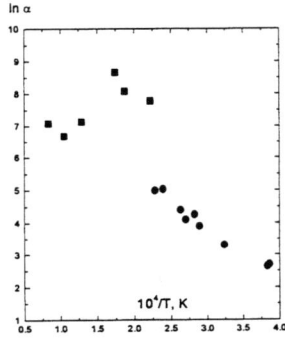

■ - aerogel (ρ_{00} = 0.15 g/cm^3),
● - silica glass (ρ_{00} = 2.205 g/cm^3)

FIGURE 6. Light absorptivity vs. temperature for aerogel and silica glass shock compression.

Light absorptivity was measured by recording radiation luminance buildup with time during the shock wave propagation through aerogel, which was related to the increase in thickness of the material layer compressed by the shock wave and its transparency [12]. The obtained results averaged for the visible spectrum (400-600nm) are presented in Fig.6 along with data for silica glass from ref. [13].

EQUATION OF STATE OF SiO$_2$. COMPARISON WITH THE EXPERIMENT

Thermodynamic properties of liquid and gaseous silicon dioxide at high pressures and temperatures were calculated including vaporization, dissociation, and ionization of constituents by the modified model of van der Waals equation of state.

Previously, the model had been successfully employed for the description of experimental data for various materials, both individual and mixed, in a wide range of states [9,10]. This is a covolume model. Variability of the covolume (intrinsic particle volume) reflects compressibility. Mixture composition is found from the condition of minimum free energy. At high temperatures, Saha equations are solved.

The calculations included the following molecules, atoms, and ions: SiO_2, SiO, Si, Si_2, O, O_2, $Si+$, $Si++$, $O+$, $O++$, etc., and electrons. When selecting the covolumes of Si and O_2, experimental data on shock compressibility of these constituents were taken into consideration. It was assumed that covolume of Si_2 was equal to that of Si, covolume of O was the same as in O_2. When selecting the covolume of molecular SiO_2 and attraction, the data on silica glass density, isothermal compressibility, and sublimation energy were used. The covolume of SiO was considered as adjustment covolume because of missing data for behavior of this constituent at high pressures. Covolume considered to be not varied during ionization. Only were considered main electronic states of particles. We treat the molecules using rigid rotator-harmonic oscillator approximation. Ionization potential was assumed constant.

The description of the experimental data by the model under discussion at relatively low pressures is presented in Fig.7.

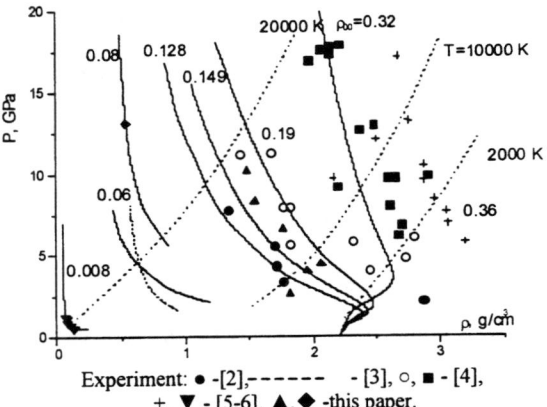

Experiment: ● -[2], ----- - [3], ○, ■ - [4], +, ▼ - [5-6], ▲, ♦ -this paper.
Initial densities of the samples are specified near the computed curves. T=2000K, 10000K, and 20000K are isotherms.
FIGURE 7. Porous quartz Hugoniots.

One can see a satisfactory agreement between the experimental and calculated data. An exception are results from refs. [6] at initial density 0.36g/cm^3. Fig. 8 depicts porous quartz Hugoniots in the pressure range up to 100GPa, and Fig. 9 gives those in the pressure range up to 3TPa. The interpretation of the results at high pressures depends on the equation of state of reference material. A displacement in some data [1, 7] when using the equation of state of aluminum from [9,10] is indicated in Fig. 9 with arrows.

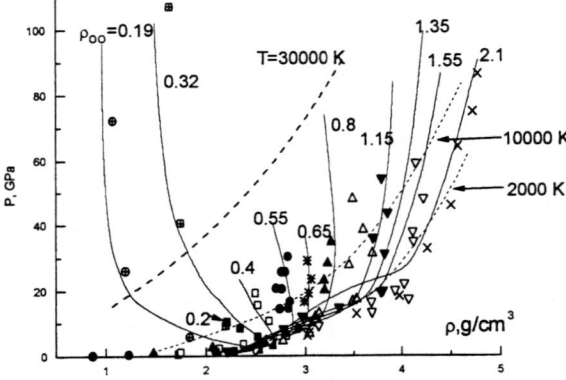

Experimental data for ρ = 0.19 and 0.32 g/cm^3 from ref. [4]. The other data is from [1]. The notations are the same as in Fig. 7

FIGURE 8. Porous quartz Hugoniots up to 100GPa pressure

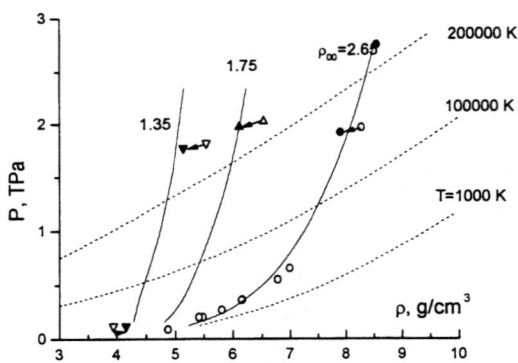

Experimental data from refs. [1,7]. The notations are the same as in Fig.7

FIGURE 9. Continuous and porous quartz Hugoniots up to 3GPa pressure

The calculated aerogel temperature as a function of shock compression energy appears in Fig.7. The measured values are seen to lie lower, than they should according to the computation. A reason for this may be equilibrium radiation shielding because of a high light absorptivity at the shock front.

This model of the equation of state of aerogel does not reproduce the temperature "shelves" depending on shock compression energy noted in Refs. [5-6].

The calculated sound speed at P = 6.59 GPa agrees with the experimental value within the measurement error (10-15)%.

REFERENCES

1. Simakov G.V., Trunin R.F. Izv. AN SSSR. Fizika Zemli, 1990, No.11, p.72
2. Holmes N.C., See E.F. In: Shock Compression of Condensed Matter -1991, pp.91-94. 1992 Elsevier Science Publishers B.V.
3. Holmes N.C. In:High-Pressure Science and Technology- 1993, pp.153-156. 1994 American Institute of Physics
4. Vildanov V.G., Gorshkov M.M. et al. In: Shock Compression of Condensed Matter- 1995, pp.121-124. American Institute of Physics. NY,1996
5. Gryaznov V.K., Nikolayev D.N., Ternovoy V.Ya., Fortov V.E. et al. Khimicheskaya Fizika, V. 17, No.2, pp. 33-37.
6. Nikolaev D.N., Fortov V.E., Filimonow A.S., Kvitov S.V., Ternovoi V.Ya. In: Shock Compression of Condensed Matter – 1999, pp. 121-124. American Institute of Physics, NY, 2000.
7. Trunin R.F., Podurets M.A., Simakov F.V. et al. ZhETF, 1995, V.108, No.3(9), pp.851-861.
8. Kormer S.B., Sinitsyn M.V., Kirillov G.A., Urlin V.D. ZhETF, 1965, V. 48, No. 4, pp. 1033-1048.
9. Medvedev A.B. Voprosy Atomnoj Nauki i Tekhniki. Ser. Teor. i Prikl. Fizika. 1992, No.1, pp. 23-29.
10. Kopyshev V.P., Medvedev A.B. Sov. Tech. Rev. B. Therm. Phys. Rev. Vol. 5. 1993. pp.37-93
11. Altshuler L.V., Trunin R.F., Krupnikov K.K., Panov N.V. UFN, 1996, V.166, No.5, pp.575-581.
12. S.B.Kormer. UFN, 1968, V. 94, p. 641.
13. Sugiura H, Kondo K, Sawaoka. J.Appl.Phys., 1982, Vol.53, No.6, pp.4512-4514.

THE HUGONIOT ELASTIC LIMIT OF ALON

James U. Cazamias[1], Peter S. Fiske[2], Stephan J. Bless[3]

[1] LLNL, L-414, PO Box 808, Livermore, CA 94551
[2] RAPT Industries, 581 Melanie Ave., Livermore CA 94550
[3] IAT, 3925 W. Braker Ln., Suite 400, Austin, TX 78759-5316

Abstract. We performed plate impact experiments on the transparent polycrystalline ceramic aluminum oxynitride (AlON, transparent alumina). From VISAR measurements, the Hugoniot Elastic Limit, σ_{HEL}, is 11.0-11.4 GPa with a corresponding yield strength of 7.2-7.5 GPa. A transverse gauge gives a yield strength of 8.7 GPa for a longitudinal stress of 13.9 GPa, which implies that the failed AlON possesses at least its σ_{HEL} strength.

INTRODUCTION

The modeling of transparent ceramics relies heavily on descriptions of failure under compressive loads which are generally derived from experience with opaque ceramics. However, the microstructural properties that are responsible for transparency also imply that sites for stress concentrations are much reduced when a medium is transparent. It is this increased homogeneity of transparent materials that is responsible for characteristic differences in shock response of transparent and opaque materials. For example, transparent materials such as glass and sapphire exhibit much higher initial spall strengths than polycrystalline ceramics.

Recently, there has been increasing interest in aluminum oxynitride (AlON, transparent alumina). AlON has spinel crystal structure (i.e., cubic symmetry) which results in isotropic optical and mechanical properties. When the ceramic is 100% dense with no voids, inclusion or grain boundary phases, it is transparent.

We have recently performed bar impact experiments which found that transparent AlON exhibited both alumina-like and glass-like behavior with a bar yield strength of 4.0 GPa [1]. We have also performed plate impact experiments which examined the σ_{HEL}, spall, and elastic properties of the material [2]. The densities for the HEL experiments in [2] were incorrect, and the corrected results are presented here.

EXPERIMENTS

25.4-mm diameter disks of AlON were subjected to planar impacts using a He single-stage gas gun at LLNL. See Table 1. The diagnostics for Shots 710 and 711 consisted of a VISAR with a fringe constant of 541 m/s/fringe. For Shot 725 (a repeat of 711), a transverse gauge (MicroMeasurements C-880113-B pulsed by a Dynasen CK1-50-300 piezoresistive power supply) was used to measure transverse stress in the sample. Impact velocities were measured using 2 pairs of piezoelectric pins.

DATA

Figure 1 shows the velocity-time histories for Shots 710 and 711. Table 2 lists derived values from the experiments.

TABLE 1. Experimental Parameters

	710	711	725
Impact Velocity (m/s)	790	684	680
Impactor	Cu	Cu	Cu
Impactor Thickness (mm)	5.08	5.08	5.07
ρ_{target} (gm/cm^3)	3.68	3.68	3.68
Target Thickness (mm)	5.11	5.11	10
c_l (km/s)	10.3	10.3	10.3
c_s (km/s)	5.91	5.91	5.91
c_o (km/s)	7.71	7.71	7.71

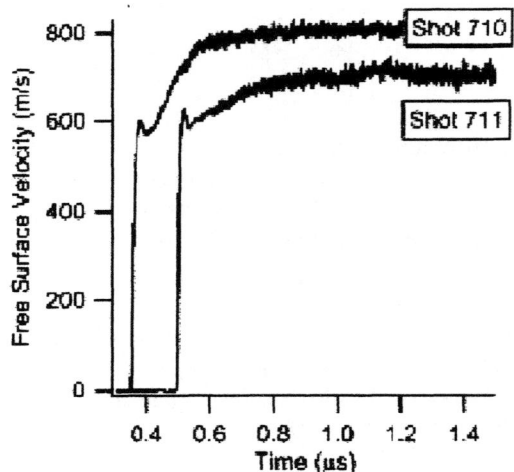

FIGURE 1. Velocity-time histories for Shot 710 and Shot 711.

TABLE 2. Experimental Measurement

	710	711
u_{HEL} (m/s)	290	301.5
σ_{HEL} (GPA)	11.0	11.4
Y (GPa)	7.2	7.5
c_{in} (km/s)	8.51	8.45
u_{in} (m/s)	428	378
σ_{in} (GPa)	15.4	13.9

Letting u_{HEL} equal one-half the velocity of the plateau after the elastic wave, we can calculate $\sigma_{HEL} = \rho_0 c_l u_{HEL}$ (11.0 GPa for Shot 710 and 11.4 GPa for Shot 711) with a corresponding yield strength of $Y = 2(c_s/c_l)^2 \sigma_{HEL}$ (7.2 GPa for Shot 710 and 7.5 GPa for Shot 711).

For further analysis, we assume that the inelastic wave is steady. With $\rho_0/\rho = 1 - u_{HEL}/c_l$ (conservation of mass), L as the plate thickness, and Δt as the difference in rear surface arrival times of the initial elastic wave and its first reflection from the inelastic wave front, the inelastic wave velocity, c_{in}, can be expressed as

$$c_{in} = c_l \frac{\rho_0}{\rho} \frac{1 - \dfrac{c_l \Delta t}{(1 + \rho_0/\rho)L}}{1 + \dfrac{c_l \Delta t}{(1 + \rho_0/\rho)L}}. \quad (1)$$

Since the magnitude of the inelastic wave is small compared to the σ_{HEL}, the inelastic wave front becomes elastic after interacting with the initial elastic release wave, which leaves a contact discontinuity at the release point. There is a step structure in the inelastic rise (Fig. 2) which is associated with this contact discontinuity, although some of the temporal thickness of the inelastic signal is associated with the thickness of the inelastic wave itself. The noise in the VISAR signal might be due to the fact that the large grain size of AlON (> 100 μm [3]) allows twinning in the grains. The subject of contact discontinuities influencing free surface velocities is discussed in more detail in [4], where it is shown that the late time free surface velocity, u_f, can be expressed as

$$u_f = 2u_{in} - \frac{\dfrac{\rho_0}{\rho} c_l - c_{in}}{c_{in}} u_{HEL} \quad (2)$$

where u_{in} is the particle velocity behind the inelastic wave. The stress behind the inelastic wave is then $\sigma_{in} = \rho_0 c_l u_{HEL} + \rho c_{in}(u_{in} - u_{HEL})$, giving peak stresses of 15.4 GPa for Shot 710 and 13.9 GPa for Shot 711.

Figure 3 shows the change in manganin gauge resistance, $\Delta R/R_0$, versus time for Shot 725. The gauge was nominally placed in the center of the target. The region of interest is the plateau which represents the stress behind the inelastic wave; here $\Delta R/R_0 = 0.174$, giving a transverse stress of $\sigma_t = 5.2 \pm 0.3$ GPa (usings Rosenberg's [5-7] analysis) with a corresponding yield strength of $Y = \sigma_l - \sigma_t = 8.7$ GPa $\pm$ 0.3 GPa. Increase of the flow stress above the σ_{HEL} is not unexpected since alumina also exhibits this behavior [8, 9].

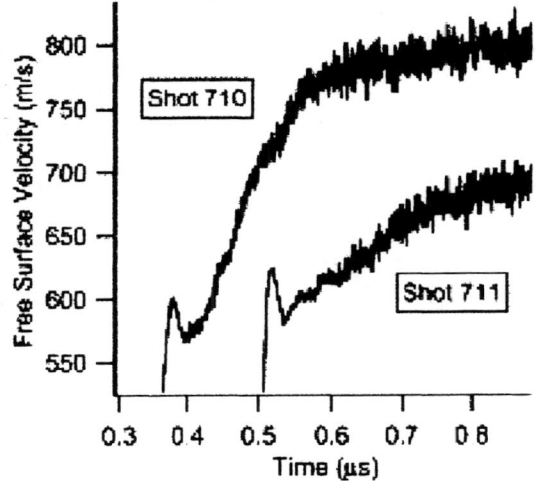

FIGURE 2. Expanded view of velocity-time histories.

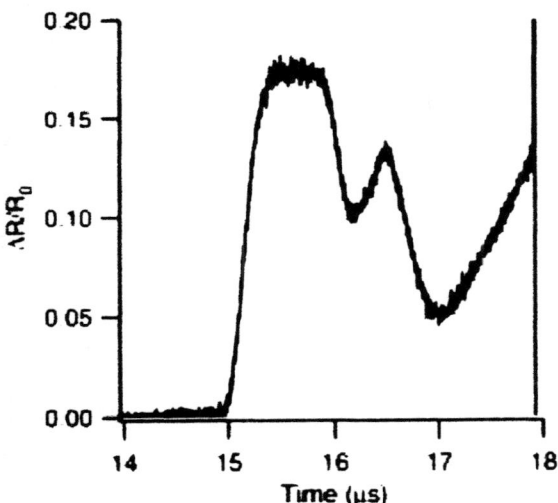

FIGURE 3. $\Delta R/R_0$-time history for Shot 725.

DISCUSSION

We do not have a good explanation for the dynamic overshoot of the elastic wave. It might be a rate effect. Dynamic overshoot is observed in B_4C, which also has extremely slow plastic waves. These behaviors are attributed to a dramatic loss of strength above the σ_{HEL} [10]. We have shown that AlON does not collapse to the hydrostat, so this is not an explanation for the overshoot behavior.

AlON's σ_{HEL} of 11.0 to 11.4 GPa compares well with those of other brittle materials (6.7 GPa for AD995 [11], 14-20 GPa for z-cut sapphire [12], 12 GPA for SiC [13], 4.2-5.8 GPa [1st σ_{HEL}] and 9-17 GPa [2nd σ_{HEL}] for TiB_2 [14], and 6 GPa for soda lime glass [9]). The strength under shock loading (7.2-7.5 GPa) is significantly higher than under bar loading (4 GPa [1]), presumably due to the suppression of microcracks by pressure.

Assuming $c_{in} = c_o + s\, u_{in}$ gives s on the order of 1.9-2.0. This agrees with sapphire (s = 1.95) [15], but is greater than polycrystalline alumina (s = 1.3) [8]. At low pressures, $(dK/dP)_T \sim 4.4$ [16]. Since $s = 0.25((dK/dP)_{0,S} + 1)$, approximating $(dK/dP)_{0,S}$ by $(dK/dP)_T$ gives s = 1.35, implying that the hydrostatic behavior of AlON should resemble that of polycrystalline alumina. Our measurements occur near the σ_{HEL} and may be affected by changes in the strength of the material [17]. Consequently, this large value of s is another indication that the strength is actually increasing.

CONCLUSION

We performed plate impact experiments on AlON using a VISAR and transverse manganin gauges. The Hugoniot Elastic Limit, σ_{HEL}, is 11.0-11.4 GPa with a corresponding yield strength of 7.2-7.5 GPa. A transverse gauge gives a yield strength of 8.7 GPa for a longitudinal stress of 13.9 GPa, which implies that the failed AlON possesses at least its σ_{HEL} strength.

ACKNOWLEDGMENTS

This work was performed under the auspices of the DOE by LLNL under contract number W-7405-ENG-48 and ARL by IAT under contract DAAA21-93-C-0101. Thanks to T. Hartnett of Raytheon Corporation for supplying samples. Thanks to C. H. M. Simha for assembling the transverse gauge.

REFERENCES

1. Cazamias, J. U., and Bless, S. J., "Bar Impact Tests on Transparent Materials," in *18th International Symposium on Ballistics*, 1999, pp. 724-730.
2. Cazamias, J. U., Fiske, P. S., and Bless, S. J., "Shock Properties of AlON," *Fundamental*

Issues and Applications of Shock-Wave and High-Strain-Rate Phenomena, 181-188 (2001).
3. Hartnett, T. M., and Gentilman, R. L., "Optical and Mechanical Properties of Highly Transparent Spinel and AlON Domes," *Advances in Optical Materials*, 15-22 (1984).
4. Grady, D. E., "Steady-Wave Risetime and Spall Measurements on Uranium (3 - 15 GPa)," *Metallugical Applications of Shock-Wave and High Strain Rate Phenomena*, 763-780 (1986).
5. Rosenberg, Z., and Partom, Y., *J. Appl. Phys.* **58** (8), 3072-3076 (1985).
6. Rosenberg, Z., and Brar, N. S., *J. Appl. Phys.* **77** (4), 1443-1448 (1995).
7. Rosenberg, Z., and Partom, Y., *J. Appl. Phys.* 57 (11), 5084-5086 (1985).
8. Simha, C. H. M., *High Rate Loading of a High Purity Ceramic - One Dimensional Stress Experiments and Constitutive Modeling*, Dissertation, The University of Texas at Austin (1998).
9. Bourne, N., Millett, J., Rosenberg, Z., and Murray, N., *J. Mech. Phys. Solids.* **46** (10), 1887-1908 (1998).
10. Grady, D. E., "Dynamic Properties of Ceramic Materials," Sandia National Lababoratories Report No. Sand94-3266 (1995).
11. Dandekar, D. P., and Bartkowski, P., "Shock Response of AD995 Alumina," *High-Pressure Science and Technology - 1993*, 1994, pp. 733-736.
12. Kanel, G. I., Razorenov, S. V., Utkin, A. V., Baumung, K., Karow, H. U., and Licht, V., "Spallations Near the Ultimate Strength of Solids," *High-Pressure Science and Technology - 1993*, 1043-1046 (1994).
13. Bartkowski, P., and Dandekar, D. P., "Spall Strength of Sintered and Hot Pressed Silicon Carbide," *Shock Compression of Condensed Matter - 1995*, 535-538 (1996).
14. Ewart, L., and Dandekar, D. P., "Relationship Between the Shock Response and Microstructural Features of Titanium Diboride (TiB_2)," *High-Pressure Science and Technology - 1993*, 1201-1204 (1994).
15. Mashimo, T., Hanaoka, Y., and Nagayama, K., *J. Appl. Phys.* **63** (2), 327-336 (1988).
16. Graham, E. K., Munly, W. C., McCauley, J. W., and Corbin, N. D., *J. Am. Ceram. Soc.* 71 (10), 807-812 (1988).
17. Carter, W. J., Marsh, S. P., Fritz, J. N., and McQueen, R. G., "The Equation of State of Selected Materials for High-Pressure References," *Accurate Characterization of the High Pressure Enviroment*, 147-158 (1968).

THE FAILURE OF ALUMINIUM NITRIDE UNDER SHOCK

I.M. Pickup, N.K. Bourne*

*Royal Military College of Science, Cranfield University, Shrivenham, Swindon, SN6 8LA, UK.

Defence Science and Technology Laboratory, Chobham Lane, Chertsey, Surrey, KT16 OEE, UK

Abstract. The shear strength of aluminium nitride has been measured over a range of impact stresses by measuring lateral stresses in plate impact experiments. The range of impact stress spanned several key shock thresholds for the material, pre and post Hugoniot elastic limit and up to values where the hexagonal to cubic phase transition starts. The shear strength measurements indicate significant inelastic damage at stress levels in excess of the HEL, but a significant recovery of strength at the highest impact stress was observed. This stress equates to the phase transition stress. The shear strength behaviour is compared to that of silicon carbide, which does not exhibit a phase change at these impact velocities.

INTRODUCTION

Over the last 10 years aluminium nitride (AlN) has been the subject of a significant number of shock and ballistic studies. It has some interesting features in both fields. The material has been the subject of ballistic trials over a wide range of kinetic energy projectile velocities [1-2] (up to 5000 ms^{-1}). It has been noted that at lower velocities (1300 ms^{-1}) the ballistic penetration resistance of AlN is significantly less than that of other non-oxide ceramics, e.g. B$_4$C, SiC and TiB$_2$. At higher velocities (~2500 m s^{-1}) it is apparently significantly greater. In shock studies several investigations have reported a wurtzite (hexagonal) to rock-salt (cubic) phase transformation initiating at pressures from 16 to 24 GPa depending on material, particularly grain-size, and measurement technique [3-6]. The work presented here describes the initial experiments in a programme which compares the deviatoric strength of AlN and SiC from relatively low impact stresses to stresses up to the phase transition in AlN with a view to determining the relative shear strength behaviour and correlating this with ballistic performance. The shear strength in the shocked ceramic was determined by embedding manganin gauges in plate impact specimens to monitor the lateral stresses.

EXPERIMENTAL

The impact experiments were conducted using a 50 mm diameter gas gun. Impact velocity was measured to an accuracy of 0.5% using a sequential pin-shorting method and tilt was fixed to be less than 1 mrad by means of an adjustable specimen mount. Impactor plates were made from lapped copper and aluminium discs and were mounted onto a polycarbonate sabot with a relieved front surface in order that the rear of the flyer plate remained unconfined. Targets were flat to less than 2 μm across the surface. Lateral stresses were measured using manganin stress gauges of type J2M-SS-580SF-025 (resistance 25 Ω). The gauges were placed at varying distances from the impact face as shown in Fig. 1. They had an active width of 240 μm and which contrasts with the 2 mm wide gauges used by Rosenberg [7].

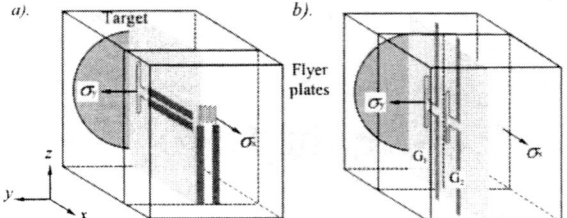

FIGURE 1. Experimental arrangement for lateral and longitudinal stress experiments. a) Longitudinal and lateral stress gauge mounting positions with rear PMMA plate. b). Sectioning for multiple lateral gauge measurements.

The lateral stress, σ_y, was used along with the longitudinal stress, σ_x, to calculate the shear strength of the material, τ, using the well-known relation

$$\tau = \frac{1}{2}(\sigma_x - \sigma_y). \quad (1)$$

This quantity has been shown to be a good indicator of the ballistic performance of the material [8]. This method of determining the shear strength has the advantage over previous calculations of being direct since no computation of the hydrostat is required.

The materials used in this study were manufactured by Cercom Inc. A linear intercept method was used to measure grain size, yielding values of 4±3 and 2±2 μm for AlN and SiC B respectively. The density was measured by a water immersion technique to be 3.19 and 3.22 g cm^{-3} for AlN and SiC B respectively.

TABLE 1. PROPERTIES OF ALUMINIUM NITRIDE

	ρ g cm^{-3}	c_L mm μs^{-1}	c_S mm μs^{-1}	ν
AlN This work	3.19	10.46	6.14	0.24
AlN Rosenberg (7)	3.23	10.72	6.27	0.24
SiC B	3.22	12.26	7.78	0.16

RESULTS

Lateral stress histories are shown in Fig. 2 for three impacts upon AlN with a 10 mm thick aluminium flyer travelling at 556 m s^{-1}, a 5 mm thick copper flyer travelling at 833 m s^{-1} and a 5 mm tungsten flyer travelling at 975 m s^{-1}. These impacts induced stresses of magnitude 5.6, 14.0 and 19.5 GPa respectively. The magnitude of the HEL for the material is around 9.5 GPa [6] whilst the phase transformation has been reported to initiate at ~19.5 to 20 GPa for AlN with a similar grain size.

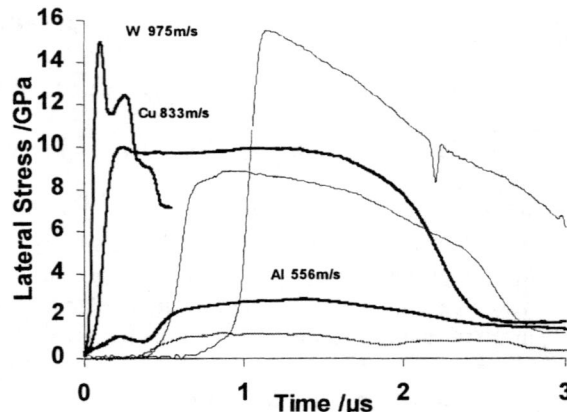

FIGURE 2. Impacts of Al alloy, copper and tungsten alloy flyers onto AlN. Lateral gauges at 2 mm (bold lines) and 6 mm (fine lines) from the impact surface.

The traces below the HEL show features typical of polycrystalline materials such as alumina. The stress rises to a plateau at the first gauge and after 400 ns rises again to a second higher lateral stress. This second rise is interpreted as the arrival of a failure wave behind which the strength reduces [8]. The gauge at 6 mm does not rise above the first level, implying that a failure wave does not penetrate to this thickness during this loading time. The trace at an intermediate stress between HEL and phase transition does not show the stepped rise described above. It may be that at this impact stress the failure wave may travel with the shock as has been seen in lead-filled glasses described elsewhere [9, 10]. The 6 mm gauge shows a lower lateral stress than that at 2 mm which may indicate that the failure does not penetrate equally from the impact face through the thickness of the tile.

The traces obtained above the phase transformation show an entirely different behaviour. Whereas the other sensors observe the lateral stress rising in a step from a lower to a higher value, these traces show the opposite behaviour. The lateral stress falls from a high to a lower value indicating a strengthening of the material. It is acknowledged that at the highest stress the manganin gauges are approaching an upper performance limit, prior to the electrical breakdown of the polymer gauge encapsulation at stresses greater than 24 GPa. Such a breakdown is mediated by the use of thicker polymer films and choice of polymers.

Lateral traces for SiC B [11], impacted at similar stresses to the 19.5 GPa AlN are shown in Fig. 3. SiC does not undergo a phase change in this stress regime and in contrast to AlN behaves as described above with the two stage trace indicating strength loss behind a failure wave. The initial and final value of the stress are used to calculate shear strengths which are nominated ahead and behind respectively in Fig. 4.

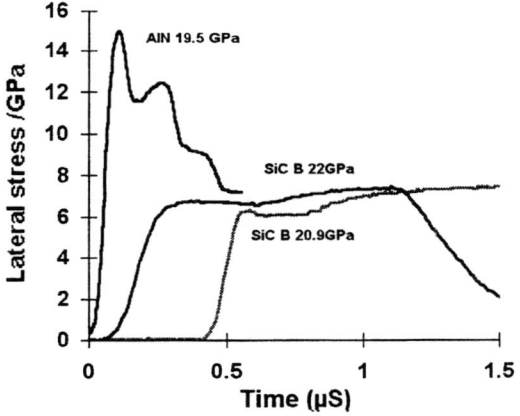

FIGURE 3. Contrasting lateral stress histories of SiC B and AlN impacted at ~20 GPa longitudinal stress (σ_x is indicated by each curve).

The calculated shear strength (using equation 1) is shown in Fig. 4. The rising diagonal lines represent the elastic values calculated using the longitudinal stress and the Poisson's ratio (solid line for SiC B and dotted for AlN). The Phase transformation initiation stress for AlN is indicated as a vertical dotted line. Two points are plotted for each experiment. The first corresponds to the value ahead of the failure front (filled symbol) and the second to that behind it (open symbol). The crosses show the data of Rosenberg [7] using gauges of greater width and a different material. His work did not exceed the phase transformation and the thickness of the sensors precluded resolution of the two components to the failure wave, but nevertheless the strength up to the phase transformation was mapped thoroughly for his material.

For the SiC material the shear strength follows the elastic line up to stresses approaching it's HEL (~ 15 GPa). Beyond this, the shear strength deviates from the elastic behaviour indicating both instantaneous damage and a further small reduction in strength behind the failure wave.

For AlN the shear strength deviates from the elastic line at approximately the HEL value (9.5 GPa) indicating significant instantaneous damage, with the initial stages of the lateral stress history suggesting inelastic behaviour. The shear strength of the AlN in the current study levels off at ~ 2 GPa, slightly lower than that of Rosenberg [7]. For the AlN specimen impacted at 19.5 GPa longitudinal stress, due to the strong reduction in the lateral stress observed in the stress history in Figs. 2 and 3, there is an apparently significant strengthening effect. The shear strength of the AlN increases to a value similar to the shear strength of SiC B rising from 3.25 GPa to 7.25 GPa. It is significant that the transition in behaviour occurs at the phase transition onset stress.

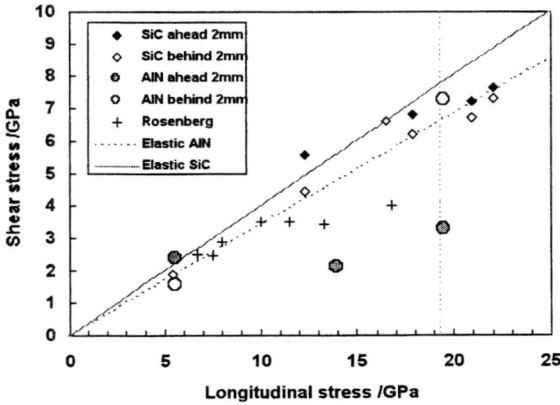

FIGURE 4. The shear strength of AlN and SiC plotted against longitudinal impact stress. The diagonal lines are the calculated strength based on elastic properties (solid for SiC and dotted for AlN).

At present neither the reasons for the strength increase nor the significance of the structure on the lateral AlN trace for the 2 mm gauge (impact stress 19.5 GPa) is understood. The phase transition which occurs at this level of stress has an associated (non-recoverable) volumetric compression of ~20% [5]. It is therefore unlikely that the apparent strength increase results from a bulking effect increasing pressure and consequently effective strength. Mashimo [6] has measured a significant increase (nearly 50%) in the bulk modulus of the high pressure, cubic phase. This may significantly affect the measured lateral stresses.

CONCLUSIONS

The deviatoric strength of AlN has been measured in shock studies from sub-HEL levels to stresses where the hexagonal to cubic phase change initiates. At the lowest stresses, failure wave characteristics were observed. At impact stresses in excess of the HEL inelastic behaviour, probably in the form of instantaneous damage travelling with the shock front was evident. Very significant shear strength recovery was apparent for impact stress at the level which initiates the phase transition. The shear strength behaviour of silicon carbide was compared over the same stress regime using exactly the same measuring techniques as a standard material which does not phase transform in the stress range. The unusual response observed in the AlN, at the highest impact stress, i.e. significant reduction in lateral stress, was not apparent in SiC at similar stresses. This tends to suggest a material response was governing the reduction rather than gauge breakdown.

REFERENCES

1. Reaugh, J. E., Holt, A. C., Wilkins, M. L., Cunningham, B.J., Hord, B.L and Kusubov, A.S., Int. J. Impact. Eng. **23**, pp. 771-782, 1999.
2. Orphal, D. L., Frantzen, R. R., Piekutowski, A. J., Cunningham, B.J., Hord, B.L and Kusubov, A.S., Int. J. Impact. Eng. **25**, pp. 221-231, 2001.
3. Vollstadt, H. Ito, E., Akaishi, S. and Fukunaga, O., Proc. Jpn. Acad., Ser. B: Phys. Biol. Sci., **66**, p.7, 1990.
4. Ueno, M., Onodera, A., Shimomura, O and Takemura, K., Phys. Rev. B **45**,10123, 1992.
5. Kipp, M.E. and Grady, D.E., 'Shock phase transformation and release properties of aluminium nitride', DYMAT 94 Internat. Conf. Mechanical and Physical Behaviour of Materials under Dynamic Loading, Journal de Physique, p. 249, 1994.
6. T. Mashimo, T., Uchino, M., Nakamura, A., Kobayashi, T., Takasawa, E., Sekine, T., Nuguchi, T., Hikosaka, H. and Fukuoka, K., J. Appl. Phys. **86**, pp.6710-6716, 1999.
7. Rosenberg, Z., Brar, N.S. and Bless, S.J., J. Appl. Phys. **70**, pp.167-171,1991.
8. Bourne, N. K. and Millett, J. C. F., 'On impact upon brittle solids', DYMAT 2000, Internat. Conf. Mechanical and Physical Behaviour of Materials under Dynamic Loading, Journal de Physique IV, pp. 281-286, 2000.
9. Bourne, N. K. and Millett, J. C. F., 'The Dynamic response of Soda-lime glass', in Shock Compression of Condensed Matter-1995, edited by S. C. Schmidt et al., AIP Press, pp.*567-572, 1996.*
10. Bourne, N.K., Millett, J.C.F., Rosenberg, Z. and Murray, N.H., J. Mech. Phys. Solids 46, pp.1887-1908, 1998.
11. Pickup, I. M. and Barker, A. K., 'Deviatoric strength of silicon carbide subject to shock,' in Shock Compression of Condensed Matter-1999, edited by M. D. Furnish et al., AIP Press, *pp.573-576, 1999.*

ON THE FAILURE OF BORON CARBIDE UNDER SHOCK

N.K. Bourne and G.T. Gray III*

Royal Military College of Science, Cranfield University, Shrivenham, Swindon, SN6 8LA, UK.
*Los Alamos National Laboratory, MS-G755, Los Alamos, NM 87545, USA.

Abstract. The failure of brittle materials during uniaxial, compressive shock-loading has been the subject of extensive recent research. For instance, the physical interpretation of the yield point, the Hugoniot elastic limit (HEL), remains poorly understood. Stress and particle velocity records show B_4C exhibits a type of behaviour different to that of other brittle materials. A number of features have been measured to investigate these features. In other ceramics, failure has been seen to occur behind a travelling boundary that follows a shock front that has been called a failure wave, across which the strength of the material is dramatically reduced. In order to elucidate whether this failure process occurs, gauges were embedded to measure the lateral stress behind the shock front in B_4C. As in other materials the stress in B_4C was seen to rise across a failure front. However, this phenomenon only occurred over certain stress ranges. More significantly, the failure penetrated further into the ceramic than has been seen in other materials. A mechanical interpretation is suggested to explain the observed behaviour of B_4C.

INTRODUCTION

Interest in boron carbide stems from its low density and substantial compressive strength and there have thus been several studies of its properties (1-5). Investigations into the dynamic compressive strengths of brittle materials have been extensive and various values for the Hugoniot elastic limits (HELs) have been reported (6). That of boron carbide has been measured to lie between 15 and 20 GPa according to the differing grain size and production routes of the materials investigated (7) but lies at 16.0 GPa for this material (5). The material displays a spall strength of *ca.* 0.35 GPa at stresses below its HEL (5), which indicates its weakness in tension. Additionally, ballistic performance data (8, 9) has shown that the material does not perform as well as may have been expected. It is know that the shear strength shows a dramatic loss of strength in post-yield flow, which may be related to its unusual crystal structure (7).

Rasorenov *et al.* (10) were the first to observe the phenomenon of delayed failure behind the elastic wave in glass, across a front that has been called a fracture, or more lately a failure wave. Later work (11) confirmed the existence of these fronts by measuring spall and shear strengths ahead of and behind them, using manganin stress gauges. The shear strength of the material may be calculated from the offset of the Hugoniot from the hydrostat, and for perfectly elastic-plastic materials this may be estimated to be $2/3\ Y$ (where Y is yield strength of the material calculated from the HEL assuming a Von Mises yield criterion). Evaluating the value for twice the shear strength in this manner gives 13 GPa using Tresca/Von Mises or 8 GPa if the Griffith's criterion is adopted (12) given the value for the HEL. This will be used later to suggest the brittle nature of the yield in this material.

Recent experiments of this type have used the measurement of the lateral stress to map the behaviour of a range of glasses and polycrystalline materials including aluminas, SiC (13) and titanium diboride. This range has been extended to include AlN (14) and, in the work presented here, B_4C. As will be shown, this measurement of the failed strength may provide a means to assess the ballistic worth of a material.

EXPERIMENTAL

Impact velocity was measured to an accuracy of 0.5% using a sequential pin-shorting method and tilt was fixed to be less than 1 mrad by means of an adjustable specimen mount. Impactor plates were made from lapped copper and aluminium discs and were mounted onto a polycarbonate sabot with a relieved front surface in order that the rear of the flyer plate remained unconfined. Targets were flat to less than 2 μm across the surface. Lateral stresses were measured using manganin stress gauges of type J2M-SS-580SF-025 (resistance 25 Ω). The gauges had an active width of 240 μm and were placed at varying distances from the impact face as shown in Fig. 1.

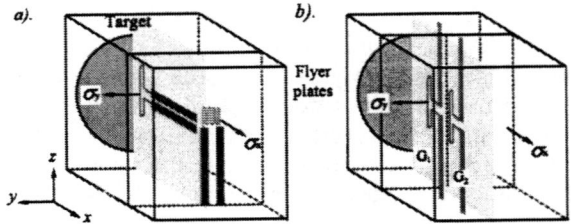

FIGURE 1. Experimental arrangement for lateral and longitudinal stress experiments. a) Longitudinal and lateral stress gauge b). Multiple lateral gauge measurements.

The lateral stress, σ_y was used along with the longitudinal stress, σ_x to calculate the shear strength of the material τ thus

$$2\tau = \sigma_x - \sigma_y. \qquad (1)$$

This quantity has been shown to be a good indicator of the ballistic performance of a material (15). This method of determining the shear strength has the advantage of being direct since no modelling and calculation of the hydrostat is required.

The material used in this study was a hot-pressed B_4C manufactured by Cercom Inc. The density of this hot-pressed material was 2.52 gm cm^{-3} or 99.5% theoretical maximum density. Materials' data are presented in Table 1.

TABLE 1. Properties of Boron Carbide tested

Purity %	ρ g cm^{-3}	c_L mm μs^{-1}	c_S mm μs^{-1}	ν	HEL GPa
99.5	2.52	13.9	8.7	0.18	16.0

RESULTS AND DISCUSSION

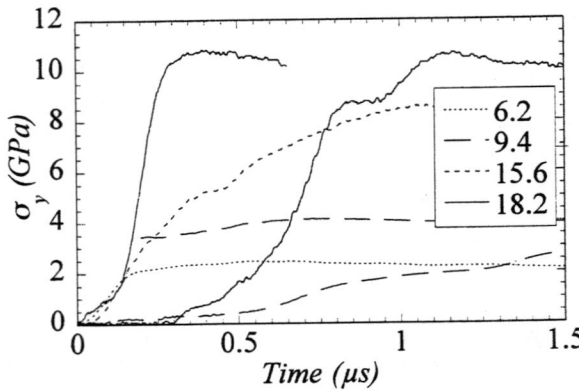

FIGURE 2. Lateral stress histories measured at 2 and 6 mm from the impact face of boron carbide targets shot to stresses of 6.2, 9.4, 15.6, and 18.2 GPa.

Figure 2 shows the lateral stress histories recorded for B_4C at longitudinal stresses of 6.2, 9.4, 15.6 and 18.2 GPa. The lowest of these histories (where the flyer plate was aluminium) show no failure wave and consist of a rise followed by a flat-topped shock pulse. The impacts, in the case of the higher impact velocities, were conducted in each case with 5 mm thick, copper and tungsten alloy flyer plates onto 25 mm thick B_4C targets with gauges positioned at 2 and 6 mm from the impact face. The longitudinal stresses for each of these impacts were calculated from the published Hugoniot (5).

All of the other histories recorded show evidence of failure waves occurring in the B_4C targets. Additionally, the gauges placed 6 mm from the impact plane show evidence of this behaviour indicating that boron carbide does not confine the failure to a surface region in the same manner as other polycrystalline ceramics previously investigated (16). This may be a consequence of the high sound speed in B_4C. The experiments conducted to date have been at relatively low stresses mainly in the elastic region so further results are necessary to define this behaviour.

The traces for the shots at a longitudinal stress of 9.4 GPa show a ramped rise and then the lateral stress at 6 mm rises further, after ca. 1 μs, to its final value. This trace is similar to that seen in glasses investigated previously (13).

There was only one gauge, at 2 mm, in the experiment conducted at a longitudinal stress of 15.6 GPa. The stress is seen to rise, again in a ramped manner, to a value of ca. 5 GPa and then in a second rise to a value of ca. 9 GPa. The corresponding longitudinal stress is calculated to be very close to that of the HEL in B_4C.

The highest stress experiments tested, above the HEL, exhibited a pair of complex traces. Both gauges at 2 and 6 mm displayed a ramped rise to ca.2 GPa and then a more rapid climb. The gauge at 2 mm reaches a plateau and then relaxes back. The gauge at 6 mm plateaus once and then rises again to a similar value to that at the first gauge. Just above the HEL the shock speed will be slow and it is thought that these later fronts may represented the shock arrival.

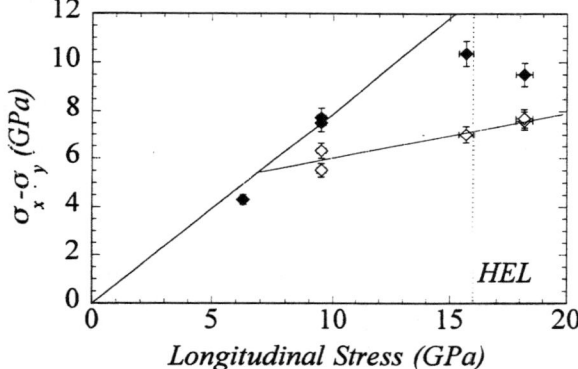

FIGURE 3. Deviatoric response ahead of and behind the failure wave for the gauges and shots discussed. The diagonal line is the calculated elastic assuming v. The HEL is indicated.

Fig. 3 shows the value of twice the shear strength plotted against the longitudinal stress and calculated from the measured values presented in Fig. 2. The diagonal line is the calculated elastic response to the induced longitudinal stresses calculated from the formula

$$2\tau = \left[\frac{1-2v}{1-v}\right]\sigma_x, \qquad (2)$$

where v is Poisson's ratio.

The dark points are measured ahead of the failure front whilst the open points are those behind. The lowest impact recorded was below the failure wave threshold. This behaviour is seen in glasses but has only been noted for TiB_2 previously in polycrystalline materials (17). The values of strength for the un-failed B_4C increase along the elastic line until near the HEL. The material above the HEL then loses strength in what has been dubbed the un-failed material, which suggests that this terminology is incorrect and that some yielding or non-linear dissipative process is occurring. This phenomenon remains to be fully explained. However, it is consistent with the findings of other workers (7).

Above this stress, ca. 8 GPa, there are two strengths for each longitudinal stress. The failed values are seen to increase with stress but the value for 2τ is ca. 7 GPa. It is assumed that by 20 GPa the material fails in compression as the shock rises.

It may be noted that assuming that the failure of boron carbide may be ascribed to brittle failure, the yield strength may be calculated from the failed strength using the Griffith's criterion. The measured failed value of 2τ at close to the HEL is 7.2 GPa which gives a calculated yield stress of 14 GPa. This is below the quoted HEL for B_4C but nevertheless suggests that brittle fracture is responsible for the elastic limit in this material.

CONCLUSIONS

The dynamic compressive strength of B_4C has been investigated by mounting piezoresistive gauges to measure the longitudinal and transverse stress components behind the shock. A failure front has been observed to propagate which can penetrate much further than is seen in other polycrystalline materials investigated. Like TiB_2, but unlike SiC and Al_2O_3, there is a stress threshold below which the failure wave does not initiate in B_4C. The measured HEL correlates with the failed strength seen by the gauges through the Griffith's brittle yield criterion.

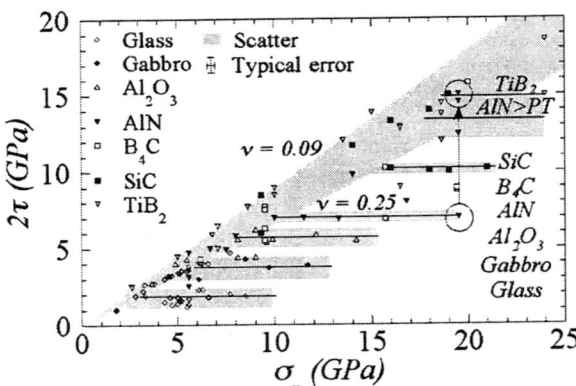

FIGURE 4. Deviatoric response at 2 mm from impact face of presented materials. Gray shaded-region of elastic response and spread of failed strengths.

An assessment of the relative response to shock loading of many common brittle materials may be assessed by plotting the shear strength data on a common plot. In Fig. 4, the known values of the longitudinal and lateral stresses are used to calculate the shear strengths of three glasses, the volcanic gabbro, and the polycrystalline ceramics SiC and TiB_2 ahead and behind the failure fronts. Now this work and another in these proceedings (14) has added B_4C and AlN to these data (13). No indication is given in this plot of the time that it takes to travel from the un-failed to the failed state which is critical in determining ballistic resistance (dwell) before penetration begins since the front of the projectile erodes at the target surface. Values of the failed state are taken from the 2 mm traces for the polycrystalline ceramics since the failure wave does not penetrate to the deeper gauge except for the B_4C presented here. It can be seen that the un-failed state of the material lies on an elastic trajectory (as constructed using equation 2) although different materials have differing Poisson's ratio as indicated by the shaded fan. B_4C shows a failed strength of the same order as AlN in the stress range looked at. However, above the latter's phase transition, it shows a declining value of lateral stress appearing to double its strength giving values equivalent to TiB_2 (14).

The failed states lie on the horizontal lines fitted through the data and represent the *initial* strength of comminuted material. It is thus the starting strength for material failed under shock that lies ahead of a penetrator. At later time this will degrade as particles flow and rotate so that it represents a maximum strength of comminuted materials. Others have shown that initial shear strength is important in the response of metals to penetration (15). The ranking of the failed strengths shown to the right of the figure represents an equivalent material property for brittle solids.

It is hoped that this work, and other observations of this type, will stimulate further consideration of the response of brittle materials to shock loading.

REFERENCES

1. Brar, N.S., Rosenberg, Z. and Bless, S.J., in *Shock Compression of Condensed Matter,* (ed. S.C. Schmidt, R.D. Dick, J.W. Forbes, and D.G. Tasker), Amsterdam: North-Holland, pp. 467-470, (1992).
2. Grady, D.E., *J. Phys. IV France* **4**, 385-391 (1994).
3. Mashimo, T. and Uchino, M., in *Shock Compression of Condensed Matter,* (ed. S.C. Schmidt and W.C. Tao), Woodbury, New York: American Institute of Physics, pp. 531-534, (1996).
4. Johnson, G.R. and Holmquist, T.J., *J. Appl. Phys.* **85**, 8060-8073 (1999).
5. Dandekar, D.P., (2001), Army Research Laboratory Report, ARL-TR-2456.
6. Gust, W.H. and Royce, E.B., *J. Appl. Phys.* **42**, 276-295 (1971).
7. Grady, D.E., in *Shock Compression of Condensed Matter,* (ed. S.C. Schmidt, R.D. Dick, J.W. Forbes, and D.G. Tasker), Amsterdam: Elsevier, pp. 455-458, (1992).
8. Orphal, D.L., Franzen, R.R., Charters, A.C., Menna, T.L. and Piekutowski, A.J., *Int. J. Impact Engng* **19**, 15-29 (1997).
9. Gooch, W.A., Burkins, M.S., Hauver, G., Netherwood, P. and Benck, R., *J. Phys. IV France* **10**, 583-588 (2000).
10. Rasorenov, S.V., Kanel, G.I., Fortov, V.E. and Abasehov, M.M., *High Press. Res.* **6**, 225-232 (1991).
11. Brar, N.S. and Bless, S.J., *High Press. Res.* **10**, 773-784 (1992).
12. Rosenberg, Z., in *Shock Compression of Condensed Matter,* (ed. S.C. Schmidt and W.C. Tao), Woodbury, New York: American Institute of Physics, pp. 543-546, (1996).
13. Bourne, N.K., Millett, J.C.F., Rosenberg, Z. and Murray, N.H., *J. Mech. Phys. Solids* **46**, 1887-1908 (1998).
14. Pickup, I.M. and Bourne, N.K., in *Shock Compression of Condensed Matter,* (ed. M.D. Furnish, N. Thadani, and Y. Horie), Melville, New York: American Institute of Physics, pp. in press, (2001).
15. Meyer, L.W., Behler, F.J., Frank, K. and Magness, L.S., in *Proc. 12th Int. Symp. Ballistics,* pp. 419-428, (1990).
16. Bourne, N.K., Rosenberg, Z. and Field, J.E., *Proc. R. Soc. Lond. A* **455**, 1267-1274 (1999).
17. Bourne, N.K. and Gray III, G.T., *Proc. R. Soc. Lond. A.* in press (2001).

SPALLATION OF HOT PRESSED BORON CARBIDE CERAMIC

Peter T. Bartkowski, Dattatraya P. Dandekar, and David J. Grove

*Army Research Laboratory, Weapons Materials Research Directorate,
Aberdeen Proving Ground, Maryland 21005-5069*

Abstract This work describes the results of plane shock wave spallation experiments conducted on Hot Pressed Boron Carbide marketed by Cercom as PAD B_4C (>99% pure). Density of the material was determined to be 2.508 ±0.016 Mg/m^3 while the longitudinal and shear wave velocities were measured at 13.49 ±0.18 km/s and 8.65 ±0.08 km/s, respectively. Spallation thresholds calculated from the measured "pull-back" velocity were determined up to an impact stress of 15.5 GPa. The values of spall threshold do not vary significantly with impact pressure but do exhibit a pulse width dependency indicating a time dependent generation of defects. The value of spall strength of boron carbide is 0.35 ± 0.07 GPa when shocked between 2 and 15.5 GPa. The values of release impedance lie between 33 and 37 Gg/m^2s and are in good agreement with the longitudinal impedance of 33.8 ±0.5 Gg/m^2s at the ambient condition measured ultrasonically. The free-surface velocity profiles obtained from these experiments were numerically simulated using Rajendran-Grove (R-G) ceramic model. The paper provides the values of material constants required by the R-G ceramic model for boron carbide.

INTRODUCTION

The present investigation is a continuation of our effort to determine the nature of deformation of ceramic materials under shock induced tension. Plane shock wave spallation experiments were conducted on Cercom Corp.'s PAD B_4C Boron Carbide Ceramic. The particle velocity records measured during the experiments were then used to determine the material constants for the Rajendran-Grove (R-G) ceramic model.

MATERIAL

PAD B_4C is a hot pressed ceramic over 99% pure. It has a density of 2.508 ±0.016 Mg/m^3 while the single crystal density of B_4C is 2.52 Mg/m^3 [1] which results in a pore volume fraction for PAD B_4C of approximately 0.5%. The average grain size, as reported by the manufacturer, is 15 μm.

Longitudinal and shear wave velocities were measured ultrasonically. The longitudinal and shear wave velocities were measured at 10 and 5 MHz to be 13.49 ±0.18 km/s and 8.65 ±0.08 km/s, respectively. Table 1 lists the elastic constants as determined from these measurements.

Specimens were cut, ground from ceramic blocks to form 40 ±1 mm diameter discs. Flyer discs were either PAD B_4C or WC and had a nominal thickness of 2 or 4 mm while the targets were nominally 4 or 6 mm thick. The disc faces were lapped and polished flat to 10 μm while the disc faces were parallel to within one in 10^4.

EXPERIMENTS

The general configuration for the shock experiments is shown in Fig. 1. Plane shock experiments were performed using ARL's 100 mm bore Light Gas Gun. Diagnostics were composed of projectile impact velocity measurement and free surface particle velocity measurement of the rear of the target specimen.

Impact velocity was determined by measuring the time intervals between the shorting of 4 sets of electrically charged pins located immediately in front of the target. Velocity is calculated using pre-measured distances between pins and the recorded

TABLE 1. PAD B_4C Material Properties

Property (units)	PAD B_4C
Density (Mg/m^3)	2.508 ±0.016
Average Grain Size (μm)	15
Void Fraction	0.005
Elastic Wave Velocity (km/s)	
Longitudinal	13.49 ±0.18
Shear	8.65 ±0.08
Bulk	9.06 ±0.22
Elastic Modulus (Gg/m^2s)	
Longitudinal	33.8
Shear	21.7
Bulk	22.7
Poisson's Ratio	0.151 ±0.014

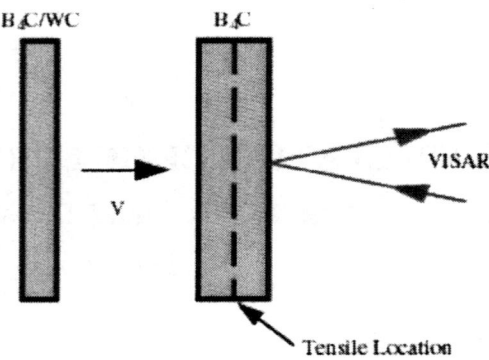

FIGURE 1. Shot configuration

transit times to an accuracy of 0.5%.

Free surface velocity measurements of the targets were made using VISAR (Velocity Interferometer System for Any Reflector) [2]. The precision of VISAR particle velocity measurements is better than 1%.

Flyer thicknesses were chosen to produce pulse widths of approximately 0.3 and 0.6 μs. Projectile velocity and flyer material were varied to produce impact stresses up to 15.5 GPa.

EXPERIMENTAL RESULTS

The reported HEL of nearly fully dense boron carbide is 15-19 GPa [3,4,5,6,7]. All experiments conducted in this work were at 15.5 GPa or below and are considered elastic. As such, the spall strength of the material can be determined by the magnitude of the pull back particle velocity change Δu_{pb} [8,9]. The spall strength is calculated using the following formula:

$$\sigma_{spall} = \tfrac{1}{2} U_{pb} Z_{el} \quad (1)$$

Where Z_{el} is the elastic impedance of the material.

A summary of the experiments conducted is given in Table 2. All experiments were symmetric impact in nature except for 826 and 846 where a WC flyer was used, instead of boron carbide, to reach a higher impact stress. Table 2 also lists the determined impact shock state, release impedance and calculated spall strength of PAD B_4C as determined from the steady free surface velocity and pull back particle velocity change measured from the VISAR waveforms. The values of release

TABLE 2. Summary of PAD B_4C shock experiments

Shot #	Flyer Matl.	Thickness Flyer (mm)	Thickness Target (mm)	Pulse Width (μs)	Impact Velocity (km/s)	Shock State Stress (GPa)	Shock State Velocity (km/s)	Shock State Density (Gg/m^3)	Release Imp. (Gg/m^2s)	Spall Strength ½ Δu_{pb} (km/s)	Spall Strength Stress (GPa)
825†	B_4C	4.052	5.956	0.60	0.6129	10.37	0.3065	2.566	32.8	0.0095	0.32±0.14
826	WC	2.033	5.957	0.59	0.6106	15.54	0.4595	2.596	35.0	0.0070	0.24±0.20
828†	B_4C	4.053	5.958	0.60	0.4086	6.91	0.2043	2.547	33.4	0.0120	0.41±0.10
829†	B_4C	4.051	5.956	0.60	0.2327	3.94	0.1164	2.530	34.6	0.0105	0.36±0.07
833-1	B_4C	4.063	5.960	0.60	0.4120	6.97	0.2060	2.547	36.3	0.0090	0.30±0.10
833-2	B_4C	2.043	5.954	0.30	0.4120	6.97	0.2060	2.547	35.7	0.0135	0.46±0.10
840	B_4C	4.053	5.956	0.60	0.1228	2.08	0.0614	2.519	35.4	0.0095	0.32±0.03
846	WC	2.028	5.954	0.59	0.5070	12.91	0.3815	2.581	33.0	0.0100	0.34±0.17
910†	B_4C	2.043	4.058	0.30	0.1229	2.08	0.0615	2.519	36.8	0.0125	0.42±0.03

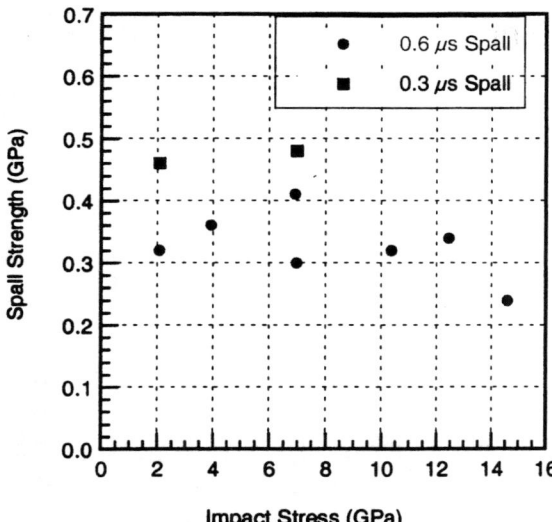

FIGURE 2. Spall stress vs. Impact stress

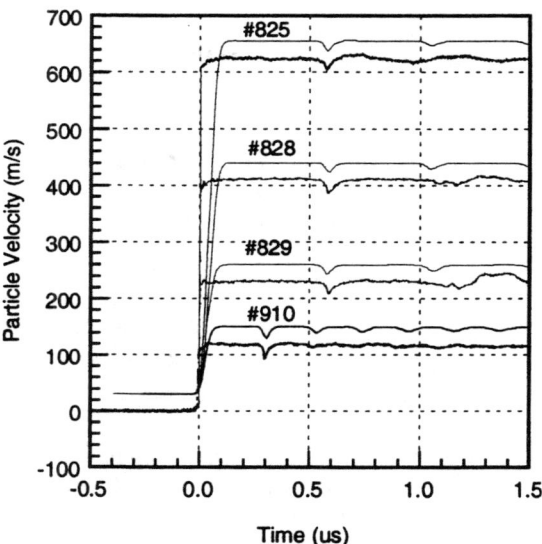

FIGURE 3. Particle Velocity Waveforms

impedance vary between 33 and 37 Gg/m^{2}s and agree favorably with the ultrasonically measured longitudinal impedance of 33.8 Gg/m^{2}s. Plotted as a function of impact stress in Fig. 2, the spall strength can be seen to decline with increasing impact stress. Two of the experiments (833-2 & 910) were conducted with a pulse width of 0.30 μs instead of 0.6 μs to investigate the effects of pulse width on the spallation process. These two experiments at first seem to indicate a time dependent generation and propagation of defects; however if one considers the calculated uncertainties given in Table 2, the difference between the two pulse widths become insignificant. Nonetheless, these values of spall strength are consistent with Grady's values for Dow Chemical's Boron Carbide Ceramic [5].

R-G MODEL

Four of the experiments listed in Table 2 (denoted by †) were simulated using the RG ceramic model [10,11,12] running in EPIC finite element code. The model is based on an elastic-plastic cracking deformation process. The scalar damage is measured by the crack density parameter γ. The number of flaws remains constant while a strain energy release based evolution law governs the crack growth, measured by the crack size parameter a. Crack orientation is not considered and assumed to be random throughout the material. Permanent strain is a function of plastic flow only and the strains from microcracking are elastic. Plastic flow occurs in the model only under compressive loading when the pressure exceeds the HEL. The strains due to pore collapse are assumed to be visco-plastic and are modeled using a pressure dependent yield function.

Microcracks are allowed to grow when the stress state satisfies a generalized Griffith criterion. This criterion uses the material fracture toughness and dynamic friction coefficient. As damage accumulates in the material from crack growth, stress relaxation occurs. The pressure in the material is a direct function of the degraded bulk modulus.

Simulations

The lowest velocity experiment (#910) was used to determine the model constants. These determined constants for PAD B$_4$C are given below in Table 3. Three other experiments (#829, 828, 825) were then simulated using these constants. The results of all four simulations can be compared to the experimentally recorded particle velocity data

Table 3. RG Model Constants for PAD B_4C

Symbol	Description	Constants
A	Static compressive strength (GPa)	12.5
C	Coefficient for strain rate dependence	0.01
K_{IC}	Static fracture toughness (MPa √m)	2.0
f_o	Initial void volume fraction	0.03
a_o	Initial microcrack size (μm)	1.0
N_o^*	Initial microcrack density (m^{-3})	5×10^{11}
μ	Dynamic friction coefficient for mode II crack growth	0.45
n_1^-	Coefficient for mode II crack growth	0.10
σ_{spall}	Spall criterion for damage evolution under high triaxial tensile loading (GPa)	0.5
γ_p	Critical crack density parameter for pulverization	0.75
β_p	Slope of linear strength-pressure relationship for pulverized material	1.5
Y_{max}	Strength "cap" for pulverized material (GPa)	3.0

in Fig. 3; the simulations are offset 30 m/s vertically above the experimental data for clarity. Clearly, the RG model does an excellent job of predicting spallation behavior of PAD B_4C.

CONCLUSION

Plane shock wave spallation experiments were conducted on Cercom Corp.'s PAD B_4C Boron Carbide Ceramic. Spallation thresholds were determined vs. impact stress for two pulse widths of 0.3 & 0.6 μs. For 0.6 μs pulse width, a spall strength of 0.35 ±0.07 GPa was determined. A pulse width of 0.3 μs resulted in a higher spall strength of approximately 0.47 GPa. Although higher spall strengths were measured with the shorter pulse width, a consideration of the errors in measurement indicates that the differences may be insignificant.

The values of release impedance were determined to be between 33 and 37 Gg/m²s which is in good agreement with the ultrasonically measured value of 33.8 Gg/m²s.

The particle velocity records measured during one of the experiments was then used to determine the material constants for the Rajendran-Grove (R-G) ceramic model. These constants were then used to accurately predict spallation behavior at higher impact stresses.

REFERENCES

1. Thevenot, F. "Boron Carbide-A Comprehensive Review," *J. Europ. Ceramic Soc.* **6**, 205-225 (1990).
2. L. M. Barker and R. E. Hollenbach, *J. Appl. Phys.* **41**, 4208-4226 (1970).
3. Wilkins, M. L. "Third Progress Report of Light Armor Program." UCRL50460, Lawrence Livermore National Laboratory, University of California, (1968).
4. Gust, W. H., and E. B. Royce, "Dynamic Yield Strengths of B4C, BeO, and Al2O3 Ceramics." *J. Appl. Phys.* **42**, 276-295 (1971).
5. Grady, D. E., "Dynamic Properties of Ceramic Materials," Sandia National Laboratory Report, SAND 94-3266 (1995).
6. Kipp, M. E., and D. E. Grady, "Shock Compression and Release in High-Strength Ceramics," Sandia National Laboratory Report, SAND 89-1461 (1989).
7. Winkler, W., and A. J. Stilp, "Spallation Behavior of TiB2, SiC, and B4C under Planar Impact Tensile Stress," in *Shock Waves in Condensed Matter-1991*, Ed. S. C. Schmidt et al., Elsevier Science, 1992, pp. 475-478.
8. Dandekar, D.P. and D.C. Benfanti, *Journal of Applied Physics* **73**, 673-679 (1993).
9. Dandekar, D.P. and P. Bartkowski, "Shock Response of AD995 Alumina," in *Proceedings of the AIP Conference 309 Part I*, 1994, pp. 733-736.
10. Rajendran, A.M., and D.J. Grove, *International Journal of Impact Engineering*, Vol. 18, No. 6, 1996, pp. 611-631.
11. Grove, D.J. & A.M. Rajendran, in *Shock Compression of Condensed Matter – 1997*, Ed. S.C. Schmidt et al., AIP, New York, 1998, pp. 255-258.
12. Grove, D.J. & A.M. Rajendran, in *Shock Compression of Condensed Matter – 1999*, Ed. M.D. Furnish et al., AIP, New York, 2000, pp 619-622.

SHOCK EQUATION OF STATE AND DYNAMIC STRENGTH OF TUNGSTEN CARBIDE

Dattatraya P. Dandekar[1] and Dennis E. Grady[2]

[1]Army Research Laboratory, Weapons Materials Research Directorate,
Aberdeen Proving Ground, Maryland 21005-5069
[2]Applied Research Associates, Inc.
4300 San Mateo Blvd. NE
Albuquerque, New Mexico 87110

Abstract Tungsten carbide ceramic is a high-density material with attractive compressive and tensile strength properties. Cercom, Inc. manufactured hot-pressed tungsten carbide ceramic was tested in the present study. The density of this ceramic varies between 15.53 and 15.56 Mg/m^3. The values of longitudinal and shear wave velocities measured ultrasonically vary between 7.04 and 7.05 km/s, and 4.30 and 4.32 km/s, respectively. Shock wave experiments were conducted at the U.S. Army Research Laboratory (ARL) and Sandia National Laboratory (SNL) to determine its shock-induced compressive behavior. The results of these experiments are summarized as: (1) the Hugoniot Elastic Limit (HEL) of this material is 6.6 ± 0.5 GPa, (2) this value of the HEL may not adequately represent the dynamic yield strength of the material because of the substantial post-yield hardening characteristics of this material shown by the pronounced slope of the precursor wave preceding the following final-state shock wave, and (3) the final shock state attained in the material indicates that the shear strength is maintained when shocked above the HEL to 80 GPa.

INTRODUCTION

The attractive mechanical properties of tungsten carbide with its high-density makes it ideally suited for use as a protective element to mitigate shock-induced effects. The current study was initiated with the goal of bringing together information on the shock-induced response of a hot-pressed tungsten carbide manufactured by Cercom, Inc. This material is, hereafter, referred to as Cercom WC. Specific properties of interest in this study are hydrodynamic equation of state, and compressive and shear strength under plane shock wave loading of this material. Shock compression data were obtained from a series of shock profile measurements conducted at ARL and SNL on Cercom WC. But the material used in the shock wave experiments performed at ARL and SNL were not from the same batch.

MATERIAL

Cercom WC is a composite consisting of two distinct materials, namely, WC (97.2% by weight) and W_2C (2.8% by weight)[1]. W_2C is a byproduct of the densification process. WC and W_2C both crystallize in hexagonal form. The theoretical densities of these carbides are 15.7 and 17.2 Mg/m^3, respectively [2]. Both melt around 3050 K and have similar thermal expansion coefficients. The only other property of W_2C reported is the value of longitudinal elastic wave velocity, which is 4.94 km/s [2]. The measured values of density, elastic longitudinal and shear wave velocities of this tungsten carbide composite are given in Table 1. The values of density and elastic wave velocities for Cercom WC measured at ARL and SNL are within the errors of measurements. Since the measured value of the density of Cercom WC is less than the

TABLE 1. Properties of Cercom WC

Properties	SNL	ARL
Density (Mg/m³)	15.56	15.53
Elastic wave velocities (km/s)		
Longitudinal	7.04	7.05
Shear	4.30	4.32
Bulk	4.96	4.98
Bulk modulus (GPa)	383	385
Poisson's ratio	0.200	0.200

reported densities of WC and W_2C, it must contain some lighter impurities with or without porosity. We have no information about these impurities but the void volume fraction is estimated to be around 0.01 [1].

EXPERIMENT

Plane shock wave experiments were conducted using the ARL 100 mm light gas gun facility and the 89 mm bore diameter single stage powder gun facility at SNL. The maximum impact velocities that can be achieved at these two facilities are 0.7 km/s and 2.5 km/s, respectively. In these experiments, a disc of either Cercom WC or 6061-T6 aluminum or C-cut single crystal sapphire or x-cut quartz (impactor) mounted in the projectile impacted another Cercom WC disc (target) with a given velocity to generate a shock compression wave of an appropriate magnitude. In SNL experiments, the impactor material was backed by a disc of polymethylmethacrylate (PMMA). The wave velocity profiles in the experiments conducted at ARL were monitored at the center of the free surface of the tungsten carbide target. At SNL wave profiles were monitored at the center of the interface of a single crystal lithium fluoride disc and a Cercom WC target. These wave profiles were recorded by employing the interferometry technique (VISAR) developed by Barker and Hollenbach [3]. The impact velocities were measured by shorting electrically charged pins located at measured distances a few millimeters ahead of the target disc. The planarity of impact was better than 0.5 mrad. The precision of free surface velocity measurements is 1%. The uncertainties in the measured values of impact velocities are 0.5%. In all, 14 experiments were performed on Cercom WC. Of these 14 experiments, 6 were performed at SNL.

RESULTS

For convenience, the experiments performed at ARL are denoted by those which begin with numerals in column 1 of Table 2. The experiments performed at SNL are denoted by those starting with letters in column 1 of Table 2. Shock compression wave profiles recorded in a few experiments are shown in Fig. 1. The results are presented in terms of elastic deformation, post-yield

TABLE 2. Shock Wave Experiment Data on Cercom WC.

Experiment/ Impactor material	Thickness (mm)		Impact Velocity (km/s)	Elastic compression			Inelastic compression			
	Impactor	Target		Stress (GPa)	Mass velocity (km/s)	Density (Mg/m³)	Plastic velocity (km/s)	Stress (GPa)	Mass velocity (km/s)	Density (Mg/m³)
504/WC	3.15	3.14	0.6130	6.79	0.0620	15.67	5.43	27.6	0.306	16.41
507/S	2.02	3.14	0.2962	7.33	0.0672	15.68	5.79	9.3	0.088	15.74
514-1/S	2.02	3.15	0.5067	6.35	0.058	15.66	5.29	15.2	0.165	15.98
514-2/WC	3.14	6.43	0.5067	7.55	0.069	15.68	5.56	23.7	0.253	16.22
516/Q	0.99	6.48	0.5004	6.47	0.0591	15.66				
517/WC	3.14	6.48	0.1848	6.84	0.0625	15.67	5.10	9.1	0.092	15.75
102/WC	4.00	6.00	0.0648	3.49	0.0319	15.60				
103/WC	4.00	6.00	0.0644	3.44	0.0314	15.60				
WC-8/WC	6.20	6.18	1.660	6.2	0.0566	15.68	6.22	81.6	0.830	17.92
WC-12/Al	1.51	6.19	0.362	4.30	0.0392	15.64	-			
WC-13/Al	1.50	6.19	0.454	5.90	0.0539	15.68	-			
WC-14/WC	6.35	6.32	1.239	6.2	0.0566	15.68	5.98	59.2	0.620	17.30
WC-15/WC	6.35	3.00	1.210	6.2	0.0566	15.68	5.93	57.5	0.605	17.27
WC-16/WC	6.30	6.34	0.824	6.2	0.0566	15.68	5.74	38.	0.412	16.69

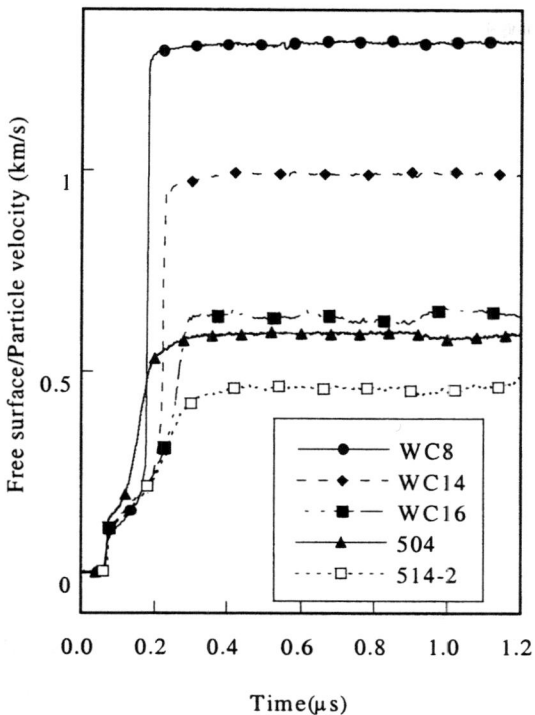

FIGURE 1. Shock wave profiles in Cercom WC

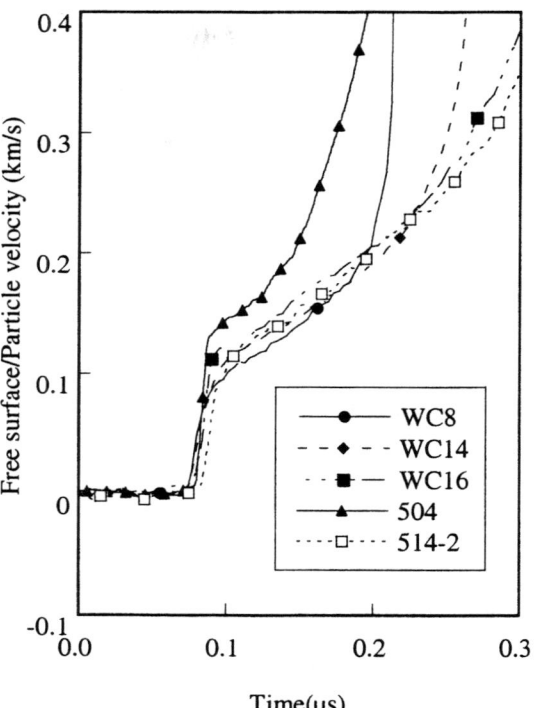

FIGURE 2. Precursor characteristics of Cercom WC

feature, inelastic deformation, and its nature.

Elastic Compression

The wave profiles showed initial break corresponding to stresses between 6.2 and 7.6 GPa (Fig. 1). This break is identified as the HEL for the Cercom WC. The decay of the elastic precursor with target thickness is not evident from the results of these experiments. The average value of the HEL is 6.6 ± 0.5 GPa. Since the precision of measurements of particle and free surface velocity is 1%, the large statistical uncertainty associated with the average value of the HEL is likely to be arising from variability in the material itself. Details of the precursor wave indicating the dynamic strength property of Cercom WC is shown in Fig. 2. This figure shows a substantial post-yield hardening in Cercom WC following the elastic precursor and preceding the attainment of final shock-induced state. These wave profiles cannot discriminate between the relative roles of pressure hardening or deformation hardening in the observed post-yield behavior of Cercom WC.

Inelastic Compression

Shock wave following the elastic precursor travels in the elastically deformed Cercom WC with velocity between 5.1 and 6.2 km/s. These values of shock velocities are larger than the bulk sound wave velocity in Cercom WC at the ambient condition, i.e., 4.96-4.98 km/s (Table 1). This suggests that the inelastic deformation proceeds plastically in this material. Since an adequate presentation of results of inelastic deformation requires knowledge of appropriate hydrodynamic compression of the material, it is dealt with next.

The hydrodynamic compression of Cercom WC is obtained from the shock compression data on tungsten carbide reported by McQueen et al. [4]. Tungsten carbide material used in their experiments contained 5 weight percent of cobalt as a binder material. The density of the material was 15.01 Mg/m^3. The values of longitudinal and shear wave velocities were reported to be 6.89 km/s and 4.18

km/s, respectively. The calculated values of bulk sound wave speed and Poisson's ratio are 4.92 km/s and 0.209, respectively. These values are not very different from the measured values of sound speeds for Cercom WC given in Table 1. Thus, the calculated hydrodynamic compression of Cercom WC based on the results of McQueen et al. [4] experiments should represent its compression faithfully. A careful analysis of their data taking into account the fact that hydrodynamic equilibrium in their tungsten carbide was not attained in their experiments until the pressures in excess of 70 GPa were reached yield the following linear relation between shock velocity (U_s) and particle velocity (u_p):

$$U_s = 4.93 + 1.309\ u_p. \qquad (1)$$

The value of correlation coefficient for the above linear relation is 0.999, and the 95% confidence intervals of the intercept (C_0) and the slope (s) in relation (1) are ±0.05 km/s and ±0.083, respectively. Further, the value of C_0 agrees well with the value of the bulk sound wave velocity, 4.92 km/s, at the ambient condition.

The hydrodynamic compression of Cercom WC calculated by using the value of its bulk modulus 384 GPa with s=1.309, and experimentally determined shock Hugoniot of Cercom WC are shown in Fig. 3. It shows that this material continues to retain shear strength under plane shock wave propagation when shocked to 80 GPa. The value of shear stress sustained by Cercom WC at the HEL obtained from the Hugoniot data and the hydrodynamic compression curve is 2.4 GPa. This compares well with the elastic value of shear stress, i.e., 2.5 ± 0.2 GPa. It appears that the magnitude of shear stress sustained by Cercom WC increases with an increase in shock-induced stress. For example, at shock stress of 82 GPa, the value of sustained shear stress is 6 GPa.

CONCLUDING REMARKS

The results of the present investigation indicate that Cercom WC deforms like an elastic-plastic solid under shock wave compression. The

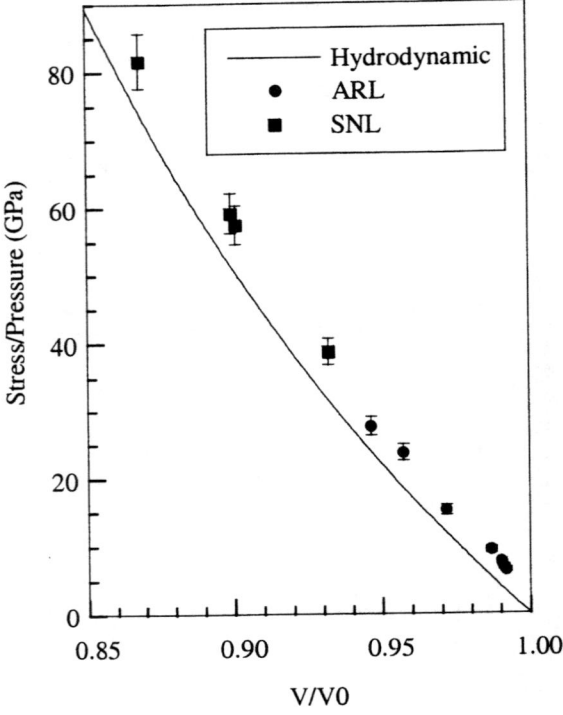

FIGURE 3. Shock compression of Cercom WC

magnitude of shear stress sustained above the HEL needs to be determined by conducting appropriate shock-reshock experiments of the type reported by Asay et al. [5] and Dandekar and Gaeta [6].

REFERENCES

1. Gooch, W. J. (Private communication).
2. Gauthier, M. M., *Engineered Materials Handbook*, ASM International, Cleveland, 1995, pp. 961-963.
3. Barker, L. M., and Hollenbach, R. E., *J. Appl. Phys.*, **43**, 4669-4675 (1972).
4. McQueen, R. G., Marsh, S. P., Taylor, J. W., Fritz, J. N., and Carter, W. J., "The Equation of State of Solids from Shock Wave Studies", in High-Velocity Impact Phenomena, edited by R. Kinslow, Academic Press, New York, 1970, pp. 293-417 and 521-568.
5. Asay, J., Chhabildas, L. C., and Dandekar, D. P., *J. Appl. Phys.* **51**, 4774-4783 (1980).
6. Dandekar, D. P., and Gaeta, P. J., "Double Shock and Release Experiments in PMMA and Z-cut Sapphire", in *Shock Wave in Condensed Matter-1987*, edited by S. C. Schimdt and N. C. Holms, North - Holland, New York, 1988, pp. 281-284.

BAR IMPACT TESTS ON ALUMINA (AD995)

James U. Cazamias[1], William D. Reinhart[2], Carl H. Konrad[3], Lalit C. Chhabildas[2], Stephan J. Bless[4]

[1] LLNL, L-414, PO Box 808, Livermore, CA 94551
[2] SNL, MS 1181, PO Box 5800, Albuquerque, NM 87185-1181
[3] Bechtel, 3900 Paradise Road, Suite 183, Las Vegas, NV 89101
[4] IAT, 3925 W. Braker Ln., Suite 400, Austin, TX 78759-5316

Abstract. Dynamic strength may be inferred from bar impact tests, although interpretation of the data is affected by the time-to-failure of the target bar. To clarify the mechanics, tests with graded density impactors were conducted on bare and confined bars, 12 and 19 mm in diameter, cut from blocks of AD995 alumina. Manganin gauge and VISAR diagnostics were employed. Larger rods displayed higher strength. In some tests the "true" yield stress of ~4.5 GPa was achieved.

INTRODUCTION

Polycrystalline alumina (Al_2O_3) is the archetypical armor ceramic. It has been studied since the 1960's, and its material and ballistic properties have been fairly well characterized experimentally. A specific high purity alumina, Coors AD995, has been characterized dynamically via a variety of experimental methods – plate impact [1-2], bar impact [3-5], spherical wave [6], and penetration [7-9].

In spite of these efforts, there has been remarkably little success at predicting impact or penetration experiments that involved large contrasts in loading state or strain rate. In this situation, it is important to focus on material property tests that mimic engineering stress states and rates. Static tests are dominated by the largest flaw in the material, and 1D shock tests are fully confined; however, 1D stress tests allow the measurement of a strength that is most relevant to ballistic applications. There have been several attempts to use the split Hopkinson bar (a 1D stress experiment) to characterize ceramics for ballistic applications; however, there are several complications in doing so: (1) ceramics typically indent the striker bar, (2) the sample undergoes several reverberations while approaching stress equilibrium, and it may fail before equilibrium is reached, and (3) the samples are small. In contrast, in the bar impact experiment, the impactor does not need to remain elastic, it is a single wave experiment, and larger specimens can be used which eliminate small scale effects and allow direct observation of the failure process.

The bar impact experiment is fairly well understood for ductile metals. A few diameters from the impact face, a 1D stress wave with a maximum amplitude of $Y_0 = \sigma_{HEL}(1-2\nu)/(1-\nu)$ travels down the bar.

In brittle materials, the picture is a bit more complicated. In the first couple of diameters, the transmitted wave is transformed from 1D strain (impact condition) to 1D stress (steady-state solution). Damage in this transition region can be compressive and/or tensile [4-5]. The maximum stress that propagates down the bar is determined by how much of the wave can get out of the impact zone before the strength of the damaged material limits the amplitude of the stress. Therefore, the compressive strength measured in a brittle bar impact experiment can possess a dependence on tensile strength and is seldom equal to the ideal unconfined compressive strength.

TABLE 1. Bar Dimensions

	Sleeve	Sleeve OD (mm)	Bar OD (mm)	Long Length (mm)	Short Length (mm)	Overall Length (mm)
Alseg-1	Foam	38.100	12.713	76.835	25.959	102.949
Alseg-2	Steel	38.100	12.713	76.810	25.959	102.705
Alseg-3	Foam	38.100	19.164	112.751	25.476	138.275
Alseg-4	Steel	38.100	19.164	114.427	26.391	140.670

TABLE 2. Projectile Properties

	Vel. (m/s)	$D_{nominal}$ (mm)	T_{TPX} (mm)	ρ_{TPX} (gm/cm^3)	T_{Al} (mm)	ρ_{Al} (gm/cm^3)	T_{Ti} (mm)	ρ_{Ti} (gm/cm^3)	T_{St} (mm)	ρ_{St} (gm/cm^3)
Alseg-1	843	50.8	0.54	0.816	0.48	2.673	0.35	4.384	19.08	7.836
Alseg-2	851	50.8	0.55	0.822	0.49	2.658	0.35	4.388	19.06	7.837
Alseg-3	859	76.2	0.99	0.829	0.95	2.705	0.85	4.529	19.19	7.837
Alseg-4	846	76.2	0.98	0.831	0.95	2.701	0.84	4.516	19.19	7.834

In the present work, three sorts of bar impact experiments have been performed: (1) normal bar impacts; (2) confined bar impacts, where the rod is completely sleeved in a material to provide a confinement pressure and the measured longitudinal stress is no longer limited to the 1D stress value and may approach the 1D strain value (σ_{HEL}); and (3) graded density impactors, where the impactor is composed of multiple thin layers to spread out the loading wave and hopefully suppress tensile damage.

The quasi-static compressive strength of AD995 is 2 GPa [10]. The compressive strength calculated from the onset of HEL-like softening in plate impact experiments is 4.3 GPa [1]. It should be noted that there is some controversy over the interpretation of HEL experiments for alumina. While we believe Y to be 4.3 GPa, some interpretations give Y values of over 8 GPa [11].

Wise and Grady [3] impacted unconfined and Ta-confined alumina (10.06 mm OD, 80 mm L) with 6061-T6 Al. For the unconfined bars, they measured a peak stress of 3.15 GPa at 1.035 km/s and 2.182 km/s. For the confined bars, they measured peak stresses of 6.32 GPa (2.14 km/s) and 5.80 GPa (1.051 km/s).

Simha [4] impacted unconfined alumina (12.5 mm OD). With 4340 steel impactors he found that the peak stress was 3.6 GPa in the 100-300 m/s impact regime. When he used alumina impactors, he measured lower peak stresses of 3.4 GPa (175 m/s) and 2.8 GPa (278 m/s) which he attributed to the fact that the impactor also fails.

Chhabildas, et al. [5] impacted confined and unconfined alumina (19 mm OD) with normal and graded density impactors. For unconfined alumina with a steel impactor at 318 m/s, they observed a two wave structure; the first wave loaded the bar to 2.1 GPa and the second to 3.4 GPa. For unconfined alumina with a graded density impactor at 318 m/s, the loading ramped to a peak stress of 3.5 GPa (300 m/s) for one shot and a peak stress of 4.2 GPa, which relaxed to 3.6 GPa for a repeat shot. For confined alumina with a graded density impactor, in a 74 mm long bar (321 m/s) they observed a peak stress of 5.1 GPa, and in a 151 mm long bar (321 m/s and 322 m/s) they observed a peak stress of 4.6 GPa.

EXPERIMENTS

A set of four experiments using an 89 mm smooth bore single stage propellant gun were performed on bare and fully sleeved rods at a nominal impact velocity of 850 m/s with graded density impactors at the Sandia STAR facility. The Coors AD995 bars were nominally 12.5 mm and 19 mm in diameter (see Table 1). The 12.5 mm rods were cut from 4 inch thick tiles. The 19 mm rods were cut from 1 inch thick tiles. The bars have a nominal density of 3.89 gm/cc, a nominal bar wave speed of 9.79 km/s, and a nominal bulk wave speed of 7.72 km/s.

The bars were instrumented with manganin gauges nominally six diameters from the impact face with a nominally two diameter long backing piece. The backing piece had W vapor deposited as a mirror surface for a VISAR. There was no window. The sleeves were made of 4340 steel. The unsleeved measurements had a foam sleeve for mounting purposes. The overall lengths differ from the sum of the long and short lengths due to gauge insertion and grinding to assure parallel faces. The graded density impactors consisted of a TPX (plastic) layer followed successively by Al and Ti layers, backed with a 4340 steel plate (see Table 2). Velocity was measured with self shorting pins.

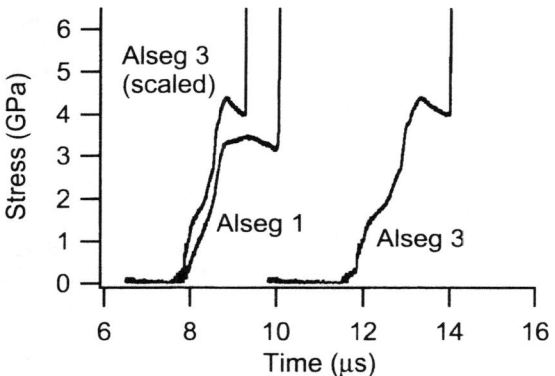

FIGURE 1. Gauge traces for unsleeved shots.

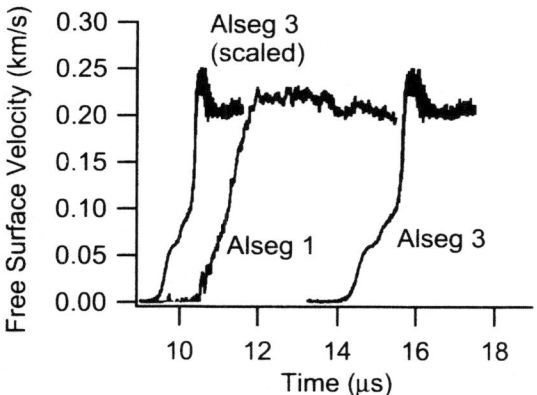

FIGURE 2. VISAR traces for unsleeved shots.

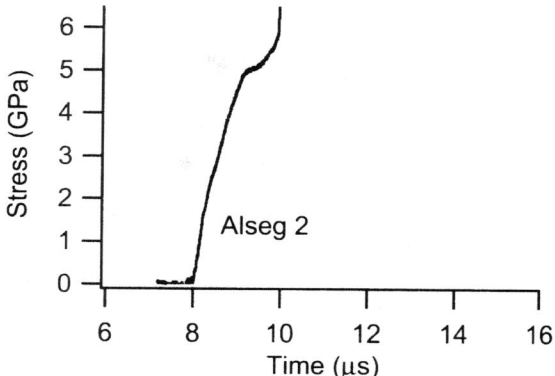

FIGURE 3. Gauge traces for sleeved shots.

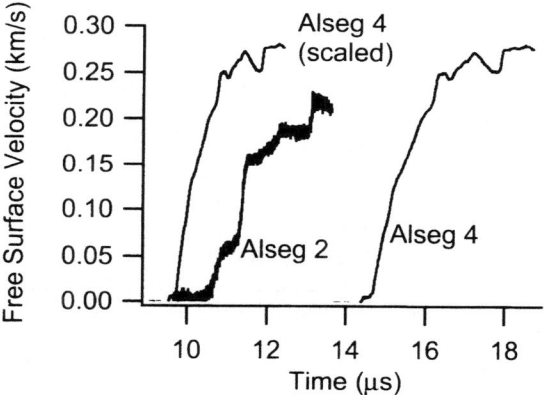

FIGURE 4. VISAR traces for sleeved shots.

The manganin gauge stress traces for unsleeved/sleeved shots are presented in Figs. 1 and 3. The VISAR free surface velocities for unsleeved/sleeved shots are presented in Figs. 2 and 4. The scaled traces use a temporal scaling determined by the diameter (d) ratio: $t^{scaled} = t_{large}d_{small}/d_{large}$.

The stress corresponding to the VISAR signal is nominally $\sigma_{max} = 0.5 \rho c_{bar} u_{max}$ giving σ_{max} = 4.6 GPa for Alseg 3 and σ_{max} = 4.2 GPa for Alseg 1. The values are higher than the manganin gauges'.

DISCUSSION

For both confined and unconfined rods, the larger diameter rods exhibit a greater strength than the smaller diameter rods. While this might be a rate effect (actually, a backward rate effect), the work of [4] and [5] at 100 - 350 m/s (discussed above) does not support this explanation since those experiments were carried out at even lower strain rates. We think that the most probable cause of the strength variation is traceable to differences in the stock material. The large scale rods came from one inch plate, while the small scale rods came from four inch plate. The greater performance of small scale ceramic armors when compared to large scale ceramic armors [12] is consistent with our strength measurements. Thus, the difference in strength might be a true scale effect rather than a rate effect. Importantly, the bar impact test appears to differentiate between the two scales. Material variability might also be attributed to the fact that the small scale rods were cut perpendicular to the face of the plate, while the large scale rods were cut parallel.

For the unconfined tests, the shapes of the rise of the stress wave are fairly consistent, with peak stresses of 4.5+-0.1 GPa for the large scale and 3.8+-0.3 GPa for the small scale. The VISAR trace for Alseg 3 exhibits a plateau before decaying. The plateau is probably not observed in the gauge trace due to the greater temporal resolution of the VISAR signal. Taking into account the lower estimate for the HEL-derived yield stress of alumina of 4.3 GPa, this is an indication that "true" yield surface has been reached in the large bar.

For the confined tests, the small scale sample exhibits a precursor wave that is not observed at large scale.

ACKNOWLEDGMENTS

This work was performed at SNL and IAT and supported by ARL under contract DAAA21-93-C-0101 and the DOE. Thanks to C.H.M. Simha.

REFERENCES

1. Grady, D. E., and Moody, R. L., *Shock Compression Profiles in Ceramics*, SAND96-0551 (1996).
2. Dandekar, D. P., and Bartkowski, P., "Shock Response of AD995 Alumina," in *High-Pressure Science and Technology – 1993*, 1994.
3. Wise, J. L., and Grady, D. E., "Dynamic, Multiaxial Impact Response of Confined and Unconfined Ceramic Rods," in *High-Pressure Science and Technology – 1993*, 1994.
4. Simha, C. H. M., *High Rate Loading of a High Purity Ceramic - 1D Stress Experiments and Constitutive Modeling*, Ph.D. Thesis, U.T. Austin, 1998.
5. Chhabildas, L. C., Furnish, M. D., and Grady, D. E., *J. de Physique. IV., Colloque C3, Suppl JPIII* **7**, 137-143 (1997).
6. Klopp, R. W., Shockey, D. A., Seaman, L., Curran, D. R., McGinn, J. T., and de Resseguier, T., " A Spherical Cavity Expansion Experiment Characterizing the Penetration Resistance of Armor Ceramics," in *ASME Winter Annual Meeting Symposium on the Mechanical Testing of Ceramics and Ceramic Composites*, 1994.
7. Subramanian, R., and Bless, S.J., *Int. J. Imp. Engng.* **17**, 807-816 (1995).
8. Anderson, C. E., and Royal-Timmons, S. A., *Int. J. Imp. Engng.* **19(8)**, 703-713 (1997).
9. Skaggs, R., "Review of PHERMEX Confined Ceramic Armor Tests," in *Proc. of the 13th Ceramics Modeling Working Group Meeting*, 1997, pp. 3-51.
10. Coors, *Ceramic Armor Products Catalog*, 1995.
11. Grady, D. E., "Dynamic Failure of Brittle Solids," *Fracture and Damage in Quasibrittle Structures* (1994).
12. Bless, S. J., Subramanian, R., Anderson, C. E., and Littlefield, D., "Prediction of Large Scale High Velocity Penetration Experiments on Ceramic Armor," in *Proc. 13th Army Symposium on Solid Mechanics*, 1993.

INVESTIGATING MULTI-DIMENSIONAL EFFECTS IN SINGLE-CRYSTAL SAPPHIRE*

W. D. Reinhart[1], L. C. Chhabildas[1], W. M. Trott[1], D. P. Dandekar[2]

[1]Sandia National Laboratories, Department 1610, Albuquerque, New Mexico
[2]U.S. Army Research Laboratory, Weapons and Materials Research Directorate
Aberdeen Proving Ground, Maryland

Abstract: Most studies in the past have focused on obtaining uni-axial strain states in shocked materials. In this study, however, results of symmetric impact gas gun experiments on single-crystal C-cut sapphire are described to *observe edge relief waves* as they propagate toward the center of the sapphire target shocked to high pressures. This is made possible by the recent development of a LINE ORVIS, which measures both spatial and time-resolved particle-velocity variations in materials. A series of experiments have been conducted over the impact velocity from ~0.25 to 0.8 km/s, and in the elastic regime (except the 0.8km/s experiment). In these experiments, a new line imaging optically recording velocity interferometer system is used over a line segment of 13mm. Edge relief waves are unmistakably visible with local variations following the edge relief wave. Heterogeneous effects following dynamic yielding is also observed.

INTRODUCTION:

A variety of established diagnostic tools are available to determine the shock Hugoniot of materials. In particular, a single point VISAR[1] (velocity interferometer) is used to measure material velocity, a variable that is necessary to relate the Hugoniot for the material (care is taken to ensure that the conditions of uniaxial-strain are met in these applications; namely, the diameter is large compared to the sample thickness). In this paper, we have attempted to investigate multi-dimensional wave propagation in single crystal C-cut sapphire (i.e crystallographic C-direction is parallel to the sample cylindrical axis). The line imaging ORVIS [2,3,4,5] (Optically Recording Velocity Interferometer System) is projected as a finite length on the target surface allowing measurements of particle velocity not only as a function of time, but also yields spatial distribution over the length of the line. In this study, a smaller diameter C-cut sapphire disc impacts a large diameter C-cut sapphire crystal target. Experiments were conducted over an impact velocity range from 0.25 to 0.8 km/s corresponding to a stress regime of 5.5 to 18.5 GPa (exceeding the elastic limit of approximately 15.5 GPa). A line with the length of over 13mm is projected on the sample. This length was specifically chosen so that multi-dimensional effects emanating from the edge of the impactor would be recorded as it traverses towards the center of the target. Three experiments having slightly different experimental configurations will be discussed. Many interesting details of this well-characterized *homogeneous* material are revealed in this study. Complex and apparently asymmetric variations in the fringe records following the edge relief waves indicate anisotropic behavior. In particular, upon dynamic yielding, the shock front is no longer planar indicative of a spatially heterogeneous yielding process. In all of these experiments, in-plane crystallographic orientation was not considered because different crystallographic orientations in the plane of the crystal, cannot be distinguished, as the longitudinal wave speeds vary as little as ½% in the orthogonal directions [6].

---

* This work was supported by the U. S. Department of Energy under contract DE-AC04-94AL8500000. Sandia is a multiprogram laboratory operated by Sandia Corporation a Lockheed Martin Company, for the United States Department of Energy.

MATERIAL DESCRIPTION:

Single crystal sapphire (Al_2O_3) discs were used for the experiments with axis parallel to the crystallographic C-axis. Sapphire was chosen for

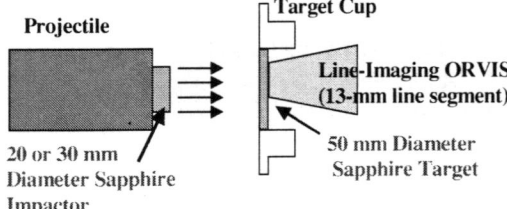

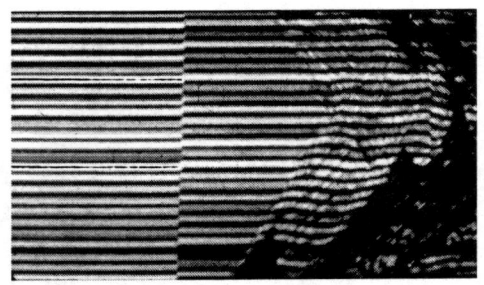

FIGURE 2. Fringe record from LVSAPH-4. Shock arrival at free surface and lateral unloading are unmistakably visible

FIGURE 1. Experimental configuration. Three or ten millimeter flier impacts five-millimeter target with ORVIS imagee on rear surface or insitu.

these experiments because of the very high Hugoniot elastic limit [7,8] and also because it is a well characterized material in the elastic regime [9].

EXPERIMENTAL TECHNIQUE:

The experimental method used in this study is indicated in Figure 1. The experiments presented were conducted on a single-stage light gas gun. A symmetric impact configuration for sapphire was used in this study where the impactor diameter for all experiments was intentionally sized smaller than that of the diameter of the target. Impact velocities were varied to induce stress states from approximately 65 to 180 GPa. Three electrical shorting pins were used to measure the velocity of the projectiles at impact and four similar pins were mounted flush with the impact plane to monitor impact planarity. The insitu and free surface velocity histories were recorded using the recently developed Line Imaging ORVIS. The ORVIS was configured to generate a low magnification image at the detector and provides a line segment of approximately 13 mm. Experimental techniques for generating line-imaging ORVIS [2,3,4,5] have been previously reported, therefore the principles of operation will not be summarized in this paper. This modification is expected to provide macroscopic spatial variations and continuous monitoring of wave fronts as they propagate through the material. Samples were coated with a thin reflective aluminum coating to allow a reflected beam to be monitored during the experiment.

LVSAPH-4: ANALYSIS:

The streak camera records obtained from this experiment performed at an impact velocity of 0.564

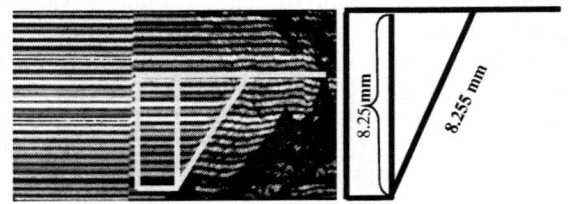

FIGURE 3. Geometric Analysis used for experiments for determination of lateral unloading.

km/s are shown in Figure 2. The ORVIS line is deliberately positioned on the rear surface of the target to observe the multi-dimensional effects from the outer diameter of the impactor as the waves propagate towards the center of the target. For this low magnification, the distance that corresponds to one fringe cycle is 550-µm and the entire line image from the target spans approximately 13 mm. The

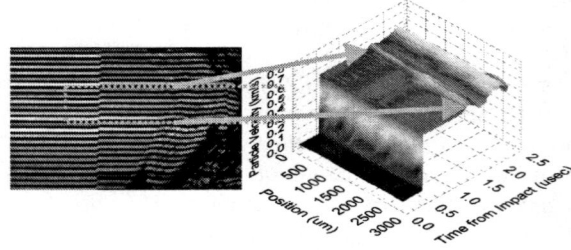

FIGURE 4. 3mm line segment near the center of the target suggests that there is a minor perturbation in the particle velocity measurements caused by the arrival of the edge release

published equation of state parameters given by the empirical relation, $U_s = C_o + su_p$, is used for shock velocity determination. Us is the shock velocity, at a given particle velocity up, and Co (11.91km/s) and s(1) are the parameters in the linear shock-velocity vs particle-velocity relations for sapphire. Simple geometric analysis is used to estimate lateral unloading wave speeds from the edge. As indicated in Figure 3, the initial fringe measurement at the free surface indicates shock arrival, and the change in particle velocity provides an accurate indication of the Hugoniot state. The lateral unloading emanating from the edges of the impactor/target interface is now traversing through a compressed medium, and is clearly visible in the fringe record at the free surface. The velocity-time profile for a 3mm line segment near the center of the target (Figure 4) suggests that there is a minor perturbation in the particle velocity measurements caused by the arrival of the edge release wave. Geometric analyses of the experiment suggest the lateral release wave velocity is estimated to be 12.7 km/s.

LVSAPH-6: ANALYSIS:

The same general experimental configuration was used on this experiment; however, the OVIS line segment was projected through the material to monitor the impact surface. This test now would eliminate most of the wave interactions that is caused by the shock front arrival at the free surface. In this experiment, a 19-mm diameter, 3-mm thick sapphire impacts a 50-mm diameter 5-mm thick target sample at an impact velocity of 0.55 km/s.

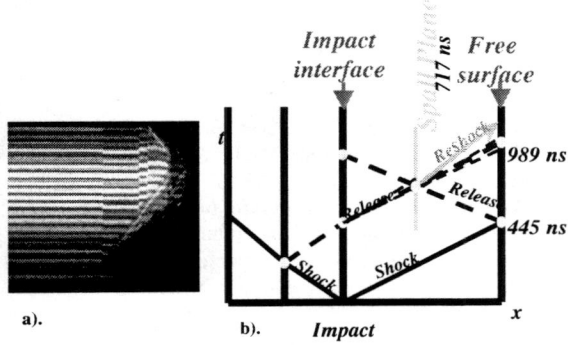

FIGURE 5. Lagrangian x-t diagram depicts the (one-dimensional) wave-interactions occurring in the impactor and the target.

The fringe records indicates impact time at t=0, and also an edge release as depicted by the slanted line. There is a perturbation in the fringe record at a time of 445 ns after impact that is caused by the arrival of the elastic shock front at the free surface. The subsequent change in particle velocity at 544 ns in the fringe record can be attributed to the release wave emanating from the impactor/TPX interface. This is shown in Figure 5b, the Lagrangian x-t diagram, depicts the (one-dimensional) wave-interactions occurring in the impactor and the target. The interaction of the two release waves at 717ns within the sample is clearly seen in the fringe record (Figure 5a). A gradual release followed by a drastic change in light intensity is observed after 717 ns. This is due to spallation of the sapphire target sample. It is interesting that the leading edge of the edge release wave appears to unperturbed in the fringe record. It is not surprising because the edge wave front is monitored at the impact interface while the interactions indicated above are occurring within the sample. An analysis of the edge release waves in this experiment yields a velocity of 11.0 km/s.

CVSAPH-5: Analysis

Using a conventional Visar system, and an experimental configuration illustrated in Figure 6, the lateral unloading was observed. The impact

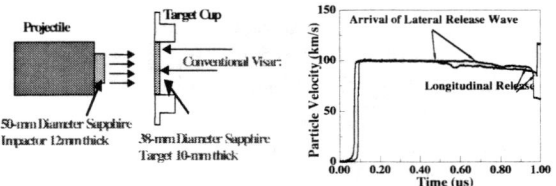

FIGURE 6. Conventional Visar experiment. Lateral unloading velocity of 12.9 km/s.

velocity for this experiment was 0.2 km/s imparting a stress of approximately 4.5 GPa. Two Visar probes were located 3-mm apart and 5 and 8-mm from the target edge respectively. The velocity profiles in Figure 4b, show that lateral unloading wave velocity to be 12.9 km/s.

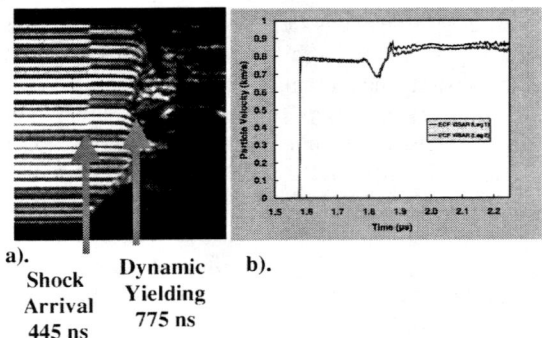

a). Shock Arrival 445 ns Dynamic Yielding 775 ns b).

FIGURE 7. Experiment conducted above the HEL of sapphire. Indication of spatial inhomgenity is apparent.

LVSAPH-3: ANALYSIS:

In this experiment, we deliberately imparted a stress above the dynamic Hugoniot elastic limit. A nominally 30mm diameter, 10mm thick impactor impacts a 50mm diameter, 5 mm thick target. The ORVIS line segment monitored the free surface of the target. In addition, a single point VISAR (Figure 7b) monitored a point (overlapping a segment of the ORVIS line) on the free surface. As seen in Figure 7a, the fringe record shows that there is an elastic wave front arriving at the target free surface at 445 ns after impact. The elastic strength is sustained for approximately 230 ns after which there is a decrease in particle velocity indicative of a precursor decay aassociated with the dynamic yielding process. It is interesting that spatial inhomogenity is observed during yielding. The edge release feature is also observed, however due to the light intensity loss it cannot be monitored once the sample has undergone the yielding behavior.

RESULTS AND CONCLUSIONS:

In this paper, results of experiments conducted to monitor multi-dimensional effects in single crystal sapphire are reported. This is an ongoing study and these are some of the first measurements of its kind in single crystal sapphire. There is, however, a wide variation in edge release wave velocities (~11-13 km/s) measurements. A similar experiment using conventional velocity interferometry gives 12.9 km/s. It appears that this wide variation is related to the anisotropic nature of the crystal. Even though it has been reported that the anisotropy in the two orthogonal crystal axis is quite small, it is quite likely that the pressure dependence of the elastic constants is quite significant. This needs to be pursued further by monitoring edge release waves in well-defined crystallographic directions. The results also indicate spatially inhomogeneous yielding process, even though the studies are performed in single crystals. This explains why there are considerable differences in the Hugoniot elastic limit measurements in the literature.

REFERENCES:

1. Barker, L. M., and Hollenback, R. E., "Laser Interferometer for Measuring High Velocities of any Reflective Surface", J. of Applied Physics 43, 4669, 1972.
2. Bloomquist, D. D. and Sheffield, S. A., "Optically recording interferometer for velocity-measurements with subnanosecond resolution," Journal of Applied Physics, vol. 54, no. 4, pp. 1717-1722, 1983
3. Bloomquist, D. D., and Sheffield, S. A.,"ORVIS, Optically Recording Velocity Interferometer System, Theory of Operation and Data Reduction Techniques, Sandia report, SAND82-2918, February 1983.
4. Trott, W. M. et al., Dispersive Velocity Measurements in Heterogeneous Materials, Sandia report, Sand2000-3082, December, 2000
5. Trott, W. M., et al., "Measurements of Spatially Resolved Velocity Variations in Shock Compressed Heterogeneous Materials using a Line Imaging Velocity Interferometer", in Shock Compression of Condensed Matter, 1999 (M.D. Furnish, L.C. Chhabildas, R.S. Hixson, eds.), part II, pp. 993-998.
6. Graham, R. A., Brooks, W. P., "Shock-Wave Compression of Sapphire from 15 to 420 kbar. The Effects of Large Anisotropic Compressions", J. Phys. Chem. Solids, 1941, Vol. 32, pp.2311-2330.
7. Brooks, W. P., and Graham, R. A., "Shock-wave compression of Sapphire," Bull. American Physical Society vol. 11, no. 3, 414, 1966.
8. Barker, L.M. "Fine Structure of Compressive and Release Wave Shapes in Aluminum Measured by the Velocity Interferometer Technique", Behavior of Dense Media Under High Dynamic Pressures, Gordon & Breach, N.Y. (1968), p. 483
9. Barker, L.M., Hollenbach, R.M., "Shock-Wave Studies of PMMA, Fused Silica, and Sapphire",J. of App. Physics, Vol 41, No. 10, 4208-4226, Sept. 1970.

EXPERIMENTAL CHARACTERIZATION OF THE DYNAMIC FAILURE RESISTANCE OF TiB_2/Al_2O_3 COMPOSITES

A. R. Keller and M. Zhou[†]

The George Woodruff School of Mechanical Engineering
Georgia Institute of Technology, Atlanta, GA 30332-0405

Abstract: The dynamic compressive strength and microscopic failure behavior of TiB_2/Al_2O_3 ceramic composites with a range of microstructural morphologies and size scales are analyzed. A split Hopkinson pressure bar is used to achieve loading rates of the order of 400 s^{-1}. The dynamic compressive strength of the materials is found to be between 4.3 and 5.3 GPa, indicating a strong dependence of strength on microstructure. Microstructures with finer phases as measured by linear intercept length (LIL) have higher strength levels. The dynamic strength levels are approximately 27% higher than the values of 3-4 GPa measured at quasi-static loading rates for these materials. These strength levels are also higher than the strength levels of monolithic TiB_2 and Al_2O_3 under similar dynamic conditions. Scanning electron microscopy (SEM) and energy dispersive spectrometry indicate that failure associated with the Al_2O_3 phase is transgranular cleavage in all microstructures and failure associated with the TiB_2 phase is a combination of transgranular cleavage and intergranular debonding and varies with the microstructures. The measured compressive strength of the materials directly correlates with the fraction of TiB_2-rich areas on fracture surfaces.

INTRODUCTION

One method to reduce brittleness and enhance failure resistance of ceramics is the development of ceramic composites (Niihara et al., 1991). Significant influence of microstructural effects on properties has been reported. For example, Niihara et al. (1991) reported that a 5% population of SiC nanoparticles increases the tensile strength of Si_3N_4 from 350 MPa to 1 GPa and improves its fracture toughness from 3.25 MPa$\sqrt{m}$ to 4.7 MPa$\sqrt{m}$. Although microstructure-induced, size-dependent toughening mechanisms at the micro and nano levels are demonstrated approaches for property enhancement, so far such effects have not been well quantified. In order to develop more advanced materials, it is necessary to characterize the influences of phase morphology, phase length scale, and interfacial behavior on fracture toughness. Recently, two-phase ceramics of titanium diboride/alumina (TiB_2/Al_2O_3) with a range of phase sizes and phase morphologies have been developed (Logan, 1992a,b). These materials have shown a wide range of fracture toughness values and some of the values are higher than those of both constituents produced separately in bulk. These materials provide an opportunity to study the correlation between microstructure and mechanical behavior.

In many applications, intensive dynamic loading occurs under normal operating conditions. Examples include particle impact on ceramic turbine blades (Bilek & Helesic, 1990), contact of a high speed cutting tools with workpiece (Komanduri, 1989), and impact of ballistic projectiles with ceramic armor (Anderson & Morris, 1991). The dynamic behavior of materials can be dramatically influenced by microstructural characteristics, including phase size, phase morphology, composition and texture (Viechnicki et al., 1991). Thus, it is crucial to develop intrinsic relationships between fabrication,

---

[†] To whom all correspondence should be addressed, 404-894-3294, min.zhou@me.gatech.edu.

resulting microstructure, and dynamic properties in order to tailor ceramic systems to the needs of applications (Bilek & Helesic, 1990). In order to establish the relationships, varying processing approaches have been used to produce two-phase TiB_2/Al_2O_3 ceramics with four different biased microstructures. These microstructures are analyzed in this research.

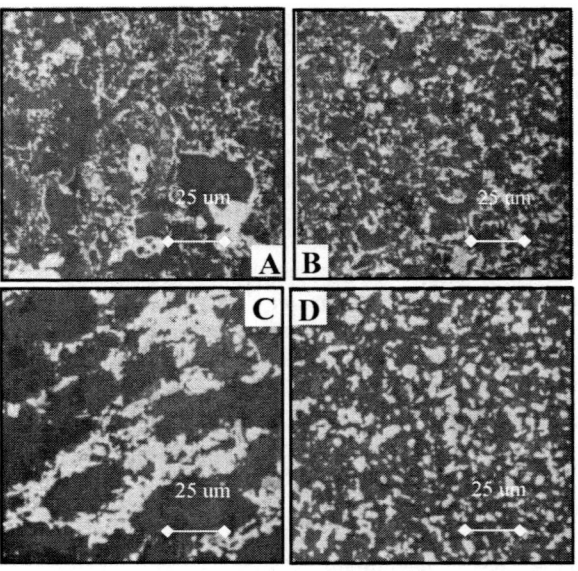

Fig. 1 Microstructures of Materials Analyzed (TiB_2 white, Al_2O_3 dark)

MATERIALS

The materials are produced through self-propagating high temperature synthesis and mechanical mixing of powders followed by hot pressing (Logan 1992b; Carney 1997). Each material has a nominal composition of 70% Al_2O_3 and 30% TiB_2 by weight. Figure 1 shows the microstructures of the four Al_2O_3/TiB_2 composites analyzed. The light areas consist of TiB_2 and the dark areas are Al_2O_3. The microstructures vary in phase distribution and length scales. This microstructure shows intertwining of alumna and titanium diboride. The microstructure is unique in that its complicated morphology precludes the identification of simple crack paths through each phase or along phase interfaces. The microstructure in Fig. 1(b) consists of particles of TiB2 embedded in an Al_2O_3 matrix. The microstructure in Fig. 1(c) consists of isolated Al_2O_3 matrix islands surrounded by a network of TiB_2 reinforcement. The isolated matrix areas are approximately 20×100 μm in size. A dispersion of small Al_2O_3 particles nanometers in size are scattered in the TiB_2 phase. Clearly, two length scales are operative in this microstructure, influencing its strength and fracture toughness. The microstructure in Fig. 1(d) is similar to that in Fig. 1(b) except for significantly larger TiB_2 particle sizes. The average TiB_2 particle size is 5-10 μm in Fig. 1(d) and at least one or two orders of magnitude smaller in Fig. 1(b). The smallest particles in Fig. 1-1(b) are of submicron or nanometer sizes. To quantify the different phase morphologies and sizes, the linear intercept length (LIL) for the two phases are measured on digital images of the microstructures. In the analysis here, the average LIL is used as a measure of phase size.

EXPERIMENTS

In order to characterize the response of the materials to dynamic compressive loading, a split Hopkinson pressure bar apparatus is used. Soft recovery of specimens is achieved through the use of a momentum trapping technique developed by Nemat-Nasser et al. (1991). This mechanism allows specimens to be recovered after loading under a single stress pulse, eliminating any unintended reloading. The specimen fragments were collected for postmortem characterization of microscopic failure.

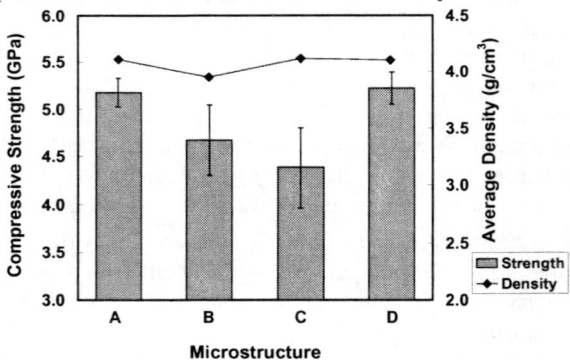

Fig. 2 Compressive Strength and Density

COMPRESSIVE STRENGTH

The measured strengths of the materials and their densities are summarized in Fig. 2. The correlation between the strength and the linear intercept length is shown in Fig. 3. Microstructures A and D show significantly higher strengths than microstructures B and C. Since material B has a much higher level of porosity than the other materials, attention is given only to materials A, C, and D. Higher strengths coincide with smaller LIL. This can be phenomenologically explained. Microstructures with smaller LIL values present a more heterogeneous media for propagating cracks, inducing more tortuous cracks paths and, therefore, increasing energy dissipation for crack growth. The micromechanical quantification of the fracture resistance of the four microstructures accounts for the actual phase distributions and arbitrary fracture patterns.

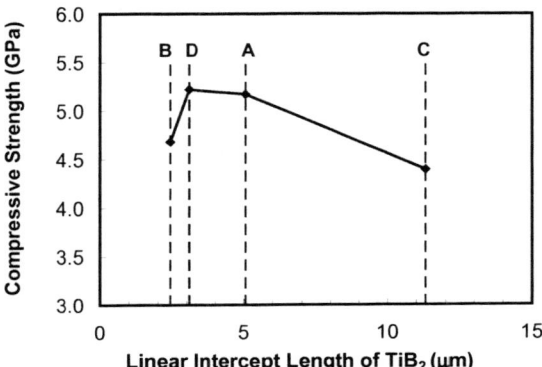

Fig. 3 Correlation between Strength and Linear Intercept Length

FRACTOGRAPHY

Coinciding energy dispersive spectrometry maps showing the distributions of Al_2O_3-rich and TiB_2-rich areas on fracture surfaces are taken and shown in Fig. 4. The use of these phase maps and fractographs greatly facilitates the identification of phases on fracture surfaces and the intergranular or intragranular nature of the fracture process. The failure process in Microstructure A shows large regions of Al_2O_3 surrounded by a continuous network of TiB_2. The Al_2O_3 areas are smooth, indicating transgranular cleavage inside Al_2O_3 grains. Failure involving the TiB_2 phase appears to be primarily intergranular pullout. The overall failure is a combination of transgranular cleavage inside Al_2O_3 and intergranular separation along Al_2O_3/TiB_2 boundaries. Approximately 74% of the fracture surfaces is cleavage planes within the Al_2O_3 and 26% are associated with the TiB_2. Note that the volume fraction of Al_2O_3 in this material is 72.6%. The fracture surfaces for microstructure B are dominated by small regions of transgranular cleavage of the Al_2O_3 phase. Failure involving the TiB_2 phase occurred primarily by intergranular pullout. TiB_2 particles are relatively evenly distributed on the fracture surfaces, suggest a fracture process which is not significantly affected by the presence of the particles. Also, numerous pores were observed in the Al_2O_3 phase, consistent with the measured low density (95% of theoretical value) of this material. Clearly, excessive porosity is detrimental to the resistance to crack propagation in this material. Approximately 84% of the fracture surfaces is Al_2O_3-rich, significantly higher than the 72.6% volume fraction of Al_2O_3 in this material. Thus, the occurrence of fracture is predominantly in the Al_2O_3 phase. Since Al_2O_3 has a lower strength and lower fracture energy compared with TiB_2, this higher ratio of fracture surfaces in Al_2O_3 appears to be the primary reason for the observed low strength reported earlier. A primarily transgranular mode of fracture inside Al_2O_3 is also observed for microstructure C. Fracture surfaces are dominated by large areas of cleavage within the Al_2O_3 phase. However, failure involving the TiB_2 phase occurred through a combination of transgranular cleavage and intergranular pullout. Most parts of the fracture surfaces associated with the TiB_2 show transgranular cleavage. The fracture surface characteristics reflect the morphology of this microstructure--large areas of cleavage in Al_2O_3 surrounded by a continuous network of TiB_2. The fact that cleavage in Al_2O_3 occurs over multiple grains contributes to its lower failure resistance, due to the fact that the effectiveness of the TiB_2 as a reinforcing phase is decreased. Approximately 77% of the fracture surfaces is Al_2O_3-rich, higher than the 72.6% volume fraction of Al_2O_3 in the material. Thus, there appears to be a slight preference for failure to occur in the Al_2O_3 phase. The failure in microstructure D is unique in that transgranular cleavage is the primary fracture mechanism for both Al_2O_3 and TiB_2. Although failure involving the TiB_2 phase occurred through a combination of transgranular cleavage and intergranular pullout, the dominant mechanism is cleavage. This is in sharp contrast to what is observed for the other microstructures. Evenly distributed TiB_2 particles across the fracture surfaces echo the microstructural morphology of fine particles embedded in an Al_2O_3 matrix. It also appears that the homogeneous distribution of TiB_2 reinforcement inhibits intergranular separation, thus forcing cracks to go through the stronger TiB_2 phase and enhancing failure resistance. The rough and ragged nature of the surface is in contrast to the relatively smooth appearance of the surfaces for microstructures A, B, and, most prominently, C. Approximately 71% of the fracture surfaces is Al_2O_3-rich, lower than the 72.6% volume fraction of Al_2O_3 in this material. Note that this is the only microstructure that shows a TiB_2-rich fracture surface fraction higher than its corresponding volume fraction in the material. This bias toward the TiB_2 points to an unusual shift of failure into the stronger phase and can be associated with significantly higher strength.

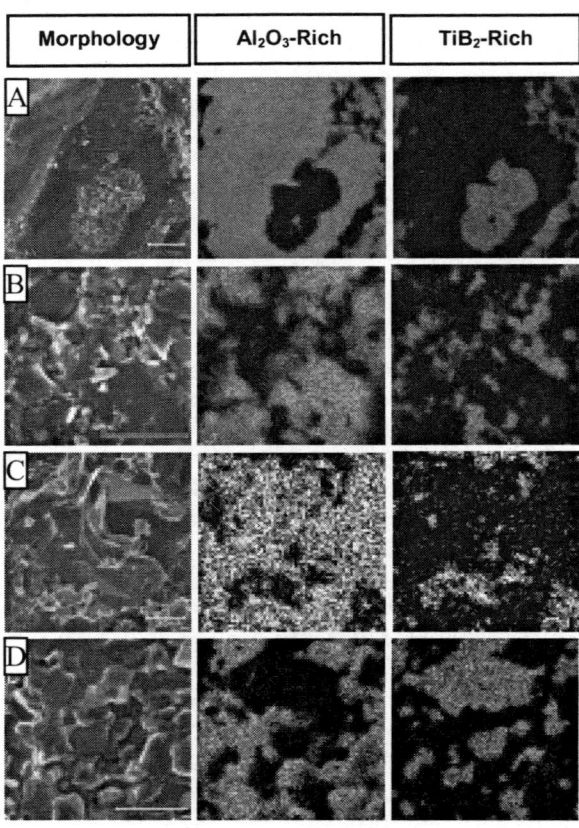

Fig. 4 Fracture Surfaces

DISCUSSION

The analysis has yielded dynamic compressive strength between 4.3 and 5.3 GPa, indicating a strong dependence of strength on microstructure. Microstructures with finer phases as indicated by smaller linear intercept length values have higher strength levels. The dynamic strength levels are approximately 27% higher than the values of 3-4 GPa measured at quasi-static loading rates. Scanning electron microscopy and energy dispersive spectrometry indicate that failure associated with the Al_2O_3 phase is transgranular cleavage in all microstructures. Failure associated with the TiB_2 phase is a combination of transgranular cleavage and intergranular debonding and varies with the microstructures. The fraction of TiB_2-rich fracture surface areas directly correlates with the measured compressive strength of the materials. The addition of titanium diboride reinforcement clearly improves the failure resistance of the materials.

ACKNOWLEDGEMENT

Support from the ARO and NSF is gratefully acknowledged. We are grateful to Dr. K. V. Logan for providing the materials and helpful discussions.

REFERENCES

1. Anderson, C.E., Morris, B.L., (1992), "The Ballistic Performance of Confined Al_2O_3 Ceramic Tiles", *International Journal of Impact Engineering*, **12**(2), 167-187;
2. Bilek, Z, Helesic, J, (1990), "Mechanical Properties of Aluminum Oxide at High Loading Rates", *Proceedings of the First European East-West Symposium on Materials and Processes*, 381-394;
3. Carney, A.F., (1997), "The Effect of Microstructure on the Mechanical Properties of a Titanium Diboride/Alumina Composite", Masters Thesis, Georgia Institute of Technology, Atlanta, GA;
4. Komanduri, R., (1989), "Advanced Ceramic Tool Materials for Machining", *Int. J. Refract. Hard. Met.*, **8**, 125-132;
5. Logan, K.V., (1992a), "Shapes Refractory Products and Method of Making the Same", U.S. Patent # 5141900;
6. Logan, K.V., (1992b), Elastic-Plastic Behavior of TiB_2/Al_2O_3 Produced by SHS, Ph.D. Dissertation, Georgia Institute of Technology;
7. Nemat-Nasser, S., (1991), New frontiers in Dynamic Recovery Testing of Advanced Composites: Tailoring Microstructures for Optimal Performance, *JSME International Journal*, series I, **34**(2), 111;
8. Niihara, K., Nakahira, A., Sekino, T., (1991), New Nanocomposite Structural Ceramics, *Mat. Research Society Symposium Proceedings*, **286**, 405-412;
9. Nordgen, A., Melander, A., (1988), "Influence of Porosity on Strength of WC-10%Co Cemented Carbide", *Powder Metallurgy*, **31**(3), 189-200;
10. Viechnicki, D.J., Slavin, M.L., Kliman, M.I., (1991), *Ceramics Bulletin*, **70**, 1035.

FRAGMENTATION OF EXPANDING CYLINDERS AND THE STATISTICAL THEORY OF N. F. MOTT

Dennis Grady

Applied Research Associates, 4300 San Mateo Blvd., A-220, Albuquerque, New Mexico 87110

Abstract. The seminal investigation of the explosive fragmentation of steel cylindrical shells by N. F. Mott in the early 1940's led to an elegant statistics-based theory of dynamic fragmentation (Mott, 1947). Experiments in which rapidly expanded metal rings undergo dynamic fragmentation provide ideal data for testing the fragment size and statistical distribution predictions of Mott's theory. In this work the theoretical development of Mott is examined and compared with expanding metal ring experimental data.

INTRODUCTION

The fragmentation of hollow metal shells subjected to rapid expansion by impulsive internal pressure loading continues to be a problem of both practical and intellectual interest. The seminal modeling of this fragmentation process by Mott (1947) continues to be a fascinating framework for investigating the consequences of differing fracture physics. The model is illustrated in Figure 1 in which a uniformly expanding metal cylinder (the Mott cylinder) has exhausted deformation hardening and the circumferential stress versus strain curve is horizontal tangent at a nominally constant dynamic tension Y. Fracture is assumed to occur within a fairly short time through a statistical fracture activation and growth process. Fractures activate at random points on the Mott cylinder. Growth is the process of stress wave propagation by which release of the tensile stress is communicated to regions away from fractures points. Density of the material is ρ while the nominal stretching rate $\dot{\varepsilon}$ is the ratio V/R of the expansion velocity and cylinder radius. In the present effort we will investigate two alternative theories underlying the fracture activation and growth process, and then offer possible explanations for reconciling them. First is the energy-based approach (Grady et al., 1984; Kipp and Grady, 1985). Second is the statistical strain-to-fracture theory of Mott (1947).

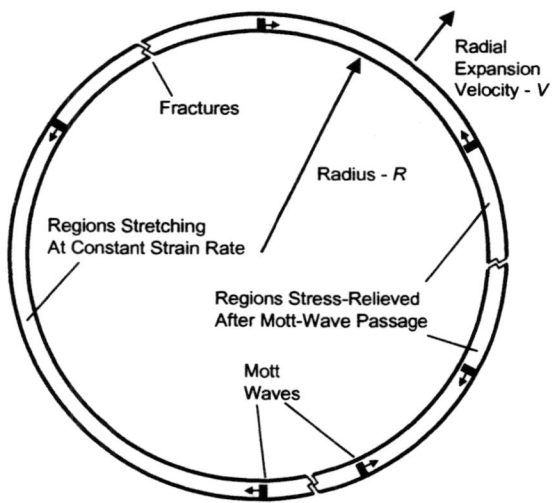

FIGURE 1. The Mott Cylinder. A model for the statistical fragmentation of an expanding cylindrical metal shell.

ENERGY-BASED FRAGMENTATION THEORY

At the heart of the energy-based fragmentation theory are details of the fracture activation process. In the 1947 paper Mott offered a solution for the propagation of stress release away from the fracture point (the Mott wave) by considering instantaneous fracture and tensile stress drop at a point and subsequent motion of a rigid region behind the propagating Mott wave through momentum considerations (Mott, 1947; Kipp and Grady, 1985). The solution,

$$x = \sqrt{(2Y/\rho\dot{\varepsilon})}t^{1/2}, \quad (1)$$

was obtained for the position of the Mott wave as a function of time t.

The solution of Mott was extended (Grady et al., 1984; Kipp and Grady, 1985) to account for dissipation at the fracture point and the time-dependent release of tensile stress. A similar solution for the position of the Mott wave, again assuming rigid-plastic properties, yields,

$$x = \frac{1}{12}\frac{Y^2}{\rho\gamma}t^2, \quad (2)$$

over the activation time τ from initial perturbation at the fracture site until fracture separation. The energy dissipated in the fracture is 2γ during the time τ.

It is readily recognized that if two fractures begin to activate within a sufficiently small linear region of the Mott cylinder than their respective Mott waves will interact before activation is complete and one or the other will arrest and not go to completion. To achieve theoretical closure it was reasoned that twice the distance traveled by the Mott wave over the time τ,

$$L_o = \left(\frac{24\gamma}{\rho\dot{\varepsilon}^2}\right)^{1/3} \quad (3)$$

should correspond to an average fragment length. A coupled statistical theory leading to prediction of the probably distribution in fragment lengths was not pursued.

STATISTICAL STRAIN-TO-FRACTURE THEORY

Mott proceeded by arguing that energy dissipated in the fracture process was unimportant. Rather, he chose to assume that fracture occurred instantaneously at points on the Mott cylinder according to a probabilistic expression which increased rapidly with the strain at that point. In particular he proposed a probabilistic hazard function of the exponential form,

$$\lambda(\varepsilon) = Ae^{\sigma\varepsilon}, \quad (4)$$

that provided the chance of fracture within unit length of the cylinder at the strain ε. It is readily shown that Equation 4 leads to an extreme value distribution of the Gumbel type where $1.28/\sigma$ is the standard deviation of the distribution (Hahn and Shapiro, 1967). Other extreme value distributions could of course be considered.

Mott (1947) then used Equation 4 for the fracture activation law and Equation 1 for the propagation of Mott waves to determine the probabilistic number of fracture sites and the probabilistic distribution in position of these sites over the fracture activation and growth duration. Through a graphical method he determined the characteristic fragment length and the distribution in lengths about this mean.

Grady (1981a, 1981b) applied statistical methods of Johnson and Mehl (1938) to provide an analytic solution to the theoretical approach proposed by Mott. It can be shown that the average fragment length is

$$L_o = \sqrt{\pi}\sqrt{\frac{2Y}{\rho\dot{\varepsilon}^2}\frac{1}{\sigma}}, \quad (5)$$

while the probability density distribution in fragment lengths is,

$$f(L) = \frac{\beta}{L_o}\left(\frac{L}{L_o}\right)^3 e^{-\frac{1}{4}(L/L_o)^3}\int_0^1 (1-y^2)e^{-\frac{3}{4}(L/L_o)^3 y^2}dy, \quad (6)$$

although a simpler strain-to-fracture hazard function than Equation 4 was assumed to carry through the analytic solution (Grady, 1981a, 1981b) for the fragment distribution. Comparison of Mott's graphical distribution and the analytic distribution in

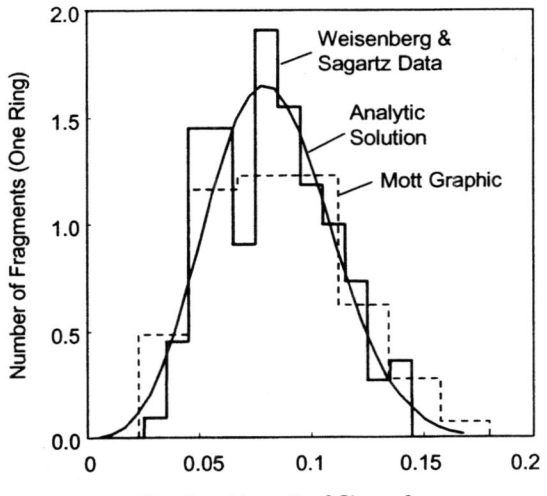

FIGURE 2. Comparisons of analytic and graphical solutions for the Mott statistical fragment size distribution and expanding ring data of Weisenberg and Sagartz (1977).

Figure 2 suggests a lack of sensitivity to the assumed hazard function but this has not been proved. Comparisons with expanding ring data of Weisenberg and Sagartz (1977) are also shown.

COMPARISON WITH EXPERIMENTAL DATA

An extensive series of fragmenting aluminum and copper ring experiments using magnetic loading (Grady and Benson, 1983) has been compared with the energy-based prediction of fragment length in Equation 3. The data are reasonably consistent with fracture energies measured by independent methods although the average fragment length dependence on strain rate is closer to an inverse first power rather than the predicted inverse two-thirds power. More recent unpublished expanding ring data on other metals have shown inverse two-thirds power dependence, however. Further, exploding cylinder fragment size data are reasonably predicted with the energy-based fragment size relation using published fracture toughness properties (e.g., Reedal et al., 1999).

Mott's theoretical prediction of fragment length based on statistical strain-to-fracture concepts is less readily compared with data due to the unavailability of the parameter σ ($1.28/\sigma$ is the standard deviation in strain-to-fracture). Calculated values of σ of a few percent to a few tens of percent for existing data are certainly reasonable, however, and the predicted inverse linear dependence on strain rate of Equation 5 is seen in at least some of the data (Grady and Bensen, 1983). Further, several comparisons of the Mott statistical size distribution (Figure 2) with experimental data have been made (Grady et al., 1984; Grady and Benson, 1983) and results are quite satisfying. Comparison with the data of Weisenberg and Sagartz (1977) shown in Figure 2 is representative.

DISCUSSION

Two theories of fragmentation following the fundamental fracture activation and growth framework of Mott have been pursued. That of Grady and Kipp is based on energy dissipated within the fracture activation process and the length scale governing the predicted fragment size contains that fracture energy. Mott's statistical strain-to-fracture theory, on the other hand, ignores fracture energy and predicted that the average fragment length is proportional to a length scale which is a unique dimensional combination of the flow stress Y, density ρ and strain rate $\dot{\varepsilon}$. Magnitude of the average fragment length is determined by the temporal standard deviation in fracture activation ($\varepsilon = \dot{\varepsilon} t$). As $1/\sigma$ goes to zero so does the average fragment size provided from Equation 5.

Both theories have attractive features and experimental data to date do not strongly favor one over the other. Several issues should be considered which would effectively merge the two. First, a statistical variation in the fracture energy γ could be considered in the energy-based theory of Grady and Kipp. This in turn would lead to a statistical variation in the activation time τ and hence the temporal variation in fracture assumed by Mott. This statistical variation in γ was tacitly assumed in the energy-based approach of Grady and Kipp anyway. Otherwise the predicted average fragment length in Equation 3 would more appropriately be a minimum fragment length. Secondly, even if each fracture energy were nominally the same, temporal variations in initial fracture perturbations should be expected.

Thus there is every reason to expect that a merging of the energy-based concepts of Grady and Kipp with the temporal fracture statistics concepts of Mott would lead to an improved predictive theory of dynamic fragmentation.

REFERENCES

1. Mott, N. F. (1947), *Proc. Royal Soc.*, A189, 300-308, January.
2. Grady, D. E., M. E. Kipp, and D. A. Benson (1984), *Inst. Phys. Conf. Ser.* **70**, 315-384.
3. Kipp, M. E. and D. E. Grady (1985), *J. Mech. Phys. Solids*, **33**, 399-415.
4. Hahn and Shapiro (1967)
5. Grady, D. E. (1981a), *J. Geophys. Res.* **86**, 1047-1054.
6. Grady, D. E. (1981b), Shock Waves and High-Strain-Rate Phenomena in Metals, M. A. Meyers and L. E. Murr, Eds., Plenum, New York, 181-191.
7. Johnson and Mehl (1938)
8. Reedal, D., L. Wilson, D. Grady, L. Chhabildas, and W. Reinhart (1999), Proceedings of the 15th U. S. Army Symposium on Solid Mechanics, 569-585, Myrtle Beach, SC, 12-14 April, 1999, Battelle Press, Columbus, OH.
9. Grady, D. E. and D. A. Benson (1983), *Exp. Mech.*, **23**, 393-400.
10. Wesenberg, D. L. and M. J. Sagartz (1977), *J. Appl. Mech.*, **44**, 643-646.

DIGITAL SPECKLE X-RAY FLASH PHOTOGRAPHY

S.G. Grantham, W.G. Proud

Cavendish Laboratory, Madingley Road, Cambridge, CB3 0HE. UK

Abstract. The new technique of digital speckle X-ray flash photography (DSXFP), which has been successfully applied to polyester and cement specimens[1], is being further developed and used to study materials in ballistic situations in a way not previously possible. The technique involves seeding the specimen with a lead layer and then taking flash X-ray images before and during an impact event. Digital cross-correlation can then be used to make measurements of the internal displacements occurring throughout the specimen. Using a stereoscopic geometry the out of plane displacements can also be determined and a full 3-dimensional displacement map constructed. In this paper these two powerful and complementary techniques of flash X-rays and DSXFP are used to study the ballistic response of a borosilicate sample to produce information that other techniques are unable to provide.

INTRODUCTION

The study of how glass fractures during a ballistic impact is of great interest given the conditions in which glass is often used. The way a glass windscreen or bullet proof glass reacts to a high velocity impact are examples of this. In this paper the processes by which a sample of glass fractures and fails under such a high velocity impact is studied.

Techniques that can be used for investigating the fragmentation of glass in high-speed ballistic events are limited. High-speed photography becomes ineffective when trying to look at the behaviour behind the damage front, as the glass comminutes and shatters leading to optical opacity. Thus, high-speed photography can give the velocity of the damage front but little else. The use of gauges is also not a viable option since this involves altering the structure of the sample quite dramatically by cutting it and inserting the gauge. If flash X-ray photography is used this can provide information on the position of the projectile without the opacity behind the damage front being a problem. This technique relies on the sample fracturing in the same manner in different experiments, as only one flash X-ray can be taken per experiment. This has been shown to be the case in previous work[2], however, and hence this technique can be applied with reasonable confidence. This still does not allow displacement measurements within the sample to be made. A dramatic improvement on flash X-ray photography is digital speckle X-ray flash photography (DSXFP). This is a relatively new technique which allows quantitative measurements of the displacement on a plane within the sample during a ballistic impact to be made. The technique requires a random sprinkling of an X-ray opaque material on a flat plane within the sample. A "before" image is taken, and then, another flash X-ray is taken at a given point during the impact event. The resulting random speckle patterns can then be correlated to find the maximum of correlation and hence where the random pattern has moved to[3]. By repeating this correlation using small subimages from the reference and deformed images, a map of displacement vectors can be calculated. The resolution of these vectors, and hence the resolution of the displacement measurements, is defined by the size of these subimages and the stepsize taken between subimages.

In the following research some traditional flash X-ray images, of borosilicate glass being impacted, have been taken which provide information on the material's failure. A specially manufactured sample is then used for DSXFP, and a comparison between these different results made.

EXPERIMENTAL

Blocks of borosilicate with a 30×15 mm^2 cross sectional area and a length of 60 mm were used for the standard flash X-ray photographs. The projectiles used were mild steel rods 9.15 ± 0.05 mm in diameter and 80 ± 0.5 mm in length with a rounded tip and were fired at a velocity of 190 ± 3 m s^{-1}. The X-ray images were taken at delays of 26µs, 79µs, 105µs and 184µs using a 150keV X-ray head with a 30 ns exposure. The X-ray film was sandwiched in a cassette between two image intensifier plates to increase the exposure and with a 2 mm lead sheet behind to prevent X-rays escaping the apparatus. The sample used for the X-ray speckle study was made by spinning a tube of glass out into two flat discs, which were then flattened. These two discs were placed in a carbon holder with a layer of tungsten filings (50 to 250 µm in size) sprinkled between them and heated to 950°C for 1 hour. The sample was then cut to 42×42 mm^2 and 10 mm thick. The tungsten layer was at a height of 5 mm through the centre of the sample. Tungsten was used for this particular experiment instead of lead because it has a thermal expansion coefficient that is better matched to glass. This prevents any cracking of the sample during cooling.

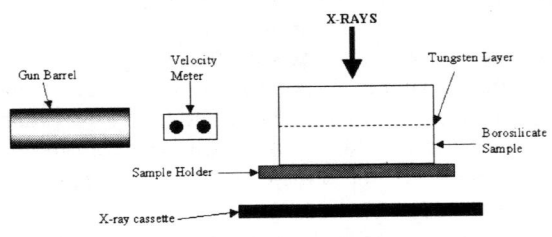

FIGURE 1. Experimental setup for DSXFP

The projectile had a velocity of 197.6 ± 3.0 m s^{-1} and a delay of 40 µs was used. The experimental configuration used for these experiments is illustrated in Fig. 1.

RESULTS

Fig. 2. through to Fig. 5. show the results from normal flash X-ray photographs of the mild steel rod impacting the block of borosilicate glass. It can be seen in Fig. 1. that the steel rod has been flattened on impact and has caused cracks to propagate away from the point of impact and into the glass. A crack also appears to be forming directly ahead of the projectile on the rear surface of the block. Fig. 2. shows the sample beginning to bulge outwards at the rear surface. The first signs of debris being ejected from this region can also be seen. The notch on this projectile is a marker which was placed 15 mm from the tip of the projectile to allow the degree of deformation in the rod to be gauged. At 105 µs following impact (Fig. 4), the rear surface of the glass is clearly starting to shatter and break into pieces. However the two sides of the borosilicate block still appear to be relatively intact. Whilst they are moving outwards and away from the projectile, the sides are not suffering from the same degree of fracturing as the main central section of the sample. The final X-ray image (Fig. 5), taken at 184 µs post impact, clearly depicts very little coherent structure left in the sample, whilst the side sections are also beginning to fail catastrophically. From these results, the expected behaviour of the glass in this velocity range has been observed. When performing the experiment on the tungsten seeded glass sample, a quantitative comparison of the structural response can be made.

The X-ray image taken during the impact is reproduced in Fig. 6. The image is a computer scan of a contact print taken from the X-ray film negative. A contact print has been used, rather than just a scan of the negative, so that more contrast can be brought up on the image, which allows a better correlation to be achieved. Two bolts can be seen behind the sample, these were used to hold the sample in place and the area behind these bolts at the top of the picture is an area of fiducial markers. These were attached to the platform that the specimen was mounted on so that rigid body motions can be eliminated. The displacement

FIGURE 2. Rod impacting glass at 190 m s⁻¹, 26 μs delay.

FIGURE 3. Rod impacting glass at 190 m s⁻¹, 79 μs delay.

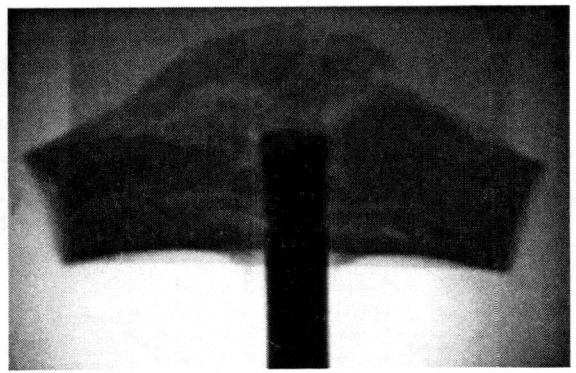

FIGURE 4. Rod impacting glass at 190 m s⁻¹, 105 μs delay.

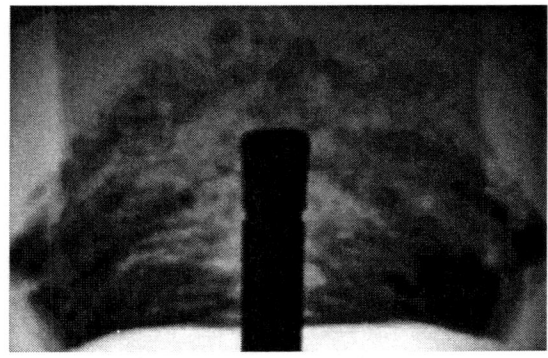

FIGURE 5. Rod impacting glass at 190 m s⁻¹, 184 μs delay.

vectors, which are overlayed, have been scaled up by a factor of 3 and one pixel represents 0.063 mm in the frame of the sample. Cracks in the glass are beginning to propagate away from the tip (represented by white dashed lines in Fig. 6.) of the projectile and, again the projectile has been flattened on impact. The displacement vectors, produced by the correlation process, appear to agree with the behaviour previously observed in the normal X-ray impacts. The central section appears to be moving backwards away from the projectile in a large section, and the regions to either side of the projectile appear to be moving outwards. These are moving to a lesser degree than the main central section however. The major fragments of the glass were retrieved after the experiment and it was found that out of 154 fragments, only 10 appeared to have tungsten filings on any of the outer surfaces. From this it would seem reasonable to infer that the glass has not preferentially fractured along the seeded layer.

CONCLUSION

We have shown that the displacement field, on a flat plane, within a glass sample can be measured during an impact event. We have also carried out standard impact tests on a non-seeded specimen to verify that the behaviour exhibited in the seeded case is reasonable, and the effect of the seeding

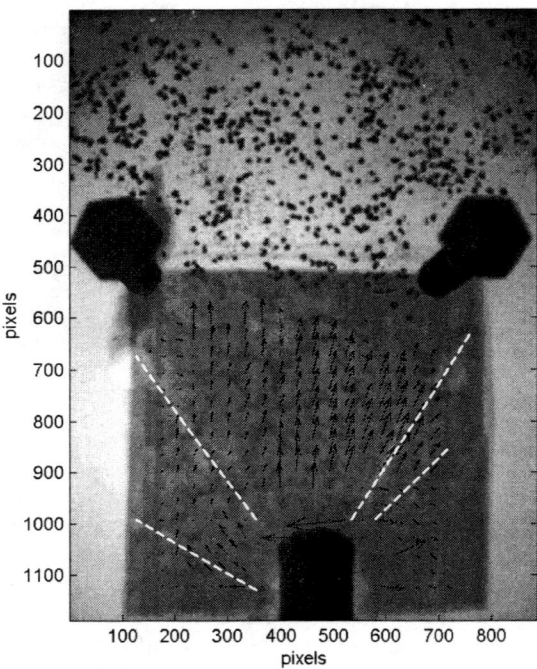

FIGURE 6. Displacement field for rod impact after 40 µs.

does not appear to have altered the structural response of the glass significantly. The natural progression now is to make thicker, and hence more realistic glass samples. These samples can then be used to measure the displacement field at varying delay times, projectile velocities and depths through the sample. The full stereoscopic X-ray speckle treatment can also then be applied to give a truly three-dimensional measure of the internal displacements[4].

ACKNOWLEDGEMENTS

The authors thank Prof. J.E. Field and Dr H.T. Goldrein, Cambridge and Dr. I.G. Cullis (DERA) for their advice and encouragement. The research is supported by the Engineering and Physical Research Council (EPSRC), and the Defence Evaluation and Research Agency (DERA). R. Smith is thanked for his technical help in preparing the samples.

REFERENCES

1. Synnergren, P., Goldrein, H.T., Proud, W.G., *Appl. Opt.* **38**, 4030-4036 (1999).
2. Bourne, N.K., Forde, L.C., Millett, J.C.F., Field, J.F., *J. Phys. IV France Colloq. C3* **7**, 157-162 (1997).
3. Sjödahl, M., Benckert, L.R., *Appl. Opt.* **32**, 2278-2284 (1993).
4. Goldrein, H.T., Synnergren, P., Proud, W.G., "Three-Dimensional Displacement Measurements Ahead of a Projectile," in *Shock Compression of Condensed Matter-1999*, edited by M.D. Furnish, L.C. Chhabildas, and R.S. Hixson, AIP Conference Proceedings 505, Snowbird, Utah, 1999, pp. 1095-1098.

THE DEVIATORIC RESPONSE OF THREE DENSE GLASSES UNDER SHOCK LOADING CONDITIONS

D.D. Radford, W.G. Proud, and J.E. Field

PCS, Cavendish Laboratory, Madingley Road, Cambridge, CB3 0HE. UK.

Abstract. In-material longitudinal and lateral stress histories in three dense, silica-based glasses were directly measured by embedded manganin stress gauges during plate impact experiments. Lateral stress profiles in all of the materials show evidence of failure fronts that behave in a similar manner to those observed in open-structured glasses. The measured stress histories were used to calculate the deviatoric responses and results indicate that ahead of the failure front the shear stress increases linearly along the estimated elastic response. Behind the failure front, however, the shear stress appears to first decrease and then increase as the pressure increases, contrary to a previous interpretation [1].

INTRODUCTION

Previous work at the Cavendish Laboratory on silica-based glass materials under shock loading has demonstrated that the material response is highly dependent upon the composition of the glass [1, 2]. The shock responses of glasses with an open structure, such as borosilicate, exhibit a ramping behaviour in the longitudinal stress histories due to structural collapse. Glass materials with a *"filled"* microstructure, as in the case of Type-**D**, **E**xtra **D**ense **F**lint (DEDF) do not exhibit a ramping behaviour and behave in a manner similar to polycrystalline ceramics [3]. Partially filled materials, such as soda-lime glass, show an intermediate response.

Although the shock response of the glasses previously tested varied considerably, one common feature was the existence of a damage front that propagated behind the initial shock at a lower velocity [4, 5]. Behind this damage front, denoted as the failure wave [6], or *failure front*, the shear strength undergoes significant reduction and the spall strength is essentially reduced to zero [7, 8]. Based on the deviatoric response in soda-lime, borosilicate and DEDF glasses, and two ceramics, it was suggested that the shear stress behind the failure front is independent of impact stress [9].

In the current investigation, results from plate impact experiments on three filled silica-based glasses are presented for a wider range of pressures than in previous studies. Contrary to the previous interpretation [1], there is now sufficient data to show that the shear stress behind the failure front initially decreases and then increases with increasing impact stress. A similar dependence of shear stress on pressure is assumed in the well-known model developed for brittle materials [10] and has recently been observed in cement paste subjected to shock loading [11].

MATERIAL

The glass materials used in this investigation consisted of two type-DEDF glasses and another filled glass designated as type-SF. The glasses are silica-based with different amounts of lead as the main filler material. One of the DEDF glasses was also tested in previous investigations [1-3, 5], and is denoted as DEDF-1 in this investigation. The new DEDF material is denoted as DEDF-2, and the type-SF material is denoted as SF-57. The wave speeds of materials were measured with quartz transducers in both the longitudinal and transverse orientations, using a Panametrics-5052PR pulse receiver operating at 5 MHz. The measured longitudinal (c_L) and shear wave (c_S) speeds, and the density at ambient conditions (ρ_0) were used to calculate the elastic properties, as shown in Table 1.

TABLE 1. Elastic Properties of Dense Glasses.

Glass Type	Density (g cm^{-3})	C_L (mm µs^{-1})	C_T (mm µs^{-1})	C_0 (mm µs^{-1})	Poissons ratio	Impedance (10^3 kg m^{-2}s^{-1})
DEDF-1	5.18±0.05	3.49±0.01	2.02±0.01	2.60±0.01	0.25±0.005	18.08
SF-57	5.53±0.02	3.42±0.01	1.98±0.01	2.53±0.02	0.25±0.005	18.91
DEDF-2	5.94±0.06	3.27±0.03	1.86±0.01	2.47±0.02	0.26±0.005	19.42

EXPERIMENTAL PROCEDURE

Plate impact experiments were performed on all three filled glasses in this investigation using the 50-mm, single stage gas gun at the University of Cambridge [12]. Longitudinal stress measurements were taken by embedding piezoresistive manganin gauges (Micromeasurements type LM-SS-210FD-050) between tiles of the target materials, or supported on the back of the target with a block of polymethylmethacrylate (PMMA). Lateral stress measurements were performed by embedding manganin gauges (Micromeasurements type J2M-SS-580SF-025) in the lateral orientation at various distances from the impact face. Figure 1 shows the specimen configurations used in the investigation. Gauge calibrations were performed for each experiment according to the work of Rosenberg [13], and the specimens were assembled using a low viscosity epoxy adhesive.

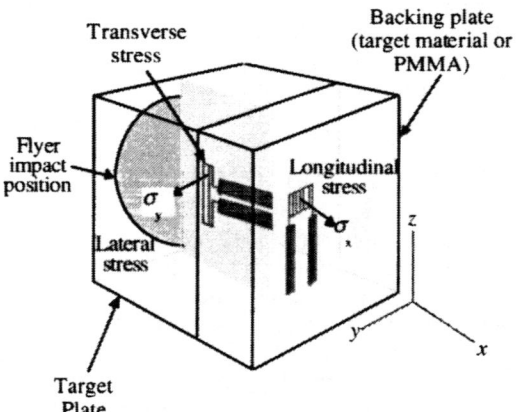

FIGURE 1. Specimen configuration.

Experiments were performed using 6, 10, and 19-mm-thick targets that had a minimum diameter of 50 mm. Copper and tungsten-alloy flyer plates (5 to 20 mm) were impacted onto the targets at velocities ranging from 0.19 to 0.84 km s^{-1}. Impact velocities were measured by the shorting of sequential pairs of pins to an accuracy of ± 0.5 %, and the specimen alignment was fixed to be less than 1 mrad by of an adjustable specimen mount.

RESULTS AND DISCUSSION

Longitudinal stress (σ_x) histories from tests performed in this investigation confirmed the HEL reported for the DEDF-1 material (4.3 GPa [3]) and extended the range of data. The HELs of the new glasses were determined to be ca. 4.4 GPa. The behaviour of the DEDF-2, however, is unique relative to the previously tested silica-based materials. In this material, there is a distinct two-wave structure that initiates at approximately the HEL. It appears that a structural densification or rearrangement of the SiO$_2$ tetrahedral is occurring that depends on material composition [14]. Work specifically related to the densification of glass materials is ongoing and will be addressed in future publications.

Results from the longitudinal stress histories were used in conjunction with the impedance matching technique to determine the principal Hugoniots, shown in Fig. 2. The Hugoniots of these filled glasses are consistent with the behaviour of other silicate glasses and follow the expected trend based on the impedance, as outlined in Table 1.

Lateral stress (σ_y) histories under the same impact conditions as the corresponding longitudinal measurements were measured to allow the deviatoric response of the materials to be characterised. Figure 3 shows the lateral response of the DEDF-2 material measured in three experiments using gauges located at 3 and 6 mm from the impact face. In these experiments longitudinal stresses of 3.3, 6.4, and 10.7 GPa, respectively, were produced. From the stress histories for the experiment in which

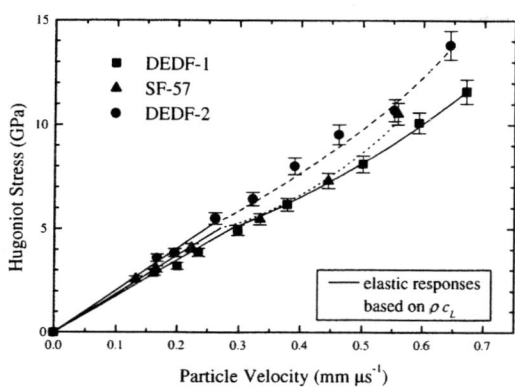

FIGURE 2. Principal Hugoniots of dense glasses in this study.

$\sigma_x = 6.4$ GPa, it is clear that a two-wave structure associated with the propagation of a failure front is present [1]. Similar results were observed in the other glasses tested and the existence of failure fronts was confirmed using high-speed photography. Comparing the lateral responses, it is seen that the delay time between the incident wave and failure front decreases as the longitudinal stress, or impact pressure, increases. This trend of decreasing delay times agrees with previous results on glasses and indicates that the failure front velocity increases with pressure [9]. It is also noted that there is a transition from the failure front lagging behind the incident shock, to the point where it is propagating with the incident shock, as the longitudinal stress is increased. In this case ($\sigma_x = 10.7$ GPa) there is no initial step in the lateral response, but an increase directly to the higher value of stress behind the failure front.

Using the lateral stress measurements ahead of, and behind the failure front, the shear stress (τ) defined by,

$$2\tau = \sigma_x - \sigma_y \qquad (1)$$

was calculated. The resulting deviatoric responses for the three glasses tested, as a function of longitudinal stress, are presented in Fig. 4. The results are based on measurements at 6 and 10 mm in the DEDF-1, and 3 and 6 mm in the SF 57 and DEDF-2. From Fig. 4, it is seen that the same basic trend occurs for all the filled glasses. That is, ahead of the failure front the shear stress increases linearly along the estimated elastic response, based on Poisson's ratio, to ca. 6 GPa. Behind the front, however, the shear stress appears to first decrease and then increases as the impact stress increases.

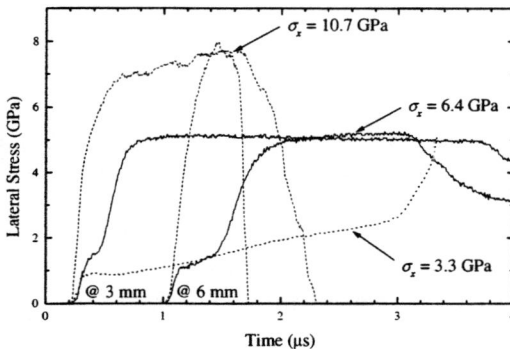

FIGURE 3. Lateral stress histories measured in DEDF-2 at 3 mm and 6 mm from impact face in three experiments.

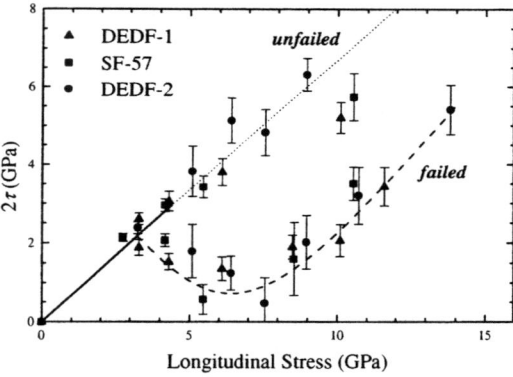

FIGURE 4. Deviatoric responses of dense glasses tested up to ca. 14 GPa longitudinal stress.

This observation modifies the previous interpretation based on longitudinal stress data to ca. 8.5 GPa, which suggested that the shear stress behind the failure front remains constant [1]. In that investigation, results from three silica-based glasses and two polycrystalline ceramics were shown to follow the same non-failed and failed curves, with the shear stress behind the failure front remaining constant within experimental error. However, study of the earlier data (Fig.4 of [1]) does suggest the trend now established in the present study. A plausible explanation is that the comminuted material behind the failure front consists of fragments that will impart force/resistance as they attempt to move past one another under a shear force. As the pressure is increased, more force is required to move the fragments, due simply to mechanical constraint. This explanation is consistent with the well-known model for brittle materials, due to Johnson and Holmquist [10]. Similar behaviour has recently been observed in cement paste [11], which also behaves in a brittle manner.

CONCLUSIONS

Results from plate impact experiments on three dense glass materials have been presented. Longitudinal stress histories were used to determine the principal Hugoniot. Lateral stress histories show the existence of failure fronts in all three materials. Based on corresponding longitudinal and lateral stress measurements, the deviatoric response was characterised. The calculated shear strengths ahead of the failure fronts were similar to previously tested silica-based glasses. Behind the failed material, however, the shear stress shows a dependency on impact stress.

ACKNOWLEDGEMENTS

We acknowledge financial support from DERA, under contract SMC/4U1348, and are grateful to T. Andrews, P.D. Church, B. Goldthorpe, and I.G. Cullis for encouragement. We also thank J.C.F Millett and N.K. Bourne for performing some of the early experiments on the DEDF-1 material. Finally, we acknowledge D.L.A. Cross and R.P. Flaxman for valuable technical support.

REFERENCES

1. Bourne, N.K., J.C.F. Millett, and J.E. Field, *On the strength of shocked glasses.* Proc. R. Soc. Lond. A, 1999. **455**: pp. 1275-1282.
2. Bourne, N.K., Z. Rosenberg, and A. Ginzburg, *The ramping of shock waves in three glasses.* Proc. R. Soc. Lond. A, 1996. **452**: pp. 1491-1496.
3. Bourne, N.K., J.C.F. Millett, and Z. Rosenberg, *The shock wave response of a filled glass.* Proc. R. Soc. Lond. A, 1996. **452**: pp. 1945-1951.
4. Bourne, N.K., Z. Rosenberg, Y. Mebar, T. Obara, and J.E. Field, *A high-speed photographic study of fracture wave propagation in glasses.* J. Phys. IV France Colloq. C8 (DYMAT 94), 1994. **4**: pp. 635-640.
5. Bourne, N.K., J.C.F. Millett, and Z. Rosenberg, *Failure in a shocked high-density glass.* J. Appl. Phys., 1996. **80**: pp. 4328-4331.
6. Rasorenov, S.V., G.I. Kanel, V.E. Fortov, and M.M. Abasehov, *The fracture of glass under high pressure impulsive loading.* High Press. Res., 1991. **6**: pp. 225-232.
7. Brar, N.S., S.J. Bless, and Z. Rosenberg, *Impact-induced failure waves in glass bars and plates.* Appl. Phys. Letts, 1991. **59**: pp. 3396-3398.
8. Brar, N.S. and S.J. Bless, *Failure waves in glass under dynamic compression.* High Press. Res., 1992. **10**: pp. 773-784.
9. Bourne, N.K., J.C.F. Millett, Z. Rosenberg, and N.H. Murray, *On the shock induced failure of brittle solids.* J. Mech. Phys. Solids, 1998. **46**: pp. 1887-1908.
10. Johnson, G.R. and T.J. Holmquist, *A computational model for brittle materials subjected to large strains, high strain rates and high pressures*, in *Shock-Wave and High-Strain-Rate Phenomena in Materials*, M.A. Meyers, L.E. Murr, and K.P. Staudhammer, Editors. 1992, Marcel-Dekker: New York. pp. 1075-1082.
11. Tsembelis, K., W.G. Proud, and J.E. Field, *The Dynamic Strength of Cement Paste Under Shock Compression.* (to appear in) Proc. of SHOCK 2001 - APS 12TH Topical Conference on Shock Compression of Condensed Matter, 2001.
12. Bourne, N.K., Z. Rosenberg, D.J. Johnson, J.E. Field, A.E. Timbs, and R.P. Flaxman, *Design and construction of the UK plate impact facility.* Meas. Sci. Technol., 1995. **6**: pp. 1462-1470.
13. Rosenberg, Z., D. Yaziv, and Y. Partom, *Calibration of foil-like manganin gauges in planar shock wave experiments.* J. Appl. Phys., 1980. **51**: pp. 3702-3705.
14. Arndt, J., H. Hornemann, and W.F. Müller, *Shock wave densification of silica glass.* Phys. Chem. Glasses, 1971. **12**: pp. 1-7.

IMPACT INDUCED FAILURE ZONES IN HOMALITE BARS

Rod Russell, Stephan J. Bless, and Tim Beno

Institute for Advanced Technology, 3925 W. Braker Ln., Ste. 400, Austin, TX 78759

Abstract. Impact tests were conducted on Homalite bars. Bars were impacted at 250 m/s with various flyer plates. Bar behavior was observed with a high-speed digital camera. Homalite bars exhibited repeatable failure modalities with little effective change coming from flyer plate or bar geometry. Failure is characterized by early, late, and intermediate morphologies. Early failure exhibits a radial damage cone near the impact event. Late damage adds a catastrophic failure zone near the bar end and multiple wave front locations along the length of the bar. Intermediate time pictures indicate that catastrophic failure starts as a series of spall-like planes in the catastrophic failure zone.

BACKGROUND

Brittle materials can be characterized by their degree of microstructure. There is a continuum, ranging from polycrystalline ceramics to homogeneous amorphous materials. In between are materials, such as cubic polycrystals (AlON, for example) and some glasses, like Pyrex, that do contain a degree of internal structure. Depending on their degree of internal structure and loading rate, brittle failure may be mainly intrinsic or extrinsic. Rapid loading is usually necessary to observe intrinsic phenomena, in which failure occurs locally and is not influenced by structural features. However, such rapid intrinsic failures do not always require impact loading; rock burst and

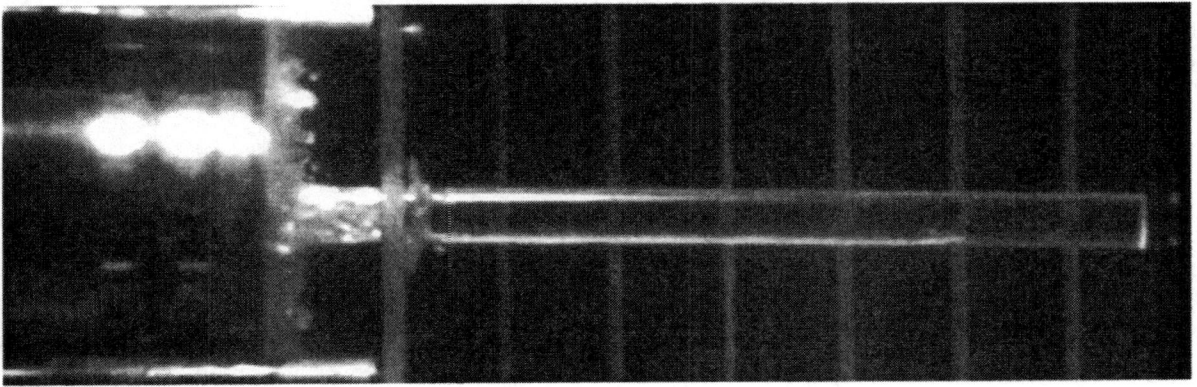

FIGURE 1. Failure zone in impacted Homalite bar typical of "early" failure mode. Homalite bar was ~200mm long by 12.7-mm diameter. Fiducial lines are spaced ~25.4 mm apart. Impacts from the left at 253 m/s. Time ~ 150 μs.

the explosion of Prince Rupert's drops are apparently examples of self-propagating intrinsic failure.

Structure is also apparently a condition for the manifestation of failure waves in brittle materials.

Across a failure wavefront, there is a loss of shear strength but little or no change in the stress component perpendicular to the wavefront.

Bar impact geometry provides a particularly useful way to steady dynamic failure of brittle materials. In this geometry, the loading near the impact face is one-dimensional strain, but in most of the sample the state is one-dimensional stress. In this way, the sample loading mimics that associated with many important engineering impact scenarios.

Use of Homalite as a Surrogate Material

Most experiments on brittle failure have been conducted with ceramics and glass. However, ceramics pose a number of experimental difficulties: they can be relatively expensive; they can be difficult to machine; they are often harder than the striker, which leads to uncertainties in impact geometry; they may have very high wavespeeds, which causes difficulty with instrumentation timing; and they are stronger and/or higher impedance than witness bars, which complicates the interpretation of measurements.

Use of a brittle plastic can circumvent many of these difficulties. Homalite has been used frequently in the past for fracture-mechanics studies, and it seems to have properties ideally suited for bar impact studies as well. Homalite is a thermoset manufactured by the Homalite Division of Brandywine Industries. According to the manufacturer, density is 1.23 g/cm^3, compressive strength is 0.18 GPa, and Young's modulus is 4.5 GPa. From these values, the bar speed, $(E/\rho)^{1/2}$, is 1.9 mm/µs.

EXPERIMENTAL DETAILS

Experiments were conducted with a 5.5-m-long, 56-mm-diameter single stage compressed gas gun. Various flyer plate geometries – 5- and 10-mm thick steel discs, 100-mm long aluminum bars, and 100- and 200-mm long Homalite bars – were used. Velocity was measured by monitoring laser light

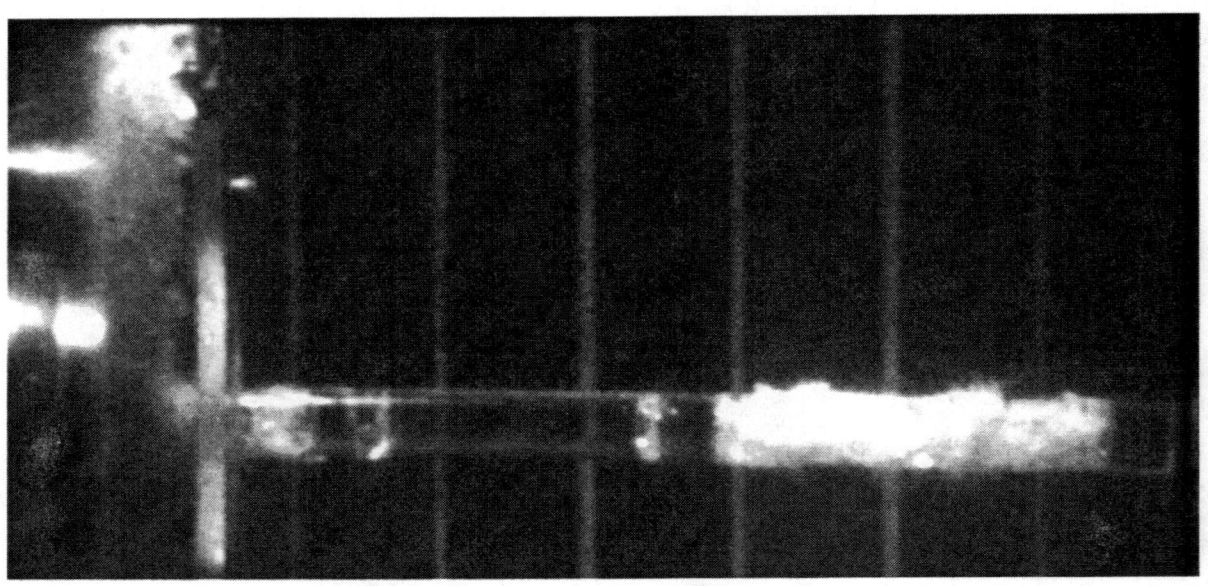

FIGURE 2. Failure zones in impacted Homalite bar typical of a "later" time failure. Note radial ejecta and numerous fracture planes. Lines are ~25.4 mm apart. Homalite bar was ~200 mm by 12.7-mm diameter. Velocity 259 m/s, time ~240 µs.

from reflecting strips on the sabot using the technique described by Simha (1).

Target bars were placed at the end of the gun barrel and positioned using an alignment system. Two bar geometries were investigated. Round bars 12.5 mm in diameter ranging in length from 170- to 200-mm, and rectangular prisms 25-mm tall and 12.5-mm thick were also used in lengths of 170 to 200 mm. Instrumentation consisted of a single frame DYCAM CCD camera manufactured by Cooke Corporation, using a 30-ns exposure. Triggering was accomplished using 2μm tungsten break-wires placed in front of the impact plane. The first wire triggers a one millisecond flash, and the second triggers the camera. Actual exposure was delayed by a preset interval. The flash lamp illuminated from above while images were taken from the side. A laser alignment system was used to maintain camera orthogonality. Various backgrounds were used; best results were obtained using a black background, against which fractured plastic appeared white. Bright fiducial lines were printed on the background for measurement.

Unlike earlier bar impact tests with glass and ceramic bars, there was no indentation of the flyer plates in the Homalite experiments.

OBSERVATION AND DISCUSSION

One-Sided Compressive Loading

Three principal morphologies were seen in bar failures. These morphologies are associated with space along the bar and time after impact. The first, or early, morphology is associated with times around 150 μs and is characterized by an impact cone extending about 25 mm from the flyer plate and no visible damage ahead. Behind the failure front, the plastic becomes opaque. The radius expands 2-3 mm, and the edges are indistinct. This appearance is consistent with a material that has been turned to rubble by the fracture. There is little additional change in radius over the next 10 mm or so, which means that radial expansion velocity behind the fracture is small or zero. Very near the flyer plate, fractured material is violently splayed out radially, as seen in Fig. 1.

Late time, ~250-μs, photographs reveal that a zone near the distal end of the bar has also been destroyed. The failed zone starts about a diameter from the rear surface, and extends about 50 mm toward the advancing projectile. The failed region also appears to be experiencing radial expansion. (See Fig. 2.) The free end of the bar has moved. Assuming that motion begins with the arrival of the 1.9-mm/μs elastic wave, average velocity is ~100 m/s. There are also several thin and stable damage zones in the midsection of the bar.

Intermediate time, ~200-μs, pictures show an expansion cone standing off from the flyer plate, similar to its appearance at early time. Fracture planes appear to be coalescing in the center of what will become the catastrophic failure zone on the distal end of the bar. (See Fig. 3.) These "intermediate" phenomena are seen in both round bars and prisms.

Sympathetic Compressive Impact

The final test included in this paper consisted of the rectangular prism geometry being loaded on

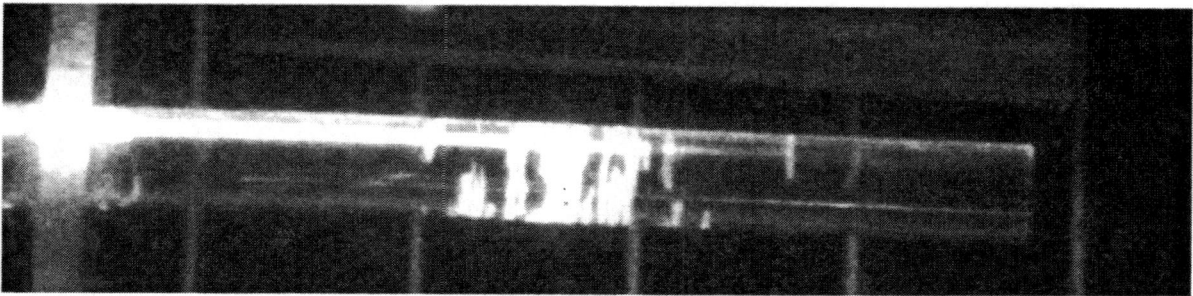

FIGURE 3. Failure zones in impacted Homalite bar showing "intermediate" failure mode. Note how multiple spall-like planes are forming along bar's length. Lines are spaced ~25.4 mm apart, Velocity 253 m/s, Time ~200 μs.

FIGURE 4. Intermediate time double ended compressive failure in sympathetically loaded prismatic bar. Projectile is impacting from the left at 261m/s. Plasticity is indicated on distal end of the bar by the apparent distortion of the ~25.4-mm fiducial line located ~100 mm to the right from the mounting plate.

both ends. A 10-mm flyer plate was glued to the distal end of the bar prior to the experiment. The bar was impacted at ~250 m/s with a 10-mm steel flyer, and an intermediate time photograph was taken. The resultant image, Fig. 4, shows the expected compressive failure cone and radial ejecta on the impact side. Compressive failure has occurred near the back end without the unfailed gap seen previously. The slight radial expansion evidences a compressive condition. The failure nucleates along the top and bottom of the bar. These machined surfaces are rougher than the sides parallel to the image plane. Unlike the free end case, the rear surface of the bar has moved only a few millimeters. It is our preliminary conjecture that the absence of "intermediate type" fracture planes suggests that they do not evolve in a compressive environment.

We conjecture that these Homalite experiments point to a complex set of wave interactions inconsistent with simple shock and release wave conditions. The nearly simultaneous formation of multiple "spall-like" planes in the unbounded fracture zone, and their suppression in the restrained case, suggest that a complex series of pressure and tension regions are evolving in the bar.

REFERENCES

1. Simha, H-M., "A Novel Technique to Measure In-Bore Velocity in a Single Stage Light Gas Gun," 46th Meeting of the Aeroballistic Range Association, Minneapolis, MN, 1995.

CHAPTER XIII

MECHANICAL PROPERTIES: REACTIVE MATERIALS

ELASTIC PRECURSOR DECAY IN HMX EXPLOSIVE CRYSTALS *

J. J. Dick and A. R. Martinez

*Group DX-1, MS P952,
Los Alamos National Laboratory, Los Alamos, New Mexico 87545,*

Abstract. VISAR experiments were performed on beta-phase HMX (cyclotetramethylene tetranitramine) crystals to measure the elastic precursor decay as a function of crystal orientation. Strong precursor decay in the first three mm of propagation was observed. There was significant dependence on impact stress, but the decay was about the same for both orientations.

INTRODUCTION

The first studies of elastic precursor decay in explosive crystals were performed by P. M. Halleck and Jerry Wackerle in pentaerythritol tetranitrate (PETN).[1] A dramatic dependence of precursor strength on crystal orientation was noted in PETN in work at Los Alamos.[2] Additional studies on PETN were performed at Institute St. Louis.[3–5] In the case of PETN the orientations with large elastic precursors also were more sensitive to shock initiation. In work reported here we present results of VISAR experiments performed on two orientations of HMX crystal to observe the nature of elastic precursor decay as a function of crystal orientation in this monoclinic explosive crystal.

EXPERIMENTAL TECHNIQUE

Plane shock experiments were performed on HMX crystals with final shock states of about 1.4 and 2.4 GPa obtained using a light-gas gun facility. Particle velocity vs time histories were recorded at the HMX/window interface using a velocity interferometer. Projectiles made of 2024 aluminum were impacted on x-cut quartz or Kel-F anvil disks. The HMX crystals were mounted on the aluminum-coated, anvil disc with a silicone elastomer.

The crystal slabs were cut from larger crystals using a low-speed, diamond saw. Identification of crystal planes was achieved by comparison of interfacet angles measured with a protractor with calculated angles until unique agreement was obtained. Lateral dimensions of the slabs were 8 to 11 mm. The crystals were polished using a series of polishing sheets consisting of alumina embedded in plastic. The crystals were water-clear and inspected with a binocular microscope at magnifications up to 50x to make sure that the interiors were free of inclusions and other visible defects larger than about 2 microns. Crystals were grown by Howard Cady by slow evaporation from acetone solution.

For VISAR studies polymethylmethacrylate (PMMA, Mil. Spec. P-5425D, preshrunk) windows 12.7 mm in both diameter and thickness were bonded to the window with elastomer. The surface of the PMMA adjacent to the crystal was coated with a diffuse aluminum mirror for VISAR particle velocity measurements. The lateral sides of the crystal and window were surrounded by an impedance-matching mixture of epoxy and 40

---

*Work performed under the auspices of the U. S. Department of Energy and partially supported by the Department of Defense/Office of Munitions under the Joint DoD/DOE Munitions Technology Program and the Explosives Technology Program at LANL.

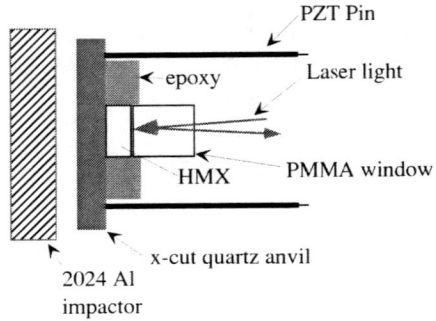

FIGURE 1. Schema of the gun impact experiment with VISAR (velocity interferometry system for any reflector) instrumentation. There is a mirror on the window at the interface.

vol% silica to keep out edge effects.

Two or four piezoelectric pins were placed on the rear surface of the anvil disc adjacent to the crystal. See Fig. 1. The average impact time for a pair of diametrically placed pins is a measure of impact time in the sample center. The time difference between impact time at the sample center measured by the pins and the shock arrival signal recorded by VISAR yields a transit time and velocity through the crystal. It is important to account for the difference in signal travel times to the digitizing oscilloscopes for the two types of signals.

The measurement system used was a dual, push-pull, VISAR system.(6) The dual VISAR with different fringe constants removes ambiguity in determining the particle-velocity jump at the shock when extra fringes must be added. The light was transported from the laser to the target and thence to the interferometer table with fiber optics.

EXPERIMENTAL RESULTS

Initial experiments were performed on samples 1.23 to 4.66 mm thick with input stresses into the crystal of about 1.4 GPa. The wave profiles obtained are shown in Figs. 2 and 3. There is an elastic shock followed by a plastic wave. Decay of the elastic shock strength with propagation distance as well as stress relaxation behind the shock are evident. The elastic shock strengths are similar for the two orientations, but the stress relaxation shows some difference with orientation. The PMMA window has a 12 ns viscoelastic relaxation time. This may distort the stress relaxation for the thinnest samples.

In addition, experiments were undertaken with about 2.4-GPa input stress into the crystal. These would detect the effect of input stress on the elastic precursor strength and decay. One experiment was performed for each orientation, Figs. 4 and 5. These show that the elastic precursor shock is stronger and has more relaxation at a given propagation distance for 2.4 GPa input stress.

The precursor decay results are summarized in Fig. 6. The points at zero thickness are the estimated elastic stresses at the input faces of the crystal. The graph shows that results are the same for both orientations. The precursor decays by about a factor of 4 in 3 mm from 1.9 GPa elastic input stress and a factor of 6 from 3.0 GPa. The decayed stresses are about 30% higher for the higher input stress. Most of the decay is in the first millimeter or so. The input elastic stress was not measured, so it remains hypothetical. For a crystal of this low symmetry the elastic wave may decompose into quasilongitudinal and quasitransverse waves. This has not been dealt with here. The VISAR experiment measures the longitudinal motion.

Shockwave Tabular Results

Table I lists the experimental conditions and measured quantities. If HMX is nonlinearly elastic, the shock slows down as the wave decays. This was the case in explosive crystals of pentaerythritol tetranitrate (PETN).(7) This appears to be so for HMX in that the measured shock velocity is higher for the thinnest samples, Shots 1180 and 1181. The shock velocity computed from the measured transit time and sample thickness is some mean value of the wave velocity over the propagation distance. The final velocity at the interface where the wave profile is measured will be less than this mean value. Halleck and Wackerle estimated the error to be about 3.5% for PETN 5 mm thick.(1) The elastic particle velocity and longitudinal stress in HMX are computed from the intersection of the Rayleigh line given by the initial density times the mean elastic shock velocity and its reflection through the HMX/PMMA interface state in the stress vs particle velocity plane. The

TABLE 1. Results of VISAR Tests for Two Orientations of HMX Crystal[a]

Shot num.	Sample type[b]	Thickness (mm)	Impactor u_I (mm/μs)	Interface u (mm/μs)	Elastic precursor			Plastic wave	
					U_S (mm/μs)	u_p (mm/μs)	P_x (MPa)	U_S (mm/μs)	Rise Time (ns)
1180	110	1.23	0.3185	0.123	4.86	0.085	790	3.48	91
1067	110	3.21	0.3170	0.080	4.16	0.058	460	3.28	122
1166	110	3.18	0.3068	0.083	4.27	0.059	480	3.19	133
1182	110	3.57	0.5209	0.094	4.07	0.068	530	3.39	74
1181	011	1.39	0.3160	0.121	4.51	0.086	740	3.26	79
1068	011	3.00	0.3140	0.086	4.07	0.062	480	3.21	123
1168	011	4.66	0.3132	0.072	4.08	0.052	400	3.26	179
1183	011	3.11	0.5204	0.109	4.31	0.078	640	3.65	47

[a] At 1.4 GPa and u_I near 0.315 mm/μs, the anvils were x-cut quartz and for the 2.44 GPa near 0.52 mm/μs they were Kel-F. The window material was polymethylmethacrylate (PMMA). Shots 1067 through 1168 used two PZT pins to measure the time of impact. Shots 1180 and 1181 used 4 pins.
[b] This is the crystallographic plane in space group $P2_1/n$ that was parallel to the impact plane.

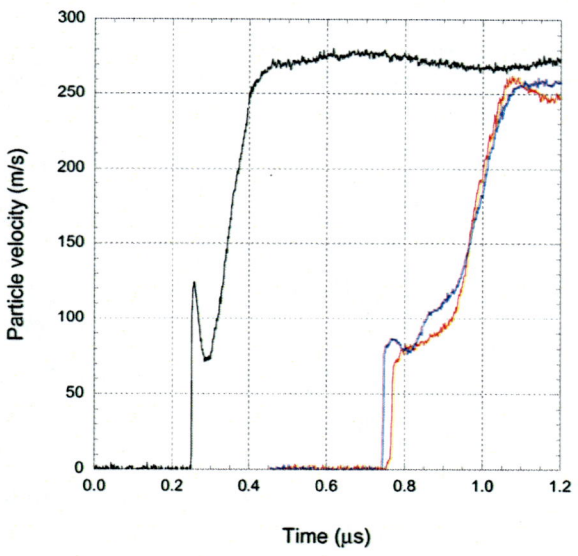

FIGURE 2. Elastic precursor decay for 1.4-GPa shocks parallel to a {110} plane. The profiles are those measured at the HMX/PMMA interface. The black, red, and blue profiles are for samples 1.23, 3.21, and 3.18 mm thick, respectively.

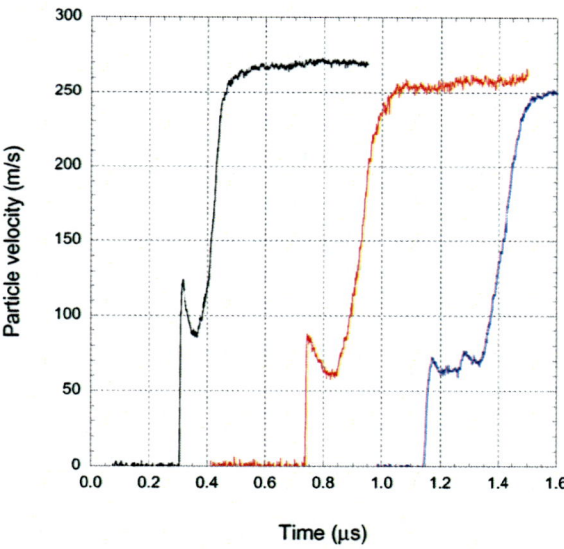

FIGURE 3. Elastic precursor decay for 1.4-GPa shocks parallel to a {011} plane. The profiles are those measured at the HMX/PMMA interface. The black, red, and blue profiles are for samples 1.39, 3.00, and 4.66 mm thick, respectively.

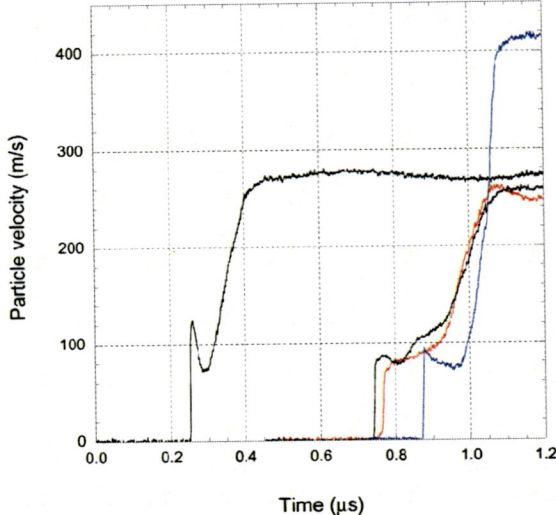

FIGURE 4. Elastic precursor decay for shocks parallel to a {110} plane at 1.4 and 2.4 GPa. The profiles are those measured at the HMX/PMMA interface. The blue profile is at 2.4 GPa, 3.57 mm thickness.

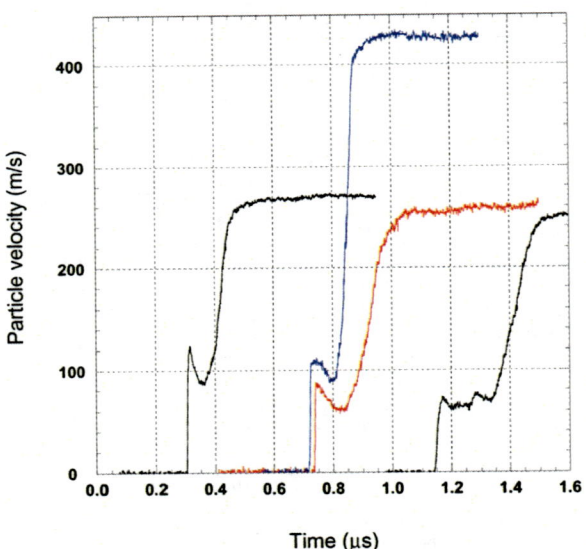

FIGURE 5. Elastic precursor decay for shocks parallel to a {011} plane at 1.4 and 2.4 GPa. The profiles are those measured at the HMX/PMMA interface. The blue profile is at 2.4 GPa, 3.11 mm thickness.

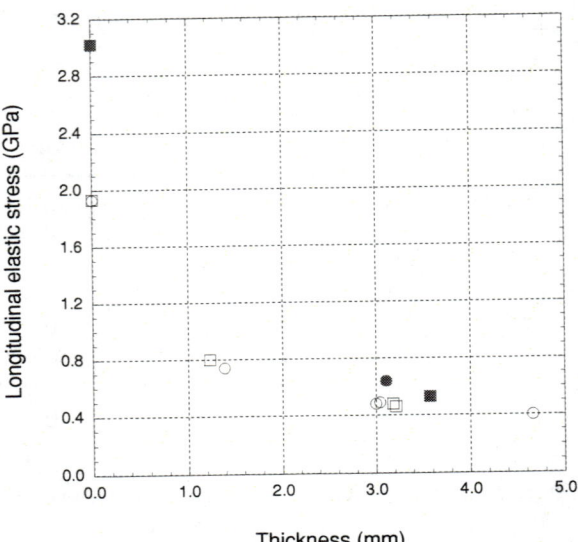

FIGURE 6. Longitudinal stress in HMX vs Lagrangian position for elastic precursor shocks in HMX parallel to {110} and {011} planes. The squares are for {110} and the circles for {011} cases. The solid symbols are for 2.4 GPa final stresses.

plastic wave speed is the Lagrangian wave velocity. The plastic wave rise time appears to increase with propagation distance.

ACKNOWLEDGMENTS

Helpful conversations with Mike Winey and Professor Y. M. Gupta on the behavior of PMMA are gratefully acknowledged.

REFERENCES

1. P. M. Halleck and J. Wackerle, J. Appl. Phys. **47**, 976 (1976).
2. J. J. Dick and J. P. Ritchie, J. Appl. Phys. **76**, 2726 (1994).
3. L. Soulard, *Etude du monocrystal de pentrite soumis a un choc plan*, PhD thesis, L'Universite de Haute Alsace, 1990.
4. L. Soulard and F. Bauer, "Applications of Standardized PVDF Shock Gauges for Shock Pressure Measurements in Explosives," in *Shock Compression of Condensed Matter – 1989*, edited by S. C. Schmidt, J. N. Johnson, and L. W. Davison, North-Holland, Amsterdam, 1990, pp. 817–820.
5. D. Spitzer, *Etude du role des defauts de taille microscopique dans la transition choc-detonation du monocristal de pentrite*, PhD thesis, Universite Louis Pasteur de Strasbourg, Strasbourg, 1993.
6. W. F. Hemsing, Rev. Scient. Instr. **50**, 73 (1979).
7. J. J. Dick, J. Appl. Phys. **81**, 601 (1997).

INFLUENCE OF POLYMER MOLECULAR WEIGHT, TEMPERATURE, AND STRAIN RATE ON THE MECHANICAL PROPERTIES OF PBX 9501

D.J. Idar,[†] D.G. Thompson,[††] G.T. Gray III,[†††] W.R. Blumenthal,[†††] C.M. Cady,[†††] P.D. Peterson,[††] E.L. Roemer,[††] W.J. Wright,[††] B.L. Jacquez[†††]

[†]MS P915, [††]MS C920, [†††]MS G755, Los Alamos National Laboratory, Los Alamos, NM 87545

Abstract. Compression and tensile measurements were conducted on newly formulated (baseline) and lower molecular weight (virtually-aged) plastic-bonded explosive PBX 9501. The PBX 9501 binder system is composed of nitroplasticized Estane 5703,™ a polyester polyurethane copolymer. The molecular weight of polyester urethanes can degrade with time as a function of hydrolysis, affecting the mechanical behavior of the polymer or a polymer composite material of high explosives, i.e. PBXs. The molecular weight of Estane 5703™ was degraded by exposure to high temperature and humidity for different periods of time, and then formulated to produce "virtually-aged" PBX 9501 specimens. Quasi-static and dynamic compression tests were conducted on the baseline and virtually-aged PBX 9501 as a function of temperature and strain rate. Quasi-static tensile tests were also conducted as a function of temperature and test rate. Rate and temperature dependence was exhibited during both compression and tensile loading. Results also show significant differences between the baseline and virtually-aged specimens for the dynamic compression tests at –15°C, and for the quasi-static compression tests at -15°C, 22°C, and 50°C.

INTRODUCTION

Understanding and modeling the response of plastic-bonded explosives (PBXs) to low amplitude stresses and strains is critical to developing an accurate predictive capability to address potential survivability and safety issues. This requires a thorough understanding of the mechanical response of PBXs as a function of a complex set of variables including test conditions and pedigree, e.g. initial formulation conditions and changes with age. PBXs are exposed to numerous environmental conditions throughout their service lifetimes, e.g. humidity and temperature gradients, which may ultimately affect the mechanical response of these materials.

PBXs have been commonly formulated with a soft phase polymeric binder system mixed with hard phase explosive to facilitate processing, and to reduce sensitivity to different stimuli. The simplest binder matrix consists of a long-chain polymer backbone or a polymer mixed with plasticizer. Some polymeric binders are susceptible to molecular weight degradation through hydrolysis, oxidation, temperature and radiation effects.

Conventional quasi-static compression and tensile methods have been used to measure mechanical properties of PBXs at low strain rates. Dynamic compression measurements were conducted using a unique split-Hopkinson pressure bar (SHPB) designed to improve the signal-to-noise level. The new design was needed to test extremely low strength materials as compared to the maraging steel bars traditionally utilized for metals as described previously (1). The mechanical responses of PBXs have been studied as a function of strain rate and temperature as reviewed previously (2,3). In general, the compressive strength and the loading modulus of polymeric materials increases with decreasing temperature and increasing strain rate with further dependence on the specimen geometry, material processing, and test method.

In the present study, we evaluate the effects of changing the Estane 5703™ (hereafter referred to as Estane) molecular weight on PBX 9501 mechanical properties. Quasi-static uniaxial compression and tensile tests were performed on baseline PBX 9501 and 3 different lots of virtually-aged PBX 9501 formulated from degraded Estane. Dynamic compression data on the same materials were also conducted. Data are compared as a function of the Estane molecular weight differences.

EXPERIMENTAL TECHNIQUES

Materials and Preparation

This investigation was performed on the plastic-bonded explosive PBX 9501. PBX 9501 is a formulation composed of 94.9/2.5/2.5/0.1 wt% HMX / Estane /a eutectic mixture of bis(2,2 dinitropropyl) acetal and bis (2,2-dinitropropyl) formal [abbreviated BDNPA-F] / and Irganox (a free radical inhibitor). Estane is the polymer constituent of the PBX 9501 binder. It is an amorphous, thermoplastic polyester polyurethane copolymer with a glass transition temperature (T_g) of $-31°C(4)$. The copolymer is composed of approximately 75% soft poly(butylene adipate) segments and 25% hard segments composed of 4,4'-methylenediphenyl 1,1'-diisocynanate and a 1,4 butanediol chain extender (5). Estane molecular weight degradation as well as the addition of the BDNPA-F plasticizer can reduce the static glass transition temperature up to 20 degrees (6), resulting in a decrease in mechanical strength and moduli.

Analyses of Estane and PBX 9501 molding powder reserves stored under uncontrolled temperature and humidity conditions indicate that the polymer molecular weight is decreasing with age as a function of hydrolysis. To mimic this aging profile the molecular weight of new Estane, in pelletized form, was degraded for 14, 23, and 36-days respectively at 70°C and 74%RH. New Estane and the 3 degraded specimens of Estane were formulated with HMX, BDNPF-A and Irganox 1010 to produce baseline and virtually-aged PBX 9501 lots (hereafter referred to as baseline, 14, 23-, and 36-day VA PBX 9501.)

Estane weight average molecular weights (M_w) were determined in tetrahydrofuran by Gel Permeation Chromatography relative to the retention times for polystyrene standards extracted from the 4 different PBX 9501 lots. M_w data were measured in triplicate for each lot, with averages of 115044±691, 88917±1127, 72068±1389, and 44919±793 daltons for the baseline, 14-, 23-, and 36-day VA PBX 9501 lots respectively.

Quasi-static compression specimens, ~9.5mm diameter by ~19.0mm long, and cylindrical dogbone tensile specimens, 76.2 mm long with ~15° tapered ends, 38.1 mm gauge length and 12.7mm diameter, were machined with the specimen loading axes parallel to the billet pressing direction. Specimen densities ranged between 1.825 to 1.831 g/cm^3. SHPB compression specimens were machined from PBX 9501 samples with dimensions of 6.35mm diameter and 3.2mm length.

Low Strain Rate Testing

Quasi-static compression tests were conducted using an Instron 5567 Materials Testing Workstation equipped with a Bemco Environmental Control Chamber at crosshead speeds of 0.5 and 5.0 in/min (strain rates approaching 0.01 and 0.1 s^{-1} respectively) at 3 different temperatures of -15 ± 1, 23 ± 1, and $50\pm1°C$ with humidity ranging from ~5 to 50%RH. Quasi-static tensile tests were performed at the same temperatures with crosshead speeds of 1.0 and 10.0 in/min using a newly modified tensile grip design and 1 to 2 contact extensometers. The new grip design reduces the number of mechanical interfaces required to seat the specimen before uniaxial tension is achieved. Strain data was averaged if two extensometers were used. Four to five tests were completed for each test condition to evaluate specimen-to-specimen variations and all specimens were loaded to failure (<10% strain). Quasi-static compression specimens were also lubricated with molybdenum disulfide.

High Strain Rate Compression Testing

Two to three duplicate dynamic tests were conducted on PBX 9501 specimens at a strain rate of ~2000 s^{-1} utilizing a modified SHPB design. Test temperatures of -15°C to +50°C were achieved during testing using a helium gas heating/cooling system described previously (1). Specimens were ramped to the desired temperature in approximately 5 minutes and equilibrated at temperature for approximately 10 minutes prior to testing. The SHPB sample faces were lubricated with a thin spray coating of boron nitride and a thin layer of molybdenum disulfide grease.

RESULTS AND DISCUSSION

Quasi-static and Dynamic Compression Tests

Selected compression curves are plotted in Fig. 1 for the baseline PBX 9501 lot. These data demonstrate a strong dependence on both strain rate and temperature. Specifically, the strength at 22°C increases by over a factor of 4 from 12 MPa at 0.5 in/min to ~59 MPa at ~2000 s^{-1}. Figure 1 also shows that the peak flow stresses for the baseline PBX 9501 consistently occur at strains between 1.5% and 2.5% before slowly decaying with further strain.

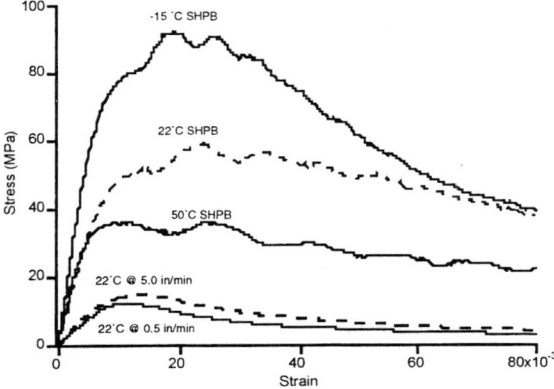

FIGURE 1. Compression behavior for baseline PBX 9501 as a function of quasi-static and dynamic strain rates and temperature.

Figures 2 and 3 are summary plots of selected compression curves for the 4 lots of PBX 9501 at quasi-static and dynamic test rates. In general the quasi-static data show a 15 to 19% increase in elastic moduli, and 18 to 23% increase in strength with an order of magnitude increase of the crosshead speed at room temperature which correlates well with previously measured PBX 9501 trends (1). Notable differences are seen in the compression strength and elastic moduli as the Estane M_w is degraded from ~115k to ~45k with a 22 to 24% loss in strength, and between 7 and 9% loss in elastic modulus. Specifically, the 36-day VA PBX 9501 data shows compressive strengths at 5.0 in/min that are weaker than the data measured for the baseline material at the slower 0.5-in/min crosshead speed.

Conversely the effect of the degraded polymer weight is not evident in the SHPB dynamic data until the test temperature begins to approach the colder dynamic glass transition temperature (T_g) of the soft segments. These differences suggest that the damage mechanism during high-rate loading is changing as a function of the molecular weight changes only at cold temperatures. The lack of a difference in the dynamic test data at 22 and 50°C as a function of molecular weight decrease, contrary to the quasi-static results, is thought to be due to the off-setting effect of the strong strain-rate sensitivity of the binder overwhelming the M_w effect.

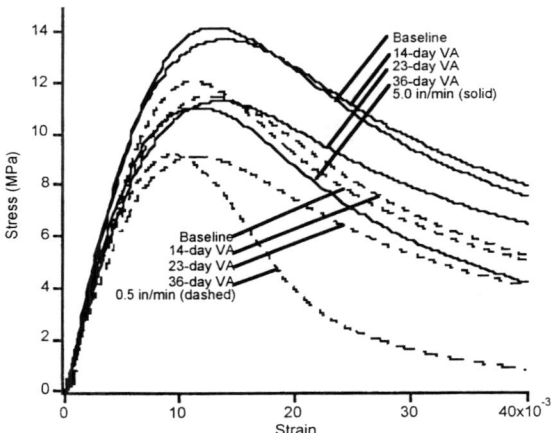

FIGURE 2. Quasi-static compression behavior for baseline and virtually-aged lots of PBX 9501 as a function of quasi-static crosshead speeds, 0.5 in/min (dashed) and 5.0-in/min (solid).

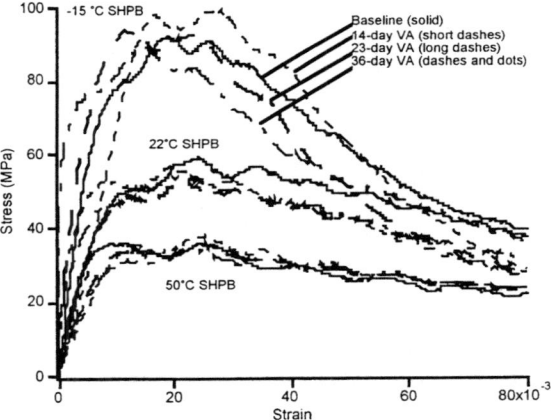

FIGURE 3. SHPB compression behavior for baseline and virtually-aged lots of PBX 9501 as a function of temperature at dynamic strain rates.

Quasi-static Tensile Tests

Figure 4 is a plot of the stress strain curves for several of the baseline PBX 9501 specimens at the three different test temperatures and two different crosshead speeds where the test data show good reproducibility from specimen-to-specimen. In general the tensile behavior exhibits rate and temperature dependence similar to the compressive behavior, but with peak flow stresses, and the strain at which the peak flow stress significantly decreased. A drop in the peak flow stress was also evident at -15°C when the quasi-static crosshead speed was increased by an order of magnitude because of increasingly brittle behavior.

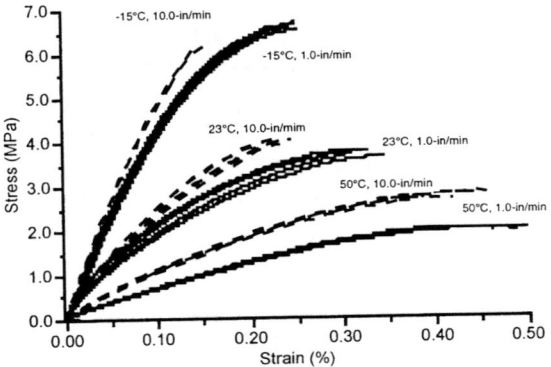

FIGURE 4. Tensile behavior of baseline PBX 9501 at 2 crosshead speeds, 1.0- and 10-in/min, and three temperatures (-15, 23, and 50°C).

Figure 5 is a summary plot of selected tensile curves at 22°C and 1.0-in/min for all 4 PBX 9501 lots (baseline and VA). Significant differences in the tensile response due to virtual aging are not immediately evident in these data until the M_w is degraded by more than 37%. These trends were also evident at the other two test temperatures, -15 and 50 °C. Similar to the baseline data, the peak flow stress at -15°C for the three virtually-aged lots also drop to a lower value at the fast crosshead speed.

SUMMARY AND CONCLUSIONS

The following conclusions can be drawn: 1) the compressive and tensile stress-strain response of PBX 9501 demonstrates strong strain rate and temperature dependence; 2) Estane molecular weight degradation significantly affects the *quasi-static* compressive behavior at *all* temperatures, 3) the *dynamic* test data at -15°C exhibits significant changes as a function of the molecular weight, suggesting changes in the failure mechanism, while no effect of M_w was seen dynamically at 22 or 50°C; and 4) the tensile response is considerably weaker and has lower flow stress with smaller differences evident in the strength, elastic moduli and flow stress as a function of molecular weight changes in comparison to compression data.

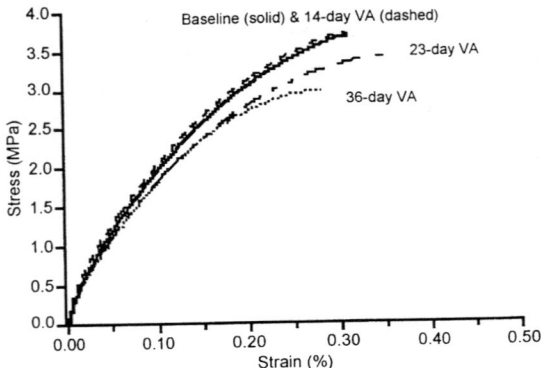

FIGURE 5. Tensile behavior of baseline, 14-, 23- and 36-day virtually-aged PBX 9501 at ~23°C, 1.0-in/min.

ACKNOWLEDGMENTS

The authors wish to thank Debra Wrobleski, Bruce Orler, Sheldon Larson, Ray Flesner, Gregg Sullivan, Robert Montoya, George Harper, Jose Archueta, and Wayne King for the preparation, density, and M_w characterization of the degraded Estane and PBX 9501 materials. This work was performed under the auspices of the U.S. Department of Energy (contract W-7405-ENG-36).

REFERENCES

1. Gray III, G. T., Blumenthal, W. R., Idar, D. J. and Cady, C. M., "Influence of Temperature on the High-Strain-Rate Mechanical Behavior of PBX 9501," in *Shock Compression of Condensed Matter-1997*, edited by S.C. Schmidt et al., AIP Conference Proceedings 429, New York, 1998, pp. 583-586.
2. Idar, D. J., Rabie, R. L., and Scott, P. D., "Quasi-Static, Low-Strain Rate Compression Measurements of Thermally Treated and Mechanically Insulted PBX 9502 Samples," Los Alamos National Laboratory report LA-UR-97-5116, Los Alamos, New Mexico, (1997).
3. Idar, D.J. and Holmes, M.D. "Quasi-static, Low Strain Rate Compression Measurements of PBX 9502 and Mock 900-24 Specimens," Los Alamos National Laboratory report LA-UR-98-5270, Los Alamos, New Mexico, (1998).
4. Goodrich, B. F., "Adhesives Technology of Estane Polyurethane", B.F. Goodrich Speciality Chemicals, TSR 76-02 TF116, March 1995.
5. Hoffman, D.M., Caley, L.E. ACS Div. Org. Coat. Plast. Chem. 44, (1981) p. 686.
6. Campbell, M.S., Garcia, D.A., Idar, D.J., "Effects of Temperature and Pressure on the Glass Transitions of Plastic Bonded Explosives", Thermochimica Acta, 357-358, (2000) pp. 89-95.

MOIRE INTERFEROMETRY STUDIES OF PBX 9501

Philip J. Rae*, H. Timothy Goldrein*, Stewart J. P. Palmer* and William Proud*

*University of Cambridge, Cavendish Laboratory, Madingley Road, Cambridge, UK. CB3 0HE

Abstract. The microstructure of polymer bonded explosives influences significantly the mechanical response to quasi-static and dynamic loading. The microstructure of PBX 9501 is examined using moiré interferometry, a sensitive optical technique useful for measuring in-plane displacement. Quasi-static deformation and fracture has been followed and the influence of the crystal microstructure is found to be significant. If moiré interferometry is to be useful at high strain rates, changes in the experimental setup are required. These alterations are outlined.

INTRODUCTION

The US composition PBX 9501 is a much studied explosive [1, 2, 3, 4]. It is manufactured from 95% by weight crystalline explosive and 5% rubbery binder [5]. An image of the microstructure, taken with polarised light, is shown in figure 1. The explosive crystals in this composition are typically angular in shape and can be extensively flawed, with growth inclusions, voids and deformation twins. Post failure optical and electron micrographs can be taken showing the failure route and nature of cracking but this method reveals no quantitative information about the material deformation. Past studies have revealed that the quality of the explosive crystals and the toughness of the binder play a key role in the mechanism of fracture [6, 7, 8, 9]. These experiments followed deformation and failure under quasi-static loading. An obvious extension of this research is into the dynamic regime.

As in many other areas of scientific investigation, researchers are seeking to create analytical and computer models of the response of PBXs to a variety of impact situations [10]. The strain-rate regimes of interest vary between creep and high intensity shock waves and such models require experimental verification. In order to understand in a quantitative manner the microscopic deformation of these materials under load, a high-resolution measurement technique is required. A number of possible techniques could be employed [11] but moiré interferometry offers a non-contact, sensitive and whole-field solution.

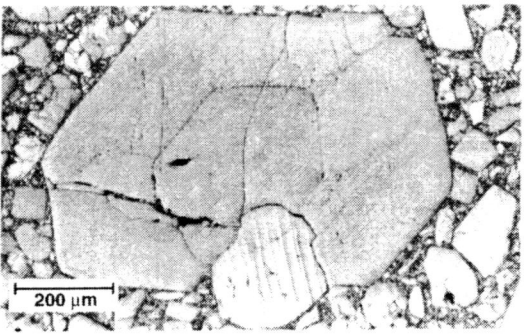

FIGURE 1. Optical micrograph of PBX 9501 showing the angular nature of the filler, growth inclusions, voidage and deformation twins.

QUASI-STATIC MOIRE INTERFEROMETRY

High Resolution moiré interferometry is a sensitive coherent optical technique which allows the measurement of in-plane displacements [12, 13, 14]. By taking white-light images of the microstructure in exact registration with laser interferograms, a direct correlation of the measured displacement field with features in the composite microstructure is made.

The technique works by using collimated laser beams falling onto a thin phase diffraction grating bonded to the test specimen surface. A single beam falling onto a phase grating surface produces a number of diffracted beams. The number and angle at which they are formed depends on the frequency of the laser light, the angle of incidence and the pitch or spatial frequency of the grating. If two collimated beams are set up so that their $+1$ and -1 diffraction orders respectivly leave normal to the grating surface, as in figure 2(left), then any change in the grating pitch produced by mechanical strain will cause a change in the angle at which the diffraction orders leave the grating, figure 2(right).

In quasi-static experiments the interference pattern created by the overlapping diffracted beams may be recorded as a two-dimensional fringe pattern on a CCD camera. It can be shown [12] that each fringe represents a local displacement of half a grating pitch, in a direction perpendicular to the grating rulings. In these experiments a He–Ne laser (wavelength 632.8 nm) is shone onto a phase grating of 1200 lines mm^{-1} cast onto the PBX surface. This produces an extra interference fringe for each 0.4167 μm of local in-plane displacement. Only local strains are measured by this system since rigid body motion of the specimen does not change the grating pitch. Out of plane motion is not measured since the path length of each beam is equally affected leading to an unchanged interference pattern. Using computer analysis and phase stepping [15, 16, 17], a sensitivity of around one hundredth of a fringe may be achieved, leading to a displacement uncertainty of approximately 10 nm. A schematic of the optical arrangement is presented in figure 3.

Phase gratings are replicated on the specimen using a low modulus epoxy resin which does not reinforce the specimen surface significantly. Thin gratings (<5 μm) are required to prevent the 'smearing' of high local strains over a larger surface area. The grating is coated with a thin layer of gold (≈ 5 nm) to enhance the diffraction efficiency, whilst allowing white light pictures to be obtained with the video camera through the grating.

Figure 4 shows a contour map of a fractured sample of PBX 9501. The specimen has been loaded at a strain rate of approximately 10^{-4} s^{-1} in the Brazilian test [18, 19]. In this biaxial test compression occurs vertically while the measurement is taken in the horizontal, tensile, direction. It can be seen that a significant vertical crack has occurred in a large filler particle (marked A). It can also be seen that the material on the left of the image has deformed uniformly and with little correlation to the underlying microstructure. Figures 5 and 6 show only the white-light micrographs obtained before and after

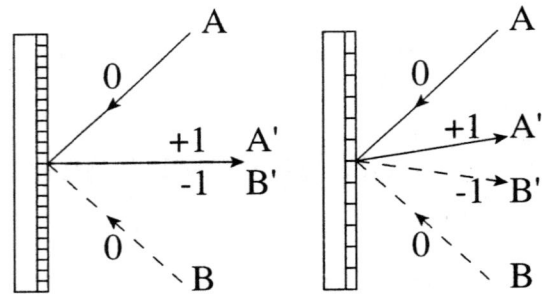

FIGURE 2. Principal of moiré interferometry. Left: undeformed phase grating with symmetric input beams. Right: deformed phase grating resulting in altered angles of diffraction.

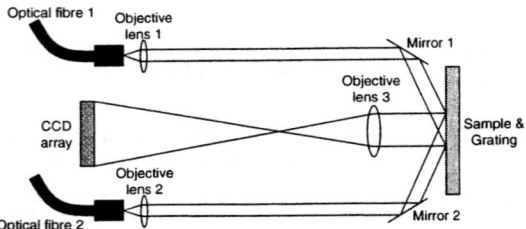

FIGURE 3. A schematic of a moiré interferometer.

failure. The cracked crystals are more obvious in these images. Only the post failure contour map is presented here. Eleven others were recorded during loading; they show incremental increases in displacement prior to total sample failure.

DYNAMIC MOIRE INTERFEROMETRY

In principal moiré interferometry is applicable to dynamic events, however some simplifications are required. Is is not possible to perform phase-stepping at more than a few hundred frames per second. For fast events the fringes patterns need to be photographed and analyzed using the 'fourier transform technique'[20]. This necessity reduces the fringe interpolation accuracy to about 1/10th of a fringe, corresponding to about 100 nm. A typical set of interference fringes showing the sinusoidal nature of the pattern produced is shown in figure 7.

In addition, powerful lasers are required. If one

FIGURE 4. Post failure contour map in PBX 9501. The applied tensile stress is horizontal and the contour displacement is 0.5 μm.

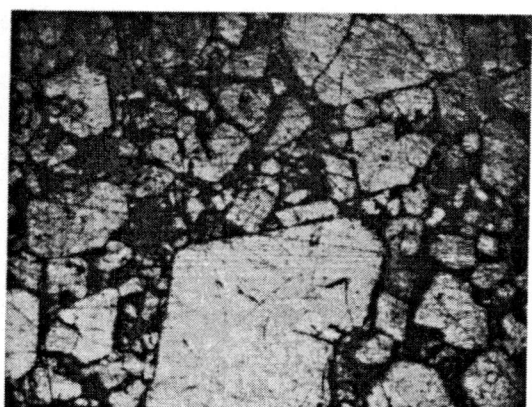

FIGURE 5. Microstructure of the sample shown in figure 4 prior to sample loading.

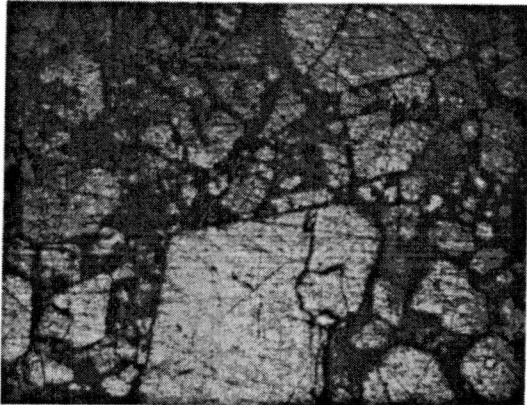

FIGURE 6. Microstructure of the sample shown in figure 4.

wishes to capture the deformation due to a shock front moving at between 2-7 km s^{-1}, exposures of less than 100 ns are required. Even with image intensified cameras a great deal of light needs to be delivered to the specimen. This high-power-density precludes the use of single-mode optical fibres in the system and forces the researcher to use bulk optics despite a considerable increase in experimental dif-

FIGURE 7. A typical set of interference fringes.

ficulty. One final limitation is that whilst white-light images may still be taken through the grating before loading it is not possible to do this during the dynamic event in addition to capturing interferograms.

ACKNOWLEDGMENTS

The authors wish to thank the Los Alamos National Laboratory (USA) for suppling samples of PBX 9501 and the Atomic Weapons Establishment, Aldermaston UK for funding the research.

REFERENCES

1. Gray III, G. T., Idar, D. J., Blumenthal, W. R., Cady, C. M., and Peterson, P. D., "High and low-strain rate compression properties of several energetic material composites as a function of strain-rate and temperature", in *11th International Detonation Symposium, Snowmass, Colorado 31 Aug.–4 Sept. 1998. ONR 33300-5*, edited by J. M. Short, 1998, pp. 76–84.
2. Idar, D. J., Lucht, R. A., Straight, J. W., Scammon, R. J., Browning, R. V., Middleditch, J., Dienes, J. K., Skidmore, C. B., and Buntain, G., "Low amplitude insult project: PBX 9501 high explosive violent reaction experiments", in *11th International Detonation Symposium, Snowmass, Colorado 31 Aug.–4 Sept. 1998. ONR 33300-5*, edited by J. M. Short, 1998, pp. 101–110.
3. Idar, D., Low amplitude impact testing of pristine, aged and damaged PBX 9501, Tech. Rep. LA-UR-00-2463, Los Alamos National Laboratory, New Mexico, USA (2000).
4. Peterson, P. D., Mortensen, K. S., Idar, D. J., Asay, B. W., and Funk, D. J., *J. Mat. Sci.*, **36**, 1395–1400 (2001).
5. Palmer, S. J. P., Field, J. E., and Huntley, J. M., *Proc. R. Soc. Lond.*, **A440**, 399–419 (1993).
6. Rae, P. J., *Quasi-static studies of the deformation, strength and failure of polymer-bonded explosives.*, Ph.D. thesis, Cavendish Laboratory, University of Cambridge, UK (2000).
7. Rae, P. J., Goldrein, H. T., Palmer, S. J. P., Field, J. E., and Lewis, A. L., *Proc. R. Soc. Lond.*, p. Submitted (2001).
8. Goldrein, H. T., Huntley, J. M., Rae, P. J., and Palmer, S. J. P., *J. Optics Lasers Eng.*, p. Submitted (2001).
9. Rae, P. J., Palmer, S. J. P., Goldrein, H. T., Field, J. E., and Lewis, A. L., *Proc. R. Soc. Lond.*, p. Submitted (2002).
10. Baer, M. R., Kipp, M. E., and van Swol, F., "Micromechanical modelling of heterogeneous energetic materials", in *11th International Detonation Symposium, Snowmass, Colorado 31 Aug.–4 Sept. 1998. ONR 33300-5*, edited by J. M. Short, 1998, pp. 788–797.
11. Ranson, W. F., Sutton, M. A., and Peters, W. H., "Holographic and laser speckle interferometry", in *Handbook on experimental mechanics*, edited by A. S. Kobayashi, Society for experimental mechanics; Prentice Hall, 1987, pp. 388–429.
12. Post, D., "Moiré interferometry", in *Handbook on experimental mechanics*, edited by A. S. Kobayashi, Society for experimental mechanics; Prentice Hall, 1987, pp. 228–313.
13. Walker, C. A., *Exp. Mech.*, **34**, 281–299 (1994).
14. Goldrein, H. T., Palmer, S. J. P., and Huntley, J. M., "Microstructural strain analysis of composites using high-magnification moiré interferometry", in *Proceedings of the Applied Optics Divisional conference of the Institute of Physics, Reading, 16-19 September 1996*, edited by K. T. V. Grattan, 1996, pp. 10–15.
15. Bruning, J. H., *Appl. Opt.*, **13**, 2693–2703 (1974).
16. Creath, K., "Temporal phase measurement methods", in *Interferogram Analysis*, edited by D. W. Robinson and G. T. Reid, Institute of Physics (IOP): Bristol, 1993.
17. Huntley, J. M., *J. Strain Analysis*, **33**, 105–125 (1998).
18. Hondros, G., *Australian Journal of Applied Science*, **10**, 243–268 (1952).
19. Awaji, H., and Sato, S., *J. Eng. Mat. Tech.*, **101**, 139–150 (1979).
20. Whitworth, M. B., Huntley, J. M., and Field, J. E., *Proc. SPIE*, pp. 667–682 (1990).

EXPERIMENTAL SIMULATIONS OF DYNAMIC STRESS BRIDGING IN PLASTIC BONDED EXPLOSIVES

Keith M. Roessig and Joseph C. Foster, Jr.

Air Force Research Laboratory/Munitions Directorate
101 W. Eglin Blvd. Ste. 135
Eglin AFB, Florida 32542

Abstract. This work investigates the role of the particle/binder interface in the formation of stress bridges within bonded particulate materials. The photoelastic technique is exploited to examine the dynamic stress states within three systems: a binderless particle bed, a particle bed with binder, and a particle bed with a binder bond strength of zero. In a binderless system, stress concentrations form readily due to the fact that the stress must be transferred through specific contact points. The particle bed with binder is shown to have a much more diffuse stress state because shear stresses are transferred at the interface between crystal and binder. In the system with bond strengths of zero, stress concentrations redevelop due to stress transfer only near the contact point between disks. Stress chains are seen to develop in front of the bulk wave in the zero bond strength condition.

INTRODUCTION

Initiation of energetic materials by mechanical loading is important for many safety and performance issues. Of specific interest in this work is the ignition behavior of cure cast plastic bonded explosives (PBXs), particulate materials consisting of explosive crystals in a plastic binder. Heat generation through bulk mechanical shear comes from the yield strength of the material, which is very low in the PBXs of interest (<10 MPa), and not thought high enough to start a reaction. But within single crystals, localization of the shear deformation within the crystals on certain slip planes can lead to an increased heat generation within the particles due to dislocation motion. Stress bridging has been shown to increase stresses in crystals loaded in a dry particle bed by localizing the volume of material loaded to a small percentage of the whole (1,2).

A micrograph of a post-test impact test specimen made of a modified PBXN-109 cure cast plastic bonded explosive is shown in Fig. 1. Energetic HMX crystals, substituted for the RDX crystals found in the standard PBXN-109 formulation, are surrounded by small aluminum particles. A plasticized binder holds all the particles together. At the mesoscale, there are three distinct components of the particulate plastic bonded explosive: the crystal, the binder, and the crystal/binder interface. The contact mechanics within the mesoscale of this material are such that under a compressive loading, shear stresses will develop. Damage to the material will include debonding of the crystals and binder, crystal fracture and binder tear. The damage accumulation is not uniform, but is concentrated along certain paths.

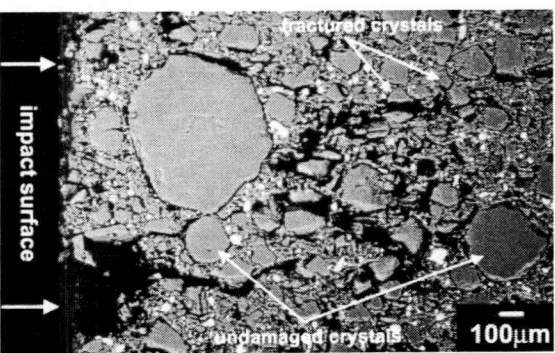

FIGURE 1. Damage pattern of a modified PBXN-109 after an unconfined impact test.

829

This work examines the effect of the particle/binder interface on the stress state within a particulate material. Of specific interest is the ability of the material to form stress concentrations along specific paths within the material. Three conditions are examined: a binderless system, a particle/binder system with an approximate acoustic wave speed ratio of one in which both compressive and shear stresses can be transferred at the boundary, and a system where shear stress is not allowed to be transferred at the boundary. This last case corresponds to a system with bond strength equal to zero.

EXPERIMENTAL METHODS

The determination of dynamic stress at the mesoscale within an explosive during an impact event is very difficult, if not impossible, with current testing methods. Therefore, both temporal and spatial scalings must be performed to understand the mechanics at this scale (3). The simulation of explosive crystals in a hard binder took the form of circular PMMA disks 6.4mm thick, 50mm diameter in an acrylic binder with similar acoustic properties to PMMA. This binder material forms a bond to the disks to allow the transfer of both compressive and shear stresses across the disk/binder interface. To examine the case where only compressive stress is transferred across the interface, silicon grease was placed on the edges of the disks to prevent bonding of the binder to the disks during the curing process. Both of the binder systems are compared to a binderless system consisting exclusively of PMMA disks.

The loading cell for the tests consists of a 4340 steel frame. Dynamic tests were conducted by placing the load frame into a compression Hopkinson bar apparatus. The loading pulse could then be controlled by the striker bar length and velocity. A 6.4 mm thick loading pin transfers the loading pulse from the Hopkinson bar to a 1018 steel bar that spans the entire specimen impact surface. In this way, planar impact conditions could be simulated. The photoelastic technique was used to generate fringes of constant maximum shear stress within the disks. High speed photographs of the fringe patterns were taken with an Imacon 460 digital camera.

RESULTS AND DISCUSSION

Dynamic fringe patterns for the particle bed without a binder are shown in Fig. 2. In this configuration, the load must be distributed through the disks due to the lack of a binder. This allows for large stress concentrations to develop at the contact points between the disks, shown by the number of fringes occurring at these points. The fringe patterns are symmetric, showing that there is little to no shear being transferred across the contact points (4). There are two distinct paths that support the most load. Other disks have a much lower loading state, while two disks remain completely unloaded.

By adding the acrylic binder, the stress state changes dramatically. The fringe patterns in Fig. 3 show a much more diffuse stress state than that of Fig. 2. The general circular wavefront of the fringe patterns is due to the loading conditions described in the experimental methods section and the fact that the binder can support the load. Each disk is loaded in this case, and the stress concentrations do not develop. Towards the front of the fringe wavefront, the contact points are seen to be supporting shear by the antisymmetry of the fringe patterns. Similar fringe shapes are seen in simpler geometries designed to promote shear deformation (3). The fringe patterns are also seen to be propagating at a much faster speed than the case without binder. This is not a change in the acoustic properties of the materials, but rather an increase in the maximum shear at points farther away from the impact at a later time. Photoelasticity is sensitive to shear stresses, and with the disk/binder interface allowing the transfer of shear, fringe patterns form earlier in disks further from the impact.

The final case with the acrylic binder and greased interfaces is shown in Fig. 4. While the distributed circular wave pattern forms as in the bonded acrylic binder case, stress concentrations form at the disk contact points as in the binderless case. The fringes are symmetric, indicating a low level of shear stress at the contact points. The stress must be transferred as compressive stress near the contact points as the disks now act as wave guides. The fringe pattern propagation speed has decreased. As shear stress cannot be transferred through the disk/binder interface, one would expect the shear stress states to develop at later times.

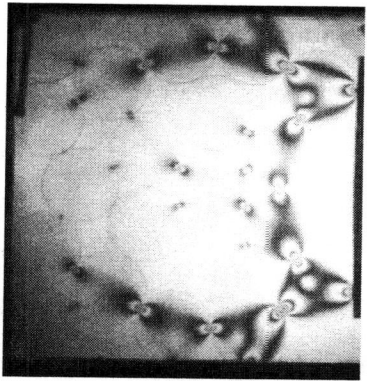 

FIGURE 2. Stress chains form in particle beds without any binder. Concentrations occur at each contact point between disks. The pictures were taken at 200 and 600µs after impact.

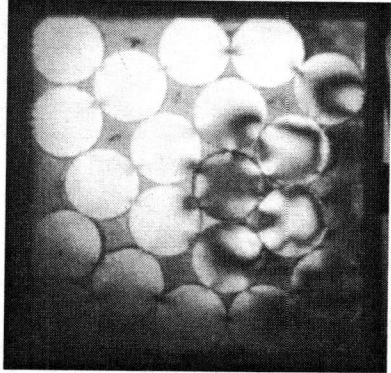 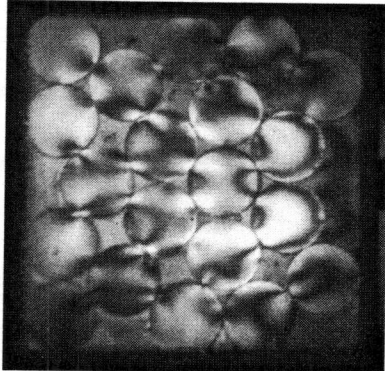

FIGURE 3. By adding an acrylic binder, the dynamic stress state changes dramaticlaly. There are very few concentrations and the load is distributed across both the binder and the disks. Pictures were taken 45 and 115µs after impact.

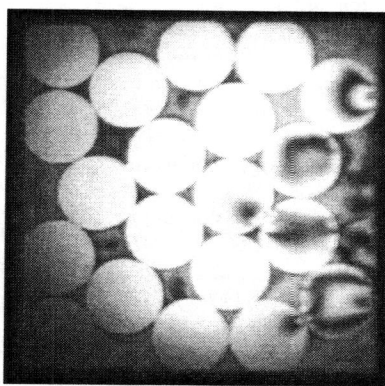 

FIGURE 4. Placing grease interfaces between the disks and binder creates a different stress state than shown in Figure 4 because shear stress is not allowed to pass across the interfaces. While a more distributed stress state develops, stress concentrations at disk contact point appear in the pictures. The pictures were taken 45 and 150µs after impact.

The interesting aspect of the second picture in Fig. 4 is the stress bridge that propagates three disks in front of the bulk fringe pattern. Figure 5 schematically shows the stress chain. This behavior has been seen in computer simulations by Bardenhagen and Brackbill (5). In the simulations, the mechanical wave speed ratio of the binder to disk material causes the stress chain. In this experiment, the chain is due to the interface condition of pure compressive stress transfer similar to the binderless case. There are two different mechanisms that both inhibit transfer of shear from one crystal to the next causing these stress chains. In the experiments described in this work, shear stress is prevented from traveling across the boundary. In the computer simulations, shear is not transmitted due to the long transit time of the shear wave within the binder. In either case, shear information is not transferred to interact with the compressive wave at the contact points between disks.

The stress chain also uses the binder in this case to propagate. The arrow in Fig. 5 goes through the binder to the last disk. There are no fringes at the contact between the last two disks in the chain, implying no contact load between the two. This behavior is seen in real explosives also. The fracture in the specimen in Fig. 1 many times propagates from crystal to crystal but also propagates along the crystal/binder interface.

CONCLUSIONS

Dynamic photoelasticity is used to examine the stress states within three systems: a binderless particle bed, a particle bed with binder, and a particle/binder system with a bond strength of zero. In the binderless system, stress concentrations form readily at the specific contact points between the crystals. The particle bed with binder is shown to have a much more diffuse stress state because shear stresses are transferred at the contact points as well as along the rest of the interface. In the system with bond strengths of zero, stress concentrations redevelop. In this case, only purely compressive stress may be transferred anywhere along the interface. Stress chains are seen to develop in front of the bulk wave in the zero bond strength condition. The transfer of shear along the interface is seen to be essential in eliminating the stress concentrations within the sample.

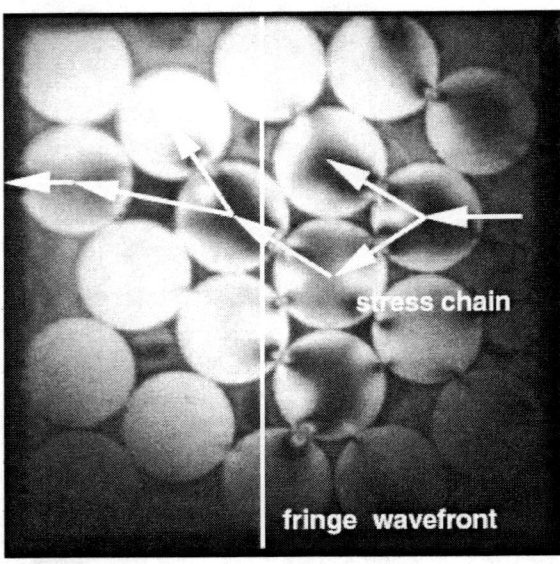

FIGURE 5. The stress chain formed in the acrylic binder system with greased interfaces extends out from the fringe wave front in the rest of the material.

REFERENCES

1. Foster, Jr., J. C., Christopher, F. R., Wilson, L. L., Osborn, J., "Mechanical Ignition Of Combustion In Condensed Phase High Explosives," *Shock Compression of Condensed Matter 1997*, edited by S.C. Schmidt et al., AIP Conference proceeding 429, pp. 389-392.
2. Roessig, K.M., and Foster, J.C., Jr., "Dynamic Stress Chain Fracture in Particle Beds," in Plastic and Viscoplastic Response of Materials and Metal Forming, edited by A.S. Khan et al., proceedings of Eighth International Symposium on Plasticity and Its Current Applications, July, 2000, pp. 437-439.
3. Roessig, K.M., "Mesoscale Mechanics of Plastic Bonded Explosives," *Shock Compression of Condensed Matter 2001*.
4. Shukla, A. and Higam, H., *Journal of Strain Analysis* **20**, 241-245 (1985)
5. Bardenhagen, S.G. and Brackbill, J.U., Journal of Applied Physics **83**, 5732-5740 (1998).

AN OPTICAL MICROSCOPY AND SMALL-ANGLE SCATTERING STUDY OF POROSITY IN THERMALLY TREATED PBX 9501

J. T. Mang[1], C. B. Skidmore[1], S. F. Son[1], R. P. Hjelm[1], and T. P. Rieker[2]

[1] Los Alamos National Laboratory, Los Alamos, NM 87545

[2] Center for Microengineered Materials, Department of Chemical and Nuclear Engineering, University of New Mexico, Albuquerque, NM 87131

Abstract. Heat transfer and combustion properties of a high explosive (HE) are influenced by the HE microstructure. The development of voids and cracks in an HE system under the conditions of thermal loading can have a strong impact on the safety and reliability of a weapon system. The optical microscopy and small-angle x-ray scattering (SAXS) techniques are useful tools for microstructural characterization. A combination of the tools allows lengthscales from hundreds of microns to tenths of nanometers to be probed, allowing a thorough description of a system's microstructure to be made. We present an optical microscopy and SAXS study of the effects of thermal loading on the microstructure of PBX 9501. Pressed pellets of PBX 9501, an HMX-based system, were heated in an oven at 180 °C for periods of 0, 15 and 30 minutes. Optical micrographs reveal the development of large pores in the microstructure with increasing thermal treatment as well as increased cracking and morphological changes of crystal grains, associated with the beta to delta phase transition in HMX. SAXS measurements were performed in order to quantify the observed porosity.

INTRODUCTION

Microstructural aspects of energetic materials are known to influence shock initiation. Similar influence is found in non-shock initiation events such as the mechanical or thermal insult that might occur in accident scenarios (1). In order to ensure the safety of a weapon system, it is essential to understand the influence different microstructural parameters have on the non-shock initiation of a weapon. Consequently, characterization of microstructural changes, resulting from insult, is necessary in order to understand and simulate the conditions that lead to initiation.

Here, we present an optical microscopy and SAXS study of the effects of thermal loading on the microstructure of the HE system, PBX 9501. The effects of thermal insult vary, depending upon the magnitude and duration of the insult and include ignition and self-sustained combustion and material sensitization (1,2). Microscopical techniques can be readily used to make qualitative, post-test assessments of microstructural damage. SAXS techniques can provide quantitative measurements of changes in microstructural parameters, such as intergranular porosity.

MICROSCOPY AND SAXS TECHNIQUES

Optical microscopy images for the current study were obtained in reflected parallel polarized light

(RPPL). This arrangement allows the return light from crystal interfaces below the plane of polish to be minimized and thus provides good grain-to-grain contrast in plastic-bonded systems. In preparation for examination, a sample is vacuum-mounted in low-viscosity epoxy for viewing in cross-section and is polished using a series of fine abrasives (3).

In a SAXS experiment, x-rays will scatter from fluctuations in the scattering length density, $\rho(\mathbf{r})$, which reflects microscale structure. The intensity of the scattered radiation, $I(Q)$, is measured as a function of the scattering vector, $\mathbf{Q}$, of magnitude $Q = (4\pi/\lambda)\sin\theta$, where λ is the wavelength of the incident radiation and θ is half of the scattering angle. $I(Q)$, for a monodisperse system of non-interacting particles, dispersed in a uniform media, can be expressed as $I(Q) = \Delta\rho^2 V \phi \langle P(\mathbf{Q}) \rangle$, where $P(\mathbf{Q})$ is the normalized, single particle form factor and is related to Fourier transform of $\rho(\mathbf{r})$, V is the particle volume, and ϕ is the volume fraction of scatterers (4). $\Delta\rho$ is the scattering length density contrast between the average scattering length density of the particle, and that of the surrounding media. In order to probe longer lengthscales, we have employed a technique known as multiple small-angle x-ray scattering or MSAXS, which can extend the range of accessible lengthscales by an order of magnitude (5). The above equation is applicable when each scattered x-ray undergoes one scattering event. Experimentally, this requires the study of thin samples, having a small ϕ. Increasing the sample thickness then, increases the probability of MSAXS. The scattered intensity can then be described as (6):

$$I_{MSAXS}(Q) = \frac{1}{2\pi} \int_0^{r_{max}} Jo(Qr)h(r)dr . \qquad (1)$$

While not shown explicitly, Eq. 1 is dependent upon the sample thickness, t, λ, ϕ, and $\Delta\rho$. The details of this technique will be discussed elsewhere (7), but the salient feature is that the measured intensity for structures undergoing multiple scattering will appear at larger values of Q than the corresponding single scattering intensity. So, scattering signals from structures, which would normally be out of the range of a given instrument, will now be accessible.

EXPERIMENTAL AND RESULTS

For the microscopy studies, samples of PBX 9501 were pressed ($\rho = 1.81$ g/cm^3), at ambient temperature, into cylinders of 1.27 cm diameter and 1.27 cm thickness. The samples were thermally treated by placing them (unconfined) in an oven at a temperature of 180 °C, corresponding to the delta phase of HMX, for 0, 15, and 30 minutes. Digital images were collected in reflective light, using a Spot camera (Diagnostic Instruments) and a DMRXA microscope (Leica). The spatial resolution of the microscope was matched with the camera pixel characteristics to nearly meet the Nyquist limit.

SAXS experiments were performed at the University of New Mexico/Sandia National Laboratory Small-Angle X-ray Scattering Laboratory, employing the Bonse-Hart camera (6). Measurements were performed on cylindrical samples, 0.9 cm in diameter, ranging in thickness from 0.05 – 0.3 cm. The samples were pressed at ambient temperature to a density of 1.79 g/cm^3. At this time only SAXS data for pristine PBX 9501 are available.

Thermal treatment of the PBX 9501 samples resulted in a 14% increase in volume for the 15-minute specimen while the volume of the 30-minute specimen increased by 16%. The theoretical volume increase for a grain of HMX having undergone the beta-delta phase transition is 7%, suggesting that additional microstructural changes have occurred.

Optical microscopy studies of the pristine and thermally insulted PBX 9501 samples were preformed in order to understand these changes. Figure 1 is an RPPL image of the pristine specimen polished on the face. Individual, large grains are easily distinguished from the surrounding matrix of fine particles mixed with binder. Most of the large grains exhibit fracture.

The heated specimens were cut along the cylindrical axis for observation of the cross-section. Figure 2 (15-minute treatment) and Fig. 3 (30-minute treatment) were taken at half the resolution of Figure 1 in order to show a larger field of view. After 15 minutes of heating (Fig. 2), numerous pockets of apparently undisturbed microstructure

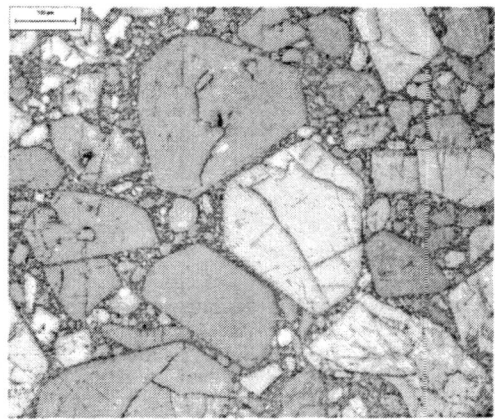

Figure 1. RPPL image of pristine PBX 9501.

and partially transformed crystals were found. The surrounding regions contain fully transformed material. In addition, we see the development of large channels in the microstructure.

Figure 3 displays the resulting microstructure after the 30-minute treatment. Significant cracking is seen, distributed throughout the sample. In contrast to the sharp crystalline facets seen in Figure 1, we see smooth, rounded edges. In some regions, individual grains are difficult to identify. In comparison to the pristine microstructure, there appears to be a loss of the fine HMX particles. The regions in Fig. 1, identified as a matrix of fine particles mixed with binder are no longer present.

These observations suggest that after fifteen minutes of heating, a partial transition from the beta to delta phase of HMX occurred. After thirty minutes, the transition appears to be complete, as no grains identifiable as beta HMX were observed. Second harmonic generation measurements of the 30-minute sample confirm this notion (2). Significant thermally-induced cracking and pore development was observed in the microstructure, which can account for the volume increase.

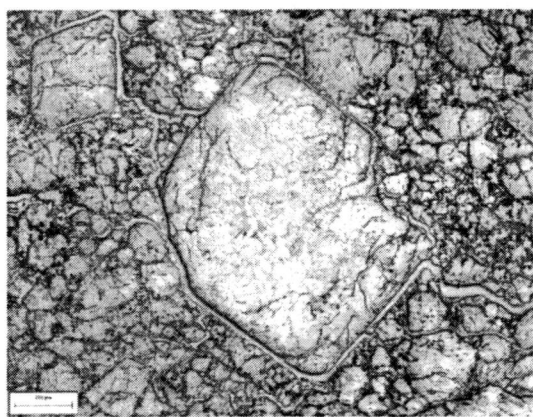

Figure 3. RPPL image of PBX 9501, 30-minute treatment.

Figure 4 shows a log-log plot of the measured SAXS lineshapes for the pristine PBX 9501 samples. Samples of different thickness were studied in order to vary the probability of multiple scattering, as discussed previously. As can be seen in the figure, the shapes of the measured curves change with increasing thickness, a hallmark of MSAXS.

Analysis of the scattering from the PBX 9501 system is complicated because of the different phases present: HMX, binder, and voids. Since HMX accounts for ~ 90% of the volume, it is considered as the continuous phase. The scattering thus arises from the HMX-binder, HMX-void and binder-void interfaces. With the current data, we cannot distinguish among the three contributions to the scattering signal. However it is possible to obtain a distribution that represents the average size distribution of the binder and void (pore) regions.

The data shown in Fig. 4 were analyzed according to Eq. 1, taking into account polydispersity. For the model calculation, we have

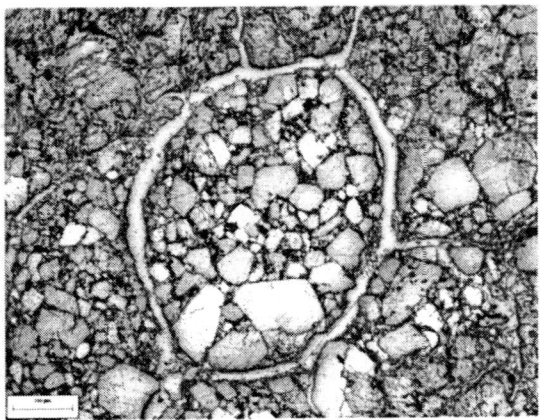

Figure 2. RPPL image of PBX 9501, 15-minute treatment.

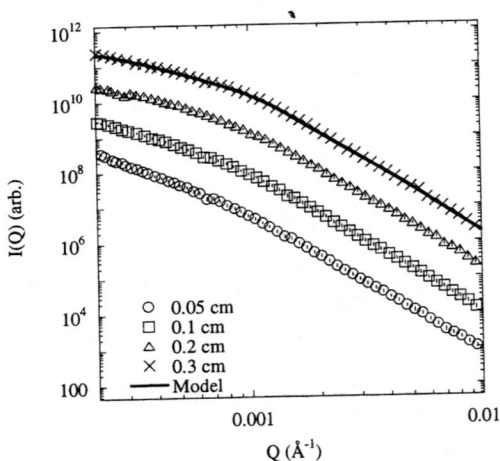

Figure 4. MSAXS lineshapes of PBX 9501.

assumed that the pores can be represented as Gaussian distributions of spherical pores. The mean size, amplitude, and width of the distributions were allowed to vary freely during the fitting process. The solid line in Fig. 4 is a representative fit of the data. Figure 5 shows the volume-weighted and number-weighted distributions of sizes obtained

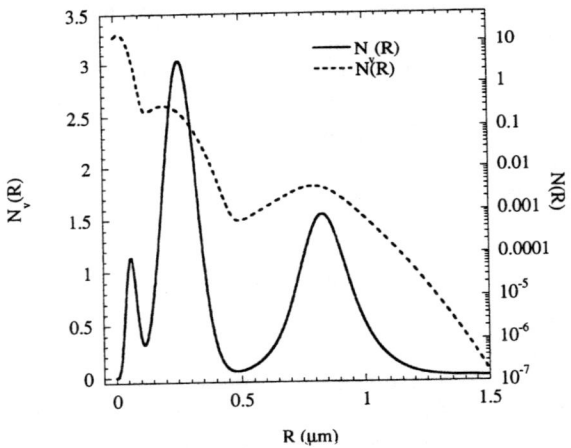

Figure 5. Pore size distributions in PBX 9501.

from the analysis. The trimodal distributions represent the probability of finding a pore of a given size in the range 0 – 1.5 µm. Averaging over these distributions, we find a mean volume-weighted pore size of 0.51 (σ = 0.32 µm) and a mean number-weighted pore size of 0.04 µm (σ = 0.05 µm). Based upon the composition of PBX 9501 and particle surface area, the average thickness of the binder region between crystals is estimated to be 0.36 µm (8). Now, from Figure 5, we see that the number-weighted distribution shows a peak at R = 0.196 µm, corresponding to a pore thickness of ~ 0.39 µm, in good agreement with the compositional calculation.

SUMMARY

We have performed optical microscopy and MSAXS studies of the microstructure of pristine and thermally damaged PBX 9501. Optical microscopy measurements show increased alteration of the PBX 9501 microstructure with increasing thermal insult. Evidence of the beta to delta phase transition was seen as well as the appearance of large channels and cracks. SAXS measurements performed on pristine PBX 9501 revealed a trimodal distribution of pores. In order to quantify the changes in porosity seen by optical microscopy, future scattering experiments will be made on thermally insulted PBX 9501. Small-angle neutron scattering measurements, employing contrast variation, will be performed in order to separate the scattering signals arising from the different interfaces found in PBX 9501.

REFERENCES

1. Idar, D. J., Lucht, R. A., et al, *Proceedings of the 11th International Detonation Symposium*, edited by J. M. Short and D. G. Tasker.
2. Son, S. F., Berghout, H. L., et al, submitted to Thermochimca ACTA.
3. Skidmore, C. B., Phillips, D. S., and Crane, N. B., Microscope, **45**, 127 (1997).
4. Brumberger, H., *Modern Aspects of Small Angle Scattering*, Kluwer Academic Publishers, 1993.
5. Schelten, J. and Schmatz, W., J. Appl. Cryst., **13**, 385 (1980).
6. Rieker, T. P., Hubbard, P. F., *Review of Scientific Instruments*, **69**, 3504 (1998).
7. Mang, J. T., Skidmore, C. B., Hjelm, R. P., and Rieker, T. P., to be submitted.
8. Kober, E., internal communication.

SUB-MOLECULAR FRACTURE STEPS IN SHOCK-SHATTERED RDX CRYSTALS AND FOLLOW-ON NANO-INDENTATION EVALUATION OF EARLY STAGE PLASTICITY

J. Sharma[1], C.S. Coffey[2], R.W. Armstrong[3], W.L. Elban[4] and S.M. Hoover[1]

[1]*Carderock Division Naval Surface Warfare Center, Bethesda, Maryland 20817*
[2]*Indian Head Division Naval Surface Warfare Center, Indian Head, Maryland 20640*
[3]*ASFRL/MNME, 2306 Perimeter Road, Eglin AFB, Florida 32542*
[4]*Loyola College, Baltimore, Maryland 21210*

Abstract. Nano-crystallites of RDX produced by aquarium shock, were examined using atomic force microscopy and found to contain sharply defined, apparent shear-type (Mode II) fracture steps having heights less than the size of an RDX molecule. The sub-molecular steps run for substantial distances along crystallite surfaces, thus indicating concerted disregistry in depth between juxtaposed molecules across the crack surfaces. The observation of locally jumbled regions of displaced/misoriented molecules suggests an obstacle barrier, not previously considered, to either shear crack propagation or dislocation pile-up release with sufficient stress intensity to cause hot spot initiation. Recent follow-on nanostructural results are reported for separate cleaved RDX crystals containing nano-indentations showing only plastic deformation with evidence of cracking.

INTRODUCTION

Aquarium-shocked RDX (cyclotrimethylene-trinitramine) crystals have shown sub-molecular high fracture steps which can potentially provide a trigger decomposition mechanism. To assess the slowly produced plastic deformation of the nanometer regime, RDX crystals were indented with ultra sharp spikes and followed with an atomic force microscope (AFM).

SURFACE STRUCTURE OF SHOCKED RDX

Application of the atomic force microscope to view surfaces of nano-crystallites (20-500 nm) of RDX resulting from aquarium shock [1] of laboratory-grown crystals revealed [2,3] that sub-molecular fracture steps had formed, running long distances along the surfaces. Step heights range from 0.05 to 0.4 nm, which are smaller than the size (0.5 nm) of an RDX molecule. The sub-molecular steps are associated with an extensive network of straight-line (planar) and sometimes non-linear cracks in bulk crystal specimens that were recovered largely intact. The steps relate geometrically to the familiar macro-scale Mode II shear fracturing. Figure 1 shows a line plot of a representative step.

The crystallographic character of the step is obvious even in a 4 nm x 4 nm area image, of Fig. 2. Further, hardly any lateral gap is discernible at the step. From the spacings of molecules in a line going perpendicular across the step, it appears that the gap is 0.1-0.2 nm wide at most. Of particular interest now is the jumbled appearance of the surface molecules shown in Fig. 2. The relative positions/orientations of molecules at and away from the two crack surfaces have been effected. This has important ramification for the chemical reactivity/decomposition of RDX. It is envisioned that conformational changes and the rubbing action of molecules at fracture surfaces could provide an important trigger mechanism. The creation of steps may involve the breaking of intramolecular bonds. The extent of such a reaction pathway would depend on how far the fracture surfaces extend into the nano-crystallites, but potentially a large number of molecules could be involved.

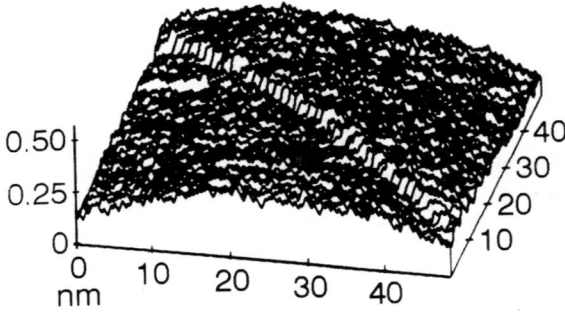

Figure 1. Line plot image of a fracture step smaller in height than an RDX molecule observed on nanocrystallites, produced by aquarium-shock.

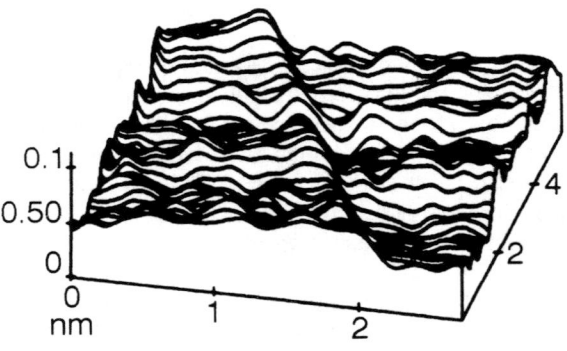

Figure 2. Line plot of a molecularly resolved AFM image, showing that the 0.25 nm high fracture step can be distinctly seen even on a 4nm x 4nm area image.

The relatively high hardness of RDX and related energetic crystals has been explained at the nanometer scale by the difficulty of dislocation motion occurring because of hindered shear displacements between the juxtaposed irregularly-shaped molecules at relatively high packing densities in the respective crystal lattices [4,5]. This consideration was subsequently used to understand [6] the formation of nitroso compounds as decomposition products in production-grade RDX that had been drop-weight impacted [7]. The possibility exists that the side nitro groups in RDX serve as obstacles needed in the dislocation pile-up avalanche model [8] for hot spot formation in energetic crystals. The foregoing discussion is based on analysis of defect-free crystal lattices containing molecules in regular positions and orientations. Given the current nano-scale observations, locally jumbled regions of displaced/misoriented molecules are proposed as potential obstacles in dislocation pile-up formation, thus providing a new nanostructural-based explanation of hot spots in crystalline materials undergoing plastic deformation.

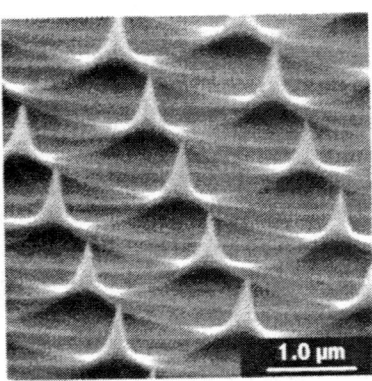

Figure 3. SEM image of the spike grating used to produce indentations in RDX crystal. The spikes were 700 nm high with tips less than 10 nm radius of curvature and 20 degree tip angle.

NANO-SCALE PLASTICITY IN INDENTED RDX

In order to study the plasticity of RDX at the nano-scale in a controlled way, a novel, inexpensive technique was devised to deform freshly cleaved surfaces at very small loads. RDX was pressed into a calibration grating, manufactured by MikroMasch˙, using a microindentation hardness tester, without affixing the usual diamond pyramid (Vickers or Knoop) indenter. A force of 25 g was applied for 10 s dwell. The grating consists of approximately 10^6 sharp spikes in square array over a 2 mm x 2 mm area on a single crystal silicon wafer, resulting in a 2 um spacing between adjacent spikes. Individual spikes have circular cross-sections, are 700 nm high and have a 20° apex angle; the utmost tip radius of curvature is <10 nm. The use of ultra sharp spikes, set in a patterned way, as seen in Fig.3, has made it possible to examine plastic deformation even when the impressions are only nanometers deep.

Figure 4. AFM image of the indented surface of RDX, showing deep and shallow indentations produced in a square pattern. The deeper ones show sharply defined 100-200 nm deep holes.

Since the RDX cleavage surfaces are not perfectly flat, the contact was not uniform causing the resultant nano-indentations to have varying sizes, which allowed successive stages of material response prior to crack formation to be followed. The distances between adjacent impressions were sufficient to allow the corresponding deformations to be independent of each other. Sometimes during the indenting process a little lateral movement of the crystal surface relative to the grating occurred, which is apparent in the AFM images. This led to streaking and very closely spaced, multiple impressions at some sites.

The indented crystal surface, indicated to be (001) as the preferred cleavage plane, also exhibited a striated appearance of hills and valleys (See Figs. 4 and 5.). The undulated surface presumably relates to an underlying internal growth sector structure first reported for RDX crystals by van der Steen and Duvalois [9]. Here, important evidence for an obstacle character of the (boundary-type) line structure is shown by the discontinuous elevation changes associated with the small indentations revealed in Fig. 5.

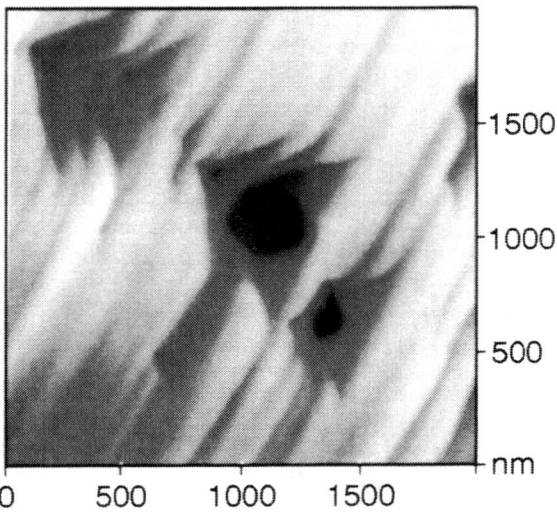

Figure 5. The indentation on the top left shows how the surface gets pressed down over a dimension of a few hundred nanometers before any hole is produced. Firmer indentations result in prominent holes.

At points of least force 3-10nm deep impressions extended over 200-300 nm are produced (See left top of Fig.5.). At higher force, a sharply defined impression is produced. Figure 5 reveals, in addition, that the indentations have polygonally-shaped edges.

The other two indentations of Fig. 5 show prominent hole formation from deeper penetration of the spike. They are approximately 100 nm deep as measured by the AFM, but the scanning tip of the AFM is too fat to probe deep and give full depth. The larger indentations were observed to have built-up regions of higher elevation distributed immediately adjacent to the residual impression. Figure 6 shows an example of these pile-up mounds.

It is instructive to relate the nanoscale indentations of various sizes to microindentation results reported for RDX crystals. First, there is the important observation that no cracking has occurred at these small indentations, also made at small effective load values, despite the fixed cone angle of the indenter points, corresponding to an otherwise large effective strain value. Elban, Armstrong and Russell [10] reported an absence of cracking for effective small strains and load values applied in a ball microindentation test. Next, there is the apparent two-step process indicated for increasing nano-indentation sizes of primary dislocation flow occurring to form the indentations at small loads, in the absence of secondary, volume conserving slip surrounding the residual nano-indentations only at larger indentation sizes [11]. Such secondary slip is very evident in Fig. 6 where the raised areas are readily identified. In Fig. 6 also, there is indication of the indentation strain field being heavily pitted, perhaps, relating to the speculative Frank suggestion of dislocations exhibiting hollow cores for large Burgers vector dislocations [12].

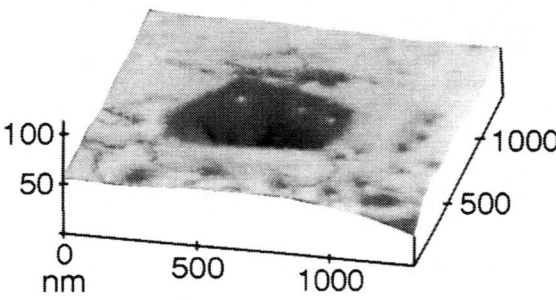

Figure 6. Surface mode image of a large indentation (approximately 700 nm in size) from which pile-up material (5-10 nm high) has been pushed out on all sides. The pile-up material has sharp borders and is full of 100-200 nm size well-defined, 3-10 nm deep, dislocation core holes.

SUMMARY

Sub-micron size indentations on RDX crystals, produced with very sharp silicon spikes, having radius of curvature smaller than 10 nm, have been reported for the first time. It appears that before the tip penetrates into the crystal, extended but shallow deformation of the surface takes place. Under higher force, sharply defined indentations are produced and mounds arise around them. The surface of the mounds shows very small craters, having a surface density of 10^8 cm^{-2}, that are interpreted as dislocation core holes.

ACKNOWLEDGEMENTS

This AFM based research has been solely supported by the U.S. Office of Naval Research (ONR). N0001401WX20909 (Program Officer Dr. Judah M. Goldwasser).

REFERENCES

1. Sandusky, H.W., Beard, B.C., Glancy, B.C., Elban, W.L. and Armstrong, R.W., "Comparison of Deformation and Shock Reactivity for Single Crystals of RDX and AP," in *Mater. Res. Soc. Symp. Proc.* **296**, 1993, pp. 93-98.
2. Sharma, J., Armstrong, R.W., Elban, W. L., Coffey, C.S. and Sandusky, H. W., *Appl. Phys. Letters* **78**, 457-459 (2001).
3. Coffey, C.S. and Sharma, J., *J. Appl. Phys.* **89**, 4797-4802 (2001).
4. Armstrong, R.W. and Elban,W.L., "Microstructural Origin of Hot Spots in RDX Crystals," in *Energetic Material Fundamentals Workshop*, Los Alamos National Laboratory, 1986, (Chemical Propulsion Information Agency [CPIA] Publication 475, 1987) pp. 177-182.
5. Dick, J.J., Mulford, R.N., Spencer, W.J., Pettit, D.R., Garcia, E. and Shaw, D.C., *J. Appl. Phys.* **70**, 3572 (1991).
6. Armstrong, R.W., *J. Physique IV*, Colloque C4, Supplement au *J. Physique III*, **5**, C4-89-102 (1995).
7. Hoffsommer, J.C., Glover, D.J. and Elban, W.L. *J. Energetic Mater.*, **3**, 149-167 (1985).
8. Armstrong, R.W., Coffey, C.S. and Elban, W.L., *Acta Metall.*, **30**, 2111-2116 (1982).
9. van der Steen, A.C. and Duvalois, W., "What Do Explosive Particles Look Like?" in *ONR/TNO Workshop on Desensitization of Explosives and Propellants*, Preprints Volume 3, Prins Maurits Laboratory, Rijswick, The Netherlands, p. 1.
10. Elban, W.L., Armstrong, R.W., and Russell, T.P., *Phil. Mag.*, **78**, 907-912 (1998).
11. Elban, W.L. and Armstrong, R.W., *Acta Mater.*, **46**, 6041-6052 (1998).
12. F.C. Frank, *Acta Cryst.*, **4**, 497-501 (1951).

* Distributed by K-TEK International, Inc, Portland, OR. 97223.

REACTION OF SHOCKED BUT UNDETONATED HMX-BASED EXPLOSIVE

P. Taylor, D.A. Salisbury, L.S. Markland, R.E. Winter and M.I. Andrew

Hydrodynamics Department, AWE, Aldermaston, Reading, Berkshire, RG7 4PR, UK

Cylindrical samples of the pressed plastic bonded HMX based explosive EDC37, backed by metal discs, were shocked through a stainless steel attenuator by an explosive donor. Reaction of the EDC37 sample was diagnosed with embedded PVDF pressure gauges and a distance to detonation for the geometry was determined. Sample length was then reduced to less than the observed detonation distance and laser interferometry was used to record the free surface velocity of the metal backing disc. The results provide data on the metal driving energy liberated by explosive which is shocked and reacting but not detonated. The results are compared with 2-D Eulerian calculations incorporating a 3-term ignition and growth reactive burn model with desensitisation. It is found that a parameter set for the reaction model which replicates the PVDF pressure profiles before reflection also gives good agreement to the metal disc velocity history at early times. The results show that an appreciable fraction of the metal driving potential of an explosive can be released without detonation being established.

INTRODUCTION

Prediction of the response of explosive containing systems to external shock stimuli is an increasingly important area of study for application to system safety assessments. A key question for any predictive tool is whether the explosive is detonated by the insult, as this is likely to lead to the highest energy responses. However, the work reported by Craig and Marshall[1] on planar shocks in explosive samples shorter than their natural run to detonation distance shows that absence of detonation conditions does not preclude significant energy release from the shocked and reacting explosive material.

Previous work by the authors of this paper[2] on the reaction of the HMX based explosive EDC37 suggested that normal reflection of strongly reacting shocks from a high impedance barrier did not lead to detonation, but did produce significant energy release as observed from indentation depths in the stainless steel reflector blocks. In order to better quantify the amount of energy liberated from the shocked but undetonated explosive, further experiments in a similar geometry have been performed with the aim of simultaneously recording the reaction profiles in the shocked explosive and the resultant velocity history of adjacent metal reflector discs. Measurement of both the reaction history and metal driving ability of the shocked explosive allows comparison with the energy release prediction of a reactive burn algorithm such as the Lee and Tarver ignition and growth model[3] to be made.

EXPERIMENTAL

The basic SI2D experimental geometry used in this study is depicted in Figure 1. An EDC29 donor charge initiated by a detonator drives a divergent shock wave through a 45 mm thick stainless steel barrier into an acceptor charge of EDC37 explosive. This sample usually consisted of a set of machined discs with PVDF pressure gauges mounted between them. The sample was terminated with a reflector disc or discs. For the reactive drive experiments, the free surface velocity of this disc was

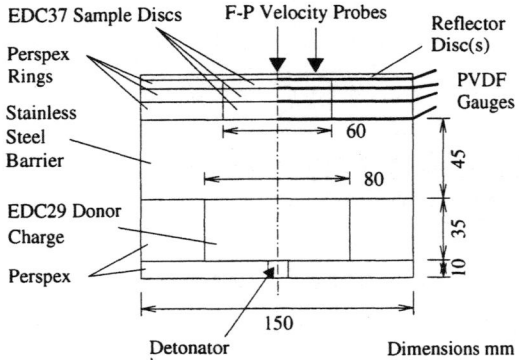

FIGURE 1. SI2D geometry with 3mm reflector disc shown.

diagnosed with Fabry –Perot velocity interferometry at locations on and off the axis. Control experiments without embedded gauges, with an inert impedance match sample and with a long sample length to determine the run to detonation distance were included in the experimental series which is detailed in Table 1.

The PVDF gauges were commercially available Dynasen gauges insulated with Kapton and having a thickness of less than 90 µm. Outside the gauge area the void between the explosive discs was filled with Sylgard184 potting compound after assembly. One experiment incorporated strain gauges in place of PVDF pressure gauges, and the strain data from this trial was used to provide a strain compensation correction for the pressure data.

The Fabry-Perot measurement was performed with a BMI Doppler Laser interferometer system, consisting of a pulsed dye laser output via fibre optic cable, with Fabry Perot analysis of the returned light. The system has a velocity sensitivity of 0.9 km/sec per fringe order (etalon spacing of 50 mm wavelength 602 nm). The dynamic fringe patterns were recorded on Thomson 506 N electro-optical streak cameras. These records were calibrated by the use of a pulsed laser diode time marker system.

RESULTS AND DISCUSSION

The first shock or detonation wave arrival time at each on axis gauge location is recorded in Table 2, where T NN.N is the arrival time (µs) of a wave at gauge location NN.N (mm), relative to the shock entering the sample. The time from detonator current zero (Io) to shock arrival in the sample was recorded as 15.6 ± 0.05 µs over the series.

SI2D/10 and SI2D/14 were control experiments to establish the experimental distance to detonation for the 45 mm barrier geometry with and without embedded gauges. From the total charge transit time alone a simple 1-D equivalent run distance can be calculated with the unreacted EDC37 Hugoniot and the Pop plot. This procedure leads to a run distance discrepancy of 5 mm between the two experiments. This suggests the effect of the embedded gauges is to reduce the effective 1-D equivalent input pressure from about 24 kb to 21 kb. The 2-D divergent nature of the geometry leads us to expect that the 1-D Pop plot will actually be

TABLE 1. 45 mm barrier SI2D Experiments

Experiment Number	Sample Material	Sample Thickness	Reflector Material(s)	Reflector Thickness	Gauge Planes mm into sample	Derived 1-D Run Distance[d]
SI2D/10	EDC37	62.5mm	St.Steel	40mm	0,10,17.5,22.5,42.5, 62.5	31.5mm
SI2D/11	EDC37	22.5mm	St.Steel	40mm	0[b],10,17.5,22.5[b]	None
SI2D/12	EDC37	22.5mm	Al.Alloy St.Steel	3mm 3mm[a]	0[b],10,17.5,22.5[b]	None
SI2D/13	KEL-F	22.5mm	Al.Alloy St.Steel	3mm 3mm[a]	0[b],10,17.5,22.5[b]	None
SI2D/14	EDC37	60mm	St.Steel	40mm	0,60	26.5mm
SI2D/15	EDC37	22.5mm	Copper	3mm[a]	0,10,17.5,22.5[c]	None

[a] The free surface velocity of these plates was monitored with F-P interferometry, on axis and 20 mm off axis.

[b] These gauge positions contained two gauges, one on axis the other 20 mm off axis.

[c] SI2D/15 contained strain gauges in identical positions to the PVDF gauges in SI2D/11-13.

[d] Derived from transit time through the complete sample length the unreacted Hugoniot and the Pop plot.

TABLE 2. PVDF Pressure gauge timing data.

Expt.	T 10 µs	T 17.5 µs	T 22.5 µs	T 42.5 µs	T 60 µs	T 62.5 µs
SI2D/10	2.73	4.84	6.24	10.86	-	13.17
SI2D/11	2.77	4.88	6.30	-	-	-
SI2D/12				-	-	-
SI2D/13	3.62	6.48	8.67	-	-	-
SI2D/14		-	-	-	11.75	-

displaced to higher run distances for a given input pressure. This is confirmed by the embedded gauge shock and detonation wave arrival times for SI2D/10 shown in Fig. 2. A simple run distance of 37 mm is suggested which corresponds to a 1-D Pop plot pressure of 19 kb. However the observed average shock velocity of 3.615 mm/µs equates to a shock pressure of 34 kb.

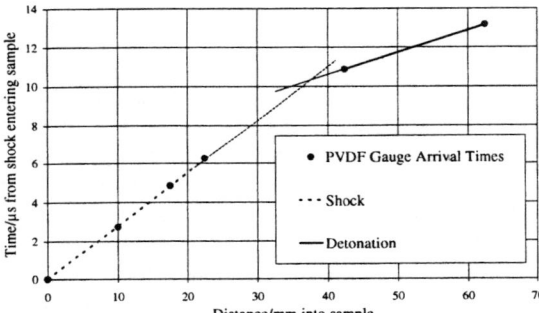

FIGURE 2. SI2D/10 x-t plot from gauge times.

The results from the two run distance experiments suggest that most of the offset from the 1-D Pop plot run distance observed in SI2D/10 is due to the 2-D divergent shock flow in the experiment, with a small proportion of the increased run distance due to the presence of the embedded gauges. As the main study involves reflection of the reactive shock well before the experimental run distance of 37 mm, the small degradation of the shock pressure and reactive growth profiles due to the embedded gauges was judged to be acceptable.

Comparison of the pressure gauge data from SI2D/10 and SI2D/11 shows the effect of reflecting the reacting shock wave from a thick high impedance barrier. Figure 3 shows that the reaction profiles before reflection are reasonably reproducible, and the reflection of the shock seems to produce an increase in reaction rate behind the reflected shock. The reflected shock is observed returning to the input face of the sample, at a typical shock velocity, well below the normal detonation propagation speed in the explosive.

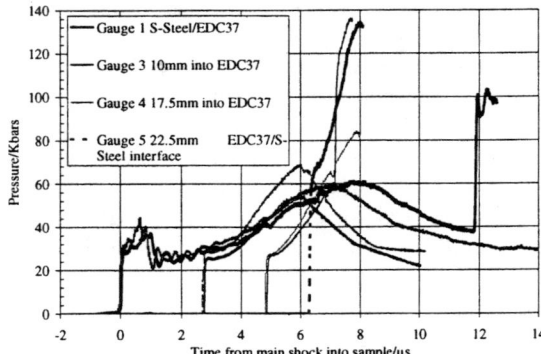

FIGURE 3. Comparison of PVDF gauge data from SI2D/11 - 22.5 mm sample (black) and SI2D/10 - 62.5 mm sample (grey).

Figure 4 shows the pressure gauge data from SI2D/12 where the reflector consisted of 3 mm thick Aluminium Alloy and stainless steel discs. The impedance mismatch between the two discs leads to a double shock reflection, and there appears to be increased reaction behind both of the reflected shocks. The short duration release wave between the two reflected shocks seen on the 22.5 mm gauge indicates a small assembly gap between the two discs. The double shock nature of the reflected pulse is still evident when the shocks return to the input face of the sample.

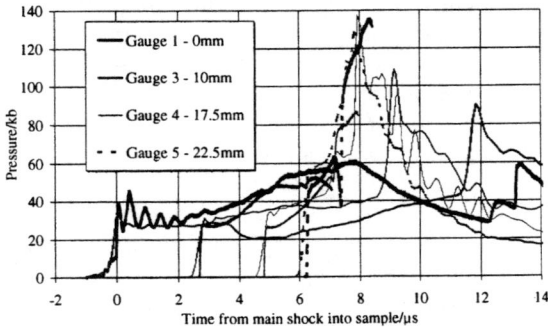

FIGURE 4. SI2D/12 pressure data against calculation in grey.

Also plotted in the figure, in grey, are pressure profiles calculated using an Eulerian hydrocode with a 0.5 mm mesh. The reactive burn model used is a three-term ignition and growth model with a shock desensitisation model. The desensitisation model is required in the calculation to prevent the explosive detonating when the shock pressure increases on reflection. The parameters for the model have been adjusted to give a reasonable fit to the experimental reaction profiles close to the reflection plane.

Figure 5 compares the experimental free surface velocity measurements from SI2D/12 with calculated histories from the same calculation shown in Fig.4. Agreement is good for the first 3-5 µs, but the ignition and growth calculation appears to be over-predicting the late time drive, giving significantly higher terminal velocities than the experimental observations.

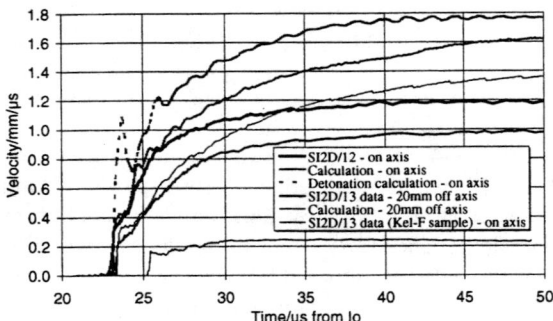

FIGURE 5. Free surface velocity measurement from 3 mm Aluminium Alloy / 3mm stainless steel reflector experiments SI2D/12 and SI2D/13 compared to calculations of SI2D/12.

A calculation was also performed on the same geometry where the EDC37 was modelled with programmed burn and a detonation point, rather than ignition and growth reactive burn treatment, and the resultant velocity profile is also plotted. It shows a terminal velocity of 1.75 mm/µs compared to 1.2 mm/µs for the experiment and 1.6 mm/µs for the reactive burn calculation. The free surface velocity trace on axis for the inert impedance match Kel-F sample experiment is also included in the figure, showing the drive given to the reflector discs from the donor charge alone. A simple velocity squared analysis suggests almost 50 % of the total detonation drive is released in the SI2D/12 experiment without detonation being established.

SI2D/15 incorporated a single 3 mm Copper reflector disc. The free surface velocity measurements for this experiment are shown in Fig. 6 for comparison with a hydrocode calculation using the same reactive burn model used to calculate SI2D/12. The early time agreement is good to around 5 µs, but at late times the hydrocode overestimates the drive and terminal velocity.

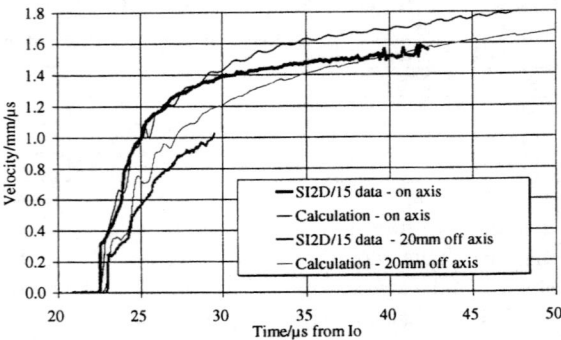

FIGURE 6. Free surface velocity measurements from 3 mm Copper reflector experiment SI2D/15 compared to calculation.

CONCLUSIONS

The data presented clearly shows significant energy release from the shocked but undetonated explosive. A 2-D Eulerian hydrocode with an ignition and growth reactive burn model and a desensitisation model provides a reasonable fit to the recorded reaction profiles recorded with embedded pressure gauges. This calculational modelling provides a good fit to the early time reflector plate velocity but tends to provide too much late time drive. A possible explanation of this discrepancy is that the calculated reaction rate remains finite at late times whereas experimentally the presence of 2-D divergence and release waves from the free surface of the reflector quench the reaction.

REFERENCES

1. Craig, B. G. and Marshall, E. F., "Decomposition of a Shocked Solid Explosive", in *5th Symposium (Int.) on Detonation*, 1970, pp. 321-329.
2. Winter, R. E., Taylor, P. and Salisbury, D. A., "Reaction of HMX Based Explosive Caused by Regular Reflection of Shocks", in *11th International Detonation Symposium*, 1998, pp. 649-656.
3. Lee, E. L. and Tarver, C. M., "Phenomenological model of shock initiation in heterogeneous explosives", *Phys. Fluids* **23**(12), pp.2362-2372 (1980).

© British Crown Copyright 2001/MOD
Published with the permission of the Controller of Her Britannic Majesty's Stationery Office.

INVESTIGATION OF DISPERSIVE WAVES IN LOW-DENSITY SUGAR AND HMX USING LINE-IMAGING VELOCITY INTERFEROMETRY[*]

Wayne M. Trott, Lalit C. Chhabildas, Melvin R. Baer, and Jaime N. Castañeda

Sandia National Laboratories, Albuquerque, NM 87185-0834

Abstract. A line-imaging optically recording velocity interferometer system (ORVIS) has been used in gas-gun impact experiments to compare the mesoscopic scale response of low-density (65% theoretical maximum density) pressings of the explosive HMX to that of an inert simulant (granulated sugar). Dispersive waves transmitted through 2.27- to 6.16-mm-thick beds of the porous sugar typically include mesoscale fluctuations that occur on length scales consistent with those seen in 3-D numerical simulations. Conditions that approximate steady wave behavior occur at a sample thickness $\geq$4 mm. Transmitted wave profiles in HMX include complex effects of chemical reaction. For coarse-grain HMX samples, reaction expands vigorously over a narrow range (0.4-0.47 km-s^{-1}) of impact velocity. Localized regions of reaction growth are evident in the spatially resolved velocity-time data.

INTRODUCTION

Current numerical simulations (1) can explore in detail the correlation between microscopic properties and the response of heterogeneous materials to shock loading at mesoscopic scales, including spatial variations in stress and thermal fields (dispersive behavior). Experimental characterization of the shock response of these materials at the requisite scale via spatially resolved measurements of transmitted wave behavior in turn provides useful information for validation and calibration of the numerical models. Validated numerical descriptions of the detailed wave fields in these materials can be analyzed to determine accurate statistical properties (1). Hence, these methods provide a promising approach for acquiring the statistical distribution information needed to develop predictive continuum level descriptions of shock-loaded materials, including heterogeneous explosives.

In this paper, we describe experiments designed to explore the mesoscopic scale response of a common secondary explosive (HMX) in comparison with that of granulated sugar (sucrose), an inert explosive simulant. These tests seek to expand on a series of magnetic gauge studies of low-density sugar and HMX performed by Sheffield et al. (2,3) Experiments on low-density, porous sugar can address mesoscopic scale thermomechanical effects in the absence of rapid reaction. The sugar experiments discussed here primarily focus on the observed dispersive wave behavior as a function of sample thickness. The tests on HMX explore the complex, additional effects of chemical reaction in the shock response.

EXPERIMENTAL

Simultaneous line-imaging ORVIS and single-point VISAR measurements have been made on waves transmitted by pressed sugar (2.27-mm to 6.16-mm thick) and HMX samples (4-mm thick) in a gas gun target design very similar to that used in the previous magnetic gauge studies (2,3). A

---
[*] Sandia is a multiprogram laboratory operated by Sandia Corporation, a Lockheed Martin Company, for the United States Department of Energy under Contract DE-AC04-94AL85000.

schematic diagram of this design is shown in Fig. 1. The target assembly consists of a Kel-F impactor and a Kel-F target cup containing sugar or HMX pressed to 65% theoretical maximum density (TMD). The porous bed is confined by a 0.225-mm-thick buffer layer of Kapton and an aluminized PMMA interferometer window. The buffer is used to mitigate the loss in reflected light intensity that typically occurs upon shock arrival at the window. To achieve consistency in preparation of the low-density, porous samples, both mass and volume of the material were carefully controlled.

Sugar samples were prepared from coarse, granulated material. A description of the particle size distribution both as received and after pressing and release has been reported previously (4). The largest weight fraction (~60%) of the granulated sugar resides in a grain size range of 250-425μm. A significant amount of grain crushing occurs even at the low pressing density used in this study. HMX samples were prepared from three different materials: [1] a coarse-grained lot of Holston batch HOL 920-32 (mean particle size near 120μm), [2] a fine-grain sieved sample (38-45μm), and [3] a coarse-grain sieved sample (212-300μm).

The line-imaging ORVIS used in this study is a compact system that combines the interferometer optics, laser source (2W NdYVO$_4$ cw laser), and streak camera/intensifier/CCD detector on a single 2' x 6' optical breadboard. Detailed discussions of the instrumentation, the optical coupling to the gas gun target chamber, and image data reduction methods are available elsewhere (4). Single-point VISAR data were obtained using a dual-delay-leg, "push-pull" assembly.

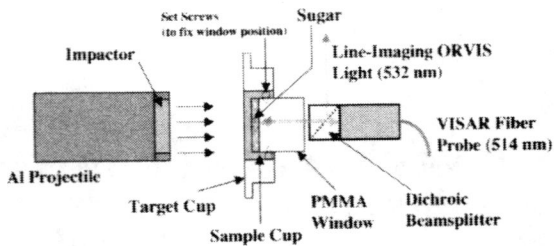

FIGURE 1. Schematic diagram of gas gun target design for measurements on low-density sugar or HMX.

RESULTS AND DISCUSSION

Low-Density Sugar Experiments

As described previously (4), results from both VISAR and line-imaging ORVIS are generally consistent with the systematically varying dispersive behavior of wave profiles vs. impact velocity reported by Sheffield et al. (2). This includes comparable measured shock and particle velocities as well as rise times in the transmitted wave that decrease from 700 ns to 200 ns as impact velocity increases from 0.3 to 0.7 km-s^{-1}, respectively. The spatially resolved ORVIS data also reveal mesoscopic scale velocity variations (both transverse and longitudinal wave structures). The length scale of the wave fluctuations can vary over a wide range (~10-200 μm) depending on the local particle size distribution.

Instructive comparisons can be made between the observed wave profiles and those generated by the 3-D simulations. A description of computational methods and results from numerical simulation of a shock-loaded 2.27-mm-thick sugar sample are given in Reference 1. The computations predict rapid deformation at material contact points in the sample. Both stress and temperature fields display large amplitude fluctuations, arising from the effects of shocks interacting with individual material surfaces and multiple crystal interactions. Figure 2 displays a time sequence of temperature fields along a midplane cross-section of the sugar bed after impact at 0.5 km-s^{-1}. Similar dispersive wave effects are evident in the stress field contours (not shown). Of particular interest is a crystal penetration event that occurs at the region marked

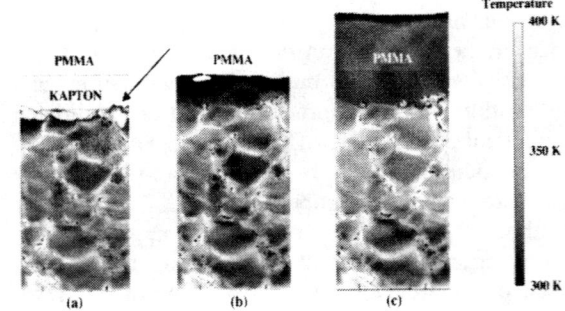

FIGURE 2. Time sequence of temperature field maps from simulation of low-density sugar under impact: (a) initial interaction with buffer layer; (b) wave approaching PMMA; (c) crystal penetration into buffer.

by an arrow. Such effects contribute to loss in reflected light intensity and drive the need to view transmitted waves through a thin buffer material.

For comparison with experimental data, particle velocity was computed at approximately 50 "tracer" points along a line segment located at the Kapton/PMMA interface. As shown in Fig. 3a, the velocity-time profile at the interface displays complex transverse mode structure (even after transmission through the buffer) in addition to a ~100-ns rise time in the wave. Representative spatially resolved velocity-time plots from line-imaging ORVIS data are also presented in Fig. 3b and 3c. In general agreement with simulation, waves transmitted through 2.27-mm and 4-mm-thick sugar beds, respectively, display significant transverse wave structure and early peaks in particle velocity (arising from a shock impedance mismatch between porous sugar and the homogeneous buffer material) followed by late-time velocities near 0.25 km·s^{-1}. For 2.27-mm-thick samples, the rise time of the transmitted wave is in good agreement with simulation. A clearly longer rise time is seen in the 4-mm case. These results point to non-steady wave behavior at the reduced sample thickness.

The issue of "steadiness" in the wave behavior vs. sample thickness is important in defining the computational domain needed to achieve representative conditions in the low-density material. Accordingly, we have extended our tests to include thicker (6.16-mm) samples. Fringe records of the wave transmitted by three different thicknesses of sugar are displayed in Fig. 4. The rise time in waves generated in the two thicker samples appear to be very similar. The transition in the 2.27-mm-thick sample is (as also indicated above) more abrupt, especially with respect to the duration of the "foot" of the ramp. A compilation of rise times (estimated as the time for wave velocity to progress from 10% to 90% of the maximum level) from both VISAR and ORVIS measurements is shown in Fig. 5. These data suggest that relatively steady wave behavior may be obtained for sample thickness ≥4 mm. It must be noted, however, that the overall wave behavior actually consists of a superposition of many wavelets, a phenomenon that is highly statistical in nature.

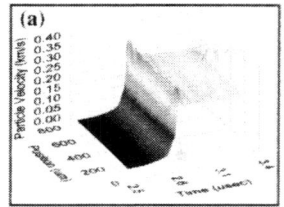
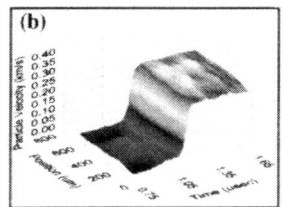
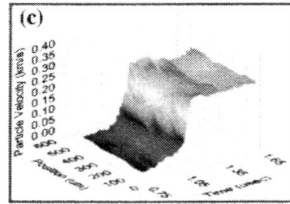

FIGURE 3. Comparison of (a) predicted and (b) measured velocity profile of wave transmitted by 2.27-mm-thick sugar sample. The measured velocity profile transmitted by a 4-mm-thick sample is shown in (c). Impact velocity near 0.5 km·s^{-1}. A 900-ns duration of time is plotted in all three cases.

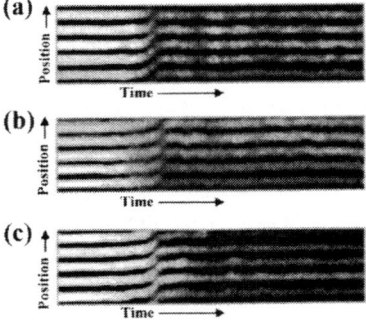

FIGURE 4. Line ORVIS image data showing transmitted wave from (a) 2.27-mm, (b) 4-mm, and (c) 6.16-mm-thick sugar, respectively; impact velocity near 0.5 km·s^{-1}.

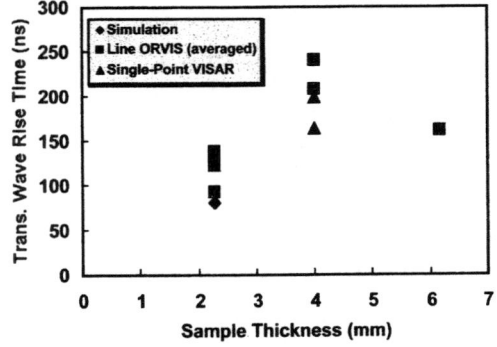

FIGURE 5. Rise time of wave transmitted by 65% TMD sugar as a function of sample thickness.

Low-Density HMX Experiments

As discussed by Sheffield et al. (3), compaction waves at levels below the threshold for reaction in porous HMX behave in a manner similar to those in sugar. At levels above the threshold for initiation, chemical reaction causes the wave to accelerate and to become steeper as it travels. Line-imaging ORVIS can reveal interesting details of the initiation and reaction growth at mesoscopic scales.

A spatially resolved velocity-time profile for 65% TMD HMX (HOL920-32) subjected to impact at 0.4 km-s^{-1} is shown in Fig. 6. The wave is diffuse with a rise time ~150-200 ns, slightly faster than that reported by Sheffield et al. (3) Also evident are localized regions of modest wave growth, reflecting the onset of exothermic reaction. This localized wave growth is a consistently observed feature of low-density HMX at impacts slightly above threshold for onset of reaction. Figure 7 provides another example, displaying a plot of the wave behavior from three different regions of the sieved 212-300μm material under impact at 0.4 km-s^{-1}. Sample particle size is also directly reflected in the amplitude and frequency of wave fluctuations seen in this plot (i.e., the sieved, coarse grained sample exhibits prominent low-frequency fluctuations in its wave profile).

The VISAR and ORVIS results also reveal that the coarse grained HMX materials (HOL920-32 and sieved 212-300μm) display a sharp increase in reactivity over a fairly narrow range of impact velocity (0.4-0.47 km-s^{-1}). In contrast, the transmitted wave profiles from the finer 38-45μm sieved HMX exhibit little evidence of exothermic reaction under the same conditions. The different response of these three materials (as reflected by the peak observed particle velocity) is summarized in Table 1. The complex phenomenology (including stochastic aspects of chemical energy release) associated with the transmitted wave profiles from low-density, porous HMX provides a substantial challenge to detailed numerical simulation.

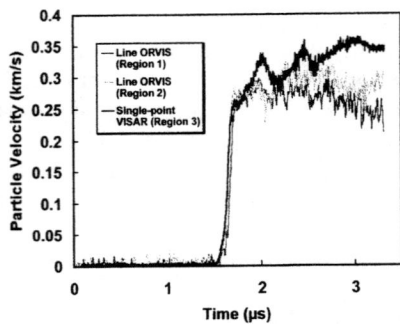

FIGURE 7. Transmitted wave profiles from different regions of HMX (212-300μm) under impact at 0.4 km-s^{-1}.

FIGURE 6. Spatially resolved transmitted wave profile from 65% TMD HMX; 0.4 km-s^{-1} impact velocity.

TABLE 1. Peak Particle Velocities of Transmitted Waves from Low-Density HMX (as determined by VISAR and line-imaging ORVIS)

Impact Velocity (km-s^{-1})	Holston Batch HOL920-32	212-300μm Sieved HMX	38-45μm Sieved HMX
0.4	0.25	0.38	0.23
0.425	>0.3	0.475	
0.47	0.53		0.275
0.53	0.97		

REFERENCES

1. Baer, M. R. and Trott, W. M., "Mesoscale Descriptions of Shock-Loaded Heterogeneous Materials" (this volume).
2. Sheffield, S. A., Gustavsen, R. L., and Alcon, R. R., "Porous HMX Initiation Studies--Sugar as an Inert Simulant," in *Shock Compression of Condensed Matter--1997*, edited by S. C. Schmidt et al., AIP Conference Proceedings 429, New York, 1998, pp. 575-578.
3. Sheffield, S. A., Gustavsen, R. L., and Anderson, M. U., "Shock Loading of Porous High Explosives," in *High-Pressure Shock Compression of Solids IV*, edited by L. Davison, et al., New York: Springer-Verlag, 1997, pp. 23-61.
4. Trott, W. M., et al., "Dispersive Velocity Measurements in Heterogeneous Materials," Sandia National Laboratories Report, SAND2000-3082, December 2000.

ISENTROPIC COMPRESSION OF LX-04 ON THE Z ACCELERATOR

D.B. Reisman[1], J.W. Forbes[1], C.M. Tarver[1], F. Garcia[1], R.C. Cauble[1], C.A. Hall[2], J.R. Asay[2], K. Struve[2], M.D. Furnish[2]

[1]Lawrence Livermore National Laboratory, P.O. Box 808, L-282, Livermore, CA 94550
[2]Sandia National Laboratories, Albuquerque, NM, 87185

Abstract. Three sets of LX-04 samples of 0.18 and 0.49 mm nominal thicknesses were all dynamically loaded by Sandia's Z-accelerator with a ramp compression wave with a 200 ns rise time and about 160 kb peak stress. The LX-04/lithium fluoride samples interface velocities were measured using VISAR's. Comparisons of experimental and computational results will be given. Compression and release isentropes both show some reaction and kinetic behavior of the LX-04. Experiments were also performed on ultra-fine TATB where PMMA windows were used. Future experiments on single crystals of HMX that are designed to measure the phase transition at high pressures will be discussed.

INTRODUCTION

Isentropic compression experiments (ICE) [1,2] were performed on LX-04 and ultra-fine TATB using the "square short" assembly. In this configuration pairs of samples of thickness 200-600 μm were placed on driver panels of aluminum of base thickness 400 μm (Fig. 1). Windows of LiF and later PMMA were bonded to the back of the HE samples to minimize wave interactions and to aid in the analysis. Pressure ramps of approximately 200 ns duration and peak pressure of 110-160 kbar were applied to the aluminum surface. The resulting HE/window interface velocities were recorded using single-point VISAR.

EXPERIMENTS AND RESULTS

Six LX-04 samples (three pairs of 300 and 500 μm thickness and 6 mm diameter) were compressed to a peak pressure of over 160 kbar. The increase in peak velocity and consequently peak pressure with increasing sample thickness suggests that the material has begun to react (Fig 2.). This made standard

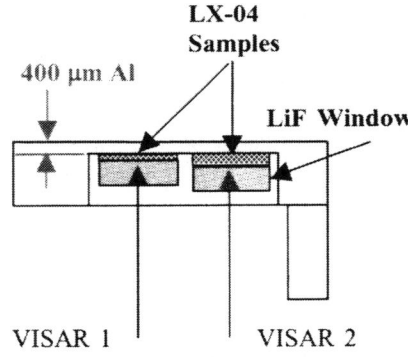

FIGURE 1. Arrangement of LX-04 samples on ICE.

characteristics analysis [3], which assumes rate independent response, difficult to perform.

Instead, the resulting waveforms were compared to hydrodynamic (HD) calculations. The HD code was run in Lagrangian mode with a Grünesien equation of state assumed for the Al

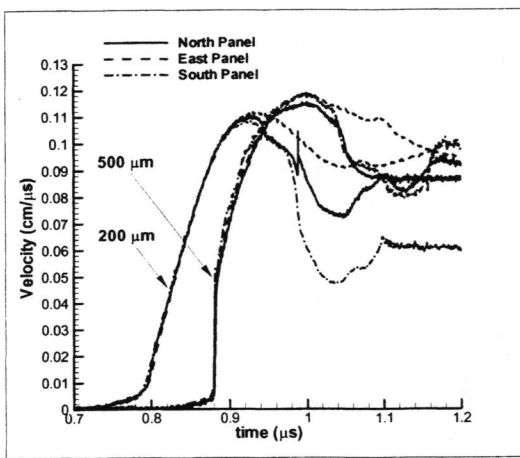

FIGURE 2. Isentropic compression of LX-04

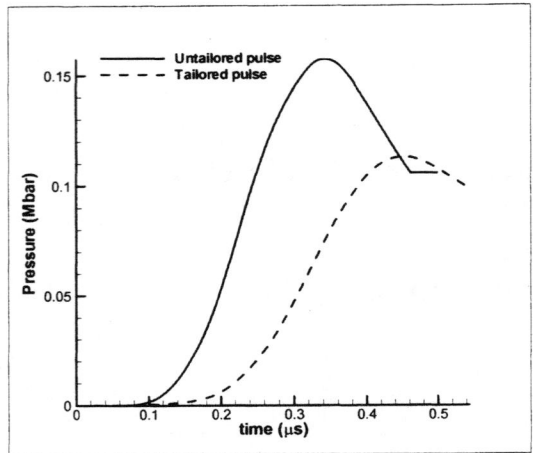

FIGURE 4. Standard pressure ramp and "tailored" ramp used in ICE.

and the LiF window. The LX-04 was modeled using the reactive flow model of Tarver [5]. The pressure drive was derived approximately from experimental current (B-dot) measurements. The calculation showed the increase in peak velocity (pressure) with propagation distance (Fig. 3). However, the calculation showed the material to have stiffer response as reflected by a steeper waveform for the 500 μm sample. This is possibly due to the fact that the LX-04 reactive flow model was calibrated to shock loading experiments which may show more

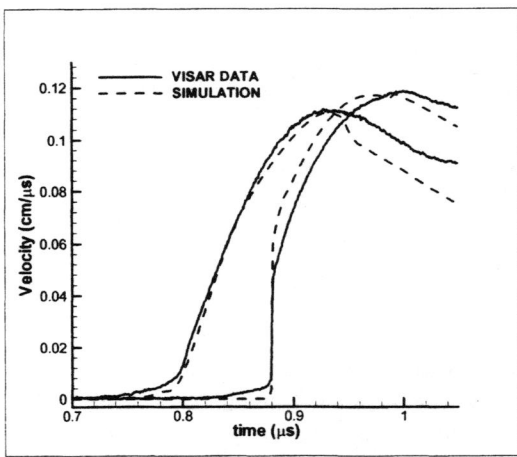

FIGURE 3. Comparison of selected VISAR data to simulation.

reaction than lower temperature isentropic loading experiments.

In the second set of experiments, which used LX-04 and fine-grained TATB, a modified or "tailored" pressure drive was applied to the sample (Fig. 4). This was accomplished by firing a level of Z's capacitor banks early, producing a 200 ns prepulse followed by the main pulse. A linear ramp will typically produce low pressure shocks that show up at the base of the velocity waveform. By gradually increasing the slope of the pressure drive with time, the low sound speed, low pressure wave is allowed propagate ahead of the higher density, higher pressure wave. The result is a pressure drive that can delay shock up and maintain isentropic compression loading over a greater sample thickness. This is of particular importance in the case of HE which has relatively low impedance and will typically shock up in only 200-300 μm under linear ramp loading.

Two experiments, each with 8 samples (consisting of 4 pairs of HE of thickness 300 and 500/600 μm), were performed on LX-04 and ultra fine-grained TATB. A peak pressure of 110 kbar was applied to the Al which produced a peak pressure of approximately 100 kbar in the HE samples. As before, simulations showed the LX-04 being driven to reaction. However, the

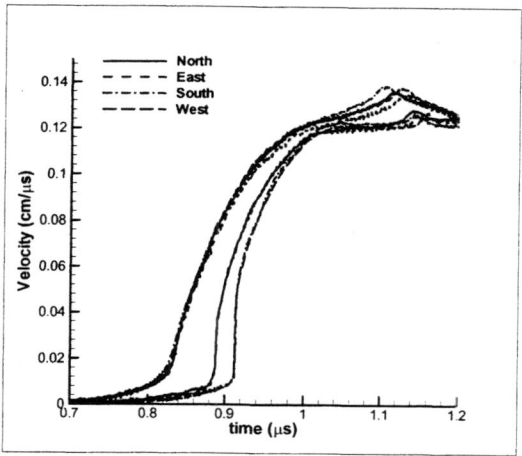

FIGURE 5. VISAR data from second LX-04 experiment.

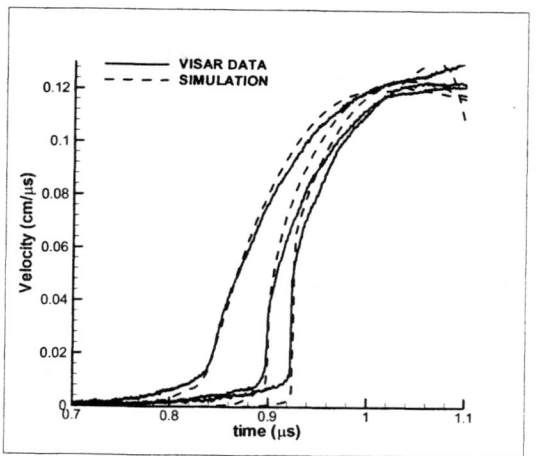

FIGURE 6. Comparison of simulation of LX-04 simulation to ICE data.

TATB, again simulated with the Tarver model, showed no such reaction.

On each panel a third VISAR probe was used to record the free surface velocity of the aluminum driver surface. By making this measurement, the input pressure drive of the aluminum can be more accurately determined. It is hoped that by using the velocity histories of the HE/window interface and the input pressure drive along with a backward integration hydrocode [6], the stress-strain relation can be accurately determined.

VISAR velocity histories from the LX-04

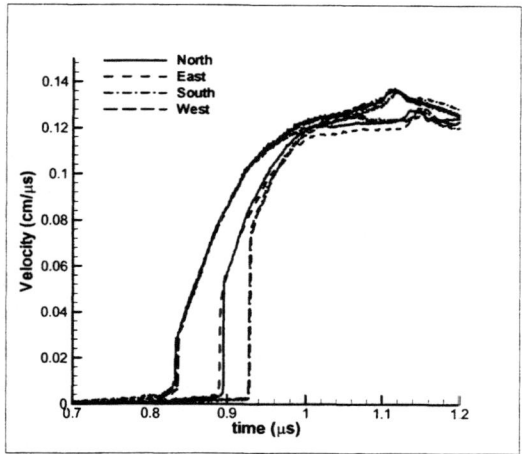

FIGURE 7. VISAR data from ultra-fine TATB experiment.

experiment suggested that the material did not undergo significant reactions (Fig. 5). Unlike the first LX-04 experiment an increase in velocity was not seen with increasing sample thickness. In fact, simulations performed with the Tarver model with reactions turned off (Fig. 6) compared reasonably well with experiment. As before, the model showed a stiffer response than the ICE data.

On the ultra-fine TATB a shock was seen even for the thinnest (300 μm) sample (Fig. 7). This, we believe, was due to the large void fraction (7%) of the material. Simulations reproduced the peak velocity and much of the higher velocity part of the waveform past the initial shock (Fig. 8). However, simulations did not reproduce the early shock formation.

In future ICE experiments we plan to compress samples of single crystal HMX to pressures of approximately 300 kbar. It is believed that HMX will undergo a phase transition at 270 kbar [4]. It should be relatively easy to detect this transition in the VISAR record since it is associated with a fairly significant (5-10%) volume change.

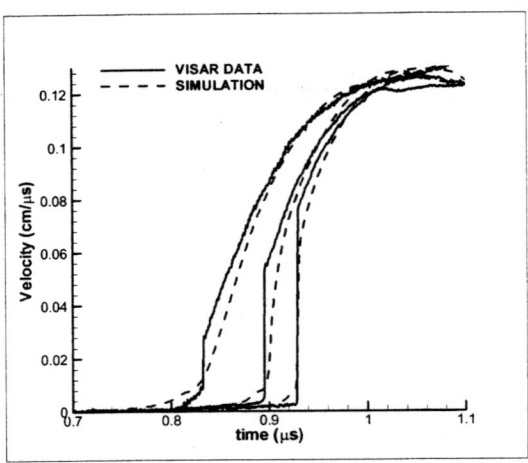

Figure 8. Comparison of simulation of ultra-fine TATB to ICE data.

ACKNOWLEDGEMENTS

This work was performed under the auspices of the U.S. Department of Energy by Lawrence Livermore National Laboratory under contract No. W-7405-Eng-48.

REFERENCES

1. Hall, C.A., *Phys. Plasmas* **7**, 2069 (2000).
2. Reisman, D.B. et. al., *J. Appl. Phys.* **89**, 1625 (2001).
3. Aidun, J.B. and Y.M. Gupta, Y.M, *J. Appl. Phys.* **69**, 6998 (1991).
4. Yoo, C.S. and Cynn, H., *J. Chem Phys.* **111**, 10229 (1999).
5. Tarver, C.M., Kury, J.W., and Breithaupt, R.D., *J. Appl. Phys.* **82**, 3771 (1997).
6. Hayes, D.B., "Backwards Integration of the Equations of Motion to Correct for Free Surface Perturbations", Sandia National Laboratories Report, SAND2001-1440 (2001).

MECHANICAL BEHAVIOR OF ENERGETIC MATERIALS DURING HIGH ACCELERATION

Y. Lanzerotti[1] and J. Sharma[2]

[1] U. S. ARMY TACOM-ARDEC, Picatinny Arsenal, NJ 07806-5000
[2] Naval Surface Warfare Center, Carderock Division, West Bethesda MD 20817-5700

Abstract. The mechanical behavior of explosives subjected to high acceleration has been studied in an ultracentrifuge at -10°C. Melt-cast TNAZ and pressed TNAZ, LX-14, Composition A3 Type II, PAX-2, and PAX-3 have been studied. Failure occurs when the shear or tensile strength of the explosive is exceeded. The fracture acceleration of melt-cast TNAZ and pressed TNAZ at -10°C was about 10% less than that at 25°C. The fracture acceleration of the plastic bonded explosives PAX-3 and Composition A3 Type II at -10°C were about twice that at 25°C.

INTRODUCTION

We have been studying the mechanical behavior of energetic materials during high acceleration by using an ultracentrifuge (1-5). Energetic materials are of significant interest for scientific and practical reasons in the extraction (mining) industry, structure demolition, space propulsion, and ordnance. In these applications the materials can be subjected to high, fluctuating and/or sustained acceleration. The nature of the fracture process of such materials under high acceleration is of particular interest, especially in ordnance and propulsion applications. For example, explosives in projectiles are subjected to setback forces as high as 50,000 g during the gun launch. These high setback forces can cause fracture and premature ignition of explosives.

Fundamental understanding of the behavior of energetic materials subjected to high acceleration is a key to better practical ordnance designs that solve the problems of abnormal propellant burning and premature ignition of explosives during gun launch. The pressure gradient that is experienced by the explosive during acceleration in the gun and under g-loading in the ultracentrifuge is unique and will produce different kinds of behavior and failure than under other material test conditions. The present work is particularly relevant to the future development of insensitive energetic materials to be used in devices with higher acceleration.

Previously, (1-5) we have used an ultracentrifuge to study the fracture behavior of TNT (trinitrotoluene), Composition B [59% cyclotrimethylenetrinitramine (RDX), 40% TNT, 1% wax], four types of Octol [70% cyclotetramethylenetetranitramine (HMX), 30% TNT; 75% HMX, 25% TNT; <75% HMX, 25% TNT, <1% HNS (hexanitrostilbene); and 83% HMX, 17% TNT], cast TNAZ (1,3,3-trinitroacetidine), pressed TNAZ, pressed plastic-bonded explosives LX-14 (95% HMX, 5% Estane), Composition A3 Type II (91% RDX, 9% polyethylene), PAX-2A [85% HMX, 9% BDNPF (bis-dinitropropyl acetal formal), 6% CAB (cellulose acetate butyrate)], and PAX-3 (85% HMX, 9% BDNPF, 6% CAB/25% aluminum). In the present work, we studied the fracture behavior at high acceleration in an ultracentrifuge at -10°C of cast TNAZ, pressed TNAZ, and pressed plastic-bonded explosives [LX-14, Composition A3 Type II, PAX-2A and PAX-3].

TECHNIQUE

A Beckman ultracentrifuge model L8-80 with swinging-bucket rotor model SW 60 Ti was used to rotate the samples up to 60,000 rpm (about 500,000 g). The distance of the specimen from the axis of rotation ranged from 6 to 12 cm.

The samples were machined into the shape of the frustrum of a cone. The large diameter was typically 11 mm and the small diameter 9 mm. The angle between the base and the side was 80°. Each sample was fitted into a 5-mm long, 11-mm o.d. aluminum cylinder. At one end the i.d. was 11 mm and at the other end 9 mm. The angle between the inner and outer sides of the sleeve was 10°. The 9-mm diameter top of the sample faced away from the axis of rotation.

For an experiment the acceleration was increased up to a maximum and then held there until the total elapsed time reached 5 min. The sample was then decelerated smoothly. In a series of experiments on a material, the initial maximum acceleration was then increased systematically in each successive 5-min run. The sample fractured when the shear or tensile strength of the material was exceeded, causing particles to break loose from the exposed surface and transfer to the closed-end of the tube. A hemispherical fracture surface resulted.

RESULTS

TNAZ

Polycrystalline melt-cast TNAZ fractured at grain boundaries at about 93 kg at -10°C. The cast had been made with 70% liquid TNAZ and 30% 40 µm maximum size TNAZ particles. This melt-cast TNAZ was 97.3% of its theoretical maximum density (TMD) of 1.84 g/cm^3. Pressed TNAZ fractured at about 55 kg at -10°C and its density was 98.4% of this TMD.

Pressed Plastic-Bonded Explosives

Table 1 compares the fracture behaviors of melt-cast TNAZ and of pressed TNAZ with those of LX-14, PAX-2A, PAX-3, and Composition A3 Type II. The fracture acceleration of LX-14 was greater than 140 kg at -10°C. This PAX-2A was pressed at 99.4% of its TMD of 1.79 g/cm^3. PAX-3 has been found to fracture at about 123 kg at -10°C. This PAX-3 was pressed at 97.4% of its TMD of 1.955 g/cm^3. Composition A3 Type II fractured at about 55 kg at -10°C. It had been pressed to 98.1% of its TMD of 1.662 g/cm^3.

The fracture acceleration of cast TNAZ was greater than that of pressed TNAZ at -10°C and 25°C. (4) Since the fracture acceleration of cast explosives is inversely related to the grain size, (2) the high fracture acceleration of this cast TNAZ may therefore be due to its very small grain size (<0.1 mm). The fracture acceleration of Composition A3 Type II was less than that of PAX-3 at -10°C and 25°C. LX-14 and PAX-2A did not fracture at -10°C in these experiments.

The fracture acceleration of the pressed plastic-bonded explosives PAX-3 and Composition A3 Type II at -10°C were about twice that at 25°C. The fracture acceleration of pressed TNAZ and melt-cast TNAZ at -10°C was about 10% less than at 25°C.

TABLE 1. Fracture Acceleration of Pressed Explosives in an Ultracentrifuge at -10°C

Explosive	Fracture Acceleration, Kg	Density, %TMD
LX-14	>140	95.9
PAX-2A	>140	99.4
PAX-3	123	97.4
TNAZ (melt-cast)	93	97.3
Composition A3 Type II	55	98.1
TNAZ	55	98.4

REFERENCES

1. Lanzerotti, Y. D., and Sharma, J., App. Phys. Letters **39**, 455-457 (1981).
2. Lanzerotti, Y. D., and Sharma, J., "Mechanical Behavior of Energetic Materials During High Acceleration in an Ultracentrifuge", in *Grain Size and Mechanical Properties - Fundamentals and Applications*, edited by M. A. Otooni et al., Materials Research Society Proceedings 362, Pittsburgh, 1995, pp. 131-135.
3. Lanzerotti, Y. D., Meisel, L. V., Johnson, M. A., Wolfe, A., and Thomson, D. J., "Fracture Surface Topography of Energetic Materials Using Atomic Force Microscopy", in *Atomic Resolution Microscopy of Surfaces and Interfaces*, edited by D. J. Smith, Materials Research Society Proceedings 466, Pittsburgh, 1997, pp. 179-184.
4. Lanzerotti, Y. D., and Sharma, J., "*Mechanical Behavior of Energetic Materials During High Acceleration in an Ultracentrifuge*", in Shock Compression in Condensed Matter - 1997, edited by S. C. Schmidt et al., AIP Conference Proceedings 429, New York, 1998, pp. 595-597.
5. Meisel, L. V., Scanlon, R. D., Johnson, M. A., and Lanzerotti, Y. D., "*Self-affine Analysis on Curved Reference Surfaces: Self-affine Fractal Characterization of a TNT Surface*", in Shock Compression in Condensed Matter - 1999, edited by M. D. Furnish et al., AIP Conference Proceedings 505, New York, 2000, pp. 727-730.

USING SIMULTANEOUS TIME-RESOLVED SHG AND XRD DIAGNOSTICS TO EXAMINE PHASE TRANSITIONS OF HMX AND TATB[*]

C. K. Saw, J. M. Zaug, D. L. Farber, B. L. Weeks and C. M. Aracne

University of California, Lawrence Livermore National Laboratory,
7000 East Ave., Livermore, CA 94550

Abstract: Simultaneous SHG (second harmonic generation) and XRD (x-ray diffraction) diagnostics have been applied to examine the phase behavior of energetic materials, HMX (octahydro-1,3,5,7-tetranitro-1,3,5,7-tetrazocine) and TATB (1,3,5-triamino-2,4,6 trinitrobenzene). This unique capability provides information about both volume and surface effects that occur during the solid-solid transformation process. This paper reports XRD results for HMX and TATB at elevated temperatures and on simultaneous SHG and XRD experiments on HMX at fixed temperature. Our results do not indicate that a solid-solid phase transformation occurs for TATB even at temperatures up to 340°C. XRD results on HMX held at 165°C and 1 bar, indicate that the β to δ transformation is incomplete after a period of 4.5 hours which do not temporally correlate with SHG. Overall information indicates that the observed SHG intensities from surface effects can, in some cases, dominate over volume generated SHG contributions. Finally, we have run *in situ* AFM scans of HMX at 180°C and 184°C that show HMX surface area increases by many orders of magnitude after the δ-phase transformation is completed.

INTRODUCTION

After several decades of study there still is not a universal set of rate laws governing solid-solid structural phase transitions of polymer bonded explosive (PBX) materials. The following parameters all affect the β to δ transformation kinetics of HMX at fixed pressure and temperature: grain size, binder content, impurity content (e.g., RDX), and compaction density of the powder. To date there is not a kinetic rate law that incorporates these critical rate-limiting parameters. Given the core mission of the stockpile safety initiative we have been motivated to develop and refine a series of experimental diagnostic tools that will allow us to derive a universal rate law. To do this we first set out to determine what type or series of diagnostic tools would be suitable for the task. X-ray diffraction was the first tool of choice as it provides accurate information concerning the lattice constants of the entire volume of a powdered sample. The limitation of our x-ray setup with the sample in the diamond anvil cell (DAC), is that it requires 5-8 minutes of exposure time to develop an interpretable pattern. This limitation led us to search for an accompanying diagnostic tool that could provide a real-time probe of solid-solid structural transitions. SHG was first demonstrated as a probe into reaction kinetics of HMX and TATB phase transitions [1,2]. In order to test the application of SHG to the study of phase transitions we developed a portable optical SHG experiment that could be put into the 10-2 x-ray beamline at the Stanford Synchrotron Linear Accelerator (SSRL). We can now conduct SHG and x-ray diffraction experiments simultaneously on PBX materials contained in DAC's.

MATERIALS ASPECT

HMX was prepared by the method of Siele et. al. [3]. This involved the treatment of octahydro-1,5-diacetyl-3,7-dinitro-1,3,5,7-tetrazocine (DADN) with 100% HNO_3 and P_2O_5 at 50°C for 50 minutes. followed by quenching in ice water. Slow recrystallization from acetone yielded HMX as colorless microcrystals. The grain size distribution is trimodal as shown in Fig.1. TATB was prepared by aqueous amination of trichlorotrinitrobenzene (TCTNB) in a water/nitrobenzene medium. The grain size distribution has a quasi-Gaussian profile centered at approximately 75 microns. Both TATB and HMX powders were introduced into a 500 microns diameter metal gasket, 100 microns thick that laterally confines samples within the DAC.

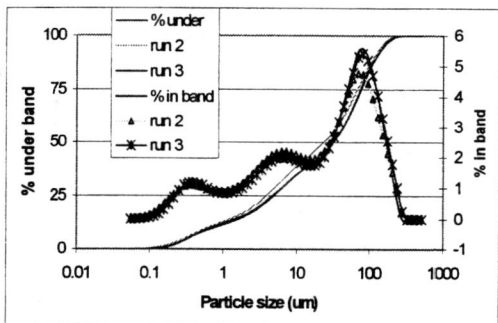

Figure 1: Grain size distribution of HMX lot # B-725.

EXPERIMENTAL

The experimental setup consists of building two separate techniques, XRD and SHG, on an optical breadboard. Schematically, the setup is shown in Fig. 2. Both systems are aligned with their beams (500 µm diameter) collinear and incident onto the sample located in the DAC. The laser operates at 1064 nm wavelength, 2-4 µJ and 20Hz PRF. The frequency doubled light from DAC is collected using a Be mirror. The experiments reported in this paper were conducted at 1 bar constant pressure in a dual heated hydrothermal DAC. Outputs from the photo-multiplier tube and photodiode were collected and recorded using a Tektronix TDS684C oscilloscope. The x-ray patterns are captured using image plates (IPs). Both experiments were performed simultaneously and were interrupted only to replace IPs after completion of 8-minute exposures. The experiment on TATB involved incrementally increasing the temperature over time. The experiment on HMX was run at a constant temperature.

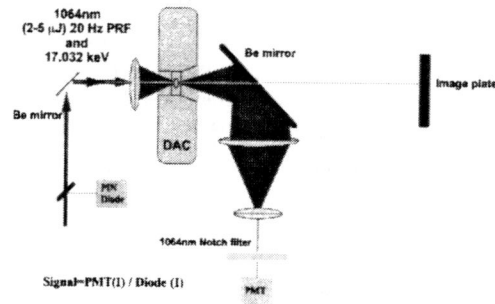

Figure 2: Schematic of the SHG/XRD experiment. Pressure measurement components are not shown as they were not used in this work.

RESULTS AND DISCUSSION

The x-ray energy is set at 17 KeV. The spectrum is then obtained by collapsing the two dimensional image. Figure 3 shows x-ray spectra for TATB with increasing temperatures at 20°C increments. Listed in the plot is the JCPDS listing for TATB (43-1708).

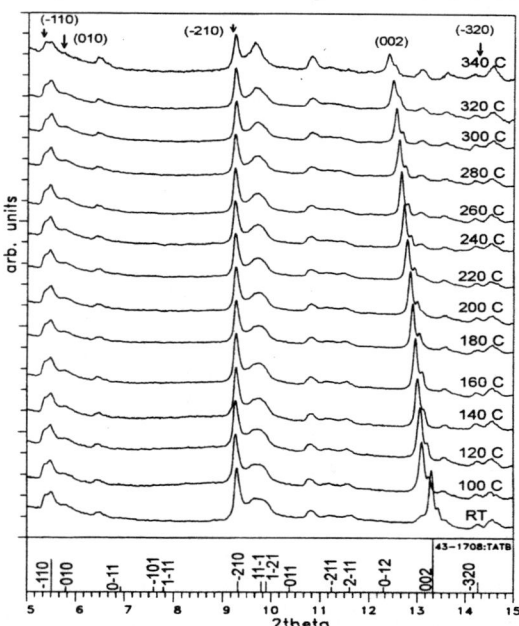

Figure 3: Truncated XRD patterns of TATB from 24° to 340°C. The JCPDS listed intensities are multiplied by ten.

TATB is triclinic with a= 9.01 b=9.028 and c=6.812 (Å), α=108.59, β=91.82 and γ=119.97. A few less intense unidentified lines can be observed, are perhaps from the background. The data do not substantiate the existence of a solid-solid phase transformation at these temperatures. However, a shift of the (002) peak position to higher d-spacings is observed with increasing temperature as shown in Fig. 4. This result suggests the opening of the intermolecular distances with temperature. The linear coefficient of thermal expansion α_c, calculated from a linear fit of the data in fig. 4 is found to be 225 x $10^{-6}/°C$. The lack of significant change in (hk0) peaks, which are related to intra-molecular arrangement, indicates no major changes in molecular structure even up to 340 °C. Hence, most of the volume expansion in TATB occurs along the c-axis due to weak inter-planar Van der Waals interactions. The (002) line shift and/or the increase in (004) peak intensity, as observed in the literature [2] cannot be interpreted as a phase transformation but merely a reorganization of the triclinic phase.

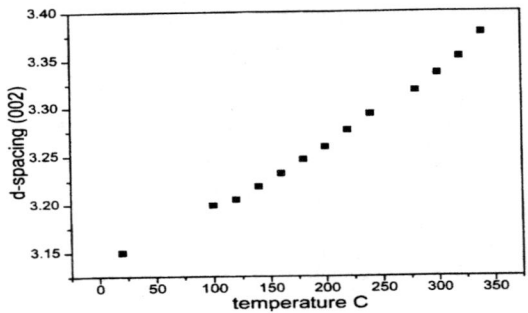

Figure 4: Changes in d-spacing (Å) for the (002) reflection of TATB.

Figure 5 shows the diffraction patterns for HMX held at 165°C as a function of time as indicated on the right side of the plot. These patterns are compared to the calculated powder pattern from single crystal results [4,5] for both β and δ phases using LAZY-PULVERIC programs. Clearly, at the start of the experiment, all the lines can be accounted for by β HMX as indicated and the δ phase emerges at ~7000 seconds into the experiment, which ran for 4.5 hours.

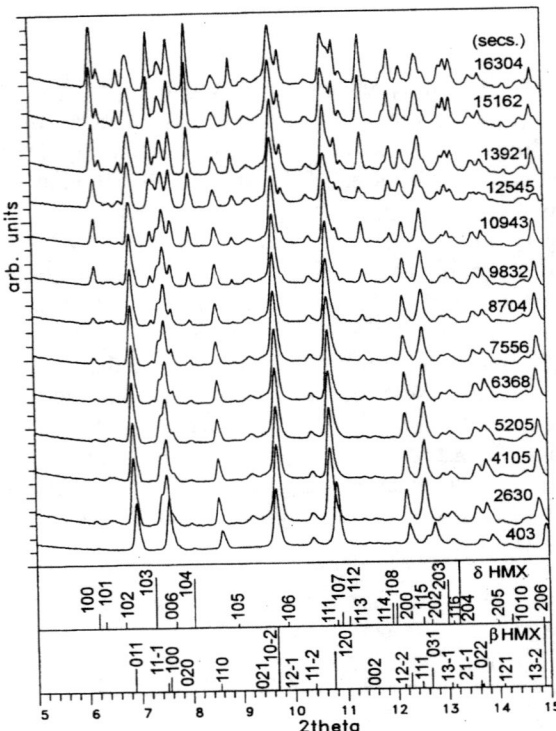

Figure 5: Truncated XRD patterns for HMX at 165°C versus time with the peak listing for the calculated powder patterns from published single crystal data.

Figure 6 shows the normalized SHG signal and four HMX diffraction peak intensities. Normalization was carried out by simply dividing the data set by the highest amplitude point. The structure in the SHG growth curve is real and is most likely related to surface energy and grain scale effects. The SHG data suggests that 80% of the HMX has converted to the delta phase after ~8000 seconds while the XRD data suggests that only ~15% of the sample volume has converted. Also note that the SHG intensity increases at the onset of the experiment even though there is no evidence of δ phase.

We have conducted *in situ* high-temperature AFM experiments on HMX single crystals [6]. The data show that the total surface area of δ-HMX is on the order of 10^3 to 10^5 times higher than the starting β-HMX material. Grain size effects and the corresponding increase in surface area can explain the incongruent contrast between our SHG and XRD data on HMX. We have conducted numerous survey XRD and SHG tests that confirm that larger grain HMX single crystals (50-200 micron length) phase convert at near instantaneous times at a given temperature whereas smaller crystals (0.1-20 microns) can take hours to convert at the same

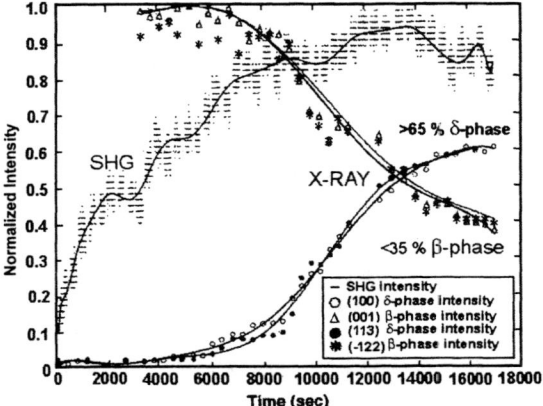

Figure 6: Normalized SHG intensity and XRD peak intensity ratios versus time at 165°C.

temperature. It is the dramatic increase in surface area that accounts for the primary contribution in SHG intensity observed. For example SHG intensity was enhanced by a factor of $\sim 10^4$ on a roughened silver surface [7]. Indeed the SHG surface effect has been used for over ten years to track protein conformational changes by generating SHG light at surface of the cell membranes [8].

CONCLUSIONS

High temperature XRD and SHG experiments have been performed. For TATB, our results do not indicate any solid-solid phase transformation occurrence, even up to 340°C, which directly conflicts with the results of Son et al. [2]. Changes in XRD peak intensity are merely due to molecular re-arrangement and annealing effects. The molecular stacking distance relating to the c- lattice parameter increases with increasing temperature. No major peak changes in the (hk0) reflections are observed suggesting that there is no change in molecular conformation. Our simultaneous SHG and XRD experiments on HMX show that SHG can give misleading results, which bring into question the rates previously derived for HMX and TATB using only the SHG diagnostic. Studying uniform grain sized HE materials (sample lot # issues should be studied too) may give SHG a foothold concerning the rigorous determination of kinetic rate determinations for polymer blended explosive materials.

ACKNOWLEDGEMENTS

The authors would like to thank D. M. Hoffman for the grain size distribution measurements and C.O. Boro, and D.G. Ruddle for assisting with the experimental setup at SSRL We thank P. Pagoria for HMX samples and F. Foltz for TATB samples.

*This work performed under the auspices of the U.S. Department of Energy by the Lawrence Livermore National Laboratory under contract number W-7405-Eng-48.

REFERENCES

1. Henson, B.F., Asay, B.W., Sander, R.K., Son, S.F., Robinson, J.M. and Dickson, P.M., Phys. Rev. Lett. **82**, 1213-1216 (1999).

2. Son, S.F., Asay, B.W., Henson, B.F., Sander, R.K., Ali, A.N., Zielinski, P.M., Philips, D.S., Schwarz, R.B. and Skidmore, C.B., J. Phys. Chem. B, **103**, 5434-5440 (1999).

3. Siele, V.I., Warman, M, Leccacorvi, J, Hutchinson, R.W., Motto, R, Gilbert, E.E, Benzinger, T.M., Corburn, M.D., Rohwer, R.K., Davey, R.K., Propell. and Explosiv., **6**, 67-73 (1981).

4. Cobbledick, R.J. and Small, R.W.H., Acta. Cryst. **B30**, 1918-1922 (1982).

5. Choi, C.S. and Boutin, H., Acta Cryst. **B26**, 1235-1240 (1970).

6. Paper submitted to Ultramicroscopy (August 2001).

7. C. K. Chen, A. R. B. de Castro, and Y. R. Shen, Phys. Rev. Lett. **46**, 145-148 (1981).

8. Campagnola, P.J., Wei, M., Lewis, A. and Loew, L.M., J. Biophys. **77**, 3341-3349 (1999).

USE OF HIGH-SPEED PHOTOGRAPHY TO AUGMENT SPLIT HOPKINSON PRESSURE BAR MEASUREMENTS OF ENERGETIC MATERIALS

Richard J. Lee and Vasant S. Joshi

*Research and Technology Department, NAVSEA Indian Head Division,
101 Strauss Ave, Indian Head MD 20640-5035*

The split Hopkinson pressure bar technique has been used successfully to characterize high strain-rate behavior (10^3 to 10^4 s^{-1}) of metals and polymeric materials in the past. Similar studies on composite energetic materials are desired to assist in developing a model that can predict the mechanical response of energetic materials. The strength of cast-cured and melt-cast explosives are far lower than metallic materials, requiring a combination of techniques to accurately resolve their deformation characteristics. High-speed-imaging techniques have been explored here and compared to conventional strain gauge data for a soft, compliant material as well as a hard, brittle one, which are distinctly different in mechanical behavior. The photographic data coupled with the strain gauge data have proven useful in determining the nature of damage that occurs during the loading cycle.

INTRODUCTION

Characterizing solid energetic materials response to mechanical stimuli is important to develop a fundamental understanding of sensitivity and ignition. Cast-cured compositions typically involve solid energetic crystals, other solids that act as fuel or oxidizers, and a plastic binder. The properties of the individual constituents and how they bond to the binder contribute to the overall mechanical-response, complicating the problem. Understanding the dominant features associated with this response provides the basis for improved models to determine the vulnerability of munitions.

The split-Hopkinson pressure bar (illustrated in Figure 1) is one of a number of tools typically used to establish constitutive relations for materials. Stress-strain data are obtained for various strain rates on the order of 10^3 s^{-1}. Cylindrical test samples are longitudinally strained in compression between two cylindrical bars by a stress wave of finite length.

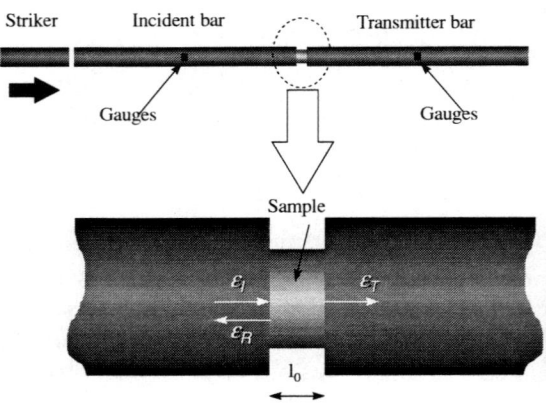

Figure 1. Hopkinson Bar Apparatus

This stress wave is induced by launching a short bar (the striker) into one of the longer bars. Stress, strain, and strain rate in the sample are determined by measuring the resulting stress waves in the two longer bars (incident and transmitted) bounding the sample. The stress waves are measured with strain gauges placed on the bars. The distance between gauges and sample is selected to keep the incident wave (established by the striker) and the reflected wave (returning from the sample) from interfering with one another at the gauge position.

Unfortunately, these types of test have proven difficult for energetic materials. Cast-cured explosives compositions have relatively low mechanical impedance (defined by low sound-speed and density) in comparison to the bars, assuring a very small transmitted wave. Signal intensities can be improved by using lower-modulus bars, e.g., aluminum, thereby reducing the disparity in mechanical impedance.[1] Additionally, typical data reduction schemes assume a linear relationship between radial expansion and axial strain of the sample, which may not always be true for the materials of interest.

Reported here is the use of high-speed photography as an augmentation to conventional strain gauge data. Two distinctly different materials were studied; (1) a cast-cured explosive based on HMX and a polymeric binder, PBXW-128, and (2) a brittle, melt-cast explosive, TNT. The photographic technique coupled with comparisons of the typical strain-gauge data (one and three-wave analysis)[1] have proven useful in validating data-integrity for PBXW-128 as well as providing insight into material failure for TNT samples.

EXPERIMENTAL

Each test sample was fabricated into right-circular cylinders with nominal diameters of 9.5-mm. Sample lengths were 4.75 and 8.55 mm for PBXW-128 and TNT, respectively.

The bars were made from 15.8-mm diameter, hardened 7075 aluminum rods. The incident and transmitted bars were 1.2 m long. Striker lengths of 508 and 308 mm provided nominal loading cycles of 200 and 120 μs for the PBXW-128 and TNT samples, respectively.

Strain gauges were placed at the mid point of the incident and transmitted bars. At each measuring location, two 350 Ω gauges were placed 180^0 from one another on the bar to compensate for any bending. Gauge-pairs were configured in a two-arm bridge with a supporting Ectron 2 MHz bandwidth amplifier. Gains of 100 and 200 (incident and transmitted bar, respectively) were used to improve signal amplitude at the oscilloscope.

A Nicolet Integra 40 digital oscilloscope recorded the data with 12 bit resolution. The gauges were 1.57 mm (0.062"), which provide a time resolution of 3.3 μs (300 kHz response).[2] Data was collected at 50 ns/point using a 300 kHz bandwidth filter to reduce signal noise.

Photographed sample deformation was taken with an Imacon 200 digital camera, which uses a series of CCD cameras to record 16 images. Individual pixels are 6.7 x 6.7 microns in a pattern of 1280 horizontal x 1024 vertical. Systematic measuring-errors combine to ±0.10 mm when using a field of view a little larger than the bar diameter. Pictures were taken every 12 μs for PBXW-128 samples and 8 μs for TNT to span the respective loading cycles. Faster framing-rates are eventually possible for probing specific periods of interest, e.g., the initial stages of compression or around an expected failure time.

Data were examined for stress equalization (a validation requirement) by comparing stress-strain curves from one and three-wave analyses.[1] (The three-wave analysis uses all three signals to determine stress and strain in the sample: Stress equalization allows reduction of the three-wave data to the one-wave, where stress is given by the transmitted signal and strain the reflected signal.) The one-wave data were also evaluated in light of the photographic data.

RESULTS

Figure 2 shows a comparison of true-stress vs. true-strain from one and three-wave analyses at a nominal strain rate of 2300 s^{-1} for the PBXW-128 sample. The strain rate actually increases over the loading cycle from 1900 to 2700 s^{-1}. It is desirable to achieve a constant true strain-rate, however, gradually increasing rates are expected for high strains in the sample. The difference between the one and three-wave plots shown here would normally indicate a loss of stress equalization in the sample resulting from some internal damage, e.g., de-wetting of solids from the binder. However the photographic data suggest otherwise. As shown in Figure 3, the sample diameter and radius obtained from the photographs as well as that calculated from the true-strain vs. time data (assuming constant volume) are comparable. (Figure 4 shows three of the sixteen photographs taken.) Moreover, volume vs. time, calculated from the photographic data, indicate that the sample volume remains constant over the loading cycle. These data argue that stress equalization was maintained during the loading cycle.

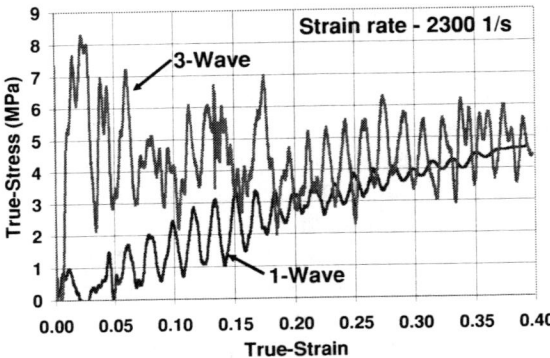

Figure 2. One and three-wave data for PBXW-128.

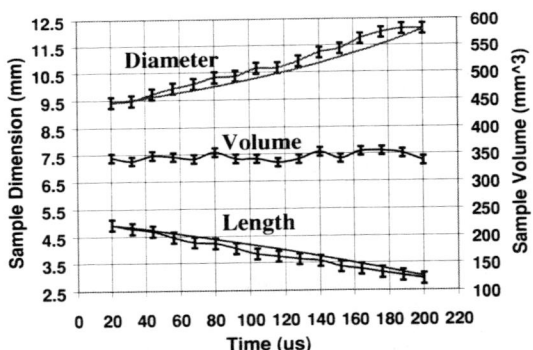

Figure 3. Photographic vs. strain gauge data for PBXW-128. Plots with error bars are measured values from photographs. Lines without error bars, are corresponding dimensions calculated from true-strain vs. time data.

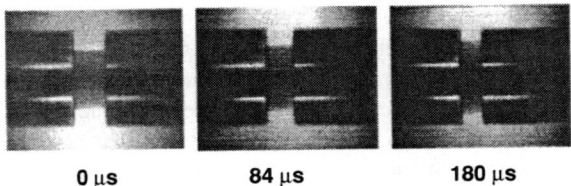

Figure 4. Selected Photographs for PBXW-128 showing smooth lateral expansion during the loading cycle.

Figure 5 shows stress versus strain at a nominal strain rate of 750 s^{-1} for a TNT sample. Strain rate varied from 650 to 870 s^{-1}. Here, the one and three-wave data are fairly comparable. These data might have been mistaken for reasonable results if the photographs had not indicated fracturing along the sample's outer surfaces as shown in lower portion of Figure 5. This fracturing begins early in the loading cycle and corresponds with the slope change from positive to negative in the stress-strain plot.

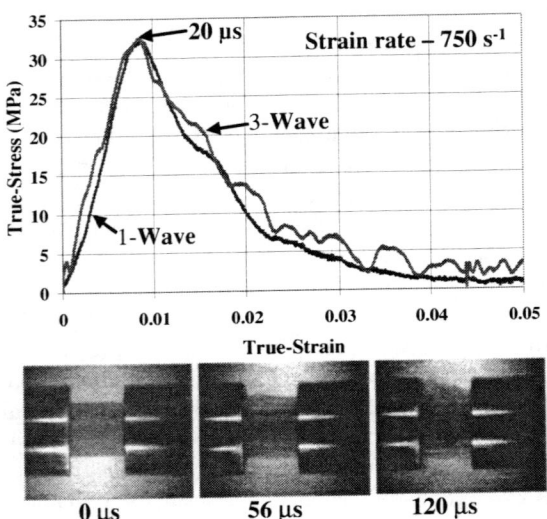

Figure 5. One and three-wave data for TNT, and selected photographs.

DISCUSSION

Photographic data indicate that PBXW-128 deforms at constant volume under compression at strain rates comparable to 2300 s^{-1}. This indicates that stress equalization occurs, on average, through the loading cycle despite observed differences between one and three-wave data reduction schemes. It must be noted that the framing rate for the photographs were relatively course with respect to oscillations observed on the stress-strain plot. A close inspection of the true-stress vs time (as indicated by the transmitted wave) suggests that there is a cyclic increase and decline of stress during the first portion of the loading pulse. Chung has observed this phenomenon in copper and attributed it to radial relaxation following a gradient buildup from the central axis to the periphery.[3]

The initial disagreement of one and three-wave data may be attributed to this phenomenon. However it is equally possible that the difference may be attributed to an inability to accurately compare the difference of incident and reflected signals to the transmitted signal, in the way that is typically done with harder material, e.g., metals. This difficulty arises from the disparity between the reflected and transmitted signal magnitudes observed in softer materials. Despite efforts to improve diagnostic precision, errors are introduced from an inability to accurately correct for dispersion introduced as the stress waves propagate down the bars. This was brought to light during recent efforts to confirm our dispersion correction routines. They are adequate for correcting data for hard materials but fall short when the material is soft.

These problems are greatly reduced for TNT because of its firmness. The one and three-wave data for TNT are similar, suggesting that stress equalization occurs in the sample despite its fracturing. The supposition here is that fracturing occurs only along the outer periphery of the sample, leaving an inner core intact. Stress equalization is maintained in this inner core allowing an opportunity to glean stress-strain data from these types of experiments. The negative slope may be attributed to a decreasing ample diameter, which may eventually be monitored and compensated for by back lit photography (not employed in this study).

CONCLUSIONS

Two very different materials have been observed in Hopkinson bar experiments using high-speed photography. Studies on PBXW-128, a very soft material, demonstrated constant volume deformation. Also, it may not be possible to use comparisons of one and three-wave data to validate for stress equalization due to the disparity in reflected and transmitted signal magnitudes typical to testing soft materials. TNT proved to be a brittle material easily fractured early in the loading cycle. However, data may be obtained if the diameter of un-fractured inner-core can be monitored.

ACKNOWLEDGEMENTS

The sponsors for this effort were the Office of Naval Research (ONR) under the Air and Surface program and the Navy Explosives Ordnance Disposal (EOD) Technology Center. We wish to also thank Chak-Pan Wong and Dennis Budd for their assistance in performing these experiments.

REFERENCES

1. Gray, G.T., and Blumenthal, W.R., *Metals Handbook., Vol 8, 10th Ed,* ASM Publications, 2000, pp 462-476 and pp 488-496.
2. Ueda, K. and Umeda, A., *Experimental Mechanics*, **38**(2), 93-98 (1998).
3. Chung, D.T., *Shock Compression in Condensed Matter - 1995*, edited by Schmidt et. al., AIP Conference Proceedings, New York 1996, pp 483-486.

MECHANICAL BEHAVIOR OF EXPLOSIVES AT HIGH PRESSURES

J. M. Kelley, V. S. Joshi, and R. H. Guirguis

Research and Technology Department
Naval Surface Warfare Center
Indian Head, MD 20640

Abstract. The mechanical and ignition behaviors of heterogeneous explosives are highly dependent on the pressure. Experimental data of the compaction of composite plastic-bonded inert samples under hydrostatic loading conditions are presented. The data reduction techniques and the procedures used to correct errors introduced by the compliance of the apparatus are addressed.

INTRODUCTION

The quasi-static compression behavior of ½" composite plastic-bonded inert samples was measured under hydrostatic loading conditions at pressures up to 8 kbar. A new apparatus was specifically built for this purpose. Unlike previous designs using a cylinder-piston arrangement, the current approach subjects the samples to quasi-static compression in a working fluid, such as to ensure hydrostatic loading conditions. The cavity pressure is recorded using a piezoelectric transducer, while its volume is determined by measuring the displacement of the piston.

APPARATUS

Various apparatus and seals to apply and maintain quasi-static high-pressure loads over an extended duration were proposed by Bridgman [1]. In these, hydrostatic loading conditions are achieved by using a working fluid as an interface between the walls of the device applying the stress and the sample. Because fluids mostly resist compression but little shear, if the vessel wall is not impulsively started, then hydrostatic loading conditions can be maintained (even under dynamic loading conditions) by selecting a working fluid with a high sound speed. In the experiments described below, glycerin was selected as the working fluid for its moderate density (1.25 g/cm^3) and high sound speed (1.9 mm/μs). By selecting the working fluid lighter than the test samples, the latter can be easily submerged, thus aiding the elimination of trapped air. Glycerin was also selected for its low toxicity (a pharmaceutical and food ingredient), minimal chemical interaction with most test samples, and because of its high freezing pressure at room temperature.

Figure 1 is a schematic of the apparatus. It was adapted from a high-pressure intensifier in which two pistons of different cross-sectional areas are rigidly mounted in line. Fluid is supplied at moderate pressure to the larger piston. In principle, this pressure is amplified by the ratio of the two areas. However, in practice, the friction introduced by the seals reduces the pressure gain below this theoretical value.

In quasi-static loading, because the piston moves slowly, a dial-indicator is adequate for recording the piston displacement in real time, from which the volume of the high-pressure cavity can be derived. Accurately measuring the initial volume of the cavity was bypassed by conducting comparative experiments and applying subtraction techniques during data reduction as described below. Three sets of experiments were performed, starting at a different piston location in each set (i. e., different initial cavity volumes). Two tests were performed in each set, one with and the other without the sample, starting each test at the same piston location.

To test explosive samples up to ½", an intensifier with a bore measuring ⅝" in diameter and 8" in length, capable of containing 200 kpsi was selected. To the original intensifier, an off-the-shelf product of Harwood Engineering Company, a small-diameter extension rod was attached to the large piston in order to measure the piston displacement. The diameter of the low-pressure piston is 2½". With the ½" extension rod attached, the area ratio was reduced from 15.5:1 to 14.88:1. The high-pressure closure plug was fitted with a T-section to accommodate: (1) a piezoelectric pressure transducer; (2) a manually-operated needle-valve to relieve any residual pressure in the high pressure section (cavity and conduits); and (3) a burst-tube designed to fail at 220 kpsi and release the pressure.

An Enerpac® 10 kpsi hydraulic hand-pump equipped with a pressure regulator and Bourdon-tube pressure gauge was selected to remotely power the low-pressure hydraulic cylinder. Test-section pressure was acquired via a Kistler Model 6213B quartz piezoelectric transducer nominally rated to 10 kbar. A Kistler Type 5010B dual-mode amplifier, recommended for measuring quasi-static pressures because of its "long time constant" charge mode, and a Nicolet Integra Model 40 12-bit digital oscilloscope were used to acquire the pressure data. Piston displacement was measured at the extension rod using a Fowler-Sylvac electronic dial indicator capable of 0.00005" (1.3 µm) resolution and a total travel of 1" and the data recorded every 1s using a PC. For safety purposes, the tests were remotely conducted and monitored via a closed-circuit television camera.

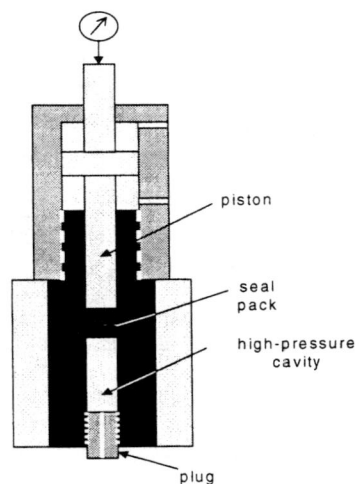

FIGURE 1. Schematic of high-pressure apparatus.

TEST SAMPLES

All the data presented in this paper are for inert samples, similar to plastic-bonded explosives, but in which the energetic crystals were replaced by a simulant. Sucrose was selected from several possible candidate simulants because of its availability in several granulations, its low toxicity, and relatively low cost. The particle-diameter specifications for several grades were provided by Tate-Lyle Company, the manufacturer of Domino® sugars.

A solid filler with a bimodal distribution was used such that the coarse-to-fine ratio was 3:1 and the ratio of the mean diameters was 10:1. Domino® "Industrial 10X Confectioner's" grade sucrose was used as the fine fraction, and a blend of larger particles sieved from various grades was used for the coarse component. Procedures representative of standard processing techniques were used to prepare a small batch of the mock-composition. The cured material was machined into ½" x ½" right circular cylinders. The volume and density of each sample were measured with a helium pycnometer.

EXPERIMENTAL PROCEDURE

Because glycerin was used as a transfer medium, special precautions were taken during the tests in order to ensure the accuracy of the measured mechanical properties:

1. Pressure-loading of the samples was done in a step-wise fashion, with sufficient dwell-time at each step to allow the materials in the cavity to equilibrate before the pressure was increased in the next step.
2. To increase the signature of the test samples versus that of the glycerin used as a working fluid, the number of samples used in each test was selected such that their total volume occupied most of the intensifier cavity.
3. Prior to placement in the apparatus, each sample was lightly coated with Fluorolube® GR-290 (Hooker Chemical Co.) to minimize intrusion of the glycerin.
4. To eliminate entrapped air, each leg of the apparatus was manually filled with glycerin.

5. The glycerin in the intensifier bore, which was replaced for each test, was evacuated in situ until no bubbles appeared.
6. After compaction, each sample was blotted free of glycerin and weighed. Weight changes were negligible in all but one instance, when a sample apparently lost weight. Removal of some of the Fluorolube® coating was determined to be the cause.

DATA REDUCTION

As explained above, in order to eliminate the errors arising from flexing of the apparatus at high pressure, three sets of experiments were performed, starting each set at a different piston location. Two tests were performed in each set, one with and the other without the sample, starting each test at the same piston location.

First, the measured pressure vs. piston displacement data describing the behavior of the neat glycerin and illustrated in Fig. 2 were reduced using

$$\left(\frac{V}{V_o}\right)_{glycerin} = \frac{x_2(p) - x_1(p)}{X_2 - X_1} \quad (1)$$

where V denotes the specific volume (per unit mass) at pressure p and V_o is the corresponding volume at atmospheric pressure. In Eq. 1, $x_1(p) - x_2(p)$ is the difference in piston locations between two glycerin tests begun at X_1 and X_2, respectively, both resulting in the same pressure p.

The calculated values of V/V_o for glycerin are illustrated in Fig. 3. The individual data sets are represented by different symbols. The solid curve is a fit of the points resulting from averaging all three sets.

In theory, V/V_o is an intrinsic material property and should not depend on the piston starting location. However, in practice, because glycerin is hygroscopic, moisture sorption varies with duration of exposure to atmospheric air, resulting in a slightly different material for each individual test, with different mechanical properties. In addition to the other possible errors committed during data acquisition and reduction, the scatter in the data illustrated in Fig. 3 is attributed to the differences in the properties of glycerin used in each test.

Next, the compaction data of mock samples and glycerin illustrated in Fig. 4 were reduced using

$$\left(\frac{V}{V_o}\right)_{sample} = \left(\frac{V}{V_o}\right)_{glycerin} + A \frac{x_s(p) - x_g(p)}{V_s} \quad (2)$$

where A is the area of the piston, and V_s is the volume of the sample(s) at atmospheric pressure. In Eq. 2, $x_s(p)$ is the piston location at pressure p, whereas $x_g(p)$ is the corresponding piston location in the test conducted in the same series but with glycerin alone.

When calculating V/V_o for the mock samples, the average fit for V/V_o of glycerin described above was used. The results are illustrated in Fig. 5. It is important to notice that the scatter in the values of $(V/V_o)_{sample}$, portraying the variation from one test to another, is smaller than the corresponding scatter in $(V/V_o)_{glycerin}$ observed in Fig. 3. That is because unlike glycerin which is sensitive to its environment, the samples made of sucrose and binder are much more stable.

Figure 6 compares the compaction properties of the test samples to the compression behavior of glycerin. In the lower pressure regimes the test samples exhibit a somewhat higher compressibility than does the glycerin, but as the pressure increases the sample becomes more rigid. This is attributed to lock-up of the filler particles.

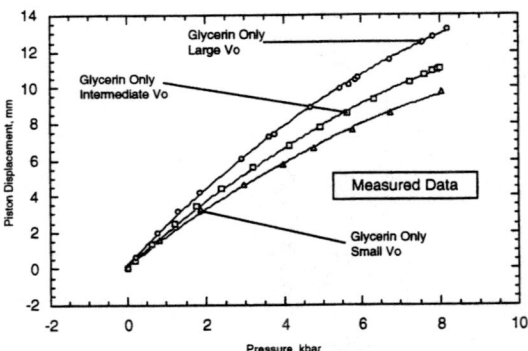

FIGURE 2. Piston displacement vs. pressure in glycerin compression tests.

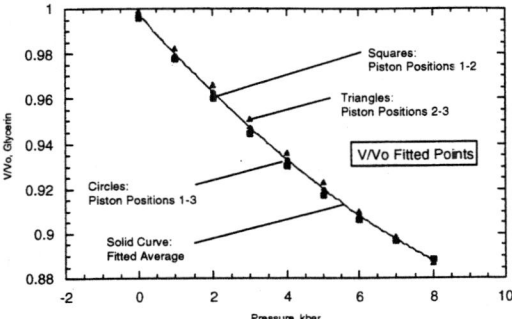

FIGURE 3. V/V_0 for glycerin.

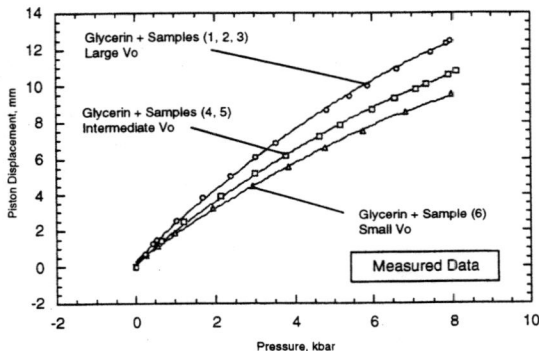

FIGURE 4. Piston displacement vs. pressure for samples compacted in glycerin.

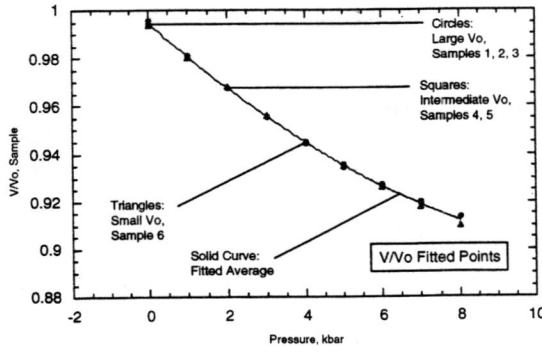

FIGURE 5. V/V_0 for test samples.

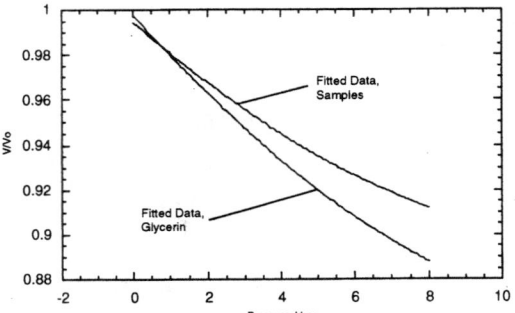

FIGURE 6. Comparison of $(V/V_0)_{sample}$ to $(V/V_0)_{glycerin}$.

CONCLUSIONS

A new apparatus capable of quasi-static compaction of explosives under hydrostatic loading conditions reaching 8 kbar was built. The errors arising from flexing of the apparatus at high pressure were eliminated by conducting comparative experiments and applying subtraction techniques during data reduction. The compaction properties of test samples were expressed in terms of the compression behavior of glycerin, selected as a working fluid for its high sound speed and moderate density. Preliminary results describing the compression of glycerin and the compaction of an inert analog of a typical plastic-bonded explosive composition are presented.

REFERENCES

1. Bridgman, P. W., *The Physics of High Pressure*, Dover Publications, Inc., New York, NY, 1970.

INVESTIGATION OF SHOCK WAVE IMPULSE INFLUENCE ON SOLID PROPELLANT COMBUSTION

Alexander Yu. Dolgoborodov and Vladimir N. Marshakov

N. Semenov Institute of Chemical Physics, Russian Academy of Sciences, 119991, Moscow, Russia

Abstract. The laboratory technique for test of solid propellant combustion under shock wave loading is described. The explosive generator was used for shock waves formation in propellant sample in pressure range 100 - 300 MPa. The experiments were conducted in the combustion chamber at pressure range 1 - 12 MPa. HE charge was initiated in 1-1.5 seconds after ignition of the propellant. For shock pressure in a sample less than 150 MPa, the experimental results have shown that the steady combustion regime is retained. The shock pressure increasing up to 230 MPa results in product pressure rise and consequent combustion chamber breakage. The analysis of possible causes of observed differences in regimes of burning was performed.

INTRODUCTION

In operation on the ground and in flight, a solid-propellant rocket engine may be subjected to shock loading caused by external shock wave, impacts of fragments generated by explosions, and directed high-intensity energy fluxes [1-3]. Duration of these loads varies from 10^{-6} to 10^{-2} s, depending on the source, and the representative pressure pulses range from 0.1 kPa∗s to 5 kPa∗s. Effect of shock wave impulse on the operating rocket engine can result in disastrous consequences. Sometimes fire bench tests are carried out for determination of safe load levels of operating engine [4]. However realisation of such tests requires significant costs. The purpose of our work was to develop a rather cheap laboratory technique, which would allow investigating the stability of propellant burning under shock wave loading in the pressure range 0.1 to 0.3 GPa.

EXPERIMENTAL

For experimental investigation of stability of composite propellant combustion we used the combustion chamber. We investigate the samples of propellant on the base of polybutadiene rubber filled by ammonium perchlorate and HMX with a characteristic size of particles of 100-300 μm and fine aluminum particles (5-10 μm). Earlier the structure of shock wave was investigated for this propellant [5]. For shock wave formation in samples we used the explosive generator similar described in this work. The scheme of combustion chamber with explosive generator (EG) is displayed in Fig.1.

The combustion camera was made by the way of thick-walled vessel with volume about 2 litres. The shock wave impulses with maximum pressure 0.1-0.3 GPa formed by EG. EG consists of a composite charge of RDX (lens by a diameter of 40 mm and weight 15 g and intermediate charge with detonator in weight 5 g) and thin-walled copper barrel with water by a diameter of 85 mm. The HE charge was installed above at centre of a barrel. The barrel was fixed on a cover of the combustion chamber. The parameters of a shock wave were adjusted by variety of height of the barrel with water. The shock wave from water passed in a propellant sample through the organic-plastic insert in a cover of the combustion chamber. The diameter of the organic-plastic insert constituted 90 mm, width of 14 mm. Propellant samples were produced as truncated cones by a thickness of 40 mm and diameters of the basis 50 and 57 mm. Samples are located inside the chamber and nestled on the organic-plastic insert by a conic cartridge clip. Ignition of samples was made from the lower end face by burning products of powder charge of weight 5 g. On a lateral area of the chamber there were landing places under the nozzle block, block of a safety valve and reducing coupling under the inductive gauge DD-10. The products of burning were assigned from the chamber through a nozzle block. The pressure in chamber was registered by gauge DD-10 with the pressure indicator ID-2I on the oscilloscope N-117. The initiation of a HE charge was made in 1-2 second after ignition of a propellant with the help of synchronization scheme.

The experiments on determination of a shock wave structure in propellant samples without burning were previously conducted. For a measurement of shock wave pressure in samples were used piezoelectric film PVDF - gauges made and calibrated in laboratory by Yakushev V.V. (IPCP RAS, Chernogolovka). PVDF gauges were placed in two planes: on the contact boundary of the organic-plastic insert of the chamber and propellant sample and in a sample on depth of 20-22 mm. The scheme of experiments is shown in Fig. 2. In order to prevent of influence of air inclusions on a structure of a wave of an irregularity between the insert and sample were filled in with an epoxy resin.

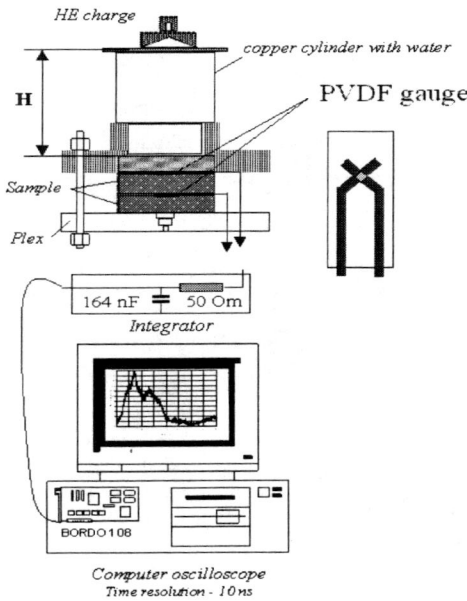

FIGURE 2. Experimental set-up for pressure measurement in propellant samples. PVDF- piezoelectric film gauges made of polarized polyvinylidene-fluoride with sensing area 5.06 mm^2.

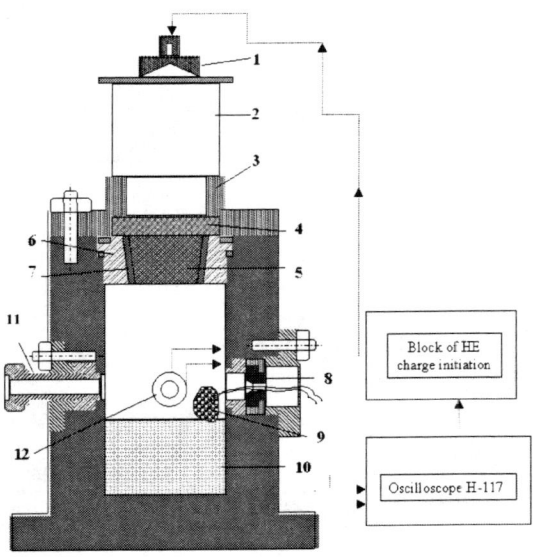

FIGURE 1. Combustion chamber with explosive generator
1 - explosive charge, 2 - copper cylinder with water, 3 - cover of chamber case, 4 - organoplastic insert, 5 - propellant sample, 6 - conical cartridge, 7 - protecting cover, 8 – nozzle, 9 - ignition powder charge, 10 - additional insert of chamber, 11 - block of safety diaphragm, 12 - orifice in the chamber to adapter and pressure gauge DD-10

EXPERIMENTAL RESULTS

The trial tests were made without combustion of samples. Samples subjected to shock compression and were visually studied after a shock loading. Fastening of samples was carried out by two ways. In the first case the sample was retained against organic-plastic insert by the plexiglas screen. The screen had the holes of different diameter. In the second case the sample was consolidated only on a lateral area by a special conical cartridge from duraluminum, thus also was given to a sample the conical form. In the latter case it was possible to achieve absence of separation of a sample from the insert after shock wave loading, and hereinafter this manner of fastening utilised for researches at combustion.

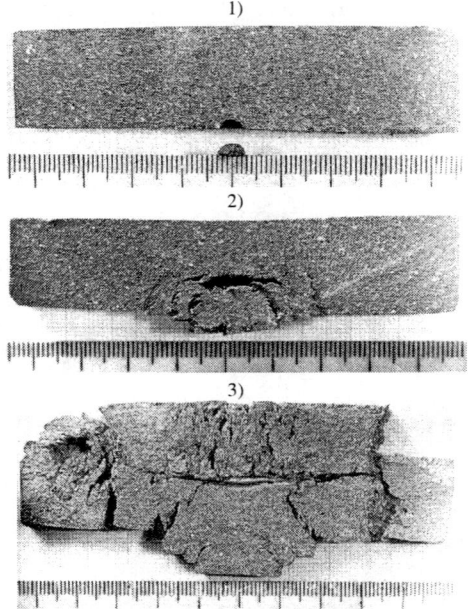

FIGURE 3. Fracture of samples after shock wave loading 1-3 - propellant samples pressed from below by plexiglas plate with a hole. 1- Plexiglas plate with 5 mm hole. $P_{sw} = 0.135$ GPa. 2- Plexiglas plate with graduated hole. $P_{sw} = 0.135$ GPa. 3- Plexiglas plate with graduated hole. $P_{sw} = 0.23$ GPa.

Visual research of samples in all cases has shown, that the large fragments of a filling material (HMX and ammonium perchlorate) from a near-surface layer take off from a sample, and the free surface of a sample gains spongy structure (see Fig. 3). In case of availability of holes in the screen in samples the destructions are observed. Thus the size of destructions decreases with growth of diameter of holes. So at diameter of a hole of 5 mm and shock wave pressure 0.135 GPa there was a full separation of spalling element by depth up to 2.5 mm. At diameter of a graduated hole of 18 mm there was a loosening of a sample in the field of a hole. Inside a propellant the spalling flaws have appeared. In case of a pressure increment till 0.23 GPa propellant lost durability and the corrupting were considerably magnified, though complete spallation does not happen.

The experiments on measurement of the profile of pressure have shown the following. At a stratum of water of height 148 mm the pressure profile on contact boundary with the insert was close to the triangular form. The maximum pressure constituted 0.18 GPa. On depth of 21 mm in a sample the main features of a structure were saved. The maximum pressure has decreased up to 0.135 GPa. At decrease of a stratum of water on 50 mm the maximum pressure on depth of 21 mm inside a sample has increased till 0.23 GPa. The pressure profiles are shown at Fig. 4.

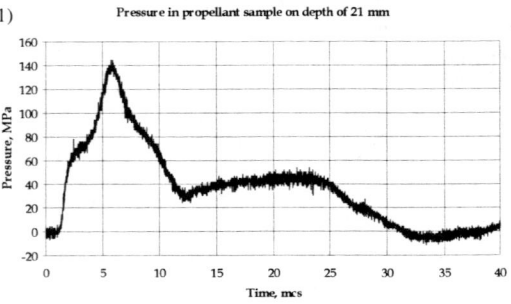

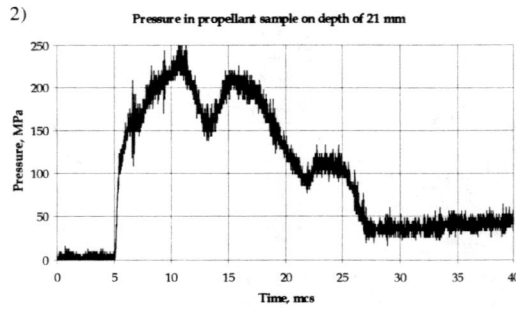

FIGURE 4. Shock pressure records in propellant samples. (H - height of a stratum of water in the explosive SW generator). 1- Pressure on the depth of 21 mm in sample, H=148 mm, $P_{max} = 0.135$ GPa, 2- Pressure on the depth of 21 mm in sample, H=88 mm, $P_{max} = 0.23$ GPa.

The experiences with burning were conducted at pressure inside the combustion chamber of 1 - 12 MPa. The pressure was regulated by selection of a nozzle diameter. In Fig. 5 two records of pressure are indicated at shock wave amplitudes - 0.135 and 0.23 GPa. The conducted experiences at a shock load by amplitude 0.135 GPa as a whole display saving stability of combustion. After shock wave exit on a shining surface happens small (approximately up to 0.2 MPa) pattern null of pressure and through 70 - 100 ms restoring of a former level. A reason of a pattern null of pressure can be break-up of a part of a warmed-over stratum from burning surface after shock wave passing.

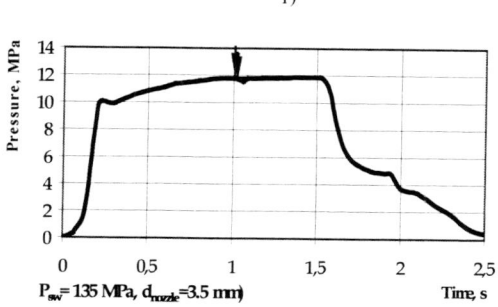

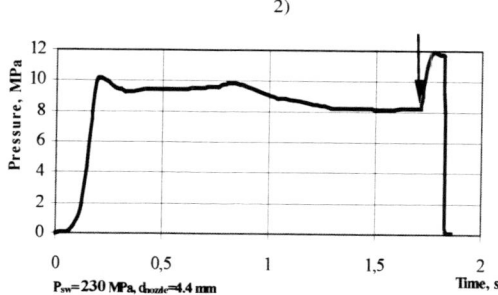

FIGURE 5. Pressure variation in combustion chamber (vertical arrow shows a moment of shock loading) 1 – shock wave pressure = 0.135 GPa, 2 - P_{sw} = 0.23 GPa.

CONCLUSION

The experimental procedure for conducting the laboratory tests on stability of propellant combustion under shock wave loading has been developed. There were manufactured and developed explosive generator, model testing unit, and combustion chamber for studying combustion of propellants under shock wave loading in the pressure range 0.1 – 0.3 GPa.

There were conducting the tests on combustion of the composite propellant in the chamber under shock wave with amplitude from 0.13 to 0.23 GPa. The experiments conducted in the combustion chamber under the pressure ranged from 1 to 12 MPa showed the following:

- The shock wave loading of the propellant sample with the amplitude 0.135 GPa did not cause the combustion failure and resulted only in insignificant (0.15 – 0.20 MPa) short-acting (20 – 70 ms) pressure decrease after which the previous pressure level restored. This means that steady state combustion regime was retained.
- Under the same conditions, shock-wave pulse with pressure amplitude of 230 MPa resulted in abrupt pressure increase and emergency conditions in the chamber.

In the last case, the abrupt pressure rise could be caused by several reasons that may be summarised as following: increasing of total burning area at the coast of surface disintegration and flame breaking into the sample; separation of heated-up layer with it consequent afterburning in the chamber; spalling fracturing of the sample etc.

The analysis of the pressure diagram suggests that, as the result of shock wave loading, separation of burning layer approximately from one third of total burning surface took place. Burning of this layer and ignition of the liberated surface caused the pressure increase in the chamber. More detail analysis of the phenomenon occurring under shock wave loading calls for further investigation.

REFERENCES

1. Sinyukov A.M., Volkov L.I.,. L'vov A.I, and Shoshkevich A.M., *Ballistic Solid-Propellant Rockets*. Voenizdat MD USSR, Moscow, 1972.
2. *Space Weapons: Dilemma of Security*. Velikhov E.P., Sagdeev R.E., and Kokoshin A.A., Eds.. Mir, Moscow, 1986.
3. Anisimov S.P., Prokhorov A.M., and Fortov V.E., *Usp. Phys. Nauk.* **142**, (3), p.395 (1984).
4. Ostrik A.V., and Petrovskii V.P., *Chemical Physics Reports*. **14** (1-3), 10-16 (1995).
5. Gafarov B.R., Utkin A.V., Razorenov C.V., Bogach A.A., and Yushkov E.S., *Applied Mechanics and Technical Physics*. **40** (3), 161-167 (1999).

CHAPTER XIV
DETONATION PHENOMENA

INVESTIGATION OF ISENTROPE FOR DETONATION PRODUCTS OF TATB-BASED COMPOSITION

Yu.A. Aminov, M.M. Gorshkov, V.T. Zaikin, G.V. Kovalenko, Yu.R. Nikitenko, G.N. Rykovanov

Russian Federal Nuclear Center – VNIITF, Snezhinsk 456770

Abstract. The modified impedance matching method was used for investigation of isentropic expansion of the detonation products of plasticized TATB-based composition. Experimental installation consists of flat aluminum flyer plate, aluminum "shield", the sample of explosive and inert barrier. The thickness of explosive sample is 15 mm. It was used 14 substances with various densities. Measured values of the shock wave velocity in the barrier were used for determination of particle velocity and pressure. Obtained results are in agreement with simulation and with experimental data for similar composition T_2.

INTRODUCTION

For build-up of empirical equations of state (EOS) for detonation products (DP) the information on DP shock compression and isentropic expansion is necessary in addition to usually determined explosive parameters (initial density, stationary detonation velocity, DP parameters in Chapman-Jouguet point, etc.). The impedance matching method is usually used for such experiments. In this method a shock wave velocity D is measured in inert barrier bordering to explosive charge. In this case, pressure P and particle velocity U in contacting substances are equal and can be found if the equation of state for barrier substance is known. Selecting different materials for barrier, it is possible to receive experimental data in, for example, P(U) form for investigated explosive. In our experiments, the modified impedance matching method was used for investigation of the plasticized TATB-based composition (PCT) with initial density ρ_0=1.91 g/cc.

EXPERIMENTAL INSTALLATION

Figure 1 shows the scheme of experimental installation used in our experiments.

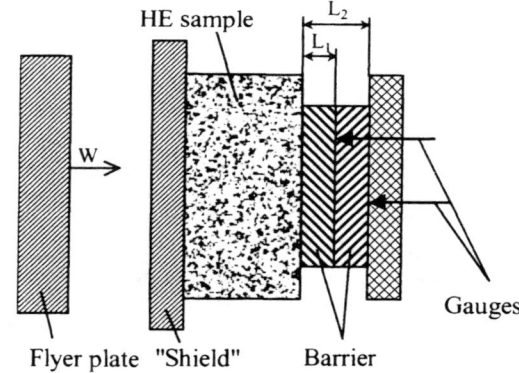

FIGURE 1. The installation for impedance matching method experiments.

A 15 mm thick explosive sample was shocked by an 8 mm aluminum flyer plate through a 4 mm aluminum "shield". The flyer velocity was W=3.6 km/s, that corresponds to the values of pressure in PCT at the leading shock front ≈28 GPa according (1). The time to detonation for similar explosive PBX-9502 at such pressure does not exceed 0.1 μs (2).

The choice of barrier materials (Table 1) was made to investigate DP in pressure range P=0.1-50 GPa. The mean shock velocity D in a barrier was measured apart from L_1=5 mm up to L_2=10 mm by electrocontact gauges to exclude the influence of chemical reaction zone.

In each measuring planes 8 gauges apart from 7 mm up to 21 mm from axis of symmetry placed, the obtained data averaged. In experiments with non-metallic substances, the gauges were covered with an aluminum foil of 10 μm width. In experiences with liquids, the gauges were positioned in handsets with brass bottom of 50 μm width.

For preliminary calculations of experimental system we used one-dimensional hydrodynamic code VOLNA (3), permitting calculations with precise fronts of shock and detonation waves. The equations of state from (4) were used for unreacted explosive and its detonation products. The reaction rate parameters were selected from the plane-wave experiments results (4). At these parameters run to detonation is near 7 mm that is in agreement with data (2) for PBX-9502 and allows us to use in experiment a 15 mm explosive charge. The calculated reaction zone length is near 1 mm. Its influence has an effect in a barrier apart less than 6 mm. According to calculation, the rarefaction wave does not perturb a constant pressure profile during measuring.

DISCUSSION OF RESULTS

Table 1 shows the measured in a barrier shock wave velocities and substances used for the barrier. For determination of the particle velocity U and the pressures P were used the law of momentum conservation on shock front and linear D(U) dependence for barrier:

$$P = \rho_0 D U, \quad D = a + bU.$$

The experimental points in P(U) form are shown in Fig. 2, where they are compared to the experimental data for similar composition T_2 (ρ_0=1.855 g/cc). Impedance matching method (5) and laser interferometry (6) were used to receive the data for T_2.

The good agreement with the laser interferometry data (6) is observed, while the lower pressures are measured in experiments (5).

In our experiments the constant pressure profile in DP was created by a thick flyer plate. If this pressure is more than Chapman-Jouguet pressure (P_{CJ}),

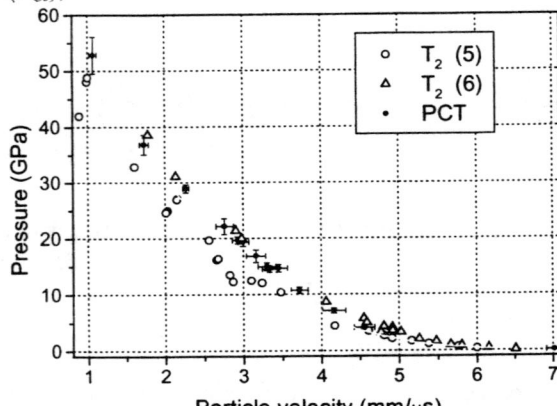

FIGURE 2. Experimental data for PCT and T_2 explosives

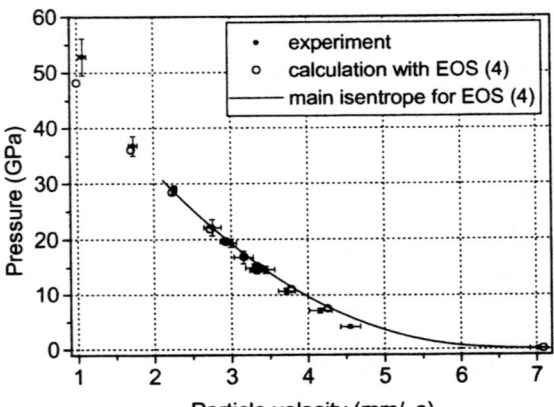

FIGURE 3. Comparison of experimental and calculated with EOS (4) curves.

we obtain an overdriven detonation. For similar compositions (with TATB/inert ratio approximately 90/10) the experimental and theoretical Chapman-Jouguet pressure spread is 26-31 GPa. To investigate influence of C-J pressure value on experimental results the VOLNA calculations were carried out with DP equations of state from (4) (P_{CJ}=30.6 GPa) and (7) (P_{CJ}=26 GPa). As it is seen from figures 3 and 4 the calculations is in the good agreement with experiments for both EOS, therefore at the used scheme of initiation it is possible the DP supracompression. Thus, the obtained experimental data can be used for EOS verification only if in the calculations the experimental set-up is accurately reproduced.

TABLE 1. Experimental Data Obtained in Impedance Matching Method Experiments

Substance	ρ_0 g/cc	D mm/µs	U mm/µs	P GPa
copper	8.92	5.528±0.068	1.07±0.05	52.8±3.3
aluminium	2.73	7.772±0.079	1.73±0.06	36.8±1.7
magnesium	1.74	7.345±0.034	2.27±0.04	29.0±0.7
PMMA	1.18	6.785±0.135	2.76±0.11	22.1±1.4
water	1.00	6.460±0.067	3.00±0.09	19.3±0.7
polyethylene	0.92	7.296±0.028	2.92±0.07	19.6±0.6
ethyl alcohol	0.80	6.634±0.155	3.16±0.12	16.8±1.1
n-hexane	0.65	6.688±0.063	3.34±0.10	14.5±0.6
polystyrene	0.77	5.881±0.031	3.30±0.12	15.0±0.7
polystyrene	0.72	5.895±0.066	3.45±0.11	14.6±0.6
polystyrene	0.51	5.647±0.109	3.72±0.11	10.7±0.5
polystyrene	0.31	5.498±0.124	4.16±0.15	7.1±0.4
polystyrene	0.16	5.583±0.068	4.55±0.13	4.1±0.2
air	0.00115	7.675±0.079	7.02±0.10	0.06±0.003

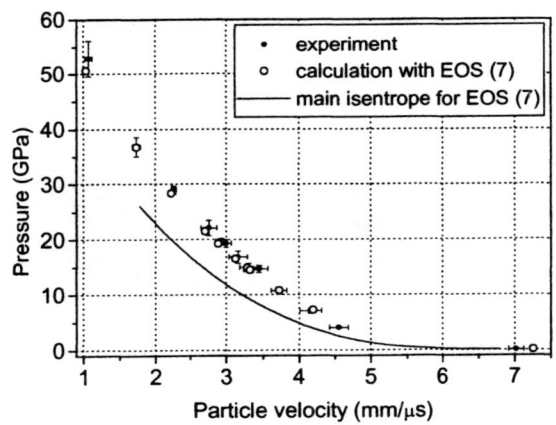

FIGURE 4. Comparison of experimental and calculated with EOS (7) curves.

REFERENCES

1. Shorokhov, E.V., and Litvinov, B.V., *Chemical Physics* **12** 722 (1993).
2. Dick, J.J., Forest, C.A., Ramsey, J.B., and Seitz, W.L., "The Hugoniot and Shock Sensitivity of a Plastic-Bonded TATB Explosive PBX 9502", *Journal of Applied Physics*
3. Kuropatenko, V.F., Kovalenko, G.V., *Problems of an atomic science and techniques. Mathematical simulation of physical processes* (1989).
4. Aminov, Yu.A., Vershinin, A.V. et al., "Research of shock-wave sensitivity of plasticized TATB-based composition", *Physics of Combustion and Explosion* **31** (1995).
5. Pinegre, M., Aveille, J., Leroy, J.C et al., "Expansion Isentropes of TATB Composition Released into Argon", *Eighth Symposium (International) on Detonation*, Albuquerque, New Mexico, 1985.
6. Chirat, R., and Baute, J., "An Extensive Application of WCA4 Equation of State for Explosives", *Eighth Symposium (International) on Detonation*, Portland, Oregon, 1989.
7. Davis, W.C., "Equation of State for Detonation Products", *Tenth International Detonation Symposium*, Boston, Massachusetts, 1993.

OBSERVATIONS ON TYPE II DEFLAGRATION-TO-DETONATION TRANSITIONS

M.J. Gifford, W.G. Proud, J.E. Field

PCS, Cavendish Laboratory, Madingley Road, Cambridge, CB3 0HE. UK.

Type II DDT has been observed in low density charges of ultrafine PETN and RDX. The compressive burning regime that mediates the final stages of type I DDT has been shown to be absent in this mechanism. Convective burning controls the propagation of reaction throughout the column and detonation breaks out at some point along the column. The exact features of the detonation vary between PETN and RDX. The study described here gives new details on the pressure and temperature regimes that operate within the column during the build-up to detonation. Velocity measurements of the waves (in particular the detonation waves) found in these systems have been made. It has been shown that the detonation wave velocities are anomalous for materials at the initial pressing densities, with PETN having an enhanced detonation velocity following a type II DDT and RDX having a retarded detonation velocity.

INTRODUCTION

Type II DDT was first observed in low density columns of tetryl[1], picric acid [2,3] and some propellants[4] by researchers in both Russia and the United States. They found that in low density columns of these materials the outbreak of detonation was preceded by a rapid build-up of pressure at the point where the detonation occurs.

Previous research at the Cavendish Laboratory[5,6] has shown that type II DDT can occur in widely used materials such as cyclotrimethylene trinitramine (RDX) and pentaerythritol tetranitrate (PETN). When columns of ultrafine material below about 50% of the theoretical maximum density (TMD) are thermally ignited, they do not exhibit the compressive burning stage that builds to the shock required for the SDT event in type I DDT. Instead, the convective burning reaches the end of the column. The burning then continues until detonation breaks out at some point along the column. The bulk of this early research into the type II DDT was carried out using high-speed photography.

This paper describes research that has been carried out to elucidate the nature of the events that occur in the build-up to the detonation in type II DDT. The use of strain gauges and thermocouples is described and the merits and problems associated with these diagnostics are discussed. Some observations made whilst reviewing previously obtained data are also discussed and possible explanations of the phenomena are presented.

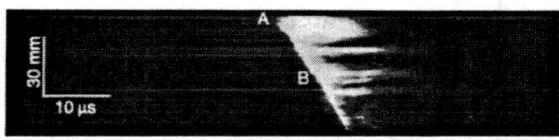

FIGURE 1. Streak record of type II DDT event in ultrafine PETN. A - initiation; B - detonation wave.

MATERIALS USED

Both PETN and RDX were supplied in ultrafine form by ICI Nobel Enterprises, Ardeer, U.K. The powders have a primary particle size of ~1 μm and are produced by a proprietary process. The loose powder densities are ~15% TMD

EXPERIMENTAL METHOD

The main tools that were used during the course of this research were thermocouples and strain gauges. Some of the discussion relates to previously reported results that were obtained using high-speed photography. This technique has been described fully in previous publications[5,6].

General

All of the experimental techniques described below relied on incrementally pressed columns in cylindrical steel confinements as shown in figure 2.

The ignition system used was based on one developed by Dickson[7]. The confinement is effectively sealed against venting by the presence of polycarbonate and aluminium plugs. The pyrotechnic igniter used was an 80% potassium dichromate, 20% boron mixture, This has few gaseous products and burns at a temperature well in excess of the ignition temperature of PETN and RDX.

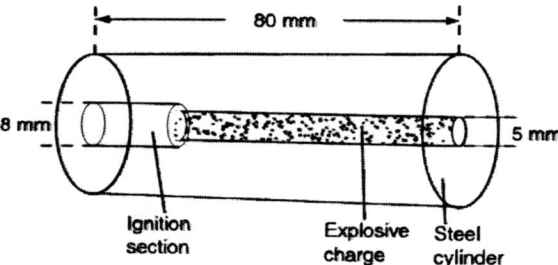

FIGURE 2. Schematic of confinements used. The thermocouples and strain gauges are not shown.

Thermocouples

Thin type K thermocouples were placed at equally spaced points along the column. They were positioned in such a way as to try and ensure that the junctions were as close to the long axis of the column as possible. The holes through which the thermocouples were led were kept as thin as possible and the thermocouples were sealed in position using epoxy.

The thermocouples were connected to high speed amplifiers the output of which was fed into a Tektronix 460A digital storage oscilloscope.

Prior to each experiment the calibration was checked and the results were subsequently corrected for variations in the level of amplification.

Strain Gauges

Micro Measurements strain gauges were mounted on the outside of the confinements to dynamically measure the hoop strain generated during the experiment. The gauges were used in a bridge configuration which enabled accurate and rapid monitoring of the changes in gauge resistance.

Due to the presence of a small amount of plastic deformation it is difficult to directly relate the strain on the cylinder surface to the pressure experienced by the column. As a result all traces arising from the use of strain gauges are given in terms of the strain actually measured.

RESULTS

Thermocouple measurements

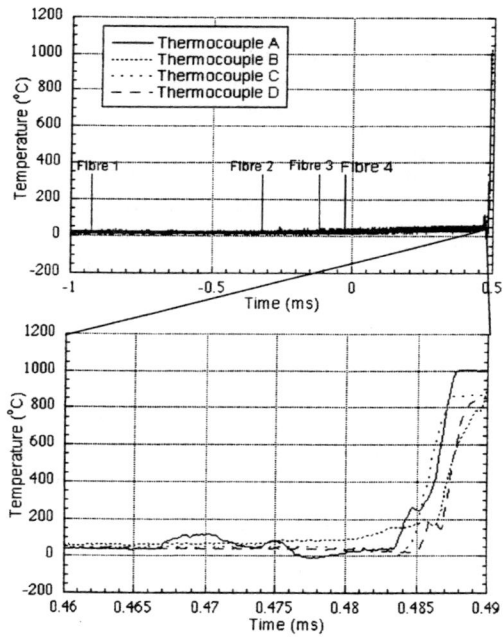

FIGURE 3. Traces from the thermocouples placed in the charge.

Figure 3 shows the traces recorded from four thermocouples positioned in a column during the build-up to a type II DDT event in PETN. Four optical fibres were also placed along the column to monitor the arrival of luminous reaction and compare it with the thermocouples' output. The oscilloscope was triggered using the optical fibre positioned at the same point as the downstream thermocouple. The time at which the light reached this fibre corresponds to the time marked as 0 on the oscilloscope record. This occurred some 930 µs after the luminous reaction passed the optical fibre positioned at the upstream thermocouple. The temperature records began to rise some 500 µs after the reaction was first detected at the position of the upstream thermocouple.

The temperature reading at the thermocouple 20 mm downstream from the ignition point rose considerably faster than the other three thermocouples reaching around 40 °C at the time when the luminous reaction passed the downstream fibre (0 s on figure 3). This thermocouple remained at a higher indicated temperature than the others as they all rose slowly until 466 µs after triggering at which point the upstream thermocouple reading jumped from 36 °C to around 120 °C in 2 µs. As the others continued to rise steadily and quite slowly this upstream thermocouple continued to go up and down erratically indicating temperatures as low as −5 °C. Between 483 and 486 µs after triggering all of the thermocouple readings rose rapidly as the detonation event occurred. Prior to that event the two downstream thermocouples only reached indicated temperatures of 40 °C while the thermocouple 20 mm from the ignition point reached nearly 200 °C.

Strain gauges

The strain gauge traces in figure 4 are from the outside of the confinement during a type II DDT event occurring in ultrafine PETN. The distances indicated on the figure refer to the distance between the position of the gauge and the ignition point of the charge.

The oscilloscope was triggered using an optical fibre located at the downstream end of the column. 0 s refers to the time at which luminous reaction was visible through the fibre

The strain gauge records show no evidence of significant pressure build-up within the column until 650 µs after the luminosity reached the optical fibre. At that point the outbreak of detonation caused all three gauges to jump to strains between 0.005 and 0.02. The traces in this region are quite noisy due to both electrical disturbance associated with the detonation and waves travelling through the confinement.

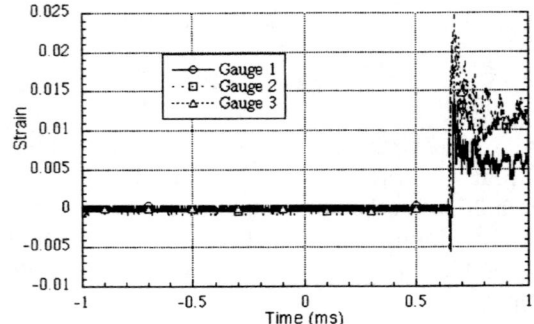

FIGURE 4. Strain gauge traces during a type II DDT event. Ignition occurred at approximately -0.9 ms

2.5 ms after the detonation event, the strain gauges settle to the strain level that remains in the confinement due to plastic deformation.

Observations from reviewing previously obtained data

As has been previously reported[5,8] the detonation velocity following a type II DDT event in PETN pressed to 30% TMD is typically 5-5.5 mm µs^{-1}. This is higher than the 3.7 mm µs^{-1} that is found when charges of this density are directly detonated.

What has also been found from reviewing data obtained prior to the recent thermocouple and strain gauge investigation is that the detonation velocity following a type II DDT event in ultrafine RDX is around 2 mm µs^{-1} which is slower than would be found in an equivalent pristine charge of that density were it directly detonated.

DISCUSSION

The thermocouple readings must be considered in the main to be only indicative of the temperatures in their vicinity. The relatively large heat capacity of the metal junctions when compared to the low density materials present within the columns will mean that there is some discrepancy between the actual temperatures and the measured temperatures[9,10]. Despite this, thermocouple readings of no more than 40 °C on three of the four thermocouples prior to the detonation breaking out seem very low.

A possible explanation of this which is consistent with the previous high-speed photographic studies is that the reaction only occurs in close proximity to the channels along which the hot product gases permeated. The photograph in figure 1 shows high luminosity reaction occurring only in some of the regions of the column in contact with the viewing window.

The higher reading given by the thermocouple 20 mm from the ignition point may well have occurred due to a reacting region being in close proximity to the thermocouple junction.

The strain gauge measurements serve to support the hypothesis that the portion of the charge that is actually reacting prior to detonation is very small when compared to the total amount of material in the charge.

CONCLUSION

The results of these studies point to the type II DDT transition being preceded by an early conductive burning that leads to a convective burn as hot gases are given off. The lack of any detectable pressure build-up within the column reinforces the previous hypothesis that no compressive burning stage such as that mediating the final stage of the type I transition is present in the type II mechanism.

It seems that convective burning is localised around the channels through which the gases are permeating and that temperature rises in other regions of the column are modest.

The enhanced detonation velocity in PETN is possible because little material is consumed in the pre-transition period and a channel is created within the column. The presence of a channel in a charge has been shown by Woodhead[11] and Bakirov[12] to lead to large increases of the detonation velocity as the shock can travel in excess of the detonation velocity through the gas filled channel and initiate material as it passes.

The retardation of the detonation velocity in the case of RDX may be due to more material being consumed in the pre-reaction part of the transition and hence less material is available to support the detonation. Further research involving columns of RDX with embedded thermocouples and strain gauges may be able to shed some light on this problem.

ACKNOWLEDGEMENTS

The authors wish to acknowledge the assistance of Mr R.P. Flaxman and Mr A.S.S.W. Johnson with the production of the confinements used.

ICI Nobel Enterprises (Ardeer) and DERA (Fort Halstead) are both thanked for their joint support of this research.

REFERENCES

[1] R. R. Bernecker, D. Price, J. O. Erkman, and A. R. Clairmont Jr., in *Proc. Sixth Symposium (Int.) on Detonation*, edited by D. J. Edwards (Office of Naval Research, Arlington, Virginia, 1976), p. 426-435.

[2] B. S. Ermolaev, A. A. Sulimov, V. A. Okunev, and V. E. Khrapovskii, *Combust. Explos. Shock Waves* **24,** 59-62 (1988).

[3] V. E. Khrapovskii, *Combust. Explos. Shock Waves* **29,** 129-134 (1993).

[4] D. Price and R. R. Bernecker, *Combust. Flame* **42,** 307-319 (1981).

[5] M. J. Gifford, P. E. Luebcke, and J. E. Field, *J. Appl. Phys.* **86,** 1749-1753 (1999).

[6] M. J. Gifford, PhD Thesis, Cambridge, 2000.

[7] P. E. Luebcke, PhD Thesis, Cambridge, 1995.

[8] M. J. Gifford, P. E. Luebcke, and J. E. Field, in *Shock Compression of Condensed Matter - 1999*, edited by M. D. Furnish, L. C. Chhabildas, and R. S. Hixson (American Institute of Physics, Melville, New York, 2000), p. 845-848.

[9] D. Rittel, *Exper. Mech.* **38,** 73-78 (1998).

[10] S. M. Walley, W. G. Proud, P. J. Rae, and J. E. Field, *Rev. Sci. Instrum.* **71,** 1766-1771 (2000).

[11] D. W. Woodhead, *Nature* **160,** 644 (1947).

[12] I. T. Bakirov and V. V. Mitrofanov, *Dokl. Akad. Nauk SSSR* **231,** 1315 (1976).

PRESSURE WAVE MEASUREMENTS FROM THERMAL COOK-OFF OF AN HMX BASED HIGH EXPLOSIVE PBX 9501

Frank Garcia, Jerry W. Forbes, Craig M. Tarver, Paul A. Urtiew, Daniel W. Greenwood, and Kevin S. Vandersall

Lawrence Livermore National Laboratory, 7000 East Avenue L-282, Livermore, CA 94550

Abstract. A better understanding of thermal cook-off is important for safe handling and storing explosive devices. A number of safety issues exist about what occurs when a cased explosive thermally cooks off. For example, violence of the events as a function of confinement is important for predictions of collateral damage. This paper demonstrates how adjacent materials can be gauged to measure the resulting pressure wave and how this wave propagates in this adjacent material. The output pulse from the thermal cook-off explosive containing fixture is of obvious interest for assessing many scenarios.

INTRODUCTION

The effects of the HMX $\beta \rightarrow \delta$ phase transition, [1-3] which at atmospheric pressure occurs near 160°C, on thermal ignition, impact sensitivity and the kinetics of the cook-off processes need to be better understood for HMX containing explosives. Questions exist on the level of violence of these events as a function of confinement and thermal heating rates. In addition, the acceleration of the metal case by this type of thermal reaction is needed to assess whether the resulting flyer can initiate detonation or reaction in a neighboring explosive item. Thus, results of cook-off events of known size, confinement, and thermal history are essential for developing and/or calibrating reactive flow computer models for calculating events that are difficult to measure experimentally.

EXPERIMENTAL PROCEDURES

Three different experiments on thermal heating of materials have been performed. Two experiments thermally exploded stainless steel encased PBX 9501 (HMX/Estane/BDNPA-F; 95/2.5/2.5 wt %) donor charges. A transmitted two-dimensional pressure wave was measured by gauges in cylinders of Teflon or PBX 9501 that were in contact with the donors' case. A third experiment measured the thermal distribution in a Teflon system of same design and hardware as the first explosive experiment with the same heating rate. A fourth experiment is currently assembled and awaiting testing.

In the two experiments containing high explosive (HE), the PBX 9501 cylindrical disc is confined by 304 Stainless Steel. The HE disc and case was designed such that the explosive would come into contact with all surfaces when the explosive was near 150°C. Some uncertainty existed as to when the HE came into contact with the donor case, because the HE was not uniformly heated (as the thermal expansion calculation assumed). The front 12.4 mm thick stainless steel plate was fastened to the 12.4 mm rear steel plate with 8 grade A hardened steel bolts tightened to 70 ft-lbs. Each 9.0 cm by 2.5 cm thick disk of PBX 9501 weighing 295 g was radially contained by a close fitting stainless steel ring with wall thickness of 34.5 mm. The ring height was slightly greater than the explosive disc to allow for the greater thermal expansion of the explosive. A flat spiral ribbon heater made of nichrome foil was placed between the steel cover plate and a 3 mm thick 6061-T6 aluminum plate. The aluminum plate, which was in contact with the explosive on the opposite side of the heater, served as a gasket for a compression seal since both steel interfaces had knife edges machined in them. This plate also aided in transferring heat from the heaters nearly uniformly to the face of the HE. Two thermocouples were included in the heater package to monitor the temperature and control the heating rate of the heater. No thermocouples were placed internal to the steel encased PBX 9501 to allow for a simple pressure seal design of the steel fixture. The same heater configuration was also placed at the back of the target assembly.

The triggering of the power supplies and the digitizers is a critical feature of this experiment. For the primary triggering system and to measure the wave arrival at the bottom steel plate surface, a series of thirteen PZT pins in a cross pattern with one pin at the center and each pin being 15 mm center to center distance apart were placed against the bottom steel plate. A back-up break wire trigger system was used to provide a trigger pulse from a circuit if any of the wires broke. These thirteen PZT pins and break wires were all summed so the first signal generated would trigger the digitizers and power supplies to allow collection of the data.

The acceptor included gauge packages with both carbon resistor and manganin pressure gauges. The carbon resistor gauge was to measure low-pressure ramp or shock waves generated. The manganin gauges were used to detect if high pressure or detonation waves were generated in the acceptors, although not accurate for large lateral strain.

Carbon resistor gauges have been used successfully in two-dimensional shock wave experiments where time resolution was sacrificed for survival of the gauge [4-6]. The constant current power supply for the carbon resistor gauges was always on driving about 16 mA through the 470 Ω resistors.

Manganin gauges have also been successfully used in numerous one-dimensional strain experiments [7]. It has also been shown to be temperature insensitive [8]. Numerous papers in the literature have discussed the calibration of this gauge, but only two [9,10] are selected here for reference.

The third experiment was a thermal simulation experiment with the Teflon discs inside the steel case replacing the PBX 9501 discs. Thermocouples were at a number of interfaces in this mock donor system. The same heating procedure was performed on this inert donor system. The thermal traces are not reported here but can be used to calibrate a thermal code, which can then provide the time and spatial history of the heated PBX 9501 donor. Recall that no thermocouples were used inside the explosive donor experiments.

One of the HE experiments (TEXT VI) is shown in Figure 1. Both manganin and carbon resistor gauges were placed at different depths in the PBX 9501 cylinder acceptor. A 10 mm thick Teflon disc is placed between the steel top plate of the confined donor system and the acceptor to provide thermal insulation for the acceptor charge. This insures that the acceptor does not cook-off. A second benefit is to keep the temperature down on the carbon resistor gauges since they are temperature sensitive and no calibration exists for this gauge at temperatures other than ambient.

The other HE experiment with a Teflon acceptor was performed previously and is not included here for brevity. Details are provided elsewhere [11].

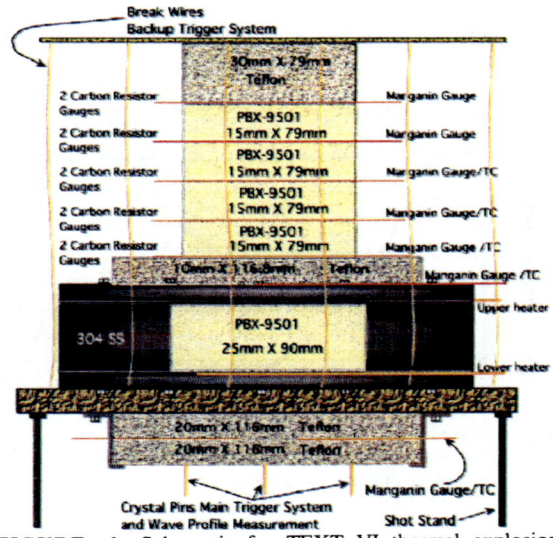

FIGURE 1. Schematic for TEXT VI thermal explosion experiment.

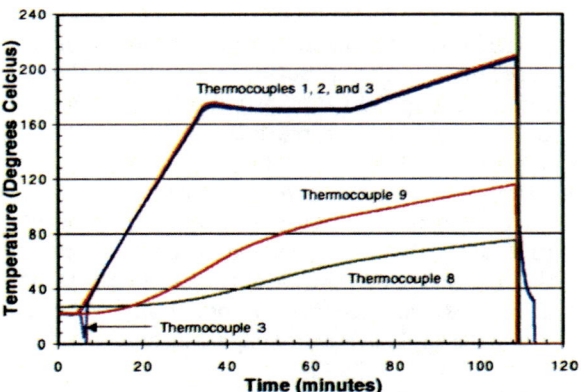

FIGURE 2. Temperature profiles of various thermocouples at various locations in the TEXT VI target.

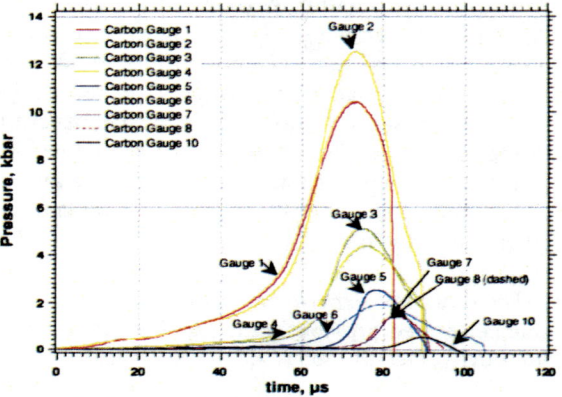

FIGURE 3. Carbon resistor pressure gauge results for TEXT VI.

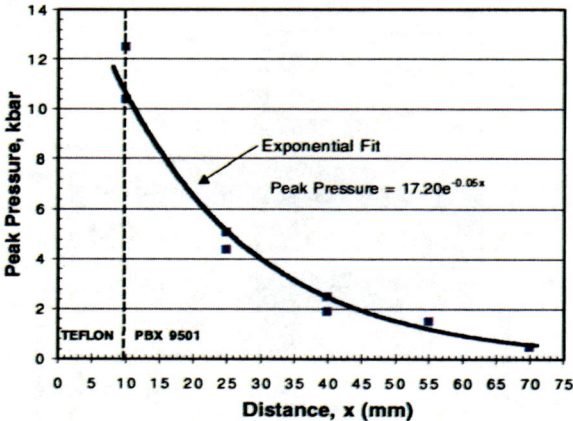

FIGURE 4. Peak pressures of the carbon resistor gauges as a function of Lagrange distance.

RESULTS

Figure 2 gives the temperature time profiles for the five thermocouples that behaved well for TEXT VI. These show that rapid explosion occurred when the thermocouples at the metal surface of the donor system reached 209°C. The initial heating rate was 5.7°C per minute up to 170°C at the metal surface of the donor. Then the temperature at this surface was held (soaked) at 170°C for 35 minutes to allow for the donor to be somewhat uniform in temperature. From the soak temperature of 170°C, the heating rate resumed at 1°C per minute until cook-off occurred. The temperatures in the acceptor did increase but at much lower rates and magnitudes. These temperatures were high enough that the carbon resistor gauge calibration will need to be done for this range of temperatures to improve the accuracy of these measurements.

The carbon resistor pressure gauge results (without temperature corrections) in Figure 3 show that a ramp wave with peak pressure of 12 kb exists at the first gauge level in the acceptor. Some variation in gauge pressure exists for gauges on the same plane which is likely due to the ramp wave not being symmetric as it propagates into the acceptor. Variation between gauges is smaller than this observed difference of 2 kb at the first gauge station. The ramp pressure wave decays very rapidly as it moves up the acceptor charge and the rise time of the ramp shortens. This decay is faster than observed in the Teflon acceptor, which is consistent with PBX 9501 being a stiffer material with faster release wave speeds. It is clear that for TEXT VI the wave did not build into a detonation, which would be a more severe safety issue. The decay of the ramp wave peak pressure is given in Figure 4. The peak pressure decay is fitted accurately to an exponential function.

Since the ramp wave did not build into a high-pressure wave or a detonation, the initial manganin records were not much above the noise level of the digitizers. Lateral gauge strain causes the records to increase significantly after a few microseconds making the gauge records of limited value and therefore not reported here.

SUMMARY AND FUTURE WORK

A multi-dimensional ramp pressure wave is transmitted to the acceptor materials (Teflon or PBX 9501) from an explosive deflagration cook-off of a confined PBX 9501 donor system with a peak pressure of around 12 kb. This ramp wave's peak pressure decays rapidly while the rise time of the ramp decreases initially and then lengthens again as the wave becomes more dispersive. These pressures are substantial and will scatter burning materials around significantly but for these experimental conditions build-up to detonation in the acceptor does not occur.

Future work in this area will include additional experiments with different heating rates and confinement. In addition, some future experiments will measure the velocity of the steel cover plate to see if a sympathetic detonation in a neighboring explosive device with a reasonable stand off is possible. Note that a ramp wave such as seen in these experiments will accelerate the cover plate of the donor system in a manner similar to the acceleration of a projectile by a powder gun. Figure 5 outlines a schematic for such an experiment that has been assembled and waiting on testing. A thermal and hydrodynamic coupled code ALE 2D will be use to model the results of these and future experiments.

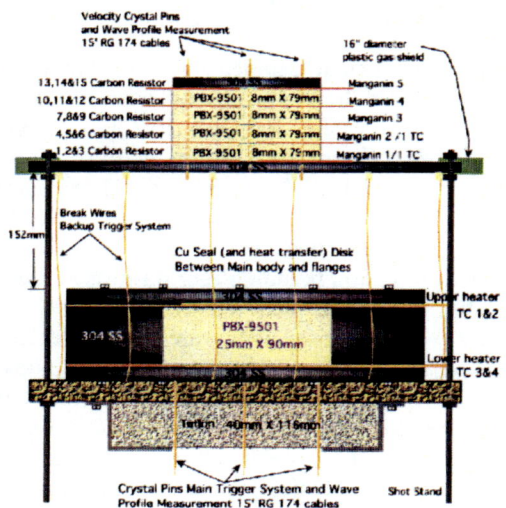

FIGURE 5. Schematic of a future thermal explosion experiment.

ACKNOWLEDGEMENTS

Jerry Dow obtained the funding for us to do this work. Pat McMaster, Ernie Urquidez, and Gary Steinhour assisted on the experiments. Douglas Tasker (LANL) and William Wilson (Eglin AFB) are acknowledged for sharing their information on the use of carbon resistor gauge. The carbon resistor gauge constant current power supply was designed by Douglas Tasker. This work was performed under the auspices of the United States Department of Energy by the Lawrence Livermore National Laboratory under Contract No. W-7405-ENG-48.

REFERENCES

1. Landers, A. and Brill, T., "Pressure-Temperature Dependence of the $\beta \rightarrow \delta$ Polymorphinterconversion in HMX,," J. Phys. Chem., 84, pp. 3573-3577 (1980).
2. Herrman, M., Endel, W., and Eisenreich, N., "Phase Transition of HMX and their Significance for the Sensitivity of Explosives," Zeitschrift fur Kristallographie 204, pp. 121-128 (1993).
3. Henson, B. F., Asay, B. W., Sander, R. K, Son, S. F., Robinson, J. M., Dickson, P., M., Phys. Rev. Ltrs, 82, No. 6, pp. 1213-1216, (1999).
4. Ginsberg, Michael J., and Asay, Blaine W., Rev. Sci. Instrum. 62 (9): 2218-2227 (1991).
5. Wilson, W. H., "Experimental Study of Reaction and Stress Growth in Projectile-Impacted Explosives," Shock Compression of Condensed Matter-1991, eds. Schmidt, Dick, Forbes, and Tasker, Elsevier Science Publishers, pp. 671-674 (1992).
6. Forbes, J. W., Tarver, C. M., Chidester, S. K., Garcia, F., Greenwood, D. W., Garza, R, "Measurement of Low Level Explosives Reaction in the Two-Dimensional Steven Impact Test" 19th Propulsion Systems Hazards Subcommittee (PSHS) Meeting, Monterey, CA. 13-17 November (2000)
7. Urtiew, P. A., Forbes, J. W., "Experimental Study of Low Amplitude, Long-Duration Mechanical Loading of Reactive Materials", 19th Propulsion Systems Hazards Subcommittee (PSHS) Meeting, Monterey, CA. 13-17 November (2000).
8. Urtiew, P.A., Forbes, J.W., Tarver, C.M. and Garcia, F., "Calibration of Manganin Gauges at 250°C", Shock Compression in Condensed Matter, Furnish, M.D., Chhabildas, L.C. and Hixson, R.S. eds., AIP Press, New York, pp. 1019 – 1022 (1999).
9. Vantine H., Chan J., Erickson L. M., Janzen J., Lee R. and Weingart R. C., Rev. Sci. Instr., 51. pp. 116-122 (1980).
10. Gupta, Y. M., J. App. Phys., 54 (11): 6094-6098 (1983).
11. J.W. Forbes, C.M. Tarver, P.A. Urtiew, F. Garcia, D.W. Greenwood, and K.S. Vandersall, "Pressure Wave Measurements from Thermal Cook-off of an HMX Based High Explosive," Paper for the 19th Propulsion systems hazards subcommittee (PSHS) meeting 13-17 November 2000 in Monterey, CA.

MEASUREMENT OF LOW LEVEL EXPLOSIVES REACTION IN GAUGED MULTI-DIMENSIONAL STEVEN IMPACT TESTS

A.M. Niles, F. Garcia, D.W. Greenwood, J.W. Forbes, C.M. Tarver, S.K. Chidester, R.G. Garza, L.L. Swizter

Lawrence Livermore National Laboratory
P.O. Box 808, L-283, Livermore, CA., 94550

Abstract. The Steven Test was developed to determine relative impact sensitivity of metal encased solid high explosives and also be amenable to two-dimensional modeling. Low level reaction thresholds occur at impact velocities below those required for shock initiation. To assist in understanding this test, multi-dimensional gauge techniques utilizing carbon foil and carbon resistor gauges were used to measure pressure and event times. Carbon resistor gauges indicated late time low level reactions 200-540 μs after projectile impact, creating 0.39-2.00 kb peak shocks centered in PBX 9501 explosives discs and a 0.60 kb peak shock in a LX-04 disk. Steven Test modeling results, based on ignition and growth criteria, are presented for two PBX 9501 scenarios: one with projectile impact velocity just under threshold (51 m/s) and one with projectile impact velocity just over threshold (55 m/s). Modeling results are presented and compared to experimental data.

INTRODUCTION

Impact sensitivity of solid high explosives is an important concern in handling, storage, and shipping procedures. Several impact tests have been developed for specific accident scenarios, but these tests are generally neither reproducible nor amenable to computer modeling. The Steven Impact test[1] was developed with these objectives in mind. Blast wave overpressure gauges and external strain gauges were initially used to measure the relative violence of the explosive reactions. High-speed film was used, in part, to obtain time to reaction data. It became clear that adding embedded gauges to the experiment would enhance understanding of the ignition of explosives in this test

Modeling efforts based on Ignition and Growth reactive flow tested several impact ignition criteria and simulated the growth of explosive reaction following ignition as the confined explosive charge produced gaseous reaction products[2-3]. The best models from these earlier works were used to model the experiments containing the embedded gauges. This paper gives details of the embedded gauge experiments and modeling results.

EXPERIMENTAL GEOMETRY

Experimental geometry for the Steven impact test is shown in Fig. 1. A 6.01 cm diameter steel projectile is accelerated via a 76.2 mm gas gun into a cylindrical explosive charge of dimension 11cm diameter and 1.285 cm thickness. The charge was confined using a 0.318cm thick steel front plate, a 1.91 cm thick steel back plate and 2.67 cm steel sides. A Teflon retaining ring positioned the charge within the confinement vessel. Up to six external blast overpressure gauges were placed ten feet from the target for direct comparison with Susan test data. A variety of embedded pressure gauges measuring the internal pressure developed during impact and the subsequent growth of reaction and induced pressure if the critical impact velocity is exceeded are depicted. To date, only carbon foil and carbon resistor embedded gauges have been used.

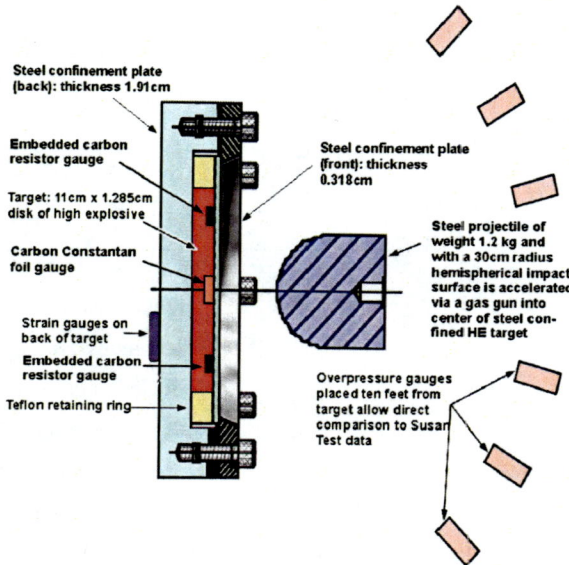

FIGURE 1. Schematic geometry of the Steven impact test.

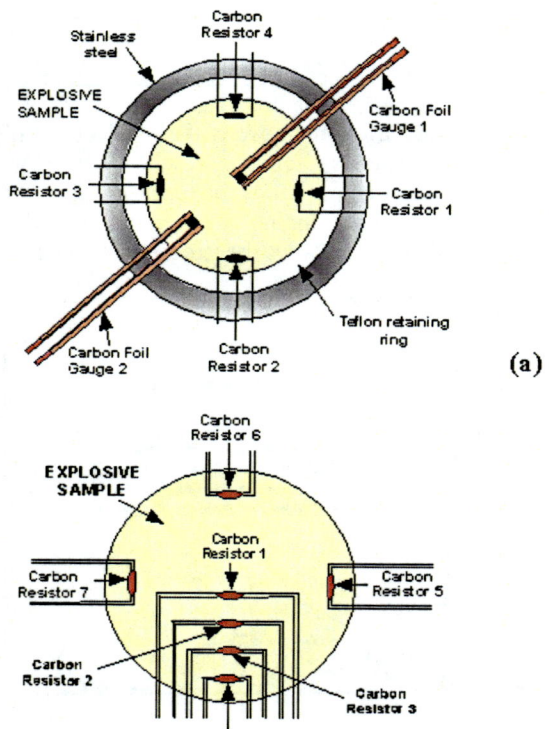

FIGURE 2. Cross-sectional view of embedded gauges inside the target for experiment #s (a) WRL 39-40, 43-47 and (b) WRL 121-122.

Figure 2 shows placement of the carbon foil and carbon resistor gauges in the targets. Two gauge layouts were used over the course of experiments. Figure 2(a) shows the resistor layout for WRL shot #s: 39-40, 43-47 and Fig. 2(b) shows the resistor layout for WRL shot #s: 122-123. The carbon resistors were placed into machined slots on the impact face of the explosive cylinder. The carbon foil gauges were sandwiched between two sheets of 0.125 mm thick Teflon. The Teflon initially extended over the entire diameter of the explosive. Later experiments eliminated the Teflon sheet and placed only a Teflon insulated gauge at the center.

The carbon foil gauge for one-dimensional longitudinal strain experiments[4-6] is good for 0-30 kb pressures with 5-10% accuracy and typical temporal resolution of 25-115 ns. Some two-dimensional flow experiments have been fired using carbon foil gauges where strain compensation on the pressure signals was attempted[6]. The carbon resistor gauge[5,7-10] is also good for one or two dimensional flow pressures of 0-30 kb with accuracy between 8-15%. The temporal resolution of the carbon resistor gauge is 1.4 µs. It is a very rugged gauge that can be used in situations where the foil gauge will not survive. Accuracy decreases for high-end pressures due to the non-linear calibration curve of the gauge. Both gauges have large hystereses on release of pressure because they are porous materials that do not behave elastically.

For the foil gauge, the lower time resolution was determined by assuming a 25 µm thick foil and the upper number assumed the foil gauge package to have insulation of 50 µm layers on both sides of it i.e. a total package thickness of 130 µm. The resistor gauge is assumed to have a 12.5 µm glue layer on both sides of it. To reach equilibrium it was assumed that the principal wave and its reflections transited the gauge element five times [roughly 4 1/2 times the package thickness] at a nominal velocity of 5 km/sec.

EXPERIMENTAL RESULTS

Experimental results for the series of gauged Steven impact tests are shown in Table 1. Impact

TABLE 1. Summary of experimental results of the gauged Steven Impact Tests

WRL Shot Number	HE type	Projectile Impact Velocity (m/s)	Projectile Impact Pressure– Carbon Foil Gauge (kb)	Late Reaction Peak Pressure - Carbon Resistor (kb)	Time of Late Reaction Peak (ms)	Comment
39	PBX 9501	81.49	-	-	-	Reaction observed
40	PBX 9501	61.06	-	-	-	Reaction observed
43	LX-04	90.60	1.52	0.60	0.200	Reaction observed
44	PBX 9501 (new)	46.59	0.59	0.15	-	No Reaction observed
44-2	PBX 9501	46.00	-	0.18	-	No Reaction observed
45	PBX 9501 (new)	51.36	0.82	0.16	-	No Reaction observed
45-2	PBX 9501	60.40	0.32	0.59	0.500	Reaction observed
46	PBX 9501 (aged)	55.40	1.17	2.10	0.540	Reaction observed
47	PBX 9501 (aged)	66.70	0.76	0.46	0.360	Reaction observed
121	PBX 9501	49.50	-	0.17	-	No Reaction observed
122	PBX 9501	55.57	-	0.39	0.315	Reaction observed

pressure histories provided by the carbon foil gauge records show no indication of fast energy release in any of the experiments. Carbon resistor gauges captured late time peak pressure data that were consistent with observed reaction/no-reaction determinations. Reactive collisions generally produced late time pressures greater than 0.35 kb, while shots with no reaction produced pressures less than 0.20 kb.

IGNITION AND GROWTH REACTIVE FLOW MODEL

Previous DYNA2D modeling[1-3] of the Steven test concentrated on its mechanical aspects, modifying the Ignition and Growth reactive flow model to calculate reaction rates under these impact conditions, normalizing these rates for various HMX-based explosives, and predicting threshold velocities for various projectile shapes. In this paper the pressures at the carbon foil and resistor gauge positions for impacts just below and above the threshold velocities for reaction in PBX 9501 are calculated and compared to the measured values. The teflon insulation on the embedded gauges reduces the friction between the steel cover plate and the explosive charge resulting in slightly higher threshold velocities for reaction. This effect is modeled by reducing the Ignition coefficient slightly. Figure 3 shows the experimental and

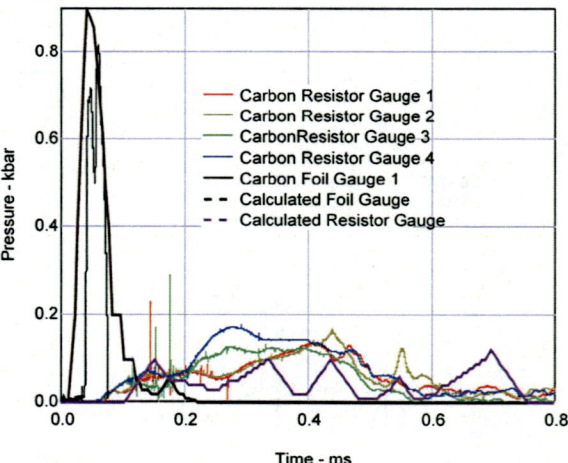

FIGURE 3. Comparison of embedded pressure gauge measurements and reactive flow calculations for WRL 45.

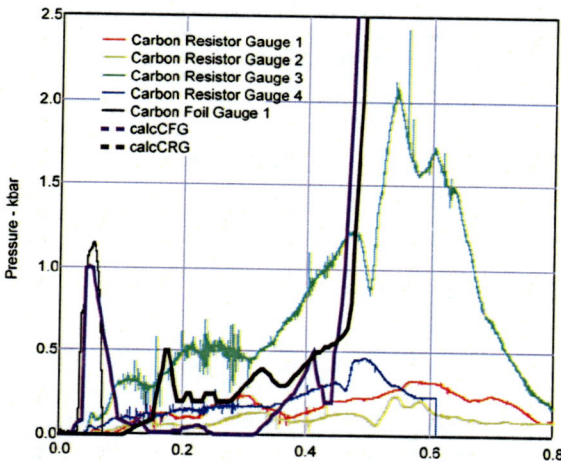

FIGURE 4. Comparison of embedded pressure gauge measurements and reactive flow calculations for WRL 46.

calculated pressure histories for an impact velocity of 51.36 m/s in experiment WRL-45, which did not cause a reaction. The calculated impact pressure and pulse duration agree closely with the carbon foil record. Figure 4 shows the comparison for reactive experiment WRL-46 impacted at 55.4 m/s. The calculated time to reaction and the pressures also agree well with this set of gauge records.

SUMMARY AND CONCLUSIONS

Both embedded carbon foil and resistor gauges gave repeatable pressure-time results in this Steven Test geometry. The carbon resistor gauge is rugged but requires several microseconds to come to equilibrium with its surrounding material. Its pressure measurements are not sensitive to the two-dimensional flow that occurs in this experiment because the gauge smoothes out the differences giving only the change in resistance. Future work includes: (1) hydrodynamic code calculations to calculate lateral strain effects; (2) lateral strain measurements with a strain gauge located near carbon foil active stress element, and (3) an analysis of carbon foil gauge response to strain.

ACKNOWLEDGMENTS

The tests were performed at Bunker 812 at Site 300 by Dave White, Tim Gates, Tom Rambur, and Don Mcdougall. Douglas Tasker (LANL) shared his constant current power supply design for the carbon resistor gauges. William Wilson (Eglin AFB) shared his file on the carbon resistor gauge technique. Frank Garcia was the originator of the gauging concept. This work was performed under the auspices of the U.S. Department of Energy by Lawrence Livermore National Laboratory (contract no. W-7405-ENG-48).

REFERENCES

1. Chidester, S. K., Green, L. G., and Lee, C. G. "A Frictional Work Predictive Method for the Initiation of Solid High Explosives from Low Pressure Impacts", *Tenth International Detonation Symposium*, ONR 33395-12, Boston, MA, 1993, pp. 785-792.
2. Chidester, S. K., Tarver, C. M., and Lee, C. G., "Impact Ignition of New and Aged Solid Explosives", *Shock Compression of Condensed Matter-1997*, edited by S.C. Schmidt et al., AIP Conference Proceedings 429, AIP Press, New York, 1998, pp. 707-710.
3. Chidester, Steven, K., Tarver, Craig, M., and Garza, Raul, "Low Amplitude Impact Testing and Analysis of Pristine and Aged Solid High Explosives", *Eleventh (International) Symposium on Detonation*, ONR 33300-5, Arlington, VA, 1998, pp. 93-100.
4. Charest, J. A, Keller, D. B., Rice, D. A., "Carbon Gauge Calibration," AFWL TR-74-207, (1972).
5. Urtiew, P. A., Forbes, J. W., "Experimental Study of Low Amplitude, Long-Duration Mechanical Loading of Reactive Materials", 19th Propulsion Systems Hazards Subcommittee (PSHS) Meeting, Monterey, CA. 13-17 November (2000).
6. Lynch, C. S., *Rev, Sci. Instrum,.* **66** (12), pp. 5582-5589, (1995).
7. Ginsberg, Michael J., and Asay, Blaine W., *Rev. Sci. Instrum.* **62** (9), pp. 2218-2227, (1991).
8. Wilson, W. H., "Experimental study of Low Amplitude, Long-Duration Mechanical Loading of Reactive materials", *Shock Compression of Condensed Matter-1991*, edited by. Schmidt, Dick, Forbes, and Tasker, Elsevier Science Publishers, 1992, pp. 671-674.
9. Austing, J. L., Tulis, A. J., Hrdina, D. J., and Baker, D. E., Propellants, Explosives, Pyrotechnics, **16**, pp. 205-215, (1991).
10. Forbes, J. W. , Tarver, C. M., Urtiew, P. A., Garcia, F., Greenwood, D. W., and Vandersall, K. S., "Pressure Wave Measurements from Thermal Cook-Off of an HMX Based High Explosive", 19th Propulsion Systems Hazards Subcommittee (PSHS) Meeting, Monterey, CA. 13-17 November (2000).

THE EFFECT OF ADDITIVES ON THE DETONATION CHARACTERISTICS OF A LIQUID EXPLOSIVE

P J Haskins, M D Cook, R I Briggs

Defence Evaluation & Research Agency, Fort Halstead, Sevenoaks, Kent TN14 7BP, England

In this paper we report new experimental results on the detonation characteristics of nitromethane containing high volume percentages of essentially inert additives. In particular, we have studied the detonation of packed beds of small spherical glass and aluminium particles saturated with pure nitromethane. These mixes are found to have reduced detonation velocities and critical diameters compared to the liquid explosive alone. We conclude with a general discussion of the propagation mechanism in such materials.

INTRODUCTION

The effect of inert additives, both solid and liquid, on the detonation of liquid explosives has been the subject of a number of previous studies (e.g. 1-5). In general, the effect of any inert additive will be to reduce the detonation velocity and pressure since some of the energy released will be used in heating and accelerating the inert material. For a miscible liquid additive, mixing will be at the molecular level and full thermal and mechanical equilibrium can be assumed. However, for solid additives the degree of equilibrium achieved will depend on the size of the additive particles. Solid additives also have the effect of introducing hot spots, thus changing the behaviour of the explosive from homogeneous to heterogeneous. This latter effect means, that despite the decrease in available energy, it is possible for solid additives to give rise to an increase in sensitivity, and a reduction in critical diameter.

The most commonly used explosive in such work has been nitromethane (NM), often sensitised by an organic amine. A systematic study of packed beds of inert spherical beads saturated with sensitised NM has been reported by Lee et al. (3, 4). In contrast, here we report some experiments carried out with packed beds of spherical glass or aluminium particles saturated with pure NM. We interpret the results of these new experiments, and draw some general conclusions about the effect of inert additives in general.

EXPERIMENTAL

The experiments were all carried out in 300mm long glass tubes, of various diameters, that were completely filled with the composition under test. The tubes were sealed at the bottom with a steel witness plate, and were initiated at the top by a booster charge. The booster had a length and diameter equal to the diameter of the glass tube.

Experiments were carried out on pure NM (to establish a baseline detonation velocity and critical diameter in the glass tubes) and on packed beds of glass and aluminium saturated with NM. The mixes were prepared by part filling the tube with NM and then slowly adding the solid until the mix just became dry at the surface, and then repeating this procedure in an incremental fashion until the tube was filled. A very small amount of excess NM was left on the top surface of the charge to assist take-over from the booster, and as an insurance against evaporation. All increments were weighed and the mass ratio of solid to liquid was calculated.

The tests were filmed using a high-speed framing camera operating at an inter-frame time of 2.1 microseconds, and the charges were front illuminated with an argon flash bomb.

RESULTS

Tests carried out on pure NM in a range of diameters showed the critical diameter in the glass tubes to lie between 20mm (no detonation) and 25mm (detonation). At 25mm diameter the detonation velocity was measured at 6.32mm/µs. The pure NM results are plotted in Fig. 1, from which the infinite diameter detonation velocity is estimated to be 6.42mm/µs. Figure 2 shows a frame from the high-speed record of a NM test at 25mm diameter.

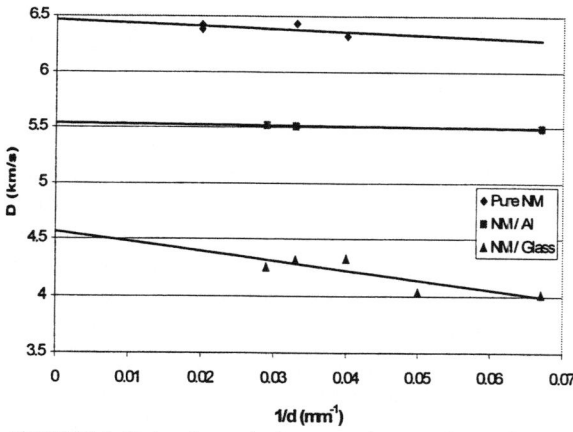

FIGURE 1. Detonation velocity versus inverse charge diameter for (a) pure NM, (b) NM/Al(10.5µm), (c) NM/glass beads.

In the bulk of the experimental work we were concerned with studying packed beds of small particles saturated with NM. The first series of tests were carried out using glass beads. The beads were spherical and 0-44µm in diameter. The compositions using these beads were approximately NM 23% / Glass 77%, by mass. It was found that the critical diameter for these mixes was reduced from that of the pure liquid. Tests were carried out down to 15mm diameter, and all exhibited stable detonations.

Smaller diameters were not tested due to the difficulty in obtaining a uniform filling in smaller diameter tubes. The detonation velocity extrapolated to infinite diameter (see Fig. 1) was estimated to be 4.64mm/µs.

The second series of tests employed aluminium particles in two grades. For both grades the compositions contained approximately NM 35% / Al 65% by mass. The first material was a spherical grade with a mean diameter of 10.5µm. As with the glass beads it was found that these mixes also detonated down to diameters of 15mm. The detonation velocity extrapolated to infinite diameter (see Fig. 1) was estimated at 5.57mm/µs.

FIGURE 2. Frame from a high-speed record showing detonation of NM in a 25mm diameter glass tube.

In addition to the tests using the 10.5µm material, one experiment was carried out using a nanometric grade of aluminium. The Argonide Corporation supplied the nanometric grade, known as Alex. This material is manufactured by an exploding wire process and has a mean particle diameter of ca. 100nm, although there are some considerably larger particles present. The composition based on this material was tested in a 15mm diameter tube, and again a stable detonation was observed. The detonation velocity, at 5.66mm/µs, was a little higher than that observed with the larger aluminium particles. It should be noted that whilst, in principle, aluminium is a reactive additive it is not believed that there is time for any significant reaction of even the Alex material within the reaction zone of NM (6). However, the after burning of the aluminium is clearly visible in Fig. 3, which shows a frame from the high-speed record of the 15mm diameter test using Alex.

FIGURE 3. Frame from a high-speed record showing detonation of NM/Alex in a 15mm diameter glass tube.

CHEETAH (version 2.0) (7) equilibrium code calculations have been carried out for these mixes assuming the aluminium to be inert or fully reactive. Table 1 gives the detonation velocities calculated using the BKWC equation of state at the experimental densities. It can be seen that the calculated velocity for pure NM is a little below (0.36mm/μs) the infinite diameter value estimated from the experiments. However, the calculated values for the NM/Al composition are considerably (>1mm/μs) below those observed experimentally, and this is discussed in the next section.

DISCUSSION

The results we have presented here show that packed beds of both glass and aluminium particles saturated with NM have smaller critical diameters, and a reduced velocity of detonation, compared with NM alone. The detonation velocities of the NM/Al compositions were higher than that for the NM/glass, but it should be noted that the NM/particle mass (and volume) ratio obtained was higher for the aluminium mixes.

TABLE 1. Calculated detonation velocities for NM and NM 35% / Al 65%.

Composition	ρ (g/cc)	D (mm/μs)
NM	1.13	6.06
NM / Reactive Al	1.698	4.46
NM / Inert Al	1.698	4.41

It is interesting to compare these results with those obtained by Lee et al. (4) with sensitised NM. By studying packed beds of spherical glass beads of different sizes they found that the critical diameter was a maximum for bead diameters of the order of the critical diameter of the liquid explosive (ca. 1-2mm for amine sensitised NM). They reasoned that there were two regimes, in which different propagation mechanisms operated, depending on the bead size. For large beads the detonation is thought to merely propagate around the obstacles. Consequently, as the bead size increases the diffraction has less effect and the critical diameter decreases towards that of the pure liquid. However, for small beads detonation cannot propagate around the beads, but shock transmission through them continues to propagate reaction of the liquid explosive in the interstitial pores. In this "small-bead" regime the critical diameter decreases as the bead size decreases, but remains above that of the liquid explosive alone. However, the smallest beads studied by Lee et al. were in the 44-88μm range, and were therefore larger than those considered here. The other, very significant, difference between the studies lies in our use of pure, as opposed to chemically sensitised, NM. Clearly, the large critical diameter of pure NM means that the "small-bead" regime would be expected to apply unless very large (ca. 20mm) beads were used.

Clearly, the addition of high percentages of inert additives represents a large potential dilution of the energy available to support a detonation, and this might be expected to lead to an increase in the critical diameter. This is certainly true when miscible liquids are added to NM. The addition of acetone, in particular, has been extensively studied (5) and shown to lead to a very rapid increase in critical diameter with increasing dilution (critical diameter > 200mm at 25% by volume acetone). However, the addition of solid particles differs in two respects from the addition of a miscible liquid. Firstly, unless the particles are extremely small they are unlikely to be in full thermal and mechanical equilibrium within the detonation reaction zone, and hence will not be fully effective as a diluent. Secondly, small particles are capable of acting as hot spots that can significantly sensitise the explosive. Since, experimentally, we observe a reduction in the critical diameter on addition of particles it seems reasonable to assume that, for the particles considered here, this effect far outweighs any dilution effects.

The CHEETAH calculations provide further evidence for lack of equilibrium between the particles and the detonation products. The CHEETAH calculations (Table 1), which assume equilibrium, are seen to predict a significantly lower detonation velocity for the NM/Al composition than that found experimentally, regardless of whether the Al is assumed reactive or inert. It would therefore appear that even for particles as small as 100nm (the Alex material) the very short reaction zone length of NM means there is insufficient time for full equilibrium.

The smaller particles used in this study are more likely to be effective as a source of hot spots (through shock interactions) than the larger ones used by Lee et al. (4). This is probably at least part of the reason that we observe a decrease in critical diameter upon addition of particles, whereas the earlier studies with larger particles see an increase over that of the liquid alone. However, it is probable that the large difference in sensitivity between pure and chemically sensitised NM also plays a part. This follows since the lower activation energy required for decomposition of the sensitised NM means there is a smaller gain in sensitivity available through the introduction of hot spots.

It is not possible to draw any firm conclusions about the differences in detonation velocity observed between the NM/glass and NM/Al mixes because we did not achieve the same NM/solid ratios. However, in view of the proposed propagation mechanism it is likely that the shock velocity in the solid particles will play a role in determining the detonation velocity. As a consequence we might expect a correlation of observed velocity with the sound speed of the additive. This would certainly be consistent with our observations with glass and aluminium, but further work is required to test this hypothesis.

CONCLUSIONS

We have shown that a packed bed of small glass or aluminium particles saturated in NM can be detonated at diameters less than that of the pure liquid. We have also observed that the propagation velocities of such mixes are less than that of NM, but are higher than would be expected if the particles and detonation products were in equilibrium within the reaction zone.

The role of particles is important with regard to understanding the important mechanisms controlling the detonation process in non-ideal explosives. Consequently, we hope to extend this work to quantify the critical diameter changes, and study the effects of different additives.

REFERENCES

1. Campbell, A. W., Davis, W. C., and Travis, J. R., *Phys. Fluids* **4**, 498-510 (1961).
2. Engelke, R., *Phys. Fluids* **26**, 2420-2424 (1983).
3. Lee, J. J., Frost, D. L., Lee, J. H. S., and Dremin, A. N., *Shock Waves* **5**, 115-119 (1995).
4. Lee, J. J., Brouillette, M., Frost, D. L., and Lee, J. H. S., *Combustion and Flame* **100**, 292-300 (1995).
5. Dremin, A. N., and Rozanov, O. K., *Dokl. Akad. Nauk. SSSR* **139(1)**, 137-139 (1961).
6. Baudin, G., Lefrancois, A., Bergues, D., Bigot, J., Champion, Y., "Combustion of nanophase aluminium in the detonation products of nitromethane", in *11th Symp. Int. on Detonation*, ONR 33300-5, Snowmass CO, 1998, pp. 989-997.
7. Fried, L. E., Howard, W. M., and Souers, P. C., "CHEETAH 2.0 User's Manual", LLNL, UCRL-MA-117541 Rev. 5, (1998).

ELECTROMAGNETIC PROPERTIES OF PRE-DETONATING EXPLOSIVES

G. P. Chambers[1], R. J. Lee[1], T. J. Oxby[2], and W. F. Perger[2]

[1]Energetic Materials Research and Technology Department, NAVSEA Indian Head Division,
101 Strauss Ave, Indian Head MD 20640-5035
[2]Department of Electrical Engineering, Michigan Technological University,
Houghton, MI 49331

ABSTRACT: Current theories of reaction processes suggest that changes in electronic band structure and radiation producing dipole oscillations occur during shock loading of an energetic crystal prior to detonation. To test these theories, a broadband antenna, capable of measuring polarization, was employed to observe shock-induced electromagnetic radiation from a crystalline explosive, RDX. The frequency spectra from these experiments were analyzed using time/frequency Fourier methods. Changes in conductivity resulting from this shock loading were also measured at the opposite end of the crystal from the shock source. A four-point-probe arrangement was used to eliminate errors involving lead resistance. This arrangement uses two leads and a fast discharge circuit to pass current through the crystal interface at the time conductivity begins to change in conjunction with the arrival of the shock wave. Also reported are corresponding light (observed with a high-speed electronic camera) and sub-microwave emission observed during the passing of the shock wave in the RDX crystal prior to detonation.

INTRODUCTION

There exists motivation from both theoretical considerations and previous experiments to pursue both the measurement of electromagnetic fields radiated from and the conductance of energetic materials under shock. Recently, work by B. Kunz's group [1] has indicated that shock energy is converted to lattice vibrations, which in turn give rise to molecular vibrations, electronic excitations and collective excitations. As the electronic energy gap of an energetic crystal destabilizes during shock induced deformation, metallization and reaction onset occur, a model advocated by J. Gilman[2]. We have therefore devised a series of experiments to measure the frequency spectrum through the radio-frequency band of electromagnetic emissions, as well as the conductance of energetic materials near the onset of shock-induced reaction.

EXPERIMENTAL ARRANGEMENT

The experimental arrangement consisted of an RDX crystal (cube, nominally 9-mm on each edge) sandwiched in a Teflon holder with a 10-mm wide groove down the middle to allow back-lit photography of the crystal. A 19-mm diameter Pentolite pellet initiated by an RP-80 detonator provided a planar shockwave into the crystal through a variable PMMA gap. CTH[3] simulations were used to confirm the planarity of the shockwave with respect to the front face of the crystal.

Four conducting pins were placed in a line against the crystal opposite from the donor charge to record voltage and current in a four-point probe arrangement. Each pin was 0.75-mm in diameter and set 1.5-mm between centers. The crystal rested on a nylon disk from which the pins protruded to make contact with the crystal surface. Silver paint was used to provide an ohmic contact between the pins and the crystal.

The circuit is similar to that used by Weir et al.[4], consisting of a 100-nF capacitor charged by 205-Volt battery. The capacitor was connected across the two outside-pins to deliver current when the crystal interface began to conduct. No current flowed through the circuit until the RDX sample changed from an insulator to a conductor upon arrival of the shock wave. The capacitor was placed just behind the nylon disk to minimize inductance and thereby insure a fast rise time for the current. Voltage was measured across the two inner pins. Both measurements were made using a LeCroy oscilloscope at 500 ps/point. The circuit was calibrated using thin wafers of copper (a good conductor) and n-doped Silicon (a typical semiconductor).

RF radiation was measured with two dipole antennae positioned nominally 38 mm from the central axis of the crystal. They were mounted on a single substrate and aligned vertically and horizontally with respect to the shock propagation, in order to measure longitudinal and transverse electromagnetic fields. Signals were recorded using separate LeCroy oscilloscopes at 125 ps/point.

For shots involving the detection of RF radiation, water was used to surround the sample and thereby prevent signals from air ionization. The detonator was fired into both Teflon and Lexan to determine the sensitivity of the antenna to noise during its operation alone.

While the antennae were found to be sensitive to the currents generated by the firing pulser, signals did not appear in the window between onset of the shockwave into the crystal (~7.5 µs) and the initiation of reaction at the crystal/nylon interface (>10.5 µs). Other control shots were fired with the detonator and pentolite donor into water, to determine the radio pickup onto the antennae as well.

RESULTS

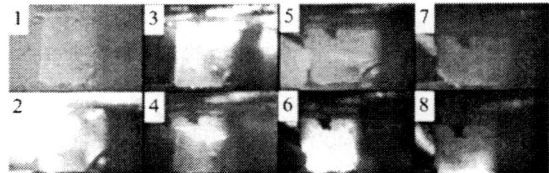

FIGURE 1. High-speed photograph of shocked RDX crystal exhibiting shock-induced luminescence.

Fig 1 shows a high-speed camera record of a 125-kbar shock wave from a pentolite pellet entering an RDX crystal. Frame 1 starts at 7.5 µs after firing of the detonator pulse. The interframe time is 400 ns, with a duration of 40 ns per frame. Light emission can be seen at 8.7 (frame 4) and again at 9.5 µs (frame 6). The light emission appears to be the result of passage of the shockwave through the sample and is occurring prior to detonation. By 8.7 µs, light from the pentolite detonation has died out, so reflection of this light through the crystal is not the source of the light emission.

Concurrent with light emission, we observed RF signals on the antenna temporally correlated with the onset of light emission. The raw voltage data from this measurement can be seen in Fig. 2.
Signals at 8.7 and 9.6 µs are temporally correlated with the light emission observed in Fig. 1.

Furthermore, the strength of the signals tends to correlate with the amount of light emission occurring at the same time. The RF signal at 8.7 µs appears weaker than the signal at 9.6 µs, consistent with the degree of light emission in the high speed photograph, which appears stronger at the later time, 9.5 µs. This agreement tends to support the view that the signals are correlated to some physical phenomena.

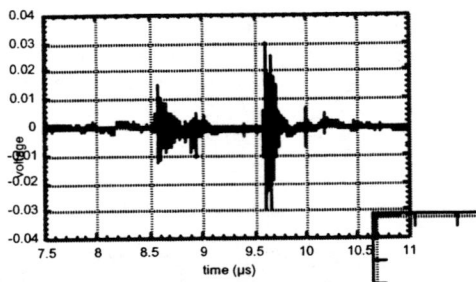

FIGURE 2. Voltage trace observed directly on the antennae as a function of time.

Fig. 3 shows spectrally analyzed signals from the RF, generated through time/frequency Fourier analysis of the antennae signals. The signal at 9.6-μs spectrally resolves into a series of discrete frequencies at 0.5, 0.43, and 0.37-GHz. At 8.7-μs, lines appear at 0.1 and 0.25-GHz. These frequencies are in the sub-microwave regime, and hence are not likely due to molecular rotational spectra which occur at microwave frequencies, in the tens or hundreds of GHz range.

A conductivity measurement is shown in Fig. 4 using the four-point probe test. This figure compares the conductivity of Cu and an n-doped Si wafer with changes in conductivity as a function of time of an RDX crystals shock loaded at 12.5 GPa.

This figure shows that the current flowing in the shocked RDX lies between that obtained by applying disks of n-doped Si and Cu across the electrodes. Hence, the conductivity associated with the shock loaded RDX crystal is comparable to that of a semiconductor. Also of interest is the timing of this event. The RDX began to conduct around 8.4 μs as a result of the shockwave arrival at the interface. (The current trace was shifted in Figure 4 to allow comparison with the two calibration trials.) Current decays quickly in each case due to the small value of capacitance used in these experiments. It is interesting to note that no reaction was observed prior to 9 μs in the camera record, i.e., 0.6 μs after the arrival of the shock front.

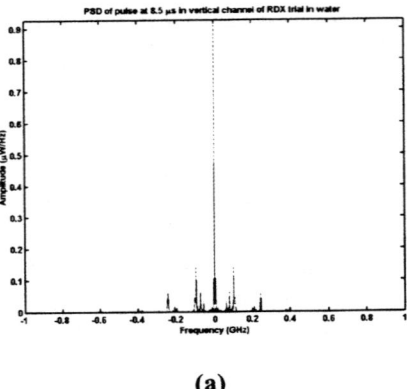

(a)

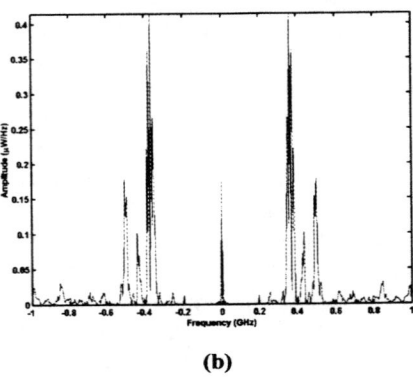

(b)

FIGURE 3. Comparison of spectrally analyzed antennae signals occurring at 8.7 μs (a) and at 9.6 μs (b).

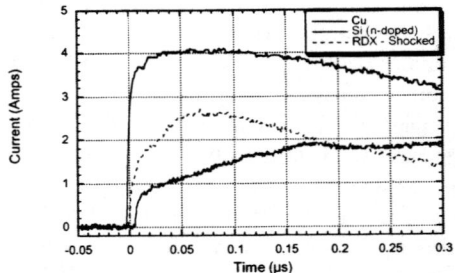

FIGURE 4. Change in RDX resistivity as a function of time for an incident shock wave.

DISCUSSION

In previous work by Dick[5], luminescent emissions from a shocked PETN crystal were interpreted as being due to shock induced decomposition products NO_2 and possibly NO_1. Visible light is known to be produced by electronic transitions, however, without spectroscopy, we cannot distinguish between the possibility that it is correlated with excited states of shock decomposition products or due to radiative emissions from the RDX molecule as a whole.

One possible explanation for the source of RF emission is excitation of plasma oscillations within the crystal[6]. Such plasma modes do not exist in insulators, but can be excited in metals or semiconductors. Since the shock wave transforms RDX into a semiconductor during its passage through the crystal, it is possible to excite such plasmon oscillations within the crystal. These oscillations may be the source of the RF emissions detected with the antennae.

The correlation of RF emissions with luminescent emissions, and the correlation of the latter with sensitivity, suggests a relationship between the RF field and reaction initiation in RDX. This mechanism may need to be considered in simulations if an accurate prediction of sensitivity is to be obtained.

Lastly, conductivity changes have been observed as a result of the passage of the shockwave through the crystal, prior to onset of reaction. This was predicted by Gilman[2]. This represents a several order of magnitude change in conductivity of a molecular RDX crystal as a result of the incident shockwave. Such changes have previously been measured in the ionic crystal KCl at 140-kbar shock pressures[7]. These changes ultimately push the organic crystal towards metallization. A metallized organic crystal therefore has electrons available to participate in intermolecular reactions, which may either facilitate or be a necessary condition for detonation chemistry to occur.

Thus, our results may provide support for both models of reaction onset. The detection of RF emissions from the crystal during shock loading directly supports theoretical work by Zwitter et al[1] while conductivity changes towards metallization support models of Gilman for intermolecular processes.

SUMMARY AND CONCLUSIONS

Shock-induced light emission from an RDX crystal prior to detonation has been observed. This luminescence correlated with shock-induced electromagnetic radiation. A change in crystal conductivity as a function of time was also observed prior to and during the detonation process. Future parametric studies, using the techniques developed here, will help broaden our understanding of the roles these phenomena play in initiation.

ACKNOWLEDGMENTS

Discussions with R. Doherty, R. Chau and funding from the IHD Internal Research Program are gratefully acknowledged.

REFERENCES

1. Zwitter, D. E., Kuklja, M. M., and Kunz, A. B., *Shock Compression in Condensed Matter*, Snowbird Utah, June 1999
2. J. Gilman, *Shock Compression in Condensed Matter*, Amherst MA, July 1997
3. R.L. Bell et al. "CTH User's Manual and Input Instructions" version 2.00 Sandia National Laboratories (Sept 1995)
4. W.J. Nellis, S.T. Weir, C Mitchell *Physical Review B,* Vol. 59. P. 3436
5. J.J. Dick, R.N. Mulford, W.J. Spencer, D.R. Pettit, E. Garcia, D.C. Shaw, J Appl. Phys. **70** p. 3572 (1991)
6. J.J. Gilman, Philos. Mag. B, 1999, Vol 79, No. 4, 643-54
7. N.K. Bourne and D. Townsend. Shock Compression of Condensed Matter, 1999 ed. M.D. Furnish, p. 109

EFFECT OF GMB ON FAILURE AND REACTION REGIME OF NM/PMMA-GMB MIXTURES

José Gois, José Campos and Igor Plaksin

Laboratory of Energetics and Detonics, Av. Universidade de Coimbra, 3150 Condeixa, Portugal
Mechanical Department, University of Coimbra, Pinhal de Marrocos, 3030 Coimbra, Portugal

Abstract. The effect of the addition of small amounts of glass microballoons (GMB) on heterogeneous explosives has been investigated with the aim of understanding mechanisms that lead to the strong reduction of its critical diameter. However, there is no clear identification of the changes on detonation wave propagation and its structural features. To obtain a better understanding of the contribution of GMB as a particular heterogeneity, the detonation failure and the re-initiation of NM/PMMA-GMB mixtures is studied. Corner turning configuration was performed in order to determine the influence of the GMB concentration and size on failure phenomena by observing the trajectories of the divergent shock waves around the corner. The shape of the printed traces on a copper witness plate, coupled with detonation velocity and front curvature measurements, was used to evaluate the evolution of the detonation reaction regime and its cellular structure. The obtained results of printed flow lines show significant changes of the original pattern of the propagation of the detonation wave. The structural colliding and divergent waves were successful identified and explained.

INTRODUCTION

The influence of the addition of a small amount of inert particles to NM has long been known to drastically change failure thickness (1). Glass microballoons (GMB) were used with success to sensitize homogeneous NM/PMMA (96/4) matrix. Increased mass fraction and decreased particle size were found to increase the shock sensitivity of NM/PMMA-GMB mixtures (2). The shock pressure to initiate detonation is lowered as manifested by the decrease of critical diameter and failure thickness (2,3).

The increase of the energy release rate due of hot spots induced by GMB collapse compensates the effect of the rarefaction, avoiding the breakdown of the reaction zone in the detonation front. As proved in previous work, a decrease of particle size and rise of concentration increases the density of the re-initiation sites and strongly reduces the critical diameter allowing the propagation of detonation front (4).

In previous work Gois et al. (4) observed for 1 percent mass fraction of GMB in NM/PMMA-GMB mixtures, significant changes in the original pattern of NM/PMMA matrix. Close to the failure thickness, the critical pressure pulse of the initiation of the shock front to induce a CJ plane wave is just critical for an initiation detonation. Dremin (5) has explained the conical failure wave generated when the detonation front (DF) reaches the section, corresponding to an abrupt increasing of cylinder charge diameter.

The purpose of the present work is to determine the effect of hot-spots induced by GMB collapse on failure phenomena and re-initiation of light heterogeneous explosives, in order to better understand the complex mechanism of propagation and detonation as described in literature. A copper witness plate and different strips of 64 optical fibres are used to measure respectively the grid of perturbations and the curvature and velocity of shock waves around the corner.

EXPERIMENTS AND RESULTS

Explosives mixtures and preparation

The heterogeneous explosives mixtures used in this study are based in a homogeneous explosive mixture of NM/PMMA (96/4 wt.). The amount of PMMA increases viscosity of the mixtures and avoids buoyancy GMB, without changing the original detonation characteristics of NM (2). GMB (QCel 520 FPS and QCel 300), supplied by Asko Inc., were selected to perform NM/PMMA-GMB mixtures. QCel 300 GMB was sieved to obtain large particle diameters. Both classes of GMB have a wall thickness of about 1µm. Table 1 shows the range and mean particle diameter of GMB's used.

TABLE 1. Range and mean diameter of GMB used in NM/PMMA-GMB mixtures.

GMB	$d_{10} - d_{90}$ [mm]	d_{50} [mm]	Effective density [kg.m^{-3}]
QCel 520FPS	16 - 79	45±1	220
QCel 300*	21 - 146	92±2	151

* Selected size distribution after sieving.

The diameter distribution of GMB's was obtained by laser diffraction spectrometry. The effective density was measured using helium picnometry. The experiments were performed for a range temperature of 15°C to 20 °C. Tab. 2 shows the mass fraction of GMB's and the density of final mixture.

TABLE 2. Mass fraction of GMB's and density of NM/PMMA-GMB mixtures.

d_{50} [mm]	GMB mass fraction [%]	Density [kg.m^{-3}]
45±1	1	1090±5
	3	1040±9
92±2	1	1088±10
	3	928±15

Corner turning tests and Results

Corner turning experiments of detonation front propagation of NM/PMMA-GMB mixtures were performed for different concentrations of GMB. The setup consists of an aluminium channel of 80 mm length with a square section of 12x12 mm. Copper plate of 5 mm thickness is used as witness plate. A strip consisting of 64 optical fibers is used in different positions to record the light induced by the detonation front around the corner. A fast electronic streak camera (THOMSON TSN 506N) is used to record the signals. Figure 1 shows the setup and configuration used to measure the curvature of shock wave propagation from the end of the channel. Two more configurations not shown were adopted to measure the radial propagation of shock front, the conical failure and the extension of the dark zone around the corner.

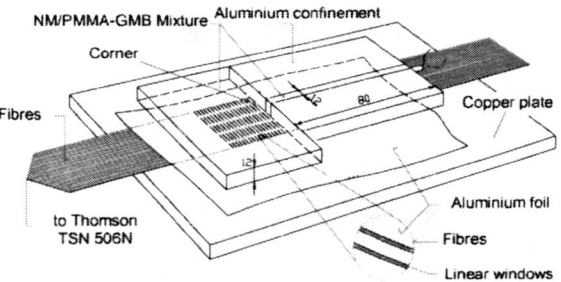

FIGURE 1. Corner turning test experimental setup.

Figure 2 shows the change on the detonation front curvature of NM/PMMA-GMB mixtures caused by the abrupt transition of detonation front between the channel and the corner. For 1% of GMB, the conical failure of shock front induced by the rarefaction waves at the corner is identified. Near the vertex of conical failure, a semi-hemispherical re-initiation point appears and curvature is strongly reduced. From the erosion in the witness copper plate (*vd.* Fig. 3) the angle corresponding to conical failure, the quenching distance at the edge of the corner, the dark zone and the interaction of divergent shock waves and the re-initiation front from the vertex of conical failure can be seen.

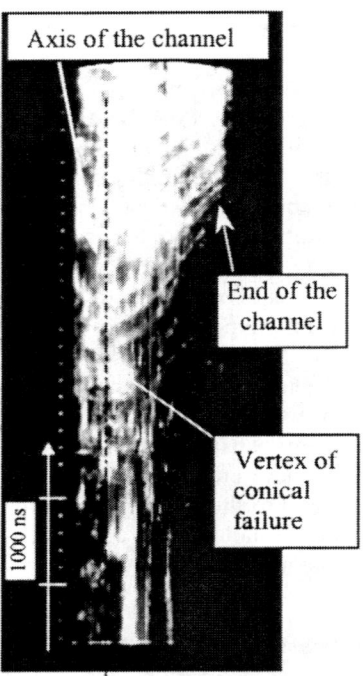

FIGURE 2. Typical streak records of detonation front curvature for NM/PMMA-GMB mixtures (1% GMB, $d_{50}=45\mu m$) obtained with the test arrangement of Fig. 1.

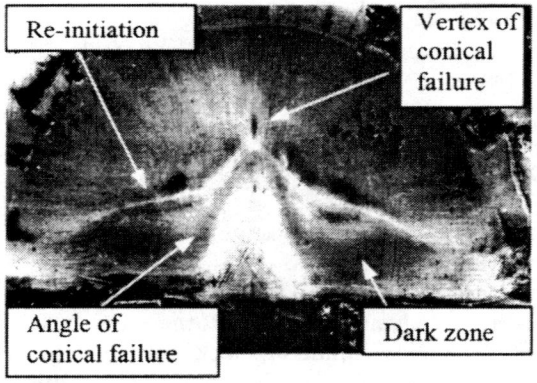

FIGURE 3. Picture of witness copper plate showing erosion around the corner induced by failure and re-initiation of NM/PMMA-GMB mixtures with 1% GMB ($d_{50}=45\mu m$).

The comparison between the results obtained for NM/PMMA-GMB mixtures and its of reference NM/PMMA(96/4) matrix allows to verify the influence of GMB on the reduction of the angle of conical failure.
The radial velocity of divergent shock wave was measured at three angles (0°, 30° and 60°) from the axis at the end of the channel (vd. Fig. 4). Previous work (4) has shown no light at 90° immediately at the corner. Results show that divergent shock wave velocity is reduced when GMB concentration increases. Shock front velocity decreases continually from the axis of initiation of divergent shock wave.

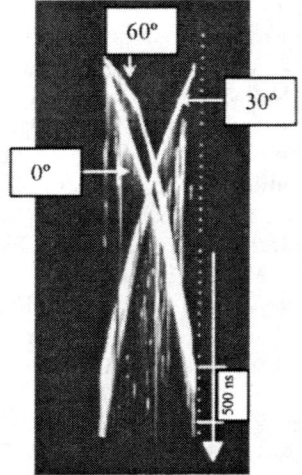

FIGURE 4. Experimental divergent shock wave velocity obtained in three directions (0°, 30° and 60°) from the axis, at the end of the channel.

Three strip of fibers, each one on them divided in three groups, and fixed at different angles from the axis of initiation channel have shown that the divergent shock wave is not circular. Fig. 5 shows the photochronogram obtained from the light transmitted from the three groups of fibers distributed in three rings around the axis, fixed at −50°, 0° and +50° from the axis of the channel around the corner. The result allows evaluates of the shape of DF as function of radial distance and time.

The following are observed in the photochronograms: (I) the preferential DF is developed in the range angle of 45° to 60°, from the explosive mixture – confinement interface; (II) the detonation front in the range angle of 60° to 90° accelerates, reducing the delay of detonation front. at the middle of the mushroom shape, characteristics of divergent detonation front.

The steady state interactions of transversal waves induced by rarefaction waves from the walls of the channel (vd. Fig. 6) are identified from the erosion profiles on the witness copper plate. The re-initiation of detonation starts at the periphery of conical failure from the collision of transverse waves. The detonation starts again at the conical failure and expands to the lateral pre-compressed zones giving rise to two hemispherical detonation fronts. The shock reflection at the collisions of the adjacent hemispherical detonation is initially regular, but as the angle of collision increases, a Mach stem appears. This Mach stem overtakes the original front and accelerates the detonation front.

reaction zone generating an inversion of two hemispherical detonation fronts created from the conical failure boundaries. The collisions of transverse waves and microdetonations, induced by hot spots, sustain a regular cellular structure proportional spaced to the reaction zone thickness. The grid of regular pattern changes with the interparticle distance of GMB.

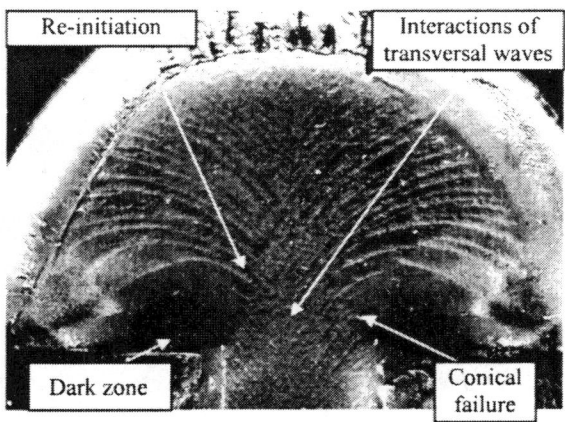

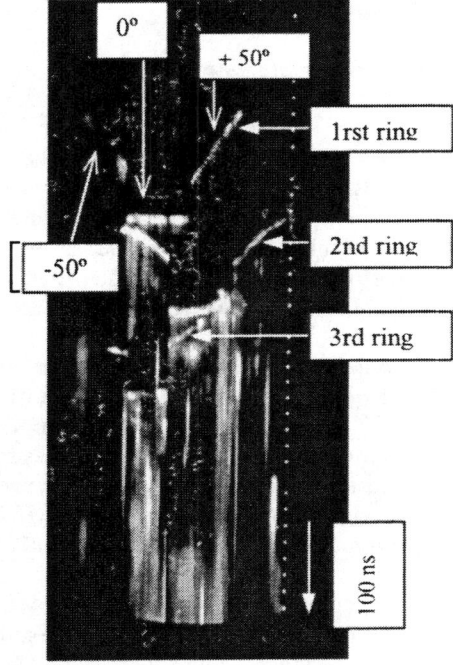

FIGURE 5. Signals of DF at the angles -50°, 0° and 50° from the axis of the channel in three equal spaced rings around the corner.

FIGURE 6. Erosion profiles on witness copper plate, around the corner, induced by propagation of detonation front of NM/PMMA-GMB mixture with 3% GMB(d_{50}=45µm).

CONCLUSIONS

The optical records obtained using an optical technique, based in several strips of fibers, and the erosion profiles on a witness copper plate are coupled to determine the influence of GMB heterogeneities in NM/PMMA-GMB mixtures. The increasing of GMB concentration reduces the angle of conical failure and the extension of dark zone. The presence of GMB induces hot spots into the

REFERENCES

1. Engelke R (1979) Effect of a physical inhomogeneity on steady-state detonation velocity. In Phys. Fluids, vol. 22, (9), pp. 1623-1630.
2. Gois JC (1995) Influência das micro esferas ocas de vidro na detonação da mistura nitrometano-polimetilmetacrilato, Ph.D. Thesis, University of Coimbra, Portugal.
3. Gois JC, Campos J, Mendes, R (1995) On extinction detonation behavior of NM-PMMA-GMB mixtures. In Proceedings of the Conference of the American Physical Society on Shock Compression of Condensed Matter, Seattle, Washington, Part. 2, pp. 827-830.
4. Gois JC, Campos J, Plaksin I, Mendes R (1997) Failure and re-initiation detonation phenomena in NM/PMMA-GMB mixtures. In Proceedings of the Conference of the American Physical Society on Shock Compression of Condensed Matter, Amherst, Massachusetts, pp. 691-694.
5. Dremin A. Razanov SD, Trofimov VS (1963) On the detonation of nitromethane. In Combustion and Flame, vol. 7, pp. 153-162.

PRESSURE WAVE MEASUREMENTS IN CYLINDERS OF DETONATING LX-17*

J. W. Forbes, P. C. Souers, P. A. Urtiew, K. S. Vandersall, F. Garcia, D.W. Greenwood, and LeRoy Green

Energetic Materials Center, Lawrence Livermore National Laboratory, P. O. Box 808, L-282, Livermore, CA 94551

Abstract: Manganin gauges with temporal resolution of less than 75 ns were used to measure the detonation wave pressure profile in a right cylinder of LX-17 (TATB/Kel-F: 92.5/7.5 wt.%). Three gauges at different Lagrange locations were on the centerline of the 5.08 cm diameter cylinder at distances greater than four times its diameter from the boosted end. At the last gauge plane, seven gauges were placed at the same Lagrange position but spaced radially across the cylinder diameter. Wave curvature and effects of lateral strain in these gauges were measured.

INTRODUCTION

In-situ gauges have the advantage over other techniques for measuring detonation properties because of the minimum disturbance to the flow. In cylinders of HE with steady detonation waves, in-situ gauges can provide 2-D hydrodynamic state data for validation of 2-D flow theories of detonation such as Wood-Kirkwood[1] and other curved front theories[2]. Another reason for this research is the general need to develop 2-D flow in-situ gauge experimentation. Improvements of the temporal resolution, survivability, and signal fidelity[3] of the manganin foil gauge in detonations has allowed us to begin this research project.

The level of difficulty of gauging 2-D flow experiments is an order of magnitude greater than difficult 1-D experiments. This is because stress and particle motion are a function of two spatial variables and lateral strain exists in the gauges.

EXPERIMENTAL DESCRIPTION

The schematic of the cylindrical charge of LX-17 boosted by a LX-10 cylindrical disc is given in Fig. 1. This experiment was designed to measure a steady detonation wave. The cylinder of LX-17 was 5.08 cm diameter and 23.4 cm long. Three gauges were placed on the centerline at distances of 20.0, 20.8, and 22.4 cm, respectively, from the originally boosted LX-17 surface. To measure the wave curvature a set of seven manganin gauges are equally spaced across the diameter of the cylinder at 22.4 cm from the boosted surface as given in Fig. 2.

The active gauge elements are different in size for the standard gauge, the mini gauge, and the multiple gauges. The standard gauge has a rectangular element of 0.7 mm wide by 2.0 mm long, the mini-gauge has an element 0.3 by 0.5 mm and the six multiple gauges each have elements of 0.3 by 0.7 mm. The temporal resolution is determined by calculating the shock transit time for 3.5 wave transits through the gauge package. The gauges 20.0 and 20.8 cm from the boosted face had 0.05 mm thick Teflon sheets in front and back of the 0.025 mm thick manganin foil while the gauges at 22.4 cm had 0.07 mm thick Teflon sheets in front and back of the foil.

RESULTS AND DISCUSSION

Manganin foils were used to measure longitudinal pressure along the axis of a steady detonating cylinder of LX-17. The lateral flow at the centerline

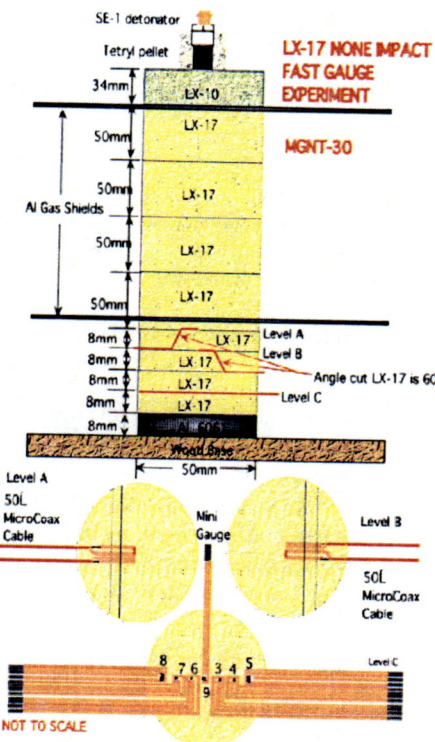

FIGURE 1. Schematic of LX-17 cylinder experiment

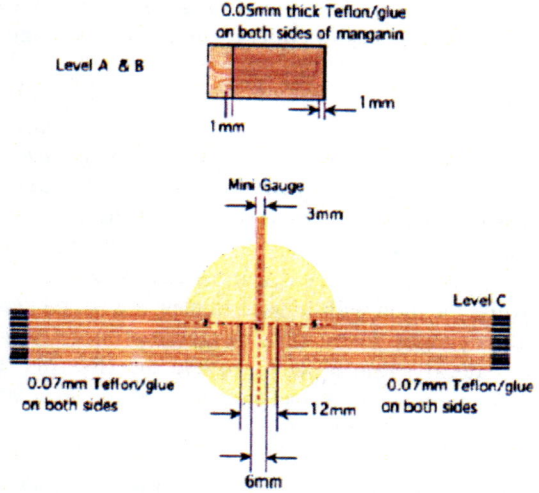

FIGURE 2. Gauge details for LX-17 cylinder experiment

of the cylinder is zero due to symmetry. Note that the four lead gauges will be affected by strain due to the lateral flow in the detonating cylinder.

Even with this 75 ns temporal resolution, the measured peak pressures were near 30 Gpa. Figure 3 gives all the gauge results as analyzed without correction for lateral strain. Two experiments were done with identical LX-17 cylinders. Micro-coaxial cables were used to eliminate gauge lead length (see Fig. 1) for the MGNT 30 experiment but not for MGNT 54.

The average steady detonation velocity from the two experiments was 7.64 mm/μs for an initial density of $\rho_o = 1.90$ g/cm^3. Note that the first gauge in MGNT 30 broke down immediately when the shock wave arrived as seen in Fig. 3. Therefore, the arrival time of this gauge was not used for determining detonation velocity. Impedance matching using initial density, steady detonation velocity, and published JWL products EOS[4,5], gives a CJ pressure of 27.0 GPa. The three gauges near the centerline of the cylinder gave nominal results of 29-30 GPa peaks and pulse widths 70 - 90 ns at P_{CJ} for experiment MGNT 30 and 28-30 GPa peaks and pulse widths 130-210 ns at P_{CJ} for MGNT 54. The reasons for the difference in the pulse widths at 27 GPa are not known. More experiments are required to determine the reasons for this difference.

Six multiple gauges were placed symmetrically from the centerline (one mini-gauge was at the

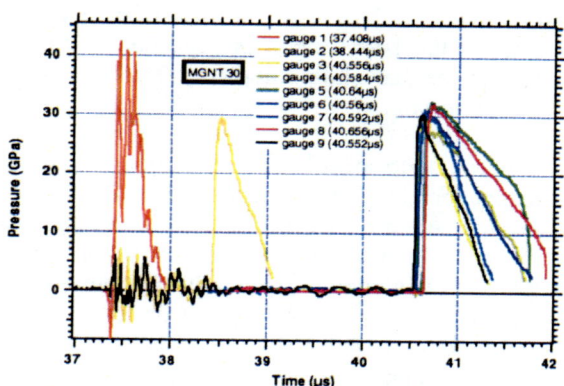

FIGURE 3. Manganin gauge records for experiment MGNT 30

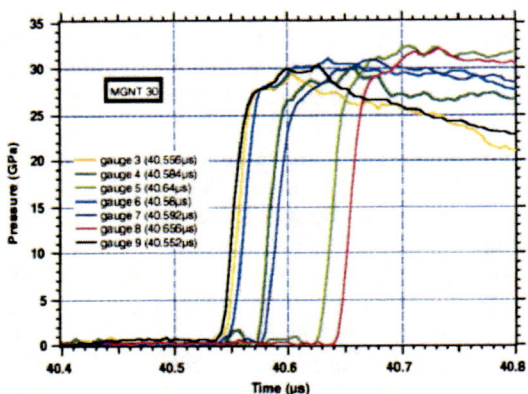

FIGURE 4. Records for gauges across diameter of plane at 22.4 cm from the boosted LX-17 surface

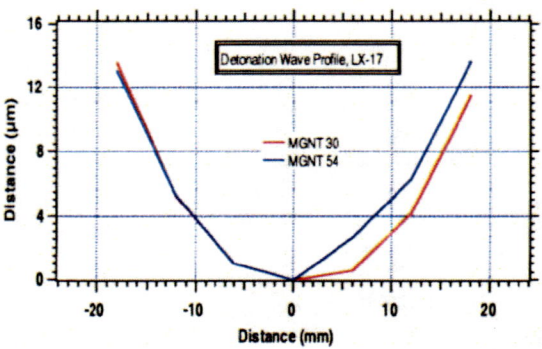

FIGURE 5. Profile of detonation waves in 5.0 cm diameter LX-17 cylinders

center) on a plane 22.4 cm from the boosted surface. Figure 4 gives the results from the seven gauges located across the diameter. The gauges not on the centerline gave voltage signals with greater peak values than those at the center. Corrections for the increased voltage signals due to lateral strain in the gauge were not done. Analyzing the data without the strain correction results in apparent higher peak pressures the further away from the centerline. This apparent result is contrary to oblique shock theory[6] which gives the highest pressure at the center and lower pressure as a function of distance from the centerline.

The lateral strain is affecting these results and needs to be compensated for. Two-dimensional code calculations could be used to determine the strain as a function of time for these gauges. However, the best solution would be to directly measure this lateral strain as a function of time and correct the manganin voltage[7]. This is a subject for future work.

Wave curvature was obtained from the detonation wave arrival times at the seven gauges located across a diameter in one Lagrange plane. The arrival times and the steady detonation velocity allowed calculation of the relative position of wave front elements along the axis of the cylinder. The wave front position along the cylinder axis versus charge diameter is given in Fig. 5. The data points are connected with straight-line segments.

The manganin gauges are accurate to 4% (i.e. twice the standard deviation of errors) for compression and release states in inert materials[8]. The accuracy of the present data has not been determined because of lack of adequate information on gauge accuracy with the gauge strain effects in the detonation environment.

CONCLUSIONS

The feasibility of using in-situ pressure gauges for determining CJ states at the center of cylinders has been demonstrated in experiments with small divergent flow. Pressures exceeding the P_{CJ} of 27 GPa in LX-17 were measured. However, accurate Von Neuman spike pressure and pulse widths were not determined because of insufficient temporal resolution and accuracy of the gauge packages used here.

In future experiments the temporal resolution of the manganin foil gauge package will be reduced to less than 30 ns. This will allow for a more accurate measurement of the spike pressure at the detonation front in cylinders of LX-17. In addition, constantan strain elements located symmetrically at the same spatial location will be used to measure lateral strain in the gauges. Correcting manganin pressure gauges for lateral strain[6] will provide the first step for obtaining accurate stress gauge records in these type of experiments with small divergent flow.

To complete an accurate gauge calibration requires applying the theory of piezoeresistance response of the manganin foil to shock loading and unloading[9] measurements of compression and release at pressures near 30 GPa in inert materials with similar mechanical properties as detonating HE.

The gauge accuracy needs to be at least 3 % for compression and release pressure states near 30 GPa for the data to be discriminating for hydrodynamic models.

The search for gauges to use in the detonation environment that are not sensitive to lateral strain and have ns time response with less than 3% pressure errors continues to be a challenge.

ACKNOWLEDGEMENTS

Mike Martin, Gary Steinhour, and Ernie Urquidez fired these experiments for us. Dave Zevely provided the machined HE parts. Paul Marples machined the inert parts. The authors acknowledge useful discussions with J. Charest on strain compensation of manganin gauges and general discussions on 2-D gauge measurement techniques with Y. M. Gupta.

REFERENCES

1. W. W. Wood and J. G. Kirkwood, "Diameter Effect in Condensed Explosives. The Relation Between Velocity and Radius of Curvature of the Detonation Wave," J. Chem. Phys., 22, p. 1920, (1954).
2. D. S. Stewart and J. B. Bdzil, "Examples of Detonation Shock Dynamics for Wave Spread Applications," Proceedings of Ninth Symposium (International) on Detonation, Coronado, CA, Aug. 24-27, pp. 773-783, 1976.
3. D. Greenwood, J. Forbes, F. Garcia, K. Vandersall, LeRoy Green, and Leroy Erickson "Improvements in the Signal Fidelity of the Manganin Stress Gauge," Proceedings of 12th Biennial APS Conference on Shock Compression of Condensed Matter, Atlanta, GA, June 24-29, 2001.
4. C. M. Tarver, J. W. Kury, and R. D. Breithaupt, "Detonation Waves in Triaminotrinitrobenzene," J. Appl. Phys. 82, (8), 1997.
5. R. L. Gustavsen, S. A. Sheffield, R. R. Alcon, J. W. Forbes, C. M. Tarver, and F. Garcia, "Embedded Electromagnetic Gauge Measurements and Modeling of Shock Initiation in the TATB Based Explosives PBX 950 and LX-17," Proceedings of 12th Biennial APS Conference on Shock Compression of Condensed Matter, Atlanta, GA, June 24-29, 2001.
6. E. R. Lemar, J. W. Forbes, and M. Cowperthwaite,"Oblique Shockwave Calculations for Detonation Waves in Brass Confined and Bare PBXN-111 Cylindrical Charges," p. 385, *Shock Compression of Condensed Matter*, eds. Schmidt, Dandekar, Forbes, AIP Conference Proceedings 429, 1977.
7. J. A. Charest, "Development of a Strain-Compensated Shock Pressure Gauge," Dynasen TR-005, February 1979.
8. H. S. Vantine, L. M. Erickson, and J. Janzen, "Hysteresis Corrected Calibration of Manganin under Shock Loading," J. Appl. Phys., 51, (4), 1980.
9. S. C. Gupta and Y. M. Gupta, "Experimental Measurements and Analysis of the Loading and Unloading Response of Longitudinal and Lateral Manganin Gauges Shocked to 90 kbar," J. Appl. Phys. 62, (7), 1987.

*This work was performed under the auspices of the U.S. Department of Energy by the University of California, Lawrence Livermore National Laboratory under Contract No.W-7405-Eng-48.

DIAMETER EFFECT CURVE AND DETONATION FRONT CURVATURE MEASUREMENTS FOR ANFO*

R. A. Catanach & L. G. Hill

Los Alamos National Laboratory
Los Alamos, New Mexico 87545 USA

Diameter effect and front curvature measurements are reported for rate stick experiments on commercially available prilled ANFO (ammonium-nitrate/fuel-oil) at ambient temperature. The shots were fired in paper tubes so as to provide minimal confinement. Diameters ranged from 77 mm (≈ failure diameter) to 205 mm, with the tube length being ten diameters in all cases. Each detonation wave shape was fit with an analytic form, from which the local normal velocity D_n, and local total curvature κ, were generated as a function of radius R, then plotted parametrically to generate a $D_n(\kappa)$ function. The observed behavior deviates substantially from that of previous explosives, for which curves for different diameters overlay well for small κ but diverge for large κ, and for which κ increases monotonically with R. For ANFO, we find that $D_n(\kappa)$ curves for individual sticks 1) show little or no overlap—with smaller sticks lying to the right of larger ones, 2) exhibit a large velocity deficit with little κ variation, and 3) reach a peak κ at an intermediate R.

INTRODUCTION

Rate stick tests on lightly confined prilled ANFO at ambient temperature were conducted to measure steady-state detonation velocity and detonation front curvature as a function of diameter. It is well known that ANFO's detonation velocity varies considerably with charge diameter. Yancik[1], Petes[2], and others have shown that several other physical properties can affect ANFO's detonation velocity. These include the type of confinement, explosive density, particle size and distribution, fuel oil content, moisture content and temperature. Furthermore, these factors can be interdependent. Differences in detonation velocity can also be attributed to slightly different physical properties of the ammonium nitrate (AN) produced by different manufacturers. Because there are no standard specifications by which AN is produced, batch-to-batch variation from a single production plant is not uncommon. Although performance data for ANFO have been studied for over four decades and become somewhat self-consistent since being introduced in the mid-1950's, variability in performance can still be expected from this highly non-ideal explosive.

Near critical diameter, ANFO exhibits a severe velocity deficit; however, with increasing diameter a slight upturn in the diameter effect curve is observed. We developed an extension of the widely used Campbell/Engelke[3] curve fit that accurately captures both behaviors together.

The detonation velocity of a given explosive is also related to the detonation front curvature[4], which is the basis of Detonation Shock Dynamics (DSD) theory[5]. Very little such data exists on ANFO. A study by Sandstrom[6] on an ANFO rate stick confined in polyvinyl chloride showed a jagged, saw-toothed detonation wave front. This is expected due to the large granular features of the ANFO prills. In this study we measured front curvature of several ANFO rate sticks of varying diameters in paper tube confinement, and discovered a similar wave front pattern. Each wave front was analyzed in a manner similar to those of previous PBX 9502[7] and nitromethane[8] tests. The generated $D_n(\kappa)$ curves indicate that ANFO exhibits a behavior that is significantly different from other explosives.

---

* Work supported by the United States Department of Energy.

EXPERIMENT

Our ANFO was composed of commercial (Titan Energy, Lot No. 30SE99C) explosive grade AN prills, with 6 wt.% diesel fuel. Separate bags were well blended prior to firing, to obtain consistent samples and to minimize batch-to-batch variability.

Twelve rate sticks were fired at ambient temperature in thin wall (6 mm) paper tubes. Seven diameters were tested between 77 and 205 mm inner diameter (ID). Three cases were successfully repeated to obtain duplicate front curvatures. The length-to-diameter (L/D) ratio was 10 in all cases. The charges were fired in a vertical position to 1) accommodate top loading, 2) record detonation front curvature at the bottom, and 3) obtain an axially symmetric density distribution to minimize wave tilt. A schematic diagram is shown in Fig. 1.

Each charge was loaded in ten separate lifts —except for the 205 mm stick, which required twenty lifts—to attain a uniform density throughout the charge. Individual lifts were weighed prior to pouring. During the pour of each lift, the outside of the tube was gently tapped. The top of each lift was lightly tamped with a flat-bottomed plunger to provide a level prill distribution between lifts. The rise height of each lift was measured and the density per lift calculated. This procedure gave a bulk density of approximately 0.90 g/cc. The shots were boosted using pressed PBX 9501 cylinders. These had the same ID as the tubes, and a L/D ratio of 1/2.

The detonation velocity was measured using eleven self-shorting capped shock pins glued into holes drilled at equal intervals along the tube. The shorting pins were inserted flush with the ID of the tube starting four diameters from the top of the rate stick. The eleven shorting pins were connected to a Los Alamos DM-11 pin board and multiplexed to a single cable so that when the detonation wave shorts the pin, it fires an RC circuit in the pin board, producing a short voltage pulse.

To measure detonation front curvature, the rate sticks were capped at the bottom with a PMMA window. A small strip of PETN paint was applied across the diameter of the inside window surface; this served as a flasher. It was applied in a thin layer and covered with copper tape to block reaction prelight from the detonation front, which scatters 20 to 30 mm ahead of the wave front.[9] Front curvature was recorded with a rotating mirror streak camera at a writing speed of 1 mm/μsec.

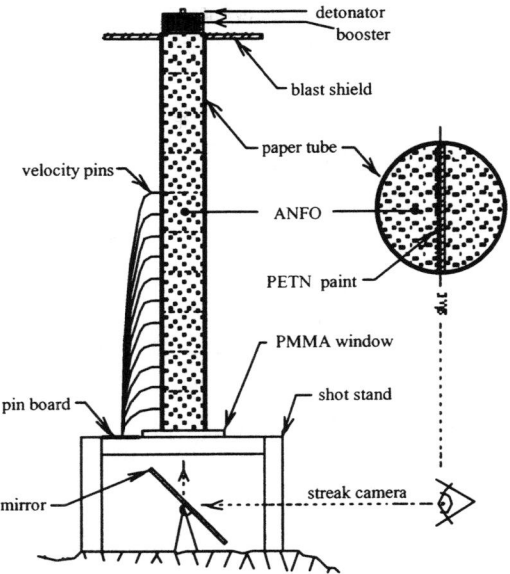

Figure 1. Schematic diagram of ANFO rate stick experiment.

ANALYSIS

A linear least squares fit was applied to the x-t pin data to obtain the axial detonation velocity, D_0, for each rate stick. Of the seven diameters tested, five were repeated to investigate velocity repeatability. For all the tests, a minimum of nine pin signals were recorded. The average standard velocity error for all the tests, excluding the test that reached failure diameter, was ± 2.7 m/s. The velocity for the test at failure diameter was recorded in the process of failing, with the first ten of eleven pins reporting.

The film records were read on an optical comparator from edge to edge. The center was then computed as the average of the left and right edges, and the two sides were partitioned. The magnification was measured from either a scale taped adjacent to the PETN strip, or a still exposure of the tube ID.

The processed film data gives breakout time t as a function of the radius r. When the two sides are overlaid the data generally do not overlay to within experimental noise. The reason is wave tilt. We applied a linear tilt correction (i.e., for the tilt component in the slit direction) based upon a linear least squares fit to the time difference, as a function of r, between the left and right data sets. When tilt-corrected records are overlaid, the data scatter from both sides combined is comparable to that for each side alone. Inferred tilt angles are shown in Table 1.

TABLE 1. Diameter Effect and Wave Front Data (Mean $T_0 \approx 19°C$ and Relative Humidity $\approx 30\%$)

Charge Diameter $\varnothing$ (mm)	Charge Density ρ (g/cc)	Detonation Velocity D_0 (mm/µs)	Fit Coefficient a_1 (mm)	Fit Coefficient η	z/R Max.	Edge Angle (degrees)	Tilt Angle (degrees)	Shot Number
77	0.91	1.68	*	*	*	*	*	3850
90	0.89	2.41	25.32	0.7126	0.4694	52.6	0.53	3849
90	0.93	2.47	—	—	—	—	—	3851
102	0.89	2.79	36.96	0.6389	0.4689	49.9	0.93	3853
102	0.88	2.82	—	—	—	—	—	3846
115	0.88	2.99	37.33	0.6522	0.4221	47.4	0.21	3847
115	0.92	3.01	40.50	0.6504	0.4573	49.6	0.26	3852
128	0.92	3.18	47.76	0.6270	0.4424	47.9	0.20	3854
128	0.92	3.22	41.64	0.6641	0.4484	49.5	0.03	3855
153	0.90	3.46	52.58	0.6425	0.4337	47.8	0.14	3857
153	0.91	3.51	54.01	0.6358	0.4314	47.5	0.18	3848
205	0.91	3.92	64.93	0.6471	0.4068	46.1	0.34	3856
5300	≈ 0.88	4.94	—	—	—	—	—	(2)
9100	≈ 0.91	5.26	—	—	—	—	—	(2)

* failed

Given that a detonation in a stick approaches a steady traveling wave after propagating a few diameters, the wave shape z as a function of r is

$$z(r) = D_0\, t(r). \quad (1)$$

The $z(r)$ data is fit to the series

$$z(r) = -\sum_{i=1}^{n} a_i \left(\ln\left[\cos\left[\eta \frac{\pi}{2} \frac{r}{R} \right] \right] \right)^i, \quad (2)$$

where R is the charge radius and a_i and η are fitting parameters, with $0 \leq \eta \leq 1$ controlling the curvature near the edge. One should not use a higher order fit than necessary, or spurious wiggles will occur. In this case (cf. 7,8), only the first term was necessary to fit wave shapes to within random scatter, as shown for Shot #3856 in Fig. 2. The granular nature of the ANFO produced some outlier points that were dropped if the standard error improved by doing so.

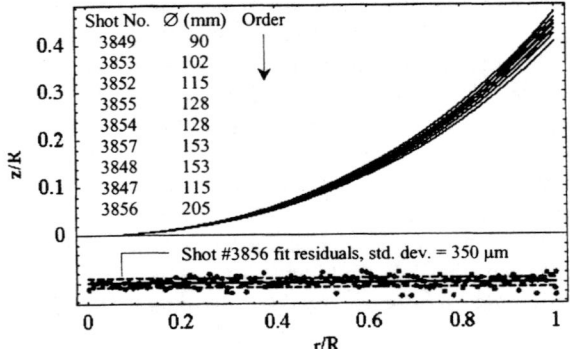

FIGURE 2. Wave shape curve fits and fit residuals (#3856).

RESULTS

Figure 2 shows that—with the exception of Shot #3847—wave profiles normalized by R are progressively flatter for larger charges than for smaller ones. This is the typical and expected behavior.

The diameter effect data are listed in Table 1. Two data points from the *Dice Throw* program of large unconfined ANFO charges that were fired at a similar density are included from Petes[2]. Diameter effect data have traditionally been fit using the form[3]

$$\frac{D_0}{D_\infty} = 1 - \frac{A}{R - R_c}, \quad (3)$$

where D_∞ is the Chapman-Jouguet velocity, and A and R_c are fitting parameters. We prefer to express Eq. 3 in "diameter effect" coordinates as

$$\frac{D_0}{D_\infty} = 1 - \frac{\beta k}{k_d - k}, \quad (4)$$

where $k = 1/R$, $k_d = 1/R_c$, and β is a dimensionless fitting parameter. Our extension involves replacing unity by the term $\dfrac{k_u - \alpha k}{k_u + k}$, so that

$$\frac{D_0}{D_\infty} = \frac{k_u - \alpha k}{k_u + k} - \frac{\beta k}{k_d - k}, \quad (5)$$

where Eq. 4 is recovered if $\alpha \to -1$. Approximately, k_u and k_d characterize the position of the upturn and downturn, respectively, while α and β are dimensionless shape factors.

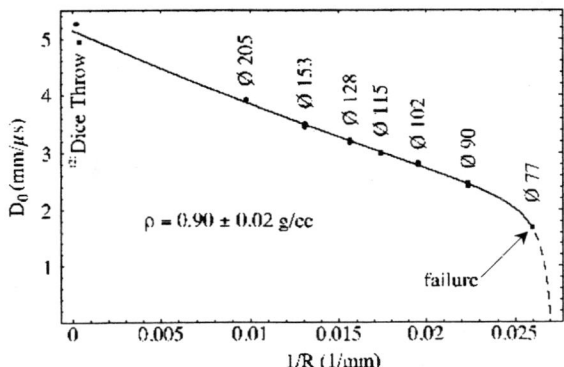

FIGURE 3. ANFO diameter effect data and curve fit.

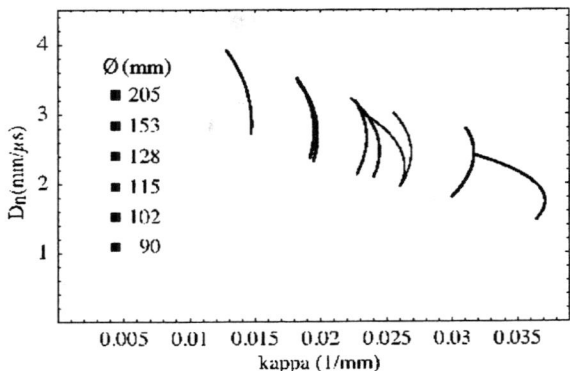

FIGURE 4. ANFO $D_n(\kappa)$ curves.

Diameter effect data, with Petes' included, and the fit to Eq. 5 are plotted in Fig. 3. Curve fit parameters are: $D_\infty = 5.1464$ mm/μsec, $k_u = 0.09526$ mm^{-1}, $k_d = 0.02745$ mm^{-1}, $\alpha = 1.6100$ and $\beta = 0.006523$.

The $D_n(\kappa)$ curve is derived from D_0 and the slope $s(r)$ of the wave profile $z(r)$:

$$s(r) = \frac{dz(r)}{dr} = \tan\theta(r), \quad (6)$$

where $\theta(r)$ is the local angle between the charge axis and the normal direction to the wave front. The normal detonation velocity, D_n, is $D_0\cos\theta(r)$, or

$$D_n(r) = \frac{D_0}{\sqrt{1+s(r)^2}}. \quad (7)$$

Total curvature, κ, is the sum of two terms:

$$\kappa(r) = \frac{s'(r)}{(1+s(r)^2)^{3/2}} + \frac{s(r)}{r\sqrt{1+s(r)^2}}. \quad (8)$$

$D_n(\kappa)$ is generated by plotting $D_n(r)$ versus $\kappa(r)$. $D_n(\kappa)$ curves for the nine charges with front curvature records are shown in Fig. 4.

Unlike $D_n(\kappa)$ curves for PBX 9502[7] and nitromethane[8], ANFO $D_n(\kappa)$ curves 1) show little or no overlap, with smaller sticks lying to the right of larger ones, 2) exhibit a large velocity deficit with little κ variation, and 3) reach a maximum κ at an intermediate R. The $D_n(\kappa)$ curve for the 205 mm charge is nearly vertical, but the 90 mm curve is much flatter—perhaps because this charge is close to failure. Comparing the three sizes for which front curvatures are repeated, the 153 mm charges show excellent repeatability, while the 128 and 115 mm charges show the same general behavior, but differ from each other in detail. Nevertheless, the three listed characteristics seem to be real and consistent features of the data that deviate substantially from those of more ideal explosives, for which curves for different diameters overlay well for small κ but diverge for large κ, and for which κ increases monotonically with R. The behavior of Fig. 4 is compatible neither with existing DSD implementation nor with current extensions[5]; however, it provides guidance for further extensions that consider the DSD consequences of disparate reaction rates[10].

ACKNOWLEDGEMENTS

We thank E. Aragon, R. Archuleta, M. Chavez, R. Critchfield, J. Keddy, D. Kennedy, and D. Murk for design/test support; R. Hopler and F. Sandstrom for data support; and R. Flesner for funding support.

REFERENCES

1. Yancik, J.J.; PhD Dissertation, Univ. of Missouri, 1960.
2. Petes, J., Miller, R., & McMullan, F.; DNA Technical Report No. DNA-TR-82-156, 1983.
3. Campbell, A.W., & Engelke, Ray; 6th International Detonation Symposium, 1976, pp. 642-652.
4. Wood, W.W., & Kirkwood, J.G.; J. Chem. Phys., 22, 1954, pp. 1920-1924.
5. Aslam, T.D.; Bdzil, J.B.; & Hill, L.G.; 11th International Detonation Symposium, 1998, pp. 21-29.
6. Sandstrom, F.W., Abernathy, R.L., Leone, M.G., & Banks, M.L.; 28th DOD Expl. Safety Board. Sem., 1998.
7. Hill, L.G., Bdzil, J.B., & Aslam, T.D.; 11th International Detonation Symposium, 1998, pp.1029-1037.
8. Hill, L.G., Bdzil, J.B., Davis, W.C., Engelke, R., & Frost, D.L.; Shock Compression of Condensed Matter, 1999, pp. 813-816.
9. Kennedy, D.L.; private communication, 2000.
10. Short, M., & Bdzil, J.B.; Submitted to J. Fluid Mech.

EXPERIMENTAL INVESTIGATION OF HETEROGENEOUS HE DECOMPOSITION MECHANISM IN DETONATION WAVE FRONT

A.V.Fedorov

Russian Federal Nuclear Center, VNIIEF, 607190, Sarov, Russia

Abstract. The mechanism of decomposition of heterogeneous HE in hot spots is considered. Experimental investigations have shown that the main reason of formation of locally heated areas is the existance high-velocity microjets, which penetrate HE layers laying ahead. Chemical reaction, thus, can begin in hot spots already before detonation wave arrival. Different structures of detonation wave and different profiles of particle velocity in the chemical reaction zone can be explained by jets penetration. There is also discussion on the relation of von Neumann spike and Chapman-Jouguet state for high-density condensed HE.

INTRODUCTION

In recent years the methods with high time resolution are used to study the detonation wave structure and chemical reactions zone (CRZ) of condensed HE [1-8]. We have performed studies of detonation front of liquid homogeneous and solid heterogeneous HE [6-8]. The model of hot spots, which explains the mechanism of decomposition of heterogeneous HE in detonation wave, is well known. Voids collapse and occurrence of hot area, as well as friction between HE particles, shifts, and dislocations are now considered as the basic reasons of hot spots formation. However, already in the 40's A. Apin wrote [9] that detonation of heterogeneous HE propagating by there is penetration of reaction products into HE layers laying ahead, and velocity of these products is higher than detonation velocity.

EXPERIMENTAL

The experimental set-up is depicted in Fig.1.
Fabry-Perot laser interferometer records particle velocity profile at the HE-LiF window interface. To obtain good time resolution, aluminum covering (≈ 1 μm) is placed between HE and window.

Shock wave passes it in ≈ 0.15 ns. Laser spot size at the surface is ≈ 100 μm.

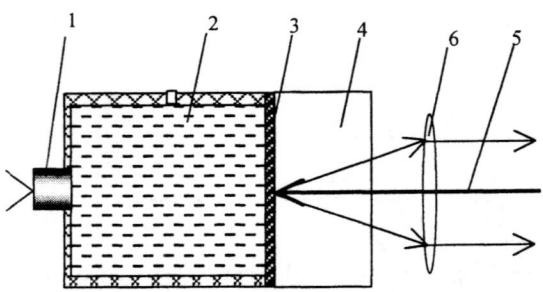

FIGURE 1. 1.-detonator; 2.-liquid HE in cylindrical cell or sample of solid HE; 3.-Al covering (0.2-1.5 μм)) or Al foil (5-10 μм); 4.- LiF crystal; 5.-laser beam (λ=694.3 nm); 6.-focusing lens.

HE is initiated by a divergent shock wave. Pressure of this wave in aluminum cap of electric detonator is 22 GPa.

Von Neumann spike of high-sensitive HE is formed in this case on thicknesses close to value of critical diameter [6, 7].

910

EXPERIMENTAL RESULTS

Earlier in [6, 7] two types of profiles of detonation wave were recorded at the HE-LiF interface: smooth concave profiles falling from von Neumann spike and protuberant profiles with decelerating oscillations of velocity. The profiles with velocity oscillations are depicted in Fig.2.

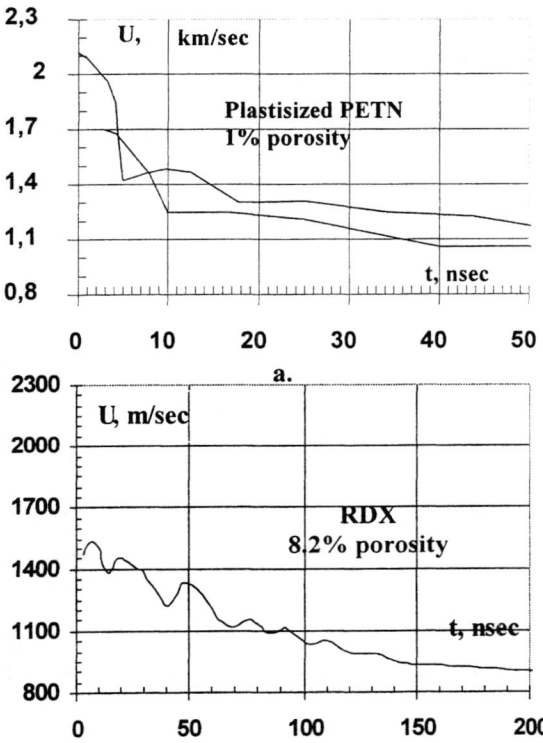

FIGURE 2 Profiles of particle velocity with decelerating oscillations

In this paper we present an experimental proof that the formation of such profiles is caused by microjets which occurred, when the detonation front interacts with voids existing in a heterogeneous HE. When such microjets outstrip detonation front and decelerate in unreacted HE in immediate vicinity of surface to be recorded (HE-LiF interface), the laser interferometer records perturbations caused by jet in profile of detonation wave.

Usually, in our experiments, we place only thin aluminum covering (≈ 1 μm) between HE and LiF window. In these experiments either different profiles of detonation wave are recorded, or the record is broken for very short time (≤ 1 ns).

The majority of experimental facts testify that Al covering and LiF layer adjacent to it are destroyed by microjets and the recording is broken. The following experimental facts are confirming that the microjets are moving with a velocity higher than the detonation velocity, penetrate into the unreacted HE and are the main reason for formation of locally heated areas (hot spots):

1. In a series of experiments (see Fig.1), laser interferometer records have shown that even before the detonation wave the arrival or before Al covering is broken, the individual HE - window interface starts to be smoothly accelerated. Time of such record is ≈ 5-7 ns, the maximum velocity before break is ≈ 100 m/s. Approaching particles of the microjets are, most probably, the reason of such an acceleration.

2. Figure 3 shows schematically the results of one of the model experiments, with conical cavities (simulating voids in HE) 0.3 - 4 mm deep and vertex angle of 30 - 60 degrees in plastisized composition of PETN. HE was in contact with to an aluminum plate. After the explosion, parameters of the cavity formed in the Al plate were measured. Sizes of the cavity formed in the Al plate are approximately equal to sizes of the conical hollow in the HE. In the case, when the cavity is filled with glue (which simulated the binder), size of the hollow in the Al plate approximately 3-4 times smaller cavity. If there was a small conical cavity in the Al plate, the size of this conical cavity was 3-4 times larger because of explosion products.

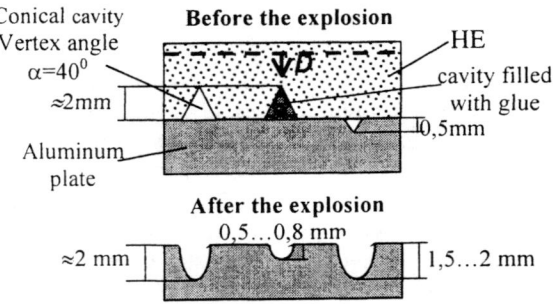

FIGURE 3 Model experiments with conical cavities.

Thus, the experiments have shown that the jets make a small cavity in the Al plate, and the subsequent effects of explosion products increase its sizes. In the case, when the jets penefiole into the unreacted HE rather that into the metal, this effect will be stronger due to less density and strength of HE.

3. In experiments with conical cavities 2-4 mm high(see Fig.3) we also made a measurement of the jet velocity using time of arrival of DW to the top of the conical cavity and time of the jet approach to Al foil 100 μm which was in contact with LiF. Under the jet effect the Al-LiF boundary is accelerated up to a velocity of about 100-400 m/s for several nanoseconds. Then the recording is broken because of the breakage Al foil. Velocity of the jet head was 9-12 km/s and higher. The detonation velocity was 7.8 km/s.

4. Tests with various HE's were performed according to the scheme presented in Fig.1 using Al covering (≈1μm). Break of interferometric recording of the HE-LiF interface velocity occurs as a result of fracture of the reflecting covering by microjets. In homogeneous liquid HE the recording breaks occur very seldom, in high-density HE (1-2% of porosity) the breaks occur in ≈30 % of the cases. Number of breaks grows sharply with the increase in HE porosity.

5. Velocity oscillations (see Fig.2) were recorded in profiles of the detonation wave Similar profiles are recorded in [1] for HMX with porosity of 35%. Average size of HE particles of ≈10 μm cause velocity oscillations of 300-400 m/s, HE particles sizes of 120 μm – cause velocity oscillation of 80-1000 m/s [1]. The recorded value of spike in window material decreases in comparison with high-density HE (≈1.5 % of porosity) from 3.6 mm/μs up to 2.2 mm/μs and lower (i.e. by 40 % and more).

6. For HE (HMX, RDX) having maximum density and minimum porosity (1-2 %), detonation wave passes the voids (of 10-20 μm) in times of about 1-2 ns [14]. Duration of chemical reactions for these HE's is about 50 ns [3]. The jets overtake the detonation wave front. HE in these jets is, probably, in the decomposition stage, since jet is formed by an HE, which is in the CRZ. On the other hand, chemical reaction is not yet completed. Hence, the jets carry forward ions, radicals, intermediate and final products of reactions. Thus, chemical reaction can start in hot spots already before detonation wave arrival. Microjets catalyze the process of HE decomposition. It is known that for liquid homogeneous HE (TNM, NM, etc.), as well as for HE monocrystals (HMX, RDX, etc.) the value of critical diameter is 10-20 mm and more. Including 1-2 % of voids in HE reduces critical diameter ten times. According to the Yu. Khariton's principle, the smaller the critical diameter, the smaller is the duration of CRZ. Thus, microjets increase the rate of chemical reactions, reduce duration of the CRZ, and catalyze the process of HE decomposition.

Relation of the von Neumann spike values and the Chapman-Jouguet state.

Last years a series of experimental works was published, where values of the von Neumann spike in condensed HE are measured by methods with high time resolution with good precision [1-8]. The values of the Chapman-Jouguet (C-J) state for these HE are also known. It is of interest to determine the relation between these values.

Many authors, for example, L. Al'tshuler [10] noted that experimental data on C-J pressures of HE is characterized by extraordinary large incoordination. So, for TNT, which is the most investigated HE, at density of 1.63-1.64 g/cm^3, the detonation pressures are found to be between 17.7 and 21.3 GPa. For PBX 9404 compositions, the C-J pressures are changed from 34.5 to 39.16. Craig [11,12] revealed that the effective C-J pressure grows in accordance with detonation wave propagation in HE. Using charges of large diameters (100-300 mm), Craig revealed that the detonation pressure at change of charge thickness from 12.7 mm up to 101.6 mm grows from 30.5 up to 37.5 GPa. The dependence curve of the effective C-J pressure on charge thickness from [12] is depicted in Fig.4. Mader writes that for thicknesses charge within 40-100 mm the C-J pressure is 35-37.5 GPa, and actual value of C-J pressure for unlimited medium is equal to ≈40 GPa. We determined the von Neumann spike pressures for four compositions close to PBX 9404 (HMX, agatized HMX, HMX with 5% and 10% binder). They are 47.8 - 49.5 GPa (average value is 49 GPa). This value of the von Neumann spike is presented in Fig.4. The fact that the value of the von Neumann spike for high-sensitive HE is formed at thickness equal to the critical diameter (d_{cr}), and then remains constant at thicknesses up to 200 d_{cr} and more, is verified by us using plastisized PETN. Thus, for a stationary detonation wave, the

Neumann spike pressure does not depend on charge thickness.

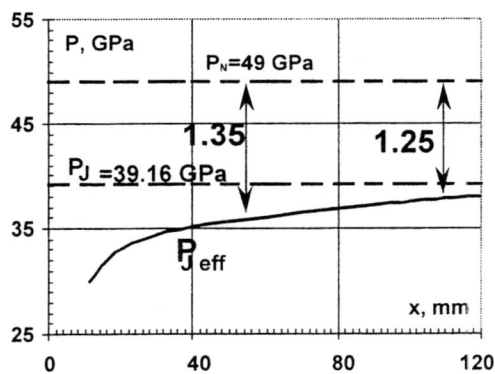

FIGURE 4. Values of the von Neumann spike pressure (P_N), C-J state (P_J), and dependence of effective pressure ($P_{J\ eff}$) on thickness of the charge for plastisized compositions of HMX of the PBX-9404 type.

Evstigneev determined the maximum value of C-J state for plastisized HMX using charges with large diameter and large thicknesses (100-200 mm). This value was 39.16 GPa. Thus, the von Neumann spike exceeds the C-J state 1.25 times. Value of C-J state, which is equal to 36±0.5 GPa, was determined using charges with thicknesses of 40-100 mm. For these charges the average exceess of the von Neumann spike is 1.35 times. The dependence of the detonation pressure on charge thickness explains the fact that for the majority of condensed HEs the excess of the value of the von Neumann spike over C-J state is usually in the interval between 1.25-1.4 and more. Shvedov analyzed in detail the change of the C-J state value versus the charge diameter. He revealed that with diameter growth, this value increases and reaches the maximum value at diameters of more than 100 d_{cr} [13].

Thus, with increase of thickness and diameter of charge, the C-J pressure grows and reaches the maximum value at $d \geq 100 d_{cr}$. With reduction of thickness and diameter of charge, the detonation pressure drops, and the value of excess of von Neumann spike over C-J state grows.

ACKNOWLEDGEMENTS.

The author greatly appreciate the help of A.L. Mikhaylov, I.P. Khabarov, V.M. Bel'sky, L.A. Gatilov, A.V. Men'shikh, D.V. Nazarov, S.A. Finyushin, V.A. Davydov in conduction of this work and for heir helpful advises and discussions.

REFERENCES

1. Gustavsen, R.L., Sheffield, S.A., Alcon, R.R. "Detonation wave profiles in HMX based explosives" Proceedings of the Int. Conf. Shock Compression of Condensed Matter, pp.739-742. Amherst, 1997.
2. Tarver, C.M., Breithaupt, R.D., Kury, J.W., *J.Appl.Phys.* **81**(11), 7193 (1997).
3. Lubyatinsky, S.N., Loboiko, B.G. "Detonation reaction zones of solid explosives," Proceedings of XI Symposium on Detonation, Snowmass, USA, 1998.
4. Tarver, C.M., Kury, J.W., Breithaupt, R.D. "Detonation Waves in Triaminotrinitrobenzene", *J. Appl. Phys.* **82**(8), 3771 (1997).
5. Green, L.G, Tarver, C.M, Erskine, D.J. Proceedings of XI Symposium on Detonation, Snowmass, USA.1998.
6. Fedorov, A.V., Menshikh, A.V., et al. "Detonation Front in Homogeneous and Heterogeneous High Explosives", *Proceedings of the Int. Conf. "Shock Compression" of Condensed Matter-1999*, Snowbird, USA, 1999.
7. Fedorov A.V., Menshikh A.V., Yagodin N.B. // Chemical Physics (Russian), №11, pp.64-68, 1999.
8. Fedorov, A.V. Paper for International Conference "Shock Waves of Condensed Matter", St. Petersburg. 2000.
9. A.Ya. Apin. On detonation and explosive combustion of HE. Papers of USSR Acad. Of Sciences, (Russian), v. **50**, pp.285-288 (1945).
10. Shock waves and extreme states of substance. (Russian) Under edition of V.E. Fortov et al. – Moscow: Nauka, 2000.
11. Mader, Ch.L., Craig, B.G. Nonsteady-state detonations in one-dimensional plane, diverging and converging geometries. - Los Alamos Sci.Lab.rep. LA-5865, 1975.
12. Mader Ch.L. Numerical modeling of detonation. – University of California Press,1979.
13. K.K. Shvedov. *Physics of combustion and explosion, (Russian)*, №4, 1988.
14. Demol, G, Goutelle, J.C., Mazel, P. // Proceedings of the Int. Conf. Shock Compression of Condensed Matter, pp.353-356. Amherst, 1997.

EXPERIMENTAL AND NUMERICAL STUDY OF TEMPERATURES IN CAVITY COLLAPSE

A.M. Milne[1] and N.K. Bourne[2]

[1] *Fluid Gravity Eng. Ltd., St. Andrews, Fife KY16 9NX, UK.*
[2] *Royal Military College of Science, Cranfield University, Shrivenham, Swindon SN6 8LA, UK.*

Abstract. The temperature of the gas enclosed in a cavity collapsing as the result of the passage of a shock wave has been the subject of much previous interest. We consider the ability of this collapse to ignite an explosive medium in which the cavity is placed. Both jet impact and hot gas ignition mechanisms are considered. A series of experiments have been conducted in which a cylindrical cavity has been collapsed under shock. This geometry has the advantage of allowing details of the interior gas to be studied. A further series of experiments has been conducted to ally this disc-shaped geometry with a spherical cavity using two novel arrangements. The development of these tests has addressed the temperature increase within the cavity. The gas content can be varied to include chemically reactive gases content so that the combustion can also be used as a temperature diagnostic. For jet impact studies we use nitromethane as the liquid and measure ignition directly. The series of experiments has been coupled with numerical modelling of the multi-material shock interactions as well as the combustion of the diagnostic gases, both in the design and validation stage.

INTRODUCTION

The collapse of a gas void within a reactive material has the potential to start local burning leading to partial reaction or run to full detonation (1). There are three main features of the collapse that provide a means for ignition. The first is the formation of the high-speed jet and elevated velocities in the convergent flow around the wall of the cavity. This gives rise to heating in viscous materials (2). The second is the shock-heated region at the point of jet impact in an asymmetric collapse (3). The third is the compression of the gaseous or vapour content of the cavity (4). These effects have been studied (5) and an experiment, showing examples of each of these modes of heating giving rise to local reaction is described below. Once ignited, a burning front may be quenched or may accelerate so that a transition to detonation may occur according to the confinement of the material.

There is evidence of sensitization of material and also of associated shock-heating of the impacted explosive. Mader has modelled shock-induced hot spots (3, 6). He has shown the process of jets impacting in a local high-pressure area, followed by compression of material followed by heating. His models have not required a gas-filled bubble to achieve the necessary ignition temperatures. This mode was considered by Kang et al. (7). Clearly, in materials with viscosity and/or strength there is also friction between and shear within moving explosives to consider. In a visco-plastic material, work is done at internal interfaces leading to heating and the formation of a melted layer at the edge of a pore at which surface burning may start (8). These effects may be summarized by stating that low viscosity and small bubbles decrease sensitivity.

The size of the cavity will determine which mechanism gives rise to heating that may ignite the matrix. Flow is favoured at the smallest scale, and gas compression and jet impact at the largest. In the following work, larger cavity diameters are considered as they are experimentally easier to visualize.

EXPERIMENTAL

The experiments described were carried out in two and three-dimensional geometries. A disc shaped cavity allowed observations of jet formation, shock and reaction within the cavities. A prepared gel slab with introduced cavity was then clamped between glass or polymethylmethacrylate (PMMA) blocks and plane shock waves were introduced into the sheets by the impact of a flyer plate from one of the RMCS gas guns. In other experiments the shock was introduced using an explosive plane wave lens and a calibrated inert gap. In some experiments a 3 mm thick layer of an ammonium nitrate (AN) emulsion explosive (described in 9) was contained between two, 25 mm thick PMMA blocks. High-speed framing photography at microsecond framing rates recorded the light emitted by the ignited sites. Schlieren was used to visualized the shock in some experiments whilst no external lighting was used for the AN experiments where reaction occurred.

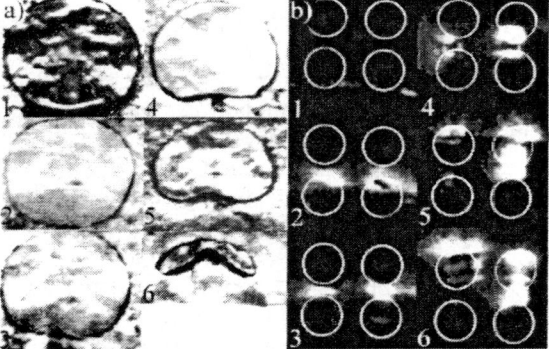

FIGURE 1. In a), a 12 mm cavity collapses in gelatine. In b), an array of four 5 mm cavities (in two rows of two) collapses under a shock of gap pressure 8 GPa.

Figure 1 a) shows six frames taken from a sequence in which a 12 mm cavity containing air is collapsed by a 0.3 GPa shock. In frame 1 the incident shock has entered the sequence from below and has crossed two thirds of its diameter. An air shock can be seen travelling away from the downstream wall at the acoustic velocity of air. The micro-jet begins to form as an instability in frame 2 and subsequently travels across the cavity to impact the downstream cavity wall. Frame 5, taken 60 µs after frame 1, is representative of an intermediate stage of the collapse where the air shock has reflected from the downstream wall, returned, reflected from the involuted upstream wall and travelled back across the cavity. After jet impact, two lobes of compressed gas are trapped in the closure. As the jet penetrates the downstream wall, a pair of linear vortices subsequently form and travel downstream in the following flow.

Figure 1 b). shows the collapse of a 2x2 square array of cavities (which are all 5 mm in diameter) by a shock of amplitude 8 GPa entering an emulsion. The initial cavity positions are shown as white circles and the interframe time is 1 µs. The shock enters from below and runs over the rear of the cavities. There is a small amount of light emission from the bubble rear walls when the shock meets the first row. The voids close between frames 1 and 2 and the material ahead of each can be seen reacting in frame 3. The sites produced by the first row still react with that to the right persisting for the longest time but no accelerating reaction occurs. In frame 4 the second row of cavities starts to collapse with the burning site to the right immediately ahead of the point of jet impact. The bulk of the explosive ahead of the cavity reacts in frame 6. The diffracted shock can clearly be seen as a dark band to the rear of the left-hand site. In the sequence presented, hot-spots have been formed, but the criterion for a propagating reaction has not been met so that the sites die. This may be due to their short duration and to the lack of adequate confinement.

NUMERICAL

This study considers low (*ca.* 0.3 GPa) and high pressure (*ca.* 3 GPa) shock impact on cavities of order 10 mm in diameter (relatively large for explosives). In the lower pressure regime the temperature of the gas in the collapsed cavity is deemed to be the main concern whilst in the higher pressure regime jet impact is seen as the more important feature when considering the likelihood of explosive ignition. The geometry of the problem is shown in fig. 2.

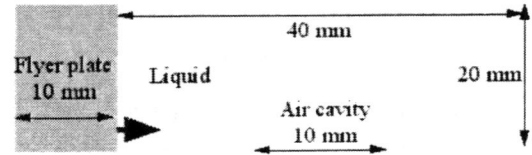

FIGURE 2. Schematic of model system.

Details of the behaviour of this system have been modelled using a 2D, multi-material Eulerian hydrocode. A variety of material models are available and these are introduced in the sections below.

FIGURE 3. a). Closure and jet penetration for the low pressure collapse of a 12 mm diameter cavity shocked to 0.3 GPa. In b). the simulation tracks both shocks and the cavity walls.

Fig. 3 a). shows details of the final stage of the collapse of the 12 mm cavity of Fig. 1 a). collapsing under a shock of magnitude 0.3 GPa. As will be seen, the code predicts the form of the walls and the interface between gas and liquid well. The position of shocks is also shown leading to regions of high pressure and temperature.

Other experimental work has shown the presence of luminescence from the enclosed interior gas soon after the jet impacts (9).

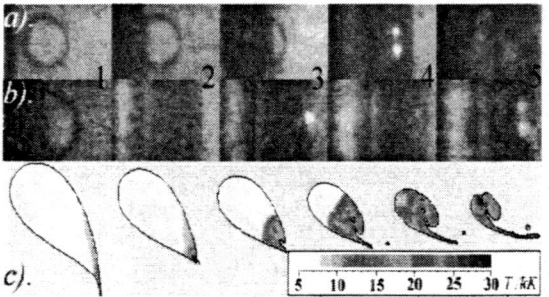

FIGURE 4. Collapse of 6 mm cavities under 2 GPa shock waves. The first frames show initial cavity positions. In a). interframe time is 1 μs, whilst it is 0.2 μs in b). In c) temperature distribution in lobes of gas is shown.

Fig. 4 shows the collapse of a 6 mm cavity by a 2 GPa shock photographed at two different framing rates (interframe times of 1 μs for a. and 0.2 μs for b.). Frame 1 of each sequence shows the initial cavity position. There are two bright flashes of light in frame 4 of the slower sequence which are shown in greater temporal detail in b). First, a single flash occurs in frame 3 that fades until the two regions of luminescence are seen in frame 4. It is interesting to note the similarity with sonoluminescence where the source of light is believed to be weak bremsstrahlung radiation.

To explain these effects, different approximations for the equation of state of air have been investigated. Ideal gas calculations show local gas temperatures peak at around 4000 K for an impact speed of 200 m s^{-1}. It is expected that air chemistry might be important under these conditions and thus the calculation is repeated with a seven-species air model including electron production.

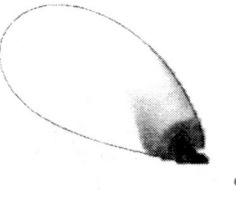

FIGURE 5. Collapsed bubble shape with electron concentration superimposed.

The distribution of electron density at the time of peak temperature is plotted in Fig. 5 and the regions of maximum intensity coincide with the observed light emission sites.

The gas chemistry model in the code was additionally employed to consider reactive gases in the cavity. Strong shocks exist within the bubble and (as Ball *et al.* (10) have recently shown) adiabatic collapse models will significantly underestimate the peak temperatures. The simulation of Fig 4 c). shows the temperature distribution in the lobes at the end of collapse showing high values at a central position at the time of jet impact followed by further elevated values in the trapped lobes at later time as seen in the sequence.

Shock waves of higher pressures will naturally produce higher gas temperatures in the collapsed cavities (*ca.* 35,000K for a steel flyer plate impacting at 1000 m s^{-1}) but these exist only for short periods of the order of 200 ns. Here, the main mode of ignition of a reactive medium is the heating at the point of jet impact in the surrounding fluid. This ignition source occurs well before peak gas temperature is achieved. For a spherical bubble a 100 m s^{-1} plate impact results in a 4000 m s^{-1} jet speed. For a cylindrical bubble this speed reduces to *ca.* 3200 m s^{-1}. The RMCS gun can impact energetic materials and it is planned to fire into nitromethane (NM) containing well-defined cavities. Simulation of these experimental studies requires an Arrhenius kinetics model for nitromethane within the hydrocode. This analysis was fitted by Cook et al. (11) from bullet impact studies and thus is well suited to jet impact.

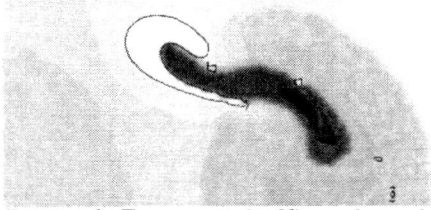

FIGURE 6. Temperature in Nitromethane showing ignition location compared with collapsed bubble.

Fig. 6 shows contours of temperature in the NM near to the state of minimum bubble volume. The darkest shading corresponds to 2600 K and shows the locus of ignition of the NM after jet impact in relation to the collapsed bubble.

Fig. 7 shows qualitative comparison of a). a 6 mm cavity collapsing under an 8 GPa shock in an AN emulsion explosive and b). a simulation of a 10 mm cavity collapsing in nitromethane under the conditions of Fig. 2.

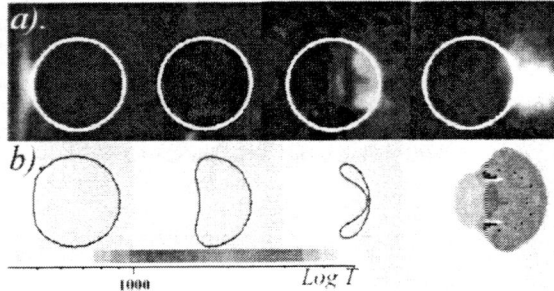

FIGURE 7. Experiment and simulation of large (6 mm) cavity collapsing in an AN matrix. Light is due to reaction. Interframe time 1 μs. b). Temperature field around a cavity collapsing in NM showing similar reaction sites.

The temperature in the cavity reaches high values only towards the end of the collapse and although there is reaction in the vapour within the cavity as the jet is close to impact, it is the temperature and pressures generated at the jet impact site itself that causes ignition of the surrounding fluid. The same feature is seen for the NM simulation. High temperatures of thousands of K are generated in the collapsing cavity but it is the jet impact that gives rise to the main ignition.

SUMMARY

A range of experiments and simulations has illustrated an approach adopted to understand the formation of hot spots in reactive media by the collapse of cavities. Experiments have illustrated features of the shock collapse of large (mm sized) cavities in spherical and cylindrical form to observe details occurring within the cavity. A numerical scheme with suitable representation of liquid and included gas has been developed and has been shown to reproduce the wall geometry, the included gas behaviour (including the electron density consistent with observed luminescence) and the temperature field within the bubble. Further experiments have been described to probe the temperature field that is present with particular regard to the ignition of energetic materials.

REFERENCES

1. Bowden, F.P. and Yoffe, A.D., *Initiation and Growth of Explosion in Liquids and Solids (republ. 1985)*, Cambridge University Press, 1952, pp. Pages.
2. Frey, R.B., in *Proc. Eighth Symposium (Int.) on Detonation*, (ed. J.M. Short), White Oak, Silver Spring, Maryland: Naval Surface Weapons Center, pp. 68-80, (1985).
3. Mader, C.L. and Kershner, J.D., in *Eighth Symposium (International) on Detonation*, (ed. Albuquerque: pp. 42-52, (1985).
4. Chaudhri, M.M. and Field, J.E., *Proc. R. Soc. Lond. A* **340**, 113-128 (1974).
5. Bourne, N.K. and Field, J.E., *Proc. R. Soc. Lond. A.* **435**, 423-435 (1991).
6. Mader, C.L. and Kershner, J.D., in *proceedings of the Ninth Symposium (International) on Detonation*, Portland, Oregon: pp. 693-700, (1989).
7. Kang, J., Butler, P.B. and Baer, M.R., *Combustion and flame* **89**, 117-139 (1992).
8. Carroll, M.M. and Holt, A.C., *J. Appl. Phys.* **43**, 1626-1636 (1972).
9. Bourne, N.K. and Field, J.E., *Phil. Trans. Proc. R. Soc. A.* **357**, 295-311 (1999).
10. Ball, G.J., Howell, B.P., Leighton, T.G. and Schofield, M.J., *Shock Waves* **10**, 265-276 (2000).
11. Cook, M.D., Haskins, P.J. and Stennett, C., in *Proc. Eleventh Symposium (Int.) on Detonation*, (ed. J.M. Short and D.G. Tasker), Arlington, Virginia: Office of the Chief of Naval Research, pp. 589-598, (1998).

DETONATION PHENOMENA OF PBX MICROSAMPLES

I. Plaksin, J. Campos, J. Ribeiro and R. Mendes

Laboratory of Energetics and Detonics, Mech. Eng. Dep., Faculty of Sciences and Technology, University of Coimbra, Polo II, 3030 Coimbra, PORTUGAL

Abstract. Detonation study of PBX micro-samples, based in HMX with an inert (HTPB, epoxy) or energetic (GAP) binder was performed on the meso-scale level, using the multifiber optical probes of 50 μm of maximum resolution, connected directly to a fast electronic streak camera with 0.6 ns resolution. The direct 2D observation of particle to particle successive transition of transmitted shock wave, through the binder, allows to analyse and to discuss, not only the cooperative formation of a multihead detonation front (DF), in the collection of particles surrounded by binder, but also the synenergetic effect, behind the DF, by the appearing of dissipative structures drawing spatial and temporal DF oscillations.

INTRODUCTION

Phenomena of pulsing detonation of PBX was originally mentioned [1-3] in the experiments, carried out on the meso-scale level, as a result of the application of high resolution multi channel optical method, based on optical fiber strip [1-3]. Phenomena of pulsing detonation, in chemically reacted media, implies that the initially smooth shock front, induced by the external source, losses its stability and it is followed by the origination of the oscillating structure [OS], behind the forward front [FF]. OS appears as a result of the inter-influence, in the strongly non equilibrium zone within the FF and chemical reaction zone [CRZ], of a few relaxation phenomena: kinetic relaxation involving release of energy (due to the exothermic reactions), thermal dissipation (due to heat/mass transfer) and stress relaxation (due to the existing limit velocity of impulse/shock waves [SW] propagation).

Theoretically it was obtained [4,5] that, independently of the physical nature of the energetic material, but according to the combination of different main relaxation times, in energy release and heat/mass transfer, and under the influence of the fluctuations), it can exist two main different kinds of detonation wave [DW] regimes, showing different unsteady spatial-temporal dissipative structures [DS].

These spatial-temporal DS are consequently the OS with the transversal SW (represented the cellular DW with the longitudinal-transversal DF oscillations) and OS without transversal waves, but with the additional longitudinal shocks (DF as a oscillating giant monocell). Characteristic sizes of DS depend on the EM physical/chemical properties and its rheology. We have detected both of these regimes in PBX [1-3].

The phenomenology of the origination and transformation of naturally unstable PBX detonation regimes has been already described [6], based on the experiments on the meso-scale level. The factor of divergence of the reactive flow behind the FF (specified by front curvature) was found to be the influent parameter determining the regimes of DW instabilities and DS. It is obvious that the increase of the divergence, in the reacting particles flow, will change the rate of relaxation, in strongly non-equilibrium temperature/stress fields, behind the FF. Micro-mechanisms of temperature-stress transfer are very important in the OS lateral phase formation. The kinetic relaxation level, in the process of propagation in crystals of shock reactive wave [SRW], constitutes the governing factor determining the initial phase of OS origination. The existing delay of the energy release, in shocked and reacting crystals (in PBX, after the FF), could imply the changing of process of energy transfer and the pattern of non equilibrium temperature and stress fields, in CRZ.

The presented study concerns the evaluation of the relaxation phenomena of energy release and of the stress fields, associated with SW propagation in explosive crystals, by the direct registration of SW within single crystals and its clusters, under the similar conditions to the PBX detonation. This objective also implies the clarification of the complex pattern of the OS depending not only by particle sizes and its compaction but also by the binder nature (inert or reactive).

EXPERIMENTS

The experiments were carried out on the meso-scale level with the micro and mini samples of PBX, based on HMX crystals surrounded by polymer binders (HTPB, epoxy, GAP) or by water. The multifiber optical probes [MFOP] of the matrix type, with 250 µm of spatial resolution (50 µm maximum, in colimation mode)[7], were connected directly to a fast electronic streak camera (THOMSON TSN 506N). The streak records, with 0.6 ns of maximum temporal resolution, allow the 2D and 3D analysis of SW and DW in PBX micro-samples and HMX crystals. This procedure allows the registration of, not only the emitted irradiation from the front surface (in SW propagation inside the µ-sample), but also the induced stress amplitudes and the front geometry (in thin layers of kapton), from the input and output SW.

RESULTS AND DISCUSSION

Non Monotonous Shock Reaction and Energy Release in Coarse HMX Crystals

Two experiments have been conducted for the direct time registration of the SW propagation in single coarse HMX crystals, surrounded HTPB binder (Figure 1) and by water (Figure 2).

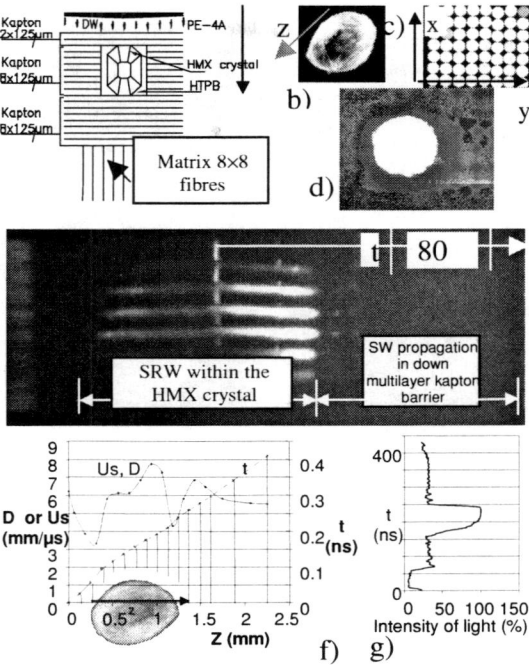

FIGURE 1. Registration on the SW propagation in HMX crystal, surrounded by HTPB binder. [a) experimental set-up; b); c); and d) micro-photos of HMX crystal, matrix MFOP and the MFOP in the background of the crystal; e) photochronogram; f) z-t diagram and velocity of SW before, during and after the crystal; g) histogram of the relative intensity of the light emitted from the central zone of the propagated SW front.

The experimental results, presented in these figures, show the significant effects of the SW propagation inside the HMX crystal: 1 – The non monotonous built up and following decrease of SW velocity D, in order of the successive increasing and decreasing of the crystal cross section, followed by the enhance of the light emission intensity, at the end of the SW run. Pike of D corresponds to 2/3 of SW total run in the crystal; 2 – The delay time of the maximum stress phase, in the down kapton barrier, corresponding to the delay of the pike of energy release.; 3 – Anisotropical effect of the SF propagated through the binder, with the enhanced stress, not in the central, but in the preferential zone of the crystal/binder interface, proved by preferential development of SW in kapton barrier (Figure 2 e)).

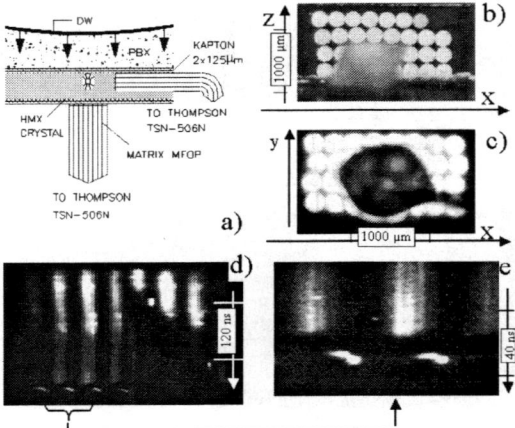

FIGURE 2. Experiment with HMX crystal, surrounded by water. a) experimental set-up, b) HMX crystal and side MFOP, c) HMX crystal and the bottom MFOP, top view d) photo-chronogram showing the pulsing SW propagation inside the crystal, e) detail of the obtained photo-chronogram (bottom MFOP), showing the surface edge effect of the enhanced SW propagation in the peripheral zone of crystal.

The obtained results clarify the mechanism of relaxation of the HMX crystal reaction, induced by strong SW, proving the existing time delay in its energy release, behind the FF. The experimental time relaxation seems to be greater than the time needed for the shock propagation inside the crystal[6]. Future developments will be focused to the quantitative evaluation of the ratio relaxation time vs. ignition delay.

Reactive Waves Propagation, Interaction and Transition in the Ensemble of HMX Crystals

These kind of experiments were performed with a collection of crystals, arranged in the vertical position, related to the initiating SW. Other results of experiments with horizontal clusters of HMX crystals, surrounded by HTPB and by GAP binders, show[3] the significant role of the binders nature in the interaction process between the induced SW in the inter-crystal space. Performed tests show colliding SW's, generating the strong reaction in GAP binder, generate a multi-head front less fluctuated (more homogeneous) that appears with the HTPB binder.

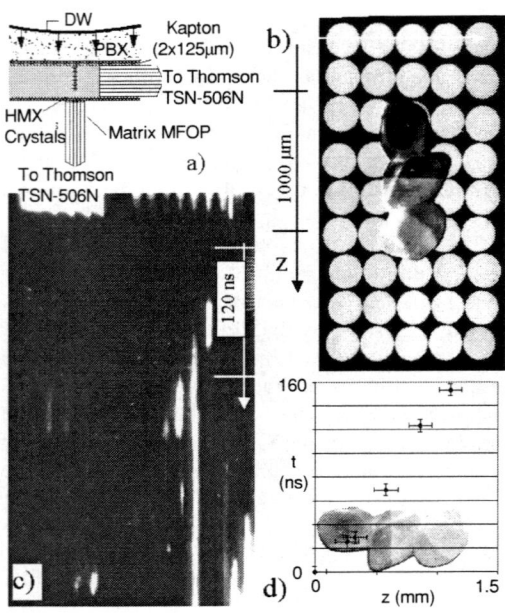

FIGURE 3. Experiments with vertically oriented group of three HMX crystals in epoxy binder. a) experimental set-up; b) HMX crystals and side MFOP; c) photo-chronogram showing the pulsing SW propagation; d) z-t diagram of SW propagation from crystal to crystal.

The results of the experiment with the vertical group of three HMX crystals, in the epoxy binder, are presented in Figure 3, showing the non monotonous (pulsing) process of the SW propagation, inside the crystals, and its transition from crystal to crystal, through the 10-20 µm thick inter-particle space. The mean velocity of SW propagation, inside the crystal's chain is estimated in 7.0 mm/µs, that exceeds in ≈1.4 times the SW velocity in the surrounding epoxy binder (but in ≈1.2 times less than the maximum velocity within the individual crystals).

From the presented results it can be concluded that the defined relaxation effect of the unstable kinetics of shock reaction of a single crystal, accompanied by the surface edge effect, shows the same essential properties that can be observed in the group of crystals, creating the conditions for high order of fluctuations of energy dissipation, behind the FF, followed by formation of the DS by the mechanism of the transversal micro shock interactions.

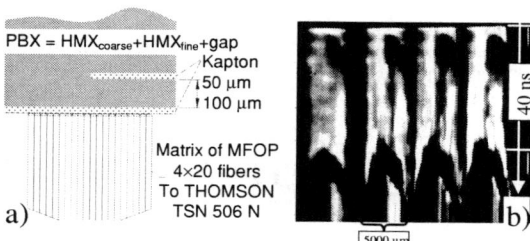

FIGURE 4. Long Channel Test (80×5×5mm, PBX charge in copper confinement) with PBX micro sample in its terminal zone. a) terminal detail of experimental setup; b) photochronogram showing the existing of the cellular structure of DF and OS.

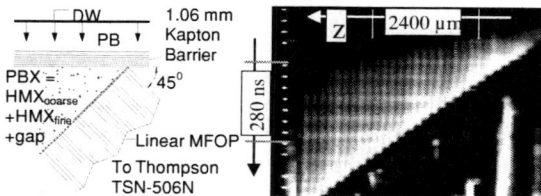

FIGURE 5. Mini Gap Test. a) experimental setup; b) photochronogram.

Experiments with micro and mini samples of PBX

The experiments, demonstrating the self-organization phenomena in PBX detonation, were carried out with µ-samples and m-samples of PBX based on 82 mass percent of HMX (80% of d_{50} = 240 µm, 20% d_{50} = 17 µm) and 18% of an energetic binder (GAP). The results of the long channel test[6] (where DF has a big positive curvature) show the existence of the DS forming the multi-head, or cellular DF, and the process of their origination and development after the DF had propagated through a kapton µ-barrier. The characteristic size of the individual cell is equal to 5-6 d_{50} of coarse HMX particles.

The results of mini-gap test[6] (Figure 5) with the quasi-plane SW input, shows the existence of the longitudinal oscillations of DF, representing in this case a giant mono-cell. Its origin is immediately after the shock run equal to $2\times d_{50}$.

These experimental results confirm the synergetic effects in PBX detonation described in the past[3,6].

CONCLUSIONS

The detonation study of PBX on the meso-scale level has been carried out applying the high-resolution multi-fiber optical technique, in original tests, with single, two and more particles, surrounded by inert and energetic binders.

The obtained results show the kinetic instability in shock reaction of coarse HMX crystals, surrounded by binder, the cooperative formation of a multihead detonation front (DF), in the collection of particles, and also the synergetic effect, behind the DF, by the appearing of dissipative structures with spatial and temporal DF oscillations.

REFERENCES

1. Plaksin, I., Campos, J., Mendes, R., and Góis, J., "Interaction of Double Corner Turning Effect in PBX", Shock Compression in Condensed Matter – 1997, edited by S. C. Schmidt, D. P. Dandekar, and J. W. Forbes, AIP CP 429, New York, 1998, p.p. 755-758
2. Plaksin, I., Campos, J., Mendes, R., Ribeiro, J., and Gois, J., "Pulsing Behaviour and Corner Turning Effect of PBX", in Eleven International Symposium on Detonation, pp. 679-685.
3. Plaksin, J. Campos, R. Mendes, J. Ribeiro and J. Góis, "Mechanism of Detonation Wave Propagation in PBX with Energetic Binder", Shock Compression in Condensed Matter – 1999, edited by M. D. Furnish, L. C. Chhabildas, and R. S. Hixon, AIP CP 505, New York, 2000, p.p. 817-820
4. Daniljenko, V. A., Kudinov, V. M., Dokladi. Akademii Nauk Ukr. SSR, 1982, N. 12, p.p. 24-27
5. Daniljenko, V. A., Dissertation. Institute of Hydrodinamics of Siberian Branch of Academy of Sciences of USSR. 1984.
6. Plaksin, I., Campos, J., Ribeiro, J., and Mendes, R., "Irregularities of Detonation Wave Structure and Propagation in PBX", 32nd International Annual Conference of ICT – Energetic Materials, Ignition Combustion and Detonation, Karlsruhe, July 3-6, 2001, pp. 31-1 to 31-14.

DETONATION MESO-SCALE TESTS FOR ENERGETIC MATERIALS

I. Plaksin, J. Campos, J. Ribeiro, R. Mendes, J. Góis, A. Portugal*, P. Simões* and L. Pedroso*

*Lab. of Energetics and Detonics, Mechanical Eng. Dept. and *Chemical Eng. Dpt., Fac. of Sciences and Technology, University of Coimbra, 3030 Coimbra, Portugal*

Abstract. The objective of the present study is to characterize, on the meso-scale level, the detonation behaviour of PBX based on HMX, based in the minimisation of the test samples of energetic materials up to 10 mg. The development of a non-intrusive, high resolution, optical metrology procedures, using multi-fibber strip, allows the testing of PBX micro-samples, formed by few crystals surrounded by binder, with the simultaneous registration of parameters as local detonation velocity and pressure, geometrical shape of detonation front and the structure of the shock-to-detonation transition zone. The enhanced information allows a better understanding of the processes of formation and propagation of detonation wave. This procedure can be applied to the study of new advanced energetic materials.

INTRODUCTION

The development of new classes of molecules of Energetic Materials [EM] needs new metrology techniques able to test small amounts of EM. Therefore, to minimise sample sizes, and at the same time, to optimise the information collected from each experiment, in order to a deeper phenomenological study, it was the main present challenges of present contribution.

The application of multi-fibber optical systems, developed in LEDAP[1-6], gave the first important results in testing EM (PBX[1-4,6] and propellants[5]) with mini-samples [m-samples] and micro-samples [μ-samples] formed by the collection of few coarse crystals surrounded by binder[3]. It allows the simultaneous registration of two/three parameters such as shock and detonation velocity [U_S,D] and pressure [P_d], detonation front [DF] curvature, sizes of the dark zones in the DF corner turning, shock-to-detonation transition [SDT] phenomena, and Mach Wave formation and attenuation processes.

Characteristic tests of m-samples and μ-samples, with typical mass (m_0) and volume (V_0) less than 2000 mg and 2 cm^3, respectively, have been developed using the multi-fibber optical technique. These characteristic tests are: Long Channel Test[1-3,6] (V_0=0.5 to 1 cm^3), with crystal scale resolution; Conical Tube Test[3] (V_0 = 0.3 to 1 cm^3); Corner Turning Test[1-3,6] (V_0 = 1.5 cm^3); Colliding Tests[4] (V_0 = 1.5 cm^3); mini-Gap Test[3,6] (V_0 = 0.5 cm^3); μ-Gap Tests[3] of crystals, particles and their clusters (V_0 < 2 mm^3 m_0 < 4 mg).

These tests allow to identify the phenomena of pulsing detonation, in PBX, as well as the two main regimes of the unstable DW propagation, in long explosive charges, namely the quasi-periodical longitudinal oscillations (of the continuous DF) and the longitudinal/transversal oscillations (of the cellular DF) and the limits of the existence of the two outlined DW regimes. The obtained results seem to show a strong evidence of the self-organisation establishment in detonation of PBX, behind the DF, where the reaction instability (in shocked energetic particles) and the small fluctuations (through the initial particleheterogeneities) generate a new regime DW propagation, rising to a bigger scale instabilities.

The results of the developed μ-Gap Test, with

the collection of HMX crystals, and the Coaxial Double Charge Test, of small PBX samples, are here presented and discussed, showing the capabilities of multi-fiber optical technique in detonation study on the meso-scale level.

EXPERIMENTS, RESULTS, DISCUSSION

Recording Optical System

A high resolution optical method[1-6] based on 64-90 fibbers strip, connected directly to a fast electronic streak camera (THOMSON-TSN 506 N) was used for registration of the DW propagation in PBX mini and µ-samples, with a maximum temporal resolution of 0.6 ns. The main element of the optical method is a polymeric 64 optic fibber ribbon, with the diameter of each fibber (as well as the inter-axis distance between two adjacent fibbers) equal to 250±1µm. Few variants of Multi-fiber Optical Probes [MFOP] were developed for the direct registration of light emitted by shock or detonation front.

Micro Gap Test

The µ-Gap Test of coarse HE crystals, surrounded by binder, was developed to study the ignition phase of DW formation, as a function of binder and size of crystals. Experimental set up is shown in Figure 1. Four PBX µ-samples (collection of the pairs of coarse crystals, with characteristic sizes between 540 and 740 µm) were placed in individual cells and then surrounded by a composition of 52.3% of the fine HMX crystals

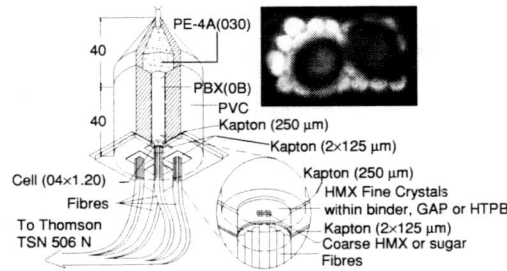

FIGURE 1 – Experimental set-up of µ-Gap-Test with the photo of the multi-fibber optical matrix (MFOP-2) with the collection of two HMX coarse crystals (µ-PBX-2) in its background.

(d_{50}=17 µm) with 47.7% of binder (respectively inert binder HTPB or energetic binder GAP) in order to simulate the typical PBX compositions. Consequently, it was tested simultaneously four different µ-samples of PBX, based on HMX, representing the collection of two coarse crystals (HMX or its inert mimic sugar) surrounded by the mentioned fine HMX crystals (d_{50} = 17 µm) within energetic (GAP) or inert (HTPB) binder (µ-PBX-1 = HMXcoarse +HMXfine + GAP; µ-PBX-2 = HMXcoarse + HMXfine + HTPB; µ-PBX-3 = Sugar+HMXfine+GAP; µ-PBX-4 = Sugar+HMXfine+HTPB).

The photo-chronogram, combined with the photos of µ-PBX-1 to µ-PBX-4 samples, is shown in Figure 2. The streak record shows successively the initial SW input and output, in the upper 250 µm Kapton barrier, the light impulses emitted by the reactive SW during the time t_1 of its propagation into the 1.20 mm thick µ-PBX specimens, and finally, the time intervals of the SW crosses of two Kapton layers placed in the contact with MFOP. The input pressure evaluated from the mean velocity, in the upper Kapton barrier is 9.2 Gpa.

The obtained results show, by the enhanced light radiation emitted by the shock front (Figure 2 and 3), the existence of more intense shock reaction in µ-PBX-1 and µ-PBX-3 (with GAP binder) than in PBX-2 and µ-PBX-4 (with HTPB binder). The time intervals of the shock run t1, for the first pair of these µ-PBX, are 11% less then for the second one, while the maximum values of shock velocity [Us] in Kapton are increased of 11%. The relatively low effect of shock reaction of HMX coarse crystals is proved by a very small difference, in t1 and Us amplitudes, between samples with coarse HMX particles and sugar. The differences between µ-PBX-1 and µ-PBX-3 (Figure 3) seem to show the contribution, in the energy release process, of HMX coarse crystals surrounded by energetic binder GAP and HMX fine crystals. The results also show the bigger effects of the HMX coarse crystals, in the energy release process, in the case of energetic binder GAP (µ-PBX-1) than in the case of inert binder HTPB (µ-PMX-2), proving previous predictions and results[3].

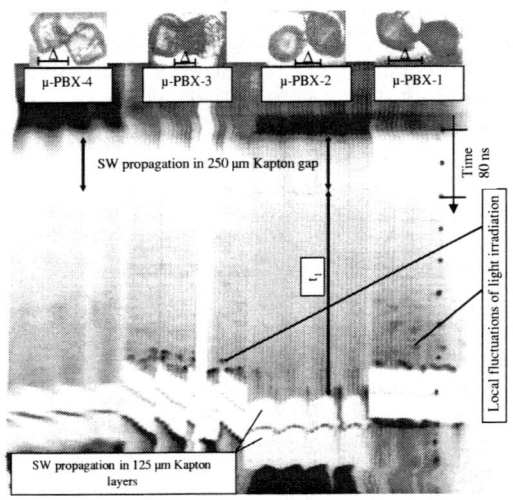

FIGURE 2 – μ-photos of PBX μ-samples (Δ=500 μm) and photo-chronogram obtained in μ-Gap-Test.

In compositions with energetic binder GAP, the local fluctuations of the light irradiation and the heterogeneities are clearly observed, showing the formation of the longitudinal/transversal oscillations in the process of SW propagation.

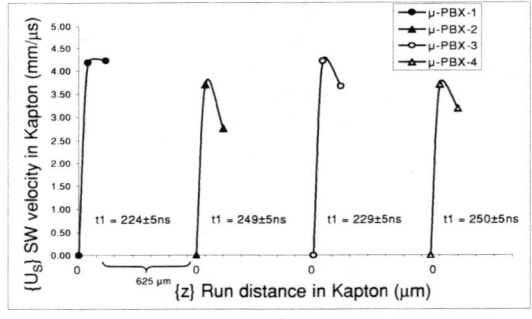

z [μm]	Us (μ-PBX-1) [mm/μs]	Us (μ-PBX-2) [mm/μs]	Us (μ-PBX-3) [mm/μs]	Us (μ-PBX-3) [mm/μs]
0.00	0.00	0.00	0.00	0.00
62.50	4.18 ± 0.10	3.70 ± 0.09	4.21 ± 0.11	3.70 ± 0.09
187.50	4.22 ± 0.11	2.76 ± 0.07	3.66 ± 0.09	3.18 ± 0.08

FIGURE 3 – SW attenuation in down Kapton (2×125 μm) barrier.

Coaxial Double Charge Test

Coaxial Double Charge Test (CDC) was developed for the simultaneous testing of μ-samples of unknown EM, with total mass on the level of 100mg, in combination with the standard PBX. The idea of CDC Test is based in the surrounding of the tested PBX μ-samples by a PBX ambient charge, that is large enough, in order to avoid the relief and decay effects (quenching layer) related to the critical diameter of detonation. Few CDC Tests configurations have been developed and applied for simultaneous registration of D, Pd and DF curvature in both PBX (PBX$_x$ μ-sample and surround standard PBX$_{St}$). An example of CDC Test is presented, with PBX composition based on HMX (82% of coarse crystals, d_{50}=204 μm, with 18% of epoxy binder) with an initial density equal 1.82g/cm^3.

Validation of CDC Test implies the correlation of the DW parameters, D and P_d, of the same PBX$_x$, measured by two independent tests, with different sizes of PBX$_x$ samples. In CDC Test (Figure 4), μ-samples have the size 1.9×5×7 mm. In the typical scale test the sample of the same PBX$_x$, of size of Ø25 × 30 mm, was initiated by the long charge (Ø25 × 100 mm) of a standard PBX (PE-4A - initial density 1.57 g/cm3, 85% of RDX with 15% of a polyurethane binder). The PBX$_{St}$ in CDC Test was the same PE-4A composition.

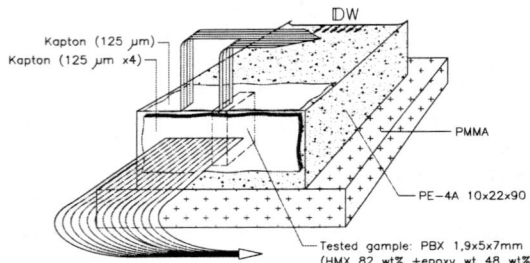

FIGURE 4 – Experimental set-up of CDC Test.

The photo-chronograms, obtained in CDC and typical scale tests, are presented in Figure 5. They show the records of the pulsing DW propagation, in both tests, fluctuation of light irradiation emitted by irregular DF and the following run, of the derived SW with the fluctuations, in the multi-layer Kapton barrier. The photo-chronogram, obtained in typical scale test of PBX$_x$, presented in Figure 5 (b), also shows the derived SW, in the multi-layer Kapton barrier (consisting of eight Kapton films of thickness of successively, 100 μm, 50 μm and 6x125 μm).

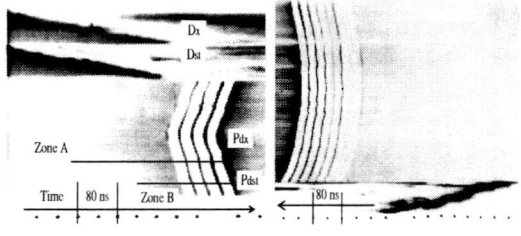

(a) (b)
FIGURE 5 – Photo-chronograms obtained in (a) CDC Test and in (b) Typical Scale Test.

In both tests, the measured mean values of D (mean velocity of pulsing DW), D_t (velocity of pulsing DW in the terminal point of its run, closely to the PBX-Kapton interface), $U_{sKapton}$ and P_{Kapton} (shock velocity and SW pressure into the Kapton barrier) and P_d (detonation pressure amplitude) are presented in Figure 6 and Table 2. Results show, for both of PBX samples (µ-sample and typical scale sample) an agreement of the measured values of D, $U_{sKapton}$ and P_d, and an inaccuracy, in the measured values of D_t, that not exceeds the experimental error of these tests.

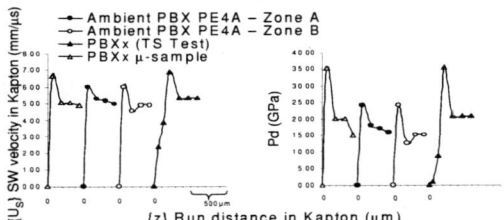

FIGURE 6 – SW attenuation ($U_{sKapton}$ vs z) in Kapton barrier and Detonation Pressure profiles. All samples are CDC test except PBXx (as noted)

TABLE 2 – Results obtained in CDC Test and in TS Test.

Test	Tested PBX	ρ_0, g/cc	D, µm/µs	D_t, µm/µs	U_{sKapto}, µm/µs	P_d, GPa
CDC	Ambient charge PE-4A	1.57	7.79 ±0.08	7.96 ±0.32	6.04 ±0.12	24.2
CDC	PBX$_x$(82% HMX+epoxy)	1.82	8.64 ±0.09	8.93 ±0.36	6.65 ±0.13	35.3
TS	PBX$_x$(82% HMX+epoxy)	1.82	8.54 ±0.09	8.30 ±0.33	6.88 ±0.10	35.4

The correlation between the main parameters of DW, detonation velocity and detonation pressure, recorded in CDC and in typical scale tests, proves the validity of CDC Test, using µ-sample mass of ≈900 times less the mass of the typical scale test.

CONCLUSIONS

The developed optical non-intrusive methods in characteristic tests, based on the application of multi-fiber optical strip, allows the registration of shock and DW in PBX on the meso-scale level, providing the observation of the phenomenological behaviour of PBX detonation. The results obtained in detonation study of mini and micro samples of PBX, allow the characterisation of the detonation process in this strongly non-equilibrium composite energetic material (coarse HE – fine HE particles – binder) as an oscillating interacting process with a tendency, in its development, to the self-organisation into more regular and a larger scale process.

REFERENCES

1. I. Plaksin, J. Campos, R. Mendes and J. Góis, Interaction of Double Corner Turning Effect in PBX, Proceedings of the Conf. on the American Phys. Soc. Topical Group on Shock Compression of Condensed Matter, Amherst, Massachusetts, July 27-August 1, 1997, pp. 755-758
2. I. Plaksin, J. Campos, R. Mendes, J. Ribeiro and J. Góis, Pulsing Behaviour and Corner Turning Effect in PBX, Proceedings of the 11th Detonation Symposium, Aspen, CO, Aug. 29 – Set. 4, 1998, pp. 658-664
3. I. Plaksin, J. Campos, R. Mendes, J. Ribeiro and J. Góis, Mechanism of Detonation Wave Propagation in PBX with Energetic Binder, Proceedings of the Conf. of the American Phys. Soc. Topical Group on Shock Compression of Condensed Matter, Snowbird, Utah, June 27-July 2, 1999, pp. 817-820
4. R. Mendes, I. Plaksin, J. Campos and J. Ribeiro, Double Slapper Initiation of PBX Proceedings of the Conf. of the American Phys. Soc. Topical Group on Shock Compression of Condensed Matter, Snowbird, Utah, June 27-July 2, 1999. pp. 915-918
5. P. Simões, L. Pedroso, I. Plaksin, J. Campos and A. Portugal, New Propellant Component: 2. Study of a PSAN/DNAM/HTPB Based Formulation, Propellants, Explosives, Pyrotechnics, (in publishing)
6. I. Plaksin, A. Portugal, L. Pedroso, P. Simões and J. Campos, Detonation Properties of HMX-DNAM-GAP Compositions, Abstracts of the International Conference in Shock Waves in Condensed Matter, St Petersburg, September 8-13, 2000. pp. 31-33

CONVECTIVE DETONATIONS

Raafat H. Guirguis[a] and Alexandra M. Landsberg[b]

[a] *Research and Technology Department* and [b] *Weapons Department*
Naval Surface Warfare Center
Indian Head, MD 20640

Abstract. Convective detonations are introduced, a novel concept whereby in some regions of the reaction zone, instead of solely depending on the shock-induced hot spots, convection significantly contributes to igniting the unreacted materials coming through the shock front. Radially-graded explosives is an application in which these detonations are likely to occur. Highly curved detonations can be sustained in these explosives, resulting in transverse pressure gradients that drive the decomposition products upstream, where these hot gases can ignite the surface of unreacted particles that crossed the shock front at a relatively weak section.

INTRODUCTION

In deflagration-to-detonation transitions (DDT) often observed in porous propellants and explosives, convective burning plays a dominant role in the transition. Specifically, the convection of the hot gas products through the pores drastically increases the flame speed beyond that of laminar propagation. However, a compression phase in which the flame-generated pressure waves coalesce ahead of the deflagration front always precedes the final stages of transition to detonation. In traditional detonations convective burning participates in the surface decomposition stage of the reaction, but it does not contribute to igniting the unreacted materials coming through the leading shock front.

This paper argues the feasibility of convective detonations in radially-graded explosives, a new type of explosives manufactured with a built-in gradual change in composition. Highly curved detonations resulting in transverse pressure gradients in the proper direction for driving the gas decomposition products upstream can be sustained in these explosives (1). Where the shock is relatively weak, instead of the hot spots, these hot products can ignite the surface of yet unreacted particles.

STRUCTURE OF REACTION ZONE IN TRADITIONAL DETONATIONS

In traditional detonations ignition is propagated by the leading shock. Figure 1 illustrates the different stages of reaction in a heterogeneous explosive composed of energetic crystals and a binder. Two different pressure gradients develop in the reaction zone – macroscopic, from one control volume to the next, and microscopic, describing changes in pressure within the same control volume over length scales comparable to the particle size. Each pressure gradient drives the decomposition gas products in a different manner.

Upon crossing the shock, the bed is compacted. As illustrated in Fig. 1a, the dissipated work is localized in a number of hot spots where ignition occurs. The resulting bulk chemical decomposition *locally* raises the pressure at the hot spots, thus introducing within the same control volume a large number of microscopic pressure gradients, each pointing in a different direction. At each of these points, the difference in pressure forces the hot gas products to burn channels around and between particles, as illustrated in Fig. 1b. These channels eventually connect the isolated pockets of decomposition gas products together, forming a network through which gases can travel macroscopic distances uninterrupted, and partially deconsolidating the bed. As the control volume moves downstream to lower pressures, the resulting dilatation also help expand the pores created at the hot spots and the channels erosively bored into the bed. When the decomposition products completely engulf the remaining solid fragments, the bed is fluidized, as illustrated in Fig. 1c, and surface decomposition becomes the dominant reaction mechanism.

All the components within the control volume are subjected to the same macroscopic pressure gradient, but being lighter, the gas products acquire a higher velocity. They infiltrate through the network of open channels and through the interstitial spaces in the fluidized bed into neighboring control volumes having lower pressures, as illustrated in Fig. 1c, same as convective burning in DDT. However, unlike this macroscopic flow, the microscopic flows described above, driven in different directions by the different microscopic pressure gradients within the control volume, do not add up to a flow with a net velocity component.

Eventually, due to the drag between the two phases, the remaining smaller fragments reach the same velocity as the gas products, and all 2-phase flow aspects cease. As illustrated in Fig. 1d, laminar surface decomposition then completes the reaction process, which in 1-D detonations and in detonations with slightly curved fronts has to end at the Chapman-Jouguet (CJ) surface.

Whether discussing DDT or convective detonations, in this paper the term "convection" specifically refers to the flow of gas products through the interstitial spaces between solid fragments. The interstitial spaces are either pre-existing, such as in porous explosive beds, or could be generated by chemical decomposition. The term "convective detonation" means that in some regions of the reaction zone, instead of solely depending on the shock-induced hot spots to start the reaction, hot gases from a partially decomposed region downstream infiltrate back and ignite some of the unreacted material upstream. In traditional detonations, the macroscopic pressure gradient is pointed in the wrong direction. However, as explained next, in detonations with highly curved fronts the transverse pressure gradient is positive ($\nabla p \cdot \underline{n} > 0$; $\underline{n}$ = unit vector normal to streamline), i.e., in the proper direction for driving the lighter gas decomposition products upstream.

MACROSCOPIC PRESSURE GRADIENT IN THE TRANSVERSE DIRECTION

Figure 2 illustrates the details of the reaction zone for three detonation waves with progressively curved fronts. A planar front is an idealization of practical detonation waves that is only applicable in the asymptotic limit, when the charge is rigidly confined or is infinitely large. In finite charges, the detonation front becomes curved in order to accommodate the divergence resulting from the lateral expansion of the high-pressure decomposition products and at the same time, satisfy the CJ condition of unit Mach number at some location within the reaction zone (2). The curvature of the streamlines, on the other hand, depends on the sign of the pressure gradient in the transverse direction (orthogonal to streamlines), the resultant of two competing factors, both tending to decrease the pressure, but one faster than the other.

In general, after crossing the shock the pressure decreases along the streamline due to the heat liberated by the exothermic chemical reactions, i.e., $p_3 < p_1$. Due to the wave curvature, the pressure also

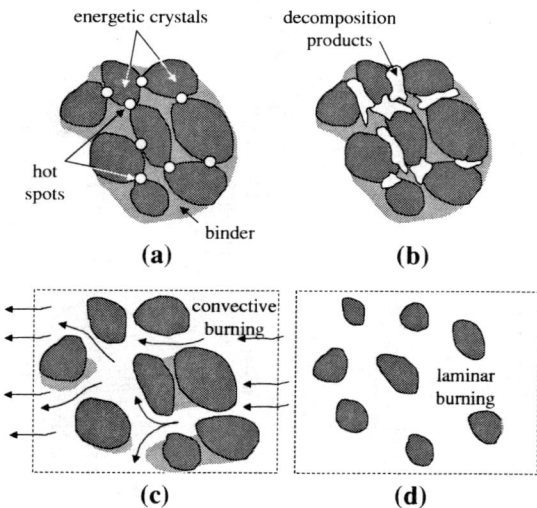

FIGURE 1. Different stages of reaction: (a) formation of hot spots in shock-compacted bed; (b) ignition of the hot spots creates a network of connected porosity; (c) convective surface decomposition in fluidized bed; (d) laminar surface decomposition of solid fragments of explosive floating in the gas products.

decreases along the detonation front and along the constant reaction-progress contours behind it, i.e., $p_2 < p_1$, but for planar detonations, $p_2 = p_3$.

If the front is slightly curved, the pressure drops along the streamline faster than it does along the detonation front, yielding $p_3 < p_2$. The inclination of the streamlines to the axis continuously decreases after the shock, but more importantly, the gas decomposition products tend to infiltrate inwards and downstream, from cooler regions closer to the front, to hotter points where the reaction has progressed further.

If the detonation front is highly curved, however, the pressure drop along the detonation front becomes larger than that along the streamline, hence $p_3 > p_2$. The gas products tend to infiltrate outwards and upstream, from regions where the reaction has progressed further and the gas products are hotter, to cooler locations. At a minimum, this convection should accelerate surface decomposition upstream. If, however, the shock is too weak to quickly ignite the incoming materials, or if the explosive is insensitive to shock initiation, these hot products may start surface decomposition first, or at least dominate the reaction process by igniting more of the unreacted particles than the shock-induced hot spots do.

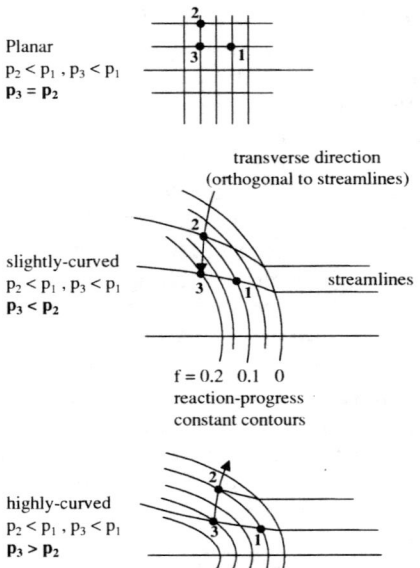

FIGURE 2. Effect of detonation front curvature on pressure gradient in the transverse direction. For slightly curved fronts, convection is pointed downstream, whereas for highly curved fronts, it is pointed upstream.

However, as explained above, ignition of these hot spots also plays a significant role in creating a network of open channels through which the gas products can travel. If ignition at the hot spots is completely eliminated, the convection of the hot gas products forced by the pressure gradient in the transverse direction will be the only remaining mechanism for boring travel channels through the bed. The shock may introduce microscopic cracks in the compacted bed, but to open such channels the hot gas products will have to push through these cracks and widen them by burning the walls.

HIGHLY CURVED DETONATIONS IN RADIALLY-GRADED EXPLOSIVES

In ideal explosive charges, the detonation front is slightly curved, but as the size of the charge approaches the critical diameter, the radius of curvature decreases. Detonation waves with significantly curved fronts are also observed in non-ideal explosives containing a considerable fraction of slow-reacting components, but in all these detonations the resulting front is not curved enough to create a strong positive pressure gradients in the transverse direction. Such highly curved detonation fronts can be created, however, in radially-graded charges.

Figure 3 compares the structure of the detonation waves resulting in two explosive charges 4 cm

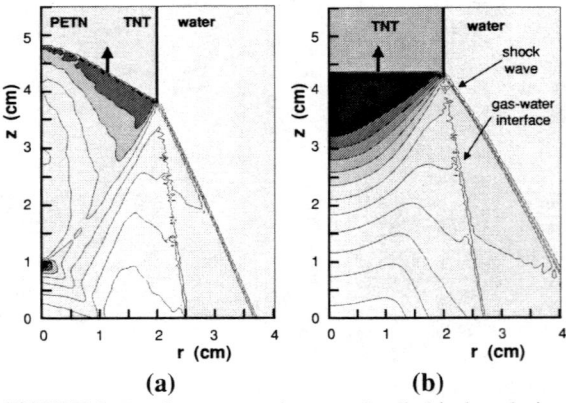

FIGURE 3. Density contours in pressed cylindrical explosive charges detonated underwater (reproduced from reference 1). Charge (a) is radially-graded - PETN at the axis, gradually changing to TNT at the outer radius, resulting in a highly curved detonation front. Charge (b) is pure TNT. The curvature of the front is imperceptible.

in diameter, both pressed to a constant uniform density = 1.63 g/cc and detonated underwater. However, charge (a) is radially-graded – PETN at the axis, gradually changing to TNT on the outer radius, whereas charge (b) is pure TNT. Because PETN has a higher detonation velocity (at 1.63 g/cc, D = 7.79 km/s) than TNT (D = 6.93 km/s), the detonation propagating axially in charge (a) has a highly curved front. Although charge (b) is not graded, the detonation front is still curved because both the charge diameter and acoustic impedance of the confining water are finite, but the curvature of the front is imperceptible.

DISCUSSION AND CONCLUSIONS

Convective detonations are introduced. They are defined in this paper as detonations in which the convection of hot decomposition products significantly contributes to igniting the unreacted materials coming through the shock front. Radially-graded explosives is an application in which these detonations are likely to occur because detonations with highly curved fronts can steadily propagate in these explosives, resulting in transverse pressure gradients in the proper direction for driving these products upstream. However, one has to wonder whether the laminar surface decomposition that starts after the gas products reach their destination is fast enough to contribute to the propagation of the front.

At the high pressures generated by detonations, the rate of surface regression driven by heat diffusion (conduction) and radiation, usually expressed as ap^n, obviously can be quite large, but a quantitative estimate is nevertheless calculated next. As explained above, even in detonation waves, laminar surface decomposition is ultimately responsible for burning the solid particles. In ideal explosives, the reaction zone is usually 0.5-1 mm thick, which takes about 50-100 ns to cross. Assuming the average particle size is 100 μm, surface regression must be proceeding at velocities at least of the order of 1-2 mm/μs (km/s) in these detonations.

In uniform explosives, if the volumetric rate of gas generation, which mainly depends on the composition, is not large enough to preserve a core of high pressure near the axis against the action of the rarefaction waves that originate at the surface of the charge, these waves will eventually kill the detonation wave by decreasing the pressure in the core, causing a concomitant decrease in the rate ap^n. That is what happens if a detonation wave is forcibly initiated by a strong booster in an explosive charge smaller than the critical diameter. However, as explained below, radially grading the explosive allows us to build into the charge a core that can autonomously support the high pressure near the axis, as well as enhance the lateral transfer of surface decomposition by seeding the outer layers with more energetic materials.

Let us assume that the cylindrical charge in Fig. 3a is replaced with one that is constructed of only two layers - a core of pressed PETN, 10 mm in diameter, and an outer layer 15 mm thick of a less energetic powder. Since the critical diameter of unconfined PETN is about 1 mm, a steady detonation can be obviously maintained in the core, independently of the outer layer. However, if the volumetric rate of gas generation in the outer layer is not large enough to keep up with the wave front propagating as fast as the detonation in the core, a non-reactive oblique shock similar to that induced in the water in Fig. 3b will result in the outer layer.

Now let us radially stratify the charge. If some of the less energetic powder is mixed in the core, the detonation will slow down. If some small fraction of PETN is mixed in the outer layer, its volumetric rate of gas generation is increased. If a significant fraction of PETN is mixed in, however, the outer layer will become similar to the core and will sustain a traditional detonation. By properly grading the fraction of PETN mixed in, it is possible to reach a condition whereby surface decomposition in the outer layer keeps up with the front, and convection, not compression, transfers ignition laterally. Thus, convective detonations are more likely to occur in radially-graded charges.

REFERENCES

1. Guirguis, R., and Landsberg, A., "Structure of Detonation Waves in Stratified/Spatially-Graded Explosives," *Proceedings of the 2000 JANNAF PSHS meeting*, 2000, CPIA, Columbia, MD.
2. Guirguis, R. H., "Streamlines Dynamics Method for Highly-Curved Detonation Waves," *Proceedings of the Tenth Symposium (International) on Detonation*, pp. 27-36, 1993, ONR, Arlington, VA.

EFFECT OF VOID SIZE ON THE DETONATION PRESSURE OF EMULSION EXPLOSIVES.

Yoshikazu Hirosaki [1], Kenji Murata [1], Yukio Kato [1] and Shigeru Itoh [2]

[1] NOF Corporation, 61-1 Kitakomatsudani, Taketoyo-cho, Chita-gun, Aichi 470-2398, JAPAN
[2] Shock Wave and Condensed Matter Research Center, Kumamoto University
2-39-1 Kurokami, Kumamoto 860-8555, JAPAN

Abstract. To study the effect of void size, detonation pressure as well as detonation velocity was measured using PVDF pressure gauge for the emulsion explosives sensitized with plastic balloons of five different size ranging from 0.05mm to 2.42mm. The experimental results were compared with the detonation pressure and velocity calculated using KHT code. The experimental results showed that the detonation pressure and velocity were strongly affected by void size, and that the fraction of ammonium nitrate reacted in the reaction zone was strongly dependent on void size.

INTRODUCTION

Explosives for civil use, such as emulsion explosives, slurry explosives and ANFO (Ammonium Nitrate-Fuel Oil), are usually ammonium nitrate (AN)-based. Those explosives are well known to show non-ideal detonation behavior due to its slow reaction rate compared to high explosives for military use. The non-ideal detonation wave propagates steadily but its characteristics are significantly affected by the conditions such as charge diameter and confinement. Detonation pressure and detonation velocity are much lower than those for infinite charge diameter. This behavior is depending on the long reaction zone length due to slow reaction rate of AN.

It is well known that detonation behavior can be widely controlled by the size and quantity of void contained in the emulsion explosives. The effect of void size and quantity on the detonation velocity, critical diameter and sensitivity of emulsion explosives were reported in many investigations [1-4]. In those studies, glass microballoons of size smaller than 0.15mm were used as void. In this paper, to study the effect of void size, detonation pressure as well as detonation velocity was measured using PVDF (polyvinylidenefluoride) pressure gauge for the emulsion explosives sensitized with plastic balloons of five different size ranging from 0.05mm to 2.42mm.

EXPERIMENTALS
Samples

The formulation of the emulsion matrix used in this study is ammonium nitrate /sodium nitrate /water /wax and emulsifier = 77.66/4.68/11.22/5.40 in weight ratio. The oxygen balance of the emulsion matrix is 0.4g/100g, and the density of it is $1.39 g/cm^3$. A certain amount of plastic balloons of mono-cell or multi-cell structure shown in Table 1 were added into the emulsion matrix to achieve the desired explosive density. Five different sizes of balloons were used. PB-1 is the smallest balloon of mono-cell structure with the average diameter of 0.05mm, and others are the balloons of multi-cell structure with average diameter ranging from 0.47 to 2.42mm.

Microscopic photographs of both structures are shown in Fig.1 as examples. The particle density of each balloon was determined from the densities of explosives that contain different amount of balloons. The size of balloon was optically measured.

Detonation Pressure Measurement

The emulsion explosive loaded into PVC pipe of 51mm in inner diameter, 60mm in outer diameter and 200mm in length was placed on a PMMA block as shown in Fig.2. A PVDF film of 10 μm in thickness and 5mm squares was sandwiched with polyimide films together with electrodes made of copper foil. The PVDF gauge was put onto a PMMA block and then covered and glued with a PMMA plate of 1mm in thickness. The emulsion explosive was initiated with an electric detonator. Additional 30grams of emulsion explosive was also used as a booster explosive, if necessary. The output of the pressure gauge was recorded with a digital oscilloscope at sampling rate of five nanoseconds. Calibration of PVDF pressure gauge was carried out by measuring electric charge created under hydraulic pressure and by comparing with the pressure measured with a manganin pressure gauge that has preliminary been calibrated. The pressure profile observed by PVDF pressure gauge is that transmitted into PMMA plate, which exists among the explosive and a PVDF gauge.

Detonation velocity was measured with ionization gaps that were placed at points 130mm and 180mm apart from the upper end of the charge.

TABLE 1. Characteristics of plastic balloons

	Average diameter (mm)	Standard deviation (mm)	Particle density (g/cm^3)	Structure	Material
PB-1	0.053	0.023	0.027	Multi-cell	Acrylonitrile / vinylidene chloride
PB-2	0.472	0.062	0.051	Multi-cell	Polystyrene
PB-3	0.795	0.129	0.077	Multi-cell	Polystyrene
PB-4	1.728	0.273	0.032	Multi-cell	Polystyrene
PB-5	2.420	0.403	0.064	Multi-cell	Polystyrene

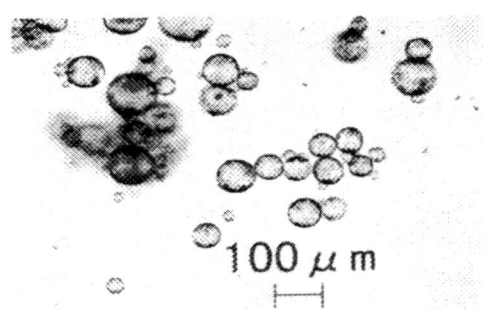 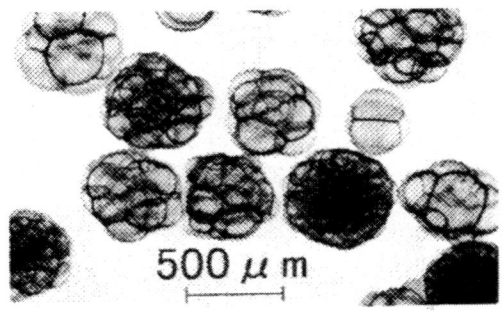

FIGURE 1. Microscopic photographs of plastic balloons PB-1 (left) and PB-2 (right)

RESULTS AND DISCUSSIONS

Fig.3 indicates pressure profiles observed for the

emulsion explosives of density 1.05g/cm³ sensitized with balloons of five different size. The pressure rises up sharply to reach its peak pressure within about 75 nanoseconds in the explosives sensitized with balloons smaller than 0.80mm in diameter: PB-1, PB-2 and PB-3. The pressure rise time of about 75 nanoseconds can be explained by the detonation front curvature measured by optical observation [5] and shock transition time in PVDF film of 10 μm thick. Pressure decrease in the reaction zone behind leading shock and following pressure decay in Taylor wave can be observed in the emulsion explosives sensitized with balloons PB-1 and PB-2. Whereas the emulsion explosives sensitized with larger balloons such as PB-4 or PB-5 require longer time to reach its peak pressure. This is due to the important irregularity of detonation front of the emulsion explosives containing large balloons, which was measured in optical observation [5]. The peak pressure is fairly low compared with that for the emulsion explosives containing smaller balloons.

The pressure observed in this experiment is that transmitted into PMMA plate of 1mm in thickness.

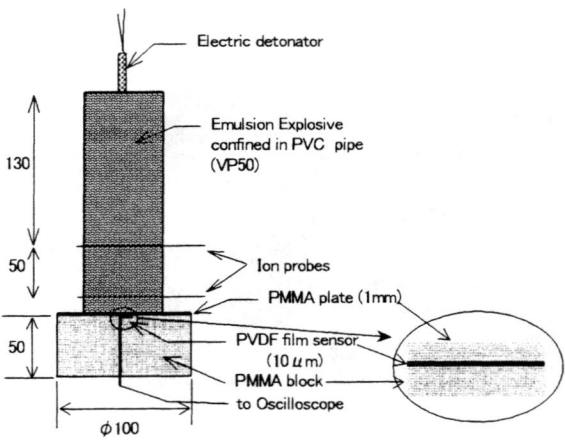

FIGUER 2. Experimental setup for detonation pressure measurements

The detonation pressure was therefore determined from the impedance match method. The Hugoniot of PMMA was based on the reference [6], and that of the unreacted emulsion explosive was supposed to be same as Universal Hugoniot [7],[8] for AN solution.

The reaction zone length estimated from the measured pressure profile is about 1.5mm for the emulsion explosives sensitized with void of 0.05mm. This value agrees well with the reaction zone length estimated from the diameter effect of detonation velocity in the same emulsion explosives [4].

The theoretical detonation pressure and velocity were calculated with a hydro-thermodynamic code KHT in which the ingredients other than AN were assumed to be completely reactive. The measured and calculated detonation pressure as well as detonation velocity were summarized in Table 2. The fraction of AN reacted in the reaction zone was evaluated based on the comparison between the measured and calculated detonation pressure and velocity. The fraction of AN reacted in the reaction zone evaluated from pressure agrees well with that estimated from detonation velocity. The fraction of AN reacted in the reaction zone is as high as 0.87 for the emulsion explosives sensitized with void of 0.05mm. On the other hand, the fraction of AN reacted is as low as 0.30 for the emulsion explosives sensitized with void of 2.42mm. This result is due to small number of void which act as hot spot, and this leads to longer reaction zone length and poor reactivity of AN. The poor reactivity of AN in the emulsion explosives containing large voids is due not only to the lateral rarefaction waves but also to the rarefaction waves from void itself.

CONCLUSIONS

To examine the effects of void size, detonation pressure and detonation velocity were measured for the emulsion explosives sensitized with plastic balloons of five different size ranging from 0.05mm to 2.42mm in average diameter.

TABLE 2. Effect of balloon size on the detonation properties of emulsion explosives.

Balloon diameter d_B(mm)	0.05	0.47	0.80	1.73	2.42
Measured detonation velocity (m/s)	5230	4480	3510	3360	2960
Measured detonation pressure P (GPa)	6.4	5.1	2.8	2.5	2.1
Calculated CJ pressure P_{CJ}(GPa)	8.21	8.21	8.21	8.21	8.21
Fraction of AN reacted at C-J state estimated from pressure P	0.84	0.69	0.42	0.39	0.33
Fraction of AN reacted at C-J state estimated from detonation velocity	0.87	0.68	0.43	0.39	0.30

When void size was increased, the difference between the measured and calculated values was increased both for detonation pressure and velocity. The fraction of AN reacted in the reaction zone was as high as 0.87 for the emulsion explosives sensitized with voids of 0.05mm. On the other hand, the fraction of AN reacted is as low as 0.30 for the large void of 2.42mm. In the case of the emulsion explosives sensitized with void smaller than 0.47mm, pressure decrease behind leading shock and following decay in Taylor wave were observed. Whereas, in the case of the emulsion explosives sensitized with void larger than 1.73mm, detonation pressure rise time is larger than 0.5 microsecond due to the important irregularity of detonation front.

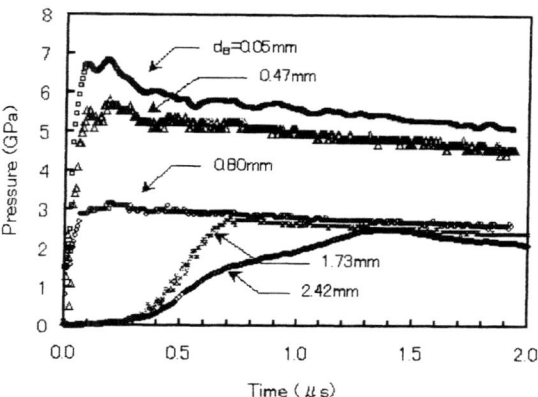

FIGUER 3. Pressure-time curves observed with PVDF gauge for the detonation of emulsion explosives of density 1.05g/cm^3 sensitized with various balloon sizes

REFERENCES

1. Hattori, K., Fukatsu, K., and Sakai, H., *J. of the Japan Explosives Society*, **45, 5**, 295-301, (1982).
2. Yoshida, M., Iida, M., Tanaka, K., Fujiwara, S., Kusakabe, M., and Shiino, K., "Detonation Behavior of Emulsion Explosives Containing Glass Microballoons," in *Eight Symposium (International) on Detonation*, NSWC MP 86-194, Naval Surface Weapon Center, White Oak, Silver Spring, Maryland, 1985, pp. 993-1000.
3. Lee, J., Sandstrom, F. W., Craig, B. G., and Persson, P. A., "Detonation and Shock Initiation Properties of Emulsion Explosives," in *Ninth Symposium (International) on Detonation*, Office of the Chief of Naval Research, Arlington, Virginia, 1989, pp. 573-584.
4. Hirosaki, Y., Takahashi, Y., Kato, Y., Hamashima, H., and Itoh, S., *J. of the Japan Explosives Society*, **61, 5**, 201-209 (2000).
5. Hirosaki, Y., Sawada, T., Kato, Y., Hamashima, H., and Itoh, S., *J. of the Japan Explosives Society*, **62, 1**, 23-32, (2001).
6. Marsh, S., P., *LASL SHOCK HUGONIOT DATA*, University of California Press, 1980.
7. Woolfork, R., W., Cowperthwaite, M., and Shaw, R., *Thermochimica Acta*, **5**, 409, (1973).
8. Hirosaki, Y., Ishida, T., Tokita, K., Mori, N., Hattori, K., and Sakai, H., *J. of the Japan Explosives Society*, **46, 6**, 376-383, (1985)

MOMENTUM TRANSFER DURING SHOCK INTERACTION WITH METAL PARTICLES IN CONDENSED EXPLOSIVES

Fan Zhang[1], Paul A. Thibault[2], Rick Link[2], and Alexander L. Gonor[3]

[1]*Defence Research Establishment Suffield, PO Box 4000, Stn Main, Medicine Hat, AB T1A 8K6 Canada*
[2]*Combustion Dynamics Ltd., Halifax, NS B3J 3J8 Canada*
[3]*University of Toronto, Toronto, ON M5S 3G8 Canada*

Abstract. Detonation propagation in a condensed explosive with compressible metal particles can result in significant momentum transfer between the explosive and the particles during the shock-particle interaction. Consequently, the classic assumption of a "non-momentum-transfer shock" used in multiphase continuum detonation initiation and propagation models may not be valid. This paper addresses this issue by performing numerical and theoretical calculations in liquid explosives and RDX with various compressible metal particles under conditions of detonation pressure. The results showed that immediately behind the shock front the velocity of particles such as Al and Mg can achieve 60 – 94 % the value of the shocked velocity of the explosive.

INTRODUCTION

Theoretical models for detonation in a fluid-solid particle system have mostly been based on multiphase fluid dynamics models taking mass, momentum and heat transfer between the phases into consideration. [1-2]. In these models, a frozen shock-particle interaction is often assumed in which the solid particles are not accelerated during crossing of the shock front. Behind the shock front, a viscous drag force is assumed to determine the momentum transfer between the fluid phase and the particles. For detonation in two-phase mixtures of gas and solid particles, the shock-particle interaction time is several orders of magnitude smaller than the velocity relaxation time related to viscous drag. Thus, the particle crosses the shock front with negligible changes in its velocity. Momentum loss behind the shock front plays an important role in the detonation velocity deficit for gas-particle systems [3-4]. However, for detonation in a condensed explosive containing metal particles of 0.1 to 1 µm, the shock-particle interaction time is about the same order as or one order of magnitude less than the velocity relaxation time. Therefore, the frozen shock assumption may fail and momentum transfer during the shock-particle interaction together with that behind the shock front could influence the detonation initiation and structure. Detonation velocity deficit was observed experimentally in condensed explosives with 0.1 µm aluminum particles [5-6]. The objective of the present paper is to calculate the momentum transfer during the shock interaction with compressible metal particles in condensed matter under conditions of detonation pressures.

SHOCK OVER A SINGLE PARTICLE

It was assumed that detonation initiation and decomposition of the condensed explosive would not start within the shock front. It was also assumed that the reaction time scale of solid particles is larger than the shock-particle interaction time so that the particles can be considered chemically frozen within the shock front. Thus, the explosive material was described using the Murnagham equation of state excluding temperature effect. The solid particle material was described using the HOM equation of state [7]. The Euler equations for mass and momentum were used to model the flow of the non-reacted explosive, and the governing equations for inviscid plastic flow were used for modeling of dynamic response in metal particles. The interaction between the explosive and the particles was solved by matching their boundary conditions of pressure and particle velocity. Numerical solution of these closure equations was obtained using the IFSAS code [8]. The resolution for the calculations corresponded to 20 cells or elements for a particle radius using a cylindrically axi-symmetrical mesh to represent a spherical particle immersed in the fluid. Calculations were conducted using a 10 μm diameter particle. Since the flow of the explosive and the metal particle is assumed inviscid and the equations of state and the constitutive model contain no rate-dependent terms, the results can be scaled to any other particle diameters using simple geometric similarity arguments.

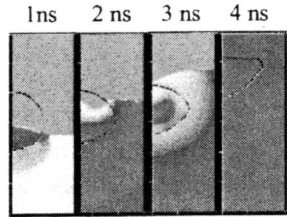

FIGURE 1. Pressure contours for an aluminum particle subjected to a 101.3 kbar shock in water.

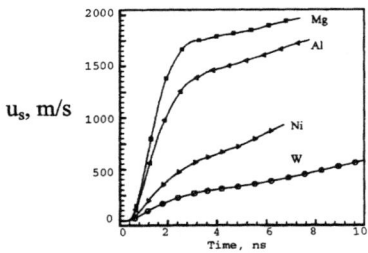

FIGURE 2. Particle velocity histories in magnesium, aluminum, nickel and tungsten subjected to a 101.3 kbar water shock.

Calculations were performed for metal particles subjected to a shock in explosive under conditions of various particle material density, particle acoustic impedance and shock strength. Explosives considered in the analysis included a liquid explosive with the same initial density and compressibility as water, and RDX with a bulk initial density of 1.4 g/cm^3 and 1.8 g/cm^3. It can be seen in Fig. 1 that an aluminum particle is severely deformed during the shock interaction process due to the fact that the incident shock pressure far exceeds the yield stress of the material. The deformation is directly related to the relative velocity between the leading and trailing edges of the particle. The early-phase of the particle acceleration is mainly controlled in the direction of motion by the transmitted shock and the rarefaction reflected off the downstream end of the particle. The lateral unloading produces a secondary effect of deceleration of the particle.

The particle velocity, based on a mass average over the particle elements, is compared in Fig. 2 for different metals for a 101.3 kbar liquid shock. The earlier time histories within about 2 ns display a shock interaction process where the particle acceleration decreases with the particle density. A velocity transmission factor α is defined as the ratio of the particle mass-averaged velocity u_s after the shock interaction time $\tau = d/D_0$ over the shocked fluid velocity u_1: $\alpha = u_s/u_1$, where d is the particle diameter and D_0 is the shock velocity. The values of α calculated for various metal particles are summarized in Table 1. Among the material properties and shock strength investigated, α is most dependent on the initial density ratio of the explosive to metal ρ_0/ρ_{s0} (see Fig. 4a). In spite of large difference of the sound speed between magnesium and beryllium and between tungsten and uranium, the velocity transmission factor remains almost the same for the same shock pressure. A curve fit of the numerical data suggests

$$\alpha = \rho_0/\rho_{s0}\,(a+b\,\rho_0/\rho_{s0})/(a+b) \quad (1)$$

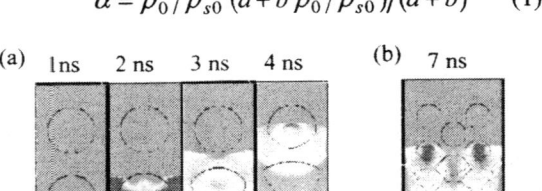

FIGURE 3. Pressure contours for cylindrical aluminum particles subjected to a 101.3 kbar water shock. a) Two particles with rigid sidewall; b) a cluster of particles.

TABLE 1. Velocity transmission factors

Material	ρ_{s0} g/cm^3	c_{s0} m/s	p_1 kbar	α
	Water, $\rho_0 = 1.0$ g/cm^3:			
Magnesium	1.770	4700	50.7	0.776
			101.3	0.790
			202.6	0.803.
Beryllium	1.870	7975	101.3	0.781
Aluminum	2.785	5350	101.3	0.600
Nickel	8.860	4646	101.3	0.227
Uranium	18.98	2540	101.3	0.108
Tungsten	19.30	4060	101.3	0.108
			202.6	0.114
	RDX $\rho_0 = 1.4$ g/cm^3:			
Magnesium	1.770	4700	101.3	0.940
Aluminum	2.785	5350	101.3	0.754
	RDX $\rho_0 = 1.8$ g/cm^3:			
Magnesium	1.770	4700	101.3	1.008
Aluminum	2.785	5350	101.3	0.802

where a = 3.947 and b = -1.951 with respect to the grid resolution used.

SHOCK OVER MULTIPLE PARTICLES

Two-dimensional calculations were conducted for a cluster of cylindrical aluminum particles 10 μm in diameter subjected to a 101.3 kbar liquid shock. Although the results of the cylindrical particle calculations cannot be directly applied to spherical particles, they do provide a qualitative trend for the effect of the particle volume fraction on the velocity transmitted to the particle. Figure 3a displays pressure contours for two particles with a 5 μm gap in the direction of motion. The effect of neighboring particles in the lateral direction is simulated by introducing a rigid wall at the symmetry plane between the particles and the lateral particles with a 5 μm gap. It can be seen that the early-phase of the particle acceleration is controlled in the direction of motion by: 1) the transmitted shock, 2) the rarefaction occurring at the downstream end of the particle, 3) the reflected rarefaction occurring at the upstream end of the particle, 4) the wave reflected off the downstream particle. Around the leading (bottom) particle, the waves reflected off the lateral particles collide and produce a secondary pressure pulse to the bottom particle. This produces an acceleration which is less significant than the deceleration effect from the wave reflected off the downstream particle in the direction of motion.

Figure 3b displays a matrix of cylindrical aluminum particles compacted by a 101.3 kbar liquid shock into a tight cluster of particles. The particles are largely deformed and coalesce due to the time lag between the acceleration of the leading (bottom) and trailing (top) particles. However, the initially non-uniform velocity in the particle quickly reaches and oscillates around an average value behind the shock front due to the multiple interactions between particles. The transmitted velocity for multiple particles, computed at a time equal to twice the shock-particle interaction time τ, is summarized in Fig. 4b. The results indicate that the transmitted particle velocity decreases with increase in particle volume fraction ϕ.

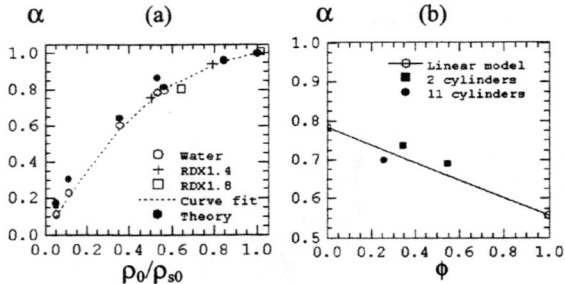

FIGURE 4. Velocity transmission factors for: a) a single particle, and b) multiple particles.

A SIMPLE MODEL

To get more insight into the mechanism, a 1D planar analytical model was established. It was assumed that the impedance matching of the explosive and particle material results in a reflected shock in the explosive and a transmitted shock in the particle. This yields an equation for the transmitted velocity in the particle u_{s3}. As the surrounding explosive shock with the velocity D_0 reaches the particle trailing edge, the transmitted shock with a velocity D_{s0} still propagates in the particle if $D_{s0} < D_0$; or is already reflected off the particle trailing edge and the reflected rarefaction runs back into the particle if $D_{s0} > D_0$. Taking into account the shock velocity difference, a momentum balance rule is assumed after the shock-particle interaction time, that is,

$$\rho_s u_s d = \rho_{s3} u_{s3} d_3, \text{ if } D_{s0} \leq D_0$$
$$\rho_s u_s d = \rho_{s3} u_{s3}(d - d_4) + \rho_{s4} u_{s4} d_4, \text{ if } D_{s0} > D_0 \quad (2)$$

where the subscript "s" denotes the mass-averaged state after the shock-particle interaction time τ, subscripts "3" and "4" represent the transmitted shock and the reflected rarefaction state respectively. Distances d_3 and d_4 are defined by the shock-particle interaction time,

$$\tau = \frac{d}{D_0} = \frac{d_3}{D_{s0}} = \frac{d}{D_{s0}} + \frac{d_4}{c_{s3} - u_{s3}} \quad (3)$$

For simplicity, the flow velocity u_{s4} behind the rarefaction wave is assumed to be $u_{s4} \sim 2u_{s3}$ and also $\rho_s \sim \rho_{s3} \sim \rho_{s4}$ is assumed. Substitution of equation (3) into equation (2) yields

$$\alpha = \frac{D_{s0}}{D_0} \alpha_s, \text{ for } D_{s0} \leq D_0$$
$$\alpha = \left[1 + \frac{c_{s3} - u_{s3}}{D_{s0}}\left(\frac{D_{s0}}{D_0} - 1\right)\right]\alpha_s, \text{ for } D_{s0} > D_0 \quad (4)$$

where $\alpha_s = u_{s3}/u_1$. The theoretical velocity transmission factors calculated from equations (4) are also displayed in Fig. 4a. Comparison of the theoretical results with the numerical calculations shows fairly good agreement, except for the theoretical values are generally larger than the numerical ones since the 1D theory does not consider the loss caused by the lateral deformation and expansion.

For multiple particles, the particle velocity calculated varies with the selection of the time for the shock interaction with the multiple particles. It may also vary with the geometric arrangement of the particles for a given particle volume fraction. For complicated cases involved in multiple particles, it may simply assume a linear model for the velocity transmission factor between the value for a single particle, α, for the particle volume fraction $\phi = 0$, and the value for the transmitted shock in the solid, α_s, at $\phi = 1$, that is,

$$\alpha_m = (1 - \phi)\alpha + \phi \alpha_s. \quad (5)$$

The results predicted using equation (5) for multiple aluminum particles are displayed in Fig. 4b and compared with the numerical calculations.

CONCLUSIONS

For a charge of a condensed explosive with metal particles, the present study indicates that the momentum transfer from the condensed explosive to the metal particles is significant during the particle crossing of the shock front. The particle velocity after the shock-particle interaction strongly depends on the initial density ratio of explosive to metal, but is relatively insensitive to the other parameters, such as the particle acoustic impedance, shock strength and bulk explosive shock Hugoniot. The transmitted particle velocity decreases with increase in the particle volume fraction. The significant momentum transfer during the particle crossing of the shock front must be taken into account when modeling the shock initiation and detonation structure for two-phase mixtures of condensed explosive and metal particles.

This work was supported under the auspices of the DND contract W7702-8-R696.

REFERENCE

1. Baer, M. R., and Nunziato J. W., *Int. J. Multiphase Flow* **12**, 861-889 (1986).
2. Powers, J. .M., Stewart, D. S., and Krier, H., *Combust. Flame* **80**, 264-279 (1990).
3. Zeldovich, Ya. B., Borisov, A. A., Gelfand, B. E., Frolov, S. M., and Maikov, A. E., *Progress in Astronautics and aeronautics* **144**, 211-231 (1988).
4. Zhang, F., and Lee, J. H. S., *Proc. R. Soc. Lond. A* **443**, 1-19 (1994).
5. Baudin, G., Lefrancois, A., Bergues, D., Bigot, J., Champion, Y., 11th Int. Detonation Symp., Snowmass, ONR 33300-5, 989-997 (1998)
6. Gogulya, M. F., Dolgoborodov, A. Yu., Brazhnikov, M. A., Baudin, G., 11th Int. Detonation Symp., Snowmass, ONR 33300-5, 979-988 (1998)
7. Mader, C., *Numerical Modeling of Explosives and Propellants*, CRC Press, Boca Raton, 1998, pp. 352.
8. *IFSAS User Manual Version 1.6*, Combustion Dynamics Ltd., 1997.

REACTION ZONE TRANSFORMATION FOR STEADY-STATE DETONATION OF HIGH EXPLOSIVES UNDER INITIAL DENSITY INCREASE

Alexander V. Utkin, Sergey A. Kolesnikov, Sergey V. Pershin, and Vladimir E. Fortov

Institute of Problems of Chemical Physics RAS, Chernogolovka, Russia, 142432

Abstract. The laser interferometric system VISAR was used to investigate the detonation waves structure of pressed RDX, HMX, TNETB, and ZOX with different initial density. The experimental results are the surface velocity profiles of foils placed at the boundary between a HE sample and a water "window". Critical initial densities ρ_c at which the reaction zone structure changes crucially were found: Von Neumann spike was recorded if the density was less than the critical value, otherwise monotone pressure increase in the reaction zone was observed. The ρ_c is equal to 1.72 g/cm^3, 1.82 g/cm^3, 1.56 g/cm^3, and 1.71 g/cm^3 for RDX, HMX, TNETB and ZOX respectively. The unusual structure of the detonation wave at high density can be explained by realization of underdriven detonation.

INTRODUCTION

According to the classical theory [1], the detonation wave consists of a shock jump and a chemical reaction zone, in which the pressure decreases and the matter expands, i.e. Von Neumann spike is shaped. Numerous experimental data confirm the validity of this model for heterogeneous high explosives (HE). However, it was found [2] that in RDX and HMX at high initial density the pressure increases in the reaction zone and the spike does not form. The detonation wave without Von Neumann spike does not correspond to the classical model. Moreover, it is not clear, whether the Chapman-Jouguet state will be reached, and what the selection rule of detonation velocity is in this case. To solve these key theoretical problems it is necessary (1) to understand if the observed phenomenon is the unique property of RDX and HMX, or it reflects the property common for powerful explosives, (2) to define the initial density at which the reaction zone structure changes crucially. For these purposes the experimental investigation of the reaction zone transformation under initial density increase in pressed RDX ($C_3H_6N_6O_6$), HMX ($C_4H_8N_8O_8$), ZOX ($C_6H_8N_{10}O_{16}$, Bis (2, 2, 2 – Trinitroethyl - N - nitro) Ethylenediamine), and TNETB ($C_6H_6N_6O_{14}$, Trinitroethyl trinitrobutyrate) was conducted.

THE SCHEME OF EXPERIMENTS

The scheme of experiments is shown in Fig. 1. The detonation in the samples was initiated by a shock wave, created by the HE plane generator (1). The diameter of charges was 30 mm, the length changed from 40 up to 80 mm. The wave profiles were registered by laser interferometer VISAR. The

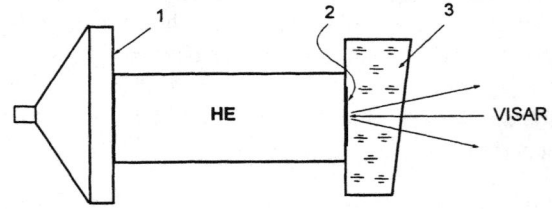

Figure 1. The scheme of experiments.

laser beam reflected from a 100 - 400 μm aluminum foil (2) placed between the charge and the water window (3). As the result of the experiment we have the velocity of the foil - water border, which represents all the details of the reaction zone structure in detonation wave.

EXPERIMENTS AND RESULTS

RDX and HMX

RDX samples of different initial density ρ_0 were pressed from powder with a mean particle diameter of 80 μm. To obtain ρ_0 in the range of 1.50 g/cm^3 to 1.75 g/cm^3, a small amount of acetone (less than 1 wt%) was added in pressed RDX (RDX$_1$). Higher density, up to 1.776 g/cm^3, was reached by increase of acetone quantity up to 10 wt% and by curing of the samples under pressure (RDX$_2$).

The experimental results for RDX$_1$ are presented in Fig. 2,3. When the initial density is less than 1.72 g/cm^3 the velocity of the foil - water boundary decreases after the shock jump. Duration and amplitude of the spike are determined by Von Neumann spike in the RDX. The subsequent velocity increase is caused by circulation of compression and rarefaction waves in Al foil.

The velocity spike duration changes insignificantly, but its amplitude notably drops as the density increases. When ρ_0 exceeds ρ_c=1.72 g/cm^3 the situation changes crucially: instead of velocity decrease after the shock jump the monotonic increase is observed (curve for ρ_0 =1.73 g/cm^3 in Fig.3). It is necessary to note, that the particle velocity of explosion products increases with the increase of initial density except for critical density, where the abnormal velocity behavior is registered. For ρ_0= 1.73 g/cm^3 the particle velocity is the same as for 1.69 g/cm^3 (Fig.3), whereas it has to increase by more than 50 m/s. It is possible to explain this by transition to underdriven detonation when Von Neumann spike disappears.

Figure 4 shows the experimental data for RDX$_2$ and the velocity spike is registered for all initial densities. This result differs essentially from the result that one would expect on the basis of the RDX$_1$ data extrapolation to higher densities. The spike amplitude drops approximately twice when the ρ_0 increases from 1.72 up to 1.776 g/cm^3, and, by analogy with RDX$_1$, it is possible to expect that for RDX$_2$ ρ_c≈1.78 g/cm^3. This conclusion agrees with the data [2]. It is also seen, that the particle velocity at 1.72 g/cm^3 density (Fig.4) is approximately 100 m/s above the velocity at 1.73 g/cm^3 density for RDX$_1$ (Fig.3), when the spike fades. That confirms the possibility of underdriven detonation existence when the density exceeds ρ_c.

The obtained results demonstrate that the ρ_c essentially depends on the sample structure and is determined not only by the RDX particle size, but also by the pressing process. It is known [3], that at pressing many of the explosive particles are cracked and sheared. Pressing with a small quantity of acetone (RDX$_1$) creates a lot of the potential centers of reaction, and the decomposition rate and explosive part reacting in a shock front increase. Therefore the detonation wave without Von Neumann spike is formed at ρ_0 > 1.72 g/cm^3. Pressing with a big quantity of acetone (RDX$_2$)

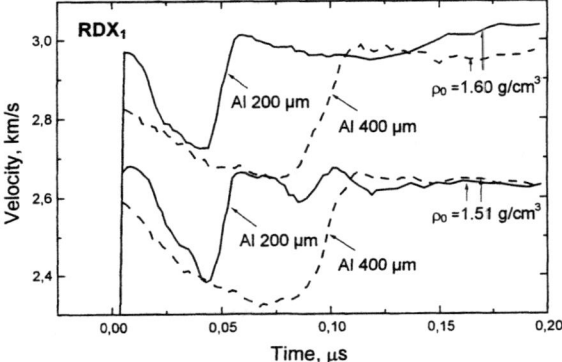

Figure 2. The velocity of Al foils-water window boundary for RDX$_1$, when the initial density is less than critical value.

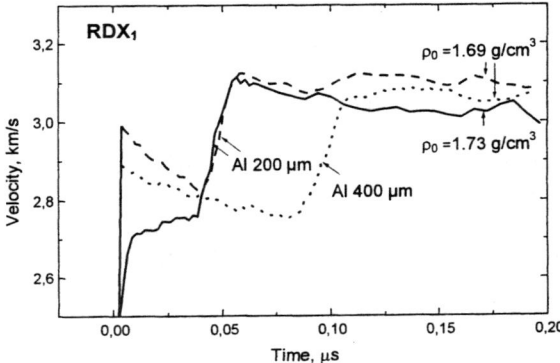

Figure 3. The velocity of Al foils-water window boundary for RDX$_1$ nearly critical density.

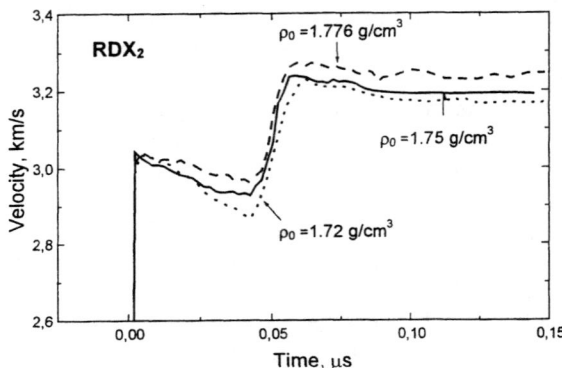

Figure 4. The velocity of Al foils-water window boundary for RDX$_2$. Thickness of Al foil is equal to 200 μm.

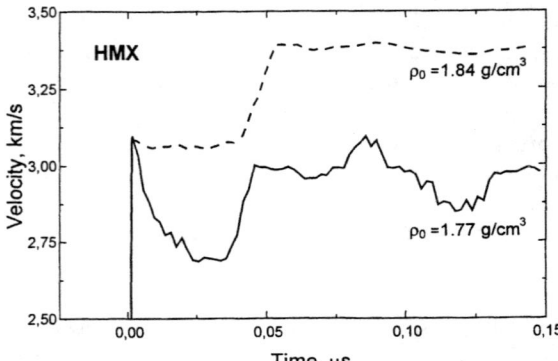

Figure 5. The velocity of Al foils-water window boundary for HMX. Thickness of Al foil is equal to 200 μm.

gives the same density at smaller damage of RDX particles, which decreases the decomposition rate and the spike is recorded.

The similar experiments were conducted for HMX. Samples were pressed from powder with a mean particle diameter of 150 μm. The quantity of acetone added in the HMX at pressing did not exceed 0.5 wt%. Figure 5 shows the experimental results. The velocity spike with the duration ~30 ns is recorded at the ρ_0=1.77 g/cm^3. When the density is equal to ρ_c=1.84 g/cm^3 the spike disappears, and the velocity remains constant after the shock jump. This result agrees well with the data given in [2], where the pressure increase in the reaction zone was found at the initial density 1.87 g/cm^3.

TNETB and ZOX

TNETB is an explosive with low negative oxygen balance (-4.15 %) and density of a single crystal 1.839 g/cm^3 [4]. The detonation heat and sensitivity of TNETB are at the same level as for RDX. The samples were pressed without any solvents.

Figures 6, 7 show experimental results. At the initial density of 1.48 g/cm^3 the spike is observed, its duration is ~50 ns. The increase of ρ_0 up to 1.51 g/cm^3 leads to the spike amplitude decrease, and at 1.56 g/cm^3 the spike disappears and after the shock jump the velocity is constant. At further increase of initial density, the velocity increases by ~100 m/s during ~30 ns after the jump. Good reproducibility of the experimental results is observed when the charge of double length is investigated (two curves in Fig.7 at 1.61g/cm^3). It means that steady-state detonation is realized in all the experiments. As well as for RDX, the abnormal change of particle velocity at critical density ρ_c=1.56 g/cm^3 (Fig.7) was found: for ρ_c the velocity is approximately 50 m/s lower than for 1.51 g/cm^3. It can be explain by transition to underdriven detonation at the disappearance of Von Neumann spike.

ZOX is a powerful explosive completely balanced on oxygen, with the density of a single crystal 1.87 g/cm^3 [5]. Figure 8 shows the experimental results for ZOX. The disappearance of Von Neumann spike is registered at ρ_c=1.71 g/cm^3: after the jump the velocity remains practically constant, whereas at smaller initial density the spike is registered. At 1.51 g/cm^3 its amplitude is equal to ~ 300 m/s, and duration is ~50 ns. The increase of initial density to 1.61 g/cm^3 reduces the amplitude and duration of the velocity spike approximately by 20 %.

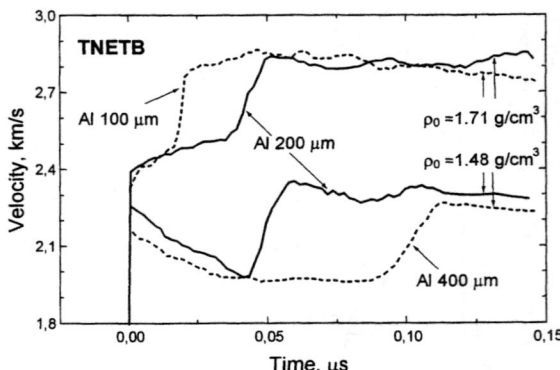

Figure 6. The velocity of Al foils-water window boundary for TNETB.

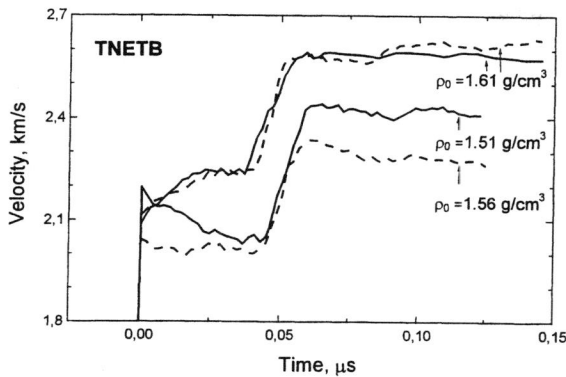

Figure 7. The velocity of Al foils-water window boundary for TNETB. Thickness of Al foil is equal to 200 μm.

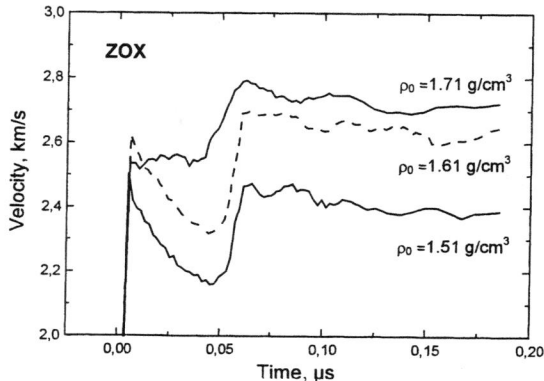

Figure 8. The velocity of Al foils-water window boundary for ZOX. Thickness of Al foil is equal to 200 μm.

DISCUSSION

The results of this work confirm the possibility of detonation wave propagation without Von Neumann spike in powerful HE. The reaction zone structure changes qualitatively at the critical initial density. It can be explained by growth of the initial decomposition rate of explosive with increase of density if it is assumed that the physicochemical transformations take place in the compression wave. The influence of RDX charge structure on the critical density agrees with this assumption. Moreover when the pressure increases in the reaction zone, the final state of detonation products can be on the weak part of detonation Huhoniot.

Steady-state detonation wave without Von Neumann spike does not correspond to the classical model which ignores the chemical reaction in the shock jump. Nevertheless, strong coupling between the shock and reaction zones was noted [6-8]. In the [6] attention is given to the fact, that after the shock jump the non-equilibrium HE Hugoniot is realized. Further there is a relaxation to an equilibrium HE Hugoniot and chemical reaction.

Mathematical model of detonation phenomenon with the reaction in shock compression is analyzed in [7,8]. It is proposed that the shock front thickness is determined by viscosity. At low reaction rate the detonation wave does not differ significantly from the classical model. The Von Neumann spike amplitude only decreases and a part of HE reacts in the shock front. However when the reaction rate increases the structure of the detonation wave changes crucially: the final state on the detonation Hugoniot is reached as a result of monotonic pressure increase. Thus underdriven detonation can be realized, and the detonation velocity will not only be determined by thermodynamics and gas dynamics, as it is in the classical detonation theory, but by the chemical reaction kinetics and shock wave front structure as well.

The work has been funded by Russian Fund for Basic Research, grant number 00-03-32308a

REFERENCES

1. Zeldovich Ya.B., *The theory of shock waves and introducing in gas dynamics*, Academy Science of USSR, Moscow, 1946, pp.45-98.
2. Ashaev V.K., Doronin G.S., Levin A.D., *Phys. of Combustion and Explosion*, **24**, N1, pp.95-101 (1988).
3. Burnside N.J., Son S.F., Asay B.W., Skidmore C.B., "Particle characterization of pressed granular HMX," in *Shock Compression of Condensed Matter–1997*, edited by S.C.Schmidt et al., AIP Conference Proceedings 429, New-York, 1998, pp.571-574.
4. Price D., *Chemical Reviews*, .**59**, pp.801-825 (1959).
5. Haishan D., "Properties of Bis(2,2,2,-Trinitroethyl-N-Nitro) Ethylenediamine and Formulations Thereof," in *X Symposium (International) on Detonation*, edited by S.C.Schmidt et al., AIP Conference Proceedings 429, New-York, 1989, pp.995-1000.
6. Dremin A.N. *Toward Detonation Theory*, Springer, New York, 1999, pp.123-203.
7. Williams F.A., *Combustion Theory*, Addison-Wesley Publishing Company, London, 1964, pp.194-213.
8. W.Fickett. *Introduction to Detonation Theory*. University of California Press, Berkely - Los Angeles - London, 1985, pp.55-61.

THE EFFECT OF VARIATION OF ALUMINUM PARTICLE SIZE AND POLYMER ON THE PERFORMANCE OF EXPLOSIVES

Diana Woody[1] and Jeffery J. Davis[2]

[1]Combustion Research Branch, Engineering Sciences Division, Research Department; and
[2]Detonation Sciences Branch, Engineering Sciences Division, Research Department,
Naval Air Warfare Center Weapons Division, 1 Administration Circle, China Lake, CA 93555-6100

Abstract. The Ballistic Impact Chamber (BIC) test was used to measure the effect of particle size upon performance of the energetic material. Two different particle sizes of aluminum, 20 micron (μm) and 150 nanometer (nm), were added to LX-17 to measure their effect upon hazard sensitivity and energy/gram output. Also, the effect of the binder on the performance of the mixtures was explored. For this paper, Viton and Kel-F were compared. The Ballistic Impact Chamber test measures the initial rate of reaction, time of reaction, the burning reaction, and the energy output during the impact of the energetic material.

INTRODUCTION

The Ballistic Impact Chamber (BIC) test is a small-scale instrumented test capable of measuring sensitivity and energy/gram output. It does so by measuring the initial rate of reaction, time of reaction, and the energy output during the impact of the energetic material. These parameters were used to discern differences in the reaction due to the particle size of the aluminum additive.

Previous studies have been performed to measure the effect upon performance of the addition of aluminum to explosives (1). The use of micron-sized aluminum additives has become the standard for explosive formulation throughout the Department of Defense (DOD) for enhancing blast.

It has since been observed that particle size plays a significant role in the exothermic output from energetic materials (2). In this paper, the effect of the size of aluminum added as well as the type of binder added to an explosive has been studied. Two aluminum particle sizes, 20 μm and 150 nm were used in the LX-17 analog compositions to compare their exothermic output upon impact. The two binders compared were Viton and Kel-F.

EXPERIMENT

The BIC test is a closed-volume version of the simple drop weight test setup. The impact machine is 2 m high and 0.30 m wide. The distance from the bottom of the drop weight to the anvil is 1.57 m. This height corresponds to the free-fall velocity of 5.5 m/s. A laser diode and light meter were used to trigger the acquisition of the data obtained from the pressure gauge.

Each sample was a cylindrical disk approximately 1.25 mm thick and 5 mm in diameter. The mass of the sample was between 35 to 50 mg. The sample to be tested was placed on garnet paper (Norton Corp. #01489 garnet A511, 180A grit paper). The garnet paper served to standardize the anvil surface friction. The sample and sandpaper were placed in the test cavity of the BIC. Figure 1 shows the BIC test chamber.

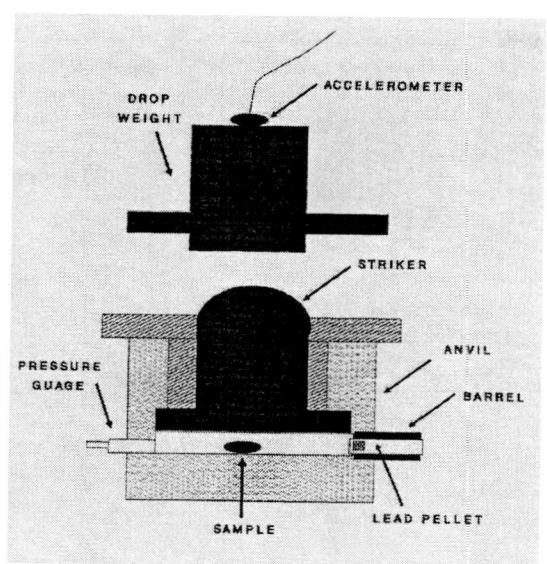

FIGURE 1. Ballistic Impact Chamber.

Two ports were located at the bottom of the wall of the cup. One port was connected to a 30-cm-long 0.177 caliber gun barrel which held a 0.177 caliber lead airgun pellet (Crossman Copperhead) weighing 0.51 gram. The second port was connected to a fast-response pressure gauge (PCB Piezotronics, Model #113). This gauge measured the pressure time development of the reacting gases from the impacted energetic material. The hot gases generated by the reacting sample accelerated the pellet along the gun barrel. Work done on the pellet by the gases was a measure of the energy released by the impacted energetic material. The integration of the pressure/time history trace was used to calculate the energy released by the sample.

The nanosized aluminum "ALEX" (Argonide Inc. Lot # A06-24-25R) had a wide particle size distribution over a range of 50 – 500 nm with a typical size of 150 – 200 nm. A major concern when evaluating the contribution of nanosized aluminum is the oxide layer that exists on the particle. This oxide layer is typically on the order of 3 nm thick for all particle sizes. It is present because of the diffusion of oxygen, which leads to passivation of the surface of the metal. For micron sized particles, the 3 nm shell does not significantly affect the amount of pure metal in the particle.

However, the oxide shell can take up an appreciable volume of the of a nanosized particle. The oxide layer of the Al used in this study was measured to be approximately 14% by weight of the particle. Thus, the total amount of Al available to react with the other materials in the nanometer aluminum composition is less than is available in the micron Aluminum composition.

Five sets of samples were tested. The baseline formulation was LX-17. The other two formulations were analogs of LX-17 containing 20 μm and 150 nm aluminum, respectively. Viton was substituted for Kel-F for two of the mixtures. Ten sample tests were conducted on each composition; the results were then averaged. The explosive compositions are given in Table 1.

RESULTS AND DISCUSSION

The averaged results of the BIC tests of this study are shown in Table 2. The total area under the pressure time curve (Equation (1)) was used to determine the average pressure released by the hot gases of the reacting sample. Typical trace of pressure vs time are shown in Fig. 2 and 3.

$$P = 1/T \int P(t)dt \quad (1)$$

where $T = t_{final} - t_0$

The total energy released from the gases of the reacting energetic material during impact was linked to the pressure via:

$$\text{Energy} = (P) \times (\text{Volume of Chamber}) \quad (2)$$

where Volume = 0.0288 cm^3

The data was standardized as Energy/gram

where

$$E/gm = \text{Energy/mass of sample (grams)} \quad (3)$$

It can be observed from the data that the addition of both sized aluminums to LX-17 increased the sensitivity measured as dp/dt and energy output measured as energy/gram. There is a

TABLE 1. Compositions Used in the BIC Tests.

Explosive	Organic	Metal	Binder	Density (g/cc)	Process
LX-17	92.5% TATB	0 %	17.5 % Kel-F	1.9	Pressed
TATB/20 μm Al/Kel-F	72.5% TATB	20% 20 μm MDX81	7.5 % Kel-F	1.97	Pressed
TATB/150 nm Al/kel-F	72.5% TATB	20% 150 nm Al - Alex	7.5 % Kel-F	1.97	Pressed
TATB/20 μm Al/Viton	72.5% TATB	20% 20 μm MDX81	7.5 % Viton	1.97	Pressed
TATB/150 nm Al/Viton	72.5% TATB	20% 150 nm Al - Alex	7.5 % Viton	1.97	Pressed

TABLE 2. Results from the BIC Test.

Material and Composition	Peak Pressure (psi)	Sensitivity dp/dt (psi/μs)	Energy (J/gm)
LX-17	67.1	0.139	8.05
TATB, Kel-F, and (20 μm aluminum)	53.4	3.45	9.65
TATB, Kel-F, and (150 nm aluminum)	146.52	6.73	20.86
TATB, Viton, and (20 μm aluminum)	63.6	0.23	10.59
TATB, Viton, and (150 nm aluminum)	147.5	0.56	17.44
PBXN-109 (cast)	245.5	30.2	37.83

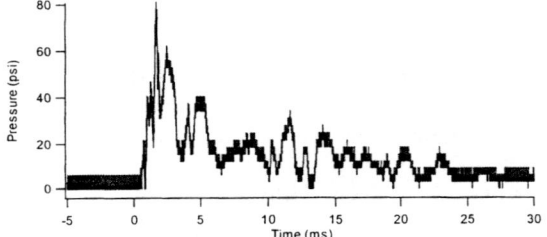

FIGURE 2. Pressure vs. time trace for LX-17.

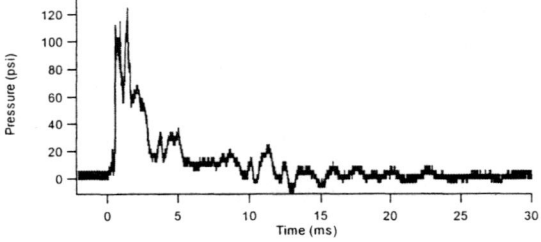

FIGURE 3. Pressure vs. time trace for LX-17 with 150 nm aluminum added.

greater increase in sensitivity and energy output for the TATB-Kel-F mixture containing the 150 nm sized aluminum than that observed for the TATB-Kel-F mixture containing the 20 μm aluminum. The increase in sensitivity is significantly lower for both sized aluminums mixed with TATB and Viton than that observed in the mixture containing Kel-F as a binder. Results from PBXN-109 are also shown in Table 2 for comparison values.

The greater sensitivity observed in the TATB mixture containing aluminum and Kel-F could be due to the contribution from the fluorine in the Kel-F polymer. Shear has been shown to be an important factor in the reactivity of energetic materials. Possibly the small size of the fluorine atom could make the potentially more polar and fluorine-rich Kel-F more reactive with the aluminum additive under shear conditions than the fluorine-deficient Viton.

CONCLUSIONS

This study has shown that the size of the aluminum added to an explosive composition can have an affect upon its energy output and sensitivity. The smaller aluminum increased the sensitivity and the energy output of the explosive. This was shown in the pressure-time traces obtained from the gases released under closed volume impact tests. The substitution of Viton for Kel-F when adding aluminum to the TATB mixture resulted in a slightly lower increase in energy output than that observed for the mixture containing Kel-F. But at the same time, the sensitivity does not show as great an increase with the TATB mixture containing Viton than it does an increase with the more polar Kel-F polymer.

Further BIC and Differential Scanning Calorimetry (DSC) studies will be performed to measure the effect that humidity may have on the different sized aluminums. This data should give an indication of which sized aluminum is more vulnerable to humidity or general cycling events that may occur upon storage of the material.

ACKNOWLEDGMENTS

This research was performed under the sponsorship of Tom Loftus, Air Weaponry Technology Program.

REFERENCES

1. Woody, D. L., Davis, J. J., and Coffey, C. S., "Comparison of Infrared Emissions from Impacted Aluminum to Non Aluminum Containing Energetic Materials," in *JANNAF Hazards Meeting Proceedings*, April 1992.
2. Woody, D. L., and Davis, J. J., "The Effect of Particle Size and Porosity on Metal/Metal Exothermic Reactions Induced by Low Velocity Impact," in *14th U.S. Army Symposium on Solid Mechanics Proceedings*, edited by K. Iyer and S. Chou, Battelle Memorial Institute, 1997.

NEAR-FIELD IMPULSE EFFECTS FROM DETONATION OF HETEROGENEOUS EXPLOSIVES

David L. Frost[1], Fan Zhang[2], Susan McCahan[3], Stephen B. Murray[2], Andrew J. Higgins[1], Marta Slanik[1], Marc Casas-Cordero[1], and Chayawat Ornthanalai[1]

[1]*McGill University, Department of Mechanical Engineering, 817 Sherbrooke St. W., Montreal, Quebec, Canada H3A 2K6*
[2]*Defence Research Establishment Suffield, PO 4000, Stn Main, Medicine Hat, Alberta T1A 8K6 Canada*
[3]*University of Toronto, Department of Mechanical and Industrial Engineering, 5 King's College Rd., Toronto, Ontario M5S 3G8 Canada*

Abstract. Particle momentum effects from the detonation of a spherical heterogeneous charge consisting of a packed bed of inert particles saturated with a liquid explosive have been investigated experimentally and numerically. When such a charge is detonated, an interesting feature of the subsequent flow field is the interplay between the decaying air blast wave and the rapidly expanding cloud of particles. Using a cantilever gauge, it is found that the particle momentum flux provides the primary contribution of the multiphase flow to the near-field impulse applied to a nearby small structure. To determine the impulse from the particle momentum flux on the structure, a novel particle streak gauge was developed to measure the rate of particle impacts at various locations. The trends of the experimental results are reproduced using an Eulerian two-fluid model for the gas-particle flow and a finite-element model for the structural response of the cantilever gauge.

INTRODUCTION

The addition of solid particles to a homogeneous explosive introduces additional length and time scales that influence both the detonation propagation within the reactive two-phase medium and the subsequent blast wave propagation and particle dispersal in the surrounding air. The detonation propagation within the heterogeneous explosive has a nonideal nature due to nonequilibrium effects associated with the acceleration, heating and fracture of the particles (1, 2). When the detonation wave reaches the surface of a spherical charge, a shock wave is transmitted into the surrounding air and the particles are accelerated radially by the expansion of the combustion product gases to supersonic speeds. For large (e.g., millimeter-sized) particles, after the rapid acceleration within the combustion products, the particle velocity remains roughly constant due to the inertia of the particles. In some cases, the particles catch up to and penetrate the leading shock wave, as predicted numerically by Lanovets et al. (3) and confirmed experimentally by Zhang et al. (4). In this case, the particles decelerate due to aerodynamic drag until eventually the shock wave overtakes the particles in the far field. The interchange of momentum and energy between the particles and the surrounding flow field changes the rate of decay of the blast wave pressure and impulse in comparison with a homogeneous explosive (5). In the near field, the impact of the particles with nearby structures can make a significant contribution to the overall structural damage. In the present paper, experimental and numerical results will be presented which investigate the relative contribution of the particles and the blast wave to the bending work done on a nearby cantilevered structure. A particle streak gauge was developed to record the time history of particles arriving at a given location. The experimental results are used to validate the predictions using a two-fluid multiphase model. The model is then used to determine the dominant mechanism responsible for the work done by the multiphase flow on nearby structures.

EXPERIMENTAL

The experiments were carried out using spherical charges consisting of packed beds of spherical steel beads saturated with nitromethane sensitized with 10% triethylamine. The heterogeneous charges tested had a diameter of 11.8 cm and contained about 10 million 463 μm steel beads (4,300 g) together with 430 g of NM + 10% TEA. Details of the experimental procedure can be found in Zhang et al. (4).

A "particle streak" gauge was developed to determine the rate of particle collisions with a surface at a given location from the charge. It consisted of an aluminum cylinder (7.5 cm dia) attached axially to the shaft of an AC motor which was operated at a speed of 3685 rpm. A thin sheet of aluminum was wrapped around the cylinder and attached with double stick tape. The motor/cylinder assembly was placed inside a steel cylinder which contained a vertical slot (5.08 cm x 0.64 cm) to allow the particles to strike the cylinder as it rotated. After a trial the steel cylinder was removed, and the aluminum foil was recovered to obtain a direct temporal history of the impact of particles with the cylinder. A photograph of a foil sample is shown in Fig. 1. The distance between the lines drawn on the foil corresponds to the width of the slot. The time for the foil to move a distance of one slot width is about 0.45 ms.

The cantilever gauges consisted of a 38.1 cm long aluminum rod (with a diameter of either 0.95 cm or 0.635 cm) with 10.16 cm of the rod clamped in a tripod stand. A steel plate (20.3 cm x 5.1 cm x 0.48 cm) was fixed to the top portion of the rod with an aluminum bracket, to serve as a witness plate for the impact of the blast wave and the particles. Strain gauges were also placed just above the bend location.

RESULTS AND DISCUSSION

When particles are added to a homogeneous liquid explosive, the peak blast wave overpressure is reduced in comparison with a homogeneous charge. For example, Frost et al. (5) showed that for the heterogeneous charge described above, the peak overpressures generated were smaller than those for a homogeneous charge (with the same amount of NM) by a factor of 2–3. In contrast, the heterogeneous charge produced substantially more bending of the cantilever gauge than a homogeneous charge, as shown in Fig. 2.

For a homogeneous charge, the load applied to the cantilever gauge will consist of the sum of i) the short-duration force applied as the blast wave diffracts around the gauge and ii) the drag force from the flow behind the blast wave.

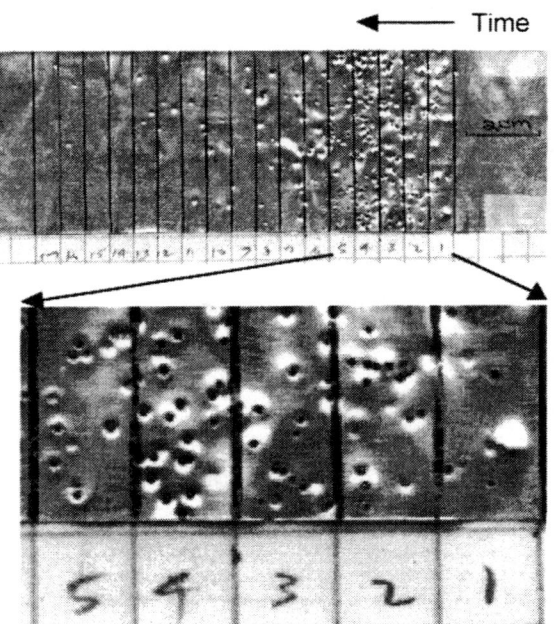

FIGURE 1. Foil recovered from particle streak gauge showing history of steel particle impacts at a distance of 90 cm from the charge. Note that some particles remain embedded in foil. Particle diameter is 463 μm.

For a heterogeneous charge, a third force is present due to the particle collisions with the cantilever gauge. To determine the relative magnitudes of the three forcing functions mentioned above, it is necessary to use the data obtained with the particle streak gauge together with overpressure data.

Experiments were carried out with the particle streak gauge located at distances of 60, 90, 156, and 200 cm away from the center of the charges. By counting the number of particle impacts in each segment of the foil (see Fig. 1), the cumulative number of impacts was determined as a function of time, and is shown in Fig. 3 for a distance of 90 cm. Note that from the particle streak gauge data alone, we cannot determine the absolute arrival time of the first particle at the gauge. This arrival time was obtained from earlier flash X-ray data of the expanding particle cloud (see Ref. 4). Also shown in Fig. 3 are the results of the Eulerian two-fluid model described in Zhang et al. (4). The model predicts that the particles will arrive at a given location slightly earlier than measured, although the arrival time profile is similar. The actual number of particle "hits" recorded on the foil was 345, which is quite close to the predicted value of 358, assuming that the particles are distributed in a spherically symmetric fashion.

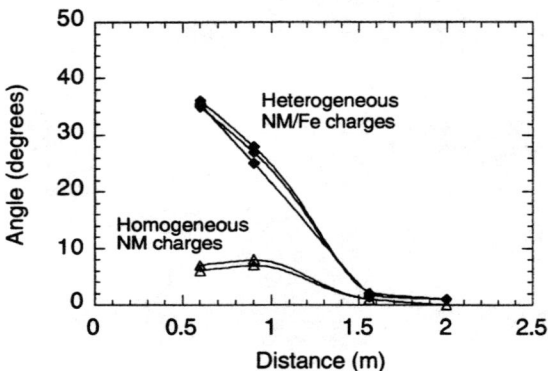

FIGURE 2. Cantilever bend angle as a function of distance from the charge for heterogeneous charges (containing 430 g of sensitized NM with 4,300 g of 463 μm steel beads) and homogeneous charges (with the same amount of NM). Cantilevers at 0.6 m had diameters of 9.5 mm whereas other cantilevers had diameters of 6.4 mm.

Using the particle streak gauge data at the four locations, an estimate of the velocities for particles arriving at a given location, as a function of time, can be made. However, considering the reasonable prediction of the particle dynamics by the two-phase model (Fig. 3), the model has been used to estimate the particle and gas velocity behind the blast wave at various distances from the charge. For example, at 90 cm from the charge, the blast wave arrives just before the particles, but the peak gas velocity behind the blast wave (about 500 m/s) is smaller by a factor of 2 than the peak particle velocity. The particle velocity also decays more slowly than the gas velocity due to the high inertia of the particles.

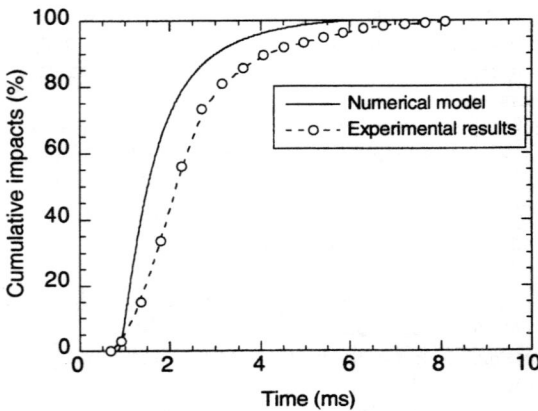

FIGURE 3. Cumulative impacts on the particle streak gauge as a function of time at a distance of 90 cm from the charge: experimental results and model predictions.

The work done on the cantilever gauge depends on the momentum fluxes ρV^2 associated with the gas and particles, which are shown in Fig. 4. From this figure it is apparent that the impulse to the cantilever gauge due to the particle collisions (which equals $(1 + e) \int \rho_p V_p^2 dt$, where $0 < e < 1$ is the coefficient of restitution) is considerably larger than the impulse due to aerodynamic drag ($C_D/2 \int \rho_g V_g^2 dt$). The reflected blast wave pressure provides the third contribution to the impulse which is generally larger than the drag force but smaller than the particle force. At a distance of 90 cm, the particle collisions provide the majority (about 70%) of the total impulse applied to the cantilever. The particle force is applied over a time of about 2 ms whereas the force due to the reflected blast wave occurs over the relatively short time (~ 200 μs) for pressure relief as the blast wave diffracts around the cantilevered plate (see Kinney & Graham, 6).

A finite element model was developed using the code ABAQUS to model the response of the cantilever to a dynamic load. The loading functions were determined assuming that the particles collide elastically with the cantilever plate and the peak reflected pressure was estimated using measured side-on pressure data and reflected pressure data from Baker (7). The calculated dynamic displacement of the end of the cantilever gauge is shown in Fig. 5 for the heterogeneous and homogeneous charges. The final bend angles agree well with the experimental results (Fig. 2) considering that the collisions are not perfectly elastic. Note that if only the drag and reflected blast wave forces for the heterogeneous charge are applied to the cantilever gauge (i.e., neglecting the particle force), then negligible bending work is done on the cantilever.

In summary, it has been found that adding inert particles to a homogeneous explosive reduces the peak overpressure in the near field (within 10 charge diameters). However, the integrated momentum flux of the particles in the near field is larger (by a factor of 3–4) than the gas momentum flux. The dynamic response (strain and final bend angle) of a cantilevered target is greater for a heterogeneous charge than for a homogeneous charge with the same energy. Further improvements to the modelling of the near-field heterogeneous blast and particle dispersion dynamics will require the development of appropriate equations of state for the dense solid-gas system, including a treatment of a large number of high-speed inelastic collisions.

ACKNOWLEDGMENTS

The authors wish to acknowledge the assistance of K. Gerrard, A. Nickel, D. Boechler and S. Trebble during the field trials. Useful information regarding cantilever gauges was also provided by Dr. A. van Netten.

REFERENCES

1) Lee, J. J., Frost, D. L., Lee, J. H. S., and Dremin, A., *Shock Waves* **5**, 115-119 (1995).
2) Lee, J. J., Brouillette, M., Frost, D. L., Lee, and J. H. S., *Combustion and Flame* **100**, 292-300 (1995).
3) Lanovets, V. S., Levin, V. A., Rogov, N. K., Tunik, Yu.-V., and Shamshev, K. N., *Fizika Goreniya i Vzryva* **29**, 88-92 (1991).
4) Zhang, F., Frost, D. L., Thibault, P. A., and Murray, S. B., *Shock Waves* **10**, 431-443 (2001).
5) Frost, D. L., Kleine, H., Slanik, M., Higgins, A. J., McCahan, S., Zhang, F., and Murray, S. B., "Blast Waves from Heterogeneous Explosives," *Proceedings of 22nd International Symposium on Shock Waves*, paper 0560, CDROM (1999).
6) Kinney, G. F. and Graham, K. J. *Explosive Shocks in Air*, 2nd Ed., Springer-Verlag, New York, (1985).
7) Baker, W. E., *Explosions in Air*, University of Texas Press (1973).

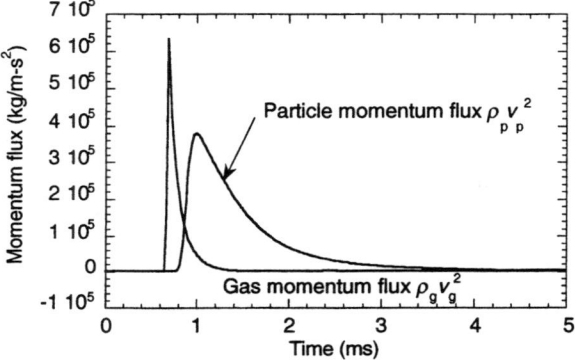

FIGURE 4. Numerical particle and gas momentum flux at a distance of 90 cm from heterogeneous charge.

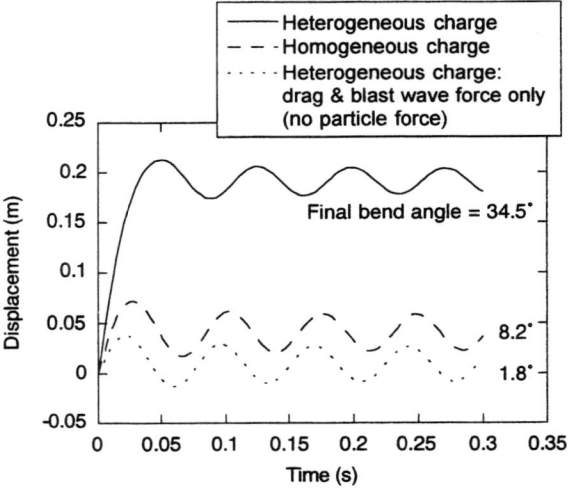

FIGURE 5. Calculated dynamic displacement of top of cantilever gauge subject to loading from heterogeneous and homogeneous charges at a distance of 90 cm from charge.

EFFECT OF METAL PARTICLE SIZE ON BLAST PERFORMANCE OF RDX-BASED EXPLOSIVES

Jeffery J. Davis and Philip J. Miller

Naval Aviation Science and Technology Office, Research and Technology Department, Code 4T4330D, Naval Air Warfare Center Weapons Division, 1 Administration Circle, China Lake, CA 93555-6100

Abstract. This paper discusses the role that aluminum particle size has on explosives blast performance. Tests were performed using a small sealed chamber and an open-ended shock tube. Three explosives were tested and the results presented. The Al particle size examined was 20 microns and 150 nanometers in a pressed PBXN-109 analog composition. (PBXN-109 was also tested.) A noticeable difference in the internal blast pressure was observed between the 20 μm and 150 nm Al in the sealed chamber but not in the shock tube. The chamber results compare favorably with modeling work performed.

INTRODUCTION

The addition of metal to energetic materials has been shown to be effective in increasing the blast characteristics of formulations. The role that metals play in affecting the reaction of energetic material has been studied for decades. The accepted reaction mechanism of the metal is that the particle reacts after the detonation wave has swept past, releasing its energy in a much longer time frame and thus creating a late time reaction and a longer pressure profile. The use of μm-sized Al additive has become the standard for explosive formulation throughout the Department of Defense (DOD). The Navy's standard explosive for blast is PBXN-109, which contains RDX and 20 μm Al.

While metals have been studied for decades, the availability of small particle metals (particularly Al) over the past couple of years has prompted researchers to examine the effect that particle size has on energetic materials (1,2,3). The dramatic results obtained by increasing the burn rate of propellants have fueled interest from the explosive community. The main metal reaction pathway is the oxidation of the metal.

$$2 Al + 3/2 O_2 \rightarrow Al_2O_3 + heat$$

This reaction is diffusion-limited if the oxide coating remains in contact with the unburned Al. As the particle size is decreased, the speed of that reaction increases and the ability for complete conversion to the oxide is increased.

The effect of using smaller sized Al should be observed in the increased pressure of the initial blast wave. As the reaction is allowed to continue, it has been surmised that the larger Al will burn late, while the smaller Al will have been consumed. Thus, the total energy is the same while the energy release profile is different.

EXPERIMENT

To test blast performance of the explosives, a closed chamber test was used to evaluate the initial blast effects from explosives. The steel cylindrical chamber used was 10 inches (inner diameter) by 16 inches (high) with 1-inch-thick walls. A pressure gauge (Kistler Model 607-C3 100,000 psi) was mounted at the top of the chamber in a Teflon mount (Fig. 1). (Teflon is used to try to isolate the gauge from the shock traveling in the chamber wall.) The amount of energetic material used was

2-3 grams and were 1/2- x 1/2-inch cylinders. An explosive pellet and a detonator (Reynolds RP-501) were held in an aluminum fixture mounted on a Plexiglas block. The fixture provided for some confinement as well as ensured that the detonator and explosive were in contact.

FIGURE 1. Internal blast chamber.

The chamber has an interior volume of 20,600 cc, this results in approximately 0.2 mole of O_2 present in the chamber under standard temperature and pressure. Typical samples have 0.02 mole of Al. Thus, we had sufficient O_2 present for complete reaction of the Al into Al_2O_3.

The second experimental setup consisted of a shock tube sealed at one end and open at the other. The steel tube was 20 feet long and 4 inches in diameter. Pressure gauges (PCB 102A) were mounted along the side of the tube at various locations away from the explosive. The gauges were at 4 inches, 15 inches, and 3, 8, 13, and 18 feet. A picture of the shock tube is given in Fig. 2.

FIGURE 2. Shock tube.

Three sets of samples were tested for this paper. The baseline formulation was PBXN-109. The two other formulations were pressed analogs to PBXN-109. One set contained 20 µm Al and the other 150 nm Al. Each composition was tested five times. The explosive compositions are given in Tables 1 and 2.

MDX-81 Al has an average particle size of 20 µm and is spherically shaped. This Al is used in PBXN-109. The nanosized Al "Alex" was from Russia (Argonide Inc. Lot # A06-24-25R). It had a wide particle size distribution with a range of 50 to 500 nm with a typical size of 150 to 200 nm.

A major concern with evaluating nanosized Al is the oxide layer that exists on the particle. This layer is typically on the order of 3-nm thick for all particle sizes. It is due to the diffusion of oxygen, which leads to passivation of the surface of the metal. For µm-sized particles, the 3 nm shell is not significant in evaluating the amount of pure metal in the particle. However, for a nanosized particle, the oxide shell can take up an appreciable volume of the particle. For the Al used in this study, the oxide layer was measured to be approximately 14% by weight of the particle. Thus, the total amount of available Al is lower in the nanosized composition.

TABLE 1. Compositions used in internal blast tests.

Explosive	Organic	Metal	Binder	Density (g/cc)	Process
PBXN-109	68% RDX	20% 20 µm – MDX-81	12% HTPB	1.69	Cast
RDX/20 µm Al/Binder	70% RDX	20% 20 µm – MDX-81	10% Zeon	1.64	Pressed
RDX/150 nm Al/Binder	70% RDX	20% 150 nm Al – Alex	10% Zeon	1.71	Pressed

TABLE 2. Compositions used in shock tube tests.

Explosive	Organic	Metal	Binder	Density (g/cc)	Process
PBXN-109	68% RDX	20% 20 µm – MDX-81	12% HTPB	1.69	Cast
RDX/20 µm Al/Binder	70% RDX	20% 20 µm – MDX-81	10% Viton	1.84	Pressed
RDX/150 nm Al/Binder	70% RDX	20% 150 nm Al – Alex	10% Viton	1.89	Pressed

RESULTS AND DISCUSSIONS

The results from the two sets of tests are presented.

Internal Blast Chamber

A typical pressure-time trace in the internal blast chamber for the 20 µm Al is shown in Fig. 3. The ringing is due to reflections in the chamber as the pressure wave cycles between the top and bottom of the chamber.

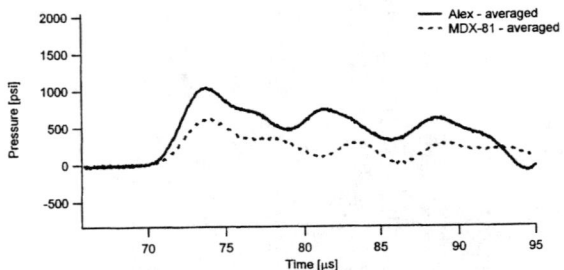

FIGURE 4. Comparison between the 20 µm Al (MDX-81) and the 150 nm Al (Alex).

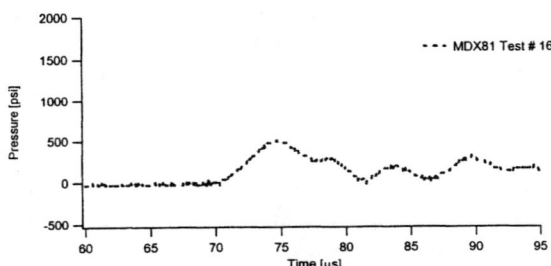

FIGURE 3. Pressure–time trace for 20 µm Al sample.

In Fig. 4, the averaged results from five 150 nm Al tests are compared with the averaged results from five 20 µm Al tests. The initial pressure of the nanosized Al is clearly higher than the µm Al (an average increase of 63%). If one determines the impulse by integrating the pressure-time curve from the initial rise in pressure for a fixed time, the impulse from the composition containing Alex is 2.75 times higher then the MDX-81 composition.

The results shown in Fig. 4 compare well with the calculated pressures from DYNA2D (4). The calculation showed the ringing as the wave is reflected inside the chamber as seen in the experiment. The model also predicted a higher peak pressure and impulse for the nanosized Al, as was observed.

Shock Tube

A comparison of the RDX/Viton/Al compositions is shown in Fig. 5.

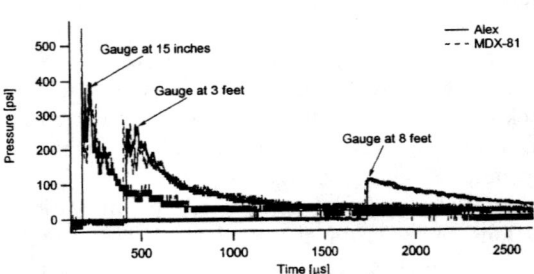

FIGURE 5. Shock tube result comparing Al particle size.

The results from the tube were not as straight forward as those observed in the chamber. First, the arrival times of the wave at the gauges were not consistent in the same test and also varied within the same material. This made analysis difficult due to the lack of a consistent time. Also, comparison was made with PBXN-109 and PBXN-5 (RDX and no Al). PBXN-5's pressure traces were very similar in structure to the pressed 109 analogs. While slight differences were noticed with the different Al, they were not as dramatic as was observed in the chamber. The impulse for the nanosized Al was slightly higher but was due, for the most part, to the mismatch of rise times. At the gauge located 4 inches from the explosive charge, the electrical noise from firing the detonator made determination of initial rise often difficult. The average from five tests on each material is presented in Figure 6. Figure 7 shows a comparison from single set of tests over a longer time base.

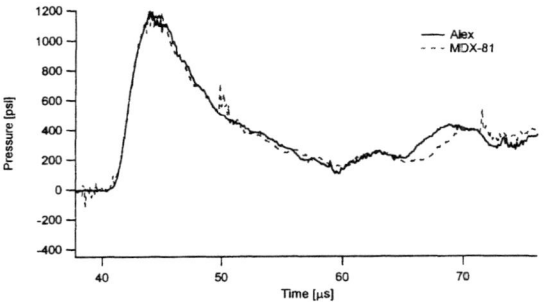

FIGURE 6. Pressure–time trace of average of five tests from gauge located 4 inches from explosive charge.

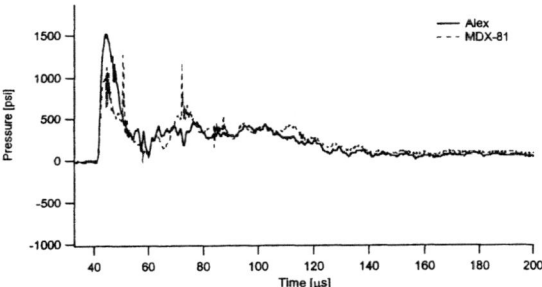

FIGURE 7. Pressure–time trace of individual tests from gauge 4 inches from charge.

Gauges mounted on the side might have caused some of the differences between the tube and the chamber that had its gauge mounted on the top in a direct path of the oncoming particles. While the gauge in the chamber was protected, it might not have been adequate. The nanometer-sized Al would be able to travel faster than the 20 µm Al and thus might be able to impact a gauge mounted in its path, increasing the observed pressure. Turbulence for mixing is known to enhance metal reaction but neither tube nor chamber should have experienced much turbulent mixing during the initial pressure rise.

CONCLUSIONS

The nanosized Al showed a significantly greater initial blast pressure in the closed chamber test but no change was observed in the shock tube. The blast chamber's experimental work has good agreement with the calculated blast pressure obtained from DYNA2D modeling. Work is necessary to develop a better test method to observe the reaction mechanisms in air. Modeling work is also continuing to develop a reaction kinetics model to account for differences in particle size of aluminum.

ACKNOWLEDGMENTS

The ONR 6.2 Air Weaponry Technology Program through Tom Loftus and the ONR 6.1 Marine Corps Research Program sponsored this work.

REFERENCES

1. Davis, J.; Miller, P. J.; and Bedford, C. "Effect of metal particle size on the detonation properties of various explosives," in the Proceedings of TTCP – WTP-4 Technical Workshop, Indian Head, MD, April 16, 1999.
2. Woody, D. L.; Davis, J. J.; and Bedford, C. D. "Comparison of the visible emissions from energetic materials containing differing particle sized aluminum," published in Proceedings of JANNAF Combustion Subcommittee and Propulsion Systems Hazards Subcommittee Joint Meeting, November 1996.
3. Davis, J.; Miller, P. J.; Bui, Q.; and Pockrandt, S., "Effect of metal particle size on internal blast explosives," published in Proceedings of JANNAF Combustion Subcommittee and Propulsion Systems Hazards Subcommittee Joint Meeting, November 2000.
4. Miller, P., "A reactive flow model with coupled reaction kinetics for detonation and combustion of non-ideal explosives," Decomposition, Combustion and Detonation Chemistry, eds. T. Brill et.al., Vol. 418, p. 413, Mater. Res. Soc. Symposium Proceedings, 1996.

EFFECT OF AN INERT MATERIAL'S THICKNESS AND PROPERTIES ON THE RATIO OF ENERGIES IMPARTED BY A DETONATION'S 1ST AND 2ND PROPULSION STAGES

Joseph E. Backofen[1] and Chris A. Weickert[2]

[1] BRIGS Co., 2668 Petersborough St., Herndon, VA 20171, USA
[2] Defence Research Establishment Suffield, P.O. 4000 Station Main, Medicine Hat, Alberta T1A 8K6, Canada

Abstract. Analysis of cylinder tests employing aluminum, steel, and copper cylinders of different thickness shows that Gurney Energy measurements have been affected by both the wall thickness and the material's dynamic properties. Experimental data for these tests and plate-push tests also show that the ratio of the initial free-surface velocity (1st propulsion stage) to the final "steady-state" velocity obtained after the explosive gases have fully expanded (2nd propulsion stage) can differ by a factor of two. The data show a clear dependence of this ratio upon the ratio of the inert material's thickness to the explosive's thickness. Phase transitions can also decrease propulsion efficiency by a significant margin.

INTRODUCTION

This paper presents observations gleaned from published experimental data while improving the BRIGS analytical package for explosive charges by separating explosive propulsion into a two step process: 1) initial motion imparted by a brisant process, and 2) subsequent acceleration by a gas-push process. As described in reference 1, the velocity imparted during the 1st propulsion stage can be described by the Energy Transference Ratio (ETR) formulas. The 2nd propulsion stage's gas-push process can be described by the familiar Gurney model wherein an explosive's gas volume expands from a "static" homogeneous "all-burned" high-pressure state into one wherein the velocities of the gases at the boundaries match those of inert boundary materials. Furthermore, due to its wide acceptance, the Gurney model also has been chosen to provide experimental data on conversion of an explosive's chemical energy to the final "steady-state" kinetic energy of the gases and inert materials. Unfortunately, when cylinder test data and plate-push data were collected and examined in order to characterize the energy that was converted during the total explosive-driven event, Gurney Energies were found to vary widely in the literature.

ANALYSIS AND DISCUSSION

Reference 2 was the first paper clearly identifying that there was a difference in the energy delivered by an explosive to different materials. This 1969 paper noted that the energy constant known as the "Gurney Energy" or the "Gurney Velocity" was significantly different when derived from US Navy, US Army, and Lawrence Livermore National Laboratory tests. For example US Navy sources found the Gurney Velocity for Composition B explosive to be 7610 or 7880 ft/sec (2320 or 2400 m/sec) while US Army sources used a value of 8800 ft/sec (2680 m/sec). Empirical data presented in refs. 3 and 4 revealed that early US Navy tests principally employed steel cylinders while US Army tests employed copper cylinders. Table 1 summarizes other examples of data found in published literature.

TABLE 1. Comparison of Gurney Velocities Derived from Experiments Using Steel Cylinders Versus Those Using Copper Cylinders

Explosive	Steel in Ref. 3 (m/sec)	Copper in Ref. 5 (m/sec)
Comp. A-3 (RDX)	2416	2630
Cyclotol (75/25 cast)	2320	2790
Comp. B	2310	2700
TNT (cast)	2040	2370
Tetryl	2209	2500

One of the problems with so-called "standard" tests, such as the copper cylinder tests, is that once a procedure is accepted as a "standard", very little other data is produced examining variations of materials or geometry. Fortunately, ref. 4 contains Gurney Velocity data, which were derived using final "steady-state" cylinder expansion velocities in the standard Gurney formula for cylinders and which show that the data were dependent upon the cylinder's material and wall thickness. Figure 1 presents a plot of these Gurney Velocities for Comp. B explosive clearly showing that experiments using steel cylinders of thickness equal to aluminum and copper cylinders yield lower Gurney Velocities.

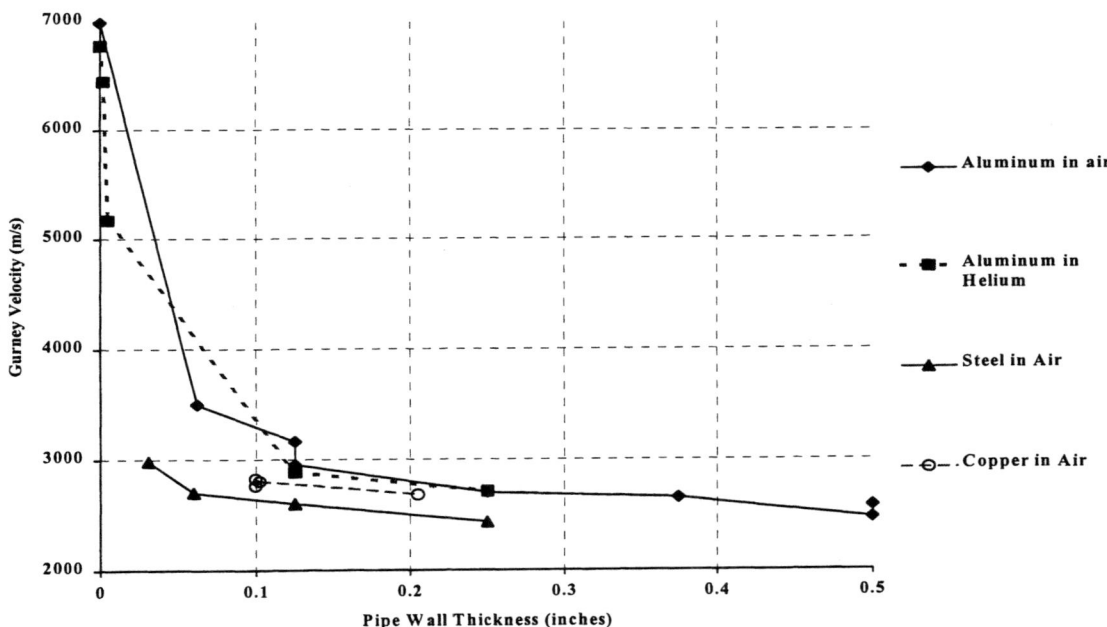

FIGURE 1. Comparison of Gurney Velocities Versus Cylinder Wall Thickness Using Comp.B Explosive

Figure 2 presents a plot of the same Gurney Velocities data versus the areal density of the cylinder wall ($t_{cyl} \times \rho_{cyl}$) in units of g/cm^2. This representation of the data shows that a Gurney Velocity derived from the final "steady-state" velocity depends upon both the cylinder material and the areal density. However, it also shows that both aluminum and steel appear to absorb more energy during explosive-driven propulsion than does copper on an areal density basis.

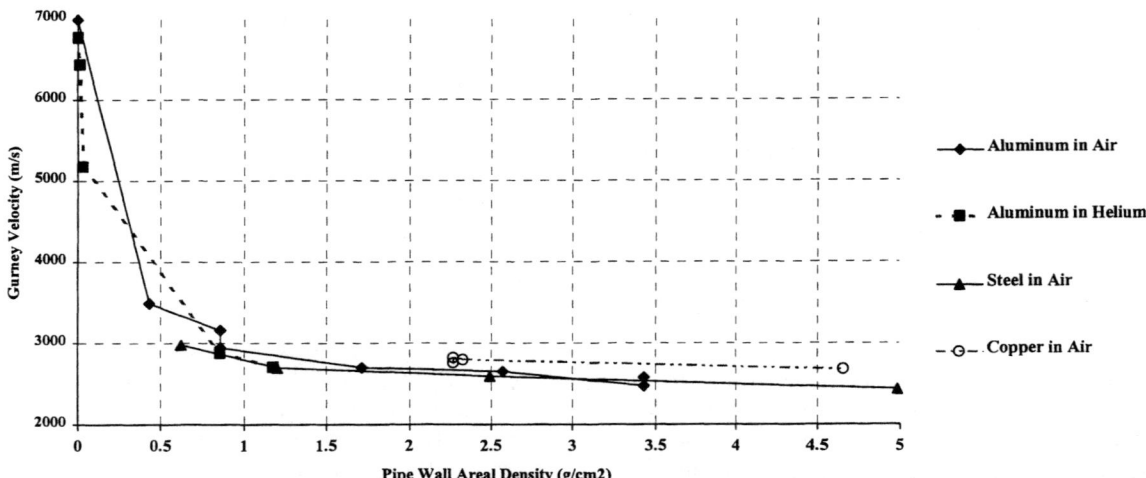

FIGURE 2. Comparison of Gurney Velocities Versus Cylinder Wall Areal Density for Comp.B Explosive

Ref.1 presented a figure plotting the ratio of initial velocity (Vi) (from ETRi) to final velocity (Vf) (from ETRf) versus the ratio of plate thickness to explosive thickness (tp/Te). The data plotted in the figure also showed that the experimentally measured initial velocity of iron plates and cylinders was significantly lower than that found for other materials. Figure 3 presents a comparable plot using an expanded data set as well as plots of Vi/Vf equations derived from ref.1's ETRi formula and Gurney equations for cylindrical and symmetric sandwich geometry modified into the same format.

According to ref.1, ETRi can be expressed as:

$ETRi = (Vi/D)(\rho_{cyl}/\rho_{ex})^{-1/2} = 0.2085 \, (t_{cyl}/R_{cyl})^{-3/40}$

Where D is the detonation velocity in km/sec, ρ_{cyl} and ρ_{ex} are the cylinder and explosive densities in g/cm^3 respectively and t_{cyl} and R_{cyl} (t_{pl} and T_{ex} for plates) represent thickness and radius in mm.

The Gurney formulas for a cylinder and a symmetric sandwich as well as a relationship for approximating the Gurney Velocity using the detonation velocity and the adiabatic expansion constant (Γ) are found in ref. 6 to be:

$Vf_{cyl} = (2Eg)^{1/2} \, [M/C + 1/2]^{-1/2}$

$Vf_{plate} = (2Eg)^{1/2} \, [M/C + 1/3]^{-1/2}$

$(2Eg)^{1/2} \cong 0.605 \, D / [\Gamma - 1]$ (Roth's formula)

Where M and C represent the masses of the inert material and the explosive.

Making the appropriate substitutions, the following formulas can be written:

$(Vi/Vf)_{cyl} = (0.2085 / 0.3457) \, (t_{cyl}/R_{cyl})^{-3/40}$
$[(t_{cyl}/R_{cyl})^2 + 2 \, (t_{cyl}/R_{cyl}) + 0.5 \, (\rho_{ex}/\rho_{cyl})]^{1/2}$

$(Vi/Vf)_{plate} = (0.2085 / 0.3457) \, (t_{pl}/T_{ex})^{-3/40}$
$[(t_{pl}/T_{ex}) + 0.333 \, (\rho_{ex}/\rho_{pl})]^{1/2}$

Data from cylinder tests used to construct Fig.1 were added to Fig.3 by using the ETRi formula to estimate the initial velocity. Since most of the data from which the ETRi formulas were derived represent experiments in which aluminum was used as the inert material, the ETRi formula calculated values were used to represent $V_{initial}$ for the aluminum cylinders. ETRi values calculated for the steel cylinders were simply divided by 2 to approximate the decreased initial velocities observed in currently available experimental data. As can be seen in Fig.3, this simple approximation for the energy lost to the iron α to ε phase transition produces "data" which nearly overlays some experimental data.

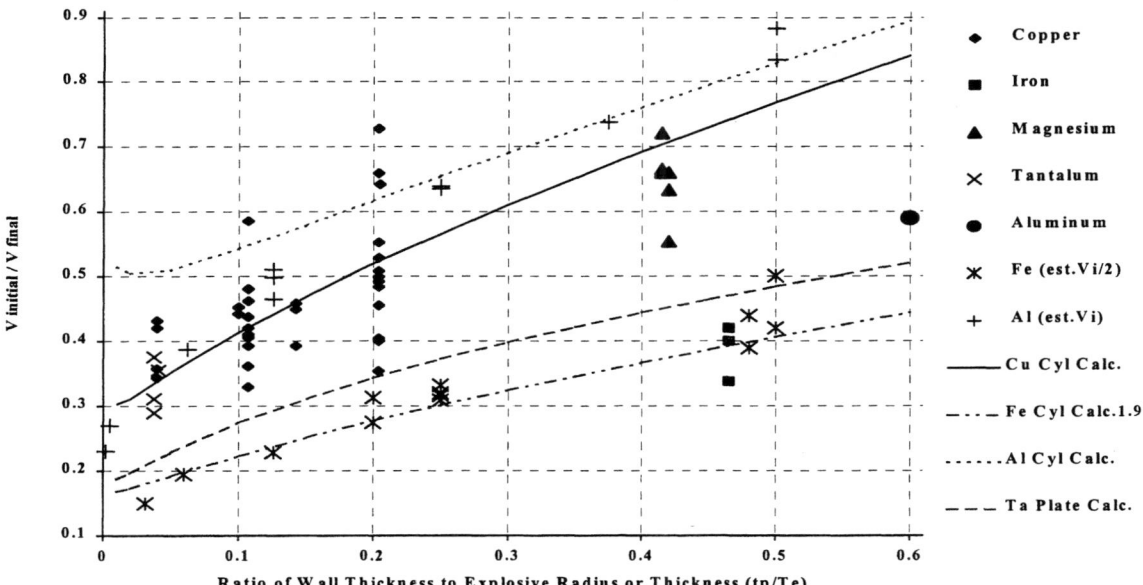

FIGURE 3. $V_{Initial} / V_{Final}$ Data Plotted for Cylinders and Plates of Various Inert Materials and Explosives

It appears that the propelled material's thickness and phase stability under pressure can have profound effects during explosive propulsion. As shown in Fig.3, the ratio of inert material thickness to the explosive radius during cylinder testing and of explosive thickness during plate-push testing can have a major effect on the partition of the energy transferred during the 1st and 2nd propulsion phases. Furthermore, the choice of material for use as a "standard" clearly affects the partitioning of energy transference as shown in Fig. 2 and 3.

CONCLUSIONS

This paper's findings underscore the caution that must be exercised when using "standard" propulsion tests to derive parameters for models which are later applied to different materials or different geometry. Detonation-driven propulsion also clearly needs to be addressed by a two-step model rather than by a single gas-push cycle model. Furthermore, the effect of phase transitions occurring during the 1st propulsion stage needs to be considered both when deriving explosive "constants" from "standard" tests and when using the "constants" during the design of explosive devices.

REFERENCES

1. J.E. Backofen and C. Weickert, "Initial Free-Surface Velocities Imparted by Grazing Detonation Waves", in *Shock Compression of Condensed Matter* – 1999, edited by M.D. Furnish, L.C. Chhabildas, and R.S. Hixon, Am. Inst. of Physics., Part 2, pp. 919 – 922
2. D.R. Kennedy, "The Elusive $(2E)^{1/2}$", 21st Annual Bomb & Warhead Section Meeting, American Ordnance Assoc., Picatinny Arsenal, 22 Oct. 1969
3. D. Price, "Dependence of Damage Effects Upon Detonation Parameters of Organic High Explosives", *Chemical Reviews* 1959, pp. 801 – 825
4. S.J. Jacobs, "The Gurney Formula: Variations of a Theme by Lagrange", NOLTR 74-86, Naval Ordnance Laboratory, Silver Spring, MD, 21 June 1974 (Public release, distribution unlimited)
5. B.M. Dobratz and P.C. Crawford, "LLNL Explosives Handbook: Properties of Chemical Explosives and Explosive Simulants", UCRL-52997, Lawrence Livermore National Lab., 31 January 1985 (National Technical Information Service, AD A272 275)
6. J.E. Kennedy, "Explosive Output for Driving Metal", in *Behavior and Utilization of Explosives in Engineering Design*, Proc. 12th Annual Symp. of the ASME New Mexico Section, edited by L. Davison, J.E. Kennedy and F. Coffey, Albuquerque, NM, 1972, pp. 109-124

OBTAINING THE GURNEY ENERGY CONSTANT FOR A TWO-STEP PROPULSION MODEL

Joseph E. Backofen[1] and Chris A. Weickert[2]

[1] BRIGS Co., 2668 Petersborough St., Herndon, VA 20171, USA
[2] Defence Research Establishment Suffield, P.O. 4000 Station Main,
Medicine Hat, Alberta T1A 8K6, Canada

Abstract. A "Gurney Energy" – reduced to represent only the propulsion imparted during the expansion of an explosive's gaseous products – is obtained so that it can be used in a two-step propulsion model which separately accounts for the initial propulsion contributed by an explosive's brisance. The derivation of this parameter from conventional Gurney Energy "constants" shows that the original "constants" are also affected by geometrical factors which previously were unknown.

INTRODUCTION

This paper presents work related to improving the BRIGS analytical package for explosive charges by separating explosive detonation-driven propulsion into a two-step process: 1) initial motion imparted by a brisant shock-dominated process, and 2) subsequent acceleration by a gas-push process. Initial motion is envisioned as being caused by the higher-pressure region of a detonation front (i.e. envision the von Neumann spike or reaction zone region as being a finite thickness of solid material squeezed at high pressure). The gas-push process is envisioned similar to that assumed by Gurney, wherein the gas volume expands from a "static" homogeneous "all-burned" high-pressure state into one wherein the velocities of the gases at the boundaries match those of inert boundary materials.

The Gurney model can be and has been used extensively with various tests, such as the cylinder expansion test and the symmetric sandwich test, to obtain a parameter representing the conversion of a portion of an explosive's chemical energy into kinetic energy of the gaseous products and propelled non-reacting materials. Historically, a single "Gurney Energy" or "Gurney Velocity" has been obtained during cylinder testing and then successfully employed to approximately describe the velocity, which could be imparted when using the same explosive in different geometrical arrangements. However, as shown in reference 1, only about 50% of the final velocity representing 75% of the final kinetic energy is imparted to the cylinder wall by the gas-push process during experiments having the geometry ratios employed during typical cylinder tests.

ANALYSIS AND DISCUSSION

Reference 2 – well known worldwide for presenting the usefulness of the Gurney model to describe explosive-driven propulsion – provides the following Gurney formulas for the final "steady-state" velocity imparted in cylindrical and symmetric sandwich geometry tests:

$$Vf_{cyl} = (2Eg)^{1/2} [M/C + 1/2]^{-1/2}$$

$$Vf_{plate} = (2Eg)^{1/2} [M/C + 1/3]^{-1/2}$$

$$(2Eg)^{1/2} \cong 0.605 \, D / [\Gamma - 1] \quad \text{(Roth's formula)}$$

Where $(2Eg)^{1/2}$ is the Gurney Velocity (km/sec) form of the Gurney Energy (Eg), D is the detonation velocity (km/sec), Γ is the adiabatic coefficient, and M and C usually represent the per unit length masses of the inert material and the explosive of an arrangement wherein the boundary losses are deemed to be un-important. When using Ro, Ri, t_{pl}, and T_{ex} to represent a cylinder's outer and inner radii, a plate's thickness and half the thickness of the explosive in a symmetric sandwich, respectively, these formulas can be rewritten as:

$Vf_{cyl} = (2Eg)^{1/2} [(\rho_{cyl} (Ro^2 - Ri^2)/\rho_{ex} Ri^2) + 1/2]^{-1/2}$

$Vf_{plate} = (2Eg)^{1/2} [(\rho_{pl} t_{pl} / \rho_{ex} T_{ex}) + 1/3]^{-1/2}$

Where ρ_{cyl}, ρ_{pl}, and ρ_{ex} are the cylinder, plate, and explosive densities in g/cm^3, respectively.

The BRIGS two-step detonation propulsion model also is based on the transfer of energy from an explosive to the propelled materials. As described in ref.1, initial motion can be represented by the following Energy Transference Ratio (ETR) for grazing (side-on) propulsion, which has been found to be approximately half the ETR for normal (head-on) impact of a detonation front with a plate.

$ETR_i = (V_i / D)(\rho_{cyl} / \rho_{ex})^{1/2}$
$= 0.2085 [3.75 / (\Gamma+1)]((Ro-Ri)/Ri)^{-3/40}$

Where V_i is the initial free-surface velocity expressed in km/sec. (For plates, t_{pl} and T_{ex} are used in the formula.)

As mentioned in ref.3, the initial velocity formulas for grazing and normal impact propulsion can be combined to account for an angle of incidence between the detonation front and the plate by using the following formula:

If $\zeta < 21°$ Then $V_i = V_{i \text{ (side-on)}}$
Else $V_i = (1 + \sin \zeta) V_{i \text{ (side-on)}}$

Where ζ is the angle between the normal to the plate or cylinder wall and the detonation wave. As described in ref.3, 21° is an approximation for the angle at which the detonation jumps to a Mach-wave / Triple-Point flow in explosives such as Comp.B, Dupont's Detasheet, and various RDX-based PBX explosives.

During cylinder tests, the detonation wave can be considered as a grazing wave at the position where measurements are usually taken. By employing the Gurney assumption both to the entire process and also to the 2nd propulsion stage of the BRIGS model, the following formula can be used to represent the total energy transfer process:

$$V_f^2 = V_i^2 + V_{gp}^2$$

Where the same geometry is used for V_f and V_{gp}, which are the final "steady-state" velocity and the velocity imparted by the gas-push process, respectively. This formula can be solved for an "equivalent geometry copper cylinder" locally in order to obtain a reduced Gurney Velocity $(2Egp)^{1/2}$ representing only the 2nd propulsion stage by substituting the initial velocity formula and by rearranging the terms to yield the following:

$(2Egp)^{1/2} = [((2Eg)^{1/2})^2 - (V_i / [\text{test geometry}])^2]^{1/2}$

This formula has been solved numerically for Comp.B explosive, in both cylindrical and symmetric sandwich geometry, to provide the Gurney Velocity values plotted in Fig.1 as functions of the ratio of inert material thickness to explosive thickness.

As shown in Fig.1, the gas-push Gurney Velocity is a decreasing function of the thickness ratio. This was anticipated from the experimental data graphed in ref.1 showing that the relative contribution of the 1st propulsion stage increased as the thickness ratio increased. However, Fig.1 also highlights a geometrical dependence in detonation-driven propulsion. For example, if the total Gurney Velocity derived by experiment is the same for both a cylinder and a symmetric sandwich having a thickness ratio of 0.205, then the gas-push Gurney Velocities for these would be 85.6% and 93.2 % of the total Gurney Velocity, respectively.

Figure 2 presents a graph of the final "steady-state" velocities calculated for Comp.B filled cylinders using the original conventional Gurney formula and the BRIGS two-step propulsion model, as well as separate calculations for the 1st and 2nd propulsion stages. The calculations for copper, aluminum, and steel cylinders were performed for the same cylinder thickness to radius ratios and used the Gurney Velocity found during copper cylinder expansion tests at a ratio of 0.205 as a baseline. A factor of 1.9 was used to decrease the steel cylinder initial velocities to approximate the energy absorbed by the α to ε phase transition.

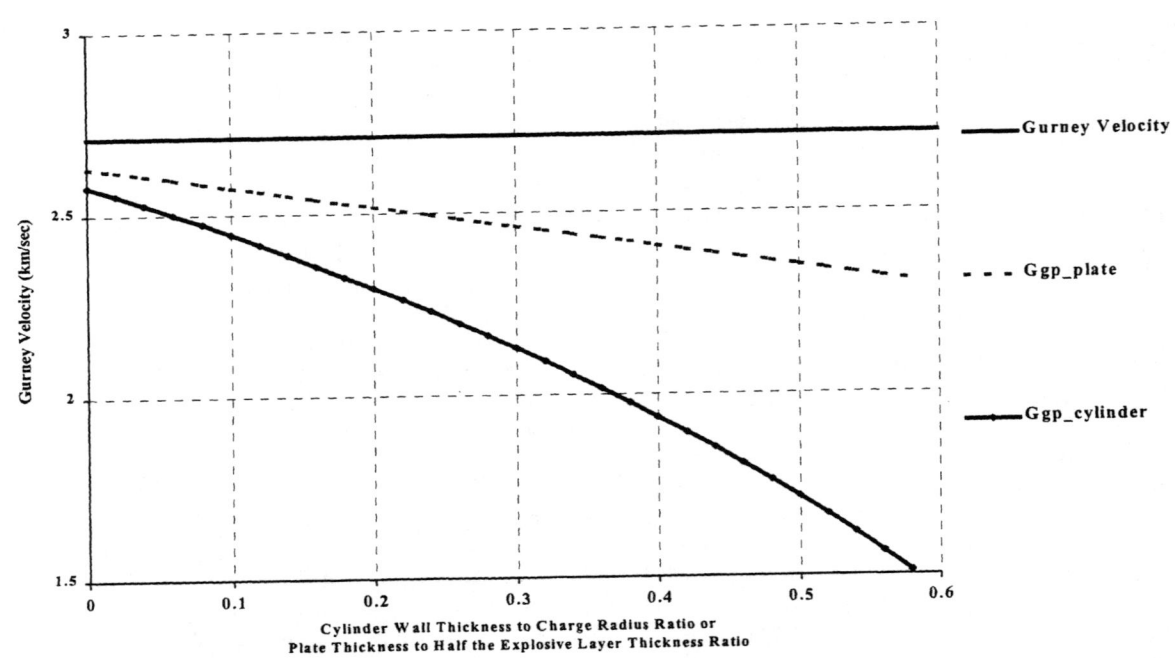

FIGURE 1. Variation of the Gurney Velocity Representing the 2nd Propulsion Stage's Gas-Push Process

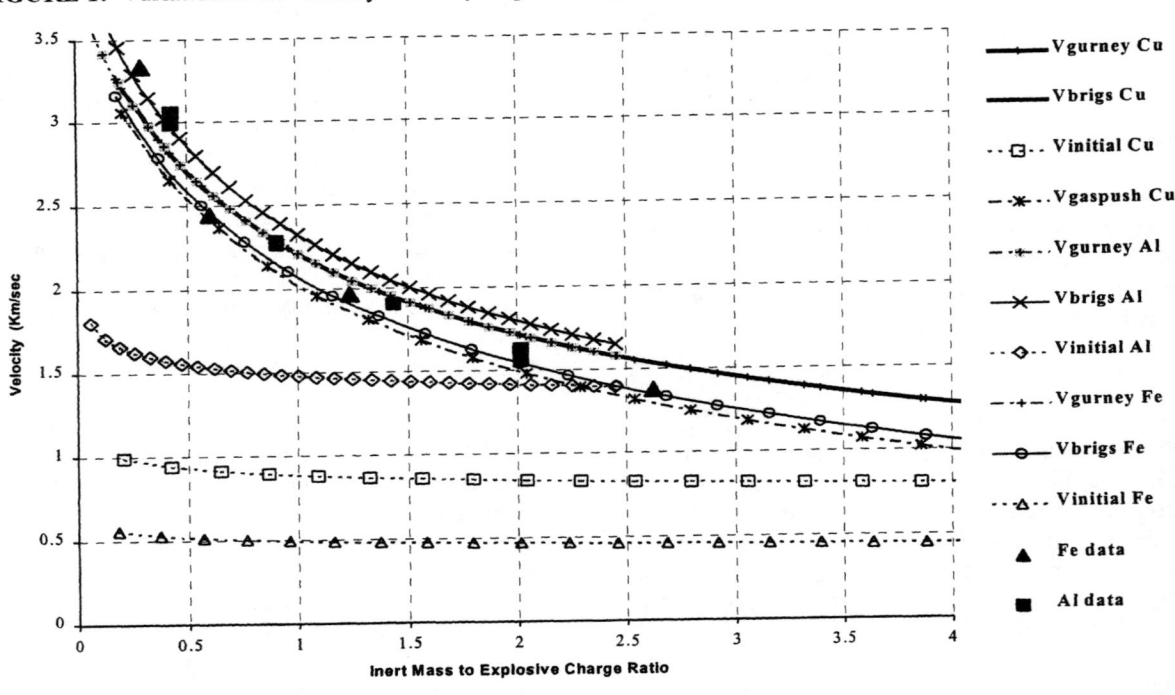

FIGURE 2. Cylinder Wall Velocities Calculated Using the Gurney Formula and the BRIGS Two-Step Model

As one can expect from ref.2, the velocities calculated using the conventional Gurney cylinder equation overlay one another for all three materials. The two-step propulsion model, however, increases the velocities for aluminum cylinders due to faster velocities from the 1st propulsion stage. However, the velocities for steel cylinders are decreased due to the effect of the α to ε phase transition which absorbs energy during the 1st propulsion stage.

Unfortunately, there is currently a paucity of experimental data for same-material / same-explosive cylinder tests or symmetric sandwich tests conducted at multiple mass-to-charge ratios and/or multiple thickness-to-radius ratios. Experimental data taken at large ratios of these parameters is also lacking. Thus, comparisons of experimental data to the two-step propulsion model's 2nd stage formulas employing reduced Gurney Velocity constants is somewhat hindered at present. The very limited data from ref.4 plotted in Fig.2 show that the steel cylinder data generally match the two-step model's calculations while the aluminum cylinder data also appear to be affected by an energy loss mechanism that reduces their values below both the conventional Gurney and two-step model calculations.

The question naturally arises *"Why have the Gurney equations generally worked so well?"*.
The answer may lay in the fact that these equations are based upon initial and end state assumptions that: 1) all the Gurney energy is released instantaneously (and uniformly) within the gaseous products, and 2) this energy only goes into the kinetic energy of the explosive products and the solid body parts being propelled during the gas-expansion-push process. Gurney constants derived under these assumptions have generally been able to hide the initial propulsion effects – some of which cancel each other out – because they are used to calculate only the final velocity imparted to similar materials. However, the Gurney equations still have been known to have problems at low and high ratios of inert mass to explosive charge mass as well as with various explosive compositions to the extent that there has always been argument over the values of the Gurney "constants" and whether they have been properly employed in practice.

CONCLUSIONS

This paper has presented a means by which an explosive-filled detonation-driven device can be modeled by a two-step system of equations. These equations can be used to account for the effects of phase changes driven by an explosive's brisance during the 1st propulsion stage as well as the effects of detonation wave-shaping. These equations also separately model the gas-push propulsion stage in a manner consistent with previous propulsion concepts. Hopefully, this paper will stimulate additional experimental work to broaden the available data base so that explosive coupling effects in both propulsion stages might be more fully revealed.

REFERENCES

1. J.E. Backofen and C. Weickert, "Initial Free-Surface Velocities Imparted by Grazing Detonation Waves", in *Shock Compression of Condensed Matter – 1999*, edited by M.D. Furnish, L.C. Chhabildas, and R.S. Hixon, American Institute of Physics., Part 2, pp. 919–922
2. J.E. Kennedy, "Explosive Output for Driving Metal", in *Behavior and Utilization of Explosives in Engineering Design*, Proc. 12th Annual Symp. of the ASME New Mexico Section, edited by L. Davison, J.E. Kennedy and F. Coffey, Albuquerque, NM, 1972, pp. 109-124
3. J.E. Backofen and C.A. Weickert, "A 'Gurney' Formula for Forward Projection From the End of an Explosive Charge", Proc. 14th *Int. Symp. Ballistics*, Quebec, American Defense Preparedness Association, 1993, Vol.2, pp.59-68
4. S.J. Jacobs, "The Gurney Formula: Variations of a Theme by Lagrange", NOLTR 74-86, Naval Ordnance Laboratory, Silver Spring, MD, 21 June 1974 (Public release, distribution unlimited)

ALUMINISED EXPLOSIVE COMPOSITIONS BASED ON NQ AND BTNEN

Michael F. Gogulya, Alexander Yu. Dolgoborodov, Michael A. Brazhnikov, Michael N. Makhov, and Vitaliy I. Arkhipov

N. Semenov Institute of Chemical Physics RAS, Kosygin st. 4, Moscow, 117334, Russia

Abstract. Aluminium containing explosive compositions based on nitroguanidine (NQ) or bistrinitro-ethylnitramine (BTNEN) were studied. The tested compositions contained Al (15% wt.) of different particles' size and particles' shape. There were measured the following explosive parameters: detonation velocity, pressure time histories and temperature time histories, velocity of accelerated metal plate, explosion heat. NQ pressure profile is of the shape predicted by ZND theory, thus C-J pressure was estimated. BTNEN detonation seems to be of more complicated nature. Effect of Al introduction into HE depends on the nature of HE and Al particles' size and shape as well.

INTRODUCTION

Aluminium is widely used as an additive enhancing detonation characteristics of HE. However, the mechanism of Al oxidation in and behind detonation wave is not well understood. This problem attracted new interest after ultra-fine Al (< 0.1 μm) became available [1]. For study, there were chosen two HE: BTNEN and NQ. The interest to BTNEN is due to its high density (1.96 g/cm^3) and positive oxygen balance (OB = +16.5%). NQ, explosive with negative OB (−30.8%), is of particular interest as HE with high hydrogen content.

PREPARATION OF HE/Al CHARGES

Five Al batches were tested including spherical particles with size <0.1{0.9}; 7{0.98}; 15{0.99}; 150{0.99} μm and flaked Al with size ≈1×20×20 μm {0.85}, containing 3.8% stearine. Here figures in parentheses indicate content of active Al. The aforementioned Al batches are referred to below as Al(0.1), Al(7), Al(15), Al(150) and Al(fl) for flaked Al. Aluminium content in mixtures was 15% wt. NQ-crystals with low loose-packed density of 0.2 g/cm^3 had needle-like shape and were about 5-10 μm of thick and about 50 μm in length. BTNEN particles had needle-like shape with diameter of 15-40 μm and length up to 500 μm. Components of the mixtures merged in hexane were mixed in a rotating drum or manually. Then it was evaporated from the mixture at its boiling point (~70^0 C). Charges were pressed to density about 0.90-0.95 TMD.

EXPERIMENTAL TECHNIQUES

Detonation velocity (D) was measured with the aid of a set of contact gauges (0.1 mm of thick) made of copper foil insulated with a plastic film. The time interval of detonation front travel was recorded by the frequency meter with an accuracy of 0.01 μs. Pressure histories and temperature ones were measured with the aid of dual-channel optical pyrometer (λ=420 and 627 nm) with time resolution about 10 ns. Indicator technique was used for pressure profile measurements [2]. Bromoform 20 mm-thick layer was used as an indicator. Temperature measurements were performed by means of window technique [3]. LiF plate served as a window. When measuring D or temperature histories or pressure ones, HE samples were initiated with a plane wave generator made of RDX-wax composition. Charges 40 mm in diameter and ~100 g in weight were tested. Plate acceleration technique [4] consists in measurements of the velocity of a 4-mm steel plate accelerated by detonation products (DP) in the direction of detonation wave propagation. NQ basic charge was 35-mm long. The length of BTNEN charge was 40 mm. The plate velocity was measured with an accuracy of ~1%. Explosion heats (EH) were measured in a bomb calorimeter made of steel vessel 5 litre in volume with an accuracy of ~1%. It is placed in a compartment with a distilled water [5]. For EH measurements, NQ and NQ/Al mixtures were pressed in charges of 30 mm in diameter and 50-60 g in weight and placed into the 10-mm thick stainless steel casing. BTNEN and BTNEN/Al mixtures were pressed in charges of 20 mm in diameter and 40–45 g in weight and placed into 7-mm thick casing of the same metal. In all aforementioned tests except EH and metal plate acceleration measurements in BTNEN mixtures, there was used an additional RDX pellet ($\rho_0 = 1.68$ g/cm^3) 10 mm of thick to reinforce the initiation impulse.

EXPERIMENTAL RESULTS

Detonation velocity data are listed in Table 1, where ρ_0 and η are the absolute and relative charge density, D_{ex} is experimental D and ρ_{0HE} is the density of HE in mixture. Basing on the relationship $D_{id} = a + b\rho_{0HE}$ for pure HE, one can recalculate D measured at different charge density to those (D*) would be measured at the same density of HE in the mixture (ρ_{0HE}^*): $D^* = D_{ex} - b(\rho_{0HE} - \rho_{0HE}^*)$. The coefficients, a and b, and ρ_{0HE}^* are given in Table 2. D* values are listed in Table 1 and they are plotted in Fig. 1. For comparison, the data for HMX/Al are also presented. The results on D are influenced by a number of factors. First one is the decreasing of the number of moles of gaseous DP caused both by the decreasing of HE amount in the mixture and by the Al reaction with carbon oxides. Second factor is the energy release caused by Al oxidation. In addition, one should concern energy losses through additive compression and its heating up. Competition of them controls D value.

TABLE 1. Detonation Velocity

HE	Al	ρ_0, g/cm^3 (η)	ρ_{0HE}, g/cm^3	D_{ex}, km/s	D*, km/s
NQ	-	1.635 (0.918)	1.635	7.94	7.94
	Al(15)	1.743 (0.929)	1.640	7.94	7.92
	Al(fl)	1.720 (0.916)	1.616	7.78	7.86
	Al(0.1)	1.785 (0.951)	1.684	8.13	7.93
BTNEN	–	1.870 (0.954)	-	8.50	8.62
	–	1.909 (0.974)	-	8.66	
	Al(150)	1.965 (0.961)	1.875	8.38	8.48
	Al(15)	1.955 (0.956)	1.864	8.30	8.44
	Al(7)	1.955 (0.956)	1.864	8.28	8.42
	Al(0.1)	1.914 (0.936)	1.820	8.04	8.35

TABLE 2. D(ρ) Relationship

HE	a, km/s	b, (km cm^3)/g s	ρ_{0HE}^*, g/cm^3	Ref.
NQ	1.44	4.015	1.635	[7]
BTNEN	1.24	3.885	1.900	[6]**

**Data of the present work are also included for D(ρ_{0HE}) relationship construction.

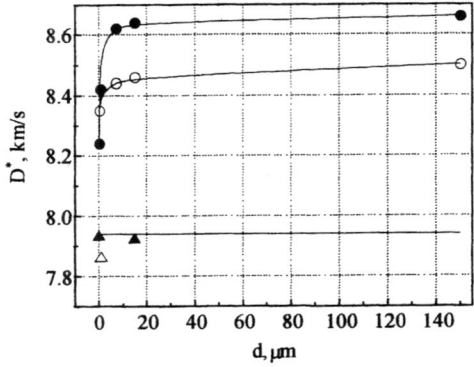

FIGURE 1. Detonation velocity versus Al particles' size: ○ - BTNEN/Al; ▲ - NQ/Al (△ - NQ/Al(fl)); ● - HMX/Al

In Fig. 2, 3 there are given pressure histories in DP for tested mixtures. For NQ, Fig. 2 demonstrates that the C-J pressure would fall in interval of (22.1÷21.6) GPa, with corresponding polytrope index of 3.66 ÷3.77 and the detonation reaction zone of (0.7÷1.0) mm. Opposite to NQ, BTNEN pressure profile is not a classical one. The peculiarities seen for BTNEN at the front during first 0.05 μs retain for BTNEN/Al mixtures. They are possibly caused by the macroscopic kinetic of BTNEN decomposition.

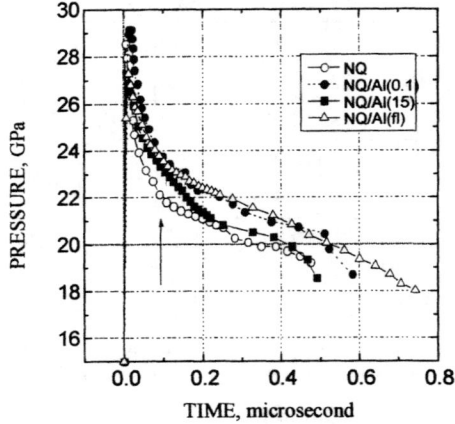

FIGURE 2. Pressure time histories in the DP for NQ and NQ/Al mixtures. The arrow shows C-J point for NQ.

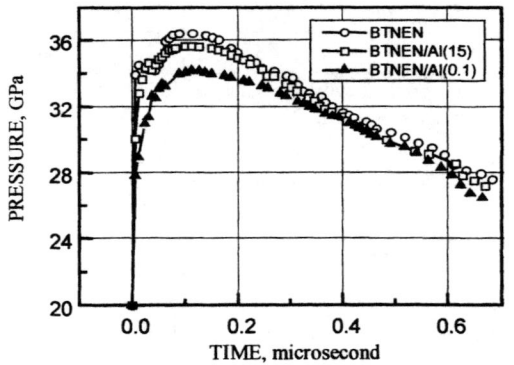

FIGURE 3. Pressure time histories in the DP for BTNEN and BTNEN/Al mixtures.

Brightness temperature histories are shown in Fig. 4. For HE/Al, rapid temperature decrease at 1.2 – 1.6 μs is caused by rarefaction entered to the observing area. NQ/Al(fl) temperature curve can be explained by peculiarities of component package in the charge. It is seen that free oxygen of BTNEN DP reacts with Al more actively.

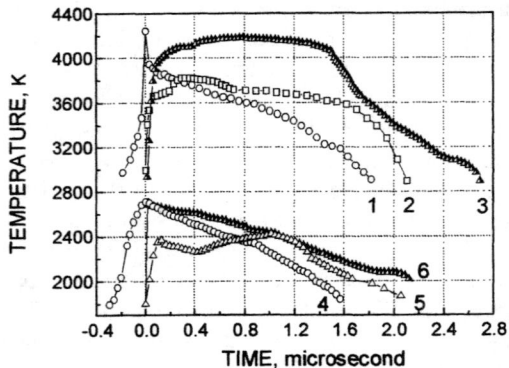

FIGURE 4. Brightness temperature time histories with LiF used as the window (λ = 627 nm). 1 - BTNEN; 2 - BTNEN/Al(15); 3 - BTNEN/Al(0.1); 4 - NQ; 5 - NQ/Al(fl); 6 - NQ/Al(0.1). The instant of time pointed as zero corresponding to the detonation wave entrance DP/LiF interface.

For NQ and NQ/Al charges, data on metal plate acceleration velocity (W) are presented in Fig. 5. For NQ, the tests were performed at two densities.

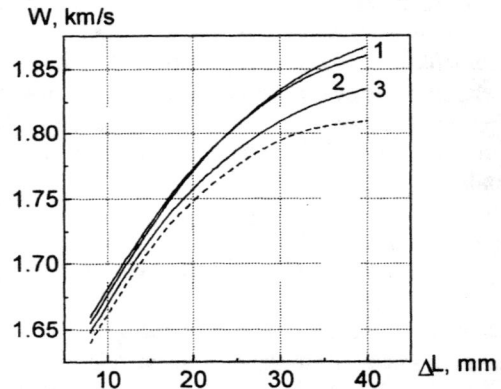

FIGURE 5. Steel plate velocity vs. distance for NQ and NQ/Al. 1 - Al(fl); 2 – Al(0.1); 3 – Al(15); Dash - pure NQ.

The increase of NQ-charge density by 0.1 g/cm^3 results in W increase by ~90 m/s. The curves for NQ/Al mixtures were recalculated from the experimental data to NQ porosity in the charge of ~ 8% basing on the W–ρ relation for NQ. Increase in metal plate velocity is of ~3.1% for NQ/Al over pure NQ at ΔL = 40 mm. The curves for BTNEN and BTNEN/Al are given in Fig. 6. For BTNEN, the tests were performed at two densities. The increase of BTNEN density by 0.1 g/cm^3 results in W increase by ~80 m/s. Trajectories for BTNEN/Al were recalculated in the same manner to BTNEN porosity in the charge ~3%.

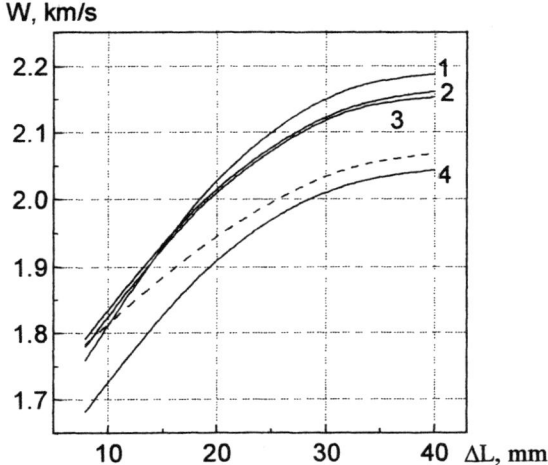

FIGURE 6. Steel plate velocity vs. distance for BTNEN and BTNEN/Al: 1 – Al(7); 2 – Al(15); 3 – Al(0.1); 4 – Al(150); dash line corresponds to pure BTNEN.

It is seen that BTNEN/Al(7) and BTNEN/Al(15) mixtures provide higher gain in metal plate velocity than do the mixtures with EH of negative OB, e.g. HMX [8]. In similar mixture with BTNEN, the plate velocity augments by nearly 6% at $\Delta L = 40$ mm.

TABLE 3. Explosion Heat

HE	Al	ρ_0, g/cm^3	Q, kJ/kg
NQ	-	1.635	3480
NQ	Al(15)	1.743	4820
NQ	Al(fl)	1.720	4930
NQ	Al(0.1)	1.785	4960
BTNEN	-	1.889	5230
BTNEN	Al(150)	1.945	8250
BTNEN	Al(15)	1.955	8450
BTNEN	Al(7)	1.945	8580
BTNEN	Al(0.1)	1.914	8420

Though plate acceleration ability of BTNEN is lower than that of HMX, the acceleration ability of BTNEN/Al approaches that of pure HMX at the same porosity. The advantages of Al(7) over Al(0.1) in metal plate velocity can be explained by higher Al(7) activity (98% over ~90%). EH data are given in Table 3. EH of BTNEN/Al(7) is the highest due to the relatively small particles and high pure Al content. On the condition of the complete Al oxidation in DP of BTNEN, one can estimate the EH as ~ 8600 kJ/kg at content of pure Al 15% and that for DP of NQ as ~5600 kJ/kg. The estimation indicates that for BTNEN (positive OB) there is complete Al oxidation in DP expanding in calorimetric bomb only for Al(7) and Al(0.1). For any tested NQ/Al mixtures, there is no complete Al oxidation by DP of NQ.

CONCLUSION

Effect of Al introduction into HE depends on the nature of HE and Al particles' size and shape as well. Ultra-fine Al manifests itself as an active powder among the tested ones. The advantages of ultra-fine Al caused by lesser particles' sizes are restricted by lower content of pure Al in powder. Al starts react with DP in detonation zone or immediately behind it, but the most part of Al oxidises in expanding DP at larger times.

REFERENCES

1. Gen M. Ya., and Miller A. V., Patents of USSR, No 814432 and No 967029.
2. Gogulya M. F., and Dolgoborodov A. Yu., *Chem. Phys. Rep.* **13**(12), 2059-2069 (1995).
3. Gogulya M. F., and Brazhnikov M. A., "Radiation of Condensed Explosives and Its Interpretation (Temperature Measurements)," in *Proceedings of the 10-th International Symposium on Detonation*, Boston-1993, Office of Naval Research, ONR 33395-12, 1995, pp. 542-548.
4. Arkhipov V. I., Makhov M. N., and Pepekin V. I., *Sov. Jnl. Chem. Phys.* **12**(12), 2395-2399 (1994).
5. Pepekin V. I., Makhov M. N., Lebedev Yu. A., *Dokl. Akad. Nauk*, **232**(4), 852-855 (1977), (in Russian).
6. Kamlet M. J., and Hurwitz H. J., *Chem. Phys.*, **48**(8), 3685-3692 (1968).
7. Price Donna and Clairmont A. R., "Explosive Behavior of Nitroguanidine," in *Proc. Twelfth Symp. (Intern.) on Combustion*, The Combustion Institute, Pittsburgh, Pennsylvania, 1969, pp. 761-770.
8. Arkhipov V. I., Makhov M. N., Pepekin V. I., et.al., *Khim. Fiz.*, **18**(12), 53-57 (1999), (in Russian).

PROTON RADIOGRAPHY EXAMINATION OF UNBURNED REGIONS IN PBX 9502 CORNER TURNING EXPERIMENTS

Eric N. Ferm, Christopher L. Morris, John P. Quintana, Peter Pazuchanic, Howard Stacy, John D. Zumbro, Gary Hogan, and Nick King

Los Alamos National Laboratory, Los Alamos, New Mexico 87545

Abstract. PBX 9502 Corner Turning Experiments have been used with various diagnostics techniques to study detonation wave propagation and the boosting of the insensitive explosive. In this work, the uninitiated region of the corner turning experiment is examined using Proton Radiography. Seven transmission radiographs obtained on the same experiment are used to map out the undetonated regions on each of three different experiments. The results show regions of high-density material, a few percent larger than initial explosive density. These regions persist at nearly this density while surrounding material, which has reacted, is released as expected. Calculations using Detonation Shock Dynamics are used to examine the situations that lead to the undetonated regions.

INTRODUCTION

The PBX 9502 corner turning experiments were one of many explosive assemblies used to study the initiation and boosting requirements of insensitive explosive. The experiment was able to identify differences in materials that had subtle particle size changes in the material. The experimental configurations shown in Fig. 1 show two of the charges used for this work. Although the charges from different experimental groups differ in details, they can be described as consisting of three sections, an initiation/booster charge, a smaller diameter donor charge, and a larger diameter acceptor charge. The most common diagnostic used on these assemblies was to place a flasher material along the edge of the acceptor charge and measure the axial distance from the face of the acceptor to the first breakout on the cylinder edge[1]. With an ideal explosive, first breakout occurs at the face of the acceptor. The further down the acceptor charge the breakout occurred, the more difficulty the donor had initiating the acceptor.

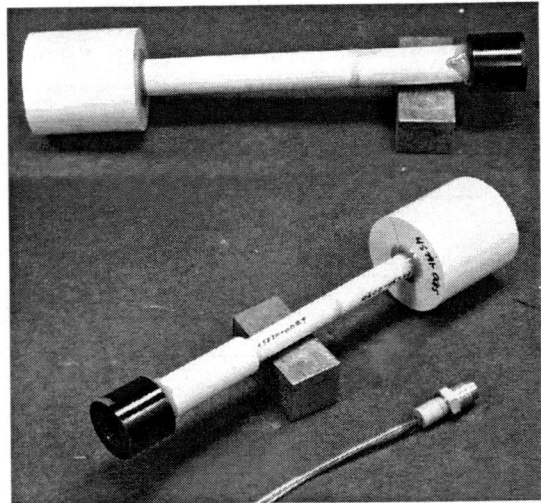

FIGURE 1. Corner turning charges used in PRad0067 and PRad0068 experiments.

Radiographs of planar corner turning experiments have been taken using the PHERMEX radiographic machine[2]. Radiographs showed regions of high-density material that did not appeared to have reacted or expanded significantly.

These have been described as "Dead Zones" or regions where the reaction rate seems to be significantly reduced.

The Proton Radiography Program at LANL has developed a radiographic facility at the Los Alamos Neutron Science Center. Multiple proton radiograph images of the same explosive experiment[3, 4] can be taken. This paper analyzes the three corner turning experiments that have been performed using this facility and allows examination of the situations that leads to the creation and persistence of the undetonated regions.

Experiment

The corner turning experiments were designed to develop a very steady detonation wave in the donor explosive. The description of the charges can be found in Table 1. The explosive was placed in a foam box which held the charge at the detonator and at the end of the acceptor charge, while the remainder of the charge was free to expand to 63.5 mm before reaching the foam wall. The charge is placed in a containment system design to allow for the magnetic imaging of the protons transmitted through the experiment at the object location to an image location where a scintillator allows cameras to photograph a signal proportional to the transmitted proton image[4]. The containment system is evacuated to minimize the blurring from atmospheric scattering of the protons in the long magnetic lens system.

The static image, the beam profile, and the camera dark current along with the dynamic images are used to obtain normalized transmission and pathlength images. The inverse Abel transforms may then be used to make density of the dynamic experiment.

Shown in Figure 2 is a ratio of the dynamic to static image of Prad0068. The image shows regions where the pathlength is thicker while surrounding areas have much thinner pathlengths.

In Figure 3 the inverse Abel transform of the image is shown. The transform raises the contrast of features like the shock front, however it also concentrates noise near the symmetry axis. The detonation front can be seen easily as well as the high-density region near the face of the acceptor charge.

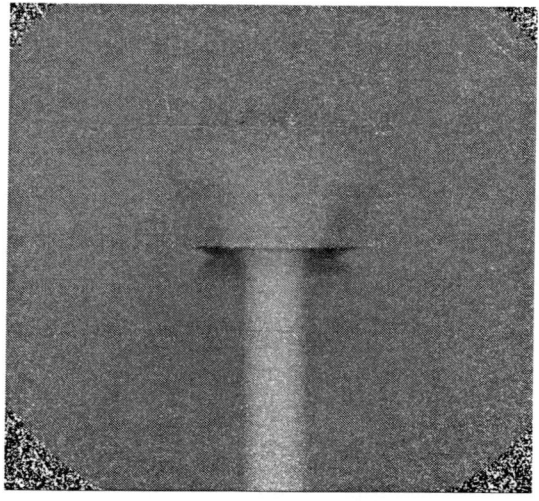

FIGURE 2. Ratio of dynamic image to static image for the 12-mm corner turner experiment at 25.3 μs after detonator breakout. The ratio shows changes in pathlength occurring since the static. Dark areas show that more material is present along the projection path than initially present.

TABLE 1. Charge description of the Proton Radiography Corner Turning Experiments

Shot	Initiation System	Donor	Acceptor
PRad 0043	SE-1 0.5"ø x 0.5" PBX 9407 17 mm ø x 50 mm PBX 9502	17 mm ø x 100 mm PBX 9502	2" ø x 2 3/16" PBX 9502
PRad 0067	SE-1 0.5"ø x 0.5" PBX 9407 18 mm ø x 50 mm PBX 9502	18 mm ø x 100 mm PBX 9502	50 mm ø x 50 mm PBX 9502
PRad 0068	SE-1 0.5"ø x 0.5" PBX 9407 18 mm ø x 50 mm PBX 9502	12 mm ø x 100 mm PBX 9502	50 mm ø x 50 mm PBX 9502

ANALYSIS

A high-density region is visible in Figure 3, which appears to emanate out of the detonation front and widen as it approaches the face of the acceptor charge. Analysis of the images indicates the average density is higher than the initial density of the explosive, which is confirmed by examining Figure 2.

The leading edge of the deadzone is radially expanding about 3.3 mm/μs while the trailing edge is expanding about 0.9 mm/μs. The attachment point to the detonation wave was estimated by extrapolating the surface leading the deadzone and the trailing edge back to where it appears to blend into the detonation front. The axial position of the attachment point is entering the charge at 3.mm/μs and then slows down to near zero. In the 12- and 18-mm donor charges, the attachment point enters acceptor as deep as 13- and 11-mm respectively. Radially growth is initially stalled, and then rapidly increases to more than 6 mm/μs. During this period it becomes clear that the detonation wave has turned the corner, i.e. the detonation wave has a point that is radially expanding to the wall, and will be the first breakout point.

MESA calculations using Detonation Shock Dynamics (DSD) [5] are compared with the experiment in Figure 4. DSD as implemented has no curvature failure criterion. Without a criterion, it can easily be seen that the deadzone is not modeled nor does the calculation reproduce the corner turning distance. Although the curvature slows the wave propagation along the face of the acceptor it is not sufficient to model the effect. The existence of the deadzone is required to model the corner turning effect.

A boundary of the dead zone can be chosen by connecting up the detonation wave attachment points found in the experiment. This region is about twice as big as the estimated mass of the dead zones from the radiographs. When this boundary is used in the calculation the corner turning distances is represented much better, although this is not surprising.

The significant overestimation of the deadzone mass indicates that there is reaction taking place in the vicinity of the attachment point. The size of dead zone region is influenced by several factors, including rarefactions coming from acceptor face, possible shock desensitization, and the length of time before rarefactions quench shocked explosive.

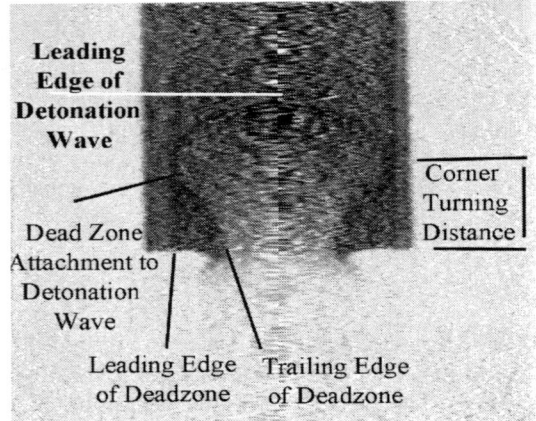

FIGURE 3. Volume density image of the 12 mm corner turner experiment 25.3 μs. The identified regions are used in describing the deadzone region.

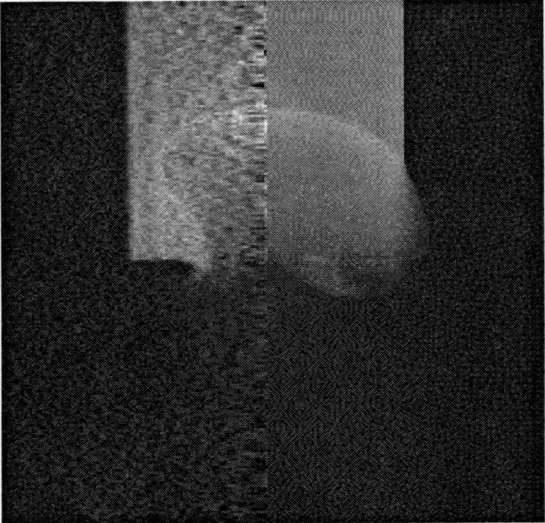

FIGURE 4. A MESA calculation of density is shown on the right half of the above image, while the left side is and average density from the two symmetric sides in figure 3.

CONCLUSIONS

The proton radiographs of "deadzones" are persistent features of corner turning experiments. The regions of material persist for more than 6 µs near initial density and typically slightly higher. The front radial edge of the deadzone appears to propagate slightly faster than sound speed in undetonated explosive. At late times this motion is stalled, either being too small of compression to measure or possibly rarefaction entering in from the side of the charge.

Regions of the deadzone are reacting near the detonation front attachment point. In the 12-mm charge it appears that the outside layer of the 50-mm charge is detonating, back from the corner turning point towards the face of the corner turning charge beyond the deadzoned.

Corner turning experiments do have regions of material that are not reacting over a time period of more than 5 µs. Other parts of the deadzone regions show indications of a slow reaction-taking place at this time scale. The deadzone region next to the attachment point is a reactive zone, which has been lengthened due to a high curvature and low amplitude shock wave, which was the stimulus for initiation. Ultimately this shock wave becomes too weak or rarefaction waves quench the reactions, leaving material behind to be observed at our latest images.

ACKNOWLEDGEMENTS

Robert Lopez, Camilo Espinoza, John Gomez helped in field testing and assembling the containment system. Larry Hill provided PBX 9502 rate stick material for the experiments, provided constants for the material and advice. John Bdzil and Tariq Aslam helped with the initial DSD calculation and provided valuable discussions on the work. This work was supported by the U. S. Department of Energy.

REFERENCES

1. M. Cox and A. W. Campbell, "Corner-Turning in TATB," presented at the Seventh International Symposium on Detonation, Annapolis, Maryland, 1981.
2. Charles L. Mader, *LASL Phermex Data Volume III* (University of California Press, Berkley and Los Angeles, California, 1980).
3. J. D. Zumbro, K. R. Alrick, R. A. Gallegos *et al.*, Technical Report Report No. LA-UR-98-1015, presented at the Eleventh Symposium on Detonation, Brekenridge, 1998.
4. N. S. P. King et *al.*, "An 800-MeV proton radiography facility for dynamic experiments", Nucl. Instru.. and Meth., **424**, 84-91, (1999).
5. L. G. Hill, John B. Bdzil, and Tariq D. Aslam, "Front Curvature Rate Stick Measurements and Detonation Shock Dynamics Calibration for PBX 9502 over a Wide Temperature Range," presented at the Eleventh Symposium on Detonation, Brekenridge, 1998.

CHAPTER XV

EXPLOSIVE AND INITIATION STUDIES

MESOSCALE MECHANICS OF PLASTIC BONDED EXPLOSIVES

Keith M. Roessig

Air Force Research Laboratory/Munitions Directorate
101 W. Eglin Blvd. Ste. 135
Eglin AFB, FL 32542

Abstract. The dynamic behavior of particulate materials is important to a wide range of problems. When dealing with energetic particulate materials, mechanical ignition is an added concern for safety and performance issues. Micrographs from unconfined impact tests show specific crystal damage paths within the matrix. Under loading conditions consistent with real world applications, these materials can be subjected to large hydrostatic pressures combined with shear deformation. Subsequent stress chain formation concentrates the compressive load into small regions, providing ignition sites within the material. A photoelastic experiment with high speed photography has been constructed to record stress state formation within PMMA disks set in different binder systems. The propagation of shear stress across disk/binder interfaces is shown to be important in the overall stress state of the particle bed. Binders with similar mechanical and acoustic properties as the PMMA disks remove stress concentrations and allow waves to propagate as if in a continuum. Softer binder with lower acoustical wave speeds and hard binders that have debonded from the disks do not allow shear stresses to be transferred. These configurations cause stress concentrations similar to a binderless system of disks.

INTRODUCTION

The initiation of reaction in energetic materials through mechanical insult is important from both a safety and performance viewpoint. Mechanical loading on the material is at much lower amplitudes, but for much longer durations, than investigated for shock initiation (1). The particulate behavior and the constituent contact mechanics at the mesoscale become important in the transfer of stress through the material. Stress concentrations develop, known as stress chains or stress bridges, which allow for localized stress and strain concentrations, and therefore subsequent heating, within the material (2,3).

A micrograph of a modified PBXN-109, a cure cast plastic bonded explosive, is shown in Fig. 1. HMX crystals 150-200μm in diameter, substituted for the RDX crystals in the standard formulation, are surrounded by small aluminum particles approximately 10-20μm in diameter. A plasticized binder holds all the particles together. At the mesoscale, there are three distinct components of

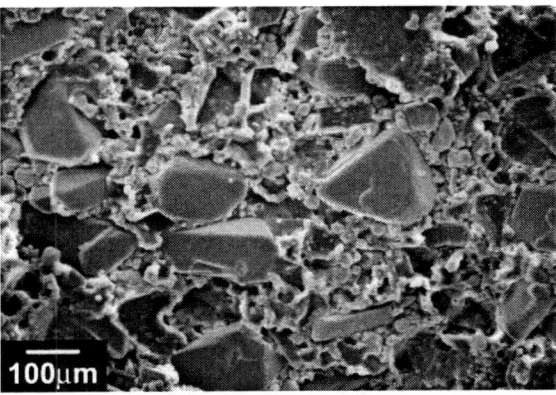

FIGURE 1. Scanning electron micrograph of the mesoscsale structure of a modified PBXN-109.

the particulate plastic bonded explosive: the crystal, the binder, and the crystal/binder interface. The transfer of stress through each is governed by the contact mechanics of the materials.

Unfortunately, it is not possible to observe the stress state of real materials at the mesoscale during these high strain rate tests. To understand how the constituents interact under a dynamic load, the process must be scaled so that meaningful data can be obtained. Therefore, photoelasticity is used on larger particles and binder amounts so that wave propagation through the material happens at a rate and length scale that is observable. Photoelasticity has been used to look at stress states in particles under dynamic loading (4-7). The purpose of this work is to use the photoelastic effect to examine the effects of binder/particle interactions at the mesoscale on the stress state that develops within the bulk material. The experimental data generated here uses only small numbers of particles, not nearly enough to simulate a real system. But numerical simulations can be calibrated against this data and then run for much larger numbers of particles. Future efforts from there could then work to homogenize this damage and its effect on bulk ignition properties.

EXPERIMENTAL METHODS

The simulation of explosive crystals in a plasticized binder took the form of circular PMMA disks, 6.4mm thick and 50mm in diameter, with different binder conditions depending on the material behavior of interest. Acrylimet, a commercial acrylic metallographic specimen mounting material, is used as a hard binder with acoustic properties similar to that of the PMMA disks. This material forms a bond to the disks to allow the transfer of both compressive and shear stresses across the disk/binder interface. To examine the case where only compressive stress could be transferred across the interface, silicon grease was placed on the edges of the disks to prevent bonding of the binder to the disks during the solidification process and reduce friction at the interface. The second binder consisted of a polyalcohol resin used in certain Air Force plastic bonded explosives. This material is much softer than PMMA, but bonds to the disks to allow the transfer of compressive and shear stresses across the interface. All of these binder systems are compared to a binder system consisting of PMMA disks with no binder.

There are two different geometries used in this work. This first consists of four disks placed in a row. Due to symmetry, there is no shear at the contact point between disks. This geometry will be referred to as the "1D" or "4 Disk" geometry. The second setup places the discs at 45° angles to each other, allowing both compressive and shear stresses at the contact point between the disks. This geometry will be labeled "45 degree contact".

The loading cell for the tests consists of a 4340 steel frame with adjustable sides that allowed for changes in frame width. Dynamic tests were conducted by placing the load frame into a compression Hopkinson bar apparatus. High speed photographs of the fringe patterns were taken with an Imacon 460 digital camera.

RESULTS & DISCUSSION

The following results show wave propagation through the two geometries mentioned in the previous section, and three different binder/interface conditions. The pictures show early and late time fringe patterns within the disks that all are close to the same location. The relative speeds at which the fringes travel are given by the time after impact shown with the pictures. All impact is on the right side of the disks with the fringe patterns traveling right to left. Reference to first, second, last disk etc. refers to the order in which the wave impinges on each disk.

4 Disk problem

The first disk/binder combination of this work repeats many previous tests conducted by Shukla et al. (4,5). The fringe patterns generated are shown in Fig. 2 and are representative of diametrical compression with Hertzian contact conditions.

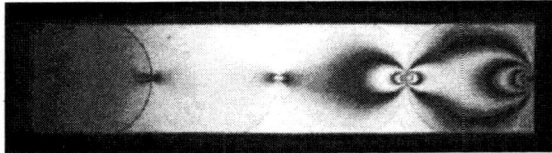

FIGURE 2. Fringe patterns in the 4 Disk geometry without binder. Pictures are taken 55 and 120 μs after impact.

The next case is the 4 Disk geometry with the acrylic binder, shown in Fig. 3. With a good impedance match between the PMMA disks and binder, an expected result occurs. There is a concentration at the first disk where impact occurs and no binder is present, but the wave expands and only a single diffuse fringe propagates smoothly down the specimen. A lower stress state develops within the specimen as the binder can support as much load as the disks themselves and shear is transferred at the boundaries. Though not shown here, this single fringe does travel back up the specimen upon reflection at the last disk. The fringes in the first disk distort due to contact with the side of the load frame.

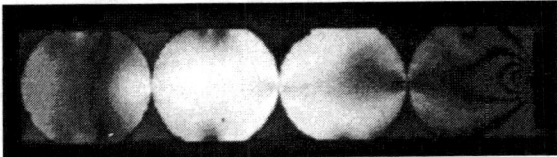

FIGURE 3. Fringe patterns in the 4 Disk geometry with an acrylic binder 45 μs and 95 μs after impact. Stress concentrations are not present in this configuration.

Figure 4 shows the third experiment, the 4 Disk problem with acrylic binder but with greased interfaces to prevent tangential loads being transferred across the interface. The stress concentrations at the disk contact points reappear in his case, as shear cannot be transmitted through the interface. The disks act as wave guides and concentrate the wave back towards the disk contact points. The contact zone between disks seems to be larger than the case with no binder at all, as there is now material to support the compressive stresses once the angle of incidence approaches 90 degrees at the end of the disk. Previous work by Sadd et al. (6) examined the effect of cementation on the propagation of stress waves in a particulate material. The fringe patterns in Fig. 4 resemble the "soft cement" case of in that work, as both of these experiments tend to reduce the stress concentration at the contact. This is in contrast to the "hard cement" case of Sadd et al. (6), which changes the position of the stress concentration but does not eliminate it.

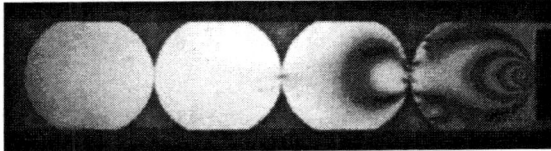

FIGURE 4. Fringe patterns generated from greased contact between the PMMA disks and the acrylic binder. Pictures were taken at 35 μs and 135 μs after impact.

The final case for the 4 Disk geometry includes the polyal binder. The contact regions for this configuration show much lower stress concentrations and wider contact areas than the binderless case, similar to the greased acrylic boundaries shown in Fig. 5. Fringe propagation speeds are also comparable to the binderless and greased acrylic interface cases. The most interesting aspect to this configuration is the ability to see the stress wave propagate through the binder itself. The binder exhibits photoelastic properties, from which a comparison of wave speeds in the two materials can be made. The wave speed in the binder is much less than the PMMA disks. This causes the same kind of stress state as in the greased interfaces because even though the compressive and shear forces can be transferred across the boundary, that information will not reach the next disk before the transfer of stress near the immediate contact point. Therefore, the contact region is increased, but only slightly as the wave speed of the binder material limits the transfer of information across the disks.

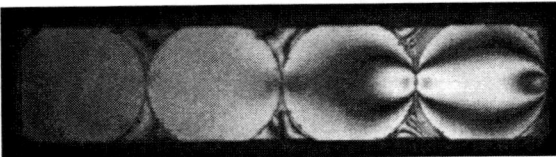

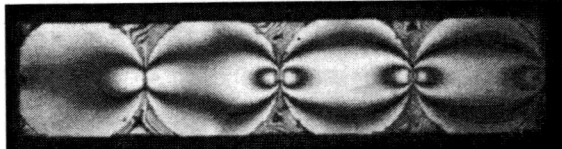

FIGURE 5. Fringe patterns of the 4 Disk problem with polyal binder taken 60 and 135 µs after impact.

45 Degree Contact Problem

The 4 Disk geometry by virtue of its symmetry did not promote shear stresses in the system. The shear stresses developed from the shapes of the disks and the reflection of the dilatational waves at the boundaries. This is the reason for the diffuse stress state in the bound acrylic case. The second geometry, the 45 degree contact, promotes shear by the disk arrangement as well.

As a reference, Fig. 6 shows the 45 degree contact geometry without any binder. Again, this is a repetition of work conducted by Rossmanith and Shukla (3). The friction coefficient between the disks is very low, exhibited by the fact that the fringe patterns are symmetric about the contact zone. With tangential loading, the fringes become antisymmertic (7). The wave propagates through the disks and attenuates due to the transfer of load into the side supports through bending.

Following the cemented particle ideas of Sadd et al. (6), the disks were bonded together with super-glue, but no other binder was added. Two things are readily apparent in Fig. 7. First, the fringe patterns exhibit high shear stresses, as they are anti-symmetric and tend to travel along the centerline of the disk formation. Only at later times do the fringe patterns between the first and second disk, exhibiting more of a normal load at the contact point. However, post test examination of the disks showed that the bond between the first and second disk fractured. This may be the cause of the fringes changing character after loading. The second difference is the speed at which the fringe patterns propagate. While the bare disks shown in Fig. 6

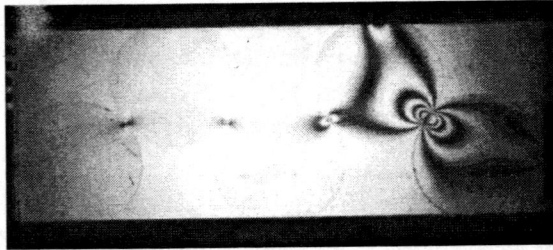

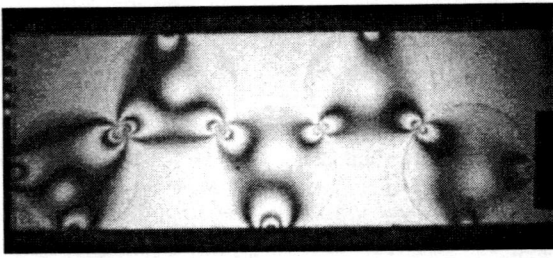

FIGURE 6. Fringe patterns at 100 and 300 µs after impact in the 45 degree contact geometry without a binder.

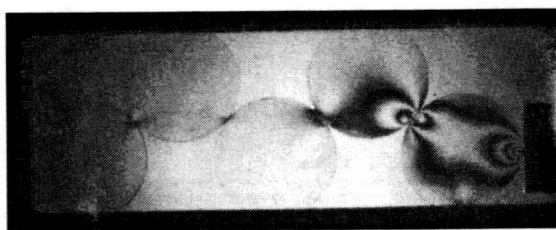

FIGURE 7. Dynamic fringe pattern for the 45 degree contact geometry with glued contact points. Pictures are taken at 60 and 135 µs after impact.

take around 300 μs to propagate from the second to the back edge of the loading frame; the same distance is traveled in about 80 μs with the glued contact. It is not the actual wave speeds of the material that are increasing. It is the state of maximum shear stress that increases at a greater rate along the centerline of the disk assembly. The fringe patterns generated are lines of constant maximum shear stress. With the glued contact, shear is transferred more readily, allowing fringes to develop at earlier times. As failure many times is determined by shear, and shearing in the crystals leads to heat generation, it is important to know how shear stresses and compressive stresses are generated in the different geometric and material conditions.

The next case is the 45 degree contact with the acrylic binder, shown in Fig. 8. Compared to the two previous scenarios, a more diffuse fringe pattern develops as in the 4 Disk problem with acrylic binder. In this case however, the fringes seem to concentrate around the centerline similar to the glued 45 degree contact problem. The ability to transfer shear along the interface exists in this case as in the glued contact, but it can now be transferred over a larger contact area. Further evidence of shearing along the centerline, there is an interesting debonding pattern in this test. Debonding occurred down the centerline, i.e. started at the first disk on top and continued to the contact point with the second disk. Debonding then occurred on the bottom of the second disk until the contact with the third disk at which point it jumps to the top of the third. This wavy pattern continues all the way through the assembly. This is a consequence of the shear forces being supported at the centerline, which are exhibited by the fringes. No debonding occurred in the 4 Disk problem where shear was not developed by the disk positions.

Figure 9 shows the greased 45 degree contact configuration with acrylic binder. The same trends are shown here as in the 4 Disk greased case. The fringe patterns from greased contact are similar to the binderless contact, but with larger contact zones between the disks. Again, as shear cannot be transferred easily across the interfaces, the stress state resembles the binderless case without glued contact.

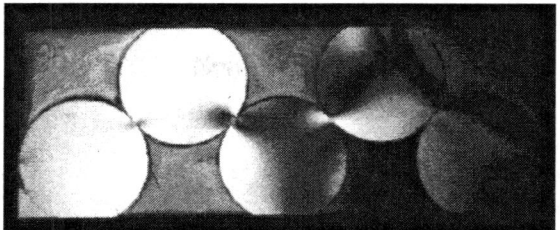

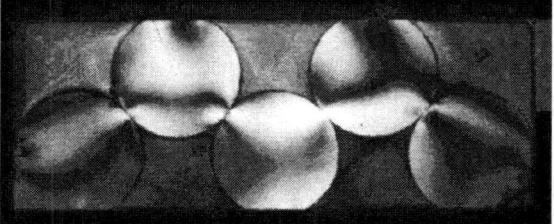

FIGURE 8. The fringe patterns generated with the acrylic binder at 75 and 150 μs after impact. The 45 degree contact with acrylic binder shows a much more diffuse pattern, but still has some shear concentratioon down the centerline.

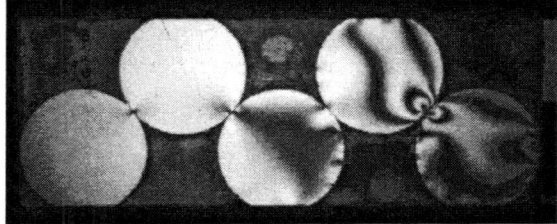

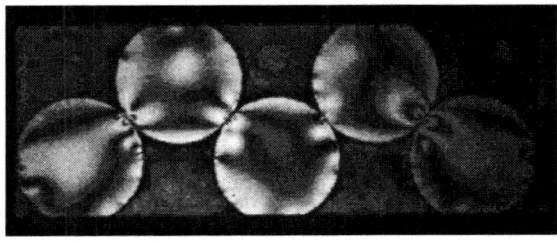

FIGURE 9. Fringe patterns developed in the 45 degree contact geometry with acrylic binder and greased interfaces. Pictures were taken 100 and 200 μs after impact.

The final configuration is polyal binder in the 45 degree contact geometry, shown in Fig. 10. The fringe patterns in the disks again are closer to the binderless case as the polyal binder has a low shear modulus. There is definitely some shear transfer though, as seen from the slight anti-symmetry in the initial fringes generated at each contact point. This effect decreases at later times once the full wave establishes itself in the disk. There is no debonding in this case, so that is not the cause of the slight change in character of the fringes. The disparate values of wave speeds are exhibited by the fringe propagation in the two materials. The average estimate of fringe propagation speed in the polyal binder is 50 m/s, compared to 1000 m/s in the PMMA disks.

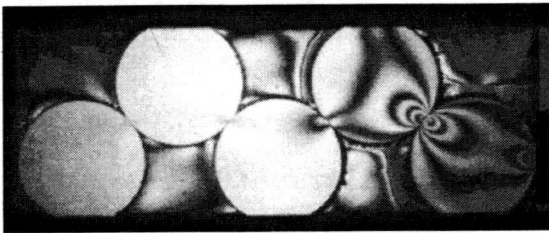

FIGURE 10. The 45 degree contact with the polyal binder produces a fringe pattern similar to the binderless case. Pictures are taken 75 and 250 μs after impact.

CONCLUSIONS

The effect of disk geometry, disk/binder materials, and disk/binder interface condition all have a large effect on the stress states developed in a particulate material under dynamic loading. This work examines the ability to transfer compressive and shear stresses through a particle and binder materials and across the interface between them. Geometries with no binder were used as test cases and also to provide repeatable results from past research. Glued contact point in the 45 degree geometry showed that the ability to transfer tangential loads at the contact point resulted in a much greater shear stress within the particles at much earlier times. Hard binders with similar mechanical properties as the particles will allow for a more diffuse stress state, but still may have shear concentrations depending upon the disk geometry.

Similar binders but with low friction interfaces prevent the transfer of shear and return the stress state to one similar to a no binder system. Stresses at the contact point are less concentrated than pure Hertzian contact conditions. Soft binders also generate a stress state similar to that of a binderless system, but low shear stresses can be supported, which allow for the diffusion of stress at the contact point.

ACKNOWLEDGEMENTS

I would like to thank Dr. Joe Foster and Dr. Scott Bardenhagen for their help and willingness to discuss technical issues at any time. I would also like to thank 1Lt Daniel Warrensford and Thomas Sprague of the Air Force Research Laboratory's High Explosive Research and Development Facility (HERD) for their help by mixing and curing the polyalcohol resin.

REFERENCES

1. Foster, Jr., J. C., Christopher, F. R., Wilson, L. L., Osborn, J., "Mechanical Ignition Of Combustion In Condensed Phase High Explosives," *Shock Compression of Condensed Matter 1997*, edited by S.C. Schmidt et al., AIP Conference proceeding 429, pp. 389-392.
2. Roessig, K.M., and Foster, J.C., Jr., "Dynamic Stress Chain Fracture in Particle Beds," in Plastic and Viscoplastic Response of Materials and Metal Forming, edited by A.S. Khan et al., proceedings of Eighth International Symposium on Plasticity and Its Current Applications, July, 2000, pp. 437-439.
3. Foster, J.C., Jr., Glenn, J.G., and Gunger, M., "Meso-Scale Origins of the Low Pressure Equation of State and High Rate Mechanical Properties of Plastic Bonded Explosives", in *Shock Compression in Condensed Matter – 1999*, edited by M.D. Furnish et al., AIP Conference Proceeding 505, 1999, pp. 703-706
4. Rossmanith, H.P. and Shukla, A., *Acta Mechanica* **42** 211-225 (1982).
5. Shukla, A. and Damania, C., *Experimental Mechanics* **44**, 268-281 (1987).
6. Sadd, M.H., Shukla, A., Sienkiewicz, F., and Gautam, A., *Wave Propagation and Emerging Technologies* **188**, 11-28 (1994).
7. Shukla, A. and Higam, H., *Journal of Strain Analysis* **20**, 241-245 (1985).

COMPACTION WAVE PROFILES IN GRANULAR HMX

Ralph Menikoff *

Theoretical Division, Los Alamos National Laboratory, Los Alamos, NM 87544

Abstract. Meso-scale simulations of a compaction wave in a granular bed of HMX have been performed. The grains are fully resolved in order that the change in porosity across the wave front is determined by the elastic-plastic response of the grains rather than an empirical law for the porosity as a function of pressure. Numerical wave profiles of the pressure and velocity are compared with data from a gas gun experiment. The experiment used an initial porosity of 36%, and the wave had a pressure comparable to the yield strength of the grains. The profiles are measured at the front and back of the granular bed. The transit time for the wave to travel between the gauges together with the Hugoniot jump conditions determines the porosity behind the wave front. In the simulations the porosity is determined by the yield strength and stress concentrations at the contact between grains. The value of the yield strength needed to match the experiment is discussed. Analysis of the impedance match of the wave at the back gauge indicates that the compaction wave triggers a small amount of burn, less than 1% mass fraction, on the micro-second time scale of the experiment.

INTRODUCTION

The sensitivity of an explosive is related to material heterogeneities. When subjected to a compressive wave, the heterogeneities generate hot spots. The hot spots dominate the overall burn rate due to the strong temperature dependence of the reaction rate. In a damaged material the heterogeneities are dominated by porosity. Consequently, granular explosives are used as a model for damaged explosives.

Compression in a granular bed results in stress concentrations at the the contact between grains. When a stress concentration exceeds the yield strength, localized plastic flow occurs. The resulting change in shape of the grains enables them to pack together more tightly. Shock waves in which the compression is dominated by the decrease in porosity are known as *compaction waves*.

Hot spots resulting from a compaction wave are sub-grain in size. To better understand the formation of hot spots, meso-scale simulations — continuum mechanics calculations in which heterogeneities are resolved — are being performed. Data from experiments provide a check on the simulations.

A series of compaction wave experiments on granular explosives performed at Los Alamos and Sandia national laboratories are summarized in an article by Sheffield, Gustavsen and Anderson (4). In these experiments, a projectile from a gas gun impacts a target consisting of a front disk of Kel-F, a granular sample of HMX, and a back disk of TPX. The dominant waves in the target are shown in figure 1. Two sets of gauges are used to measure either velocity or stress. The front gauge, located at the Kel-F/HMX interface, records the wave profiles of the incident shock as it is transmitted into the granular bed, and of the return shock in the compressed bed. The back gauge, located at the HMX/TPX interface, records the wave profile as the compaction wave reflects from the back disk.

The simulations reported here correspond to experiments with a low impact velocity (280 m/s); LANL shot #912 and SANDIA shot #2477. This case is chosen in order to test that the meso-scale simulations describe quantitatively the mechanical behavior of a granular bed before proceeding to the more interesting cases of higher impact velocities

---

* This work supported by the U.S. Department of Energy.

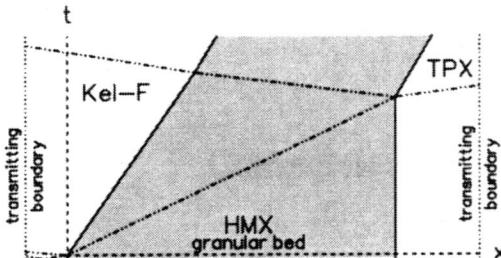

FIGURE 1. Wave diagram for gas gun experiments.

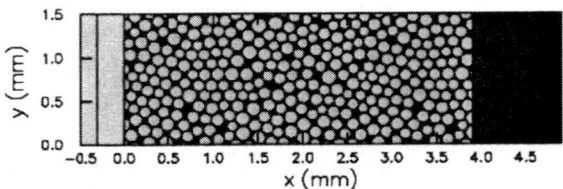

FIGURE 2. Initial configuration for simulations. The two light gray region at the left are Kel-F. The first region is given a velocity corresponding to the projectile. A transmitting condition at the left boundary is used to obtain the effect of a larger front disk. For the granular bed the HMX grains are shown in gray and the pores (voids) in black. The dark gray region on the right is TPX.

that result in significant burn on the micro-second time scale of the experiments.

Similar experiments have been simulated previously. Baer (1) used a two-phase (coarse grain) model. This type of model averages over the heterogeneities, and allows for a much coarser resolution than the grain diameter. However, an empirical compaction law is needed to account for the evolution of the porosity. In the meso-scale simulations the evolution of the porosity is determined by the plasticity model for pure HMX and the structure of the granular bed. In addition, burn models used in two-phase simulations are heuristic in nature and generally are accurate only when applied to experiments similar to the ones used to calibrate model parameters. In contrast, meso-scale simulations aim to determine the burn rate based on the measured chemical reaction rate, and the distribution of hot spots.

Horie and collaborators (5) simulated a compaction wave experiment with the discrete mesodynamics method. Though very general force laws between elements can be employed, convergence with increased resolution (number of elements per grain) and the continuum limit of the underlying model have not been studied. In addition, the focus has been on mechanical properties. Thermal quantities, such as temperature needed for reaction rates, have been neglected.

Finally, we note that 3-dimensional mesoscale simulations of reactive flow have been reported by Baer (2). Even on a super-computer, his simulation with 5 μm resolution is limited to a 1 mm cube and to a time interval of 50 ns. In order to observe reaction on this short time scale, a high impact velocity (1000 m/s) is needed. This is in the regime of a shock-to-detonation transition.

The mesoscale simulation reported here are in the regime of a deflagration-to-detonation transition. A two-dimensional simulation with 10 μm resolution is used for a sample 3.9 mm × 1.5 mm over a time interval of 8 μs. On the current generation of personal computers, a simulation takes well under a day.

SIMULATIONS

The initial configuration used for the meso-scale simulations is shown in figure 2. The porosity of the granular bed is 36%. The grains are randomly distributed and have an average diameter of 120 μm with a uniform variation of ±10%. A Mie-Grüneisen equation of state based on a linear u_s-u_p relation is used for the Kel-F front disk and TPX back disk. An elastic-plastic constitutive model is used for the HMX grains. It consists of a Mie-Grüneisen equation of state for the hydrostatic component of the stress, an elastic shear stress, and a von Mises yield condition. The strength model is isotropic, perfectly plastic and rate independent. The material parameters are specified in (3, table 1).

Columns of Lagrangian tracer particles are placed just in front and just behind the granular bed. The column averages of the velocities and normal stresses of the tracer particles are shown along with the gauge data in figure 3. The data and simulation of the front records indicate that after a transient, the velocity and stress behind the compaction wave are in good agreement. As discussed in (3), the transient response of the tracer particles as compared to the gauges explains the initial overshoot in the simulated record. The remainder of the front records shows a slow in-

(a) Velocity gauge data, LANL Shot-912

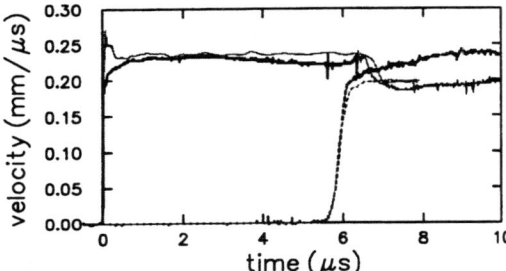

(b) Pressure gauge data, SANDIA Shot-2477

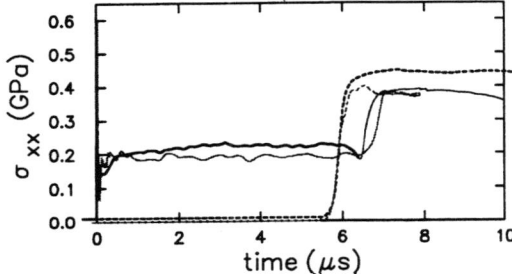

FIGURE 3. Comparison of gauge data with simulations. Black lines are gauge data taken from (4, figure 2.7). Thin line ($t > 6.4\,\mu s$) indicates possible contamination from side rarefactions. Gray lines are simulated results.

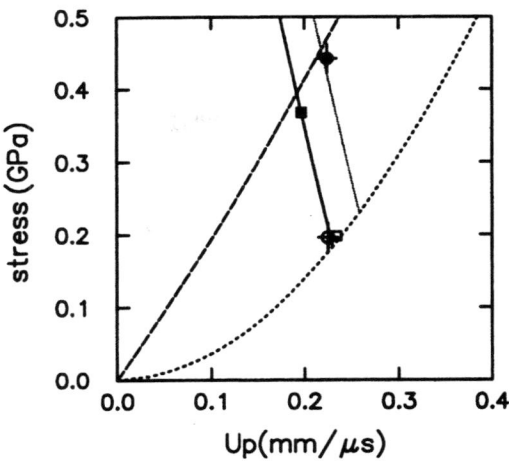

FIGURE 4. Impedance match at back gauge. Short and long dashed lines are the shock loci for the granular HMX and the TPX, respectively. Solid line is the reflected shock locus in compacted HMX. The symbols denote the states of the incident wave (from the front record) and the reflected wave (from the back record) from the experiment (squares) and simulation (circles). The crosses on the circles are an estimate of the experimental error. The gray line is the reflected shock locus assuming that burn increases the stress behind the compaction wave by 10 % as it propagates through the granular HMX.

crease of the stress and a corresponding decrease of the velocity of the gauge data as compared to the simulations. Burning behind the wave front would have this effect. In (3) it is estimated that a small amount of burn, mass fraction of less than 1 %, is sufficient to increase the stress by the observed 10 %.

At the back records, the agreement in the arrival times indicates that the compaction wave speed is correctly determined by the simulations. Using the jump condition for mass conservation, we find that the solid volume fraction ($\phi = 1 - $ porosity) behind the wave front

$$\phi = \frac{\phi_0}{1 - u_p/u_s}. \qquad (1)$$

is nearly 1. The porosity is determined by the grain distribution and yield strength. To obtain the nearly zero porosity needed to match the wave speed, the yield strength had to be lowered to 0.15 GPa from the value of 0.26 GPa inferred from the measure elastic precursor in single crystal HMX. For comparison, we note that the average normal stress component (σ_{xx}) behind the compaction wave is 0.2 GPa. The yield strength is discussed further in (3).

The most notable discrepancies at the back records are the high values of stress and velocity at the back gauge as compared to the simulations. The difference is due to the lack of burning in the simulation. To verify this we consider the impedance match of the compaction wave impacting the back disk.

The wave curves for the impedance match are shown in figure 4. In addition, for both the experiment and the simulation the states are shown corresponding to the incident shock (as determined from the front record) and to the reflected shock (as determined from the back record). The simulation is in agreement with the impedance match.

A property of the impedance match is that the change in stress must have the opposite sign from the change in velocity. This property is violated for the data if it is assumed that the compaction wave propagates with constant strength. On the other hand, if the stress behind the compaction wave increases by

10%, as indicated by the front gauge, then the data is consistent with the result of the impedance match.

Finally, we note that the arrival of the reflected shock at the front of the granular bed is slightly delayed in the simulation as compared to the gauge data. This is probably due to the small amount of burn occurring in the experiment and absence in the simulation.

BURNING ISSUES

The average state behind the compaction front is determined largely by the jump conditions derived from the conservation laws. The state from mesoscale simulations compares well with experimental data since it is an insensitive quantity. Because reaction rates are strongly temperature sensitive, the overall reaction rate is dominated by hot spots. Hot spots represent fluctuations, the tail of the temperature distribution, and as such are a more sensitive quantity to compute. The temperature field and the average profile are shown in figure 5. Even though the simulations do show that 'hot spots' are generated by a compaction wave, the hot-spot temperature is low and would result in a negligible amount of burn, even if reactions were included in the simulations.

Hot spots are determined by the grain distribution and dissipative mechanisms. The simulations include three dissipative mechanisms: (i) plastic work, (ii) shear heating, and (iii) artificial bulk viscosity. Conspicuously absent is frictional heating at grain interfaces. The computational problem is that algorithms that can handle the large distortion of the grains are inaccurate at interfaces and vice versa. Hence, dissipative mechanism that take place predominantly at interfaces are difficult to simulate accurately. Moreover, the peak hot-spot temperature, which is critical for reaction, requires fine resolution. Presently, on supercomputers it is possible to do simulations with sufficiently high resolution in two-dimensions but not yet in three-dimensions.

ACKNOWLEDGEMENTS

The author thanks Prof. David Benson, Univ. of Calif. at San Diego, for providing the code used for the simulations, and Dr. Richard Gustavsen and DR. Stephen Sheffield for providing the data from their experiments.

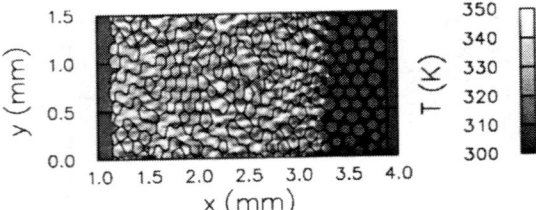
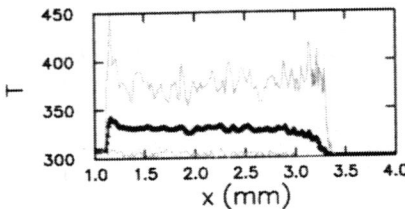

FIGURE 5. Temperature behind wave front from simulations, at time of $5\,\mu s$. Profile is average parallel to wave front (y-direction). Gray lines correspond to maximum and minimum values.

REFERENCES

1. Baer, M. R., in *High-pressure shock compression of solids IV: response of highly porous solids to shock compression*, edited by L. Davison, Y. Horie, and M. Shahinpoor, Springer-Verlag, New York, 1997 pp. 63–82.

2. Baer, M. R., in *Shock Compression in Condensed Matter–1999*, AIP, Melville, NY, 1999 pp. 27–33.

3. Menikoff, R., *J. Appl. Phys.* **(to be published)**.

4. Sheffield, S. A., Gustavsen, R. L., and Anderson, M. U., in *High-pressure shock compression of solids IV: response of highly porous solids to shock compression*, edited by L. Davison, Y. Horie, and M. Shahinpoor, Springer-Verlag, New York, 1997 pp. 23–61.

5. Tang, Z. P., Horie, Y., and Psakhie, S. G., in *High-pressure shock compression of solids IV: response of highly porous solids to shock compression*, edited by L. Davison, Y. Horie, and M. Shahinpoor, Springer-Verlag, New York, 1997 pp. 143–175.

MECHANISTIC MODEL OF HOT-SPOT: A UNIFYING FRAMEWORK

K. Yano, Y. Horie, and D. Greening

Applied Physics Division, Los Alamos National Laboratory, Los Alamos, NM 87545

Abstract. A one dimensional hot-spot model is proposed to consider wide range of ignition scenarios for solid high explosives. Model components commonly found in typical mechanistic hot-spot models are included in the model. They are a finite region of enhanced heating in the solid matrix, a space occupied by gases, heat flow between solid and gas phases, and chemical reaction in the gas phase, at the solid-gas interface, or both. The model also contains advanced features of spherical pore-collapse models. Test calculations based on cyclonite (RDX) data show that the model behavior is comparable to that of spherical models under conditions indicative of shock initiation. Critical conditions for ignition is summarized in a single chart in terms of total energy dissipation, the rate of energy dissipation, and the size of hot-spot.

INTRODUCTION

In high explosive (HE) science, the existence of "hot-spot," or the region where energy dissipation is highly localized is generally accepted as essential to the ignition of explosives. Various mechanisms for hot-spot have been proposed, such as pre collapse, shear banding, friction, and fracture (1-5). Currently, there is no agreement as to the mechanism(s) by which the dissipation is localized (6).

To model hot-spot initiation of HE, one typically choose a mechanism that describes mechanical energy dissipation. The mechanism tends to imply a specific geometry that may not be typical of the actual localization. In addition, any single-mechanism approach quickly becomes problematic in a complex system such as polymer-binder explosives (PBX).

If one examines the various models, setting aside the mechanisms, a common modeling structures become clear. They are a region of enhanced heating, creation or existence of the region occupied by gas, heat flow between the regions, and chemical reaction in the gas, at the solid-gas interface, or both. In this study, we consider a model that incorporates these features without introducing the mechanism-specific traits.

MODEL DESCRIPTION

Figure 1 shows the simplified one dimensional model that incorporates the components commonly found in pore collapse, shear band, friction models of hot-spot. The model consists of solid mass, gas cavity, and the interface. The solid phase is assumed to be incompressible. Thus the particle velocity is common, and it is determined by the force balance on the solid phase through the equation of motion. Temperature profile is calculated using analytic solution for a fixed length bar with heat source. The energy localization is represented parametrically by the size of the heating region, δ, with a known heating rate, ϕ_s. In this way, the very significant issue of how to consider the localization as dependent on the state of the material is set aside for the moment. Instead, attention is focused on whether a model constructed on this framework produces results that are consistent with other

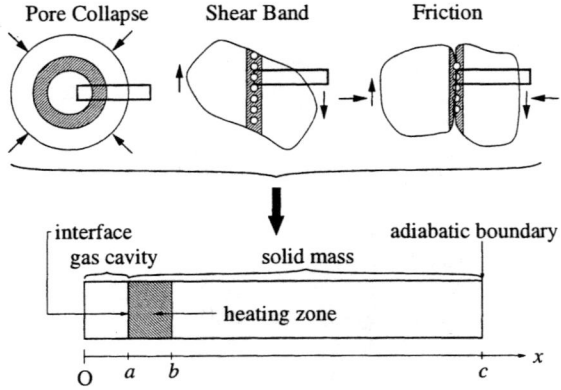

FIGURE 1. A schematic of the one dimensional model structure that consists of the essential components commonly found in typical mechanistic hot-spot models. Three major regions are identified: gas cavity, solid-gas interface, and solid mass with a region of localized heating.

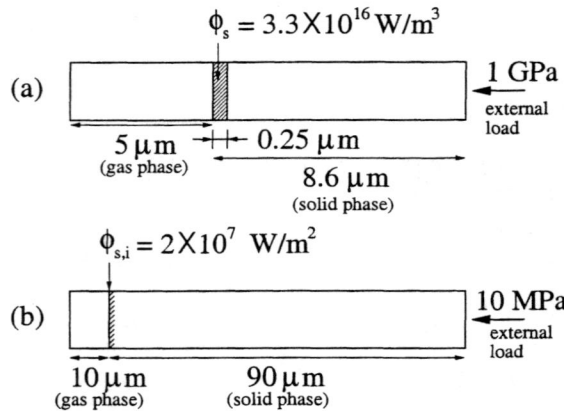

FIGURE 2. Conditions for ignition simulations. (a) The heat is deposited over a finite thickness region at the rate ϕ_s. (b) The heat is deposited at the solid-gas interface at the rate $\phi_{s,i}$.

mechanistic models by ranging over the parameter space.

At the solid-gas interface, conditions are imposed for mass, momentum, and energy conservation. There is mass flux, $\dot{m}$, into the gas phase, which is caused by the localized heating in the solid phase. The flux is determined by the interface temperature. Part of the mass flux may go through an instantaneous chemical reaction. Heating at the solid-gas interface can also be considered as a known heating rate, $\phi_{s,i}$, depending on the modeling needs such as friction.

In the gas phase, time evolution of the average density, temperature, and chemical species are described by a set of ordinary differential equations. The equations are obtained through integral form of conservation equations. Heat flow is considered between the gas phase and the interface. The equation of state is a modified form of the ideal gas equation of state (Abel equation of state). The gas-phase reaction is described by a single-step reaction with the Arrhenius kinetics model.

Mathematically, the system consists of 7 ordinary differential equations with 10 algebraic equations. Runge-Kutta scheme was used for time integration. The model contains essentially the same features found in the spherical hot-spot model by Kang, et al. (2), except that in Kang's model the heating rate and zone width is coupled with the geometry instead of free parameters as in this paper. Details of the model are found in (7).

MODEL CALCULATIONS

Three simulation results are presented. The first two are concerned with ignition simulations for different types of heat deposition as depicted in Figs. 2 (a) and (b). The third result presents a parametric study of critical ignition conditions.

Figure 2 (a) shows simulation conditions where heat is deposited within a finite volume in the solid matrix. At time $t = 0$, a confining pressure of 1 GPa was applied to the solid matrix. At the same time, heat was added over the thickness of 0.25 μm at a constant rate of 3.3×10^{16} W/m^3. The heat deposition was terminated at 150 ns. These conditions are comparable to those in Kang et al. (2), and indicative of shock initiation. The rate of heat flux was deduced from the results by Kang et al. (2), Massoni et al. (3), and Bonnet and Butler (8). Our value is close to the average of the three. Lastly, the duration of heat deposition was selected so

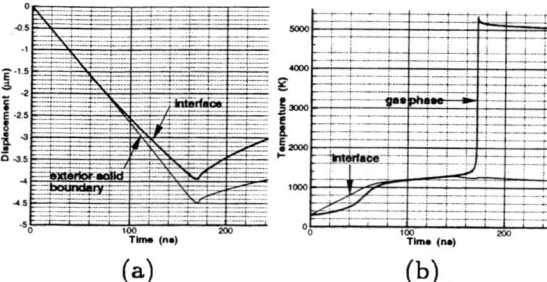
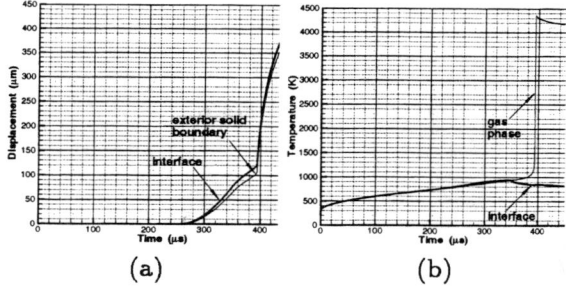

FIGURE 3. Results of simulation where heat is added over a finite region as shown in Fig. 2 (a). Time evolutions are shown for (a) displacement of the interface and the adiabatic solid boundary measured from their initial positions, and (b) temperatures of the gas phase and the interface.

FIGURE 4. Results of simulation where heat is added at the interface as shown in Fig. 2 (b). Time evolutions are shown for (a) displacement of the interface and the adiabatic solid boundary measured from their initial positions, and (b) temperatures of the gas phase and the interface.

that the system results in ignition after removal of the heat. Figures 3 (a) and (b) show time histories of the interface and solid outer boundary displacements, and temperatures of gas phase and the interface, respectively. Upon the application of external pressure, the pore starts to collapse at almost constant velocity during the period $0 < t < 170$ ns. The deviation of interface displacement from that of outer solid boundary reflects the effect of mass flux at the interface. The interface temperature rises steadily until $t = 150$ ns, when the heat addition is turned off. When the temperature exceeds approximately 940 K, the rate of temperature increase slows down because the heat consumed for sublimation of the solid exceeds the heat evolved from surface chemical reaction. In the gas phase, the temperature increases steadily, mainly because of heat conduction from the interface.

At about 170 ns, the conditions in the gas phase reach a critical state, and the accumulated reactant gas reacts in a near discontinuous manner. Because of high density and high temperature after the reaction, the pressure of the gas reaches well above the confining pressure (about 3 GPa) and expands the gas pore (Fig. 3(a)). Once the expansion occurs, both the gas temperature and pressure start to drop. Although there are some minor differences, the observed model behaviors and those presented by Kang et al. (2) are very similar in general.

In the second simulation, the heat was deposited to the interface, as shown in Fig. 2 (b). The heating rate was made comparable, in estimate, to that caused by sliding solid explosive over a rough surface at 10 m/s under the imposed normal stress of 10 MPa. The corresponding energy flux is 2×10^7 W/m². The deposition of energy was maintained for the duration of 350 μs.

Figures 4 (a) and (b) show time evolutions of the interface and solid outer boundary displacements, and temperatures of gas phase and the interface, respectively. The maximum compression is reached almost instantaneously and the solid plate remains virtually at rest for $0 < t < 150$ μs. During this time period, the interface and gas-phase temperatures rise steadily. Also, the temperatures are nearly in equilibrium (Fig. 4(b)). As the interface temperature rises, it starts to deviate from the gas temperature at about 150 μs as a result of exothermic reaction at the interface. At the same time, mass flux increases rapidly and effects a pressure rise in the gas phase. At about 270 μs, the gas-phase pressure exceeds the imposed stress, and the pore starts to expand, as seen in Fig. 2(a). Finally, at about 395 μs, the reactant accumulated in the gas phase reaches a critical temperature and undergoes an explosive reaction, causing an almost instantaneous jump in temperature and pressure. These jumps are followed by rapid drops because of pore expansion.

Lastly, Fig. 5 shows the results of critical

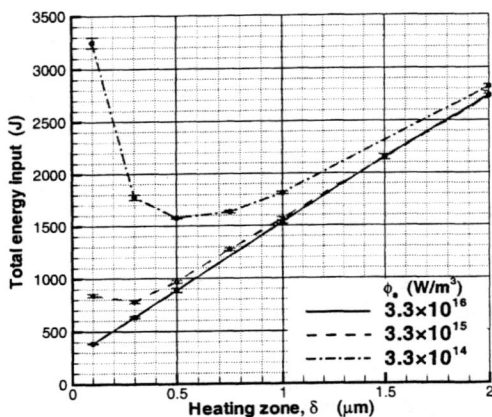

FIGURE 5. Parametric study of critical ignition conditions in terms of the total energy input to the hot-spot as a function of the heat deposition rate ϕ_s and the size of the heating zone δ.

conditions for ignition in terms of total energy deposition as a function of hot-spot size δ and rate of energy dissipation ϕ_s. The simulation condition is essentially the same as those shown in Fig. 2 (a), except that δ and ϕ_s are changed for the parametric study. For a given δ and ϕ_s, any energy above the profile leads to ignition and any energy below does not lead to ignition. The profiles are characterized by the existence of minima and convergence for large δ. In the left branch of the minimum point, less energy is required for larger δ. This is because heat loss at the interface resulting from conduction becomes smaller as the size of the hot-spot increases. On the other hand, in the right branch, more energy is required for larger δ simply because of the volume increase. The curve approaches a linear profile for large values of δ as the effect of thermal conduction becomes negligible. The convergence behavior for large δ indicates that the ignition occurs as the interface temperature reaches the critical value (thermal ignition).

CONCLUSIONS

Two distinct conclusions are highlighted. Firstly, the model was tested well under conditions indicative of a shock-load/pore-collapse ignition regime and a frictional ignition regime. The specific calculations, based on the RDX data, show ignition behaviors consistent with those published for a comparable mechanism-specific model (2,5).

Secondly, an initial parametric study was presented. The results, in the form of an ignition chart, are compatible with observations on the size range and duration of heating required for an effective hot-spot ignition source, 0.1 to 10 μm and 10^{-5} to 10^{-3} s, respectively (6). The results also show that ignition thresholds depend not only on total energy deposition, but also on the rate of deposition and the size of the zone where the energy is localized.

REFERENCES

1. Davis, W.C, *Los Alamos Science* **2**, 48-75 (1981).
2. Kang, J., Butler, P.B., Baer, M.R, *Combustion and Flame* **89**, 117-139 (1992).
3. Dienes, J.K., "A unified theory of flow, hot spots, and fragmentation, with an application to explosive sensitivity," in *High-Pressure Shock Compression of Solids II*, edited by L. Davison, D.E. Grady, and M. Shahinpoor, Springer, New York, 1995, pp. 366-398.
4. Bennett, J.G, Haberman, K.S., Johnson, J.N., Assay, B.W., and Henson, B.F., *J. Mech. Phys. Solids* **46**, 2303-2322 (1998).
5. Massoni, J., Saurel, R., Baudin, B., and Demol, G., *Phys. Fluids* **11**, 710-736 (1999).
6. Field, J.E., Bourne, N.K., Palmer, S.J.P., and Walley, S.M., *Phil. Trans. R. Soc. Lond.* **A339**, 269-283 (1992).
7. Yano, K, Horie, Y., and Greening, D.R., "A unifying framework for hot spots and the ignition of energetic materials," Los Alamos National Laboratory Report (in process).
8. Bonnet, D.L. and Butler, P.B., *J. Propul. Power* **12**, 680-690 (1996).

MICROSTRUCTURAL MODEL OF IGNITION FOR TIME VARYING LOADING CONDITIONS

Richard V. Browning and Richard J. Scammon

Los Alamos National Laboratory, PO Box 1663, Los Alamos, NM 87545

Abstract. A micro-mechanical based model of ignition was developed about five years ago based on a simple inter-granular friction model of mechanical dissipation coupled with a fit to extensive direct numerical simulations of the resulting thermally induced decomposition. The chemical model used was the McGuire-Tarver ODTX based model for HMX decomposition. The resulting power law type model has been reasonably successful in predicting threshold conditions for Steven type experiments. The final power law form was obtained by assuming a constant time history for both the pressure and shear strain rate, resulting in time independent loading conditions for the chemical model.

Here we propose to extend the model to handle time varying loading conditions. This is done using a linear operator that models reactive heat transfer simulations done for a wide variety of loading conditions. The linear operator is represented by a convolution integral with Prony series kernel form for efficient numerical implementation. To complete the model the same inter-granular friction model used previously is employed. Comparisons are made with results of numerical simulations and experiments.

The technique used here is based on the notion of linearizing the reactive heat transfer problem. Although the chemical model involves four reactions and is highly nonlinear, we effectively linearize the problem around ignition conditions with a linear operator fit. We use a simple power law approximation that gives useful accuracy over at least 4 orders of magnitude in time and fluence. A non-dimensional scaling method is used to determine the final form. We believe the techniques used here could also be used with more detailed chemical models and with other types of mechanical dissipation models.

INTRODUCTION

Models for ignition of reactive materials are needed for safety studies. They need to incorporate an energy dissipation mechanism, heat conduction, and chemical reaction. In this paper we focus on modeling the heat conduction and chemical reaction mechanisms in a concise form. Browning (1) developed a microstructural based model of ignition based on intergranular friction coupled with fits to numerical simulations of thermal ignition based on McGuire-Tarver (2) kinetics and finite element numerical techniques in Chemical Topaz (3). The result was a very simple power law relation between time of ignition, pressure and shear strain rate, however a basic assumption was that the pressure and shear strain rate was constant up to the time of ignition. The simplicity of the model allowed easy implementation and use compared with using direct numerical simulation of sub-scale reactive heat transfer models (4), and produced reasonable agreement with results from certain classes of experiments (5,6,7).

In this paper we investigate a technique for dealing with non-constant time history loading by using a linear operator to model the heat conduction and chemistry involved in the ignition process. Effectively, we linearize about the ignition conditions, incorporating the nearly linear heat conduction process and effectively hiding the severe non-linearities of the detailed chemistry.

THERMO-CHEMICAL MODELING

We start with the thermo-chemical process as defined by the McGuire-Tarver kinetics model implemented in Chemical Topaz. By running the code on simple axi-symmetric problems for varying spot sizes we can calculate ignition conditions for varying spot sizes and flux time histories. To transform the computed results into a more compact numerical model we first develop a non-dimensional representation of the results following Barenblatt(8).

We assume seven dimensional variables, the critical flux φ_{cr}, critical time t_{cr}, spot radius R, heat capacity c_p, thermal conductivity k, density ρ, and a heat of reaction q. This is simplified from the actual system, where we have three reactions in the chemical model each with separate q values, for example. These variables have five dimensional units, so we can choose two independent non-dimensional parameters. We choose a flux related parameter $\pi_1 = \varphi_{cr} c_p R / qk$ and a time related parameter $\pi_2 = t_{cr} \lambda / R^2$ where we used thermal diffusivity $\lambda = k / \rho c_p$ as a derived quantity.

Our basic assumption is that the non-dimensional parameters are related by a linear operator, $F_{cr} = L(\pi_1(\pi_2))$. We implement the linear operator as a convolution integral eventually using a Prony series representation for the kernel function. Considering π_1 a function of π_2 we can write,

$$L(\pi_1(\pi_2)) = \int_0^{\pi_2} K(\pi_2 - \tau) \pi_1(\tau) d\tau,$$

and then $F_{cr} = L(\pi_1(\pi_2))$. This assumption is valid for linear heat transfer problems as the temperature at any point of the body can be obtained from the response of a linear operator acting on the time dependent boundary conditions. In the case of non-linear reactive heat transfer, the ignition condition is not exactly defined by a linear operator, but we fit computational results to the model, and then verify the accuracy by comparisons with the results.

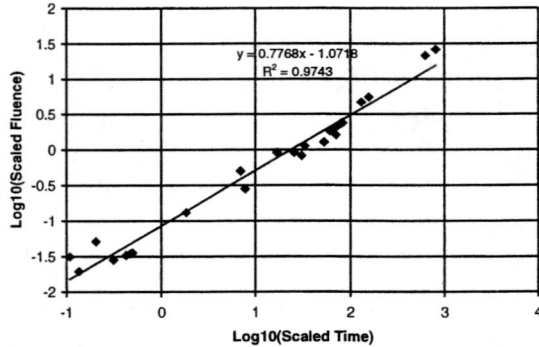

Figure 1. Scaled results from original 2-D calculations. A constant flux is applied over spot sizes 0.25 to 250 micron diameter. The linear fit to logarithmic values is equivalent to a power law fit.

To express the relation in terms of real time, we substitute the definitions of the non-dimensional parameters back into the integral representation, and after changing the integration variable obtain a representation,

$$F_2(t) = \int_0^t K(s_2 t - s_2 x) s_1 \varphi(x) s_2 dx,$$

where we have used scaling factors defined by $\pi_1(\pi_2) = s_1 \varphi(t)$ and $\pi_2 = s_2 t$. Figure 1 shows the scaled results of a series of calculations on spot sizes 0.25, 2.5, 25 and 250 μm. These are given in terms of scaled fluence $\pi_{1b} = s_1 s_2 \Phi = \Phi / \rho qR$. On this log-log plot the straight line fit is equivalent to a power law function and so we take $\pi_{1b}(x) = ax^n$ and then the kernel becomes $K(x) = x^{-n}(1-n)/a$. Given the power law form for the kernel, we can factor out the scaling terms to obtain the thermo-chemical model in real time variables,

$$F_2(t) = \int_0^t s_1 s_2^{1-n} (t - x)^{-n} \varphi(x) dx.$$

The critical time is then defined by $F_{cr} = F_2(t_{cr})$, where F_{cr} is a fitting constant.

We use a Prony series kernel in the convolution integral to simplify the integration process. The Prony terms are taken as constant ratios, with the ratios selected to cover the range of times involved, and to generate a good fit to the power law function. Fig. 2 shows a comparison of the Prony model results and the original calculations for various spot radii. The scaling gives a good approximation for the various spot sizes, and the slope of the fluence time curve is accurate, however we do miss the time-to-ignition depending on the distance to the calibration point. This is basically the result of the scatter in the original scaled time-to-ignition data. Some of this scatter is inherent in the approximation, however some also results from numerical effects in the Chemical Topaz calculations.

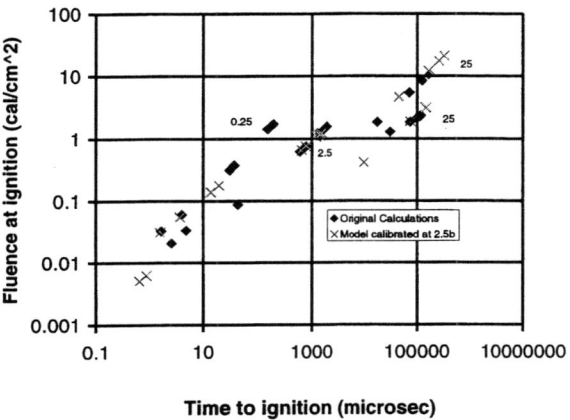

Figure 2. Comparison of Prony model with results from original model.

COMPARISON WITH OTHER HISTORIES

To verify the linear operator technique we ran a series of problems with non-constant flux conditions. The numerical reactive heat transfer results are compared with the predictions from the linear operator model. We first ran a series of calculations on a one-dimensional problem. In this series the time history was an on-off-on sequence where the total on time was split equally between the first and second heating pulse. The results are compared in Fig. 3 in terms of the reduction of the ignition criterion. In this case the agreement is excellent. Other more complex problems are in progress, with good results emerging.

OVERALL MODEL

To create an overall model we use our previous results(1) on inter-granular friction to complete the model. The spot radius and flux are expressed in terms of the macroscopic pressure P and shear strain rate $\dot{\gamma}$, as well as the material properties grain radius R_G, coefficient of friction β and grain elastic compliance $C = 2(1-\nu)/E$, where ν is Poisson's ratio and E is Young's modulus. We use,

$R = 0.721 \cdot 2^{1/6} \cdot R_G C_E P^{1/3} = C_2 P^{1/3}$ and $\varphi = C_1 P^{1/3} \dot{\gamma}$, where $C_1 = 1.031 \beta C_E^{-2/3} R_G^{-1}$.

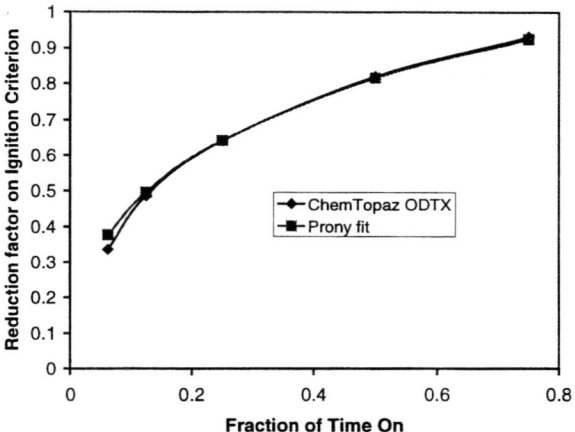

Figure 3. Comparison of Prony model and Chemical Topaz calculations for one-dimensional geometry. The flux is on for an interval, then off, then back on. The total on time is split evenly between the first and second segment. The agreement is very good indicating the ability to deal with widely separated flux pulses.

After combining all these expressions, most of the constants are merged into the fitting constant F_{cr}, but we retain the explicit dependence of R within the integral, to give the final expression

$$F(t) = \int_0^t (t-\tau)^{-n} P^{(2n-1)/3}(\tau) \dot{\gamma}(\tau) d\tau$$

We have tried fitting this with n=0.7768, and the critical value of F taken at several different times. The fit works well in the vicinity of the selected time and radius, with increasing discrepancies further from the fitting point. The overall model results are shown in Fig. 4 compared with our previous results and experimental results from the Steven experiments done at Los Alamos National Laboratory and Lawrence Livermore National Laboratory(9).

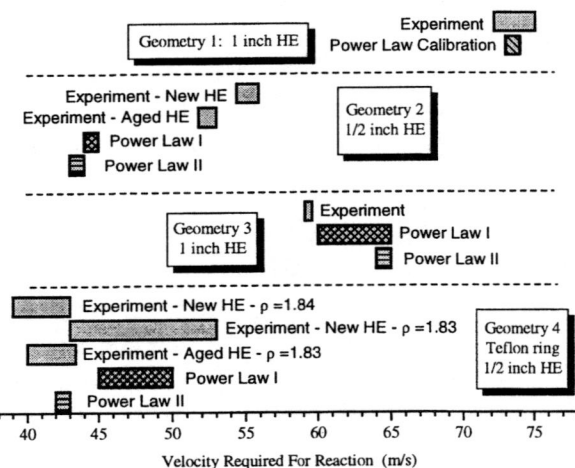

Figure 4. Comparison of models with experimental results from Steven tests. The horizontal bars show the difference between the highest no-go test and the lowest go condition for both the experiments and the models. Geometry 1 is used to calibrate the model. Geometry 2 involved a thin section of HE and there was evidence of contact between the cover plate and back support plate. The Geometry 4 experiments were done at LLNL by Steve Chidester.

CONCLUSIONS

The incorporation of the linear operator gives improved consistency in applying the ignition model, even though it does not appear to have a major effect on the results for the Steven test. We hope to expand on the model by including other dissipation mechanisms such as binder flow and void collapse. The most important need is to develop better constitutive equations for the macroscale behavior of the material. The localization behavior is key to correctly predicting the ignition behavior of real HEs. None of the currently used constitutive models correctly capture the localization effects at failure conditions.

ACKNOWLEDGEMENTS

The patient and enthusiastic experimental work of Deanne Idar and Jim Straight was critical to the continued modeling. We thank Steve Chidester for sharing experimental details and results from his tests. This work is supported by the United States Department of Energy under contract W-7405-ENG-36.

REFERENCES

1. Browning, R. V., "Microstructural Model of Mechanical Initiation of Energetic Materials," in *Shock Compression of Condensed Mater - 1995*, edited by S. C. Schmidt et al., AIP Conference Proceedings 370, New York, 1996, pp. 405-408.
2. McGuire, R.R. and Tarver, C.M., "Chemical Decomposition Models for the Thermal Explosion of Confined HMX, TATB, RDX, and TNT," *Seventh (International) Symposium on Detonation*, June 1981, pp. 56-64.
3. Nichols, A.L. III, *Chemical-TOPAZ - An Addendum to the TOPAZ manual*, UCRL-ID-104558 Add. 1, Lawrence Livermore National Laboratory, Jan. 1993.
4. Dienes, J.K. and Kershner, J.D., "Multiple-Shock Initiation via Statistical Crack Mechanics", Eleventh International Detonation Symposium, August 1998, pp. 717-724.
5. Idar, D.J., et al., "Low Amplitude Insult Project: PBX 9501 High Explosive Violent Reaction Experiments," Eleventh International Detonation Symposium, August 1998, pp. 101-110.
6. Idar, D.J. et.al.,"Low Amplitude Impact of Damaged PBX 9501," in *Shock Compression of Condensed Matter - 1999*, edited by M. D. Furnish et al., AIP Conference Proceedings 505, New York, 2000.
7. Scammon, R.J. et al. "Low Amplitude Insult Project: Structural Analysis and Prediction of Low Order Reaction," Eleventh International Detonation Symposium, August 1998, pp. 111-118.
8. Barenblatt, G. I., *Scaling, self-similarity, and intermediate asymptotics*, Cambridge University Press, Cambridge, 1996.
9. Chidester, S.K., Tarver, C.M., DePiero, A.H., and Garza, R.G., "Single and Multiple Impact Ignition of New and Aged High Explosives in the Steven Impact Test", in Shock Compression in Condensed Matter - 1999, edited by M. D. Furnish et al., AIP Conference Proceedings 505, New York, 2000, pp. 663-666.

DEVELOPMENT OF A SIMPLE MODEL OF "HOT-SPOT" INITIATION IN HETEROGENEOUS SOLID EXPLOSIVES

N. J. Whitworth

AWE, Aldermaston, Reading, United Kingdom

Abstract. Previously we numerically studied "hot-spot" formation in an explosive material as a result of shock induced pore collapse via microscale one-dimensional hydrocode simulations. Following this work, a simple model of the shock compaction process, leading to the formation and subsequent ignition of "hot-spots", has been developed for use in macroscale simulations of shock initiation problems of interest. The simple model is presented, where "hot-spots" are formed as a result of elastic-plastic and viscous stresses generated in the solid explosive during pore collapse. Results from the model are compared with corresponding results from the hydrocode simulations to illustrate how well, or otherwise, the simple model is performing. The model has also been used to help analyse data obtained from single and double shock initiation experiments on the HMX-based explosive PBX-9404.

INTRODUCTION

Work is in progress to develop a new, physics-based, explosive ignition and burn model. The requirements of the model are that it should explicity describe the important physical and chemical processes involved in explosive shock initiation, yet still be simple enough for use within a hydrodynamics code to simulate real problems of interest.

In this paper a simple model to explicitly describe the formation and subsequent ignition of "hot-spots" is presented. It is assumed that the solid heterogeneous explosive contains pores, and that "hot-spots" are formed as a result of elastic-plastic and viscous stresses generated in the solid explosive during the collapse of the pores under shock wave loading. The described model is based on very similar assumptions and equations to other published "hot-spot" ignition models [1,2,3].

In a previous paper [4], we numerically studied the dynamic formation of "hot-spots" in an explosive material as a result of shock induced pore collapse using a hydrocode which incorporated an elastic-viscoplastic constitutive model. This hydrocode modelling work is used as a reference to test the validity of the various assumptions in the simplified model, and results obtained from the simple "hot-spots" model are compared with the corresponding hydrocode calculations to illustrate how well, or otherwise, the simple model is performing.

SIMPLE "HOT-SPOT" INITIATION MODEL

Physical Model and Assumptions

The explosive is represented by the 1D hollow sphere pore collapse model developed by Carroll and Holt [5] for compaction of inert porous materials, see Figure 1. The initial inner radius, a_0, represents the average pore radius in the explosive, and the external radius, b_0, is chosen such that the initial porosity and the measured overall porosity of the explosive material are equal. P_s is the applied stress. The initial distention ratio, α_0, of the material is the ratio of the total volume of the porous material to the volume of the solid material, and Φ_0 is the initial porosity.

In conjunction with the physical model of the explosive, it is assumed that: *(i)* pore collapse and flow of the solid material is treated as 1D, spherically symmetric, *(ii)* the solid explosive material is incom-

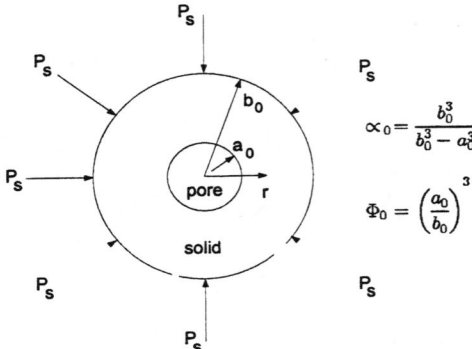

FIGURE 1. Hollow sphere configuration used to model "hot-spot" formation.

pressible during its radial motion, *(iii)* the solid explosive is an isotropic elastic-viscoplastic material, where the shear modulus, μ, yield strength, Y, and viscosity, η, are constant, and *(iv)* the pore is a void.

Model Equations

The time evolution of the pore radius is given by,

$$-\rho\left(a\ddot{a} + 1.5\dot{a}^2\right) = P_s + P_v - P_y \quad (1)$$

where ρ is the density of the solid explosive, a is the pore radius, P_v is the viscous (strain-rate dependent) stress, P_y is the elastic-plastic (strain dependent) stress, and a dot above a symbol denotes a derivative with respect to time.

Following [5], the deformation of the pore occurs in three distinct phases: *(i)* an initial (elastic) phase ($\alpha_0 \geq \alpha \geq \alpha_1$), *(ii)* a transitional elastic-viscoplastic phase ($\alpha_1 \geq \alpha \geq \alpha_2$), and *(iii)* a viscoplastic phase ($\alpha_2 \geq \alpha \geq 1$), where α_1 and α_2 define the distention ratios at which the transition from one state of stress to the next occurs. Other "hot-spot" ignition models based on viscoplastic pore collapse [1,2,3] assume that volume changes during the first two phases of collapse are negligible, and pore collapse occurs by virtue of viscoplastic flow only in the material. An important difference between our model and the other "hot-spot" models, is that all three possible phases of pore collapse are modelled as hydrocode calculations [4] have shown that, for weak and moderate shock waves, the spherical shell is usually in the transitional elastic-viscoplastic state. The viscous and elastic-plastic stresses, which act together to resist the pore collapse, are given by,

(i) Elastic phase ($\alpha_0 \geq \alpha \geq \alpha_1$);

$$P_v = 0, \quad P_y = \frac{4\mu(\alpha_0 - \alpha)}{3\alpha(\alpha - 1)}, \quad \alpha_1 = \frac{2\mu\alpha_0 + Y}{2\mu + Y} \quad (2)$$

(ii) Elastic-viscoplastic phase ($\alpha_1 \geq \alpha \geq \alpha_2$);

$$P_v = 12\eta\, a^2\, \dot{a} \int_a^c \frac{1}{r^4}\, dr, \quad \alpha_2 = \frac{2\mu\alpha_0}{2\mu + Y}$$

$$P_y = \frac{2}{3}Y\left(1 - \frac{2\mu(\alpha_0 - \alpha)}{Y\alpha} + \ln\left\{\frac{2\mu(\alpha_0 - \alpha)}{Y(\alpha - 1)}\right\}\right) \quad (3)$$

where the time-dependent interface between elastic and viscoplastic flow, c, is given by,

$$c = \sqrt[3]{\frac{2\mu B}{Y}}, \quad \text{where } B = \frac{a_0^3(\alpha_0 - \alpha)}{\alpha_0 - 1} \quad (4)$$

(iii) Viscoplastic phase ($\alpha_2 \geq \alpha \geq 1$);

$$P_v = 12\eta\, a^2\, \dot{a} \int_a^b \frac{1}{r^4}\, dr, \quad P_y = \frac{2}{3}Y\ln\left(\frac{\alpha}{\alpha - 1}\right) \quad (5)$$

The hydrocode modelling work [4] has also shown that the size of a "hot-spot" created in the vicinity of a collapsing pore is roughly equal to the size of the pore before it collapsed. These findings are used to define a domain for the "hot-spot" which is then defined by a number of Lagrangian points, where the local temperature, reaction rate, and mass fraction reacted are computed at each step.

The temperature increase at the Lagrangian positions as a result of the mechanical deformation during pore radial motion is given by,

$$\rho C_v \dot{T} = 12\eta\left(\frac{u^2}{r^2}\right) + 2Y\left(\frac{|u|}{r}\right), \quad u = \frac{\dot{a}a^2}{r^2} \quad (6)$$

where T is the local temperature, C_v is the specfic heat capacity at constant volume, r is the radial (Lagrangian) position, and u is the local velocity in the solid shell. The local reaction rates are calculated using the Arrhenius rate law,

$$\dot{F} = (1-F)\, Z\, e^{-\frac{E^*}{RT}} \quad (7)$$

where F is the local mass fraction of explosive that has reacted, Z is the frequency factor, E^* is the activation energy, and R is the universal gas constant.

MODEL CALCULATIONS

To initially test the model, the HMX-based explosive PBX-9404 was chosen as the majority of its material parameters, as required for input to the model, are to be readily found in the literature, see Table 1.

Sample results from the described simplified "hot-spots" model, in terms of the time evolution of the pore radius and pore surface temperature, are compared with the corresponding hydrocode calculations in Figures 2 and 3 respectively. Good agreement with the hydrocode results are obtained. The observed small difference in pore surface motion at low shock pressures is due to the assumption of incompressibility of the solid material in the model.

Also shown in Figures 2 and 3 are the corresponding model results assuming that pore collapse occurs by virtue of viscoplastic flow only in the material, as assumed in other "hot-spot" ignition models. Significant differences in pore response are observed between the different simplified modelling approaches, particularly for weak shocks, with the results from the model described in this paper being in closer agreement to the hydrocode results where we model the problem as accurately as possible. At higher pressures, very similar results are obtained.

Empirical ignition and burn models cannot adequately describe the response of an explosive to multiple shock loading. Experimentally it is known that preshocked explosives are less sensitive than virgin material [6]. Here, the response of the hollow sphere model to a planar double shock input has been cal-

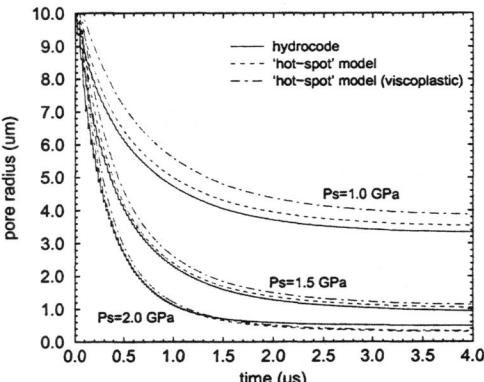

FIGURE 2. Time evolution of pore radius.

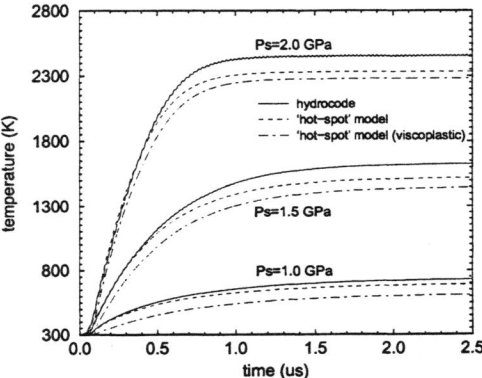

FIGURE 3. Temperature at surface of collapsing pore.

culated. The double shock loading consisted of a preshock of 1.0 GPa followed a given time later by a second shock of 2.0 GPa. Computed results from the "hot-spot" model and hydrocode, corresponding to a time delay of 0.5 μs between the two shocks are compared in Figures 4 and 5, where the single shock results are also shown. Good agreement with the hydrocode results are again obtained, and it is seen that the temperature at the pore surface in a double shock process is less than in a single shock at the pressure of the second (main) shock alone. Preshocking results in a lower temperature "hot-spot" due to the reduction in pore size before arrival of the main shock, thus making the material less sensitive. In addition, the calculated "hot-spot" temperatures are also dependent on the time delay between the precursor and main shocks, with increasing preshock duration resulting in lower temperature "hot-spots".

TABLE 1. Material parameters for PBX-9404.

Material parameter	Value
Initial density, ρ_0 (g/cc)	1.84
Yield strength, Y (GPa)	0.2
Shear modulus, μ (GPa)	4.54
Viscosity, η (GPa μs)	0.1
Initial temperature, T_0 (°K)	300.0
Specific heat, C_V (GPa cc/g/°K)	1.512e-03
Frequency factor, Z (μs^{-1})	1.81e+19
Activation energy, E^* (GPa cc/mole)	220.5
Initial pore radius, a_0 (μm)	10.0
Initial outer radius, b_0 (μm)	46.416
Initial porosity, Φ_0 (%)	1.0
Rise time of shock, τ (μs)	0.1

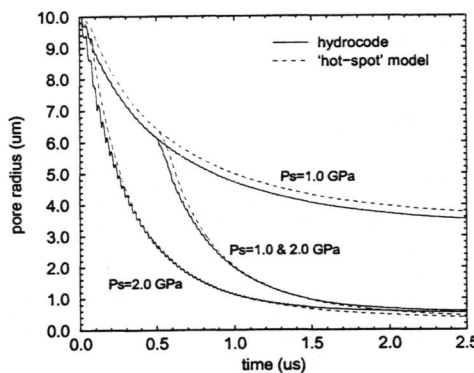

FIGURE 4. Time evolution of pore radius for a double shock.

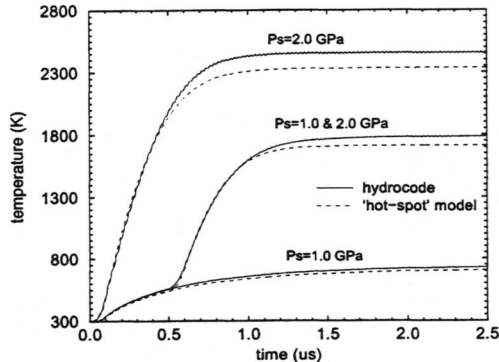

FIGURE 5. Temperature at pore surface for a double shock.

MODEL IMPLEMENTATION

The described simple "hot-spot" ignition model has been implemented in a 2D Lagrangian hydrocode to enable modelling of shock initiation problems of interest. The model has been applied to both single and double shock experimental data on PBX-9404, where the growth of reaction from the "hot-spots" is modelled using the Lee and Tarver model growth terms with published parameters for PBX-9404 [7].

Calculations of sustained single shock inputs show very similar results to Lee and Tarver ignition and growth model calculations which are in agreement with experiment eg [7]. The PBX-9404 double shock experiments of Mulford et al. [6], where a 2.3 GPa preshock was followed 0.65 μs later by a 5.6 GPa shock, have also been calculated. The experimental and calculated particle velocity histories are compared in Figure 6. Concentrating on the early

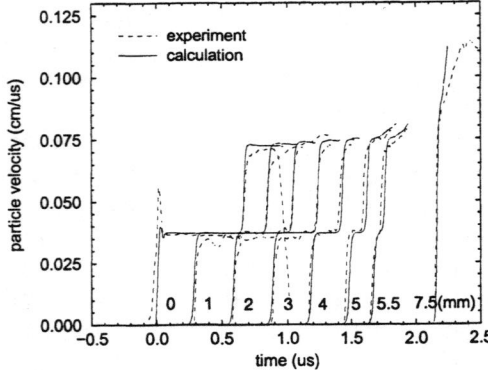

FIGURE 6. PBX-9404 double shock experiments vs calculation.

stages of reaction build-up corresponding to the ignition phase, reasonable agreement with experiment is obtained.

CONCLUSIONS

Overall, the performance of the simple "hot-spots" model is encouraging, and provides a useful starting point for further development of the physics-based reactive burn model. Future enhancements will include taking account of the effects of heat conduction on "hot-spot" initiation. Consideration will also be given to improving the growth of reaction from the "hot-spots" as the bulk of the explosive is consumed.

REFERENCES

1. Kang, K., Butler, P. B., and Baer, M. R., *Combustion and Flame* **89**, 117-139 (1992).
2. Bonnett, D. L., and Butler, P. B., *Journal of Propulsion and Power* **12**, 680-690 (1996).
3. Massoni, J., Saurel, R., Baudin, G., and Demol, G, *Physics of Fluids* **11**(3), 710-736 (1999).
4. Whitworth, N. J., and Maw, J. R., *Shock Compression of Condensed Matter-1999*, AIP Conference Proceedings 505, New York, 2000, pp. 887-890.
5. Carroll, M. M., and Holt, A. C., *Journal of Applied Physics* **43**, 1626-1635 (1972).
6. Mulford, R. N., Sheffield, S. A., and Alcon, R. R., *Proceedings of the Tenth Symposium (International) on Detonation*, 1993, pp. 459-467.
7. Tarver, C. M., and Hallquist, J. O., *Proceedings of the Seventh Symposium (International) on Detonation*, 1981, pp. 488-497.

©British Crown Copyright 2001/MOD

INITIATION OF PETN POWDER BY PULSE LASER ABLATION

Kunihito Nagayama*, Kazunari Inou*, and Motonao Nakahara**

*Department of Aeronautics and Astronautics, Faculty of Engineering,
Kyushu University, Hakozaki, Fukuoka 812-8581, JAPAN
** Department of Computer and Communication Engineering, Faculty of Engineering,
Fukuoka Institute of Technology, Fukuoka 811-0295 Japan

Abstract We developed a laser-driven method of initiating a secondary high explosive charge, based on the energy conversion from the pulsed laser energy to the shock wave energy. Enhanced energy conversion was found from optical to mechanical by an intentionally roughened surface of a transparent medium through which laser beam propagates. In this study, a PMMA plate is used, one of whose surface was intentionally roughened. Aluminum coating of less than 1 μm thickness was made to further enhance the laser absorption at the surface. Observed air shock wave indicates that it is very strong with Mach number larger than 30. Nd:YAG laser of 4 ns duration, 1064 nm wavelength and of up to 200 mJ was used as an energy source. Detonation emission of PETN powder was recorded by an image converter streak camera. Delay of 200-300 ns in the initiation was detected.

INTRODUCTION

By focusing very intense laser beam onto solid surface, laser energy absorption at the surface layer causes extreme high temperature and high pressure which drives explosive burst of plasma and particles called laser ablation. Such process can be applied to various physics experiments as well as engineering and medical applications. Laser pulse energy is one of the promising source of energy to be used to initiate high explosive charge without danger of accidental explosion by electrical noise. Accurate timing of initiation can be expected by the proper choice of energy deposition method. This is a simple extension of the impact test. Various attempts have been made by focusing the laser energy directly on the explosive charge. In this application, infrared and ns duration laser is preferable. Since most of the explosives are not a good absorber of light, direct absorption of laser energy by explosives is not expected to be very efficient. Watson et al have studied an experimental feasibility of initiating PETN powder by pulse laser driven thin metal foil.[1] They claimed that at least 1 μm grain PETN powder can be initiated by a extremely thin foil of 3-5 μm accelerated up to 8 km/s.

In this study, we propose a new energy deposition method on explosive through the generation of high-temperature metal plasma due to the pulse laser ablation of vacuum deposited thin metal layer on a roughened polymer surface.[2,3] We have found an enhanced absorption of pulse laser energy by a roughened surface.[2,3] High-pressure shock wave generation was observed in an ambient medium. This is evidenced by the laser shadowgraphy technique. Such effects may find several applications. One application is medicine as a well controllable tool of short pressure pulse for various kinds of microsurgeries.

Strength of shock wave produced in these conditions is further enhanced, if a thin metal layer is deposited on the roughened surface. In such cases, laser energy transfer to the desired position might be made through use of plastic or glass optical fiber. End surface of the optical fiber is intentionally roughened and then metallized. Deposition of the laser energy through the fiber causes an intense ablation of the metal layer. Then, we have studied initiation of high explosive charge by this process.

In this report, we will give an evidence of the effect of the surface roughness to absorb laser energy at the surface resulting in the generation of high-pressure shock wave in air. Preliminary explosive initiation test results are described, and discussed.

SHOCK WAVE GENERATION BY LASER ABLATED HIGH-TEMPERATURE METAL PLASMA

To apply this phenomena to the explosive initiation, laser beam energy is provided through the transparent fiber medium to the roughened and metallized surface. This condition is different from the common situation encountered in pulsed laser deposition of thin films. The most important difference is that the laser fluence should be less than the ablation threshold fluence of the laser transmitting medium, but be larger than the ablation threshold fluence of the roughened and metallized surface layer.

We first observed the shock wave propagation in air produced by the laser ablation of the roughened layer. Figure 1 (a) and (b) shows the laser shadowgraphs of air shock waves by focusing a Nd:YAG laser beam through PMMA plate to air. Air breakdown-induced shock front by laser focus can be seen in Fig. 1(a). In this case, laser beam transmits through smooth PMMA surface, although a small portion of laser energy is absorbed at the surface causing a weak pressure wave in air and in PMMA shown clearly in Fig. 1. When the PMMA surface is roughened, one notices that a very strong absorption of laser energy causes a much stronger shock wave in air and in PMMA emanating from the roughened interface, as shown in Fig. 1(b). In this case, a small amount of laser beam transmitted through the interface causes a weak air breakdown and shock wave front. This is also seen in Fig. 1(b).

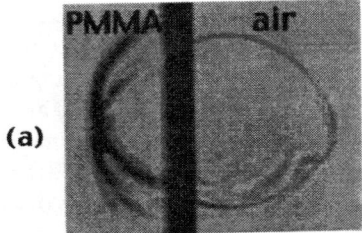

(a)

Al thickness 100nm   delay time = 536 ns

(b)

Al thickness 300nm   delay time = 524 ns

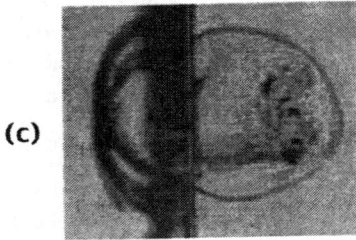

(c)

Al thickness 420nm   delay time = 560 ns

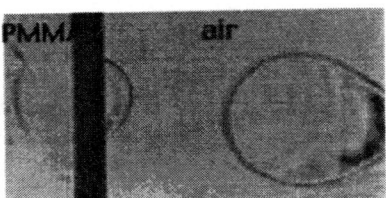

(a) Mirror surface, delay time=520ns

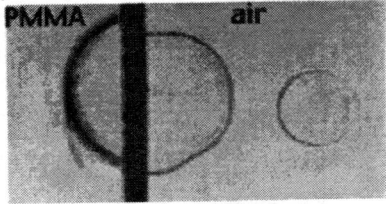

(b) Roughness #800, delay time=542ns

FIGURE 1 Shadowgraphs of shock waves generated at the PMMA-air interface. (a) smooth interface, (b) intentionally roughened interface. Nd:YAG laser pulse of 4 ns duration, 180 mJ/pulse is focused on the interface with a spot radius of about 2 mm.

FIGURE 2 Shadowgraphs of shock waves generated at the PMMA-air interface with thin aluminum coating. (a) Al thickness 100 nm, (b) 300 nm, and (c) 420 nm, respectively.

Further enhancement of laser absorption is realized by depositing a thin metal layer on the roughened interface of transparent medium with air. In this study, aluminum layer is deposited. A series of shadowgraphs are taken with a thickness of alumi-

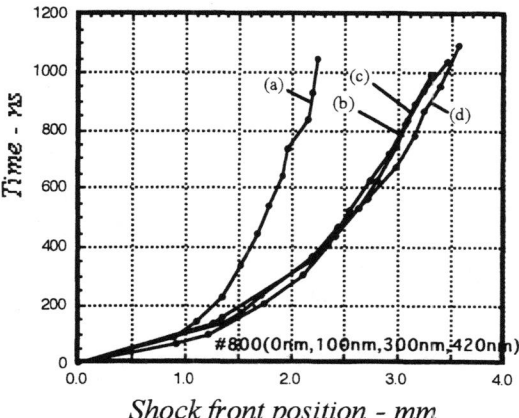

FIGURE 3 The x-t diagram of shock fronts in air for four cases. (a) Rough surface, no Al layer, (b) Rough surface, 100 nm Al thickness, (c) Rough surface, 300 nm Al thickness, and (d) Rough surface, 420 nm Al thickness, respectively.

num layer as a parameter. Figure 2 shows comparison of resultant shock waves with different aluminum layer thickness. In these experiments, PMMA surface is intentionally roughened and aluminized. Since the figures in Fig. 2 are different only in the Al thickness, all the figures in Fig. 2 can be compared with Fig. 1(b). Much stronger shock wave generation is seen compared with the picture in Fig. 1(b). Air shock velocity in these cases exceeds 10 km/s. This is evidenced by the difference in the shock front position at similar delay time. As seen in Fig. 2, difference in Al layer thickness has no appreciable effects on the shock front position. A dark undulation observed inside the shock wave front in air is created due to a small aperture inserted in optics of shadowgraphy experimentation. Comparing three pictures in Fig. 2 with different Al thickness, quite different feature in an area inside the air wave front is seen. In Fig. 2(c) of 420 nm Al thickness, granular density distribution strongly suggests an incomplete evaporation of Al layer by the present illumination condition. Even so, the shock front position in three cases is found to be almost identical.

Figure 3 shows the x-t diagrams of four different cases with different Al layer thickness and no layer. Roughness effect is apparent, but the Al layer thickness has almost no effects on the x-t diagram of air shock propagation. Shock velocity in air exceeds over that in PMMA in an early stage, but decays very fast. Shock in PMMA decays very fast as well especially at an early stage approaching rather steady propagation. No appreciable difference is seen in the x-t diagram for shock front in PMMA.

INITIATION OF VERY SLENDER PETN POWDER EXPLOSIVE COLUMN

Experimental results in the previous section reveals that high-temperature metal plasma can be generated by pulse laser ablation of thin metal layer on a roughened surface. This high-energy density flow is used to initiate high explosive charge of extremely small amount. In the present situation, scenario of the explosive initiation by pulse laser energy is described as follows: (i) laser energy pulse is converted to the high energy plasma flow by the absorption of laser energy by metal layer on a roughened surface; (ii) high-temperature high-pressure metal plasma compresses and heats up the high explosive powder grains; (iii) some kind of local hot spot formation takes place, that leads to the initiation of high explosive layer.

We have tested the initiation test of PETN powder charged into very thin layer of 0.5 mm to 1 mm. Total amount of PETN charge is less than 10-20 mg per shot. This extremely small amount of explosive is supposed to be far less than the limiting size of DDT to steady detonation. Previous data on the impact initiation tests suggests the high sensitivity of powdered PETN. Impact test using very thin foil, however, showed dependence of grain size of PETN due to the finite duration of high pressure exerted on the grain.

Figure 4 shows the streak photographs of self emission due to laser induced reaction of PETN thin layer. Although thickness of the explosive layer is very thin, explosion is observed for the laser focus onto less than 2 mm diameter. Explosion is determined by the streak record of self luminous front propagation, sound, smell, and broken assemblies. In some cases, incomplete detonation, half detonation was observed. All the evidences in this case is intermediate between complete detonation and no detonation. As seen in Fig. 4, detonation reaction is found to be delayed 200-300 ns after laser ablation. In the photographs, intense flash due to laser ablation is recorded in the streak photographs through the opaque powder explosive layer. This is due to the extreme intensity of light flash of ablation. Light intensity of ablation is brighter than the self emission of detonation. This is attributed to be the extreme high temperature of ablated metal plasma. It is shown in Fig. 4 that detonation front proceeds almost at constant velocity, although the cross section of explosive is only 1mm^2. This result is surprising in the sense of existence of limiting diameter of steady detonation. Detonation velocity estimated by the slope of the streak photography is 4 km/s, while the published data of density dependence of detonation velocity in powdery PETN is 4.5-4.8 km/s. Agreement of these two values suggests that the emission observed in this study seems to be due to almost steady detonation wave front even in an extremely slender explosive column.

We have tested sensitivity to study the laser energy dependence of initiation. In this experiment, laser pulse of 200 mJ was focused to 0.5 mm thick PETN layer with different diameter. Very simple scaling suggest the correspondence of the conditions with the same value of laser fluence, but detonation sensitivity of the present tests cannot be described by the threshold value of laser fluence.

CONCLUSION

We have proposed a new energy conversion mechanism from the laser pulse energy to shock wave energy by the intentional roughening of the energy deposition surface and thin metal coating. It is found that surface roughness and thin metal layer is very effective for the enhanced absorption of laser energy and produces a very strong shock wave in ambient media including a transparent medium through which laser pulse transmits. We have succeeded in the detonation of very slender PETN powder column almost in the steady propagation mode.

ACKNOWLEDGEMENTS

Authors wish to thank Ms. S. Hatano and Mr. Y. Mori of Kyushu University, and Mr. Murakami for their help of experiments. They also wish to thank Asahi Chemical Industry for providing PETN explosives.

REFERENCES

1. S. Watson, M.J. Gifford and J.E. Field, J. Appl. Phys., **88**, 65 (2000).
2. Nakahara M, Nagayama K, Proc. 21st Shock Wave Symp. In Great Keppel, Australia. 1997, vol. 2, (1998) p. 801.
3. M. Nakahara, and K. Nagayam, J. Materials Processing Technology, **85**, 20 (1999).

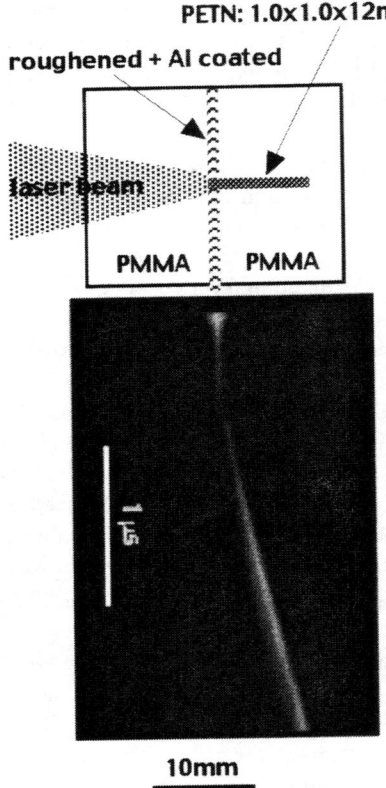

FIGURE 4 Streak photograph of delayed detonation of PETN powder of 10 mg. Nd:YAG laser pulse of 180 mJ is focused to about 1.5 mm diameter.

DOUBLE SHOCK INITIATION OF THE HMX BASED EXPLOSIVE EDC-37[†]

R. L. Gustavsen, S. A. Sheffield, R. R. Alcon[1] R. E. Winter, P. Taylor, and D. A. Salisbury[2]

[1]Los Alamos National Laboratory, Los Alamos, NM 87545
[2]Atomic Weapons Establishment, Aldermaston, Reading, Berks, RG7-4PR, U.K.

We have completed a series of double shock initiation experiments on EDC-37. EDC-37 is a unique HMX based explosive because it has an energetic liquid binder, is composed primarily of fine (< 40 μm) particles, and has a very low void content (< 0.3%) versus 1.5 – 2.0% for other HMX based explosives. It is also considerably less sensitive to initiation by shocks than other HMX based explosives such as PBX 9501. Double shocks were created by impacting the EDC-37 with gas gun launched sapphire impactors faced with a 1 to 1.5 mm thick layer of Kel-F. Varying the impact velocity controlled the magnitude of the shocks, and varying the thickness of the Kel-F layer controlled the duration of the first shock. Wave profiles were measured using embedded electromagnetic particle velocity gauges. Results show buildup to detonation commencing after the first and second waves coalesce into a single shock, provided there is not significant reaction in the first wave. That is, in the doubly shocked region, the explosive is completely desensitized by the first shock. If there is significant reaction in the first wave, the explosive is only partially desensitized.

INTRODUCTION

At least as early as the work of Gittings[1], people have been interested in multiple shock initiation, or equivalently, shock desensitization, of pressed granular explosives.[1-8] This work has involved experiments, theory and modeling. Initiation of granular explosives is widely understood to involve "hot spots", small regions of explosive which have been heated more than the rest by the passage of the shock wave. Chemical reaction in these hot spots determines the reaction rate of the explosive. Successful numerical models of multiple shock phenomenon have understood that hot spots are generated only in the first shock.[2,4,6] The second shock or rarefaction does not generate more hot spots, nor does it destroy them, so it has little effect on the rate.

In the present study we will be discussing experiments in which a weak wave is followed by a stronger second wave. Because the second wave is traveling in pre-compressed material it will travel faster than the first wave, and will eventually overtake and coalesce with the first wave, forming a single shock.

Combining this wave pattern with ideas about hot spots in multiple shocks predicts that, if the first shock is weak enough that minimal hot-spot reaction is generated, buildup to detonation does not start until the two waves coalesce forming a strong shock with high reaction rates. This is the basis for the often quoted rule of thumb that buildup to detonation does not begin until the waves coalesce.[5] In this paper we show that if the first shock generates significant hot-spot reaction, this simple rule no longer applies.

EXPERIMENTAL DETAILS

We have recently done a great deal of work to characterize the Hugoniot and initiation characteristics of EDC-37.[8-10] This work merely extends that work to include multiple shock initiation.

The overall configuration for the experiments is shown in Figure 1. This is the same configuration described in Refs. 2, 5, and 11. A gas gun projectile is faced with an impactor disk made of a thin disk of Kel-F[12], over a thick disk of sapphire[13].

[†] Work performed under the auspices of the U.S. Dept. of Energy. Work funded by the U.K. Atomic Weapons Establishment.

The Kel-F launches a weak wave into the explosive, and the sapphire launches a following stronger wave into the explosive. The strength of the waves is controlled by the impact velocity, and the width of the pulse is tailored by using different thicknesses of Kel-F. Impact velocities, wave strengths and Kel-F thicknesses are given in Table 1 below.

Electromagnetic particle velocity gauges are embedded in the sample at 10 – 12 different depths. Two experiments are done for each stress level/wave width combination. In the first experiment the explosive sample is made without a front disk (a disk is shown in Fig. 1) and the gauges cover depths of 0.0 and 2.5 through 7.0 mm. In the second experiment, a 6 mm thick disk of explosive is used on the front of the sample. The gauges are at depths of 0.0, 6.0, and 6.5 through 11.0 mm.

In addition to the 10 – 12 particle velocity gauges, these experiments also used the "shock tracker" gauges.[11] Outputs from these gauges can be used to construct $x - t$ plots of the shock front position with time, similar to those obtained in optical or pinned wedge tests. If a transition to detona-

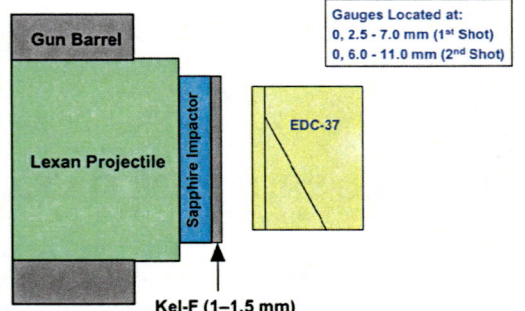

FIGURE 1. Experimental setup. Gauges are located along the inclined line. A disk of explosive is placed on front of the sample to put the disk farther into the explosive on some experiments.

tion occurs with depth spanned by the shock tracker, the time and distance of the shock-to-detonation transition can be determined.[11]

RESULTS AND DISCUSSION

Wave profiles of particle velocity vs. time and x-t plots of the shock trajectories were obtained for two pairs of experiments with wave stresses of 2.9 and 6.2 GPa, and 3.9 and 8.6 GPa. Figure 2 pre-

Table 1. Summary information for double – shock experiments in EDC-37.

Shot #	Impact Vel. (km/s)	Kel-F Thickness (mm)	1st shock Pressure (GPa)	2nd shock pressure (GPa)	Run (after coalescence) (mm)
1175	0.921	1.033 ± 0.005	2.87	6.20	
1176	0.925	1.033 ± 0.005	2.87	6.20	12.3 (6.5)
1194	1.170	1.428 ± 0.005	3.92	8.56	9.43 (2.40)
1195	1.165	1.428 ± 0.005	3.92	8.56	8.4 (2.0)

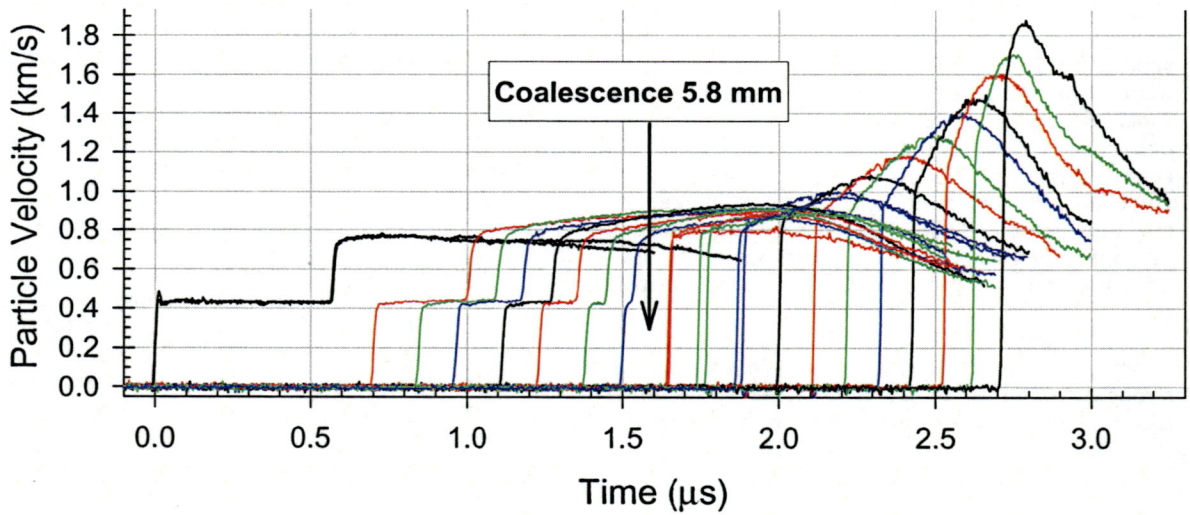

FIGURE 2. Results of Shots 1175 and 1176 with shock strengths of 2.9 and 6.2 GPa in the first and second waves. Gauges are located at 0.0 and 2.5 through 11 mm on ½ mm intervals.

sents data from the pair of lower pressure experiments. Note that there are duplicate gauges at the impact surface as well as at 6.0, 6.5, and 7.0 mm. The first wave appears to have no reaction whatsoever in it. The second wave appears to be reacting a little; however, it is not until the waves coalesce at 5.8 mm and 1.6 µs, that the reaction really gets under-way.

Figure 3 shows as-measured wave arrival times for experiments 1175 and 1176. This is also equivalent to a Lagrangian position - time or *x-t* plot. From these plots, we can determine *x-t* points for wave coalescence (5.8 mm, 1.6 µs), and more importantly, the transition to detonation (12.3 mm, 2.9 µs). This is about 1.3 mm beyond the last gauge, which is located at 11.0 mm, and 6.5 mm beyond the position where the first and second shocks coalesce. From Figure 2, we can also determine the Lagrange velocities of the first and second waves to be 3.6 and 5.6 km/s respectively. For this pair of experiments, the run distance to detonation after coalescence falls on the single shock Pop – plot. This will be further discussed later in the paper.

In summary, for the first set of experiments we saw no reaction in the first wave. While there was considerable reaction in the following wave, it did not contribute to the final buildup, and the run distance after coalescence fell on the single shock Pop – plot.

In the next set of experiments we wanted to get reaction in the first wave and even more reaction in

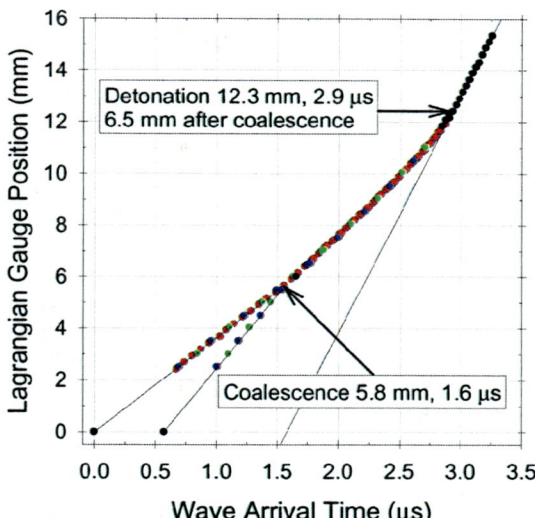

FIGURE 3. *x-t* plot obtained from shock arrival at shock tracker elements and particle velocity gauge elements for the first pair of shots. The x-t points for coalescence and detonation are shown..

the second wave. We also wanted to increase the pulse length by using a thicker layer of Kel-F, thereby further allowing the reaction to run longer.

Figure 4 shows the results of these experiments. The stresses of the first and second waves were 3.9 and 8.6 GPa, corresponding to single shock run distances of 13 and 4 mm respectively. Clearly there is a lot of reaction in the first and second waves. This is revealed by the positive slopes in particle

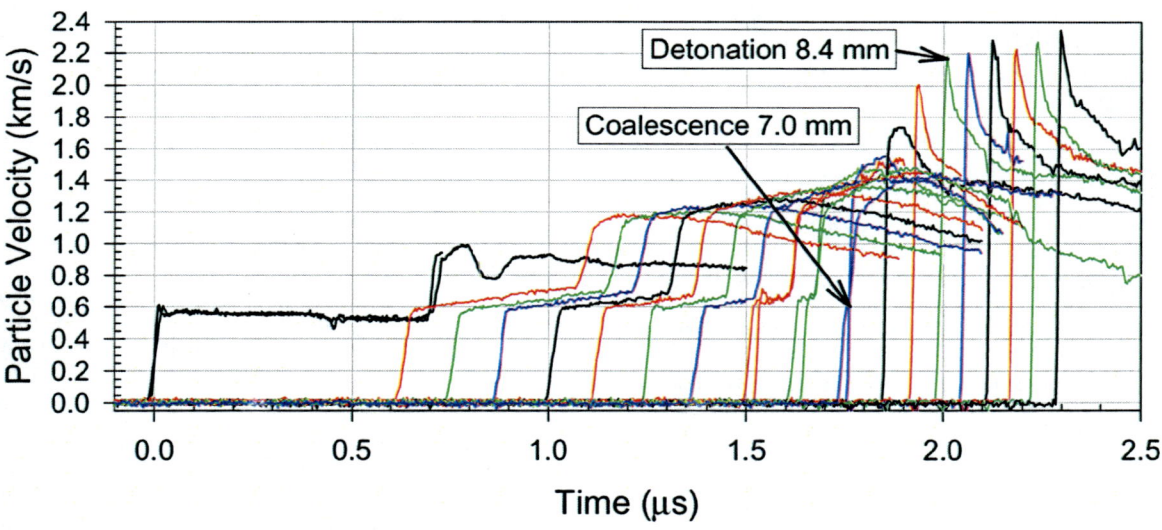

FIGURE 4. Particle velocity wave profiles from Shots 1194 and 1195. The strengths of the two waves are 3.9 and 8.6 GPa. Gauges are spaced at 0.0 and 2.4 through 10.9 mm on ½ mm intervals.

velocity following the initial shock as well as the shock amplitude increasing with depth. Coalescence of the waves occurs at 7.0 mm. From the two experiments we got x-t plots which didn't overlay quite as well as those of Figure 2. We also got slightly different run distances to detonation of 8.4 mm and 9.4 mm.

Figure 5 shows the Pop – plots for single[9] and double shocked EDC-37 (this paper). For the double shock experiments, the pressure of the second or the coalesced wave is shown. The low pressure point is from Shots 1175 and 1176. The run distance *after* coalescence is plotted and it lies on the single shock Pop – plot.[9] This pair of shots follows the rule of thumb for double shock experiments: "Buildup to detonation does not begin until after the two waves coalesce."[5]

The higher pressure experiments, 1194 and 1195, are also shown in Figure 4. Run distances from these experiments lie significantly off the Pop – plot. The run distance after coalescence is significantly shorter than that predicted by the Pop – plot. The total run distance (as shown circled) does not fall on the Pop – plot either. Apparently if there is significant reaction in the first wave, the rule about buildup to detonation starting after wave coalescence no longer holds.

Finally, if we use the actual pressure level of the wave at coalescence, we get about the right distance to detonation. Furthermore, it appears that a more sophisticated reactive burn model than those currently in use will be needed to accurately replicate these experiments.

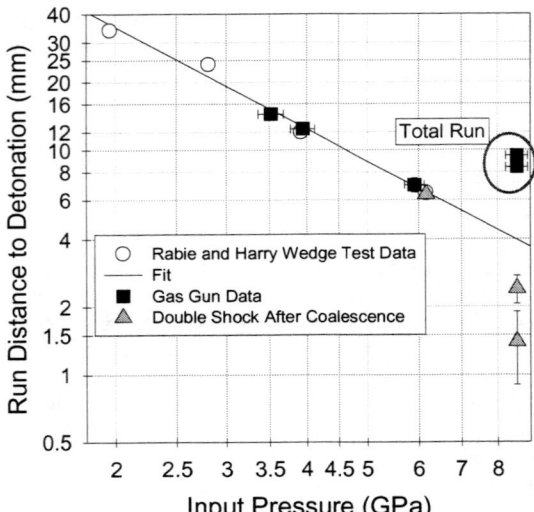

FIGURE 5. Pop – plots for EDC-37 and double shocked EDC-37.

ACKNOWLEDGEMENTS

Robert Medina operated the gas gun for all of the experiments.

REFERENCES

1. E.F. Gittings, "Initiation of a Solid Explosive by a Short Duration Shock", in *Proceedings of the Fourth Symposium (International) on Detonation*, Office of Naval Research Report ACR-126, p. 373, 1965.
2. J. E. Vorthman, and J. Wackerle "Multiple – Wave Effects on Explosive Decomposition Rates" in *Shock Waves in Condensed Matter – 1983*, Elsevier Science Publishers (1984), p. 613.
3. A.W. Campbell, and J.R. Travis, "The Shock Desensitization of PBX 9404 and Composition B-3", in *Proceedings of the Eighth Symposium (International) on Detonation*, Report NSWC-MP- 86-194, p. 1057, 1986.
4. C.L. Mader, *Numerical Modeling of Explosives and Propellants, 2nd Ed.*, CRC Press, 1998, p. 213 – 220.
5. R.N. Mulford, S.A. Sheffield, and R.R. Alcon, "Initiation of Preshocked High Explosives PBX 9404 PBX 9501, and PBX 9502", in *Proceedings of the Tenth Symposium (International) on Detonation*, Office of Naval Research Report ONR 33395-12, p. 459 1993.
6. C.M. Tarver, T.M. Cook, P.A. Urtiew, and W.C. Tao, "Multiple Shock Initiation of LX-17", in *Proceedings of the Tenth Symposium (International) on Detonation*, Office of Naval Research Report ONR 33395-12 p. 696 1993.
7. R. E. Winter, P. Taylor, and D. A. Salisbury, "Reaction of HMX Based Explosives Caused by the Regular Reflection of Shocks" in *Proceedings of the Eleventh Symposium (International) on Detonation*, Office of Naval Research Report ONR 3300-5 p. 649 1998.
8. R. L. Rabie and H. H. Harry, *Characterization of the British Explosives FD16, EDC29, EDC35 and EDC37*, Los Alamos National Laboratory Report # LA-UR-92-1928.
9. R.L. Gustavsen, S.A. Sheffield, R.R. Alcon, L.G. Hill, R.E. Winter, D.A. Salisbury, and P. Taylor, in *Shock Compression of Condensed Matter – 1999*, American Institute of Physics (AIP) Conference Proceedings 505 (1999), p. 879.
10. R. E. Winter, Lee Markland, and S. D. Prior, in *Shock Compression of Condensed Matter – 1999*, American Institute of Physics (AIP) Conference Proceedings 505 (1999), p. 883.
11. S.A. Sheffield, R.L. Gustavsen and R.R. Alcon, in *Shock Compression of Condensed Matter – 1999*, American Institute of Physics (AIP) Conference Proceedings 505 (1999), p. 1043
12. $\rho_0 = 2.14$ g/cm^3, $U_S = 1.99 + 1.76 u_p$ km/s S.A. Sheffield and R.R. Alcon, in *Shock Compression of Condensed Matter – 1991*, S.C. Schmidt, R.D. Dick, J.W. Forbes, D.G. Tasker (editors), Elsevier Science Publishers (1992), p. 909.
13. L. M. Barker and R. E. Hollenbach, "Shock Wave Studies of PMMA, Fused Silica and Sapphire," *J. Appl. Phys.*, **41**, 4208, 1970.

PLASTIC DEFORMATION RATE AND INITIATION OF CRYSTALLINE EXPLOSIVES

J. Namkung[1] and C. S. Coffey[2]

[1]*Naval Air Warfare Center, Pax River, Maryland 20640*
[2]*Indian Head Division, Naval Surface Warfare Center, Indian Head, Maryland 20640-5035*

Abstract. Recent theoretical calculations have demonstrated a relationship between the rate of energy dissipation and the rate of plastic deformation in crystalline solids subjected to plastic flow due to shock or impact. In the case of explosive crystals the energy dissipated locally within the crystals during plastic deformation forms the hot spots from which chemical reaction can be initiated. Prompted by this prediction relating the plastic deformation rate with initiation, a series of experiments were undertaken to measure the plastic deformation rate at the initiation site at the moment of initiation for a number of polycrystalline explosives when subjected to impact or mild shock. The experiment and the results will be reviewed here.

INTRODUCTION

Recent calculations have demonstrated a relationship between the rate of energy dissipation in a deforming crystal and the rate of plastic deformation that the crystal experiences.[1,2,3] The energy dissipation required to raise the temperature of HMX crystals to their initiation temperature during mild impact has been determined. The plastic deformation rate associated with this energy dissipation has also been determined and found to be about 10^4 s^{-1}. Similarly, the plastic deformation rate at the impact initiation threshold of RDX was determined to be about 10^4 s^{-1}. The plastic deformation rates at the impact initiation threshold of TNT and TATB were estimated to be about 2×10^5 s^{-1} and $> 2 \times 10^5$ s^{-1} respectively. The uncertainty in these latter calculations is mainly associated with the value chosen for the shear modulus.[4] At the current time it is not possible to predict the plastic deformation rate at the moment of initiation of polymer and explosive crystal compounds.

The experiment to be discussed here attempts to measure the plastic deformation rate at the initiation site at the moment of initiation. Experimental results will be presented that appear to substantiate the above mentioned predictions for HMX, RDX, TNT and TATB. Results will also be presented for several PBX materials.

THE CONCEPT

It has been observed repeatedly that when explosive or propellant samples were impacted between two hardened steel anvils chemical reactions were always first initiated in the high shear region near the edge of the expanding sample.[5] Here, advantage is taken of these observations to estimate the plastic deformation rate at the moment of initiation. By choosing the sample geometry to be a right circular cylinder it is possible to measure the radial velocity at the edge of the expanding sample disc at the moment of initiation. The moment of initiation is determined by fast photo diodes that detect the first light due to reaction initiation in the sample since reaction always occurs at or near the perimeter of the expanding sample disc. The photo-diodes were positioned to monitor the entire circumference of the expanding sample disc.

Because the sample is radially symmetric it does not matter on which radius initiation takes place since all radii are equivalent. For the mild impacts typical of a drop weight impact machine the loading force levels are low and the assumption can be made that the sample has a constant volume during the impact. This permits a simple relation between the velocity of the drop weight and the radial velocity at the outer edge of the cylindrical sample. Since the sample volume during impact is assumed constant, the time derivative of the volume is zero, $d(\pi r^2 h)/dt = 0$ and $r_0^2 h_0 = r^2 h$ where r_0 and h_0 are the initial radius and initial height of the sample while r and h are the radius and height of the sample at the moment of initiation. Combining the above relations gives following expression for the radial velocity at the perimeter of the sample disc, dr/dt, in terms of the vertical velocity of the impactor, dh/dt,

$$\frac{dr}{dt} = -\frac{r_0}{2h}\sqrt{\frac{h_0}{h}}\frac{dh}{dt} \quad . \quad (1)$$

The negative sign above is canceled by the negative sign associated with the velocity of the impactor which is responsible for decreasing the height of the sample.

Equation 1. permits the evaluation of the radial expansion velocity of the edge of the sample disc at the moment of initiation. To accurately measure the plastic deformation rate requires an accurate description of the radial flow throughout the sample disc. Among other things, this requires specifying the coefficient of friction between the disc and the anvil and striker surfaces which is likely to be an impossible task. Here, the plastic deformation rate will be approximated by assuming that the sample completely adheres to the anvil and striker surfaces and on these surface the radial velocity is zero. The maximum velocity is assumed to occur mid way up the height of the sample at $h/2$, so that the plastic deformation rate can be approximated by

$$\frac{d\gamma}{dt} \approx \frac{r_0}{h^2}\sqrt{\frac{h_0}{h}}\frac{dh}{dt} \quad . \quad (2)$$

THE EXPERIMENT

While there are a number of ways that the radial velocity could be measured among the simplest and quickest to implement is to measure the deceleration of the impactor as it encounters and crushes the sample. The Ballistic Impact Chamber (BIC) Test apparatus was used as the test vehicle.[6] The sample size was typical of that of the BIC Test and consisted of a right circular cylinder 5 mm in diameter and about 2mm high, with a mass of approximately 80 to 100 mg. The walls of the BIC impact chamber were modified to accommodate four photo diodes as shown in Fig.1.

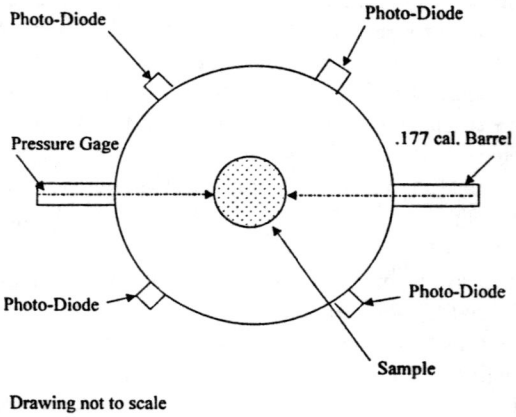

Figure 1. Schematic of modified BIC Test

The moment of initiation was detected by one or several of these photo-diodes whose outputs were summed and recorded on a single channel of a multi-channel digital recorder. An accelerometer mounted on the impactor provided a measure of the deceleration of the impactor as it encountered the sample. This also was recorded on the multi-channel recorder as was the pressure-time record of the reaction gases confined in the BIC Test chamber. The multi-channel recorder provided a common time-base for all of the recorded data so that it was possible to determine the accelerometer data from initial impact to the moment of initiation. Integrating the accelerometer record provided the velocity, v, and the displacement, h, of the impactor necessary to evaluate the plastic deformation rate, Equation (2), at the moment of initiation.

RESULTS

The plastic deformation rates at the moment of initiation for several different materials are listed in the following table.

TABLE. Plastic Deformation Rate at Initiation, s^{-1}

HMX(125 μ)	$.7 \times 10^4$
HMX(5 μ)	$.8 \times 10^4$
HMX(5 μ, calculated)	1×10^4
RDX(calculated)	1×10^4
IH-H7-D	2×10^4
IH-H7-D2	2×10^4
IH-H7-F	7×10^4
Comp B	7×10^4
TNT	$> 2 \times 10^5$
TNT(calculated)	2×10^5
PBXN-109(heated)	1.4×10^5
PBXN-109	1.7×10^5
PBXW-128	2×10^5
TATB(calculated)	$> 2 \times 10^5$
PBX-9502	$> 3 \times 10^5$
Detonation (All Materials, calculated)	a few times 10^6

The compositions IH-H7-D and IH-H7-D2 are mainly composed of ammonium perchlorate and aluminum and IH-H7-F was mainly potassium perchlorate and aluminum.

DISCUSSION

The predicted plastic deformation rates required for initiation of pure crystalline materials and the measured plastic deformation rates for these same materials are in reasonable agreement. This is true generally, but particularly so for the case of hard crystals of sensitive materials. For the softer materials, TNT and TATB, the agreement between prediction and experiment is still good. But to obtain the predicted plastic deformation rates required much smaller sample thickness at initiation and for TATB a higher impact velocity, 20 m/s. The sample thickness needed to achieve a plastic deformation rate of a few times 10^5 s^{-1} required for initiation is less than 100 μ so that the spatial resolution of the twice integrated accelerometer data must approach 10 μ. This represented an instrumentation challenge as does the survival and calibration of the accelerometer at 20 m/s impacts. It is possible to calibrate the accelerometer on every experiment by integrating the acceleration to determine the velocity change at the moment the impactor stops and comparing that velocity with the independently measured velocity of the impactor at the moment of impact.

The measured plastic deformation rates required to initiate the plastic bonded explosives fall in the expected order. To exploit and explore this ordering represents both experimental and fundamental physics challenges. Experimentally, it was very difficult to prepare and measure cylindrical samples of extremely soft materials such as PBXW-128. It maybe that a non-intrusive means can be used to obtain the initial sample thickness. However, to quickly prepare a sample pellet of a soft material like PBXW-128 will be a much more difficult task.

Recall that this experimental effort was prompted by theoretical calculations relating the plastic deformation rate with the energy dissipation rate in shocked or impacted solids and the numerical prediction of the plastic deformation rate of several crystalline explosives at the moment of initiation. The agreement between the predicted and the measured plastic deformation rates for crystalline materials is gratifying. The plastic deformation rate data obtained for the PBX materials is encouraging and suggests an underlying regular behavior. In recent work one of us (CSC) has shown that all crystalline solids and liquids approach a maximum plastic deformation rate of a few times 10^5 s^{-1} to about 5×10^6 s^{-1} for shock wave amplitudes ranging from about 5 GPa to in excess of 200 GPa.[7,8] The viscosity of all liquids and solids are shown to approach a few time 10^4 poise over this shock wave pressure range. These predictions are in good agreement with experiment. Similar calculations are the basis of the final entry in the above Table.[2]

ACKNOWLEDGEMENTS

The authors want to thank Dr. C. W. Anderson and the Office of Naval Research for their encouragement and support. They also want to thank their colleagues both for their insights and for supplying many of the materials used in the test series. In particular they want to thank P. A. Thomas, F. J. Zerilli, R. H. Guirguis, N. Jones and J. M. Kelley.

REFERENCES

1. Coffey, C. S., Phys. Rev. B **24**, 6984 (1981).
2. Coffey, C. S. and Sharma, J., Phys. Rev. B **60**, 9365 (1999).
3. Coffey, C. S. and Sharma, J., J. Appl. Phys. **89**, 4794 (2001).
4. The following values were used for the shear modulus, G_{HMX} = 4.3 GPa., G_{RDX} = 4.0 GPa., $G_{TATB} \approx G_{TNT} \approx 1$ GPa. The shear modulus for TATB was suggested by H. Cady, private communication.
5. Coffey, C. S., Frankel, M. J., Liddiard, T. P., and Jacobs, S, J, in Seventh Detonation Symposium, p. 970, (1981).
6. Coffey, C. S., DeVost, V. F. and Woody, D. L., in Ninth Detonation Symposium, p. 1234, (1989).
7. Coffey, C. S., Phys. Rev. B **49**, 208 (1994).
8. Coffey, C. S., Submitted to Phys. Rev. B June 2001

FACTORS AFFECTING SHOCK SENSITIVITY OF ENERGETIC MATERIALS

A. Chakravarty, M.J. Gifford, M.W. Greenaway, W.G. Proud, J.E. Field

PCS, Cavendish Laboratory, Madingley Road, Cambridge, CB3 0HE. UK.

Abstract. An extensive study has been carried out into the relationships between the particle size of a charge, the density to which it is packed, the presence of inert additives and the sensitivity of the charge to different initiating shocks. The critical parameters for two different shock regimes have been found. The long duration shocks are provided by a commercial detonator and the short duration shocks are imparted using laser-driven flyer plates. It has been shown that the order of sensitivity of charges to different shock regimes varies. In particular, ultrafine materials have been shown to be relatively insensitive to long duration low pressure shocks and sensitive to short duration high pressure shocks. The materials that have been studied include HNS, RDX and PETN.

INTRODUCTION

When a shock-wave is incident on an energetic charge, a number of parameters must be considered when determining whether detonation is likely to result. The nature of both the charge and the shock-wave are important.

In a very simplistic way a shock can be described by its pressure and duration (ignoring shock profile at this stage). For a shock to cause initiation it must be capable of creating sufficient chemical reaction to sustain it. Acting against this chemical reaction, to weaken the shock, are rarefactions due to the expansion of the material which, due to the subsonic flow of the material following the shock, will eventually reach the front. The relationship between the required pressure and duration for initiation is such that the shock level must be high enough to cause sufficient reaction to sustain the shock before the initial shock decays. If this criterion is met then a detonation will propagate in the charge.

The magnitude and duration of a shock required for a particular charge to be initiated are dependent on the microstructure and chemistry of the charge. The microstructure is crucial in determining the nature of hot-spots that are created in the charge and the chemistry is important in determining the response of the material to the presence of the hot-spots.

A large number of researchers have attempted to elucidate the role of hot-spots in the shock initiation of detonation. The reviews of the field given by Khasainov et al.[1] and Dremin[2] give a very complete account of the state of the literature on this subject.

The present study has focussed on varying the density and grain size of the charges and the nature of the imparted shock in an attempt to alter the hot-spot parameters and so determine the critical factors associated with them.

MATERIALS USED

Both the pentaerythritol tetranitrate (PETN) and cyclotrimethylene trinitramine (RDX) were supplied in ultrafine and conventional forms by ICI Nobel Enterprises, Ardeer, U.K. The ultrafine powders have a primary particle size of ~1 µm and are produced by a proprietary process. The conventional grain material has a particle size of about 180 µm.

The hexanitrostilbene (HNS) used in these studies was supplied by DERA, Fort Halstead and came originally from Bofors AB in Sweden. The ultrafine form is known as HNS IV and has a grain size of less than a micron. The HNS IV was supplied both in a pure form and with pressing additives. In the case where zinc stearate and graphite were added to act as pressing agents, the additives contributed approximately 1% to the total mass of the material. The coarse grain HNS (known as HNS II) had a grain size that was typically of the order of 25 µm.

EXPERIMENTAL METHOD

Two principal experimental methods were used during the course of the research described here. For the imparting of relatively long duration shocks, a gap testing geometry was used. When short high pressure shocks were required a system for generating laser-driven flyer plates was used.

Long Duration Shocks

The charges used in these experiments were incrementally pressed columns of either RDX or PETN. The confinements used were 25 mm long 25 mm diameter PMMA cylinders. The explosive columns were 5 mm in diameter.

The donor charge used during the experiments was a PETN boosted C8 detonator which was found to have a reliable output in terms of the shock pressure produced.

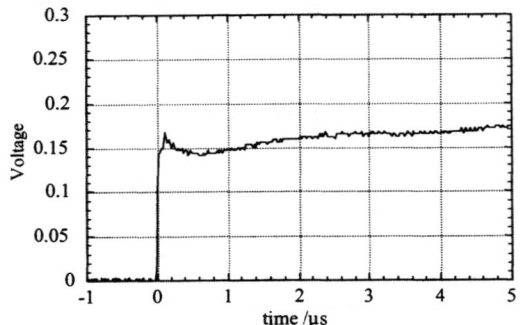

FIGURE 1. Typical trace from a PVDF gauge.

The gaps that were used to mediate the shock pressure were discs of PMMA placed between the detonator and the surface of the column. A thin layer of silicone grease was used between all three components of the test in order to aid the reproducibility of the testing. PVDF gauges placed between the PMMA gap and another piece of PMMA in the charge position were used to obtain an indication of the shock pressure during a test. A typical trace from a PVDF gauge is shown in figure 1.

Both photographic streak recording and brass witness plates were used to determine whether a detonation event had occurred during a test.

Short Duration Shocks

The HNS charges used for the short-duration shocks were 5 mm long, 5mm diameter cylinders contained within 25 mm diameter PMMA confinements. The charges were incrementally pressed into the confinements. The surface of the charges was polished with 2500 grade SiC paper to provide a consistent surface finish. The quality of the surface finish was checked using a Sloan DekTak II surface profilometer.

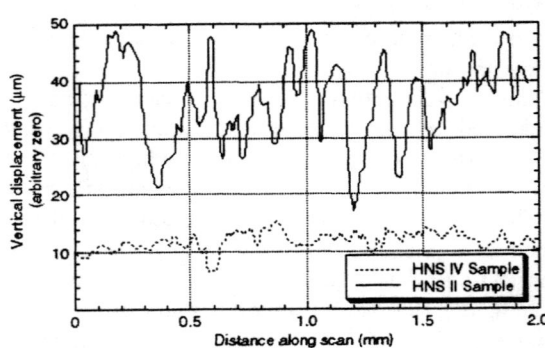

FIGURE 2. Profilometer traces from the Sloan DekTak II.

The laser-driven flyer launching system is described fully in previous publications from this laboratory[3-6] and details can also be found in the paper by Greenaway et al. in these proceedings.

The system uses a Nd:YAG laser to accelerate flyers 1 mm in diameter and 5 µm thick to velocities up to 8 mm µs^{-1}. On impact these flyers provide intense shocks lasting approximately 1 ns. The energy of the pulse imparted to the flyer is controlled in order to determine the velocity of the flyer. Energies between 50 and 400 mJ were accessed during this study.

A Hadland Imacon 790 high speed image converter camera was used to provide streak photographs of the initiation events. The camera was triggered from the signal that fired the laser with a suitable delay added. These photos allowed calculation of the position of the initiation event within the column.

RESULTS

Long duration shocks

Figure 3 shows the results of the experiments which used long duration shocks in a gap test geometry. These experiments were carried out on PETN and RDX in both ultrafine and conventional grain sizes. As can be seen the density was also varied in the RDX study in order to determine the effect that increased porosity has on the sensitivity of the charges. Although there is some overlap in the go/no go gaps for some of the densities, in general the experimental reproducibility was extremely high.

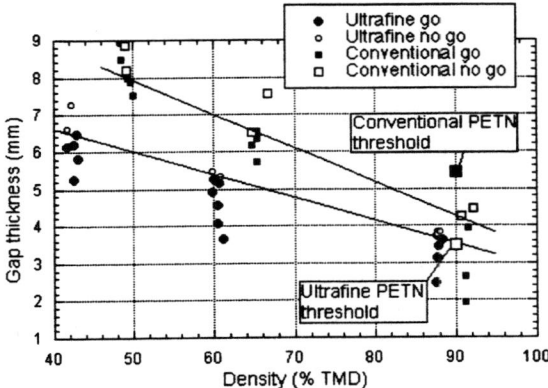

FIGURE 3. Results of gap testing on RDX. Thresholds for PETN are also indicated.

The ultrafine PETN at a density of 90% TMD had a critical gap of 3.68 ± 0.01 mm compared with a gap of 5.57 ± 0.02 mm for the conventional grain size material. These were shown to correspond to shock pressures of approximately 4.1 and 2.1 GPa as measured using the PVDF gauges described previously.

The gap required to prevent initiation of the RDX charges increased significantly in both the ultrafine and conventional materials as the porosity increased, but the ultrafine material was consistently less sensitive to this form of initiation.

Short Duration Shocks

The findings of this study into initiation by short duration shocks have been explained in some detail in the paper by Greenaway et al. within these proceedings. The results of this study involving laser-driven flyer plates are that HNS II could not be initiated with very short duration shocks at the energies available in that system, but that the HNS IV could be readily initiated with a go/no go threshold of about 250 mJ of laser pulse energy.

The presence of zinc stearate and graphite as additives in some of the HNS IV acted to increase the flyer energy required for initiation of the charges to approximately 350 mJ.

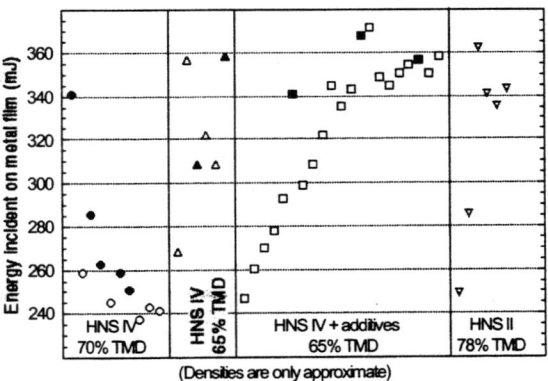

FIGURE 4. Results of the laser-driven flyer tests. Filled objects denote a "go" result.

The results of this study indicate a strong correlation between pressing density and sensitivity. Unfortunately due to the nature of the pressing technique employed and the powder, it was difficult to accurately reproduce a given density of charge. Within the limits of the study, it can be said that the charges pressed to a density of 65% TMD appear to be less sensitive than those pressed to 70% TMD.

Without performing a larger study, however, it is impossible to say what the exact nature of the dependence on density of the sensitivity is for this form of initiation.

DISCUSSION

This study together with previous studies carried out within this group has shown that simple orderings of materials by sensitivity cannot be done. It is not even possible to do this for sensitivity to initiation by shock as has been demonstrated here.

The results of this study have shown that for a given situation, the sensitivity of the material is dependent on the chemistry, the grain size, the density of the charge and the nature of the shock itself. The way in which all of these variables determine the likely response of a charge to an insult can be linked to their effect on the distribution, nature and form of the hot-spots that are caused by the shock.

It has been shown[7-9] that the effect of increasing the shock pressure is to change the relative importance of the jetting and the gas compression in the process of pore collapse. As small pores are more effective for the rapid formation of hot-spots by jetting and large pores are more effective in the case of gas compression it seems that it may be the pore size rather than the grain size that is critical. In the case of the high pressure short duration shocks imparted by laser-driven flyers, the incident shock is not sufficiently large for it to cover an entire pore in the coarse material and so the releases will act to hinder collapse. These short shocks are, however, of sufficiently high pressure for jetting to be significant and this may well be the dominant mechanism in the ultrafine charges where the pores are extremely small.

With the longer duration, lower pressure detonator-supplied shocks, the shock is sufficiently large to encompass whole gas spaces in both the ultrafine and the conventional powders. Due to the reduced pressure, jetting is a less important mechanism for hot-spot production than gas compression and as a result the larger pores that are found in the conventional charges are more conducive to the creation of hot-spots capable of causing reaction.

The effect of density on the sensitivity of the charges is caused by the change in the relative density of hot-spot nucleation sites compared to the density of material available for reaction. It appears from the results of the gap testing of the RDX that it is the number of available sites for hot-spots that determines the sensitivity (at least down to the density of 40% TMD that was used in this study).

It is not so clear from the laser-driven flyer study that the same is true in this regime. The importance of good coupling between the energetic material and the hot-spot is more pronounced due to the short duration of the shock, so this may account for what appears to be a higher sensitivity of the more densely packed charge. Further research would have to be carried out with more emphasis on density in order to determine the optimum density for charge sensitivity in this shock regime.

ACKNOWLEDGEMENTS

The authors would like to acknowledge ICI Nobel Enterprises (Ardeer), U.K. and DERA, Fort Halstead for their support of this research. Dr. M. Cook of DERA is particularly thanked for useful comments that he has made.

REFERENCES

[1] B. A. Khasainov, A. V. Attetkov, and A. A. Borisov, *Chem. Phys. Rep.*, **15**, 987-1062 (1996).

[2] A. N. Dremin, *Toward Detonation Theory* Springer-Verlag, Berlin, 1999.

[3] S. Watson, PhD Thesis, University of Cambridge, 1998.

[4] S. Watson and J. E. Field, *J. Phys. D: Appl. Phys.* **33**, 170-174 (2000).

[5] S. Watson, M. J. Gifford, and J. E. Field, *J. Appl. Phys.* **88**, 65-69 (2000).

[6] S. Watson and J. E. Field, *J. Appl. Phys.*, **88**, 3859 (2000).

[7] J. P. Dear, J. E. Field, and A. J. Walton, *Nature* **332**, 505-508 (1988).

[8] N. K. Bourne and J. E. Field, *Proc. R. Soc. Lond. A,* **435**, 423-435 (1991).

[9] J. E. Field, *Accounts Chem. Res.*, **25**, 489-496 (1992).

THE BURNING RATE OF ALUMINIUM PARTICLES IN CYLINDER TESTS.

David J. Evans[1], Alec M. Milne[1] and Ian Softley[2]

[1]*Fluid Gravity Engineering Ltd, 83 Market Street, St. Andrews, Fife, KY16 9NX, U.K.*
[2]*Defence Evaluation Research Agency, Porton Down, Salisbury, Wiltshire, SP4 0JQ, U.K.*

Abstract. Aluminium is a common fuel component in propellants and explosives. There is a wealth of literature on Aluminium combustion in gases at relatively low pressure but limited data on combustion at high pressure (as in explosive detonation products). In this work we have carried out and analysed cylinder tests with Aluminium loaded explosives with a view to assessing the applicability of low pressure burning rates in this regime. The analysis makes use of detailed numerical two phase flow modelling and a range of experiments used to validate other relevant aspects of the physics, such as drag laws. We conclude that the burning rate is significantly faster than that implied by extrapolating laws applicable at lower pressures.

INTRODUCTION

Aluminium particles burning in a low pressure environment, e.g. air, typically exhibit a dependence of the burn time on particle size of the form [1, 2, 3],

$$t_b = \alpha d_0^2 \qquad (1)$$

where t_b is the burn time, d_0 is the particle diameter and α is a constant. Typically the value of α is $4 \times 10^6 sm^{-2}$. In the course of studying propellants and explosives it is of interest to study the burning of such particles in a high pressure environment such as occurs detonation. There is limited data available on this subject so a coupled modelling (FGE) and experimental (DERA) programme was instigated. We have also made use of cylinder test data published by Baudin et al. [4].

The experimental data for this work was provided from Standard Cylinder Tests. The experimental arrangement for our cylinder tests is shown in Fig. 1. The cylinder is 300mm long with inner diameter 25mm and outer diameter 30mm. The Debrix pellet is 25mm long and is centrally initiated by an RP80 detonator. The mixtures considered here are Nitro-Methane combined with Aluminium particles at varied loading densities and particle sizes. The standard diagnostic is to measure the radius of the cylinder at 200mm from the point of initiation.

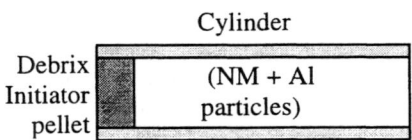

FIGURE 1. Standard Cylinder Test

These experiments have been modelled using a two-dimensional, axi-symmetric hydrocode previously used in two-phase modelling of Aluminium Detonation [5]. The mixture is modelled using two-phase flow where the NM uses a "programmed burn" approach: the burn front moves at a constant, pre-determined, speed releasing energy as it moves through the explosive. In order to infer the effects of particle burning rate From cylinder tests it is important to ensure first that the model accurately represents cylinder tests with no particles and also with inert particles.

MODEL VERIFICATION

First, consider cylinder tests with no particles present. Data from Baudin and Softley are available to compare with and Fig. 2 shows the results obtained. The plots show the cylinder radius at 200mm from the initiator. The programmed burn model for the NM uses parameters obtained from the CHEETAH code [6]. The model fits the Baudin data better in the early stages but later the match is better for the terminal speed of the Softley data.

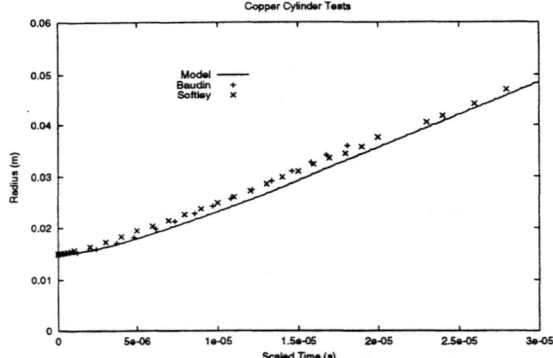

FIGURE 2. Comparison of programmed burn model with experimental data

To assess the dynamic effects of the particles an experiment was carried out using 1mm diameter steel particles, which do not burn, in the Nitro-Methane at a volume fraction of 62%. It can be seen that the agreement is good though there is some discrepancy in the early stages when the radius variation is small. The photographic measurement technique employed in these experiments is probably less reliable than Baudin's VISAR data in the early stages

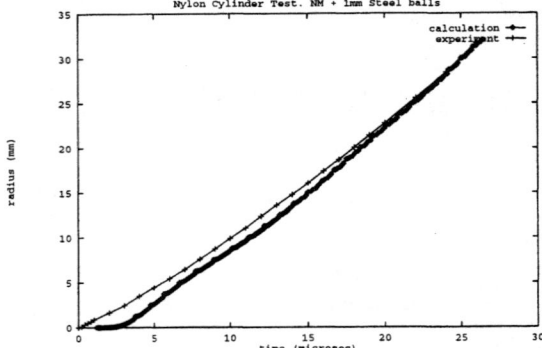

FIGURE 3. 1mm steel particles (not burning) in Nitro-Methane: comparison of experimental and numerical data

BURNING PARTICLES

Baudin's paper presents results using 5μm and 100nm diameter particles at 20% and 40% loading densities by weight. The experimental data shown for the 100nm AL at 40% loading indicates radial motion much slower than the 5μm case. This is difficult to understand in that all other work has indicated that the smaller particles should burn more quickly and yet this result moves more slowly than even the simulation of inert particles. Comparison with the 20% experimental data shows that this is a unique result and so calls into question the validity of this particular experiment and for t this reason this case is neglected here.

In order to assess the particle burning it is necessary for the the numerical model to incorporate a burn law. Usually the EDEN code uses a constant radial burn velocity law. Assuming the particles are spherical and burning occurs all over the surface simultaneously, the radius of the particle reduces at a constant rate. The rate of mass burnt per second is then proportional to the rate of change of the particle volume:

$$\frac{dV}{dt} = 4\pi r^2 \frac{dr}{dt} \qquad (2)$$

An alternative law often used in these models is to assume that the rate of change of the surface area of the particle is a constant. In this case the rate of change of volume is:

$$\frac{dV}{dt} = 2\pi r \frac{dr^2}{dt} \quad (3)$$

By varying the particle burn time until a best fit is obtained with the 40%, 5μm Al data it is possible to get an estimate of the burn time. For this case a burn time of 30μs was found. The fit is good but there is a discernible difference in the second derivative with the simulation producing more motion at late time. One possible solution to this is to reduce the burn energy of the particles. This is reasonable since the given energy release is only valid if the Aluminium is completely burnt. It seems plausible that this will not be the case since an oxide layer will be present initially and complete combustion requires carefully controlled amounts of the fuel and oxidant. Therefore a second fit was attempted with the particle burn energy reduced from 1.33×10^7J/kg to 1.0×10^7J/kg. The resulting fit is shown in Fig. 4 and was achieved using a burn time of 24μs. This is a better fit to the data than the full burn model.

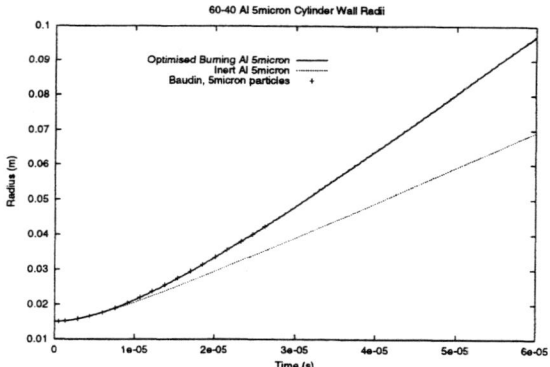

FIGURE 4. Best fit of 5μm particle burn simulation to Baudin data

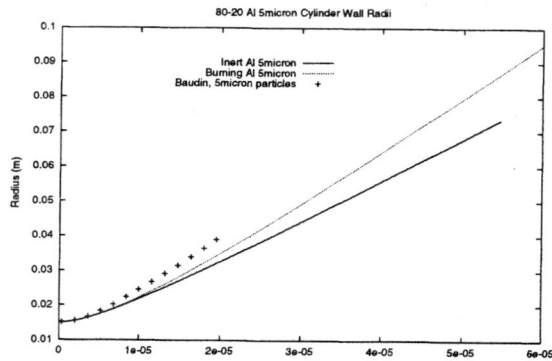

FIGURE 5. Best fit of 5μm particle burn simulation with reduced burn energy

The 20%, 5μm particle case was simulated using the same burn energy and time parameters derived for the 40% Al runs, as would be expected if the d-squared law held. The resulting simulation shows that in the early time the cylinder wall motion is slower than with no particles for the first few μs, disagreeing with the experimental data as can be seen in Fig. 5. Even with the burn energy returned to the full value of 1.33×10^7J/kg the radial motion is still too slow as can be seen in Fig. 6. In this case it seems that a more complex relationship between burn time and particle size than the simple square law discussed earlier may apply.

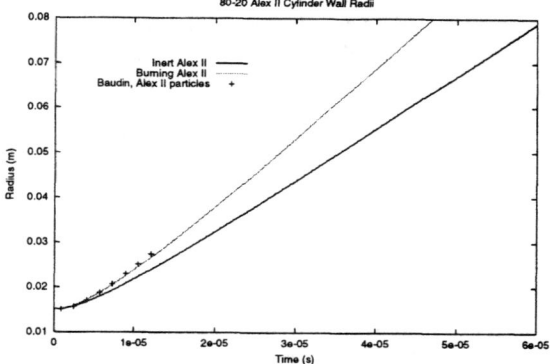

FIGURE 6. Summary of results for 20% by weight of 5μm Al particles

Using the *d*-squared law and extrapolating from the optimised case above (40%, 5mm) gives a burn time of 9.6ns for 100nm particles. Using this in the simulations gives a good fit to the experimental data as can be seen in Fig 6.

These models used the constant radial burn velocity law. To consider whether there is any dependence on the form of burn law the first run was repeated with the constant rate of area change law. The optimisation procedure for the 40%, 5μm case was repeated using such a burn law and this yielded a burn time of 25μs, very close to the previous value. There appears to be little dependence on the nature of the law, only the speed. If the burn times were much longer this may not remain true.

CONCLUSIONS

With a limited set of experimental data available it has been possible to begin testing the applicability of relationships between particle burn time and particle diameter in high pressure environments. In all cases it has been found that the burn time is significantly shorter than would be predicted by the d-squared law with the a value appropriate for combustion in low pressure environments. The burn times are reduced by a factor of 4 in general. However not all results are consistent with this reduction: one result in particular showed an even shorter burn time. It seems likely that the d-squared law is still applicable in high pressure environments but with different coefficients and possibly dependence on other factors such as loading density.

Improved modelling capabilities for the Nitro-Methane and the particle elements have been developed and these will be applied to this problem in the near future so that a better understanding of the burn laws may be developed.

REFERENCES

1 Khasainov, B. A. and Veyssiere, B., *Archivium Combustionis*, **7**, 333-352, 1987
2 Bouriannes, R., Ph.D. Thesis, Univ. of Poitiers, 1971
3 Marion, M., Chauveau, C. and Gokalp, L., "Studies on the Ignition and Burning of Levitated Aluminium Particles", *Combust. Sci. and Tech.*, 1996, **115**, 369-390
4 Baudin, G. et al., "Combustion of Nanophase Aluminium in the Detonation Products of Nitromethane, 11th Detonation Symposium, Snowmass, 1998
5 Milne, A. M. and Evans, D. J., "Numerical Modelling of Two-Phase Reactive Flow and Aluminised Explosives", in *Shock Wave Processes in Condensed Media*, I. G. Cameron (Ed.), Hunting Brae, U.K., 1997
6 Fried, L. E., Cheetah 1.39 User Manual, 1996

FIRST RESULTS OF REACTION PROPAGATION RATES IN HMX AT HIGH PRESSURE

Daniel L. Farber, Anthony P. Esposito, Joseph M. Zaug,
John E. Reaugh, and Chantel M. Aracne

*University of California, Lawrence Livermore National Laboratory,
7000 East Ave., Livermore, CA 94550*

Abstract. We have measured the reaction propagation rate (RPR) in octahydro-1,3,5,7-tetranitro-1,3,5,7-tetrazocine (HMX) powder in a diamond anvil cell over the pressure range 0.7-35 GPa. In order to have a cross-comparison of our experiments, we conducted RPR experiments on nitromethane (NM) up to 15 GPa. Our results on NM are indistinguishable from previous measurements of Rice and Foltz. In comparison to high-pressure NM, the burn rates in solid HMX are 5-10 times faster at pressures above 10 GPa. Numerical simulations of the burn rate of pressurized HMX were also performed for comparison to the results obtained. The simulated burn rates closely approximate the observed rates at pressures up to 3 GPa. However, further refinement to the computational model is required for the calculated burn rates to approach those observed at higher pressures.

INTRODUCTION

Presently there is a strong interest in first-principles modeling of chemical reactions in high explosive (HE) materials. However, the validation of these models requires experimental data at the appropriate pressure and temperature conditions of the reactions of interest. Because these reactions occur over time scales of microseconds, they have resisted experimental characterization of the fundamental processes governing combustion and detonation. The diamond anvil cell (DAC) is well suited for studying these reactions because it provides a high-pressure, variable-temperature sample environment, as well as a window for spectroscopic study of reactions within the DAC. The reaction propagation rate (RPR) of an HE material can be studied directly by confining the material within the DAC and initiating combustion with a focused laser pulse. Our experimental approach is a modification of the earlier work of Rice and Foltz (1,2). Here we report the first results of the RPR measurements on octahydro-1,3,5,7-tetranitro-1,3,5,7-tetrazocine (HMX) over the pressure range 0.7-35 GPa. In addition, we report calculated RPR values for HMX from 0.1 to 30 GPa.

EXPERIMENTAL

The experimental setup is presented in Figure 1. The apparatus and procedure employed were similar to those used by Rice and Foltz (1,2). Samples were contained in a DAC consisting of two opposed 0.25 carat diamonds with culet diameters of 0.5-1.0 mm. Lateral confinement of the ~50 μm thick samples was achieved using Inconel 718 or rhenium gaskets with 150-400 μm hole diameters. Ruby powder was deposited onto the surface of the back diamond for determination of the initial pressure using the ruby fluorescence pressure scale.

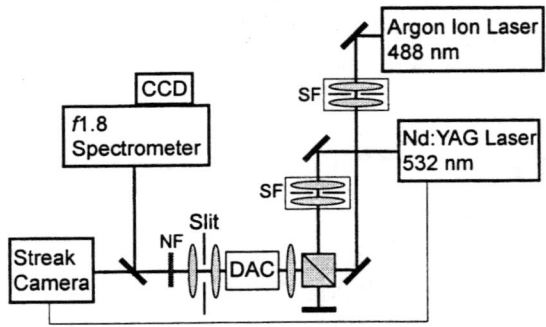

FIGURE 1. Schematic layout of RPR experimental apparatus as described in the text. The following abbreviations have been used: DAC: diamond anvil cell; NF: holographic notch filter; SF: spatial filter.

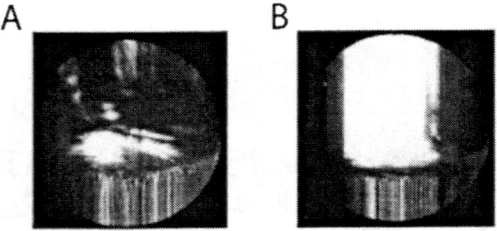

FIGURE 2. Streak camera records of reacting HMX. The vertical dimension is time, where the edge-to-edge length is 3.2 μs, and the horizontal dimension is distance, where the edge-to-edge length is 353 μm. The parallel vertical lines are due to the undisturbed laser speckle pattern being streaked in time. Deflagration within the sample disturbs the speckle pattern from the point where the ignition pulse strikes the sample. The disturbance moves outward from this point, resulting in the pattern shown here. (A) HMX sample at 21 GPa. The corresponding RPR is 223 m/s. (B) HMX sample at 35 GPa recorded with the same streak duration as in Fig. 3A. The corresponding RPR is 641 m/s.

Both sample illumination and excitation for ruby fluorescence were provided by an argon ion laser operating at 488 nm and ~3 W (Lexel model 95). The cw 488 nm beam was passed through a spatial filter and defocused into the DAC in order to fully illuminate the sample. Sample ignition was provided by a Q-switched Nd:YAG laser (New Wave MiniLase II-20), frequency-doubled to 532 nm, with pulses of ~9 ns duration. During optical alignment, the 532 nm pulse energies were kept below 0.1 μJ to prevent accidental ignition. The 532 nm beam was made collinear with the 488 nm beam using a holographic bandpass filter, and was focused to a ~5 μm spot size in the center of the sample region illuminated by the 488 nm beam. Transmitted light from the sample was magnified (~10x) and focused onto a 10-50 μm-wide slit, and was then magnified (9.5x) after the slit. For ruby fluorescence measurements, the emission was focused onto the entrance slit of a f1.8 spectrograph (Kaiser Optics), and detected by a liquid-nitrogen cooled CCD camera (Princeton Instruments).

After pressure measurements, the laser speckle pattern from the DAC, due to illumination by the 488 nm beam, was directed to an EG&G L-CA-20 electronic streak camera (Polaroid film type 57, 3000 speed) operating at streak durations between 1.8 and 10 μs. Ignition pulse energies were determined by a Molectron EPM 2000 energy meter, and were in the range of 1-10 μJ. A holographic notch filter placed before the streak camera slit was used to attenuate the 532 nm light to prevent over-exposure of the streak image. Typical streak images are shown in Figure 2.

Samples consisted of ultrafine HMX containing less than 0.7% RDX, with a uniform grain size of ~3 μm. In addition, RPR studies of nitromethane (NM) were conducted for comparison with the previous data of Rice and Foltz (1). Figure 3 presents a comparison of NM burn rates obtained by Rice and Foltz with those obtained in our laboratory. Our results are essentially indistinguishable from those previously obtained.

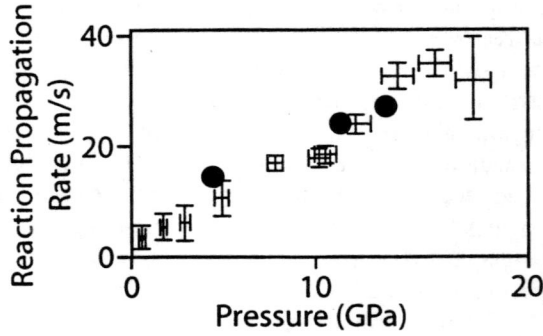

FIGURE 3. Pressure dependence of the RPR for NM. The data obtained in our laboratory are shown in large circles over the RPR data obtained by Rice and Foltz (1).

RESULTS AND DISCUSSION

Table 1 presents the experimental RPR values obtained in the present study on HMX in the pressure range 0.7-35 GPa, and those calculated for pressures between 0.1 and 30 GPa (see below). The data presented in Table 1 have been plotted in Figure 4; also depicted in Fig. 4 is a best-fit curve to strand burner data on LX-04 (85 wt% HMX, 15 wt% Viton-A) obtained by Maienschein et al. (3). There is good agreement between the experimental and calculated values for pressures up to 3 GPa; however, at higher pressures the data sets diverge.

Direct numerical simulations of a propagating planar flame front were made using ALE-3D, an arbitrary Lagrange-Eulerian computer simulation program under development at this laboratory. In these simulations, the program was exercised as though it were a one-dimensional Lagrange program with plane symmetry. The simulations include heat transfer by conduction, and a simplified global reaction scheme fitted to one-dimensional time-to-explosion experiments (4). The first of the three reactions is endothermic, the second moderately exothermic, and the third

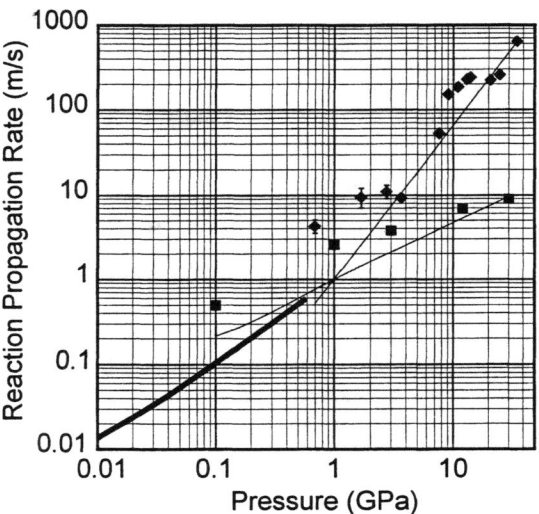

FIGURE 4. Pressure dependence of the reaction propagation rate for HMX and LX-04. Values obtained in this study on pure HMX are indicated by the points (diamonds: experimental; squares: calculated), while the thick curve is the best fit to strand burner data on LX-04 from Maienschein et al. (3). The thin curve fits to the data are present to guide the eye, and are described by the following equations: fit to experiment: RPR = (pressure)^1.81; to calculation: RPR = (pressure)^0.67.

exothermic. Four equations of state are required: the solid unreacted material; the solid endothermic product; moderately exothermic gas, described as a relatively high molecular weight gas; and the final products of HMX decomposition. The latter two gas equations of state were described by interpolating tables constructed using CHEQ, a thermochemical equilibrium computer program (5).

The numerical simulation is ignited by raising the temperature of one face of the one-dimensional slab to the approximate flame temperature. That face is also maintained as a constant pressure boundary, and the opposite face has no normal displacement. After an initial transient the flame propagates as a steady, constant velocity process through the initially cold but pressurized HMX reactant. This steady velocity is then recorded. Repeating the calculations with different initial pressures provides the simulation results of flame speed as a function of pressure. The spatial resolution (zone-size) needed to resolve the flame

TABLE 1. Experimental and Calculated Reaction Propagation Rates for HMX.

Pressure (GPa)	Rate (m/s) Experimental	Calculated
0.1		0.5
0.7	4.3 ± 0.8	
1.0		2.6
1.7	9.5 ± 2.5	
2.8	11.0 ± 2.0	
3.0		3.8
3.7	9.3 ± 0.9	
7.7	53.0 ± 5.3	
9.1	152 ± 5.5	
11	186 ± 19	
12		7.0
13	228 ± 18	
14	242 ± 23	
21	222 ± 23	
25	257 ± 26	
30		9.0
35	641 ± 70	

front depends on pressure and flame speed. Higher flame speed requires finer mesh resolution to capture the spatial gradient. Lower pressure requires finer mesh resolution to capture the spatial gradient in the product gas. For these Lagrange simulations, the gas products, and so the finite-difference zones, expand much more at low pressure than at high pressure. In our simulations the zone-size was typically a few nanometers, and the flame thickness a few hundred nanometers.

The Arrhenius chemical reaction rates used in these simulations are temperature-dependent, but not pressure-dependent. The observed pressure dependence in the simulations comes from the separation of hot gas products from the cold surface that is large for the low-density, low-pressure products, and is small for the high-density, high-pressure products. A second-order effect is the change of thermal conductivity with density.

Although there are many parameters used to describe the mechanical, chemical, and thermal properties of the four species used in these simulations, none are specifically fit to or determined by flame propagation. They are obtained independently. The factor of two or three difference between theory and experiment at low pressure is undoubtedly due to deficiencies in the thermal and chemical properties used, which were taken from near-atmospheric pressure experiments. Nevertheless, we consider the agreement with experiment to be surprisingly good. The substantial deviation from the measured flame propagation speed and our simulations at high pressure is apparently due to the inadequacies of our simplified and global chemical reactions that were determined by ODTX experiments at roughly 0.1 GPa. We anticipate further research in this area.

The significance of the experimental results is that they may constrain the physics and chemistry behind our canonical model of initiation and growth of reaction in explosives from hot spots. We feel that time-resolved temperature, and, to a lesser degree, pressure measurements under the RPR experimental conditions are required to confidently guide future computational developments. In all such models, it is postulated that the hot spots, once formed, link up by the mechanism of laminar flame spreading. Knowing the ratio of the flame spread velocity at high pressure, on the order of 35 to 50 GPa, to the detonation velocity is the necessary coupling between the average separation of hot spots and the thickness of the reaction zone of a quasi-planar detonation front.

ACKNOWLEDGEMENTS

*This work was performed under the auspices of the U.S. Department of Energy by the Lawrence Livermore National Laboratory under contract number W-7405-Eng-48. We thank C. M. Tarver for his support of this work.

REFERENCES

1. Rice, S.F., and Foltz, M. F., *Combustion and Flame* **87**, 109-122 (1991).
2. Foltz, M. F., *Propellants, Explosives, Pyrotechnics* **18**, 210-216 (1993).
3. Maienschein, J. L., and Chandler, J. B., "Burn Rates of Pristine and Degraded Explosives at Elevated Temperatures and Pressures," Eleventh International Detonation Symposium, Snowmass, CO, 1998.
4. McGuire, R. R., and Tarver, C. M., "Chemical Decomposition Models for the Thermal Explosion of Confined HMX, TATB, RDX, and TNT Explosives," Seventh Symposium (International) on Detonation, Annapolis, MD, 1981.
5. Ree, F. H., *J. Chem. Phys.* **81**, 1251-1263 (1984).

EMBEDDED ELECTROMAGNETIC GAUGE MEASUREMENTS AND MODELING OF SHOCK INITIATION IN THE TATB BASED EXPLOSIVES LX-17 AND PBX 9502[†]

R. L. Gustavsen, S. A. Sheffield, R. R. Alcon[1] J.W. Forbes, C.M. Tarver, and F. Garcia[2]

[1]*Los Alamos National Laboratory, Los Alamos, NM 87545*
[2]*Lawrence Livermore National Laboratory, Livermore, CA 94550*

We have completed a series of shock initiation experiments on PBX 9502 (95 weight % dry aminated TATB explosive, 5 weight % Kel-F 800 binder) and LX-17 (92.% wet aminated TATB, 7.5 % Kel-F 800). These experiments were performed on the gas/gas two stage gun at Los Alamos. Samples were prepared with ten or eleven embedded electromagnetic particle velocity gauges to measure the evolution of the wave leading up to a detonation. Additionally, one to three shock tracker gauges were used to track the position of the shock front with time and determine the point where detonation was achieved. Wave profiles indicate little delay between formation of hot-spots in the shock front and release of hot-spot energy. In other words, a great deal of the buildup occurs in the shock front, rather than behind it. Run distances and times to detonation as a function of initial pressure are consistent with published data. The Ignition and Growth model with published parameters for LX-17 replicate the data very well.

INTRODUCTION

The TATB based explosives LX-17 and PBX 9502 are among the most thoroughly studied high explosives in current use. References 1-6 provide examples of work on these explosives, which we will refer to throughout this paper. The reasons for interest in these explosives are their insensitivity to initiation by shock and heat, their high power, and their manufacturability. By manufacturability, we mean that uniform density pressings can be made and from these, parts are easily and safely machined

One of the finest methods for measuring the shock initiation of detonation in explosives is the embedded multiple electromagnetic particle velocity gauge method developed and used at Los Alamos.[7] Until recently, this method could not be applied to these insensitive TATB based explosives. The reason for this is that very high shock pressures are needed to initiate these explosives. These pressures can be achieved with ~ 2 km/s projectile velocities achievable with powder guns and high impedance metallic impactors such as copper or stainless steel. However, electromagnetic particle velocity gauge measurements cannot be made using metallic (conducting) impactors or projectiles because of the eddy currents produced as the metal moves through the magnetic field.

About 5 years ago, we started doing shots on the Los Alamos 50 mm bore gas driven two-stage gun. This gun is capable of the 2 – 3 km/s projectile velocities needed to initiate TATB based explosives using plastic impactors. Unfortunately, being the only gun of its kind, it has had a lengthy shakedown period. We have encountered projectiles which disintegrated when launched, huge impact tilts, gouged barrels, and projectile velocity measuring systems that refused to work. Fortunately, this is now behind us. Hereafter, we present our progress on making embedded magnetic particle velocity gauge measurements in these insensitive explosives. Throughout, we will try to show that our work is in accord with, and complementary to, the fine previous studies.

EXPERIMENTAL DETAILS

The overall configuration for the initiation experiments is shown in Figure 1. This is the same configuration described in great detail in Ref. 7. A

[†] Work performed under the auspices of the U.S. Dept. of Energy.

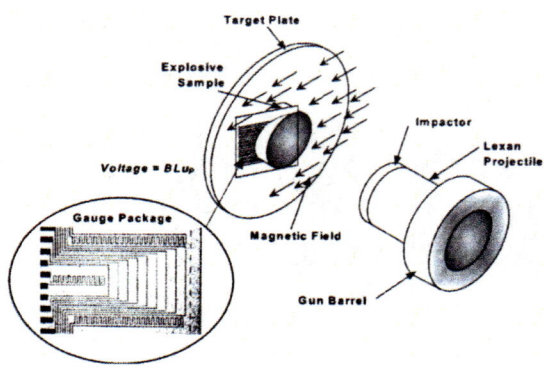

FIGURE 1. Experimental setup. Explosive sample installed in gun target chamber and magnetic field.

Lexan projectile is faced with an impactor disk made of Kel-F, a high-density plastic, and coincidentally, the binder material for the explosives studied. Because of our problem with projectiles disintegrating when launched, we have till now been reluctant to use impactors made of anything but plastic.

In all but one of the experiments, the projectile velocity was measured using optical methods. We must confess, however, that the optical method changed from one experiment to the next as "improvements" were made. Nevertheless, all the methods we used gave standard errors in the velocity of < 10 m/s.

When the impactor strikes the explosive sample, a planar shock wave is generated which begins the initiation process. The stress of the initial shock is determined using the impact velocity, the Hugoniot of the Kel-F impactor,[8] the Hugoniot of the explosive,[9] and standard impedance matching techniques.

Electromagnetic particle velocity gauges are embedded in the explosive sample at ten or eleven different depths. These vary from the impact surface to about 11 mm into the sample. These, of course, produce voltages proportional to the local mass (particle) velocity at the Lagrangian position of the gauge. Three "shock trackers" also allow construction of distance - time (x-t) plots of the position of the shock front with time as it moves through the sample. These x-t trajectories are used to determine the position and time where detonation is achieved.

EXPERIMENTAL RESULTS

Wave profiles of particle velocity vs. time and x-t plots of the shock trajectories were successfully obtained for eight experiments on PBX 9502 and three experiments on LX-17. In addition, there were many additional experiments that failed for one reason or another. Impact stresses of 10.8 to 15.4 Gpa were created with projectile velocities of 2.4 to 3 km/s. This produced run distances of 4.4 to over 14 mm.

Figure 2 shows wave-profiles from shot 2S-47 where LX-17 was impacted with a Kel-F impactor at a velocity of 2.951 ± 0.004 km/s producing an input of 14.96 GPa. The data quality is seen to be exceptional in this figure. Surprisingly, this quality was typical for experiments that worked. Experiments that didn't work failed in an unmistakable fashion. Nine good wave profiles from gauges located at depths of 0.0 through 6.7 mm were recorded. The first profile is from a gauge on the front of the sample and the remaining 8 from embedded gauges. The input particle velocity is 1.45 km/s and this grows to a full detonation well before the wave reaches the last gauge.

All wave profiles other than the first show that the amplitude of the wave at the shock front is increasing as the wave travels into the sample. This is observed in all heterogeneous explosives. Additionally, there is a small hump behind the shock front. This is consistent with some delay between the shock passing through un-shocked material and the release of energy.

Additionally, as observed by Wackerle, Stacy and Seitz,[6] the slope of the particle velocity immediately behind the shock front is positive. This per-

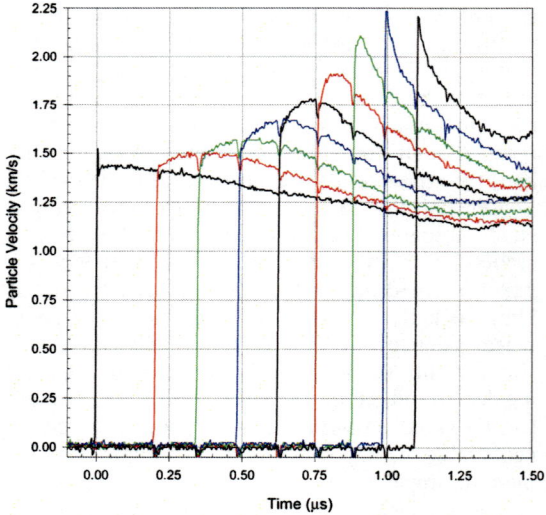

FIGURE 2. Particle velocity wave profiles from shot 2S-47 on LX-17. The input is 14.96 GPa and was created by impacting Kel-F on the LX-17 at 2.951 km/s.

sists and is observable up until detonation is achieved, and only then can it not be resolved. This is indicative variously of a slow initial reaction or an induction time before reaction begins. The models of Tarver and also of Pier Tang account for this using a weak or slow initial reaction rate.

When one accounts for the ~ 40 ns time response of the embedded gauges, the detonation wave profiles are consistent with those observed in Refs. 4 and 6 using Fabry-Perot Velocimetry. That is, after the initial shock, the time to go from ~ 2.2 km/s to 1.5 km/s is roughly the same for both types of experiments. Additionally, in Refs. 4 and 5, the time to drop from the spike at ~ 2.65 km/s to ~ 2.2 km/s, where our records begin, is very short ~ 40 – 50 ns which is roughly the time response of our gauges.

The x-t plot showing the position of the shock front with time is shown in Figure 3. This again is for shot 2S-47 in LX-17. Points shown are from all three shock tracker gauges and the particle velocity gauges. The slope of the line through the last few data points indicates the detonation velocity.

From plots such as those shown in Figure 3, there are a number of ways to determine the run distance and time to detonation. Because the approach to detonation in these TATB based explosives is so gradual, we have found it useful to use the point where the data closely approach the detonation line as the transition point. Even so, it is often helpful to have the guidance of corresponding wave profiles such as those in Figure 2.

Figure 4 shows "Pop – Plots" of distance to detonation as a function of the initial shock pressure. Historic work for PBX 9502 (Ref. 2) and LX-17 (Refs. 1,3,5) is shown as well as data from the present study. Note first that PBX 9502 is slightly more sensitive than LX-17. This is due to; 1) slightly more explosive and less binder in the PBX 9502; 2) For the case of 1.900 g/cm^3 LX-17, PBX 9502 and LX-17 have nearly identical void content of ~ 2.6 – 2.7%. The reason for the decreased sensitivity in LX-17 in the work of Dallman and Wackerle is that the explosive they used had a lower void content ~ 2.0% (density 1.913 g/cm^3).

PBX 9502 data shown in the Pop – Plot of Figure 4 is from three powder lots. Dick et. al.[2] used explosive pressed from "Virgin" PBX 9502 molding powder. Our work used two different lots of "Recycled" PBX 9502 molding powder. Recycled molding powder is made from 50% Virgin molding powder and 50% ground up machining scraps. All sets of data lie on the same curve. The Pop – Plot for PBX 9502 is thus,

$$\log(X) = 4.26 \pm 0.09 - (3.06 \pm 0.08)\log(P). \quad (1)$$

This work and Jackson et. al.[1] used LX-17 pressed to 1.900 g/cm^3 from two different molding powder lots. The Pop – Plots for this density of LX – 17 is

$$\log(X) = 4.53 \pm 0.12 + (3.22 \pm 0.10)\log(P). \quad (2)$$

Dallman and Wackerle[7] used LX-17 pressed at Pantex to a density of 1.913 g/cm^3. The Pop – Plot for this material is

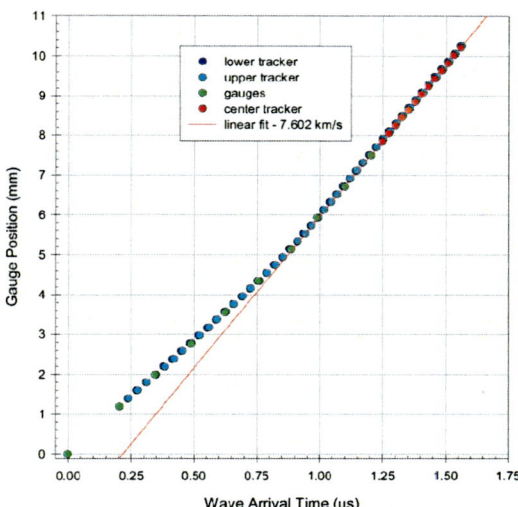

FIGURE 3. x-t plot obtained from shock arrival at shock tracker elements and particle velocity gauge elements. The shot is #2S-47 on LX-17.

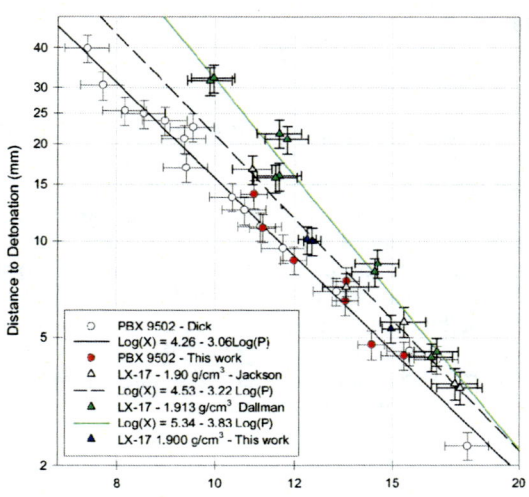

FIGURE 4. Pop – plots for TATB based explosives. Names of first authors identify data sets.

$$\log(X) = 5.34 \pm 0.26 - (3.83 \pm 0.23)\log(P). \quad (3)$$

NUMERICAL MODELING

We used the Ignition and Growth "Reactive Burn" model to simulate these experiments. As with all reactive burn models, equations of state are required for the unreacted explosive and for the reaction products. We used the JWL model for these equations of state

$$P = A\exp(-R_1 V/V_0) + B\exp(-R_2 V/V_0) \\ + \omega C_V T V_0 / V \quad (4)$$

Table 1 gives EOS parameters for LX-17 and PBX 9502 products and unreacted explosive. Note that aside from the initial volume, both explosives have identical Equations of State.

Table 1. EOS Parameters for LX-17 and PBX 9502

Parameter Eq.(4)	Unreacted EOS (PBX 9502)	Product EOS
A (Mbar)	632.07	13.454
B (Mbar)	-0.04472	0.6727
R_1	11.3	6.2
R_2	1.13	2.2
ω	0.8938	0.5
$C_V * 10^{-5}$ (Mbar/K)	2.487	1.0
V_0 (cm^3/g)	0.525 (0.531)	
E_0 (Mbar)		0.069
D (cm/µs)		0.7596
P_{CJ} (Mbar)		0.2714

The Ignition and Growth reactive burn model was used, where λ is the fraction reacted is

$$d\lambda/dt = I_0(1-\lambda)^{2/3}(\rho/\rho_0 - 1 - a)^X \\ + G_1(1-\lambda)^{2/3}\lambda^{0.11}P^1 + G_2(1-\lambda)^{1/3}\lambda^1 P^Z \quad (5)$$

Table 2 lists Parameters used for the reaction modeling. Note again that most parameters were the same for both explosives. The lower value of a for PBX 9502 indicates that this explosive starts reacting at a slightly lower pressure or compression. There was not room in this paper to include comparison between the measured and calculated wave profiles, but these parameters have been found to model all of the available data very well. A more complete paper including all the data and these comparisons will be published later.

Table 2. Ignition and Growth reactive burn parameters.

Parameter Eq. (5)	LX-17	PBX 9502
I_0(1/µs)	4.4(10^5)	4.4(10^5)
G_1(Mbar^{-1} µs^{-1})	0.6	0.6
G_2(Mbar^{-Z} µs^{-1})	400	400
a	0.22	0.214
X	7.0	7.0
Z	3.0	3.0

ACKNOWLEDGEMENTS

Robert Medina operated the gas gun for all of the experiments presented here.

REFERENCES

1. R.K. Jackson, L.G. Green, R.H. Barlett, W.W. Hofer, P.E. Kramer, R.S. Lee, E.J. Nidick, Jr., L.L. Shaw, and R.C. Weingart in *Proceedings of the Sixth Symposium (International) on Detonation*, Office of Naval Research, Report ACR-221, p. 755, 1976

2. Jerry J. Dick, C.A. Forest, J.B. Ramsay, and W.L. Seitz,, *J. Appl. Phys.*, **63**, 4881, (1988)

3. K. Bahl, G. Bloom, L. Erickson, R. Lee, C. Tarver, W. Von Holle, and R. Weingart in *Proceedings of the Eighth Symposium (International) on Detonation*, Office of Naval Research, Report NSWC MP 86-194, p. 1045, 1986

4. W.L. Seitz, H.L. Stacy, Ray Engelke, P.K. Tang, and Jerry Wackerle in *Proceedings of the Ninth Symposium (International) on Detonation*, Office of Naval Research, Report OCNR 113291-7, p. 657, 1989

5. J.C. Dallman and Jerry Wackerle in *Proceedings of the Tenth Symposium (International) on Detonation*, Office of Naval Research, Report ONR 33395-12, p. 130, 1995

6. Jerry Wackerle, H.L. Stacy, and W.L. Seitz in *Proceedings of the Tenth Symposium (International) on Detonation*, Office of Naval Research, Report ONR 33395-12, p. 468, 1995

7. S.A. Sheffield, R.L. Gustavsen and R,R. Alcon, in *Shock Compression of Condensed Matter – 1999*, American Institute of Physics (AIP) Conference Proceedings 505 (1999), p. 1043.

8. $\rho_0 = 2.14$ g/cm^3, $U_S = 1.99 + 1.76 u_p$ km/s

   S.A. Sheffield and R.R. Alcon, in *Shock Compression of Condensed Matter – 1991*, S.C. Schmidt, R.D. Dick, J.W. Forbes, D.G. Tasker (editors), Elsevier Science Publishers (1992), p. 909.

9. $U_S = 1.90 + 3.00 u_p : u_p \leq 0.82$

   $U_S = 2.90 + 1.78 u_p : u_p \geq 0.82$

   C.A Forest, unpublished (1995). This Hugoniot for PBX 9502 is based primarily on the data of Ref. 2.

DETONATION INITIATION IN PRESHOCKED LIQUID EXPLOSIVES

Andrew J. Higgins[1], François X. Jetté[1], Akio C. Yoshinaka[1], John H.S. Lee[1], and Fan Zhang[2]

[1]*McGill University, Department of Mechanical Engineering, 817 Sherbrooke St. W, Montreal, Quebec, Canada*
[2]*Defence Research Establishment Suffield, P.O. 4000 Stn. Main, Medicine Hat, Alberta, Canada*

Abstract. The initiation of detonation in a homogenous liquid explosive by the reflection of a strong shock from a high impedance anvil is investigated. By transmitting a sub-critical shock through a test sample of sensitized nitromethane and then reflecting it normally off a steel plate bounding the explosive, detonation can be initiated in the pre-shocked medium. The initiation of detonation is observed via fiber optics monitored by photodiodes and by manganin pressure gauges mounted on the steel plate. The initiation of detonation by the reflected shock is inferred from the appearance of intense luminosity and an increase in pressure at the explosive/steel interface, both appearing about 1 μs after shock reflection. The manganin gauge measurements indicate that the critical pressure for incident initiation by a 100 mm diameter shock is 4-5 GPa, while the critical pressure for reflected shock initiation is 7 GPa.

INTRODUCTION

Since the early work of Chaiken [1] and Campbell et al. [2], the initiation of detonation in homogenous explosives by a planar shock has been intensively studied. However, the actual mechanism of initiation in scenarios such as gap tests can be considerably more complex, involving multiple shock interactions and reflections. The use of a stiff anvil to reflect the incident shock back into the test explosive, for example, may result in initiation for conditions where the incident shock by itself would not.

Recent experiments by Winter et al. [3] and Tarver et al. [4] examined the reflection of shock waves transmitted through plastic-bonded explosives off high impedance backing plates. Initiation of reaction and energy release without detonation as well as detonation initiation by shock reflection were reported. As for homogeneous explosives, attempts by Presles et al. [5] to initiate reaction by reflecting a sub-critical shock off an aluminum plate back into pure nitromethane did not result in any measurable decomposition reaction upon reflection. This was attributed to the relatively low reflected-shock temperature (750 K). Extensive work in recent years by Gruzdkov, Winey, and Gupta [6] involved multiple shock reverberations between two stiff anvils to compress nitromethane to pressures of up to 19 GPa. The onset of reaction is observed at temperatures above 940 K, suggesting a thermal mechanism of initiation in nitromethane. While the reverberating shock wave technique provides a means to study the onset of chemical reaction in off-Hugoniot (quasi-isentropically compressed) states, the samples used are too small to permit the initiation and propagation of detonation to be observed.

The present study examines the initiation of detonation in a homogeneous liquid explosive by reflecting a subcritical incident shock off a high impedance plate located underneath the test explosive. The charge is much larger than the critical diameter of the test explosive, permitting detonation initiation and propagation in a shock-compressed explosive to be observed. These experiments seek to identify and compare the critical pressure to initiate detonation in the incident versus reflected mode of initiation, with an ultimate goal of understanding the limits to shock compression of homogenous energetic materials.

EXPERIMENTAL DETAILS

The experimental charge configuration is similar to a conventional gap test, with a point-initiated donor charge of 100 mm diameter and 200 mm length (Fig. 1). The donor explosive is nitromethane sensitized with 10% diethylenetriamine (DETA). The donor explosive transmits a shock though an attenuator of inert material (gray PVC plastic). PVC was used rather than the more typical PMMA attenuator because of its compatibility with nitromethane. The thickness of the attenuator is varied to control the strength of the shock transmitted into the test charge. The test charge is contained in a PVC capsule of a larger diameter (200 mm) than the donor. This configuration was used to eliminate shock interactions with the capsule walls that can result in initiation at anomalously low pressures as compared to ideal shock initiation [7-9]. For the experiments examining incident initiation, the

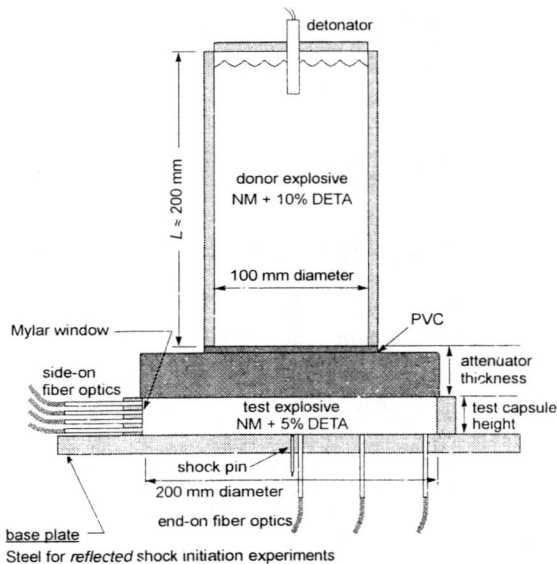

FIGURE 1. Experimental charge configuration.

bottom of the charge was sealed with a sheet of 0.25 mm Mylar mounted on a medium density fiber board (MDF). For experiments with reflected shocks, the test capsule was prepared directly on a 12.7-mm-thick mild steel plate. The capsule height was varied between 12.5 and 50 mm. This range of heights was chosen to ensure that the shock reflection from the bottom plate occurred before the incident shock reached the capsule side walls, so that initiation would not occur due to interaction with the side walls and the incident shock would not interfere with the side-on fiber optics observing the reflection through a Mylar window.

The luminosity generated by detonation was observed via fiber optics connected to photodiodes. For incident initiation, both end-on and side-on fiber optics, mounted in brass light pipes, were used. For reflected initiation, only side-on fiber optics were used so as not to interfere with the shock reflection off the bottom steel plate. As discussed below, the photodiodes were only sensitive enough to determine the onset of detonation but not chemical reaction. The arrival time of the shock at the test explosive/steel interface was determined by a shock pin centered on the plate. Select experiments were performed with manganin gauges (Dynasen MN4-50-EK) embedded in the attenuator to determine the critical incident pressure or mounted on the steel plate to determine the critical reflected shock pressure for initiation.

The nitromethane used for the test mixture was commercial grade sensitized with 5% DETA by mass. The experiments were always performed within a few hours of mixture preparation and were fired at ambient conditions (5-22 °C). Considerable care was taken in

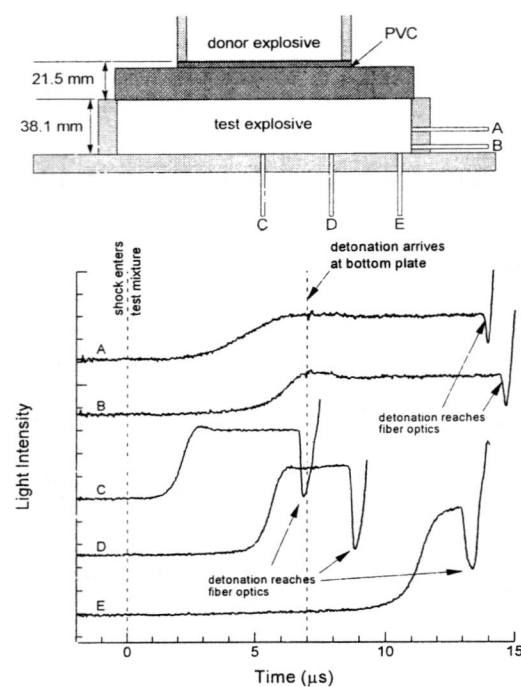

FIGURE 2. Luminosity signals for incident initiation.

final charge assembly to ensure no air bubbles were trapped on the attenuator/test explosive interface or on the steel plate.

RESULTS

Incident Initiation

Before experiments to examine the initiation of detonation by shock reflection were performed, the critical shock pressure for incident shock initiation for this scale of experiment (100-mm-diameter donor) was identified. The results with a 21.5-mm-thick PVC attenuator are shown in Fig. 2. The entry of the shock into the test mixture is time zero. Note the photodiodes observing the charge via fiber optics detect luminosity beginning about 1 µs after shock entry. The side-on fiber optics clearly show a detonation propagating down the charge, and the end-on fibers show that initiation occurred on the central axis of the charge.

If the attenuator thickness is increased to 35 mm, no luminosity is observed at all and only a decaying shock propagates through the test charge. The fact that the photodiodes detect no luminosity in a slightly subcritical case (where reactions are present prior to being quenched by lateral and rear-generated rarefactions) suggests that they are not sensitive enough to observe shock-initiated reaction and can only be used to determine the presence of detonation. Repeating these experiments with different thicknesses of attenuator showed that the critical gap thickness for incident initiation was in the range of 30-35 mm. Based on manganin measurements of shock pressure in the attenuator and the average shock velocity over this range, the critical shock pressure for incident initiation is estimated to be 4-5 GPa from impedance matching calculations.

Reflected Initiation

If the experiments described above were repeated with a steel plate on the bottom of the test charge, initiation of detonation could be observed upon reflection. Shown in Fig. 3a are the photodiode and manganin gauge traces of a subcritical shock (as transmitted by a 42-mm-thick attenuator) propagating through a 22.5-mm-thick capsule and then reflecting off the steel bottom plate. The time of shock reflection was 7 µs, as indicated by the shock pin and manganin gauge. Within 1.5 µs after shock reflection, the appearance of intense luminosity was detected by the photodiodes. A pressure signal as measured by a manganin gauge mounted on the center of the bottom steel plate is also shown in Fig. 3a. The post-reflected-shock pressure of 7.0 GPa is nearly constant in amplitude, until the appearance of a "hump" 1.2 µs after shock reflection. The appearance of this hump is simultaneous with the start of luminosity as detected by the fiber optics/photodiodes and is apparently associated with the onset of detonation. The amplitude of this hump (8.5 GPa) is below the CJ pressure, so it is unclear if this is the record of the establishment of a self-sustained detonation (retonation wave) or a result of reaction without complete detonation. The manganin gauge used is also only rated to 12.5 GPa, so the fact that the signal did not exceed this value cannot be taken as a conclusive indication of the pressures reached during reflected initiation. It is clear from the photodiode traces, however, that a detonation did propagate from the charge axis to the wall of the test capsule after shock reflection.

Shown in Fig. 3b are manganin traces of an experiment similar to that in Fig. 3a, except that pure nitromethane as opposed to sensitized nitromethane was used. The photodiode traces (not shown here) in this case did not detect any luminosity, and the explosive is believed to have remained inert throughout the experiment. The signal of manganin gauge A mounted on the steel plate can be compared to the signal from Fig. 3a. Manganin gauge B (suspended in the liquid 3 mm above the steel plate) clearly shows the record of incident and, as the shock passes the gauge a second time, reflected shock as well. The amplitude of the incident shock (3.5 GPa) and the reflected shock (7.0 GPa) are in good agreement with impedance matching calculations, and also match the pressures in the experiment with sensitized

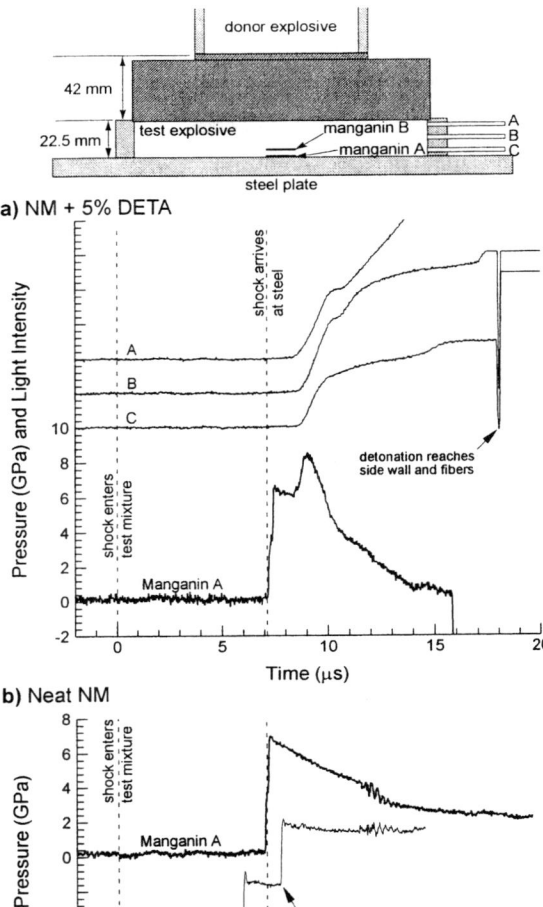

FIGURE 3. (a) Reflected shock pressure and luminosity for reflected initiation in sensitized NM (b) Incident and reflected shock pressure in pure (nonreacting) NM.

nitromethane in Fig. 3a. This agreement suggests that the reflected shock in sensitized nitromethane is initially nonreacting. The results in pure nitromethane show a slow decay in pressure and lack the distinctive "hump" associated with the reflected initiation of Fig. 3a.

To prevent initiation upon reflection, one can increase the attenuator thickness (thus lowering the pressure of the incident shock) or increase the height of the test capsule (thus further attenuating the shock in the test liquid before shock reflection occurs). In this study, both approaches were taken. Figure 4 shows the results of approximately 20 such experiments with shock reflection in NM+5% DETA. Experiments with attenuators thinner than 35 mm (the shaded region in

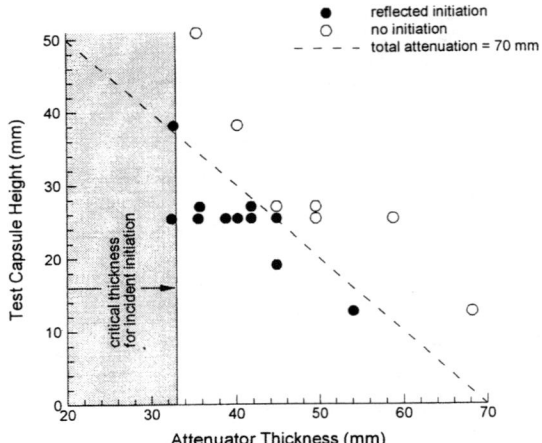

FIGURE 4. Results for shock reflection off a steel plate as a function of the attenuator thickness and the test capsule height.

Fig. 4) will typically result in initiation on incident shock, and therefore are not plotted on this figure. Experiments in which luminosity traces identical to those shown in Fig. 3a were obtained are shown as solid symbols. Experiments in which no luminosity was observed at all during the test time are shown as open symbols. The reflected initiation events were very reproducible, with the onset of luminosity consistently occurring 1.2-1.5 μs after shock reflection in the case of initiation.

Since the acoustic speed, shock impedance, and Hugoniots of nitromethane and PVC are similar, it would be expected that the strength of the reflected shock is determined by the total distance of shock travel in either material prior to reflection. Indeed, a line denoting a total propagation distance of 70 mm appears to bound the region of "initiation upon reflection" from "no initiation" in Fig. 4. Based on the manganin gauge measurement shown in Fig. 3 and additional pressure measurements made 70 mm in PVC, the pressure of the shock in sensitized nitromethane prior to reflection in the critical case is 3.5 GPa, giving a reflected shock pressure of 7.0 GPa.

DISCUSSION

The reflection of a strong subcritical shock off a steel plate and back into the test explosive can result in the initiation of detonation. The results obtained here suggest that the critical pressure for incident initiation of detonation in NM + 5% DETA at the 100 mm scale is 4-5 GPa, while the pressure obtained upon reflection must be 7 GPa for initiation. The higher pressure required for reflected initiation is likely a result of the fact that a single shock compression results in greater shock heating than for the case of two (or more) successive shocks to the same final pressure.

The critical pressures reported here refer only to peak pressure. Since the use of point-initiated charges and attenuators involving lateral and rear-generated rarefactions will result in a shock followed by an expansion gradient, these values of critical pressure cannot be directly compared to square-wave loading. Nonetheless, the fact that unambiguous initiation of detonation in a pre-shock-compressed explosive has been obtained provides a new means to examine the mechanisms of shock initiation in homogeneous explosives.

ACKNOWLEDGEMENTS

The authors would like to thank Massimiliano Romano, Leo Nikkinen, and David Hanna for their expertise with the photodiagnostics and Oren Petel for assistance in conducting the field trials. Steve Kacani and Charles Dolan are thanked for timely fabrication of the charges. The dedicated assistance of the technical staff of the Defense Research Establishment Suffield is generously acknowledged.

REFERENCES

1. Chaiken, R.F, *J. Chem. Phys.* **33**, 760-761 (1960).
2. Campbell, A. W., Davis, W. C., and Travis, J. R., *Phys. Fluids* **4**, 498-510 (1961).
3. Winter, R.E., Taylor, P., and Salisbury, D.A., "Reaction of HMX-Based Explosive Caused by Regular Reflection of Shocks," *11th Symp. (Int.) on Detonation*, 1998, pp. 649-656.
4. Tarver, C.M., Cook, T.M,. Urtiew, P.A., Tao, W.C., "Multiple Shock Initiation of LX-17," *10th Symp. (Int.) on Detonation*, 1993, pp. 696-703.
5. Presles, H.N., Fisson, F., and Brochet, C., *Acta Astro.* **7**, 1361-1377 (1980).
6. Gruzdkov, Y.A., Winey, J.M., and Gupta, Y.M, "Use of Time-Resolved Optical Spectroscopy to Understand Shock-Induced Decomposition in Nitromethane," *11th Symp. (Int.) on Detonation*, 1998, pp. 521-524.
7. Travis, J.R., "Experimental Observations of Initiation of Nitromethane by Shock Interactions at Discontinuities," *4th Symp. (Int.) on Detonation*, 1965, pp. 386-393.
8. Seely, L.B., Berke, J.G., and Evans, M.W., *AIAA J.* **5**, 2179-2181 (1967).
9. Jetté, F.X., Yoshinaka, A.C., Romano, M.,. Higgins, A.J., Lee, J.H.S., Zhang, F., "Investigation of Lateral Effects on Shock Initiation of a Cylindrical Charge of Homogeneous Nitromethane," *18th Int. Colloquium on the Dynamics of Explosions and Reactive Systems*, Seattle, WA, 2001.

LAGRANGIAN ANALYSIS OF EDC37 SHOCK INITIATION DATA

J. R. Maw

AWE Aldermaston, Reading, RG7 4PR, U.K.

Previous papers have presented experimental and theoretical studies of shock initiation in EDC37. Multiple embedded electromagnetic gauges were used to provide particle velocity histories over a range locations in the explosive but attempts to simulate these data by adjusting reaction rate parameters in an ignition and growth model were only partially successful. In an attempt to improve the modelling, the Lagrangian Analysis technique has been applied to the data to obtain direct information on the reaction history in EDC37. Results are presented illustrating the sensitivity of the inferred pressure, density and energy fields to uncertainties introduced in the application of the technique. Reaction histories are derived corresponding to different assumptions in the treatment of the Equation of State of the partially reacted mixture and these are compared with those obtained from a number of commonly used reaction rate formulations.

INTRODUCTION

Two earlier papers (1,2) described experimental and theoretical studies of shock initiation in EDC37, an HMX based explosive. In (1) planar shock wave initiation of EDC37 was studied using embedded electromagnetic gauges to measure particle velocities over a range of depths. These data, together with measurements of the trajectory of the leading shock, clearly demonstrated the build up to detonation. In (2) the experiments were modelled using the Lee and Tarver ignition and growth model (3). Reasonable agreement with the data was obtained by suitable adjustment of the model parameters but it was recognised that there was room for improvement in both the predicted run distances to detonation and the detailed modelling of the particle velocity data.

The process of refining the model by parameter adjustment is time consuming and there is also some doubt that the true reaction rates can indeed be represented within the confines of the reaction rate formulation. An alternative approach is to extract information on the reaction rates directly from the data using the Lagrangian analysis technique. This paper describes an attempt to follow this approach.

LAGRANGIAN ANALYSIS

There is extensive literature on the theory of the Lagrangian Analysis and its application to inert and reacting materials (e.g. 4-6). In this paper we present only the basis of the technique restricting attention to the analysis of particle velocity gauge data. Integrated equations for the conservation of the mass, momentum and energy are written in the form

$$v(h,t) = v(h,t_0) + v_0 \int_{t_0}^{t} \left(\frac{\partial u}{\partial h}\right)_t dt \quad (1)$$

$$p(h,t) = p(h_0,t) - \frac{1}{v_0} \int_{h_0}^{h} \left(\frac{\partial u}{\partial t}\right)_h dh \quad (2)$$

$$e(h,t) = e(h,t_0) - v_0 \int_{t_0}^{t} p\left(\frac{\partial u}{\partial h}\right)_t dt \quad (3)$$

Where v is the specific volume, p the pressure and u the particle velocity. t and h denote time and Lagrangian position and subscript 0 denotes an initial condition. Equations (1) and (3) give the specific volume and internal energy as functions of time at a given Lagrangian position, while equation (2) gives the pressure as a function of Lagrangian position at a given time. Ideally the integration of the latter equation should start from initial pressure conditions at a specified Lagrangian position provided by pressure gauge data. In the absence of such data the starting point of the integration is on the shock front along which all conditions must be evaluated.

APPLICATION TO EDC37 DATA

The Shock Front

Figure 1 shows the shock trajectory obtained from the gauge records and a "shock tracker" gauge (1) in an experiment where the HE was subjected to a 3.5 Gpa shock.

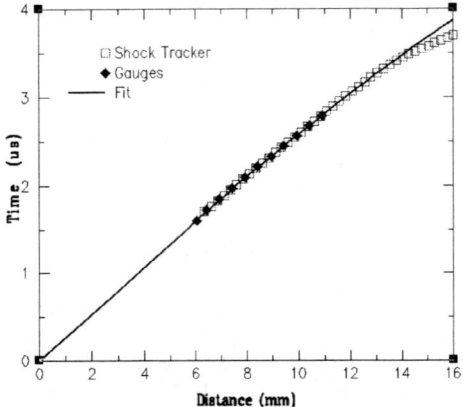

FIGURE 1. Shock trajectory.

In applying the technique to these data we have fitted the shock arrival time at position h by

$$t_s = b_1 h + b_2 h^2 + b_3 h^3 \quad (4)$$

from which the shock velocity follows as

$$U_s = \frac{dh}{dt_s} = 1/(b_1 + 2b_2 h + 3b_3 h^2) \quad (5)$$

Figure 1 shows that the fit is good up to 13 mm where the shock develops into a detonation wave.

Assuming a linear shock velocity-particle velocity relation for the unreacted explosive

$$U_s = C_0 + s u_p \quad (6)$$

enables the particle velocity to be calculated. The pressure, specific volume and internal energy are then calculated using the Rankine Hugoniot relations.

Particle velocity

Figure 2 shows the corresponding particle velocity records. Measurements were made at ~0.5 mm intervals at depths between 6 and 11 mm but for clarity the figure shows only the records at 6,7,8,9 and 10 mm depths.

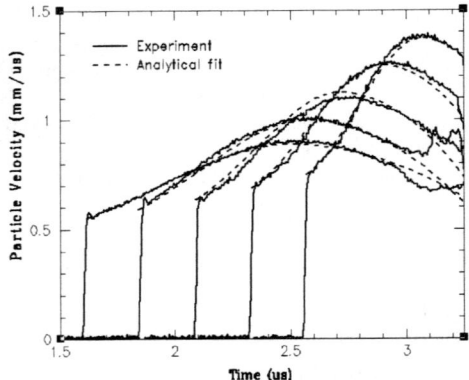

FIGURE 2. Particle velocity data.

These data were fitted by an equation of the form

$$u = u_s(h) + [u_{max}(h) - u_s(h)]\overline{u}(\tau) \quad (7)$$

where u_s is the particle velocity at the shock front and u_{max} is the maximum particle which is fitted by a polynomial in h. The non dimensional time co-ordinate τ is

$$\tau = \frac{(t - t_s(h))}{(t_{max}(h) - t_s(h))} \quad (8)$$

Where $t_{max}(h)$ is the time of maximum velocity at each gauge location also fitted by a polynomial in h. The non-dimensional velocity $\overline{u}$, is represented by a polynomial with coefficients which are also functions of h.

The fitted velocities obtained using this functional form are compared with those measured in fig 2. No attempt has been made to precisely fit individual records but it can be seen that the overall fit is good and the main features of the data are well reproduced.

RESULTS

Pressure – Volume Paths

Pressure specific volume and internal energy have been calculated using equations (1)-(3) together with the above representations of the experimental data.

Figure 3 shows the inferred pressure – volume paths at a number of Lagrangian positions. Also shown is the unreacted shock Hugoniot. The general behaviour is as expected with the HE initially undergoing some compression behind the shock. At the smaller depths the paths lie above the Hugoniot indicating that significant reaction is occurring. At greater depths the paths lie initially below the Hugoniot indicating isentropic compression with little initial reaction. At later stages the HE begins to expand and the pressure starts to decrease due to further reaction.

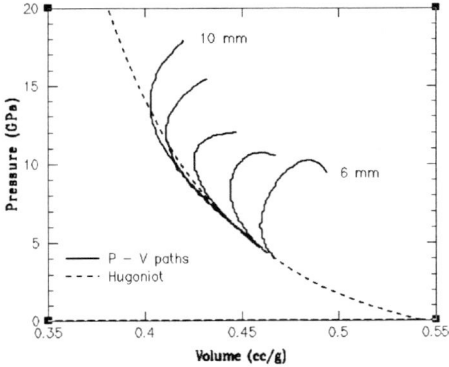

FIGURE 3. Pressure – Volume paths.

Degree of Reaction

Two methods have been used to calculate the degree of reaction, λ, based on different treatments of the EOS of the partially reacted mixture of unreacted HE and reaction products. In the first the unreacted HE is assumed to lie on the Hugoniot while the products lie on the isentrope passing through the Chapman Jouguet state. Writing the mixture equation for the volume in the form

$$v = (1-\lambda)v_s(p) + \lambda v_p(p) \quad (9)$$

Gives the following relation for λ as a function of the pressure and volume

$$\lambda = \frac{v - v_s}{v_p - v_s} \quad (10)$$

In the second method the unreacted HE lies on the isentrope through the shocked state and the products are assumed to satisfy an EOS of the form

$$p = F(v_p, e_p + Q) \quad (11)$$

Note that the chemical energy Q is added to e_p since the latter, deduced from the Lagrangian analysis, is only the energy of compression. Equation (9) together with the EOS and the energy of the mixture in the form

$$e = (1-\lambda)e_s(p) + \lambda e_p(p) \quad (12)$$

enables λ to be obtained by an iterative procedure.

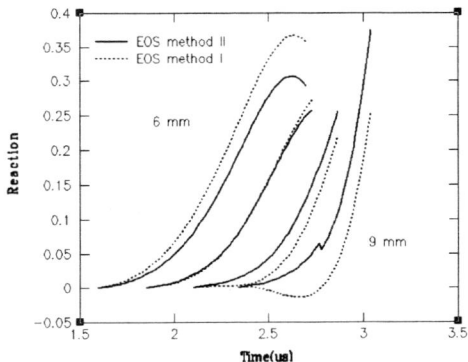

FIGURE 4. Reaction histories

Figure 4 plots the variation of λ with time obtained by each method at depths 6,7 and 8 mm. The two methods give very similar results at 6 and 7 mm but there is some disagreement at 8 mm and

9mm. Closer inspection shows that at these depths method I gives negative reaction initially, a consequence of the assumption that the unreacted material compresses along the Hugoniot. Note that it has only been possible to calculate the degree of reaction up to about 30%.

Reaction Rates

Figure 5 shows reaction rates at the 6,7 and 8mm locations obtained by differentiating the reaction calculated using the second method. These appear reasonably smooth apart from some jitter due to the numerical differentiation at the 8 mm depth. This figure also shows reaction rates calculated using the Lee and Tarver model with the parameters for EDC37 obtained in (2). The overall agreement is surprisingly good bearing in mind the very different ways in which the reaction rates were obtained. More detailed examination shows that the Lagrangian Analysis gives reaction rates higher than those of the model at the 6 and 7mm depths but somewhat lower at 8 mm. Further work is now needed to determine whether the model parameters can be further adjusted to give an improved fit to the reaction rates deduced from this analysis.

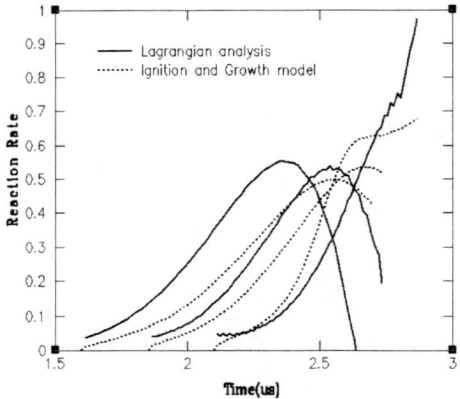

FIGURE 5. Reaction rates

CONCLUSIONS

We have shown that the Lagrangian Analysis technique can be successfully applied to deduce sensible reaction rates from shock initiation data. This offers the prospect in the short term of improving the modelling of experimental data using existing reactive burn models and, in the longer term, will allow assessment of the value of more physically based models. It should be noted that with particle velocity data alone the method gives information only on the early stages of the reaction. It would be useful to have additional pressure gauge data to extend the technique to later times

ACKNOWLEDGEMENTS

My thanks to Rick Gustavsen for providing the experimental data in digital form.

REFERENCES.

1. Gustavsen, R.L. et al., "Initiation of EDC37 Measured with Embedded Electromagnetic Particle Velocity Gauges," in *Shock Compression of Condensed Matter-1999*, edited by M. D. Furnish et al., AIP Conference Proceedings 505, New York, 2000, pp. 879-882.
2. Winter, R.E. et al., "Modelling Shock Initiation of HMX-Based Explosive," in *Shock Compression of Condensed Matter-1999*, edited by M. D. Furnish et al., AIP Conference Proceedings 505, New York, 2000, pp. 883-886.
3. Lee, E. L. and Tarver, C. M., "Phenomenological Model of Shock Initiation in Heterogeneous Explosives", *Physics of Fluids*, **23(12)**, 1980.
4. Fowles, F. and Williams R. F., "Plane Stress Wave Propagation in Solids", *J. Appl. Phys.*, **41(1)**, 1970.
5. Cowperthwaite, M. and Rosenberg, J. T., "Lagrange Gage Studies in Ideal and Non-Ideal Explosives", *Seventh Symposium on Detonation*.
6. Forest, C.A., "Lagrangian Analysis, Data Covariance, and the Impulse Time Integral," in *Shock Compression of Condensed Matter-1991*, edited by S. C. Schmidt et al., North-Holland, 1992, pp. 317-324.

© British Crown Copyright 2001/MOD
Published with the permission of the Controller of Her Britannic Majesty's Stationery Office.

TRANSIENT DETONATION PROCESSES IN A PLASTIC BONDED EXPLOSIVE

Keith A. Thomas[1], Eric S. Martin[1], James E. Kennedy[1], Ismael A. Garcia[1], and Joseph C. Foster Jr[2]

[1]Los Alamos National Laboratory, Los Alamos, NM 87545
[2]Air Force Research Laboratory, Eglin AFB, FL 32542

Abstract. Experiments involving the transfer of detonation from small booster charges of PBXN-5 (95% HMX and 5% Viton A) into larger charges of various plastic-bonded explosives (PBXs) have produced some surprising results and have stimulated investigation into the factors governing observed responses. To understand these results, we conducted a series of tests with different miniature detonator-booster configurations using laser velocimetry to quantify the pressure pulse that is transmitted from the PBXN-5 booster. Models were used to determine the ideal explosive behavior for comparison with the measured results. The differences are interpreted as being due to transient behavior and late-time energy release from the booster charge. We characterize these behaviors as evidence of microdetonics, where we define microdetonics as the study of less-than-CJ detonation performance due to curvature and/or transient behavior. This provides useful insights into the fundamentals of the detonation process that can feed into advanced modeling approaches such as Detonation Shock Dynamics (DSD).[1,2]

INTRODUCTION

Research on the detonation of condensed phase explosives spans many decades in both length and time scales. Early work was limited primarily by the instrumentation response. Improvements in the instrumentation have revealed many aspects of the detonation process that are best studied during the shock-to-detonation transition phase under conditions which represent high geometrical divergence. An asymptotic solution to the classical Zel'dovich, Von Neuman, Doering (ZND) theory of detonation, called Detonation Shock Dynamics (DSD), has emerged in recent years.[1,2] Continuum variables emerge out of DSD for the influence of curvature of the detonation front and the reaction-zone chemistry on the detonation wave propagation. It has been expanded to include transient effects of accelerating detonation waves.[3,4] Experimental calibrations of DSD to date have addressed only the curvature effects, through measurements of the curved front in steady-detonation experiments in cylindrical columns only slightly larger than the failure diameter[5] and with high convergence.[6] Those efforts have focused on explosives that are used for the large-volume, main charge of an explosive assembly. However, an essential element of the detonation process is initiation, which is usually triggered by a detonator and booster system that occupies a small volume in comparison to the main charge.

Historically, very little attention has been paid to the performance of real initiation systems. It is generally assumed that the initiation-system explosives operate at steady-state C-J detonation conditions as they are used to transfer detonation into the main explosive of interest. However, depending on the experimental configuration, the miniature detonator-booster system's small size implies a highly divergent, transient region of detonation physics, which we call *microdetonics*. Microdetonics refers to the combination of behaviors dominated by transient effects in detonation, such as detonation acceleration and spreading, or curvature effects associated with initiation by small sources. This paper is an attempt to determine whether these dynamics manifest themselves in miniature detonation systems with highly localized, short-duration, shock pulse outputs. Thus we will focus our efforts on the behavior of the second element in the explosive train commonly referred to as the booster.

EXPERIMENTS

To determine the presence of transient effects in the output of charges typically used in detonators or boosters, we consider a few different experimental configurations. For most of our experiments we use a typical detonator arrangement with an electrically driven exploding foil initiator (EFI) detonating a pellet of pentaerythritol tetranitrate (PETN). The PETN output drives detonation into a booster pellet of PBXN-5, which is a commonly used pressed explosive booster composed of 95% HMX with 5% Viton A binder. The PBXN-5 in all of our tests was pressed to a density of 1.77 g/cm^3. Since most real initiation systems are sealed in cups, we observe the outputs through a barrier material. Although interaction of this barrier with the detonation front complicates the shock waveform, we feel that it represents a real-world constraint in designing or analyzing explosive initiation systems.

To quantify the performance of these miniature detonator-booster systems, we used a VISAR system to measure the particle velocity history of an experiment driven into a coated lithium fluoride (LiF) window. The impedance of the Al barrier and the LiF crystal are closely match to minimize any reflections at this interface. The VISAR system for the experiments had a dual-leg arrangement utilizing two interferometer cavities with different fringe constants. This arrangement enables unambiguous recognition of and correction for missed fringes in the data trace, and thus provides more accurate data.

In order to determine deviations from ideal behavior, we simulated the experimental arrangements with the CTH hydrocode.[7] The one-dimensional model we used applies a constant energy release rate across the detonation front traveling at full detonation velocity to achieve proper C-J pressure. Thus, it represents an ideal steady-state explosive behavior and serves as a basis for comparison for the measured VISAR records. Any observed deviations from the predicted CTH models should indicate the presence of transient or nonideal behavior.

Case 1: Aluminum Barrier

The first experimental setup consisted of an exploding foil initiator detonating a 4-mm-diameter x 2-mm-long PETN pellet in contact with a 5-mm-diameter PBXN-5 booster pellet. A 0.254-mm-thick aluminum barrier plate between the PBXN-5 and the LiF crystal simulated the booster cup. VISAR measurements were then made of the particle velocity at the aluminum-LiF interface.

Figure 1 shows the measured particle velocities in the LiF for two lengths of PBXN-5 pellets, 6mm and 2.2 mm, as well as the CTH simulations for both lengths. There is a large discrepancy between the calculated and measured velocity histories for the 2.2-mm-long pellet. The lower peak amplitude

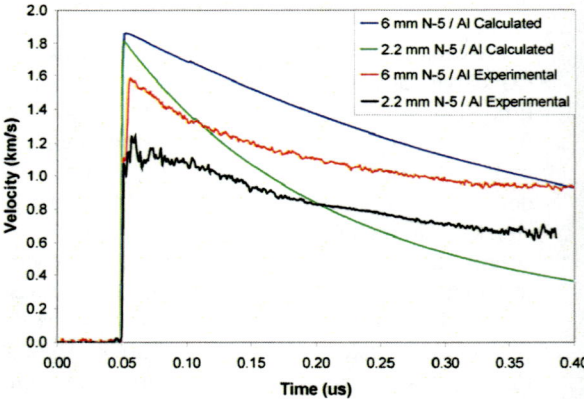

FIGURE 1. Particle velocity for the 6-mm and 2.2-mm long PBXN-5 pellets with an aluminum barrier. CTH calculations representing an ideal explosive behavior are shown for comparison.

in the measured waveform is approximately half of the C-J pressure. The shapes of the waveforms behind the front are also considerably different. The measured waveform displays a relatively flat segment just behind the front, and its subsequent decline in velocity is slower than that in the calculation. We interpret this to indicate the presence of an extended reaction zone that extends through the supersonic and the subsonic portions of the flow behind the front.

The measured velocity history for the 6-mm pellet is higher than that for the 2.2-mm pellet, but it is still lower than the calculated curve for 6 mm. Also, the slope of the wave immediately behind the front more closely resembles the CTH model. This indicates a buildup of detonation has occurred in the interval between the 2.2 and 6 mm lengths, but the detonation front has still not completely attained a full C-J detonation.

Case 2: Initiation Methods

In case 1 the initiation of the booster was accomplished by placing the output of a detonator pellet in contact with the PBXN-5 booster. To determine what effect the form of initiation has on the booster performance, we conducted tests with a configuration designed to initiate the booster via a flying plate that is believed to overdrive detonation. A PETN pellet, initiated using an EFI, propelled a 0.127-mm-thick, 2-mm diameter stainless-steel flyer across a 0.91-mm gap. The booster pellet for this case was a 2.2-mm PBXN-5 similar to that used in case 1. The output of the PBXN-5 is in direct contact with the LiF crystal, i.e. there is no barrier for this case.

Figure 2 shows the output from this experiment along with the CTH estimate, and the experimental results are also shown again for the 2.2-mm-long pellet initiated by direct contact with a similar PETN detonator pellet. The particle velocity data from the flyer-initiated PBXN-5 pellet is more representative of a C-J detonation. The first peak jumps to a velocity of 1810 m/s corresponding to a measured pressure of 36 GPa, which agrees with the accepted C-J value for PBXN-5.[8] The steep velocity drop behind the front resembles a Taylor decay following a more ideal detonation. The differences between these two experimental measurements have been seen in repeated tests and must be the result of the difference in initiation in the two cases.

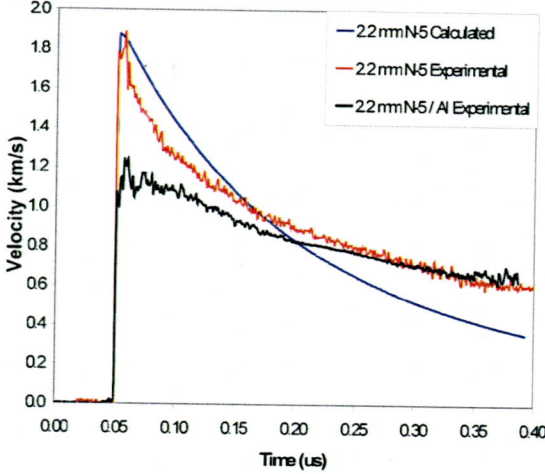

FIGURE 2. Particle velocity for a 2.2-mm long N-5 pellets at the LiF interface. A CTH calculation without the aluminum plate is shown for comparison.

DISCUSSION

The experimental results from the 6-mm and 2.2-mm boosters shown in Figure 1 reveal very different performances for the two different column lengths of explosive. The fact that the peak of the longer booster pellet more closely approaches the CTH estimate is not surprising since the longer column of explosive may be expected to produce a more "ideal" detonation. At first glance one might conclude that the 2.2-mm pellet is still running up to detonation as is observed in a wedge test. While the VISAR data is not particularly clean, it appears the energy is being released over a period of at least 50 ns; implying a very extended reaction region that is much longer than the 6-mm PBXN-5 pellet.

The presence of this extended reaction region indicates that DSD theory may not be well suited to describe these waves since it may violate one of the fundamental assumptions of the DSD approach, i.e. the reaction zone length is smaller than the radius of curvature.[1,2] Perhaps these miniature systems will have to be modeled using direct numerical simulation explicitly addressing the chemical reactions and their associated energy release. This may provide insight to clarify whether this type of waveform represents a feeble detonation or a re-initiation of detonation following detonation failure.

The 6-mm PBXN-5 pellet has a more classical wave shape with a negative slope immediately behind the initial peak indicative of a detonating explosive. However, the peak amplitude is well below the C-J level implying that even at this longer column length the PBXN-5 may be undergoing a buildup *of* detonation. This behavior is evidence for the presence of a transient (dD_n/dt) term.[3,4] Further experiments with PBXN-5 columns of even longer lengths leading up to a full C-J particle velocity are needed to confirm this observation.

Comparing the 2.2-mm-long booster performance in Figures 2, we again see dramatically different behaviors. The peak pressure of the experimental arrangement for case 1 is 40% lower than that of case 2. Given the slopes behind the shock fronts, this disparity cannot be attributed to differences in attenuation losses in the barrier. We attribute the disparity to the difference in initiation methods. Late energy release from the case 1 initiation scheme does not appear to be present in the data from case 2. This not only indicates that the initiation mechanism matters, but appears to be evidence of a transient state in the detonation of case 1, i.e. a dD_n/dt behavior as described in the DSD theory.[3,4]

CONCLUSIONS

While the results shown here are preliminary, they indicate a few trends that are significant considerations for explosive train designers. For short explosive booster columns, the initiation method affects the explosive output, as does the length of the booster column. We have presented evidence indicating that there is delayed energy release in short columns of PBXN-5 that corresponds to nonideal detonation. The data implies a transient detonation behavior building toward C-J pressure with increased column length. While further experiments are needed in this area, we believe that these results indicate that microdetonics, i.e. transient effects, occurs in detonator-booster systems.

ACKNOWLEDGEMENTS

The authors wish to thank Willard Hemsing and Michael Shinas for their assistance with the VISAR system and data reduction.

REFERENCES

1. J.B. Bdzil, W. Fickett, & D.S. Stewart, Ninth Symposium (Intl.) on Detonation, pages 730-742, OCNR 113291-7 (1989).
2. J.B. Bdzil & D.S. Stewart, Phys. Fluids A **1**, 1261 (1989).
3. J. Yao & D.S. Stewart, J. Fluid Mech. **309**, 225 (1996).
4. T.D. Aslam & D.S. Stewart, Combst. Theory Modeling **3**, 77 (1999).
5. L.G. Hill, J.B. Bdzil, & T.D. Aslam, Eleventh Symposium (Int.) on Detonation, pages 1029-1037, Office of Naval Research, ONR 33300-5 (1998).
6. L.M. Hull, Tenth Symposium (Intl.) on Detonation, pages 11-18, ONR 33395-12 (1993).
7. J.M. McGlaun, S.L. Thompson, L.N. Kmetyk, & M.G. Elrick, Intl. J. Impact Eng. **10**, 351, (1990).
8. T.N. Hall & J.R. Holden, Navy Explosives Handbook, NSWC MP 88-116, Naval Surface Warfare Center, Dahlgren, VA.

AN INVESTIGATION INTO THE INITIATION OF HEXANITROSTILBENE BY LASER-DRIVEN FLYER PLATES

M. W. Greenaway[a], M. J. Gifford[a], W. G. Proud[a], J. E. Field[a], S. G. Goveas[b]

[a] Physics and Chemistry of Solids, Cavendish Laboratory, Madingley Road, Cambridge, CB3 0HE, United Kingdom
[b] AWE, Aldermaston, Reading, RG7 4PR, United Kingdom

Abstract. An investigation into the shock sensitivity of hexanitrostilbene (HNS) has been carried out. A Q-switched Nd:YAG laser was used to launch miniature flyer plates from substrate-backed aluminium films. The impact produces a shock with duration of the order of 1 ns and pressure of the order of 10 GPa. The explosive samples were pressed into PMMA cylinders to 65-78 % theoretical maximum density. The threshold laser pulse energy required to produce a flyer with sufficient velocity to cause detonation was found. A high-speed camera was used to record the entire event. Initial curvature of the streak record, for impacts just below the detonation threshold, showed that reaction started inside the column. This feature was not seen in a previous study[1]. It was found that conventional HNS, with a mean particle size of approximately 25 μm, could not be detonated while fine grained HNS (sub-micron particle size) would detonate.

INTRODUCTION

A technique for imparting very short duration, high pressure shock waves into energetic materials has been developed.

The system uses a high power laser to drive minature plates of aluminium at velocities up to 8 km/s[2]. The primary motivation for the development of this technique has been the initiation of explosives. Other authors have reported similar systems for high strain rate material testing and for the ground-based simulation of micrometeorite impact[3].

This study was a preliminary investigation to determine whether this system could be used to initiate hexanitrostilbene (HNS). The main aim was to determine the influential parameters and compare with the findings from a study on the explosive pentaerythritol tetranitrate (PETN)[1]. For these reasons no statistical methods, for finding thresholds and suchlike, were employed.

The desire to enhance safety provides the main motivation behind the development of techniques such as the laser-driven flyer plate. More sophisticated detonators, like this one, can use more insensitive explosives while retaining similar output characteristics.

The shock-to-detonation transition (SDT)

The laser-driven flyer plate is really an extension of the exploding foil initiator (EFI) and the slapper detonator. Here, a metallic plasma is formed by depositing a large current through a small copper bridge. The exploding bridge fires a kapton flyer in a similar manner to the laser-driven flyer. A barrel is often used to give directionality to the flyer. The attraction of shock initiation is its promptness and reproducibility. The flyer is a means for imparting a shock wave into the explosive. The otherwise decaying shock is supported by the chemical reaction of the detonation. In this manner, a detonation could be defined as a reaction supported shock wave. Thus reaction must start within the temporal shock width in order for the support to remain. This is where a large dependence on the grain size of the energetic is introduced.

The pressure of the shock wave must also be sufficient to start the reaction. The pressure threshold is dependant on the production of ignition hot spots within the material. The key mechanisms are granular friction, shear deformations and the compression of gas spaces. These hot spot mechanisms must occur during the shock compression in order for the reaction to be supported.

Since the technique was first reported[4], much interest has surrounded its development including research in this laboratory[5-7] and elsewhere[8-10].

EXPERIMENTAL

A Nd:YAG laser operating at the fundamental wavelength is used to deliver horizontally polarised pulses of optical radiation up to 1 J in approximately 9 ns (FWHM). Energy modulation is achieved by incorporating a variable angle half-wave plate and polarising beamsplitter.

A plano-convex lens is used to focus this optical energy onto a substrate-backed aluminium film. This energy is readily absorbed in the metal, which generates a hot plasma at the substrate-film interface. Rapid expansion of this plasma blasts off the remaining depth of film as a miniature flyer plate. This mechanism is illustrated in Fig. 1. For the purpose of this research, the flyers were 1 mm in diameter and approximately 5 μm thick. The key properties of these flyers have been measured and the results are reported elsewhere[2].

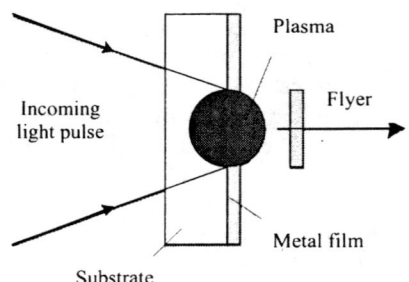

FIGURE 1. A schematic of the flyer launch mechanism.

A preliminary investigation into the sensitivity of PETN to this type initiator has been carried out[1]. This study showed the importance of grain size on the initiation threshold by comparing a fine grain and coarse grain variant.

Hexanitrostilbene (HNS) was selected for this study with a view to enhancing our understanding of the initiation mechanisms involved. Its importance as a secondary explosive and availability as a fine and coarse grain material made it an ideal choice.

Three forms of the energetic material were tested with this system. HNS IV is the fine grain form of the material, supplied in its pure form and with pressing additives. The pressing additives were zinc stearate and graphite and contributed approximately 1% of the total mass. HNS II is the coarse grain version and has a grain size of the order of 25 μm.

The charges were pressed into PMMA confinements 5 mm deep and 5 mm diameter. The columns were polished with 2500 grade SiC paper to provide a consistent surface finish. The charges were backed with a brass witness plate and held 100 μm from the metal film on the impact (front) side. A schematic of this set-up is given in Fig. 2.

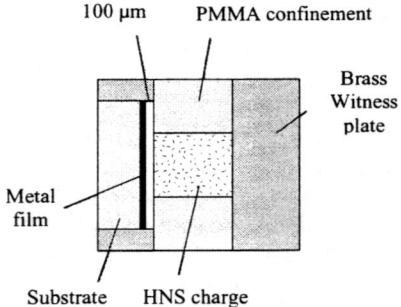

FIGURE 2. A schematic of the substrate-charge arrangement.

The initiation event was recorded by a Hadland Imacon 790 electronic image converter camera.

RESULTS

The aim of these experiments was to determine the minimum amount of optical energy required to produce a flyer of sufficient velocity to induce detonation.

The results are given in Fig. 3 where a solid marker indicates a go (i.e. successful initiation) and a hollow marker a no go.

Pure HNS IV at 70% theoretical maximum density (TMD) showed the greatest sensitivity, in this study, with a threshold in the region of 250 mJ. A small drop in TMD to 65% yields a substantial rise in the threshold, as can be seen in comparing the first and second columns of data. HNS IV with the additives and pressed to approximately 65% TMD shows a threshold of about 350 mJ. The contribution of the additives on the sensitivity cannot be extracted from this data but it does not appear to be hugely influential.

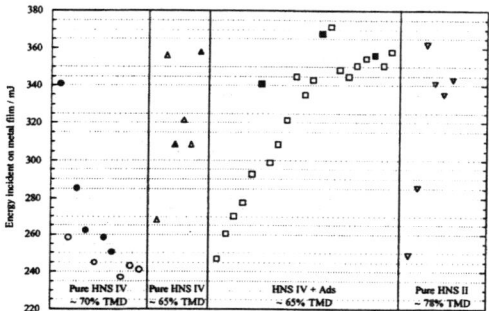

FIGURE 3. A scatter plot of the results for four different variations of HNS.

Perhaps the most interesting result comes from the photographic record. Non-linear streak records occurred for low density samples. An example is given in Fig. 4. This hooking nature indicates that detonation is breaking out inside the column rather than on the surface at the impact site. This was not seen in the previous study with PETN. Higher density charges did not show this effect.

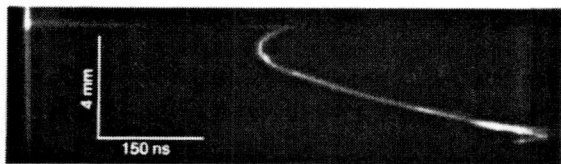

FIGURE 4. Streak record of the detonation of HNS IV.

Using a simple curve fitting macro, the depth at which reaction broke out is found. This macro assumes reaction breaks out at a single point and propagates outward as a spherical wave of constant velocity.

Using this model, the detonation in Fig. 4 is found to have broken out 750 µm inside the column. This is in good agreement with the impact site found on charges that did not react just below the threshold. A clear correlation between the depth at which detonation breaks out and the charge density is evident. Higher density charges showed less hooking, indicating that reaction has begun on or just below the surface. In some cases, shots just below the threshold showed some evidence of limited reaction.

CONCLUSIONS

This laser-driven flyer plate system was found to be capable of initiating fine grain HNS. A key aim of this study was to compare HNS with another important secondary explosive, PETN. HNS was found to be more insensitive to this type of initiator. In both cases, the sensitivity of the material was heavily dependent on particle size. The coarse grain HNS II could not be initiated, just like the PETN equivalent.

The initiation process for this type of loading has to be by a hot spot mechanism which can respond extremely rapidly. In general, this suggests that a mechanism associated with a rapidly collapsed gas space is more likely than a sheer or friction process.

The large dependence on particle size is explained by the size of the critical hot spots. The shock width is of the order of a few microns, this is comparable to the grain size (and hence gas spaces) of the fine materials but is much shorter than the grain size of the coarse grain materials. Thus, in the case of the latter, the gas spaces cannot be entirely collapsed during the short duration of the shock. This is illustrated in Fig. 5. In addition, we have better mixing of the hot spots and explosive particles with a fine grain material. These differences offer an explanation as to why the fine grain materials are more sensitive to these very short shocks.

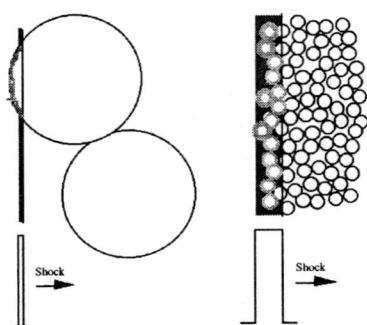

FIGURE 5. The shock induced reaction in a relatively large grain explosive (left) and a relatively small grain explosive (right).

A hooking of the streak record indicated that reaction breaks out inside the column. This was not seen with PETN. The degree of hooking being heavily dependant on the density of the charge. Shots carried out at high density showed much less prominent hooking.

Improvements to the flyer generation process are under investigation. The use of fibres to deliver the optical energy to the launch substrate is one such idea currently under research in this laboratory. The attraction of this technique is the improved spatial profile that can be achieved using a multimode fibre. The main difficulty is reliably carrying such large power densities down the fibre. The preparation of the fibre and the coupling method and procedure are the key issues. Such a system has been successfully developed in this laboratory[11].

ACKNOWLEDGMENTS

The equipment was purchased on grants from AWE Aldermaston and the Engineering and Physical Sciences Research Council (EPSRC).

Thanks also go to Dr. S. Watson, Dr. R. C. Drake and Dr. J. Andrew for some useful discussions.

REFERENCES

1. Watson, S., Gifford, M. J. and Field, J. E., *J. Appl. Phys.* **88**, 65-69 (2000).
2. Watson, S., *The production and study of laser-driven flyer plates*, PhD Thesis, University of Cambridge, 1998.
3. Tighe, A., *Ground based simulation of orbital debris using laser driven flyer plates*, PhD Thesis, University of Southampton, 2000.
4. Sheffield, S. A. and Fisk, G. A., "Particle velocity measurements in laser irradiated foils using ORVIS", *Shock Waves in Condensed Matter - 1983*, edited by J. R. Asay, R. A. Graham and G. K. Straub, pp. 243-246, North-Holland, Amsterdam, 1984.
5. Dickson, P. E. and Field, J. E., *Laser initiation of fast reactions*, 1994, Cavendish Laboratory internal report (unpublished).
6. Watson, S. and Field, J. E., *J. Phys. D: Appl. Phys.* **33**, 170-174 (2000).
7. Watson, S. and Field, J. E., *J. Appl. Phys.* **88**, 3859-3864 (2000).
8. Paisley, D. L., "Laser-driven miniature flyer plates for shock initiation of secondary explosives", *Shock Compression of Condensed Matter - 1989*, edited by S. C. Schmidt, J. N. Johnson and L. W. Davidson, pp. 733-736, Elsevier, Amsterdam, 1990.
9. Trott, W. M. and Meeks, K. D., *J. Appl. Phys.* **67**, 3297-3301 (1990).
10. Frank, A. M. and Trott, W. M., "Investigation of thin laser-driven flyer plates using streak imaging and stop motion microphotography", *Shock Compression of Condensed Matter - 1995*, edited by S. C. Schmidt and W. C. Tao, pp. 1209-1212, American Institute of Physics, Woodbury, New York, 1995.
11. Greenaway, M. W., Proud, W. G., Field, J. E., Goveas, S. G. and Drake, R. C., "The high power transmission characteristics of fused-silica optical fibres" in *Laser-Induced Damage in Optical Materials*, edited by G. J. Exarhos, A. H. Guenther, M. R. Kozlowski, K. L. Lewis and M. J. Soileau, SPIE 4347, Boulder CO, 2000, pp. 599-607.

(c) British Crown Copyright 2001 /MOD

This document is of United Kingdom origin and contains proprietary information which is the property of the Secretary of State for Defence. It is furnished in confidence and may not be copied, used or disclosed in whole or in part without prior written consent of the Director Commercial 2, Defence Procurement Agency, Ash 2b, MailPoint 88, Ministry of Defence, Abbey Wood, Bristol, BS34 8JH, United Kingdom.

pressure in the silicone fluid (Dow 200 silicone oil, 20 centistokes viscosity [17]). The calibrated PCB Piezotronics model 136A [18] quartz reference transducer and carbon resistor pressure gauge sample were placed equal distances from the center of the fluid cell. A PCB model 443A101/443A102 dual mode amplifier was used to amplify the calibrated gauge signal and a Tektronix TDS 784D oscilloscope was used to measure the carbon resistor gauge output during the experiment. In the analysis, the peak value from the calibrated pressure gauge was correlated to the peak resistance change from the carbon resistor gauge. Both signal peaks correlated well in time.

observed that there was little difference in relative resistance change between individual resistors. Note that small jogs are seen at 0.5 GPa increments where the increase in gas pressure was held constant to enable equilibration of the carbon resistor gauge. The split-Hopkinson bar data is shown as open squares and the drop tower data is shown as open circles. It should be noted that with the drop tower data, some of the points were obtained from multiple drops on the same resistor with no observable deviation. This is an indication of the rugged nature of the carbon resistor gauge.

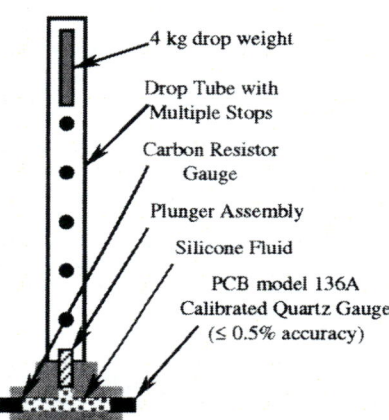

FIGURE 3. Schematic showing the drop weight carbon resistor gauge calibration set-up.

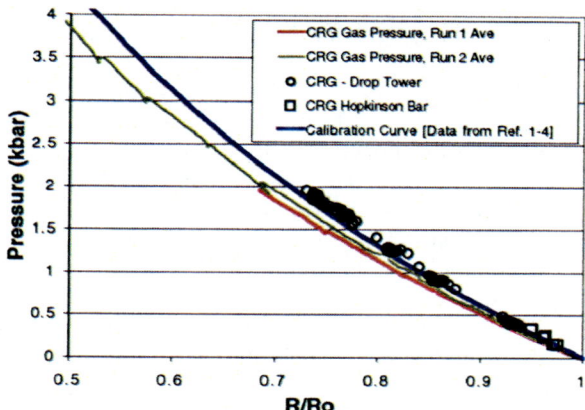

FIGURE 4. Summary plot of the carbon resistor gauge calibration compared with the calibration curve fit to previously published data.

Investigation of Heating at Atmospheric Pressure

The effect of heating at atmospheric pressure on the carbon resistor pressure gauge was investigated by placing an array of resistors in an oven and heating them in air to 160°C. During the heating the carbon resistor gauge resistance was continuously scanned on a Keithley digital ohmmeter and output to a computer that recorded the resistance value of the gauges and the thermocouple voltage.

DISCUSSION

A summary plot of the results compared with the calibration curve fit to previously published data is given in Figure 4. Two runs of the gas pressure chamber data to 0.4 and 0.2 GPa are shown as the average value from the array of 5 resistors. It was

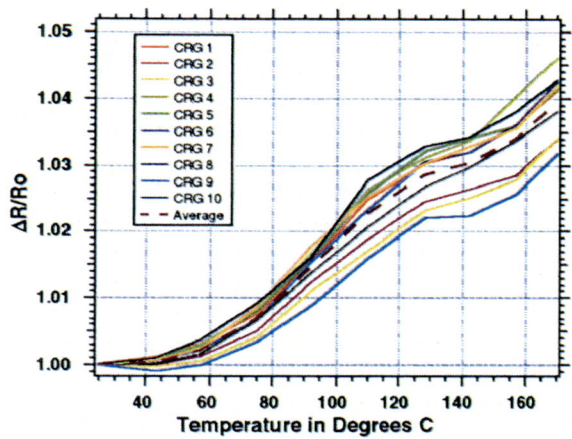

FIGURE 5. Change in resistance as a function of temperature (at atmospheric pressure) for 10 carbon resistor gauges. Note the dashed line is the average value.

In this summary plot (Figure 4), the gas pressure chamber data shown falls slightly below the dynamic calibration curve, which is expected due to the static nature of the loading. The current dynamic data calibration points follow the previously fitted calibration curve reasonably well.

Figure 5 displays the relationship between temperature and resistance change of the carbon resistor gauge. From this plot it can be seen that there is only a 4% change in resistance on average, and a 1.4% difference among groups when gauges are heated from ambient to 160°C.

SUMMARY AND FUTURE WORK

Calibration experiments were performed at low stresses (<0.4 GPa) to compare results with a calibration curve fit to previously published calibration data. The new experiments used: a split-Hopkinson pressure bar, a drop tower apparatus, and gas pressure chamber. The gas pressure chamber data obtained falls slightly below the dynamic curve as might be expected due to differences in static and dynamic loading. The calibration experiments using dynamic loading agree with the calibration curve fit to previously published data. Resistance of the carbon resistor gauge was shown to only vary 4%, on average, when heated from ambient to 160°C at atmospheric pressure.

Gas gun experiments are in progress to extend the scope of investigation from the current low-pressure region to ~1.6 GPa. Experiments are also being conducted at elevated temperatures (~80°C).

ACKNOWLEDGEMENTS

Jerry Dow is thanked for obtaining the funding for the research. Assistance by Jeff Wardell on the static gas pressure chamber experiments is greatly appreciated. Douglas Tasker (LANL) is acknowledged for sharing information on his use of the carbon resistor gauge and his design of a constant current power supply. This work was performed under the auspices of the United States Department of Energy by the Lawrence Livermore National Laboratory under Contract No. W-7405-ENG-48.

REFERENCES

1. R.W. Watson, "Gauge for Determining Shock Pressures," Rev. Sci. Instrum. 38, 978 (1967).
2. J.A. Charest and M.D. Lilly, "A Rugged Disposable Pressure Transducer," 45th ARA Meeting, Huntsville, Alabama, Oct. 10-14, 1994.
3. Ginsberg, Michael J., and Asay, Blaine W., "Commercial Carbon Composition Resistors as Dynamic Stress Gauges in Difficult Environments," Rev. Sci. Instrum. 62 (9): 2218-2227 (1991).
4. F. Scholz, "Uber die Druckbeeinflussung von Sprengladungen durch die Schwaden fruher detonierender Nachbarladungen beim Sprengen mit Millsekundenzudern im Karbongestein," Ber. Versuchs mbH, Heft 16, Versuchsgruben Gessellschaft mbH, Dortmund, FRG (1981).
5. J. Stankewicz and R.L. White, Rev. Sci. Instrum. 42, 1067 (1971).
6. Wilson, W. H., "Experimental Study of Reaction and Stress Growth in Projectile-Impacted Explosives," Shock Compression of Condensed Matter-1991, eds. Schmidt, Dick, Forbes, and Tasker, Elsevier Science Publishers, pp. 671-674 (1992).
7. W.M. Wilson, D.C. Holloway, and G. Bjarnholt, Techniques and Theory of Stress Wave Measurements for Shock Wave Applications (American Society of Mechanical Engineers, New York, 1987), pp. 97-108.
8. Austing, J. L., Tulis, A. J., Hrdina, D. J., and Baker, D. E., "Carbon Resistor Gauges for Measuring Shock and Detonation Pressures I. Principles of Functioning and Calibration", Propellants, Explosives, Pyrotechnics 16, pp. 205-215 (1991).
9. A.J. Tulis, J.L. Austing, D.E. Baker, and D.J. Hrdina, Propellants, Explosives, Pyrotechnics, 16, 216-220 (1991).
10. J.L. Austing, A.J. Tulis, R.P. Joyce, C.E. Foxx, D.J. Hrdina, and T.J. Bajzek, Propellants, Explosives, Pyrotechnics, 20, 159-169 (1995).
11. Frank Garcia, Jerry W. Forbes, Craig M. Tarver, Paul A. Urtiew, Daniel W. Greenwood, and Kevin S. Vandersall, "Pressure Wave Measurements from Thermal Cook-off of an HMX Based High Explosive PBX 9501," 12th APS Conference on Shock Compression of Condensed Matter, Atlanta, GA, June 245-29, 2001, this proceedings.
12. A.M. Niles, F. Garcia, D.W. Greenwood, J.W. Forbes, C.M. Tarver, S.K. Chidester, R.G. Garza, and L.L. Switzer, "Measurement of Low Level Explosives Reaction in Gauged Multi-dimensional Steven Impact Tests," 12trh APS Conference on Shock Compression of Condensed Matter, Atlanta, GA, June 245-29, 2001, this proceedings.
13. J.L. Maienschein and J.B. Chandler, "Burn Rates of Pristine and Degraded Explosives at Elevated Pressures and Temperatures," 11th International Symposium on Detonation, p. 872.
14. Kistler Instrument Corporation, 75 John Glenn Drive, Amherst, NY 14228-2171.
15. Follansbee, P.S., ASM Metals Handbook, Volume 8 Mechanical Testing, American Society of Metals, 1992, pp. 190-207.
16. R.M. Davies, "A critical Study of the Hopkinson Pressure Bar," *Phil. Trans. A*, vol. 240, 1948, pp. 375.
17. Dow Corning Corporation, Midland, Michigan 48686-0994.
18. PCB Piezotronics, Inc., Buffalo, NY, 14225.

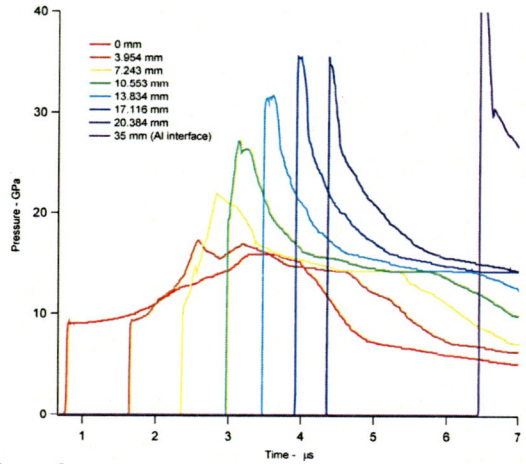

Figure 3. Calculated pressure histories for ambient UF-TATB impacted at 1.688 mm/μs by an aluminum flyer

250°C UF-TATB RESULTS

Figures 4 – 6 show the pressure histories measured in 250°C UF-TATB impacted by aluminum flyer plates at 1.1, 1.343, and 1.482 mm/μs, respectively. In Fig. 4 for an initial pressure of 4.5 GPa, gauge records at 0, 5, 7, and 9 mm show buildup of reaction behind the leading shock but no transition to detonation. In

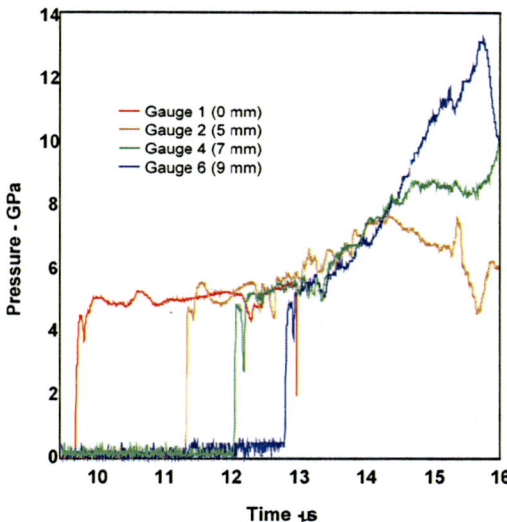

Figure 4. Pressure histories for 250°C UF-TATB impacted by an aluminum flyer at 1.1 mm/μs

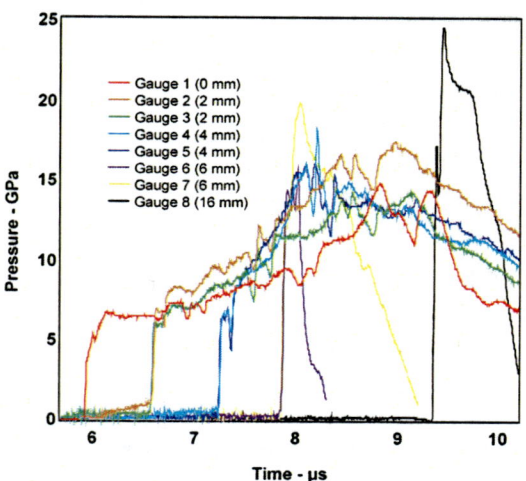

Figure 5. Pressure histories for 250°C UF-TATB impacted by an aluminum flyer at 1.343 mm/μs

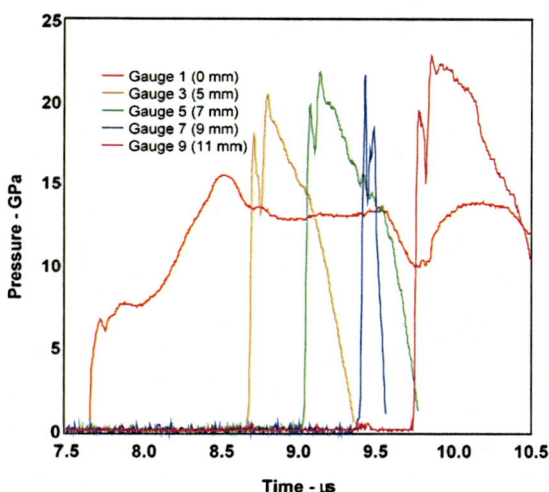

Figure 6. Pressure histories for 250°C UF-TATB impacted by an aluminum flyer at 1.482 mm/μs

Fig. 5 for a pressure of 6 GPa, gauge records at 0, 2, 4 and 6 mm show rapid buildup and the last gauge at 16 mm bordering the aluminum back plate is close to detonation. In Fig. 6, transition to detonation occurs in less than 5 mm, and for the highest velocity impact 1.711 mm/μs, detonation transition occurs in less than 2 mm.

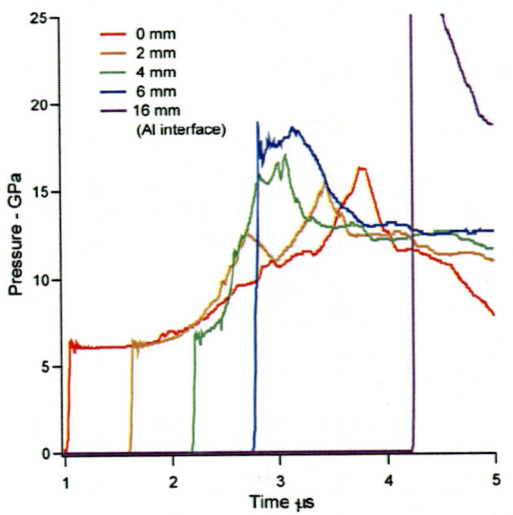

Figure 7. Calculated pressure histories for 250 °C UF-TATB impacted by an aluminum flyer at 1.343mm/μs

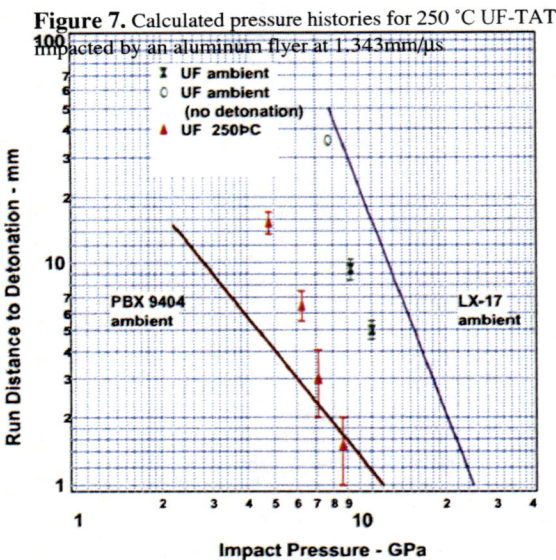

Figure 8. Pop Plots for ambient and 250°C UF-TATB compared to other TATB and HMX-based explosives

Figure 7 shows the calculated pressure histories for the experiment shown in Fig. 5. For 250°C UF-TATB, the maximum fraction ignited Figmax and the G_i coefficient in Eq. (2) are twice those for UF-TATB at ambient temperature. Figure 8 contains the Pop Plot data for ambient and 250°C UF-TATB compared to LX-17 and PBX 9404, an HMX-based explosive, at ambient temperature.

CONCLUSIONS

The shock sensitivity of unconfined charges of UF-TATB is reported at ambient temperature and at 250°C. UF-TATB has steep slopes in its Pop Plot results, which are similar to other TATB-based explosives. At high pressures, the shock sensitivity of 250°C UF-TATB is similar to that of PBX 9404, the most sensitive HMX-based plastic bonded explosive. Ignition and Growth reactive flow models for ambient and 250°C UF-TATB yield good agreement with the embedded gauge data and thus can be used to simulate other shock initiation scenarios that can not be tested.

REFERENCES

1. Urtiew, P.A., Tarver, C.M., Maienschein, J.L. and Tao, W.C., Combustion and Flame **105**, 43-53 (1996).
2. Urtiew, P.A., Cook, T.M., Maienschein, J.L., and Tarver, C.M., Tenth International Detonation Symposium, ONR 33395-12, Boston, MA, 1993, pp. 139-147.
3. Dallman, J.C. and Wackerle, J., Tenth International Detonation Symposium, ONR 33395-12, Boston, MA, 1993, pp.130-138.
4. Urtiew, P.A., Tarver, C.M., Forbes, J.W., and Garcia, F., in Shock Compression in Condensed Matter-1997, Schmidt, S.C., Dandekar, D.P., and Forbes, J.W., eds., AIP Press, New York, 1997, pp. 727-730.
5. Forbes, J. W., Tarver, C. M., Urtiew, P. A., and Garcia, F., Eleventh International Detonation Symposium, Office of Naval Research ONR 33300-5, Snowmass, CO, 1998, pp. 145-152.
6. Tarver, C. M., Forbes, J. W., Urtiew, P.A., and Garcia, F. in Shock Compression in Condensed Matter-1999, Furnish, M. D., Chhabildas, L. C., and Hixson, R. S., eds., AIP Press, New York, 2000, pp. 891-894.
7. Urtiew, P. A., Forbes, J. W., Tarver, C,. M., and Garcia, F., in Shock Compression in Condensed Matter-1999, Furnish, M. D., Chhabildas, L. C., and Hixson, R. S., eds., AIP Press, New York, 2000, pp. 1019-1022.

MANGANIN GAUGE AND REACTIVE FLOW MODELING STUDY OF THE SHOCK INITIATION OF PBX 9501*

C. M. Tarver, J. W. Forbes, F. Garcia and P. A. Urtiew

*Lawrence Livermore National Laboratory,
P.O. Box 808, L-282, Livermore, CA 94551*

A series of 101mm diameter gas gun experiments was fired using manganin pressure gauges embedded in the HMX-based explosive PBX 9501 at initial temperatures of 20°C and 50°C. Flyer plate impact velocities were chosen to produce impact pressure levels in PBX 9501 at which the growth of explosive reaction preceding detonation was measured on most of the gauges and detonation pressure profiles were recorded on some of the gauges placed deepest into the explosive targets. All measured pressure histories for initial temperatures of 25°C and 50°C were essentially identical. Measured run distances to detonation at three input shock pressures agreed with previous results. An existing Ignition and Growth reactive flow computer model for shock initiation and detonation of PBX 9501, which was developed based on LANL embedded particle velocity gauge data, was tested on these pressure gauge results. The agreement was excellent, indicating that the embedded pressure and particle velocity gauge techniques yielded consistent results.

INTRODUCTION

The relative safety of high energy materials based on octahydro-1,3,5,7-tetranitro-1,3,5,7-tetrazocine (HMX) is very important. PBX 9501, which contains 95 weight % HMX, 2.5 weight % estane binder, and 2.5 weight % BDNPA/F, is a widely used HMX-based plastic bonded explosive. Its shock sensitivity has previously been studied using embedded particle velocity gauges (1,2) and VISAR at low input shock pressures (3,4). In this paper, the shock sensitivity of PBX 9501 at 25°C and 50°C was measured using embedded manganin pressure gauges to determine whether particle velocity measurement techniques agreed with pressure measurement techniques. The experimental records were compared through the use of the Ignition and Growth reactive flow model for PBX 9501, which had been previously normalized to particle velocity gauge data in the same input shock pressure regime (3). If this PBX 9501 Ignition and Growth model can calculate manganin pressure gauge records accurately with no adjustments, then the two experimental techniques are producing equivlanet shock initiation data. Thus they can be used interchangeably or in combination.

EXPERIMENTAL

The experimental geometry for the PBX 9501 embedded gauge experiments is identical to those in previous studies (5-8). A 100 mm diameter, 12.5 mm thick aluminum flyer plate impacts a target consisting of: a 90 mm diameter, 6 mm thick aluminum plate; a 90 mm diameter, 20 mm thick PBX 9501 charge; and a 90 mm diameter, 6 mm thick aluminum back plate. In the heated experiments, the heaters were placed within the aluminum plates, and the PBX 9501 was heated to approximately 50°C at a rate of 1.6°C/minute. A total of three shots were fired. One 25°C experiment was fired with an aluminum flyer velocity of 0.697 mm/µs producing a shock pressure of approximately 3.2 GPa. Two shots were fired at 50°C with aluminum flyer plate velocities of 0.649 and 0.8005 mm/µs, imparting pressures of 3.1 GPa and 4 GPa, respectively. An initial temperature of 50°C for PBX 9501 was used. It is known that the shock sensitivity of HMX-based explosives increases for temperatures exceeding 150°C (8). However, no data exists for 50°C, to which explosives may be subjected in hot climates and certain applications.

REACTIVE FLOW MODELING

The Ignition and Growth reactive flow model uses two Jones-Wilkins-Lee (JWL) equations of state, one for the unreacted explosive and another one for the reaction products, in the temperature dependent form:

$$p = A e^{-R_1 V} + B e^{-R_2 V} + \omega C_v T/V \quad (1)$$

where p is pressure in Megabars, V is relative volume, T is temperature, ω is the Gruneisen coefficient, C_v is the average heat capacity, and A, B, R_1 and R_2 are constants. The unreacted equation of state is fitted to the available shock Hugoniot data, and the product equation of state is fitted to cylinder test and other metal acceleration data. The reaction rate equation is:

$$dF/dt = I(1-F)^b (\rho/\rho_o - 1 - a)^x + G_1(1-F)^c F^d p^y$$
$$0 < F < F_{igmax} \quad\quad 0 < F < F_{G1max}$$
$$+ G_2(1-F)^e F^g p^z \quad (2)$$
$$F_{G2min} < F < 1$$

where F is the fraction reacted, t is time in μs, ρ is the current density in g/cm^3, ρ_o is the initial density, p is pressure in Mbars, and I, G_1, G_2, a, b, c, d, e, g, x, y, and z are constants. This three term reaction rate law models the three stages of reaction generally observed during shock initiation of pressed solid explosives (6). The equation of state parameters for PBX 9501, aluminum, and Teflon, and the Ignition and Growth rate law parameters for PBX 9501 are listed in Table 1. These parameters were previously normalized to several particle velocity gauge experiments fired at Los Alamos National Laboratory. Sheffield et al. (1) did a careful study of the effect of initial density on the shock sensitivity of new and aged PBX 9501. The average density of the three PBX 9501 charges used in this study was 1.838 g/cm^3. A change in temperature from 25°C to 50°C does not change the density significantly. The only change in the 50°C PBX 9501 parameters from the ambient parameters is a lower B value in the unreacted JWL equation of state so that p=0 at V=1 and T_o = 323°K.

TABLE 1. Equation of State and Reaction Rate Parameters

1. Ignition and Growth Model Parameters for PBX 9501

A. T_o=298°K; ρ_o =1.838 g/cm^3; Shear Modulus=0.0354 Mbar; Yield Strength=0.002 Mbar

Unreacted JWL	Product JWL	Reaction Rate Parameters	
A=7320 Mbar	A=16.689 Mbar	I=1.4e+11	G_2=400
B=-0.052654 Mbar	B=0.5969 Mbar	a=0.0	e=0.333
R_1=14.1	R_1=5.9	b=0.667	g=1.0
R_2=1.41	R_2=2.1	x=20.0	z=2.0
ω=0.8867	ω=0.45	G_1=130	F_{igmax} = 0.3
C_v=2.7806e-5 Mbar/°K	C_v=1.0e-5 Mbar/°K	y=2.0	F_{G1max}=0.5
	E_o=0.095 Mbar	c=0.667, d=0.277	F_{G2min}=0.5
B. T_o=50°C=323°K	B=-0.055179 Mbar (this is the only adjustment)		

2. Gruneisen Parameters for Inert Materials

$$p = \rho_o c^2 \mu [1+(1-\gamma_o/2)\mu - a/2\mu^2] / [1-(S_1-1)\mu - S_2\mu^2/(\mu+1) - S_3\mu^3/(\mu+1)^2]^2 + (\gamma_o + a\mu)E$$

where $\mu = (\rho/\rho_o) - 1$ and E is thermal energy

Inert	ρ_o(g/cm^3)	c(mm/μs)	S_1	S_2	S_3	γ_o	a
6061-T6 Al	2.703	5.24	1.4	0.0	0.0	1.97	0.48
Teflon	2.15	1.68	1.123	3.98	-5.8	0.59	0.0

TABLE 2. Experimental flyer velocities, impact pressures, and run distances to detonation

Flyer Velocity (mm/μs)	Impact Pressure (GPa)	PBX 9501 Temperature (°C)	Experimental Run to Detonation Results	
			Distance(mm)	Time(μs)
0.697	3.4	25	11	3.4
0.649	3.0	50	13	4.3
0.801	4.0	50	8	2.4

EXPERIMENTAL RESULTS

Table 2 contains the experimental flyer velocities, impact pressures, and run distances to detonation for the three PBX 9501 shots. Figure 1 shows the measured pressure histories at the six gauge locations (0,5,7,9,12, and 15 mm) in PBX 9501 at 25°C impacted by an aluminum flyer plate at 0.697 mm/μs. The gauge records and calculations show that this input shock pressure of 3.4 GPa causes detonation to occur just before the 12 mm gauge depth. Two other comparisons for 50°C PBX 9501 are shown in Figs. 2 and 3. Figure 2 contains six gauge records at 0,5,10,12,14, and 17 mm for the 0.649 mm/μs aluminum flyer impact velocity experiment. The transition to detonation occurs between 12 and 14 mm. Figure 3 contains the manganin gauge records at four depths (0,3,7, and 10 mm) for the 50°C PBX 9501 shot which had an aluminum flyer velocity of 0.801 mm/μs. The transition to detonation occurred just beyond the 7 mm deep gauge. For all three experiments, the run distances and times to detonation agree exactly with those measured by Sheffield et al. (1,2).

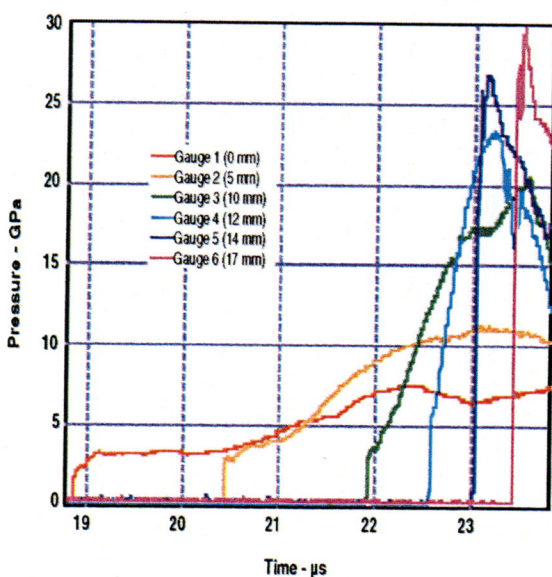

FIGURE 2. Pressure histories for 50°C PBX 9501 shock initiated by an aluminum flyer at 0.649 mm/μs

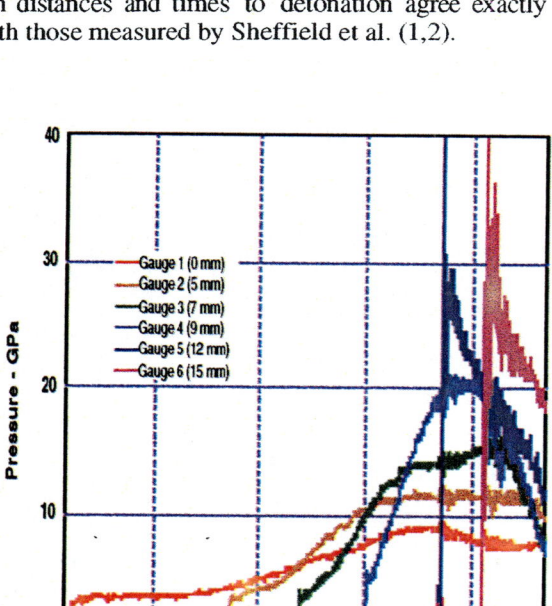

FIGURE 1. Pressure histories for 25°C PBX 9501 shock initiated by an aluminum flyer at 0.697 mm/μs

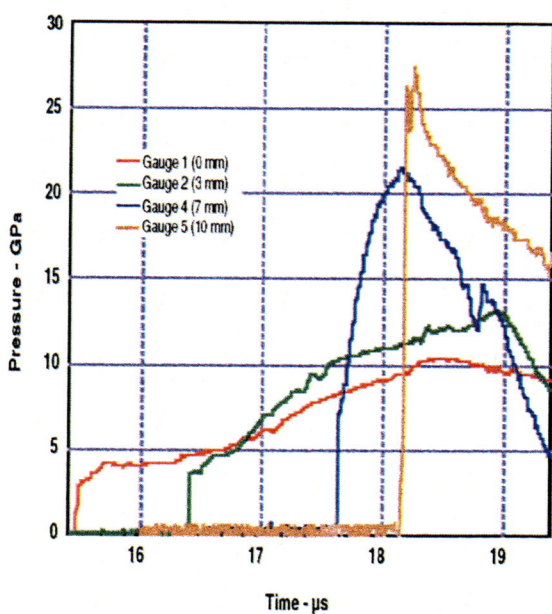

FIGURE 3. Pressure histories for 50°C PBX 9501 shock initiated by an aluminum flyer at 0.801 mm/μs

MODELING RESULTS

The Ignition and Growth reactive flow modeling results for the three PBX 9501 experiments in Figs. 1 – 3 are shown in Figs. 4 – 6, respectively. Ignition and Growth reactive flow modeling shows that manganin pressure gauges are yielding equilvalent results to particle velocity gauges.

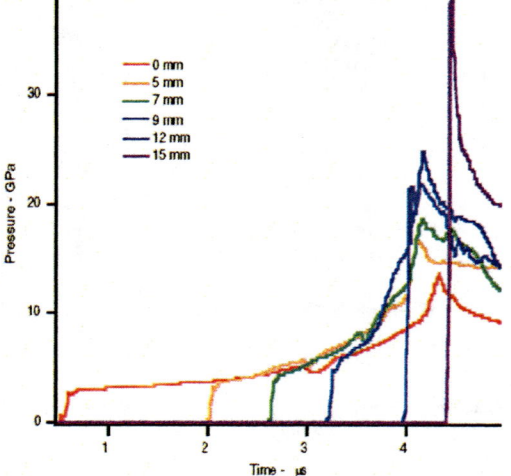

Figure 4. Calculated pressure histories for 25°C PBX 9501 impacted by an aluminum flyer at 0.697 mm/μs

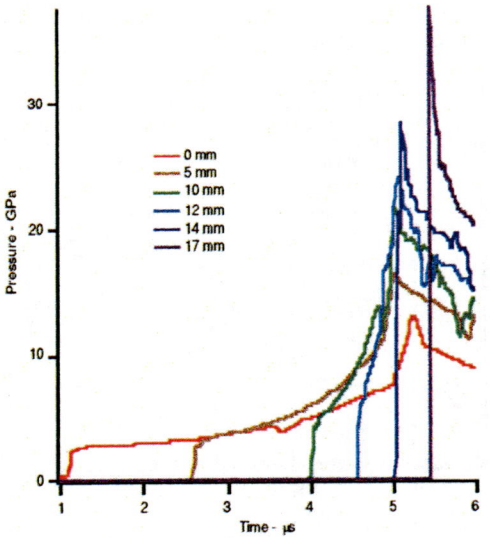

Figure 5. Calculated pressure histories for 50°C PBX 9501 impacted by an aluminum flyer at 0.649 mm/μs

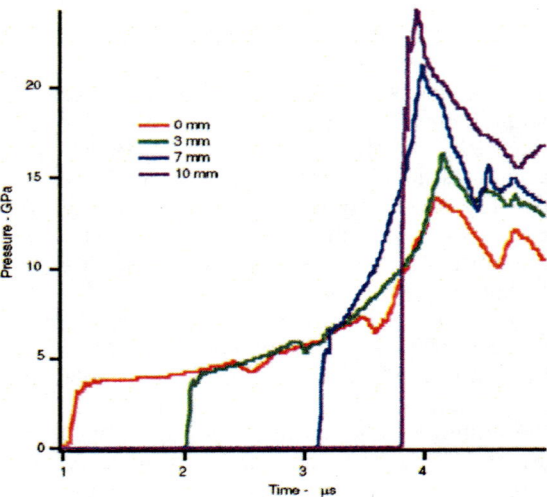

Figure 6. Calculated pressure histories for 50°C PBX 9501 impacted by an aluminum flyer at 0.801 mm/μs

ACKNOWLEDGMENTS

*This work was performed under the auspices of the U.S. Department of Energy by LLNL (contract number W-7405-ENG-48).

REFERENCES

1. Sheffield, S. A., Gustavsen, R. L., Hill, L. G., and Alcon, R. R., *Eleventh International Detonation Symposium*, ONR 33300-5, Snommass, CO, 1998, pp. 451-458.
2. Sheffield, S. A., Gustavsen, R. L., and Alcon, R. R., Shock Compression of Condensed Matter-1999, M. D. Furnish, L. C. Chhabildas, and R. S. Hixson, eds., AIP Conference Proceedings 505, Snowbird, UT, 1999, pp. 1043-1048.
3. Dick, J. J., Shock Compression of Condensed Matter-1999, Furnish, M. D Chhabildas, L. C., and Hixson, R. S.,eds., AIP Conference Proceedings 505, Snowbird, UT, 1999, pp. 683-686.
4. Dick, J, J., Martinez, A. R., and Hixson, R. S., *Eleventh International Detonation Symposium*, ONR 33300-5, Snommass, CO, 1998, pp. 317-324.
5. Urtiew, P. A., Tarver, C. M., Forbes, J. W., and Garcia, F., *Shock Compression of Condensed Matter-1997*, Schmidt, S. C., Dandekar, D. P., and Forbes, J. W., eds. AIP Press, New York, 1998, pp. 727-730.
6. Forbes, J. W., Tarver, C. M., Uritew, P. A., and Garcia, F., *Eleventh International Detonation Symposium* ONR 33300-5, Snommass, CO, 1998, pp. 145-152.
7. Tarver, C. M., Forbes, J. W., Urtiew, P.A., and Garcia, F. Shock Compression of Condensed Matter-1999, Furnish, M. D., Chhabildas, L. C., and Hixson, R. S., eds., AIP Press, New York, 2000, pp. 891-894.
8. Urtiew, P. A., Forbes, J. W., Tarver, C,. M., and Garcia, F., Shock Compression of Condensed Matter-1999, Furnish, M. D., Chhabildas, L. C., and Hixson, R. S., eds., AIP Press, New York, 2000, pp. 1019-1022.

FRAGMENT IMPACT CHARACTERIZATION OF MELT-CAST AND PBX EXPLOSIVES

Malcolm D. Cook[1], Peter J. Haskins[1], Richard I. Briggs[1], Chris Stennett[1], Justin Fellows[1] and Phil J. Cheese[2]

[1]*Defence Evaluation and Research Agency, Fort Halstead, Sevenoaks, Kent TN14 7BP, England*
[2]*Defence Ordnance Safety Group, Walnut 2c #67, MoD Abbey Wood, Bristol, BS34 8JH, England*

Abstract. In this paper we report new experimental results on the shock to detonation transition characteristics of the melt-cast explosive RDX/TNT 60:40, and two PBX explosives, one containing RDX, and the other HMX, with HTPB as the binder in both cases. These experiments employed right-regular cylindrical steel projectiles impacting charges covered by either steel or aluminium barrier plates. Response curves were generated giving the threshold impact velocity for prompt shock initiation as a function of barrier thickness. The results of these experiments showed some general trends. Firstly, the melt-cast explosive was generally more shock sensitive than the PBX formulations. The PBX compositions showed similar shock sensitivities; despite the RDX based material having a higher percentage of nitramine (88%) compared to the HMX material (85%). All the response curves appeared to have at least one discontinuity. For the melt-cast explosive this appeared at thicker barriers than for the PBX formulations. The results of these experiments are discussed in terms of the mechanisms that may be responsible for the observed shape of these response curves.

INTRODUCTION

The effects of fragment impact on explosives have been investigated in some detail during recent years. At this laboratory we have investigated the effects of a variety of steel fragment sizes and shapes impacting explosive charges both bare and covered by materials of differing shock transmission properties, including aluminium and steel (1-4). As a result of these studies we have observed a number of mechanisms occurring depending on the explosive type, confinement, fragment velocity and geometry. These include the shock to detonation transition (SDT), deflagration to detonation transition (DDT) and unknown to detonation transition (XDT) (3).

More recently, we have concentrated on determining the shock to detonation transition (SDT) characteristics of several explosive compositions in some detail. Data of this type are essential for the validation and verification of any model purporting to predict the response of explosives to fragment impact, and have immediate application in hazard assessments of any munition.

In this paper we report the results of a comprehensive study of both a melt-cast and two PBX explosives.

EXPERIMENTAL

In the experiments described in this paper three explosive formulations have been employed. One melt-cast and two PBX compositions.

RDX/TNT (60:40) was melt-cast into right regular cylindrical moulds 57mm in diameter and 100mm long. One end of each charge was machined flat and perpendicular to the axis of the cylinder and was designated the target face. The average density was 1.690g/cm³. A PBX explosive consisting of 85% RDX and 15% HTPB binder was vacuum cast and machined to the same dimensions as the RDX/TNT charges. The average density was 1.560g/cm³. For comparison, a PBX consisting of 88% HMX and 12% HTPB binder was also, vacuum cast but this time into plastic tubes of 65mm internal diameter and 100mm length for convenience. The average density of this formulation was 1.692g/cm³.

Flat-ended cylindrical steel projectiles were used in two diameters: 13.15mm and 20mm. The 20mm diameter projectiles were somewhat shorter so as to be of equal mass (27g) to the 13.15mm projectile.

The sub-calibre projectiles were housed in nylon sabots and fired from a 30mm RARDEN gun. The projectiles were propelled using a standard 30mm RARDEN percussion cartridge filled with a known quantity of NRN41 propellant. The precise quantity of propellant in the cartridge was varied from round to round to produce fragment velocities in the range 700-2000m/s.

The experiments were all back-lit by a bank of flash bulbs and filmed using a quarter-height Fastax camera. The camera was operated at ca. 30,000 pictures/s, and was used to observe the projectile before impact, and determine its velocity. The film record was also used to reveal the projectile orientation at the moment of impact, and to provide visual confirmation of the degree of reaction of the target store. Projectile velocities measured from the film record are estimated to be accurate to ±5%.

RESULTS

A large number of experiments have been carried out in order to obtain tight error bounds between the highest velocity non-detonations and the lowest velocity detonations. The results are summarised in Figs. 1-4, which show plots of projectile velocity against barrier thickness covering the explosive. Only the lowest velocity detonations are shown for clarity. It should be noted that the lowest velocity detonations have been used here because they are positive results that can be directly attributed to an SDT process. In contrast there can always be some doubt over the validity of a non-detonative event. Note that in the experiments using the 13.15mm projectiles it was sometimes impossible to induce detonation even at the highest velocity attainable with our gun system.

Some general observation can be made. Firstly, the projectile threshold velocity required to produce a detonation in the target generally increases as the

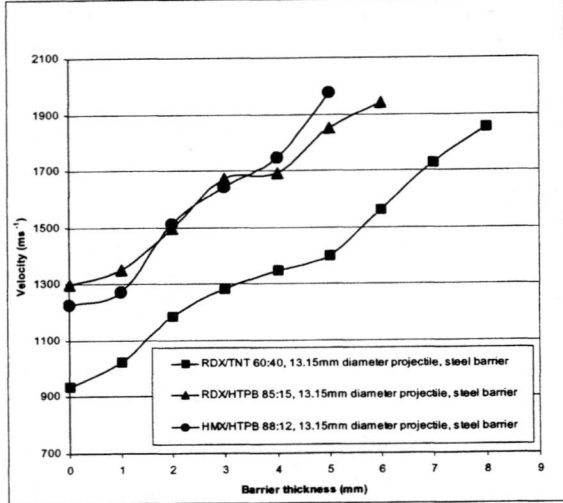

FIGURE 1. Summary of firings of 13.15mm diameter projectiles at charges covered by steel barriers. Only the lowest velocity detonations are shown for clarity.

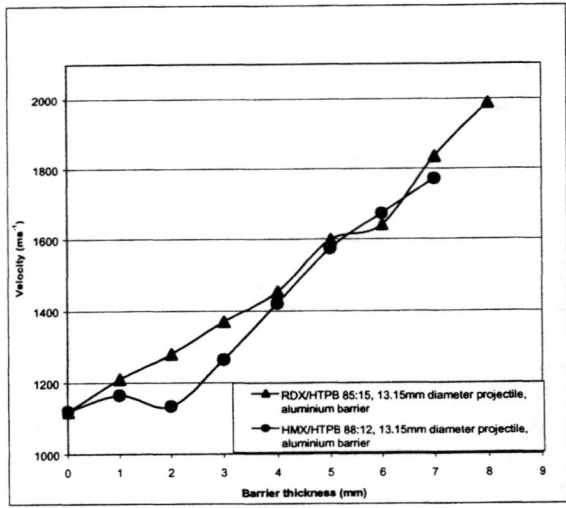

FIGURE 2. Summary of firings of 13.15mm diameter projectiles at charges covered by aluminium barriers. Only the lowest velocity detonations are shown for clarity.

barrier plate thickness increases for a given barrier plate thickness. Secondly, the projectile threshold velocity required to produce a detonation in the target decreases with increasing projectile diameter for flat faced projectiles; larger impact-area projectiles are more efficient at causing SDT in explosives.

The curve for RDX/TNT impacted with 13.15mm projectiles through steel barriers (Fig. 1) shows a distinct flattish region at barriers of 3-5mm thickness. In view of the high number of firings made against RDX/TNT, we have reasonable confidence that this curve feature is not due to experimental error, and represents a mechanistic effect. For RDX/TNT impacted with 20mm diameter projectiles through steel barriers (Fig. 3), there is a flat region at barriers of 3-4mm thickness. Comparatively fewer firings were made using the 20mm projectiles, and it should be noted that the kinks in this curve are within experimental error.

The curve for RDX/HTPB impacted by 13.15mm projectiles through steel barriers (Fig. 1) displays a distinct increase in slope as the barrier thickness increases from 2 to 3mm. A further slight change appears as the barrier thickness increases from 5 to 6mm. By contrast, the RDX/HTPB targets impacted by 13.15mm projectiles through aluminium barriers result in a smoothly increasing threshold velocity curve (Fig. 2). This suggests that the shock impedance match between the barrier material and the projectile, or between the barrier material and the target composition plays some part in governing the SDT threshold velocity. The curve for RDX/HTPB impacted by 20mm projectiles through aluminium barriers (Fig. 4) shows a series of slope changes as the barrier thickness increases. Most notably, the curve is relatively flat for barriers thinner than 4mm. Other slope changes are evident at the 7mm and 12mm barrier thicknesses, although these are not so marked as for other compositions. It is difficult to see how such behaviour can be modelled by a simple critical energy criterion. The presence of one 'kink' to produce an S-shaped curve has previously been explained (2) as resulting from the elimination of the pseudo-1D shock by thicker barriers. The more complex behaviour seen here suggests an additional mechanistic change. This is most likely as a result of an intricate interaction between the chemistry and effective hot spots.

The curve for HMX/HTPB impacted by 13.15mm projectiles though steel barriers (Fig. 1) shows a shallower slope at barriers of 0-2mm thickness than for the thicker barriers. The curve for the same composition and projectile, but using aluminium

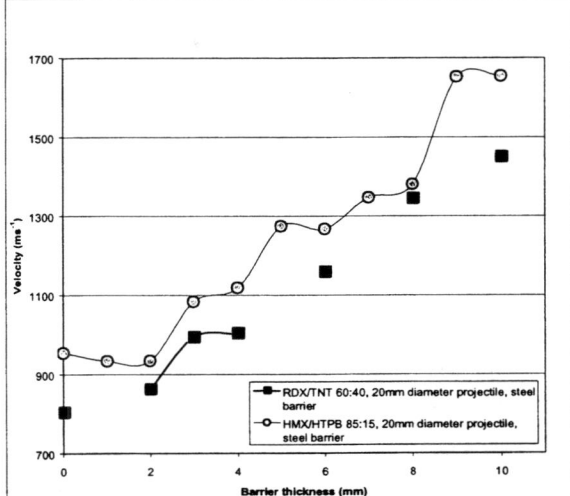

FIGURE 3. Summary of firings of 20mm diameter projectiles at charges covered by steel barriers. Only the lowest velocity detonations are shown for clarity.

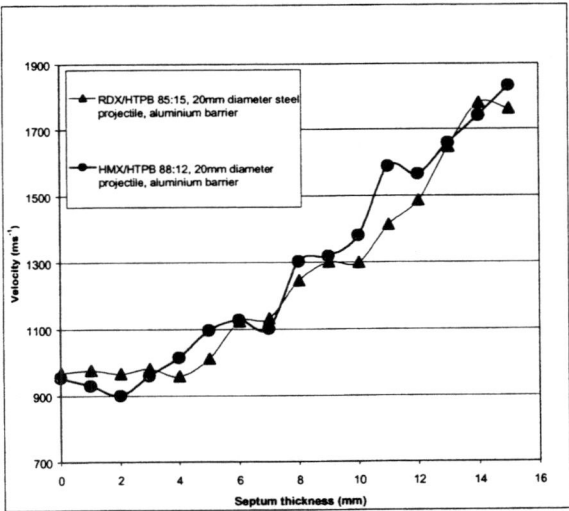

FIGURE 4. Summary of firings of 20mm diameter projectiles at charges covered by aluminium barriers. Only the lowest velocity detonations are shown for clarity.

barriers (Fig. 2) again shows a flattish response for barriers of 0-2mm thickness, followed by a smooth increase in threshold velocity as the barrier thickness increases. The curve for HMX/HTPB impacted by 20mm projectiles through steel barriers (Fig. 3) seems to display several slope changes. However comparatively few firings were made with this combination so that the GO/NOGO bounds are looser, and the size of these structures may appear exaggerated. The curve for the same composition attacked by 20mm projectiles through aluminium barriers (Fig. 4) shows several slope changes in a similar manner to that for RDX/HTPB under the same conditions. Again, a flattish response for barriers of 0-3mm thickness is evident, followed by a slope change at the 7mm barrier thickness.

CONCLUSIONS

The RDX/TNT curve in Figure 2 shows the 'S' shape characteristic of heterogeneous explosives, with the inflection occurring at barrier thicknesses of 4 to 5mm. Although this study fired relatively few shots at this composition, there is reasonable confidence that the curve shape is due to some real phenomenon, and is not the result of experimental error. There are several possible explanations for the structure in the RDX/TNT curve (1, 2, 3).

Both the RDX/HTPB and HMX/HTPB compositions display very similar shock sensitivity for 13.15mm diameter projectiles fired through both steel and aluminium barriers (Figures 2 and 3), and show similar shock sensitivity for 20mm projectiles fired through aluminium barriers. No firings were made against RDX/HTPB using 20mm projectiles and steel barriers, and so no comparison can be made. However, the shots that were fired provide reasonably strong evidence that these compositions display similar shock sensitivity.

The steel barrier plate experiments with RDX/TNT and HMX/HTPB show that these compositions have similar shock sensitivities when attacked with a 20mm projectile (Figs. 2 and 4). However the shots fired with 13.15mm projectiles show that the two compositions differ significantly in sensitivity; in this case HMX/HTPB is much less sensitive to shock. It is clear that the diameter of the projectile is an important factor in the assessment of the shock sensitivity of explosives using this type of test. In addition, changes to the projectile geometry can invalidate assumptions about the apparent sensitivity of different explosives. The curves for HMX/HTPB attacked with 20mm projectiles through both types of barrier show structure, although the precise structure differs depending on the barrier material chosen. This suggests that the structure is at least partially related to the shock impedance matching, either between the projectile and the barrier, or between the barrier and the explosive. RDX/HTPB was only tested using the 20mm projectile against aluminium barriers, so it is not yet possible to assess whether this effect is duplicated in the RDX-based composition.

The RDX/HTPB and HMX/HTPB threshold velocity curves are almost linear with barrier thickness when attacked with a 13.15mm projectile through aluminium barriers (Fig. 3), although the HMX/HTPB curve shows a distinct 'flat' region for barriers thinner than 3mm. At barrier thicknesses greater than 3mm, both curves are almost coincident, and are free from structure. This is believed to be a result of critical diameter effects (4): the 13.15mm projectile is similar in diameter to the critical diameter of both HMX/HTPB and RDX/HTPB, so that 'hotspot' mechanisms play a less significant role in the initiation process in these materials.

REFERENCES

1. James, H.R., *Propellants, Explosives and Pyrotechnics*, **13**, 35, (1988).
2. James, H.R., Grixti, M.A., Cook, M.D., Haskins, P.J., and Stuart Smith, K., "The Dependence of the Response of Heavily-Confined Explosives on the Degree of Projectile Penetration" *in The Tenth Symposium (International) on Detonation*, ONR 33395-12, Boston, MA, 1993, pp 89-93 and refs therein.
3. Haskins, P.J., Cook, M.D., "An Investigation of XDT Events in the Projectile Impact of a Secondary Explosive" *in The Tenth Symposium (International) on Detonation Symposium*, ONR 33395-12, Boston, MA, 1993, pp 148-154.
4. James, H.R., Cook, M.D., and Haskins, P.J., "The Response of Homogeneous Explosives to Projectile Attack", *in The Eleventh Symposium (International) on Detonation*, ONR 33300-5, Snowmass, CO, 1998, pp 581-588.

HUGONIOT AND SHOCK INITIATION STUDIES OF ISOPROPYL NITRATE[†]

S. A. Sheffield*, L.L. Davis*, M. R. Baer[+], Ray Engelke*, R. R. Alcon*, and A. M. Renlund[+]

*Los Alamos National Laboratory, Los Alamos, NM 87545
[+]Sandia National Laboratories, Albuquerque, NM 87185

Isopropyl nitrate (IPN) is a liquid explosive of rather low energy. We have measured the sound speed and used it in the universal liquid Hugoniot to produce an estimated Hugoniot for this material. Gas-gun-driven, multiple-magnetic-gauge measurements were made to measure a Hugoniot state at 6 GPa; it was in good agreement with the prediction. Two similar experiments were conducted at higher pressure inputs to study the shock-to-detonation transition in IPN; the high inputs required for initiation necessitated the use of a two-stage gun. One experiment with an input of 9.0 GPa into the IPN produced a run to detonation of about 3 mm and the in-situ particle velocity profiles showed the expected homogeneous initiation behavior of a growing wave behind the shock front that overtakes the front and decays to a steady detonation. The reactive wave in the shocked IPN appears to have achieved a steady superdetonation in both of the initiation experiments. This is the first time a steady superdetonation has been measured with in-situ gauges.

INTRODUCTION

Liquid explosives have been of interest for many years because they offer the opportunity to experiment with a homogeneous material and determine the properties of the material without the influence of other phenomena that might lead to hot spots (heterogeneous behavior). Nitromethane (NM) has typically been the explosive of choice because it is easy to obtain and has been extensively studied. Most other studies of liquid explosives have concentrated on critical diameter, gap test sensitivity, detonation velocity, initial temperature effects, etc., and have not studied in detail the shock initiation properties. In this study we have concentrated on isopropyl nitrate to more carefully study its shock and reaction properties. This information has led to increased understanding of IPN as well as adding important new information on the homogeneous initiation process.

Liquid isopropyl nitrate [$(CH_3)_2CHONO_2$] (IPN) (see Fig. 1) is a rather low energy explosive because it does not have a good oxygen balance. It has been used in propellants or as a monopropellant. A rather large study of the detonation (failure) properties of IPN in steel and glass tubes as a function of temperature was reported in Ref. 1. Unfortunately, the accuracy of the detonation velocity measurements was only 3 to 5%, less than one would hope because NM has been shown to have a velocity change (velocity deficit) of only 1% from infinite diameter to failure diameter in glass.[2] We have taken the liberty of replotting some of the room temperature data from Ref. 1 to get some idea of what the diameter effect curve is for IPN in steel tubes; the data are shown in Fig. 2. The detonation velocity in steel tubes is between 5.3 and 5.4 mm/µs, depending on the tube diameter. The line stops at

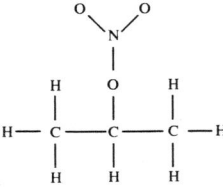

FIGURE 1. Chemical structure of isopropyl nitrate.

[†] Work performed under the auspices of the U.S. Dept. of Energy.

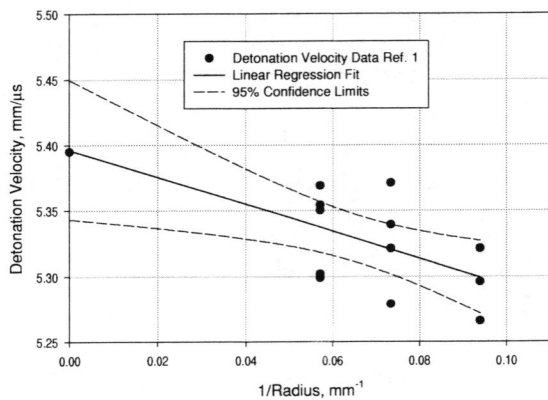

FIGURE 2. Diameter effect curve for IPN in steel tubes at room temperature. Data are from Brochet in Ref. 1.

roughly the critical diameter in steel, about 10 mm radius (20 mm diameter). The velocity deficit for IPN appears to be about 2%, somewhat larger than that for NM but comparable.

In Ref. 1, the IPN detonation velocity was measured to drop about 0.2 mm/μs when the temperature increased from room temperature to 350K; about −3.3 m/s/K. It was stated that the change was due to density changes as a function of temperature. NM detonation velocity drops with increased temperature at a rate of −3.7 m/s/K.(Ref. 3) This is thought to result mostly from density change but there is also a contribution due to internal energy changes with temperature.[2]

EXPERIMENTAL DETAILS AND DATA

Sound speed measurements at ambient conditions were made on liquid IPN to provide the information necessary to estimate the Hugoniot. Magnetic gauge experiments were completed using both our single-stage and two-stage guns to measure the Hugoniot state and provide shock initiation information. Information about and data from these experiments are presented below.

Material – Liquid IPN was obtained from Aldrich Chemicals with a purity of 99 wt% IPN. It was used as received. IPN has an initial density at room conditions of 1.036 g/cm^3, a boiling point of 101–102 °C and a freezing point of 12 °C. IPN is a colorless liquid with a viscosity about like water.

Sound Speed and Estimated Hugoniot – An estimate of the Hugoniot for this material was obtained using the universal liquid Hugoniot.[4] This empirical equation has the form

$$U_s = C_0\{1.37 - 0.37\exp(-2u_p/C_0)\} + 1.62u_p$$

where U_s is the shock velocity, u_p is the particle velocity, and C_0 is the room condition sound speed – the only required parameter. It has been shown to provide reliable Hugoniot estimates for essentially all the liquids for which shock data are available. The room condition sound speed for IPN was measured (see Ref. 5) to be 1.10 mm/μs.

Gun Experiments/Liquid Cell Design – In the gun experiments of this study, the input shock was produced by impacting an impactor-faced projectile on a plastic cell containing the liquid IPN. The diameter and depth of the liquid sample was such that edge effects did not complicate the measurements, i.e., they were designed to be 1-D.

The cell body was made from two pieces of PMMA which were machined to fit together with the gauge membrane epoxied between them. The design was such that the suspended membrane was a plane inclined 30° with respect to the cell front (impact plane). The cell front was made from either PMMA or Kel-F, depending on the input desired into the IPN. An epoxy coating was put on the inside of the cell to isolate the IPN from the PMMA. The cell front was epoxied and screwed (with nylon screws) to the cell body. Fill holes were located on the side of the cell. Details of how these experiments were done are contained in Refs. 6 and 7.

In each experiment the gauge membrane included ten gauges and a "shock tracker." Another gauge (called a "stirrup" gauge) was epoxied on the back of the cell front to measure the input to the IPN. This setup provided a total of eleven in-situ particle velocity gauges and the shock tracker, which provides data to use in constructing a distance vs. time (x-t) plot of the shock front as it moves through the IPN. The particle velocity gauges provide dynamic information relating to the state of the shocked IPN (which may be reacting) at specific Lagrangian positions.

There are some difficulties in using an inclined gauge membrane suspended in a liquid. The membrane is composed of FEP Teflon. It has been shown that if the liquid is the same shock impedance as the gauge, the gauges provide an accurate measurement of the particle velocity. If the liquid is a higher impedance, the gauge measures high and if the liquid is lower impedance, it measures low.[8] The errors can be up to ± 10%, depending on the impedance difference. IPN is a lower impedance than the membrane so the gauges read low in this material. These errors in the measurement are due to slippage at the gauge plane; this happens in experiments with liquids but not those involving solid materials.

IPN Gas Gun Experimental Data – Three multiple-magnetic-gauge gun experiments were completed. One was done on the single-stage gun (Shot 1129) below the condition where reaction was initiated to confirm that the estimated Hugoniot was correct. Two higher pressure input experiments were completed on the two-stage gun (Shots 2s-28 and 2s-29) to measure the details of the shock-to-detonation transition in IPN.

Shot 1129 provided Hugoniot a point to compare to the estimated Hugoniot. This shot involved a Z-cut single-crystal sapphire impactor hitting a Kel-F cell front at a velocity of 1.408 mm/μs. A stirrup gauge on the cell front, in contact with the IPN, measured a particle velocity input of 1.50 mm/μs to the IPN. The membrane gauges measured about 8.5 % lower than this. The shock tracker provided a value for the shock velocity of 3.896 mm/μs. Using these values, the IPN input pressure was 6.05 GPa. This Hugoniot point is plotted on Fig. 3 along with the universal liquid Hugoniot prediction. It is obvious that the predicted Hugoniot is accurate.

On Shot 2s-28 the input shock to the IPN was generated by a Kel-F impactor on the projectile hitting a PMMA cell front. Unfortunately, the projectile velocity measurement failed so we can only estimate the impact velocity to be between 2.75 and 2.85 mm/μs. Particle velocity measurements from this experiment include the stirrup gauge at the beginning of the IPN and ten gauges in the IPN. The particle velocity waveforms from all eleven gauges are shown in Fig. 4. In addition, a shock tracker measured the progress of the shock as it moved through the IPN. The initial shock velocity was 4.8 mm/μs, the particle velocity (as shown in Fig.4) was about 1.6 mm/μs, so the initial shock pressure was about 8 GPa. The waveforms in Fig. 4 show that the superdetonation achieved a steady state and, using

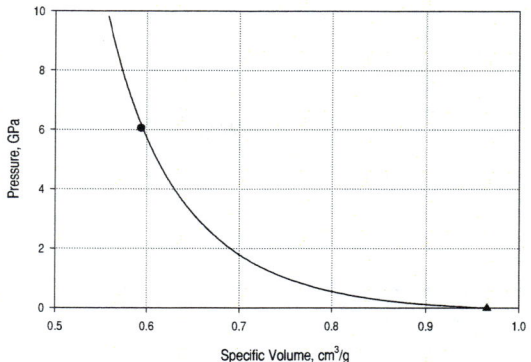

FIGURE 3. P-V Hugoniot plot for IPN. The curve is obtained from the universal liquid Hugoniot and the data point is that measured in Shot 1129. The initial specific volume is shown as a triangle.

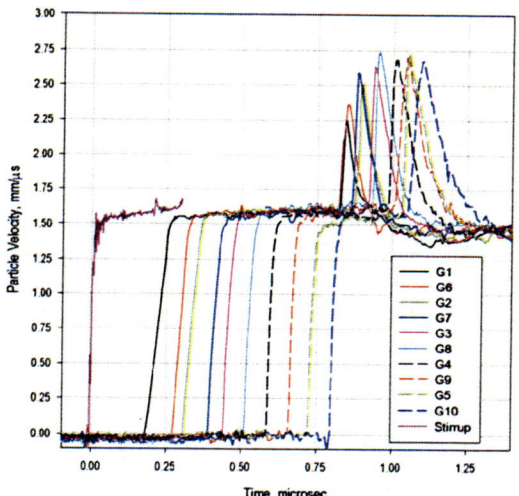

FIGURE 4. Particle velocity waveforms obtained from Shot 2s-28. Eleven gauges are shown – the first gauge is the stirrup gauge at the input face of the IPN and the other ten gauges are in-situ in the IPN at positions from 0.8 to 4 mm deep.

the gauge positions and arrival times, a velocity of about 10 mm/μs was measured for this wave as it moved through the initial state. Data from the shock tracker was used to determine that the detonation immediately after overtake had a velocity of 6.9 mm/μs, which decreased to a steady 5.34 mm/μs by the end of the shock tracker data. The position of overtake was 5.7 mm into the IPN.

Shot 2s-29 produced the particle velocity waveforms shown in Fig. 5. The projectile velocity was 2.97 mm/μs with a Kel-F impactor hitting a

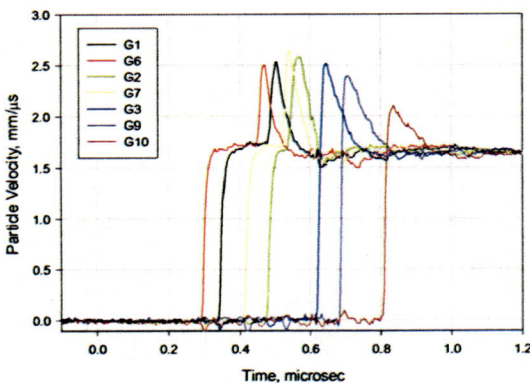

FIGURE 5. Particle velocity waveforms obtained from Shot 2s-29. Seven gauges are shown, all of them are in-situ in the IPN at positions from 1.3 to 4.5 mm deep. The stirrup gauge broke so the input to the IPN was not recorded.

Kel-F cell front and providing a shock of 8.9 GPa into the IPN. Several gauges failed in this experiment so they are missing from the figure. However, the shock tracker provided information about the shock front propagation. The input particle velocity was measured at 1.73 mm/μs; it was calculated to be 1.89 mm/μs, about 8.3% low. The input shock velocity was 4.5 mm/μs. The waveforms show the superdetonation achieved a steady velocity of about 8 mm/μs. After overtake of the initial wave, the overdriven velocity was 6.3 mm/μs which decreased to a steady 5.34 mm/μs by the end of the shock tracker. The position of overtake was 2.7 mm into the IPN.

DISCUSSION OF RESULTS

From the low pressure experiment, it is obvious that the universal liquid Hugoniot is a good estimate of the IPN unreacted Hugoniot. The shock initiation experiments both showed the same initiation behavior. A reactive wave builds behind the initial input shock to a steady superdetonation (steady velocity 8 to 10 mm/μs) which overtakes the initial shock, producing an overdriven detonation (6.3 to 6.9 mm/μs) that decreases to a steady detonation of 5.34 mm/μs. This agrees with the process proposed earlier for NM.[9] In this study of IPN, the superdetonation reached a steady velocity, the first time this has been measured with in-situ gauges. The steady detonation velocity of 5.34 mm/μs would be expected to be the infinite diameter velocity because the experiments are 1-D. This value agrees reasonably well with the earlier data of Fig. 2.

When the three experiments are considered as a group, there are some inconsistencies. Shots 1129 and 2s-29 agree with each other in that the input particle velocities measured by the gauges are about 8 % low, when compared to what was expected; this is as it should be for liquid IPN. However, Shot 2s-28 was different. The particle velocity measured by the in-situ gauges agreed with the stirrup gauge as one would expect for a solid and the measured particle velocity and shock velocity were not what was expected. We have conjectured that this may be the result of freezing in the initial wave on Shot 2s-28. This means the initiation occurred in the solid state rather than the liquid state, producing a faster superdetonation. However, this is highly speculative and should be considered suspect until additional experiments are completed.

ACKNOWLEDGMENTS

We appreciate the help of Rick Gustavsen and Bob Medina with this work.

REFERENCES

1. Brochet, C., Astronautica Acta, **15**, 419 (1970).
2. Engelke, R., Phys. Fluids **23**, 875 (1980).
3. Campbell, A. W., Malin, M. E., and Holland, T. E. Proceedings of the Second Symposium on Detonation, Naval Surface Warfare Center Report NSWC MP 87-194, pp. 454-471 (Reprinted in 1987, Short, J., Editor).
4. Woolfolk, R. W., Cowperthwaite, M., and Shaw, R., Thermochimica Acta **5**, 409 (1973).
5. Engelke, R., Sheffield, S. A., and Davis, L. L., J. Phys. Chem. A **104**, 6894 (2000).
6. Sheffield, S. A., Gustavsen, R. L., and Alcon. R. R., Shock Compression of Condensed Matter – 1999, Eds. Furnish, M. D., Chhabildas, L. C., and Hixson, R. S., AIP Proceedings 505, p. 1043 (2000).
7. Sheffield, S. A., and Alcon. R. R., Shock Compression of Condensed Matter – 1989, Eds. Schmidt, S. C., Johnson, J. N., Davison, L. W., North-Holland, Elsevier Science Publishers B. V., Amsterdam, The Netherlands, p. 683 (1990).
8. Gustavsen, R. L., Sheffield, S. A., and Alcon, R. R., High Pressure Science and Technology – 1993, Eds. Schmidt, S. C., Shaner, J. W., Samara, G. A., Ross, M., AIP Conference Proceedings 309, p. 1703 (1994).
9. Sheffield, S. A, Engelke, R. and Alcon. R. R., Proceedings of the Ninth Symposium (Intl.) on Detonation, Office of Naval Research OCNR 113291-7, pp. 39-49 (1989).

STRESS GROWTH MEASUREMENTS FOR THE EXPLOSIVE IRX-4

Gerrit T. Sutherland

Naval Surface Warfare Center, Indian Head Division, Indian Head, MD 20640

Embedded gauge experiments were performed to measure the shock reactivity of IRX-4, a plastic-bonded explosive that contains HMX, Al, AP and HTPB binder. The pressure-time profiles obtained are similar to those obtained for similar composite explosives. Hugoniot points obtained are in agreement with those obtained with mixture theory.

INTRODUCTION

A series of mono-modal research explosives denoted IRX-1 (independent research explosive one), IRX-3A, and IRX-4 were formulated to elucidate the roles of HMX, aluminum (Al), and ammonium perchlorate (AP) in shock reactivity and detonation property experiments. All three explosives contain mono-modal HMX and a polyurethane binder (HTPB). IRX-3A also contains aluminum, and IRX-4 contains both Al and AP.

Shock reactivity and detonation property experiments for IRX-1 and IRX-3A have been published.[1-4] In this paper, we present shock reactivity results for IRX-4.

The shock reactivity experiments performed were embedded gauge experiments, which will be used to obtain reaction kinetics. Pressure-time or particle velocity-time records are used to adjust parameters in reactive rate models such as the Lee-Tarver [5] model. In this model, the rate at which unreacted explosive is converted into detonation products is expressed as a function of the: mass fraction of explosive reacted, pressure and density. When inserted in a hydrocode, the Lee-Tarver model has been successful in simulating [6-7] the results of shock sensitivity and detonation property experiments. Our intent is to further refine the Lee-Tarver model parameters until computer simulations predict the results of tests such as modified gap, wave curvature, detonation velocity decrement and the embedded gauge experiments. When reactive rate model parameters are obtained for all of the IRX series explosives, the differences in global reaction kinetics between the explosives will be determined. The roles that each constituent (HMX, Al, AP) plays in shock reactivity and detonation properties will be quantified.

EXPERIMENT

IRX-4 consists (by weight) of 30% HMX, 16% Al, 24% AP and 30% HTPB binder. The HMX was sieved and had a average particle size of 124 µm. The Al and AP particle sizes were estimated to be about 3 and 200 µm respectively. In our experiments, the sample density was taken as the casting density of 1.46 g cm^{-3}.

A schematic representation of our embedded gauge experiments appears in Fig. 1 and experimental parameters appear in Table 1. The nominal diameters of the explosive disks were 70 mm. Impactors and cover plates were constructed from OFHC copper (Cu) for experiments 3 and 5, and from 6061T6 Aluminum for the remaining experiments. The Dynasen Corporation manufactured Manganin gauges from manganin foil supplied by NSWC. The gauges (MN10-.050 type S) [8] had an electrical impedance of 0.050 ohms, used a four-wire terminal configuration and incorporated Teflon® encapsulation resulting in gauges of either 0.25 or 0.50 mm nominal total thickness. All experiments were performed using a 100-mm diameter light gas gun.

TABLE 1. Experimental Parameters

Exp.	Thickness Layer 1 (mm)	Thickness Layer 2 (mm)	Impact Velocity (mm/μsec)	Calculated Impact Pressure (GPa)
1	5.982	6.007	0.639	2.18
2	5.009	5.994	0.756	2.68
3	7.010	7.023	0.965	4.44
4	5.982	5.956	1.066	4.12
5	7.023	7.008	0.881	3.92

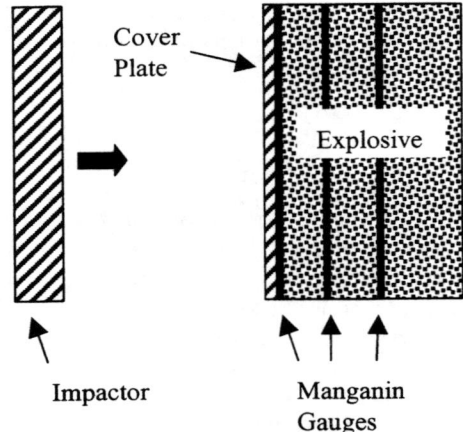

Figure 1. Schematic representation of an embedded gauge experiment.

RESULTS

Pressure -Time Profiles

Experimental results for the IRX-4 experiments appear in Figs. 2–4. In these records, considerable electrical noise, which will be discussed later, was observed, so the records were smoothed with a smoothing routine with a 150 nsec time span. For each record, the relative resistance change ($\Delta R/R_o$) was calculated and the stress determined from an empirical calibration of Vantine et al.[9]

The pressure-time profiles show a response similar to a composite explosive [3] containing a nitramine (ie. HMX, RDX), Al, AP and binder. The similarity includes the square-wave shape in pres-

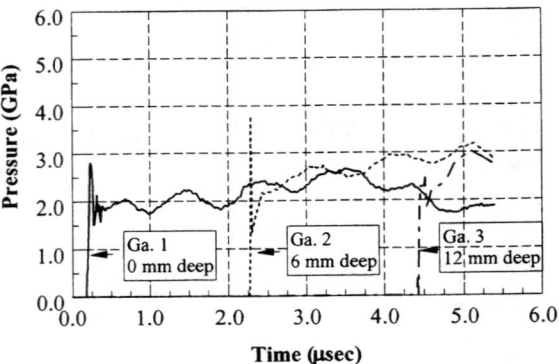

Figure 2. Pressure-time profiles for experiment 1. The calculated impact pressure is 2.18 GPa.

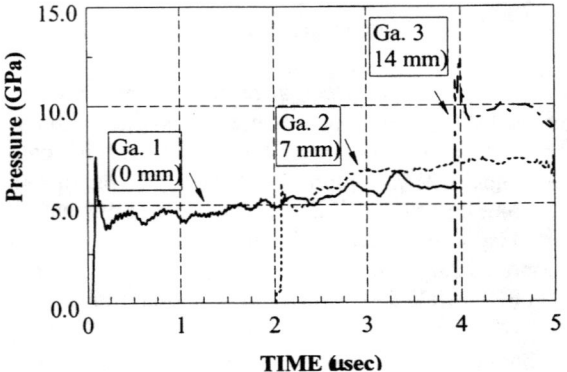

Figure 3. Pressure-time profiles for experiment 5. The calculated impact pressure is 3.92 GPa.

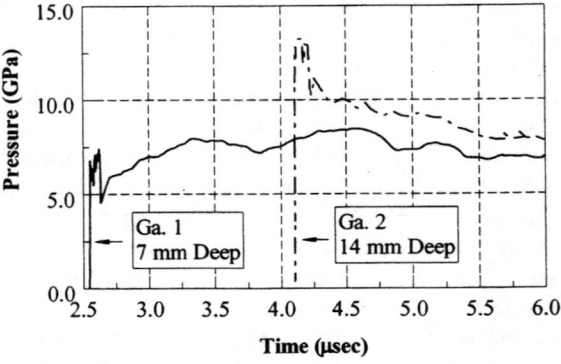

Figure 4. Pressure-time profiles for experiment 3. The calculated impact pressure is 4.44 GPa.

sure-time profile for gauge 1 on experiment 3 and gauge 3 of experiment 5. The record for our lowest pressure experiment indicated a modest level of reaction, something not seen for a similar composite explosive.[3] This reaction is likely due to the fact that the composition, although a composite explosive, contains almost 24% volume percent of coarse HMX. A detonation wave was observed at gauge 2 in the pressure-time profile in Fig. 4. It is believed that the peak pressure was slightly higher as the original voltage digitizer record was clipped. Comparison of Figs.3-4 shows that IRX-4 is an insensitive explosive; it takes 3.9-4.5 GPa impact pressure to produce a detonation at a run distance of 14 mm.

Hugoniot measurements

Reactive rate modeling of our experiments requires a Hugoniot for IRX-4. Minimal reaction at the shock front allowed Hugoniot points to be obtained from transit time measurements between the two gauge planes. A correction was made for the transit time through the Teflon® encapsulation of the gauge planes. The Hugoniot points obtained are plotted with a Hugoniot obtained from mixture theory in Figure 5. For our mixture theory calculation, we used Hugoniots for the HTPB binder from Grady,[10] and for all other components Hugoniots from Bernecker[11-12] were used. The agreement between the experimental points and mixture theory implies that mixture theory Hugoniot can be used in the computer simulations. Hugoniot points were not determined from front surface gauge pressures; the gauge calibration is not accurate for low pressures.[13]

Electrical Noise Observed

Figure 6 shows the substantial amount of electrical noise that we observed for experiment 1. Other experiments showed similar noise levels. We have observed the high frequency noise component for another composite explosive but we have not always observed the lower frequency noise component ($\approx$ 1Mhz).[3] The gauges are susceptible to noise since a pressure change of 2 GPa represents a voltage change of only 20 millivolts. The lower frequency noise may be present due to a different method of connecting the gauge leads to the coax cable leading to the instrumentation. Thin copper strips, twisted hookup wire, and circuit boards in-

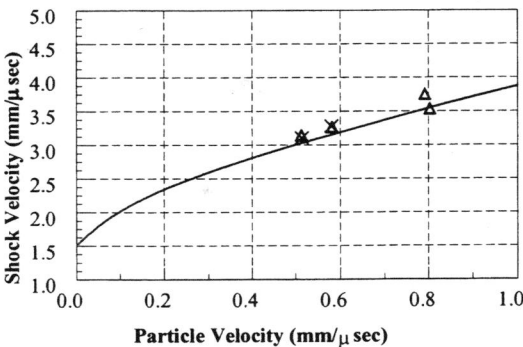

Figure 5. Hugoniot points plotted against the results of the mixture theory. The open triangles represent shock velocities obtained between gauges 1 and 2; the crosses represent shock velocities obtained between gauges 2 and 3.

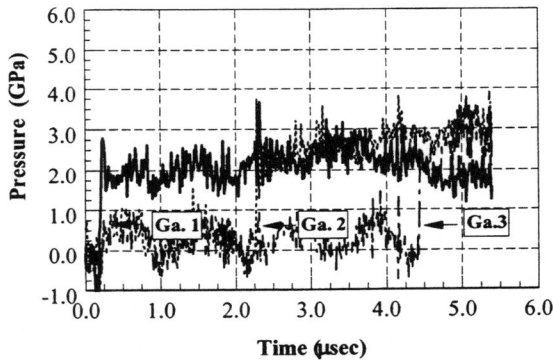

Figure 6. Pressure-time profiles showing electrical noise observed for experiment 1.

corporating strip lines have each been used to make this connection. Circuit board type connections were most likely used for these experiments.

FUTURE WORK

The pressure-time profiles will be used to determine parameters for the Lee-Tarver reactive rate model. The parameter set will be refined by further adjusting the parameters until the results of all IRX-4 experiments such as wave curvature and detonation decrement are predicted.

To reduce the 1MHz electrical noise observed for the gauge records, the method used to make the electrical connection between the gauge leads and the coax cable will be optimized.

ACKNOWLEDGEMENTS

John O'Connor, Paul Gustavson, Bob Baker, Carl Groves, Dale Ashwell, and Ray Lemar are thanked for their help in constructing and performing the experiments. Jerry Forbes and Charles Dickinson are thanked for his support and encouragement of this work. This work was supported by Independent Research Program of the Naval Surface Warfare Center.

REFERENCES

1. Sutherland, G.T., Lemar, E.R., Forbes, J.W., Anderson, E.W., Ashwell, K.D., and Baker, R.N., in Proceedings of the JANNAF PSHS Meeting, May 11-13, 1993, Fort Lewis, WA.
2. Sutherland, G.T., Forbes, J.W., Lemar, E.R., Ashwell, K.D, and Baker, R.N., and Liddiard, T.P., "Shock Wave and Detonation Wave Response of Selected HMX Based Research Explosives with HTPB binder systems" in High Pressure Science and Technology-1993, Woodbury, NY, AIP, 1994, pp. 1413-1416.
3. Sutherland, G.T., Ashwell, K.D., O'Connor, J.H., Barker, R.N., and Lemar, E.R., "Shock Response of Several Plastic Bonded Explosives, in Shock Compression of Condensed Matter-1995, edited by S.C. Schmidt et al., AIP Conference Proceedings 370, New York, 1996.pp. 763-766.
4. Lemar, E.R., Forbes, J.W., Sutherland, G.T., Detonation Wave Curvature of IRX-4 and PBXN-110", in Shock Compression of Condensed Matter-1995, edited by S.C. Schmidt et al., AIP Conference Proceedings 370, New York, 1996.pp. 791-794.
5. Lee E.L., Tarver, C.M., *Phys. Fluids* **23**, 2362-2372 (1980).
6. Miller, P.J., Sutherland, G.T., Reaction Rate Modeling of PBXN-110, Shock Waves in Condensed Matter-1995, editors Schmidt, S.C., and Tao, W.C., AIP, Woodbury, NY
7. Murphy, M.J., Simpson, R.L., Breithaupt, R D., Tarver, C.M., "Reactive Flow Measurements and Calculations for ZrH_2 Based Composite Explosives, in Proceedings of Ninth Symposium (International) on Detonation, Office of the Chief of Naval Research, OCNR 113291-7, pg. 525.
8. Information of the Dynasin pressure gauges can be found at www.dynasin.com.
9. Vantine, H.C, Erickson. L.M., Janzen, J.A., J. Appl. Phys. **51** (1979) 1957.
10. Grady, D.E., Private Communication
11. Bernecker R.R.., "Observations on the Hugoniot for HMX", in Shock Compression of Condensed Matter-1995, edited by S.C. Schmidt et al., AIP Conference Proceedings 370, New York, 1996.pp. 137-140.
12. Bernecker R.R.., "Observations on the Hugoniot for HMX", in Shock Compression of Condensed Matter-1995, edited by S.C. Schmidt et al., AIP Conference Proceedings 370, New York, 1996.pp. 137-140.
13. Forbes, J.W. Private Communication. Work on a revised manganin gauge calibration for low pressures is underway at the Lawrence Livermore National Laboratory.

THE COMBUSTION OF EXPLOSIVES

S. F. Son

Los Alamos National Laboratory, Los Alamos, NM 87545

Abstract. The safe use of energetic materials has been scientifically studied for over 100 years. Even with this long history of scientific inquiry, the level of understanding of the important deflagration phenomena in accidental initiations of high explosives remains inadequate to predict the response to possible thermal and mechanical (impact) scenarios. The search also continues for improved explosives and propellants that perform well, yet are insensitive. Currently, the most significant uncertainties are in the processes immediately following ignition. Once ignition occurs in an explosive, the question then becomes what the resulting violence will be. The classical view is that simple wave propagation proceeds from the ignition point. Recently, several experiments have elucidated the importance of reactive cracks involved in reaction violence in both thermally ignited experiments and impacted explosives, in contrast to classical assumptions. This paper presents a view of reaction violence, in both thermal and mechanical insults, that argues for the importance of reactive cracks, rather than simple wave propagation processes. Recent work in this area will be reviewed and presented. Initial results involving novel energetic materials will also be discussed. Novel materials may yield insight into the mechanisms involved with rapid deflagration processes.

INTRODUCTION

Energetic materials, including propellants, explosives, and pyrotechnics, are used in applications such as rocket motors, guns, explosive bolts, weapon systems, air bags, and of course fireworks. Since the first energetic materials were discovered, the safe use of these materials has been of interest. For example, in 1864 a major explosion at the Nobel factory in Stockholm claimed the lives of Alfred's brother Emil and four other people. This accident, in part, drove Alfred Nobel to the invention of dynamite that was much safer than nitroglycerine. The safe use of energetic materials remains a concern today. In this paper, an overview of recent efforts at Los Alamos to obtain an improved understanding of the mechanisms of accidental initiation of energetic materials will be presented. The importance of the interaction of combustion with cracks in both thermally and mechanically insulted explosives will be presented.

Ongoing work using new classes of explosives and propellants, specifically high nitrogen materials and nanoscale thermites (also called known as Metastable Interstitial Composites, or MIC), will be presented also. High-nitrogen (HN) compounds may

Figure 1. This is an image of the long confined crack experiment. A Plexiglas window insert was used. PBX 9501 was mounted to brass inserts and shimmed to a gap of 80 μm. The two parts shown are bolted together. Pressure ports were placed in the crack region and in the breach area.

be key to meeting the advanced performance objectives of next-generation energetic materials (1-3). High-nitrogen solids offer the possibility of high performance (both as propellants and explosives), reduced emissions, and lower plume signature (low temperature and no HCl) than materials used in current systems.

Nanoscale thermites offer the possibility of tunable energy release rates, high density, high energy, high temperatures, and low toxicity (4). These materials have been shown to have very high propagation rates in loose powders. In sharp contrast to HN materials, the Al/MoO$_3$ system considered produces little gas, but yields high temperatures. In this paper, we present some initial results and argue that by studying these novel materials we gain insight into the physics of unusual combustion, and therefore gain understanding, and confidence in the modeling, of accident scenarios involving more

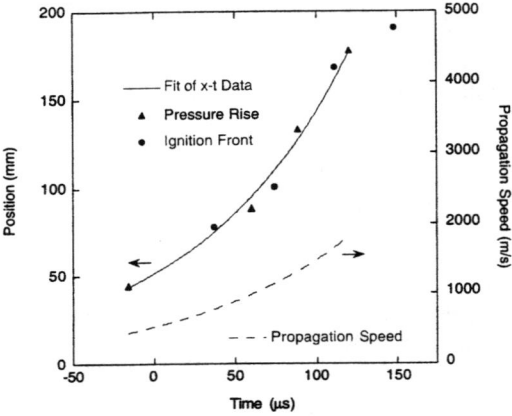

Figure 2. Flame front location in long confined slot experiment.

typical materials.

In accident scenarios, relatively low-speed deflagration processes play a critical role in determining the ultimate reaction violence. Reaction propagation may begin as conductively driven (normal) combustion, transit to convective combustion (advective energy transport), induce compaction-initiated combustion (mechanical dissipation), and possibly lead to detonation (propagation involving shock waves). Normal deflagration and detonation are relatively well studied. Convective and compaction driven combustion are much less well understood. The objective of this paper is to present an overview of recent and current work proceeding at Los Alamos in the area of convective burning.

BACKGROUND

Convective burning can be an important step in the deflagration-to-detonation transition in explosives and other energetic materials (5-7). Normal deflagration involves primarily conductive heat transfer from the gas-phase flame region to the surface, and to a lesser extent, radiation transport from the gas to the solid. In contrast, convective burning involves heat transfer via mass flow. Defects increase the available surface area where combustion can occur and are necessary for convective burning in energetic materials. The effect of defects on combustion has major implications for the safety and reliability of energetic materials.

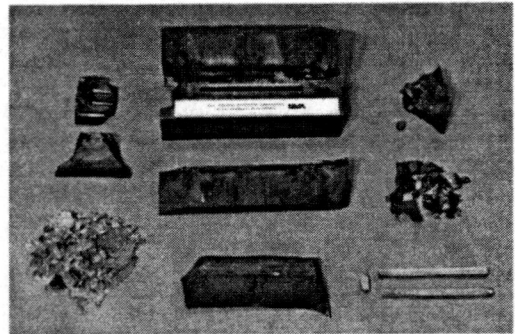

Figure 3. Damaged long confined slot experiment.

Voids and cracks in explosives may result from numerous environmental and physical factors. Impact, aging, and variations in temperature and pressure associated with combustion can produce defects. At sufficiently high pressures, the surface area of a defect becomes accessible to deflagration. Defects can trap the hot reaction products, creating the necessary pressure gradient for convective burning. The pressure from this burning may induce further cracking. A few studies exist on the effects of voids and cracks on the combustion of some common propellants (7-8), but relatively few studies exist of the effects of voids and cracks on the combustion of high explosives (HE) (7). One example is the study by Ramaswamy and Field who studied hot spot and crack propagation in single crystals of RDX (9). Formulated explosives, using HMX, (octahydro-1,3,5,7-tetranitro-1,3,5,7-tetrazocine), typically include a binder that makes it possible to desensitize and shape the explosive formulation. Binder affects the number, shape, and size of voids, as well as influences somewhat the combustion.

Recent experiments highlight the importance of

cracks and voids in the ignition, combustion, and reaction violence of PBX 9501. The Steven Test determines the critical impact velocity of a lightly confined energetic material to the low-speed impact of a blunt steel projectile. Radial cracks emanating from the impact point are apparent for a test where no sustained reaction occurred (10). Idar *et al.* (10) find that damaged PBX 9501 has a significantly lower impact threshold for violent reaction than pristine material.

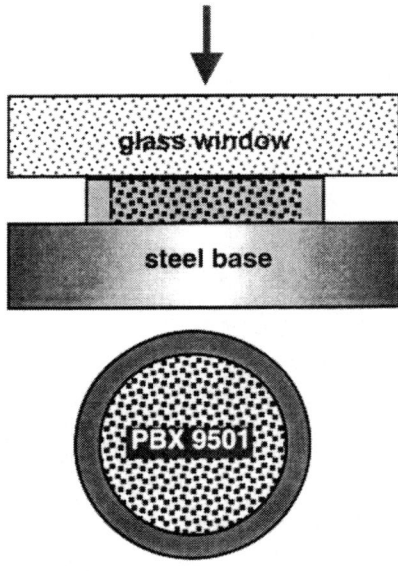

Figure 4. Schematic of radial burn experiment. Sample is ignited in center using a coiled nichrome wire. Pressure is measured in the center of the sample.

Henson *et al.* have conducted shear impact experiments using thin samples of PBX 9501 (11). A rectangular steel plunger is driven into the lightly confined sample at about 100 m/s. Plunger intrusion causes both shear and non-shear fracturing with reaction initiated along fracture zones. Skidmore *et al.* (12) have used microscopy to study damaged samples recovered from the shear impact experiments and find that the HMX along the fracture zones shows clear signs of heating and quenched reaction (12).

Evidence of the importance of crack-sustained combustion also appears in elevated-temperature experiments, such as the Mechanically Coupled Cookoff (MCCO). Dickson *et al.* slowly heat a confined sample of PBX 9501 to a well-defined temperature field, then ignite the center of the sample. They detect reaction, indicated by luminous emission, throughout cracks that are caused by pressurization due to production of reactive gases (13). The fast reactive waves, indicated by the luminosity, propagate through the cracks at velocities on the order of 500 m/s. An interesting question is whether reaction is spread into these cracks via convective processes, or whether crack tip dissipation ignites the material. Crack tip dissipation is expected to provide only a small amount of energy. However, in the MCCO experiments the material is already at an elevated temperature so very little energy is needed for reaction to occur. Recent initial experiments have been performed on unheated pristine materials that show similar reactive cracks. These results are shown below.

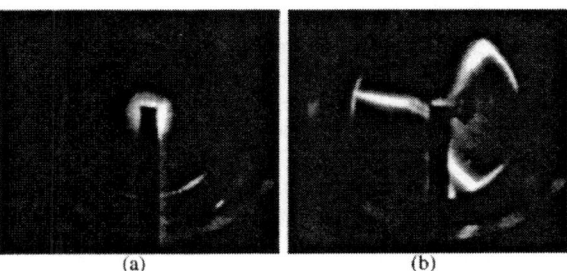

Figure 5. Images from radial burn experiment. Part (a) shows the initial ignition in the center and part (b), 0.5 ms later, shows illuminated cracks that extend from the center ignition site to the confinement ring.

In recent carefully heated cook-off (thermal explosion) experiments involving larger-scale explosive charges in an annular configuration (14), nearly symmetric compression of the inner wall was observed, although thermocouple records indicate ignition occurred asymmetrically. Consequently, it is surmised that some physical mechanism must spread reaction around the annular explosive charge, followed by violent reaction yielding a nearly symmetric compression of the inner walls. Spreading of ignition sites must occur at a rate on the order of 1000 m/s. Connected porosity is expected in this material because it is heated for several hours at elevated temperatures. Consequently, convective burning is one possible mechanism to provide this spreading of ignition. Convective burning in this material has not been observed before at speeds near 1000 m/s in PBX 9501. Recent experiments have shown that these rates are achievable in PBX 9501. Initial results are presented below.

COMBUSTION IN MICROCHANNELS

Precisely machined slots in PBX 9501 have allowed us to examine the propagation of fast reactive waves in psuedo-cracks of PBX 9501, focusing on the reactive wave velocity and on the interplay of pressure and crack size in PBX 9501 (15). In other words we try to eliminate the mechanics in the problem to study the combustion in isolation. Initial experiments were performed with 4 cm slots that were open to a pressurized large-volume environment. Experiments at initial pressures of 6.0 MPa reveal monotonic reactive wave propagation velocities around 7 m/s for a 100-μm slot. Reactive wave velocities as high as 100 m/s are observed in experiments at initial pressures of 17.2 MPa and various slot widths. Similar experiments at lower pressure sometimes exhibit oscillatory reactive wave propagation in the slot with periodic oscillations whose frequencies vary with combustion vessel pressure. Although the propagation rates achieved were impressive considering the length of the channel and that an end of the gap was open to a large volume, the question remained concerning the ultimate ignition spread rates attainable.

Long Highly Confined Gap Experiment

More highly confined experiments, with longer channels were designed and initial experiments have been performed. Figure 1 shows the experimental setup used. PBX 9501 was mounted on brass and shimmed to produce an 80 μm gap, 19 cm long. The end of the gap was ignited in a small breach area with an equal mixture (by weight) of AP and DHT. Pressure was measured in the breach area and at four locations down the length of the crack. A Plexiglas window provided optical access to the experiment. For low violence events it was expected that this window would release.

In the experiment, the ignition front spread rapidly through the slot, approaching speeds of 1500 m/s, followed by failure of the cell. The speeds were obtained from both the visual and pressure data (see Fig. 2). Figure 3 shows the resulting damage to the cell. There were no clear indications of detonation, although the damage was severe. An additional series of experiments in this configuration is planned and a more complete description of this series of experiments will appear in a later publication. However, this initial experiment clearly shows that convective burning in slots in PBX 9501 can reach ignition spread rates over 1000 m/s, which was a primary aim of this experiment.

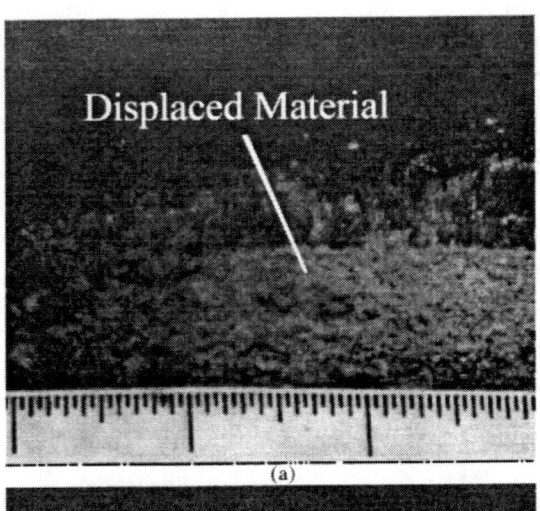
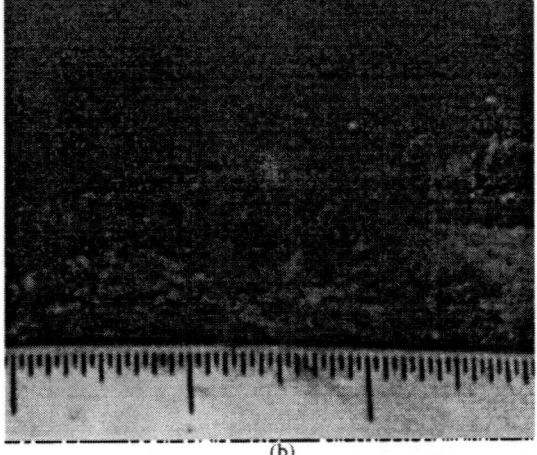

Figure 6. High speed video image of a mound of BTATZ. Part (a) and part (b) are taken 6 ms apart. In part (a) movement is evident ahead of the main front in the mound of material.

RADIAL BURN EXPERIMENT

Figure 4 shows a schematic of the radial burn experiment, or Cindy test. The radial burn experiment is nearly the same as the MCCO experiment, except that it is designed to fit into our pressure vessel, pressure is measured in the center ignition hole, and pristine material is considered (it is not heated). We found that we could achieve ignition in pristine materials if the pressure was initially sufficiently high or a good seal is achieved (using thin Teflon sheets) such that the nichrome wire could pyrolyze enough of the sample to raise the pressure sufficiently.

Figures 5 shows some initial observations from this experiment. The pressure rises very fast as ignition occurs in the small center hole (1 mm diameter). This quickly creates radial cracks that are illuminated by reaction, which is very similar to the MCCO experiments. The aluminum confinement rings were quickly breached by the hot reactive gases, in these initial experiments. Additional experiments will use copper rings that may provide more robust confinement. Since the material was not preheated, the energy dissipated by the crack tip cannot explain the ignition of the reaction in the cracks. Consequently, the most probable explanation is that the cracks are opened by the pressurization from the burning, and convective burning spreads the reaction through the crack.

NOVEL ENERGETIC MATERIALS

Recently at Los Alamos we have studied the combustion of several novel energetic materials. The two materials considered here exhibit rapid, presumably convective, burning in a small mound of material. The first material is BTATZ (3,6-bis(1H-1,2,3,4-tetrazol-5-amino)-s-tetrazine), which is a high nitrogen compound. This material burns rapidly, and produces lots of gas at moderate temperatures. The second is a nano-scale thermite (Al/MoO_3) system that reacts very quickly at high temperatures, and produces solid products. It is interesting to contrast and compare the behavior of these materials in similar experiments.

High Nitrogen (HN) Materials

Some HN materials, such as BTATZ, burn rapidly with very little luminosity. Burning BTATZ in an open configuration (simple mound of material) is shown in Fig. 6. BTATZ burns on the order of centimeters per second in normal burning. However, in this configuration the ignition spread rate is two orders of magnitude faster than the normal burning rate. In part (a) of Fig. 6 material on the surface of the mound can be seen to move ahead of where the material is clearly reacting. This appears to be a self-confined convective spread of ignition within the mound.

Nanoscale Thermite

Similar unconfined experiments were performed using nanoscale thermites (MIC). Since reaction of these materials yield solid phase products, it might be assumed that rapid convective burning would not occur, yet propagation rates exceed 100 m/s. This far exceeds BTATZ in a similar configuration. With the small size of the reactants the barrier to reaction is small in the MIC material. Furthermore, initial air is likely heated and behaves as a working fluid, and gaseous intermediates are probable for this system because of the high temperatures. Figure 7 shows an image from a high-speed video record. A large plume is evident, with the leading edge in the mound of material. The resulting propagation rate is nearly constant except initial and final transients.

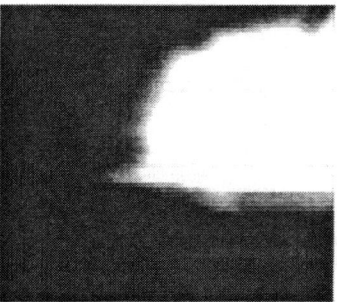

Figure 7. This figure shows an image of nanoscale Al/MoO3 burning from right to left.

Figure 8 shows the same material burning in a Pyrex tube. The spread of reaction can be seen to accelerate rapidly. Speeds exceeding 800 m/s are obtained, although the tube is essentially undamaged. This affect of confinement indicates a mechanism such as convective burning is dominating, although other possible mechanisms are currently being investigated. In contrast, BTATZ burning in a similar tube propagated reaction exceeding 1000 m/s and pulverized the tube. By changing the particle size of these MIC materials the reaction rate can be adjusted. This could potentially be a valuable tool in the study of the propagation physics involved in these materials, and could provide unique model validation data. Classical energetic materials, of course, provide no means of adjusting the reaction rate. Visual access is also easier to obtain because confinement requirements are less difficult.

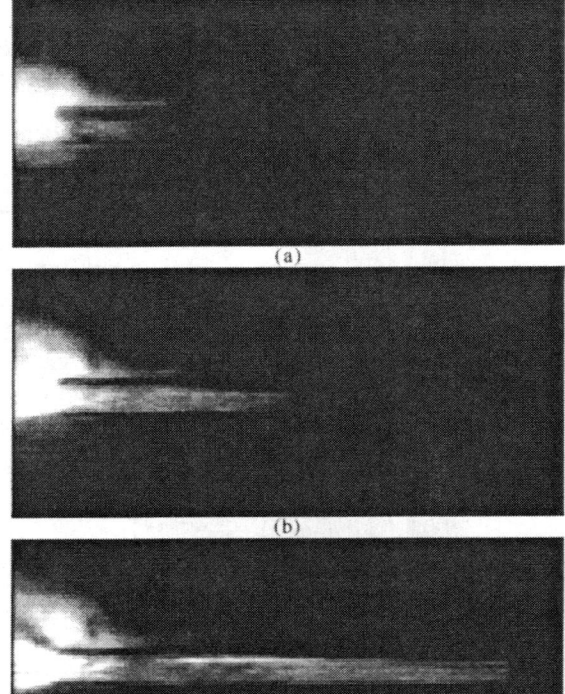

Figure 8. Burning nanoscale Al/MoO$_3$ in 3.8 mm inner diameter Pyrex tube. Part (a) is at t=0, part (b) at t=37 µs, part (c) at t=74 µs, part (d) at t=111 µs.

SUMMARY

This paper examines the role of reactive cracks in reaction violence. Recent results are presented that illustrate the importance of convective burning in accident scenarios. Recent work in this area was reviewed and presented. Initial work involving novel energetic materials was presented. These materials create conditions that are in a different parameter space than classical materials and therefore offer a unique perspective of rapid combustion in energertic materials.

Acknowlegments

These experiments were designed, performed and discussed in close collaboration with several co-workers, including Blaine Asay, Larry Hill, Laine Berghout, Cindy Bolme, Mike Hiskey, Darren Naud, and Bryan Bockmon. Without their efforts this paper would not have been possible.

REFERENCES

1. S. F. Son, H. L. Berghout, C. A. Bolme, D. E. Chavez, D. Naud, and M. A. Hiskey, *Proc. Combust. Inst.*, 28, 919-924 (2000).
2. Hiskey, M. A., Goldman, N., and Stine, J. R., *J. Energetic Mat.*, 16, 119-127 (1998).
3. Chavez, D. E., and Hiskey, M. A., *J. Pyrotechnics*, 7, 11-14 (1998).
4. Aumann, C.E., Skofronick, G.L., and Martin, J.A., *J. of Vac. Sci. Tech.* B 13(2): 1178-1183 (1995).
5. B. W. Asay, S. F. Son, and J. B. Bdzil, *Int. J.of Multiphase Flow* 22, 92-952 (1996).
6. A. F. Belyaev and V. K. Bobolev, *Transition From Deflagration to Detonation in Condensed Phases*, 1975 translation ed. (National Technical Information Service, Springfield, VA, 1973).
7. H. H. Bradley and T. L. Boggs, *Convective Burning in Propellant Defects: A Literature Review*, Naval Weapons Center, China Lake, CA Report No. NWC-TP-6007, 1978.
8. M. Kumar and K.K. Kuo, in *Fundamentals of Solid Propellant Combustion*, (K. Kuo and M. Summerfield, Eds.) AIAA, Inc., New York, 1984, Vol. 90, pp. 339-350.
9. A. L. Ramaswamy and J. E. Field, *J. of Applied Phys.* 79, 3842-3847, (1996).
10. D. J. Idar, R. A. Lucht, J. W. Straight *et al.*, *11th International Detonation Symposium*, Snowmass, Colorado, 101-110, (1998).
11. B. F. Henson, B. W Asay, P. M. Dickson *et al.*, *11th International Detonation Symposium*, Snowmass, Colorado, 325-331, (1998).
12. C. B. Skidmore, D. S. Phillips, B. W Asay *et al.*, *Shock Compression of Condensed Matter -- 1999*, American Institute of Physics, Snowbird, Utah, 1999, pp. 659-662.
13. P. M. Dickson, B. W Asay, B. F. Henson *et al.*, *11th International Detonation Symposium*, Snowmass, Colorado, 606-611, (1998).
14. B. W Asay, P. M. Dickson, B. F. Henson, *et al.* "Large-Scale Annular Cookoff Experiment, JANNAF PSHS meeting, Cocoa Beach, FL, October 1999.
15. H. L. Berghout, S. F. Son, and B. W. Asay, "Convective Burning in Gaps of PBX 9501," *Proc. Combust. Inst.*, 28, 911-918, 2000.

EFFECT OF TEMPERATURE PROFILE ON REACTION VIOLENCE IN HEATED AND SELF-IGNITED PBX 9501

Blaine Asay, Peter Dickson, Bryan Henson, Laura Smilowitz, and Larry Tellier

Los Alamos National Laboratory, Los Alamos, NM 87545

Abstract. Historically, the location of ignition in heated explosives has been implicated in the violence of subsequent reactions. This is based on the observation that typically, when an explosive is heated quickly, ignition occurs at the surface, leading to premature failure of confinement, a precipitous drop in pressure, and failure of the reaction. During slow heating, reaction usually occurs near the center of the charge, and more violent reactions are observed. Many safety protocols use these global results in determining safety envelopes and procedures. We are conducting instrumented experiments with cylindrical symmetry and precise thermal boundary conditions which are beginning to show that the temperature profile in the explosive, along with the time spent at critical temperatures, and not the location of ignition, are responsible for the level of violence observed. Microwave interferometry was used to measure case expansion velocities which can be considered a measure of reaction violence. We are using the data in a companion study to develop better kinetic models for HMX and PBX 9501. Additionally, the spatially- and temporally-resolved temperature data are being made available for those who would like to use them.

INTRODUCTION

Thermal studies of explosives have traditionally measured the time to ignition as a function of an isothermal temperature as the figure of merit. Kinetics for the process have been derived for these measurements. For example, the one-dimensional time to explosion experiment (ODTX) has been used for a number of years, and has produced the best results to date [1]. In this experiment, a sphere of explosive is placed between two heated anvils and the time required for significant reaction is recorded. The results are then plotted as 1/T vs. time. The location of the ignition is not measured and is not known.

Over many years of slow and fast heating tests, the conclusion has been drawn that the location of ignition governs the violence of the explosive response. A slow heat scenario will allow the center of the HE to heat up sufficiently so that once self-heating begins, the reaction runs away at the center. A rapid heating produces high temperatures first at a region near the boundary, and runaway thus occurs there. It has been noted that fast cookoff usually results in a less violent reaction than does slow cookoff, and since the location of ignition is so different between the two cases, this has been thought of as being paramount.

We have conducted a series of tests wherein we measure a temporally and spatially resolved thermal profile [2]. This allows us to ascertain not only at what time the reaction begins, but the location as well. Using these two measures, we have been able to modify the classic ODTX kinetics as well as investigate the effects of location of ignition on violence, separate from the heating rate. This study was designed to investigate the role of thermally induced damage on the violence as measured by wall velocity.

EXPERIMENTAL

The experiments have been previously described [Peter Dickson, 1999 #2] but have been slightly modified for the current study. We have used PBX 9501 which is 95% HMX and 5% binder. The explosive is encased in 2 copper half-cylinders split down the middle (each 10.3 mm in length x 19.4 diameter, with 3.7 mm-thick wall). Heat is supplied to heating wire wrapped around the Cu cylinders via two circuits, one on the top and one on the bottom, each individually controlled. Five thermocouples are placed at the union of the two cylinders at different radial positions (see Fig. 1). The experiment is designed so that the centerline will be the hottest point with heat flow occurring out of each end cap. The system is held together by threaded rod, but is not sealed against gas loss.

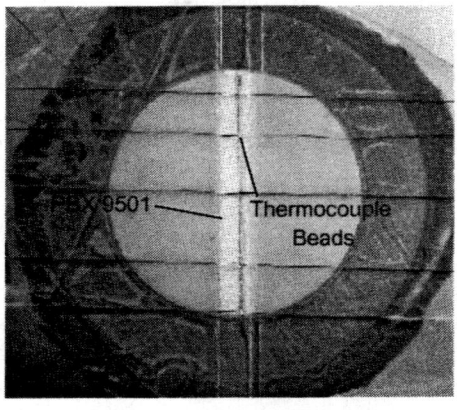

FIGURE 1. Photograph of one-half of experiment showing copper cylinder, explosive and thermocouples

We measure the wall velocity using a 35 GHz microwave interferometer (λ= 8.45 mm). The interferometer is coupled to a PTFE waveguide that is inserted into the armored box containing the explosive. There is a small disk attached to the wall of the experiment that reflects the microwaves (see Fig. 2). We have also reflected the microwaves from the wall of the cylinder with good results. The quadrature signal is then analyzed and position as a function of time is extracted. The data are differentiated to obtain velocities.

For this set of experiments we applied a steady heating ramp until the temperature reached 185C. This is slightly above the phase transition temperature, but below the temperature at which rapid reaction begins. This was then maintained for a predetermined amount of time after which a second ramp was imposed, and self-ignition resulted. The experiments were designed to provide a varying amount of time during which thermal damage could occur. The second ramp accelerated the self-ignition time for convenience.

FIGURE 2. Photograph of assembled experiment showing optional microwave reflector.

RESULTS

Soak times of 1, 2, 3, 4, 6, and 8 hours were used, with two experiments performed using 1 hour soak times to examine reproducibility. For presentation here, the time at which the second ramp was started for each experiment was shifted so that each experiment had a common point. This was done to compare self-ignition times. Figure 3 shows the entire temperature histories while Fig. 4 shows an enlarged view of the ignition region.

The thermocouple records shown are from the regions of the charge that showed the fastest temperature rise. Although each experiment had uniform heating that was well controlled, they did not all ignite at the same radial position.

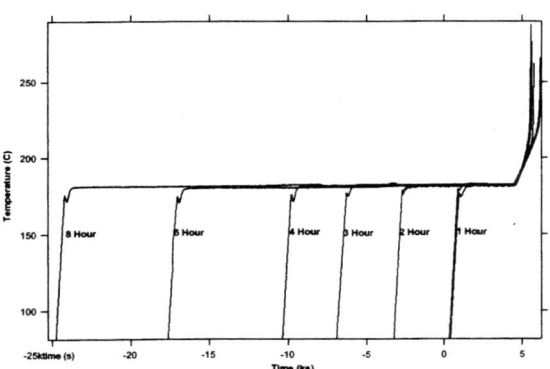

FIGURE 3. Temperature traces showing effect of soak time on self-ignition time. Times shifted for comparison.

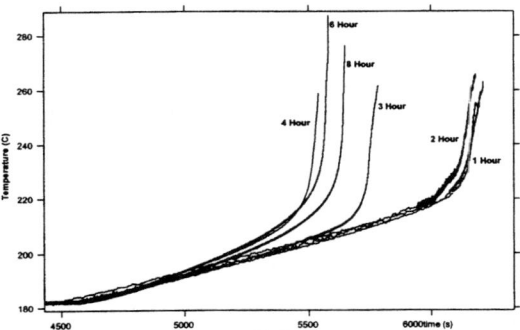

FIGURE 4. Temperature traces showing effect of soak time on self-ignition time. Times shifted for comparison. Time scale enlarged for clarity.

The experiments with the 1 hour soak showed good reproducibility. The successive shots followed a predictable pattern up to and including the 4-hour soak, with each one igniting at earlier times (see Fig. 4). This was expected because with increasing soak time, more damage (e.g., thermal decomposition and fracture) is occurring and the chemical reactions are further advanced. However, the experiment with the 6-hour soak showed a slightly increased ignition time and the test with the 8-hour soak showed a much longer time to ignition. This most likely results from an increasingly significant loss of gaseous products from the system. As heating time increases, porosity of the explosive sample also increases. At some point, the porosity will become interconnected, permitting wholesale loss of products. Because the major exothermic reactions occur in the gas phase, loss of these products results in an increase in ignition time. It is widely known that gas loss can have a profound effect on reaction times. We have maintained however that for relatively short experiments, or very large ones, that the permeation of the gas out of the system is slow enough so as not to affect the outcome. Permeabilities of pressed unreacted systems are very low. This set of experiments demonstrates that fact. We have observed in each experiment that near the end of the exponential temperature rise there is an inflection point in the temperature record. This could be the signature of the melt, or another, unidentified chemical step. We took the derivative of temperature with respect to time at that point. Those results are presented in Fig. 5 along with the measurement of the self-ignition time. The derivative results are linear with heating time. This could illustrate that while globally the time to ignition decreases and then increases because of gas loss, the chemistry occurring at the point of maximum heat release maintains a memory of the thermal history. This measurement could provide a quantitative indicator of reaction violence in that it demonstrates a large increase in heat release (at least three-fold in this example) as a function of damage state.

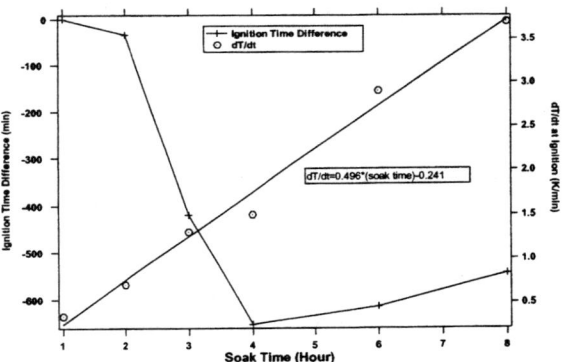

FIGURE 5. Plot showing the self-ignition time difference as a function of soak time and the derivative of the temperature at the endotherm occurring at runaway.

The experiments have been modeled using the 4-step kinetics reported earlier [Peter Dickson, 1999 #2]. The parameters were chosen so as to provide a best fit for the 1-hour case, and then held constant for the remaining cases. The results are presented in Fig. 6. The overall agreement is good across the entire spectrum, but there are some differences. The rise in the calculated baseline temperature for the case with the longest soak time is consistent with the notion that there is appreciable gas loss for this particular experiment. The code predicts heating that is not present in the data. The source of this heating would arise from gas phase reactions.

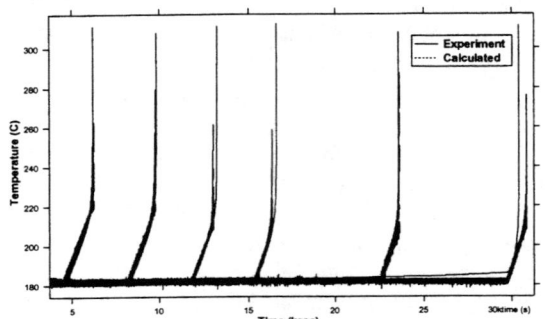

FIGURE 6. Comparison of model prediction with experimental values.

The case expansion velocities were measured using a microwave interferometer. We performed one experiment wherein we purposely detonated the encased explosive. The measured wall velocity was approximately 500 m/s, which compares favorably to a calculation using the Gurney method. The velocities for each of the cookoff experiments have been tabulated and in every case the values have been approximately 100-150 m/s, well below that of a detonation. No distinguishing differences in velocity were noted for the experiments performed under different thermal conditions. One interpretation of this result is that there was no marked difference in reaction violence with the large changes in damage state. However, the difference in reaction rate measured at late times (see Fig. 5) demonstrates that there is a significant difference in the HE behavior at during ignition. The case expansion velocity is an integrated measurement of reaction history, and as such, may not be a reliable indicator of reaction violence in this geometry or at this scale.

Fragments were collected after each experiment. Their size and distribution corroborated the velocity measurements, in that no evidence of detonation was found, and no major differences in fragment distribution with experimental change were noted. However, further analysis is required.

DISCUSSION AND CONCLUSIONS

These experiments were designed to examine the violence of reaction when the explosive was subjected to different degrees of thermal damage. We found that, with the prescribed heating profiles, no differences in violence were obtained. We also noted that the location of ignition varied between the experiments from near the case to the center of the charge. Thus, we have shown that the ignition location is not a primary factor in the determination of violence for a given reaction. The experiments all had different soak times, so the thermal conditions were somewhat different. However, we do not believe that this changes this basic conclusion.

We believe that the second temperature ramp may have masked subtle reactive behavior, in that it overdrove ignition somewhat. Our next series of experiments are designed to modify this part of the profile in an effort to drive the reaction to different levels of violence.

We have also shown that the definition of reaction violence needs to be clarified, in that two separate measures provided different conclusions. We need to better understand the differences in measured reaction rates as a function of damage state, and the reasons that these differences were not reflected in case expansion velocities.

The thermal profiles from these experiments are available in text format to facilitate comparison with various computer models. They can be secured from the authors.

ACKNOWLEDGEMENTS

We acknowledge the support of the LANL HE Science and Surety programs that made this work possible.

REFERENCES

1. McGuire, R.R. and C.M. Tarver. in *Seventh Symposium (International) on Detonation*. 1981. Annapolis, MD.
2. Peter Dickson, et al. *Measurement of Phase Change and Thermal Decomposition Kinetics During cookoff of PBX 9501*. in *Shock Compression of Condensed Matter, American Physical Society Topical Conference*. 1999. Snowbird, UT.

IGNITION CHEMISTRY IN HMX FROM THERMAL EXPLOSION TO DETONATION

Bryan F. Henson, Blaine W. Asay, Laura B. Smilowitz and Peter M. Dickson

Los Alamos National Laboratory, Los Alamos, NM 87545

Abstract. We present a global chemical decomposition model for HMX based materials. The model contains three component processes, the initial beta to delta phase transition, solid to gas decomposition and gas phase ignition, for which all kinetic and thermodynamic parameters are fixed by independent measurement. We present an isothermal ignition calculation over the range of temperatures from thermal explosion to detonation. The calculation is performed for a sphere of material and the critical diameter and time for ignition are determined. The sample diameter, and thus the balance of heat generation and dissipation, is the only degree of freedom in the calculation. The results of the calculation are in good agreement with data with respect to both the ignition times and length scales over the full temperature range of energetic response in HMX.

INTRODUCTION

It has long been known that HMX (octahydro-1,3,5,7-tetranitro-1,3,5,7-tetrazocine) exhibits a wide variety of behaviors when subjected to various thermal fields. Temperatures just above 450 K induce explosion after an induction time from 10^5 to 10^1 s (1). IR laser irradiation generates surface temperatures from 500-700 K and ignition at 10^{-3} s to 10^1 s (2). Shear or frictional heating of pressed solids to 700-900 K results in ignition in 10^{-4} s (3). Planar shocks from 10 GPa to 30 GPa result in detonation over times of 10^{-7} s to 10^{-6} s (4). Despite considerable progress in understanding the chemistry governing these processes, which span twelve decades in time, no comprehensive model of decomposition linking the separate regimes currently exists.

We first present a model independent compilation of ignition data from experiments on HMX based explosive formulations which indicates that a single decomposition model may be sufficient to reproduce the ignition chemistry of this material over the full range of energetic response, from thermal explosion to detonation. We then present a kinetic model of thermal decomposition based on the known chemistry of HMX and a simplified calculation of ignition that is in agreement with observation.

COMPILATION OF IGNITION DATA

We have collected data from experiments under conditions ranging from thermal explosion (1,5-7) and fast pyrolysis (8) to laser ignition (2) and impact induced shear and frictional heating (3). In addition, we include a classic set of detonation initiation experiments (9) where the heating profile from the confined surface of an HMX plastic bonded explosive was measured radiometrically subsequent to shock passage. In a separate publication we have extracted an experimentally constrained chemical rate constant for energy release from these experiments (10). Finally, using recent calculations of mean temperature in the detonation reaction zone of HMX we include the measured reaction zone time as an ignition time in stable detonation (10,11,12).

The set of ignition data for HMX based samples is shown in Fig. 1. The data are compiled, with the exception of the detonation data, as the measured time to ignition as a function of the inverse of the applied temperature. The data set is to our knowledge inclusive, with the following exceptions. Some unconfined thermal explosion experiments have been neglected from the data set as the loss of reactant gases on long timescales can lead to spuriously long ignition times and even ignition in the gas phase away from the sample. In the fast pyrolysis data set of Brill and Brush (8) on thin films we have shown only the subset of data where reaction time was attributed to chemistry rather than thermal transport.

The simplest interpretation of such a linear fit in an inverse temperature plot is presence of a single rate limiting chemical step in the sequence of reactions leading to ignition. We fit the ignition data set of Fig. 1 using the classic Arrhenius form for the temperature dependence of the canonical rate constant $k = A \exp(-E/RT)$, where k is a first order rate constant, A a prefactor, E the activation energy, R the gas constant and T the temperature. We assume the measured ignition times compiled in Fig. 1 to reflect a characteristic reaction time to ignition of $t = 1/k$, which is equivalent to assuming a first order rate law with ignition at a fractional decomposition of 1/e. The solid line of Fig. 1 is the result of a linear regression of the data set according to

$$\ln(t) = (\frac{E}{R})\frac{1}{T} - \ln(A) \quad (1)$$

which yields values of $E = 149 \pm 1.1$ kJ/mole and $A = e^{29.35 \pm 0.26}$ s^{-1}.

CHEMICAL DECOMPOSITION MODEL

The chemical decomposition model presented here is based solely on independently measured rate constants and thermochemical parameters. The model involves first a nucleation and growth mechanism describing the β–δ phase transition, followed by the formation of gas phase species approximately first order in the concentration of δ-HMX and ending in a highly exothermic gas phase reaction mechanism initiated by bimolecular reaction of NO_2 and CH_2O.

The set of coupled differential equations describing the evolution of the key chemical species β and δ HMX, NO_2, CH_2O and HCO are given by

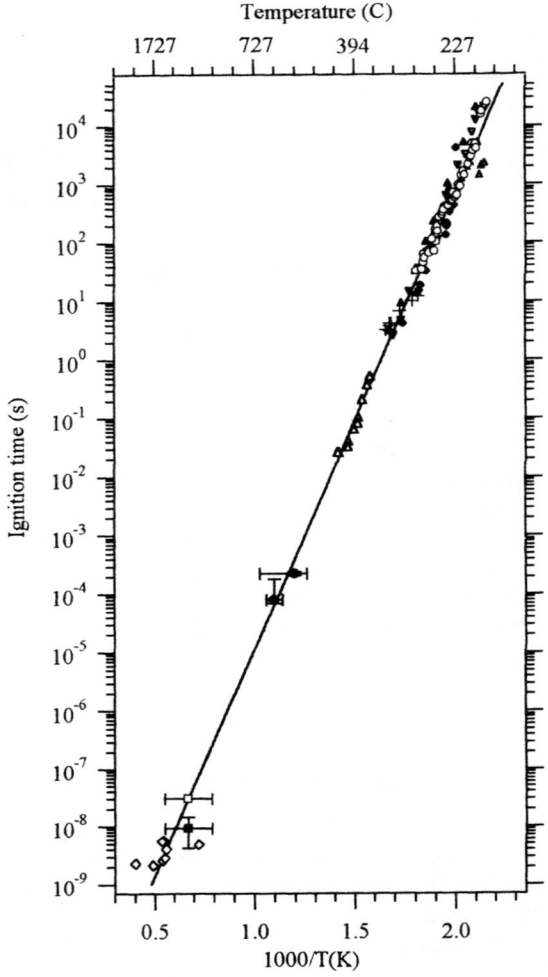

FIGURE 1. Ignition data from HMX compiled as follows: confined thermal explosion studies, filled daimonds (1), filled triangles and inverted triangles (5), open triangles and inverted triangles (6), and open circles (7), fast pyrolysis studies, crosses (8), laser ignition studies, open triangles (2), shear and frictional heating studies, filled circles (3), and detonation experiments, open diamonds (9), solid square (11) and open square (12). The solid line is the fit to the data using Eq. (1)

$$\frac{\partial[\beta]}{\partial t} = -k_1[\beta] + k_{-1}[\delta] - (k_2 - k_{-2})[\beta][\delta] \quad (2)$$

$$\frac{\partial[\delta]}{\partial t} = k_1[\beta] - k_{-1}[\delta] + (k_2 - k_{-2})[\beta][\delta] \quad (3)$$

$$\frac{\partial [NO_2]}{\partial t} = 2\frac{k_{3a}}{k_{3a}+k_{3b}}k_3[\delta] - k_4[NO_2][CH_2O]$$
$$- (k_{5a}+k_{5b})[NO_2][HCO] \quad (4)$$

$$\frac{\partial [CH_2O]}{\partial t} = 4\frac{k_{3b}}{k_{3a}+k_{3b}}k_3[\delta]$$
$$- k_4[NO_2][CH_2O] \quad (5)$$

$$\frac{\partial [HCO]}{\partial t} = k_4[NO_2][CH_2O]$$
$$- (k_{5a}+k_{5b})[NO_2][HCO] \quad (6)$$

The rate constants are given by

$$k_i(T) = e\frac{k_B T}{h}Q\exp(\frac{TDS^* - DH^*}{RT}) \quad (7)$$

where $k_i(T)$ is a canonical rate constant in s^{-1} or cm^3/mole s, depending on reaction order, T the absolute temperature, k_B and h are Boltzman's and Plank's constants respectively, H_i^* and S_i^* are the activation enthalpy and entropy of the activated state, R is the gas constant and e is the base of the natural logarithm. Q is an equilibrium constant relating the concentrations of the activated species to reagents. The parameter values used in the calculation are given in Table 1.

IGNITION CALCULATION

The calculation of ignition time, which will be detailed in a future publication, was performed as follows. The coupled set of differential equations describing the chemical model, Eqs. 2-6, was solved starting from an initial temperature T_o and density, ρ_i = 0.0063 moles/cm^3. At each time step the amount of heat liberated or consumed was determined from the incremental concentration change and corresponding heat of reaction, taken from the literature. The multiphase character of the system and the relevant length scale for thermal diffusion were incorporated into the calculation by introducing a thermal loss term

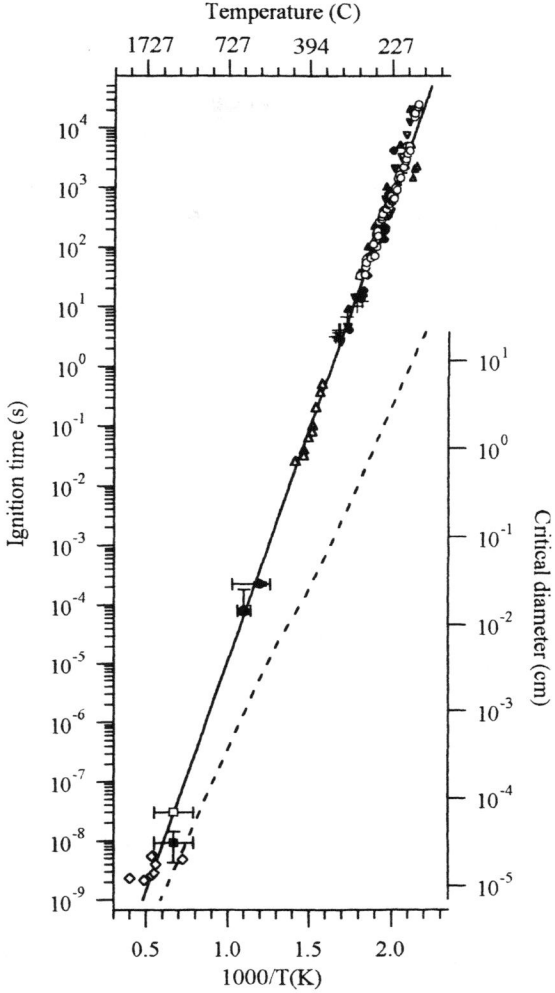

FIGURE 2. Ignition data from HMX as in Fig. 1. The solid line is the calculated time to ignition and the dashed line the critical diameter.

$$\frac{1}{\pi d^2 \left(1 - \frac{[\beta]+[\delta]}{\rho_i}\right)}\frac{6l}{C_p\rho} \quad (8)$$

where d is diameter, ρ the density, C_p the system heat capacity and l the thermal diffusivity. This term applies a diameter to the problem which defines the length scale for the calculation. The term $1-([\beta]+[\delta])/\rho_i$ is the fraction of the initial density that has converted to gaseous species. The heat capacity of the system is calculated as a weighted average of solid and gaseous components from this fraction.

The calculation is run at the initial T_o and ρ_i as a function of d until a critical diameter, d_c, is determined for ignition (i.e. for $d < d_c$ no ignition is observed). The time to ignition is then determined for this diameter. The resulting ignition time as a function of temperature is shown as the solid line in Fig. 2. Also shown in Fig. 2 is the critical diameter determined for each temperature.

TABLE 1. Rate constant parameterization of Eq. 7

	Q	ΔS^* J/mole K	ΔH^* kJ/mole
k_1	1	159.57	213.8
k_{-1}	1	136.81	204.0
k_2	10^{-9}	149.34	79.7
k_{-2}	10^{-9}	126.58	69.9
k_3	1	-0.84	141.0
k_{3a}	1	38.47	178.0
k_{3b}	1	-79.18	106.0
k_4	1	-24.29	57.0
k_{5a}	1	-0.51	0.0
k_{5b}	1	-1.18	0.0

CONCLUSIONS

The primary result presented here is the model independent existence of a single linear relationship in the ignition literature for HMX when compiled in a ln t vs. 1/T Arrhenius plot. Such a linear relationship is strong evidence for a single rate limiting chemical step in the ignition mechanism that appears to persist over the entire range of energetic response. Such a result indicates the possiblility of constructing chemical decomposition models of HMX which may of sufficient simplicity to be of use in large scale calculations and at the same time reproduce the behavior of the material with some accuracy.

In addition, the broad temperature range encompassed by this relationship provides severe constraints on models of decomposition. For instance, although the phase transition kinetics (13) and ignition mechanism involving NO_2 and CH_2O and HCO (14) utilized in this model have been studied in detail, the kinetics of delta phase decomposition to gas phase species is still uncertain. In particular, the observation that gas phase species appear in two channels, involving either NO_2 or CH_2O and characterized by two rate constants (here k_{3a} and k_{3b}) (15) leads naturally to a competitive mechanism of formation from δ-HMX. Such a mechanism, however, exhibits two limiting slopes in an Arrhenius plot and is not consistent with observation. We have utilized a first order decomposition of delta here, given by k_3, and modified the appearance ratio of the two channels with ratios of k_{3a} and k_{3b}. It is an interesting algebraic feature of the work in the literature that k_3, which corresponds to the first order rate constant controlling ignition, may be set equal to $\sqrt{k_{3a}k_{3b}}$ to within the precision of the measurements. The details of the intermediate steps in this mechanism will be the subject of future work.

REFERENCES

1. J. Zinn and R. N. Rogers, J. Phys. Chem. **66**, 2646 (1962).
2. G. Lengelle, A. Bizot, J. Duterque and J.- C. Amiot, Rech. Aerosp. **2**, 1 (1991).
3. B. F. Henson, B. W Asay, P. M. Dickson, C. Fugard, and D. J. Funk, *Eleventh Symposium (International) on Detonation*, (1998).
4. *LASL Explosive Property Data*, Edited by T. R. Gibbs and A. Popolato (University of California Press, 1980).
5. C. M. Tarver, R. R. McGuire, E. L. Lee, E. W. Wren and K. R. Brein, *Seventeenth Symposium (International) on Combustion*, (1978).
6. R. R. McGuire and C. M. Tarver, *Seventh Symposium (International) on Detonation*, (1981).
7. C. M. Tarver, S. K. Chidester and A. L. Nichols, J. Phys. Chem. **100**, 5794 (1996).
8. T. B. Brill and P. J. Brush, *Seventh Symposium (International) on Detonation*, (1981).
9. W. G. Von Holle and C. M. Tarver, *Seventh Symposium (International) on Detonation*, (1981).
10. *Occam's razor and detonation: Evidence for thermal equilibrium in the detonation of HMX*, B. F. Henson, L. Smilowitz, B. W Asay, P. M. Dickson and P. M. Howe, submitted to Phys. Rev. Lett.
11. R. L. Gustavsen, S. A. Sheffield, and R. R. Alcon, *Eleventh Symposium (International) on Detonation*, (1998).
12. L. G. Green and E. James Jr., *Fourth Symposium (International) on Detonation*, (1965).
13. L. Smilowitz, B. F. Henson, B. W Asay and P. M. Dickson, to be submitted to J. Chem. Phys.
14. C.-Y. Lin. H.-T.Wang, M. C. Lin and C. F. Melius, Int. J. Chem. Kin.**22**, 455 (1990).
15. T. B. Brill, H. Arisawa, P. J. Brush, P. E. Gongwer and G. K. Williams, J. Phys. Chem. **99**, 1384 (1995).

INSTRUMENTATION OF SLOW COOK-OFF EVENTS

H. W. Sandusky and G. P. Chambers

Energetic Materials Research and Technology Department
NAVSEA Indian Head Division, 101 Strauss Ave., Indian Head MD 20640-5035

Abstract. An arrangement was developed for validating models of slow cook-off. Experiments were conducted on the explosive PBXN-109 with measurements of temperature, pressure, and volume until the onset of reaction; and measurements of case velocity and blast overpressure during reaction. The goal is to relate changes in the energetic material during heating with time and position for onset of reaction plus reaction violence as a function of sample size, confinement, gas sealing, and heating profile. A mild range of reactions occurred as evidenced by fragmentation of the confinement into mostly large pieces; however, at the highest confinement no sample was recovered.

INTRODUCTION

Cook-off is both complex and quite dependent on a variety of environmental factors, such as the rate of heating, which is fast in a fuel fire and orders of magnitude slower from indirect heating. In addition to heating rate, the variables include sample size (diameter and length-to-diameter ratio), radial and axial confinement, initial ullage, and sealing of pyrolysis products during heating. Since only a limited number of full-scale tests can be conducted, which limits the number of variations in environmental factors, it would be advantageous to predict cook-off response to different hazard scenarios with computer models. Models at the Sandia National Laboratory, Albuquerque (SNLA) (1,2) and the Lawrence Livermore National Laboratory (3) were evaluated against small-scale screening tests, such as the Variable Confinement Cook-off Test (VCCT). It was recognized that model validation requires better controls on the tests and more measurements in each test. In addition, the properties of heated explosives are being characterized (4,5) to support the modeling effort.

Suitable metrics for comparing models and experiments include the rate of expansion of the energetic material, temperature at various locations within the energetic material and at various points on and in the apparatus, strain in the confinement and pressure buildup within the confinement as a function of time, evolution of gaseous decomposition products, time at which cook-off occurs, and the violence of that reaction. In some experiments, it would be advantageous to stop the heating at some point in the cycle and remove the explosive for evaluation of thermal damage and decomposition. To meet these requirements, an experimental arrangement was developed that is different than the usual closed pipe in most small-scale cook-off tests. An initial series of experiments was conducted on the explosive PBXN-109, which is RDX and aluminum in a rubber binder. A companion program (6) with a closed pipe is being conducted at the Naval Air Warfare Center/China Lake (NAWC/CL).

EXPERIMENTAL ARRANGEMENT

The apparatus shown in Figure 1 consists of a test cell mounted between flat springs in a load frame, which is simply two pieces of 152 mm wide steel channel connected by 25 mm threaded rods. Between each spring and base there can be a spring stop in which is mounted a potentiometer-based displacement transducer. Spring displacement has also been measured by a strain gage mounted on the spring, which is calibrated by replacing the explosive samples with hydraulic oil and pressurizing the test cell. There is a clear field of view around the test cell for measurements of rapid expansion and fragmentation of the confinement with flash radiography and high-speed photography. The other measurement of reaction violence is blast overpressure by a transducer within a meter of the apparatus.

Details of the test cell are shown in Figure 2. A cylindrical sample of the same dimensions as that in the VCCT, 25.4 mm diameter by 63.5 mm long, is radially confined in a seamless mechanical tube of 1018 steel with variable thickness and axially with spring-loaded rams. Confinement is dependent on the thickness of the tube and the strength of the springs. The springs reduce the internal pressure buildup – expansion from heating and damage in the sample and pyrolysis products – so that the seals on the rams are not breached. The rams have axial ports for instrumentation within the samples, which to date have been thermocouples sealed by epoxy. Each ram also has an O-ring seal with the confinement tube. The tube is heated by resistance wire with minimal insulation so that the field of view is not obscured. The rams are somewhat thermally isolated from the springs to reduce the heat loss from the ends of the sample and thereby maintain more uniform temperatures over the sample length.

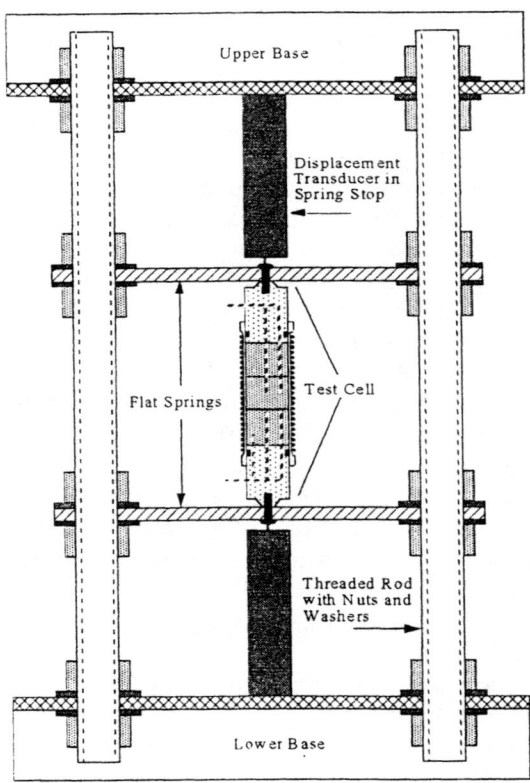

FIGURE 1. Overall experimental apparatus.

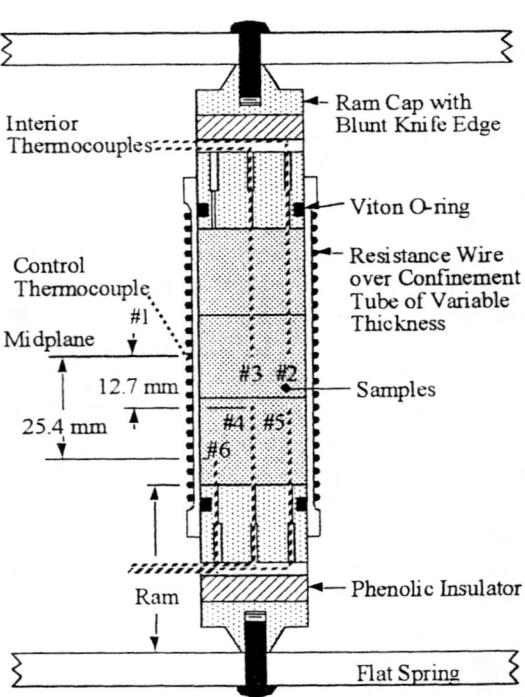

FIGURE 2. Details of test cell.

TABLE 1. Summary of PBXN-109 Cook-off Experiments

Tube Wall Thickness (mm)	Heating Profile, Start & Ramp	Ullage (%)	Gradient (°C/25 mm)	Cook-off Temp. (°C)	% Explosive Recovered	No. of Tube Fragments
1.27	130 °C, then 6 °C/hr	~3.6	18	172	35	4
1.90	150 °C, then 3 °C/hr	~3.6	11	170	6	3
2.54	150 °C, then 3 °C/hr	~1.0	11	165	0	6

The displacement of the springs was measured either by a potentiometer-based displacement transducer, as shown in Figure 1, or with a strain gage on each spring calibrated for displacement. For both techniques, spring displacement was related to pressure on the rams by hydraulically pressurizing the tube before each experiment. Temperatures on the confinement and within the sample were measured by copper-constantan thermocouples from 0.25-mm diameter wire with Teflon insulation. The tube typically had a strain gage at the midplane and a break-wire, both circumferentially mounted. During heating, tube strain is a second measure of interior pressure, calibrated by hydraulically pressurizing the tube. If a high-elongation strain gage of annealed constantan is used, this gage can also follow the initial tube expansion during the onset of cook-off. Along with the break-wire, the strain gage also serves as trigger probe for the dynamic diagnostics during cook-off. The apparatus and the associated instrumentation were evaluated experimentally and computationally (7) with inert samples of Teflon.

The PBXN-109 is from the same batch as that used in the NAWC/CL tests. (6) The sample, with a total mass of ~52 g, is in three pieces with 1-mm diameter holes drilled for the thermocouples.

RESULTS AND DISCUSSION

The conditions and results from three experiments are summarized in Table 1. The major input condition varied was the wall thickness of the confinement tube. Heating profiles replicated those at NAWC/CL (6). The relatively high starting temperatures prior to the slow heating permitted completion of each experiment in 8 hrs. The 3.6% ullage in the first two experiments was from the sample being 0.33 mm smaller in diameter than the tube and slightly oversized holes for thermocouples. The 1% ullage in the last experiment was achieved by eliminating the clearance between the sample and tube.

With the ~3.6% ullage, there was no significant spring deflection from thermal expansion; however, during the three hours prior to cook-off there was an exponential increase from thermal damage and pyrolysis. This is illustrated in Figure 3 for the second experiment, where the spring deflection at cook-off corresponds to 1.9% increase in sample length and a 2.7 kpsi (18.7 Mpa) sample pressure. With the minimal ullage in the last experiment, thermal expansion was recorded as shown in Figure 4; however, the confinement leaked at 4 kpsi. The last pressure drop was probably from seepage around a thermocouple whose signal was lost.

Heat losses through the rams caused the thermal gradients listed in Table 1 between the midplane thermocouples and the one 25.4 mm from the midplane. The gradient is significant in that the levels of thermal damage and self-heating

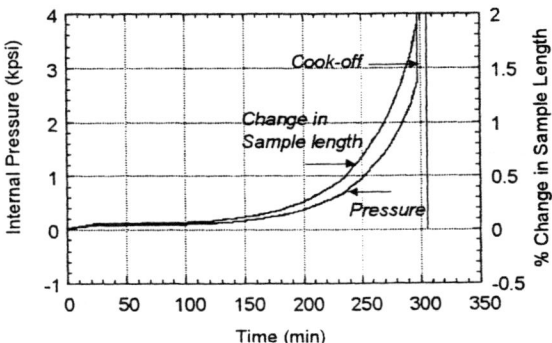

FIGURE 3. Spring response in second experiment.

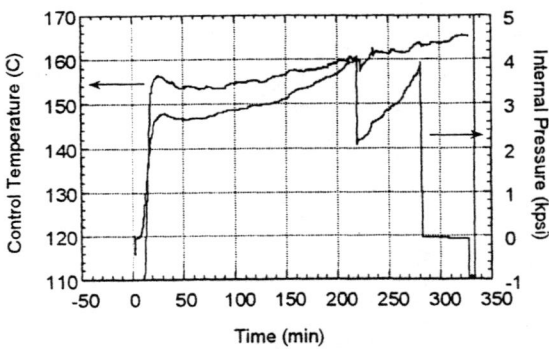

FIGURE 4. Control temperature and spring response in third experiment.

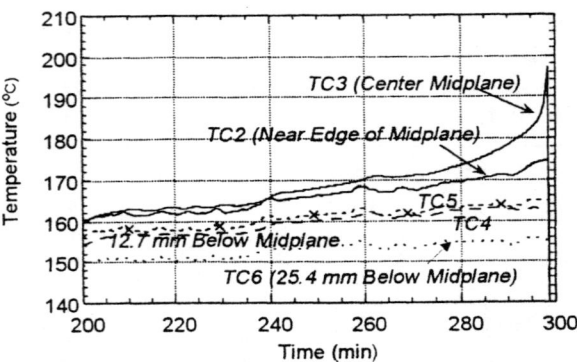

FIGURE 5. Interior temperatures in second experiment.

in the samples are reduced near the rams. Internal temperatures near cook-off are shown in Figure 5 for the second experiment, with the thermocouple locations shown in Figure 2. Despite the uniformity of temperature for the two midplane locations up to one hour before cook-off, self-heating only appeared on the axis. At 12.7 mm below the midplane where temperatures are 6 °C lower, no self-heating appeared even on the axis.

The amount of explosive recovered decreased with increasing tube wall thickness. The explosive recovered in the second experiment was only millimeter size pieces, indicating an axial reaction that fractured the surrounding annulus once the confinement failed. In the experiment with the thickest wall, an overpressure of 12.6 psi was measured 0.60 m away; there were several small tube fragments; and at a circumferential wall strain of 8%, the wall velocity was 42 m/s. Perhaps this was an explosion, whereas the reactions in the previous experiments were deflagrations.

SUMMARY AND CONCLUSIONS

Simultaneous mechanical and thermal measurements were made during slow cook-off of an explosive designed to be insensitive to various hazardous stimuli. Cook-off violence for PBXN-109 was similar to that observed in other small- and full-scale tests. There is significant pressure from thermal expansion when ullage is essentially eliminated, but only expansion without a pressure increase for several percent of ullage. At a heating rate of 3 °C/hr, self-heating occurs on the axis in a small, <10 mm zone.

ACKNOWLEDGEMENTS

This program was promoted by a number of individuals, most notably Art Ratzel at SNLA, Alice Atwood at NAWC/CL, and Ruth Doherty at this Center. Ken Schebella when visiting this Center from DSTO in Australia, Kevin Gibson, Richard Lee, and Vasant Joshi assisted various aspects of the experimental development. The Office of Naval Research provided funding.

REFERENCES

1. Baer, M. R. et al., in *Proceedings of Eleventh International Detonation Symposium*, Office of Naval Research, ONR 33300-5, 1998, pp. 852-861.
2. Schmitt, R. G. et al., op. cit., pp. 434-442.
3. Nichols, A. L. III et al., op. cit., pp. 862-871.
4. Maienschein, J. and Chandler, J. B., op. cit., pp. 872-879.
5. Renlund, A. M. et al., op. cit., pp. 127-134.
6. Atwood, A. I. Et al., in *Proceedings of JANNAF 19th Propulsion Systems Hazards Subcommittee Meeting*, CPIA Publ. 704, Vol. I, 2000, pp. 205-220.
7. Bill Erikson, SNLA, private communication.

KINETICS OF THE β–δ PHASE TRANSITION IN PBX 9501

L. B. Smilowitz, B. F. Henson, B. W. Asay, P. M. Dickson, J. M. Robinson

Los Alamos National Laboratory, Los Alamos, NM 87545

Abstract. The initial step in the thermal decomposition of HMX is the solid state phase transition from the centrosymmetric beta form to the noncentrosymmetric delta form. The symmetry change makes the phase transition amenable to the application of second harmonic generation (SHG) as a probe of transition kinetics. We have used SHG to study the temperature dependence of the kinetics for unconfined PBX9501 and HMX. Spatially resolved SHG measurements have shown a nucleation and growth mechanism for the solid state phase transition. We have measured the transition rate as a function of temperature in order to obtain the activation energy and entropy of transition, which determine the phase transition kinetics. Additionally, we have observed temperature dependent reversion of the delta phase to beta phase and have found that we can control the reversion rate by controlling the cooling.

INTRODUCTION

PBX9501 is based on the energetic material HMX, which is found to have four polymorphs, three of which are stable under various conditions of pressure and temperature[1]. The solid state phase of the material which is most stable at ambient temperatures and pressures is the relatively insensitive β phase [1]. The phase that becomes stable at higher temperatures and is the sensitive phase of the material is the δ phase [1]. Due to a change in the symmetry of the material during this phase transition, second harmonic generation (SHG) is a physical observable which is strongly correlated with the δ phase of the material and virtually nonexistent with the β phase[1, 2]. Thus, SHG provides a virtually instantaneous probe of the δ fraction with a nearly zero background generated by the β phase. This allows us to use SHG to measure the kinetics of the solid state phase transition as a function of temperature history of the material. Additionally, we have used spatially resolved SHG images to identify the transition mechanism as consistent with nucleation and growth of the new phase. By measuring the conversion rates as a function of temperature and using the rate equations for a nucleation and growth process, we are able to model the full temperature dependent behavior of the transition kinetics. Using the transition state theory formulation for rates, we are able to fully define the kinetics of the transition based on independently measured thermodynamic parameters for the β–δ phase transition in HMX. The thermodynamic behavior shows how well we are able to predict the dependence on temperature for the transition time. The kinetics show how well the model predicts the details of the process. The qualitative dependencies of the reverse phase transition from δ back to β will be discussed.

EXPERIMENTAL

The experimental work presented here is based on an application of SHG as a probe for the phase fraction of δ-HMX. We have demonstrated the utility of SHG in this application in a previous letter[2]. We have shown the signal measured from HMX samples to have the narrow spectral distribution and square intensity dependence of an elastic χ^2 process[2]. We have not measured an absolute χ^2 efficiency. In these experiments, the SHG intensity is normalized to that from the fully converted sample and we use the progress from zero to one as a measure of the transition from zero to 100% delta. The SHG is interpreted as being proportional to the square of the delta fraction present.

Materials and Methods

For this study, we have used PBX 9501 samples obtained by pressing molding powder made from 95% Holsten HMX with 2.5% estane and 2.5% nitroplasticizer by weight.

The experiments were conducted in a sample oven with optical access to allow uniform heating of the samples. Thermocouples are placed on the sample and oven to monitor and control the sample temperature. We have found that for heating rates of 5 K/minute or slower, radial gradients across the samples of less than ½ K can be maintained. The extreme sensitivity of the reaction kinetics on temperature makes this temperature uniformity essential. Heating profiles used are a 5 K/minute ramp to a temperature at which the phase transition can occur. The sample is held at that temperature to allow the transition to occur. Cooling is done at approximately 3 K/minute to 293C, or to a temperature below the delta phase stability temperature to allow phase reversion to occur.

The SHG microscope consists of a pulsed Nd:YAG pump laser with approximately 50 ps pulses at 1064 nm and approximately 10 mJ per pulse at 10 Hz repetition rate. The beam is very mildly focused onto the sample using a 1 m focal length lens. The sample is imaged by a microscope that is directly coupled to a low noise CCD.. Filters are used between the sample and the microscope to insure that imaging shows only the 532 nm second harmonic light generated by the sample. Typically, SHG microscopy is performed in transmission through the sample. A magnification of 4.5x is used and the pixels are 24 micron squares yielding a pixel image size of ~5 microns on a side. Collection time per image varies from hundreds of milliseconds to tens of seconds depending on light level.

The total second harmonic light generated by the sample is monitored by a photomultiplier tube which measures the backscattered SHG from the sample. A small percentage of the laser intensity is split off and monitored to normalize out laser intensity fluctuations.

RESULTS AND DISCUSSION

An important result of this work comes from spatially resolved SHG imaging of the transition which indicates a nucleation and growth mechanism for the phase transition. The qualitative imaging observations are that the delta phase grows via nucleation and growth in HMX crystals. Initial nucleation occurs at one or two sites within a crystal and then growth of the delta phase propagates within the crystal from those sites. The spatially resolved nucleation and growth is not as clearly observable in PBX9501 due to its highly scattering nature, but the integrated SHG signal from PBX9501 has the recognizable sigmoid shape representative of a nucleation and growth process.

We have mapped out the kinetics of the phase transition as a function of the temperature history of the samples for a range of temperatures and heating rates. The rate equation for the reversible nucleation and growth of a new phase is:

$$\frac{\partial \delta}{\partial t} = k_1 \beta + k_2 \beta \delta - k_{-1} \delta - k_{-2} \beta \delta \quad (1)$$

We have assumed a fully reversible nucleation and growth mechanism based on our observations of the reversion from δ to β phase upon controlled cooling of the samples. The rate constants from Eq. (1) are given by [2]

$$k_i(T) = e \frac{k_B T}{h} Q \exp(\frac{TS_i^* - (H_i^* + PV_i^*)}{RT}) \quad (2)$$

where $k_i(T)$ is a canonical rate constant in s^{-1} or cm^3/mole s, depending on reaction order, T the absolute temperature, k_B and h are Boltzman's and Plank's constants respectively, E_i^*, S_i^* and V_i^* are the activation energy, entropy and volume of the activated state, R is the gas constant and e is the base of the natural logarithm. Q is an equilibrium constant relating the concentrations of the activated complex to those of the stable reagent(s). The units of Q determine the concentration units in the problem, and vary according to the order of the reaction By taking the transition state to be the melt state of HMX in accordance with experimental observations, we can base the rate constants for the growth step entirely on thermodynamic parameters for HMX that have been independently measured.

The nucleation rates have been determined by starting with values within the range cited in the literature for the thermodynamic parameters E_a and S and optimizing the best fit for the integrated rate law to a series of conversions at different temperatures [3].

The times for conversion as a function of conversion temperature are shown in Fig. 1. This figure compares the expected half time (time for 50% conversion) with the experimentally observed half times for approximately isothermally heated samples. The times are measured with time zero being the time at which the sample reaches 158C. At temperatures below 158C, the degree of conversion to delta phase would be negligible.

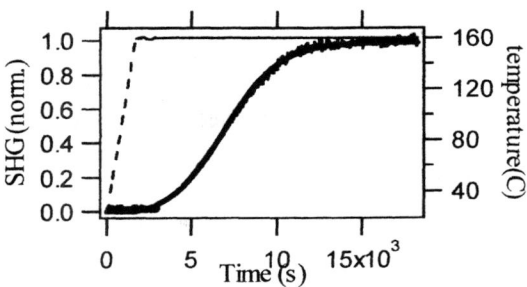

FIGURE 2: Fit of conversion data to model.. The integrated SHG data are shown as points. The solid lines are optimized fits to the data using the integrated form the of the rate law, and the dashed line is the temperature history.

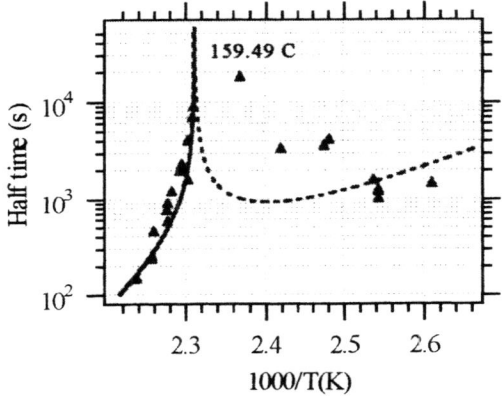

FIGURE 1. Compilation of measurements. All points are from measurements made by second harmonic generation. The data are plotted as the time to half conversion as a function of 1000/T(K), where T(K) is the isothermal temperature. Both β-δ conversion and δ-β reversion data are shown (triangles).

As shown in Fig. 1, the model predicts the half time for conversion. What can be seen in Figure 2 is that it is also able to predict the detailed kinetics for the phase transitions quite well. We have made several simplifying assumptions in this model. Additional parameters which affect the kinetics but have not yet been included in our model include pressure affects and the third thermally accessible phase of HMX which is the noncentrosymmetric alpha phase. Other limitations of our model are that we have not identified the exact reversion mechanism and have made the simplifying assumption that the SHG is quadratically proportional to the delta fraction.

We have also addressed the question of the stability of the delta phase. We have found that we can control the reversion to beta phase by controlling the cooling rate of the material. Figure 3 demonstrates the difference in phase behavior between a rapid cooling and a slowly stepped cooling process. FTIR spectra were taken to confirm the identifications of the final states as beta and delta. This demonstrates that the final phase of the material upon cooling back to room temperature is dependent on the cooling rate. From previous work on the phase diagram for HMX, we know that the thermodynamically stable phase at standard temperature and pressure is the beta phase [1, 4]. However, by rapidly cooling samples, we have kinetically trapped the material in its delta phase. This dependence on temperature history explains the wide range of previously reported stability times for the metastable delta phase upon cooling to room temperature.

The exact mechanism for the phase reversion is still not known nor is its dependence on physical parameters understood. However, the assumption of a reversible nucleation and growth mechanism allows us to predict forward going kinetics as well as to approximately predict the half times for reversion as a function of cooling. Understanding the reversion mechanism is important for addressing safety issues for explosives that have been heated to temperatures below ignition and then quenched.

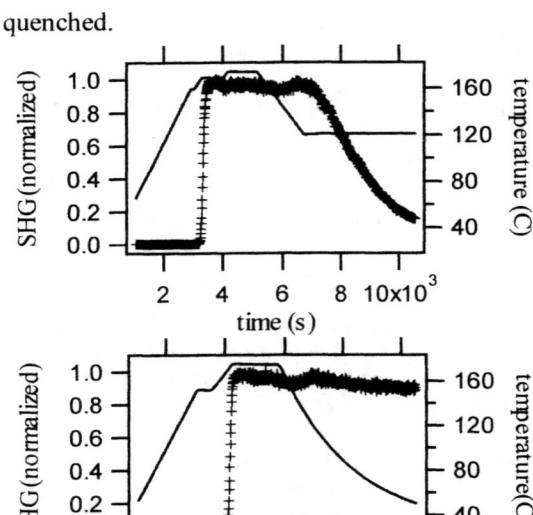

FIGURE 3. The dependence of reversion on cooling rate. SHG data (points, left axis) and temperature (solid curve, right axis) are plotted. In the top panel the temperature is raised to 174 C to generate conversion and then lowered to 121 C, leading to reversion. In the bottom panel the temperature is allowed to relax to room temperature.

By cycling a sample through the β–δ conversion, reversion back to β phase, and then a second conversion to δ phase, we can test the affects of irreversible work done during the initial conversion process on the phase transition kinetics. We found that we could fully revert the sample to β phase, confirming the source of the SHG as a volumetric phase transition and showing that the contribution to the SHG from the surface area generated by cracking the material does not have a significant contribution to the SHG signal. Also, the kinetics of the second conversion process closely match the original onset behavior implying that the nucleation and growth rates are insensitive to the degree of cracking of the sample as well.

CONCLUSIONS

We conclude that the beta to delta phase transition in PBX 9501 proceeds via nucleation and growth of the delta phase. It is a thermodynamically first order phase transition whose kinetics are governed by a mixed second order reaction mechanism. Using the experimental observation of the transition state being the melt state, we are able to determine the governing rates based almost entirely on independently obtained entropies and energies [5]. These rate equations provide an excellent prediction for transition times as a function of temperature over a broad range of times and temperatures. The model additionally is able to predict the detailed kinetic behavior of the phase transition at each temperature. Observation of reverse delta to beta phase transitions upon cooling to temperatures below the delta phase stability point led us to use a fully reversible nucleation and growth model. While the thermodynamics predicted by this model approximately capture the reversion process, the details of the behavior are not captured. Further work is underway to understand the detailed mechanism for the reverse phase transitions. All work to date has been performed at atmospheric pressures. Future work will address the question of how pressure will impact the dynamics of the transition through the volume change between the phases.

ACKNOWLEDGEMENTS

Funding for this work came from the HE Basic Sciences and Surety programs.

REFERENCES

1. Karpowicz, R.J., L.S. Gelfand, and B. T.B., *Application of solid-phase transition kinetics to the properties of HMX.* AIAA Journal, 1983. **21**(2): p. 310-311.
2. Henson, B.F., et al., *Dynamic measurement of the HMX beta-delta phase transition by second harmonic generation.* Physical Review Letters, 1999. **82**(6): p. 1213-1216.
3. Henson, B.F., et al., *The β – δ phase transition in PBX 9501: Thermodynamics.* J. Chem. Phys., 2001: p. submitted.
4. Cady, H.H., A.C. Larson, and D.T. Kromer, Acta Cryst. Sect. B, 1963. **16**: p. 617.
5. Cady, H.H.. 2000- private communication.

THE MEASUREMENT OF HOT-SPOTS IN GRANULATED AMMONIUM NITRATE

W.G. Proud

Physics and Chemistry of Solids Group, Cavendish Laboratory, Madingley Road, Cambridge, CB3 0HE, United Kingdom

Abstract. Ammonium Nitrate (AN) is one of the components of the most widely used explosive in the world namely, ammonium nitrate: fuel oil mixtures (ANFO). By itself, it is an oxygen positive explosive with a large critical diameter. Hot-spots are produced in explosives by various means including gas space collapse, localised shear or friction. If these hot-spots reach critical conditions of size, temperature and duration reaction can grow. This deflagration stage may eventually transition to detonation. This paper describes high-speed image-intensified photography study in which the number and growth of hot spots in granular AN are monitored for a range of different impact pressures. The results can be used in detonation codes to provide a more accurate and realistic description of the initiation process.

INTRODUCTION

The ignition of energetic materials is important for studies of initiation of explosives and their safe handling and use. Calculations carried out in the middle of the 20th century indicated that ignition did not usually occur due to bulk heating of material but following local concentration into "hot-spots" [1,2]. If heat losses by conduction, convection, radiation and self-heating dominate, the "hot spot" is quenched. However, if the energy produced by chemical reaction exceeds the losses reaction will build up and spread through the material [3]. Under suitable conditions there may be a transition from deflagration to detonation.

In liquids, the conditions for critical hot spot formation have been extensively studied [4-8] and the size, duration and temperature required have been reasonably well-quantified. One advantage of a liquid or a solid crystal of explosive is the inherent homogeneity of the system and relative ease of visualisation of the reaction front. Many explosive systems are heterogeneous and energy concentration relies on an interplay between the components. In the ammonium nitrate system studied here, some of the causes of energy concentration would be compression of gases in the pores of the granular bed, fracture of the crystals, jetting of fragmented material, friction between the explosive grains or shear banding in either single crystals or compacted material [9-11].

Many of these processes occur on a very short time-scale and the diagnostics used need to have sufficient time resolution and sensitivity. The present study uses visualisation of light emission from granular beds of ammonium nitrate (AN) impacted by copper flier plates at velocities of 200 - 700 m s^{-1}. High-speed photodiodes are used to track the light emission throughout the impact process and a DRS Hadland Ultra-8 camera to take high-speed images of the system. This camera has a maximum framing rate of 10^8 f.p.s. and takes eight images. At such high capture rates, the light levels of many systems would be low, however, this camera is fitted with an image intensifier.

The experimental study was conducted with several aims; to assess the contribution of adiabatic pore collapse, to measure the hot-spot density for different impact conditions and to determine if the kinetics of reaction are affected by the induced shock strength.

EXPERIMENTAL

An impact cell, figure 1, contains a bed of AN 2 mm thick. The front plate is 2 mm thick and the rear glass window is a 25 mm thick. The optic fibre is fed into a photodiode sensitive to visible wavelengths and with a time resolution of 1 ns. A mirror, at 45° to the rear of the cell, directs light into the high-speed camera.

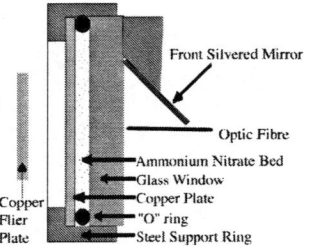

FIGURE 1. Ammonium Nitrate Cell. Copper flier plate included to indicate the 12 mm impact diameter.

The cell is mounted in at the end of a 19 mm bore gas gun. The projectile was a 2mm thick copper plate mounted on a plastic sabot. The pressure in the impact chamber can be left at atmospheric pressure or pumped to a vacuum of 10^{-3} bar. The flier plate was aligned to the cell to a tolerance better than 0.05°.

IMPACT EXPERIMENTS

Experiments were performed using framing photography to observe the fracture of the glass window. These showed that a ring crack formed which developed into a cone crack around the contact area, The cone was formed within ~4 µs of impact. The centre of the window remained transparent beyond 10 µs after impact. Given the shock speeds in the system the high-pressure state induced by the impact lasts 1-2 µs.

A test sample of sugar with grain size 150 - 210 µm size was impacted at a velocity of 700 m s^{-1}. Sugar is a material noted for its triboluminescence and hence any light seen in this image is due to sugar fracturing, the compression of the gas in the pores of the bed or fracture of the cell materials. No signal was detected on the photodiode output or the Ultra-8 framing picture shown in figure 2, which is the *negative* of the captured image, and records the integrated light output for 10 µs. A scale on the image indicates that the field of view encompasses the whole impact area. There was limited light emission and the conclusion is that there is no significant contribution from these effects on a timescale *twenty* times that used in the studies with AN.

FIGURE 2. Impact on sugar at 700 m s^{-1}, negative image, exposure time 10 µs, gain 100%.

A series of tests were then conducted on AN beds of 78% theoretical maximum density (TMD). These were impacted at velocities ranging from 200 to 700 m s^{-1}. The results of some of these impacts are shown in figure 3, the vertical column corresponds to the impact velocity and the horizontal rows to the time after impact. Each frame had an exposure time of 500 ns. One advantage of the Ultra-8 system is that the frames are independent of each other so images can be taken with no interframe time. The images have undergone the same data manipulation and the frames have been adjusted for variations in camera sensitivity. In the images 12 mm corresponds to 600 pixels: each pixel corresponds to the area of 2 or 3 AN crystals. A long focal length microscope, such as a Parfocal K2, would magnify the system so that individual grains could be resolved. This research is not reported here.

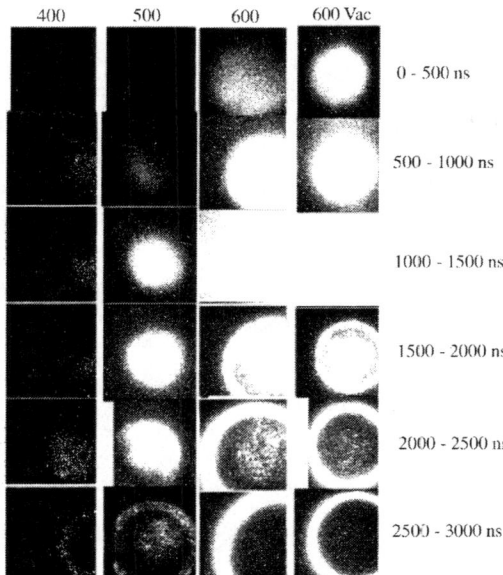

FIGURE 3. The high-speed sequences of impact on a bed of 78% TMD AN. Vertical columns show the intensity of light at the impact velocity indicated at the top of the column. The rows show images at the same capture time.

Below 400 m s^{-1} no light was detected either with the photodiode or with the camera. At 400 m s^{-1} there is limited light output and spots of light are recorded which rapidly fade. At 500 m s^{-1} no light is seen in the first 500 μs but intensity starts to build up from 1 μs onwards. The annular structure that can be seen towards the end of most of the sequences is due to interaction between the edge of the impactor acting on the impacted copper plate and the glass anvil i.e. a region of intense shear. At 600 m s^{-1} two sequences are shown "600" was performed at atmospheric pressure while "600 Vac" was performed at a reduced pressure of 1 mbar. Interestingly there is very little difference between the two images except they are at slightly different magnifications.

The output from the photodiode gives an indication of the intensity of the light output figure 4.

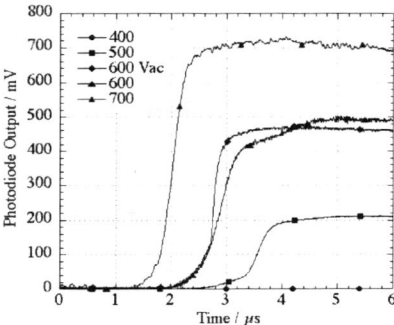

FIGURE 4. The photodiode output from experiments conducted at velocities between 400 and 700 m s^{-1}. The legend indicates the impactor velocity in m s^{-1} and "Vac" indicates an impact performed under vacuum.

The intensity is higher and the rise time shorter with increased impact velocity. The traces could also indicate that reaction spreads faster at higher velocities and is more intense due to grain fragmentation and friction between the fragments.

In figure 5 the traces are normalised according to maximum intensity while the time axis is normalised with respect to the velocity of impact after the sohck pulse has entered the AN bed.

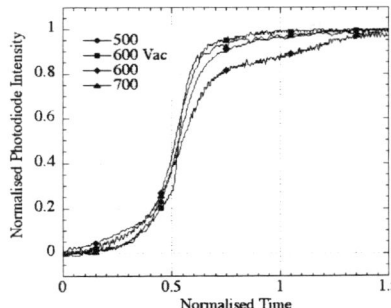

FIGURE 5 Normalised Photodiode Intensity traces.

There is remarkable similarity between these traces. The possible exception being "600" though "600 Vac" matches well with the atmospheric data from both 500 and 700 m s^{-1}. This is indicative that the kinetics of the process are self-similar under the range of impact conditions applied.

DISCUSSION AND CONCLUSIONS

A system has been developed capable of allowing hot-spots to be visualised in granular beds.

Interestingly, ignition for oxygen-positive AN, is independent of adiabatic heating of the initial gas content in the pores of the material, though gas produced by fracture and frictional rubbing may have a role.

The reaction proceeds in a self-similar fashion once the ignition threshold at ~400 m s^{-1} impact velocity has been exceeded.

The light output in figure 3 for AN compared with in figure 2 for sugar confirms that hot-spot ignition has taken place.

The similarities between "600" and "600 Vac" suggest that crystal fracture produes some gas as suggested by Chaudhri and Field [14]. Comparison with earlier work on slurries and emulsions of AN to which microballoons have been added suggests that for the shock process involved (several GPa) jetting could be a significant hot-spot mechanism. Crystal fragmentation would help ignition break out. Future research will aim to identify the precise hot-spot mechanism. Studies are now underway to look at the effect of bed density, additives which increase or deacrease friction, different grain sizes and composition of the gaseous phase in the bed.

ACKNOWLEDGEMENTS

J.E.Field is acknowledged for his encouragement. J. Gilbert of DERA Fort Halstead and I. Kirby are acknowledged for their support. C. Granstrom, E. Molin and P. Kalafatis are acknowledged for their help. R. Marrah of the Cavendish Laboratory is thanked for his technical aid. EPSRC is acknowledged for funding of the high-speed camera.

REFERENCES

1 F. P. Bowden, M. A. Stone, and G. K. Tudor, Proc. R. Soc. Lond. A **188,** 329-349 (1947).
2 F. P. Bowden and A. D. Yoffe, *Initiation and Growth of Explosion in Liquids and Solids (republ. 1985)* (Cambridge University Press, 1952).
3 J. E. Field, Accounts Chem. Res. **25,** 489-496 (1992).
4 F. P. Bowden and M. P. McOnie, Proc. R. Soc. Lond. A **298,** 38-50 (1966).
5 G. D. Coley and J. E. Field, Proc. R. Soc. Lond. A **335,** 67-86 (1973).
6 N. K. Bourne and J. E. Field, Proc. R. Soc. Lond. A **435,** 423-435 (1991).
7 E. Wlodarczyk, J. Tech. Phys. **33,** 35-61 (1992).
8 E. Wlodarczyk, J. Tech. Phys. **33,** 133-166 (1992).
9 N. K. Bourne and J. E. Field, Proc. R. Soc. Lond. A **455,** 2411-2426 (1999).
10 J. E. Field, N. K. Bourne, S. J. P. Palmer, and S. M. Walley, Phil. Trans. R. Soc. Lond. A **339,** 269-283 (1992).
11 J. E. Field, G. M. Swallowe, and S. N. Heavens, Proc. R. Soc. Lond. A **382,** 231-244 (1982).

CHAPTER XVI

SHOCK-INDUCED MODIFICATIONS AND MATERIAL SYNTHESIS

COMPUTATIONAL MODELING OF THE SHOCK COMPRESSION OF POWDERS

David J. Benson*, Ian Do† and Marc A. Meyers*

*University of California, San Diego, CA, U. S. A.
†Livermore Software Technology Corporation, Livermore, CA, U. S..A.

Abstract. The modeling of both inert and reactive materials at the meso-scale with an Eulerian finite element program is discussed. Issues that have an effect on the calculated response, including mixture theory and interface tracking, are briefly presented with recent calculations modeling shock initiated chemical reactions (SICR).

INTRODUCTION

Research on the shock compression of powders is very challenging because it is currently impossible to experimentally monitor the response of individual particles. Experimentalists can only measure the bulk response of the powder and characterize the specimen before and after the experiment. The details of the transient response on the particle level are therefore largely inferred. Analytical models which generate closed-form solutions must necessarily introduce simplifying assumptions so that the powder is treated as a continuum or as a periodic structure. Computational modeling, which permits arbitrarily complicated models limited only by the available computer resources, has therefore become increasingly important as researchers try to bridge the gap between experimental investigations and our understanding of the response on the mesoscopic level.

Hydrocodes have been the most productive computational tool for investigating the shock compression of powders. For periodic arrays with closest packing at modest pressures, the Lagrangian formulation is both accurate and effective. However, real powders have random packing and are usually at a density significantly below their theoretical packing density. Large scale jetting occurs with localized melting, and the Lagrangian mesh can no longer track the flow of the material.

The Eulerian formulation has been popular for many years for ballistic impact and shaped charge calculations [1, 2, 3]. Eulerian and arbitrary Lagrangian-Eulerian (ALE) formulations permit material to move relative to the mesh, and therfore permit arbitrarily large material flows. Individual elements, or zones, may contain several materials, and numerical interface reconstruction methods [4, 5, 6] calculate the location and orientation of the material interfaces within the elements. Phase changes, chemical reactions, and material failure are relatively easy to include in an Eulerian formulation by simply changing the composition within the elements. Williamson [7] pioneered the discrete modeling of particles in powders with the Eulerian hydrocode CTH [2], and the Eulerian formulatinon has proven effective for simulating a variety of powder processes, e.g., [8, 9, 10, 11, 12, 13, 14, 15, 16, 17].

THE EULERIAN FINITE DIFFERENCE/ELEMENT METHOD

Eulerian hydrocodes were originally developed by the finite difference community. In many instances, including the research presented here, the same algorithm can be described with either a finite difference or finite element formalism. A brief overview of the Eulerian finite element method is presented in this section to introduce the essential computational concepts. The detailed description [18] of the numerical methods used in this paper, and the review of the methods currently in general use [19], provide additional information on multi-material Eulerian formulations.

Operator splitting permits the sequential solution in two steps of the Eulerian conservation equations,

$$\frac{\partial \rho}{\partial t} + \nabla \cdot (\rho u) = 0 \qquad (1)$$

$$\frac{\partial \rho u}{\partial t} + \nabla \cdot (\rho u \otimes u) = \nabla \cdot \sigma + \rho b \qquad (2)$$

$$\frac{\partial \rho e}{\partial t} + \nabla \cdot (\rho e u) = \sigma : \dot{\varepsilon} \qquad (3)$$

where ρ is the density, u is the velocity, x is the spatial coordinate, σ is the Cauchy stress tensor, $\dot{\varepsilon}$ is the strain rate tensor, b is the body force, and e is the internal energy. The Lagrangian step, performed first, advances the solution in time, while the Eulerian step accounts for the transport between the elements. An Eulerian formulation uses a spatially fixed mesh, while an arbitrary Lagrangian-Eulerian (ALE) adds the option of the mesh evolving with time, but in a manner that is independent of the material motion.

The Eulerian and ALE formulations include algorithms that aren't in the Lagrangian formulation and which have a potential to affect the accuracy of a mesoscale analysis of powders:

- The interface reconstruction defines the material boundaries.
- The mixture theory for elements containing a mixture of materials governs the distribution of strain between the materials, and therefore the material response between adjacent powder particles.
- Both the mixture theory [20] and the interface reconstruction may require information about the ordering of the material interfaces [21].

INTERFACE RECONSTRUCTION

Modern Eulerian hydrocodes use either volume of fluid (VOF) [22] or level set [23] methods to calculate the location of the material interfaces. For problems in high pressure physics, volume of fluid methods are currently more popular. Most codes use Young's method [4], or a method derived from it, e.g., [5, 6, 24]. On a conceptual level, the approach is very simple, however, the algorithmic details are very complicated. Within an element, the material interface is approximated as a straight line or plane. The orientation of the interface is determined by a difference stencil or least squares algorithm, and the location of the interface is determined by volume conservation.

Adjacent particles must have individual material numbers to prevent them from blending into each other. When the material interfaces of particles blend together, the particles have effectively bonded mechanically, which may adversely affect the accuracy of the calculation.

The spatial resolution of the interface reconstruction scheme may limit the resolution of material jetting or exhibit break-up in the jet which is not physically real. The orientation of the material interfaces may enter into the physics of the calculation, e.g., contact mixture theories [20]. At the high pressure associated with shocks, the shear stress at a contact point is limited by the material flow stress, which is likely to be very low relative to the shock pressure. Therefore, the precise orientation of the material interface has little influence on the particle interactions, but it is important for the transport.

MIXTURE THEORIES

The mixture theory in an element containing multiple materals partitions the strain rate during each time step among the materials, and calculates the mean stress in the element from the stresses in the individual materials. The mixture theory therefore has a large potential for affecting the accuracy of the solution since it governs the interactions of the material interfaces of adjacent particles.

The mean stress, $\bar{\sigma}$ is calculated as the volume-weighted average of the individual stresses in the materials, σ_m,

$$\bar{\sigma} = \sum_m V_m^f \sigma_m \qquad (4)$$

where V_m^f is the volume fraction of material m in the element. This average, which can be justified by homogenization theory, is used by virtually all mixture theories.

The two most popular mixture theories for hydrocodes are the mean strain rate ("springs in parallel") and the mean pressure ("springs in series") mixture theories. In the former, the mean strain rate of the element is used to update the stress in every material. The latter mixture theory equilibrates the pressure between the materials by adjusting the partitioning of the volumetric strain rate. These two mixture theories bound the volumetric response of the mixture, with the mean strain rate giving the stiffer response. One serious flaw with the mean strain rate mixture theory is that empty space, usually referred to as the "void material", is compressed at the same rate as all the other materials in the element, and the void between two material interfaces is never completely closed. The mixture theory is therefore usually modified to preferentially compress out void before compressing the remaining materials.

In a series of computational experiments [12], the shock compression of a powder was simulated with both mixture theories. To explore the interaction of the material models with the mixture theories, the simulations were run with and without a viscous term in the material model. Contrary to expectations, the simulations with the viscous term were more sensitive to the mixture theory than the inviscid simulations in terms of the temperature profiles and the collapse of the voids between the particles. Overall, the pressure equilibration mixture theory gives better results.

THE MODEL BOUNDARY VALUE PROBLEM

A typical model for a metallic powder [14], idealized as having spherical particles, contains approximately 50 to 250 particles, see Figure 1. The number of particles is chosen so that the particle size distribution is represented with a reasonable statistical accuracy and to eliminate boundary effects.

The compression of the powder is modeled by imposing a velocity on one boundary while leaving the remainder fixed. These boundary conditions idealize the sample holder and piston as rigid bodies, but they are typically stiffer than the material being compressed. Boundary effects are observed in the calculations, namely a different packing geometry due to the rigid walls, and for strong shocks, there are some temperature anomalies at the piston face due to the shock viscosity.

Although three-dimensional calculations would be best, two-dimensional calculations show excellent qualitative correlation with experiments. For example,

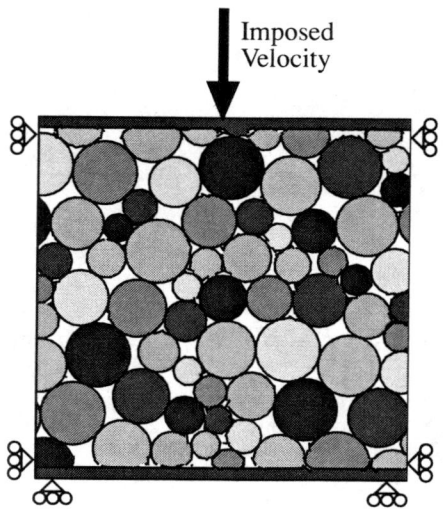

FIGURE 1. The initial state of a metallic powder subjected to shock compression.

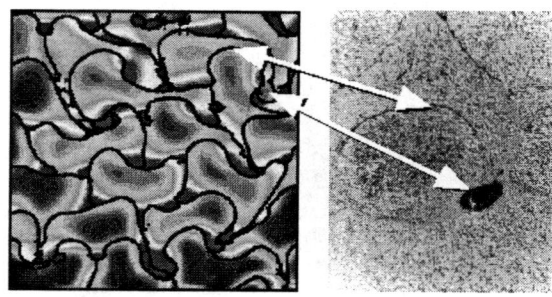

FIGURE 2. Comparison of the predicted morphology of a steel powder with the experiment. The contours in the solution represent temperature, with the light regions being the hottest.

in Figure 2, the calculation predicts the characteristic shapes of the deformed particles and regions of localized melting.

Accurate microstructural models are required for accurately modeling the response of a powder. Many metallic powder particles are adequately approximated by circles in two dimensions. Other powders, however, have particles with irregular shapes, and it is difficult to generate realistic model microstructures. To circumvent this problem, digitized micrographs have been used to generate the computational model.

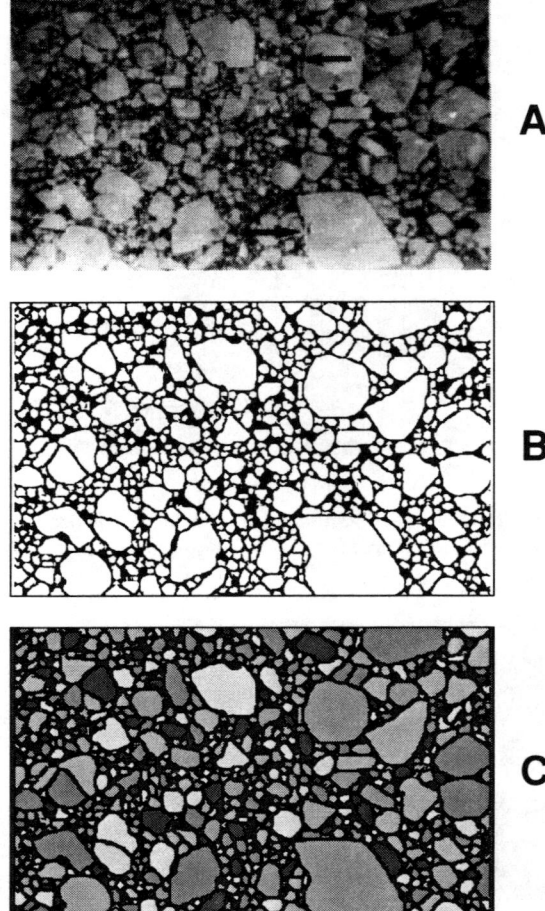

FIGURE 3. The three stages of image processing for importing an experimentally acquired microstructure: A) original micrograph, B) after enhancing the contrast, and C) after the gray scale levels are assigned to indicate the material number.

The steps for building a model from an experimentally acquired microstructure, shown in Figure 3, are:

1. Generate the digital image of the microstructure. CCD cameras directly provide a digitized image, while photographs must be scanned.
2. Use image processing techniques so that each material element (ME) has a unique gray level. When MEs of the same gray level are adjacent, one of them is assigned a new gray level to preserve their individual identities.
3. The microstructure is mapped to the computational mesh pixel by pixel. Based on a pixel's gray level, a material number is associated with it, and the pixel's volume is distributed to the appropriate overlapping elements.

SHOCK INITIATED CHEMICAL REACTIONS

The roles of the material properties, the geometry, and the configuration on the shock-induced reaction threshold (SICR) and the extent of reaction (EOR) in reactions of the type $A + B \rightleftharpoons C$ are investigated in [25, 26, 27]. The model reaction, **Nb + 2Si $\rightleftharpoons$ NbSi$_2$**, was chosen for the research because it has been studied extensively. Modeling the process turned out to be very challenging because of the binary reaction. The material models are independently evaluated for every element in the computational mesh. Using this strategy in our early calculations, when a multi-material element exhausted one of its reactants, the reaction was quenched in the element, thereby limiting the reaction zone to less than one element width. An operator splitting method was introduced to permit nonlocal reactions to occur [25, 27]. Further development of this approach is planned.

The chemical kinetics are modeled with an Arrhenius rate equation. There are many methods for constructing the species evolution equations for chemical reactions. The method chosen here closely follows the approach taken by [28] in CHEMKIN, a software package for treating chemical kinetics, developed at Sandia National Laboratory.

The general species balance equation for the k-th reaction can be written as

$$\sum_{i=1}^{I} \nu'_{ik}[x_i] \rightleftharpoons \sum_{i=1}^{I} \nu''_{ik}[x_i], \qquad (5)$$

where $[x_i]$ is the molar concentration (mole per unit volume) of the i-th species; and ν'_{ik} and ν''_{ik} are, respectively, the forward and backward stoichiometric coefficients of the i-th species and the k-th reaction. The corresponding molar production (or destruction) rate of the i-th species (concentration per unit time, summing over K reactions) is given as

$$[\dot{x}_i] = \sum_{k=1}^{K} [\nu''_{ik} - \nu'_{ik}] r_k, \qquad (6)$$

where r_k is the reaction rate for the k-th reaction, which can be expressed as

$$r_k = \left\{ k_{f_k} \prod_{i=1}^{I} [x_i]^{v'_{ik}} \right\} - \left\{ k_{b_k} \prod_{i=1}^{I} [x_i]^{v''_{ik}} \right\}. \quad (7)$$

As indicated, r_k is a function of the species concentrations and the instantaneous thermodynamic state of the reacting species. The forward and backward reaction rate coefficients, k_{f_k} and k_{b_k}, respectively, contain the physical state information. The general forms of the Arrhenius rate coefficients are

$$\begin{aligned} k_{f_k} &= A_{fk} \exp\left\{\frac{-E_{fk}}{RT}\right\}, \\ k_{b_k} &= A_{bk} \exp\left\{\frac{-E_{bk}}{RT}\right\}. \end{aligned} \quad (8)$$

A_{fk} and A_{bk} are the forward and backward frequency factors, respectively, which represent the collision probabilities among the species. E_{fk} and E_{bk} are the forward and backward activation energies which represent the energy barriers that the system must overcome in order for the reaction to occur. R is the universal constant and T is the reacting temperature.

The morphology evolution in Figure 4 shows that the reaction front essentially coincides with the shock front. Severe plastic flow and jetting of the solid particles can be seen. The silicon reacts almost completely within this duration, and the unreacted material is the excess niobium in the non-stoichiometric powder. Experiments [29, 30, 31, 32] have shown that shock-induced chemical reactions can occur over a time range from hundreds of nanoseconds to microseconds for various solid powder mixtures. Some results support the hypothesis that the reactions initiated during the shock compression are dominated by processes occurring during the stress pulse rise time or throughout the high pressure state before the expansion release, i.e. on the scale of the mechanical equilibration time. The present model produces results that approximate those observations. The pressure pulse first arrived at the right hand side boundary at about 0.34 μs, and all the voids have collapsed by 0.36 μs.

Comparing the left one-third of the domain at times of 0.24 and 0.36 μs, it is observed that reactions in this region went to completion before the shock had passed through the specimen. For example, notice the deformed particle shapes in the lower left hand corner

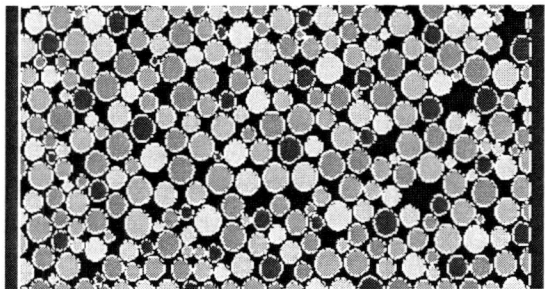
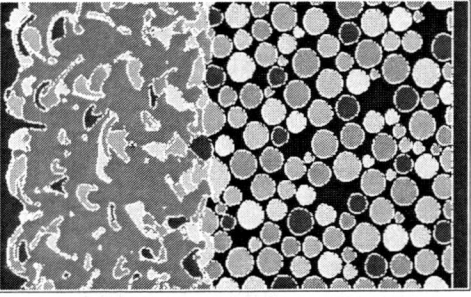
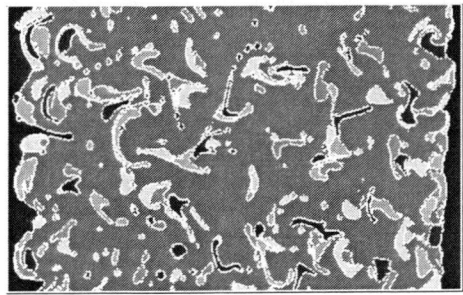

FIGURE 4. Material evolution of a baseline model during a SICR simulation at 0.00, 0.24, and 0.36 μs.

of the computational domain. The reaction consumes the reactant particles quickly. The unreacted niobium particle shapes deform further, but the niobium content stays the same, indicating that the local reaction rate is very high, and that possibly the entire reaction occurs during the pressure rise time.

CONCLUSIONS

The multi-material Eulerian finite element method has proven to be an effective tool for modeling the shock processing of powders, including processes with chemical reactions. Although the present calculations are limited to two dimensions, they compare well in a qualitative sense to the available experimental data, and in some cases, the quantitative agreement is also very good.

ACKNOWLEDGEMENTS

Funding for this research was provided by the NSF DMI grant 9612017, LANL grant UC94610017-3L, and the ARO MURI grant for lightweight armor.

REFERENCES

1. Holian, K. S., Mandell, D. A., Adams, T. F., Addessio, F. L., Baumgardner, J. R., and Mosso, S. J., "MESA: A 3-D Computer Code for Armor/Anti-Armor Applications", in *Proc. of the Supercomputing World Conference*, 1989.
2. McGlaun, J. M., Thompson, S. L., and Elrick, M. G., "CTH: A Three-Dimensional Shock Wave Physics Code", in *Proceedings of the 1989 Hypervelocity Impact Symposium*, 1989.
3. Hancock, S., PISCES–2DELK theoretical manual, Tech. rep., Physics International (1985).
4. Youngs, D. L., "Time Dependent Multi-Material Flow with Large Fluid Distortion", in *Numerical Methods for Fluid Dynamics*, edited by K. W. Morton and M. J. Baines, 1982, pp. 273–285.
5. Johnson, N., personal communication, Los Alamos National Laboratory (1990).
6. Bell, R. L., and Hertel Jr., E. S., An improved material interface reconstruction algorithm for Eulerian codes, Sandia Report SAND92-1716-UC-410, Sandia National Laboratories, Albuquerque, NM (1992).
7. Williamson, R. L., and Berry, R. A., "Microlevel numerical modeling of the shock wave induced consolidation of metal powders", in *Proceedings of the Fourth American Physical Society Topical Conference on Shock Waves in Condensed Matter*, 1986, pp. 341–346.
8. Williamson, R. L., *J. Appl. Phys.*, **68**, 1287–1296 (1990).
9. Benson, D. J., and Nellis, W. J., *Applied Physics Letters*, **65**, 418–420 (1994).
10. Benson, D. J., and Nellis, W. J., "Numerical Simulation of the Shock Compaction of Copper Powder", in *Proceedings of the AIRAPT/APS High Pressure Science and Technology Conference*, American Physical Society, Colorado Springs, 1993.
11. Meyers, M. A., Benson, D. J., and Shang, S. S., "Energy Expenditure and Limitations in Shock Consolidation", in *Proceedings of the AIRAPT/APS High Pressure Science and Technology Conference*, American Physical Society, Colorado Springs, 1993.
12. Benson, D. J., *Modelling and Simulation in Materials Science and Engineering*, **7**, 333–354 (1999).
13. Howe, P. M., Conley, P., and Benson, D. J., "Microstructural Effects in Shock Initiation", in *Proceedings of the Physics of Explosives Technical Exchange Meeting*, Paris, France, 1996.
14. Benson, D. J., Nesterenko, V. F., Jonsdottir, F., and Meyers, M. A., *Journal of the Mechanics and Physics of Solids*, **45**, 1955–1999 (1997).
15. Benson, D. J., *Modelling and Simulations in Materials Science and Engineering*, **2**, 535–550 (1994).
16. Benson, D. J., *Wave Motion*, **21**, 85–99 (1995).
17. Benson, D. J., Tong, W., and Ravichandran, G., *Modelling and Simulations in Materials Science and Engineering*, **3**, 771–796 (1995).
18. Benson, D. J., *Computational Mechanics*, **15**, 558–571 (1995).
19. Benson, D. J., *Computer Methods in Applied Mechanics and Engineering*, **99**, 235–394 (1992).
20. Benson, D. J., *Computer Methods in Applied Mechanics and Engineering*, **140**, 59–86 (1997).
21. Benson, D. J., *Computer Methods in Applied Mechanics and Engineering*, **151**, 343–360 (1998).
22. Hirt, C. W., and Nichols, B. D., *Journal of Computational Physics*, **39**, 201–225 (1981).
23. Chang, Y. C., Hou, T. Y., Merriman, B., and Osher, S., *Journal of Computational Physics*, **124**, 449–464 (1996).
24. Rider, W. J., and Kothe, D. B., *Journal of Computational Physics*, **141**, 112–152 (1998).
25. Do, I. P. H., and Benson, D. J., *International Journal of Computational Engineering Science*, **1**, 61–80 (2000).
26. Do, I. P. H., and Benson, D. J., *International Journal of Plasticity*, **17**, 641–668 (2001).
27. Do, I. P. H., *Shock Induced Chemical Reactions of Multi-Material Powder Mixtures – An Eulerian Finite Element Computational Analysis*, Ph.D. thesis, University of California, San Diego (1999).
28. Kee, R. J., Rupley, F. M., Meeks, E., and Miller, J. A., CHEMKIN III: A Fortran chemical kinetics package for the analysis of gas phase chemical and plasma kinetics, Tech. Rep. SAND96-8216, Sandia National Laboratory, University of California (1996).
29. Boslough, M. B., and Graham, R. A., *Chemical Physics Letters*, **121**, 446–452 (1985).
30. Batsanov, S. S., Doronin, G. S., Klochkov, S. V., and Tuet, A. I., *Combustion, Explosion and Shock Waves - Fizika Goreniya I Vzryva*, **22**, 765–768 (1986).
31. Boslough, M. B., *International Journal of Impact Engineering*, **5**, 173–180 (1987).
32. Thadhani, N. N., Graham, R. A., Royal, T., Dunbar, E., Anderson, M. U., and Holman, G. T., *Journal of Applied Physics*, **82**, 1113–1128 (1997).

THREE-SCALE MODEL FOR NUMERICAL SIMULATION OF MECHANO-CHEMICAL PROCESSES IN SHOCK-COMPRESSED POWDER BODIES

Vladimir N. Leitsin, Vladimir A. Skripnyak, and Maria A. Dmitrieva

Department of Mechanics of Solids, Tomsk State University, Lenin ave., 36, Tomsk 634050, Russia

Abstract. The shock-assisted and shock-induced chemical reactions in powder systems are investigated. Three-scale model of a reacting powder mixture representing physicochemical processes of shock synthesis of materials at micro-, meso-, and macroscopic levels is used. The element of macroscopic structure of concentration inhomogeneity of powder mixture is considered as a representative volume of heterogeneous medium. The model takes into account the modification of the structure of a powder material and changing the reactivity of powder mixture under shock compression. The model is used for computer simulation of the synthesis process in Ni-Al powder mixtures.

INTRODUCTION

The shock synthesis of materials is a perspective direction in the development of techniques of powder metallurgy. This method combines the technological stages of the activation of components, the compaction of a powder mixture, and initialization of chemical reactions. The experimental investigations show different regimes of mechanochemical conversions in powder mixtures under shock loading [1, 2]. It was shown that the solid-phase chemical reactions in shock compressed powder mixtures with given duration and amplitudes of shock pulse can be observed in the narrow range of the powder grain sizes, and porosity [3].

The creation of a computer model taking into consideration the structure of powder material is necessary for prediction of mechanical behavior and chemical processes in the reacting powder mixtures under shock loading.

The objective of this article is the development of the three-scale model defining principal parameters controlling the regimes of processes of structural and chemical conversions in powder mixtures under shock loading.

The simulation of mechanochemical processes of shock synthesis includes the simulation of processes of shock modification of powder mixture, mass and heat transfer in the reacting layer, and chemical reaction.

PHYSICAL MODEL OF A POWDER BODY

The model of a reacting powder mixture takes into account the macroscopic structure of concentration inhomogeneity. The initial macroscopic structure is formed in the stage of preparation of the powder materials due to creation of the particle conglomerates, and forming a porous structure under compaction [4]. These processes are connected with self-organizing in discrete powder system and produce the formation of a macroscopic interior structure of a powder body.

The reacting powder mixture can be represented by a set of elements with determined macroscopic structure of the concentration inhomogeneity.

For example, the Ni-Al powder mixtures consist of the reacting components, voids, and inert nickel aluminides.

An element of the Ni-Al powder material with macroscopic structure of concentration inhomogeneity is shown in Fig. 1. The characteristic size b depends on the powder grain size.

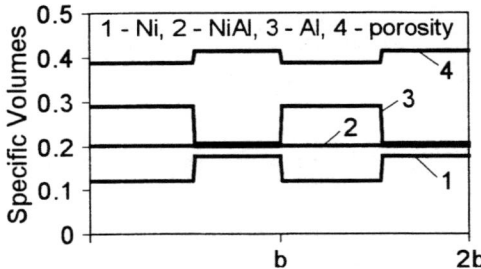

FIGURE 1. The reacting layer of powder mixture consists of two elements of the macrostructure of concentration inhomogeneity. Here b is the characteristic size of the element.

The maximum of aluminum concentration is at the left side of each element of the macrostructure of the concentration inhomogeneity. The maximum of nickel concentration is at the right side of each element. In the reacting layer the distribution of porosity (curve 4) and concentration of a product of chemical reaction (curve 2) changes during the process of mechanochemical transformations.

MODEL OF A COMPRESSED POWDER MIXTURE

The simulation mechanical behavior of a powder mixture during shock compression is based on the mechanical models of porous elastic-plastic medium [5-7]. The conservation laws of mass, momentum, and energy during loading were considered in [6] at the macroscopic level neglecting the formal value of average density of a porous medium. At the mesoscopic level in compressed powder mixtures the concentration inhomogeneity causes the high local temperatures, convection heat and mass transfer.

The dissipation of the kinetic energy activates the components of mixture and forms the hot spots. These heat sources are taken into account in the equation of energy balance.

Shock compression of powder mixture at the microscopic level was described in [5]. It was shown that when the size of pores is less than 10 μm the inertial effect is negligible. Then the mechanical behavior of powder mixture at the front of the shock pulse substantially depends on the process of pore collapse.

The modified model [7] of the mechanical behavior of porous medium was used in the present work. The modification of the model consists of taking into account the exothermic effect of chemical conversion in a reacting mixture. In this case heat source $Q\dot{z}$ appears in the equations of energy balance. Here $\dot{z}$ is the rate of chemical transformation of mixture components, and Q is the heat of chemical reaction.

HEAT AND MASS TRANSFER IN POWDER BODY

Usually reactions run in a wide range of temperatures that may result in phase changes in the components of the powder mixture. In this case the fusible component and product of chemical transformation can be transformed into liquid phase. When the structure of porosity exist in powder mixture, the liquid phase will move into the porous skeleton, and provide the convective heat and mass transfer. The model of processes of heat and mass transfer in a reacting powder body was suggested in [8]. The energy conservation law is considered for the reacting layer. When convective heat and mass transfer takes place, the laws of conservation of energy are described by the two-temperature equations of heat transfer with variable coefficients, and heat sources. All thermal parameters of the powder medium are defined as effective functions of concentration of components, porosity, and temperature for microvolumes of the reacting layer of a powder body. These functions are defined using the approach of the mechanics of the reacting granular layer [9].

The exothermic effects of chemical conversions determine heat sources in heat conduction equations, and the endothermic effects of the phase transitions determine heat losses. In the model, the heating is considered only in temperature region lying between T_0 and T_1. Here, T_0 is initial temperature of the mixture, and T_1 is the temperature of the mixture, at which the porous powder skeleton loses its strength.

SIMULATION OF PROCESSES OF SHOCK MODIFICATION OF A MIXTURE

At shock loading of powder mixtures plastic strain of powder grains, destruction of oxide and adsorbed layers on the surface of grains are observed. These processes influence on the reactivity of the powder mixture [2].

In the presented model the velocity of chemical conversion ($\dot{z}$) is defined by the kinetic equation

$$\dot{z} = k_0 \varphi(z) \exp(-E_a/RT), \quad (1)$$

where k_0 is a constant, $\varphi(z)$ is a function, which depends on the nature of conversion [10], E_a is the activation energy of chemical reaction, and R is the absolute gas constant.

The parameters k_0, $\varphi(z)$, and E_a are the characteristics of the kinetics of chemical conversion at the mesoscopic level.

Coefficient k_0 depends on the size of considered microvolumes of powder mixture. Therefore, the kinetics of chemical conversions will be different for same mixtures with various grains size.

When the parabolic law approximates the width of the reacting layer, k_0 can be defined by formula

$$k_0 = k/b^2, \quad (2)$$

where k corresponds to the element of unit width.

Shock compression decreases the actual size of the element of concentration inhomogeneity of the porous powder mixture. In this case, the parameter k_0 also changes.

In the model the changes of the reactivity of the powder mixture under shock compression are described by the parameter of activation energy

$$E_a = E_0 - H(P - P_i^*) \alpha_i A_i, \quad (3)$$

where E_0 is the energy of thermal activation of the chemical transformations in the reacting mixture, H is the Heaviside function, P_i^* are the critical values of shock pressure specifying the mechanisms of the activation of chemical reaction, A_i are the parameters specifying the contributions of activation energy for the different mechanisms caused by the work of shock compression, and α_i are the changes in powders mixture reactivity caused by different mechanisms.

The critical values of pressure P_i^* and portions of dissipated works A_i are defined by the method [7] in a points of a porous medium. The possibility of plastic deformation and destruction of surface layers of powder grains is considered for each component separately.

NUMERICAL REALIZATION OF THE MODEL

Shock compression of volume of the powder body consisting of the several reacting layers is numerically simulated.

The problems of shock compression, heat transfer, and macrokinetics of chemical conversions are solved for the set of reacting layers.

In the case of shock-assisted chemical processes, the nonlinear boundary problem of thermal conductivity is additionally solved by the finite-difference method using the implicit central-difference scheme [11, 12].

RESULTS AND DISCUSSION

The computer experiment was carried out for Ni-Al powder mixtures, in which mechanochemical conversions were initiated by shock pulse loading with different duration, amplitudes, and initial temperatures of the powder sample. Various structures of powder samples were investigated. The thermal, mechanical and macrokinetic parameters of the model were borrowed from the literature. The parameters of mechanical activation of powder mixtures were estimated using the experimental data on the mechanically activated synthesis of nickel aluminides. The computational experiment has shown the possibility of realization of different processes of synthesis depending on the inhomogeneity, porosity, and intensity of shock loading.

The possibility of transformation of chemical transition similar to the combustion into volumetric thermal explosion was detected.

The presence of solid inert fillers in compressed powder mixture allows to increase degree of its mechanical activation and to decrease the maximum temperature of the initiation of chemical conversions.

The results of estimation of characteristic times of mechanochemical processes at chemical conversions are shown in Fig. 2.

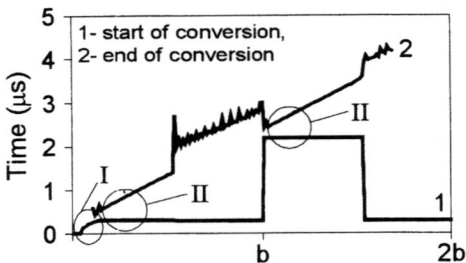

FIGURE 2. Characteristic times of mechanochemical processes in the reacting layer.

Regime of chemical conversion in the Ni-Al powder mixture similar to the "layer-by-layer combustion" is transformed to volumetric thermal explosion at the region I in Fig.2. The ultrafast chemical reactions were obtained in the region II where there was the maximum gradient of the concentration inhomogeneity of components. The calculations show that ultrafast solid phase reactions can take place in fine-grained mixtures under intensive shock loading.

CONCLUSION

A new approach to the investigation of behavior of a reacting powder mixture under compression has been suggested.

The obtained results demonstrate that the macrostructure of concentration inhomogeneity in powder mixture play important role in the initiation of mechanochemical conversions. The presented model taking into consideration the macrostructure of concentration inhomogeneity can predict different regimes of mechanochemical conversions in shock-compressed powder mixtures.

REFERENCES

1. Thadhani, N. N., *J. Appl. Phys.* **76**, 2129-2138 (1994).
2. Batsanov, S. S., *Effects of Explosions on Materials*, New York: Springer-Verlag, 1994.
3. Aizawa, T., Kamenosono, S., Tanaka, K., Kihara, J. "Shock Reactive Mechanisms for Direct Synthesis of Titanium Aluminides from Element Powder Mixture" in *Metallurgical and Materials Application of Shock Waves and High –Strain-Rate Phenomena* editted by L.E. Murr, K.P. Staudhammer and M.A. Meyers, Elsevier Science B.V. 1995,pp. 645-652.
4. Balshin, M. Yu., Kiparisov, S. S., *Fundamentals of Powder Metallurgy*, Metallurgy, Moscow, 1978.
5. Buzyurkin, A. E., Kiselev, S. P., *J. Appl. Mech. and Tech. Physics* **41**, 192-197 (2000).
6. Schetinin, V. G., "Shock Compression and Heating of Porous Media" in *Shock Waves in Condensed Matter*, Int. Conf. Proc., Saint Petersburg, 1998, pp.186-187.
7. Nesterenko, V. F., *Pulse Loading of Heterogeneous Materials*, Nauka. Sib. Branch, Novosibirsk, 1992.
8. Dmitrieva, M. A., Leitsin, V. N., *Izvestia VUZov. Fizika (Rus. Phys. J.)* **3**, 57-62 (1999).
9. Goldshtik, M. A., *Transfer Processes in Granular Layer*, Institute of Thermophysics SB AS USSR, Novosibirsk, 1984.
10. Merzhanov, A. G., *Phys. Chemistry* **3**, 6-45 (1983).
11. Leitsin, V. N., Dmitrieva, M. A., Kobral, I. V., *Phys. Mesomechanics* **4**, 43-49 (2001).
12. Leitsin, V. N., Skripnyak, V. A., Dmitrieva, M. A., "Simulation of Processes of Shock Synthesis of Aluminides" in *Shock Waves in Condensed Matter*, Int. Conf. Proc., Saint-Petersburg, 2000, pp.107-110.

EFFECT OF SHOCK-ACTIVATION ON POST-SHOCK REACTION SYNTHESIS OF TERNARY CERAMICS

Jennifer L. Jordan and Naresh N. Thadhani

School of Materials Science and Engineering, Georgia Institute of Technology, Atlanta, GA 30332-0245

Abstract. The effects of shock-compression of powder precursors on subsequent reaction synthesis and formation of Ti_3SiC_2 and Ti_2AlN ternary ceramics were investigated in this study. Mixtures of the powder precursors were shock-densified at different pressures using an 80-mm diameter gas gun and the double tube cylindrical implosion technique. Characterization of the shock-densified compacts showed an intimately mixed state of powders with high retained strain in unreacted compacts and little or no reaction at low pressures. The high pressure compact showed formation of non-stoichiometric TiC. The subsequent reaction behavior of the shock-densified compacts resulting in the formation of Ti_3SiC_2 was studied via heat treatments and differential thermal analysis (DTA). A non-stoichiometric TiC_x phase was observed as an intermediate phase prior to the formation of Ti_3SiC_2. This paper will present the results of the effects of shock compression on the the reaction mechanisms and kinetics of reactions leading to the formation of the ternary ceramics in the shock-densified precursor powders.

INTRODUCTION

Ti-based complex ternary ceramics called 312 and H-phases, including Ti_3SiC_2 and Ti_2AlN, are unique materials having high stiffness (~320 GPa for Ti_3SiC_2) but low hardness (4.5 GPa for Ti_3SiC_2 and 4.3 GPa for Ti_2AlN) [1,2]. They possess metal-like properties including electrical conductivity, thermal conductivity, and easy machinability, while demonstrating oxidation resistance, refractory behavior, and insusceptibility to thermal shock, typical of ceramics. Recent equation of state studies on Ti_3SiC_2 have shown that its bulk modulus is ~ 64% of the elastic modulus [3]. Hence, while its Young's modulus is similar to molybdenum metal, it is more compressible than Mo, but less than Si and α-Ti [3]. The compressibility and deformation response of these ternary ceramics makes them interesting candidates for damage tolerant armor applications.

Barsoum, et al. [1,4] have developed a hot pressing process for producing pure, bulk Ti_3SiC_2 starting with a mixture of Ti, SiC, and graphite. A controlled rate of heating, which ensures reaction in the solid state, has been found to be essential for the synthesis of pure Ti_3SiC_2 and Ti_2AlN [1,2].

The rationale for the proposed work was therefore to use shock compression to activate powder precursors for subsequent reactions occurring in the solid state and with activated kinetics. Hence, in this work, the reaction behavior of precursor powders was investigated by characterizing the as shock-compacted state of powder mixtures and determining their reaction kinetics. The high strain rate deformation behavior of the ceramics is currently being studied and will be presented in a later publication.

EXPERIMENTAL PROCEDURE

The precursor powders used for shock densification and subsequent reaction synthesis for Ti_3SiC_2 were titanium (Alfa Aesar), silicon carbide (Superior Graphite Company and Performance Ceramics Company), and graphite (Cerac, Inc. and Aldrich Chemical Company). One batch of the precursor powder ("as blended") was prepared by combining Ti, SiC, and graphite in the stoichiometric ratio and mixing in a V blender. The other batch of precursor powders was prepared by ball milling small quantities Ti and SiC powders for 2 hours in a Spex mill ("ball milled") or roller milling larger quantities for 6 hours ("roller milled"). Graphite was then added to both ball and roller milled mixtures, and the final mixing of the powders was performed using a V-blender which was run overnight.

The precursor powders used for shock densification and reaction synthesis for Ti_2AlN formation were titanium (-325 mesh, Alfa Aesar) and AlN (Alfa Aesar). The powders were blended in a V blender overnight.

The shock compression experiments were performed using a three capsule recovery fixture with the single-stage 80-mm diameter gas gun at Georgia Tech and the double-tube cylindrical implosion fixtures at the Energetic Materials Research and Testing Center in Socorro, NM. For the three capsule fixture, the powders were pressed in the steel capsules at ~65% theoretical maximum density and shock compressed at calculated peak pressures of 5 and 9 GPa. For the double-tube cylindrical implosion experiments, powders were packed to ~55% TMD in 1 inch diameter tubes. The powder containment fixtures were placed in 6 inch diameter PVC, which was packed with ANFO or ANFOIL explosive. The corresponding calculated maximum peak pressures are ~ 4 and 6 GPa, respectively.

RESULTS AND DISCUSSION

Table 1 summarizes the results of the compacts in the as shocked state and that following reaction synthesis of shocked precursors for Ti_3SiC_2 experiments.

Shock Densified State

The recovered shock compressed compacts from low pressure (< 6 GPa) experiments showed retention of reactants in both Ti + SiC + graphite mixtures. XRD line broadening analysis showed extensive residual microstrain ($\varepsilon \approx 10^{-2}$) retained in all of the precursor powders. The magnitude of

TABLE 1. Characteristics of recovered shock compressed Ti + SiC + Graphite samples before and after reaction synthesis

Sample	As Shocked	10 °C/min to 1600 °C Hold 4 hours
As Blended Powder	N/A	Ti_3SiC_2 TiC (0.430)*
590 m/s, ~ 5 GPa** As Blended	Reactants (Ti, SiC, and graphite)	TiC (0.431)*
870 m/s, 9 GPa** As Blended	Reactants (Ti, SiC, and graphite) and TiC_x (0.432)*	TiC (0.433)*
As Ball Milled	N/A	TiC (0.428)* Ti_3SiC_2
Implosion Cylinder 4 GPa** Rolling Mill 6 h	Reactants (Ti, SiC, graphite)	Ti_3SiC_2 TiC (0.432)*
870 m/s, 9 GPa** Ball Milled 2 h	TiC (0.430)*	Ti_3SiC_2 TiC (0.429)*

** Pressures calculated from Autodyn-2D [5]; * lattice parameter in parenthesis

retained strain in the shock densified compacts is of the same order of magnitude as the powders ball milled in a Spex or roller mill. At higher pressures (~9 GPa), the recovered powder compact of the as blended Ti + SiC + graphite powder showed evidence of partial reaction, forming TiC_x, while the compact of the ball milled precursor showed almost complete reaction, forming TiC. For the reaction product observed in the reacted compact, the TiC has a lattice parameter of 0.430 nm corresponding to a non-stoichiometric TiC_x phase.

No reaction was observed in either the low (~ 5 GPa) or higher (9 GPa) pressure experiments in the case of Ti + AlN precursors.

Reaction Behavior of Shock-Densified Compacts

<u>Titanium – Silicon Carbide, Ti_3SiC_2</u>

The as blended and as Spex milled powders and sections of the recovered, shock densified compacts were heat treated in a tube furnace to 1600 °C at 10 °C/minute, with a hold time of four hours, and subsequently characterized by XRD analysis. Reaction synthesis of the as blended, shock densified precursor powder compacts showed formation of TiC with a lattice parameter the same as that of the stoichiometric compound for the 9 GPa sample and non-stoichiometric compound for the 5 GPa sample. No Ti$_3$SiC$_2$ phase was observed to be found in either shocked sample. Formation of the stoichiometric TiC phase appears to be due to a self-sustained SHS-type combustion reaction in this as blended powder mixture, which in turn inhibits the formation of the ternary carbide phase.

Reaction synthesis of the unshocked, as milled powder and the Spex and roller milled, shock densified powder compacts showed the formation of both Ti$_3$SiC$_2$ and TiC products. The TiC formed in conjunction has a lattice parameter less than that of the stoichiometric (0.433 nm) value.

Thus, while reaction synthesis of the as blended, shock densified compacts reveals a tendency to form stoichiometric TiC and no ternary carbide, the milled and shock densified compacts yield the ternary phase along with TiC. Furthermore, the non-stoichiometric TiC$_x$ phase formed during the reaction synthesis of milled, shocked compacts appears to be a TiC – Si solid solution having possibly formed by silicon diffusion into TiC. The TiC – Si solid solution could be an intermediate phase prior to the formation of the ternary phase. The intimate mixing during milling and the dense packed, highly activated state attained during shock compaction, appear to aid the solid state diffusion of carbon into titanium and subsequently silicon into TiC$_x$, thereby resulting in Ti$_3$SiC$_2$ formation.

To further evaluate the effect of shock compression on reaction synthesis and formation of Ti$_3$SiC$_2$, kinetic studies were conducted to determine the activation energy for reaction by heating the powders in a DTA. At low heating rates (10 °C/min), a single broad DTA peak was observed characteristic of a solid state diffusion reaction. At higher heating rates (40 °C/min), two DTA peaks became obvious – a low temperature peak corresponding to solid state diffusion and a higher temperature peak from an SHS reaction, which initiates as the rate of heat release exceeds the rate of heat dissipation. Hence, depending on the activation induced by the shock compression process, changes in both the peak reaction temperature and the degree of reaction by solid state and SHS mechanisms were manifested by the exotherms observed in the DTA traces. The peak

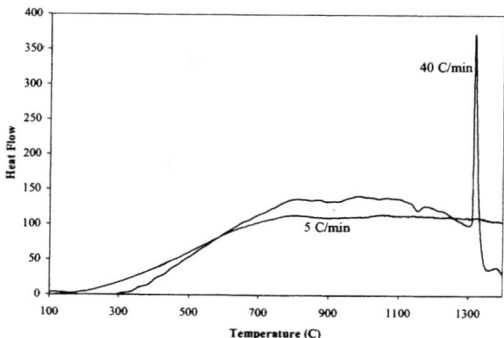

FIGURE 1. Sample DTA trace, from the Ti + SiC + graphite compact showing one peak at low heating rates and two peaks at high heating rates.

temperatures corresponding to the solid state reaction ranged from 700 °C – 920 °C in as blended, 680 °C – 930 °C in as ball milled and 800 °C – 1024 °C in the 4 GPa imploding cylinder sample. The corresponding peak temperature for the self-sustained combustion type reaction remained constant over a much narrower range of 1320 °C – 1485 °C in all samples. The 9 GPa shocked sample, which contained TiC product that had formed during shock compression, showed further solid state reaction in the temperature range of 720 °C – 1010 °C, but with no subsequent combustion reaction.

The modified Kissinger method [6, 7] was used to determine the activation energy from the peak temperature (T$_t$) of the solid state reaction exotherm obtained from the DTA traces. The activation energy, E, is obtained from the slope of the linear plot of heating rate (ϕ) over peak temperature according to the following equation:

$$\frac{d\left(\ln\frac{\phi}{T_t^2}\right)}{d\left(\frac{1}{T_t}\right)} = -\frac{E}{R}$$

where R is the gas constant. The results illustrate that the activation energy of the solid state reaction decreases from 80 kJ/mole for the as blended state to 71 kJ/mole for the ball milled state, and 56 kJ/mole in the case of the powder shock compressed using the cylindrical implosion geometry, and 68 kJ/mole for the 9 GPa gas gun sample.

Titanium-aluminum Nitride, Ti_2AlN

Reaction synthesis experiments on Ti + AlN have also been performed using a heat treatment similar to that used for the formation of Ti_3SiC_2. Reaction heat treatment of the 4 GPa samples showed higher XRD peak intensities of Ti_2AlN phase compared to the as blended, reacted sample. However, the peak intensities of the Ti_2AlN phase decreased from the 4 GPa to the 6 GPa sample indicating that there might be an optimum pressure window for shock activation. In the as blended and 4 and 9 GPa shocked samples, TiN was present in addition to Ti_2AlN. Reaction synthesis of the high pressure (9 GPa) sample showed formation of TiN and $Ti_3Al_2N_2$, which is a ternary phase with a stacking sequence of ABABACBC and is typically observed to be formed in a narrow temperature range (1200-1300 °C) [8].

Kinetic studies were, also, performed on the Ti + AlN compacts. These samples showed evidence of only a single broad peak indicative of a solid state reaction. The activation energy for this solid state reaction for the as blended sample was 97 kJ/mole, for the shocked, 4 GPa sample was 16 kJ/mole, and the 6 GPa and 9 GPa samples had activation energies of 58 kJ/mole. The activation energy decrease for the 4 GPa samples agrees with the increased amount of Ti_2AlN formed during reaction synthesis.

DISCUSSION AND SUMMARY

Shock compression of Ti, SiC, and graphite powders at ~ 4 – 5 GPa showed formation of dense packed highly activated state of reactants, while compression at a high pressure (~9 GPa) resulted in TiC formation in the recovered shock-densified compacts. Reaction heat treatment showed the formation of Ti_3SiC_2 and a TiC phase. The lattice parameter of the TiC phase was different from that of the stoichiometric value, suggesting that the TiC phase may be a TiC_x + Si solid solution, i.e. an intermediate state prior to Ti_3SiC_2 formation. In addition, the fraction of Ti_3SiC_2 in the compacts was higher in those densified at a higher shock pressure. These results, along with those of the reaction kinetic studies to determine the activation energies of solid state and combustion type reactions, illustrate that shock compression activates powder precursors and promotes the formation of the Ti_3SiC_2 phase via solid state diffusion.

Shock activated reaction synthesis of Ti + AlN also showed the formation of Ti_2AlN. However, the amount of phase formed decreased with increasing shock pressure indicating that there is an optimum window of shock compression pressure in which shock activation can be beneficially used for the formation of Ti_2AlN.

In general, the results of the present work on reaction synthesis of shock densified powder precursors illustrate that shock compression does activate the powder precursors thereby favoring the formation of Ti_3SiC_2 and Ti_2AlN ternary ceramics. However, formation of the ternary phases is not complete with the reaction treatments investigated. In most cases, formation of TiC (in the case of the ternary carbide) or TiN (in the case of the ternary nitride) by combustion-type reactions inhibits the completion of reaction via solid-state mechanisms. Modeling of the reaction behavior of shock-densified precursor powders will be conducted to predict the reaction treatment coupled with the degree of shock activation desired to ensure complete formation of the ternary phases by solid state diffusion reactions.

ACKNOWLEDGEMENTS

This work is funded by DOD/ASSERT program through the Army Research Office, contract number DAAG55-98-1-0161.

REFERENCES

1. Barsoum, M.W. and T. El-Raghy, *J. of the Am. Cer. Soc.*, **79** [7], 1953-1956 (1996).
2. Barsoum, M.W., D. Brodkin, and T. El-Raghy, *Scripta Materialia*, **36** [5], 535-541 (1997).
3. El-Raghy, T. and M.W. Barsoum, *J. of the Am. Cer. Soc.*, **82** [10], 2849-2854 (1999).
4. Onodera, A., H. Hirano, T. Yuasa, N.F. Gao, and Y. Miyamoto, *Sci. and Tech. of High Pressure: Proc. of AIRAPT-17*, Ed. M.H. Manghnani, et al., Universities Press, Hyderabad, India, p. 918-920 (2000).
5. "Autodyn-2D," Century Dynamics Inc., Oakland, CA, 1995.
6. Kissinger, H.E., *Analytical Chemistry*, **29**, 1702-1706 (1957).
7. Boswell, P.G., *J. of Thermal Anal.*, **18**, 353-358 (1980).
8. Schuster, J.C. and J. Bauer, *J. of Solid State Chem.*, **53**, 260-265 (1984).

SYNTHESIS OF FUNCTIONAL CERAMICS LAYERS USING NOVEL METHOD BASED ON IMPACT OF ULTRA-FINE PARTICLES.

J. Akedo[1] and M. Lebedev[2]

[1]Advanced Technology Process Mechanism Group and [2]Digital Manufacturing Research Center, National Institute of Advanced Industrial Science and Technology, Tsukuba East, Tsukuba, Ibaraki 305-8564 Japan

Abstract. A novel method of shock wave ceramics synthesis is reported. 0.3 μm in diameter ultrafine ceramics particles were accelerated by gas flow up to velocity of 100 - 500 m/s. During interaction with substrate, these particles formed dense, uniform and hard ceramics layers. Experiments were fulfilled at room temperature. No additional procedure for synthesis was required. The results of syntheses of piezoceramics oxide materials (Lead Zirconate Titanate) are presented. The density of material over 95% of the bulk density and hardness of as synthesized ceramics over 400 Hv were achieved. The microstructure and elemental composition were investigated. Applications of functional ceramics fabricated by reported method are discussed, as well.

INTRODUCTION

The synthesis of ceramics by shock compression has a long history. The attractive point of such synthesis is the high speed of the process. In the "conventional" shock synthesis concept the primary powder is compressed at one time. But, unfortunately, unloading processes, which are followed after compression, drastically destroy the ceramic, i.e. some cracks were appeared.

Other concept of synthesis is to excite a shock wave in a local area of primary powder material to be synthesized. In this case to compress row powder, individual particles of this powder are accelerated to a velocity of a few meters per second and impact onto the substrate. As a result of impaction area of shock compression does not exceeded a few diameters of particles and not destroyed other parts of material.

Several deposition methods based on the principle of particle impaction have already been investigated. This family of methods include depositing ultrafine particles via electrical field acceleration (Electrostatic Particle-Impact Deposition (EPID), which was originally developed by Ide et al. [1], or via acceleration by mixing with high-speed gas flow (Gas Deposition Method (GDM) [2], which was originally developed in ERATO UFP-Project of Japan [3]. For two-dimensional pattern formation of the metal, Cold Spray Method (CSM) [4, 5], developed by Institute of Theoretical and Applied Mechanics in Russia and Sandia National Laboratory for coating by metal material, and Hypersonic Plasma Particle Deposition (HPPD), originally developed by Minnesota University group [6] for Si, SiC, ceramics coating are used. Fundamentally, these methods are based on shock loading consolidation with or without thermal or plasma energy assistance. EPID is appropriate only to conductive material, for example metals or carbon, due to necessity of charging up the particles. Thick film formation (over 1 μm) in EPID has not been reported. GDM is applicable to metal and ceramics material using ultrafine particles, which has small diameter under 100 nm and has highly activated surface. In CSM large size particles with diameter over 1 μm are accelerated by hot gas. This method is very similar to GDM and conventional thermal spray coating, but for ceramics material coating has not been success. In HPPD active ultrafine particles are also used. These particles are produced under the high pressure after condensation from the gas phase in the nozzle. Deposition efficiency of EPID seems very low. On the other hand, GDM, CSM and HPPD have high potential in deposition rate.

In introduced Aerosol Deposition Method (ADM), ultra fine (UFP) submicron particles were accelerated by a gas flow in the nozzle and ejected onto substrate. It is suggested that during interaction of UFP with substrate and UFP with each other a part of kinetic energy is transformed into thermal energy in a local area to promote bonding between particles. But real mechanism of deposition is not clarified yet.

In this paper we report the result of deposition of PZT (Lead Zirconate Titanate) ceramics by ADM.

EXPERIMENT

Our ADM apparatus had two vacuum chambers connecting each other through a gas pipe. The first was a deposition chamber for the formation of ceramics. Deposition chamber contained the nozzle, substrate holder with or without heating system and window for diagnostic. This chamber was vacuumed during the deposition by a rotary vacuum pump and by mechanical booster pump. The second chamber was an aerosol chamber for generation of UFP aerosol. It had the accelerating gas introducing system and vibration system for powder mixing with accelerating gas. Aerosol flow from aerosol chamber was transported to deposition chamber by pressure difference between two chambers. The UFP ceramics powder was continuously ejected through the micro orifice nozzle and deposited onto the substrate. The orifice size of nozzle had rectangular shape. To get ceramics with uniform thickness, the nozzle was continuously scanning along the substrate. Schematic of ADM is presented in Fig. 1. Gas flow, which was controlled by mass flow controller, determined velocity of ejected particles. Table 1 shows the typical parameters of deposition condition for ADM. The details of apparatus were described elsewhere [7,8]. As a PZT powder, commercially available raw-material powder (PZT- LQ; Sakai Chemical Ind. Japan) with dry-milling process to improve the deposition rate was used.

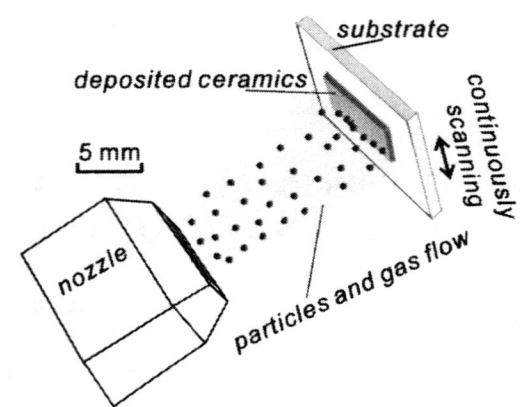

FIGURE 1. Schematic diagram of Aerosol Deposition Method.

The PZT powder had the perovskite structure and a composition of $Pb(Zr_{0.52}, Ti_{0.48})O_3$ which was close to the morphotropic phase boundary. According to Scanning Electron Microscopy (SEM) observations, the particle size of the powder varied through the $0.08 \sim 0.5$ μm range.

Velocity of particle flow was measured by time-of-flight method [9], in which some part of particle flow was mechanically cut from the total flow and deposited onto moving substrate, the deflection of deposited pattern from the axis, geometrical dimensions and speed of substrate provide data of particle flow velocity. The values of PZT particle flow velocity were varied from 100 up to 500 m/s in these experiments.

TABLE 1. Experimental parameters.

Pressure in deposition chamber	0.4 ~2 Torr
Pressure in aerosol chamber	80 ~ 600 Torr
Size of nozzle orifice	5 x 0.3 mm^2 10 x 0.4 mm^2
Accelerating gas	He, N$_2$, air
Consumption of accelerating gas	1 ~ 10 l/min
Maintained substrate temperature during deposition	300 K
Scanning area (area of deposition)	40 x 40 mm^2
Scanning speed of the nozzle motion along substrate	0.125 ~ 1.25 mm/sec
Distance between the nozzle and substrate	1 mm ~ 20 mm

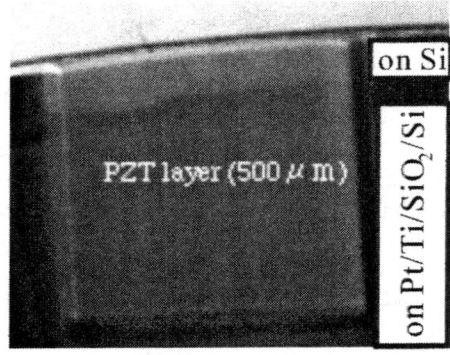

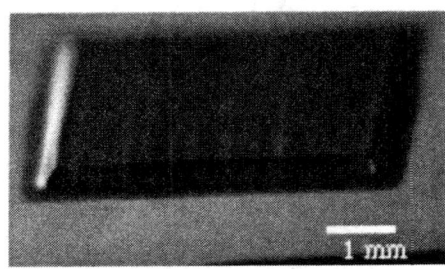

FIGURE 2. Optical image of PZT ceramics synthesized by ADM on Si and on Pt/Ti/SiO$_2$/Si substrates. Substrate temperature during experiment was maintained at 300K.

RESULTS AND DISCUSSION

The results of deposition of PZT on Si and Si coated by Pt layer substrates are shown in Fig. 2. Volume and weight of PZT film were measured using a three-dimensional stylus profiler and a precise weight balance with resolution of 0.1 μm and 10 μg, respectively. The bulk density of the PZT film was estimated as 7.76 g/cm^3, which is more than 95% of the theoretical density (8.10 g/cm^3). Although, the interaction time of particles with substrate and/or particles with each other was a few nanoseconds, the ceramics formation is a continuous process. It took 15 min to fabricate a 500-μm-thick PZT with the area 5x5 mm^2.

Adhesion force of the PZT deposited films on stainless steel and Si substrates was measured by a tensile testing machine and was higher than 50 MPa.

Crystal structures of the deposited films have been observed by X-ray diffraction (XRD).

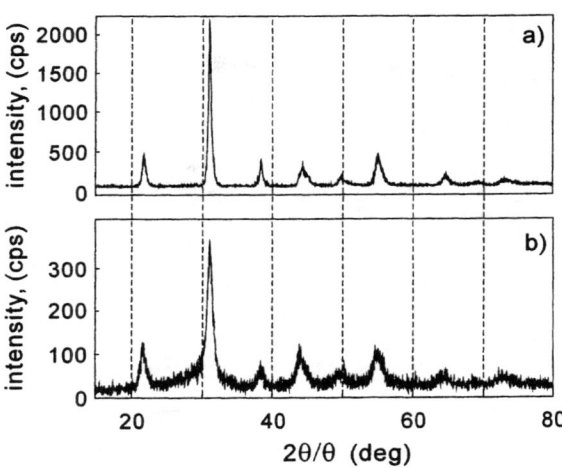

FIGURE 3. XRD patterns of PZT: a) - primary powder; b) - as synthesized film (thickness 25 μm) at room temperature; (XRD: Cu K-alpha, 40 kV/120 mA)

Figure 3 shows the results of XRD observations for primary PZT powder and PZT deposited by ADM without any additional external energy assistance. The deposited films have randomly oriented polycrystalline structures and have the spectra phases similar to raw-powder. A rhombohedral perovskite structure was retained before and after deposition. Pyrochlor and amorphous phase were not observed in the as-deposited PZT film. However, broadening of the spectra and slight shifting of the spectra angle in a higher degree were observed. The reason of the changing between the raw-powder and the deposited film spectra may be due to reducing of the films' crystallite size or their distortion during the deposition.

Structural characterization of the PZT films was carried out using transmission electron microscopy (TEM) (H-9000UHR, 300 kV). According to the TEM image (Fig. 4), the PZT films have a dense polycrystalline structure.

Elemental composition of PZT films was measured by an energy-dispersive X-ray microanalyzer (EDX). The measurement confirmed that the PZT films after deposition had a stoichiometric composition with a Pb/(Zr+Ti) ratio of about 1/1 and with Zr/Ti ratio of about 52/48 [10]. Elemental composition was the same as that of the primary powder and that of the bulk PZT material.

Results of measurements of micro Vickers hardness (Hv) (DUH-W201, Shimazu Co) of deposited films are presented in Fig. 5. The hardness of the deposited film does not increase if velocity of particle i.e. shock wave pressure, was increased. This result indicates that synthesis of ceramics was complete. For piezoelectronics applications an additional heat treatment procedure to improve ferroelectric properties is required. PZT layers made by ADM after annealing have no cracks and did not peel from substrate. PZT films deposited by ADM have high potential for producing microactuators and other applications of piezoelectric ceramics [11,12].

CONCLUSION

1) Thick, dense ceramic layers with thickness up to 1 mm were obtained during interaction of a ceramics particle flow with the substrate; 2) No external energy is required for synthesis; 3) Layers have a polycrystalline structure with strong bonding between the crystallites; 4) Chemical compositions of ceramics did not change; 5) Layers demonstrated high hardness, and good adhesion with the substrates.

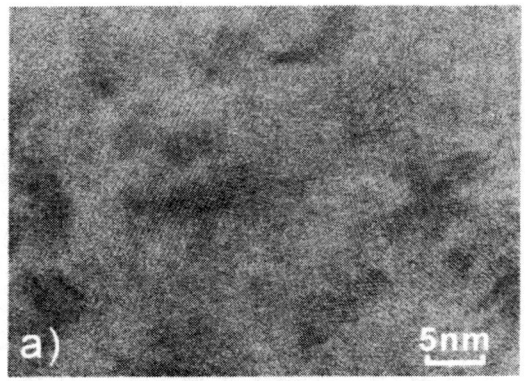
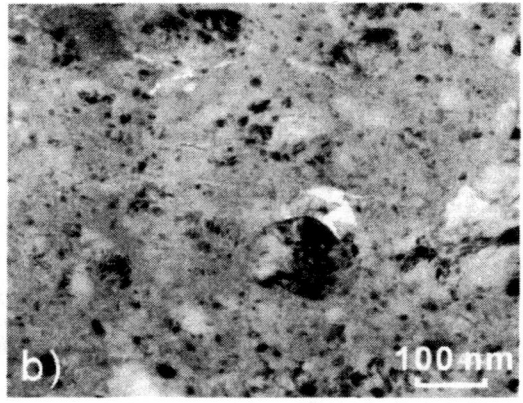

FIGURE 4. TEM images of PZT ceramics synthesized by ADM.

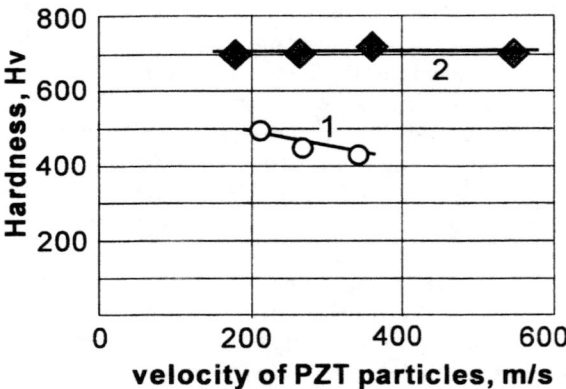

FIGURE 5. Micro Vickers Hardness of PZT for different velocities of ultrafine particles during experiment: 1) - As synthesized using oxygen as accelerative gas; 2) after heat treatment at 600°C during 1 hour in air atmosphere. Indentation force is 50 gf; Dwell time is 15 s

REFERENCES

1. T. Ide, Y. Mori, I. Konda, N. Ikawa and H. Yagi, *J. Jpn. Soc. Prec. Eng.* **57**, 143 (1991) [in Japanese].
2. S. Kasyu, E.Fuchita, T. Manabe, and C. Hayashi, *Jpn. J. Appl. Phys.* **23**, L910 (1984)
3. C. Hayashi, S. Kashu, M. Oda and F. Naruse, *Mater. Sci. Eng.* **A163**, 157 (1993)
4. P. Alkhimov, V.F. Kosarev and A.N. Papyrin, *Sov. Phys. Dokl.* **315**, 1062-1065 (1990)
5. R.C.Dykhuizen and M.F.Smith, *J. Therm Spray Technol.* **7**, 205 (1998)
6. Rao, N. et al., *J. Aerosol Sci.*, **29**, 707 (1998)
7. J. Akedo, *Oyo Ruturi* **68**, 44 (1999) [in Japanese]
8. J. Akedo, N. Minami, K. Fukuda, M. Ichiki and R. Maeda, Ferroelectrics **231**, 285 (1999)
9. M. Lebedev, J. Akedo, K. Mori and T. Eiju, *J. Vac. Sci. & Technol. A*. **18**, 563 (2000).
10. J. Akedo and M. Lebedev, *Jpn. J. Appl. Phys.*, **38**, 5397 (1999)
11. J. Akedo and M. Lebedev, *Appl. Phys. Lett.*, **77**, 11, 1710 (2000)
12. M. Lebedev, J.Akedo and Y. Akiyama, *Jpn. J. Appl. Phys.* **39**, 5600 (2000)

THE STUDY OF INTERNAL DEFORMATION FIELDS IN GRANULAR MATERIALS USING 3D DIGITAL SPECKLE X-RAY FLASH PHOTOGRAPHY

H.T. Goldrein, S.G. Grantham, W.G. Proud, J.E. Field

Cavendish Laboratory, Madingley Road, Cambridge, CB3 0HE. UK

Abstract. Digital Speckle X-ray Flash Photography is a technique which combines Digital Speckle Photography with Flash X-Ray Photography to measure 2- and 3-D displacement fields within dynamically deforming specimens. Measurements are made throughout a plane within the specimen seeded with X-ray opaque particles. This technique has already been successfully applied to the study of polyester, cement[1] and sand[2], and is used here to study the influence of water on a sand bed under impact from a hemispherical-tipped copper rod travelling at 100 m s^{-1}. Significant differences in the response of dry and wet sand beds were detected, and examples of the deformation fields measured are illustrated here. These results may be applicable in many spheres, for example, in the design of mechanisms to destroy buried ordnance.

INTRODUCTION

Being able to measure the internal response of sand to shock and ballistic impacts is of great use in many fields. One possible application is in the destruction of ordnance which are often buried in sand. There have been investigations into the penetration behaviour of shaped charges travelling through sand[3], but it is very difficult to obtain dynamic measurements of the sand's response. The technique described here is known as digital speckle X-ray flash photography which uses digital speckle photography (DSP) combined with flash X-rays. Standard DSP algorithms are used[4], but instead of producing the speckle pattern by white light or laser methods we instead seed a specimen with a sprinkled layer of X-ray opaque filings. A specimen such as sand lends itself naturally to this technique because of the granular nature of the material, and the relative ease of producing a specimen that incorporates a layer of lead filings at a chosen depth in the sample. The difficulty associated with mounting conventional gauges in a granular material also demonstrates the advantageous nature of this technique. In this case the "gauge" (the lead filings) is of the same dimension as the sand grains, introduces an insignificant perturbation in the sample, and no leads or power supply are required. The method provides very data rich results with x and y components of displacement obtained at every point on the plane to a resolution defined by the fourier sampling size chosen in the analysis. Here sand in both dry and saturated states has been subjected to an impact from a copper rod and the effect of the moisture in the sand studied.

EXPERIMENTAL

The sand used for this study was fraction C (David Ball Group plc, Huntingdon Road, Bar Hill, Cambridge, CB3 8HN, England), which has a grain size ranging from 300 to 600 μm. The projectile was a round ended copper (XM) rod with a diameter of 5.0 ± 0.1 mm, a length of 50.0 ± 0.5 mm and a mass of 8.6 ± 0.1g. To find how much water was needed to achieve 100 % saturation of the sand, a volume of sand was placed in a measuring cylinder and a known volume of water

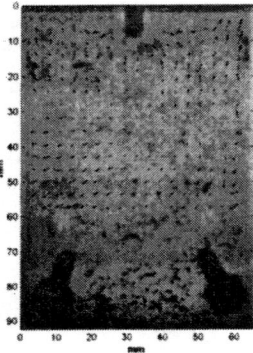

Figure. 1a. 120 µs delay, dry (displacement vectors ×3).

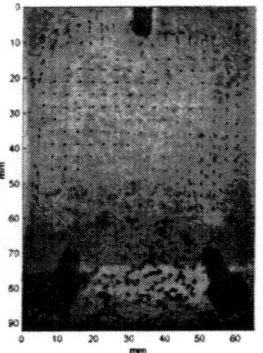

Figure.1b. 120 µs delay, saturated (displacement vectors ×3).

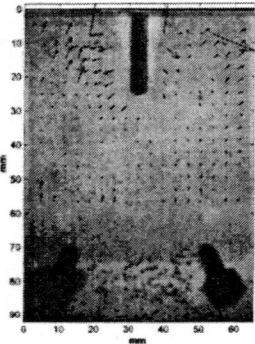

Figure. 2a. 240 µs delay, dry (displacement vectors ×3).

Figure. 2b. 240 µs delay, saturated (displacement vectors ×3).

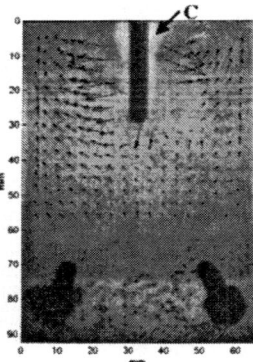

Figure. 3a. 360 µs delay, dry (displacement vectors ×3).

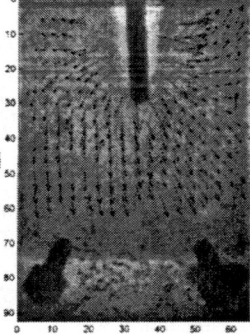

Figure. 3b. 360 µs delay, saturated (displacement vectors ×3).

added and left to soak through the sand. The volume of supernatant water could then be subtracted away from the total added, and the water/sand ratio and its density could then be calculated.

The samples were prepared in PMMA (polymethylmethacrylate) containers, with dimensions of $60 \times 70 \times 30$ mm^3, (the 60×30 mm^2 face being the impact face) filled to a depth of 10 mm with sand. The lead layer was introduced at a depth of 5 mm. A sand depth of 10 mm was chosen to give the optimal exposure of the film using our 150keV X-ray heads. The X-ray flashes had a duration of 30 ns. The projectile was fired at 100 ± 3 m s^{-1}, and X-ray photographs were taken at delays of 120 µs, 240 µs, and 360 µs after impact, for both the dry and saturated samples.

RESULTS

The results of these experiments can be seen in figures 1a to 3b. In each of these pictures, the bolts used to stop the sample container from moving backwards during the impact can be clearly seen, as can the fiducial marker region at the bottom of each picture below the bolts. This region enables the rigid body motions introduced by the scanning process to be calculated and subtracted. The displacement vectors in these images have been scaled up by a factor of three to make them more visible. A comparison of the 120 µs impacts shows that there is some movement away from the tip of the projectile in the dry case, but very little obvious movement in the saturated case with the displacements appearing quite noisy. This noisy displacement map is probably due to out-of-plane motions being comparable in magnitude to the in-plane motion. By 240 µs after impact larger displacements occur ahead of the projectile in the saturated case than in the dry case. After 360 µs it is clear that there is more bulk motion occurring ahead of the projectile in the saturated case than in the dry case. In all of these images, the response of the sand to the impact is to flow away from the rod, both forward and to the sides, in a manner that is more hydrodynamic in behaviour than the response a solid would exhibit. It is also possible to see the effect of cavitation to the sides of the rod in all these results (indicated by "C" in Fig. 3a). It causes a change in density in the sand which is apparent as a lighter shade in the X-ray image.

To investigate more quantitatively the displacement ahead of the projectile, graphs have been plotted of the y-component of displacement

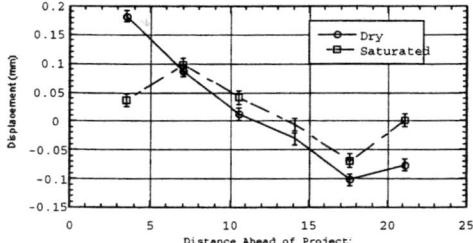

Fig. 4a. Displacements ahead of projectile, 120 µs.

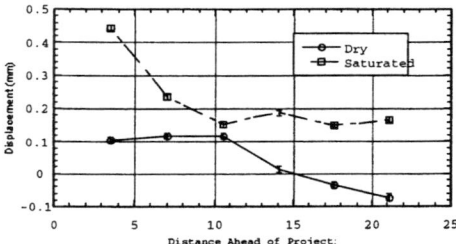

Fig. 4b. Displacements ahead of projectile, 240 µs.

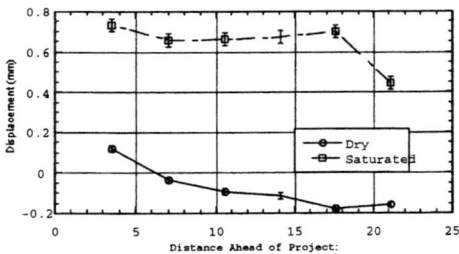

Fig 4c. Displacements ahead of projectile, 360 µs.

ahead of the projectile. In figure 4a to 4c, the result of averaging the 4 columns of the displacement vectors in the centre of the image, 21 mm in front of the projectile can be seen. The dry sample is represented by the solid line, and the saturated sample by the dashed line. From these graphs it is clear that in the 120 µs case there is very little difference between the two graphs, with the displacement reducing quite gradually further from the projectile tip. At 240 µs, the displacements in the dry sand have changed little, whereas in saturated sand the displacements are larger directly ahead of the projectile, and remain relatively constant at approximately 0.2 mm further into the target. By 360 µs the disparity between the two cases is quite pronounced. The dry sand has remained very similar to the 240 µs case with very little change, whereas the saturated sand has moved significantly further, the graph being relatively flat throughout. The behaviour seen in these plots can be explained by considering the effect that saturation has on the sand. The more open structure of the dry sand enables it to compact and move away from the projectile to the sides and even slightly backwards, this reduces the overall effect of bulk forward motion. In the saturated case, however, the sand cannot compact in the same way since the cavities are already filled with water, effectively an incompressible fluid at these pressures. Consequently the sand appears to move more like a rigid body ahead of the projectile.

CONCLUSION

A series of experiments have been carried out on sand utilising the technique of digital speckle X-ray flash photography. The measurements which have been made of the internal displacements of dry and moist sand would not have been possible using any other existing technique. It was shown that there was a measurable difference between the response of the dry and saturated sand when subjected to identical impacts. The purpose of this research was to validate the use of this technique when applied to granular materials, and its ability to make useful comparative studies. With this proven, the technique can now be used for more interesting and realistic situations, and we are currently in the process of scaling up the experiment to study impacts from shaped charge jets on sand beds and using a stereoscopic geometry with two X-ray heads[5]. This will allow out of plane motion to be measured as well, reducing the errors in the in-plane measurements.

ACKNOWLEDGEMENTS

The authors would like to thank Dr. I. G. Cullis (Defence Evaluation and Research Agency (DERA), UK, for his advice and encouragement. The research is supported, in part, by the Engineering and Physical Sciences Research Council.

REFERENCES

1. Synnergren, P., Goldrein, H.T., Proud, W.G., *Appl. Opt.* **38**, 4030-4036 (1999).
2. Grantham, S.G., Proud, W.G., Goldrein, H.T., Field, J.E., "The Study of Internal Deformation Fields in Granular Materials Using 3-D Digital Speckle X-Ray Flash Photography," in *Laser Interferometry X: Applications-2000*, edited by G.M. Brown, W.P.O. Juptner, and R.J. Pryputniewicz, Proceedings of SPIE 4101, San Diego, USA, 2000, pp. 321-328.
3. Resnyansky, A.D., Wildegger-Gaissmaier, A.E., "Hydrocode Modelling of High-Velocity Jet Penetration Into Sand," in *Proceedings 19th International Symposium of Ballistics-200*, edited by I.R. Crewther IBS Conference Proceedings, Interlaken, Switzerland, 2001, pp. 1561-1567.
4. Sjödahl, M., Benckert, L.R., *Appl. Opt.* **32**, 2278-2284 (1993).
5. Goldrein, H.T., Synnergren, P., Proud, W.G., "Three-Dimensional Displacement Measurements Ahead of a Projectile," in *Shock Compression of Condensed Matter-1999*, edited by M.D. Furnish, L.C. Chhabildas, and R.S. Hixson, AIP Conference Proceedings 505, Snowbird, Utah, 1999, pp. 1095-1098.

INVESTIGATION OF SHOCK-INDUCED CHEMICAL REACTIONS IN Mo-Si POWDER MIXTURES USING INSTRUEMTED EXPERIMENTS WITH PVDF STRESS GAUGES

Kevin S. Vandersall[1] and Naresh N. Thadhani[2]

[1]*Lawrence Livermore National Laboratory, 7000 East Avenue, L-282, Livermore, CA 94550*
[2]*School of Materials Science and Engineering, Georgia Institute of Technology, Atlanta, GA 30332-0245*

Abstract. Shock-induced chemical reactions in ~58% dense Mo+2Si powder mixtures were investigated using time-resolved instrumented experiments employing PVDF-piezoelectric stress gauges placed at the front and rear surfaces of the powders to measure the input and propagated stresses and wave speed through the powder mixture. Experiments performed on the powders at input stresses less than 4 GPa showed characteristics of powder densification and dispersed propagated wave stress profiles with rise time >~40 nanoseconds. At input stresses between 4-6 GPa, the powder mixtures showed a sharp rise time (<~10 ns) of propagated wave profiles and an expanded state of products revealing evidence of shock-induced chemical reaction. At input stresses greater than 6 GPa, the powder mixtures showed a slower propagated-stress-wave rise time and transition to a low-compressibility (melt) state indicating lack of shock-induced reaction. The results illustrate that premature melting of Si, at input stresses less than the crush-strength of the powder mixtures, restricts mixing between reactants and inhibits "shock-induced" reaction initiation.

INTRODUCTION

The Mo-Si intermetallic-forming system contains constituents having large differences in properties (e.g., density, sound speed, yield strength, and melt temperature), and a high heat of reaction. Prior shock synthesis studies on Mo-Si have been performed by Meyers et. al. [1,2], Marquis and Batsanov [3], Montilla [4], Aizawa et. al. [5], and Vandersall and Thadhani [6], in which reaction products having a variety of microstructures have been observed. While the microstructures observed via post-mortem analysis provide possible evidence of how the product phase may have been formed, it is difficult to ascertain whether the phases formed due to "shock-induced" reactions in time scales of pressure equilibrium, or subsequent to the shock event in time scales of temperature equilibrium via "shock-assisted" processes [7]. Inference of "shock-induced" chemical reactions can only be obtained via in-situ measurements of shock-properties using time-resolved experiments [8]. In the present work, instrumented experiments employing PVDF stress gauges were used to study the reaction behavior during shock compression of Mo+2Si powder mixtures.

EXPERIMENTAL PROCEDURE

Mo (Cerac No. M2000) and Si (Cerac No. S1053) powders (-325 mesh, <44 µm) were mixed in a stoichiometric ratio corresponding to $MoSi_2$ (using a mechanical V-blender) and pressed into fixtures at a density of ~58% of theoretical maximum density (TMD). The setup for instrumented experiments is similar to that used in prior work [8]. PVDF piezoelectric stress gauges were placed in intimate contact with front and back powder-capsule planar surfaces to monitor both "input-shock" and "propagated-wave" characteristics. The propagation of shock wave sensed by the "input" and "propagated" gauges at their respective locations provided the precise transit time through the ~3mm thick (50.8 mm diameter)

TABLE I. Summary of experimental results.

Expt. No.	Packing Density (g/cm³, %TMD)	Projectile Velocity (km/s)	Input Stress (GPa)	Input Risetime (ns) (10%-90%)	Equilibrium Propagated Stress (GPa)	Propagated Risetime (ns) (10%-90%)	Wave Speed (km/s) toe-to-toe-10%, ½ max	Relative Volume (toe-to-toe 10%, ½ max)
9806	2.59, 57	0.507	1.52	11.5	1.82	186	1.28, 1.23	1.58, 1.50
9818	2.50, 55	0.700	2.09	7.5	2.36	86	1.43, 1.41	1.43, 1.39
9902	2.69, 59	0.964	3.15	6.5	3.95	76.5	1.66, 1.61	1.13, 1.08
9910	2.70, 59	0.851	4.36	8.5	4.29	10, 14 †	1.87, 1.87	1.46, 1.46
9908	2.71, 59	0.940	5.4	4.5	5.18	6	2.10, 2.10	1.55, 1.55
9907	2.71, 59	0.966	6.3*	‡	6.16	8, 7 †	1.99*	1.13*
9913	2.50, 55	0.967	6.3*	‡	5.07	22.5	2.07 +	1.02
9919	2.51, 55	0.914	6.65	5	4.74	25.5	2.17, 2.15	1.10, 1.08

Symbols: * indicates calculated value, ‡ no measurement obtained, † two slope wave structure, + toe-to-toe at shock arrival instead of 10%.

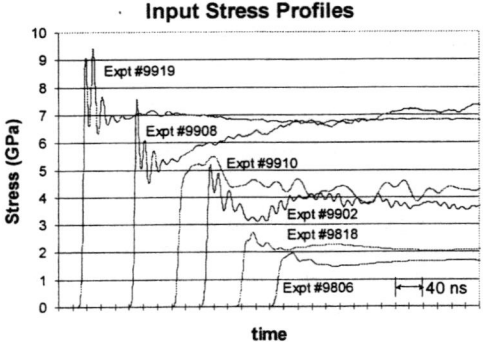

FIGURE 1. Combined plot of input stress traces.

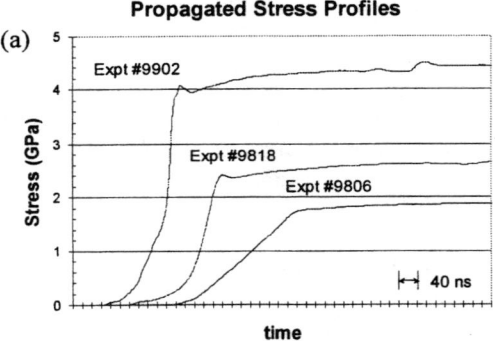

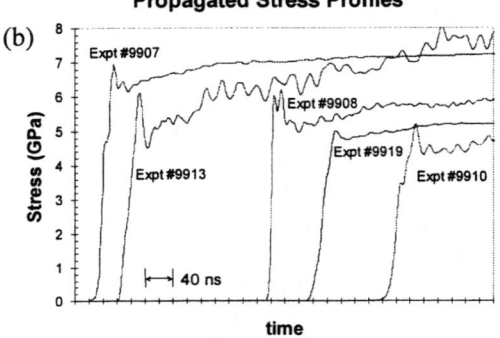

FIGURE 2. Propagated stress traces of (a) three low velocity experiments and (b) higher velocity experiments.

powder-mixture samples. OFHC-copper flyer plates were used for experiments #9806, #9818, and #9902 and a tungsten alloy (MIL-T-2014) flyer plate was used in all other experiments for generating higher pressures. The projectile velocity was measured using shorting pins, spaced 12.7 mm apart and the digital oscilloscopes were triggered from standoff pins placed 6.35 mm from the impact surface.

RESULTS AND DISCUSSION

A summary of the experimentally determined parameters obtained from the instrumented experiments is listed in Table 1. These include: the input stress and the input pulse rise-time (from 10% to 90% of peak) measured by the input shock gauge; the equilibrated propagated stress and propagated pulse rise-time recorded by the propagated stress gauge; wave speed determined using both the toe-to-toe and half-max values of input- and propagated-wave profiles; the relative volume calculated using the values of initial powder density, measured input stress, shock wave speed (both toe-to-toe and ½ max), and shock jump conditions for conservation of mass and momentum.

A plot displaying the measured input-stress profiles from all experiments is included in Figure 1. The propagated stress traces for the low and high velocity experiments are provided in Figures 2 (a) and (b) respectively. The varying amplitude of the different input-stress profiles corresponds to experiments performed at different impact velocities. The rise time of the input stress pulse is less than 10 ns, while the propagated stress wave is dispersed (rise time varying between 6 to 186 ns). Lower amplitude stress waves show the longest rise time, due to the behavior being dominated by powder densification.

Densification of the powder mixture from an initial to final solid density was considered using the P-α pore collapse model [9]. A thermodynamic consideration was also used to generate the pressure-volume (Hugoniot) curve of a fully reacted $MoSi_2$ product, based on the model recently developed by Bennett and Horie [10]. The important concept of this analysis is that it actually determines a calculated Hugoniot of the *products* formed via "shock-induced" reaction in a powder mixture. Details of both the P-α and thermodynamic curves as applied to this plot are described elsewhere [11].

Figure 3 (a) shows the pressure-volume space with the calculated curves representing the P-α densification behavior, the pressure-volume data points obtained from the PVDF gauge experiments, and the calculated compressibility curve of the fully reacted $MoSi_2$ product formed from Mo+2Si reactants at ~58% TMD. The calculated mixture Hugoniot is also displayed and considers densification of the Mo+2Si powder from $V_{TMD}/V=1.78$ to $V_{TMD}/V=1$, occurring at practically zero stress.

It can be seen that while the cluster of the three data points at pressures less than 3.1 GPa follow the trend representing the P-α densification behavior, the two data points at 4.3 and 5.3 GPa show significant expansion as they approach the fully reacted powder Hugoniot (forming $MoSi_2$ product) curve. Hence it can be reasoned that the 5.3 GPa data point corresponds to almost 100% shock-induced reaction occurring in the 58% dense powder mixture and the 4.3 GPa data point represents a shock pressure state in which the 58% dense Mo+2Si powder mixture undergoes an appreciable shock-induced reaction. From Figure 3 (a), if the 5.3 and 4.3 GPa data points are respectively considered to represent evidence of complete and partial shock-induced chemical reaction, then the cluster of data points corresponding to the three experiments at 6.2-6.6 GPa, which show minimal expansion and remain close to the inert Mo+2Si mixture Hugoniot, can be considered to reveal very limited or practically no shock-induced reaction.

The experimentally obtained data points of wave speed versus input stress are plotted in Figure 3 (b) along with the calculated curve corresponding to the 58% dense inert Mo+2Si powder mixture (illustrated as a dashed line). It can be seen that the experimental data points appear to follow the inert Hugoniot curve at stresses <6 GPa. At higher stresses, the data points actually show lower wave speed corresponding to that of low-compressibility melt phase of Si (obtained from [12]). As shown in Figure 3 (b), it can be seen that the Hugoniot of melted Si (dashed-dot line) intersects the Mo+2Si inert Hugoniot curve at P_m (~5 GPa) which represents the stress at which Si in the ~58% dense Mo+2Si powder mixture undergoes melting. The data points corresponding to higher pressures are found to lie more closely on the silicon melt Hugoniot than on the Hugoniot of the Mo+2Si powder mixture or its product.

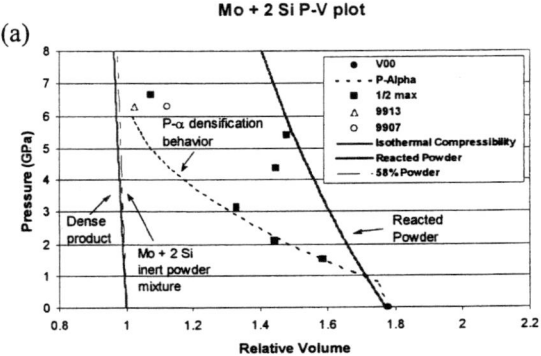

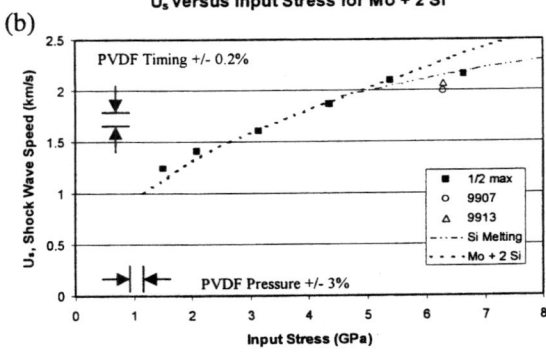

FIGURE 3. Plots of (a) measured input stress versus calculated relative volume (based on half-max values) plotted with the isothermal compressibility curves of dense $MoSi_2$ alloy, Mo+2Si inert mixture with zero crush strength, P-α densification behavior, and reacted product Hugoniot (data for experiment #9907 based on calculated values of wave speed and input stress, and that for Experiment #9913 based on calculated value of input stress and measured wave speed) and (b) measured input stress as a function of powder wave speed, for Mo+2Si powder mixture with silicon behavior.

A similar result has also been observed in the prior work on ~55% dense Nb-Si powder mixtures [12], in which the authors observed that in experiments performed at the same shock pressure (and thus particle velocity) the data points fell on either of two branches. Hence, premature melting of silicon (in some cases) was considered to inhibit shock-induced reaction, while in other cases under similar conditions, lack of melting of silicon led to shock-induced reaction in Nb+Si powder mixtures at stresses in the vicinity of the crush strength. Recent modeling work by Tamura and Horie [13] on reaction initiation in Nb+Si powder mixtures inside regions of an adiabatic shear band, also observed that a higher shear rate resulted in greater degree of deformation of reactants and consequently both mixing and increased propensity for shear-induced solid-state reaction initiation. In contrast, presence of a melt phase decreased the shear rate and resulted in inhibiting the reaction.

CONCLUSIONS

Time-resolved experiments performed on ~58% dense Mo+2Si powder mixtures at input stresses less than the crush strength (i.e. < 4 GPa), show characteristics of densification represented by the P-α behavior and measured propagated wave stress profiles showing wave dispersion with rise time >~40 nanoseconds. In experiments at input stresses between 4-6 GPa, the powder mixtures show evidence of shock-induced reaction, based on propagated wave profiles showing a sharp rise time (<~10 ns), and the data points of shock states revealing expansion that approaches pressure-volume compressibility curve of a thermodynamically determined reacted powder Hugoniot. Experiments on Mo+2Si powder mixtures at input stresses greater than 6 GPa showed lack of shock-induced reaction. The lack of reaction is inferred based on facts that the propagated stress profiles show a slower rise-time, the wave speeds recorded are lower, and the data points on the pressure-volume compressibility curve are closer to the un-reacted (inert) Mo+2Si powder mixture. Likewise, in the plot of wave speed versus input stress, the measured data points fall along the melted silicon Hugoniot curve, suggesting that melting of Si inhibits shock-induced reactions in Mo-Si powders.

ACKNOWLEDGEMENTS

This research was funded by the Army Research Office under Grants DAAH0495-1-0235 and DAAG55=97-1-0163. We also gratefully acknowledge the partial stipend (for KSV) support provided by the Georgia Institute of Technology Molecular Design Institute, under prime contract N00014-95-1-1116 from ONR.

REFERENCES

1. Meyers, M.A., Yu Li-Hsing, and Vecchio, K.S., *Acta metall. mater.* **42**, 701-714 (1994).
2. Vecchio, K.S., Yu Li-Hsing, and Meyers, M.A., *Acta metall. mater.* **42**, 715-729 (1994).
3. Marquis, F. D. S. and Batsanov, S. S., "Advances in Shock-induced Synthesis and Densification of Metal Silicides", in *Powder material: Current Research and Industrial Practices*, TMS, 113-128 (1999).
4. Montilla, K., PhD. Thesis, California Institute of Technology, (1997).
5. Aizawa, T. and Yen, B. K., "Shock-induced Reaction Mechanism to Synthesize Refractory Metal Silicides" *Shock Waves in Condensed Matter – 1997*, AIP Conf. Proc., pp. 651-654, (1998).
6. Vandersall, K.S., and Thadhani, N.N., in Molybdenum and Molybdenum Alloys Symposium held at the 126th TMS Annual Meeting and Exhibition, in San Antonio, TX, February 16-19, TMS Press, pp. 61-69, (1997).
7. Thadhani, N.N., *J. Appl. Phys.* **76**, 2129-2138 (1994).
8. Thadhani, N.N., Graham, R.A., Royal, T., Dunbar, E., Anderson, M.U., and Holman, G.T., J. Appl. Phys, **82** (3), pp. 1113-1128, (1997).
9. Carroll, M. M. and Holt, A. C., *J. Appl. Phys.*, **43**, No.4, April (1972).
10. Bennett, L.S. and Horie, Y., *Shock Waves*, **4**, pp. 127-136, (1994).
11. Vandersall, Kevin S., PhD. Thesis, Georgia Institute of Technology, (1999).
12. Yoshida, M., and Thadhani, N. N., "Study of Shock-induced Solid State Reaction by Recovery Experiments and Measurements of Hugoniot and Sound Velocity;" in Shock Waves in Condensed Matter-1991, edited by S.C. Schmidt, R.D. Dick, J.W. Forbes, and D.G. Tasker, Elsevier Science Publishers, B. V., pp. 586-592, (1992).
13. Tamura, S., and Horie, Y., J. Appl. Phys., Vol. **84**, No. 7, pp. 3574-3580, (1998).

SHOCK-INDUCED CUBIC SILICON NITRIDE AND ITS PROPERTIES

Toshimori Sekine

Advanced Materials Laboratory, National Institute for Materials Science,

Namiki 1-1, Tsukuba 305-0044, Japan

Abstract. The shock-induced phase transition of Si_3N_4 has been investigated through recovery technique and Hugoniot measurements. Recovered samples indicates the formation of a cubic spinel (c-Si_3N_4) and the yield of c-Si_3N_4 increases with increasing the shock pressure and reaches 100% at 63 GPa. Hugoniot measurements have revealed a phase transition above 36 GPa and the associated volume change of about 25%. The shock synthesized c-Si_3N_4 is nano crystalline and displays a high-temperature metastability up to about 1620 K. c-Si_3N_4 is one of hard materials based on the measured equation of state. Shock-synthesized c-Si_3N_4 has been characterized by electron microscopy and ^{29}Si magic angle spinning NMR spectroscopy.

INTRODUCTION

Silicon nitride ceramics exhibit high strength at high temperature, good thermal stress resistance and good resistance to oxidation. Previous studies [1-4] on shock-loaded Si_3N_4 have centered on processing techniques. Hugoniot measurements on sintered Si_3N_4 with considerable amounts of additives have been carried out and they indicated no phase transition up to about 40 GPa [5, 6]. It has recently found that the low-pressure phases transform into a high-pressure phase of Si_3N_4, c-Si_3N_4 [7]. There are three methods for high-pressure synthesis of c-Si_3N_4. The first one is a reaction of Si and N_2 fluid at 15-30 GPa and 2000 K in diamond anvil cells coupled with laser heating [7]. The second one is a shock transformation from the low-pressure phases, α-Si_3N_4 and β-Si_3N_4, above 20 GPa [8]. The third one is a solid-solid transformation from the low-pressure phases, at pressures of 18-20 GPa in multi-anvil high-pressure cells [9]. c-Si_3N_4 is the first known nitride spinel in which one third and two thirds of the Si atoms are tetrahedrally and octahedrally coordinated to nitrogens, respectively. It has been generally accepted that high-pressure phases display better performances in strength and resistance, because the chemical bond is stronger in the high-pressure phases than in the low-pressure ones. Therefore we need to know properties of c-Si_3N_4 for industrial applications.

We have already announced a massive production method of c-Si_3N_4 from β-Si_3N_4 [8]. In this paper, shock synthesis of c-Si_3N_4 is reviewed as well as a method for purifying and separating c-Si_3N_4. Some properties of shock synthesized c-Si_3N_4 powders and the Hugoniot of sintered β-Si_3N_4 with less amount of additive are summarized. The shock-induced transformation of β-sialon to a cubic spinel also is investigated in recovery experiments. The results indicate that the oxynitride spinel makes a solid solution with c-Si_3N_4.

EXPERIMENTAL METHODS

For the shock synthesis, we used a propellant gun to generate shock wave by hypervelocity impact. A projectile with a flyer metal plate is accelerated up to a velocity of ~ 2 km/sec and impacts a container target. Samples are stored in copper containers to protect them from the destruction by the rarefaction wave after the shock compression. A detailed description has been published elsewhere [10]. Preliminary shock recovery experiments [11] indicated the chemical reactions between the steel containers and the silicon nitride powder. Flyers are copper and tungsten, dependent on the required pressure. Pressure is estimated by the impedance match method and temperature is calculated on the thermodynamic ground. After successful recovery of the container, a sample was taken out from the container and immersed into a nitric acid solution to remove copper pressure media. Thus obtained powders were investigated by X-ray diffraction (XRD), electron microscopy, thermal analysis, and NMR spectroscopy.

For Hugoniot measurements, we used a two-stage light-gas gun to increase the impact velocity up to about 6 km/sec. Inclined mirror method coupled with an electrical streak camera has been employed to measure shock velocity and particle velocity simultaneously. The details of the techniques have been reported elsewhere [12]. Pressure and density at the compressed state have been calculated with aid of the Rankine-Hugoniot equations.

Two kinds of Si_3N_4 starting powders were used for recovery experiments: a pure β-Si_3N_4 and a mixture of 96% α-Si_3N_4 and 4% β-Si_3N_4 powders. The powders contain 0.5 wt% and 1.3 wt% oxygen, respectively. These Si_3N_4 powders were mixed with a large amount of copper powders (9 times by weight) to increase the shock pressure and temperature. Two kinds of sintered β-Si_3N_4 were used for the Hugoniot measurements. One is black and has about 2wt% Nd_2O_3 and Y_2O_3 as additions and the other contains no additive and is light gray. These sintered bodies were cut into the pieces of about 10 mm x 12 mm x 2.5 mm.

We have developed a method of purifying and separating shock-synthesized c-Si_3N_4. By applying a heated concentrated hydrofluoric acid only low-pressure Si_3N_4 phase as well as oxygen-rich, amorphous parts could be dissolved out. c-Si_3N_4 powder survived for a limited duration. This treatment allowed us to purify and separate as the residual powders of c-Si_3N_4.

Thermal properties were investigated by thermogravimetry and differential thermal analysis up to about 1800 K in Ar flow with a range of heating rates of 5 to 20 K/min. The phases present in the heated and quenched samples were identified by the XRD method. ^{29}Si magic angle spinning NMR spectroscopy was investigated to characterize the local structures around Si atoms in c-Si_3N_4.

RESULTS AND DISCUSSION

A. Formation of c-Si_3N_4 from β-Si_3N_4 and α-Si_3N_4

Figure 1 summarizes a series of XRD patterns of initial β-Si_3N_4 and post-shock samples quenched at pressures of 33 GPa to 63 GPa. The yield of c-Si_3N_4 increases with increasing peak shock pressure, and it reaches almost 100% at 63 GPa [11]. The shock temperature increases with increasing peak shock pressure when the density of initial mixtures is kept nearly constant. We compare the XRD patterns of post-shock samples quenched from β-Si_3N_4 at 49 GPa and from 96% α-Si_3N_4 and 4% β-Si_3N_4 at 43 GPa. The shock temperatures were not much different. According to a comparison of the relative yield of c-Si_3N_4, the starting material of α-Si_3N_4 is preferred in the shock transformation. A theoretical consideration [13] also indicates a lower pressure required for the phase transition from α-Si_3N_4 than β-Si_3N_4 because β-Si_3N_4 is slightly denser than α-Si_3N_4.

Shock-synthesized c-Si_3N_4 powders were subjected to transmission electron microscopy

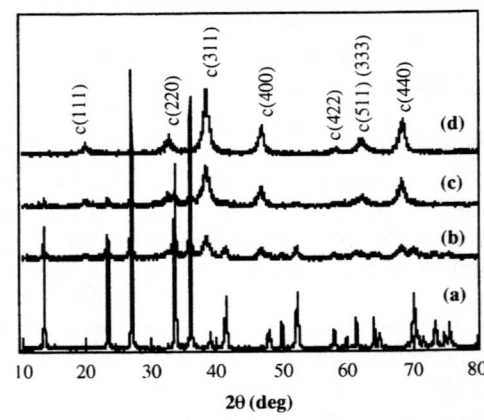

FIGURE 1. XRD patterns of starting material and its post-shock samples. (a) Starting β-Si_3N_4, (b) post-shock sample quenched at 33 GPa and 1770 K, (c) post-shock sample quenched at 49 GPa and 2400 K, and (d) post-shock sample quenched at 63 GPa and 3300 K

(TEM) observations [8, 14]. The grain sizes range between 7 and 35 nm. The electron diffraction pattern and high-resolution TEM image also were taken and analyzed. They are consistent with a cubic spinel structure. Some small c-Si_3N_4 grains have been observed to be embedded in the amorphous oxygen-rich phase. The composition of c-Si_3N_4 has been checked by EELS and EDX analyses during the TEM observations, and indicates no significant change from the initial Si_3N_4. The electron energy loss near edge spectroscopy method has been applied for structural identification and the measured results have been compared with the pattern predicted by theoretical simulation. The results has been reported separately [15].

B. Hugoniot measurements

We have measured Hugoniot of β-Si_3N_4 up to pressures of 150 GPa [12]. It is illustrated in the planes of shock velocity (Us)-particle velocity (Up) (Fig. 2) and pressure-density (Fig. 3). According to the Us-Up relation, elastic and elasto-plastic regions are seen in the low pressure phase. The HEL value is about 16 GPa, which is close to the previous data [5]. A phase transition has been detected at about Up=1.1 km/sec and last up to about 2.5 km/sec, The shock velocity for the high-pressure-induced phase appears above Up =2.5 km/sec. The Us-Up relation for the high-pressure phase can be approximated by a linear equation of Us = 4.59 + 1.49 Up.

The pressure-density relation also indicates a phase transition starting at about 36 GPa. The high-pressure region can be fitted to derive an isentropic compression curve of the high pressure phase through the third-order Birch-Murnaghan equation of state [12]. The results indicate a zero-pressure bulk modulus of 300 ± 10 GPa and its pressure derivative of 3.0 ± 0.1, as seen in Fig.3. We assumed that the high-pressure phase is a cubic spinel with a zero-pressure density of 4.01 g/cm³, based on the recovery experimental results.

There seems to be a slight difference in the onset pressure for the phase transition between Hugoniot measurements and the recovery experiments. It should be noted that the temperature in the Hugoniot experiments is significantly lower than that in recovery experiments because of the presence of pores in the powder mixtures. The estimated bulk modulus of c-Si_3N_4 is in good agreement with the estimations by the theoretical calculations [7, 11, 17], and is greater by about 30 % than that of β-Si_3N_4. Vogelgesang et al. [18]

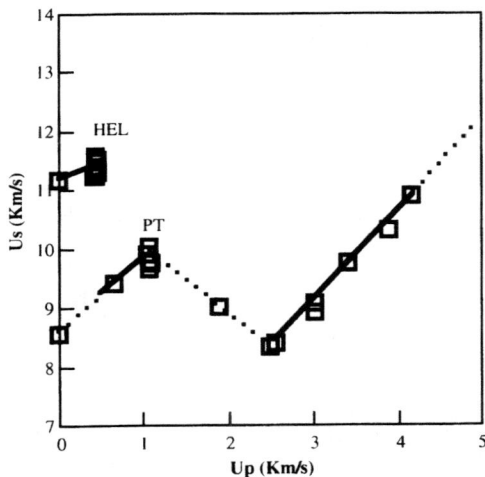

FIGURE 2. A relationship between shock velocity (Us, Km/s) and particle velocity (Up, Km/s) of β-Si_3N_4, measured by the inclined mirror method. HEL = Hugoniot elastic limit, and PT = phase transition.

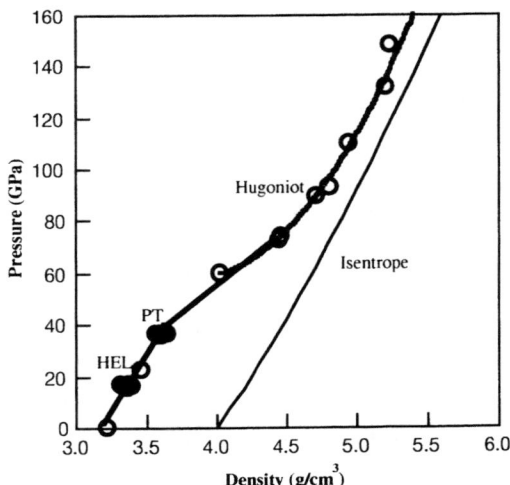

FIGURE 3. Hugoniot pressure-density relation of β-Si_3N_4 and a calculated isentrope of c-Si_3N_4 with a bulk modulus of 300 GPa and its pressure derivative of 3.0.

measured the elastic constants of single crystal β-Si_3N_4 using Brillouin scattering. He et al. [12] determined the elastic constants of sintered β-Si_3N_4 with no additives by the resonance sphere technique. These two data are significantly smaller than that based on the P-V data of β-Si_3N_4 in DAC by Li et al [19].

C. Metastability of c-Si_3N_4 at high temperature

In order to check the metastability of shock-synthesized c-Si_3N_4 powders, DTA analysis up to 1770 K has been investigated in Ar atmosphere. The DTA curve obtained for c-Si_3N_4, quenched from 63 GPa indicate a broad, exothermic reaction at temperatures over 1370 K and a relatively sharp exothermic reaction at temperatures 1620 K - 1770 K when temperature increases at a rate of 20 K /min. XRD patterns for quenched samples from temperatures of 1073 K, 1473 K and 1773 K are shown in Fig. 4. Oxidation starts at about 1470 K and a significant amount of SiO_2 cristobalite is identified in the sample quenched from 1773 K. The weigh gain indicate a formation of about 7 wt% SiO_2. c-Si_3N_4 is stable at 1473 K and transforms mostly into β-Si_3N_4 as well as a small amount of α-Si_3N_4 at 1773 K. These DTA and XRD data indicate that c-Si_3N_4 is thermally metastable at least up to 1620 K at one atmosphere.

High-resolution TEM image has indicated oxygen-rich amorphous phase in shock-synthesized c-Si_3N_4 sample. The pressure required for the complete transition to c-Si_3N_4 is too high to be achieved by conventional explosive methods used in industries. Shock recovered samples subjected to pressures of 30-50 GPa contain c-Si_3N_4 and low-pressure phases of Si_3N_4. Therefore it is necessary to develop a method to separate c-Si_3N_4 from the samples recovered below the complete transition pressure. We have developed a chemical method to purify and to separate c-Si_3N_4 from a mixture containing c-Si_3N_4, low-pressure phase Si_3N_4 and oxygen-rich amorphous phase. After using this method to purify shock-synthesized c-Si_3N_4 powders, we have obtained DTA data again and compared with that for c-Si_3N_4 sample without purification. The results have indicated a greater metastability. XRD data of heated and quenched samples are shown in Figure 5. c-Si_3N_4 survived partially in a heat treatment up to 1793 K. The DTA result indicate that the transition from c-Si_3N_4 to β-Si_3N_4 occurs at about 1670 K with an enthalpy change of 29.2±3.5 kJ/mol.

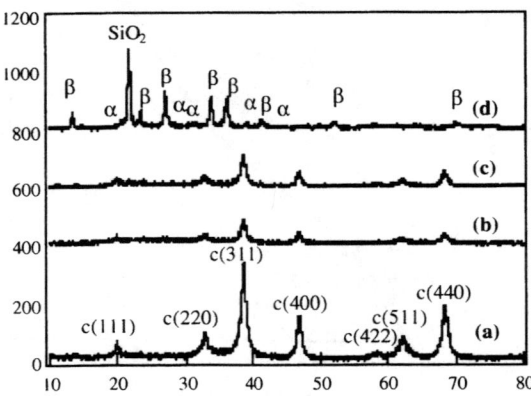

FIGURE 4. XRD patterns of c-Si_3N_4 powders (a) quenched from 63 GPa and its heated samples at 1073 K (b), 1473 K (c), and 1773 K (d). α and β are Si_3N_4 polymorphs. SiO_2 is cristobalite.

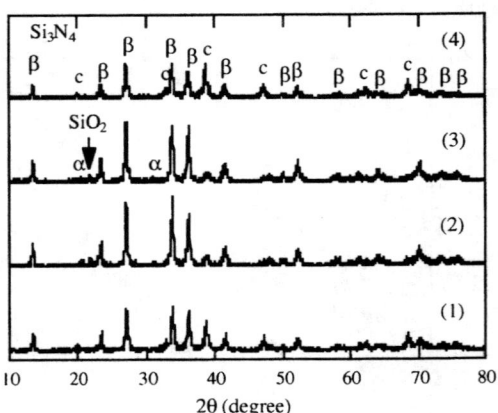

FIGURE 5. XRD patterns of c-Si_3N_4 powders shock-synthesized at about 40 GPa and purified by high-temperature HF solution and its heated samples. (1) sample heated at a rate of 5 K/min and quenched from 1753 K. (2) sample heated at a rate of 5 K/min and quenched from 1793 K. (3) sample heated at a rate of 10 K/min and quenched from 1793 K. (4) sample heated at a rate of 20 K/min and quenched from 1793 K. α, β, c and SiO_2 indicate peaks from phases α-Si_3N_4, β-Si_3N_4, c-Si_3N_4, and cristabalite, respectively.

D. ^{29}Si MAS NMR data of c-Si$_3$N$_4$

^{29}Si MAS NMR spectroscopy for c-Si$_3$N$_4$ powders has been performed to characterize the local structure of Si atoms in the spinel. The details have been published elsewhere [20]. The chemical shift has been reported to be -46.1 and -48.2 ppm for α-Si$_3$N$_4$ and -48.8 ppm for β-Si$_3$N$_4$ [21, 22]. The measured values for c-Si$_3$N$_4$ have two sharp peaks of -50.0 ± 0.2 ppm and -225.0 ± 0.2 ppm. These correspond to SiN$_4$ and SiN$_6$ units and an integration of the spectrum of SiN$_4$/SiN$_6$ is about 1/2, that of the spinel structure. The chemical shift for SiN$_6$ is more negative than that for SiO$_6$ and provides a useful information to determine the units in oxynitrides.

E. Formation of sialon spinel

β-sialon, Si$_{6-z}$Al$_z$O$_z$N$_{8-z}$ (0<z≤4.2), is known as oxynitride with a similar structure to β-Si$_3$N$_4$. Compositionally the sialon extends toward a spinel formula Al$_3$O$_3$N (Al$_2$O$_3$/AlN =1) which is not known yet. γ-Al$_2$O$_3$ exists as a spinel with defects. It is generally considered that Al may prefer the octahedral site to the tetrahedral site in spinel, and it is quite interesting to investigate whether sialon spinel can form like in Si$_3$N$_4$ spinel. We have carried out some recovery experiments using crystalline sialon powders with z = 1.8 and 2.8. XRD data of recovered sialon from 30 to 50 GPa indicate the formation of sialon spinels [23]. However the sample with z =2.8 quenched from 60 GPa becomes amorphous. The amorphization occurs in a wide range of the shock conditions, and a detailed investigation of shock condition reveals that the amorphization is due to a sluggish transformation of β-sialon into spinel but not due to melting. We are carrying out further study to obtain more information.

CONCLUDING REMARKS

Shock synthesis of c-Si$_3$N$_4$ was summarized from α-Si$_3$N$_4$ and β-Si$_3$N$_4$ in the presence of copper powders. The yield of c-Si$_3$N$_4$ increases with increasing shock pressure, and a higher yield was indicated from α-Si$_3$N$_4$ than β-Si$_3$N$_4$. The shock-synthesized c-Si$_3$N$_4$ powders are nano crystals and displays a high-temperature metastability up to about 1620 K. c-Si$_3$N$_4$ was characterized by ^{29}Si MAS NMR spectroscopy. The determined equation of state indicates that c-Si$_3$N$_4$ is one of hard materials. Our recent experimental results indicate that sialon spinel exists as well.

ACKNOWLEDGEMENTS

The author thanks H. He, F. Mitsuhashi, M. Tansho, I. Tanaka, K. Kimoto, Y. Yajima, T. Kobayashi, M. Kanzaki, and M. Mitomo for collaboration research and discussion.

REFERENCES

1. Beauchamp, E.K., Loehman, R.E., Graham, R.A., Moroshin, B., and Venturini, E.L., *Emergent Proc. Metha. High-Rech. Cerami.* **17**, 735-748 (1984).
2. Hirai, H. and Kondo, K., *J. Am. Cer. Soc.* **77**, 487-492 (1994).
3. Lurner-Adornatis, B.L., and Thadhani, N.N., *Mater. Sci. Eng.* **A256**, 298-300 (1998).
4. Tomoshige, R., Chiba, A., Nishida, M., Imamura, K., and Fujita, M., *J. Ceram. Soc. Jpn.* **100**, 1209-1214

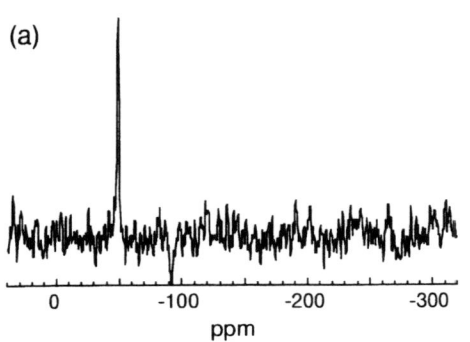

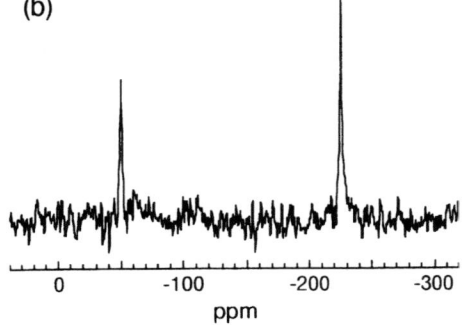

FIGURE 6. ^{29}Si MAS NMR spectra of (a) β-Si$_3$N$_4$ consisting of SiN$_4$ units and (b) c-Si$_3$N$_4$ consisting of SiN$_4$ and SiN$_6$ units with a ratio 1:2. The MAS spinning speed was about 4 KHz.

(1992).
5. Yamakawa, A., Nishioka, T., Miyake, M., Wakamori, K., Nakamura, A., and Mashimo, T., *J. Cer. Soc. Jap. Int. Ed.* **101**, 1322-26 (1993).
6. Grady, D.H., and Moody, R.L., *Sandia Report* SAND96-0551 (1996).
7. Zerr, A., Michie, G., Serghiou, G., Schwarz, M., Kroke, E., Riedel, R., Fueb, H., Kroll, P., and Boehler, R., *Nature* **400**, 340-42 (1999).
8. Sekine, T., He, H.L., Kobayashi,T., Zhang, M., and Xu, F., *Appl. Phys. Lett.* **76**, 3706-3708 (2000).
9. Jiang, J.Z., Stahl, K., Berg, R.W., Frost, D.J., Zhou, T.J., and Shi, P.X., *Europhys. Lett.* **51** [1] 62-67 (2000).
10. Sekine, T., *Eur. J. Solid State Inorg. Chem.* **34**, 823-833 (1997)
11. Sekine, T., He, H.L., and Kobayashi, T., "Shock Synthesis of Spinel-Type High-Pressure Phase of Si_3N_4"; in *Fundamental Issues and Applications of Shock-wave and High-Strain-Rate Phenomena*, Edited by K.P. Staudhammer, L.E. Murr and M.A. Meyers, Elsevier Sci. Ltd., Oxford, 2001 pp. 245-48.
12. He, H. L., Sekine, T., Kobayashi, T., Hirosaki, H., and Suzuki, I., *Phy. Rev.* **B 62** [17] 11412-17 (2000).
13. Ching, W.Y., Ouyang, L., and Gale, J.D., *Phys. Rev.* **B61**, 8696-8700 (2000).
14. Zhang, M., He, H.L., Xu, F.F., Sekine, T., Kobayashi, T., and Bando,Y., *J. Appl. Phys.* **88** [5] 3070-72 (2000).
15. Tanaka, I., Mizoguchi, T., Sekine, T., He, H.L., Kimoto, K., Kobayashi, T., Mo, S.D., and Ching, W.Y., *Appl. Phys. Lett.* **78** [15] 2134-2136 (2001).
16. Mo, S.D., Ouyang, L., Ching, W.Y., Tanaka, I., Koyama, Y., and Riedel, R., *Phys. Rev. Lett.* **83** [24] 5046-49 (1999).
17. Soignard, E., Somayazulu, M., Dong, J., Sankey, O.F., and McMillan, P.F., *J. Phys. Cond. Mat.* **13**, 557-63 (2001).
18. Vogelgesang, R., Grimsditch, M., and Wallace, J.S., *Appl. Phys. Lett.* **76** [8] 982-84 (2000).
19. Li, Y.M., Kruger, M.B., Nguyen, J.H., Caldwell, W.A., and Jeanloz, R., *Solid St. Common.* **103** [2] 107-12 (1997).
20. Sekine, T., Tansho, M., and Kanzaki, M., *Appl. Phys. Lett.* **78** [20], 3050-3051 (2001).
21. Carduner, K.R., Blackwell, C.S., Hmmond, W.B., Reidinger, F. and Hatfield, G.R., *J. Am. Chem. Soc.* **112** [12] 4676-79 (1990).
22. Kohn, S., Hoftbauer, W., Jansen, M., Franke, R., and Bender, S., *J. Non-Cryst. Solids* **224**, 232-42 (1998).
23. Sekine, T., He, H., Kobayashi, T., Tansho, M., and Kimoto, K., *Chem. Phys. Lett.* (in press).

DYNAMIC RESPONSE OF TITANIUM CARBIDE-STEEL, CERAMIC-METAL COMPOSITES.

B. Klein[1], N. Frage[1], E. Zaretsky[2] and M.P. Dariel[1]

[1]*Department of Material Engineering*, [2]*Department of Mechanical Engineering,*
Ben-Gurion University of the Negev, P.O.Box 653, Beer-Sheva 84105, Israel

Abstract. The dynamic response of a titanium carbide (TiC)-carbon steel, ceramic-metal composite, was studied in planar impact experiments, using a copper impactor with velocity in the 80 to 450 m/sec range. The composites were prepared by pressureless infiltration of TiC ceramic preforms with molten steel. The metallic component had either a pearlitic or a martensitic microstructure, determined by an appropriate heat treatment. Fully dense composites, consisting of TiC and 1060 steel, in pearlitic and martensitic states, were used as reference samples. The values of the HEL and of the spall strength were derived from the VISAR records of the free surface velocity of the impacted samples. The results indicate that the confining stress, produced by metallic matrix on the TiC particles, changes drastically the dynamic response of the composite.

INTRODUCTION

Ceramic-metal composites (cermets) have a potential as armor plates but require an in-depth understanding of their dynamic response. The relevant available information on cermets is very scarce and the available data are non-systematic. In particular, the influence of the state of the metallic component on the dynamic response of cermets was never studied.

The present work is an attempt to provide some information regarding any such influence. TiC-carbon steel composite was chosen for the present study. Although the dynamic properties of dense TiC ceramics are not high, they present some definite advantages. Ceramic TiC matrices with controlled open porosity (preforms) may be manufactured easily. Moreover, owing to the good wetting of TiC by molten steel, the porous preforms can be completely infiltrated by the molten metal. By varying the heat treatment applied to the cermet, the state of the metallic component can be changed while that of the ceramic matrix is kept constant.

MATERIALS

The composites were prepared by pressureless infiltration of TiC ceramic preforms with 0.6%C molten steel, followed by furnace cooling. The infiltrated ceramic-metal pieces (presamples) contained of about 30 vol.% of steel. These presamples were divided into three groups: The samples, referred henceforth as (a), were directly machined from the presamples into 3-4-mm thick, 20-mm diameter disks. A second group of presamples, (q), was heated to 870°C, water-quenched and machined into disks. A third group of presamples, (qa), was tempered after the quenching for one hour at 250°C, furnace cooled, and machined into the disks. Similar size disks were cut from a rod of commercial, normalized 1060 steel. Part of the steel discs underwent the same heat treatments as the cermet presamples, and are also referred as (a), (q) and (qa), respectively. Finally, disk-shape samples made of fully dense TiC were also prepared and tested.

The various heat-treatments were carried out in order to determine the effect of the microstructure of the metallic component on the dynamic response of the composite material. Prior to the impact experiments the surfaces of the disk samples were lapped to better than 0.005 mm parallelism, the density, ρ_0, determined by liquid displacement in distilled water and the longitudinal C_l and transversal C_t sound velocities, measured by the ultrasonic pulse-echo method. The average results of these measurements and the corresponding Poisson's ratio are shown in Table 1.

RESULTS AND DISCUSSION

The prepared samples were studied by planar impact experiments, using 1-mm thick copper impactors, accelerated by a 25-mm gas gun to velocities ranging from 80 to 450 m/sec. The impactor-sample misalignment did not exceed 0.5 mrad in all the experiments. The velocity w of the free surface of the samples was continuously monitored by VISAR [1]. Part of the recorded velocity profiles is shown in Fig.1. The profiles corresponding to the materials that underwent heat treatments are shifted upward for the sake of clarity and the dimensional-less time is $\tau = tC_l/\delta$, where δ is the sample thickness and t is the time after impact. The estimations of the Hugoniot Elastic Limit, σ_H, and the spall strength, σ_{spall}, listed in Tab. 1, were obtained from the VISAR records of the sample free surface velocity w using the formulas $\sigma_H = 0.5\rho_0 C_l w_{HEL}$ and $\sigma_{spall} = 0.5\rho_0 C_l \Delta w_{spall}$.

Since the sample made of dense TiC was completely destroyed by the compressive stress of the HEL level (see Fig. 2), a weaker impact experiment was performed in order to determine the spall strength of TiC. The detected velocity pullback, Fig.2, yields the TiC spall strength $\sigma_{spall}^{TiC} = 0.29$ GPa. The weak (below HEL) impact experiments were also performed with cermet samples that had undergone different heat treatments, Fig. 1c. The values of the corresponding spall strengths, σ_{sp}^*, are also given in the Tab. 1.

The velocity profile of the stronger shot with fully dense TiC, Fig. 2, reveals that any compressive deformation above HEL leads to the complete fracture of the ceramics (arrow I in Fig. 2). In order to evaluate the yield stress, Y_{TiC}, corresponding to the fracture of the brittle material in compression, we used the expression derived by Rosenberg [2] from Griffith's yield criterion

$$\sigma_H^{TiC} = \frac{1-\nu}{(1-2\nu)^2} Y_{TiC}, \qquad (1)$$

where ν is the Poisson's ratio. The data of Tab. 1 yield a value of $Y_{TiC} = 2.41$ GPa for the $\sigma_H = $ =5.87GPa. According to the Griffith's biaxial-stress yield criterion, $Y_{TiC} = 8\sigma_0^{TiC}$, where σ_0^{TiC} is the material tensile strength under uniaxial stress tension. By taking the spall strength, σ_{spall}^{TiC} of the

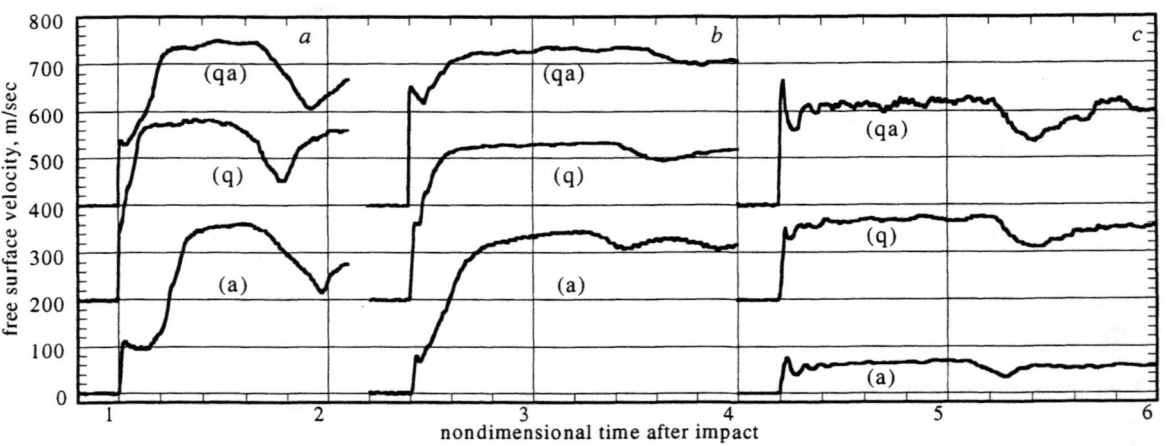

Figure 1. VISAR records of the sample free surface velocity profiles obtained in the shots with steel, *a*, and cermet, *b* and *c*, samples after different heat treatment (shown near the profiles).

TABLE 1 Static and dynamic properties of the studied materials.

Materials and treatment	density, g/cm³	C_l km/sec	C_b km/sec	Poisson's ratio	σ_{HEL}, GPa	σ_{spall}, GPa	σ^*_{spall}, GPa	Y, GPa (Griffith)	Y, GPa (von Mises)
TiC	4.78	10.12	7.280	0.216	5.87	0	0.29	2.42	4.25
Composite (a)	5.50	8.92	6.540	0.235	1.98	0.97	0.84	0.73	1.37
Composite (q)	5.43	8.76	6.370	0.226	4.47	1.04	1.48	1.73	3.16
Composite (qa)	5.47	8.72	6.280	0.217	5.68	0.91	2.14	2.32	4.11
Steel 1060 (a)	7.84	5.94	4.610	0.288	2.21	3.37	-	-	1.32
Steel 1060 (q)	7.76	5.87	4.590	0.294	3.64	2.79	-	-	2.12
Steel 1060 (qa)	7.72	5.93	4.650	0.296	3.46	3.16	-	-	2.01

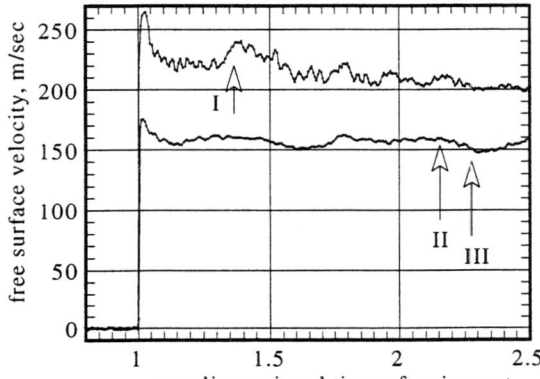

Figure 2. VISAR records of the sample free surface velocity profiles obtained in different shots with TiC samples. The arrows mark the arrival of the plastic wave (I), the unloading wave (II) and the spall signal (III) to the TiC samples free surface.

material as σ_0^{TiC}, we expect to get $Y_{TiC} = 8\sigma_{spall}^{TiC}$, while according to the data of Tab. 1, $Y_{TiC} / \sigma_{spall}^{TiC} = 2.41/0.29 = 8.3$. Within the accuracy of the experimental data, the agreement allows to conclude that dense TiC is a typical brittle ceramic.

The spall strength of the cermets in the stronger (over-HEL) shots is almost independent of the state of the metallic matrix. The spall strength of the cermets varies between 0.91 and 1.04 GPa, while the spall strength of the corresponding steel samples changes from 2.79 to 3.37 GPa. The cermet-to-steel spall strength ratio varies between 29 to 35% and is close to the volume fraction of the metal within the cermet. During the tensile path of the over-HEL shots, the effective cross-section of the cermet is actually the cross section of the metallic matrix, because the ceramic component has been destroyed in the course of the compressive path of the loading cycle. In the sub-HEL shots, see Fig. 1c, the matrix thermal treatment results in the increase, with respect to the over-HEL shots, of the σ_{spall} of the (q) and (qa) samples, by a factor of 1.42 and 2.35 times, respectively, while the σ_{spall} of the (a) sample stays almost unchanged. The increase may be attributed to some hydrostatic compression p^*, generated within the spheroid ceramic particle by the steel envelope under tension. This compression prevents the premature opening of the cracks in the ceramic particle and, thus, maintains the integrity of the sample cross section. The 30/70 matrix-to-particles volume ratio results in an average ratio of the envelope thickness, δ, to the particle radius, R, of about $\delta/R = 0.09$. We assume that the excess spall strength of the ceramic particle is due to this compression, $p^* = \Delta\sigma_{spall} = \sigma^*_{spall} - \sigma_{spall}^{TiC}$. In this case the tension in the steel matrix is $\sigma_T = \delta p^*/R$ and is equal to 0.10 and 0.16 GPa for (q) and (qa) materials, respectively. Since the response of the steel matrix under such a low tensile stress is purely elastic, it stays unchanged during the compression. The fracture of the sample starts when this excess pressure is cancelled in tension. The ceramic particles in the un-treated, (a), cermet sample seems free of this excess pressure.

The presence of the compression p^* may explain the striking difference revealed between the values of σ_{HEL} of the steel and cermet samples, tempered after quenching. The σ_{HEL} of both the steel (q) and cermet (q) samples increased with respect to the σ_{HEL} of the untreated materials (a). The tempering (qa) of the cermet sample results in a substantial increase of the σ_{HEL} while the σ_{HEL}

of the steel sample decreased by a few percent after tempering, see Tab.1.

It was shown that the yield behavior of pure dense TiC follows the Griffith's criterion, Eq. (1). Metals, and, in particular, steel follow the von Mises yield criterion, which, in the case of 1-D compression, can be written:

$$\sigma_{HEL} = \frac{1-v}{(1-2v)} Y. \quad (2)$$

The values of the yield stress, calculated for steel samples according to Eq. (2), and for cermet samples, according to both Eqs. (1) and (2), are listed in the two right columns of Tab. 1. It is apparent from the data in these columns, that for the same value of σ_{HEL}, yielding according to Eq. (1) occurs at a lower stress, and, consequently, yielding of the cermets starts in their ceramic component. Since three cermets were made from similar TiC performs and differed only in the state of the metal component, the difference in the values of the σ_{HEL} of the cermets is due to the mechanical interaction of the steel matrix with TiC inclusions. In the presence of the hydrostatic compression p^*, in the Griffith criterion

$$(\sigma_1 - \sigma_2)^2 = Y(\sigma_1 + \sigma_2) \quad (3)$$

the stress σ_1 and σ_2 have to be replaced by $\sigma_1 + p^*$ and $\sigma_2 + p^*$, respectively. Accounting in that for 1-D strain conditions $\sigma_2 = \sigma_1 v/(1-v)$ the Eq. (3) the Eq. (3) yields

$$[\sigma_{HEL}(1-2v) - p^*v]^2 = \\ = Y_c(1-v)[\sigma_{HEL} + p^*(2-v)]. \quad (4)$$

The yield stress of the ceramic Y_c is assumed to be constant. Equation (4) yields the dependence of σ_{HEL} on the excess pressure p^*:

$$\sigma_{HEL} = \frac{Y_c(1-v) + 2p^*v(1-2v)}{2(1-2v)^2} + \\ + \frac{\sqrt{Y_c^2(1-v)^2 + 8Y_c p^*(1-v)^3(1-2v)}}{2(1-2v)^2}. \quad (5)$$

Notice that (5) is the larger root of Eq. (4) and expression (5) coincides with (1) when $p^* = 0$. The dependence (5) of the σ_{HEL} on the hydrostatic pressure p^* is shown in Fig. (3) for different values of the yield stress Y_c.

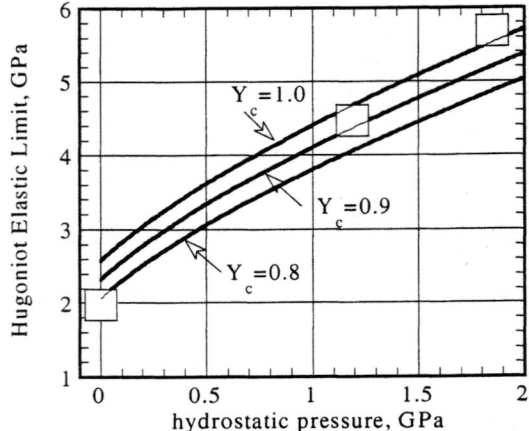

Figure 3. Hugoniot Elastic Limit of cermet samples that had undergone different heat treatments as function of the excess pressure p^* (squares), and pressure dependence of the HEL according to Eq. (5) calculated for different values of the constant yield stress Y_c.

The experimentally measured values of σ_{HEL} of the cermets that had undergone different heat treatments are shown in the same figure as function of the calculated pressure surplus. Accounting in the error margins of the experimentally obtained figures and the assumptions made ($v = const$, $Y_c = const$), the agreement seems reasonable.

The results obtained allow us to conclude that the dynamic response of the ceramic-metal composite is governed by the mechanical interaction between the metallic matrix and the ceramic particles and, thus, may be controlled by choosing a proper thermal treatment of the cermet.

REFERENCES

1. Barker, L. M., and Hollenbach, R. E., *J.Appl.Phys.*, **45**, 4872, (1974).
2. Rosenberg, Z., *J. Appl. Phys.* **74**, 752, (1993)

INVESTIGATION OF SHOCK-INDUCED CHEMICAL REACTIONS IN Ni-Ti POWDER MIXTURES USING INSTRUMENTED EXPERIMENTS

Xiao Xu and Naresh N. Thadhani

School of Materials Science and Engineering, Georgia Institute of Technology, Atlanta, GA 30332-0245

Abstract. Instrumented experiments using PVDF stress gauges were employed to investigate the occurrence of shock-induced chemical reactions in ~50% dense Ni+Ti powder mixtures. At low input stresses (~1 GPa), the as-blended powder mixture showed characteristics of powder densification and dispersed propagated-wave stress profiles with rise-time of 36 nanoseconds. At input pressure as high as 3.22 GPa, the as-blended mixture showed a sharp rise-time (< 15 ns) of the propagated-wave profile and an expanded state of products revealing evidence of shock-induced chemical reaction. Experiments performed on the ball-milled Ni+Ti powder mixtures showed that while the powder ball-milled to the state of becoming fully alloyed, mechanically amorphized, remained inert and showed no expansion, those powder mixtures ball-milled for intermittent times underwent shock-induced reaction. The expansion due to the resulting shock-induced reaction increased with decrease in ball-milling time. These results support previous studies on other intermetallic forming systems that show similar volume expansion.

INTRODUCTION

Shock compression of elemental powder mixtures produces a unique configuration of dense-packed highly activated state of material that can result in "shock-induced" chemical reaction (due to the effects of high pressure in the microsecond-duration time scale of pressure equilibration), or "shock-assisted" reaction (due to residual *post-shock* thermal effects in the time scale of thermal equilibration) [1-4]. Nickel-titanium represents an exothermic intermetallic forming system, which can undergo reaction forming intermetallics by shock compression of the elemental powder mixtures under certain conditions [5,6]. Without the help of in-situ measurements of shock states, post-mortem analysis of the product microstructures may not ascertain whether the observed reaction products formed via "shock-induced" reactions or post-shock "shock-assisted" process [7]. In the present work, instrumented experiments using PVDF piezoelectric stress gauges were used to study the reaction behavior during shock compression of as-blended and ball-milled Ni+Ti powder mixtures.

EXPERIMENTAL PROCEDURE

Elemental Ni (Cerac) and Ti (Alfa Aesar) powders of -325 mesh (<40 μm) were mixed in an equiatomic ratio, either using a slow speed V-blender or by ball-milling. Ball-milling was performed using Spex 8000 Mixer/Mill for different time periods in Ar atmosphere. The powder mixtures were then pressed into a copper capsule to form a ~ 3 mm thick disk of 50.8 mm in diameter at a density of ~50% of the theoretical maximum density (TMD). The setup for instrumented experiments was similar to that used in prior work [7,8]. Two PVDF stress gauges were placed in intimate contact with front and back powder-capsule planar surfaces, respectively, to monitor the input-shock and propagated-wave characteristics as well as the transit time of shock wave through the powder layer. Tektronix TDS 784A digital oscilloscope was used to capture the current signals generated from PVDF gauges. OFHC-copper flyer plates were used in all experiments. The projectile velocity was measured using three in-line shorting pins, and standoff pins were used to trigger the digital oscilloscopes. The ball-milled mixtures were also reacted in the Perkin-Elmer DTA 7, to determine the reaction heat evolved, as a function of ball-milling time.

TABLE 1. Summary of Experimental Results

Expt. No.	Packing Density (%TMD)	Projectile Velocity (m/s)	Input Stress (GPa)	Input Rise-time (ns) (10%-90%)	Equilibrium Propagated Stress (GPa)	Propagated Rise-time (ns) (10%-90%)	Wave Speed (mm/µs) (toe-toe-10%, ½ max)	Relative Volume (toe-toe-10%, ½ max)
0109	50	522	1.12	26.5	1.84	36	0.92, 0.92	1.15, 1.14
0104	47	930	2.71	4.5	3.39	8.5	1.50, 1.50	1.27, 1.27
0105	49	1046	3.22	4	3.88	14.5	1.77, 1.77	1.36, 1.35
9923	52	918	3.67	6.5	0.40	33	1.79, 1.77	1.24, 1.23
0101	53	930	3.82	6.5	0.22	26	1.68, 1.67	1.11, 1.10
0108	46	940	3.75	7	0.44	40.5	1.58, 1.57	1.03, 1.01

RESULTS AND DISCUSSION

A summary of the experimental results obtained from the aforementioned instrumented experimental measurements is listed in Table 1. The parameters include, the input stress and input pulse rise-time (from 10% to 90% of peak) measured by the input shock gauge; the equilibrated propagated-stress and propagated-pulse rise-time recorded by the propagated-stress gauge; wave speed determined using both the toe-to-toe (10%) and half-max values of input- and propagated-wave profiles; the relative volume calculated using the values of initial powder density, measured input stress, shock-wave speed (both toe-to-toe and half-max), and shock jump conditions for conservation of mass and momentum.

A. Reactions in As-Blended Powder Mixture

Figure 1 shows the measured input stress profiles from experiments performed on the as-blended powder mixture. The propagated-stress traces for those three experiments are shown in Figure 2. With increasing impact velocities, the amplitude of the corresponding input stress increases, while the rise-time of the input stress pulse decreases. In addition, at lower stress level, propagated-stress profile shows longer rise-time revealing characteristics dominated by powder densification instead of reaction. In all three cases, the equilibrium propagated-stress is slightly higher than the input stress.

Densification of the powder mixtures from an initial porous to final solid density is considered using the P-α pore collapse model [9]. A thermodynamic consideration is also used to calculate the pressure-volume (Hugoniot) curve of a fully reacted NiTi product, based on Bennett and Horie's model [10], which implements a constant pressure adjustment of the reference state. This model provides a method to determine a calculated Hugoniot of the reaction products formed via "shock-induced" chemical reaction in a powder mixture.

Figure 3 shows the pressure-volume relationship calculated based on the measurements of the three PVDF gauge experiments (#0109, #0104 and #0105) performed on the as-blended powder mixture. Also shown in the P-V plot is the P-α densification behavior of as-blended powder mixture

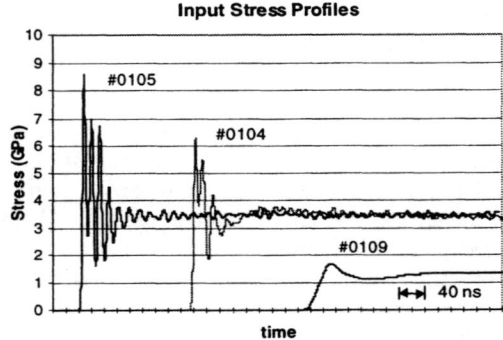

FIGURE 1. Combined plot of input stress profiles from PVDF gauge experiments on as-blended powder mixture.

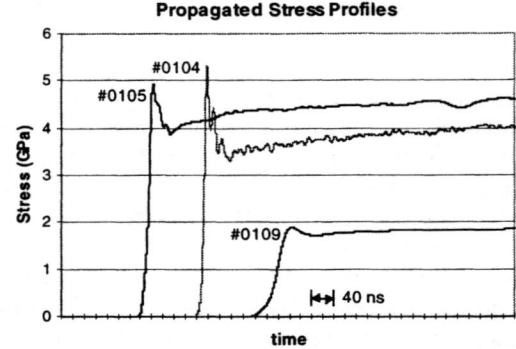

FIGURE 2. Combined plot of propagated-stress profiles from PVDF gauge experiments on as-blended powder mixture.

drawn to pass through the datum for Expt. #0109, which illustrates crush strength of about 1.6 GPa for the as-blended Ni+Ti mixture. The calculated solid product NiTi Hugoniot, the porous Hugoniot considering densification of inert powder mixture from $V/V_0=2$ to $V/V_0=1$ with zero crush strength, and the calculated compressibility curve (obtained based on Bennett and Horie's model [10]) of the fully reacted NiTi product formed from Ni+Ti reactants at ~50% TMD, are also indicated.

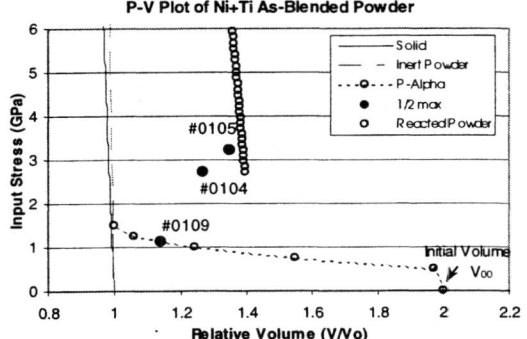

FIGURE 3. Plots of measured input stress versus calculated relative volume (based on half-maximum values), isothermal compressibility curves of dense NiTi, Ni+Ti inert mixture with zero crush strength, P-α densification curve and Hugoniot of reacted powder forming product.

It can be seen that if the data point at 1.12 GPa is considered to fall on the curve representing the P-α densification behavior, then the two data points above 2.5 GPa show significant volume expansion, and approach the calculated compressibility curve of the fully reacted NiTi product formed from Ni+Ti powder reactants. Hence, it can be reasoned that the 3.22 GPa data point represents almost complete shock-induced reaction in the powder mixture, and 2.71 GPa data point indicates an incomplete but appreciable shock-induced reaction.

B. Effect of Ball-Milling

The effect of ball-milling on the reaction behavior of Ni+Ti powder mixtures was also investigated using the PVDF gauge experiments. Ball-milling was used to change the configuration of powder mixtures. The obvious effect of ball-milling is the improved intimate mixing of the powder reactants. In present work, Ni+Ti powder mixtures were ball-milled for 4, 8 and 18 hours (designated as BM 4hr, 8hr and 18hr, respectively). The BM 4hr and 8hr powders were prepared in a sealed steel miller vial filled under Ar gas, with dripping liquid nitrogen as coolant outside the vial. The BM 18hr powder was prepared using hexane as lubricant. The reaction behaviors of those ball-milled powders as well as the as-blended powder mixture were also examined using a Perkin-Elmer DTA 7 [6]. It was found that the reaction heat evolved decreased with increasing ball-milling time, indicating occurrence of partial reaction during ball-milling, also known as mechanical alloying. Reaction behavior of BM 18hr powder was different from those of the as-blended and BM 4hr and 8hr powders in that the heat evolved was dominantly due to the crystallization of the amorphous phase.

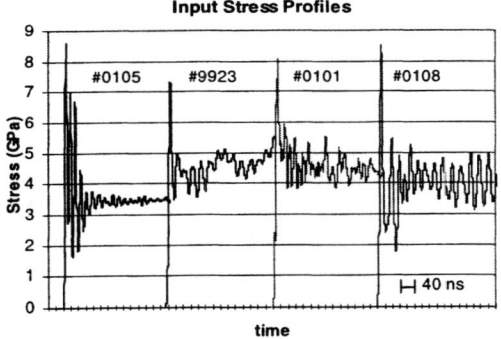

FIGURE 4. Combined plot of input stress profiles from PVDF gauge experiments at high velocity.

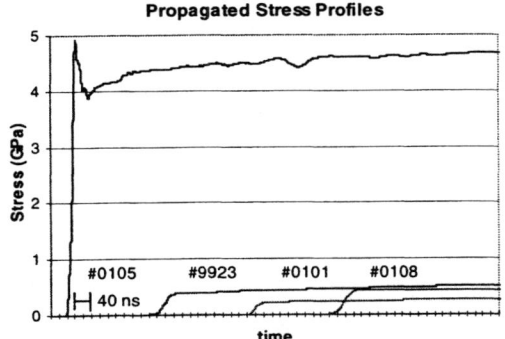

FIGURE 5. Combined plot of propagated-stress profiles from PVDF gauge experiments at high velocity.

Figure 4 compares the measured input stress profiles from experiments performed on the three ball-milled powders with the as-blended powder mixture at high velocities. The corresponding propagated-stress profiles are included in Figure 5.

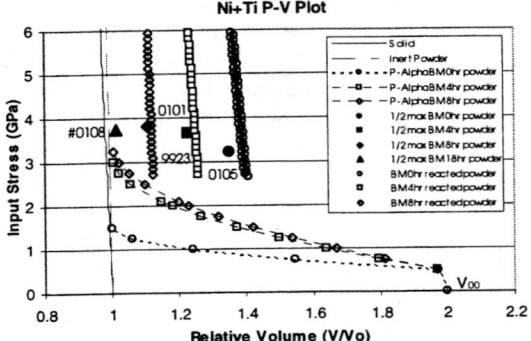

FIGURE 6. Plots of measured input stress versus calculated relative volume (based on half-maximum values), isothermal compressibility curves of dense NiTi, Ni+Ti inert mixture with zero crush strength, P-α densification curves and Hugoniots of reacted powders forming product.

It can be seen that both the input stress and input rise-time for the ball-milled powder mixture are slightly higher than those of the as-blended powder. The higher input stress can be attributed to the work hardening of the ball-milled powders. However, as shown in Figure 5, the equilibrium propagated-stress level of the ball-milled powders is one order of magnitude less than that of as-blended powder mixture, indicating that while the shock compression of ball-milled powders is dominated by shock attenuation, the behavior of the as-blended powder is dominated by shock induced reaction. Furthermore, the propagated-stress rise-time for the ball-milled powders is also higher than that of as-blended powder mixture.

Figure 6 plots the pressure-volume relationship, showing results of the measurements obtained from the above four PVDF gauge experiments at higher velocity. Also shown in the P-V plot are P-α densification behaviors of each of the as-blended and BM 4hr and 8hr powder mixtures, considering different values of crush strengths. The calculated compressibility curves corresponding to possible NiTi reaction products formed in as-blended, BM 4hr, and BM 8hr Ni+Ti powder reactants, are also illustrated. It can be seen that reactions in BM 4hr and 8hr powder mixtures show less volume expansion due to less exothermic heat (as revealed by the DTA analysis).

CONCLUSIONS

Instrumented experiments performed using PVDF gauges on ~50% dense as-blended Ni+Ti powder mixture at input stress less than the crush strength (~1.6 GPa), show characteristics of densification represented by the P-α behavior. The measured propagated-wave stress profiles show characteristics of wave dispersion with rise-time of 36 nanoseconds. In experiments at input stress higher than 2.7 GPa, the powder mixtures show evidence of shock-induced reaction, based on propagated-wave profiles showing a sharp rise-time (<~5 ns), and the data points of shock states revealing expansion and approaching the pressure-volume compressibility curve of thermodynamically determined Hugoniot of reacted powders. Experiments on Ni+Ti powder mixtures ball-milled for 4, 8 and 18 hours at input stresses greater than 3 GPa, show partial reaction or complete lack of shock-induced reaction. Their propagated-stress profiles show slower rise-time and significantly reduced stress amplitude due to shock attenuation. The recorded wave speeds in these experiments for ball-milled powders, performed at even higher pressures, are also reduced. Finally, the data points falling on the corresponding pressure-volume compressibility curves for ball-milled powder mixtures show decreasing evolution of reaction heat with increasing ball-milling time.

ACKNOWLEDGEMENTS

Funding for this research was provided by the Army Research Office under Grant DAAG55-97-1-0163. (Dr. W. Mullins, program monitor)

REFERENCES

1. Horie, Y. and Kipp, M.E., *J.Appl.Phys.* **63**, 5718-5727 (1988).
2. Graham, R.A., Morosin, B., Venturini, E.L. and Carr, M.J., *Annu.Rev.Mater.Sci.* **16**, 315-341 (1986).
3. Thadhani, N.N., *Prog. in Mater.Sci.* **37**, 118-224 (1993).
4. Thadhani, N.N., *J.Appl.Phys.* **76**, 2129-2138 (1994).
5. Zhu, Y.L., Li, T.C., Liu, J.T., Han, X.D. and Yang, D.Z., *Scripta Metall. et Mater.* **30**, 775-780 (1994).
6. Xu, X. and Thadhani, N.N., "Synthesis and Characterization of Nanocrystalline NiTi Shape Memory Alloy by Shock-Compression," in proceedings of Explomet 2000 Meeting, June 19-22, 2000, Albuquerque, NM.
7. Thadhani, N.N., Graham, R.A., Royal, T., Dunbar, E., Anderson, M.U., and G.T. Holman, *J.Appl.Phys.* **82**, 1113-1128 (1997).
8. Vandersall, K.S., PhD. Thesis, Georgia Institute of Technology, 1999.
9. Carroll, M.M. and Holt, A.C., *J. Appl. Phys.* **43**, 1626-1636 (1972).
10. Bennett, L.S. and Horie, Y., *Shock Waves: Int.J.* **4**, 127-136 (1994).

TiC BY SHS AND DYNAMIC COMPACTION

E.P. Carton, M. Stuivinga, A. Boluijt

TNO Prins Maurits Laboratory, P.O. Box 45, 2280 AA, Rijswijk, The Netherlands

Abstract. By ball-milling the Ti/C powder mixture before their Self-sustained High-temperature Synthesis (SHS) to TiC, the propagation velocity of the SHS process has been increased from 7 to 22 mm/s. The reaction of the milled powder mixture was accompanied by the release of a large volume of gas.
The micro-strain and the crystallite size of the milled reactants have been determined using x-ray diffraction line-broadening analysis. The graphite did not change during the milling process, but a large line-broadening effect was measured in the x-ray diffraction peaks of the milled titanium particles. The TiC fabricated by hot shock compaction of the porous SHS-product (SHS/DC) contained cracks and an axial hole in the center. It is believed that the large gas release rather than a Mach-stem was responsible for the formation of the axial hole.

INTRODUCTION

Dynamic or explosive compaction of ceramic powders to a high final density is difficult, since micro and macro cracks occur due to the brittle fracture behavior of these materials at room temperature. Above the Ductile-Brittle Transition Temperature (DBTT) materials show a remarkable increase in fracture energy due to a more ductile fracture behavior. The hot explosive compaction of TiC by combining the Self-sustained High-temperature Synthesis (SHS) and dynamic compaction (DC) processes was demonstrated in [1, 2, 3].

The combination of both processes enables one to compact TiC above its DBTT without the need of expensive equipment. The use of the indirect cylindrical configuration has the benefit of having an intrinsic thermal insulation layer between the metal tube in which the SHS occurs and the explosive layer surrounding the flyer tube, see Figure 1. However, the length of the SHS-tube is restricted by the rather slow propagation velocity of the SHS process (7 mm/s). The detonation can only be initiated when the SHS has been completed. For a 10 cm long SHS-tube it takes 14 seconds to complete the SHS process. In this time heat is leaking away to the surroundings; the SHS-tube, the flyer-tube, and the explosive layer. In order to decrease the SHS reaction time, and therefore reduce the heat loss, the propagation velocity of the SHS should be higher. In this work efforts to increase the SHS propagation velocity by ball-milling the reactants and its use in the SHS/DC combination are reported.

THEORY

A simple energy balance for a one-dimensional interpretation of the SHS process was used by Merzhanov [4] to determine its propagation velocity, V_{SHS}:

$$V^2_{SHS} = A(T_{ad}) \exp(-E_a/RT_{ad}) \qquad (1)$$

Here, A is a constant, R the gas constant, T_{ad} is the adiabatic temperature, and E_a is the activation energy. Apart from the temperature, as the most important rate determining parameter, the propagation velocity can be influenced by the activation energy of the reaction (E_a). The lower this energy, the higher the propagation velocity.

Lee [5] introduces the work of Benderskii et al. [6], who advanced the hypothesis that the global activation energy of a reaction (E_a) is a combination of a thermal energy (E_0) and an elastic compression energy (π):

$$E_a = E_0 - k.\pi \qquad (2)$$

Here, k is a constant of the order of unity. The elastic compression energy can be increased by elastic lattice deformation of the reactants. The elastic lattice deformations can, for example, be introduced by ball milling the SHS reactants. Ball milling the powder mixture for several hours results in an intense cold plastic deformation of the particles. Ball-milling also distributes the reactants more homogeneously and brings the reactants into a more intimate contact with each other. This will reduce the amount of mass transport during the reaction.

The average elastic strain in the lattice can be determined by analysis of line broadening in X-ray diffraction peaks, for example using the Hall-Williamson plot [7]. In these figures the 2θ-position and the width of all x-ray diffraction peaks of a crystalline material are plotted in a graph with special axis. When a straight line is drawn through these points, the average micro-strain is determined by the slope, while the intercept gives the crystallite size (average size of the coherently diffracting regions in the material).

EXPERIMENTS

Elastic lattice deformations have been generated by ball milling the reactants, an equiatomic mixture of Ti (<45 micron) and C (<50 micron) powder, in a planetary ball mill, see Figure 2. The powder mixture (100 gram) was milled for 24 hours in ethanol as a liquid coolant using 150 alumina balls with a diameter of 5 mm. After the milling process, the liquid was removed by decantation and further evaporated under vacuum at 40 °C. Then the x-ray line-broadening analysis was performed. The SHS propagation velocity was measured by timing the response of thermocouples placed in the reactants (with a relative density of 65 %TMD) at known axial positions. For comparison, also the SHS propagation velocity of a Ti/C mixture, with only the Ti-particles had been ball-milled was measured.

Finally, the ball-milled powder was used in SHS/DC experiments and the cross-section of the hot shock compacted TiC was analyzed by light microscopy. For comparison, an at room temperature shock compacted Ti/C powder mixture was analyzed. The experimental parameters for the SHS/DC experiments was the same as described earlier [1]. The set-up is schematically shown in Figure 1.

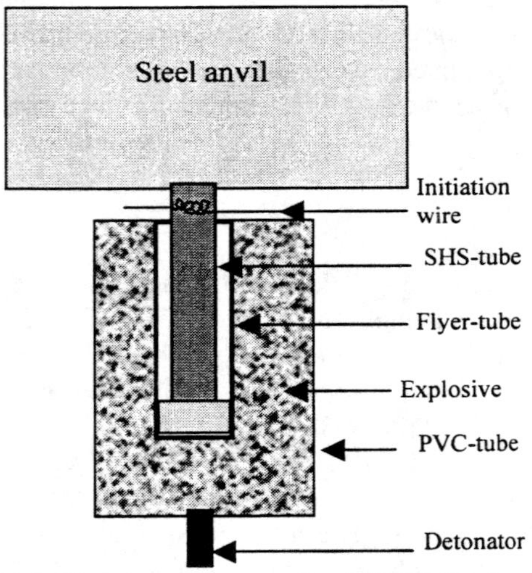

Figure 1: Experimental set-up for SHS/DC in the indirect cylindrical configuration.

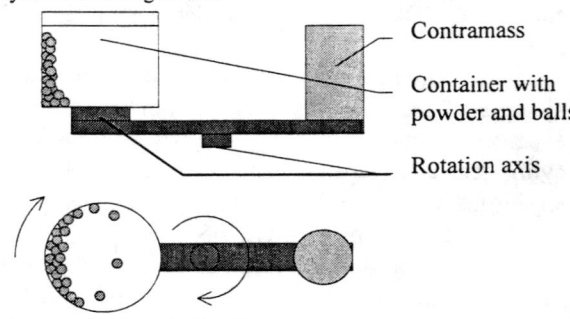

Figure 2. Planetary ball-mill.

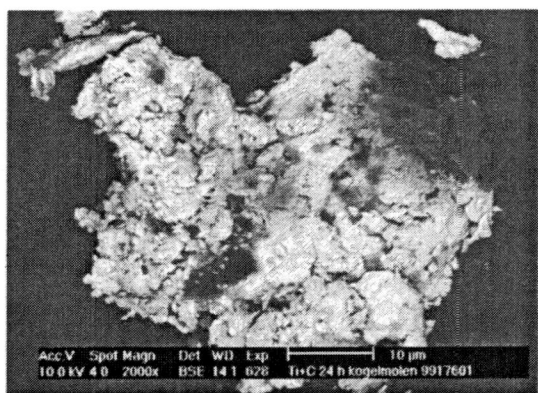

Figure 3: Particle after 24 hours ball-milling in a Ti/C powder mixture.

RESULTS AND DISCUSSION

Figure 3 shows a SEM image of a 24 hours ball-milled Ti/C powder particle. Graphite plates (the black spots) have been crushed into the Ti-particle, clearly indicating a closer contact between the two reactants. XRD analysis further indicated that no TiC had formed during the ball-milling process.

The measurement of the SHS propagation velocity indicated an increase from 7 mm/s for the starting Ti/C powder mixture, to 22 mm/s for the 24 hours ball milled powder mixture (all at a relative density of 65 %TMD). During the reaction of the latter, a much greater volume of gas escaped from the SHS-tube compared to the starting powder mixture. This indicates that the powder has not been properly dried.

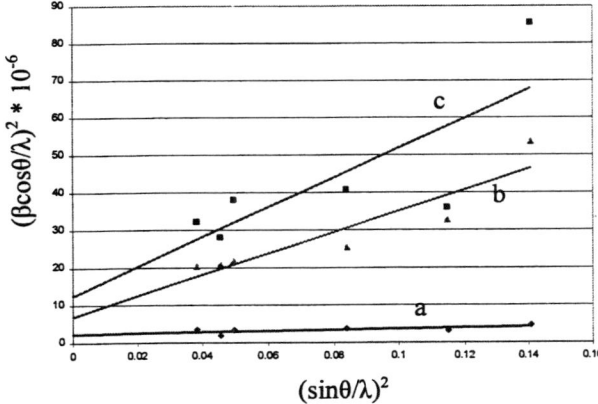

Figure 4: Hall-Williamson plot of three Ti-powder samples.

The graphite probably acts a storage material for ethanol during and after ball-milling, preventing a complete drying of the reactants in vacuum at 40 °C. The SHS propagation velocity was not higher then normal, when the Ti/C powder mixture was used with only the Ti-particles being ball milled. Also the gas release was at its "normal" volume.

Ball-milling did not lead to broadening in the x-ray diffraction peaks of graphite. However, the milling broadened the x-ray diffraction peaks in Ti considerably. Figure 4 shows Hall-Williamson plots for three Ti-powder samples. Sample (a) is the starting Ti-powder (as atomized), it has a low micro-strain value of 0.09% and an average crystallite size of 657 Å. Sample (c) is the Ti after 24 h ball-milling as a Ti/C mixture. The average crystallite size of the Ti had decreased to 283 Å, while the micro strain had increased to 0.49%. Sample (b) is the starting Ti/C mixture after shock compaction at room temperature. During this only micro-seconds lasting powder treatment the micro-strain in the Ti-particles has increased to 0.42%, and the crystallite size was reduced to 380 Å.

Morosin and Graham in [8] have done research on the differences and similarities between the defects formed in TiC powders that were treated by shock wave and ball-milling, respectively. Although both processes can lead to the same amount of micro-strain and crystallite size, the anisotropy in residual strain is large in ball-milled powder and more homogeneous in shock-modified powder. This can also be seen in Figure 4, since the scatter around line c (ball-milled) is larger than that for the line b (shock-modified).

Combined SHS/DC experiments with the ball-milled powder have been performed using the same parameters as described in [1]. The compacts showed an increase in plastic deformability of the hot TiC, due to the absence of spiral cracks that typically occur at the shock compaction of brittle material at low starting density.

The density of the TiC after SHS is only 47 %TMD, but increased to 98 %TMD due to the hot shock compaction process. However, an axial hole did form at the center of the compact as well as some cracks. The hole did not form due to a Mach-stem, since no loss of mass was detected in a cylindrical segment of the sample.

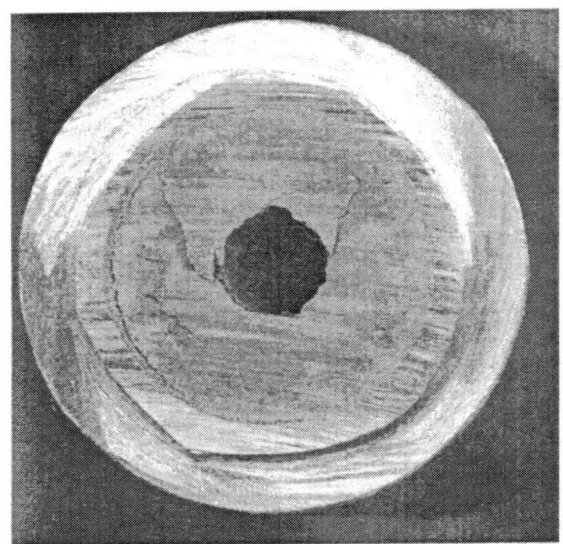

Figure 5: Cross-section of ball-milled sample after SHS/DC.

In a Mach-stem material normally escapes from the tube, due to its high kinetic energy and liquid or even gaseous state. The hole formed during this SHS/DC, probably formed by the accumulation of gasses at the line of symmetry of the cylindrical configuration.

CONCLUSIONS

Line broadening analysis of X-ray diffraction peaks of the milled Ti/C powder mixtures indicated no change for graphite, and an increase in microstrain, as well as a decrease in crystallite size for the Ti-particles.

Ball-milling the reactants in a planetary ball-mill did increase the SHS propagation velocity from 7 to 22 mm/s. This made it possible to reduce the time for the SHS process to complete, and therefore to shock compact the TiC at a higher temperature. However, the large volume of gas escaping from the powder during SHS of ball milled powder produced a central hole in the compacts. This central (axial) hole was not formed by a Mach-stem. Probably the ethanol, that was used as a coolant during the milling process, was adsorbed by the graphite. Several hours of drying in vacuum, did not dry the powder enough to prevent the large gas escape during SHS.

No increase in SHS propagation velocity was measured if only the Ti-particles had been ball-milled. This indicates that the cause of the increase in SHS velocity in the ball-milled Ti/C powder mixture, is primarily the better distribution and contact between the reactants.

ACKNOWLEDGEMENT

The authors thank E. Wilken for performing the X-ray line-broadening analyses.

REFERENCES

[1] Carton, E.P., Stuivinga, M. and Verbeek, H.J.,"Shock compaction of combustion synthesized ceramics in the cylindrical configuration", in *Shock Compression of Condensed Matter-1999*, edited by M.D. Furnish et al., AIP Conference Proceedings 741, New York, 1999, pp. 549-552.

[2] Grebe, H.A., Advani, A., Thadhani, N.N., Kottke, T., "Combustion synthesis and subsequent explosive densification of titanium carbide ceramics", *Metall. Trans. A*, Vol 23A, pp. 2365-2372 (1992).

[3] Grebe, H.A. and Thadhani, N.N.,"High-rate chemical reaction and high pressure processing of bulk titanium-carbide ceramics", *Processing and fabrication of advanced materials for high temperature applications*, TMS Proceedings, eds. Srivatsan T.S. and Ravi, V.A. (1992).

[4] Merzhanov, A.G.,"Pyrotechnical aspects of Self-Propagating High-Temperature Synthesis", in *XX Intern. Pyrotechnics Seminar*, Colorado Springs, 1994 NSWCCR/RDTN-94/004.

[5] Lee, J.H.S., Goroshin, S., et al., "Attempts to initiate detonations in metal-sulphur mixtures", in *Shock Compression of Condensed Matter-1999*, edited by M.D. Furnish et al., AIP Conference Proceedings 741, New York, 1999, pp. 775-778.

[6] Benderskii, V.A., Fillipov, D.G., Ovchimikov, "Ratio of thermal and deformation ignition in low temperature solid phase reactions:, *Doklady Akad. Nauk. SSR*, 308 (2), 401 (1989).

[7] Morosin, B. and Graham, R.A., "X-ray diffraction line-broadening studies on Shock-modified Rutile and Alumina", *Materials Science and Engineering*, Vol. 66, pp. 73-87 (1984).

[8] Morosin, B. and Graham, R.A., "X-ray diffraction line-broadening of shock modified titanium carbide", in *Materials Letters*, Vol. 3(3), pp. 119-123 (1985).

COOLING RATE THRESHOLD IN TRANSFORMATION OF C_{60} FULLERENE TO AMORPHOUS DIAMOND AND HIGHLY DISORDERED CARBON IN SCARQ EXPERIMENTS

Tomotaka Homae[1], Atsushi Okamoto[1], Kazutaka G. Nakamura[1], Ken-ichi Kondo[1], Masatake Yoshida[2], Keiji Hirabayashi[3], and Keisuke Niwase[4]

[1]*Materials and Structures Laboratory, Tokyo Institute of Technology, 4259 Nagatsuta, Midori, Yokohama 226-8503, Japan*
[2]*National Institute of Advanced Industrial Science and Technology, 1-1 Higashi, Tsukuba, Ibaraki 305-8565, Japan*
[3]*Canon Inc., 3-30-2 Shimomaruko, Ohta, Tokyo 146-8501, Japan*
[4]*Hyogo University of Teacher Education, 942-1 Shimokume, Yashiro, Hyogo 673-1494, Japan*

Abstract. C_{60} films on gold substrates (film thickness of 6-20 μm) were prepared. These films were shock compressed to 48 GPa and recovered using "shock compression and rapid quenching (SCARQ)" technique. The recovered samples were amorphous diamond, when the initial thickness of the sample was less than 10 μm, and disordered carbon, when the initial thickness was 20 μm. The temperature history of the sample was estimated by one-dimensional thermal diffusion analysis. It was revealed that there was a lower limit of cooling rate for recovery of amorphous diamond. The chemical bond change of carbon after shock compression was also discussed.

INTRODUCTION

Shock compression and recovery technique is very useful for exploring new carbon phases, because it brings the sample to extremely high pressure and temperature conditions, although the duration of shock compression is short. But, if the high shock temperature continues after arrival of a release wave, the transformed carbon phase retransforms to sp^2 bonded carbon phase. A technique, which allows rapid cooling the sample during the compression followed by quenching the metastable phase, was developed to deal with this problem. If a sample is in contact with another material, whose compressibility is lower than that of the sample and the heat conductivity is high, will be shock-compressed, the maximum shock temperature of the sample and the another material should be different. The low-compressible material work as a heat sink and the sample is cooled rapidly by heat conduction. For example, diamond synthesis from graphite powder mixed with copper powder have already been developed [1]. Hirai et al. improved this technique and called it "SCARQ" (Shock Compression And Rapid Quenching) technique [2]. In this improved technique, thin sample is sandwiched by metal disks and shock compressed. New carbon phases such as n-diamond [3], amorphous diamond [4], and nanocrystalline diamond ceramics [5] were successfully obtained. In the case of the thin-film uniform sample, the temperature history of the sample may be estimated easily with aid of one-dimensional thermal diffusion analysis.

C_{60} fullerene is considered as the promising initial material for exploring the new phase of carbon. When tapped C_{60} fullerene powder was shock compressed to more than 27 GPa, highly disordered carbon was recovered [6]. On the other hand, when C_{60} fullerene was shock compressed to 55 GPa and cooled rapidly using SCARQ technique [2], amorphous diamond [4] and/or nanocrystalline diamond ceramics [5] were obtained. These results suggest that the cooling rate of the shocked sample may dictate what kind of carbon phase will be recovered, but the details are unknown. The transition paths from C_{60} fullerene to amorphous diamond and to highly disordered carbon are also of interest.

In the present work, C_{60} fullerene films (thickness of 6-20 μm) sandwiched by gold disks were shock compressed to 48 GPa. Temperature-history of the sample was estimated by the one-dimensional thermal diffusion analysis. The relation between the cooling rate of the sample and the recovered samples and the change of the carbon chemical bond after the shock compression was discussed.

EXPERIMENTAL METHOD

A commercial grade C_{60} fullerene purified to 99.9% was used as starting material. The C_{60} films for the present investigation were prepared by vacuum deposition on gold disks (12 mm in diameter and 100 μm in thickness). C_{60} was heated to 400 °C under vacuum (1 x 10^5 torr). Sublimed C_{60} was cooled by the gold disk, which was placed over the crucible, and deposited. The deposition rate was about 1 mm/h and the prepared films were approximately 3 μm, 5 μm, and 10 μm thick. Spectrum of deposited film shows only a peak at around 1469 cm^{-1}, which is assigned to Ag_2 pentagonal pinching mode of C_{60} fullerene crystal (Figure 1). The gold disks were used as heat sinks in this experiments because gold has relatively low reactivity with carbon and the gold lattice constant differs from that of known carbon phase such as diamond. A sandwich of C_{60} films, made by superimposing the disks face to face, was inserted into a capsule made of stainless steel. Thus, the initial thickness of the samples were 6 μm, 10 μm, and 20 μm. The capsule was put into a protective assembly and subjected to shock loading with a 3 mm-thick stainless steel flyer accelerated by an explosive plane-wave generator. The flyer velocity was estimated to be approximately 2.0 km/s. The shock pressure and a duration were estimated to be 48 GPa and 800 ns, respectively.

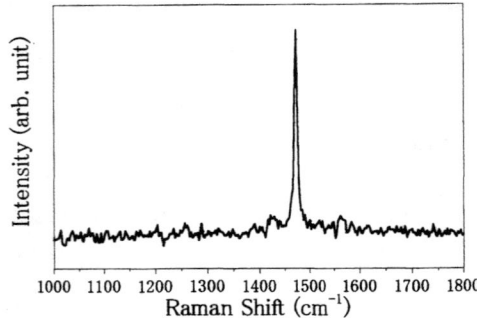

Figure 1. Raman spectrum of deposited film.

TEMPERATURE-HISTORY CALCULATION

The shock-compressed state was calculated first on the basis of the conservation of mass, momentum, and energy, the Hugoniot equations of state of gold and carbon (initial density of 1.77 g/cm^3, as substitute for C_{60} fullerene) [7], and the Mie-Grüneisen equation of state involving Debye's theory. Single shock was assumed and the openings between the films were neglected. Since the heat conduction during the shock compression was neglected because the rate of pressure increase is significantly higher than the rate of heat conduction for actual sample thickness of micrometers and tens of micrometers, the two-step process can be assumed. The changes of the sample temperature after the shock compression were calculated on the basis of one-dimensional thermal diffusion analysis. Thermodynamic parameters of gold and diamond at room pressure and room temperature were used for calculation. Figure 2 shows the calculated results for the temperature history at the center of various thickness (3-40 μm) diamond sample sandwiched by 100 μm-thick gold disks. As the density of the

C_{60} fullerene crystal is 1.65 g/cm^3 and the density of the diamond crystal is 3.51 g/cm^3, if initial C_{60} film transforms to diamond under shock compression, the thickness reduces to one-half of its initial value. Thus, the initial thickness of 6 μm, 10 μm, and 20 μm in this experiment correspond to 3 μm, 5 μm, and 10 μm in Figure 2, respectively. It is obvious that the thickness of the sample is thinner, the cooling rate is higher. For example, cooling rate of first 100 ns after shock compression is 3.2 x 10^8 K/s, 5.8 x 10^9 K/s, 1.5 x 10^{10} K/s in the case of thickness of 3, 5, and 10 mm, respectively. At 800 ns after shock compression, corresponding to the arrival of the release wave, the temperature of the diamond is estimated to be 2590 K, 1510 K, and 1160 K, in the case of thickness of 3, 5, and 10 mm, respectively.

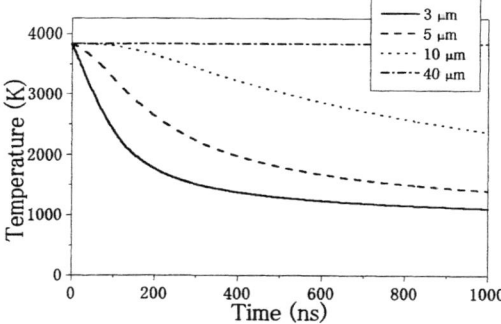

Figure 2. Temperature history calculation based on one-dimensional thermal diffusion analysis.

RESULTS

The recovered samples were studied using a optical microscope. When the initial thickness of the sample was 6 μm and 10 μm, the recovered sample was transparent and were tile-like fragments whose size was less than 100 μm (Figure 3 (a)). The gold substrate can be seen through the sample. In the case of 20 μm thickness, the recovered sample was black, unsettled shape, and size of a couple of mm (Figure 3 (b)). Transparent recovered samples show no Raman peaks and a broad photoluminescence peak (Figure 4 (a)). Since these characteristics are in agreement with that of the amorphous diamond reported previously [4], these samples are identified as amorphous

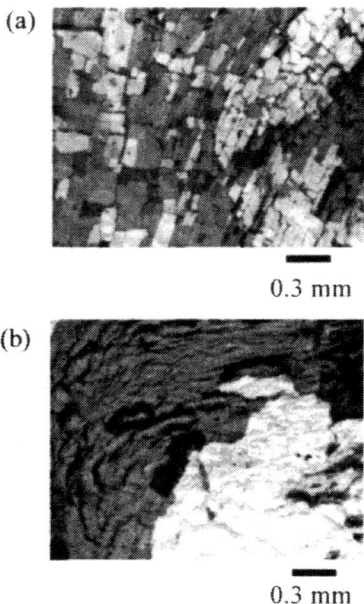

Figure 3. Recovered sample (a) initial thickness of 10 μm (b) initial thickness of 20 μm. Some of the sample exfoliated from the gold substrate when the recovery capsule was opened (looks white in these pictures).

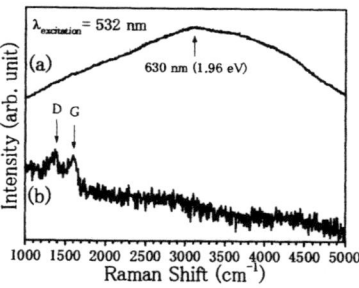

Figure 4. Raman spectra of recovered samples. (a) initial thickness of 6 μm and (b) initial thickness of 20 μm.

diamond.

In contrast, black recovered samples show G (1580 cm^{-1}) and D (1360 cm^{-1}) Raman peaks of sp^2 bonded carbon (Figure 4 (b)). Since the intensity of D peak is comparable to that of G peak, these recovered sample are identified as sp^2 bonded highly disordered carbon [8]. All of the recovered samples show no Raman peak around 1469 cm^{-1}, which is assigned to Ag_2 pentagonal pinching mode of C_{60} fullerene crystal and observed before the shock compression (Figure 1 (b)).

DISCUSSION

The samples of three initial thickness were shock compressed, but the maximum pressure and the temperature were almost identical, because initial thickness of the sample does not affect the maximum pressure and temperature. As shown in the model calculation (Figure 2), the dominant difference is the cooling rate of the sample.

As these samples were shock compressed to diamond stable region (up to 48 GPa and 2000 K), it is supposed that during the shock compression, all of these samples transformed into sp^3 bonded carbon. If the cooling rate was high enough and the sample was cooled enough before the release wave arrived, this sp^3 bonded carbon was recovered as amorphous diamond without crystal growth. In contrast, if the cooling rate was lower and the sample was exposed to high temperature and normal pressure after the release wave arrived, this sp^3 bonded carbon transformed into sp^2 bonded carbon without crystal growth, as graphite, because the duration of high temperature acting was not long enough for the crystal growth. In this case, the recovered sample was highly disordered carbon (Figure 5).

It is obvious that the lower limit of the cooling rate and upper limit of the initial thickness of the sample to recover amorphous diamond are exist. This threshold must depend on the shock duration. But when the shock duration is about 1 ms, which is typical value for shock compression apparatus of many laboratories, the upper limit of the initial sample thickness is between 10 and 20 μm. In general, this fact implies that one cannot recover quenched metastable materials thicker than 20 μm using the shock-compression and recovery technique.

The way of the designing temperature-history of the shock compression was established. But it is not revealed what is more effective for quenching and recovering the metastable phase: the cooling rate of first several tens of ns or the temperature of the sample at the time of the arrival of the release wave. Further experiments are required to deal with this problem.

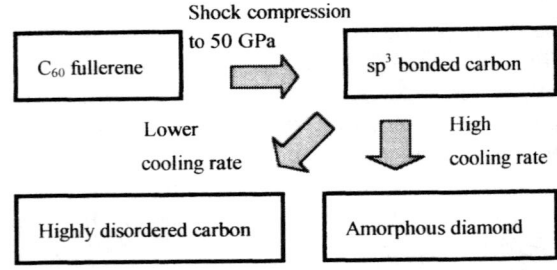

Figure 5. Transition path from C_{60} fullerene to amorphous diamond or highly disordered carbon.

ACKNOWLEDGEMENTS

This work was supported by Core Research for Evolutional Science and Technology (CREST) program of Japan Science and Technology Corporation (JST). The authors thank M. Hasegawa for his experimental help.

REFERENCES

1. Trueb, L. F., *J. Appl. Phys.* **39**, 4707-4716 (1968).
2. Hirai, H., and Kondo, K., *Science* **253**, 772-774 (1991).
3. Hirai, H., and Kondo, K., *Proc. Jpn. Acd.* **67(B)**, 22-26 (1991).
4. Hirai, H., and Kondo, K., *Appl. Phys. Lett.* **64** (1994) 1797-1799.
5. Hirai, H., Kondo, K., Kim, M., Koinuma, H., Kurashima, K., and Bando, Y., *Appl. Phys. Lett.* **71**, 3016-3018 (1997).
6. Yoo, C. S., Nellis, W. J., *Science* **254**, 1489-1491 (1991).
7. Marsh, S. P., *LASL Shock Hugoniot Data*, University of California Press, Berkeley, 1980.
8. Knight, D. S., and White, W. B., *J. Mater. Res.* **4**, 385-393 (1989).

CHAPTER XVII

INSTRUMENTATION

CARBON RESISTOR PRESSURE GAUGE CALIBRATION AT LOW STRESSES

Bruce Cunningham[1], Kevin S. Vandersall[1], Angela M. Niles[1], Daniel W. Greenwood[1], Frank Garcia[1], Jerry W. Forbes[1], and William H. Wilson[2]

[1]Lawrence Livermore National Laboratory, Energetic Materials Center, 7000 East Avenue L-282, Livermore, CA 94550
[2]AFRL-Munitions Directorate, Eglin AFB, FL 32524

Abstract. The 470 Ohm carbon resistor gauge has been used in the stress range up to 4-5 GPa for highly heterogeneous materials and/or divergent flow experiments. The attractiveness of the gauge is its rugged nature, simple construction, low cost, reproducibility, and survivability in dynamic events. Gauge drawbacks are the long time response to pressure equilibration and gauge resistance hysteresis. In the regime below 0.4 GPa, gauge calibration has been extrapolated. Because of the need for calibration data within this low stress regime, calibration experiments were performed using a split-Hopkinson bar, drop tower apparatus, and gas pressure chamber. Since the performance of the gauge at elevated temperatures is a concern, the change in resistance due to heating at atmospheric pressure was also investigated. Details of the various calibration arrangements and the results are discussed and compared to a calibration curve fit to previously published calibration data.

INTRODUCTION

The carbon resistor gauge has previously been studied by numerous researchers [1-10] for several different initial resistance values. This gauge is a simple carbon composition resistor that can be used as a pressure gauge with little or no modification. The equipment needed is a power supply to provide a small amount of constant current or alternatively a Wheatstone bridge could be used. Because of ease of use, ability to measure pressures up to 3 GPa, and survivability in harsh environments, it can be used in cases where no other gauge would survive. Because the gauge is manufactured to be a resistor and not a pressure gauge, an empirical calibration is required.

Recent experiments at Lawrence Livermore National Laboratory (LLNL) [11,12] have incorporated the 470 Ω carbon resistor gauge in energetic materials. Measurements from 0.4 to 3 GPa were made. A calibration curve was fit by one of the authors (WHW) to data from several publications [1-4]. This curve is extrapolated into the low-pressure regime from higher-pressure data. The goal of this work is to characterize the calibration of the 470 Ω carbon resistor gauge at low pressures (<0.6 GPa) using a static gas pressure chamber (argon environment), a split-Hopkinson bar, and a drop tower apparatus. Some experiments require heating of the experimental assembly, so the behavior of the resistor at ambient pressure and at elevated initial temperatures was investigated.

EXPERIMENTAL PROCEDURE

The resistors used in this work were standard 1/8 W, 470 Ω carbon composition resistors made by Allen-Bradley Corporation. The nominal dimensions of the resistor are 1.7 mm diameter and 4 mm long, with wire leads extending from each end of the cylinder. The details of the procedure for each calibration as well as the results for that calibration, are in the respective sections below. During each experiment the constant current power supply for the carbon resistor gauges remained on at all times and supplied ~16 mA of constant current through the 470 Ω resistor gauges. This simplifies the experiment by requiring only the digitizers to be triggered when the event occurs.

Static Gas Pressure Chamber

The static gas pressure calibrations of the carbon resistor gauge were performed in a pressure chamber that is usually used to measure burn rates of enegetic materials at elevated temperatures and pressures. Further details of the apparatus are included elsewhere [13]. Figure 1 outlines a general schematic of the assembly used. The chamber has the capability to be pressurized to 0.4 GPa with a gas (in this case argon). As indicated in Figure 1, a calibrated Kistler model 6213B quartz gauge [14] is located at one end of the chamber to measure the pressure in the chamber and the carbon resistor gauge array (5 gauges) is located at the other end. During the loading in the experiment, the output from the gauges are continuously scanned on a Keithly digital voltmeter and saved on a computer.

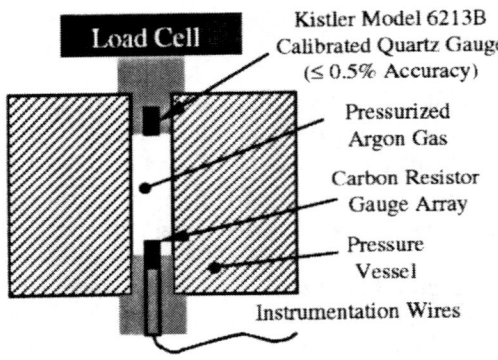

FIGURE 1. Schematic outlining the gas pressure chamber assembly.

Split-Hopkinson Bar

Calibration experiments were performed using a split-Hopkinson bar apparatus. Details of the operation of the split-Hopkinson pressure bar can be found in associated references [15,16]. Figure 2 (a) shows a schematic of the apparatus which consists of a striker bar impacting an incident bar that is adjacent to the sample and backed by a transmitter bar. The bars are operated in the elastic regime and strain gauges are placed on the input and transmitter bars to measure the strains. The input and transmitted stresses can be calculated using the elastic modulus of the bar material. Bars made from 9.5 mm diameter 6061-T6 aluminum were used in these calibration experiments. Figure 2 (b) outlines the schematic carbon resistor gauge arrangement that consisted of two sample halves that are 7.6 mm diameter by 2.5 mm and 5 mm thick respectively. The carbon resistor gauge was placed in grooves in the larger sample half, and then the sample was joined together with Dow Corning 3145 RTV sealant. For the analysis, the sample stress was calculated by using the transmitted stress in the transmitter bar (calculated from the peak strain) multiplied by the ratio of the aluminum bar diameter to the final Teflon sample (with embedded gauge) diameter. A Tektronix TDS 784D oscilloscope was used to measure the carbon resistor gauge output during the experiment.

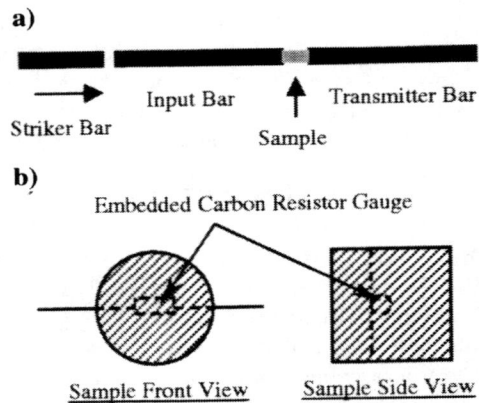

FIGURE 2. Schematic showing (a) split-Hopkinson bar arrangement and (b) general illustration of carbon resistor gauge (~4mm long by 1.7 mm diameter) inside Teflon sample.

Drop Weight Apparatus

A commercially available drop weight apparatus, model 913B02 Hydraulic Impulse Calibrator, obtained from PCB, Piezotronics [18] was used. Figure 3 shows a schematic of the apparatus with a 4 kg drop weight dropped from different heights (0.1, 0.41, 0.71, 1.18, and 1.49 meters) which impacts a plunger assembly that creates hydraulic

SHOCK INITIATION OF UF-TATB AT 250°C*

Paul A. Urtiew, Jerry W. Forbes, Frank Garcia and Craig M. Tarver

Lawrence Livermore National Laboratory
P.O.Box 808, L-282
Livermore, CA 94551

The shock initiation properties of pure ultrafine grade triaminotrinitrobenzene (UF-TATB) pressed to an initial density of 1.80 g/cm^3 and fired at ambient temperature and 250°C are reported. Embedded manganin pressure gauges are used to measure the pressure histories during the buildup to detonation at several input pressures. The ambient temperature results confirm previous run distance to detonation versus shock pressure results. UF-TATB at 250°C is shown to be much more shock sensitive than it is at ambient temperature. At high impact pressures, the shock sensitivity of UF-TATB at 250°C approaches that of HMX-based explosives under ambient conditions. Ignition and Growth reactive flow models are developed for UF-TATB at both temperatures to allow predictions to be made for other scenarios.

INTRODUCTION

One of the concerns in today's work with energetic materials is their safety when they are exposed to extreme environmental conditions. Hazard scenarios can involve multiple stimuli, such as heating to temperatures close to the thermal explosion conditions followed by fragment impact, producing a shock in the hot explosive. This scenario has been studied for triaminotrinitrobenzene(TATB)-based insensitive explosives LX-17 and PBX 9502 under various thermal and confinement conditions (1-3) and for LX-04, an HMX (octahydro-1,3,5,7-tetranitro-1,3,5,7-tetrazocine)-based solid high explosive (4-6). In this paper, embedded manganin pressure gauges and reactive flow calculations are employed to study the shock sensitivity of pure ultrafine (6 micron) TATB (UF-TATB) at ambient (25°C) and high (250°C) temperatures.

Manganin gauges have been shown to perform normally at 250°C using Teflon armor and 6061-T6 aluminum targets (7).

EXPERIMENTAL

The experiments were fired using a 100mm diameter propellant driven gas gun, capable of accelerating a 1kg projectile to a velocity of 2.5 km/s. The target assembly is identical to those used in previous studies (1-6). 12.5 mm thick, 90 mm diameter aluminum plates impact targets consisting of 6 mm thick aluminum buffer plates, 16-31 mm thick UF-TATB charges and 6 mm thick aluminum back plates. For the hot experiments, heaters are placed within the aluminum plates, and the UF-TATB is heated at approximately 1.6°C per minute to 250°C. Two ambient temperature shots were fired with aluminum flyer plate velocities of 1.436 and 1.688 mm/μs, respectively. Four 250°C shots were fired with aluminum flyer plate velocities of 1.1, 1.343, 1.482, and 1.711 mm/μs, respectively.

---

*This work was performed under the auspices of the United States Department of Energy by the Lawrence Livermore National Laboratory under Contract No. W-7405-ENG-48.

REACTIVE FLOW MODELING

The Ignition and Growth reactive flow model uses two Jones-Wilkins-Lee (JWL) equations of state, one for the unreacted explosive and another one for the reaction products, in the form:

$$p = A e^{-R_1 V} + B e^{-R_2 V} + \omega C_v T/V \qquad (1)$$

where p is pressure in Megabars, V is relative volume, T is temperature, ω is the Gruneisen coefficient, C_v is the average heat capacity, and A, B, R_1 and R_2 are constants. The equations of state are fitted to the available shock Hugoniot data. The reaction rate equation is:

$$dF/dt = I(1-F)^b(\rho/\rho_0-1-a)^x + G_1(1-F)^c F^d p^y$$
$$0<F<F_{igmax} \qquad 0<F<F_{G1max}$$
$$+ G_2(1-F)^e F^g p^z \qquad (2)$$
$$F_{G2min}<F<1$$

where F is the fraction reacted, t is time in μs, ρ is the current density in g/cm^3, ρ_0 is the initial density, p is pressure in Mbars, and I, G_1, G_2, a, b, c, d, e, g, x, y, and z are constants. This reaction rate law models the three stages of reaction generally observed during shock initiation of solid explosives (5). Equation of state parameters for ambient or hot Al and Teflon are used. Ignition and Growth rate law parameters are found for ambient and hot UF-TATB. Based on the measured density changes for LX-17 and PBX 9502, the density of UF-TATB decreases from 1.80 g/cm^3 at 25°C to 1.686 g/cm^3 at 250°C.

AMBIENT TEMPERATURE RESULTS

Figure 1 shows the measured pressure histories at 0, 4.11, 7.38, 10.64, 13.92, 17.2, 20.43, and 35 mm deep manganin gauges in UF-TATB impacted at 1.436 mm/μs. At this initial pressure of 7 GPa, the UF-TATB exhibits very little growth of reaction behind the shock front. LX-17 and PBX 9502 also show little growth of reaction below 8 GPa (1-3). The 35 mm gauge shows a higher pressure, because it is between UF-TATB and the Al back plate. Figure 2 shows the gauge records for ambient UF-TATB impacted by Al at 1.688 mm/μs. The gauges are at 0, 3.95, 7.24, 10.55, 13.83, 17.116, 20.38, and 35 mm. For this 9.5 GPa input shock pressure, transition to detonation occurs near the 17mm deep gauge. Figure 3 shows the calculated pressures for the UF-TATB shot shown in Fig. 2.

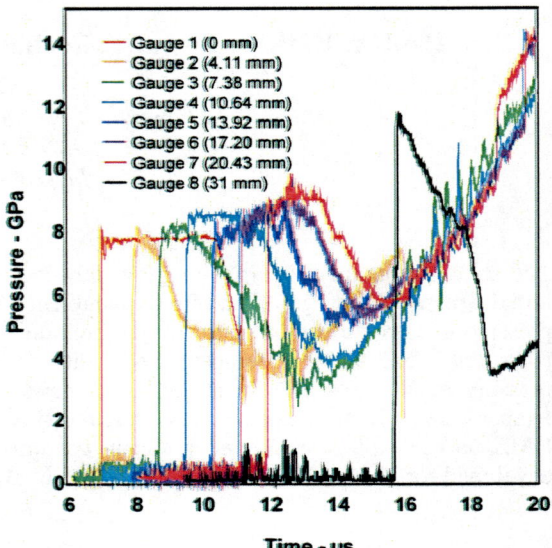

Figure 1. Pressure histories for ambient UF-TATB impacted at 1.436 mm/μs by an aluminum flyer

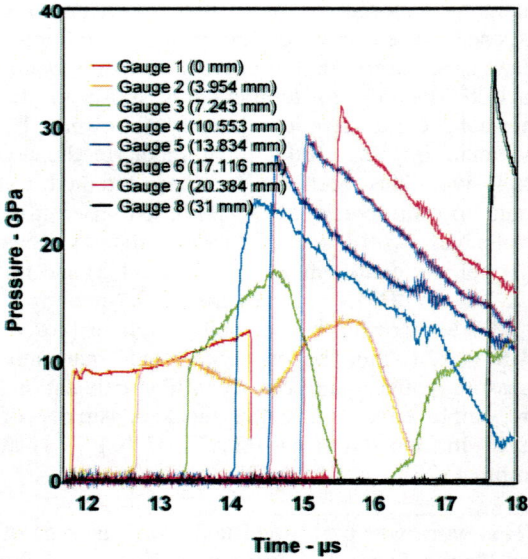

Figure 2. Pressure histories for ambient UF-TATB impacted at 1.688 mm/μs by an aluminum flyer

ADVANCED CRYOGENIC SYSTEM CAPABILITIES FOR PRECISION SHOCK PHYSICS MEASUREMENTS ON Z

D. L. Hanson, R. R. Johnston, M. D. Knudson, J. R. Asay,
C. A. Hall, J. E. Bailey, and R. J. Hickman

Sandia National Laboratories, Albuquerque, NM 87185 USA

Abstract. We have developed a general purpose cryogenic target system for precision EOS studies with the Sandia Z accelerator short-circuit current drive. Condensation of large-area cryogenic liquid samples is accomplished using a cryostat evaporation refrigerator, shielded for survivability in the Z blast environment and connected by a thermal link to an expendable sample holder with active temperature control. Accurate shock physics measurements are performed using multiple, thermally isolated, fiber-optic-coupled VISAR interferometry, active shock breakout, and time-resolved optical spectroscopy diagnostics. We describe variations of this cryogenic system currently under development for three EOS measurement applications: (1) liquid D_2 shock Hugoniot experiments using magnetically driven flyer plates; (2) liquid N_2 isentropic compression experiments; and (3) EOS measurements on liquid ^{4}He at 1.5 K.

INTRODUCTION

Recently we have used the fast pulsed power technology of the Sandia Z accelerator (20 MA peak current in the long-pulse, short-circuit mode) to develop new experimental techniques for accurate EOS measurements at high pressure. Magnetic pressure loading of material samples in high current density electrode geometries on Z allows the generation of continuous isentropic compression curves previously unavailable at Mbar pressures [1] and the launching of magnetically driven flyer plates at velocities exceeding 20 km/s for accurate shock Hugoniot measurements [2].

To extend the range of materials that may be studied using these new precision EOS techniques, we are developing a general purpose cryogenic target system for Z. Cryogenic samples are cooled with a cryostat evaporation refrigerator, shielded for survivability in the Z blast environment. The version of the system currently in use is capable of condensing large area cryogenic liquid samples from a variety of permanent gases with normal boiling points above 15 K, including H_2, D_2, Ne, N_2, O_2, and Ar. In this paper, we will describe variations of this cryogenic system currently under development for three EOS measurement applications on Z: (1) liquid deuterium (LD_2) shock Hugoniot experiments using magnetically driven flyer plates; (2) liquid nitrogen (LN_2) isentropic compression experiments; and (3) liquid ^{4}He (LHe) EOS measurements.

LD_2 HUGONIOT MEASUREMENTS

We have recently performed Hugoniot measurements of LD_2 in the range of 25–70 GPa on Z [2]. The results of these experiments challenge the controversial NOVA laser data [3,4] which show significantly increased compressibility compared to predictions of first principles, *ab initio* EOS models for the hydrogen isotopes [5]. These Z measurements use the technique of magnetically driven planar flyer plate impact on relatively large area, thick samples, with multiple redundant diagnostics to improve accuracy and address many of the concerns over the experimental conditions of the laser measurements.

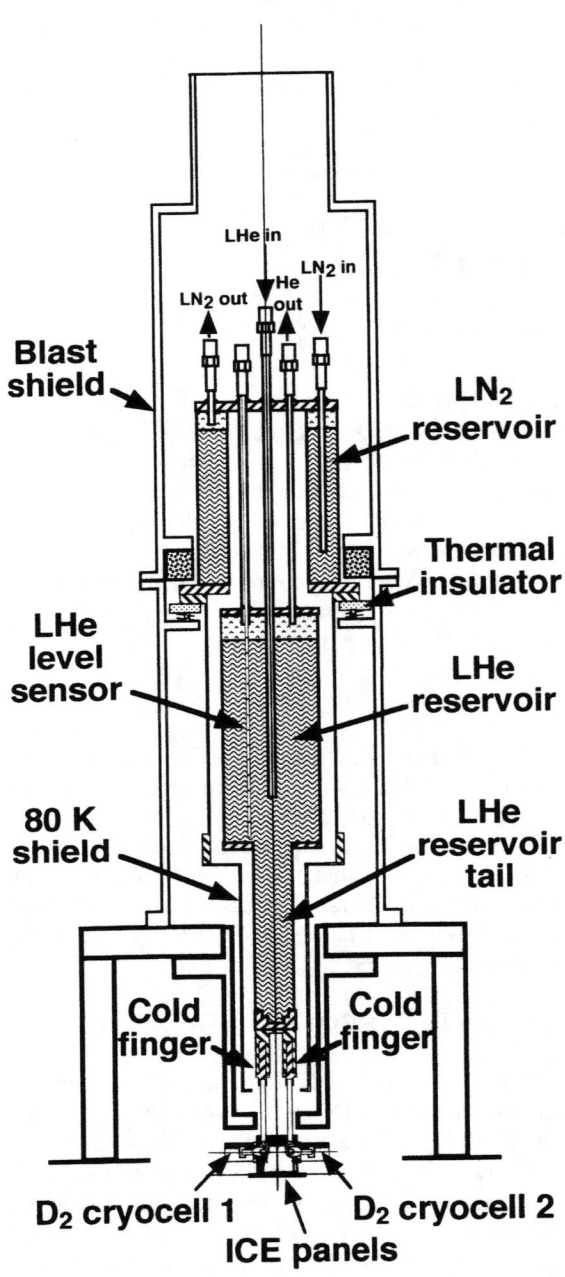

FIGURE 1. Static-fill LHe cryostat with double cold finger for simultaneous cooling of two LD$_2$ cryocells.

Figure 1 shows the arrangement of the LHe cryostat used to simultaneously cool two separate LD$_2$ samples on the Z experiments. Survivability of high value cryogenic components is a major issue for cryogenic system design on Z. Following the current pulse, more than 1 MJ of energy is dissipated in the form of hot plasma, molten metal, and shrapnel ejected from the load region. The LHe cryostat is shock-mounted inside a robust stainless steel blast shield. Cryostats in this arrangement have survived many Z shots without significant damage. The design of the cryocell for LD$_2$ shock Hugoniot measurements is shown in Fig. 2. The LD$_2$ sample is condensed from high purity D$_2$ gas at 18 psi in a cavity formed by a stepped Al pusher plate and a sapphire rear window. This window provides optical access for fiber-optic-coupled VISAR, active shock breakout, and time-resolved visible light spectroscopy diagnostic probes. The cryocell is thermally isolated from the anode mounting panel by a thin Nylon thermal break. Each cryocell is cooled to about 19 K and then warmed to 22.5 K by a combination of heater energy and absorbed diagnostic laser light to produce a quiescent (bubble-free) LD$_2$ sample with a boiling point of about 24.7 K. Spatially uniform, constant pressure shock loading of the Al pusher is produced by impact of the flyer plate accelerated by magnetic pressure from a ramped current drive.

ISENTROPIC COMPRESSION OF LN$_2$

We have recently demonstrated in studies of Cu and Fe that magnetic pressure loading with the ramped current pulse of the Z accelerator can be used to accurately determine the off-Hugoniot isentropic response of materials at high pressures [1]. The experimental arrangement being developed to extend this technique to an LN$_2$ sample is shown in Fig. 3. A quiescent large area LN$_2$ sample is condensed from high purity N$_2$ gas at about 78 K in a cryocell similar to that shown in Fig. 2. Cooling is provided by the cryostat shown in Fig. 1, with each reservoir now filled with LN$_2$. In an isentropic compression experiment, the Al pusher plate at the front of the cryocell forms part of the current path of the short circuit load and is gently accelerated by the magnetic pressure from a ramped, shaped current pulse, quasi-isentropically compressing the LN$_2$.

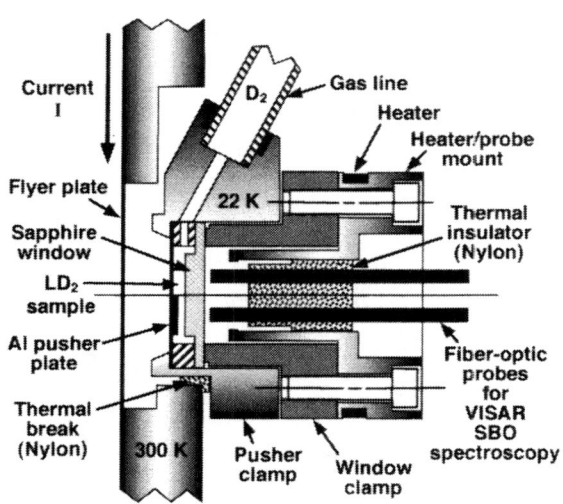

FIGURE 2. Cryocell in magnetically driven flyer plate configuration for LD$_2$ Hugoniot measurements.

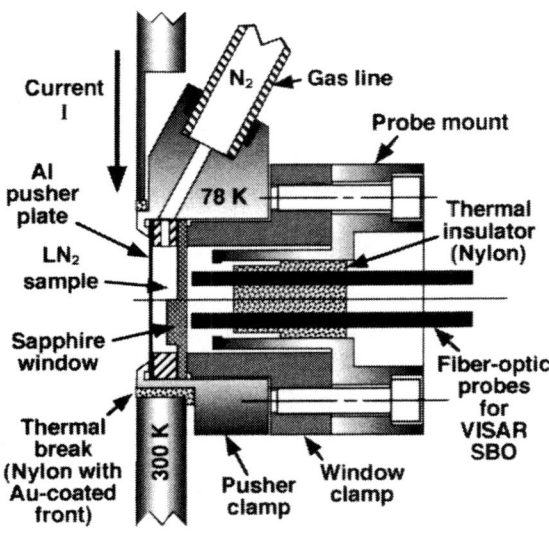

FIGURE 3. Cryocell design for isentropic compression of a LN$_2$ sample.

LHe EOS MEASUREMENTS

^{4}He is the most fundamental of the permanent gases. This isotope has the smallest closed-shell atom with an extremely weak intermolecular potential and exhibits unique quantum properties in the liquid state. It is a component of many systems of astrophysical, condensed matter, and ICF interest. Very little EOS data exist for ^{4}He at high pressures because of the difficulty of condensing quiescent liquid ^{4}He samples at less than 4 K in an environment suitable for shock physics measurements.

We are currently developing a variation of the Z cryogenic target system to condense superfluid ^{4}He-II samples at 1.5 K for He EOS measurements. The cryostat for this system is shown in Fig. 4. To minimize the heat load and maintain a stable sample environment at 1.5 K, the ^{4}He sample holder must be surrounded by a double thermal radiation shield consisting of an outer layer cooled by LN$_2$ to 80 K and an inner shield cooled by LHe to 5 K. These shields are segmented into overlapping sections, some of which serve as part of the cryostat blast shield. This acts to mechanically decouple the He sample holder from the main LHe cryostat and allow the survival of the most expensive cryogenic system components in the Z blast environment. The design of the sample holder assembly for LHe shock Hugoniot measurements with magnetically driven flyer plates is shown in Fig. 5. The sample holder is operated as a continuously fed ^{4}He evaporation refrigerator [6]. The sample holder consists of a small LHe bath, which controls the cooling of the sample holder body, and a cavity at the front in which high purity ^{4}He gas is condensed into a quiescent LHe sample for the experiment. LHe from the main cryostat at 4.2 K and 760 Torr is drawn through a flow impedance and partially evaporated to cool the LHe bath at 3.6 Torr to 1.5 K.

SUMMARY

We have briefly described several of the cryogenic sample capabilities being developed to complement new precision EOS Hugoniot and off-Hugoniot measurement techniques on Z.

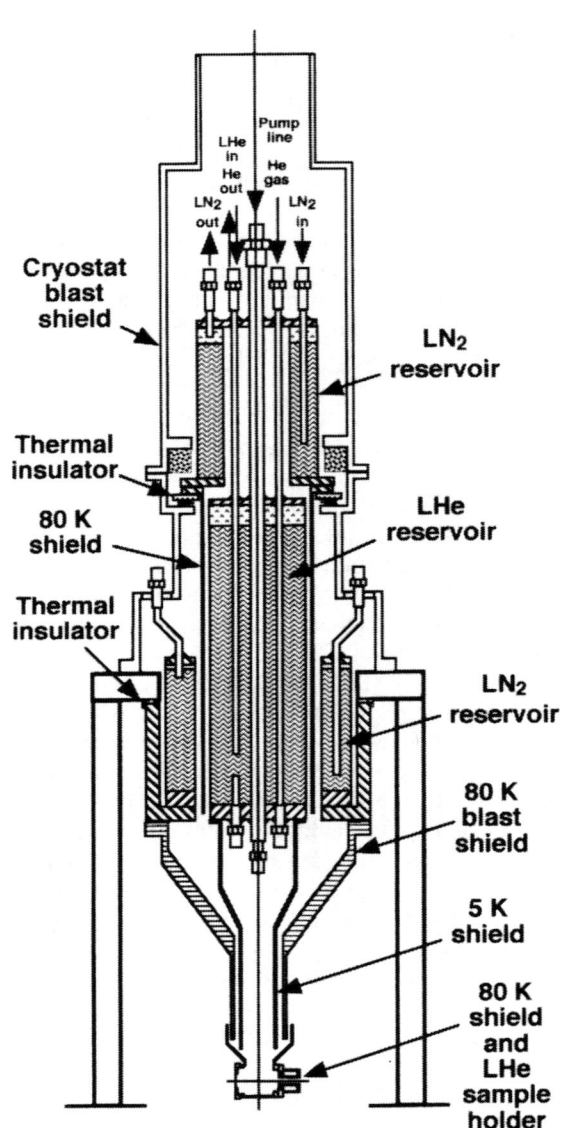

FIGURE 4. Cryostat and radiation shielding for cooling the LHe sample holder assembly.

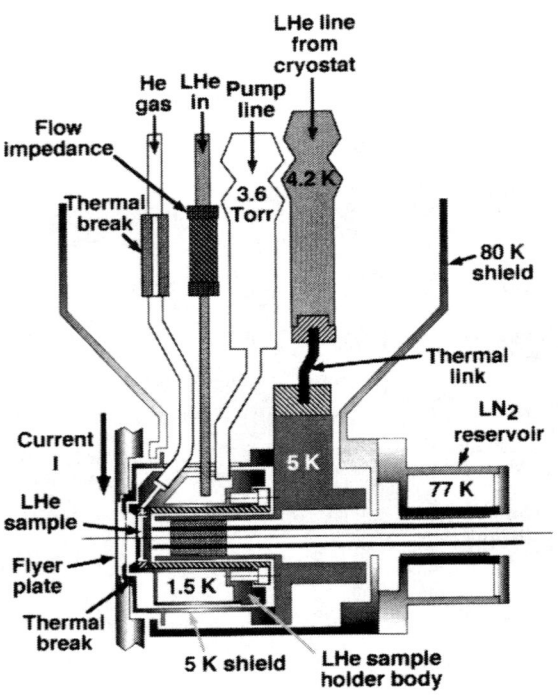

FIGURE 5. Sample holder assembly for condensing a superfluid ^{4}He-II sample at 1.5 K for LHe Hugoniot measurements.

ACKNOWLEDGMENTS

Sandia is a multiprogram laboratory operated by Sandia Corporation, a Lockheed Martin Company, for the United States Department of Energy under Contract DE-AC04-94AL85000.

REFERENCES

1. J. R. Asay, "Isentropic Compression Experiments on the Z Accelerator," in *Shock Compression of Condensed Matter-1999*, edited by M. D. Furnish *et al.*, AIP Conference Proceedings 505, New York, 2000, pp. 261–266.
2. M. D. Knudson, *et al.*, "Equation of State Measurements of Liquid Deuterium Subject to Magnetically Driven Hypervelocity Plate Impact," submitted to Phys. Rev. Lett.
3. L. DaSilva, *et al.*, *Phys. Rev. Lett.* **78**, 483 (1997).
4. G. W. Collins, *et al.*, *Science* **281**, 1178 (1998).
5. T. J. Lenosky, *et al.*, *Phys. Rev.* **B56**, 5164 (1997).
6. L, E, DeLong, *et al.*, Rev. Sci. Instrum. **42**, 147 (1971).

TEMPERATURE CONTROLLED VESSEL FOR EQUATION OF STATE MEASUREMENTS

Ted D. Rupp[1], Russell J. Gehr[1], David B. Stahl[2], Stephen A. Sheffield[2], and David L. Robbins[2]

[1]Honeywell Federal Manufacturing & Technologies/New Mexico, Los Alamos, NM 87544*
[2]Dynamic Experimentation Division, Los Alamos National Laboratory, Los Alamos, NM 87545

Abstract. We have designed and constructed a vessel capable of heating and cooling hazardous samples used in the laser-driven miniflyer experiments. For cooling, either liquid or gaseous nitrogen may be used. For heating, an electric element is used. The accessible temperature range is −100 °C to 300 °C. O-ring containment seals in the inner containment vessel establish temperature limits. The outer level of containment uses copper gaskets and commercial vacuum components. The vessel may be operated with a gas atmosphere or a vacuum. Temperature is monitored using two thermocouples, one on the heater and one on the inner containment vessel. A controller module monitors one thermocouple to reach and maintain the desired temperature. Using this vessel we can perform equation of state or spall strength measurements on hazardous materials in different phases or near solid-solid or solid-liquid phase transitions. Initial data taken with this system will be presented.

INTRODUCTION

The dynamic behavior of materials can be strongly dependent on the material temperature. [1-4] In experiments involving shock waves, such as spall strength and equation of state measurements, this is especially true near phase transformations. [1,2] The pressure change due to the shock can induce a phase change, altering the material properties as the shock wave passes. For example, the spall strength drops precipitously at temperatures above 90% of a metal's melt temperature. [1,2] To witness this and other effects it is necessary to control the material temperature.

Hazardous materials such as actinides typically have one or more solid-solid phase transitions. [5] At atmospheric pressure most of these transitions occur at very high temperatures. However, for several actinides a transition is accessible at high pressure and moderate temperatures. In these cases a shock wave propagating through the heated material can induce a phase change. Investigation of these material properties requires a temperature control vessel that can safely contain the hazardous material.

FIGURE 1. Picture of the temperature control vessel.

---
* This work was supported by the NNSA Enhanced Surveillance Campaign through contract DE-ACO4-01AL66850.

We have developed a temperature controlled vessel for use with the laser-driven miniflyer setup. [6] (See Fig. 1.) In the miniflyer system a 3 mm diameter, 0.05 to 0.10 mm thick flyer is launched at high velocity by a single shot laser. The flyer impacts a target foil, inducing a shock wave. The front surface velocity of the target is monitored using dual velocity interferometers to determine spall strength or equation of state. Given the small size of the flyer, the target material is typically 0.10 to 0.20 mm thick and 15 mm in diameter.

EXPERIMENTAL

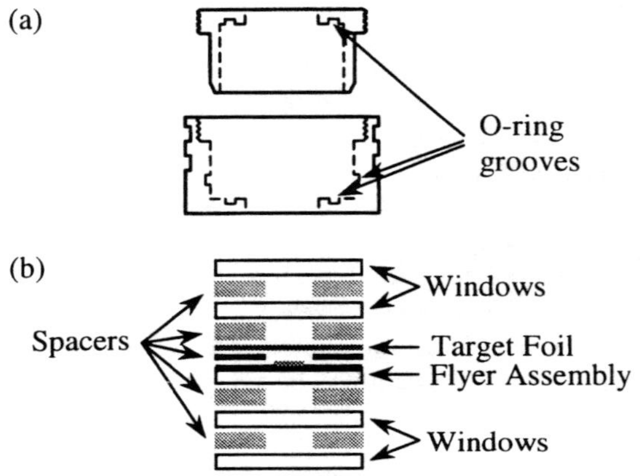

FIGURE 2. a) Inner containment vessel. b) Target stacking arrangement.

To safely confine hazardous materials during a dynamic experiment, the temperature controlled vessel requires two levels of containment. Primary containment is accomplished with an inner vessel. (See Fig. 2a.) This chamber is made of copper for efficient heat conduction. It consists of two pieces that screw together. To provide adequate seals, there are three o-rings: two o-rings seal the front and rear windows to the chamber and one o-ring seals the two pieces of the chamber together. Glass or thick PMMA windows are used to contain debris generated during the experiment. The pump laser is sufficiently energetic to pulverize the flyer substrate window, creating glass dust. In addition, it is possible to achieve flyer velocities that break through the target material. Thus the probe side windows may be required to stop the flyer while maintaining the seal. For these reasons two windows are used on both sides of the vessel. (See Fig. 2b.)

Secondary containment is accomplished with commercial vacuum components. A 6-way cube for 4 ½ inch conflat flanges is used as the main body of the vessel. The temperature control unit, manufactured by Thermionics Northwest Inc., attaches to the top opening of the cube. Two sapphire viewports, one each for the pump and probe beams, cover opposite side openings. These viewports are specified for both vacuum and overpressure up to 100 psig. Thus, the sample may be heated with an atmosphere present in the vessel without concerns of breaking the viewport seal. This situation is necessary for hazardous samples; contamination could occur if primary containment failed and a vacuum pump pulled the atmosphere out of the vessel. Therefore experiments on hazardous samples are done in an atmosphere. The remaining cube openings are used for vacuum connections, sample insertion, or are blanked off.

The temperature controller heats the sample using an electric heater and cools the sample using liquid or gaseous nitrogen flowing through the unit. Heat transfer between the temperature controller and the inner vessel is achieved at one face only. Therefore there are two thermocouples available to monitor the temperature, one at the heater interface and one on the side of the inner vessel. The temperature controller unit monitors one thermocouple and a user-selectable program drives the electric heater so that the approach to the preset temperature level does not significantly overshoot or undershoot.

Since the built-in thermocouples do not come into direct contact with the target material, it was necessary to investigate the relation between the thermocouple readings and the actual target foil temperature. For these tests, an extra thermocouple was mounted against the target foil. The windows on the probe side of the target had holes drilled through them so the thermocouple could be epoxied in place. A vacuum flange with thermocouple wire throughputs was used to connect the thermocouple to an external meter. The temperature of the target was compared with the temperatures recorded by the two thermocouples during heating and cooling cycles to determine the time required to reach equilibrium. Tests were performed in vacuum and at atmospheric pressure.

Two target materials were used for the tests: copper and bismuth, each in a foil approximately 100 micrometers thick. Copper was chosen for its high thermal conductivity, 401 W/m°K, and as one of the materials commonly used in our setup. Bismuth was chosen for its extremely low thermal conductivity: 7.9 W/m°K. In addition to being a worst-case scenario, this conductivity is similar to those of uranium and other actinides.

RESULTS AND DISCUSSION

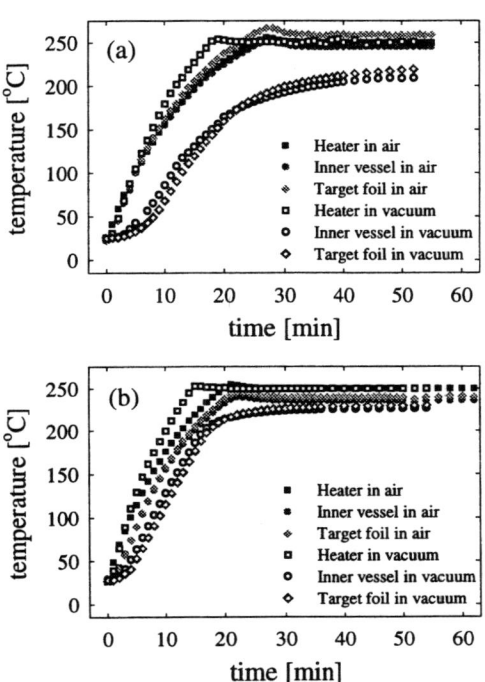

FIGURE 3. Measured temperature of the heater, the side-wall of the inner vessel, and the target foil for (a) copper and (b) bismuth foils, with a temperature goal of 250 °C.

The results of the heating tests for copper and bismuth targets are shown in Fig. 3. From room temperature the targets were heated to 250 °C. This temperature was chosen based on the melting temperature of bismuth, 271 °C. The maximum current used to drive the electric heater was set between 6.0 and 6.5 amps. This causes the temperature to rise quickly, but not rapidly enough to damage the system.

From Fig. 3 it is evident that there are significant differences between heating in a vacuum and heating in an atmosphere. In a vacuum, the heater element rapidly attains the desired temperature, but the inner vessel and the target foil temperatures rise much more slowly and do not reach their goal. In an atmosphere the heater element reaches temperature more slowly, and the inner vessel and target foil temperatures closely follow the heater temperature. The probable cause of this difference is the addition of conductive heating due to the atmosphere.

In an atmosphere both the copper and the bismuth targets achieved thermal equilibrium at the 250 °C setting in approximately 35 minutes. For the copper foil the actual temperature attained was 258 °C, which is slightly higher than the desired temperature. For the bismuth target the actual temperature attained was 240 °C. This is lower than the copper target due to the low thermal conductivity of bismuth (7.9 W/m°K). In both cases the rate of heating is sufficient to perform several experiments in one day.

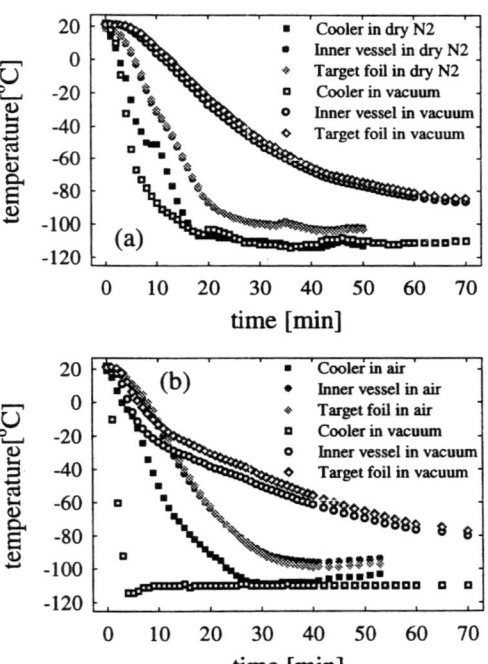

FIGURE 4. Measured temperature of the heater, the side-wall of the inner vessel, and the target foil for (a) copper and (b) bismuth foils, with a temperature goal of -100 °C.

The results for cooling the copper and bismuth foils, shown in Fig. 4, are similar to the heating tests: the cooling element achieves temperature more rapidly in vacuum, but the inner vessel and target foil do not achieve the desired temperature. In a dry nitrogen atmosphere the cooling element cools more slowly, but the inner vessel and target foil cool more rapidly and reach equilibrium in under 40 minutes. Note in this case that the inner vessel and target foil maintain similar temperatures, but do not follow the cooling element temperature. In addition, they do not reach the same temperature as the cooler, but stabilize approximately 10 degrees warmer. With the cooling element at −110 °C, the target foil reaches −100 °C.

Once again the bismuth foil takes longer to achieve equilibrium than does the copper foil. The time difference is relatively small, and in each case the foil achieves equilibrium in well under one hour.

It is important to note that the difference in reported temperature between the thermocouple on the side of the inner vessel and the thermocouple on the target foil is less than 2%. Therefore it is possible to know the target foil temperature by monitoring the inner vessel thermocouple.

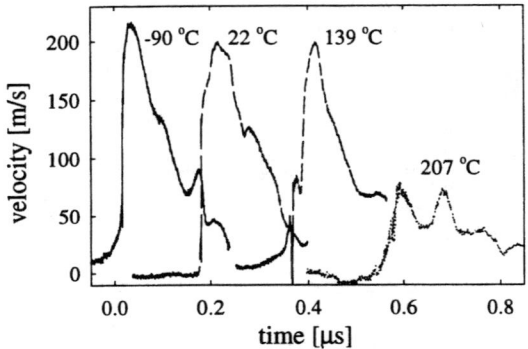

FIGURE 5. Equation of state measurements as a function of temperature on bismuth foils.

Equation of state measurements were performed on bismuth foils at temperatures between −90 °C and 230 °C. At these temperatures and at pressures below 4 GPa, bismuth has four different solid phases, as well as the liquid phase. Analysis of the initial data shows at least one solid-solid phase transition in all of the low temperature shots. The pressure at which this transition occurs decreases slowly with increasing temperature, in agreement with the bismuth phase diagram. [5] This is the solid I to solid III phase transition, which is weakly dependent on temperature. In addition, a second, lower-pressure phase transition is evident in the shots at 22 °C and 139 °C. This is the solid I to solid II phase transition, which is strongly temperature dependent.

In the two highest temperature experiments, at 207 °C and 230 °C, the measured shock did not attain high pressures. At these temperatures, the solid-liquid phase transition is accessible. [7] Our preliminary conclusion is that the shock wave propagates much slower in the liquid and therefore is overtaken by the relief wave.

CONCLUSIONS

We have demonstrated a new temperature control vessel for dynamic experimentation on hazardous samples. The two levels of containment designed into the vessel ensure that hazardous material cannot escape to create contamination. The temperature control system is capable of setting the sample temperature between −100 °C and 300 °C. Starting at room temperature, the controller can achieve the desired temperature in less than 1 hour. Including time to return the sample to near room temperature after the test, this implies that up to four experiments can be performed in a single day.

We have performed temperature-dependent equation of state measurements on bismuth foils. The preliminary data shows the boundaries of the solid I, II, and III phases, and the solid-liquid transition, in agreement with the known phase diagram of bismuth.

REFERENCES

1. Kanel G.I., Razorenov S.V., Bogatch A., Utkin A.V., Fortov V.E., and Grady D.E., *J. Appl. Phys.* **79**, 8310-8317 (1996).
2. Razorenov S.V., Bogatch A.A., Kanel G.I., Utkin A.V., Fortov V.E., and Grady D.E, in *Shock Compression of Condensed Matter – 1997*, AIP, New York, 1997, pp. 447-450.
3. Gu Z. and Jin X., in *Shock Compression of Condensed Matter – 1997*, AIP, New York, 1997, pp. 467-470.
4. Gu Z., Jin X., and Gao G., *J. Matl. Sci.* **35**, 2347-2351 (2000).
5. Young D.A., *Phase Diagrams of the Elements*, University of California Press, Los Angeles, 1991.
6. Stahl D.B., Gehr R.J., Harper R.W., Rupp T.D., Sheffield S.A., and Robbins D.L., in *Shock Compression of Condensed Matter – 1999*, AIP, New York, 1999, pp. 1087-1090.
7. Asay J.R., *J. Appl. Phys.* **45**, 4441-4452 (1974).

PVDF GAUGE PIEZOELECTRIC RESPONSE UNDER TWO-STAGE LIGHT GAS GUN IMPACT LOADING

François Bauer

Institut Franco-Allemand de Recherches de Saint-Louis, (ISL), 68301 Saint-Louis, France

Abstract. Stress gauges based on ferroelectric polymer (PVDF) studies under very high pressure shock compression have shown that the piezoelectric response exhibits a precise reproducible behavior up to 25 GPa. Shock pressure profiles obtained with "in situ" PVDF gauges in porous H.E. (Formex) in a detonation regime have been achieved. Observations of a fast superpressure of a few nanoseconds followed by a pressure release have raised the question of the loading path dependence of the piezoelectric response of PVDF at high shock pressure levels. Consequently, studies of the piezoelectric behavior of PVDF gauges under impact loading using a two-stage light gas gun have been conducted recently. Symmetric impact as well as non symmetric impact and reverse impact techniques have been achieved. Strong viscoplastic behavior of some materials is observed. In typical experiments, the piezoelectric response of PVDF at shock equilibrium could be determined. These results show that the PVDF response appears independent of the loading path up to 30 GPa. Accurate measurements in situ H.E. are also reported with very low inductance PVDF gauges.

INTRODUCTION

Piezoelectric materials are widely used as stress gauges to provide nanosecond, time-resolved stress measurements of rapid impulsive stress pulses produced by impact, explosion or rapid deposition of radiation. In the earliest work the gauges used crystalline sensors made of either X-cut quartz or various cuts of lithium niobate with thicknesses of many millimeters. The wave transit times through such a sensor range from many tens of nanoseconds to a few microseconds. The upper response of the crystalline sensors is limited by either, or both, mechanical or electrical properties: dynamic yielding of the sensors or dielectric breakdown due to the large internal fields produced by the piezoelectric effect. Early work has shown that highly reproducible poled 25 µm polymer film PVDF (Poly(vinylidene fluoride)) can be reliably used in a wide range of precise stress and stress-rate measurements (1, 2). Reliable behavior under the various extreme shock conditions requires specific sample preparations according to the Bauer process (1, 2). The 25-micron thickness of such PVDF sensors allows to place the gauges unobtrusively in a variety of locations within samples. Their direct stress-derivative or stress-rate signals with a few nanosecond and higher operating stress limits offer capabilities not available with any other technique and move experimental capability to a new sensitivity. The behavior of PVDF studied over a wide range of pressures using high-pressure shock loading has yielded well-behaved reproducible data up to 25 GPa in inert materials. Two years ago (1) solutions using a new appropriate shielding were identified and were applied to shock measurements of polar materials. This shielding is achieved via a deposition of thin metallic layers sputtered on samples to be studied and then connected to the ground. It has also been shown that this shielding technique can be applied to shock pressure profile measurements "in situ" porous H.E. in a detonation regime. The measured shock pressure profiles with the PVDF gauge show a fast superpressure of a few nanoseconds followed by a pressure release down to a plateau level and then by a pressure decay. These observations raise naturally the question of the loading path dependence of the piezoelectric response of PVDF at such high shock pressure levels.

The present paper summarizes the recent studies of the piezoelectric behavior of PVDF gauges under impact loading using a two-stage light gas gun up to 35 GPa. Symmetric impact as well as non symmetric impact and reverse impact techniques on PVDF gauges are presented. Accurate measurements in situ H.E. are also reported with very low inductance PVDF gauges.

PVDF RESPONSE UNDER TWO-STAGE LIGHT GAS GUN IMPACT LOADING: EXPERIMENTS AND RESULTS

The experimental measurements of the electrical response of a shock-compressed PVDF film are carried out on the ISL two-stage light gas gun facility. A range of impact velocities from 0.85 km/s to 3 km/s is achieved with a two-stage light gas gun which accelerates the projectile to a predetermined velocity. The planarity control and accuracy of velocities (0.5%) are comparable to those of our powder gun (1). In the impact experiment, symmetric impact as well as non symmetric impact and reverse impact techniques (3) on PVDF gauges are used. PVDF gauges are placed on the impact surface of either Z-cut sapphire crystals, aluminum (Al 2024) or OFHC copper which serve as standard materials to define the stress (Fig. 1). The PVDF gauge response is determined by recording the short-circuited current during the time when the shock waves reverberate within the samples until the mechanical equilibrium is reached, corresponding to the longitudinal stress in the standard material. In the electrical measurement circuit a carbon resistance replaces the expensive current viewing resistor "CVR" (1). The electrical charge is determined by the numerical integration of the recorded current (2).

Non Symmetric Impact Experiments

In the non symmetric impact experiments, 3 mm thick aluminum (Al 2024) or 2 mm thick copper samples were mounted on the projectile. The target consisted of a 10 mm thick electrically shielded Kel-F polymer. This electrical shielding was achieved via a deposition of thin metallic layers sputtered on polymer samples and then connected to the ground material (1). The PVDF gauge sandwiched between a 125 µm thick film and a 25 µm thick film of PFA –Teflon is bonded on the Kel-F target. The 125 µm thick layer faces the projectile. A typical record of the piezoelectric charge versus time obtained in such experiments is depicted in Fig. 2. As can be seen on the figure, the negative jump of the electrical charge, which corresponds to the early shock, is followed by a rapid decrease of the charge corresponding to a rapid relaxation to higher velocities. Then, the charge continues to decrease with a slope corresponding to a slow

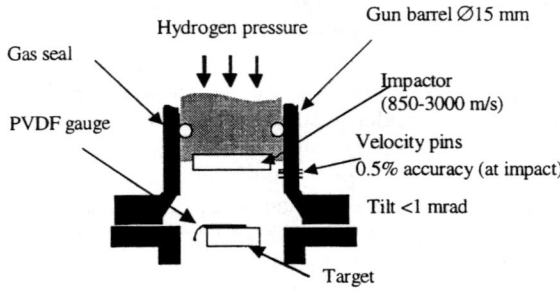

FIGURE 1. The piezoelectric response of the PVDF film is studied with the impact of materials with samples placed either on the impact surface or "in-situ".

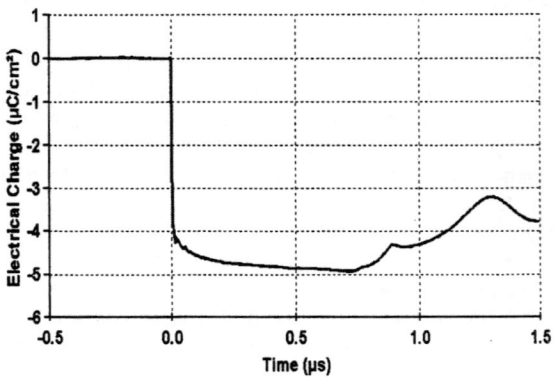

FIGURE 2. The figure shows the electrical charge released from shock-compressed PVDF bonded on a Kel-F polymer. The Al impactor velocity is equal to 2958 m/s.

relaxation to higher velocities (2), until the release wave from the impactor comes. Looking at the Kel-F relaxation, we cannot attain in our experiment the stress equilibrium, and consequently the charge released cannot be correlated with the induced stress.

Reverse Impact Technique

The reverse impact configuration (3) consists of the inverse of the above-described experiment. Projectiles are made of Kel-F polymer. PVDF gauges are bonded on sapphire or Al or Cu targets (Fig. 1). In a first series of experiments, single crystals of Z-cut sapphire, which are elastic to stresses in excess of 12 GPa, were used as the target. The shielding of the Kel-F projectile was obtained via the sputtering of thin metallic layers (1500 Å) and then connected to the ground. In both cases, (Fig. 3), a plateau follows the early shock. In the case of the PVDF embedded between two pieces of 1 and 2 mm thick sapphire, the equilibrium is reached after 0.10 µs. For the test with the PVDF bonded between a 125 µm thick PFA-Teflon and a 3 mm thick sapphire, we can observe after 0.15 µs a small change in the charge. At a time of 0.52 µs, the change in the shape of the charge is due to the arrival of the released wave coming from the 3 mm thick sapphire backer. At this time, the leads of the PVDF gauge are destroyed. In both cases, the piezoelectric charge can be correlated with the induced stress. In a second series of experiments, shielded Kel-F projectiles impacted PVDF bonded on a Cu target. The PVDF gauge was

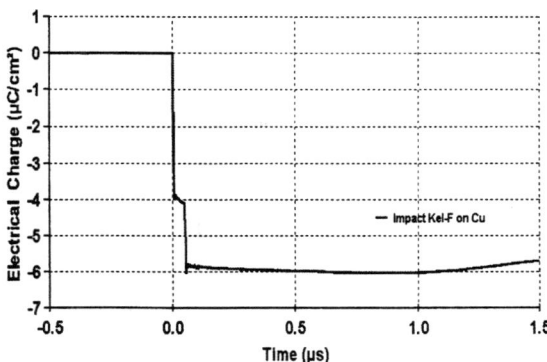

FIGURE 4. The figure shows the electrical charge released from shock-compressed PVDF bonded on a Cu target. The Kel-F impactor velocity is equal to 2561 m/s.

sandwiched between two 125 µm thick films of PFA-Teflon. Figure 4 gives a typical record of the piezoelectric charge released. The equilibrium is reached after two compression steps. Even a small relaxation is observed, the charge can be correlated with the equilibrium-induced stress.

Symmetric Impact Technique

Impactor and sample materials are identical. One half of the velocity at impact is imparted to the sample. PVDF gauges between two 125 µm thick films of PFA-Teflon are bonded on the impact surface. Figure 5 shows the piezoelectric response of the PVDF gauge subjected to the impact of a Cu projectile at a velocity of 1588 m/s. The loading of the film proceeds as a series of reverberations with increasing stress until equilibrium is reached.

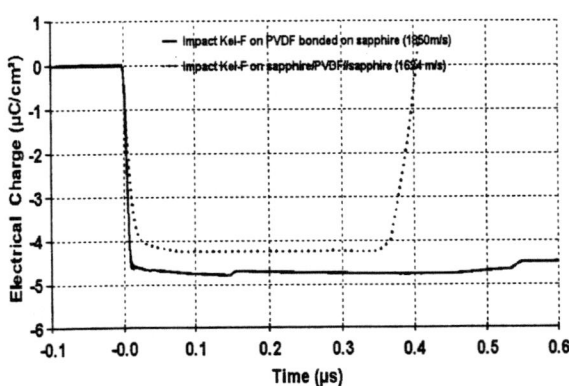

FIGURE 3. The continuous curve corresponds to the test with PVDF bonded between a 125 µm thick PFA-Teflon and a 3 mm thick sapphire. The dotted curve corresponds to the test with PVDF bonded inside two pieces of 1 and 2 mm thick sapphire.

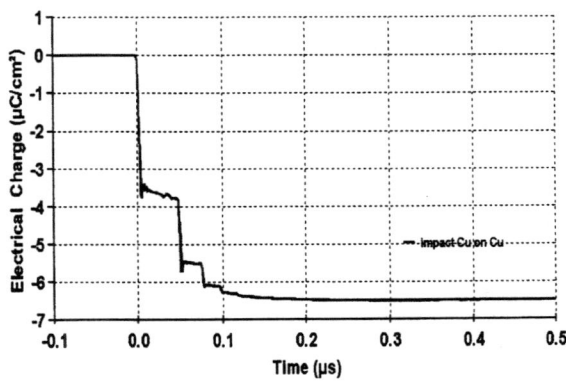

FIGURE 5. The electrical charge versus time is given for a Cu symmetric impact on PVDF. The projectile velocity is equal to 1588 m/s.

TABLE 1. Table of the valid results

Type of experiment	Impact velocity (m/s)	Stress (GPa)	Charge (µC/cm²)
Kel-F→Sa	1624	12.2	4.26
Kel-F→Sa	1850	14.7	4.79
Kel-F→Al	2695	17.8	4.89
Kel-F→Cu	2561	22.9	5.75
Cu→Cu	1357	30.0	6.16
Cu→Cu	1588	36.3	6.50

The charge is correlated with the equilibrium stress (for example, Figure 5 at 0.25 µs). Table 1 recapitulates the results obtained for reliable experiments.

Figure 6 shows the experimental and computed charges versus stress including these results. The observed charge versus stress data for these various loading paths show that the final charge seems to be independent of the loading path.

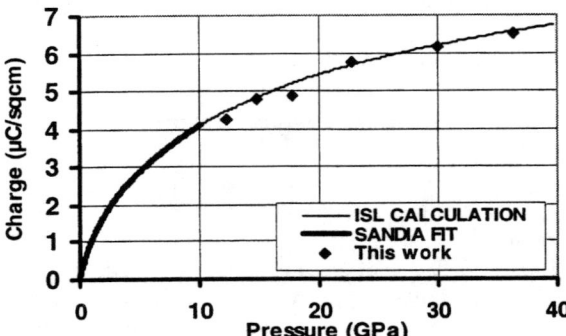

FIGURE 6. The electrical charge observed at various loading path pressures is shown for PVDF. The experimental and computed data are on the same curve.

SHOCK DETONATION PRESSURE PROFILES

Shock pressure profiles in situ porous H.E. in a detonation regime have been repeated (1). The overall features of H.E. studies under precise impact loading were described earlier (1). We have studied the influence of the value of load resistance on the charge released as well as on the pressure profile obtained. Figure 7 gives an example of detonation pressure profiles obtained with low-inductance 1 mm²-gauges "in situ" H.E. When the detonation wave propagates parallel to the PVDF gauge, the shock pressure profile shows a fast superpressure of a few nanoseconds followed by a pressure release down

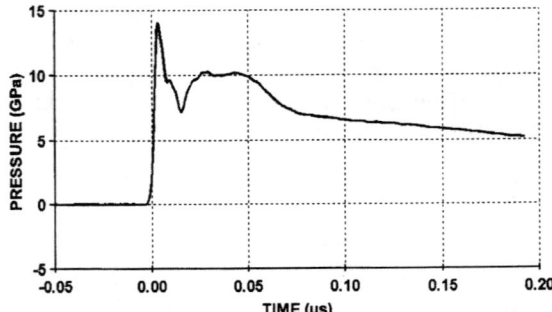

FIGURE 7. Zoom in a PVDF record of the detonation pressure "in situ" H.E. "Formex".

to a plateau level and then by a pressure decay (4). The changing resistance value of the CVR (between 0.4 and 1 ohm) does not affect significantly the shape of the profile. The values of the superpressure range between 15 and 20 GPa.

CONCLUSIONS

Studies of the piezoelectric behavior of PVDF gauges under impact loading using a two-stage light gas gun have been presented. In spite of the difficulties encountered to obtain reliable measurements, the valid data obtained show the reproducible response of our ISL PVDF, independently of the loading path. PVDF in a low-inductance configuration can measure detonation waves of some specific porous H.E.

REFERENCES

1. Bauer, F., *IEEE Transactions on Ultrasonics, Ferroelectrics, and Frequency Control* **47**, 1448-1458 (2000) and references therein.
2. Graham, R.A., *Solids under High Pressure Shock Compression*, Springer Verlag, New York, 1993, pp. 44-47 and pp. 103-113 and references therein.
3. Anderson, M., Setchell, R. E., Cox, D. E., "Shock and Release Behavior of Filled and Unfilled Epoxies," *Shock Compression of Condensed Matter-1999*, edited by M.D. Furnish et al., AIP Conference Proceedings 505, New York, 2000, pp. 551-554.
4. Sheffield, S. et al. "Detonation Properties of Nitromethane, Deuterated Nitromethane and Bromonitromethane" , " *Shock Compression of Condensed Matter-1999*, edited by M.D. Furnish et al., AIP Conference Proceedings 505, New York, 2000, pp. 789-792.

OUTPUTS OF SHOCK-LOADED SMALL PIEZOCERAMIC DISKS

Jacques A, Charest [1] and Jonathan Lee Mace [2]

[1] President of Dynasen Inc, 20 Arnold Place, Goleta, CA 93117
[2] DX-2, Los Alamos National Laboratory, Los Alamos, NM 87545

Abstract. Thin small-diameter polycrystalline Lead-Zirconate-Titanate piezoceramic disks were shock loaded in the D_{33} orientation over a stress range of 0.1-30 GPa. Their electrical outputs were discharged into 50 Ω viewing resistors, producing typically 0.15 μs quasi-triangular impulses ranging from 50-700 V. The gas gun flat plate impact approach and the high explosives (HE) plane wave lens approach were used to load piezoceramic elements. These piezoceramic elements consisted of 0.25 mm thick and 1.32 mm diameter disks that were ultrasonically machined from 25 mm piezocrystal disks of type APC 850, commercially produced by American Piezo Ceramic Inc. To facilitate our experiments, the piezoceramic elements were coaxially mounted at the tip of a 2.35 mm diameter brass tube, an arrangement that is commercialized by Dynasen, Inc. under the name Piezopin of model CA-1136. Simple calculations on the electrical outputs produced by these piezoceramic disks reveal electrical outputs in excess of 3000 W. Such short bursts of electrical energy have the potential for numerous applications where critical timing is needed to observe fast transient events.

INTRODUCTION

The use of piezoceramic materials has virtually exploded over the past 20 years. Indeed, to describe all their applications would require numerous publications. This company, being a manufacturer and developer of shock sensors, has of course utilized various types of piezoelectric materials to construct many of its sensors. As several experimenters know, we have supplied our sensors to nearly all US DOE and DOD laboratories and to several oversea research organizations engaged in fundamental research of shock physics or ammunition testing. One of our particular products, which we call the Piezopin, has been used intensively by shock wave physicists over the past thirty years because of its low cost and high output performance reliability.

A Piezopin is a small and simple coaxial device that utilizes a thin piezoceramic disk as its detector. It is capable of producing a short-duration high-output electrical signal when impacted by a fast moving object, a shock wave or a strong blast.

To date, Piezopins have been utilized primarily for event timing and triggering of instrumentation. Despite their great popularity at many research laboratories, relatively little quantitative information has been published on their output performance under shock wave loading or on potential applications.

This particular study is concerned with the high-output short-duration impulse capability of small piezoceramic elements. The experiments presented in this paper focus on results obtained by utilizing Dynasen s Piezopin configuration when subjected to a wide range of dynamic loads. Two series of controlled tests were conducted to investigate the output of these Piezopins. The first series of tests was conducted at Dynasen s impact facility using gas gun-driven plane wave impact tests of 0-10

GPa, whereas the second series of tests, 10-30 GPa, was conducted at Los Alamos National Laboratory (LANL) using HE lenses. The experimental approach for these tests is described next, followed by a brief data summary and some conclusions.

EXPERIMENTAL APPROACH

The intent of these experimental series was to develop well characterized small, passive gauge elements that can be used in applications where critical timing information from fast transient events is transmitted to a receiver from an antenna or wave guide. These experiments demonstrated that short, high power bursts of electrical energy from small dipolar piezoceramic disks potentially satisfied these requirements. Also, they provided Hugoniot data up to about 80 $kbar$. In the experiments reported Piezopins were oriented such that the surface of a small dipolar piezoceramic disk was parallel to an incident shock. When shock loaded, this dipolar separation of charge is compressed on the time scale of shock dynamics, thereby producing a large electrical impulse across 50 Ω that is recorded onto various digitizers via coaxial cable.

Figure 1 shows the basic construction of Dynasen's Piezopin, whereas Figures 2-3 illustrate the gas gun and HE testing approaches.

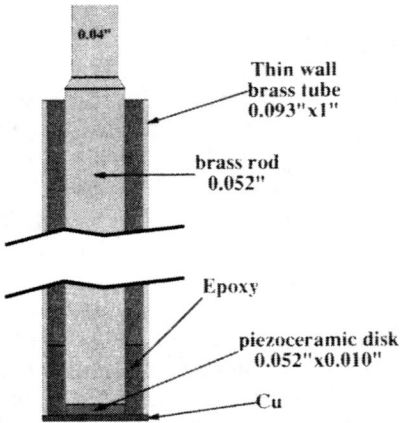

FIGURE 1. Construction of the Dynasen Piezopin. The piezoceramic disk contains a dipolar charge separation and is the active element. The surface of this element is grounded to the brass tube via a thin Cu disk, and the backside is connected to the brass rod, thus providing a convenient arrangement for coaxial transmission of the electrical impulse induced by the incident shock.

Eleven gas gun tests and six HE plane wave lens tests were performed, each consisting of six to eight Piezopins, plus several of Dynasen's Carbon Film or Low Impedance Manganin gauges, depending on the pressure range, for accurate recording of shock pressure and shock arrival times. Standard shock matching techniques were used to calculate pressure in Teflon gauge packages.

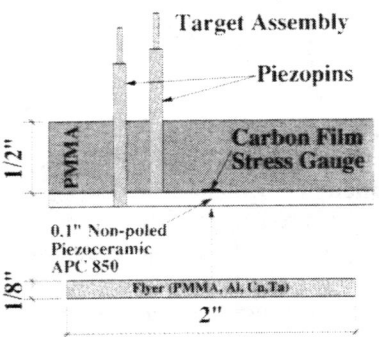

FIGURE 2. Typical gas gun assembly; 0-10 GPa. A flyer, mounted on the front of a projectile, is fired from a gas gun into the target assembly at normal incidence to the piezopins. Each target assembly typically consisted of multiple Piezopins and Carbon Film gauges.

Figure 2 illustrates the typical gas gun flyer and target assemblies utilized in experiments at Dynasen's impact facility. During initial experiments we began to observe large variations in (P,V) data points as successive tests in this experimental series produced higher impact pressures; where P is pressure and V is voltage across 50 Ω.

Our initial guess as to why this variation occurred had to do with the fact that a small copper film was vapor deposited over the front surface of the Piezopin in order to form a complete circuit (Fig. 1). The large voltage spikes that developed over the circuit (3000 W average power) caused us to believe that the integrity of this thin film of copper was questionable over the duration of the pulse. Therefore, this copper film was replaced by a thin copper disk. This resulted in only slightly better data.

We now realize that this measured scatter in data (see Fig. 6) probably results from dipolar variations between individual piezoceramic disks. These small disks were ultrasonically machined from much larger piezoceramic crystals with no control over variations in dipolar properties. Nevertheless, the significant measured outputs of these small passive elements should be noted.

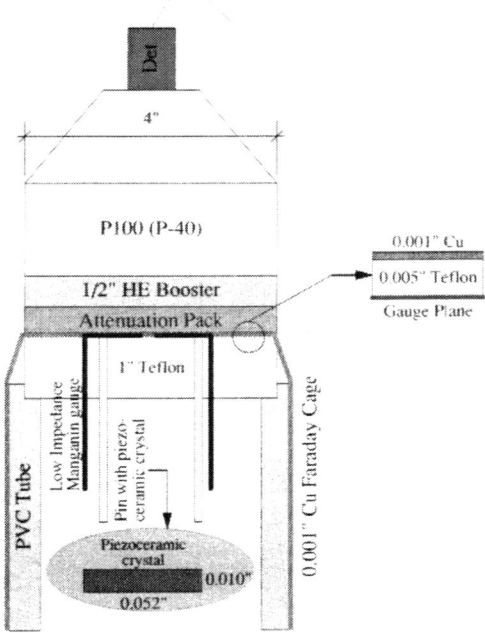

FIGURE 3. HE plane wave assembly; 10-30 *GPa*. A 4 inch plane wave lens produces a flat detonation wave in the HE booster material, which in turn produces a flat shock in the attenuation package that becomes incident on the Teflon. The gauge plane consists of an array of piezoceramic elements and low impedance Manganin gauges. A Faraday cage is used to isolate the gauge plane and cables from electrical noise inherent in the detonation front.

Figure 3 illustrates the HE plane wave lens experimental arrangement. In each experiment both the 1/2 *inch* booster and the attenuation pack were chosen such that the shock matched pressure transmitted into the Teflon gauge package varied from 10 to 30 *GPa*. This pressure was verified using low impedance Manganin gauges.

Figures 4-5 illustrate a typical pulse whereas Figures 6-7 provide a summary of measured data. Notice that the duration of the pulse illustrated in Figure 4 is on the order of $0.1\text{-}\mu s$. This time is related to shock transit time in the piezoceramic material.

A simple Gaussian is fit to this data in order to illustrate the average power and frequency content (Fig. 5) of a typical pulse. These frequencies are such that electromagnetic transmission through air and time resolved reception should be viable.

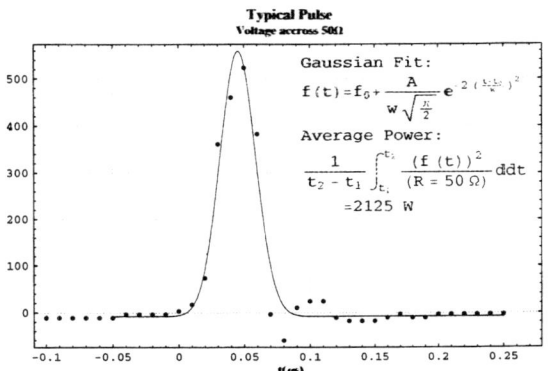

FIGURE 4. Typical pulse from a shocked 0.052 *inch* x 0.010 *inch* piezoceramic element. A Gaussian is fit to the data and is used to illustrate the average power and frequency content of a typical pulse.

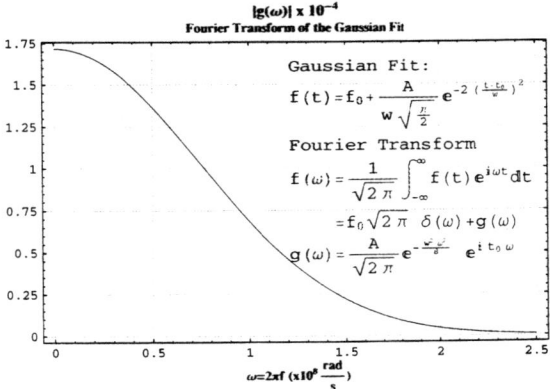

FIGURE 5. Frequency content of a typical pulse from a shocked piezoceramic element.

Figures 6-7 provide a summary of measured peak output voltage across 50 Ω and measured Hugoniot data. The data of Figure 6 demonstrate that peak pulse voltages from 300 to 700 *V* can be expected over a range of about 20 to 350 *kbar*. Notice the large scatter in data. We think this scatter is largely due to dipolar variations between piezoceramic elements. We expect that this scatter is due to polycrystaline nature of material, a property that is not detectable by the Hugoniot measurements of Figure 7.

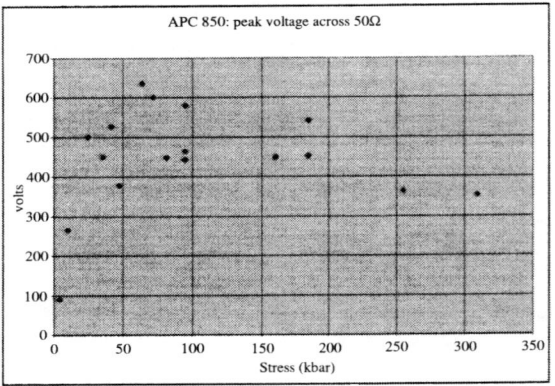

FIGURE 6. Peak voltage output of shock loaded 0.052 x 0.010 piezoceramic elements. Each point represents the output average of 6 or 8 piezopins.

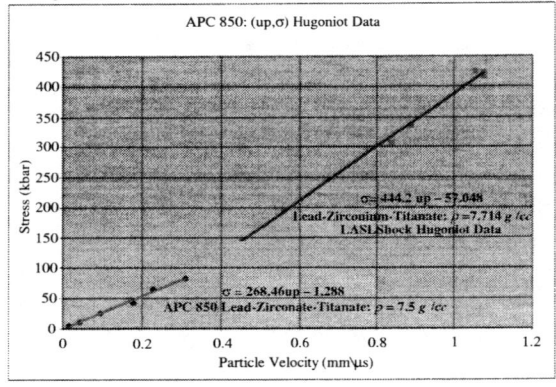

FIGURE 7. APC 850 Lead-Zirconate-Titanate Hugoniot measurements (ρ=7.5 g/cc) and Lead-Zirconium-Titanate (ρ=7.714 g/cc) data from the LASL Shock Hugoniot Data handbook.

Figure 7 provides six measured Hugoniot points of APC 850 over a range of 0-80 $kbar$. The slope of the resulting line is about 268 $kbar/(mm/s)$. These results were obtained from gas gun tests performed at Dyansen, and are the first published results for APC 850 that we are aware of (1).

CONCLUSIONS

These experiments demonstrate that a small, passive piezoceramic element can produce a high-power pulse over frequencies of up to 1-200 MHz when shock loaded. There are, of course, many potential applications that can be imagined for such small, non-intrusive, passive elements. In particular, electromagnetic transmission through air should be viable.

Furthermore, we suspect that the development of a method for producing consistent dipolar characteristics between individual piezoceramic elements should eliminate much of the data scatter seen on Figure 6. Further investigations on the subject are expected to yield more insights on the process of shock loading APC 850 polycrystal piezopceramic.

ACKNOWLEDGEMENTS

This work was funded under Los Alamos National Laboratory contract numbers 100554-001, 100554-009X and 16326-001-009X.

REFERENCES

1. Stanley P. Marsh, Editor. LASL Shock Hugoniot Data, Publisher: University of California Press, 1980

IMPROVEMENTS IN THE SIGNAL FIDELITY OF THE MANGANIN STRESS GAUGE

Dan Greenwood, Jerry Forbes, Frank Garcia, Kevin Vandersall, Paul Urtiew, LeRoy Green and Leroy Erickson

*Lawrence Livermore National Laboratory,
P.O. Box 808, L-283 Livermore CA 94551.*

Abstract: The manganin stress gauge has been and still is the primary diagnostic tool for measuring longitudinal stresses in materials shocked from 10 to 400 kb in one-dimensional (1D) uniaxial strain experiments [1]. Its simple and robust design allows this gauge to survive in harsh environments. The manganin gauge has several limitations. For example, in the eventual failure mode, the manganin gauge has a reputation of being a noise generator to the remaining functioning manganin gauges at different lagrangian positions in the experiment. The manganin gauge also demonstrates undesirable signal effects when the front edge of the incoming shock first makes contact. These two limitations and the experiments for the mediation of these effects on shock experiments will be presented in this paper. Our ultimate goal is to provide practical manganin gauging that has true fast rise time and little or no noise generation on failure in explosive detonation waves. A device was found that mitigates the noise generation without compromising the integrity of the pressure data.

INTRODUCTION

One goal is to provide nose mitigation due to the gauges failing with minimum gauge redesign. This minimum gauge redesign is necessary because of the extensive history and data base generated using this gauge design [1]. The standard Livermore manganin gauge [2] is shown in Fig. 1. It has a nominal thickness of 0.025 mm. The gauge is a peizoresistive device with the active element being nominal 50 mΩ. It uses a four-wire (Kelvin) method for measuring resistance change in the element. A pulsed constant current, typically 50 A, power supply is used to excite the gauge element. The voltage change due to the piezoresistive effect is measured on separate sense leads using high-speed digitizers.

Figure 2 reveals where potential noise problems exist. The manganin gauge leads form two loops and these loops are susceptible to pickup from otherloops (i.e. manganin gauges) in the experiment. These loops have a nominal 50 nH (calculated and measured) of inductance along with a small amount of stray capacitance 4.25e-14 F (calculated) for a 55 mm long gauge.

The RG 58 coax cable has an inductance of 1.2 µH. At 50 A excitation current the stored energy in the RG 58 and gauge current loop is 1.51 mJ. The excitation power supply has an open circuit voltage of 350 V. During the experiment the current leg often opens in the nanosec time frame. This causes the small magnetic field in the gauge to collapse and

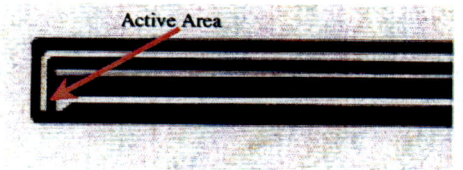

FIGURE 1: LLNL Manganin Gauge

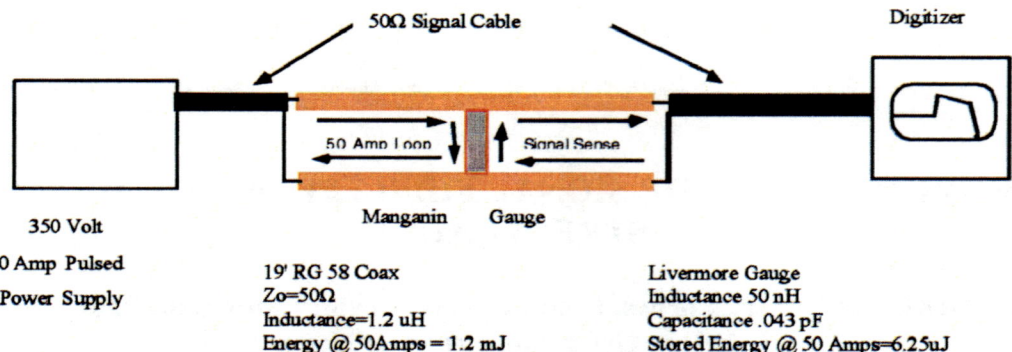

FIGURE 2. Manganin Gauge Electrical Circuit

the RG 58 transmission line to go from a shorted condition to open and reflect a large amplitude wave down both legs of the manganin gauge. This wave is not terminated at the power supply and gets reflected back to the gauge. This mismatch along with the open end of the manganin gauge can cause data disturbance in the other functioning gauges all from a few hundred nanosecs to over a microsec time duration. For the coax/current loop to dissipate the energy in that period, it would be between 1.5 to 7.5 kW.

EXPERIMENTAL PROCEDURE

To economically test various devices for this work, small scale explosive tests were used (see Fig. 3). These had 2.54 cm diameter right cylindrical pellets of explosives which sent strong shock waves into 2.54 cm diameter by 5 to 10 mm thick Lucite disks with manganin gauges between these disks. Gauges were placed at two distances from the explosive/Lucite interface.

Our testing sequence started with a base line experiment of the noise for a typical or standard gauge setup without any noise abatement devices (i e. circuit protection). We then conducted experiments to reduce or eliminate the noise sources (i e. ground loops, capacitance in the power supply and common mode rejection). These experiments provided with good noise records but very little in the way of noise reduction when gauge breakage occurred. Figure 4 shows a set of experiments that have varying gauge cross talk as the gauges break. The red traces show noise pickup from the previous gauges (noise donors not shown). The blue and green traces show noise reduction from the breakage of the noise donor gauges.

Looking for a suitable shunting device was a parametric search. The shunting device had to be fast to match the rapid opening of the gauge and have low parasitic capacitance so the signal integrity was not affected. It also had to absorb the fault energy.

Current shunting devices were tried. These devices are attached to the current leg of the manganin gauge and act as an open circuit until the gauge breaks. The shunt device conducts in the nanosecond time frame and carries the current intended for the gauge. The drawback on most semiconductor shunt devices is there high parasitic capacitance, which causes the signal to ring.

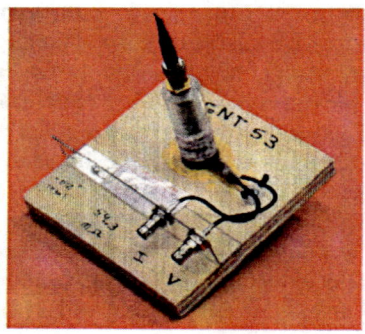

FIGURE 3. Typical Noise Test Bed Experiment

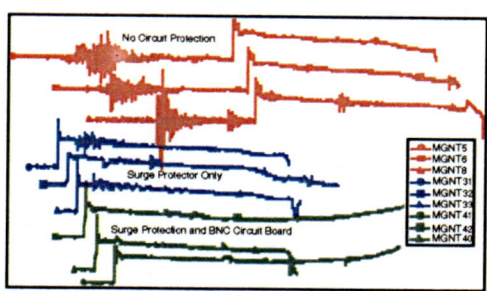

FIGURE 4. Normalized Data. Red Traces without Shunt, Blue and Green Traces with Shunt

FIGURE 5. Transient Suppressor Wired to Manganin Gauge

Semiconductor and polymer based devices with small junction capacitance were also tried but they could not absorb the energy without failing in the time frame of importance. In addition, a low voltage gas discharge tube from CP Clare has low capacitance but it is noisy when it turns on.

A small electrostatic discharge protection device was found that fit the requirements. It is normally used in protection of Input Output (I/O) lines on computers (high speed, low capacitance). The device is made by Semtech and is called LCDA15. It is a four-device array of transient suppression diodes.

One of the four diodes was used per gauge in our experiments. The results show major reduction of noise from gauge breakage. The blue and green traces in Figure 4 shows a typical LCDA15 coupled with a manganin gauges. The red traces also show gauge records without shunt devices. Figure 5 is a photograph of the transient device soldered to the manganin gauge leads

SUMMARY

We have examined the cause of the noise disturbance due to gauge breakage in shock wave experiments. A device was found that mitigates the noise generation without compromising the integrity of the gauge response. It was also discovered that the gauge leads are susceptible to the incoming shock and will disturb the front edge of the manganin foil gauge record. Although not shown in this paper, the noise-mitigating device has worked well on gun experiments using multiple gauges at different depths in the explosive targets.

ACKNOWLEDGEMENTS

Gary Steinhour, Ernie Urquidez and Denise Grimsley assisted on the experiments. Douglas Tasker (LANL) gave advise on noise mitigation for gauge circuits. We are saddened to announce the loss of our colleague Leroy Erickson. This work was performed under the auspices of the United States Department of Energy by the Lawrence Livermore National Laboratory under contract No. W-7405-ENG-48.

REFERENCES

1. Vantine, H. C., Erickson, L.M., and Janzen, J., " Hysteresis-Corrected Calibration of Manganin under Shock Loading," J. Appl. Phys., 51(4), pp. 1957-1962. (1980).
2. Vantine H., Chan J., Erickson L. M., Janzen J., Lee R. and Weingart R. C., "Precision Stress Measurements in Severe Shock-Wave Enviromnents with Low Impedance Manganin Gauges," Rev. Sci. Instr., 51: 116-122, (1980).

CHAPTER XVIII

EXPERIMENTAL TECHNIQUES

RECENT ADVANCES IN QUASI-ISENTROPIC COMPRESSION EXPERIMENTS (ICE) ON THE SANDIA Z ACCELERATOR

C. A. Hall, J.R. Asay, M.D. Knudson, D.B. Hayes, R.L. Lemke, J.P. Davis, C. Deeney

Sandia National Laboratory, Albuquerque, NM 87185-1181*

Abstract. The Z Accelerator is a pulsed power machine capable of delivering currents to loads of ~20 MA over times of 100-300 ns. This current produces smoothly increasing, time dependant magnetic pressures that can be applied to specimens allowing quasi-isentropes for these materials to be inferred. A new load design has been developed that allows this pressure to be uniformly applied to as many as 8 samples simultaneously. Diagnostics have recently been fielded that have resulted in an increased understanding of the magneto-hydrodynamic effects and our confidence in the utility of this experimental configuration for EOS measurements. Efforts are also underway on Z to provide a capability for shaping the pressure profile applied to the samples which should increase useful sample thicknesses to > 1mm by eliminating the formation of low-level shocks. In addition to direct measurements of quai-isentropic material response, the impulse from this loading technique has been demonstrated to launch macroscopic flyer plates to velocities of ~21 km/s for high-pressure Hugoniot studies. Results of ICE measurements on 6061-T6 aluminum to ~1 Mbar will be discussed.

INTRODUCTION

A principal goal of the Sandia shock physics program is to establish a capability to make accurate equation of state (EOS) measurements using the Z pulsed power machine. Early efforts centered on use of high energy x-rays from Z-pinches to produce ablatively driven shocks. Difficulties with steady shock durations, fiber-coupled diagnostics, and the ability to achieve 1-D loading hindered our efforts. However, recognition by Asay that pressure acting on samples from the smoothly increasing magnetic fields generated as the machine discharged could produce high magnitude, shockless ramp waves led to the development of isentropic compresion experiments (ICE) on Z[1]. Recent advancements in this technology have enhanced the quality of EOS information that can be obtained through improvements to loading conditions, our understanding of MHD effects, and the ability to field accurate, complex targets as required. This capability has recently been used to study aluminum to a stress state of ~1 Mbar, constituitive properties[2], optical effects in interferometer windows[3], phase changes[4,5], and energetic material response[6]. Strictly speaking, compressive isentropic material response implies that the compression is both adiabatic and reversible. Ramp loading of solids over the timescales of these experiments (typically 100–300 ns) is adiabatic but not perfectly reversible. Viscosity, strength effects, and phase transition kinetics, if present, are dissapative terms that produce entropy and are present to some degree in all solids. Therefore, ICE is actually quasi-isentropic, but ramp loading of solids has historically been referred to as isentropic and this convention will be used throughout the remainder of the paper.

The ramp loading capability on Z can also be used to make high-pressure Hugoniot measurements with accuracies approaching that of conventional gas gun data. Macroscopic aluminum flyers of ~12 mm width and ~0.5 mm thickness have been launched to velocities of 21 km/s allowing symmetric impact studies of aluminum to pressures approaching 5 Mbar. These flyers have also been used as a high-pressure, constant drive input for the Sandia National Laboratories effort to obtain the EOS of liquid deuterium.

SQUARE PANEL GEOMETRY

The Z Accelerator[7] is a low inductance pulsed power generator capable of capacitively storing 11.6 MJ of electrical energy. The accelerator uses a combination of fast switches and transmission lines to deliver a 20-MA, 100 - 300 ns risetime current pulse to generate time-varying magnetic fields between the anode and cathode that continuously load planar specimens under study. Low inductance loads designed for the Z accelerator are fielded at the center of Z's radially converging magnetically-insulated transmission lines.

When first attempted on Z, ICE samples were limited to material disks pressed into stainless steel conductors[1]. This limited the ability to assemble multiple material layers either before or after the pressing process due to sample distortions. In addition, magnetic pressure was not constant across the samples due to the radial, converging geometry or varying vacuum gap between the current carrying conductors at the point where they were mounted.

These limitations led to the development of the "square panel geometry" as shown in Figure 1 which is now the standard experimental configuration used for ICE on Z[8]. In this arrangement, four seperate panels are manufactured independently, with one side used to carry the current during the experiment (power flow surface), and the other containing one or more counterbores with prescribed conductor thicknesses for gathering experimental data. Both the powerflow surface and counterbore floors are diamond machined to tightly constrain both surface figure and parallelism. Because magnetic pressure scales with the square of current density, and current density at the sample location is simply the current divided by the current carrying perimeter, loading pressure can easily be affected through changes in panel width. Several panel widths have been successfully used to produce loading from a peak of ~500 kbar (26 mm wide panels) to a peak of ~1.5 Mbar (15 mm wide panels).

A major advance in the panel arrangement is the ability to create 1-D planar loading over a large portion of the panel face, thus allowing large, nonconductive samples to be studied. Static electromagnetic simulations using the code QUICKSILVER[9] indicate that for the 26 mm wide panels, an 11 mm region with 1% horizontal loading uniformity exists in the center of the panel, and covers its entire height. When the panel is decreased to 15 mm in width, the 1% uniformity region is reduced to ~6 mm. These results have been experimentally validated using a spatially resolved line imaging interferometer[10] to the light-limited resolution of the diagnostic. Results indicated that uniformity over this region was in agreement with the simulations.

FIGURE 1: *The newly developed "square panel geometry" for conducting ICE on Z.*

Symmetry of loading between panels has also been characterized[8]. VISAR[11] was used to record surface motion at the same points on both the upper and lower portions of counterbores on multiple panels. Results show that, in general, loading is the same to approximately 1.5-2% between panels. Efforts are underway to improve this with more careful panel placement relative to the cathode.

MHD EFFECTS

When obtaining material response information from observing ramp wave evolution, it is essential that the material be at a known initial state. This is an area of concern when the ramp waves are generated through pulsed magnetic loading because magnetic field can potentially diffuse through the sample under investigation causing joule heating and density variations. This effect could lead to systematic errors in the resulting inferred pressure-volume material response.

An effort to characterize, and be able to accurately model, this field diffusion was undertaken. Data was taken at specific material depths in both copper and aluminum where diffusion was predicted to occur (using Lee-Moore conductivity models in an MHD code) during the ramp compression. Both time resolved rear surface velocity measurements using VISAR and time resolved field intensity measurements using Bdot probes[12] were taken. The resulting records were overlayed to define a minimum material thickness where diffusion does not occur during compression for both copper and aluminum at given current densities and profiles useful for ICE.

In addition, this and other data was modeled with ALEGRA, an MHD code under development at Sandia National Laboratories, to compare predicted diffusion rates to data[13]. In general, predictions were consistently overestimating diffusion rates by approximately 35%. It was determined that the existing conductivity models were inadequate over the range of conditions encountered on the power flow surface in a typical Z firing. Modifications to the model were made in the description of conductivity at the onset of melt. Comparisons between simulations and data after the modifications were made indicate good agreement.

1.0 Mbar ICE ON ALUMINUM

Previous ICE studies on aluminum and copper were limited to stress states of ~300kbar[8,2] In these materials, as with many, there is little difference in P-v response between the isentrope and Hugoniot. This is useful when validating the technique, but it was desired to extend ICE to a stress level where deviation between the two curves could easily be seen. Aluminum was chosen to be investigated because it is a standard panel material for ICE, it is reasonably well understood, and it has been accurately characterized along the Hugoniot to stresses greater than 1 Mbar[14].

The experiment, shot Z575, used 15 mm wide 6061-T6 aluminum panels with 10 mm diameter counterbores with 400 μm, 498μm, and 851μm floor thicknesses, backed by nominally 1 mm thick, 6 mm diameter LiF interferometer windows. Epoxy was used to bond the windows to the aluminum with a typical thickness of ~ 2 μm. The interface surface of each window was coated with 1μm of aluminum prior to bonding to provide a specular reflector for the interferometer (VISAR). Velocity was recorded for each of the material thicknesses at the aluminum-LiF interface with resulting wave profiles shown in Figure 2.

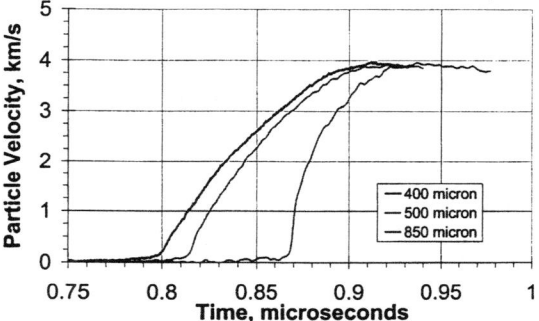

FIGURE 2: *Velocity profiles (shot Z575) from which the aluminum isentrope to ~1 Mbar was inferred*

Z575 was analyzed using the method of characteristics[15,8]. For this approach, the Lagrangian sound speed (C_L) was obtained as a function of particle velocity (u_p) at 0.01 km/s increments. Stress and specific volume were then inferred from the relations $dp = \rho_0(C_L)du_p$ and $dv = 1/v_0(du_p/C_L)$. The material must be assumed to be rate independent for these relationships to apply. To validate this assumption, arrival times for characteristics at 0.5 km/s increments were plotted versus initial material thickness for each of the three samples and found to be straight within experimental error. The stress-volume curve that resulted from this analysis is shown in Figure 3 compared to published Hugoniot data and a prediction for the isentrope using a simple model with a linear U_s-u_p relationship and γ/v held constant.

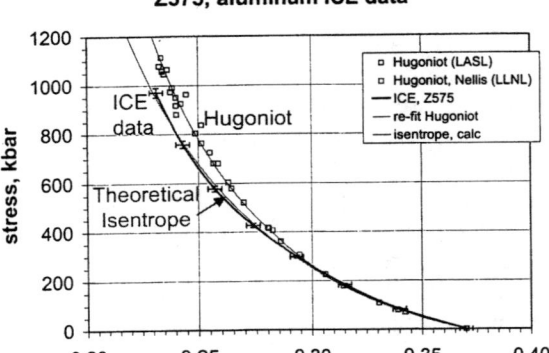

FIGURE 3: *the inferred aluminum isentrope to ~1 Mbar from shot Z575 compared to the Hugoniot and a predicted isentrope using γ/v as a constant*

It appears there is good agreement between current data and the simple theory used to predict the isentropic response of 6061-T6 aluminum to this stress level. Error bars on the 1Mbar data are determined from accuracy in relative timing between velocity profiles, accuracy in mass velocity measurements from VISAR, and sample characterization.

With all VISAR measurements made through LiF interferometer windows in these experiments, effects of ramp loading on the strain-dependant index of refraction window calibration must be considered. Recent theory reported by Hayes[16] and supported by data in sapphire[3] indicate that when a linear index of refraction vs. density relationship describes a window material, the calibration factor used to infer interface velocity information through the window is the same for ramp loading as for shock loading. Recent reanalysis of published shock data on LiF[17] indicate that a linear relationship does represent the material to within experimental error. Therefore, it is assumed that the window correction used to obtain the velocity data in these experiments is adequate. Experiments are currently underway to validate this assumption.

PULSE SHAPING

As has been discussed, there is a minimum material thickness required to prevent magnetic field diffusion from influencing the data during compression. This requirement, in combination with the need to keep wave reflections from influencing the results when samples are bonded to the aluminum or copper counterbore floors, defines the minimum sample thickness for ICE. For accuracy in material response inferences using characteristic analysis, a large seperation (ΔX) in sample thickness is needed for accuracy in determineing C_L. Therefore, it is desirable to have an ability to shocklessly load thicker samples.

In many materials, Lagrangian sound speed (C_L) varies more for a given increment of stress where compression is the greatest (i.e. before the material response stiffens in the P-v plane). When the pressure characteristics associated with their respective Lagrangian sound speeds are plotted in a material thickness vs. time plot (X-t plot), it can be seen that intersection of the early characteristics occurs before those arriving later in time for a linearly increasing input pressure profile. This is shown schematically in Figure 4. The result of the intersection of these characteristics is a low level shock that forms early in time and grows, as was typically seen when the "standard" pressure profile available on Z was used early in ICE development.

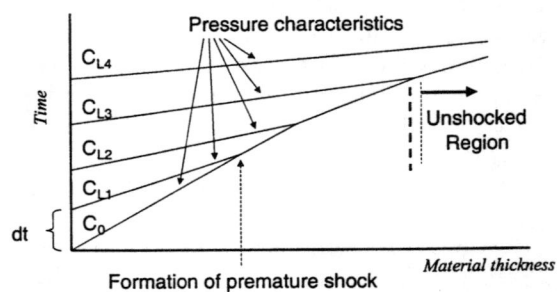

FIGURE 4: *A figurative X-t plot showing how pressure characteristics intersect early in time for a nonlinear material and linear ramp loading*

Figure 5 shows this effect when LiF was loaded to ~1Mbar using ICE on Z with 0.25, 0.5, and 1.0

mm thick LiF samples backed by LiF interferometer windows. Each sample/window assembly was mounted on a 400μm aluminum conductor. The interface velocity between the common conductor and a LiF window is labeled as "input".

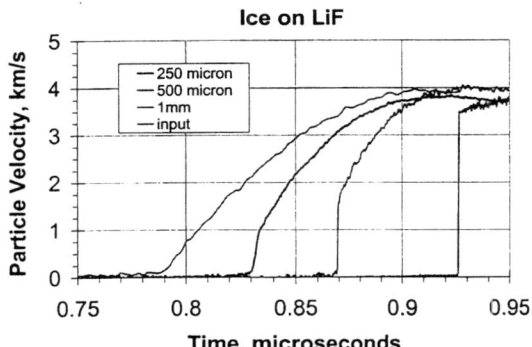

FIGURE 5: *Shot Z576, ~1 Mbar 0.25, 0.5, and 1.0 mm thick LiF backed by LiF windows showing formation and growth of shock.*

To extend shockless loading using ICE to higher stresses in many materials, or to investigate more compressible materials at lower stresses, the time-dependent loading rate for a given nonlinear material needed to be controlled. An effort was undertaken on Z to institute a method for shaping the input magnetic pressure pulse. This was accomplished by discharging one forth of the machine's stored energy 100-200 ns prior to the main pulse, and has been demonstrated on multiple firings. In addition to the prepulse, the risetime of the main discharge can be extended to ~240 ns. A combination of these two capabilities was used to investigate aluminum to ~500kbar. The resulting velocity profiles for the 0.8 and 1.5 mm thick samples are shown in figure 6. As can be seen, there is no evidence of a shock in the thicker profile, but it appears that a large amplitude shock would form in the velocity range of 0.2 – 1.75 km/s range if the wave were propagated into a thicker sample. This is in stark contrast to shots where pulse shaping was not used and smaller shocks formed much earlier in time.

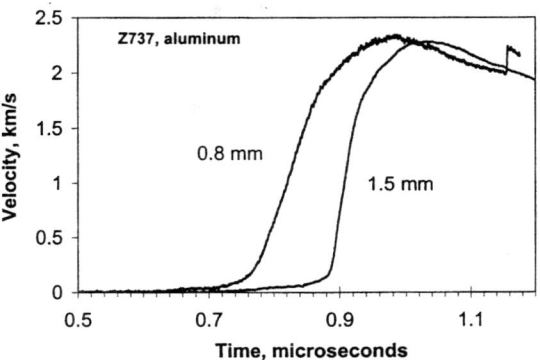

FIGURE 6: *Shot Z737, ~500 kbar aluminum with pulse shaping applied. Shockless loading seen at a thickness of 1.5 mm*

FLYER PLATES

In addition to studies of material properties along the isentrope, the impulse from ramp loading can be used to impart momentum to plates for impact studies giving material response along the Hugoniot. Initial attempts at launching plates on Z[18] of ~10 mm diameter and on order 0.5 mm thickness resulted in velocities of 10 km/s for copper, 12 km/s for titanium, and 13 km/s for aluminum. Plate motion was continuously monitored throught the launch cycle with VISAR. In addition, diagnostics were fielded to investigate plate integrity and state. For these launches, the 1.5 Mbar panel configuration was used with a partial machine charge.

Recently, an aluminum flyer was succesfully launched to a terminal velocity of 21 km/s[19] on Z using a similar configuration, but with an increased current density acting on the panels. Two 1.5 Mbar square load panels (15 mm wide) were seperated by two 8 mm wide spacers forming a rectangle instead of a square. In addition, the entire panel was machined to an initial thickness of 725 μm except for a 2.0 mm perimeter which gives the panel structural integrity. Because of the current density increase, the loading pressure increased to ~2.5 Mbar. The standard, essentially linear current profile was used causing a high loading rate during launch. The result is an ~700 kbar shock causing initial plate motion as shown in the velocity profile of Figure 7. This launch capability has been used to investigate the Hugoniot response of aluminum to, and release from, ~5 Mbar[20].

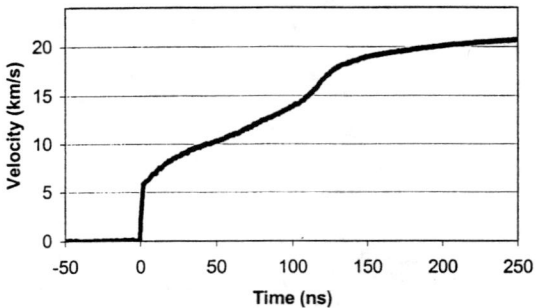

FIGURE 7: *Time resolved velocity profile of a 21 km/s aluminum flyer launched on Z.*

SUMMARY

Ramp loading experiments to obtain material response measurements along the quasi-isentrope to ~1 Mbar have been demonstrated in aluminum using pulsed magnetic loading on the Z accelerator. These results were obtained with improvements in experimental configuration over earlier attempts. These improvements provided uniform pressure loading over large diameter samples with attached interferometer windows. Results show deviation from the Hugoniot in agreement with theoretical predictions using a simple material model.

The ability to shape the input loading profile to minimize low-level shock formation and an increased understanding of MHD effects will lead to the design of more accurate experiments in the future.

Macroscopic aluminum flyer plates have also been launched on Z to velocities of 21 km/s with dimensions that are useful for accurate material response measurements.

REFERENCES

[1] J. R. Asay, in *Shock Compression of Condensed matter-1999*, edited by M.D. Furnish, L.C. Chhabildas, and R.S. Hixon, AIP Conference Proceedings, Melville, NY, pp. 261-266.

[2] D.B. Reisman, A. Toor, R.C. Cauble, C.A. Hall, J.R. Asay, M.D. Knudson, and M.D. Furnish, *J. Appl. Phys.*, **89**, 3, pp 1625-1633, (2001).

[3] C.A. Hall, D.B. Hayes et.al., to be published.

[4] J.P. Davis, D.B. Hayes, "Investigation of Liquid–Solid Phase Transition Using Isentropic Compression Experiments (ICE)", *this proceedings*.

[5] J. R. Asay, C.A. Hall, et.al., in *Shock Compression of Condensed matter-1999*, edited by M.D. Furnish, L.C. Chhabildas, and R.S. Hixon, AIP Conference Proceedings, Melville, NY, pp. 1151-1154.

[6] D.B. Reisman, J.W. Forbes, et.al., "Isentropic Compression of LX-04 on the Z Accelerator", *this proceedings*.

[7] M.K. Matzen, *Phys. Plasmas.* **4** (5), 1519- 1527 (1996).

[8] C.A. Hall, J.R. Asay, et.al., *Rev. Sci. Instrum.*, **72**, 9, Sept. 2001

[9] D.B Seidel, M.L. kiefer, R.S. Coats, T.D. Pointoin, J.P. Quintenz, and W.A. Johnson, *Proceedings of CP90 Europhysics Conference on Computational Physics*, edited by A. Tenner (World Scientific, Singapore), pp. 475-482 (1991)

[10] W.M. Trott, M.D. Knudson, et.al., in *Shock Compression of Condensed matter-1999*, edited by M.D. Furnish, L.C. Chhabildas, and R.S. Hixon, AIP Conference Proceedings, Melville, NY, pp. 993-998.

[11] L.M. Barker and R.E. Hollenbach, *J. Appl. Phys.* **43**, 4669 (1972).

[12] G. Sharp, Dissertation, Univ. New Mex., Oct. 2001

[13] Lemke, this proceedings

[14] A.C. Mitchell and W.J. Nellis, *J. Appl. Phys.*, 52, 5, pp. 3363-3374, (1981).

[15] J.B. Aidun and Y.M. Gupta, *J. Appl. Phys.* **69**, 6998-7014 (1991).

[16] D.B. Hayes, *J. Appl. Phys.*, **89**, 11, pp. 6484-6486, (2001)

[17] J.L. Wise, and L.C. Chhabildas, in *Shock Waves in Condensed Matter*, Edited by Y.M. Gupta (plenum, New York, 1986), P. 441.

[18] C.A. Hall, M.D. Knudson, et.al., *Intl. J. Impct. Eng*, 2001.

[19] M.D. Knudson, to be published.

[20] M.D. Knudson, to be published.

TEMPERATURE MEASUREMENT OF ISENTROPICALLY ACCELERATED FLYER PLATES

Thomas Bergstresser and Steven Becker[1]

Sandia National Laboratories, P O Box 5800, Albuquerque, NM 87185-1168*
[1]*Bechtel Nevada, P O Box 98521, Las Vegas, NV 89193-8521*

Abstract. Two frequently-used methods to accelerate flyer plates to extreme velocity (>10 km/s) are magnetic acceleration and impact of the flyer with a high velocity, layered impactor. In either case the temperature of the flyer is not definitely known, either because of diffusion of the magnetic field into the flyer or the quasi-isentropic nature of the impactor's acceleration. We have measured the temperature of flyers in both methods using radiation thermometry. Our pyrometer has four channels in the range 1.3 to 5.1 microns. It can resolve a 20 ns rise time, a necessary feature in these experiments where the available time can be as short as 200 ns. We have also seen the large temperature rise due to magnetic diffusion in plates that are too thin.

An optical pyrometer is an instrument for measuring temperature by the intensity of light radiated by an incandescent body. The theory and practice of such an instrument has been presented in detail[1]. We have constructed an infrared pyrometer and have used it to estimate the temperature of hypervelocity flyer plates. The pyrometer is outlined in Fig. 1. Radiation is brought to the pyrometer by a single optical fiber. This radiation is divided in the pyrometer by dichroic beamsplitters into four spectral bands centered at these wavelengths: 1.3, 2.4, 3.5 and 5.1 microns, representing channels 1 through 4 respectively. The beamsplitters were chosen to make use of an infrared transmissive, chalcogenide (As_2S_3) glass optical fiber. The wavelength bands are almost overlapping in order to increase the pyrometer output. The split beams are focussed with coated zinc selenide lenses onto photovoltaic detectors composed of mercury-cadmium telluride for channels 2 to 4 and indium-gallium arsenide for channel 1. The (Hg,Cd)Te detectors have thermoelectric coolers. Channels 2 to 4 can resolve a risetime of 20 ns, channel 1 is somewhat faster. This speed is required because of the short timescale of hypervelocity experiments. The pyrometer is calibrated while the fiber is viewing a cavity blackbody set at a sequence of temperatures. In this it is important to use the fiber that is to be used for the experiment because the fiber is somewhat lossy and also variable from batch to batch. Sometimes a lens has been used at the sample end of the fiber to focus the viewing onto a 1 mm diameter spot. When it is used this lens must also be a part of the calibration. The spectral responsivity of each individual channel is also measured, and this information is used to check the quality of the blackbody calibration as well as to extend it to higher temperatures than the blackbody can attain. The blackbody used currently has a maximum of 1273 K.

The first use of the pyrometer on a flyer plate occurred at Sandia's hypervelocity launcher, a three-stage gun. This work has been reported[2] and will

---

* This work was supported by the U.S. DOE under contract DE-AC04-94AL85000. Sandia is a multi-program laboratory operated by Sandia Corporation, a Lockheed Martin Company, for the U.S. DOE.

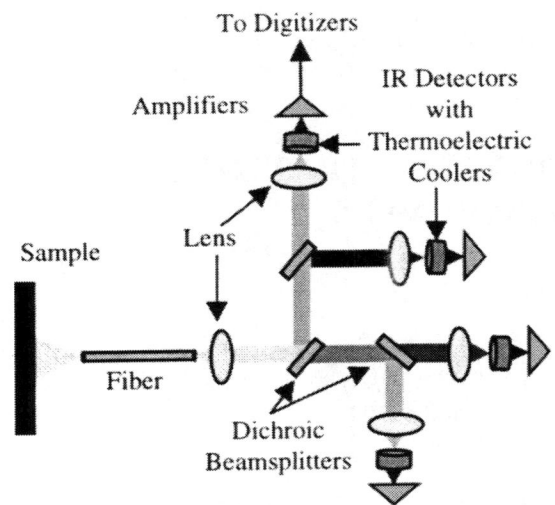

FIGURE 1. Schematic of the Pyrometer

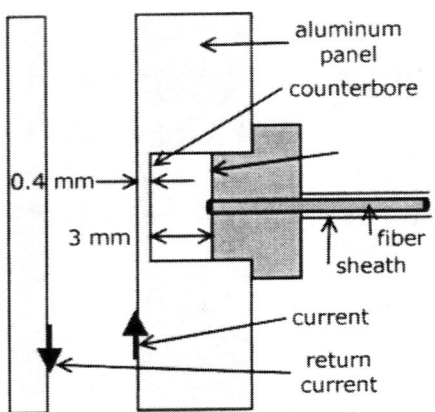

FIGURE 2. Magnetic Acceleration

be summarized here. In this type of experiment a composite, layered projectile strikes the third-stage flyer plate, accelerating it to high velocity without causing it to break up. The acceleration proceeds by a sequence of small shocks so that the flyer remains nearly isentropic. In two experiments the flyer-plate velocity was 9.8 and 10.8 km/s. The temperature was 605 ± 35 and 750 ± 70 Kelvin, respectively. There was good agreement with later calculations. Evidence of some stray light was apparent, possibly from jetting at the edge of the flyer where it met the inside portion of the target fixture. This is reflected in the uncertainty of the temperature.

Experiments have been done using Sandia's Z-machine as a source of intense electrical current with a short rise time, less than 200 ns. The current and resultant magnetic field produce a magnetic pressure which accelerates the conductor carrying the current, see Fig. 2 and refer to Ref. 3 for an extended discussion. The compression wave will ultimately shock up, but a counterbore with a thin floor provides the opportunity to observe the effects of rapid, shockless compression. The counterbore floor can be accelerated to high velocity, in the present case 12 km/s, and then used as an impactor onto a target plate. In this application the density of the impactor must be known. A high, unknown temperature with its associated density decrease would cause errors. An unacceptable temperature rise could be caused by cold working, near-surface porosity, unanticipated effects of the rapid compression, magnetic field ("B-field") penetration to the surface with its accompanying current density, etc. Calculations indicated that B-field penetration was sufficiently delayed and that the other effects were not consequential, but measurements had not been done.

To remove the pyrometer from the intense electromagnetic noise nearby to the Z-machine, a long (41 m) optical fiber carried the signal to the pyrometer stationed in a screen room. The required length of fiber made impossible the use of the lossy chalcogenide fiber for which the pyrometer was designed. Accordingly, we used a low-OH silica fiber fabricated for use in the infrared. This fiber entirely blocked channels 3 and 4, leaving the shortest wavelength channel 1 and part of channel 2.

The measured response of the pyrometer is shown in Fig. 3. Starting at -0.4 μs is the noise caused by the operation of Z, picked up mainly by channel 2. At 0.589 μs there is a fiducial time-mark on channel 2. Centered at 0.95 μs is a spike due to fiber fluorescence, discussed in the next paragraph. Judging from visar measurements on similar experiments, first motion of the counterbore surface occurs at about 1 μs. There was no visar measurement on this particular counterbore. After the fluorescence spike there is a short time interval before the temperature rise due to B-field penetration at 1.288 μs.

The optical fiber used in the experiment was sheathed in a plastic tube, providing protection from light but not x-rays or charged particles. On two subsequent runs at Z similar fibers were installed

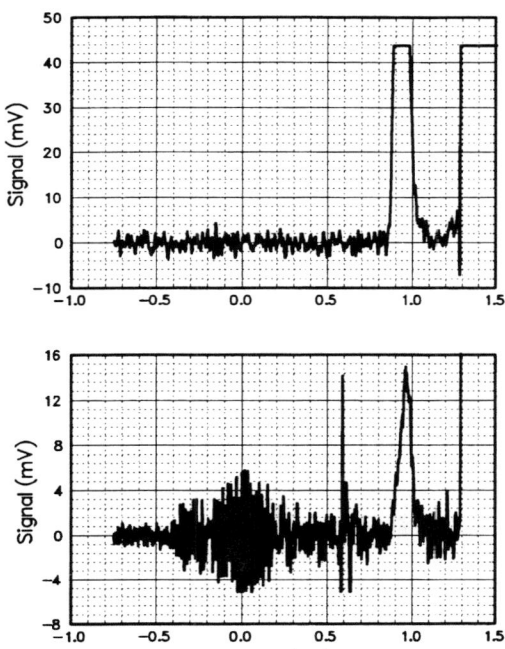

FIGURE 3. Pyrometer output from the bare floor of an aluminum counterbore. Top: channel 1, Bottom: channel 2

but the tips were blocked off. The same spike was seen in these experiments, demonstrating that the light did not come from the counterbore surface. These tests also revealed that there was a long-lived, low-level fluorescence that continued after the spike, visible only in the shorter-wavelength channel 1. In yet another test the fiber was sheathed in a light-weight hydraulic tubing with a thin, flexible steel layer. this completely eliminated the fluorescence spike. That such insubstantial protection eliminated the problem suggests that the fluorescence was due to low-energy electrons rather than x-rays.

Data from the short time interval between the fluorescence spike and field penetration is available for estimating the temperature of the counterbore. The average signal and standard deviation for channel 1 between 1.096 and 1.248 µs is 1.4 ± 1.6 mV. The corresponding information for channel 2 is -0.2 ± 1.4 mV. It is not possible to find a temperature from these results, but it is possible to find an upper bound. Take as an upper bound for the pyrometer output the average value plus the standard deviation. This is to be compared to the blackbody calibration result multiplied by the emissivity. To increase the emissivity, the counterbore surface had been prepared with a 32 finish rather than the usual diamond-turned, 20 nm rms finish. The emissivity of similar pieces was measured, yielding an emissivity of 0.255 for channel 1 and 0.22 for channel 2. This could be changed by the acceleration process, but we cannot estimate this effect. The resulting temperature upper bounds are 790 Kelvin for channel 1 and 695 K for channel 2.

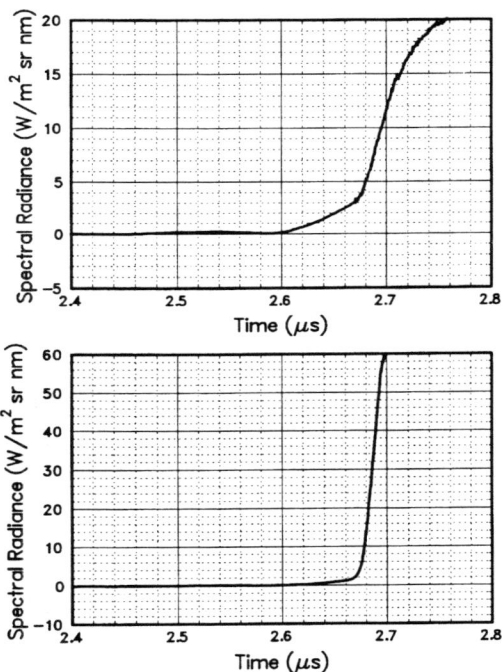

FIGURE 4. Pyrometer output from the aluminized coating on a LiF anvil glued to a copper counterbore surface. Top: channel 1, Bottom: channel 2

Recently three experiments were completed at Z, one with an Al panel, two with Cu, one of which is discussed here. The purpose was to measure B-field penetration; pyrometry was an add-on. The B-field was measured[4] by a Faraday-law loop, and to slow down the destruction of the loop by the advancing counterbore surface a 1 mm thick LiF anvil was glued to the counterbore. The glued side was aluminized, 1 μm thick, so that visar measurements could be made. This thermally complex system is not ideal for pyrometry. The pyrometer output, converted to spectral radiance, is presented in Fig. 4. The output shows three features: a low radiance from 2.45 to 2.59 μs, an almost-linear rise to 2.67 μs, and thereafter a more rapid rise. These features are present for channel 2 although the figure hardly shows it, and these features are qualitatively present in all three experiments. The output of channel 1 after 2.67 μs can not be used. Although this detector can resolve better than a 20 ns rise time, it is subject to a maximum slue rate, and the output is at that rate from 2.67 μs until the detector has begun to saturate after 2.7 μs. Only channel 2 remains for estimating the temperature.

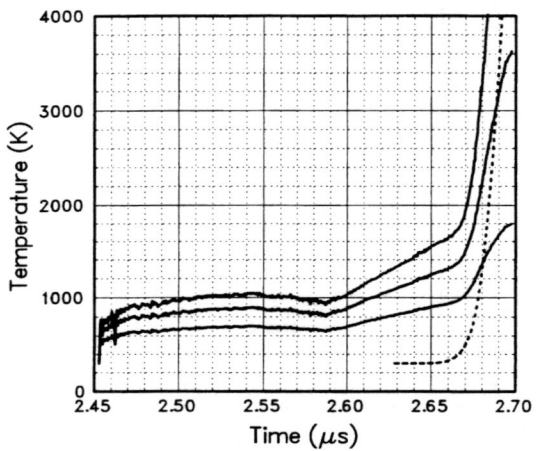

FIGURE 5. Temperature corresponding to Fig. 4. Middle solid line is the estimated temperature, and minimum and maximum values are shown (but see text). Dotted line is an approximate temperature due to magnetic field penetration.

The estimated temperature appears in Fig. 5, prepared assuming that the aluminum surface had an emissivity of 0.1. Minimum and "maximum" temperatures are produced using an emissivity of 1. and 0.03 respectively. An approximate temperature due to the B-field penetration is shown by the dotted line. The third of the mentioned three features is due to the B-field. The other two remain unexplained. The pressure history at the Al-LiF interface is known; using visar records of the velocity. The Hugoniot or isentrope of glue or Al are incapable of explaining the temperature. Channels 1 and 2 together imply that the radiance is from a small hot spot or stray light from a hot object. In this case the "maximum" temperature does not apply and the minimum applies to the hot spot, not the aluminum surface. The dotted line is an estimate of the effect of the B-field on the temperature. It is $T = 0.5 B^2 / (\mu_0 c)$, where T is the temperature, B is the B-field, μ_0 is the permeability of free space and c is the heat capacity of aluminum. This can be given an heuristic justification as a approximation, but suffice it to say here that it is dimensionally correct.

It is desirable to improve these results. More responsive detectors are available. Metal surfaces can be coated to increase the emissivity[5]. A more sophisticated use of the LiF anvil[6] can eliminate the possibility of stray light as well as suppress ejecta in a shock experiment. Other techniques may be possible.

REFERENCES

1. D. P. DeWitt and G. D Nutter, eds., *Theory and Practice of Radiation Thermometry*, Wiley-Interscience, New York, 1988
2. W. D. Reinhart, L. C. Chhabildas, D. E. Carroll, T. Bergstresser, T. F. Thornhill and N. A. Winfree, "Equation of State Measurements of Materials using a Three-Stage Gun to Impact Velocities of 11 km/s," in *Int. J. Impact Eng. Vol. 26, Hypervelocity Impact, Proc. 2000 Symp., Galveston, TX, 6-11 Nov. 2000*, in press.
3. C. A. Hall, J. R. Asay, M. D. Knudson, W. A. Stygar, R. B. Spielman, T. D. Pointon, D. B. Reisman, A. Toor and R. C. Cauble, *Rev. Sci. Instr.* **72**(9), 1 – 9 (2001).
4. Greg Sharp, Sandia National Laboratories, Private Communication.
5. M. Perez, "Residual Temperature Measurements of Shocked Copper and Iron Plates by Infrared Pyrometry," in *Shock Compression of Condensed Matter 1991*, edited by S. C. Schmidt, R. D. Dick, J. W. Forbes and D. G. Tasker, Elsevier, Amsterdam, 1992, pp. 73.-740.
6. D. Partouche-Sebban, D. B. Holtkamp, J. L. Pelissier, J. Taboury and A. Rouyer, "An investigation of shock induced temperature rise and melting of Bismuth using high-speed optical pyrometry", to be published.

SYRINX PROJECT : HPP GENERATORS DEVOTED TO ISENTROPIC COMPRESSION EXPERIMENTS.

Ch. Mangeant[1], F. Lassalle[1], P. L'Eplattenier[1], P-L. Héreil[1], D. Bergues[1]
G. Avrillaud[2]

[1]*Centre d'Etudes de Gramat, 46500 Gramat, France*
[2]*ITHPP, 46500 Thégra, France*

Abstract. Some compact pulsed-current generators are described here. They allow the generation of isentropic compression loading of metals and other materials. The range of pressures achievable is 30-300 kbar for the Explosive Switch Compact Generator (ESCG). This generator consists of a RLC circuit that discharges in a strip (also called plate)-line insulated by dielectric foils. Placed on the end of this strip-line, some material samples can be studied under dynamical loading. Typical dimensions of the tested samples are 1 mm thick, 8mm diameter. A current of 700 kA to 1.6 MA allows some 30-300 kbar ramp pressures generated with a 500ns rise time. The switch used in the ESCG is a linear-wave-explosive switch.
0D circuit and MHD simulations are discussed and compared with experimental results. Diagnostics based on current-voltage and free-surface velocity measurements are presented. Finally, the generation of isentropic compression principles and simulations discussed above are used to analyse the potential of a new compact generator which development is contracted to ITHPP, a French company. This generator should allow us to explore the 100kbar–1Mbar isentropic compression regime in order to study material behavior under dynamic loading for a large range of material and geometry samples.

INTRODUCTION

Isentropic compression waves have been extensively studied in the 70's. The most widely developed techniques were quasi-isentropic impact experiments using pillows projected by gas-guns (1), and magneto-explosive generators (2). These two techniques were hampered by some problems or restrictions. Since a few years, there has been a new interest in pulsed-current-generators for generation of quasi-isentropic compression experiments (ICE) (3-4). One of the goals of SYRINX project during the past two years -related in the present article- is to study the opportunity to use and/or develop high pulsed power (HPP) generators for ICE.

When a conductor is carrying a current, the magnetic pressure P related to the magnetic field H is $P=\mu_0.H^2/2$, where μ_0 is the magnetic permitivity of vacuum. When the magnetic field is generated by a current I flowing into a perfect plate-line of width W, the current-pressure relation becomes:

$$P=(\mu_0/2).(I/W)^2 \qquad (1)$$

Theoretically a current of 1MA flowing into a strip-line of 1cm width leads to a pressure of 63 kbar. A generator delivering 5 MA into this same strip-line would create a pressure up to 1.5 Mbar ! Although equation (1) must be revised for practical cases, this promising approach has led us to study this type of configuration.

EXPERIMENTAL SETUP

We have attempted to experimentally study two versions of an ESCG which has already delivered a 0.8 MA peak current in its first version and a 1.6 MA peak current in its second version. In its first version, this generator consists of a single 3.95 µF capacitor charged to 70 kV, a strip-line connexion ended by an explosive switch, and a load (Fig.1). In its second version, the number of capacitors is tripled. The energies stored are respectively 9.6 kJ and 29 kJ for the two versions. The whole system occupies a volume of less than 2 m^3.

The switch is a linear-wave-explosive switch allowing the current to be discharged in a 8 cm-centimeters-long line with a closing time of a few tens of nanoseconds. Currents up to 1.6 MA have been commuted by this switch. The destruction following the detonation is limited to the consumable parts thanks to an appropriate setup and to the small quantity of explosive involved (a few tens of grams only).

The load is a strip-line whose electrodes are made of copper and are insulated by thin dielectric foils. Typical widths W of these electrodes are 12 mm to 4 mm. Typical gap separating the two load-electrodes (insulated by dielectric foils) are 0.4 to 0.1 mm. Two VISAR free-surface-velocity measurements are done during each shot on one side of one of the two load-electrodes. This number could be increased to four by using both sides of the load. No material sample has been set on the copper electrode in order to be studied. Thus, in the following lines, what we will call "sample" is at present the electrode itself.

Both current and voltage are measured just before the explosive switch with dedicated B-dot and D-dot sensors built at CEG. As they were not properly calibrated in a strip-line configuration, a calibration shot was done for the 3-caps-version of the ESCG, where the reference sensor was a Rogowski coil inserted between the load-electrodes. Temporal resolution and synchronisation of all the sensors and VISAR is about 1ns.

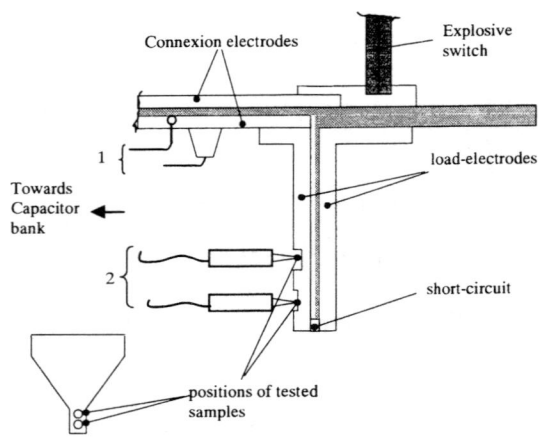

FIGURE 1. Load region and measurement setup.
1- B-dot and D-dot sensors, 2- VISAR heads.

The width of the strip-line (load) electrodes are chosen so that magnetic field distribution can be considered as uniform on the sample surface to better than 3 % and that edge effects don't affect the experiment. This allows VISAR samples to be studied under 1-D planar compression waves conditions (no edge effects) during the risetime of the first main-pulse of the current.

RESULTS AND ANALYSIS

Nine experiments have been performed with the first version of the ESCG. Only seven experiments have been done till now with the 3 caps-version. The first goal of these shots was to demonstrate the validity of the concepts used in this type of generators: dielectric insulation, explosive switch, and performance (in term of pressure achievable) of the strip-line load-electrodes. They have shown a very good reproducibility of the switch behavior and thus have led to a 0D electrical modelization. The inductances and resistances have been calculated with simple geometrical considerations and the results are in very good agreement with the experimental data (Fig. 2).

The load impedance is supposed variable during the shot. The key points for calculating this variation are the diffusion of magnetic field in the copper of the load-electrodes and the increase of the gap due to the pressure wave during the shot.

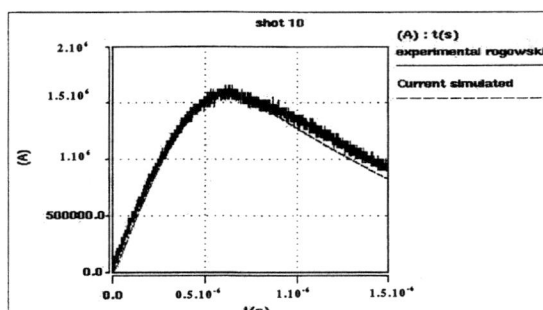

FIGURE 2. Comparison between electrical simulation and current measurement in the case of calibration-shot#10 (3caps. version).

We define what we call an "effective gap" $gap_{effective}$ by equations (2):

$$gap_{effective} = gap_{(t=0)} + f[d(t), \delta(t)] \quad (2\text{-}a)$$

where $\delta(t)$ is a skin depth and $d(t)$ is the displacement of inner faces of load-electrodes. If one supposes that there is no effective displacement of electrodes during the shot, one has:

$$f[d(t), \delta(t)] = \delta(t) \quad (2\text{-}b)$$

On the other, one can suppose that electrodes stay solid and are pushed by magnetic pressure, hence:

$$f[d(t), \delta(t)] = 2.d(t) + \delta(t) \quad (2\text{-}c)$$

In order to evaluate $\delta(t)$ some MHD calculations have been performed at linear current densities and typical timing scale of our experiments, i.e. between 1 and 2 MA/cm and 500 ns respectively. Different resistivity models have been used: TAPP, RRCK and Burgess (SESAME tables are planned).

Analytical evaluations of inductance and resistance give some values that differs by less than 20 % from the value obtained by 2D magnetodynamic simulations. Notice that ESCG performances are not so affected by the load impedance variations, which justifies the use of the very simple analytical relations.

In Eqns. 2 the displacement of inner faces of load-electrodes is easily calculated by integration of velocity of these electrodes. This integration is performed at each time step of the 0D electrical simulation. The code used is SABER and allows to couple electrical simulations to physical models (flux penetration, velocity increase of inner faces of electrodes, etc...) (5). Figure 3 shows the VISAR data of shot#14 where the thickness of the samples were 1.51mm±6µm and 2.09mm±6µm respectively and the width of the load-electrode was 6mm±20µm.

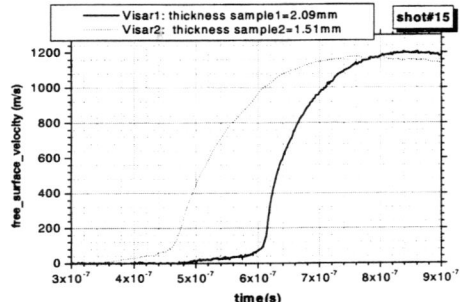

FIGURE 3. VISAR free-surface-velocity histories due to a 275kbar ICE wave in copper samples of thicknesses 1.51mm and 2.09mm (shot #15)

In order to evaluate the maximum pressure achieved, the deconvolution of one VISAR record was performed with the method of characteristics for the temporal aspect coupled to the equations (3) and (4) of conservation of momentum and mass:

$$d\sigma = \rho_0.D.du_p \quad (3)$$

$$dV = -(V_0/D).du_p \quad (4)$$

where

σ = stress in the direction of wave propagation
V (resp. V_0) = specific volume (resp. initial specific volume)
ρ_0 = original density (=$1/V_0$)
D = Lagrangian wave speed
u_p = particle velocity $\approx u_{sl}/2$ (5)

Eqn. (5) can be used as a first approximation in this deconvolution procedure. Some more accurate relations were chosen here: they were based on analysis of hydrodynamic simulations in which the elasto-plastic behavior of copper was taken into account. The typical error introduced by equation (5) on the pressure result is about 10%.

One can notice that Eqn. (1) needs to be corrected to explain the pressure achieved (250 kbar instead of 450 kbar predicted by Eqn. 1 with a maximum experimental current of 1.6 MA and a plate line of width 6 cm). The corrected equation is:

$$P = (\mu_0/2).(I/W)^2 / k_p \quad (6)$$

where k_p is a factor >1 function of the effective gap given by Eqn. 2-a. The relation $k_p=k_p(gap_{effective})$ is evaluated by 2D magneto-dynamic simulations and finally, one obtains the very simple relation:

$$k_p = a.gap_{effective}+1 \qquad (7)$$

with a is a constant depending only on the initial geometry of electrodes of the strip line (width and thickness).

Finally, the compression loading path of copper can be obtained through equations (3) to (4) and Lagrangian analysis. Figure 4 shows the curve $P=P(\eta)$ where η is the compression rate ($\eta=\rho_0/\rho$) compared to experimental datas of Sandia National Laboratories (Isentropic) and McQueen datas (4 and 6):

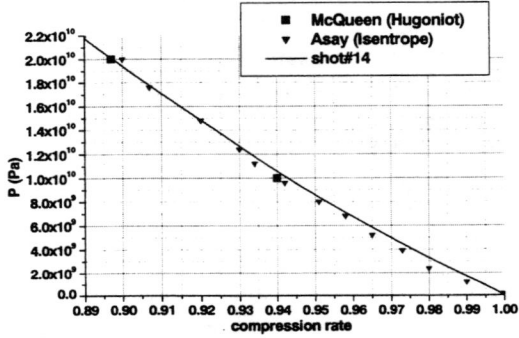

FIGURE 4. Pressure vs. compression rate ($=\rho_0/\rho$) obtained from the VISAR data of Fig. 3 compared to experimental datas.

Hugoniot and isentropic data are in good agreement, except for low pressures. This still needs to be analyzed.

FUTURE

The ESCG is a "first step generator": it allows us to study the opportunity to use strip-line for ICE. The results obtained are very encouraging. Areas like material properties (EOS, phase transitions, polymorphic transitions, etc...) can now be studied by this way. The ESCG is actually a very versatile tool for studying both materials under quasi-isentropic loading and high current density physics. Another application of this generator could be the test of new sensors in a harsh environment (linear current density or magnetic field up to 2 MA/cm). Examples of such sensors are µB-dots or Faraday rotation sensors for current measurement. Another great advantage of this generator is its easiness of use. Thus it is possible to use it for series of experimental campaigns that would mobilize bigger HPP generators elsewhere. For instance, we plan to measure magnetic field diffusion in copper thanks to this ESCG.

A new generator is now being built at CEG: it is based on the same plate-line principle but should deliver a current of about 4 MA in a load of 10 mm width. It should allow us to achieve pressures up to 1 Mbar. It should allow us to study material behavior under dynamic loading for a large range of material and geometry samples.

ACKNOWLEDGMENTS

The authors are very indebted to CEG technical teams (LPE, GSI and MAGIX) for their support. We wish to thank W. Farabolini too for his devotion.

REFERENCES

1. L.M. Barker, "High-Pressure Quasi-Isentropic Impact Experiments", in "Shock Waves in Condensed Matter" edited by J. R. Asay and R.A. Graham, 1983, pp. 217-224.

2. R.S. Hawke and R.N. Keeler, J. Appl. Phys.,**43**, 6, 2734-2741 (1972).

3. S.I. Krivosheev, "Pulsed Current Generator for Microsecond Duration Pressure Pulse Generation" in 12th Symposium on High Current Electronics. Tomsk, 24-29 Sept. 2000.

4. J.R. Asay, "Isentropic Compression of Iron with the Z Accelerator", in Shock Compression in Condensed Matter-1999, edited by M.D. Furnish et al., AIP Conference Proceedings 505, New-York, 1999, pp. 1151-1154.

5. P. L'Eplattenier, G. Avrillaud., "0D Numerical Modelisation and Optimization of the Magnetic Flux Compression Scheme for Isentropic Compression Experiments", Published in this conference.

6. R.G. Mc Queen in High-Velocity Impact Phenomena, Academic Press, NY 1970, Appendix E.

CORRECTING FREE SURFACE EFFECTS BY INTEGRATING THE EQUATIONS OF MOTION BACKWARD IN SPACE*

Dennis Hayes and Clint Hall

Sandia National Laboratories, Albuquerque NM 87185-1181 USA

Abstract. Free surface and window interfaces perturb the flow in compression wave experiments. The velocity of these interfaces is routinely measured in shock-compression experiments using interferometry (i.e., VISAR) and the perturbations must be accounted for before meaningful material property results can be obtained. Using the VISAR results as "initial conditions" we integrate the flow fields backward in space to the interior of the specimen where the VISAR interface has not perturbed the flow at earlier times and results can be interpreted as if the interface had not been present. This provides a rather exact correction for free surface perturbations. The method can also be applied to window interfaces by selecting the appropriate initial conditions. Applications include interpreting Z-accelerator ramp wave experiments. The method can be applied to elastic-plastic and quasi-elastic materials for experiments with multiple layers and multiple reverberations.

BACKGROUND AND EXAMPLE

The backward integration technique (Backward) was motivated by the need to analyze experimental results from the Sandia Z-accelerator. These experiments(1) measure free-surface velocity for two or more specimen thicknesses using VISAR(2) interferometry. The desire is to interpret these measurements by assuming the particle velocity characteristics are straight lines in space-time and thus infer the Lagrangian wave-speed as a function of particle velocity. This is sufficient to determine the stress-strain behavior of the specimen material through Reimann invariants. The central problem is as follows: as early parts of the ramp wave arrive at the free surface, they reflect and interact with the later parts of the oncoming ramp wave. This interaction bends the later oncoming characteristics negating the assumption of straight-line-characteristic behavior required for conventional analysis techniques. Since ramp waves steepen with propagation distance, perturbations are different for each specimen thickness.

For isentropic compression waves, one might erroneously expect that one-half the free surface velocity would exactly equal the *in situ* particle velocity (the velocity at that same location if the specimen were thick and no free surface perturbations oc-

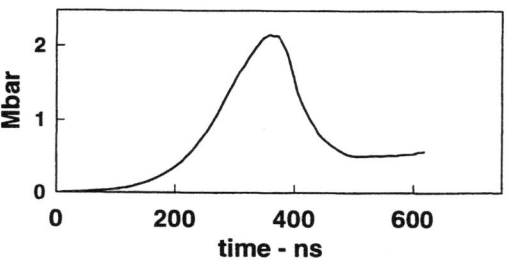

FIGURE 1. The pressure load applied to the front surface of a 0.8 mm copper specimen.

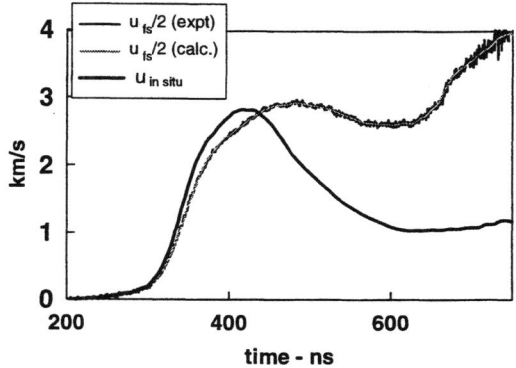

FIGURE 2. The particle velocity at depth 0.8 mm in a thick plate is compared with one-half the free-surface velocity of a 0.8 mm-thick plate. Experimental and calculated free-surface results agree almost exactly.

---

*This work supported by the US Department of Energy.

curred). That would be true for an isentropic step load but is not the case for a ramp load as explained above. For example, Fig. 1 shows a 2 Mbar ramp load that is applied to a 0.8 mm sample of copper. Figure 2 shows a dramatic difference between $u_{fs}/2$ and u_{insitu} for the ramp load of Fig. 1. Using the former as an approximation for the latter introduces errors of 10's of ns in time and more than 10% in velocity for an experimental situation that requires more than an order of magnitude better accuracy to be useful.

CORRECTING FOR FREE SURFACES

This paper treats the above situation. Actually, in the above example, we started from the experimental free-surface velocity history of Fig. 2 and the pressure history of Fig. 1 was deduced by the backward integration technique described below. This deduced pressure history was subsequently used as a boundary condition in a traditional forward WONDY(3) calculation to produce the simulated *in situ* velocity. In order to validate the accuracy of the Backward procedure, a separate WONDY simulation used this same pressure boundary condition to calculate the free-surface velocity. The experimental and calculated free-surface motion agree almost exactly (see Fig. 2) verifying the accuracy of the entire procedure.

BACKWARD METHOD

The one-dimensional Lagrangian equations of motion,

$$\sigma_{,x} = -\rho_0 u_{,t}, \quad (1)$$

$$V = F(\sigma), \quad (2)$$

$$u_{,x} = \rho_0 V_{,t}, \quad (3)$$

are numerically integrated from the free surface backward in space to the interior of the sample to a location where the relevant part of the wave has not been perturbed by the free surface. The "initial" conditions are the VISAR free surface velocity history and zero stress.

Integration can be done backward in space to any convenient location, provided it is far enough back to avoid perturbations from the free surface at times of interest. In the two-sample experiment described in example 1 below, the same stress-strain behavior would be obtained if the integration were carried back to say 0.1 mm, 0 mm as was done in the analysis, or even to -0.1 mm! The position x=0 is usually chosen although the inferred loading profile at the 0 mm location is not the one actually experienced by the copper whenever the magnetic field has penetrated the copper during loading. However, the important results obtained by Backward are unaffected by this uncertainty.

When a Backward-determined stress history at some interior location is used in a subsequent forward calculation, the starting simulated VISAR history is always replicated, even if the material model is incorrect, provided the integrations were done accurately. Because the same equations are being integrated back in space and then forward in time, the backward/forward procedure is just a mathematical mapping of the VISAR record onto itself. That is why another constraint must be found to make use of this method. For Z experiments, this constraint is usually that two specimen thicknesses must both infer the same load. This constrains the specimen stress-strain relation. For the spall experiment that is given as another example, stress after spallation must remain zero at the spall plane, *etc.*

At present, the method cannot be used if strong shocks are present anywhere in the flow. Since shocks produce entropy, which destroys information, there are simply regions of the x-t plane that are inaccessible by information obtained at the measurement surface. As a practical matter, if a shock is present in the VISAR record, it immediately begins to spread into an isentropic compression wave as the integration proceeds back into the interior of the specimen. Thus for large amplitude shocks, significant entropy generation will be ignored by Backward. But for low amplitude shocks like the 20 kbar shock present in the spall experiment, the entropy generated by the shock is completely negligible and apparently, as evidenced by the good comparison of the starting VISAR record with the subsequent forward calculation, solutions are accurate enough for our purposes. Thus if we use the method for experiments with weak shocks, a case by case evaluation must be made to assess if errors introduced are significant.

The general method is also extended to hysteretic elastic-plastic(4) and quasielastic materials.(5) These extensions can change the governing equations from hyperbolic to parabolic. However in the spall example below, the growth of expected instabilities is "managed" well enough to get an accurate solution for the stress history at the spall plane.

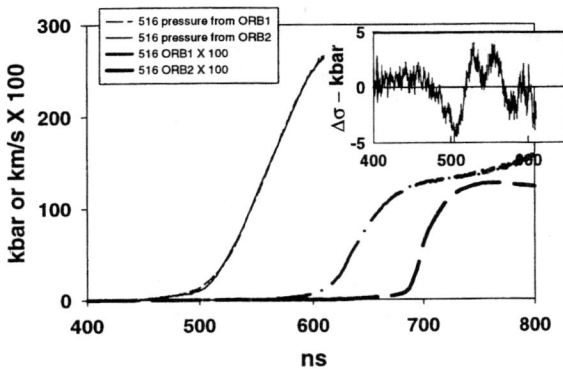

FIGURE 3. Analysis of Z-516. Two VISAR records from copper samples of two different thicknesses were obtained during a single Z experiment. Each free surface record was integrated backward in space to $x = 0$. The stress-strain relation was varied systematically until the two calculated load histories at $x = 0$ were the same. They agree to about 1%. The deduced stress-strain agrees with the known behavior of copper. See Fig. (4).

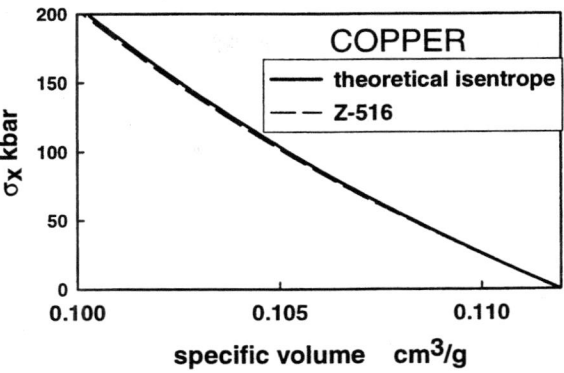

FIGURE 4. The deduced stress-strain behavior of copper for experiment Z-516 compared with the known behavior, our best result to date.

WINDOWED VISAR EXPERIMENTS

The method is not restricted to analyzing free-surface experiments. If the measurement interface has a VISAR window, the initial conditions at the measurement plane for Backward are chosen along the time axis as: VISAR velocity, stress in the window at that particle velocity and specific volume in the specimen. Care must be taken to ensure the window's stress-strain is calibrated for ramps and that its optical properties (VISAR correction) is measured for isentropic compression.(6) Furthermore, the VISAR technique itself can give different results for ramps than for steady waves.(7) A further complication arises for windowed experiments in which the stress-strain relation for the specimen is being studied by using more than one sample thickness described in example 2 below. Backward must use the "answer" to initialize the integration. This problem is tractable(6) but the implicit nature of this problem magnifies the errors in the deduced stress-strain relation for the sample.

EXAMPLES

Backward is presently being used to analyze a variety of experiments at Sandia, Los Alamos and Lawrence Livermore National Labs. Below are selected examples that show the breadth of application:

1. Starting from two free surface measurements on the Z accelerator,(6) determine the stress-strain relationship for the copper specimen. (Figs. 3,4)

2. Starting from two simulated LiF window velocity histories,(8) determine the stress-strain relationship for the specimen and the pressure load originally applied in the simulations. (Fig. 5)

3. Starting from a free-surface velocity history from a spall pullback experiment in aluminum,(5) determine the location of the spall plane and the stress history of the failure process at that location. Also determine parameters for the time-dependent quasielastic model of plasticity. (Fig. 6)

Each of these examples is part of an ongoing experimental study in which Backward is finding application.

CONCLUSION

There are many ways to combine forward and backward calculations. For instance, if one specimen of a two-specimen Z experiment has a shock, the shockless result can be integrated backward and the specimen with a shock integrated forward with a code like WONDY. By constraining the calculated and measured shock results to agree, the equation of state could be determined. Or the free-surface acceleration of a high velocity flyer plate can be used to calculate the various gradients in the flyer at impact, information needed to quantify the character of the shock generated at impact.(9) Recent experiments have combined free-surface and window measurements in a unique way to define window optical properties.(6) Backward is an essential part of this investigation. There is seemingly an endless vari-

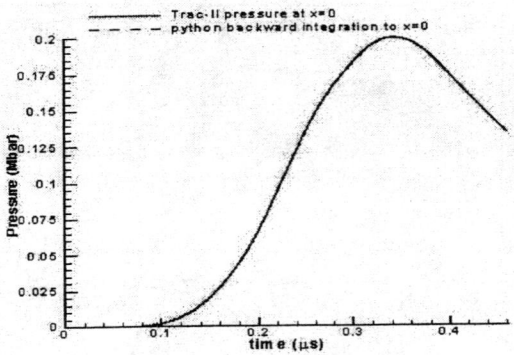

FIGURE 5. Pressure load. Simulated VISAR records for two windowed copper specimens(8) (not shown) were used by Backward to deduce the stress-strain relation used for the simulation to ≈1/4% (not shown). Backward also extracts the load (shown) accurately. For this idealized example Backward makes essentially an exact correction for the window interface perturbations.

ety of ways to put forward and backward calculations together to tease new results from experiments, provided we can shed our habit of viewing the x-t plane from only one direction.

ACKNOWLEDGMENTS

The authors are indebted to J. Asay, J. Fritz, M. Knudson, and R. Menikoff for useful discussions. Sandia is a operated by Sandia Corporation, a Lockheed Martin Company, for the Department of Energy under contract DE-AC04-94AL85000.

REFERENCES

1. C. Hall et al., submitted to Rev. Sci. Instr., 2001.
2. L. M. Barker and R. E. Hollenbach, J. Appl. Phys. **43**, 4669 (1972).
3. M. E. Kipp and R. J. Lawrence, WONDY V - a one-dimensional finite-difference wave propagation code, Technical Report SAND81-0930, Sandia National Laboratories, 1982.
4. D. Hayes, Backward integration of the equations of motion to correct for free surface perturbations, Technical Report SAND2001-1440, Sandia National Laboratories, 2001.
5. D. Hayes, J. Vorthman, and J. Fritz, Backward integration of a VISAR record: Free surface to the spall plane, Technical Report LA-13830-MS, Los Alamos National Laboratory, 2001.
6. C. Hall et al., to be published.
7. D. Hayes, J. Appl. Phys. **89** (2001).
8. D. Reisman, LLNL, private communication.
9. M. Knudson, SNL, private communication.
10. J. N. Johnson, J. Phys. Chem. Solids **54**, 691 (1993).

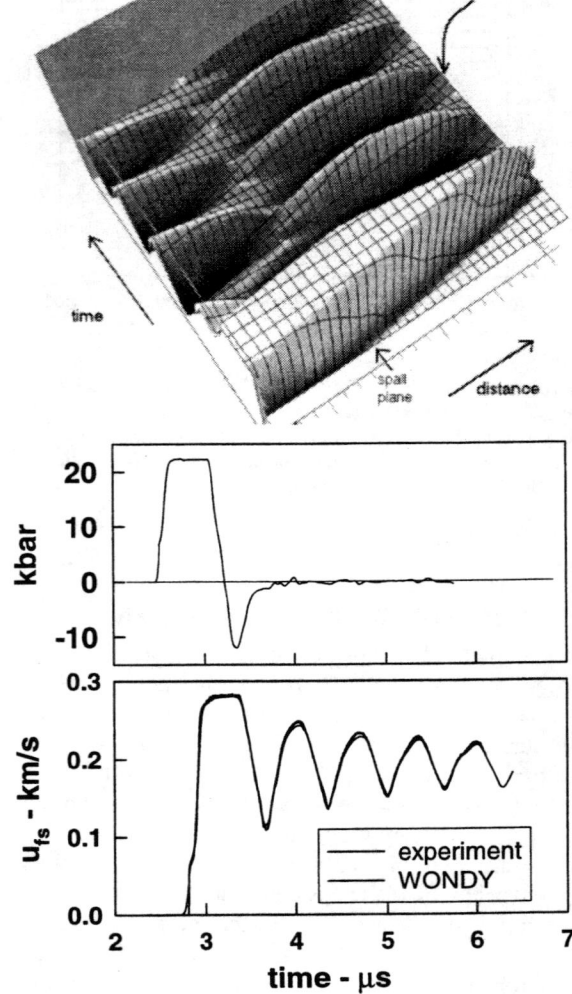

FIGURE 6. Backward integration of a spall pullback free surface VISAR record on 6061-T6 aluminum.(5) The VISAR free surface record (bottom) was integrated backward in space to ≈150% of the estimated scab thickness. The 3-D graph (top) shows the calculated stress as a function of time and Lagrangian distance from the free surface. Note the slice marked "spall plane". That is the plane where the stress stays at zero after spall occurs. Stress at that plane is shown in the middle graph. Backward sought best values for the two quasielastic parameters and for the position where the RMS of the late time stress was a minimum. Two quasielastic parameters deduced for 6061-T6 Al agree well with those determined by Johnson.(10) The position of the deduced spall plane agrees with experiment. The two-step failure seen in the middle graph displays secondary spall resistance behavior seen previously in tantalum. The calculated free-surface velocity (bottom) was obtained with WONDY where the Backward-determined load (middle) was applied to an aluminum layer with thickness equal to the Backward-determined scab thickness.

PICOSECOND TIME-RESOLVED X-RAY DIFFRACTION : ESTIMATION OF LOCAL PRESSURE

Yoichiro Hironaka, Fumikazu Saito, Akio Yazaki, Kazutaka G. Nakamura, and Ken-ichi Kondo

Materials and Structures Laboratory, Tokyo Institute of Technology, 4259 Nagatsuta, Midori, Yokohama 226-8503, Japan

Abstract. We have performed time resolved X-ray diffraction experiments with picosecond time resolution on Si single crystal compressed by laser irradiation. From the measured diffraction profiles, temporal and spatial distribution of the strain in the sample have been estimated using the direct search of optimization method based on the dynamical X-ray diffraction theory. The maximum compression of 1.05% was measured at the irradiation power density of 4.7×10^9 W/cm^2. We discussed pressure distribution analyzing observed data.

INTRODUCTION

The time resolved X-ray diffraction from shock compressed samples can give important information on the dynamics of transient phenomena such as phase transition and shock induced plasticity. Transient X-ray diffraction for shock compressed material has been studied using plate impact and flash X-ray technique [1-3]. However in order to investigate mechanisms of atomic motion induced by the shock wave front, more severe temporal and spatial resolutions on measurement will be required[4]. In particular, time-resolved recording on experiment is needed for detail of the dynamics in transient phenomena.

Recently, ultra short pulse X-rays can be generated by high intense femtosecond laser irradiation on metals. This enable to perform ultra fast X-ray diffraction of shocked material by combining with laser shock technique. In this paper, we perform pump and probe X-ray diffraction experiment for investigating the dynamics of a laser irradiated Si (111) single crystal[5]. Strain profiles and EOS are also obtained.

EXPERIMENT AND RESULTS

The pulsed X-rays (Fe $K\alpha 1$ and $K\alpha 2$) were generated by irradiation of femtosecond laser on the Fe target. The pulse width of the X-rays were measured to be 10 ps. Shock was generated by picosecond pulse (300 ps, 780 nm) was focused on the Si (111) wafer with the power density of 4.7×10^9 W/cm^2. The Si wafer was slightly translated to ignore the effect of the damage during the accumulation of signal (600 shots). Figure 1 shows the results of diffraction patterns obtained at every 60 ps. At the early time of laser irradiation (0~300 ps) the new peak is grown up at the larger angle. This corresponds to shock-compression of Si.

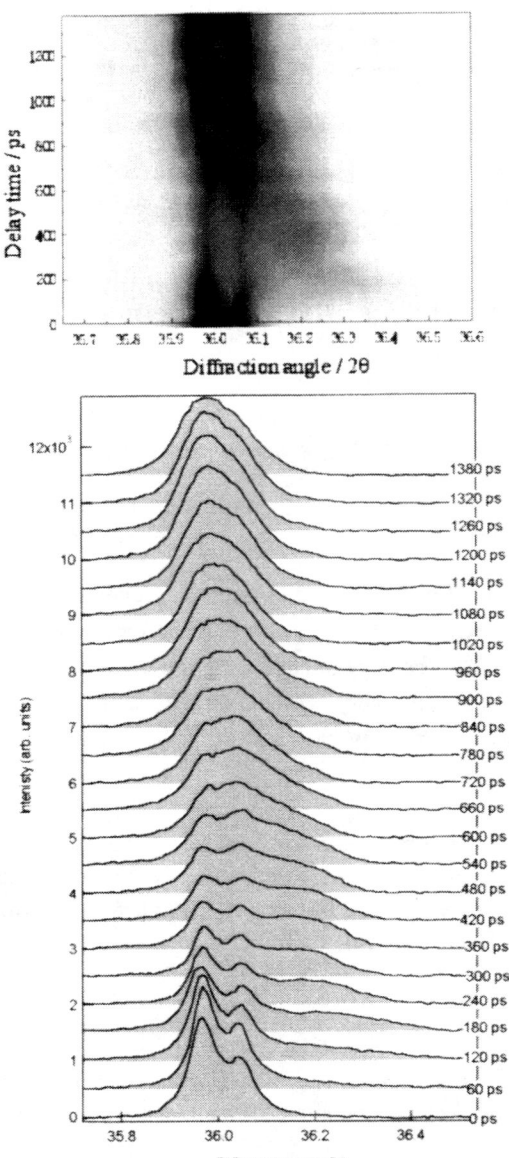

FIGURE 1. The results of Time resolved X-ray diffraction profile for laser perturbed Si(111) single crystal at every 60ps.

DISCUSSIONS
Strain Profile Analysis

We used DSOM (Direct Search of Optimization Method) based on the dynamical X-ray diffraction theory for the analysis of Figure 1, by assuming uniaxial strain perpendicular to the surface. According to the dynamical X-ray diffraction theory, the depth dependent scattering amplitude from a strained crystal is expressed as formula (1)[6].

$$i\frac{dX}{dA} = (1+ik)X^2 - 2(y+ig)X + 1 + ik$$

$$A = r_e \lambda \left| f'_g \right| t / V \sin\theta_B, \quad g = \frac{f''_o}{\left|f'_g\right|}, \quad k = \frac{f''_g}{f'_g} \quad (1)$$

$$y = \frac{\pi V \sin(2\theta_B)}{\lambda^2 r_e \left|f'_g\right|}(\theta - \theta_B + \varepsilon(t)\tan\theta_B) - \frac{f'_o}{\left|f'_g\right|}$$

A is therefore a dimensionless measure of the depth t relative to the crystal surface. r_e and λ are classical electron radius and X-ray wave length, respectively. V is the unit cell volume, and structure factor is $f = f'_{o,g} + if''_{o,g}$, $\varepsilon(t)$ means strain at depth t. The reflecting power on the surface of strained crystal is calculated by analytical solution of equation (1) and layered assumption[7]. We separated 100 layers within 10μm in depth. Information of each layer which has constant strain is expressed by the 4×4 real matrix (equation(2)).

$$\xi_{ij} = \begin{pmatrix} \xi_{11} & \xi_{12} & \xi_{13} & \xi_{14} \\ -\xi_{12} & \xi_{11} & -\xi_{14} & \xi_{13} \\ -\xi_{13} & -\xi_{14} & \xi_{33} & \xi_{34} \\ \xi_{14} & -\xi_{13} & -\xi_{34} & \xi_{33} \end{pmatrix} \quad (2)$$

$\xi_{11} = SR - yTI - gTR, \quad \xi_{12} = SI + yTR - gTI,$
$\xi_{13} = kTR + TI, \quad \xi_{14} = kTI - TR$
$\xi_{33} = SR + yTI + gTR, \quad \xi_{34} = SI - yTR + gTI$

$$AR = -\frac{\lambda r_e \left|f'_g\right|}{V\sin\theta}\delta_j SR, \quad AI = -\frac{\lambda r_e \left|f'_g\right|}{V\sin\theta}\delta_j SI$$

$$TR = \frac{2\sin(2AR)}{e^{2AI} + e^{-2AI} + 2\cos(2AR)},$$

$$TI = \frac{e^{2AI} - e^{-2AI}}{e^{2AI} + e^{-2AI} + 2\cos(2AR)}$$

where δ_j means thickness of j-th layer. SR and SI are real and imaginary part of $\sqrt{(y+ig)^2 - (1+ik)^2}$, respectively. The reflecting amplitude on the surface of crystal is calculated by equation (3).

$$\gamma = D_{total}\chi = \xi_{ij}^1 \xi_{ij}^2 \xi_{ij}^3 \cdots \xi_{ij}^{N-1} \xi_{ij}^N \chi$$

$$R(\theta) = \frac{\gamma_1^2 + \gamma_2^2}{\gamma_3^2 + \gamma_4^2} \quad here \quad \gamma = \begin{pmatrix} \gamma_1 \\ \gamma_2 \\ \gamma_3 \\ \gamma_4 \end{pmatrix} \quad (3)$$

$R(\theta)$ is the reflecting intensity at diffracted angle of θ. The vector χ is the boundary condition at the bottom of layers. The components of χ are (0,0,0,1) for the case the rear surface of crystal contact with vacuums. For the case of infinite crystal, boundary condition of (XI, XR, 0, 1) are used. XI and XR mean imaginary and real part of reflecting amplitude from the perfect crystal.

In the equation (3), we optimized value of each matrix of ξ_{ij} using DSOM to the observed data Thus, we obtain strain distribution inside of the crystal.

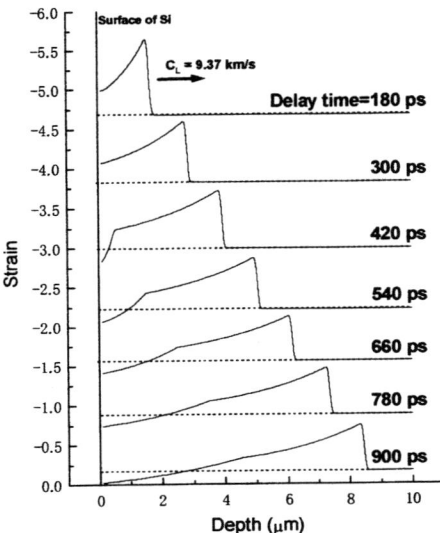

FIGURE 2. Estimated strain distribution inside of the Si crystal by DSOM computation based on the dynamical X-ray diffraction experiment.

Figure 2 shows the time-evolution of the obtained strain profile. The maximum compression of 1.05% is obtained at the delay time of 180ps. The induced wave has no steady state because the temporal profile of irradiated laser is Gaussian. After the end of laser irradiation (~300ps), expansion wave starts to propagate from the surface and caches up the wave.

EOS Determination

Here, we can translate strain distribution to the density distribution under the uniaxial assumption directory. Usually, we use three conservation laws and equation of state for the calculation of flow. The conservation of mass and momentum gives new particle velocity and density. The energy conservation law and EOS give P-V relation. Thus we can calculate new pressure value using new density value. Here, we note, the form of conservation of mass and momentum are independent from the ensemble (equation (4), the form of energy conservation law depends on the ensemble).

$$\left(\frac{\partial \rho}{\partial t}\right) + u\left(\frac{\partial \rho}{\partial x}\right) + \rho\left(\frac{\partial u}{\partial x}\right) = 0 \quad (mass)$$
$$\rho\left[\left(\frac{\partial u}{\partial t}\right) + u\left(\frac{\partial u}{\partial x}\right)\right] + \left(\frac{\partial P}{\partial x}\right) = 0 \quad (momentum)$$
(4)

Using finite differential method on first order assumption, equation (4) are developed to forms of equation (5) and equation (6).

$$u_i = \frac{1}{2\rho_{i+1} - \rho_i}\left[\frac{\Delta x}{\Delta t}(\rho_i^{n+1} - \rho_i^n) + \rho_i u_{i+1}\right] \quad (5)$$

$$P_i = \frac{\Delta x}{\Delta t}\rho_i(u_i^{n+1} - u_i^n) + \rho_i u_i(u_{i+1} - u_i) + P_{i+1} \quad (6)$$

The notation of i and n means position (depth) inside of the crystal and time, respectively. At the infinite crystal, the particle velocity of u_{i+1} will be zero. In the equation of (5), we have already estimated the value of spatial and temporal distribution of density. Thus, the particle velocity of u_i is calculated by equation (5). In this sense, the particle velocity distribution will be calculated by repetitively applying the formula (5) from deep inside of the crystal (infinite crystal) to the surface. Then the pressure distribution will be calculated using formula (6) in the same sense. Figure 3 shows the results of pressure distribution just after the laser irradiation using analysis mentioned above, and we plot analyzed value as the P-V relation in Figure 4 with the line of known EOS of Si(111). Our solution shows good agreement with known EOS of Si crystal.

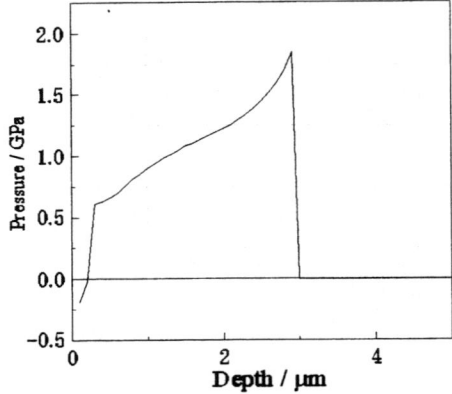

FIGURE 3. Estimated pressure distribution just after the laser irradiation.

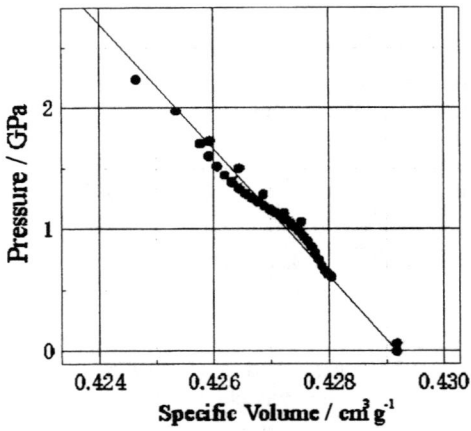

FIGURE 4. The relation between pressure and specific volume estimated using analysis mentioned in the text.

CONCLUSION

We successfully obtained the signal of time resolved X-ray diffraction for the laser perturbed Si single crystal with 60ps time step. We performed DSOM for the estimation of strain distribution based on the dynamical diffraction theory and we confirmed maximum compression of 1.05% at the irradiation power density of $4.7 \times 10^9 \text{W/cm}^2$. According to the analysis using conservation laws, the estimated value shows good agreement with the known value of Si single crystal.

ACKNOWLEDGEMENT

This work was supported by Core Research for Evolutional Science and Technology (CREST) program of Japan Science and Technology Corporation (JST).

REFERENCES

1. Jhonson, Q., Michell, A., Keeler, R. N., and Evans, L., Phys. Rev. Lett. **25**, 1099 (1970).
2. Kondo, K., Sawaoka, A., and Saito, S., Proc. 4th International Conference on High Pressure Kyoto1974, 845 (1974).
3. D'Almeida, T. and Gupta, Y. M., Phys. Rev. Lett. **85**, 330 (2000).
4. Wark, J. S., Whitlock, R. R., Hauer, A., Swain, J. E., and Solone, P. J., Phys. Rev. B, **35**, 9391 (1987).
5. Hironaka, Y., Yazaki, A., Saito, F., Nakamura, K. G., Kondo, K., Takenaka, H., Yoshida, M., Appl. Phys. Lett. **77**, 1967 (2000).
6. Klar, B. and Rustichelli, F., Nuovo Cimento B 13, 249 (1973).
7. Wie, C. R., Tombrello, T. A. and Vreeland, J., Appl. Phys., **59**, 3743 (1986).

LASER TRIGGERED SYNCHRONIZABLE X-RAY SYSTEM FOR REAL TIME STUDY OF SHOCK WAVES IN CONDENSED MATERIALS

J. Paul Farrell, K. Batchelor, V. Dudnikov, T. Srinivasan-Rao[*], J. Smedley[*] and J. McDonald[**]

Brookhaven Technology Group, Inc. 120 Lake Ave. South, Nesconset, NY 11767
[*]*Brookhaven National Laboratory* Upton, NY 11973*
[**]*Pacific Northwest National Laboratory, OPO Box 999 Richland, WA 99352*

Abstract A laser excited, sub-nanosecond, pulsed, electron beam system is described. The system consists of a high voltage pulser and a coaxial laser triggered gas or liquid spark gap. The spark gap discharges into a pulse forming line designed to produce and maintain a flat voltage pulse for 1 ns or greater duration on the cathode of a photodiode. A synchronized pulsed laser is used to illuminate the photo-cathode to produce an electron beam with very high brightness, short duration and current at or near the space charge limit. The system can be configured to operate at energies from less than 500 keV to 1 MeV and pulse width from less than 10 ps to 1000 ps and higher. This laser controlled electron beam system can be used to produce synchronizable monochromatic fluorescent or broad spectrum Bremsstrahlung x-rays for shock wave studies.

INTRODUCTION

Dynamic studies of shock compression utilize fluorescent x-rays from flash x-ray sources to characterize the change in lattice constant and other transient properties (1). These real time x-ray diffraction measurements require close synchronization of the x-ray source with the transient shock wave. The times of interest for these studies are a few nanoseconds to sub-nanosecond and lower. The laser excited photo-diode system described here produces synchronizable short pulse electron and photon beams in this time interval. Since the system uses laser excitation to generate an axial electron beam, the x-ray pulse duration, photon number and source dimension are determined by the corresponding characteristics of the laser. Since the cathode and anode are not damaged in the pulse discharge, they can be used for thousands of shots without the need for replacement.

TECHNICAL DESCRIPTION

The system described here was initially designed as an electron gun for advanced high-energy electron accelerators (2). Figure 1 is a block diagram of the complete system. It is comprised of a master timer, a laser system that includes a laser amplifier and optical pulse compressor, a high voltage pulse power supply, a pulse forming line and a photo-diode electron gun. In this high voltage (2 MV) variant, the output of the high voltage power supply is terminated in a spark gap that discharges into the pulse forming line. In a low voltage (300 kV to 500 kV) system, all spark gaps would be replaced by solid-state switches.

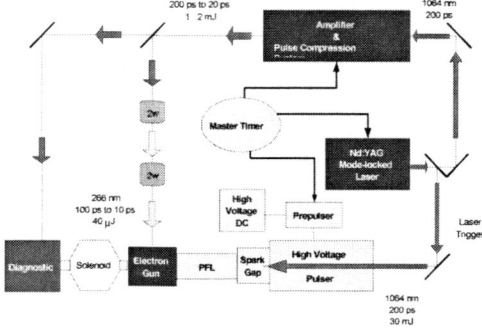

Figure 1. Block diagram of the laser controlled x-ray system,

In operation, the master timer sends trigger pulses with appropriate delays to the laser and the high voltage pulser. The laser output is split into two components; one (100 ps at 1064 nm) travels up the axis of the high voltage pulser and is timed to arrive at the triggered spark gap at or near the peak of the high voltage output waveform. The second component is amplified, time compressed and its frequency is adjusted (~ 255 nm) to optimise photoemission from the cathode of the photodiode. The pulse forming line (PFL) is terminated with a low inductance characteristic impedance to prevent reflection. In the 2 MV high voltage pulser, the PFL is designed to produce a sustained flat top (< 5%) output voltage of ~ 1 ns duration. The second laser component is time delayed to arrive at the photocathode during the 1 ns pulse on time. By triggering the output spark gap with the same laser that is used to photo excite the cathode, a very high level of synchronization (< ~ 100 ps jitter) in emission of electron current and arrival of the voltage pulse is achieved.

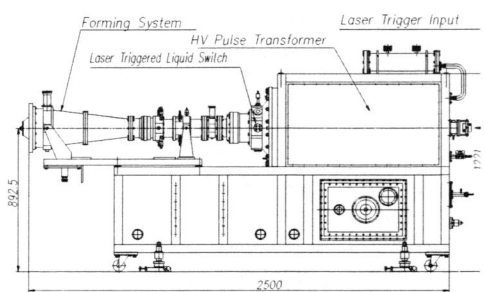

Figure 2 General view of the 5 MV Pulsed Power Supply System

When operated as a synchronizable flash x-ray source, electrons emitted from the photo cathode travel 2 to 3 mm to the anode, which is also the x-ray producing target. Since the cathode does not dump *all* its charge into the anode, the electron energy remains nearly constant during the current pulse. This results in increased fluorescent photon yield and virtually eliminates destruction of the anode and cathode surfaces that is observed in standard flash x-ray devices.

A general view showing the shape and dimensions of the 2 MV pulsed power supply is shown in Figure 2. The pulser is an integral unit comprising the following components:

- A metal casing
- A pulse generator (100 kV) for exciting the primary winding of the pulse transformer
- A pulse transformer
- A pulse forming line for generating the short (~1 ns) high-voltage pulse

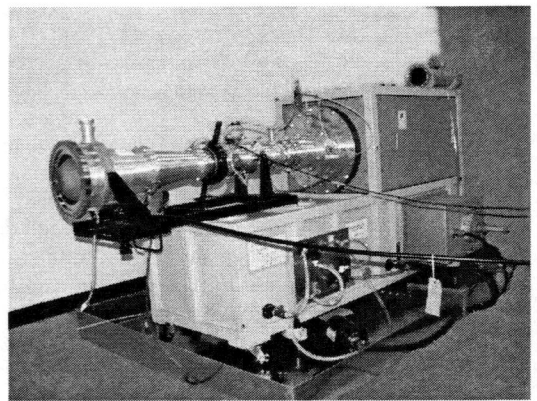

Figure 3, Photograph of the pulsed power supply system.

The 2 MV system shown here is 2.5 meters long and 1.22 meters high by ~ 1 meter wide. The welded casing forms the framework onto which all of the other parts of the pulser are mounted. The upper section, which is not sealed, houses a solid dielectric cylinder that forms a support for the pulse transformer winding. The side and end plates of this section of the casing are detachable to allow access to the components of the low voltage 100 kV pulse generator that drives the high voltage pulse transformer in the upper section.

FLUORESCENT X-RAY SOURCE FOR DIAGNOSTIC OF SHOCK WAVES

The pulse length and spot size of this beam based x-ray source is determined by the corresponding pulse length and spot size of the laser beam on the photo-cathode. A typical source spot size would be ~ 1 mm to 2 mm in diameter. This small spot size provides a very high photon flux density at the source.

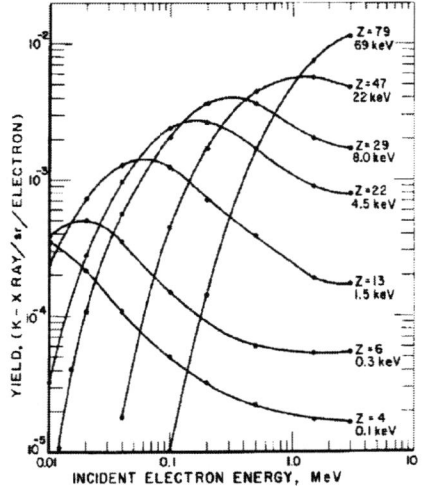

Figure 3. Dependence of K x-ray yield from thick targets of Z = 4 to 79 on incident electron energy. (From F. H. Attix 1986.)

Using a conventional solid-state laser system, an electron charge of ~ 50 nC can be drawn from the cathode without significant voltage droop. The K x-ray yield from thick targets of Z = 4 to 79 are shown in Figure 4 (3). From the figure it can be seen that the yield of 4.5 keV photons peaks at ~ 2 × 10^{-3} x-rays/sr/electron at an electron energy of ~ 150 keV. This corresponds to ~ 0.6 × 10^9 photons/sr/pulse for a 50 nC electron bunch.

Experiments are needed to determine if the total photon yield from this short pulse high brightness photon source is sufficient signal for real time x-ray diffraction studies.

An advantage of this laser synchronized x-ray source is that the system is easily adapted to include both plate impact and laser induced shocks. The same laser that is used to induce a shock in the sample could be used to photo-excite the cathode of the electron gun To produce laser induced shocks. This approach has the possibility of achieving a very high degree of synchronization between the arrival of the shock wave and the diagnostic x-ray pulse.

CONCLUSION

The same basic design concepts of laser excited electron emission and synchronized high voltage pulsed systems that are used in this state-of-art high voltage electron gun can be used to produce synchronized x-ray pulses for real time study of shock waves in condensed materials. A complete x-ray system operating at 300 kV to 500 kV uses all solid-state components.

ACKNOWLEDGEMENT

This work is supported, in part, by the U.S. Department of Energy in the following contracts: DE-FG02-97ER8233, DE-AC02-98CH10886 and DE-AC06-76RLO-1830.

REFERENCES

1. Y.M. Gupta, K.A. Zimmerman, P.A. Rigg, E.B. Zaretsky, D.M. Savage and P.M. Bellamy, Experimental developments to obtain real-time x-ray diffraction measurements in plate impact experiments. Rev. Sci. Instr. 70, No. 10, p 4008-4013 (1999).

2. Kenneth Batchelor, J. Paul Farrell, R. Conde, T. Srinivasan-Rao and J. Smedley, A Laser Triggered Synchronizable, Sub-Nanosecond Pulsed Electron Source, Proc. of International Conf. on Future Accelerators, Stony Brook, NY (June 2001). To be published.

3. J.H. Sparow and C.D. Dick, The development and application of monoenergetic x-ray sources. Report NBS SP456 (1976) and reproduced in Introduction to Radiological Physics and Radiation Dosimetry, F.H. Attix, John Wiley & Sons, New York p. 209 (1986).

0D MODELISATION OF THE MAGNETIC FLUX COMPRESSION SCHEME FOR ISENTROPIC COMPRESSION EXPERIMENTS

P. L'Eplattenier [1], G. Avrillaud [2], J. Vanpoperynghe [3]

[1] Centre d'Etudes de Gramat, 46500 Gramat, France
[2] ITHPP, 46500 Thegra, France
[3] CEA/DAM Bruyères le Chatel, France

Abstract. This paper deals with the advantages of using a flux compression scheme in High Pulsed Power (HPP) generators in order to produce isentropic magnetic high pressure ramps. Our field of interest are pressures in the range of several Mbars. For isentropic compression above 1 Mbar, the main advantage of the flux compression is to give considerable freedom to shape the current waveform and thus the magnetic pressure waveform. The optimizations of the shape of this magnetic pressure waveform are performed using improved 0D codes. Some physical models have been added to the initial 0D circuit code to reproduce the experimental results obtained on Z at Sandia National Laboratories, and on ECF1 and ECF2 at Centre d'Etudes de Gramat. Moreover, a simple hydrodynamic model is used to determine the isentropicity of the compression in a given material. We present some comparisons between experimental results obtained on the Z generator and our models. We will also give potential improvement of those results based on the use of a coil as a stator in order to have a better shape of the current waveform.

INTRODUCTION

Isentropic compression loading in a sample allows to determine the isentrope of the material $P_{S0}(\rho)$ issued from the initial condition P_0, T_0, ρ_0.

Since a few years, there has been a new interest in High Pulsed Power (HPP) generators for the generation of quasi isentropic compression experiments (ICE)[1]. In those experiments, the pressure is generated by a magnetic field created by a current flowing on one side of a sample. In order to get quasi-isentropic compression above 1Mbar, the shape of the pressure waveform and thus of the current waveform is critical.

The magnetic flux compression scheme has first been introduced on pulsed power generators as a power amplification scheme[2]. As shown in the following, it can also be used as a current waveform shaper with no changes on the generator itself.

Section I exposes the hydro techniques that we use in our 0D models to judge the isentropicity of a given pressure waveform. Section II presents the magnetic flux compression scheme as an intermediate stage to improve ICE experiments. It also introduces the numerical tools developed at CEG to optimize this stage. Finally, section III shows an example of such an optimization on the Z generator at Sandia National Laboratories (SNL).

I. HYDRO TECHNICS FOR PRESSURE OPTMISATION

A current I(t) generates a pressure P(t) on one side of a sample (referenced as the loaded side). At the back side of the sample, the free surface velocity can then be measured by the VISAR technique. The simultaneous measurement of the free surface velocity on 2 samples with 2 different thicknesses allows to compute the pressure versus the density along the isentrope, $P_{S0}(\rho)$, for $0 \leq P_{S0} \leq P_{expl\_isen}$, where P_{expl_isen} is the maximal exploitable pressure in isentropic compression. This pressure is extracted from the so called characteristics method well known in the hydrodynamic field (fig.1).

When a pressure is applied on a sample, the information propagates from the loaded side to the back side on characteristics. The first characteristic generated from the loaded side is called C_0^+. It reflects on the back side and creates the C_0^-. In the area below C_0^-, the characteristics are straight lines with a slope given by the Lagrangian sound velocity, $C_l = C_s(\rho)\frac{\rho}{\rho_0}$.

Using $U_S = C_0 + S \cdot U_P$
and by assuming $\left(\frac{\delta P}{\delta U}\right)_S = \left(\frac{\delta P}{\delta U}\right)_H$,

one can derive the following equation for C_l:

$$C_l = C_0 \cdot \sqrt{1 + \frac{P}{P^*}} \text{ with } P^* = \frac{\rho_0 C_0^2}{4S},$$

with C_S the sound velocity, ρ_0 the density at $P=P_0$, U_S the shock velocity, U_P the particle velocity, C_0 the sound velocity at $P=P_0$ and S a material constant. In the area above C_0^-, the characteristics are not straight lines any more. However, they can be well approximated by straight lines or parabolas. The flow can thus be analytically resolved for all the characteristics coming from the loaded side before the C_0^- reaches it. A shock forms in the sample when 2 characteristics intersect before the back side.

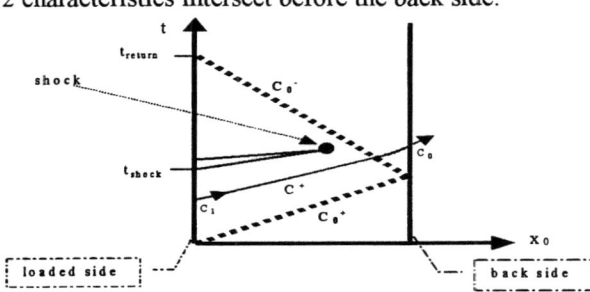

FIGURE 1. Illustration of the characteristics method, in lagrangian coordinates

Two VISAR diagnostics can be used to simultaneously measure the free surface velocity on 2 samples with different thicknesses $T_1 < T_2$ for which the same pressure waveform $P(t)$ has been applied on the loaded side. The maximal exploitable pressure in isentropic compression P_{expl_isen} is not necessarily the maximal value P_{max} of $P(t)$. Depending on the shape of $P(t)$, there could be a formation of a shock in the thicker sample limiting P_{expl_isen} to $P_{shock}=P(t_{shock})$ at time t_{shock} corresponding to the first characteristics where the shock occurs. The C_0^- could also reach the loaded side of the thinner sample at a time t_{return} limiting P_{expl_isen} to $P_{return}=P(t_{return})$. Finally, the free surface velocity measured with the VISAR does not allow too fast a variation of this velocity (and thus of the pressure on the loaded side) limiting P_{expl_isen} to $P_{mes}(t_{mes})$.

The hydro model allows, for a given shape $P(t)$ of the pressure, to calculate, as functions of the thickness of a sample, P_{shock}, P_{return} and P_{mes}, as well as $P_{isen}=\min(P_{shock},P_{return},P_{mes})$. The model then finds two thicknesses T_1 and T_2 with $\Delta T=T_2-T_1$ large enough, at least 50µm so that the VISAR method can be practically exploitable, and such that $P_{expl\_isen} = \min(P_{isen}(T_1), P_{isen}(T_2))$ is as high as possible. This is shown on figure 2, where P_{mes} has been omitted for clarity reasons.

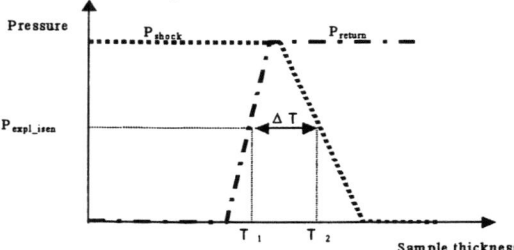

FIGURE 2. Schematic of the determination of P_{expl_isen}

In the frame of this model, for a given thickness, one can find the ideal pressure waveform to get isentropic compression as high as possible. An ideal pressure waveform is computed so that all the characteristics converge on the same point at a distance from the loaded side x_0 larger than T_2. One is then ensured there is no shock formation in the sample. The highest value of the pressure is then limited by the time t_{return} when the C_0^- in a sample with a thickness $T_1=T_2-\Delta T$ reaches the loaded side. One thus have with such a pressure waveform $P_{expl\_isen} = P_{T2}(t_{return})$ in the thinner sample.

II. THE FLUX COMPRESSION SCHEME

The magnetic flux compression scheme (fig.3) is usually used as a power amplification scheme on HPP generators as an alternative to pulse forming water lines or plasma opening switches. It can also be used as an intermediate stage in between the main generator and the load that allows to shape the current waveform.

Two different currents are needed. The first one is generated using typically ¾ of the whole stored energy and is called the primary current. It is used to implode a primary liner called armature, usually made with an aluminum wire array for conductivity reasons. The armatures already tested experimentally were between 6 and 15 cm long with an initial radius between 4 and 5 cm and made with 108 to 428 wires. The other current, generated with the rest of the stored energy is then injected inside the armature through a secondary injection gap. It thus flows in the central rod, usually called stator, on the load and on the inner part of the armature. It is called the secondary circuit. When the liner passes through the injection gap, the flux injected in the secondary circuit is trapped and since the inductance between the armature and the stator decreases with the radius of the armature, the current in the secondary circuit will rises at the same time.

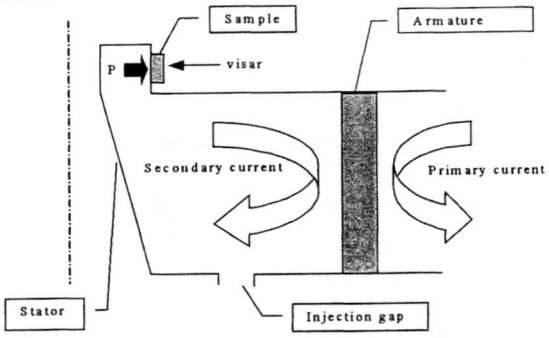

FIGURE 3. Schematic of the flux compression scheme

The advantage of that scheme is that it introduces many parameters that allow to shape the current waveform in the load: the armature initial radius, length, mass, the radius of the injection gap, the shape of the stator that can be a cylinder, a cone or a coil, and, depending on the generator, the levels of primary and secondary currents. 2D MHD codes can be used to simulate the flux compression scheme. However, their long CPU times make them practically not usable for the optimization all the parameters of the flux compression. For the purpose of optimizing those parameters, we have thus developed at CEG improved 0D codes. Those are circuit codes coupled with some physical models reproducing the main feature of the flux compression. They are fast running codes that give promising results in good agreement with the experimental ones, as shown on figure 4. This figure shows the experimental secondary current compared to the one given by the 0D code for different flux compression shots on generator Z at SNL.

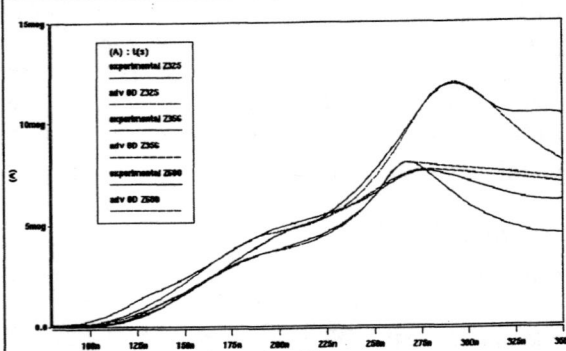

FIGURE 4. Comparison of experimental and numerical secondary currents for different shots on generator Z

We then have developed optimization procedures in multi dimensional spaces based on the gradient method driving the 0D codes, allowing to complete our basic analytical optimization of the scheme.

III. APPLICATION: OPTIMIZATION OF AN ISENTROPIC COMPRESSION EXPERIMENT

Those procedures can be used to optimize the flux compression parameters in order to get P_{expl_isen} as high as possible for a given generator. Once such an optimization is done, one can compare the obtained pressure waveform with the ideal pressure waveforms presented in section I. We now present an example of optimization of the magnetic pressure on a sample with isentropic compression on the Z generator at SNL. We started with a set of parameters corresponding to the next flux compression shot on Z (scheduled for July 2001) and which will be called case 1 in the following. For that shot, for experimental reasons, we decided to keep the central rod as a full cone. We thus have optimized the liner mass, the small and big radius of the cone, and the gap of the secondary area. We numerically obtained for that shot P_{expl_isen}=2.7Mbars with 2 sample thicknesses T_1=555µm and T_2=605µm. figure 5 shows how the pressure waveform compares to an ideal pressure waveform for a 500 µm thick sample. The general shape is already pretty good, but the first part corresponding to the injection of the secondary current and the time before the amplification is far from being optimal yet.

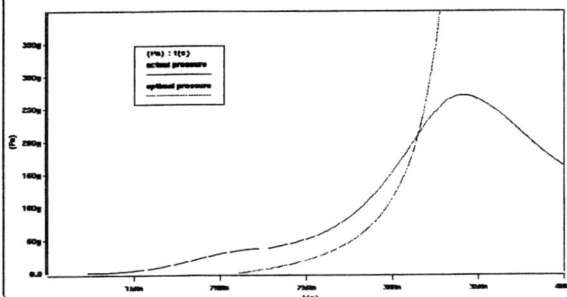

FIGURE 5. comparison of actual and optimal pressure waveforms for case 1

We then introduced a coil as the stator to show how case 1 could be improved in further experiments. The dependence of the inductance in the secondary circuit with the radius of the armature is then completely different than with a solid conical stator. There is now a large inductance at crowbar time, and a very fast varying one when the armature gets closer to the stator. We first have chosen a coil with a constant pitch and made an optimization on the flux compression parameters, including the pitch. This case will be referenced as case 2 in the following. We obtained after optimization $P_{expl\_isen}=3.21$Mbars, with $T_1=420\mu m$ and $T_2=470\mu m$. After that we made a 3rd case where the coil has a linearly varying pitch and made again the optimization. We obtained then $P_{expl\_isen}=3.55$Mbars with $T_1=430\mu m$ and $T_2=480\mu m$. Figure 6 shows the actual and ideal pressure waveforms for both cases 2 and 3.

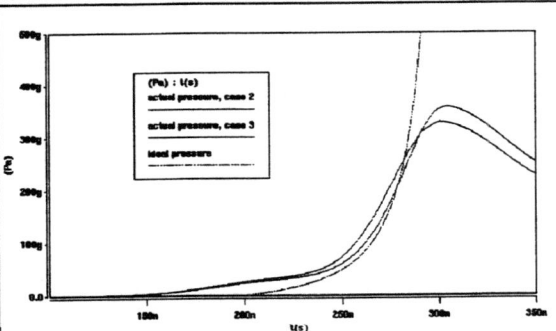

FIGURE 6. Comparison of actual and optimal pressure waveforms for cases 2 and 3

This figure shows that we get pressure waveforms much closer to the optimal ones than with a solid stator. The concept of a coil as a stator is still to be tested experimentally, in particular for magnetic insulation problems at the secondary injection gap, but this example shows how efficient it could be to use such a stator.

CONCLUSION

We have presented in this paper the use of HPP generator coupled with the flux compression scheme for isentropic loading on material samples. The flux compression is an intermediate stage with many degrees of freedom allowing to shape the pressure waveform in order to get the highest possible pressure in isentropic compression. We also presented the numerical tools developed at CEG in order to optimize those degrees of freedom.

The first numerical studies gave very promising results. However, they have to be confronted with more experimental results and 2D MHD simulations. In particular, the magnetic field diffusion in the sample have to be checked carefully. The use of a coil as a stator should be very efficient but could also present possible experimental issues that have to be checked.

The ECF2 generator will be on line at CEG by the fall of 2001[3]. This 3.6 MJ energy stored generator using a flux compression stage will allow to get many experimental results on isentropic compression loading.

ACKNOWLEDGMENTS

The authors are very indebted to J.F. Leon and the ECF1/ECF2 team at CEG; and also to R.B. Spielman, M. Mazarakis, and the Z Team at SNL for their support.

REFERENCES

[1] S.I. Krivosheev, « *Pulsed Current Generator for Microsecond Duration Pressure Pulse Generation* », *12th Symposium on High Current Electronics, Tomsk, 24-29 Sept. 2000*.

[2] J.F. Leon, « *Flux Compression Experiments on Z accelerator* », in *Pulsed Power Conference- 1999*

[3] P. Monjaux, in *Pulsed Power Conference- 2001*.

SIMULTANEOUS VISAR AND TXD MEASUREMENTS ON SHOCKS IN BERYLLIUM CRYSTALS

D. C. Swift, D. L. Paisley, G. A. Kyrala and A. Hauer

Los Alamos National Laboratory,* NM 87545, USA

Shock waves were induced in single crystals of beryllium, by direct illumination using the TRIDENT laser at Los Alamos. The velocity history at the surface was measured using a line-imaging VISAR, and transient X-ray diffraction (TXD) records were obtained with a plasma backlighter and X-ray streak cameras. At lower pressures, the VISAR records exhibited an elastic precursor followed by a plastic wave and spall. At higher pressure, the velocity records showed a two-wave structure suggesting a phase change, then at the highest pressure a single broad wave suggesting a shock directly into the high pressure phase. The rocking curves of the crystals were typically about 2 degrees wide, so analysis of the TXD records is complicated by the relatively large amount of blurring. However, the Bragg record of the shocked 0002 peak clearly indicates a smaller lattice parameter at higher pressure. In the shots where polymorphism seemed to appear in the VISAR record, additional lines appeared in the Bragg record, and new lines also appeared within the field of view of the Laue camera. These results are consistent with a new quantum mechanical equation of state for beryllium, which suggests that the hexagonal to body-centered cubic transition occurs at ~40 GPa on the principal Hugoniot.

INTRODUCTION

Beryllium capsules are among the designs being developed to contain the fuel for inertially-confined fusion (ICF). It is important to understand the properties of beryllium to underwrite capsule designs, such as specifications for the grain size so that the capsule implosion meets symmetry and uniformity requirements.

The relevant properties of beryllium include its strength as a function of orientation, melt, and solid-solid phase transformations. Beryllium adopts the hexagonal (hex) structure at STP, but transforms to body-centered cubic (bcc) on heating.[1] Gas gun experiments have provided data at slower deformation rates.[2]

We have performed a series of experiments using the Trident laser at Los Alamos to investigate the dynamic response of beryllium, and compared these with quantum mechanical equations of state incorporating polymorphism.

---
*This work was performed under the auspices of the U.S. Department of Energy under contract # W-7405-ENG-36.

EXPERIMENTAL METHOD

Shock waves were induced in foils and single crystals of beryllium by direct illumination with a beam from the Trident laser.[3] The pulse duration was 1.0 to 3.6 ns over a spot 4 mm in diameter. The spatially averaged intensity history of the drive beam was measured using a photodiode. The shock conditions in the sample were measured using a line-imaging VISAR recording on a pair of optical streak cameras[4] and – for the single crystal samples – transient X-ray diffraction recorded on time-integrating film and on X-ray streak cameras. (Fig. 1.)

The X-rays were generated by focusing a second Trident beam at full energy in the green (~200 to 250 J) tightly (~100 μm spot) onto a manganese foil 6 μm thick. The resulting hot plasma emitted predominantly helium-like line radiation, of wavelength 2.755 Å. The wavelength was verified with a transmission spectrometer.

The single crystal samples were 30 to 40 μm thick, and cut with the drive and VISAR sur-

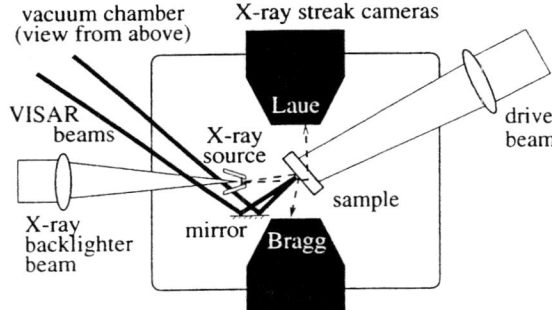

FIGURE 1. Schematic of experimental set-up.

faces parallel to the 0001 planes. The samples were cut from a single crystal boule made by zone refining. The boule dates from the 1960s, and only a little remains from the neck. As a result, there was a considerable distribution of orientations within each sample, thought to be in the form of low-angle grain boundaries, giving crystal rocking curves (distribution of scattered intensity about the nominal Bragg angle) about 2° wide.

The TXD records were interpreted in terms of lattice compression by deconvolving them using the rocking curve for the appropriate crystal as the kernel. One objective of this program is to identify melt on the Hugoniot from the TXD signal. It is likely that the hex/bcc phase transition takes place before melt occurs, so an additional challenge is to follow Bragg reflections in the bcc phase. It is not known whether the bcc phase would always form in the same orientation with respect to the hex, and hence whether the position of a bcc reflections can be predicted in advance, allowing experiments to be performed with a limited TXD field of view.

The Bragg angle for the unshocked beryllium was 46.7°. In each experiment, the sample was oriented so that the unshocked reflection would enter the 'Bragg' (south) camera, leaving space in the field of view for the reflection from shocked material. The position of the 'Laue' (north) camera was constrained by the design of the target chamber, and it was not possible to position the target so that an unshocked Laue line would appear in the field of view.

The line VISAR imaged 3 mm of the free surface across the center of the sample.

EXPERIMENTAL RESULTS

We performed experiments on four single crystals of beryllium. No attempt was made to recover shocked samples, to avoid compromising the dynamic diagnostics.

In one shot (12196) the sample was not shocked, to verify that a TXD signal could be obtained. Shielding was added to protect the sample from the X-ray source; the sample was recovered and subsequently re-used with a shock wave. Time-integrated and time-resolved TXD records were obtained. The profile of the 0002 Bragg reflection matched the rocking curve measured for that sample extremely well, so in these experiments the width of the TXD lines was dominated by the rocking curve and not by the spectrum of the X-ray source.

Subsequent shots (12198, 12199, 12202 and 12206) were performed with increasing intensity in the drive beam. (Fig. 2.)

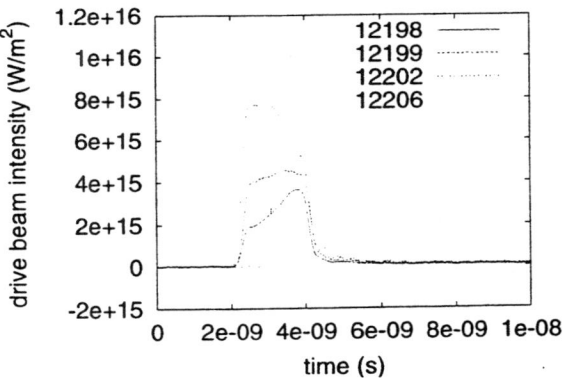

FIGURE 2. Drive intensity history for single crystal shots.

The line VISAR records showed no spatial variation in the velocity history. The records in the higher energy shots were quite noisy because the plasma / hot electron shield was omitted to make TXD alignment more accurate. There was no evidence that the sample was preheated before the arrival of the shock wave. The records from the lower intensity shots showed a precursor at ~1 km/s followed by a second wave. It seems likely that this is an elastic precursor followed by a plastic wave. There was evidence of spall and ringing after the maximum velocity was reached. The records from the higher intensity shots showed evidence of a multiple wave structure with a stronger precursor. (Fig. 3.)

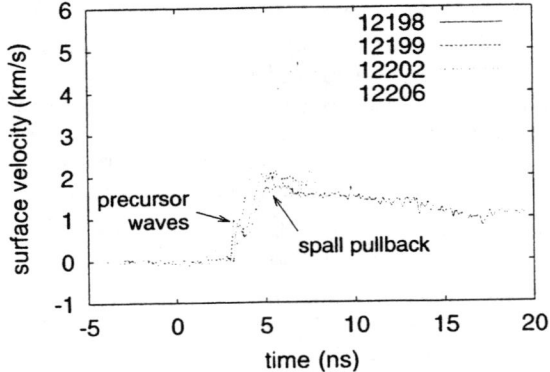

FIGURE 3. Line VISAR records.

The magnitude of the elastic precursor was considerably larger than one would expect from models calibrated to μs-scale experiments. This probably reflects rate-dependent or orientation-dependent contributions to the strength.

The Bragg TXD records showed a peak from the unshocked material, and also one or more peaks corresponding to compressed material. Because of the wide rocking curves, the noise was quite large on the time-resolved records, so we have concentrated so far on analyzing the time-integrated results (static films and time-average of the streak records). These will be used to guide a Bayesian analysis of the time-resolved signals. The records could be reproduced quite accurately by fitting a set of discrete lines (each contributing a distribution given by the rocking curve) at different deviations to the unshocked Bragg peak. The weaker lines seem to be statistically significant. The principal compressed peak deviated more from the unshocked peak at higher intensities. (Fig. 4.)

In the Bragg records, additional lines appeared at the highest intensities. In the same experiments, noisy lines appeared in the Laue records, one line remaining roughly constant and

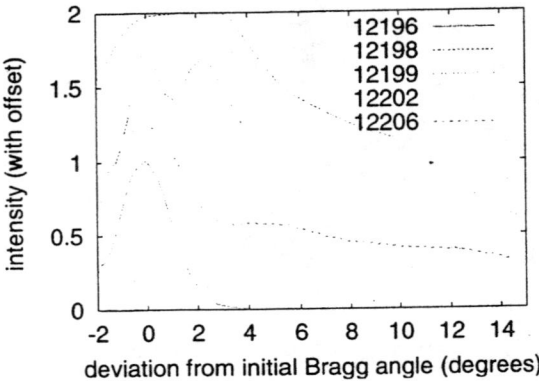

FIGURE 4. TXD records (south camera).

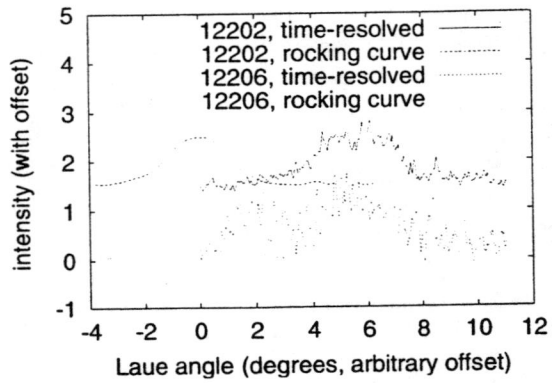

FIGURE 5. TXD records (north camera). Rocking curve for each crystal is shown for comparison.

the other moving (Fig. 5). Beryllium has a phase transformation from the hexagonal to the body-centered cubic (bcc) structure: it is possible that these additional lines are from the bcc. If so, it seems likely (albeit from only one comparison) that the bcc forms in the same orientation with respect to the hex, and thus that the position of a bcc line can be predicted in advance. The angles were used to calculate lattice spacings, assuming the orientation of the planes (Figs 6 and 7).

COMPARISON WITH THEORY

A two phase equation of state was calculated using *ab initio* quantum mechanics.(5) The equation of state matched the hex/bcc boundary

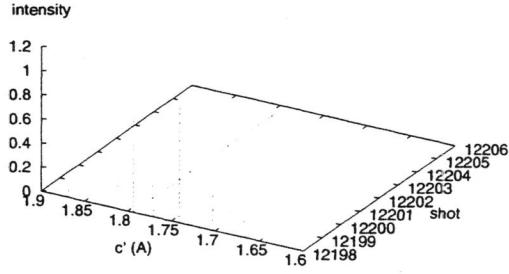

FIGURE 6. Lattice parameters deduced from south camera records.

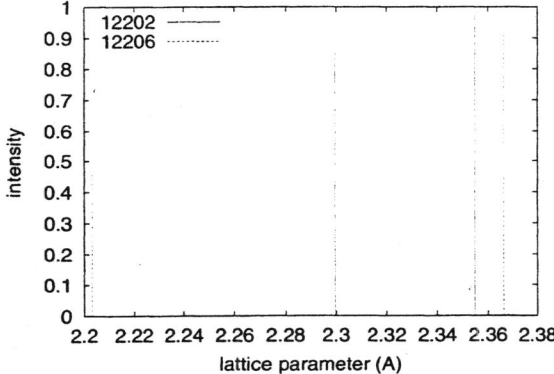

FIGURE 7. Lattice parameters deduced from north camera records.

observed at low pressures quite well. The principal Hugoniot was predicted to cross the hex/bcc boundary at ~40 GPa.

Radiation hydrodynamics simulations were performed using the measured drive profiles but with a simpler equation of state for beryllium. Velocity histories and density profiles were extracted from the simulations. The velocity histories were similar to the VISAR records but were not in good agreement. This could be the result of using an inaccurate equation of state, or because of uncertainties in the models used for transport properties.

Assuming the extra Bragg and Laue lines indicate the occurrence of bcc material, the transformation pressure deduced from the *ab initio* equation of state is consistent with the experiments.

CONCLUSIONS

Shock waves were induced in samples of beryllium by irradiation with the Trident laser, and measured using line-imaging VISAR and transient X-ray diffraction (TXD). No undesirable hydrodynamic effects of a laser drive, such as preheat, were observed, though some problems were encountered with plasma or hot electrons reducing the VISAR signal level. It was possible to analyze the TXD records, even though the rocking curve of the samples was relatively large at 2°, giving broad diffraction lines.

At the lower drive intensities, the VISAR records exhibited an elastic precursor followed by a plastic wave, with evidence of spall and ringing behind. At the higher drive intensities, there was some sign of a two-wave structure tentatively identified with the hex/bcc phase transition.

The TXD records showed the 0002 Bragg line moving with drive intensity. At the higher drive intensities, additional lines appeared in the Bragg and Laue records, again suggesting a phase transition. The location of the lines was reasonably reproducible between shots, indicating that the bcc structure formed at the same orientation to the hex in each experiment.

ACKNOWLEDGEMENTS

We would like to acknowledge the contributions of Dan Thoma and Jason Cooley for preparing the crystal samples, Bob Springer for measuring the rockin curves, Roger Kopp for performing LASNEX calculations, and the staff of Trident for their help in performing the experiments.

REFERENCES

1. Young, D.A., "Phase diagrams of the elements," University of California (1991).
2. S P Marsh (Ed), "LASL Shock Hugoniot Data", University of California (1980).
3. Paisley, D.L., Swift, D.C., Johnson, R.P., Kopp, R.A. and Kyrala, G.A., this conference.
4. Forsman, A.C. and Kyrala, G.A., Phys. Rev. E vol. 63, 056402 (2001).
5. Swift, D.C. and Ackland, G.J., *First principles equations of state for simulations of shock waves*, submitted to Phys. Rev. B (2001).

EXPERIMENT TO CAPTURE GASEOUS PRODUCTS FROM SHOCK-DECOMPOSED MATERIALS

W. H. Holt, W. Mock, Jr., F. Santiago, and R. M. Gamache

Naval Surface Warfare Center, Dahlgren Division, 17320 Dahlgren Rd., Dahlgren, VA 22448-5100

Abstract. Vacuum-tight steel containers have been designed and impact tested for the purpose of capturing gaseous products from the shock decomposition of porous polymer materials. This work extends earlier gas gun experiments in which initially-porous polytetrafluoroethylene (PTFE) powder specimens were shock compressed inside closed non-vacuum-tight steel containers and soft recovered. Although a powder decomposition residue was produced in the containers and analyzed *in situ*, there was no attempt to recover any gaseous decomposition products for analysis. A series of gas gun impact experiments has now been performed to develop a gas-tight specimen container that can survive impact shock loading. The container is evacuated prior to the experiment. The impact of a gas-gun-accelerated metal disk produces a shock wave that passes through the container wall and into the specimen material. If gaseous products are formed, they can be collected in a sample cylinder for subsequent chemical analyses. Initial results are presented for PTFE powder specimens.

INTRODUCTION

Improved understanding of the response of materials to mechanical shock loading, and in particular, the shock-induced decomposition of materials, can lead to extensions of existing models for more realistic simulations of material response to a wide range of shock loading conditions.

In our earlier experiments, initially-porous (44-47% porosity) polytetrafluoroethylene (PTFE) powder specimens were shock compressed at initial planar stresses of 1.5 and 0.72 GPa, respectively, inside closed but unsealed steel containers, and soft recovered (1,2). X-ray photoelectron spectroscopy analysis of the solid residue indicated significant unbound carbon and a variety of other chemical species. No attempt was made to capture gaseous products, although visual evidence of the expulsion of solid products from the containers suggested the presence of gaseous products. Morris, et al., (3) in experiments at ~50 GPa, have reported evidence of shock-induced dissociation of initially-solid PTFE, leading to CF_4 gas and a residue of amorphous carbon. Gas recovery experiments have also been performed on serpentine rock (4).

The purpose of the current experiments is to extend our earlier work to the include capture of the gaseous products.

EXPERIMENTAL TECHNIQUE

Figure 1(a) is a schematic of the experiment before impact. Figure 1(b) shows the experiment at the time of impact. The Naval Surface Warfare Center Dahlgren Division Research Gas Gun Facility (5) was used for this work. The specimen container is attached to a steel support block, located 0.8 m from the muzzle of the 40.0 mm smoothbore barrel. The impactor disk is carried on the end of a sabot. A thin mica cover on the muzzle permits evacuation of the barrel prior to firing the gun. The sabot and impactor disk pass through the mica cover and through a hole in the support block to impact the specimen container. A steel cover disk and momentum trap disks were used to reduce damage to the container.

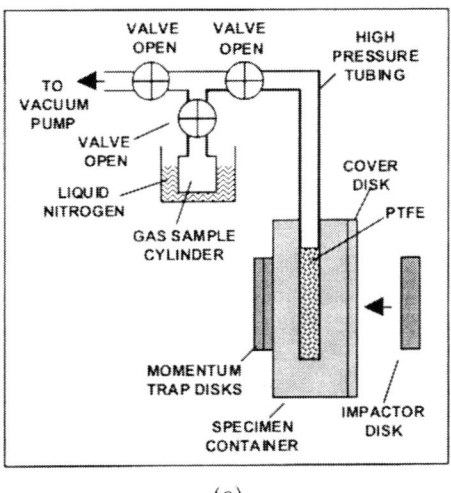

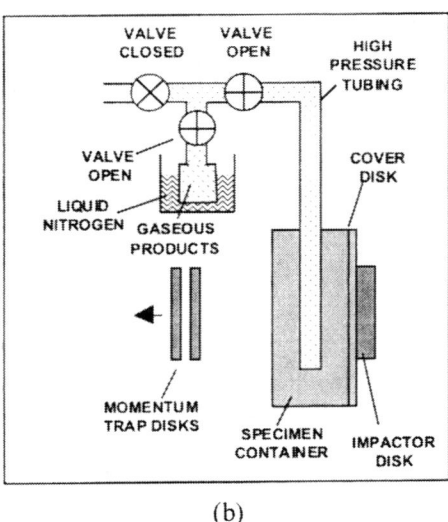

FIGURE 1. (a) Schematic of gas collection experiment before impact. The impactor disk is accelerated on a sabot in a gas gun. Impact of the disk produces a shock wave that passes through the cover disk and container wall and into the specimen material. The cover disk receives the primary damage of impact. The system is evacuated prior to impact, and the sample cylinder is maintained at liquid nitrogen temperature to minimize any subsequent reactions of collected gaseous products. All the tubing, fittings, and valves are made of 316 stainless steel. (b) Schematic of gas collection experiment at time of impact. The valve connected to the sample cylinder is closed after impact so that the captured gas sample can be removed for analysis.

The PTFE specimen material (6) was characterized with respect to initial particle morphology via optical and scanning electron microscopy, and was in the form of nearly spherical white particles having obvious surface substructures. The average particle size was 534 microns (7). In all the experiments, each specimen was pressed to have an initial porosity in the 44-47% range.

RESULTS

A series of experiments was performed with different specimen container designs to assess container survivability prior to attempting to capture product gases. The first container was a pair of 11.5-mm-thick stainless steel vacuum flanges (8) bolted together with a copper gasket seal and having a porous PTFE specimen disk (approximately 37mm diameter and 2.3mm thick) sandwiched between the flanges. A 4.6-mm-thick 6061-T6 aluminum disk impacted this container directly (no cover disk was used). Figure 2 shows one-half of the container before and after shock loading, with visual evidence of shock-induced chemical reaction

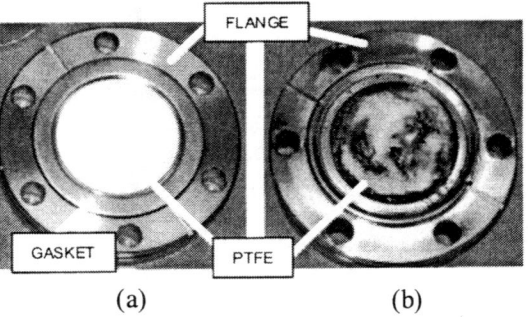

FIGURE 2. Specimen disk in one-half of container. (a) Before impact. (b) After impact. The dark regions indicate shock-induced chemical changes. Impact speed was 986 m/s.

in the specimen (dark regions). Because of damage to the bolts and to the copper gasket, this configuration did not retain a gas-tight seal.

Figure 3 shows parts for an experiment with an alternate specimen container design. To fabricate the container, a 4.6-mm-diameter hole was drilled radially into the center of a 76.1-mm-diameter,

28.6-mm-thick solid cylinder of 4130 steel. A stainless steel tube fitting was welded into the hole to provide a gas connection. PTFE powder was pressed into the hole. For this experiment a 3.1-mm-thick, 35.6-mm-diameter OFHC copper disk impacted the 9.53-mm-thick steel cover disk at 976 m/s. This container survived impact and appeared to be free of cracks affecting the specimen cavity. A residue of partially-reacted PTFE was found in the cavity.

Another container design consisted of a 316 stainless steel tube (9.7 mm outside diameter, 7.5 mm inside diameter) that fits through a centered diametric hole in a solid cylinder of 4130 steel. A 9.5-mm-thick steel cover disk was placed against the flat end of the 31.9-mm-thick solid cylinder. The tube is deformed inside the solid cylinder, compressing the specimen material.

Figure 4 shows this container design mounted near the gas gun muzzle for impact. The gas collection system (Figure 1) was attached to this configuration in an attempt to capture gaseous products. The pressed porous specimen filled a

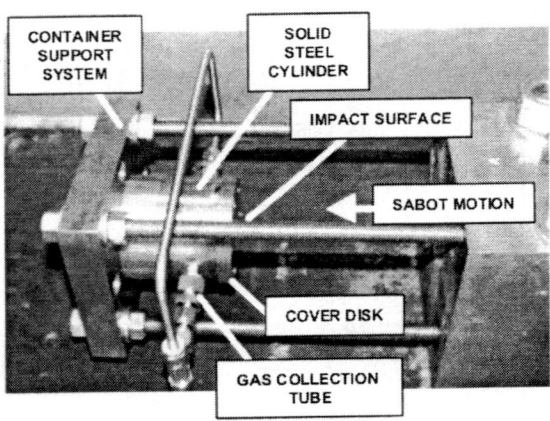

FIGURE 4. Cylinder and tube container mounted on the support system for connection to the gas collection system and impact. All the tubing and fittings were made of 316 stainless steel.

25-mm-long region of the gas collection tube and was centered with respect to the axis of the gas gun barrel. Connection was made to both ends of the straight tube to facilitate evacuation of the specimen region. The container was evacuated by a turbomolecular pumping system to 1.5×10^{-6} Torr, measured by an ion gauge near the pump. The gas collection system was baked at ~125°C, while the specimen container was kept at ~25°C to minimize thermally-induced changes in the specimen. A 3.1-mm-thick OFHC copper disk impacted the configuration at 977 m/s. As shown in Figure 1, product gases were collected in a sample cylinder (9) that was immersed in liquid nitrogen.

Figure 5 shows the container after impact. The impact side was deformed but had no visible fractures. The post-impact pressure in the system was in the 10^{-4} Torr range. The tube contained a residue that was partially-reacted PTFE.

Preliminary mass spectrometric analyses of the captured gas from this experiment have been

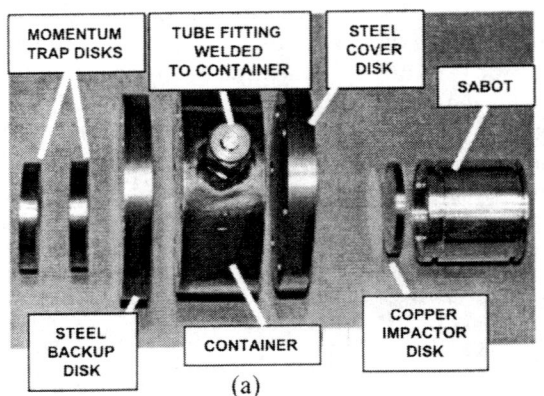

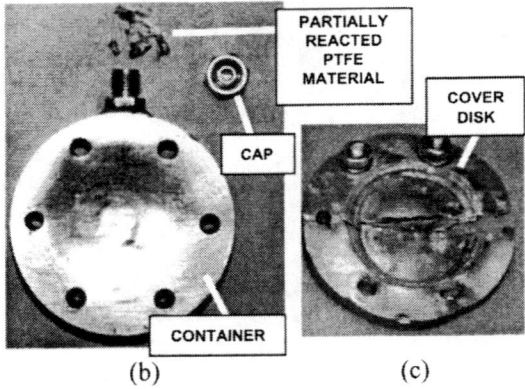

FIGURE 3. (a) Parts for experiment before impact. (b) Deformed container after impact. This container had one visible crack near the edge that did not affect the specimen cavity. (c) Deformed and fractured cover disk for container. Impact speed was 976 m/s.

performed using a residual gas analyzer (RGA). A background spectrum was first obtained for the RGA at 10^{-6} Torr. When small samples of the captured gas were leaked into the RGA, the spectra showed a variety of additional peaks, mostly corresponding to higher masses than were observed in the background spectrum. These measurements were repeated eight times with consistent results. One of the strong peaks in each of the sample spectra corresponded to the mass for trifluoromethane (CHF_3), which has a boiling point of -84°C (1 atm) (10). These results are considered preliminary since the post-impact solid residue appeared to be only partially reacted. Additional experiments are planned to include higher impact stresses and longer stress durations, along with analyses using a higher resolution mass spectrometer.

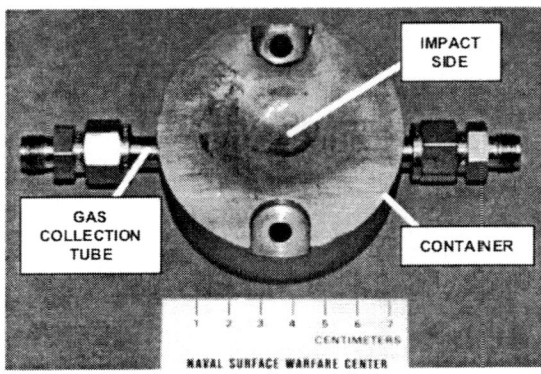

FIGURE 5. Cylinder and tube container after impact. The tube was deformed but not fractured and remained vacuum-tight. Impact speed was 977 m/s.

SUMMARY

A series of impact experiments has been performed using different specimen container designs to develop a configuration that would permit capture of product gases from shock-decomposed materials. For one of the experiments a gas collection system was connected to the specimen container. Gaseous products were collected in an evacuated sample cylinder. Preliminary mass spectrometric analyses were performed using a residual gas analyzer. One of the observed strong spectral lines corresponded to the mass of a gaseous fluorine compound, trifluoromethane (CHF_3).

ACKNOWLEDGEMENT

This work was supported by the In-House Laboratory Independent Research Office of the Naval Surface Warfare Center, Dahlgren Division.

REFERENCES

1. Mock, W. Jr., Holt, W. H., and Kerley, G. I., "Shock and Recovery of PTFE Powder," in *Shock Compression of Condensed Matter-1997*, edited by S. C. Schmidt et al., AIP Conference Proceedings 429, New York, 1998, pp. 671-674.
2. Holt, W. H., Mock, W. Jr., and Santiago, F., *J. Appl. Phys.* **88**, 5485-5486 (2000).
3. Morris, C. E., Fritz, J. N., and McQueen, R. G., *J. Chem. Phys.* **80**, 5203-5217 (1984).
4. Tyburczy, J. A., Krishnamurthy, Epstein, S. and Ahrens, T. J., *Earth Planet. Sci. Lett.* **98**, 245-260 (1990).
5. W. Mock, Jr. and W. H. Holt, *The NSWC Gas Gun Facility for Shock Effects in Materials*, NSWC/DL TR-3473, Naval Surface Weapons Center, Dahlgren, VA 22448, July 1976.
6. The material was DuPont Type 8B Teflon PTFE powder, purchased from E. I. du Pont de Nemours and Company, Wilmington, DE 19898. Teflon is a registered trademark of DuPont.
7. Average particle size information provided by Dupont Certification Office, P.O. Box 1217, Parkersburg, WV 26102-1217.
8. Varian Conflat® Flange, 316 stainless steel, Part No. F02750000NC6. Varian Vacuum Technologies, 121 Hartwell Avenue, Lexington, MA 02421-3133.
9. Whitey Miniature Sample Cylinder, Single Ended, No. SS-4CS-TW-25. Whitey Co., Highland Heights, OH 44143-1490.
10. *The Condensed Chemical Dictionary, 9th Ed.*, edited by G. G. Hawley, Van Nostrand Reinhold Company, New York, 1977, p.391.

SOUND VELOCITY DOPPLER LASER INTERFEROMETRY MEASUREMENTS ON TIN

E. Martinez - J-M. Servas

Commissariat à l'Energie Atomique
Centre d'Etudes de BRUYERES-LE-CHÂTEL
B.P. 12 - 91 680 - BRUYERES LE CHÂTEL – France

Abstract. Measurement of the sound velocity behind shock waves can be used as a sensitive probe for detecting phase transition. In order to validate an experimental set-up and to verify the evolution of the sound velocity versus pressure, we performed an experimental program on tin. By use of Doppler Laser Interferometry (D.L.I.) measurements, we have measured the sound velocity in shocked tin at twelve points along its Hugoniot in the pressure range 2 to 33GPa. The shock generator is a 60mm diameter powder gun.

INTRODUCTION

The sound velocity is an important physical parameter which evolves with changes in structural properties. The velocity increases with pressure until shock heating causes phase transition. The break occurring in the sound velocity versus pressure is attributed to structural change. Thus, phase transitions along the Hugoniot can be observed through sound wave velocity measurements.

In order to validate an experimental set-up and to verify the evolution of sound velocity versus pressure, we performed an experimental program on tin. By use of D.L.I. measurements, we have measured the sound velocity in shocked tin at twelve points along its Hugoniot in the pressure range 2 to 33GPa.

In this paper, we describe our experimental set-up using D.L.I. measurements and we give the experimental results on the tin sample.

PRINCIPLE

To generate the shock, a flyer plate, accelerated in a powder gun to velocities up to 1600m/s, hits a tin target. After the plate impact, a shock propagates forward in the target and backward in the flyer. A rarefaction wave originating at the back surface of the projectile overtakes the shock wave. The sound velocity C_L is defined as the velocity of the leading release wave (Fig. 1).

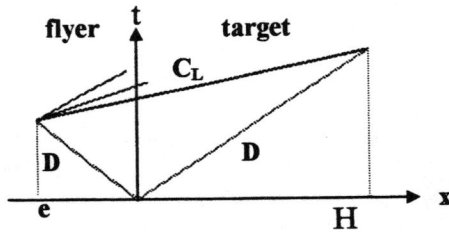

Figure 1 : Lagrangian diagram. The impactor and the target are of the same material

For a symmetrical impact, the impactor and the target are of the same material, and the sound velocity C_L is calculated taking into account:

- the flyer thickness e,
- the shock velocity D,
- and the overtake distance H.

$$C_L = D \cdot (H+e)/(H-e)$$
and $$C_{EL} = C_L \cdot \rho/\rho_0$$

where D, shock velocity,
e, flyer thickness
ρ_o, initial density,
ρ, density after impact.

From this expression, we note that the determination of the sound velocity relies on the overtake distance measurement. Then, to determine the overtake distance H, we have adapted and applied a shock and rarefaction overtake method described by McQUEEN, FRITZ, MORRIS (1.) and ASAY, CHHABILDAS (2.). The time between the shock and the arrival of the release wave is proportional to the target thickness (Fig.2). Then, using a stepped-target, the determination of the overtake distance can be realised (Fig.3).

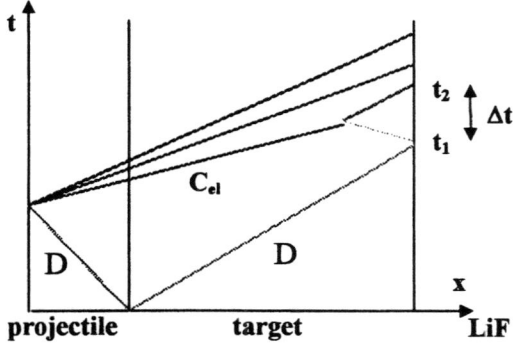

Figure 2: Lagrangian diagram

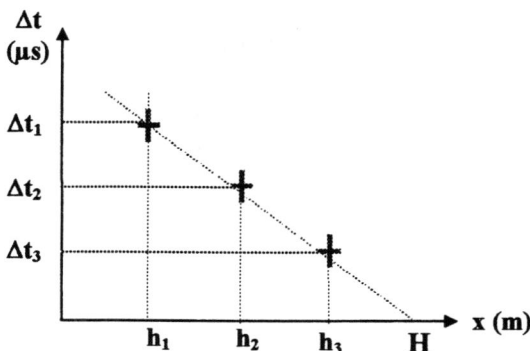

Fig. 3: Determination of the overtake distance H.

EXPERIMENTAL SET-UP AND DIAGNOSTICS

Considering that tin presents a phase transition solid-solid under shock loading in the range 5 to 10GPa (3.) and the maximum velocity of the flyer is 1600m/s, we have developed two set-ups to fulfil our requirements:
- a symmetrical set-up: both impactor and target are of the same material. It enables us to reach pressures up to 20GPa. (Fig. 4)
- an unsymmetrical set-up: to reach pressures up to 33GPa, we used tantalum impactors.(Fig. 5, 6)

The launcher diameter is 60mm and the target has three steps of different thicknesses. A window material, the Lithium fluoride (LiF) is attached to these steps.

The measured experimental parameters are:

- the particle velocity,
- the impactor velocity,
- and the shock velocity.

The velocity histories at the interface between the sample and the transparent window are measured using D.L.I. technique. Shock pressure is established by measuring the projectile velocity with D.L.I., and piezoelectrics gauges. Shock velocity is determined using piezoelectric gauges

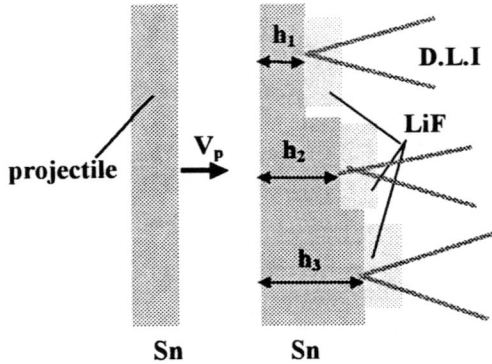

Figure 4 : Schematic representation of the symmetrical set-up.

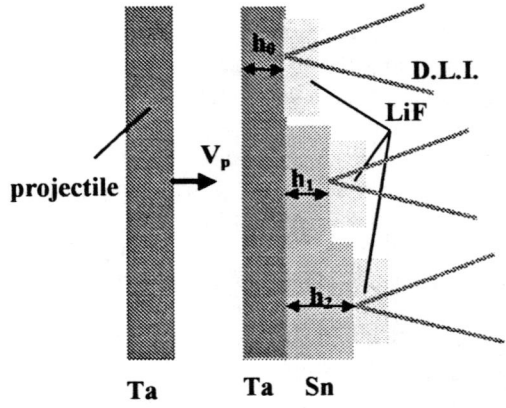

Figure 5 : Schematic representation of unsymmetrical set-up.

EXPERIMENTAL RESULTS

Twelve experiments were performed in the range pressure 2 to 33GPa: seven, using symmetrical impact and five, using unsymmetrical set-up.

Calculations were performed using RANKINE-HUGONIOT relations.
- For symmetrical impact, H is determined from overtake method, D is measured and the particle velocity is deduced from projectile velocity measurement.

- For unsymmetrical impact, the particle velocity is calculated from measurement of shock velocity in tin and HUGONIOT of impactors. the sound velocity is calculated as follows :

$$C_{LSn} = \frac{D_{Sn}.H}{H - D_{Sn}.\Delta t_0}$$

and

$$C_{el} = \frac{(D_{Sn} - U_{Sn}).H}{H - D_{Sn}\Delta t_0}$$

where D_{Sn}, shock velocity in tin
U_{Sn}, particle velocity
Δt_0, transit time on the base plate
H, overtake distance in tin.

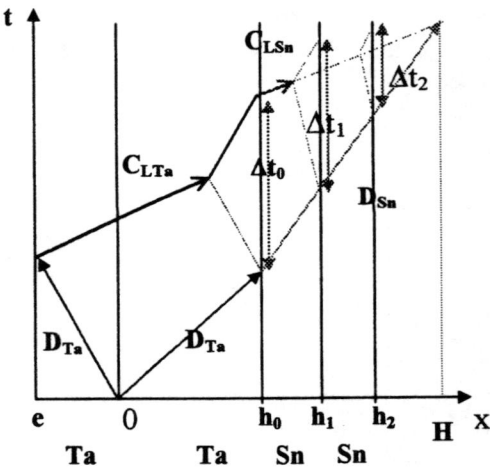

Figure 6 : Lagrangian diagram using unsymmetrical set-up

Tables 1 and 2 summarizes the results obtained.

Shot	Vp (m/s)	e (mm)	H (mm)	P (GPa)	Cel (m/s)
S1	225	2	15,2	2,4	3605
S2	434	2	15,2	5,0	3861
S3	574	2	13,4	6,6	3775
S4	620	2	17,0	6,9	3658
S5	972	2	15,8	11,3	3963
S6	1180	2	11,2	14,2	3939
S7	1490	2	11,1	19,8	4207

Table 1 : Results obtained using symmetrical set-up.

The experimental results are presented in figure 7.

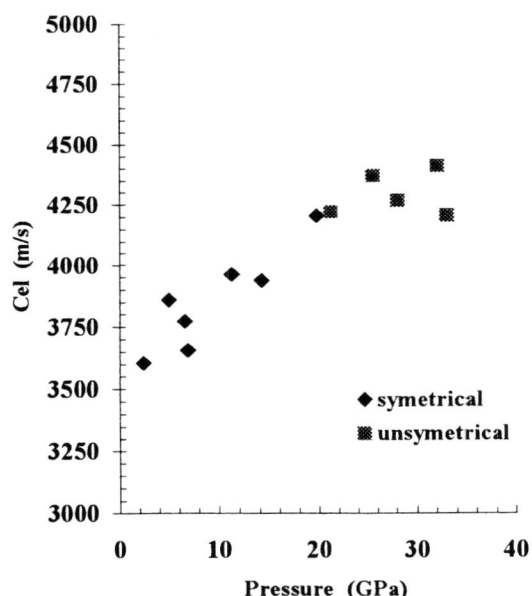

Figure 7 : Diagram (P_{choc}, C_{el})

Shot	Vp (m/s)	E (mm)	H (mm)	P (GPa)	Cel (m/s)
U1 Ta/Ta	1145	2	8,4	21,3	4222
U2 Ta/Ta	1270	2	4,7	25,6	4372
U3 Ta/Ta	1458	2	6,8	28,1	4269
U4 Ta/Ta	1524	2	10,6	32,1	4411
U5 W/W	1476	2	9,46	33,1	4207

Table 2 : Results obtained using unsymmetrical set-up.

CONCLUSION

The realized measurements validate the experimental method. We observed a break in the sound velocity versus pressure in the 5-10GPa pressure range. The accuracy of the sound velocity measurements is approximately 5%. This precision depends on the number of step and on the uncertainly in determining transit times (overtake distance H). In order to determine the phase transition, this experimental program requires some more experiments, especially in the pressure range (5-10 GPa).

REFERENCES

1. Mc Queen, R.G., Fritz, J.N. and Morris, C.E.
« The velocity of sound behind strong shock in 2024 Al » *in Shock waves in condensed matter - 1983*

2. Asay, J.R., and Chhabildas, L.C.
Shock waves and high strain rate phenomena in metals. *Meyer and Murr Eds, Plenum Publishers 1981*

3. Elias, P., Chapron, P., Laurent, B.
« Detection of melting in release for a shock-loaded tin sample using the reflectivity measurement method. ». *Opt. Comm., Vol.66, Nber 2,3 (15 April 1988), pp. 100-106.*

PROJECTILE ACCELERATION AIMING AT VELOCITIES ABOVE 9 km/s BY A COMPACT GAS GUN

Tatsumi Moritoh, Nobuaki Kawai, Kazutaka G. Nakamura and Ken-ichi Kondo

*Materials and Structures Laboratory, Tokyo Institute of Technology,
Nagatsuta, Midori, Yokohama, Japan*

Abstract. Optimization conditions were studied for the conventional two-stage light-gas gun, which is a part of a three-stage light-gas gun having a preheating and filling stage. A firing test achieved a velocity of 6.2 km/s using helium as driver gas with a projectile weighing 0.6 gram. In the case of firing test using hydrogen with the same projectile mass and a piston weighing 430 gram, a velocity of 8.1 km/s was obtained. Moreover, it was found that higher velocity can be achieved using a lighter piston to increase the piston velocity. A velocity of 8.9 km/s was achieved using a piston weighing 290 gram.

INTRODUCTION

A two-stage light-gas gun is one of the most important tools for shock compression studies[1,2] because of its abilities to generate plane shock-waves. However, conventional two-stage light-gas guns are limited to launch heavy projectile at velocities of 7-8 km/s. In order to perform shock compression studies in the region of terapascals pressure by using this gun, a projectile velocity more than 9 km/s is necessary even for a symmetric impact of platinum.[3] Recent developments of laser-shock technologies have made it possible to generate pressures above terapascals.[4] However, a gas gun launching at low acceleration without changing the initial state of flyer materials will be required even in future. In 1995, we proposed a non-destructive type of three-stage light-gas gun having an additional preheating and filling stage which allows us to regulate initial temperature of driver gas in order to obtain higher projectile velocities.[5] In preliminary tests, helium and hydrogen were preheated to 1080 K and 500 K, respectively by using a 9-kg pre-pump piston. In order to examine effects of preheating on projectile velocity, it is necessary to determine the performance of this gun as a normal two-stage light gas gun and to know the relationship among operating parameters in the non-destructive experiments. In this paper, we report optimization

TABLE 1. Specification of The Compact Two-Stage Light-Gas Gun Used In This Work as Compared With Those of NASA Gun and LLNL Gun.

	LLNL gun	NASA gun	TITECH gun
Pump tube length (m)	10	15.18	4.2
Accelerating reservoir length (mm)	445	343	370
Launch tube length (m)	9	3.87	3.58
Pump tube diameter (mm)	90	64.4	50
Launch tube diameter (mm)	28	12.7	11.8
Piston mass (g)	4540	888	290, 430

studies of the compact two-stage light-gas gun which is a part of the three-stage light-gas gun. We examined the following five operating parameters: (1) powder mass, (2) kinds of light gas used as driver gas, (3) initial pressure of driver gas, (4) projectile mass, (5) piston mass. In particular, this paper describes effects of piston mass on projectile velocities and importance of optimizing initial gas pressure.

EXPERIMENTAL

Specifications of the compact two-stage light-gas gun developed in this work are given in Table 1 as compared with those of NASA gun and LLNL gun.[6] All the units of the gun are mounted on a considerably heavy steel base (10.5-m long and 0.9-m wide). This gun is very compact for a two-stage light-gas gun, which can be placed in a small experimental room. In particular, the length of the pump tube is much less than the others listed in Table 1, so that the purpose of this experiment is to raise the temperature of driver gas in such a short pump tube. Pistons are made of high-density polyethylene, having a diameter of 50-mm. Two types of piston with different weights were examined: one was 430 g with a 200 mm length, and the other was 290 g and 165 mm length. Smokeless powder (Nippon Oil & Fat Co, NY-500) was used as propellant for driving the piston. Projectile velocity was measured by Magnetoflyer method.[7] Polycarbonate was used as the sabot material, and a small magnet was embedded in the sabot for projectile velocity measurement. The sabot is 11.85 mm in diameter, and its mass was varied from 0.6 to 2.6 g by varying its length from 4 mm to 12 mm, respectively.

RESULTS AND DISCUSSION

Firing Tests Using Hydrogen As A Driver Gas

The maximum velocity achieved in tests using helium as a driver gas was 6.2 km/s, but erosion of the gun barrel was so severe. In order to obtain higher projectile velocities, we performed firing tests using hydrogen, whose sound velocity is higher than that of helium at the same temperature. Initial pressure of hydrogen was initially fixed at 5 atm, and mass of powder was increased gradually from 40 g to prevent the apparatus from being damaged. A firing test achieved a velocity of 6.6 km/s using a projectile weighing 0.8 g and 95 g of powder. This velocity is higher than the maximum velocity in the case of helium, and gun erosion was not detected.

Consequently, we performed tests to obtain higher velocities by changing initial gas pressure and powder mass, monitoring projectile velocity and energy efficiency (E), which is the ratio of kinematic energy of projectile to combustion energy of the powder,

$$E = \frac{\frac{1}{2}M_{pro}v^2}{M_{pow}Q}, \quad (1)$$

where M_{pro}, M_{pow}, Q, and v are projectile mass (kg), powder mass (kg), combustion energy per unit mass of powder (3.38×10^6 J/kg), and projectile velocity (m/s), respectively.

Results of the firing test using hydrogen at a piston weight of 430 g are shown in Fig. 1. Energy efficiency was better at the condition of lower gas pressure. Although the projectile velocity is higher than that of heavy projectiles, energy efficiency decreases in the case of a light projectile. In these tests, a projectile velocity of 8.1 km/s was achieved by optimization to an initial hydrogen pressure with a 0.6 g projectile.

When changing powder mass and initial gas pressure, we were careful to prevent the piston from invading into the launch tube in all firing tests by checking the stop position of the piston after firing. The stop position after the firing test that we obtained a velocity of 8.1 km/s was so close to the launch tube that we chose to stop increasing powder or decreasing initial gas pressure, in order to prevent excess momentum of the piston resulting in damage to the launch tube.

Effect of Piston Mass on Velocities of Projectile

In order to decrease the momentum of the piston but increase the kinetic energy of the piston, we carried out several firing tests using a lighter piston to optimize both powder mass and initial gas pressure. Pistons weighing 290 g were also prepared. Piston velocities were measured by three strain gauges fixed on the outside wall of the pump tube at the same intervals, as shown in Fig. 2. Figure 2 shows that the rise in the velocities of 290 g pistons is much greater than those of 430 g pistons. Figure 3 shows energy efficiency as a function of

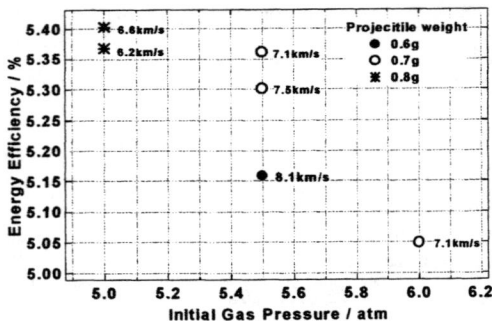

FIGURE 1. Changes in energy efficiency at three different projectiles vs initial hydrogen pressure. Projectile velocities are also shown in the figure.

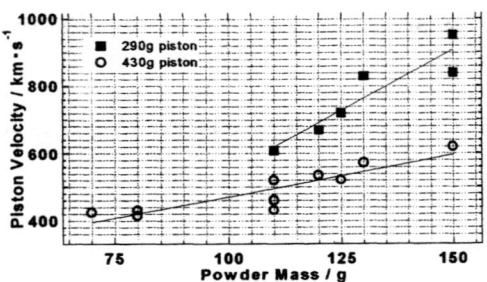

FIGURE 2. Piston velocities measured vs powder mass as a function of piston mass.

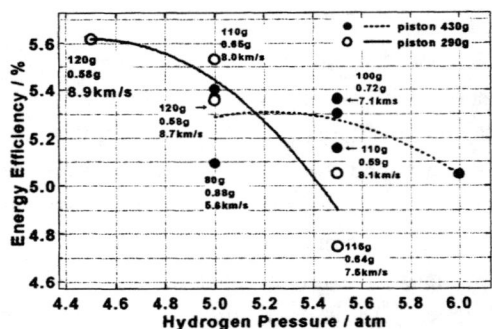

FIGURE 3. Energy efficiency vs initial hydrogen pressure as a function of piston mass. Powder mass, projectile mass, and projectile velocity are shown from the top to the bottom in the vicinity of each symbol, respectively.

hydrogen pressure and piston mass. It was possible to use a larger powder mass than those for a 430 g piston without any damage. When the initial gas pressure was lower, energy efficiency increased with 290 g pistons. Moreover, higher velocities were achieved under conditions of lower initial gas pressure and energy efficiency was better. It is necessary to decrease the initial pressure of driver gas in order to accelerate the projectile efficiently in the case of a light piston. The maximum projectile velocity achieved so far was 8.9 km/s using piston weighing 290 g.

SUMMARY

We performed optimization studies on the compact two-stage light-gas gun having a short pump tube. It was found that velocities up to 8.9 km/s were obtained by using hydrogen as driver gas and a light piston for such a short launcher. The use of a light piston makes it possible to increase powder mass without any damage of apparatus. The projectile is accelerated efficiently to higher velocities. We believe that this is because the lighter piston with higher velocity creates stronger shock-waves in the driver gas, resulting in higher gas temperature. It is necessary to optimize initial gas pressure, which strongly affects the energy efficiency. It will be possible to achieve higher velocities by further optimization of parameters such as thickness of the diaphragm and inner diameter of the launcher.

ACKNOWLEDGEMENT

This work has been supported by CREST (Core Research for Evolutional Science and Technology) program organized by Japan Science and Technology Corporation (JST). We thank M. Hasegawa for his experimental help.

REFERENCES

1) A. H. Jones, W. M. Isbell, and C. J. Maiden, J. Appl. Phys. **37**, 3493(1966).
2) N. C. Holmes, J. A. Moriarty, G. R. Gathers, and W. J. Nellis, J. Appl. Phys. **66**, 2962(1989).
3) S. P. Marsh, LASL Shock Hugoniot Data, Univ. Of Calif. (1980).
4) D. Batani, A. Balducci, D. Beretta, and A. Bernardinello, Phys. Rev. B **61**, 9287(2000).
5) K. Kondo, O. V. Fat'yanov, Y. Hironaka, T. Moritoh, and S. Ozaki, *Shock Compression of Condensed Matter-1999*, edited by M. D. Furnish, L. C. Chhabildas, and R. S. Hixson, (AIP, New York, 2000) pp. 1167.
6) L. A. Glenn, *Shock Waves in Condensed Matter-1987*, edited by S. C. Schmidt and N. C. Holmes, (Elsevoir, Amsterdam, 1988) pp. 653.
7) K. Kondo, A. Sawaoka, and S. Saito, Rev. Sci. Instrum. **48**, 1581(1977).

CHARACTERIZATION OF IMPACT IN COMPOSITE LAMINATES

Karel Minnaar and Min Zhou[†]

School of Mechanical Engineering Georgia, Institute of Technology, Atlanta, GA 30332-0405

Abstract. A new experimental technique is developed to determine the onset and evolution of delamination in fiber-reinforced composites. The configuration uses a split-Hopkinson bar for low-velocity impact loading and two Polytec laser vibrometer systems for real-time monitoring of the initiation and progression of delamination. The experiment allows the histories of load, displacement, and velocity of impacted specimens to be recorded and analyzed. Numerical simulations are conducted using a cohesive finite element method. The method employs a cohesive zone model to simulate in-ply cracking and interlaminar delamination and a transversely isotropic, elastic model to characterize the bulk behavior of each ply. The simulations provide time-resolved characterization of damage during the impact loading. The time at which delamination is detected decreases as the impact velocity is increased, and delamination is detected at similar surface displacements. The progression of damage changes as the bonding strength between plies is increased. The speed of delamination decreases as the bonding strength is increased.

INTRODUCTION

Fiber-reinforced polymeric (FRP) composite laminates are susceptible to damage due to transverse impact. Such damage can lead to significant reduction in the strength and stiffness of the material. While mechanisms of impact damage are relatively well understood, there is still a lack of experimental tools that allow time-resolved analysis of damage initiation and progression. Recently, Hallett (2000) used a modified Split Hopkinson bar apparatus and high-speed photography to correlate the failure of impacted beams to abrupt changes in measured deflection. Also, acoustic emission techniques have been combined with microscopic observations to continuously monitor damage growth (Benmedakhene et al., 1999).

The first objective of the current investigation is to develop an experimental approach for time-resolved analysis of impact response and damage development. To this end, a split Hopkinson bar apparatus is used for loading and for the determination of histories of applied load, contact point velocity displacement, and mechanical work. The Hopkinson bar setup used an incident bar for loading and no transmitter bar is used. Further, a system of two laser interferometers is used to obtain differential surface velocity and displacement measurements at opposites sides on an impacted specimen. This combination of diagnostics offers a novel capability that allows real-time detection of the onset and progression of interlaminar delamination along with time-resolved analysis of full impact response.

The cohesive finite element method (CFEM) provides a unique and powerful tool for analyzing damage and fracture in a material through explicit account of fracture. Since the discrete model possesses the attributes of both deformation inside elements and separation along embedded cohesive

---

[†] To whom all correspondence should be addressed, 404-894-3294, min.zhou@me.gatech.edu.
Citation: Minnaar, K. and Zhou, M., 2001, " Characterization of Impact in Composite Laminates", Conference Proceedings of the APS Topical Conference on Shock Compression of Condensed Matter, Atlanta, GA, June 24-29.

surfaces, fracture is an inherent attribute of the mode and this approach does not require any crack initiation and propagation criteria.

The cohesive zone formulation allows fracture to evolve as a natural outcome of the combined effects of bulk constituent response, interfacial behavior, and applied loading. The second objective of this paper is to use the CFEM to develop a framework for analyzing impact response and impact-induced damage for composite laminates.

EXPERIMENTAL SETUP

The material tested is a [0°/90°/0°] laminate made from NCT-301-1G150(50K) epoxy impregnated tape. The specimen is a rectangular strip 75 mm × 18 mm in size and 3 mm in thickness. Its two ends are placed on two roller pins. The specimen is impacted at the center by a shaped indentor. A starter crack with a width of 0.3 mm is machined across the center of the longitudinal ply and extends to the transverse ply. Two Polytec laser vibrometers are used to measure the surface velocities at one point on the front surface (impact side) and one point on the back surface of the impacted specimen. The lasers are aligned such that the two beams are co-linear and perpendicular to the specimen surface before impact.

The lasers are aligned at a distance L from the center of the specimen. The difference between the surface displacements, $\delta = D_2 - D_1$, is used to detect delamination, see Fig. 1. Where D_1 is the displacement of the impacted surface and D_2 is the displacement of the rear surface. Positive displacement is defined to be in the direction of impact for both impacted and rear surfaces. Upon impact, the starter crack initiates a matrix crack in the transverse ply that propagates in a direction perpendicular to the ply interfaces. The matrix crack propagates towards the impacted surface until it reaches the 0°/90° ply interface where delamination initiates. The delamination propagates as a mode I crack away from the impact site. The response from an experiment with an impact velocity of approximately 6.7 ms^{-1} is shown in Fig. 2. The interferometer was placed 11.71 mm from the impact site. The surface displacement and differential displacement histories are shown. There is a significant non-zero signal in the relative displacement at 0.8 ms indicating that the

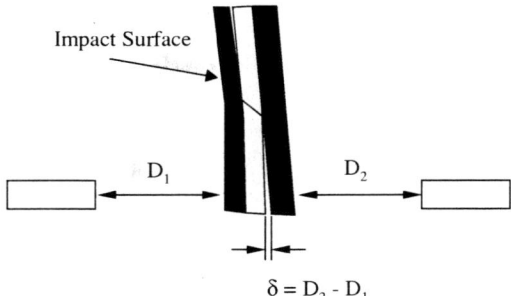

FIGURE 1. Detection of Delamination

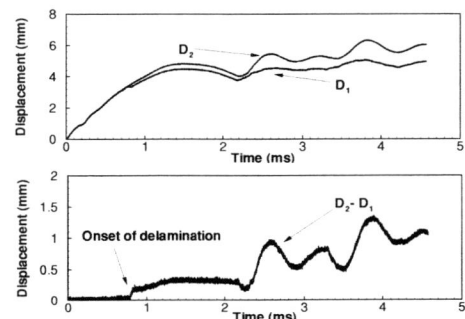

FIGURE 2. Surface displacement

delamination front has reached the point of measurement. An increase in impact velocity decreases the time at which delamination is detected at a constant distance away from the impact site. The time of detection decreases from 3.0 ms to 0.8 ms when the impact velocity is increased from 3.5 ms^{-1} to 6.7 ms^{-1}, see Fig. 3. This indicates that the average delamination speed increases as the impact velocity is increased. From the response of a speci-

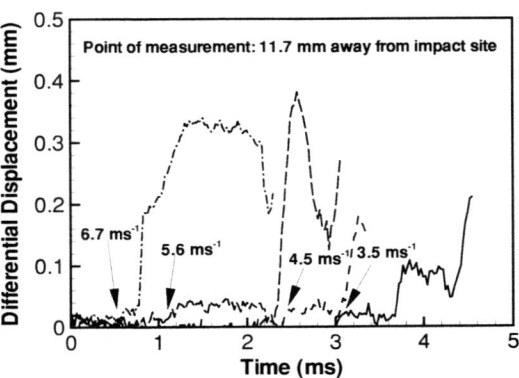

FIGURE 3. Time of onset of delamination

imen, it is possible to record D_1 at the time at which delamination is detected. In Fig. 4, D_1 at the time of detection is plotted as a function of impact velocity for various values of L. It can be seen that delamination is detected at similar surface displacements for the impact velocities shown.

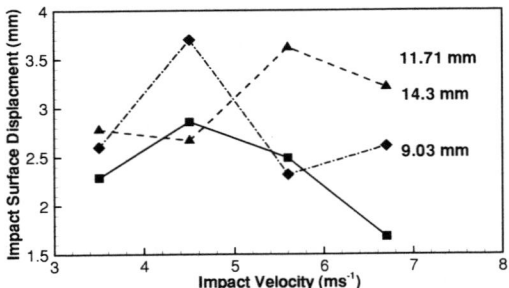

FIGURE 4. Surface Displacement at onset of delamination

NUMERICAL SIMULATION

Numerical simulations of the experiment are carried out using a CFEM. Each laminate is considered to be homogeneous and transversely isotropic. The fracture model is based on a cohesive surface formulation of Xu and Needleman (1994) and represents a phenomenological characterization for atomic forces on potential crack surfaces. Damage evolution is characterized using the time history of crack length. In the current study, a bilinear cohesive law implemented by Zhai and Zhou (2000) is used to describe the constitutive traction-separation relationship. Four material parameters are required to describe the cohesive relation: the original tensile strength of the interface (T_{max}^n), the original shear strength of the interface (T_{max}^t), critical normal separation (Δ_{nc}), and the critical tangential separation (Δ_{tc}). Discretization of the specimen is based on triangular elements arranged in uniform "cross-triangular" quadrilaterals. Cohesive elements are placed between all element boundaries to model fracture in the forms of interply cracking and delamination. The cohesive parameters assigned to a cohesive surface pair depend on its location and orientation. There is also a distinction between cohesive surfaces located on the interfaces between adjacent laminates and cohesive elements located within a particular laminate layer. It is assumed that T_{max}^n varies between $E_{22}/5$ to $E_{11}/10$ depending on the orientation of the ply, where E_{22} and E_{11} are the Young's Modulus of the composite in the transverse and longitudinal directions respectively. A nominal strength value of $E_{22}/20$ is assumed for the interface between two plies. Reported values of mode I and mode II energy release rates (G_{Ic} and G_{IIc}) are used to estimate the characteristic length Δ_{nc}. It is assumed that G_{Ic}/G_{IIc} is constant and $\Delta_{nc} = \Delta_{tc}$.

PROBLEM ANALYZED

Figure 5 shows the configuration used. This configuration contains a small starter crack in the lower longitudinal ply at the center of the beam that extends to the transverse ply. This configuration is chosen to study Mode I (opening mode) delamination growth. A constant impact velocity of 6.7 ms^{-1} is applied at the center of the upper surface of the beam. A 2D plain strain formulation is used.

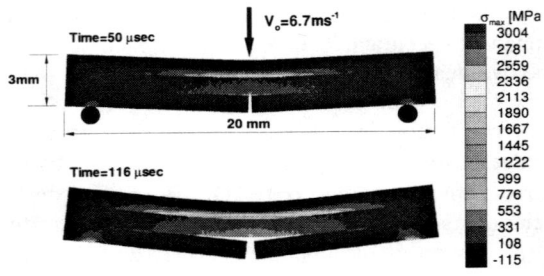

FIGURE 5. Weak interface damage progression

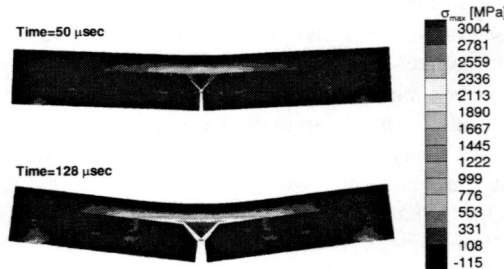

FIGURE 6. Strong interface damage progression

RESULTS AND DISCUSSION

Currently, there exists no direct measurement technique to determine the cohesive model parameters. The interface strengths are varied to understand how the strength values influence the simulation. The evolution of deformation and failure is compared to the experiment. Figures 5 and 6 show the progress of failure in an impacted specimen with the distribution of σ_{max} superimposed on the deformed configurations. The progression of damage changes as the interface strength is increased by 50% from the nominal values of $T^n_{max} = 400 MPa$ and $T^t_{max} = 500 MPa$. The results show that, delamination along the lower interface starts at the tip of the starter crack and growths away from the point of impact. In the case of the stronger interface (Fig. 6), the inply crack begins at the tip of the starter crack and propagates to the upper interface. Delamination along the upper interface starts at this point and there is no delamination in the lower interface. This progression of damage at the higher interface strength value compares well with the experiments. Figure 7 shows the peak maximum delamination speed reduces from 450 ms^{-1} to 230 ms^{-1} as the interface strength is increased.

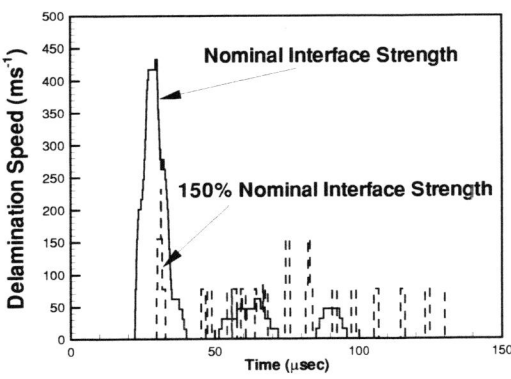

FIGURE 7. Delamination Speed

CONCLUSION

Experimental and numerical studies on the deformation and failure of fiber-reinforced structural composites subjected to low-velocity impact are being conducted. An experimental technique for real-time monitoring of delamination progression in composite laminates has been developed. A framework for the simulation of the impact deformation and damage based on a cohesive finite element method is presented. The model allows fracture in the forms of interply cracking and delamination to be tracked individually. Calculated damaged modes and progression of failure agree qualitatively with experimental observations. The damage patterns are controlled by the interface bonding strength and delamination speed decreases as the bonding strength increases. The onset and progression of delamination is not sensitive to change in impact velocity in the range tested.

ACKNOWLEDGEMENT

Support of this work by the Office of Naval Research through grant N00014-99-1-0799 to Georgia Tech (Scientific Officer: Yapa D. S. Rajapakse) is gratefully acknowledged.

REFERENCES

Benmedakhene, S., et al., 1999, Comps. Sci. and Tech., Vol. 59, pp. 201-208.

Hallett, S.R., 2000, Comps. Sci. and Tech., Vol. 60, pp. 115-124.

Xu, X.-P. and Needleman, A., 1994, J. Mech. Phys. Solids, Vol. 42, pp. 1397-1434.

Zhai, J. and Zhou, M., 2000, Intl. Journal of Fracture, pp. 161-180.

Erratum:
FRICTION IN HIGH-SPEED IMPACT EXPERIMENTS

Robert A. Pelak, Paul Rightley, and J. E. Hammerberg

Los Alamos National Laboratory, Los Alamos NM 87545 USA

The torsional strains calculated in our paper are incorrect in that the results presented are too small by a factor of two. The surface strain shown in Figure 3 and the coefficients of friction shown in Table 1 should therefore be multiplied by a factor of two. The corrected versions are reproduced below.

The original paper appeared in *Shock Compression of Condensed Matter – 1999*, M. D. Furnish, L. C. Chhabildas, and R. S. Hixson (eds), AIP Conference Proceedings Volume 505, American Institute of Physics, Melville, New York, 2000, pp. 1221-1224.

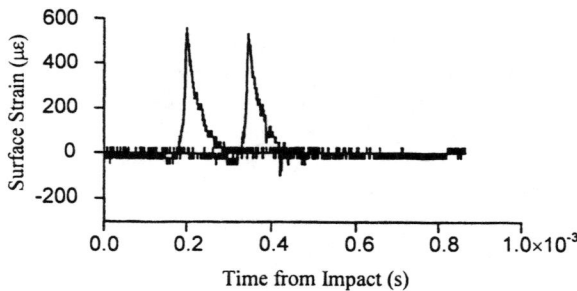

Figure 3. Torsional strain traces from impact, forward and rear gauges.

TABLE 1. Measured estimates of the coefficient of friction.

Impact Vel.	Max. Sliding. Vel.	μ
1095 m/s	148 m/s	0.110
1023 m/s	138 m/s	0.146
897 m/s	121 m/s	0.046

CHAPTER XIX

OPTICAL AND ELECTRICAL MEASUREMENTS

SHOCK TEMPERATURE OF NaCl MEASURED WITH WIDE-BAND OPTICAL RADIOMETRY

Toshiyuki Ogura[1], Kazutaka G. Nakamura[1], Hisataka Takenaka[2], and Ken-ichi Kondo[1]

[1]*Materials and Structures Laboratory, Tokyo Institute of Technology, 4259 Nagatsuta, Midori, Yokohama 226-8503, Japan*
[2]*NTT Advanced Technology Corporation, 3-9-11, Midori, Musashino, Tokyo 180-8585, Japan*

Abstract. Shock temperature of NaCl was measured with time-resolved (3-nanosecond resolution) wide-band optical radiometry observing the radiation ranging from 0.6 to 13 μm in the pressure range between 17 and 43 GPa. In case of samples above 2000 K, the emitted radiation was measured with a visible to near-IR radiometer (0.6-1.6 μm). This radiation associated with a phase transition which arises at the defect sites such as dislocations. An IR radiometer sensitive to 9-13 μm was used to measure the bulk temperature below 1000 K. In the mixed-phase region between 23 and 33 GPa, thermal heterogeneity was observed in shock-loaded NaCl. In the low-pressure phase ($B1$) between 17 and 21 GPa, the shock temperature obtained with IR radiometer was almost 400 K lower than the values obtained by Fritz et al.[1]

INTRODUCTION

Mechanical properties of sodium chloride (NaCl) under high pressure have been pursued by various methods, x-ray analysis, shock and particle velocity measurements, and ultrasonic measurements. These experimental devotions were succeeded in the construction of high pressure EOS of NaCl by Decker,[2] which is utilized as a pressure standard below 30 GPa. The shock wave data by Fritz et al.[1] play an important role on this pressure standard. In their treatment, the shock temperature, however, was derived from Hugoniot with assumptions on thermodynamic parameters. It is desirable to determine the shock temperature experimentally without any assumption.

Optical radiometry has been usually used for shock temperature measurement. It has, however, a problem for measuring low shock temperature below 2000 K because of an insufficient emission in a visible wavelength range. For NaCl in the pressure range below 45 GPa, while the radiometric technique revealed the nature of local hot spot,[3] it was impossible to determine the bulk temperature. The thermal information of the bulk is indispensable not only for constructing EOS but also for understanding the mechanism of the phase transition under dynamic high pressure loading. In a previous study, the infrared radiometry was developed for the purpose of observing low shock temperature.[4]

In this study, we measured the shock temperature of NaCl by observing shock-induced emission at the wavelength between 0.6 and 13 μm. An isothermal compression curve was derived from experimentally determined shock temperature data and compared with the available NaCl pressure standards.

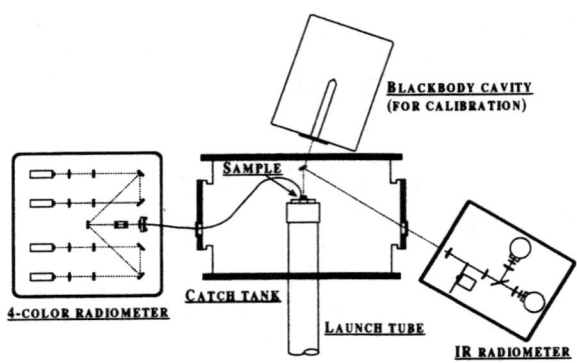

FIGURE 1. The schematic drawing of the experimental setup for shock temperature measurement.

EXPERIMENT AND DATA ANALYSIS

Shock wave was generated by a plate-impact method with a flyer accelerated with 20mm-bore, double-stage light-gas gun at Tokyo Institute of Technology. The copper impactor 1.7-3.7 km/s brought NaCl single crystal ([100] or [111] perpendicular to the shock wave front) to the pressure 17-43 GPa.

Emission between 0.6 and 13 μm from shock-loaded NaCl was observed using wide-band optical radiometric system shown in Fig. 1. The system consisted of the 4-channel visible – near IR radiometer (3-ns temporal resolution, silicon and InGaAs optical device) sensitive to 0.6-1.6 μm emission and the 1-channel IR radiometer (HgCdTe device cooled to 76 K) sensitive to 9-13 μm.

The temporal change of the IR (9-13 μm) emission was analyzed in order to derive the temperature with the equation;

$$I(\lambda,t) = \varepsilon I_\infty(\lambda,T)\left(1 - e^{-a_S(U_S - u_P)t}\right)e^{-a_U(d - U_S t)}, \quad (1)$$

where ε is the emissivity and I_∞ is the corresponding blackbody radiation. a_S and a_U are absorption coefficients of the shocked and the unshocked NaCl, respectively. d is the initial thickness of the sample. The experimental record is shown in Fig. 2 (a) along with the least square fit to Eq. (1). In order to determine the value of ε, the reflectivity of the shock wave front was measured and Kirchhoff's relation was applied;

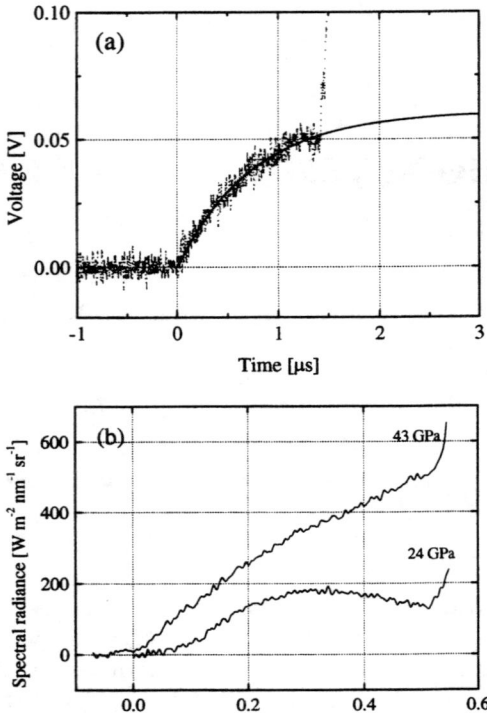

FIGURE 2. The temporal change of the emission from shock-loaded NaCl. (a) infrared (9-13 μm) at 21 GPa, (b) near IR (1.1 μm) at 24 and 43 GPa.

$$A + R + T = 1, \quad (2)$$

where A, R, T are absorption, reflection, and transmission of the sample, respectively.

The temporal profile of 0.6-1.6 μm emission changed with the shock pressure as shown in Fig. 2 (b). While the emission obeyed Eq. (1) for the pressure above 40 GPa, it peaked during the shock duration for the pressure below 35 GPa. The 4-color 0.6-1.6 μm spectrum was fitted to the graybody spectrum;

$$I(\lambda,T,\varepsilon) = \varepsilon \frac{C_1}{\lambda^5}\left(e^{C_2/\lambda T} - 1\right)^{-1} \quad (3)$$

to determine the temperature, T, and the emissivity, ε, where λ is wavelength and C_1 and C_2 are constants ($C_1 = 1.191 \times 10^{-16}$ W·m^2·sr^{-1}, $C_2 = 1.439 \times 10^{-2}$ m·K).

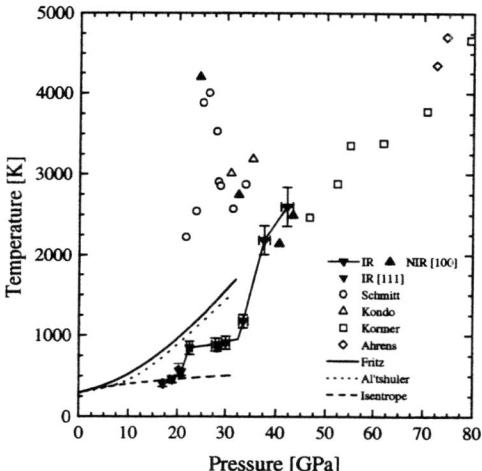

FIGURE 3. The shock temperatures of NaCl below 80 GPa determined with the wide-band optical radiometry along with the data in the literatures.

RESULTS

The experimental results are shown in Fig. 3. Above 35 GPa, the temperatures measured by the IR (9-13 µm) observations are comparable with those from the near-IR (0.6-1.6 µm). These agree well with the results by Kormer.[5]

Between 23 and 33 GPa, where the crystal is in the mixed phase state of $B1$ (NaCl) and $B2$ (CsCl) structures, the results obtained by the two methods differ from each other. The IR temperatures keep constant value around 800 K. In contrast, the near-IR temperatures are significantly higher values from 2000 to 4000 K and the brightness of the near-IR emission is small (the emissivity is in the order of 0.01). All the values of temperatures differ from the expected Hugoniot temperature obtained by Fritz *et al.* (solid line in Fig. 3).

The IR temperatures become lower and approach to the temperatures for the isentropic compression (dashed line) below 23 GPa. The 0.6-1.6 µm emission becomes weaker abruptly and is below the sensitivity of the radiometer.

The reflectivity of the shock wave front is 44% at maximum at 19 GPa for the IR emission range.

DISCUSSIONS

The temperature obtained with the IR radiometer is significantly lower than the expected continuum Hugoniot temperature below the phase transition pressure 23 GPa. This is explained as follows; the expected temperature is obtained by separating the increase in internal energy induced by shock wave into lattice compression part and thermal part,

$$\Delta E_{shock} = \Delta E_{compression} + \Delta E_{thermal} . \quad (4)$$

On the other hand, it is reported that a large number of defects such as slip surfaces, dislocations and point defects are generated in the shock-compressed solids. In considering dislocation, formation of dislocations will consume the shock energy as the form of strain energy. The new term should be added to Eq. (4),

$$\Delta E_{shock} = \Delta E_{compression} + \Delta E_{thermal} + \Delta E_{defect} . \quad (5)$$

Comparing with Eq. (4), the thermal part of the internal energy can decrease the shock temperature. Assuming the formation of the screw dislocation as an example, the elastic strain energy per unit length is written as;

$$E^S = \frac{Gb^2}{4\pi} \ln\left(\frac{R}{r_0}\right) , \quad (6)$$

where G, b, R, r_0 are shear modulus, Burgers vector, grain size including dislocation considered, and radius of dislocation core (almost equal to $5b$), respectively. Using Eq. (6) and the deviation of the temperature from the expected Hugoniot temperature, the dislocation density in shocked NaCl is estimated to be 10^{17} m^{-2}. This value is larger than the typical density of $10^9 - 10^{16}$ m^{-2}. The point defect-dislocation and the dislocation-dislocation interactions may decrease the density.

In the mixed phase region between 23 and 33 GPa, the measured temperatures, both the IR and the near-IR measurements, abruptly increase. This will be explained by the disappearance of dislocations and the release of the strain energy. It was reported that the dislocation of the order of 10^{16} m^{-2} existed in the shock-loaded steel recovered from 10.4 GPa and that the density decreased when the shock pressure exceeded the transition pressure because of the

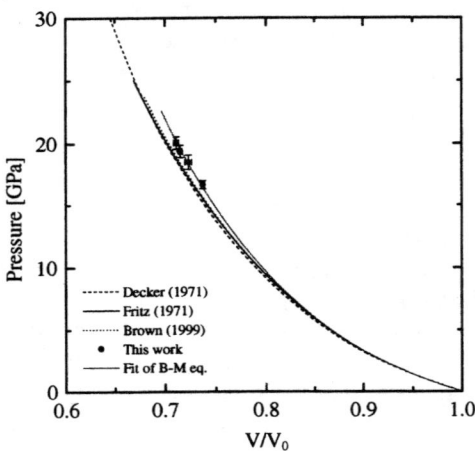

FIGURE 4. The isothermal compression curve at room temperature obtained from Eq. (8) along with the available NaCl pressure standards.

rearrangement of atoms.[6] The energy released from dislocations will be converted to the thermal energy and increase the temperature. Therefore, the disappearance of dislocations and the release of the elastic strain energy can generate the high temperature regions heterogeneously scattered in shocked solid. The extremely high temperature ranging 2000-4000 K observed with the near-IR (0.6-1.6 µm) radiometer is considered to be due to these hot spots.

DERIVATION OF THE 293 K ISOTHERMAL COMPRESSION CURVE FOR NaCl

Using the temperature measurement data of NaCl below the transition pressure, 293 K isothermal compression curve was derived and compared with the available NaCl pressure scale, i.e., the calculation from atomic model by Decker[2] and Brown,[7] and the shock wave data by Fritz et al.[1] The measured shock temperature was utilized to convert the shock wave EOS to the isotherm under room temperature using the equation,

$$\Delta P \mid_V = C_V \left(\frac{\partial P}{\partial E} \right)_V \Delta T \mid_V = \frac{\gamma C_V}{V} \Delta T \mid_V, \quad (7)$$

with the Debye model of heat capacity and the constant γ/V. The result is shown in Fig. 4.

It is revealed that the EOS obtained from shock temperature measurements is above the others by 5.9% in the pressure range 16-20 GPa when the reflectivity of the shock front is negligible. Even if the reflectivity of the shock front is assumed to be 44%, which is the maximum value measured at 19 GPa, our EOS is still higher by 4.9%. This value is larger than the experimental error. It is, therefore, concluded that the isotherm may be shifted to higher pressure in the pressure range of 16-20 GPa.

By fitting the Birch-Murnaghan equation;[8]

$$P = 3K_0 f (1+2f)^{5/2} (1 + af + bf^2),$$

(8)

$$\frac{V}{V_0} = (1+2f)^{-3/2},$$

it is possible to estimate the pressure derivative of bulk modulus, K_0''. Using $K_0 = 23.84$ GPa and $K_0' = 5.352$ (obtained from ultrasonic measurement below 3 GPa[9]), K_0'' becomes -0.30 GPa^{-1}. This value is smaller than that obtained from ultrasonic measurements, -0.68 GPa^{-1}. Using the EOS by Fritz et al., K_0'' becomes -0.46 GPa^{-1}.

ACKNOWLEDGEMENT

This work was supported by Core Research for Evolutional Science and Technology (CREST) program of Japan Science and Technology Corporation (JST).

REFERENCES

1. Fritz, J. N., Marsh, S. P., Carter, W. J., and McQueen, R. G., in *Accurate Characterization of the High Pressure Environment*, edited by E. C. Lloyd, National Bureau of Standards, Washington DC, 1972, Vol. 326, pp. 201-208.
2. Decker, D. L., *J. Appl. Phys.* **42**, 3239 (1971).
3. Schmitt, D. R., Ahrens, T. J., and Svendsen, B., *J. Appl. Phys.* **63**, 99 (1988).
4. Fat'yanov, O. V., Ogura, T., Nicol, M. F., Nakamura, K. G., and Kondo, K., *Appl. Phys. Letters* **77**, 960 (2000).
5. Kormer, S. B., *Sov. Phys. Uspekhi* **11**, 229 (1968).
6. Huo, D. T. C., and Ma, C. H., *J. Appl. Phys.* **46**, 699 (1975).
7. Brown, J. M., *J. Appl. Phys.* **86**, 5801 (1999).
8. Birch, F., *J. Geophys. Res.* **83**, 1257 (1978).
9. Spetzler, H., Sammis, C. G., and O'Connell, R. J., *J. Phys. Chem. Solids* **33**, 1727 (1972).

ULTRAFAST SPECTROSCOPIC INVESTIGATION OF SHOCK COMPRESSED GLYCIDYL AZIDE POLYMER AND NITROCELLULOSE FILMS

J. H. Reho, D. S. Moore, David J. Funk, G. L. Fisher, and R. L. Rabie

Dynamic Experimentation Division, Los Alamos National Laboratory, Los Alamos, NM 87545

Abstract. As a first experiment to observe the initial chemical reactions induced by sub-picosecond laser driven shocks in energetic materials, we have investigated thin films of glycidyl azide polymer (GAP) and nitrocellulose (NC). The GAP and NC films were spin coated onto thin film metal layers that had been vapor coated onto transparent substrates. Sub-picosecond laser pulses were used to launch planar shock waves through the metal layer and into the GAP and NC films. Time-resolved displacement measurements of both GAP and NC films suggest that shock-induced chemical reaction may be occurring. While the GAP film velocity falls as expected between that of the Al free surface and the particle velocity (using the free surface approximation), the displacement measurements in NC films yield velocities lower than the particle velocity. While data modeling discounting reaction has been unsuccessful, FTIR microscopy confirms the presence of unreacted GAP and NC in the shocked regions of the films.

INTRODUCTION

A full description of shock-induced chemical reactivity requires a microscopic and ultrafast (sub-picosecond) understanding of the interaction of a shock wave with the energetic material in question. The present experiments represent initial steps toward the extraction of such information in energetic thin film environments that are informative of not only the molecular but also the bulk behavior of energetics under shock loading. The ability to generate flat shocks by means of ultrafast laser pulses combined with spatial interferometry and various analytical tools is proving quite promising in our study of shock-induced chemistry. The experiments on thin films of NC and GAP presented here illustrate the initial strides that have been made in preparation, analysis, and shock loading of thin films of energetics.

SAMPLE PREPARATION

Our substrates consist of a 250 nm Al layer that is vapor coated onto standard borosilicate glass microscope cover slips of 100-150 μm thickness by CVI, Inc. The surface RMS deviation of these Al-coated substrates has been measured to be approximately 5 nm using atomic force microscopy (AFM). These metal-coated glass cover slips serve as the substrate for our spin-coating process in which 150-300 μL of a solution in methyl ethyl ketone, 2-pentanone, cyclobutanone, and n-butyl acetate ("magic solvent") of either GAP or NC (12.6% N) is dropped onto the center of an Al-coated cover slip and then spun for 25s at 2500 rpm.

Film thickness is varied by varying the concentration in "magic solvent" of the energetic polymer under study, yielding final NC or GAP thicknesses ranging from 50 nm to several microns. Film thickness is measured by means of a Filmetrics reflectance spectrometer, both at the center of the sample and at several other points radiating outward toward the edge in order to get a sense of thickness distribution across the sample. Good samples exhibit less than a 10% deviation in thickness across the entire area.

The film surface is also measured using the AFM in order to determine the RMS roughness across surface areas as large as 80 μm x 80 μm. Typical samples exhibit a 10 nm dispersion in height and good samples exhibit less than half of this value. We have observed the tendency of such spun films of GAP to pool toward the center over long (hours, days) timescales, causing uneven dispersion of the GAP on Al and possible error in initial film thickness measurements. However, exposure of such films to 365 nm light for one hour after spinning encourages wetting, thus inhibiting this pooling and producing film thicknesses that remain even and constant for weeks. This exposure to 365 nm light has been found to cause a ~15% reduction in intensity of the azide band of the GAP film as measured by FTIR. However, as the azide absorption in GAP is particularly strong, this reduction does not pose a serious hindrance.

LASER SHOCK GENERATION

The apparatus used to generate the planar shocks in the GAP and NC films consists of a chirped pulse amplified Ti:Sapphire laser system (Spectra Physics) which produces 110 fs laser pulses at 800 nm with approximately 10 mJ per pulse at 10 Hz. These pulses are split so that ca. 500 μJ is directed at the borosilicate side of the target and focused to a nominal 75 μm spot size. These Gaussian incident laser pulses are reproducibly flattened to better than 0.7 nm RMS over the entire 75 μm spot size after passage through the borosilicate cover slip[1], allowing for planar shock generation into the Al/GAP and Al/NC systems. The aftermath of shock break out from a single laser shot driven into a BK7-Al-NC target is shown in Figure 1.

A relatively weak portion of the laser output is employed in spatial interferometric detection[2] of optical phase and reflectivity change due to the shock pulse, and is thus temporally correlated with the shock-generating pulse. The weak portion is itself split into two relatively equal beams, one of which is thus directed to the "front" (i.e., the NC or GAP) side of the target, probing a large (300 μm) spot size inclusive of the "shocked" 75 μm spot.

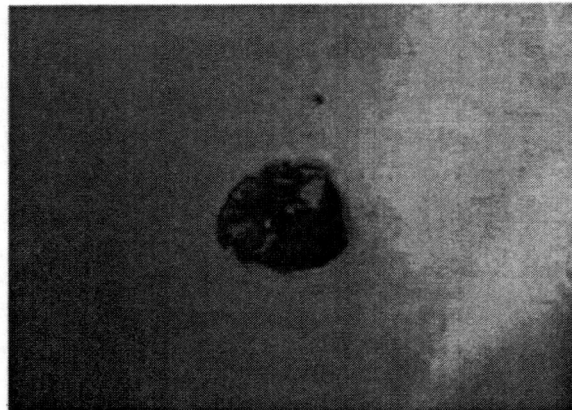

Figure 1. Aftermath of shock breakout in GAP on Al.

The other half of the weak beam is recombined with this probe both spatially and temporally after being appropriately delayed in order to generate a fringe pattern that forms the crux of our data analysis. Measurement of change in optical phase ($\Delta\phi$) with time can be converted to surface displacement (Δx) through the relation

$$\Delta x = \Delta\phi\, \lambda\, (4\pi n \cos(\theta))^{-1}$$

in which λ is the wavelength, n is the refractive index (of GAP or NC or of air in the case of bare Al), and θ is the half angle between the incident and reflected probe pulse.

A small amount of the initial 800 nm beam is also diverted from the amplifier into an optical parametric amplifier (Spectra Physics OPA800) in order to generate (after sum-difference mixing) tunable infrared sub-picosecond light pulses. These pulses, when tuned, for example, to the azide transition of GAP (~2100 cm^{-1}) or to the O_2NO stretch of NC (~ 1625 cm^{-1}), form a spectroscopic diagnostic for probing the initial decomposition pathways taken by GAP or NC upon shock-induced reaction.

The use of the infrared OPA, together with the temporal measurement of shock breakout, will allow for a fuller understanding of detonation chemistry than has been possible thus far. Using the OPA concomitantly with shock displacement measurements, we are able to study shock-induced chemistry spectroscopically and with sub-picosecond resolution. The incorporation of the infrared probe(s) is currently a work-in-progress, and while data presented here will concentrate only on results from surface displacement measurements, it is important to see the current results in the matrix of the eventual incorporation of the spectroscopic probe.

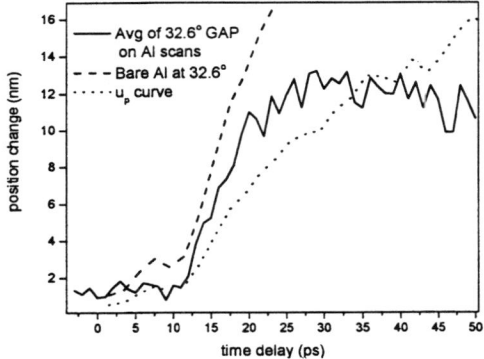

Figure 2. Al free surface velocity, particle velocity, and GAP surface displacement.

SHOCK-COMPRESSED GAP FILMS

Surface displacement due to shock propagation through a BK7-Al-GAP target has been measured using our spatial interferometric technique described above. Figure 2 shows the change in displacement measured for such a target from the "front" (GAP) side (solid curves) in comparison with the change measured at the free surface of a pure Al target (dotted line) due to a planar shock driven by our pump laser. Included in this figure is a curve representing particle velocity (u_p) based on the free surface approximation. The target employed in the collection of these data is composed of an 880 nm layer of GAP over 250 nm of Al vapor coated onto a 100-150 μm thick borosilicate microscope cover slip.

It can be found from the figure that the slope of the initial displacement v. time signal for the GAP sample lies between that of the free Al surface and the particle velocity for the shocked free Al. This result is sensible in consideration of the impedance mismatch between GAP ($\rho_o = 1.3$) and Al ($\rho_o = 2.8$). As the P-U Hugoniot of Al is steeper than that of GAP, the crossing point of the GAP Hugoniot with the reflected Hugoniot of Al (corresponding to the rarefaction wave from the free surface) is found to between the Al particle and free surface velocities.

One can also find from the figure that the optical phase change (and correspondingly the displacement[3])

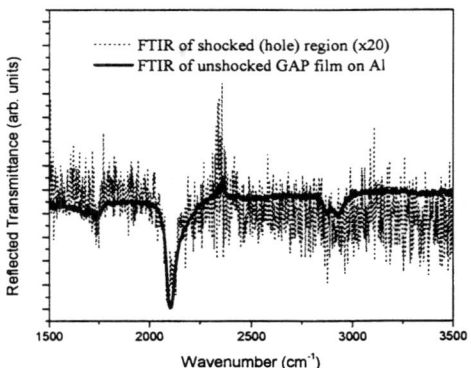

Figure 3. FTIR microscopy in region of azide band for shocked (dotted) and unshocked (solid) GAP on Al.

seems to level off and even drop at later times (~30 ps after the initial rise). This surprising result may be indicative of chemical reactivity due to shock propagation through the GAP, although presently film density variability and/or the presence of the rarefaction in the Al driver cannot be discounted. It has been postulated[4] that the initial step in GAP decomposition involves the endothermic formation of an ionic species with a concomitant volume decrease of ~20%, which would correspond well with experimental findings. Further, modeling using a simple nonreactive model is unable to reproduce the data, and our initial work with reactive models seems to promise more success.

FTIR microscopy, however, has shown minimal changes in the GAP azide intensity in the area through which our shock has propagated in comparison with the unshocked film (Figure 3) when scaled by a factor of twenty which brings the "baseline" reflectivities to the same level. While this may constitute a signature of non-reactivity, there is still a question as to how best to define the background to be subtracted from the shocked holes and thus the interpretation of the FTIR results remain open.

Thus while shock-initiated chemical reaction may already be occurring in our GAP thin films, we are presently working on increasing our shock pressure in order to unequivocally provide spectra from shock-induced chemistry of thin GAP films.

SHOCK-COMPRESSED NC FILMS

Experiments similar to those described for GAP have been done using a nitrocellulose layer of 550 nm thickness on our Al-BK7 substrates. Figure 4 shows the change in optical phase for a NC thin

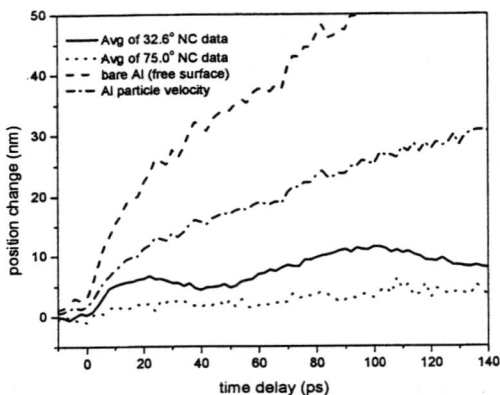

Figure 4. NC displacement at 2 angles, Al free surface velocity, and Al particle velocity.

film sample against the Al free surface velocity curve and a curve representing Al particle velocity. It is seen from the Figure that in the case of NC the initial phase change closely follows the particle velocity, which would indicate a more substantial "push back" (possibly indicative of reactivity) than has been found for GAP. It is also seen from the Figure that phase data taken at different angles of incidence differ, which is indicative of an optical component in the phase signal.

While FTIR microscopy has shown that NC is present in the regions through which our shocks have been launched, analysis of the velocity data is unsuccessful using a non-reactive model, and the decrease in displacement at longer times is likewise indicative of reactivity. It would be expected that the NC would react under milder loading conditions than would GAP, and one possible explanation reconciling our data analysis with the FTIR results would be that NC from the unshocked regions could "creep" into the reacted hole on long time scales. This possibility is currently being investigated.

We have also fabricated samples in which a thin (50 nm) layer of Al has been coated on top of NC layers of variable (50-200 nm) thicknesses, forming BK7-Al-NC-Al samples. Recent investigations of the breakout times through these samples have been undertaken in light of past results through bare Al order to measure the shock velocity though the NC layer.

CONCLUSIONS AND FUTURE DIRECTIONS

These initial experiments are paving the way toward a microscopic understanding of shock-initiated chemistry on sub-picosecond timescales. Incorporating spatial interferometry with short pulse lasers, "fast" infrared spectroscopy, FTIR microscopy, AFM, and other analytical methods, we are building an exacting and multifaceted methodology by which shock-induced chemistry can be probed and understood. We are presently working on increasing our shock pressures to ensure reaction of our thin film samples by tailoring the shock profile of the impinging pulse. Such work has led us to the study of electron-phonon coupling in metals and of shock propagation through bare metals as foundational to our thin film work with high energetics. We are furthering our understanding of our microscopic sample systems through data modeling and through the incorporation of several complimentary analytical tools.

ACKNOWLEDGEMENTS

We would like to acknowledge discussions with T. Lippert concerning sample preparation. We also thank K. Laintz for obtaining the FTIR microscopy data. This work has been carried out under the auspices of the Department of Energy (DOE).

1 D. S. Moore, K. T. Gahagan, J. H. Reho, David J. Funk, S. J. Buelow, R. L. Rabie, and T. Lippert. *App. Phys. Lett.* **78**, 40 (2001).
2 K. T. Gahagan, J. H. Reho, D. S. Moore, D. J. Funk, R. L. Rabie, in this volume.
3 Although the presence of an optical component, which would also affect the optical phase signal, cannot be ruled out and is certainly present in NC.
4 L. L. Davis, private communication, June 2001.

EMISSION SPECTROSCOPY APPLIED TO SHOCK TO DETONATION TRANSITION IN NITROMETHANE

Viviane Bouyer[1,2], Gérard Baudin[2], Christian Le Gallic[2], Philippe Hervé[1]

[1]*Laboratoire d'Energétique et d'Economie d'Energie, Paris X University, 1 Chemin Desvallières, 92410 Ville d'Avray, France, viviane.bouyer@cva.u-paris10.fr*
[2]*DGA/DCE/Centre d'Etudes de Gramat(CEG), 46500 Gramat, France, bauding@cegramat.fr*

Abstract. The objective of this work is to clarify the mechanism of shock to detonation of nitromethane by time-resolved emission spectroscopy. This paper presents the experiments performed in the spectral range 0.3-0.85 μm. Experimental results provide several radiance values depending on wavelength and time that we compared to measurements of pyrometry previously performed. Determination of the temperature profile is treated from the resolution of the equation of radiative transfer by an inversion method. The case of the steady state detonation is considered here.

INTRODUCTION

To improve knowledge of shock to detonation transition (SDT) of homogeneous high explosives, especially nitromethane (NM), temperature profiles are one of the parameters required. Optical techniques as pyrometry are often used to determine the NM detonation temperature (1). One of the most successful techniques today is the six-wavelength pyrometer developed at CEG (2)(3)(4). These means are based on the assumption of a surface emissivity. To obtain more information on the optical properties of condensed matter, spectroscopy techniques are widely used (5)(6)(7). This context and the fact that the different media involved in the SDT are likely semitransparent brings us to set up a time-resolved emission spectroscopy technique that enables a continuous measurement of the radiation emitted during the detonation of NM with a 1 ns time resolution. This paper presents the study of the visible range, where the NM in liquid state is transparent. From the radiance values obtained versus time and wavelength, we aim at determining profiles temperature by resolving the equation of radiative transfer. The fact that emissive media are made of gaseous compressed species at temperature around 3000-4000K and pressure between 10-30 GPa complicates the problem.

EXPERIMENTAL CONFIGURATION

One interesting property of NM for optical studies is its semi-transparence. Figure 1 shows the NM spectrum in the 0.6-3 μm range. In the visible range 0.4-0.6 μm, NM is transparent (2).

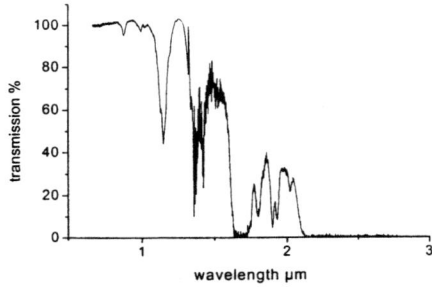

FIGURE 1. Transmission spectrum of the NM for a thickness of 15 mm measured by a FT-IR Fourier Spectrometer

The experiments consist in plane shock impacts on explosive targets at 8.6 GPa, under conditions of one-dimensional strain. A single stage powder gun propels the projectile on the target at a velocity of 1940 m/s to initiate the detonation (Fig. 2). The NM is in a polyethylene chamber of 15 to 25 mm depth, closed by a copper transfer plate. An optical probe collects the thermal radiation emitted during the detonation through a lithium fluoride (LiF) window. This radiation is transmitted to the spectroscopy system by an optical fiber. A Jobin Yvon Triax 180 spectrometer spectrally disperses the light beam. The dispersion is 33.5 nm/mm at 565 nm. A multianode photomultiplier tube (16 channels/ Hamamatsu R5900U-01-L16) is used to detect the signal in the spectral range [0.3-0.85 µm]. Its response time is 0.6 ns. Complementary measurement techniques are used: a polarization electrode records the shock entrance, the superdetonation and the detonation, piezo-electric pins measure the shock and detonation velocities.

that propagates until the interaction with the LiF window at 4.5 µs.

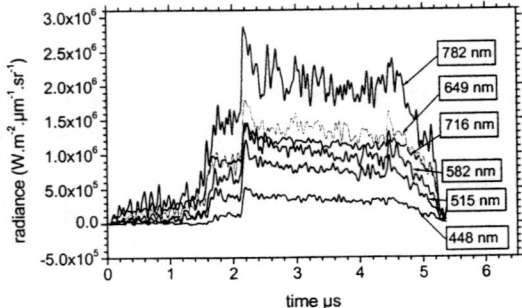

FIGURE 3. Results of one experiment at 8.6 GPa on a 25 mm NM target

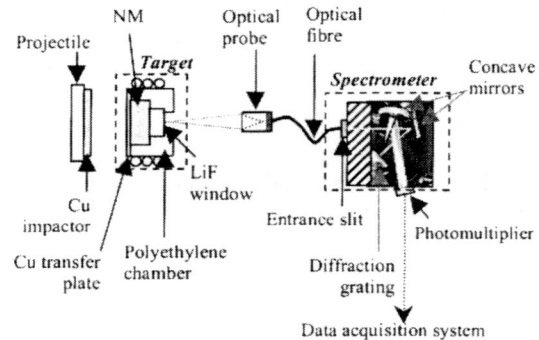

FIGURE 2. Experimental device.

EXPERIMENTAL RESULTS

We analyzed one of our experiments carried out at 8.6 GPa on a 25 mm thick NM target. Figure 3 represents the radiance signals for different wavelengths versus time. Only 6 measurements among 16 are represented. The spectral resolution is 28 nm and time resolution is 20 ns after filtering through noise. These measurements, with the piezo-electric pins and electrode signals clearly show the different phases of the SDT as described by Chaiken (8): a first jump at 1.65 µs characterizes the formation of the superdetonation, which overtakes the initial shock wave at a second jump at 2.15 µs. A strong detonation is then formed and gradually decays into an overdriven steady state detonation

RADIANCE PROFILE ANALYSIS

Analysis of one experiment at 8.6 GPa

Changes in radiance depending on wavelength have been studied for different typical moments of the SDT (Fig. 4). From the formation of the superdetonation, a discontinuity appears between 0.65 and 0.75 µm. It remains until the end of the propagation of the detonation wave.

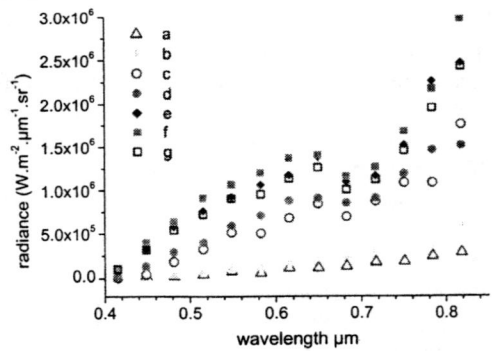

FIGURE 4. Radiance profiles at different times: a) after shock entrance (0.3 µs), b) before the superdetonation formation (1.2 µs), c) superdetonation formation (1.65 µs), d) 1.9 µs, e) catch up of the shock wave (2.15 µs), f) strong detonation (2.45 µs), g) overdriven steady state detonation (3.8 to 4.4 µs). Radiance values were averaged around the given time value.

One interpretation for this discontinuity may be gaseous H_2O produced during the SDT: the H_2O spectrum shows absorption lines between 0.65 and 0.75 µm. Another interpretation could be given from Gruzdkov and Gupta works (6). They measured the emission spectrum between 0.4 and 0.75 µm of NM shocked at 16.7 GPa under a stepwise loading process, and a peak appeared at 0.65 µm. They explained it as luminescence from reaction products, maybe NO_2.

Comparison with previous pyrometry measurements

In 1998, a time-resolved six-wavelength pyrometer was developed at CEG (4). The same experiments on NM had been carried out with this device so we can compare our results to those obtained by pyrometry (Fig. 5). However, the comparison is limited to the overdriven detonation phase because we do not obtain the same time of apparition of the superdetonation and detonation.

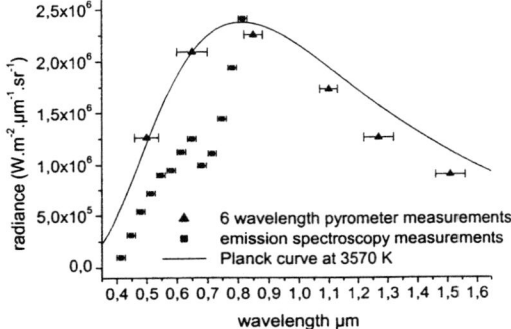

FIGURE 5. Radiance profiles obtained by pyrometry and spectroscopy during the steady state detonation. x error bars represent the spectral resolution.

The spectroscopy technique gives more information than the pyrometry technique because of the greater number of detectors and of the better spectral resolution (28 nm against 40 to 100 nm). Thus, the discontinuity can not be recorded by pyrometry. Consequently, the exploitation of the only six radiance values can give approached results, in particular for the temperature. Moreover, the black or grey body assumption often used when determining the temperature of the overdriven detonation may be questioned from these results. As shown on Fig. 5, emission spectroscopy measurements do not match with the Planck curve. Further to this, we analyze the influence of the number and kind of radiance values on the steady state detonation temperature calculation.

Determination of the temperature from the equation of radiative transfer

By using inversion methods, it should be possible to determine the temperature profiles during the SDT from the equation of radiative transfer (ERT). In this paragraph, we will only present the ERT and compare the detonation temperatures obtained from pyrometry and spectroscopy results.

Radiative transfer for a semitransparent medium

Given a semitransparent medium at a temperature T, with an absorption coefficient K_λ and a thickness (x^*-x_0), the ERT gives the spectral intensity transmitted and emitted by the medium along the detonation axis as:

$$L_\lambda(x^*) = L_\lambda(x_0)exp\left(-\int_{x_0}^{x^*}K_\lambda(x)dx\right) \\ + \int_{x_0}^{x^*}K_\lambda(x')L_\lambda^0(T)exp\left(\int_{x'}^{x^*}-K_\lambda(x)dx\right)dx' \quad (1)$$

Applying (1) to the SDT, different media are involved: neat NM, shocked NM, reaction products and detonation products. The two parameters T and K_λ are unknown except for neat NM. The difficulty lies in the knowledge of the media involved. If thermochemical codes give the main present chemical species, a model giving their absorption properties at the pressure and temperature encountered during the SDT does not exist. Therefore, the resolution of (1) by mathematical methods will require suitable absorption models.

Case of the overdriven steady state detonation

To get onto the resolution problem and to improve the comparison between the pyrometry and the spectroscopy techniques, it could be interesting to approach the case of the determination of the temperature of the overdriven steady state detonation. T is uniform (according to the Jouguet model) and equal to the semitransparent detonation products temperature T_J. K_λ^J is only depending on the wavelength λ. Then, as the detonation wave interacts with the window, the intensity is (ℓ is the depth of the cell):

$$L_\lambda(\ell) = L_\lambda^0(T_J)\left(1 - e^{-K_\lambda^J(\ell - x_0)}\right) \quad (2)$$

No accurate model for K_λ^J is available yet that could consider the absorption window. Except for this discontinuity, calculation results can point out relevant remarks. Table 1 represents calculations of T_J and of the transmissivity $\tau_\lambda = exp(-K_\lambda^\ell \cdot \ell)$ from the pyrometry measurements and from 5 and 7 values of spectroscopy measurements.

TABLE 1. Determination of temperature T_J

	T_J (K)	$\tau_{\lambda=0.5}$	$\tau_{\lambda=0.6}$	$\tau_{\lambda=0.8}$
pyrometry	3570±16	0.01	0.02	0.06
5 measurements of spectroscopy	3533±21	0.66	0.38	0.04
7 measurements of spectroscopy	3604±36	0.7	0.47	0.14

The three calculations give similar values. However, transmissivity values obtained from pyrometry measurements justify the hypothesis of a surface emissivity, close to black body's, whereas spectroscopy results don't confirm it. Consequently, the choice of the wavelength affects the estimation of the optical characteristic of the studied medium.

CONCLUSION

Time-resolved emission spectroscopy performed during the detonation of NM gives radiance measurements versus time between 0.3 and 0.85 µm, with a 28 nm spectral resolution. This technique brings out more information than time-resolved six-wavelength pyrometry, which was previously used to study the SDT of the NM at CEG. Indeed, our results showed a discontinuity between 0.65 and 0.75 µm in the radiance profile, appearing from the formation of the superdetonation. We still have to explain the presence of this discontinuity. It affects the determination of the temperature; however, we resolved the equation of radiative transfer outside the discontinuity in the case of an overdriven detonation. Detonation temperatures calculated are close to the temperature obtained by pyrometry but there is a significant difference between the transmissivity. Therefore, black or grey body assumption commonly used seems doubtful and the choice of the pyrometer wavelength is critical. Afterwards, emission spectroscopy will be performed in the infrared range 0.8-5 µm to characterize the functional groups of gaseous species.

The determination of temperature profiles by an inversion method is not so obvious because of the absorption coefficient. Chemical species produced, N_2, H_2O, CO_2, CO and C(s), are in such a thermodynamic state (high pressure and temperature) that it is not represented by existing absorption models (like Hitemp or Hitran data bases). Thermochemical calculations, existing detonation models will be then required to initiate the ETR resolution. But first, it is necessary to resolve the direct problem to analyze the sensitivity of the solution $L(\lambda,t)$ to the parameters of the chosen absorption model.

ACKNOWLEDGMENTS

The work described here was carried out with financial support from DGA/SPNuc, for the interest of CEA. Each impact experience was performed at Physics Explosive Laboratory of CEG with the assistance of ARES and Metrology staff.

REFERENCES

1. Kato Y., Mori N., Sakai H., Detonation temperature of nitromethane and some solid high explosives, in *8th Symposium on detonation*, Albuquerque, NM, 1985, pp. 558-566.
2. Léal Crouzet B., Ph. D. Dissertation, University of Poitiers, France, 1998.
3. Léal Crouzet B., Baudin G., Presles H.N., *Combustion and Flame* **122**, 463-473 (2000).
4. Léal Crouzet, Bouriannes R., Baudin G., Goutelle J.-C., *EPJ Applied Physics* **8**, 189-194 (1999).
5. Winey J. M., Gupta Y. M., *J. Phys. Chem. A* **101**, 9333-9340 (1997).
6. Gruzdkov Y. A., Gupta Y. M., *J. Phys. Chem. A* **102**, 2322-2331 (1998).
7. Piermarini G. J., Block S., Miller P. J., *J. Phys. Chem.* **93**, 457-462 (1989).
8. Chaiken R. F., The kinetic theory of detonation of high explosives, M.S Thesis, Polytechnic Inst. of Brooklyn, 1957.

ULTRAFAST MEASUREMENT OF THE OPTICAL PROPERTIES OF SHOCKED NICKEL AND LASER HEATED GOLD

David J. Funk, D. S. Moore, J. H. Reho, K. T. Gahagan[1], S. D. McGrane, and R. L. Rabie

Dynamic Experimentation Division
Los Alamos National Laboratory
Los Alamos, NM 87545
[1]*Present Address, Corning Incorporated, Corning, NY*

Abstract. We have used high-resolution Frequency Domain Interferometry (FDI) to make the first ultrafast measurement of shock-induced changes in the optical properties of thin nickel (~500 nm) targets. Data taken at several angles of incidence allowed the separation of optical effects from material motion, yielding an effective complex index for the shocked material. In contrast to our previous studies of aluminum, measurements with an 800 nm probe wavelength found a phase shift attributable to optical property changes with the same sign as that due to surface motion, during an 11.5 GPa shock breakout. A similar experiment was attempted with thin gold films (~180 nm) using Ultrafast Spatial Interferometry (USI). However, since the electron-phonon coupling in gold is extremely weak, a shock is observed as it "forms". Ballistic electrons and electron-electron equilibrium cause fast heating of the electrons in the entire thickness of the thin film, followed by lattice excitation through electron-phonon coupling, eventually leading to melt and frustrated thermal expansion yielding the observed surface motion. We suggest that these experiments offer a new path for observation of phase changes or for temperature measurements, by allowing a determination of the complex index under dynamic loading conditions and comparing the measured values to those obtained under static conditions.

INTRODUCTION

Development of optical diagnostics that are complementary to X-rays or pyrometric measurements, i.e., diagnostics that are sensitive to phase changes in shocked metallic systems (i.e. melt or solid-solid), or that yield values for the complex index of shocked systems, will help revolutionize the field of shock physics. In addition to the volumetric changes that accompany first-order phase changes, discontinuities occur in the complex index of refraction (which describes the linear interaction of the material with electromagnetic fields), for both first- and second-order phase changes. Using interferometric techniques, we are able to measure the changes in the complex index during, and for hundreds of ps after, shock wave breakout at free or windowed surfaces. Thus, we suggest that the changes in complex conductivity be used as an indicator of phase or to dynamically measure emissivity.

We have recently used such ultrafast interferometric techniques to characterize shock wave breakout in thin aluminum films [1,2]. The interferometric techniques are capable of resolving *sub-nanometer* surface deformations with ~100 femtosecond time resolution and are also sensitive to the dynamic optical properties resulting from shock wave or thermal loading [1,2]. We have applied the same techniques to studies of shocked nickel and have examined the dynamics of pulse heated gold. In both experiments, changes in the optical properties of the materials are observed and in the case of gold, we attribute the changes to melt, hinting at the possibility of observing shock-induced phase changes with these techniques.

EXPERIMENTAL

Thin nickel (in-house) and gold (CVI) films of ~500 nm and ~180 nm thickness, respectively, were vapor plated onto microscope slide cover slips or white glass. The shock or heating pulses were created in the films by the deposition of laser light from a chirped pulse amplified Ti:sapphire laser. The pulses from this laser are ca. 100-200 fs (FWHM) long at a wavelength of 800 nm with between 250-500 µJ of energy per pulse. The illuminated spot diameter is nominally 75 µm. Two types of interferometry have been used. FDI was used to acquire the nickel data as in [1,2]. USI was used to acquire the gold data and the

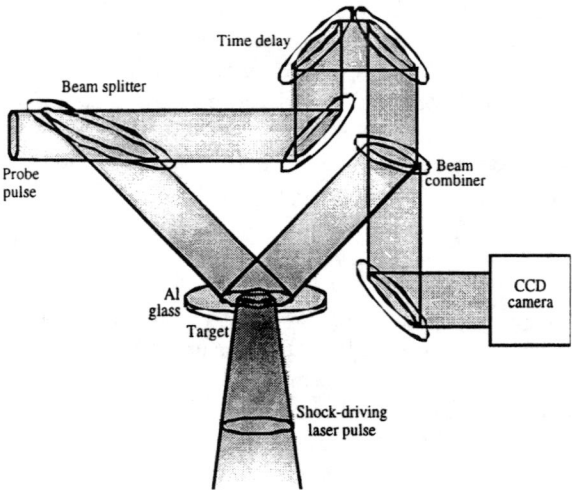

FIGURE 1. Schematic diagram of the femtosecond laser-driven shock/spatial interferometry experiment on thin film aluminum samples.

technique is discussed elsewhere in these proceedings [3]. In both cases, phase and reflectivity data are acquired dynamically.

The experimental arrangement for USI is shown schematically in Fig. 1. The thin film targets were mounted on a computer-controlled x-y translation stage. The target was rastered about 4 laser diameters between experiments so that each laser-driven shock would propagate into undisturbed material. Experiments have been performed in the above sample orientation, with the shock-driving laser focussed through the glass substrate onto the nickel and gold films, as well as in the reverse orientation with gold, with the shock-driving laser focussed onto the gold film side. The probe pulses observed the free gold/nickel surface in the orientation shown in Fig. 1 and the gold/glass interface in the reverse orientation. The time delay between the probe pulses and the the shock-driving laser pulse were adjusted with better than 100 fs precision using computer controlled precision translation stages (Newport Corp.). The two probe pulses interfere either in the exit plane of a spectrometer (FDI) or directly in the focal plane of the CCD (USI). The resulting CCD image for each experiment was analyzed using Fourier transform methods, as in [2,3,4] to extract the difference in phase between the two probe pulses caused by changes in the film surface position or the optical properties of the film. Data were taken at two or three angles to separate out the optical effects from material motion.

RESULTS

Shown in Fig. 2 are the FDI data and fits obtained for the nickel thin film. In contrast to the FDI results obtained previously for aluminum [1], the data taken near the quasi-polarizing angle (78.3°) exhibit a positive phase shift attributable to changes in the optical properties, in the same direction as that due to surface motion. Thus, if one naively had taken data only at 32.6°, one could have incorrectly determined the particle velocity and resulting shock state, by not accounting for the contribution due to optical properties. Shown in Figure 3 and 4 are the reflectivity and phase data for the laser-heated gold taken at two angles of incidence and pumping the bare gold (Fig. 3) and through-glass (Fig. 4). It both cases, note the early negative phase shift in the 75° data. Also note that the first changes in phase and/or reflectivity occur near –42 ps. Time zero in these experiments was determined by pumping a standard 250 nm aluminum film through glass and using "breakout" of the shock as the zero. Given that the shock velocity in aluminum is ~6nm/ps, it would take ~42 ps for the shock to transit the thin film. Thus, the first signals in the gold occur nearly simultaneously with the pump, which would at first appear contrary, but can be explained as resulting from the weak electron-phonon coupling in gold.

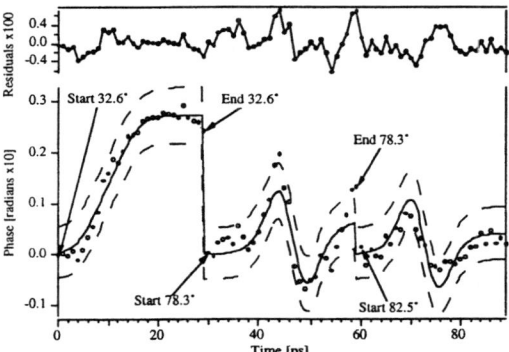

FIGURE 2. Plot of measured phase difference between the probe beams vs. relative delay (to the pump laser) for shock wave breakout from a 500 nm thick Ni thin film. The data were obtained using p-polarized 800 nm light and taken at three angles of incidence (which are offset in time for clarity). Note the large positive phase difference in the data taken near the quasi-polarizing angle in nickel (78.3°), indicating significant complex index changes leading to phase differences of the same sign as those from surface motion. The dashed lines are 95% confidence limit prediction bands.

DISCUSSION

To extract a particle velocity, rise time, and the complex index for the shocked nickel data, the data was fit as previously [1], using the following expression for the measured phase shift for each pulse as a function of time:

$$\phi_j(t) = \Delta\phi_{n_{\text{eff}},k_{\text{eff}}}\text{sech}^2(\frac{t - t_j + \delta_{t_j}}{\tau}) + \frac{4\pi\cos(\theta_o)}{\lambda_o}\int_{t_j}^{t_j} u_p(1 + \tanh(\frac{t - t_j + \delta_{t_j}}{\tau}))dt$$

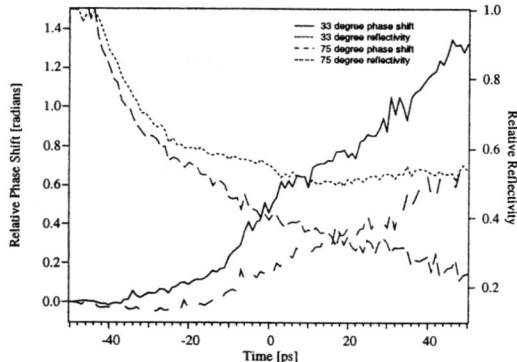

FIGURE 3. Plot of measured phase difference and relative reflectivity between the probe beams vs. delay (relative to the pump laser) at the glass interface of an 180 nm thick Au thin film. The data were obtained using p-polarized 800 nm light and taken at two angles of incidence.

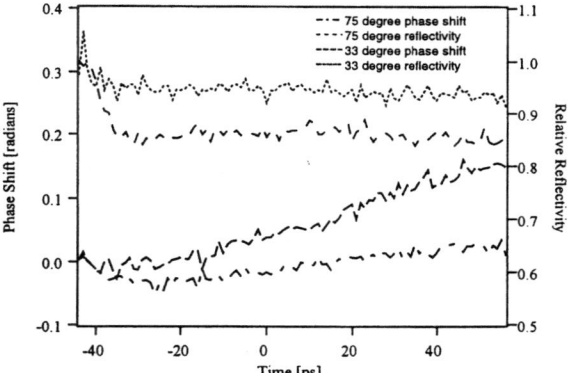

FIGURE 4. Plot of measured phase difference and relative reflectivity between the probe beams vs. delay (relative to the pump laser) at the air interface of an 180 nm thick Au thin film. The data were obtained using p-polarized 800 nm light and taken at two angles of incidence.

The functional assumes that the velocity of the free surface has the form of the hyperbolic tangent. It also assumes that the complex index change is a maximum at the peak acceleration, and varies linearly and smoothly from the ambient to the shocked values. The data for the three angles are fit simultaneously by varying the following parameters: τ, u_p, t_j, n_{eff}, k_{eff}, where τ is the hyperbolic tangent time constant, u_p is the shock state particle velocity, t_j are offsets for the data from each set. As can be seen in Figure 2, the fits are satisfactory and yield n=1.56 and k=3.35 for the shocked material's complex index, changing from the ambient values of the nickel (n_o=1.6, k_o=3.28) assuming literature values [5]. However, the initial values are likely different from the literature values since Atomic Force Microscope analysis of the films indicate that the RMS roughness of the nickel films is ~7 nm. Finally, the direction of the change in the optical properties is consistent with that of a Drude material under volume compression, which one expects should well represent nickel optical properties, since nickel has no interband transitions near 800 nm.

Concerning the gold data and as noted previously, the first air/Au or glass/Au interface motion occurs nearly simultaneously with the pump laser. Many studies of laser-heated gold have been conducted [6], and have conclusively shown that the electron-phonon coupling in gold is extremely weak, and ca.. 18 times smaller than that in nickel. This results in the following qualitative picture of the laser-material interaction: laser light excites electrons, which, prior to equilibration, have a ballistic penetration depth of ~ 100 nm at veocities of 1nm/fs. Thus, some fraction of the electrons will penetrate the entire 180 nm film on a timescale slightly greater than that of the pump laser. The hot electrons will equilibrate with each other through electron-electron collisions, and subsequently with the lattice, through electron-phonon coupling. The final temperature of the lattice will depend on the total energy absorbed and the heat capacity of the gold film, with the timescale for equilibrium determined by the electron phonon coupling. The timescales and onset of reflectivity changes, observed in Fig. 2, are consistent with this picture [6].

Finally we would like to make one final observation regarding the data obtained by pumping the bare gold directly. In another study, we believe we have pumped gold to melt using 400 nm laser light [7]. In that case, the reflectivity at 800 nm was measured to be approximately 65% of the original value at an incident angle of 33 degrees. In the experiments we report here, the reflectivity drops to approximately 55% of the original value at an incident angle of 33 degrees. Though the quantitative extraction of the complex index has not been completed, we believe that, based on the reflectivity measurement and estimates of the gold temperature [8], that, as previously, we have pumped the gold to melt.

CONCLUSIONS

We have measured changes in the optical properties of shocked nickel and have observed changes in the optical properties of gold while simultaneously measuring the interface position. We believe that these studies pave the way for new investigations of the measurement of temperature and shock-induced phase changes, through correlation of the dynamic complex index with thermodynamic states measured under static conditions.

ACKNOWLEDGEMENT

This research was performed under the auspices of the US Department of Energy.

REFERENCES

1. Funk, David J., et al., Phys. Rev. B, in press (2001).
2. Gahagan, K.T., et al., Phys. Rev. Lett. **85**(15), 3205-3208 (2000).
3. Gahagan K.T., et al., in this volume p.
4. Evans R., Badger A.D., Faillès F., Mahdieh M., Hall T.A., Audebert P., Geindre J.-P., Gauthier J.-C., Mysyrowicz A., Grillon G., and Antonetti A., Phys. Rev. Lett. 77, 3359 (1996).
5. Smith, D.Y., Shiles, E., and Inokuti, M., in "Handbook of Optical Constants of Solids," E. D. Palik, Ed., p. 374 (Academic Press, San Diego, 1985).
6. Hohlfeld, J., et al., Chem. Phys. 251, 237-258 (2000).
7. Funk David J., unpublished results.
8. Rabie, R. L., unpublished results. The two-temperature model was assumed [7] and a lattice temperature of ~2200 K obtained, well in excess of melt (1338 K).

OPTICAL EXTINCTION OF SAPPHIRE SHOCK-LOADED TO 250–260 GPa

D.E. Hare[1], D.J. Webb[2], S-H. Lee[2,3], and N.C. Holmes[1]

[1]*Shock Physics, Lawrence Livermore National Laboratory, Livermore CA 94551*
[2]*Physics Dept., University of California, Davis CA 95616*
[3]*Present address: Rudolph Technologies, Inc., Flanders, NJ 07863*

Abstract. Sapphire, a common optical window material used in shock-compression studies, displays significant shock-induced optical emission and extinction. It is desirable to quantify such non-ideal window behavior to enhance the usefulness of sapphire in optical studies of opaque shock-compressed samples, such as metals. At the highest stresses we can achieve with a two-stage gas gun it is technically very difficult to study the optical properties of sapphire without the aid of some opaque backing material, hence one is invariably compelled to deconvolve the optical effects of the opaque surface and the sapphire. In an effort to optimize this deconvolution process, we have constructed sapphire/thin-film/sapphire samples using two basic types of thin films: one optimized to emit copious optical radiation (the hot-film sample), the other designed to yield minimal emission (the cold-film sample). This sample geometry makes it easy to maintain the same steady shock-stress in the sapphire window (255 GPa in our case) while varying the window/film interface temperature. A six-channel time-resolved optical pyrometer is used to measure the emission from the sample assemblies. Two different sapphire crystal orientations were evaluated. We also comment on finite thermal conductivity effects of the thin-film geometry on the interpretation of our data.

INTRODUCTION

Sapphire is an important window material for optical studies of shock-compressed materials [1,2]. Sapphire has an important advantage over the (100) orientation of lithium fluoride (hereafter simply LiF); namely it has a significantly greater shock impedance. On the other hand, the optical transparency of sapphire under shock-compression conditions is not as good as that of LiF [3,4] and so some care does need to be exercised when using sapphire in quantitative optical studies.

Both Urtiew [5] and McQueen and Isaak [6] investigated sapphire as a shock-compression window above 80 GPa. At such high shock stresses one is compelled for technical reasons to use an opaque backing material which is then viewed through the window. Both the backing and the window are shock-compressed and emitting optical radiation and it is very difficult to deconvolve the emission due to the window from that due to the backing.

In this paper we use a novel thin film sample construction to facilitate the deconvolution of the window emission component from that due to the opaque backing. We present preliminary results of an attempt to quantify some aspects of sapphire's non-ideal window behavior for single shock loading of sapphire in the stress range 250 – 260 GPa. This is the upper limit of stress we can achieve in sapphire in a single shock wave at our two-stage light gas gun facility using a tantalum impactor.

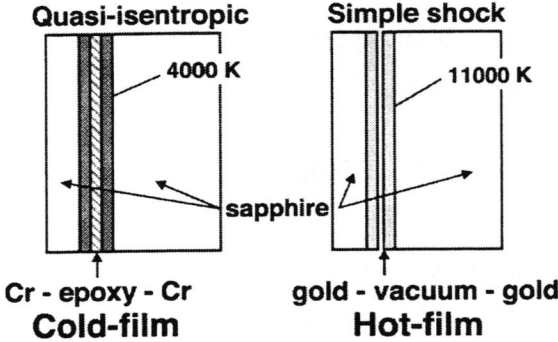

Figure 1. Our two types of samples. The sample on the left is designed to emit relatively less and we call it the "cold-film" sample. The one on the right is designed for copious optical emission and we call it the "hot-film" sample. The impactor generates the shock from the left, the shock wave propagates from left to right, and optical emission is viewed by the pyrometer in the thicker sapphire from the right. All chromium and gold films are 500 nm thick.

EXPERIMENTAL PROCEDURE

The sapphire used was single-crystal synthetic material of typical purity better than 99.996 %. Two window surface orientations were used: (0001) c-plane, and (1,-1,0,2) r-plane orientations. The indices refer to the x-ray (also known as "structural") hexagonal cell [7].

Sample construction is outlined in figure 1. A thinner sapphire of 1.00 mm thickness was always used for the impacted side of the sample, while a thicker sapphire of 3.00 mm was used for the optical observation side. The thin and thick sapphires were always matched in surface orientation within a given sample. That is, each sample contains two sapphires and they were either both r-plane or both c-plane, but not mixed.

The cold-film sample uses chromium, a relatively incompressible metal that therefore shocks to relatively low temperatures. Furthermore, a thin layer of epoxy (on the order of ten microns) is sandwiched between the chromium coatings and causes it to ring-up to the final steady stress (determined by the sapphire), which further keeps down the temperature of the chromium film. The hot-film sample is of gold, with a vacuum gap of about three microns. The higher-than-sapphire shock impedance of gold together with the absence of the low impedance layer of epoxy cause the initial shock in the downstream gold layer to overshoot the final stress. Gold was chosen because its shock temperature under the same conditions is much higher than chromium. Also, gold is not a good reflector for the UV and bluer visible wavelengths so that it should have a large spectral emissivity at these same wavelengths. All these things add up to copious emission from the gold backing of the hot-film sample.

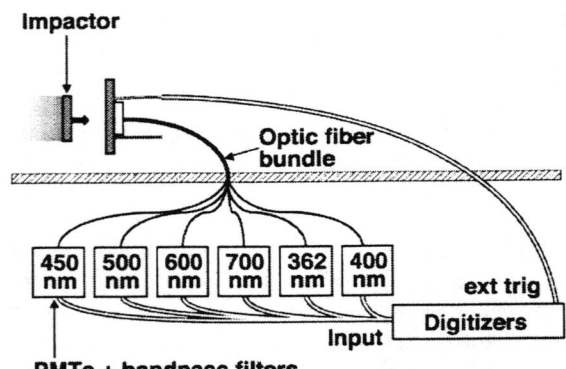

Figure 2. Optical pyrometer used for these experiments.

Shock stress was achieved by using a 1.5 mm tantalum impactor in our two stage light gas gun at Lawrence Livermore National Lab [8]. While we strove to make the sapphire shock stress identical for each experiment, small variations in gun performance produced the finite stress range 250 – 260 GPa. The emission was collected by our six-channel time-resolved optical pyrometer system [9]. A diagram of the collection scheme is shown in fig. 2. The pyrometer was calibrated by a tungsten-halogen lamp standard of spectral irradiance.

An effort was made to keep signal levels at the 100 to 200 mV level, requiring significantly more neutral density filtering for the hot-film samples than for the cold-film samples. To obtain the correct relative intensities between the relevant hot-film and cold-film sample data, the latter were reduced in amplitude by the transmittance of the excess of filters used for the hot-film sample data.

RESULTS

Figure 3 displays idealized data traces.

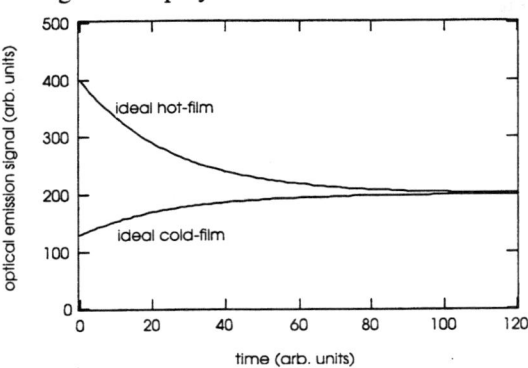

Figure 3. Idealized trace showing film-as-sole-emitter to window-as-sole-emitter transition.

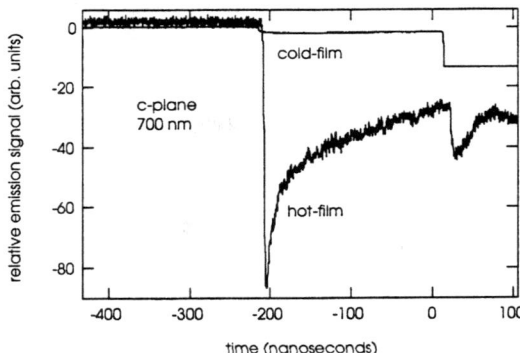

Figure 4. Hot-film and cold-film samples, c-plane sample, 700 nm wavelength. Because these are PMT output the signal is negative. Shock-compression of the thin film occurs at about −200 ns. Shock wave reaches rear sample surface at about +10 ns.

Time zero is the signal immediately after the shock wave has passed through the thin film but has not progressed far into the thicker "viewing" sapphire. The hot-film signal is much greater than the cold-film signal because the film signal is relatively unobstructed by the mostly unshocked, (and thus transparent) sapphire. Eventually the shocked sapphire becomes optically thick enough to be opaque. The pyrometer will only see opaque sapphire at the temperature resulting from the shock loading to 255 GPa. This is represented by the two different sample signals eventually merging to the same result with time. The rate at which the signal makes the transition from film-as-sole-source to sapphire-as-sole-source is determined in both cases by the optical attenuation characteristics of sapphire. The analysis depends also on the spectral emissivity of the film sample. If the backing films are perfect absorbers/emitters (i.e. blackbody) then the transition from film-as-sole-emitter to sapphire-as-sole-emitter is single exponential with the sapphire attenuation coefficient as the exponent. We acknowledge that this assumption is imperfect but in the present analysis we adopted it as an expedient in order to obtain approximate, but reasonable, results.

The pyrometry traces from the hot and cold films for the longest wavelength channel (700 nm center, 36 nm bandwidth) are displayed together for c-plane sapphire in fig. 4.

We see that the hot-film sample signal is much more intense than the cold-film sample.

The shortest wavelength (362 nm center, 10 nm bandwidth) c-plane data is shown in fig 5. As with 700 nm, the hot-film sample signal is much more intense than the cold-film sample signal. However, the hot-film sample signal appears to decay more rapidly for 362 nm than for 700 nm.

A very similar set of results is found for the r-plane orientation of the sapphire at both 362 and 700 nm (not displayed).

The analysis of the time rate of change of the signal yields an upper bound to the optical attenuation coefficient (i.e. worst-case scenario attenuation: the material can be more transparent but not less so). This is because the cooling of the hotter film by heat transport into the surrounding sapphire will also give the appearance of a signal that decays with time. Since both the optical and heat transport effect combine together to give the total apparent attenuation coefficients that we measure, we claim that our measurements reported are upper bounds on the optical attenuation. For c-plane sapphire we compute upper bounds to the optical attenuation coefficients of 3.5 cm^{-1} and 9.2 cm^{-1} for 700 and 362 nm respectively. For r-plane sapphire we compute 3.2 and 9.4 cm^{-1} for 700 and 362 nm respectively.

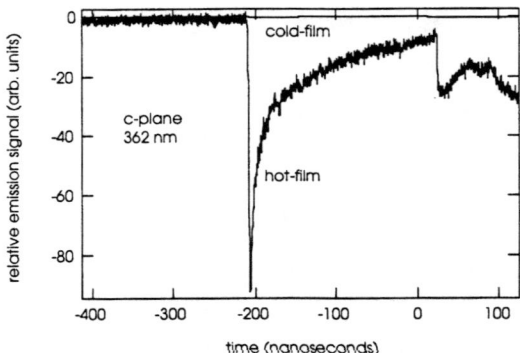

Figure 5. Hot-film and cold-film samples, c-plane sample, bluest wavelength (362 nm).

DISCUSSION

Perhaps the most important result of this work is that the cold-film sample signal level allows us to make a reasonable assessment of the window's emission contribution to the case of the hot-film signal. This is because the window emission contribution to both the cold-film and hot-film signal should be nearly identical. For the case of sapphire between 250 and 260 GPa, we see that the emission from sapphire contributes very little to the overall emission in the hot-film case. This in turn puts a tighter constraint on the arithmetic process of determining a total attenuation coefficient from the hot-film sample data.

We advise the reader that the attenuation coefficient results reported here are most appropriate to the use of sapphire with an optical pyrometer similar to ours. Light scattering contributes to optical attenuation but its impact varies with system design. For example, the results we presented here may not be as appropriate for sapphire as a VISAR window.

There does not seem to be much difference between our r-plane and c-plane result at 255 GPa. Lower stress optical studies showed substantial differences in c-plane versus r-plane optical and mechanical behavior [10,11].

The upper bound for attenuation coefficient does seem to be significantly greater for 362 nm relative to 700 nm. However, thermal relaxation of the emitting films also would cause a qualitatively similar effect.

ACKNOWLEDGMENTS

We would like to thank David A. Young of Lawrence Livermore Lab for supplying calculated values of the sample expected bulk temperatures. Keith Stickle, Leon C. Raper, Steven J. Caldwell, James G. Van Lewen, Erikk A. Ojala, and Jeffrey D. Van Lue provided invaluable technical support. DEH would like to thank Oleg Fat'yanov of Washington State University for insightful comments regarding this work.

This work was performed under the auspices of the U.S. Dept. of Energy at the University of California/Lawrence Livermore National Laboratory under contract No. W-7405-Eng-48.

REFERENCES

1. W.J. Nellis, M. Ross, and N.C. Holmes, Science, **269**, 1249 (1995).
2. J.D. Bass, B. Svendsen, and T.J. Ahrens, in High-pressure research in Mineral Physics, ed. M.H. Manghnani and Y. Syono, pp 393-402 (Terra, Tokyo, 1987).
3. L.M. Barker and R.E. Hollenbach, J. Appl. Phys. **41**, 4208 (1970).
4. J.L. Wise and L.C. Chhabildas, in Shock Waves in Condensed Matter-1985 (Proceedings of the Fourth APS Topical Conference on Shock Waves in Condensed Matter, Spokane, WA, July 22-25, 1985), edited by Y.M. Gupta (Plenum, New York, 1986), p. 441.
5. P.A. Urtiew, J.Appl. Phys. **45**, 3490 (1974).
6. R.G. McQueen and D.G. Isaak, J. Geophys. Res. **95**, 21753 (1990).
7. M.L. Kronberg, Acta Metallurgica, **5**, 507 (1957).
8. A.H. Jones, W.M. Isbell, and C.J. Maiden, J. Appl. Phys. **37**, 3493 (1966).
9. N.C. Holmes, Rev. Sci. Instrum. **66**, 2615 (1995).
10. J. Hyun, S.M. Sharma, and Y.M. Gupta, J. Appl. Phys. **84**, 1947 (1998).
11. D.E. Hare, D.J. Webb, and N.C. Holmes, in Shock Compression of Condensed Matter-1999, edited by M.D. Furnish, L.C. Chhabildas, and R.S. Hixson (American Institute of Physics, New York, 2000), p. 637.

TEMPERATURE MEASUREMENT OF TIN UNDER SHOCK COMPRESSION

Pierre-Louis Héreil, Catherine Mabire

Centre d'Etudes de Gramat - 46500 GRAMAT - FRANCE

Abstract. The results of pyrometric measurements performed at the interface of a tin target with a LiF window material are presented for stresses ranging from 38 to 55 GPa. The purpose of the study is to analyze the part of the interface in the temperature measurement by a multi-channel pyrometric device. The results show that the glue used at target/window interface remains transparent under shock. The values of temperature measured at the tin/LiF interface are consistent with the behavior of tin under shock.

INTRODUCTION

The experimental determination of temperature of materials under shock has been the subject of numerous studies since the works of Kormer [1] in 1968. The measurement of temperature under shock has been obtained for transparent materials, rocks, metallic materials and explosives [2-13]. Temperature measurement, which is difficult to perform, is essential to understand the behavior of materials under shock and, notably, to identify phase changes of the matter.

The most common technique used to measure the temperature under shock is to record the radiation evolution (or luminance) of the target with an optical pyrometer, and to determine its temperature through Planck's law. While the luminance measurement technique is well characterized, the calculation of the actual temperature of the material is not : the main difficulty concerns the emissivity evaluation of the measured surface. The experimental configura- tion used by numerous authors consists in depositing a metallic material on a transparent window to obtain the most ideal interface. Blanco [14] proposed an alternative which consists in gluing the window material on the metallic target to insulate the metallic target from the window material. If the glue used remains transparent under shock, the measured interface temperature is close to the temperature of the metallic target. The final difficulty is to evaluate the emissivity of the measurement surface.

This paper presents the temperature measurements obtained at the interface between a tin target and a LiF window for stresses ranging from 38 to 55 GPa. The purpose of this study is to analyze the part of this interface composed of glue.

EXPERIMENTS

The configuration of the plate impact experiment is depicted in Figure 1 and the parameters of the three experiments performed on tin are presented in Table I.

The experiments were performed in CEG with the single-stage powder gun ARES. The relative error of impact velocity measurement is 1%. For the three experiments performed, the impact tilt varies from 1 to 2 mrad. The impactor disks, buffer plate, target and window all present a 1/100 mm flatness on their two faces and the parallelism defect between two faces is lower than 2/100 mm. The face of the tin disk aimed by the pyrometer presents a polished surface with a rugosity of 0.01. The faces of window material are optically flat.

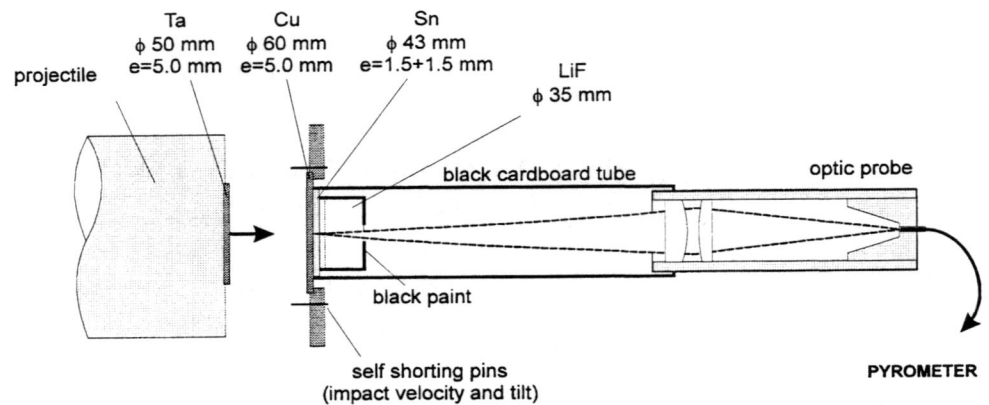

FIGURE 1. Experimental configuration of plate impact tests for temperature measurement at tin/LiF interface

The targets consisted of two tin disks because of a supplying problem. The window and the tin target were glued together with L358 glue (UV-hardened LOCTITE 358). The three experiments have different interface conditions.

In the first experiment (#A25), the glue thickness of each interface was less than 10 μm and in the second one (#A31) this thickness was 50 μm. In these two cases, the window is composed of two LiF disks glued together with L358. In the third experiment (#A30), the window is composed of a single LiF disk which is 15-mm thick and the target surface is coated with a 1-μm thick graphite film by cathodic sputtering. The glue thickness between the window and the target is less than 10 μm as in the first experiment. The goal of this experiment was to increase the emissivity of the measurement surface by using a highly emissive material (graphite).

The optical pyrometer used in this study has been exhaustively described by Léal [15]. The pyrometer uses six measurement channels with different wavelengths : 500 nm, 650 nm, 850 nm, 1100 nm, 1270 nm and 1510 nm. Its response time is 5 ns and its operating conditions are limited to a low temperature of 1500 K. The relative error associated with the luminance measurement is 3 %.

Figure 2 presents the luminance profiles obtained in the experiment A25 for the six wavelengths of the pyrometer. These signals exhibit an initial ramp when the shock reaches the Sn/LiF interface and, after 0.4 μs, a singularity which corresponds to the reflection of compression and release waves from the different interfaces (Sn/Sn and LiF/LiF). The large increase in signals after 2.2 μs corresponds to the emergence of the shock wave at the free surface of LiF.

This shows that the propagation of the shock wave through the 10 μm thick glue interface, at 2 mm in depth in the LiF window, does not modify the pyrometric signal. The L358 glue remains

Table 1. Plate impact parameters for temperature measurement at Sn/LiF interface

Shot number	Window1	Window2	Target / window interface	Projectile velocity (m/s)
A25	LiF (2.0 mm)	LiF (15.0 mm)	L358 (< 10 μm)	2258 ± 25
A31	LiF (2.0 mm)	LiF (15.0 mm)	L358 (50 μm)	2240 ± 20
A30	-	LiF (15.0 mm)	graphite (1 μm) +L358 (<10 μm)	2253 ± 25

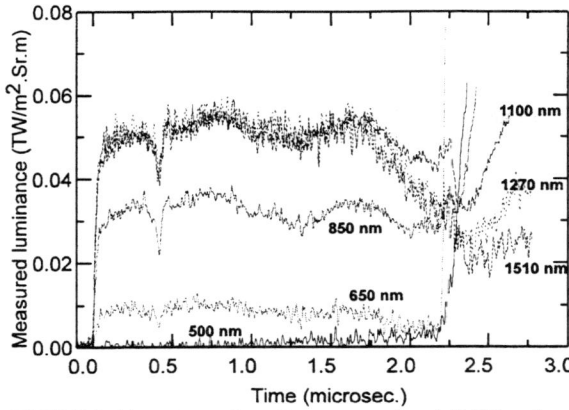

FIGURE 2 : Experimental profiles measured at tin/LiF interface for shot #A25.

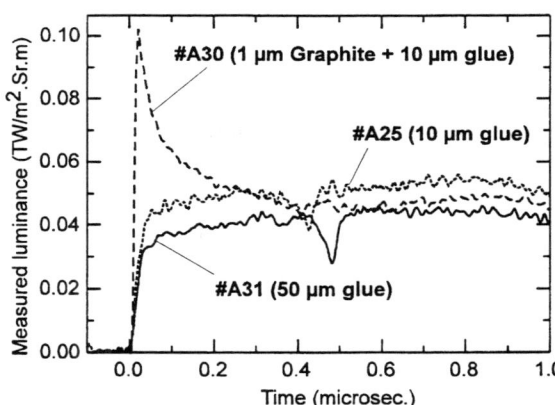

FIGURE 3. Comparison beween the experimental profiles measured during the three experiments for the 1100 nm wavelength.

transparent under shock at this stress and temperature. Otherwise, a substantial drop in the luminance would have been observed when the shock reached the LiF/LiF interface.

In Figure 3, the luminances measured during the three experiments for the 1100-nm wavelength are compared. It appears that a glue thickness of 50 μm does not modify significantly the shape of the luminance profile whereas a 1-μm thick graphite layer highly influences the profile. The essential information provided by these results is that the glue remains transparent under shock for these stresses and temperatures. If the glue became opaque under shock, the influence of graphite on the luminance profile would not be observed since the graphite deposit is in direct contact with the tin sample. Moreover, it is noticeable that the rise times of the first shock are similar for 10-μm and 50-μm thickness of glue. This implies that the measured radiation comes from the rear side of the tin sample and that the glue remains transparent under shock.

Concerning the signal shape measured in the experiment with a graphite layer, this phenomenon can be linked either to a progressive loss of emissivity of graphite, or to a modification of the conduction phenomena resulting from the presence of graphite at the interface.

The values reached in stress and temperature for the three experiments are listed in Table 2 [16]. The stresses in tin and, after release, in the LiF window are estimated from the known properties under shock of the material used and from the impact conditions.

The as-calculated values of temperature are compared in Figure 4 with the phase diagram of tin. The points obtained under shock and symbolized by circles have been calculated from the phase change model of tin [17]. The first shock results in a mixed solid-liquid state and the release occurs along the melting curve. Thus the points calculated from

Table 2. Values of stress, temperature and emissivity calculated at Sn/LiF interface

Shot number	Impact stress (GPa)	Interface stress (GPa)	Grey body temperature (K)	Grey body emissivity	Calculated * température (K)	Calculated emissivity
A25	55.3	38.8	2180	0.26	2125 ± 235	0.33 ± 0.15
A30	55.2	38.7	2180	0.23	2120 ± 230	0.28 ± 0.18
A31	54.7	38.3	2220	0.18	2078 ± 219	0.29 ± 0.14

* temperature calculated with emissivity bounded between 0.1 and 1.0

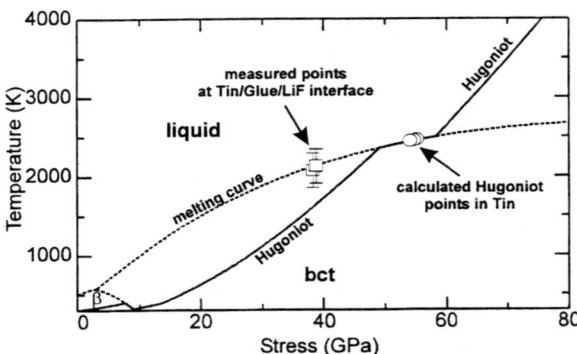

FIGURE 4 : Comparison between experimental point and phase diagram of tin

temperature measurements at the tin/LiF interface are coherent with the previous modelling.

CONCLUSION

The purpose of this paper was to analyze the influence of an interface in an optical pyrometry temperature measurement. The stresses obtained in three plate impact experiments ranges from 38 to 55 GPa in tin.

The main lessons drawn from these experiments are the following :
- the LOCTITE 358 glue used for interface joining remains transparent under shock,
- the tested glue thickness (from 10 µm to 50 µm) does not result in significant difference of the measured signal,
- the existence of a 1-µm thick graphite deposit on the tin target modifies considerably the shape of the signals.

These results confirm those obtained on bismuth [18], particularly the part of the glue interface under shock. The temperature measurements are consistent with the results obtained in a previous study on the phase change of tin under shock. If these initial results are confirmed in subsequent experiments, the experimental technique consisting of gluing a window on a metallic target would be an original and attractive alternative to the deposition technique due to its ease of use.

ACKNOWLEDGMENT

This work was supported by the French Ministry of Defense (Délégation Générale pour l'Armement). The autors are grateful to Pascal Bouinot and Yannick Sarrant for shock experiments.

REFERENCES

1. Kormer S. B., *Sov. Phys. Usp.* **11, 2** (1968) 229-254
2. Urtiew P. A. and Grover R. G.., *J. Appl. Phys.* **48, 3** (1977) 1122-1126
3. Lyzenga G. A. and Ahrens T. J., *Geophys. Res. Lett.* **7** (1980) 141-144
4. Kondo K. and Ahrens T. J., *Phys. Chem. Minerals.* **9** (1983) 173-181
5. Boslough M.B., Ahrens T. J. and Mitchell A.C., *Geophys. J. R. Astr. Soc.* **84** (1986) 475-489
6. Svendsen B. and Ahrens T. J., *High-Pressure Research in Mineral Physics.* (1987) 403-423
7. Costeraste J. and Pérez M., rapport CEG n° S87-0003 (1987)
8. Svendsen B., Ahrens T. J. and Bass J. D., *High-Pressure Research in Mineral Physics.* (1987) 403-423
9. Ahrens T. J., Tan H. and Bass J. D., *High Press. Res.* **2** (1990) 145-157
10. Tan H. and Ahrens T. J., *High Press. Res.* **2** (1990) 159-182
11. McQueen R. G. and Isaak D. G., *J. Geoph. Res.* **95, B13** (1990) 753-765
12. Yoo C. S, Holmes N. C., Ross M., Webb D. J. and Pike C., *Phys. Rev. Letters* **70, 25** (1993) 3931-3934
13. Mondot M., *Thèse du Conservatoire National des Arts et Métiers*, Paris (1993)
14. Blanco E., *Thèse de Doctorat*, Université Paris X - Nanterre (1997)
15. Léal B., *Thèse de Doctorat*, Université de Poitiers (1998)
16. Héreil P.L. and Mabire C., Conférence DYMAT, Cracow (2000)
17. Mabire C. and Héreil P.L., *Shock Compression of Condensed Matter*, Snowbird (1999)
18. Partouche D., Pelissier J.L. and Wetta N., rapport CEA/DAM (1998)

GATED IR IMAGES OF SHOCKED SURFACES

Stephen S. Lutz[1], W. Dale Turley[1], Paul M. Rightley[2], and Lori E. Primas[2]

[1]Bechtel Nevada (BN), Special Technologies Laboratory, 5520 Ekwill St., Suite B, Goleta, CA 93117
[2]Los Alamos National Laboratory (LANL), Los Alamos, NM 87545

Abstract. Gated infrared (IR) images have been taken of a series of shocked surface geometries in tin. Metal coupons machined with steps and flats were mounted directly to the high explosive. The explosive was point-initiated and 500-ns to 1-microsecond-wide gated images of the target were taken immediately following shock breakout using a Santa Barbara Focalplane InSb camera (SBF-134). Spatial distributions of surface radiance were extracted from the images of the shocked samples and found to be non-single-valued. Several surfaces were modeled using CTH, a 2- or 3-dimensional Eulerian hydrocode.

INTRODUCTION

Experimental methods used to determine the state of a shocked material are extensive and in many cases return data with a reported accuracy of a few percent(1). Although significant progress has occurred in the advancement of diagnostic methods in shock experiments, one opportunity for improvement is the extension of a single-spatial-point measurement to a high-resolution multipoint image. A number of single-spatial-point studies have been reported describing the measurement of time-resolved temperature change on the surface of a metal subjected to shocks up to 100 GPa (2-7). These researchers report the ability to resolve a 100°K temperature rise above ambient to a few percent accuracy, with time resolution on the order of 100 ns.

With recent advances in single-pixel detectors have come significant improvements in gated thermal cameras. One such example is the SBF-134. This camera is based on an indium antimonide (InSb) detector, bump bonded to a silicon CID array readout structure. With a gate time of 500 ns, we found that ~500 photons/pixel were required to produce an image with 2:1 signal-to-noise ratio. In practical terms, this camera is able to resolve distinct facial features of a human using a 500 ns gate width. We used this camera to capture free-surface thermal images of metal samples shocked to pressures on the order of 20 GPa. By modifying surface finish and surface structure we were able to affect the apparent residual temperature of the shocked materials. We modeled shock-induced temperature rise and compared the results with the experimental measurements.

EXPERIMENTAL

Experiments were carried out in a RISI explosive test chamber. Signal from the shocked coupon was reflected off a gold-coated mirror, through a sapphire viewing port and imaged onto the camera focal plane. A sketch of a typical high explosive experiment package is shown in Figure 1. A RP80 detonator single point initiated a two gram charge of Detasheet (Dupont C2-EL-506). The Detasheet charge directly contacted the metal coupon under study. The whole assembly was housed in a black Delrin package.

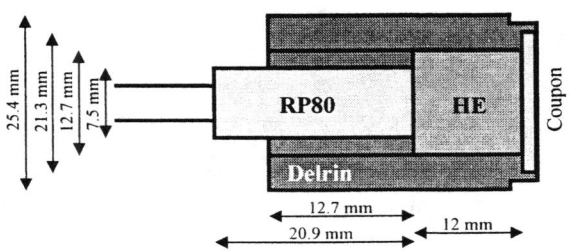

FIGURE 1. Shot package design.

TABLE 1. Physical Characteristics of Tin Targets and Camera Gate Times; t_o is Shock Breakout Time

#	Geometry	Surface Finish	Step Height (mm)	Camera open and close gate times relative to shock breakout (ns)		
				Thin Step (open,close)	Thick Step (open,close)	Flat (open,close)
1	Step	Machine finish (#8)	1.0	t_0+40, t_0+540	t_0-230, t_0+270	
2	Flat	Polished w/3 μm Al_2O_3				t_0-230, t_0+270

Metal samples were prepared with three different surfaces: single step, periodically spaced grooves, and flat polish. Although a variety of surfaces were studied, only single step and flat polished tin (99.9%) data are presented in this paper. The details of sample preparation and IR camera gate times (open and close), relative to shock breakout, are summarized in Table 1. All metal samples were nominally 2 mm thick and 21 mm in diameter.

Camera (SBF-134) Setup and Calibration

General performance characteristics of the SBF-134 are summarized in Table 2. A detailed temperature calibration was performed using a blackbody source placed in the explosive chamber at the target position. However, for this publication radiance data was not temperature converted. We intend to present temperature calibrated data in a future report.

Pressure Calibration and Timing Measurements

The pressure generated by the explosive in the metal coupon was calculated from Hugoniot and measured shock velocity data. The shock velocity was obtained by measuring the time required for the shock wave to travel across a 1 mm step, machined in a test sample. The step was coated with a thin (<25 μm) film sensor that produces a prompt (<10 ns rise) burst of light when shocked. This thin film is a mixture of cerium doped lutetium orthosilicate (LSO:Ce) phosphor powder suspended in a silica glass binder. An optical fiber positioned 2 mm from the sample surface simultaneously views the high and low surfaces of the step. When the shock arrives the sensor emits a burst of light that is detected with a photomultiplier tube and recorded with a transient digitizer. Using this method, a pressure of 20±3 GPa was determined for the tin experiments.

For each gated imaging experiment, a 3-mm-diameter spot of the previously-discussed shock arrival sensor was applied approximately 5 mm from the target center. Shock-induced light from this sensor was recorded simultaneously with the high explosive and camera triggers. Such cross timing is essential to interpret the acquired IR image. The jitter between the trigger used to initiate the high explosive and the time of arrival of the shock at the target surface was found to be on the order of 200 ns.

RESULTS

Figures 2 and 3 show a set of images of samples 1 and 2 in Table 1. In each of these figures, a gated IR image of the shocked target surface (a) is displayed next to a white light image of the pre-shocked sample (b) taken with a digital camera. A "log counts" grey scale map is used to display the 14-bit digitizer range and to highlight the subtle features in the shocked metal surface.

Figure 2 (a) shows the results from the stepped tin. Notable features include high radiance along the step boundary, and an offset in radiance between the thin and thick steps. The bright circular feature is the high-emissivity phosphor LSO:Ce sensor. The semi-circular feature marking the outer edge of the lower half of the coupon results from motion of the thinner step, presumably with hot gasses escaping from around the edge.

TABLE 2. Camera Performance Characteristics

Detector	Array Size	Pixel Size (μm)	Quantum Efficiency (3-5 μm)	Bit Depth	Photons/Pixel Required for 2:1 SNR	Minimum Gate Width (ns)
InSb	256 x 256	30	~85%	14	500	140

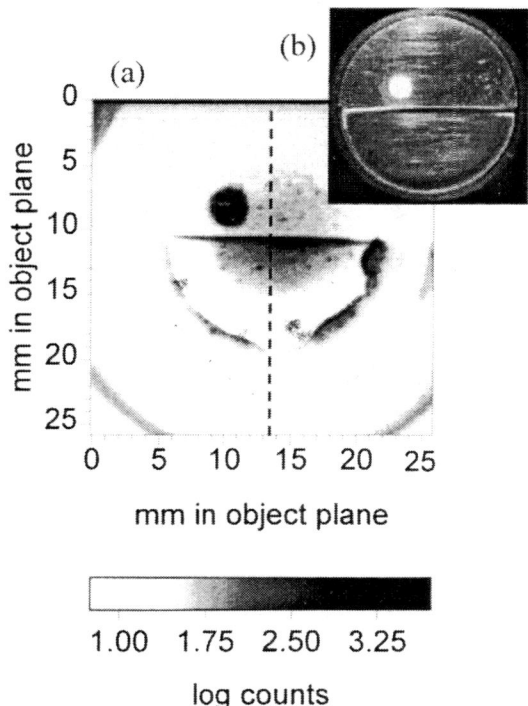

FIGURE 2. (a) IR image of shocked stepped tin coupon; (b) preshot picture of test coupon.

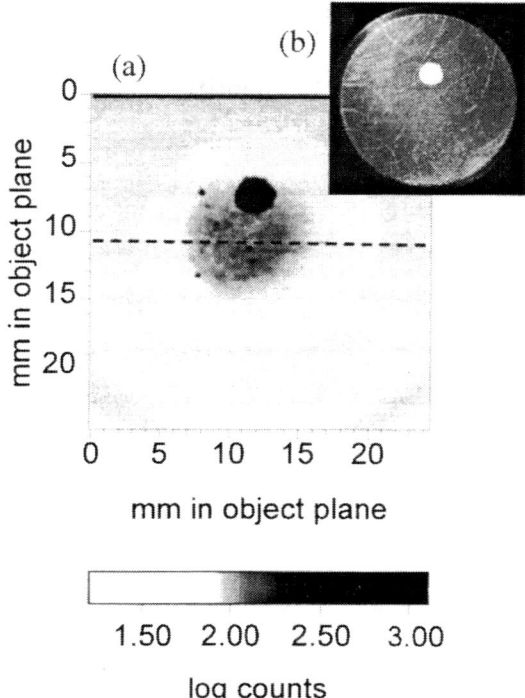

FIGURE 3. (a) IR image of shocked polished tin coupon; (b) preshot picture of test coupon.

Figure 3 (a) shows the results from polished tin. Most notable is the variance in the radiance field. Distinct circular features, some hot and some cold, relative to the mean, are clearly evident. The variance is far in excess of the statistical limit imposed by the detection system. Figures 4 and 5 show normalized radiance values taken from locations marked by dashed lines in Figures 2(a) and 3(a).

MODELING

We have simulated the tin experiments using Sandia National Laboratories' Eulerian mesh CTH code. Our rectangular, 2-dimensional modeling included the Detasheet, Delrin and tin, but did not include the actual detonator geometry (Figure 1). A detonation wave was started at a point corresponding to the center of the output area of the detonator. The equation of state for the explosive products was approximated using JWL parameters for PETN (the principle component of Detasheet). Mie-Grüneisen equation of state data were used for Delrin and tin. Delrin was modeled as an elastic, perfectly plastic material while a Steinberg-Guinan-Lund strength model was used for tin.

Lagrangian tracer particles were placed into the simulation at equal intervals along the radius just under the surface of the coupon. The depth of the tracers was approximately one mesh cell below the surface for both samples. A point approximately one third of the way from the center of the coupon (approximating the location of the breakout sensor) was used to determine breakout timing. Breakout was considered to occur when the temperature at this point rose from ambient to 300°K. Time histories of temperature at each tracer were produced using the experimental timing of the camera gate relative to break-out. To compare the modeled temperature with the measured radiance data it was necessary to integrate the radiance estimated from the modeled temperatures over the duration of the camera gate. Figures 4 and 5 show normalized experimental results over-plotted with the corresponding CTH data.

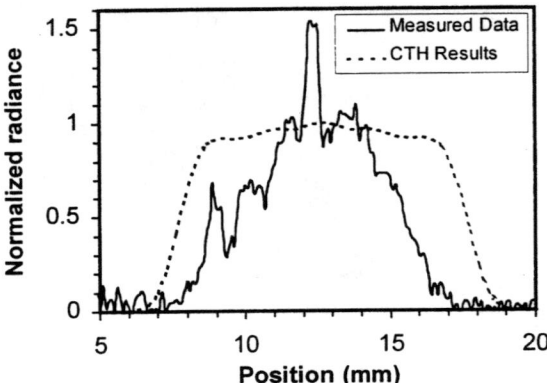

FIGURE 4. Sample 1, polished tin, measured and modeled results.

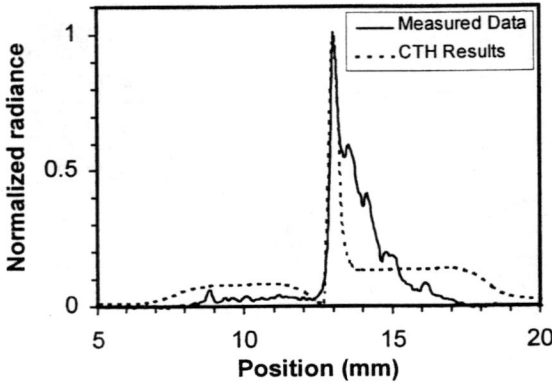

FIGURE 5. Sample 2, stepped tin, measured and modeled results.

CONCLUSION

The large-scale features of the CTH calculations are in general agreement with the radiance data. However, CTH calculations do not account for the magnitude of the variations seen in the radiance data on a sub-millimeter scale. At this scale, differences between the modeled and measured results may be due to grain structure or other features not accounted for in the model. The wealth of data obtained from two-dimensional radiance measurements over a wide range of spatial scales is important to understanding the accuracy and precision limitations of single-point pyrometry measurements. Further, this technique may provide impetus for creation and validation of new constitutive models.

ACKNOWLEDGEMENTS

We would like to thank Arn Adams of Santa Barbara Focalplane for loan of a SBF-134 camera, Mike Grover and Guy Leach of BN for technical support, Ron Boat of LANL for design of the explosive package, and Dave Holtkamp and Mark Wilke of LANL for review and critique of the overall experiment.

This work was supported by the Department of Energy, Nevada Operations Office, under Contract No. DE-AC08-96NV11718. DOE/NV/11718–542. LA-UR-01-3319. Reference herein to any specific commercial product, process, or service by trade name, trademark, manufacturer, or otherwise, does not necessarily constitute or imply its endorsement, recommendation, or favoring by the United States Government or any agency thereof or its contractors or subcontractors. The views and opinions of authors expressed herein do not necessarily state or reflect those of the United States Government or any agency thereof.

REFERENCES

1. Asay, J.R., and Shahinpoor, M., *High-Pressure Shock Compression of Solids*, Springer-Verlag, New York, 1993, pp 43–70.
2. Taylor, J.W., "Residual Temperatures of Shocked Copper," J. Appl. Phys. **43(9)**, 2727-2731 (1963).
3. Von Holle, W.G., and Trimble, J.J., "Temperature Measurement of Shocked Copper Plates and Shaped Charge Jets by Two-Color IR Radiometry," J.Appl. Phys. **47(6)**, 2391-2394 (1976).
4. Von Holle, W. G., "Shock Wave Diagnostics by Time-Resolved Infrared Radiometry and Non-Linear Raman Spectroscopy," in *Shock Waves in Condensed Matter*-1983, edited by J.R. Asay, et. al., Elsevier Science B.V., 1984, pp. 283-291.
5. Perez, M., "Residual Temperature Measurements of Shocked Copper and Iron Plates by Infrared Pyrometry", in *Shock Compression of Condensed Matter*-1991, edited by S.C. Schmidt, et. al., Elsevier Science B.V., 1992, pp. 737-740.
6. Blanco, E., Remiot, C., Mexmain, J., Herve, P., "Temperature of Shocked Materials Measured with an Infrared Pyrometer through a Window," in *Infrared Technology and Applications XXII*-1996, edited by B.F. Andresen et. al., SPIE proceedings #2744, pp. 677-683.
7. Mabire, C., Hereil, P., "Shock Induced Polymorphic Transition and Melting of Tin," in *Shock Compression of Condensed Matter*-1999, edited by M.D. Furnish, et. al., AIP Conference Proceedings, 2000, pp.93-96.

OPTICAL PROBING OF THE ELECTRON TEMPERATURE GRADIENT

T. Ao, I. Vollrath, A. Ng

*Department of Physics & Astronomy, University of British Columbia,
Vancouver, British Columbia, Canada V6T 1Z1*

Abstract. Previous pyrometric measurements on shock waves in silicon have revealed the existence of different electron and ion temperature gradients in the shock front. This was attributed to the relatively low energy equilibration rate between electrons and ions in the strongly coupled plasma that exists in a shock wave. In this paper, we will describe a new approach to assess this effect based on the reflectivity and change in phase of a P- and S-polarized probe reflected from a shock front in-flight in silicon. Predictions from numerical simulations will be presented and discussed.

INTRODUCTION

Ideally, a shock wave is treated as a propagating discontinuity in all of the thermodynamic variables. However, finite gradients can exist due to relaxation processes occurring in the shock compressed material; namely the energy exchange or equilibration between electrons and ions. In this paper, the effect of the equilibration rate between electrons and ions on the electron thermal gradient at the shock front is discussed. In addition, a new approach using the reflectivity and reflected phase of P- and S-polarized light to measure the strength of electron-ion coupling is described.

BACKGROUND

Initially, shock compression of a material leads to heating of the ions. The electrons are subsequently heated via electron-ion Coulomb collisions. Thermal equilibrium between electrons and ions would only be established somewhere behind the shock front. However, at the shock front the temperature of the electrons may deviate from that of the ions depending on the rate of energy exchange between the electrons and ions. In this paper, the process of equilibration is described using a phenomenological electron-ion coupling coefficient g. Since no simple analytical expression for g is available in the regime of interest, it is left as a free parameter. The energy equations for the electrons (1) and ions (2) are given by the following equations,

$$\frac{\partial}{\partial t}(\rho E_e) = -\frac{\partial}{\partial x}\left[\rho u\left(E_e + \frac{P_e}{\rho}\right) - \kappa\frac{\partial T_e}{\partial x}\right] + u\frac{\partial P_e}{\partial x} - g\frac{\rho}{\rho_o}(T_e - T_i) \quad (1)$$

$$\frac{\partial}{\partial t}\left(\rho E_i + \frac{\rho u^2}{2}\right) = -\frac{\partial}{\partial x}\left[\rho u\left(E_i + \frac{u^2}{2} + \frac{P_i}{\rho}\right)\right] - u\frac{\partial P_i}{\partial x} + g\frac{\rho}{\rho_o}(T_e - T_i) \quad (2)$$

The difference between the electron and ion temperatures in a shock wave due to a finite electron-ion energy exchange rate is shown in Fig. 1. It is evident that at the shock front the electron and ion temperatures differ substantially. In addition, there is a noticeable difference in the temperature gradients between the electrons and ions. At sufficiently high electron-ion coupling coefficient values, these temperature profiles will collapse together and the shocked regime may be treated as a single-temperature plasma.

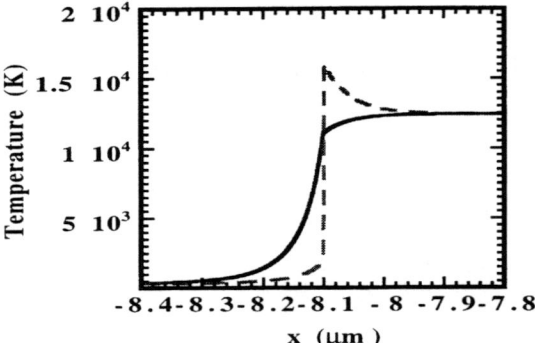

FIGURE 1. Electron temperature (solid line) and ion temperature (dashed line) profiles of a 10 km/s shock in Si for $g=10^{17}$ W/m³-K.

So far, observations of the electron-ion equilibration rate within a shock wave have relied upon emittance (spectral radiance) and absorbance (emissitivity) measurements. The first emittance measurement was obtained by Celliers et al. in 1992 on laser-driven shock waves in-flight in silicon.[1] For shock speeds of 15 to 20 km/s, the observed emittance were found to be a factor of 20 to 50 lower than that predicted for an equilibrium shock state. These results gave the first assessment of the electron-ion coupling coefficient in a shock wave in silicon to be about 10^{16} W/m³-K.

More recently, Lower et al.[2,3] were able to measure simultaneously the emittance and absorbance of x-ray driven shock waves in silicon and aluminum. Their results were also consistent with that of low g values of 10^{16} and 10^{17} W/m³-K in silicon and aluminum, respectively.

Meanwhile, theoretical predictions of electron-ion equilibration rates in strongly coupled plasmas show large variations. The results based on the traditional simple Spitzer model,[4] and the Fermi Golden Rule and coupled mode calculations of Dharma-wardana and Perrot,[5] are presented in Table 1. Also included is the value given by More[6] who extended the Spitzer formulation by using a maximum scattering cross-section which is consistent with the minimum electron mean-free path.

g (W/m³-K)	Method
2×10^{18}	Spitzer[5]
6×10^{17}	Fermi Golden Rule[5]
5×10^{16}	Coupled Mode[5]
2×10^{17}	Modified Spitzer[6]

TABLE 1. Theoretical calculations of the energy relaxation rate of 3 eV electrons in Al at solid density with the ions kept at the melting point.

NEW APPROACH

The following is a description of an alternative approach for studying the electron-ion equilibration rate within a shock wave. First of all, at the shock front a gradient in electron temperature leads to a gradient in the electrical conductivity, σ_ω, which in turn would lead to a gradient in the dielectric function, ε_ω. Now, consider an electromagnetic wave reflected from a shock wave in-flight in silicon. The amplitude and change in phase of the reflected electromagnetic wave would depend upon not only the dielectric value, but also the gradient of ε_ω at the shock front. In addition, the motion of the shock front would affect the amplitude and phase of the reflected electromagnetic wave.

An electromagnetic wave incident upon a surface may be in either P- or S-polarization. It is known that P- and S-polarized light have different sensitivities to the gradient in ε_ω because of resonance absorption that occurs in P-polarized light. Thus, examinations of the reflected amplitude and phase of both P- and S- polarized light may provide insight into the electron-ion equilibration rate in a shock wave. A plausible experimental scheme using this alternative approach is shown in Fig. 2.

FIGURE 2. Experimental schematic of new approach.

Silicon is chosen to be the material of study since unperturbed silicon is nearly transparent at near-infrared wavelengths. The target would consist of a sample of silicon sandwiched between an aluminum pusher layer and anti-reflective layer. A steady shock wave is launched into the sample layer, while a probe laser is incident obliquely on the AR-coated free surface of the sample layer.

To illustrate this approach, a 1-D hydrodynamic code is used to simulate a desired shock wave. In the 1-D hydrodynamic code, the shock is treated as a two-temperature electron-ion fluid. The equation of state used was based on the QEOS model of More et al.[7] The electrical and thermal conductivities used were based on the dense plasma conductivity model of Lee and More.[8] As stated earlier, a phenomenological coupling coefficient g was used to describe the electron-ion equilibration rate.

The interaction of an electromagnetic wave with the shock wave is treated by solving the Helmholtz equations for P- and S-polarizations.[9] The dielectric function of the shock material is obtained using the Drude model, where the collision frequency, ν_{ei}, is related to the DC conductivity, σ_0.

Shown in Fig. 3 are the reflectivities as a function of time for the P- and S-polarized probes. Due to the propagation of the shock wave towards the free surface, the reflectivities steadily increase as less and less unperturbed silicon remains ahead of the shock front.

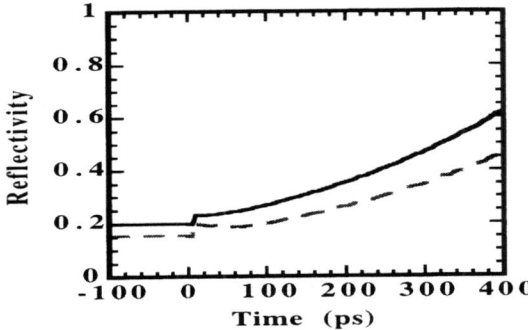

FIGURE 3. Reflectivity of P- (dashed line) and S-polarized (solid line) 800 nm probes @ 45° off a 10 km/s shock in-flight in Si. Time zero refers to the point when the shock reaches the Al/Si interface. ($g=10^{16}$ W/m^3-K)

The effect of shock propagation is suppressed by comparing the ratio of the reflectivities, (R_P/R_S), which remains nearly constant after an initial transient behavior, as shown in Fig 4. This initial jump in (R_P/R_S) is due to thermal conduction across the Al/Si interface. However, the reflectivity ratio exhibits an interesting dependence on the electron-ion coupling coefficient, as shown in Fig. 5.

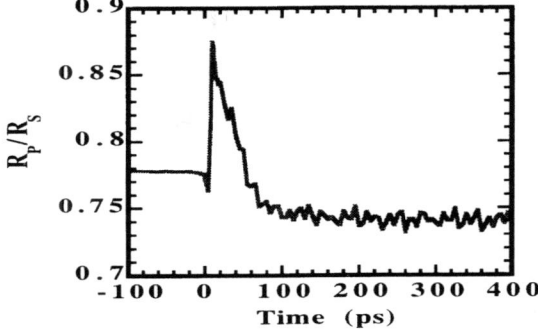

FIGURE 4. Ratio of reflectivity of P- and S-polarized 800 nm probes @ 45° off a 10 km/s shock in-flight in Si as a function of time. ($g=10^{16}$ W/m^3-K)

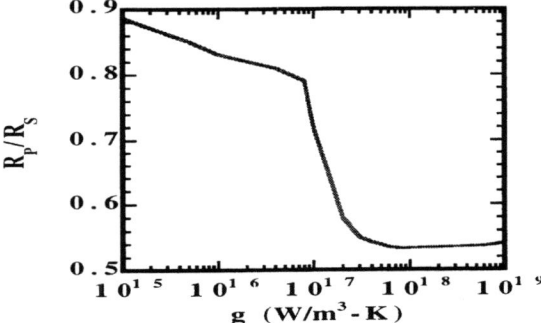

FIGURE 5. Ratio of reflectivity of P- and S-polarized 800 nm probes @ 45° off a 10 km/s shock in-flight in Si at various electron-ion coupling.

Fig. 6 shows the reflected phases for the P- and S-polarized probes. No phase change occurs until the shock reaches the Al/Si interface, but after that the motion of the shock dominates the change in the both of the reflected phases.

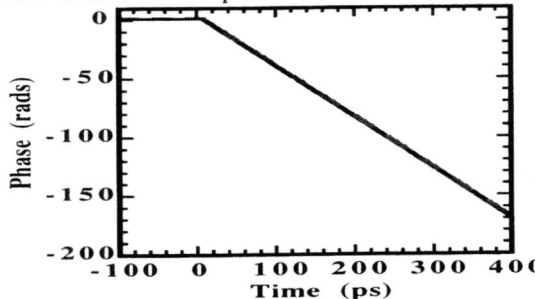

FIGURE 6. Reflected phase of P- (dashed line) and S-polarized (solid line) 800 nm probes @ 45° off a 10 km/s shock in-flight in Si. Time zero refers to the point when the shock reaches the Al/Si interface. ($g=10^{16}$ W/m^3-K)

However, Fig. 7 shows that the difference between the reflected phases, ($\delta_P - \delta_S$), after the initial transient effect, remains constant during the shock propagation.

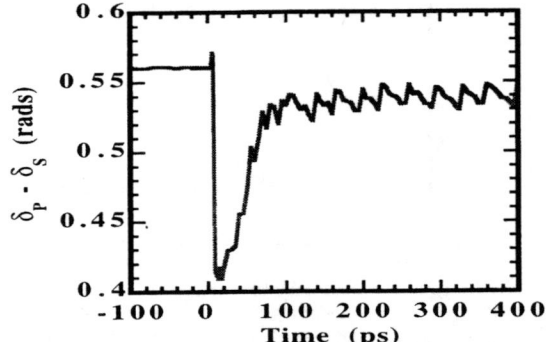

FIGURE 7. Differential phase change of P- and S-polarized 800 nm probes @ 45° off a 10 km/s shock in-flight in Si as a function of time. ($g=10^{16}$ W/m^3-K)

A comparison of this differential phase change between P- and S-polarizations for various g values is shown in Fig. 8.

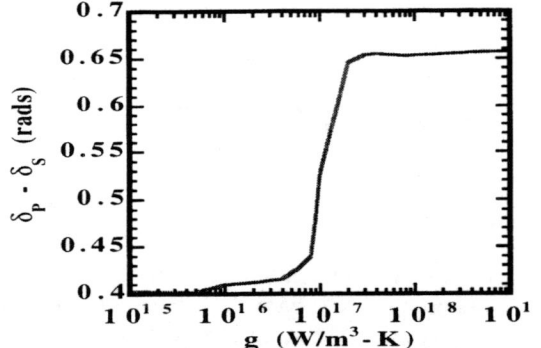

FIGURE 8. Differential phase change of P- and S-polarized 800 nm probes @ 45° off a 10 km/s shock in-flight in Si at various electron-ion coupling.

DISCUSSION AND CONCLUSIONS

It should be noted that temporal variations in the reflectivities of the P- and S-polarized probes are governed by the motion of the shock wave due to the change in the amount of unperturbed material ahead of the shock front attenuating the probe signals. On the other hand, temporal variations in the reflected phases of the P- and S-polarized probes are dominated by the Doppler motion of the shock front. Both of the observations can be used to provide a direct measurement of the shock speed, thus giving a unique assessment of the shock state.

Measurements of the ratio of reflectivities of P- to S-polarized probes to examine the strength of electron-ion coupling in strongly coupled plasmas should be viable using impact generated shock waves. Meanwhile, a high sensitivity diagnostic such as frequency domain interferometry may be used to assess the value of g using measurements of reflected phases of P- and S-polarized probes. The latter method would be more suitable in laser generated shock wave experiments.

In conclusion, a new approach for an independent assessment of the thermal equilibration rate in a shock wave has been presented. This approach is applicable to other samples, provided they have the following properties. First, the unperturbed sample must be characterized by low photo-absorption to allow the probe light to reach the shock front with little attenuation. Second, a high reflectance in the shocked state is required for high signal levels for reflected probe measurements. Accordingly, an interesting class of samples to study would be ionic crystals.

REFERENCES

1. Celliers, P., Ng, A., Xu, G., and Forsman, A., *Phys. Rev. Lett.* **68**, 2305-2308 (1992).
2. Lower, T., Kondrashov, V.N., Basko, M., Kendl, A., Meyer-ter-Vehn, J., and Sigel, R., *Phys. Rev. Lett.* **80**, 4000-4003 (1998).
3. Basko, M., Lower, T., Kondrashov, V.W., Kendl, A., Sigel, R., Meyer-ter-Vehn, J., *Phys. Rev. E* **56**, 1019-1031 (1997).
4. Spitzer, L., *Physics of Fully Ionized Gases*, Interscience, New York, 1962, pp. 131-136.
5. Dharma-wardana, M.W.C., and Perrot, F., *Phys. Rev. E* **58**, 3705-3718 (1998).
6. More, R. (private communication).
7. More, R.M., Warren, K. H., Young, D. A., and Zimmerman, G. B., *Phys. Fluids* **31**, 3059-3078 (1988).
8. Lee, Y. T., and More, R. M., *Phys. Fluids* **27**, 1273-1286 (1984).
9. Celliers, P., and Ng, A., *Phys. Rev. E* **47**, 3547-3565 (1993).

ELLIPSOMETRY IN THE STUDY OF DYNAMIC MATERIAL PROPERTIES

Andrew W. Obst[1], Keith R. Alrick[1], William W. Anderson[2], Konstantinos Boboridis[1], William T. Buttler[1], Steve K. Lamoreaux[1], Bruce R. Marshall[3], Stefanie L. Montgomery[1], Jeremy R. Payton[1], and Mark D. Wilke[1]

[1]*Physics Division, MS H803, Los Alamos National Laboratory, Los Alamos, NM 87545*
[2]*DX Division, MS P952, Los Alamos National Laboratory, Los Alamos, NM 87545*
[3]*Bechtel Nevada, Special Technologies Laboratory, 5520 Ekwill Street, Suite B, Santa Barbara, CA 93111*

Abstract. Measurements of the time-dependent absolute temperature of high-explosive (HE) shocked surfaces provide valuable constraints on the equations-of-state (EOS) of materials and of the state of ejecta from those surfaces. In support of these dynamic surface temperature measurements, we are developing techniques for measuring the dynamic surface emissivity of shocked metals in the near-infrared (IR). These consist of time-dependent laser polarimetric measurements, using several approaches. We include here a discussion of these polarimeter techniques. Polarimetry permits an accurate determination of the dynamic emissivity at a given wavelength, and may also provide a signature of melt in shocked metals.

INTRODUCTION

HE-shocked surface temperatures lie in the range of 0.04 to 0.2 eV, corresponding to shock-heating of surfaces to temperatures from about 400 K to 2000 K. This is equivalent to IR wavelengths from about 1 to 6 microns. The dynamic (and in some cases static) emissivities over these wavelength ranges are not well known, and can only be approximately inferred from multichannel pyrometer measurements, limiting the accuracy of these temperature measurements. Polarimetry or ellipsometry (1)(2), illustrated in Figure 1, permits a complementary determination of the dynamic emissivity by measuring the real and imaginary parts of the index of refraction. These emissivity measurements can be used with pyrometer data at similar wavelengths to calculate true temperatures. Dynamic emissivity data may also provide an indication of phase change. Once the emissivity is derived at one wavelength, a multichannel pyrometer permits determination of the emissivity at all the other pyrometer wavelengths, since the temperature is independent of wavelength.

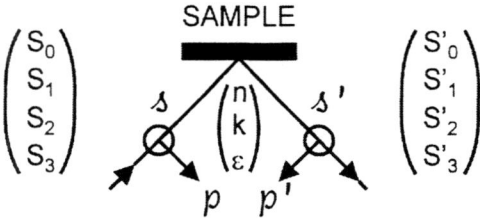

FIGURE 1. Graphic depiction of laser ellipsometry. A laser beam in a known polarization state **S** from a polarizer is reflected off a surface. The polarization state **S'** of the scattered or reflected beam is measured in the analyzer. Knowledge of both **S** and **S'** permits an inference of the optical properties (n,k,ε) of the sample surface at the laser wavelength.

EMISSIVITY MEASUREMENTS

Typical spectral behaviour of the emissivity in the near IR is illustrated in Figure 2. Multichannel pyrometric temperature studies of dynamically-shocked surfaces utilize emissivities bounded at the lower end by the static value and at the higher end by 1, the maximum possible value. Depending on the material and the speed of the shock, this may or may not be valid at a phase transition. For example, in the case of a non-magnetic niobium wire, explosively driven by high current, the emissivity drops through melt. This can be seen in Figure 3. In the case of a high-current driven magnetic nickel wire, on the other hand, at melt the emissivity is seen to rise, as shown in Figure 4. Surface roughness, for example, will certainly increase the emissivity, in the absense of melt. The dynamic emissivity may therefore be a more sensitive indicator of melt than the radiance temperature.

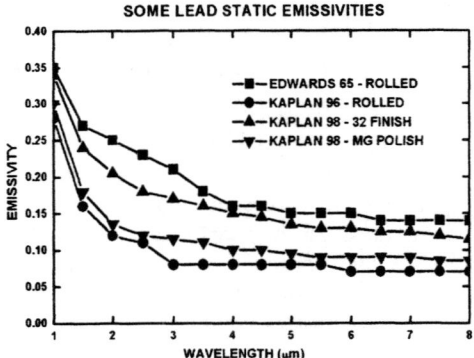

FIGURE 2. Note the large variation of absolute values, depending on the surface finish. Data are private communication.

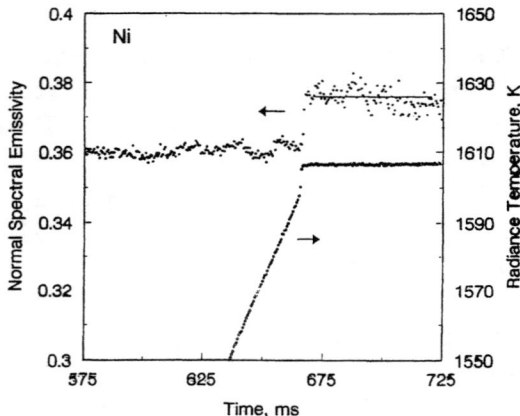

FIGURE 4. Increase in emissivity at the melt transition in an exploding magnetic nickel wire. Data are private communication.

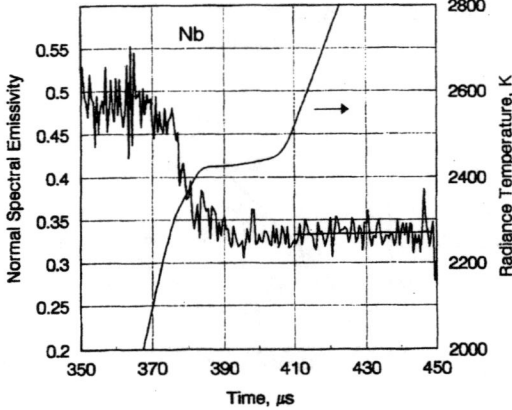

FIGURE 3. Reduction of emissivity at the melt transition in an exploding non-magnetic niobium wire. The radiance temperature assumes unity emissivity. Data are private communication.

EXPERIMENTAL TECHNIQUES

Three ellipsometers are being developed using three somewhat different approaches. All three are currently operating at 1550 nanometers, which is about the longest wavelength at which both visible-optics work and an industry-standard laser is available. This relatively short wavelength should still permit some overlap with pyrometric temperature measurements below 1000 K. It is hoped that once this wavelength regime has been explored, we will develop infrared ellipsometers up to about 6 microns. However these will pose some technological challenges. In the presence of shock background light, an ultimate modulation-frequency goal is 1 gigahertz with 12-bit recording for 1000:1 effective dynamic range. However again it will be some time before this speed will be realized.

A commercial reference instrument has been purchased from Containerless Research (CRI) in Evanston, Illinois. This instrument has been upgraded to meet our demanding field requirements and is being tested. A conservatively-built 4-channel instrument, some of its features are as follows. Laser power is currently 200 milliwatts, upgradeable to 2000 milliwatts. Rougher surface finishes will require this power. Modulation frequency is currently 50 MHz, and is limited by the following considerations. The 4-channel sampling frequency, always twice the modulation frequency, is currently limited to 100 MHz at 12 bits. However also the existing four 2-mm diameter InGaAs detectors are limited to 100 MHz. Faster speeds can be realized by underfilling, say, 1-mm or even 0.5-mm diameter detectors. This instrument is shown schematically in Figure 5.

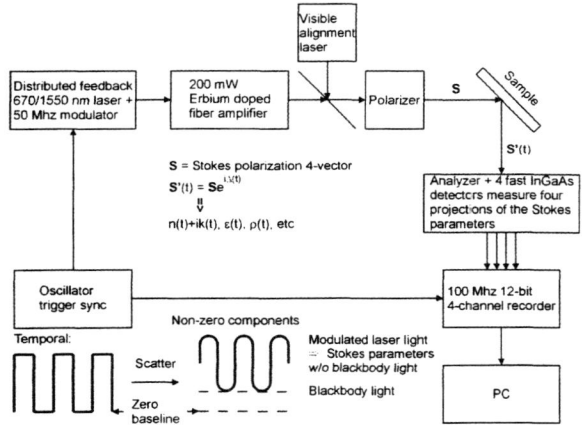

FIGURE 5. Schematic layout of CRI ellipsometer.

A second field instrument is being assembled with slightly different features. Instead of 4 channels it will have 6 balanced channels, offering some redundancy in remote field operations under adverse conditions. Using smaller fiber-coupled pin-diode detectors, a modulation frequency of 200 MHz has already been tested with stacked 8-bit recording. The laser power output is 1200 milliwatts at 1550 nonometers.

A third instrument is being assembled at Bechtel Nevada's Special Technology Laboratory (BN/STL) in Santa Barbara, California. This instrument also uses three channel-pairs, for a total of six channels. The BN/STL instrument will have several noteworthy features. Polarization effects in beamsplitters will be avoided by using near-normal incidence. As in the second instrument, complementary outputs provide background-light and common-mode noise suppression. Minimizing the number of surfaces avoids unwanted reflections. Equal coupling is only required within each pair of fibers, not between all pairs. Sensitivity to angular tilt should be reduced. Finally, the construction will be simple and stable to avoid adjustments in the field. Laser power at 1550 nanometers will be 2000 milliwatts. As in the second instrument, stacked 8-bit recording will be used.

ERRORS AND UNCERTAINTIES

Sources of error include alignment errors and polarization impurities as well as noise sources. Also there may be incomplete or incorrect polarization of the probe beam, probe beam laser noise, calibration errors, digitizer clocking errors and jitter, electronics noise, poor sample reflectivity, and depolarization of the probe beam by the sample roughness so that the analytical form of the Mueller matrix is not applicable. Furthermore noisy signals may result from the geometry of the measurement, as in the case of exploding wires or loss of surface reflectivity at shock breakout in the case of explosively shocked surfaces. Nevertheless the ongoing program of modeling the CRI ellipsometer, for example, indicates that typical errors, noise and imprecision produce tolerable errors in the output.

GOALS AND SUMMARY

Near and long-term objectives include acquisition of emissivity data at near-IR wavelengths for materials of interest. These should improve the accuracy of pyrometric temperature measurements. Also, variation of emissivity at melt may provide a signature of phase transitions, spall, and ejecta characteristics. The spatial resolution of this technique is on the order of 1 to 10 mm.

Comparison of diamond-anvil measurements at static high pressure and temperature with shock conditions at the same high pressure and temperature will be interesting, since the crystal structure may be different in the two cases. A wide variety of diagnostics are anticipated for these instruments. Local testing will include co-operative efforts in neutron resonance spectroscopy (NRS), laser shock-physics, gas-gun shock experiments, HE-shock tests at BN/STL, and exploding wire work. Shown in Figure 6 is the CRI instrument setup at the local single-stage gas gun.

REFERENCES

1. Krishnan, S., *J. Opt. Soc. Am. A*, **9,** 1615-1622 (1992).

2. Kliger, D.S., Lewis, J.W., and Randall, C.E., *Polarized Light in Optics and Spectroscopy*, Academic Press, San Diego, 1990.

ELLIPSOMETRY AT THE GAS GUN

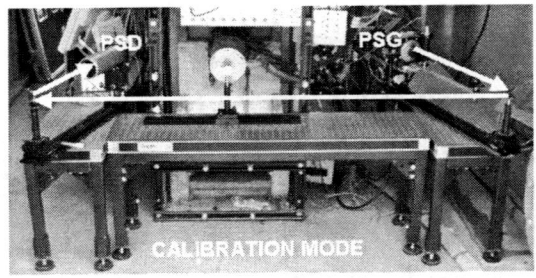

FIGURE 6. CRI reference ellipsometer at the single-stage gas gun. Here PSG and PSD refer to the polarization state generator and detector, respectively. After calibration and before the shot, the two rails are toed back out. Nothing is touched above the rails.

SHOCK INDUCED BIREFRINGENCE IN LITHIUM FLUORIDE

Jeffrey H. Nguyen and Neil C. Holmes

Physics and Advanced Technology, Lawrence Livermore National Laboratory, Livermore CA 94551

Abstract. We have used an ellipsometer to measure the birefringence of lithium fluoride in shock compression experiments. In previous x-ray diffraction experiments, single crystal [100] LiF has been reported to remain cubic at moderate pressures.

INTRODUCTION

Lithium fluoride is a commonly used window in shock compression experiments. It has been reported to remain transparent at megabar pressures [1]. Since data are collected through the window, it is important that its optical properties are well characterized. In fact, its opacity and emission are some of the properties examined by various researchers [2]. For studies where polarization of light is measured, there is currently no data on its birefringence except at very low pressures [3-10].

Under stress of less than 1 kilobar, lithium fluoride exhibits birefringence [3-10]. Yet, under large shock compression along the [100] direction, LiF unit cell is compressed isotropically [11-14]. Shock along the [111] direction, however, yields uniaxial compression.

EXPERIMENTAL PROCEDURE

We have developed a miniaturized ellipsometer to look at the optical properties of the window under shock compression. Ellipsometry is well developed for thin film measurements as well as for emissivity studies [15]. The technique essentially measures the change in polarization of a reflected laser beam off a surface under study.

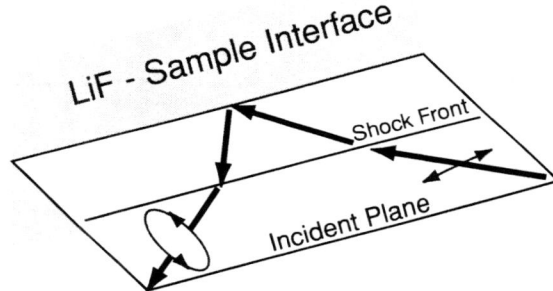

Figure 1. Path taken by the polarized light. Incident plane perpendicularly intersects the LiF-Sample interface. The shock front forms a boundary between the compressed and uncompressed sections of LiF window.

We used a LiF window in order to study the optical properties of the metal surface at high pressure. The window allows observations at the interface, and keeps the surface from releasing to ambient pressure. The target is made of an iron or aluminum disk and a [100] direction LiF window (see fig. 1). The window is coated with a 7000 Å layer of aluminum or iron, matching that of the disk. The LiF window is a circular disk with 25-mm diameter and 2 to 5 mm in thickness. We optimize the window thickness to allow sufficient time for observation and at the same time keep collecting optical fibers close to the metal-window interface.

Shock compression is a violent process for which sensitive optical measurements such as ellipsometry need a robust design. We accomplished that by enclosing the entire ellipsometer in a fiber bundle (fig. 2). The bundle is composed of seven optical fibers. One centrally located and the other six are arranged along the perimeter of the bundle. Polarizers cover the entrances of the six outer fibers. The polarizers are oriented at 30 degrees apart, covering a span from 0 to 150 degrees. The center fiber is left unpolarized. It is primarily used for intensity normalization.

As the shock front crosses the metal-window interface, it causes the interface to move and tilt slightly. This slight perturbation of the reflecting surface is sufficient to for a precise optical system to lose signal. We compensate by expanding the incoming 532 nm laser beam to cover the whole surface of the reflecting surface being shocked. The incoming and outgoing light typically enters or exits the window at roughly 45-degree angles.

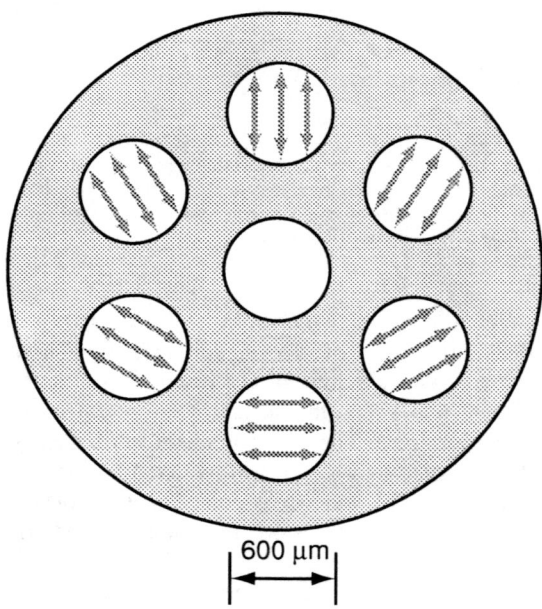

Figure 2. Front end of fiber bundle. Polarizers oriented at 30 degrees apart cover the six outer optical fibers. Central fiber is not polarized.

RESULTS

Propagation of the light through the air-LiF interface, LiF window, and reflection off the LiF-metal interface can be characterized a priori. As the shock front travels through the LiF window, two additional components become important to the light path: the shocked LiF window and the shocked and unshocked LiF-metal interface. Thicknesses of the LiF windows – both shocked and unshocked – change with the movement of the shock front, resulting in an optical system that changes with time. Moreover, the metal-window interface goes from an ambient pressure state abruptly to a high temperature and pressure state, forcing a change in the reflectivity of the polarized light. Polarization prior to shock compression can be described mathematically as

$$P = ARA. \qquad (1)$$

Polarization is affected by the LiF window, A, and the metal-window interface, R. We have ignored the air-LiF window interface in our calculation since it does not change during the relevant duration of the experiment.

As the shock front travels through the window, mathematical representation of the polarization state becomes

$$P(t) = A(t)IA'(t)R'[\theta(t)]A'(t)IA(t). \qquad (2)$$

The polarization state is affected by the distance it has to travel through the shocked and unshocked windows, A'(t) and A(t). I is the possible effect of traversing the shocked and unshocked window interface. Reflectivity at the metal-window changes abruptly as it undergoes compression. As the polarization of the light arriving at the interface changes as it travels through the windows, reflectivity at the interface changes with the polarization state R'[θ(t)].

Change in thickness of the windows result in a sinusoidal change in the polarization due to stress-induced birefringence in the window. However, a change in the reflectivity of the sample-window interface due to changing polarization is more complex.

DISCUSSION

Prior to the introduction of the shock wave at the interface and window, the output has become elliptically polarized since the light traverses the window and reflects off the interface. Polarization, and thus ellipticity, remains constant prior to the arrival of the shock front. After being shocked, the interface remains at a shocked steady state. Changes in the optical properties at the interface thus result in a discontinuous change in the polarization data. Optical properties of the shocked LiF window are only observed after the shock front enters the window.

In figure 3, we present light intensity vs. time at various polarization angles. The fibers measure output light at fixed polarization angles. As the distance the light has to travel through the sections of the window changes linearly with time, differences in thickness of the window sections contribute to the changes in the polarization, which manifest itself as the rotation of polarization.

As polarization changes with the growth of the shocked region of the window, the reflectivity of the metal-window interface changes with the state of polarization of the incoming light source. This results in a non-sinusoidal change in the output intensity. We note that if the shocked window material is isotropic in its optical character, i. e. no birefringence, the polarization in the window remains constant in time.

We performed an additional experiment to positively identify the source of the sinusoidal changes in the output signal. We replaced LiF with water as the window material. Water is transparent to visible light at pressures up to 25 GPa. Since it remains in the liquid phase for the duration of the experiment, shear strength is essentially nonexistent. Water at high pressure is thus not expected to change polarization of light traversing the window.

In figure 4, we present data taken with water as a window. At similar pressure and temperature, data taken with water exhibit no sinusoidal property. The data show changes at the water-sample interface as expected. The results clearly indicate that LiF is responsible for the sinusoidal change in polarization of the light beam.

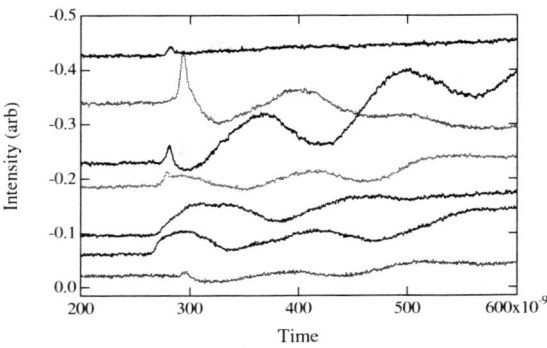

Figure 3. LiF window. Output of polarized light recorded at various polarizations. The topmost line is from the output of the unpolarized optical fiber. From the second line to the bottom are outputs of light polarized at 150, 90, 0, 60, 30, and 120 degrees. For clarity, these lines have been offset vertically. Input light is polarized at 45 degrees.

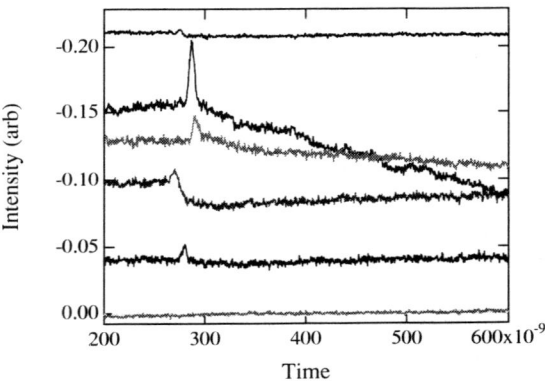

Figure 4. Water window. Output of polarized light recorded at various polarizations. The topmost line is from the output of the unpolarized optical fiber. From the second line to the bottom are outputs of light polarized at 150, 120, 0, 90, and 30 degrees. The last line is actually an overlap of 30 and 60 degrees lines. For clarity, these lines have been offset vertically. Input light is polarized at 45 degrees.

ACKNOWLEDGMENTS

We gratefully acknowledge the invaluable technical support provided by Samuel B. Weaver, William J. Metz, Jr., Steven J. Caldwell, James G. Van Lewen, Erikk A. Ojala and Jeff van Lue. We also benefited greatly from discussions with Dr. Dave Hare.

This work was performed under the auspices of the U.S. Department of Energy by Lawrence Livermore National Lab under contract No. W-7405-ENG-48

REFERENCES

1. J. E. Bailey, J. Asay, M. Bernard, A. L. carlson, G. A. Chandler, C. A. Hall, D. Hanson, R. Johnston, P. Lake and J. Lawrence, *J. Quant. Spect. Radiat. Transfer* **65**, 31 (2000).
2. J. L. Wise and L. C. Chhabildas, in *Shock Waves in Condensed Matter-1985* (Proceedings of the Fourth APS Topical Conference on Shock Waves in Condensed Matter, Spokane, WA, July 22-25, 1985), edited by Y.M. Gupta (Plenum, New York, 1986), p. 441.
3. I. I. Afanas'ev, L. K. Andrianova, T. V. Gracheva, and G. P. Zueva, *Sov. J. Opt. Techol.* **53**, 217 (1986).
4. A. F. Konstantinova, A. N. Stepanov, B. N. Grechushnikov, and I. T. Ulukhanov, *Sov. Phys. Crystallogr.* **35**, 247 (1990).
5. C. S. Chen, J. P. Szczesniak, and J. C. Corelli, *J. Appl. Phys.* **46**, 303 (1975).
6. J. P. Szczesniak, D. Cuddeback, and J. C. Corelli, J. Appl. Phys. 47, 5356 (1976).
7. K. V. Rao and T. S. Narasimhamurty, *Appl. Opt.* **9**, 155 (1970).
8. K. G. Bansigir and K. S. Iyengar, *Proc. Phys. Soc.* **71**, 225 (1958).
9. R. Srinivasan, *Zeitschrift für Physik* **155**, 281 (1959).
10. A. Rahman and K. S. Iyengar, *Acta Crsyt.* **20**, 144 (1966).
11. P. A. Rigg and Y. M. Gupta, *Appl. Phys. Lett.* **73**, 1655 (1998).
12. Q. Johnson, A. Mitchell, and L. Evans, *Appl. Phys. Lett.* **21**, 29 (1972).
13. K. Kondo, A. Sawaoka, and S. Saito, *in High Pressure Science and Technology*, edited by K. D. Timmerhaus and M. S. Barber (plenum, New York, 1979), p. 905.
14. R. R. Whitlock and J. S. Wark, *Phys. Rev. B* **52**, 8 (1995).
15. S. Krishnan, P. C. Nordine, *J. Appl. Phys.* **80**, 1735 (1996).

VIBRATIONAL SPECTRA OF NITRO COMPOUNDS UNDER SHOCK COMPRESSION

Takamichi Kobayashi, Toshimori Sekine, and Hongliang He

*Advance Materials Laboratory, National Institute for Materials Science,
1-1 Namiki, Tsukuba, Ibaraki 305-0044, Japan*

Abstract. Real-time vibrational spectra of shock-compressed nitro compounds have been measured using a single-pulse laser Raman spectrometer in conjunction with a propellant gun and vibrational mode-dependent behavior has been examined. The NO_2 stretching mode shows smaller frequency shift compared to other stretching modes, which may be attributed to increased intermolecular interaction under pressure. Pressure-induced shift of nitromethane-d_3 shows monotonic increase up to ~5.0 GPa. However, above this pressure, the monotonic increase no longer exists and a more complicated behavior is observed. Above ~ 8.5 GPa, a strong background emerges over the whole spectral range (500 ~ 2600 cm^{-1}) and Raman bands are not detectable. A chemical reaction induced by a single shock may be initiated at ~ 8.5 GPa.

INTRODUCTION

In situ vibrational spectroscopy provides essential information on shock-induced phenomena such as chemical reactions, intermolecular interactions, and phase transitions. Especially, observation of vibrational mode-dependent behavior is important because it can give a clear picture on how molecular structure or crystal structure changes under shock compression.

In the experiments described here, we focused on the vibrational mode-dependent behavior of shock-compressed nitro compounds (nitrobenzene and nitromethane-d_3) to obtain information on inter-molecular interaction and shock-induced reaction in molecular liquids. Increased intermolecular interaction under pressure and shock-induced initiation of a chemical reaction in nitromethane-d_3 are discussed.

EXPERIMENT

Shock experiments were performed using a single stage propellant gun (30 mm in bore diameter). Aluminum impactors and aluminum driver plates (base plates) were used with the impact velocity up to ~2.0 km/s to obtain single shock pressure up to ~8 GPa. For higher pressure experiments, a stainless steel flyer was impacted on an aluminum driver plate. Nitrobenzene and nitromethane-d_3 were chosen as our initial samples. They are relatively strong Raman scatterers and shock experiments on nitromethane have been performed by several researchers.[1-4]

In this experiment, deuterated nitromethane (CD_3NO_2) was used because, in normal nitromethane (CH_3NO_2), the NO_2 stretching mode overlaps with the CH_3 bending mode. Liquid samples were confined between a driver plate and a glass window. Typical sample thickness was ~5 mm. Figure 1 shows a schematic diagram of the experimental setup. The second harmonic of a Nd:YAG laser (532 nm, 8 ns) was used as an excitation light. The excitation laser pulse was introduced into the sample just before the shock wave reached the rear surface of the sample. Raman frequency shifts of stretching modes were measured against single shock pressure. The uncertainties in the measured peak shifts were about ± 3 cm^{-1}. Since the Hugoniot for nitromethane-d_3 is not known, that for normal nitromethane was used to estimate shock pressures of nitromethane-d_3 by the shock impedance matching method.

RESULTS AND DISCUSSION

Typical Raman spectra of nitrobenzene and nitromethane-d_3 under ambient and shock pressure are shown in Fig. 2. Only totally symmetric stretching modes were selected for investigation because they are well separated from other bands and also generally more intense. They are the NO$_2$ stretching mode (1346 cm^{-1}) and the C-H stretching mode (3082 cm^{-1}) of nitrobenzene and the C-N stretching mode (895 cm^{-1}), the NO$_2$ stretching mode (1390 cm^{-1}), and the CD$_3$ stretching mode (2283 cm^{-1}) of nitromethane-d_3.[5]

It is seen in Fig. 2 that all Raman bands mentioned above are blue shifted under shock compression but the magnitude of the shift depends on vibrational mode. In general vibrational frequency of a stretching mode increases with pressure because bond length is reduced and effective force constant at the new equilibrium position is usually larger than that at the original equilibrium position. [6] However, in the case where relatively strong intermolecular interaction such as hydrogen bonding exists, the situation can become quite different.

Mode-dependent behavior of pressure-induced vibrational frequency shift

Observed Raman frequency shifts vs. single shock pressure are summarized in Fig. 3. It is seen that the NO$_2$ stretching mode shows significantly smaller blue shifts in both molecules. In the case of nitromethane-d_3, the C-N and the CD$_3$ stretching modes show similar blue shifts up to ~5 GPa and they are much larger than that of the NO$_2$ stretching mode. Also we reported previously that the NO$_2$ stretching mode of nitrobenzene (1436 cm^{-1}) shows much

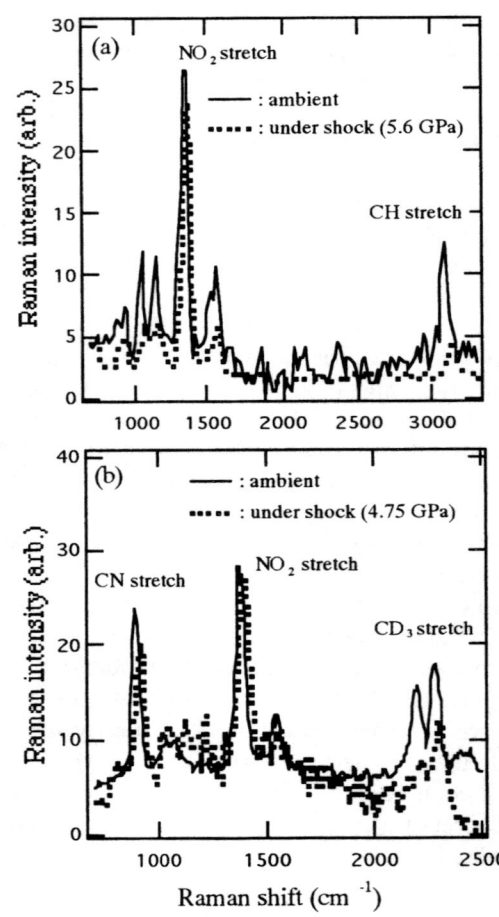

FIGURE 2. Raman spectra of (a) nitrobenzene and (b) nitromethane-d_3 under ambient and shock pressure.

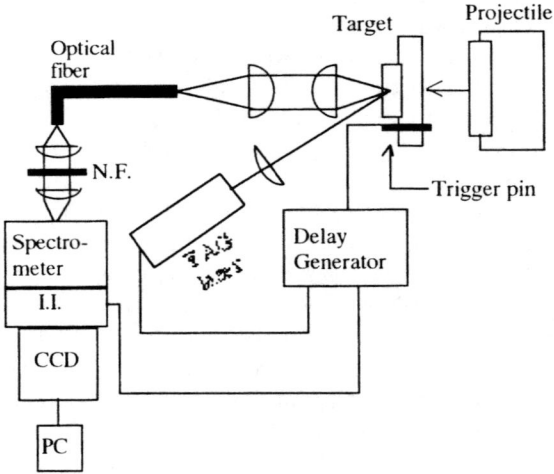

FIGURE 1. Schematic diagram of single-pulse laser Raman arrangement.

smaller blue shift than the C-C stretching mode (992 cm^{-1}) and the C-H stretching mode (3080 cm^{-1}) of benzene.[7,8] There seems to exist some kind of softening mechanism to account for the small blue shift of the NO_2 stretching mode under compression.

It is noted here that, in general, vibrational bands become broader and peak positions may shift with temperature due to hot bands.[9] In our spectra, it is difficult to see this hot band effect because of low resolution. One of the other possible explanations for the softening mechanism of the NO_2 stretching mode may be an increased intermolecular interaction

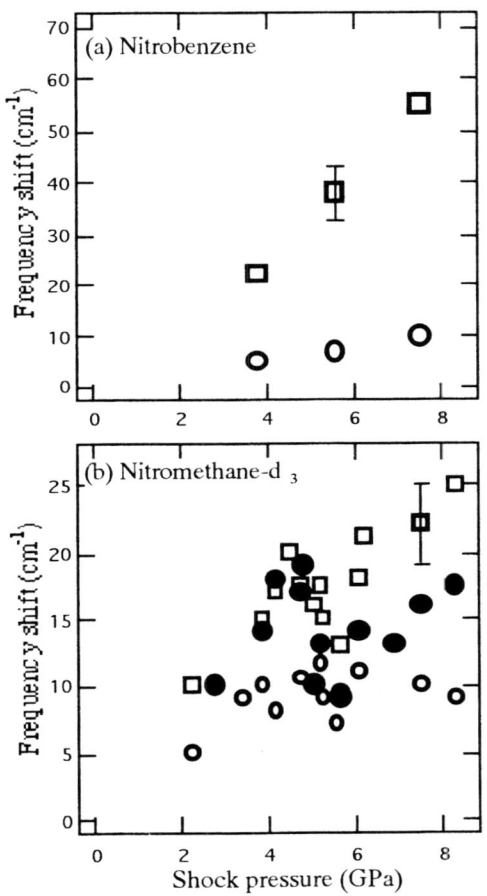

FIGURE 3. Raman frequency shift vs. shock pressure. (a) □: C-H stretching mode (3082 cm^{-1}), ○: NO_2 stretching mode (1436 cm^{-1}). (b) □: CD_3 stretching mode (2283 cm^{-1}), ○: NO_2 stretching mode (1390 cm^{-1}), ●: C-N stretching mode (895 cm^{-1}). Experimental uncertainties in frequency shift measurements are ~±5 cm^{-1} for nitrobenzene and ~±3 cm^{-1} for nitromethane.

under compression. In static high-pressure experiments of some hydrogen-bonded solids, softening of vibrational modes with pressure has been observed.[10-11] An example is the softening of the O-H stretching mode in H_2O ice. The vibrational frequency of this band decreases with pressure until the band disappears at ~60 GPa.[11] Above this pressure it is reported that nonmolecular, symmetric hydrogen-bonded state is formed, where the proton is delocalized along O-O directions. A Raman study of shock-compressed liquid water by Holmes et.al. indicated that hydrogen bonding diminishes with increasing shock pressure,[12] which may be due to high temperature effect. There are some reports on weak hydrogen bonding in ambient nitromethane [13] and hydrogen bond formation at high pressures [4,14]. If the situation of nitro compounds is similar to that of H_2O ice, softening of the NO_2 stretching mode may take place under shock compression while the O--H or O--D bonds between molecules become stronger.

Small Raman frequency shifts under compression observed for the NO_2 stretching mode may be explained as a result of two factors, i. e., (1) the pressure-induced softening mechanism in hydrogen-bonded materials which decreases the vibrational frequency and (2) the general pressure-induced hardening mechanism which increases vibrational frequencies of stretching modes. The cancellation of these two effects may be responsible for the observed small frequency shifts of the NO_2 stretching mode. This seems to explain above mentioned fact that the CD_3 stretching mode shows similar blue shifts to those of the C-N stretching mode. It is inferred from analogy with C-H stretching modes of other molecules with little intermolecular interaction[8] that the CD_3 stretching mode would show larger frequency shifts than observed unless influenced by some kind of softening mechanism such as pressure-induced intermoleculer interaction.

Shock-induced Reaction in nitromethane-d_3

Shock-induced initiation of chemical reaction in nitromethane-h_3 has been studied by several researchers. In single shock experiments, Renlund et.al.[2] suggested the initial stage of reaction near 6.0

GPa by *in situ* Raman measurements and Von Holle[3] suggested a reaction at above 7.0 GPa by time-resolved infrared radiometry. In this study, deuterated nitromethane was used and the results are somewhat different from those of normal nitromethane. Under static pressure, different reactivity between normal nitromethane and deuterated nitromethane has been reported.[15]

In Fig. 3, it is seen that pressure-induced blue shift of the C-N stretching mode suddenly drops down at ~5.0 GPa and starts increasing again at higher pressures. The CD_3 stretching mode displays similar behavior but the drop at ~5.0 GPa is not as large. A similar drop in Raman peak shift for the C-N stretching mode at ~6 GPa was observed in normal nitromethane.[2] Pressure-induced frequency shift of the NO_2 stretching mode appears to level off at around this pressure. Up to 8.3 GPa, Raman bands of nitromethane-d_3 are observable but above 8.5 GPa strong non-resonant emission suddenly emerges throughout the observed spectral range (500 ~ 2600 cm^{-1}) and the background jumps up by nearly two orders of magnitude and Raman bands are no longer detectable. This may be due to emission from reaction products and thus a chemical reaction by single shock may be initiated at ~8.5 GPa.

ACKNOWLEDGMENTS

The authors would like to thank David S. Moore, Los Alamos National Laboratory, for his helpful comments and discussions. We also thank Harumi Otsuka for preparing the manuscript.

REFERENCES

1. Moore.D. S., Schmidt.S. C., Shaner.J. W., Shampine D. L., and Holt.W. T., in *Shock Waves in Condensed Matter - 1985*, edited by Y. M. Gupta, Plenum Press, 1986, pp. 207-210.
2. Relund.A. M., and Trott.W. M., in *Shock Compression of Condensed Matter - 1989*, edited by S. C. Schmidt, J. N. Johnson, and L. W. Davidson, Elsevier Science Publishers B. V., 1990, pp. 875-878.
3. Von Holle.W. G., in *Shock Waves in Condensed Matter - 1981*, edited by W. J. Nellis, L. Seaman, and R. A. Graham, AIP Conference Proceedings 78, New York, 1982, pp. 287-291.
4. Winey.J. M., and Gupta.Y. M., *J. Phys. Chem.* B **101**, 10733 (1997).
5. Hill.J. R., Moore.D. S., Schmidt.S. C., and Storm.C. B., *J. Phys. Chem.* **95**, 3037 (1991).
6. M. R. Zakin and D. R. Herschback, *J. Chem. Phys.* **85**, 2376 (1986).
7. Kobayashi. T., and Sekine, T., in *Shock Compression of Condensed Matter -1999*, edited by M. D. Furnish, L. C. Chhabildas, and R. S. Hixson, AIP Conference Proceedings 505, New York, 2000, pp. 951-954.
8. Kobayashi. T., and Sekine. T., *Phys. Rev.* B **62**, 5281 (2000).
9. D. S. Moore, *J. Phys. Chem.* A, **105**, 4660 (2001).
10. Wolanin. Ph. Pruzan, E., Gauthier. M., Chervin.J. C., Canny.B., Hausermann. D., and Hanfland. M., *J.Phys. Chem.* B, **101**, 6230 (1997).
11. Goncharov. F., Struzhkin.V. V., Mao. H., and Hemley.R. J., *Phys. Rev. Lett.* **83**, 1998 (1999).
12. Holmes,. N. C., Nellis,. W. J., and Graham,. W. B., *Phys. Rev. Lett.* **55**, 2433 (1985).
13. E. Knoezinger, H. Kollhoff, and R. Wittenbeck, *Ber. Bunsenges. Phys. Chem.* **86**, 929 (1982).
14. D. M. Adams and J. Haines, *J. Phys.: Condens. Matter* 3, 9503 (1991).
15. Piermarini G. J., Block S., and Miller P. L., *J. Phys. Chem.* **93**, 457 (1989).

TRANSIENT BOND SCISSION OF POLYTETRAFLUOROETHYLENE UNDER LASER-INDUCED SHOCK COMPRESSION STUDIED BY NANOSECOND TIME-RESOLVED RAMAN SPECTROSCOPY

Kazutaka G. Nakamura[1], Kunihiko Wakabayashi[2], Ken-ichi Kondo[1]

[1]*Materials and Structures Laboratory, Tokyo Institute of Technology, 4259 Nagatsuta, Midori-ku, Yokohama, Kanagawa 226-8503, JAPAN*

[2]National Institute of Advanced Industrial Science and Technology, 1-1-1 Higashi, Tsukuba 305-8565, JAPAN

Abstract. Nanosecond time-resolved Raman spectroscopy has been performed to study polymer films, polytetrafluoroethylene (PTFE), under laser driven shock compression at laser power density of 4.0 GW/cm^2. The CF$_2$ stretching mode line of PTFE showed a higher shift (18 cm^{-1}) at delay time of 9.3 ns due to the shock compression and corresponding pressure was estimated to be approximately 2.3 GPa. A new vibrational line at 1900 cm^{-1} appeared only under shock compression and was assigned to the C=C stretching in transient species such as a monomer (C$_2$F$_4$) produced by the shock-induced bond scission. Intensity of the new line increased with increasing delay time along propagation of the shock compression.

INTRODUCTION

Dynamic and microscopic behavior of materials is required to investigate in order to specify a transient state of excitation and relaxation processes under shock-compression. Raman spectroscopy is used to investigate structure and bond strength of molecules and crystals under shock compression [1]. Using laser-shock generation and a pump and probe technique, it is possible to investigate molecular vibrations under shock-compression in a time domain of nanosecond or much shorter [2]. In this paper, we performed nanosecond time resolved Raman spectroscopy of laser-shocked poly-tetrafluoroethylene ((C$_2$F$_2$)$_n$, PTFE) using the pump and probe technique.

EXPERIMENTS

The laser used is a Q-switched Nd:YAG laser (Continuum Powerlite Plus) with maximum output of 3 J/pulse at wavelength of 1064 nm. The second harmonic light (512 nm) was generated by using a KD*P crystal. Pulse widths of the fundamental and second harmonic lights are 10 and 8 ns at FWHM, respectively. Stability of the output energy was within ±2.5 %. In pump and probe experiments, the fundamental beam was used for a pump beam, which generates shock wave, and the second harmonic beam was used for a probe beam, which excites Raman scattering [3]. The second harmonic beam was separated from the fundamental beam by a dichroic mirror and introduced into an optical delay line. Delay time between the pump and probe beams can be controlled within 25 ns, since the length of the delay line is about 8 m.

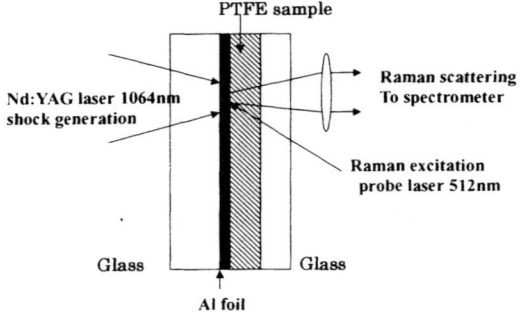

FIGURE 1. This is a schematic drawing of the

The target assembly has a glass confinement geometry (shown in Fig. 1), which consists of a back-up glass (100x100x5 mm³), an aluminum foil (25-μm thick), a PTFE film sheet (Dupont-Mitsui Fluorochemicals Co., 80-μm thick) and a cover glass (2-mm thick). The aluminum foil was stick to the back-up glass with an adhesive. The target assembly was mounted on an X-Z stage, which was computer-controlled for synchronizing to the laser pulses. The pump beam was focused on the aluminum foil with a spot (1.0 mmφ). Laser induced plasma was generated between the sample and the backup glass and drove a shock wave through the aluminum foil into the PTFE sample. The peak pressure driven in the aluminum foil by laser irradiation in the glass confinement geometry was estimated by an empirical equation proposed by Devaux et al. [4]. The peak pressure in the PTFE sample in the present irradiation condition (4.0 GW/cm²) was calculated to be 2.3 GPa by using impedance matching.

The probe beam was focused on the rear side of the target with a diameter of 500 μm after passing through the optical delay line. The energy of the probe beam was 10 mJ/pulse. Raman scattering was collected with a camera lens, spectrally resolved by a polychromater (Kaiser Co.) with a 2400 lines/mm grating and a notch filter (bandwidth of 350 cm⁻¹), and detected by a CCD camera. Each Raman spectrum was obtained by accumulating 1000 laser shots. Raman spectroscopy with excitation light of 512-nm detect a whole volume of the 80-μm PTFE film, since a penetration depth is larger than 500 μm.

RESULTS AND DISCUSSIONS

Raman measurements

Figure 2 shows a typical example of a Raman spectrum of the pristine sample (PTFE).

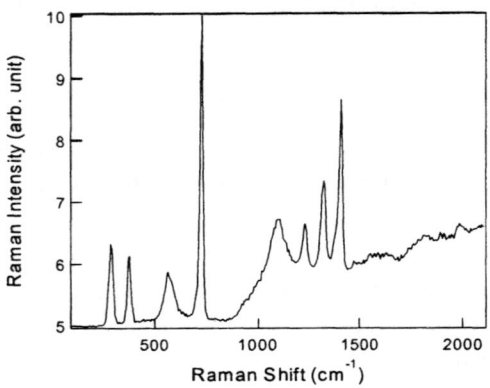

FIGURE 2. This is a typical example of a Raman spectrum of pristine PTFE.

The observed lines at 291, 381, and 729 cm⁻¹ are assigned to twisting, bending and symmetric stretching modes of CF_2, respectively. The line at 1379 cm⁻¹ is assigned to a C-C stretching mode. Overtone and combination modes are observed at 1215 and 1295 cm⁻¹. Broad lines at 550 and 1100 cm⁻¹ are due to the glass substrate.

The observed lines at 291, 381, and 729 cm⁻¹ are assigned to twisting, bending and symmetric stretching modes of CF_2, respectively. The line at 1379 cm⁻¹ is assigned to a C-C stretching mode. Overtone and combination modes are observed at 1215 and 1295 cm⁻¹. Broad lines at 550 and 1100 cm⁻¹ are due to the glass substrate.

Figure 3 shows a typical example of the nanosecond time-resolved Raman spectra detected at delay times of 9.3, 14.7, 17.6 and 20.6 ns after the pump beam irradiation. After the shock generation, a new peak appeared at around 1900 cm^{-1} and its intensity increases as increase of delay time. However this new peak is not observed in the Raman spectrum of the recovered sample, which is identical to that of the pristine PTFE. The new peak, therefore, is due to the transient species generated under the shock compression.

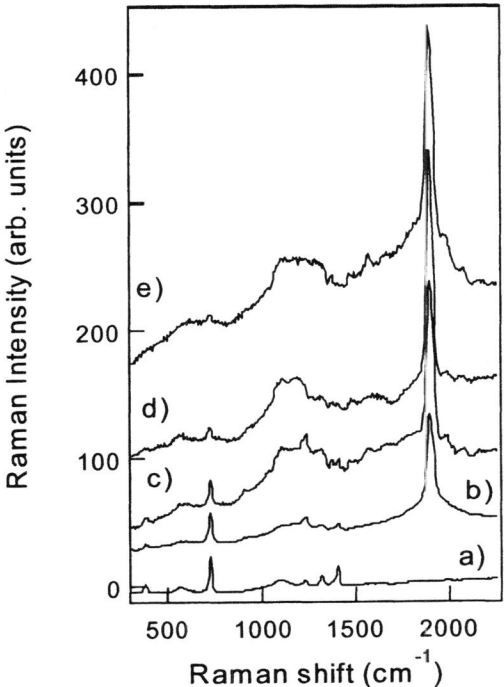

FIGURE 3. This a time-resolved Raman spectra of laser-shocked PTFE: a) 0 ns, b) 9.3 ns, c) 14.7 ns, d) 17.6 ns, and e) 20.6 ns.

The increase of the new peak can be explained in connection with the propagation of shock waves. Since Raman spectroscopy probes a whole volume of the PTFE, the observed Raman spectrum is made up by the superposition of the scattering from both volumes under shock compression and in front of the shock wave. By propagation of the shock wave inside the film along with the delay time, the volume under the shock compression increases and the intensity of the new peak increases.

The CF_2 stretching peak at delay time of 9.3 ns shows higher shift of 18 cm^{-1}. The pressure-induced shift has been reported in experiments of hydrostatic compression [5]. The relationship between the pressure-induced shift (S [cm^{-1}]) and the pressure (P [GPa]) is well fitted by linear equation: S=6.97P+1.93. Using this equation, the present shift (18 cm-1) corresponds to pressure of 2. 3GPa, which is comparable to the ablation-pressure.

Ab initio Calculation

In order to assign the new Raman line (1900 cm^{-1}), *ab initio* calculation was performed. The most conceivable candidate is a stretching mode of carbon double bonds (C=C), because the sample consists of C and F atoms. Ab initio calculation was performed for C_nF_m (n<11, m<23) molecules and radicals containing the C=C bond. The calculation was performed with a Gaussian98W program [6] and the Hartree-Fock (RHF and UHF) level and 6-31G basis sets were used. PTFE was modeled with a $C_{10}F_{22}$ molecule. The frequency was calculated after the full geometry optimization

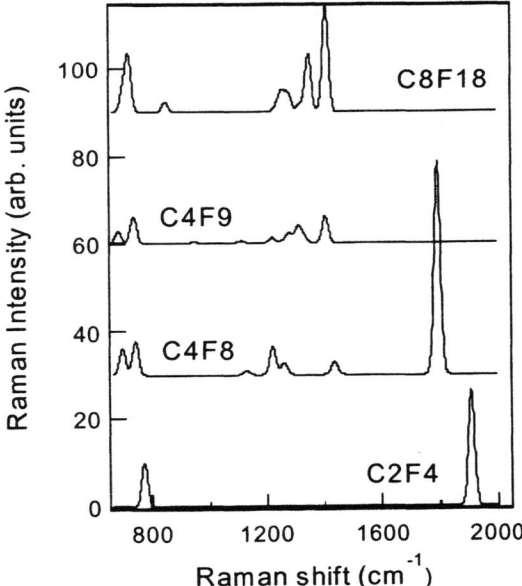

FIGURE 4. These are calculated Raman spectra obtained by *ab initio* calculation (HF 6-31G).

and the scale factor of 0.89, which is systematically used for Hartree-Fock level calculations [7]. The C_8F_{18} is not so big for a model of polymer, but the calculated Raman lines represent well the experimentally obtained ones and the Raman spectrum of a larger molecule such as $C_{10}F_{22}$ is comparable with that of C_8F_{18}.

Figure 4 shows the calculated Raman spectra. The pristine PTFE is well reproduced by C_8F_{18}.

The highest Raman frequency is a C-C stretching mode at 1409 cm^{-1}. In radicals such as C_4F_9, the highest peak is also the C-C stretching. In the long-chain radicals with a C=C bond such as C_4F_8, the highest frequency is the C=C stretching (1792 cm^{-1}) but this is smaller than the observed frequency. However, the C_2F_4 has the C=C stretching at 1902 cm^{-1}. The transiently observed Raman line at 1900 cm^{-1} is assigned to the C=C stretching of C_2F_4 monomer, which is generated under shock-compression.

It is known that the degradation of PTFE leads to the formation of C_2F_4 monomer with high yield. The bond scission of C_8F_{18} was also calculated by modeling: $C_8F_{18} \rightarrow 2(C_4F_9) \rightarrow 2(C_2F_5+C_2F_4)$. Since the bond energy of C-F is higher than C-C, the C-C bond is broken. The calculation shows that the first step is endothermic (3.04 eV) but the second step is exothermic (2.99 eV). If a C-C bond scission is induced, the successive scission may occur. Graham proposed the mechanical chemical reaction and shock-induced bond scission [8]. Therefore the observed Raman line at 1900 cm^{-1} may be due to the C_2F_4 monomer transiently generated by shock-induced bond scission.

CONCLUSION

Nanosecond time-resolved Raman spectroscopy has been performed to study PTFE polymer, under laser-driven shock compression at 2.3 GPa (laser power density of 4.0 GW/cm^2). A new vibrational line at 1900 cm^{-1} appeared only under shock compression and was assigned to the C=C stretching in transient species such as a monomer (C_2F_4) produced by the shock-induced bond scission.

ACKNOWLEDGEMENTS

This work has been supported by the CREST (Core Research for Evolutional Science and Technology) program organized by the Japan Science and Technology Corporation (JST).

REFERENCES

1. G.I. Pangilinan and Y.M. Gupta, J. Phys. Chem. **98**, 4522 (1994).
2. S.A. Hambir, J. Franken, D.E. Hare, E.L. Chronister, B.J. Baer, and D.D. Dlott, J. Appl. Phys. **81**, 2157 (1997).
3. K. Wakabayashi, K.G. Nakamura, K. Kondo, and M. Yoshida, Appl. Phys. Lett. **75**, 947 (1999).
4. D. Devaux, R. Fabbro, L. Tollier, and E. Bartnicki, J. Appl. Phys. **74**, 2268 (1993).
5. C. Wu and M. Nicol, Chem. Phys. Lett. **21**, 153 (1973).
6. Gaussian 98, Revision A.6, M. J. Frisch et al., Gaussian, Inc., Pittsburgh PA, 1998.
7. J. B. Foresman and A. Frisch, in *Exploring Chemistry with Electronic Structure Methods* (Gaussian Inc. 1993).
8. R.A. Graham, J. Phys. Chem. **83**, 3048 (1979).

SHOCK-INDUCED ORIENTATION OF BENZENE MOLECULES STUDIED BY NANOSECOND TIME-RESOLVED RAMAN SPECTROSCOPY

Kunihiko Wakabayashi[1], Kazutaka G. Nakamura[2], Ken-ichi Kondo[2]

[1]*National Institute of Advanced Industrial Science and Technology, 1-1-1 Higashi, Tsukuba 305-8565, JAPAN*
[2]*Materials and Structures Laboratory, Tokyo Institute of Technology, 4259 Nagatsuta, Midori-ku, Yokohama, Kanagawa 226-8503, JAPAN*

Abstract. Nanosecond time-resolved Raman spectroscopy has been used to study a molecular response of benzene under shock compression in the pressure range up to 1.6 GPa. S-polarized light against the target surface was used for excitation of Raman scattering. Shock wave was generated by pulsed-laser irradiation with repetition rate at 1.25 Hz. Here we used a multi-lens array for smoothing and homogenizing the intensity distribution of a shock driver laser on the surface of a target. This technique made it possible to generate a spatially uniform shock wave. The change in vibrational frequency of 993 cm^{-1} (breathing vibrational mode) has been observed with respect to applied shock pressure, and a vibrational frequency shift increased with increasing shock pressure. Temporal changes of the integrated intensity of Raman band at 993 cm^{-1} indicated the existence of mechanical processes, which were associated with uniaxial nature of shock compression. By comparing the Raman spectra under shock compression to that under ambient state, it is concluded that the temporal change of its intensity implies the occurrence of the shock-induced orientation of benzene molecules.

INTRODUCTION

Shock wave experiments have provided reliable information on mechanical and thermo-physical properties of materials under high-pressure and high-temperature conditions. However, in order to understand the transient states of excitation and relaxation processes of materials under shock compression, it is important to investigate the microscopic[1,2] and dynamic behavior[3] of materials. Vibrational spectroscopy has been recognized as a powerful tool for extracting microscopic information of molecular dynamics. In particular, Raman or infrared spectroscopy is well suitable for examination of molecular changes. In this work, we performed nanosecond time-resolved Raman spectroscopy of Benzene, using a pump-probe technique. Shock wave was generated repetitively by laser irradiation.

EXPERIMENTS

The repetitive experimental system[4] for shock compression with Table-Top laser was used, and applied it to pump-probe Raman spectroscopy. The Q-switched Nd[3+]:YAG laser (Continuum Co., Powerlite Plus) is operated at repetition rate of 10 Hz. The maximum outputs of fundamental (1064 nm) and second harmonic (532 nm) radiations are 3 and 1.5 J/pulse, respectively. A temporal profile of the laser beam is Gaussian. The full widths at half maximum (FWHM) of the fundamental and second harmonic are 10 and 8 ns, respectively. The output energies in different shots were constant within 2.5 %. The fundamental radiation was directed through a multi-lens array (MLA) and a focusing lens, and used to irradiate a target for shock wave generation. We used a MLA for smoothing and homogenizing the intensity distribution of a shock driver laser on the target surface, by its overlapping effect of many beamlets. This technique made it possible to obtain a flat-top beam, so that it could be supposed to genetare a spatially uniform shock wave. The second harmonic light was directed through a variable optical delay line and a polarized beam splitter. Only S-polarized component of the second harmonic light against a target surface was used for excitation of Raman scattering.

Fig. 1 shows a target assembly, which is called

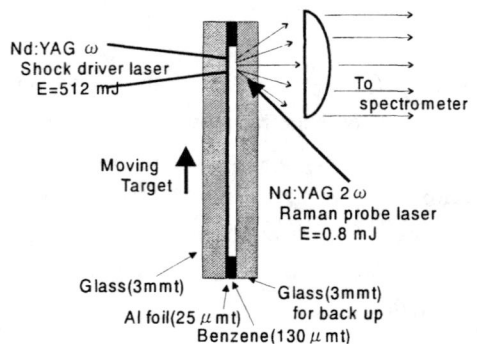

Fig. 1 Schematic drawing of the target assembly. The sample is liquid-Benzene filled between both glasses.

glass-confinement geometry[5]. The target assembly in glass-confinement geometry (Fig.2) was fabricated with a back-up glass substrate (FL3, 100 $\times 100 \times 3$ mm^3) on which aluminum foil (25 μm thick) was glued with epoxy adhesive dissolved in Toluene, a spacer made of PTFE sheet (130 μm thick), and a cover glass substrate (FL3, 100×100 ×3 mm thick). The PTFE spacer was sandwiched between glass substrates. Liquid-Benzene was filled in the space held by the PTFE spacer.

The target assembly was mounted on computer-controlled X-Z stage. The movement of X-Z stage was synchronized to the laser pulse reduced the repetition rate at 1.25 Hz with mechanical shutter, so that fresh target's surface was exposed after each shot. The driver laser homogenized with the MLA was focused on the aluminum foil at normal incidence through a glass window with laser energy up to 512 mJ/pulse, with a 1.25 mm spot diameter. Confined plasma was generated near the aluminum-glass interface, which drove a shock wave through the aluminum foil into the Benzene sample.

Optically delayed S-polarized Raman probe laser (λ = 532 nm, 0.8 mJ/pulse) was focused on the rear side of the Benzene sample with a diameter of 1000 μm at the center of the region irradiated by the shock driver laser. Raman scattered light was collected with a camera lens, and led to a polychromator (Kaiser Co.,) with optical-fiber coupling (400 μm core). Two Raman notch filters (bandwidth of 350 cm^{-1}) were placed in front of the fiber, and inside a polychrometer for reducing the Raman scattered-light from fiber and Raman exciting laser. The spectrally resolved signal was detected with an intensified charge coupled device (ICCD) camera (ANDOR). Spectral resolution was about 3 cm^{-1}. Raman spectrum was obtained by accumulating data of 60 laser shots.

RESULTS AND DISCUSSIONS

Fig. 2(a) shows a Raman peak of Benzene (ring-breathing mode) under ambient state. Fig. 3 (b)-(f) show time-resolved Raman spectra of shocked Benzene at delay time of 23.4, 28.2, 34.2, 40.2, 46.2 ns, respectively. The zero delay time (t_c=0.0 ns) denoted an arrival of the driver laser pulse at the glass-aluminum interface.

by time-resolved techniques were made up by the superposition of Raman scattered light from both the compressed area (behind the shock front) and the uncompressed area (in front of the shock front). By assuming that the frequency distribution of Raman spectrum scattered from both areas is Lorentzian shape, temporal change of integrated intensity of both spectrum (Fig. 3) was obtained. Fig. 3 showed that integrated intensity of Raman peak from the compressed area was increasing with

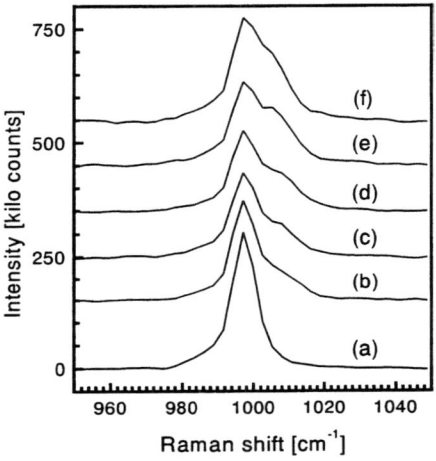

Fig. 2 Time-resolved Raman spectra of Benzene under laser induced shock compression.

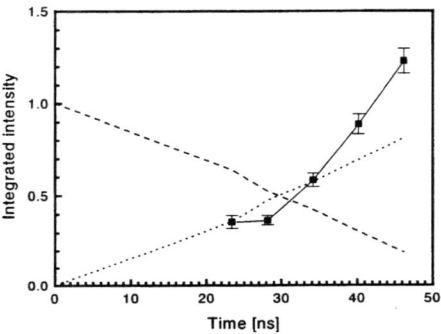

Fig. 3 The integrated intensity of the peak at 993 cm^{-1} was plotted against the delay time.

As shown in Fig. 2, vibrational frequency shift which supposed to be due to the effects of compression was observed. Under present experimental condition, temporally and spatially resolved particle velocity measurements of Benzene has been also performed with the use of the line-ORVIS[6] system. As a result of the line-ORVIS measurements, it was revealed that spatial and temporal profile of shock wave generated by laser irradiation with confined geometry target was relatively flat (800μm ϕ) and steady (about 40 ns). Generated shock pressure was about 1.6 GPa. Based on the results of line-ORVIS measurements, it has been supposed that Raman spectra obtained

the increase of the delay time. It has been considered that this temporal change of intensity has well associated with a propagation of shock wave. However, this behavior could not been explained on the basis of the intensity change which accounted for increase of the number of Benzene molecules inside the compressed area (dotted line in Fig. 3). It was considered that there were three factors which caused the increase of the intensity, as follows,

1. Geometrical configuration of experimental set up.
2. Changes of the polarizability of Benzene molecules.
3. The effect between the polarization of probe laser and the orientation of the Benzene molecules.

Taking into account of the present experimental condition, it was reasonable to suppose that the reason why the intensity increased was the results of orientation of Benzene molecules along with a propagation of shock wave. Because rotational energy of Benzene molecules was relatively small (0.005 eV), then it was excited easily at room temperature (0.026 eV). This effect could be expected under shock condition. Up to this time, it has been reported that Benzene molecules were tend to orient to the particular direction under static compression. Based on this assumption that intensity change depends solely on the molecular orientation, Raman intensity I_0 could be described as a function of the angle θ between the polarization vector of probe laser and the polarization vector of the polarizability of Benzene ring-stretching vibrational mode by following expression;

$$I_0 = I_L \cdot P_0 \cos\theta \cdot \rho_0 \cdot l \cdot A ,$$

where I_L is probe laser intensity, P_0 the Raman cross section of Liquid-Benzene, ρ_0 the initial density. A is a probed area, $l\,[\mu\mathrm{m}]$ the sample thickness.

Fig. 4 showed the temporal change of the molecular-angle inside the compressed area with the use of both the temporal change of Raman intensity (Fig. 3) and the equation. Molecular angle of Benzene tends to small with the increase of the delay time. This results indicated that molecular rotation were depressed against the particular direction. Temporal change of the molecular orientation and the Raman shift indicated that there were something mechanical dynamics that the vibrational energy state of Benzene molecules were led to lower state with relaxing the Raman shift inside the compressed area.

CONCLUSION

Nanosecond time-resolved Raman spectroscopy has been performed to investigate the microscopic behavior of Benzene shocked about 1.6 GPa. Peak intensity observed were increasing with the increase of the delay time. It was supposed that the temporal change of the peak intensity was due to the orientation of Benzene molecules. It was concluded that the temporal change of its intensity implied the occurrence of the shock-induced orientation of Benzene molecules.

ACKNOWLEDGEMENTS

This work has been supported by the CREST (Core Research for Evolutional Science and Technology) program organized by the Japan Science and Technology Corporation (JST).

REFERENCES

1. S. C. Schmidt, D. S. Moore, D. Schiferl and J. W. Shaner, Phys. Rev. Lett. **50**(1983)661.
2. T. Kobayashi and T. Sekine, Phys. Rev. B **62**(2000)5821.
3. S. A. Hambir, J. Franken, D. E. Hare, E. L. Chronister, B. J. Baer and D. D. Dlott, J. Appl. Phys. **81**(1997)2157.
4. K. Wakabayashi, K. G. Nakamura, Ken-ichi Kondo and M. Yoshida, Appl.Phys. Lett. **75**(1999)947.
5. D. Devaux, R. Fabbro, L. Tollier and E. Bartnicki, J. Appl. Phys. **74**(1993)2268.
6. W. F. Hemsing et. al., Shock compression of condensed matter (1991)767.

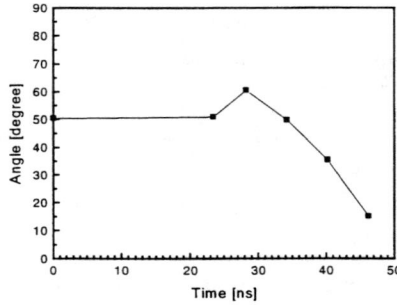

Fig. 4 The changes of the mean molecular angle was plotted as a function of the delay time.

MEASUREMENTS OF THE CONDUCTIVITY OF SHOCKED POLYMETHYLMETHACRYLATE

D. Townsend and N.K. Bourne*

Sowerby Research Centre, British Aerospace, FPC 30, PO Box 5, Filton, Bristol, BS34 7QW, UK.
*Royal Military College of Science, Cranfield University, Shrivenham, Swindon, SN6 8LA, UK.

Abstract. Shock polarization and changes in conductivity have been observed in previous electrical investigations of the shock behaviour of crystalline materials. For this reason a differential system has been designed by LLNL to separate these effects in their investigations of the conductivity of sapphire under shock. The measurement removes voltages produced in the shock electrical field allowing determination of those induced by resistance changes. Polymers are, at present, under mechanical investigation at RMCS and experiments are reported in which the thermoplastic polymethylmethacrylate (PMMA) is shocked in the range up to 10 GPa in order to observe its electrical response. The variations in induced field and in the material conductivity are reported.

INTRODUCTION

The response of polymethylmethacrylate (PMMA) to shock loading has excited interest as an example of a thermoplastic below its glass transition temperature. Barker and Hollenbach showed that up to 2.2 GPa, PMMA displays nonlinear behavior, and quote a Hugoniot elastic limit (HEL) of 0.7 GPa (1). This value was determined since the shock velocity became dependent upon the thickness of the sample, (reducing by a maximum of 2%). They suggested that this was due to the strain-rate sensitivity of the material in its plastic response above 0.7 GPa. No obvious change in slope in the particle velocity–time curves was noted, although there was a significant rounding of the upper part of those curves as the maximum particle velocity was reached. This would likely be the result of the nonlinear nature of the shock velocity-particle velocity curve due to the viscoelastic behavior of PMMA. Schuler and Nunziato (2) also report an HEL of 0.75 GPa, derived from analysis of shock–reshock experiments. Below the estimated HEL, the rise of the reshock part of the pulse was fast, indicating that loading was elastic in nature. Above 0.75 GPa, the slope of the reshock was shallower, suggesting that inelastic deformation was taking place.

Shear strength measurements performed in PMMA show an initial increase in shear strength to 7.5 GPa before dropping to near zero (3, 4). Gupta has also detected a drop in shear strength, although he has placed it somewhat lower, around the HEL (5). It is interesting to note that temperature measurements using either thermocouples or thin copper film gauges show a marked increase in temperature at 2.0 GPa (6). Here, results were explained in terms of the onset of a shock-induced, exothermic reaction, possibly due to bond breaking. A similar threshold has been noted by Hauver during measurements of the shock-induced electrical polarization of PMMA (7). Rosenberg and Partom (8) used strain gauges in PMMA to deduce residual temperatures after complete release from shock loading and the tensile strains induced at the spall plane. Here they demonstrated that the measured strain (after complete release) could be related to the residual temperature.

The classic study of Carter and Marsh (9) shows breaks in the shock velocity-particle velocity curves which (they suggested) constituted a phase transformation in the polymers they tested. This is also observed to occur in PMMA and thus it may be of interest to probe the electrical behaviour around this value which is at 26 GPa (9). This is one of the goals of this work.

There are many questions still unanswered concerning the response of the material. One of these relates to the induction of electrical fields across the front when shocked. Polarization has proved a feature of previous electrical investigations of shock behaviour. The deformation process introduces defects which can induce electronic states in the band gap of an insulator. Such changes have been observed in shocked sapphire (10).

To differentiate polarization from resistance change behind the shock, a differential circuit (fig. 1) has been designed for investigations into the conductivity of sapphire under shock (10). The measurement removes voltages produced in the induced shock field allowing determination of those induced by resistance changes. This method thereby avoids the additive measurements of polarization and resistance change previously reported by other groups.

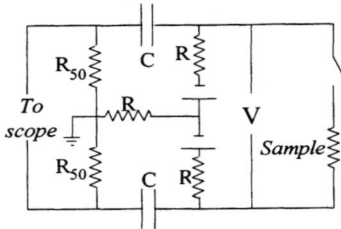

FIGURE 1. Circuitry for measuring resistance of target. After Weir *et al.* (10).

However, Lysne (11) has developed models for dielectric relaxation in polymers that suggest that interpretation of signals attributed to polarisation or to conduction are frequently confused. It will be seen that this may be the case in this investigation.

EXPERIMENTAL

The experiments were carried out using 50 mm and a 75 mm bore diameter, single stage guns and a 35 mm bore, (final stage) two stage gun. Impact velocity was measured to an accuracy of 0.5% using a sequential pin-shorting method and tilt was fixed to be less than 1 mrad by means of an adjustable specimen mount. Impactor plates were made from lapped tungsten alloy, copper and aluminium discs and were mounted onto a polycarbonate sabot with a relieved front surface in order that the rear of the flyer plate remained unconfined. Targets were flat to better than 5 μm across the surface. Stress profiles were measured with commercial manganin stress gauges both embedded into the specimens (Micro-measurements type LM-SS-125CH-048).

For the measurement of conductivity change in the sample, the method used for the experiments on sapphire (10) was adapted.

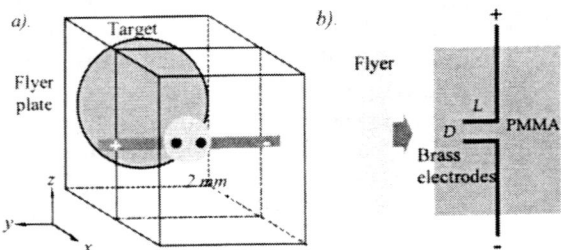

FIGURE 2. Experimental arrangement for experiments. a) Longitudinal stress gauge mounting positions with rear PMMA plate. b) Flyer hits target. Electrodes penetrate distance L and are D apart.

Each test was conducted by drilling two 1 mm diameter holes into the surface of the sample. These holes were 3 mm apart and 2 mm from the surface at which a conducting flyer would impact. They were thus located within the region that the lateral releases from the edge of the sample could not penetrate during the loading time of interest. Equally, the electrodes did not penetrate the entire sample. The holes were drilled 2 mm from the impact face so that they did not short to the conducting flyer plate. The drills used for this task were flat-bottomed. Care was necessary to avoid cracks opening up as the brittle polymer was drilled.

The arrangement of the sample and flyer are shown schematically in figure 2. The electrodes were formed by placing flat-ended brass rods into the drilled holes with a low viscosity epoxy around them. By this means the hole is completely filled if correct degassing techniques are used to eliminate bubbles. Two electrodes were placed in each sample which were connected to a differential power supply.

The electrodes were then soldered to 10 mm thick copper shims which were then taken out laterally for connection to the positive and negative poles of the differential power source. The surface of the tile has a copper electrode of thickness 25 µm bonded to it which is earthed to remove any possibility of any stray charge on the flying plate. The earth plane is defined to be a point on the metal gun framework adjacent to the sample mounting.

In all experiments, the shock wave would thus travel for a distance after impact and then sweep the length of the electrode. Clearly, any damage introduced by drilling would be encountered by the shock but this was discounted as shot to shot reproducibility was good. The PMMA target was machined and lapped on its front surface to remove any surface inhomogeneities and to ensure a flat impact face. There we no voids visible in the microstructure under optical examination.

RESULTS AND DISCUSSION

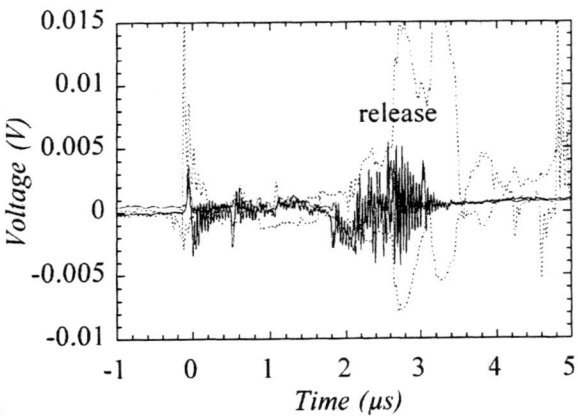

FIGURE 3. Two pairs of traces for the positive and negative electrodes. There are two experiments, at nominally the same impact stress, shown for comparison. One pair is dark and the other dotted. Note similar features illustrating reproducibility.

Figure 3 shows a pair of experiments showing induced voltages on the electrodes under nominally the same conditions. The two sets of traces are shown to emphasise the reproducible nature of the experiments. In both cases a copper impactor hits a PMMA target at a velocity of $ca.$ 550 m s^{-1} inducing stresses of $ca.$ 2 GPa. This represents one of the thresholds in behaviour noted earlier (for in this case, the onset of temperature rise in the material). The voltages induced on the conducting electrodes at these stresses are the order of mV. Note that there is a rise and then fall at the start of the trace (of $ca.$ 1 µs) which is believed to be induced before the arrival of the wave at the electrodes. The voltage then drops to a plateau for 2 µs until a release enters from the rear of the flyer plate perturbing the system. The other traces observed in the following study are typical of this form.

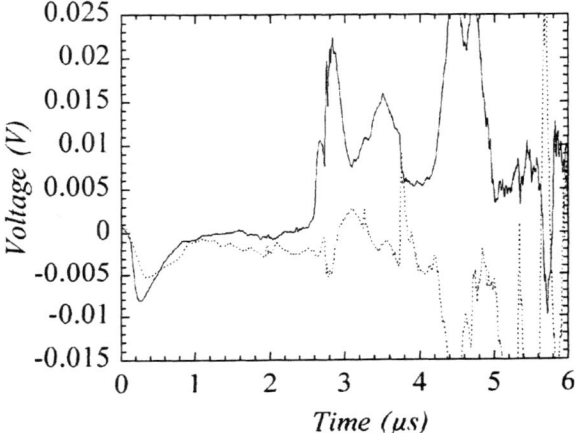

FIGURE 4. Traces on positive and negative electrodes for the impact of a 3 mm thick tungsten alloy flyer plate travelling at 805 m s^{-1}. The shock induces a negative spike of $ca.$ 10 mV before it reaches its base. The positive electrode is dark whilst the negative is dotted.

The histories of fig. 4 show a similar form. This time the stress has been increased to a value $ca.$ 3.5 GPa. The initial negative polarisation signal is increased in magnitude to its value at the lower stress. As the wave encounters the end of the electrode, the trace steady and both remain at zero until around 2 µs. At this time releases from the rear of the flyer plate enter the target interacting with the electrodes and the compressed material around them.

It is interesting to note that here, and in fig. 3, the signals show diverging positive and negative branches at $ca.$ 2.8 µs, reminiscent of what is expected to occur should conduction begin. This time corresponds to the arrival of the compression wave at the free rear surface of the target and as the wave front reflects, the signal magnitude increases at both electrodes. This may reflect increased shocked area as the legs have been reached.

The next shots showed similar results at higher stress to those shown earlier. No conduction was observed to occur but the polarisation signal increased in value although keeping the same sign.

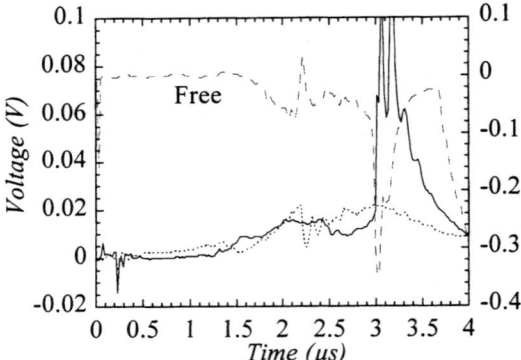

FIGURE 5. Traces on positive and negative electrodes for the impact of a 2 mm thick tungsten alloy flyer plate travelling at 1787 m s^{-1}. The shock induces a negative spike of *ca.* 15 mV before it reaches its base. The positive electrode is dark whilst the negative is dotted. The signals both rise as the shock sweeps the electrode. A dashed free electrode was also included to investigate the polarisation signal.

Fig. 5 shows results of traces taken from one of the higher stress experiments conducted. This was at 10.1 GPa. The shock induces a negative spike of *ca.* 15 mV before it reaches the electrodes' base. The positive electrode is dark whilst the negative is dotted. Traces are seen to rise slowly over the microsecond of loading but then the signals digress under action from release interactions. The signals both rise as the shock sweeps the base of the electrodes. A feature of fig. 5 is a signal recorded on a free, unbiased electrode introduced into the sample along with the other two to track the polarisation. Note the magnitude of the voltages induced which are found on the right-hand scale. The signal at this stress is of order 100 mV and is negative. Thus the magnitude of the induced signal measured has increased an order of magnitude in the series of experiments reported.

Other experiments were done at higher stresses but none more than 10.8 GPa. Thus the goal of reaching the break in U_s-u_p (9) was not achieved. At these stresses, the polarisation signal was smaller and the stress plateaux were at the level of the very first shots done.

CONCLUSIONS

A series of experiments have been described in which the method of Weir *et al.* (10) has been adapted for measurements of the conductivity of ionic solids which behave as insulators under ambient conditions.

PMMA was chosen since previous electrical measurements had tracked its dielectric properties (7). In none of the experiments conducted (up to a value of 10.8 GPa) did the material conduct. It was only in the very highest stresses achieved that the signals at the electrodes rose at all. This suggests that the material is indeed a good insulator. It was not possible to reach some of the higher mechanical thresholds but further work is planned to address these issues.

Induced electrical fields across the shock front were measured at all levels studied and these were not observed to changed direction with increasing stress in the range considered although their magnitude increased by an order of magnitude.

REFERENCES

1. Barker, L.M. and Hollenbach, R.E., *J. Appl. Phys.* **41**, 4208-4226 (1970).
2. Schuler, K.W. and Nunziato, J.W., *J. Appl. Phys.* **47**, 2995-2998 (1976).
3. Rosenberg, Z. and Partom, Y., *J. Appl. Phys.* **58**, 3072-3076 (1985).
4. Millett, J.C.F. and Bourne, N.K., *Journal of Applied Physics* **88**, 7037-7040 (2000).
5. Gupta, Y.M., Keough, D.D., Henley, D. and Walter, D.F., *Appl. Phys. Letts* **37**, 395-397 (1980).
6. Rosenberg, Z. and Partom, Y., in *Shock Waves in Condensed Matter – 1983*, (ed. J.R. Asay, R.A. Graham, and G.K. Straub), Amsterdam: North-Holland, pp. 251-254, (1984).
7. Hauver, G.E., in *Proc. Fifth Symp. (Int.) on Detonation*, Arlington, Virginia: Office of Naval research, pp. 387-398, (1970).
8. Rosenberg, Z. and Partom, Y., *J. Phys. D. Appl. Phys.* **16**, 1195-1200 (1983).
9. Carter, W.J. and Marsh, S.P., (1995), *Hugoniot equation of state of polymers*, Los Alamos Report, LA-12006-MS.
10. Weir, S.T., Mitchell, A.C. and Nellis, W.J., *J. Appl. Phys.* **80**, 1522-1525 (1996).
11. Lysne, P.C., *J. Appl. Phys.* **49**, 4186-4190 (1978).

CHAPTER XX

IMPACT PHENOMENA, BALLISTICS, HYPERVELOCITY STUDIES, AND EXOTIC SHOCK CONFIGURATIONS

NEW DIRECTIONS AND NEW CHALLENGES IN ANALYTICAL MODELING OF PENETRATION MECHANICS

James D. Walker

Southwest Research Institute, P.O. Drawer 28510, San Antonio, TX, 78228

Abstract. With the development of plasticity theory in the 1940s and 1950s, the modeling of penetration has become increasingly analytical and accurate. Currently, analytic penetration models are able to accurately predict depths of penetration for simple penetration geometries where the target and projectile are metals. The most accurate models use sophisticated plasticity analysis to obtain target resistances, but they usually end up expressible as relatively simple explicit differential equations. The most promising development in recent years is the centerline momentum balance, and this technique will be reviewed with some discussion of the meaning of terms by way of first principle physics. A recent model that successfully predicts ballistic limits of fabrics will also be discussed. In addition to addressing what is known, the most pressing questions that need to be answered and what is currently known on those problems will be discussed. Questions include: the transition from eroding to rigid penetration; the stress state transition for eroding projectiles; calculation of impact crater diameter; fracture time for ceramic targets; modeling targets comprised of anisotropic composites and fabrics; analytical approaches to projectile yaw; and modeling back surface bulging, failure and perforation.

INTRODUCTION

Analytic penetration models are models that predict the detailed penetration history of a target based upon first principles assumptions about mechanics. Penetration of a target essentially divides into two parts: how the target resists penetration, and what happens to the projectile as penetration occurs. In cases where these parts can be dealt with separately, good models currently exist for penetration when the target is a metal. Where the specifics of the target and projectile together matter a great deal (as it can with crater radius, the transition of eroding to rigid penetration, and yawed penetration), there is still work to be done towards the goal of an explicit, accurate, analytical model of penetration.

As with any overview paper of this sort, personal biases appear. In particular, this paper relies heavily on the centerline momentum balance, which in the previous decade yielded excellent penetration models as well as insight into the penetration process [1].

One of the results of the last 50 years of penetration modeling is that, in a generic sense, the target resistance can be written in the form

$$\frac{1}{2}\rho_t u^2 + R_t(u) \quad (1)$$

where u is the penetration velocity, ρ_t is the target density, and R_t is a term with units of strength (the R_t notation is due to Tate [2]). This resistance explicitly develops in both the centerline momentum balance [1] and cavity expansion techniques [3]. The first term in Eq. (1) reflects the inertia of the target and the second term reflects the strength of the target. $R_t(u)$ shows a weak dependence on penetration velocity. The calculation of $R_t(u)$ depends on plasticity theory, and major advances towards its calculation occurred in the 1980s and 1990s based on plasticity theory as developed in the 1940s and 1950s. In an energy context, the first term of Eq. (1) represents the energy that is temporarily being stored in the target as kinetic and elastic compressive energy, and the second term represents energy that is immediately being dissipated through plastic flow. This understanding of the various terms is recent, and represents a large step forward in the understanding of the penetration process [4]. Before, there was considerable discussion on the topic of "energy vs. momentum," but now it is fairly clear how the two approaches in modeling relate to each other. One result of this understanding is the recognition that in order to model from an energy perspective, it is necessary to include the intermediate energy transfer mechanisms. Including such terms makes the modeling very tedious and therefore this paper will address the topic of penetration modeling from the momentum perspective.

The understanding of penetration described above is in large part due to analytic models and large scale numerical simulations (hydrocodes). One of the tools available to today's analytic modeler is large scale numerical simulation. Simulations allow the stresses and velocities within the target to be examined. This numerical ability to look within the targets and projectiles during the penetration event provides ideas for better models as well as provides direct comparisons for the analytic model results. Modern analytic models provide not only final depth of penetration results, but provide depth vs. time information and velocity and stress states within the target and projectile during the penetration event. The model development [1] was greatly aided by results from hydrocode calculations.

With the hydrocode, *verification* of the analytic model can occur, meaning that when the same constitutive models are used in the large scale numerical simulations as are used in the analytic models, then nearly identical agreement between the simulations and analytic model confirms that the mechanics in the model have been implemented as intended. Today's analytic models can use quite sophisticated constitutive models for materials, and so such verification can be extremely useful. For example, such a verification of a penetration model using a pressure depended flow stress for the target with a cutoff (thus producing an interior boundary in the target flow region that had to be determined by the model) was performed in [5].

There is, of course, still the additional step of *validating* the model against experimental results. Currently, analytic penetration models are able to accurately predict depths of penetration for simple penetration geometries where the target and projectile are metals. Part of the reason for producing analytic models in the days of relatively successful large scale numerical simulations is that the analytic models are fast. Analytic models allow the parameter studies necessary for design and optimization. Also, when an analytic model agrees with experiment, there is confidence that the event being studied is understood from a fundamental physics viewpoint. This paper primarily discusses analytic models where a central axis of symmetry is assumed, but for those interested in 3D problems lacking traditional symmetries, the potential time savings through use of analytic models is enormous.

In order to clarify some of the discussion, it will be helpful to explicitly write down a penetration model so that the various terms can be identified. In particular, the method of the centerline momentum balance involves taking the momentum balance,

$$\rho \frac{du_i}{dt} = \frac{\partial \sigma_{ji}}{\partial x_j} \qquad (2)$$

and then integrating along the centerline to obtain

$$\rho_p \int_{z_p}^{z_i} \frac{\partial u_z}{\partial t} dz + \rho_t \int_{z_i}^{+\infty} \frac{\partial u_z}{\partial t} dz + \frac{1}{2}\rho_p u_z^2 \Big|_{z_p}^{z_i}$$
$$+ \frac{1}{2}\rho_t u_z^2 \Big|_{z_i}^{+\infty} - \sigma_{zz} \Big|_{z_p}^{+\infty} - 2\int_{z_p}^{+\infty} \frac{\partial \sigma_{xz}}{\partial x} dz = 0 \qquad (3)$$

The u^2 terms appear because we are in an Eulerian framework. For the specific model described in [1], three primary assumptions are made to produce the equations of motion. First, a velocity profile is assumed along the centerline in both the target and the projectile. Second, the rear of the projectile is assumed to decelerate by elastic waves traveling up and down the length of the projectile. Third, a hemispherical velocity field is assumed within the target that, combined with rigid plastic assumptions, provides a stress field making it possible to compute the derivative of the shear stress along the centerline, as required by the last term of Eq. (3). The velocity field is derived from a potential and describes the plastically flowing target material. In addition to a velocity field describing behavior deep within a target, using a multiplicative blending of potentials a velocity field has been developed that describes target material motion near the back surface as the target bulges [6]. This flow field produces back surface bulges that agree very well with computer simulations and experimental data. With the above three assumptions, the centerline momentum balance equation becomes

$$\rho_p \dot{v}(L-s) + \dot{u}\left\{\rho_p s + \rho_t R \frac{\alpha-1}{\alpha+1}\right\}$$
$$+ \rho_p \frac{d}{dt}\left(\frac{v-u}{s}\right)\frac{s^2}{2} + \rho_t \dot{\alpha} \frac{2Ru}{(\alpha+1)^2} \quad (4)$$
$$= \frac{1}{2}\rho_p(v-u)^2 - \left\{\frac{1}{2}\rho_t u^2 + \frac{7}{3}Y_t \ln\alpha\right\}$$

where v is the velocity of rear of the projectile and u is the penetration velocity, L is the length of the projectile, s is the plastically flowing zone in the projectile, $\alpha(u)$ is the extent of the plastic zone within the target, R is the crater radius and Y_t is the flow stress of the target. The term in brackets on the right hand side is the target resistance. The deceleration of the rear of the projectile is

$$\dot{v} = -\frac{\sigma_p}{\rho_p(L-s)}\left\{1 + \frac{v-u}{c} + \frac{\dot{s}}{c}\right\} \quad (5)$$

where σ_p is the projectile flow stress and c is the bar wave speed in the projectile. Projectile erosion is

$$\dot{L} = -(v-u) \quad (6)$$

These three equations are the central part of what has become known as the Walker-Anderson model, and the full development can be found in [1]. There are also additional assumptions required, one to determine the extent of the plastic zone within the projectile, one to determine the extent of the plastic zone within the target and one to determine the crater radius. The last two are important topics and will be discussed further below. This model agrees very well with experiment for larger L/Ds (D is projectile diameter) and models using the same ideas have been produced that do well predicting penetration into thick ceramics and glasses [5].

Recently a model that predicts well the ballistic limit of fabrics has been produced [7]. This model begins by assuming the fabric is comprised of elastic springs connected where the fabric yarns cross. Next the static deflection problem is solved. The static solution allows an explicit calculation of the strains. The strains give a force vs. deflection curve. Next, the longitudinal wave speed is calculated. These pieces combine to form a momentum balance where the deceleration of the impacting projectile and inertially involved fabric is balanced against the force versus deflection curve. When the state of the projectile just coming to rest is set to occur when the strain in the fabric equals the fiber breakage strain, the following equation for ballistic limit results [6]:

$$V_{bl} = \frac{9}{2}(1+\beta X)c_f \varepsilon_f \left\{\left(\frac{R_{bl}}{R_p}\right)^{2/3} - 2\left(\frac{R_{bl}}{R_p}\right)^{1/3} + 3\right\}^{-1} \quad (7)$$

where $\quad R_{bl}/R_p = \sqrt{\frac{9\pi}{8}\left(\frac{1}{X}+\beta\right)}$

X is the areal density of the fabric divided by the areal density of the projectile, $\beta=(1.6)^2$ a constant determining how much fabric material is inertially involved in the impact, c_f is the fiber wave speed and ε_f is the fiber breaking strain. This model predicts ballistic limit extremely well. Though not a centerline momentum balance but rather a Lagranian model, the model indicates a successful approach that may be applied to composites and other nonflowing materials.

Though analytic models have become very accurate (predictions to within 5% on depth of penetration for simple metal projectiles into monolithic metal targets), there are still problems. For example, no one model predicts the full L/D effect, that is, the observation that low L/D projectiles penetrate deeper in terms of L than larger L/D projectiles. Experimentally, the effect is surprisingly large [8].

PROJECTILE MODELING PROBLEMS

Three problems at present arise in the modeling of projectiles.

1. Transition in stress state. When the projectile nearly completely erodes during the penetration event, as it approaches L/D=1 the projectile stress state changes from uniaxal stress to uniaxial strain. The change in stress state allows larger decelerations of the back of the projectile and results in larger recovered residual projectiles than the model predicts. When and how the transition occurs is not well known. In [9] a relatively smooth transition based upon remaining L/D was used.

2. Modeling complex projectiles with centerline approximations. Choosing to go the route of the centerline momentum balance, the problem arises of modeling projectiles that are not simple rods with hemispherical noses but have complex 3D structure, such as jackets. Work recently performed in this area modeled a 0.30" APM2 projectile in the context of the centerline momentum balance [10]. Various approaches of allocating the projectile material were examined. The conclusion for that work was to model the projectile as a length of lead followed by a length of steel. However, a straightforward, consistent algorithm for defining complex projectiles in the context of a centerline momentum balance would be useful.

3. *Projectile side loading*. The models discussed above all assume the projectile load occurs on the nose of the projectile. However, if the projectile impacts with obliquity or yaw, or the target plate is in motion, the side of the projectile could be impacted by the crater wall. Such a collision leads to questions of how the projectile will deform and break. Accurately modeling these behaviors probably will require discretizing the projectile.

TARGET MODELING PROBLEMS

There are still a number of open problems with respect to target modeling. Here are six of them.

4. *Extent of plastic flow in the target $\alpha(u)$*. All penetration models that rely on an assumed flow field require the calculation of the extent of the flow field. This topic is dealt with in the centerline momentum balance models through the cavity expansion technique. In penetration models that use the cavity expansion technique directly the production of an extent of the flow field within the target is implicit. To describe what is currently done in [1], a cavity expansion is calculated where the penetration velocity u is the assumed driving velocity for the inner surface of the cavity. The subsequent velocity of the interface between the elastic and elastic/plastic response region $c(u)$ is then used to obtain a scaling factor $\alpha(u)=c(u)/u$ that is then used to calculate the flow field extent αR (R is the crater radius). This approach seems to work fairly well, but success in part could be due to the fact that in the model the flow field extent appears as a logarithmic term in the target resistance (Eq. 4). However, there is still work to be done here. For example, at higher velocities a fix is required to keep α greater than 1, since the equation of state used for the compressible metal is linear and does not have the higher order terms that become important at higher compressions. Also, the cavity expansion solution gives a sharp decrease in α as the velocity increases. Though it has been shown that α does indeed decrease with increasing velocity [11], such a large change for small velocities seems surprising. There are other methods that have been proposed based on theorems from plasticity, and for a careful discussion see [12].

5. *Different projectile nose shapes*. It may be thought that this topic should be under projectile problems, but in fact the issue of projectile nose shape is a target modeling issue, since it assumes the projectile is penetrating in such a fashion as to maintain its nose shape (eroding penetration always tends towards a hemispherical eroding nose on the projectile, regardless of the initial nose shape of the projectile [13]. Also, topics such as "self sharpening" effectively fall under the crater-radius problem, discussed later). It should be straightforward, assuming that the approach of producing a flow field leads to terms that can be inserted in the momentum balance, to model the target material flow around a different nose shape. However, this has proven to be difficult for the following reason. Hemispherical flow has only one length scale, based on the radius of the crater. Thus, given the crater radius and an extent of the flow field, the flow pattern is defined. For flow fields that lack this spherical symmetry, there are at least two length scales. Thus, producing a good flow field for a pointed or ogival projectile also requires calculating flow field extents in terms of at least two variables. Perhaps the flow field extents can be related in terms of a constant, but an argument must be put forward for doing so.

6. *Breakout models*. When a target of finite thickness is impacted, the projectile may perforate the target. When the thickness of the target is relatively large (at least a couple of projectile diameters) and the target is a ductile metal, then a flow field has been developed for the target that reproduces the shape of the back of the plate [6], but the next step is the target failure. Breakout has been addressed through assumptions about internal failure and geometric failure modes in [14] based on experimentally-based observations and empiricism, but a first principles approach is needed. Most failure modes involve fracture of the material or the localization of shear, two notoriously difficult subjects in applied mechanics, particularly in a predictive context.

The next problems address different target materials. In each case the material does not engage in ductile flow, as does a metal. Because the target response is so "nice" for a metal, it can be accurately addressed through analytical modeling. However, these next target materials do not behave in the same fashion, and in many ways are not "nice."

7. *Failure time for thin ceramic tiles*. As a simplification, ceramics in analytic penetration models have two modes: breaking and broken (or perhaps better wording is failing and failed). For thick ceramics, the penetration by large L/D projectiles is dominated by the response of the failed ceramic material. This failed material is usually modeled, in both analytic models and large scale hydrocode simulations, as a pressure dependent yield with a cut-off, also referred to as a Drucker-Prager yield surface with cutoff. Such a constitutive model seems to model well failed ceramic material, and an analytic model has been developed that uses this constitutive model [5]. Once the ceramic is ground up (comminuted), it essentially flows, thus allowing the modeling of the failed target material with the same flow fields as seen in flowing metals. Because of the success in using a pressure dependent constitutive model for failed ceramic, the behavior of the failed ceramic is not considered an open problem.

What is an open problem is modeling the time it takes the ceramic to break. The kinematics of fracture is an important question for light armors, meaning armors where the ceramic thickness is on the order of the projectile diameter. In these armors, the ceramic exerts significant stopping power to the projectile while it is in the process of breaking – producing dwell, where the projectile erodes against an essentially anvil-like ceramic face – and the amount of time the ceramic takes to break (or, more accurately, grind up into small pieces) is a large determining factor in the performance of the ceramic target. Currently, analytical models taking into account ceramic fracture usually just include a fracture time, defined as the time that it takes for the ceramic to break and begin behaving like a failed (pressure-dependent) material. There is to date no explicit calculation of the fracture time based upon the properties of the ceramic. Ref. [9] used hydrocode calculations to calibrate a curve fit to the fracture time for different ceramic thicknesses for a given impact velocity. Since every ceramic is different, and the fracture time depends on impact velocity as well as tile thickness, this approach is not reasonable for analytic modeling.

A first principles model to arrive at this fracture time is needed, but where to begin? It is still not known what material parameters are important in ceramic fracture, and the fact that for some ceramics (e.g. boron carbide) the microstructure properties seem to have little influence on the final ballistic performance leaves considerable room for concern as to the relevant properties.

8. Nonflowing target resistance calculations. Composite materials, such as carbon and glass fiber reinforced composites and fabrics with resin, introduce new complexities because they do not nicely flow. Modeling them in the context of the centerline momentum balance requires the ability to compute a stress gradient term for use in the centerline momentum balance, or it may require the use of a Lagrangian framework for the target linked with the projectile treated in an Eulerian framework. This approach would allow for the computation of elastic in-plane stresses within the target. (This approach may also be best for very thin metal plates.) Dealing with these materials is proving difficult, not just for analytic models but also for large-scale numerical simulations. Once successful constitutive models are developed, it should be possible to apply them in the analytic modeling framework. Damage is also more complicated, since, in addition to deformation, fiber breakage can occur.

9. h/R ratio for fabrics during impact. When a fabric is impacted with a projectile, a pyramid is formed in the fabric as it deflects and deforms. The edges of the pyramid run in the direction of the fibers. An outstanding problem in modeling fabrics is determining, from first principles, the ratio of the height to the radius of the base of the pyramid (h/R). It appears from experiments that this ratio is constant during the penetration, but that fact has yet to be shown through modeling. The model described in [7] assumes a constant h/R value. The model is very successful at predicting the ballistic limit of a fabric based on its fiber density, Youngs modulus, and failure strain. However, the model predicts smaller h/R than are seen in experiments; experimentally, $h/R \sim 2/3$, while the model predicts $\sim 1/3$ or less. The reason for the difference is most likely due to looseness and crimping in the fabric, but this has yet to be demonstrated.

COUPLED PROJECTILE/TARGET MODELING PROBLEMS

Finally, the last two problems require the simultaneous consideration of the projectile and the target.

10. Crater diameter. An explicit equation for crater diameter has proven elusive. In [1] an experimental curve fit was assumed universal for all materials to provide a crater diameter based on impact velocity. However, experimentally it is observed that the crater diameter decreases with penetration depth. It is straightforward to include in the analytic model a time (or depth) dependent crater diameter. The equations for crater diameter clearly will depend on both the strengths and densities of the target and projectile. Failure strain in the projectile also comes into play in the diameter of the formed crater. There is an additional level of need in crater diameter modeling. In order to analytically model certain projectile/target interactions, it is important to have a dynamic crater growth model – that is, the model should include the transient motion of the crater wall as it moves outwards to its final diameter.

11. Projectile rigid/eroding transition. A holy grail of penetration modeling: When does a projectile penetrate as a rigid body, and when does it erode, and what is the velocity that demarcates the two regimes? To be able to determine this analytically would be a major achievement. Current models do a transition from eroding to rigid penetration based on their calculations of front and tail velocities, and when the equations produce a larger value for the front velocity u than for the back velocity v, then it is assumed the penetration is rigid. However, such an approach does not accurately predict whether a striking projectile will initially penetrate in a rigid fashion or will erode. Currently, a different model is used from the outset if it is known the projectile does not erode throughout the whole penetration event.

CONCLUSIONS

There are of course more problems, but these ones are central. For example, it is likely that a solution to the crater diameter problem (#10) and the transition in stress state within the projectile problem (#1) will solve the L/D effect problem within the current model. Also, solution of the transient crater diameter problem (#10) and the projectile side loading problem (#3) will make it possible to solve complicated, 3D projectile/target interaction problem. For example, with that additional information, it should be possible to model oblique and yawed impacts. Adding a good breakout model (#6) will provide all the pieces required for the interaction of rods with armors comprised of dynamically moving plates. Such modeling will allow examination and optimization of modern armors employing plates at angle and in motion. Finally, solution of the ceramic failure time problem (#7), the nonflowing-target resistance problem (#8) and the fabric problem (#9) coupled with modeling complex projectiles on the centerline (#2) would allow a detailed examination and optimization of today's light armors comprised of ceramic tiles attached to composite plates backed by loose fabric.

In conclusion, analytic penetration models can be expected to become more capable and address more complicated targets, projectiles, and impact geometries in the future. Solution of the problems presented in this paper will allow more detailed examination and optimization of armors.

ACKNOWLEDGMENTS The author thanks those he has worked with in terminal ballistics over the years, particularly Charles Anderson of Southwest Research Institute.

REFERENCES

1. Walker, J.D. and Anderson, Jr., C.E., *Int. J. Impact Engng* **16**, pp. 19-48 (1995).
2. Tate, A., *J. Mech. Phys. Solids* **15**, pp. 387-399 (1967).
3. Forrestal, M., Okajima, K. and Luk, V.K., *J. Appl. Mech.* 55(4), pp. 755-760 (1988).
4. Walker, J.D., "Hypervelocity Penetration Modeling: Momentum vs. Energy and Energy Transfer Mechanisms," *Int. J. Impact Engng* to appear (2001).
5. Walker, J.D., and Anderson, Jr., C.E., "An Analytic Penetration Model for a Drucker-Prager Yield Surface with Cutoff," in *Shock Compression in Condensed Matter-1997*, AIP Conference Proc. 429, New York, pp. 897-900 (1998).
6. Walker, J.D., "An Analytic Velocity Field for Back Surface Bulging," *Proc., 18th Int. Ballistic Symp.*, pp. 1239-1246 (1999).
7. Walker, J.D., "Constitutive Model for Fabrics with Explicit Static Solution and Ballistic Limit," *Proc., 18th Int. Ballistic Symp.*, pp. 1231-1238 (1999).
8. Anderson, Jr., C.E., Walker, J.D., Bless, S.J., and Partom, Y., *Int. J. Impact Engng* **18**, pp. 247-264 (1996).
9. Walker, J.D., and Anderson, Jr., C.E., "An Analytical Model for Ceramic Faced Light Armors," *Proc., 16th Int. Ballistic Symp.* **3**, pp. 289-298 (1996).
10. Chocron, S., Grosch, D.J., and Anderson, Jr., C.E., "DOP and V_{50} Predictions for the 0.30-Cal APM2 Projectile," *Proc., 18th Int. Ballistic Symp.*, pp. 769-776 (1999).
11. Anderson, Jr., C.E., Littlefield, D.L., and Walker, J.D., *Int. J. Impact Engng* **14**, pp. 1-12 (1993).
12. Walker, J.D., "On Maximum Dissipation for Dynamic Plastic Flow," *Proc., 15th Int. Ballistic Symp.* **1**, pp. 67-74 (1995).
13. Walker, J.D. and Anderson, Jr., C.E., *Int. J. Impact Engng* **15**(2), pp. 139-148 (1994).
14. Ravid, M., and Bodner, S.R., *Int. J. Engng Sci.* **21**(6), pp. 577-591 (1983).

BALLISTIC RESPONSE OF FABRICS: MODEL AND EXPERIMENTS

Dennis L. Orphal[*], James D. Walker[+] and Charles E. Anderson, Jr.[+]

[*]*International Research Associates, 4450 Black Ave., Suite E, Pleasanton, CA 94566, USA*
[+]*Southwest Research Institute, Engineering Dynamics Department*
P.O. Drawer 28510, San Antonio, TX 78228-0510, USA

Walker [1] developed an analytical model for the dynamic response of fabrics to ballistic impact. In the model the force on the bullet is a function of fabric displacement (h) along the axis of impact and the radius (R) of the fabric deformation or 'bulge". Ballistic tests against Zylon™ fabric have been performed to measure h and R as a function of time. The results of these experiments are presented and analyzed in the context of the Walker model.

INTRODUCTION

Walker [1] developed a model describing the response of fabrics to ballistic impacts. A result of this model is that for normal impacts the force on the fabric, F, is related to the axial displacement of the fabric, h, by

$$F = -(8/9) E_f T^* h^3/R^2 \qquad (1)$$

where E_f is the Young's modulus of the fabric, R is the radius of the "bulge" in the displaced fabric, and T^* is the "effective thickness" of the fabric. T^* is equal to the ρ'/ρ_f where ρ' is the areal density of the fabric, and ρ_f is the density of the fabric fiber. This equation agrees very well with static force-deflection data for Kevlar™ 129 sheets [1].

It is observed, at least for some fabrics, that the ratio of h/R is approximately constant during fabric deformation prior to perforation by the projectile [2]. Walker finds an approximate solution for the transverse wave speed (c) at which the bulge radius increases with time:

$$c = (c_f V)^{1/2} \qquad (2)$$

where V = dh/dt = the velocity of the axial fabric deformation; and $c_f = (E/\rho_f)^{1/2}$, which is the bar velocity of a fabric fiber. Then self similarity is assumed, leading to:

$$h/R = \text{constant} = (V/c) = (V/c_f)^{1/2} \qquad (3)$$

This assumption results in an analytical solution for the force in terms of the single variable h, as can be seen from Eqn. (1).

Using Eqns. (1) and (2) along with an estimate of the amount of fabric that is "carried along" with a projectile as the fabric deforms, Walker's model successfully produces V_{50} data for Kevlar™ KM2 fabric for a wide range of projectile sizes and fabric areal densities, as shown in Fig. 1. In Fig. 1 V_{50} data have been scaled by the parameter $V^* = c_f \varepsilon_f^{2/3}/2^{1/3}$ where ε_f is the failure strain of the fiber,

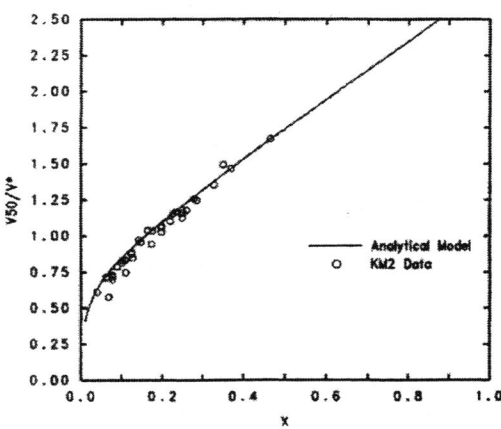

FIGURE 1. Kevlar KM2™ V_{50} data and Walker model fabric.

and the abscissa-parameter $X = \rho' A_p / m_p$ where A_p is the presented area of the projectile and m_p is the mass of the projectile. These scaling parameters are derived naturally from the model; however, it is noted that Cunniff first suggested these nondimensional parameters on the basis of dimensional arguments [3]. The model has also been successfully applied to Kevlar™ 129 [4].

TESTS

A program is in progress that involves the ballistic response of a fabric made of a relatively new fiber material. This new fiber material is PBO, poly (p-phenylene-2,6-benzobisoxazole); the specific fabric being used is called Zylon™, made by Toyobo of Japan. Key nominal physical and mechanical properties for the PBO fibers are: $\rho_f = 1560$ kg/m^3, $E = 169$ GPa, tensile strength = 5.2 MPa, and $\varepsilon_f = 3.10\%$ [3]. The Zylon™ fabrics being used in testing are composed of multiple layers or plies to obtain various areal densities. The individual plies are 30 by 30 weaves of 500 denier fibers.

For the specific experiments here, the fabric is placed behind a ceramic/metal target. The target and fabric are forced into a steel fixture, holding the fabric tightly around the edge of the fixture but allowing the fabric to deform through a 101.6-mm (4-inch) diameter hole in the fixture. (Some tests were performed with a 152.4-mm (6-inch) diameter hole and no significant differences were noted). The impacting projectile in these tests is the 7.62-mm APM2 bullet that impacts the ceramic face of the target at nominally 850 m/s.

The deformation of the fabric as a function of time following impact of the projectile against the ceramic/metal target is recorded using an Imacon 468 digital camera system. A mirror is used so that the axial and radial deformation of the fabric can be recorded simultaneously. As used the Imacon camera recorded eight images of the deforming fabric at different preset times.

TEST DATA

Figure 2 (next page) shows the eight Imacon images from a typical test (Test 77). From these images, a spatial fiducial, and the known exposure times, the axial displacement and the radial deformation of the fabric as a function of time are determined. Test data are quite repeatable. Although tens of tests have been conducted, for clarity, Figs. 3 and 4 show the measured axial ("height") and radial fabric deformation for Test 77 along with the measurements from four additional, nominally identical, tests. In these tests, the ceramic/metal target causes the bullet to "dwell" on the target surface. This dwell, combined with the time of interaction of the bullet with the ceramic/metal target, means that the fabric does not begin to deform for several 10's of microseconds after initial impact (time zero). Additionally there is some obscuration due to the edge of the target/fabric fixture, thus the deformation of the fabric is not immediately visible when it begins. Therefore, it is approximately 45 μs after impact before fabric deflections are recorded. As can be seen, except at late times when the deformation is arresting, the speed of both the axial and radial deformations are

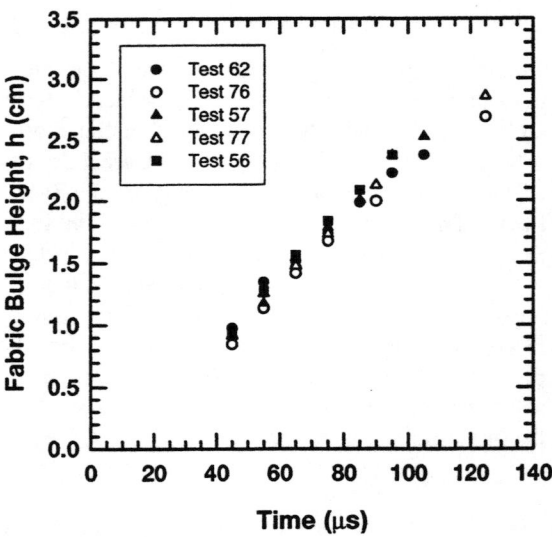

FIGURE 3. Axial deformation for Zylon™ fabric.

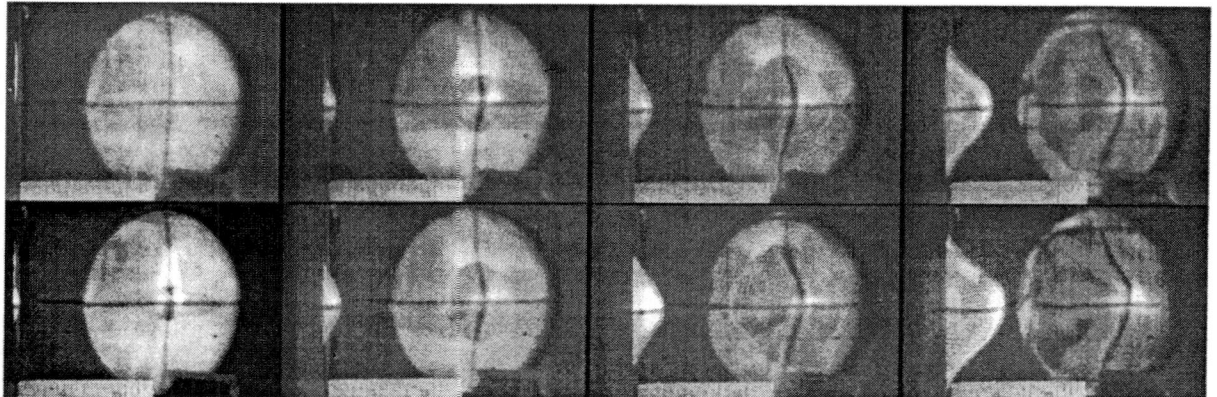

FIGURE 2. Imacon digital camera images showing both axial and radial Zylon™ deformation at eight times after impact (Test 77) (Times are: 35 μs, 45 μs, 55 μs, 65 μs, 75 μs, 90 μs, 125 μs, and 165 μs).

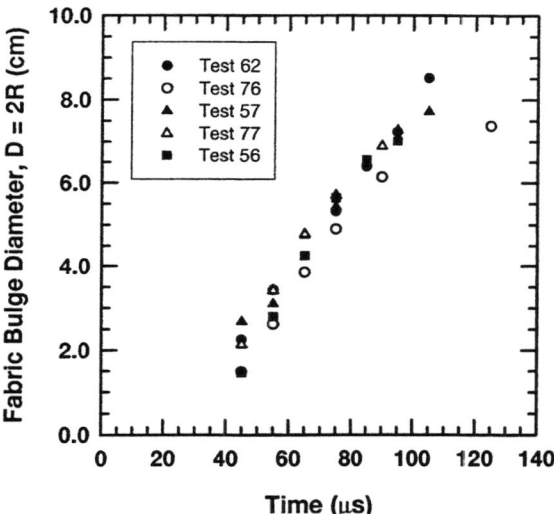

FIGURE 4. Diameter of Zylon™ fabric deformation versus time.

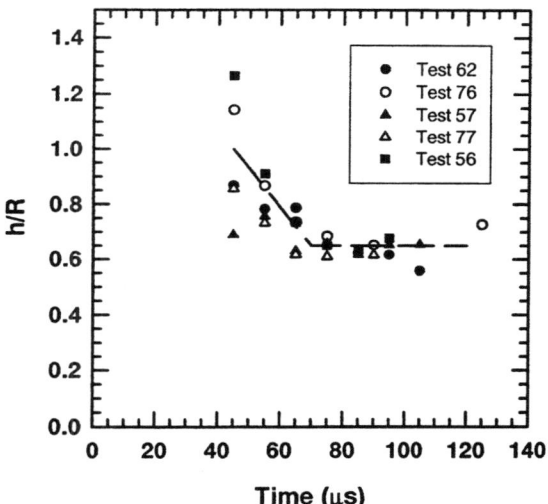

FIGURE 5. The ratio of deformation height to radius versus time for Zylon™.

about constant. The slopes for the axial and radial deformation, respectively, are

$$dh/dt = V \approx 300 \text{ m/s} \quad (4a)$$

$$dR/dt = c \approx 500 \text{ m/s}. \quad (4b)$$

Figure 5 shows the ratio h/R calculated from the measured data shown in Figs. 3 and 4. For these tests with Zylon™ fabric, the ratio h/R is not constant during the entire deformation response. As indicated by the dashed line, the ratio h/R from 45 μs to ~70 μs appears to decrease from about 1, or a little more, and then achieves an approximately constant value of about 0.6-0.7 for times between 60 μs and 110 μs.

DISCUSSION

Fabric deformation, as measured by h/R, is not constant during early portions of the deformation history, although somewhat later h/R does become constant. Walker [1] modeled the fabric as an extended system of orthogonal springs. As initially formulated the model has no provision for the crimp or "slack" associated with a woven fabric, such as tested here. It is hypothesized that inclusion of this slack would result in a non-constant and lower transverse wave velocity, c, at early times until the slack is removed. By Eqn. (3) this would in turn result in a non-constant h/R. Thus, if this hypothesis is correct, the ratio h/R would decrease from some value associated with the removal of the slack and then achieve a constant, or nearly constant, value after the slack has been removed.

There is much yet to be learned about the response of fabrics to ballistic impacts. The work reported here is intended as a contribution to this effort. Additional experiments and analysis are in progress. Walker has already extended his model to include fabrics with resin, with excellent results [4]. Work is also in progress to include the effects of slack in Walker's model.

ACKNOWLEDGMENTS

This work was performed under U.S. Army Contract No. DAAD16-00-C-9260. The authors would like to thank Janet Ward and Phil Cunniff of the U.S. Army Soldier Biological and Chemical Defense Command and Steve Wax of DARPA for their guidance and support. The efforts of Don Grosch and the ballistic testing crew at Southwest Research Institute are also gratefully acknowledged.

REFERENCES

1. Walker, J. D., "Constitutive Model for Fabrics with Explicit Static Solution and Ballistic Limit", *Proc. 18th Int. Symp. on Ballistics*, Vol. 2, pp. 1231-1238, Technomic Publishing Co., Lancaster, PA, 1999.
2. Cunniff, P. M., "An Analysis of the System Effects in Woven Fabrics Under Ballistic Impact," *Textile Res. J.*, **62**(9), 495-509, 1992.
3. Cunniff, P. M., "Dimensionless Parameters for Optimization of Textile-Based Body Armor Systems," *Proc. 18th Int. Symp on Ballistics*, Vol. 2, pp. 1303-1310, Technomic Publishing Co., Lancaster, PA, 1999.
4. Walker, J. D., "Ballistic Limit of Fabrics with Resin", *Proc. 19th Int. Symp. on Ballistics*, Vol. III, pp. 1409-1414, Interlaken, Switzerland, 7-11 May, 2001.

LONG-ROD MOVING-PLATE INTERACTION

Y. Partom[*]

Rafael, P.O. Box 2250, Haifa 31021, Israel

Understanding the mechanics of interaction of a long rod projectile with a forward moving plate at an angle is essential to understanding long rod interaction with an explosive reactive armor cassette. To investigate the mechanics of such an interaction we use AUTODIN2D/EULER in plane geometry, although the problem is 3D. We assume that this is a satisfactory approximation, as we're only interested in the main features, and are not comparing fine details to experimental results. From the simulations we learn that the interaction never reaches steady state. Initially each material splits into two streams, and the interaction plane is perpendicular to the rod. But with time the interaction plane rotates slowly, until it becomes parallel to the rod, which is then able to continue moving forward without interruption. During this process interacting rod material of length ΔL is diverted at an angle and becomes ineffective for penetrating the main target. We made many such runs to determine the dependence of ΔL on the parameters of the problem. This dependence makes it possible to predict ΔL for a variety of rod-plate situations.

INTRODUCTION

The outcome of the interaction of a long rod projectile with an explosive reactive armor cassette depends mainly on its interaction with the forward moving plate (FMP) of the cassette. The outcome of this interaction depends on many parameters such as: rod velocity, diameter, density and strength; obliquity angle; plate velocity, thickness, density and strength.

In what follows we use computer simulations to study the physical picture of the interaction, and to perform an extensive parameter study of the problem. From the results we formulate a procedure to predict the performance of the FMP on degrading the penetration capacity of the rod.

SIMULATIONS

We use AUTODYN2D/EULER in plane geometry. The problem is 3D, but AUTODYN3D/EULER was not fully debugged when we did this work. WE know that plane geometry is sometimes only a crude approximation to a 3D problem, but we believe that it can reproduce the correct physical picture. Using an Euler grid, it is more efficient to apply to the problem a Galilean velocity transformation, so that the interaction region is almost stationary and doesn't move out of the grid. Referring to Fig. 1, V_p is the rod velocity, V_t the plate velocity, and θ is the angle between the rod and its projection on the plate.

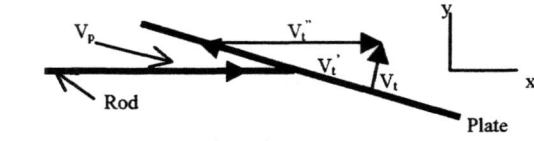

FIGURE 1. Velocity transformation.

---

[*] Work done while on sabbatical leave at SwRI, San Antonio, TX.

We first decompose V_t into V_t' and V_t'' and then apply to the system the velocity $-V_t''$. The transformed velocities V_p' and V_t' are shown in Fig. 2 and given in (1).

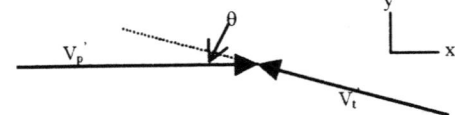

FIGURE 2. Transformed velocities.

$$V_t' = V_t/\tan\theta \; ; \; V_t'' = V_t/\sin\theta$$
$$V_p' = V_p - V_t'' = V_p - V_t/\sin\theta \quad (1)$$

We see that the transformed problem is that of two interacting streams.

In Fig. 3 we show material location plots every 100 μs from an AUTODYN run. The grid cells are 1×1 mm×mm throughout. The initial projectile is H_p=10 mm thick tungsten alloy with ρ_p=17.0 g/cm^3, Y_p(flow stress)=1.2 GPa, G_p(shear modulus)=140 GPa, STF$_p$(strain to failure)=0.1 and V_p'=0.75 km/s. More projectile material is fed in from the boundary at the same velocity up to a total length of 300 mm (L/H=30). The plate is at an angle of 30°, is H_t=10 mm thick, is made of steel with ρ_t=7.83 g/cm^3, Y_t=1.0 GPa, G_t=80 GPa, STF$_t$=0.5 and V_t'=0.75 km/s. More plate material is fed in from the boundary at the same angle and velocity.

From Fig. 3 we see that both rod and plate split into two streams of unequal thickness. For the rod the upper branch is thicker, while for the plate the lower branch is thicker. The interaction surface rotates slowly clockwise until it finally breaks up, and the two flows slide past each other. In this run the length of the rod that stays approximately straight and aligned with the x direction is 34 mm, so that ΔL=266 mm of the length is diverted and becomes ineffective for penetrating the main target.

We repeated the run with different values of V_p', and in Fig. 4 we show results for ΔL as a function of V_p' for runs with V_t'=0.5 km/s. We see from Fig. 4 that for V_p'≤0.73 km/s, the entire rod (300 mm) is diverted away from its original direction, most of it upwards. For V_p'>0.73 km/s only part of the rod is diverted. We also see that ΔL is quite sensitive to

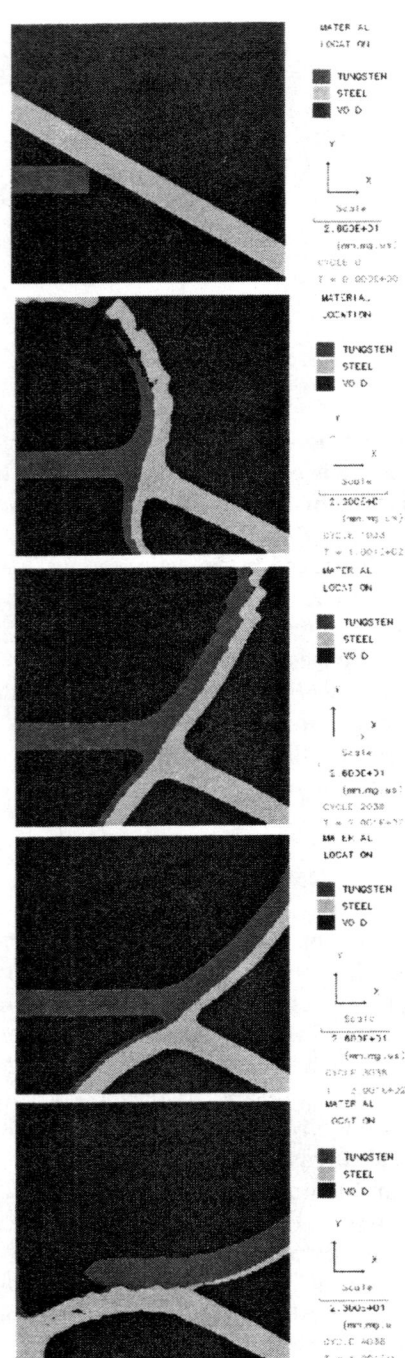

FIGURE 3. Material location plots for a run with V_p'=V_t'=0.75 km/s.

projectile velocity. For the above plate velocity (0.5 km/s at 30°), the range $0.5 \leq V_p' \leq 0.8$ km/s is equivalent to $1.5 \leq V_p \leq 1.8$ km/s (also shown in Fig. 4), which is the ordnance range for long rod projectiles. We see that for the high end of this range the FMP is only partly effective in diverting the rod. Beyond $V_p=2$ km/s only a small part of the rod would be diverted.

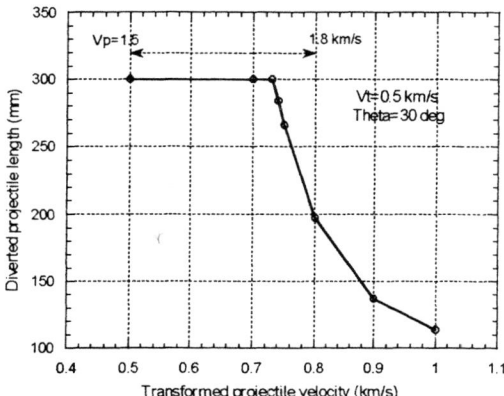

FIGURE 4. Diverted projectile length versus transformed projectile velocity for $H_p/H_t=1$, $V_t=0.5$ km/s and $\theta=30°$.

PARAMETER STUDY

Performing many runs like that described in the previous section, but with different values of the parameters, we find that for all the cases that we checked, the $\Delta L(V_p')$ relation is always a hyperbola-like curve that can be fitted with:

$$(\Delta L - a_1)(V_p' - a_2) = a_3 \qquad (2)$$

where a_i (i=1,3) depend on the material and kinematics parameters. Denoting the parameters by x_j we have:

$$a_i = a_i(x_j) \quad i=1,3 \; ; \; j=1,10 \qquad (3)$$

To first order approximation we can express $a_i(x_j)$ by:

$$a_i = (a_i)_{nominal} + \sum_j \frac{\partial a_i}{\partial x_j}(x_j - (x_j)_{nominal}) \qquad (4)$$

relative to a nominal set of parameters x_j.

The nominal set of parameters we use is $V_t=0.866$ km/s, $\theta=30°$, $H_p=H_t=10$ mm, $Y_t=1.0$ GPa, $Y_p=1.2$ GPa, $STF_t=0.5$, $STF_p=0.1$, $\rho_t=7.83$ g/cm³, $\rho_p=17.0$ g/cm³. Around the nominal set we evaluate the partial derivatives $\partial a_i/\partial x_j$ by giving each parameter a displacement Δx_j and running the simulation. We then evaluate the derivatives by:

$$\frac{\partial a_i}{\partial x_j} = \frac{(a_i)_{displaced} - (a_i)_{nominal}}{(x_j)_{displaced} - (x_j)_{nominal}} \qquad (5)$$

Two examples of the displaced and nominal hyperbolas are shown in Figs. 5 and 6.

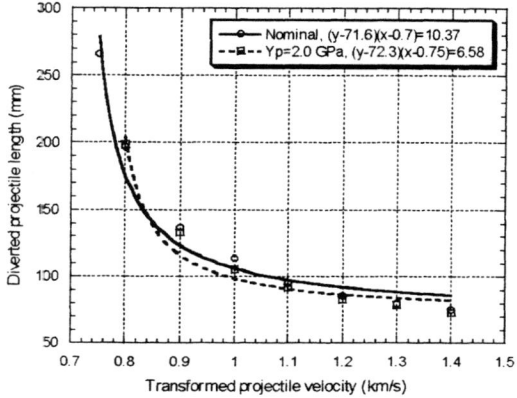

FIGURE 5. ΔL versus V_p'. Nominal runs (circles) and runs with displaced Y_t (squares). Curves are the fitted hyperbolas.

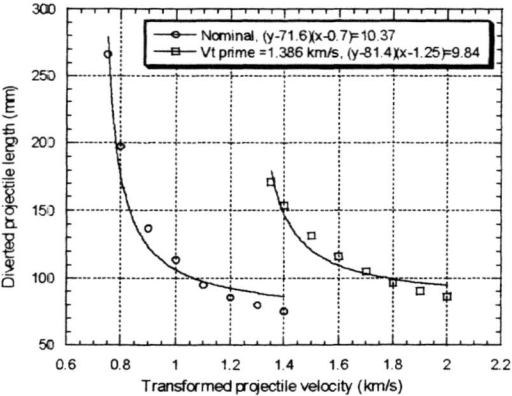

FIGURE 6. Same as Fig. 5 but with V_t' displaced to 1.386 km/s.

The partial derivatives extracted in this way are shown in Table 1.

To check our prediction procedure using these partial derivatives we performed additional simulation runs in which we used the nominal set of parameters, but with the rod and plate materials interchanged. In Fig. 7 we compare the results obtained from the runs to those predicted from the partial derivatives.

TABLE 1. Partial derivatives of parameters a_i in (2) with respect to material parameters x_j.

x_j displacement		i=1	i=2	i=3
	Nominal a_i	71.6 mm	0.7 km/s	10.37 mm²/μs
0.5 GPa	Displ. a_i	67.4	0.75	8.13
	$\partial a_i/\partial Y_t$	-8.40	0.10	-4.48
0.8 GPa	Displ. a_i	72.3	0.75	8.58
	$\partial a_i/\partial Y_p$	0.875	0.0625	-2.24
- 0.4	Displ. a_i	70.8	0.75	7.80
	$\partial a_i/\partial STF_t$	2.0	- 0.125	6.425
0.4	Displ. a_i	64.1	0.7	10.14
	$\partial a_i/\partial STF_t$	-18.75	0	- 0.575
5 g/cc	Displ. a_i	111.8	0.90	11.86
	$\partial a_i/\partial \rho_t$	8.04	0.04	0.298
- 5 g/cc	Displ. a_i	88.7	0.85	11.97
	$\partial a_i/\partial \rho_p$	-3.42	- 0.03	- 0.32
5 mm	Displ. a_i	74.1	0.7	14.02
	$\partial a_i/\partial H_p$	-0.5	0	0.73
0.52 km/s	Displ. a_i	81.4	1.25	9.84
	$\partial a_i/\partial V_t'$	18.85	1.058	-1.019
5 mm	Displ. a_I	65.7	0.75	22.51
	$\partial a_i/\partial H_t$	-1.18	0.01	2.428
5 degrees	Displ. a_i	56.0	0.65	10.12
	$\partial a_i/\partial \theta$	-3.12	-0.01	-0.05

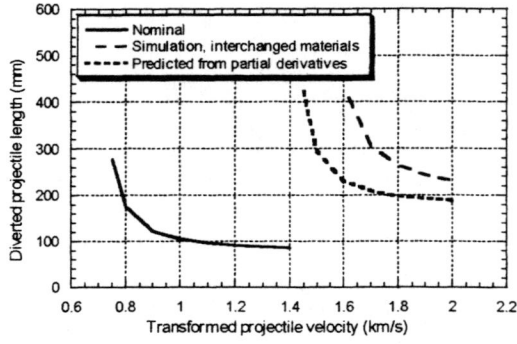

FIGURE 7. Results of check runs.

We see from Fig. 7 that the prediction is not ideal, but in view of the large change in density relative to the nominal case, it is satisfactory.

In subsequent work use this approach in a model to predict the performance of explosive reactive cassettes against long rod projectiles.

SUMMARY

We use AUTODYN2D/EULER in plane geometry to study the phenomenology of long rod interaction with a forward moving plate. The essence of the phenomenology is that part or the entire projectile is diverted from the original direction, and thereby becomes less effective in penetrating the main target. In reality, the diverted projectile is bound to break into several fragments.

Performing many simulation runs with different sets of parameters we conclude that, the relation $\Delta L(V_t')$ can always be fitted by an hyperbola with three parameters a_i that depend on the problem parameters x_j. We evaluate the partial derivatives $\partial a_i/\partial x_j$ numerically, and are able to predict $a_i(x_j)$ using a first order approximation.

Applying this approach to realistic situations of long rod interaction with an explosive reactive cassette (not reported here), we find that potential benefit of such cassettes against long rods cannot be fully exploited because of practical geometrical constrains.

REFERENCES

This reference list is empty, as we couldn't find anything relevant to the subject in the open literature.

CONVERSION OF FINITE ELEMENTS INTO MESHLESS PARTICLES FOR PENETRATION COMPUTATIONS INVOLVING CERAMIC TARGETS

Gordon R. Johnson[1,2], Robert A. Stryk[1], Stephen R. Beissel[2] and Timothy J. Holmquist[2]

[1]*Alliant Techsystems, 600 2nd St. N.E., Hopkins, MN 55343*
[2]*Network CS, 1200 Washington Ave. S., Minneapolis, MN 55415*

Abstract. This paper presents a new computational algorithm to automatically convert distorted finite elements into meshless particles during dynamic deformation. It also presents computed results for projectiles impacting ceramic targets. The new computational algorithm, together with an appropriate ceramic model, provides computed results that are in good agreement with test data. Included are problems involving dwell and penetration. This computational approach is especially well-suited for brittle materials such as ceramics, because the conversion from elements into particles generally occurs after the material has failed. The result is that the particles represent only failed material, which does not produce any tensile stresses. For some particle algorithms it is possible to introduce tensile instabilities, but this is not a concern if the particles represent only failed material.

INTRODUCTION

Lagrangian meshless methods (or particle methods) have been recently developed and applied to solid mechanics problems. An important characteristic of meshless methods is that they can be used to represent severe distortions in a Lagrangian framework. Although the accuracy and efficiency of meshless methods are not generally as good as finite element methods for dynamics problems with mild distortions, the meshless particle methods can be more accurate, more efficient and more robust for dynamics problems involving severe distortions. Therefore, a logical computational approach would be to use finite elements for the mildly distorted regions and meshless particles for the highly distorted regions.

This paper presents an explicit Lagrangian algorithm to automatically convert distorted elements to meshless particles during the course of the computation. This is an extension of previous work [1], where the general approach was demonstrated with a simplified algorithm. For brittle ceramic materials the conversion occurs after failure such that there are no tensile stresses (and therefore no tensile instabilities) in the particle nodes. The finite element formulation is provided in Reference 2 and the Generalized Particle Algorithm (GPA) is presented in References 3 and 4.

CONVERSION ALGORITHM

Figure 1 shows a finite element grid with a surface defined by nodes $a \ldots j$. Three elements on the surface (A, B, C) are designated as candidates for conversion to particles. An element is converted to a particle when the element has at least one side on a surface or interface, and the equivalent plastic strain exceeds a user-specified value (in the range of 0.3 to 0.6). Criteria other than plastic strain could also be used.

Interface (nodes a ... j) before conversion

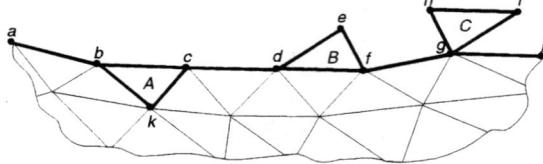

Interface after conversion of elements to particles

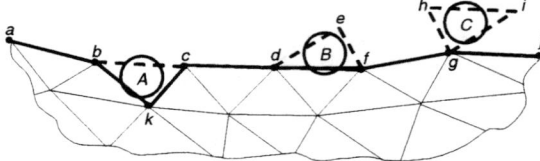

FIGURE 1. Conversion of Finite Elements Into Particles

If element A exceeds the plastic strain criterion the following steps are taken:
- Element A is removed from the finite element grid.
- Particle A is added as a GPA node.
- All of the element variables (stress, strain, internal energy, damage, etc.) are transferred to the GPA node.
- The mass, velocity and center of gravity of the GPA node are set to those of the replaced element.
- The GPA nodal diameters, (initial and current) are determined from $d = \sqrt{A}$, where A is the cross-sectional area of the element.
- The masses of nodes b, c and k are reduced by the removal of element A.
- Line segment b–c is removed from the list of interface (master surface) segments, and line segments b–k and k–c are added to the list.

The conversion of element B (with two sides on the surface) to GPA node B, and the conversion of element C (with three sides on the surface) to GPA node C, follows in a similar manner.

When a finite element is converted into a GPA node (under the conditions described for GPA nodes A and B in Fig. 1) the newly generated GPA node must be attached to the adjacent master line segment. Unlike a sliding algorithm, the attached algorithm does not allow the GPA node to separate from the master line segment or to slide along the master line segment. The details of this algorithm are provided in Reference 5.

EXAMPLES

The first example is shown in Fig. 2. It is an Armco iron cylinder that is impacted against a rigid surface at a velocity of 305 m/s. This is not a problem that requires conversion, but it demonstrates the algorithm for a problem that can be compared to a finite element computation.

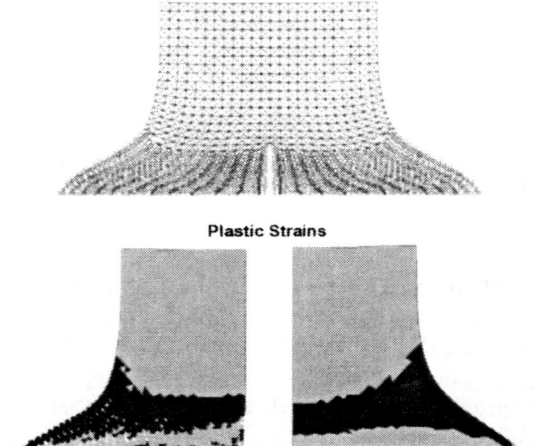

FIGURE 2. Impact of an Armco Iron Cylinder onto a Rigid Surface at 305 m/s

The upper position of Fig. 2 shows the remaining finite elements and the generated GPA nodes at 40 μs after impact. The lower left position shows the plastic strain contours for the finite element/GPA solution and the lower right provides the plastic strain distributions for a finite element solution without any conversion to particles. There is good general agreement between the strain distributions. The deformed lengths and diameters of the two results are essentially identical.

The next example involves a tungsten rod impacting a steel target at 1500 m/s, as shown in Fig. 3. For this problem both the projectile and target materials are converted to elements. The interfaces between the particles of different materials are represented by springs and dashpots, determined from the characteristics of the materials and the artificial viscosity coefficients [4]. With this interface treatment the GPA algorithm includes only those neighbors which are of the same material.

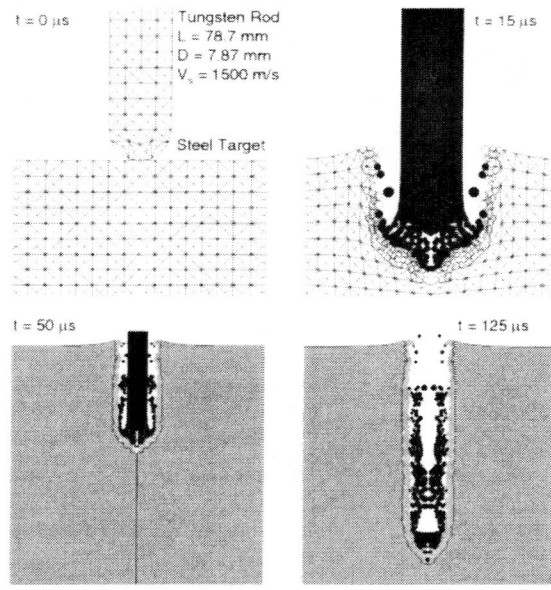

FIGURE 3. Impact of a Tungsten Rod onto a Steel Target at 1500 *m/s*

The lower portion of Fig. 3 shows the penetration process at 50 and 125 μs after impact. Here the tungsten is represented by darkened elements and particles, but the steel target is represented by light gray elements and particles. The black line in the target represents the outline of the elements such that all of the gray target material on the highly strained side of the line has been converted to particles. The final penetration is very close to comparable experimental data [6].

The final example is shown in Figs. 4 and 5. It involves a very long tungsten rod impacting a complex target composed of boron carbide ceramic surrounded by a steel case. The left side of Fig. 4 shows the configuration as provided by Lundberg *et al* [7]. The penetration in Fig. 4 is measured from the top surface of the boron carbide prior to impact.

The right side of Fig. 4 shows computed results and selected test data (penetration versus time) for four impact velocities. The test data are provided by Lundberg *et al* [7]. For the lowest velocity (1427 *m/s*) the tungsten rod does not penetrate the ceramic. When the velocity is increased slightly (1480 *m/s*), the tungsten rod dwells on the top surface of the ceramic until 25 μs after impact, and then it penetrates at a significant velocity. The two higher impact velocities penetrate in an expected manner. It can be seen that the computed results are in excellent agreement with the test data.

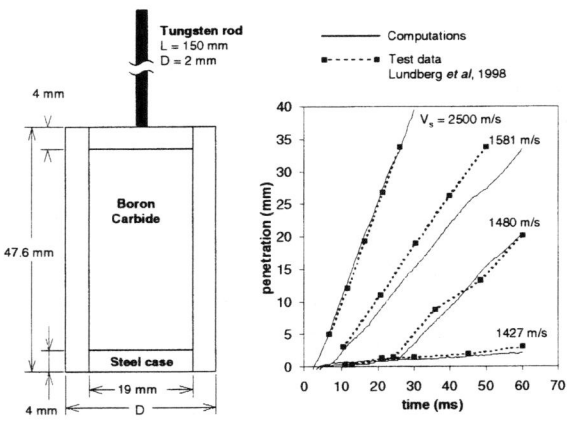

FIGURE 4. Impact of a Tungsten Rod onto a Boron Carbide Target at Various Velocities

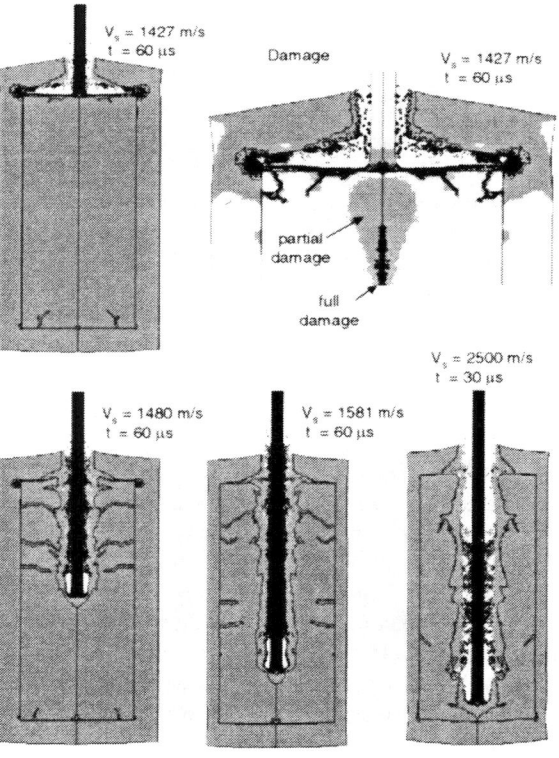

FIGURE 5. Computed Results for Impact of a Tungsten Rod onto a Boron Carbide Target

Figure 5 shows the computed results. For the four geometry plots (not including the damage plot in the upper right), the tungsten rod is represented by the darkened elements and particles, while the steel and ceramic are represented by light gray elements and particles. The black lines define the outlines of the elements. In some instances the lines represent the outlines of the two materials (case and ceramic), and in other instances they represent an interface between the elements and the particles.

The upper left of Fig. 5 shows the response for an impact velocity of $V_s = 1427$ m/s. Here the ceramic remains intact, while the defeated rod moves radially outward along the top surface of the ceramic until it is contained by the steel case. The distribution of damage is shown in the upper right and it can be seen that there is a thin region of undamaged ceramic directly under the impacting rod, and this enables the ceramic to remain intact and to defeat the rod.

The lower left of Fig. 5 is for a slightly higher impact velocity of $V_s = 1480$ m/s. The response for this case is similar to the slightly lower impact velocity ($V_s = 1427$ m/s) for the first 25 μs, as shown in Fig. 4. The higher impact velocity causes the defeated rod to push the case wall outward more than the lower velocity, and this in turn reduces the confining pressure in the ceramic thus allowing it to be more damaged and to fail. After the ceramic directly under the rod is fully damaged the rod penetrates the ceramic. For the two higher velocities the dwell is limited and the rod penetrates in a normal manner.

The boron carbide ceramic model used for these computations is similar to that presented by Johnson and Holmquist [8], although some changes have been made for damaged material.

A final comment concerns an important advantage of converting distorted elements into particles rather than simply eroding (or removing) the distorted elements. When an element is eroded it introduces a void which allows surrounding material to expand into the void and to lose pressure as it expands. If the material strength or failure characteristics are pressure dependent (as they are for brittle materials) then the pressure drop can lead to lower strength and/or more damage. If an erosion algorithm (rather than a conversion algorithm) is used for the impact conditions ($V_s = 1427$ m/s) in the upper left of Figure 1, premature failure occurs.

The highly distorted rod elements erode and the adjacent ceramic elements expand and fail, thus allowing the rod to penetrate the failed ceramic rather than being defeated by the intact ceramic. Replacing the distorted elements with particles does not introduce a void and/or pressure drop.

ACKNOWLEDGEMENTS

This work was funded by the U.S. Army Tank-Automotive Research, Development and Engineering Center (TARDEC), the U.S. Army Soldier and Biological Chemical Command (SBCCOM), the Defense Advanced Research Projects Agency (DARPA), the Army High Performance Computing Research Center (AHPCRC), and Southwest Research Institute (SwRI), under contracts DAAN02-98-C-4039, DAAD16-00-C-9260, and DASW01-01-C-0015. The content does not necessarily reflect the position or policy of the government, and no official endorsement should be inferred. The authors would also like to thank D.W. Templeton (TARDEC), P.M. Cunniff (SBCCOM), J.E. Ward (SBCCOM), and C.E. Anderson (SwRI) for their contributions.

REFERENCES

1. Johnson, G.R., Peterson, E.H., and Stryk, R.A., *International Journal of Impact Engineering*, **14**, 373–383 (1992).
2. Johnson, G.R., Stryk, R.A., Holmquist, T.J., and Beissel, S.R., *Numerical Algorithms in a Lagrangian Hydrocode*, Report WL-TR-1997-7039, Wright Laboratory, U.S. Air Force, July 1997.
3. Johnson, G.R., Beissel, S.R., and Stryk, R.A., *Computational Mechanics*, **25**, 245–256 (2000).
4. Johnson, G.R., Beissel, S.R., and Stryk, R.A., *International Journal for Numerical Methods in Engineering*, to appear (2001).
5. Johnson, G.R., Stryk, R.A., Beissel, S.R., and Holmquist, T.J., "An Algorithm to Automatically Convert Distorted Finite Elements into Meshless Particles During Dynamic Deformation," Submitted for publication.
6. Anderson, C.E. and Walker, J.D., *International Journal of Impact Engineering*, **11**, 481-501 (1991).
7. Lundberg, P., Holmberg, L., and Janson, B., "An Experimental Study of Long Rod Penetration into Boron Carbide at Ordnance and Hypervelocities," *Proceedings of the 17th International Symposium on Ballistics*, Midrand, South Africa, 1998.
8. Johnson, G.R. and Holmquist, T.J., *Journal of Applied Physics*, **85**, 8060-8073 (1999).

USING THE PENETRATION-VELOCITY RELATIONSHIP TO CORRECT FOR VARIATIONS IN TARGET HARDNESS

S. J. Bless[1] and J. Cazamias[1,2]

[1]*Institute for Advanced Technology, The University of Texas at Austin, 3925 W. Braker Ln., Suite 400, Austin, TX 78759*
[2]*currently at Lawrence Livermore National Laboratory, L-414, PO Box 808, Livermore, CA 94551*

Abstract. There is a correlation between the variation of penetration with impact velocity and the variation with target strength. This is because penetration is dependent on the ratio of inertial to strength forces. By using a Taylor series expansion, one can use penetration-velocity data to predict penetration-strength relationships.

BACKGROUND

It frequently happens that penetration data must be adjusted for relatively small changes in target hardness. Sometimes corrections for variations in penetrator density are also required. Past treatment of this problem include references [1] and [2].

Many penetration equations take the form

$$P/L = f(\rho, R, v) = C\sqrt{\rho}\, g(x) \qquad (1)$$

$$x = \frac{2R}{\rho v^2} \qquad (2)$$

where P is penetration, L is projectile length, ρ is penetrator density, v is impact velocity, R is target cavity expansion stress, C is an empirical constant, and g(x) is a function to be determined. Examples of formulae with this form include the Poncelet equation and high velocity forms of the Tate equations and Odermatt equation.

That penetration must depend primarily on x also follows from dimensional arguments, for x represents the target strength non-dimensionalized by the projectile stagnation pressure. We assume that we know the variation of P/L as an empirical or analytical function of velocity around the velocity of interest.

THEORY

We can find the effects of hardness, H, and density variations by Taylor expansion. First we work out the velocity partial derivative.

$$\frac{\partial f}{\partial v} = -\frac{4CR}{\sqrt{\rho v^3}} \frac{\partial g}{\partial x} \qquad (3)$$

Rewriting this as

$$\frac{\partial g}{\partial x} = \frac{\partial f}{\partial v}\left[\frac{-\rho v^3}{4C\sqrt{\rho}R}\right] \qquad (4)$$

and assuming that we know $\partial f/\partial v$ from empirical data, we can then write the other partial derivatives as

$$\frac{\partial f}{\partial \rho} = \frac{f}{2\rho} + \frac{v}{2\rho}\frac{\partial f}{\partial v} \qquad (5)$$

1291

$$\frac{\partial f}{\partial R} = -\frac{v}{2R}\frac{\partial f}{\partial v} \qquad (6)$$

The 1st order Taylor expansion at constant v is

$$\frac{P}{L} = f\big|_0 + \frac{\partial f}{\partial \rho}\bigg|_0 \Delta\rho + \frac{\partial f}{\partial R}\bigg|_0 \Delta R \qquad (7)$$

In ductile cavity expansion theory, R is proportional to strength. Strength is proportional to hardness. Hence, R is proportional to hardness, and $\Delta R/R_o = \Delta H/H_0$. After substitution and rearranging, this becomes

$$\frac{P}{L}(\rho_o + \Delta\rho, H_o + \Delta H, v)$$

$$= f(\rho_o, H_o, v) \cdot \left(1 + \frac{\Delta\rho}{2\rho_o}\right)$$

$$+ \frac{\partial f(\rho_o, H_o, v)}{\partial v} \cdot \left(\frac{v}{2}\right) \cdot \left(\frac{\Delta\rho}{\rho_o} - \frac{\Delta H}{H_o}\right) \qquad (8)$$

For extremely hard projectiles in which Y and R are commensurate, R should probably be replaced with R-Y in equation (2) and $\Delta H/H$ with $\Delta(R-Y)/(R_o-Y)$ in equation (8). However, for the rigid projectile case, one would again expect equations (1) and (2) to be valid.

COMPARISON WITH DATA

In order to evaluate this technique, we used the data from Hohler and Stilp for L/D=10 tungsten rods. They have data for $\rho = 17$ and 17.6 g/cm^3 tungsten alloys penetrating into steel of BHN hardness 180, 255, 295, and 388 (taken from [3], also quoted in [2]). We fit the BHN 255 and 388 data between 1.4 and 2.6 km/s with an Odermatt type equation,

$$f = A\exp(-B/v)^2 \qquad (9)$$

The correction for density was negligible. The best fit parameters are given in Table 1. The data are plotted in Fig. 1.

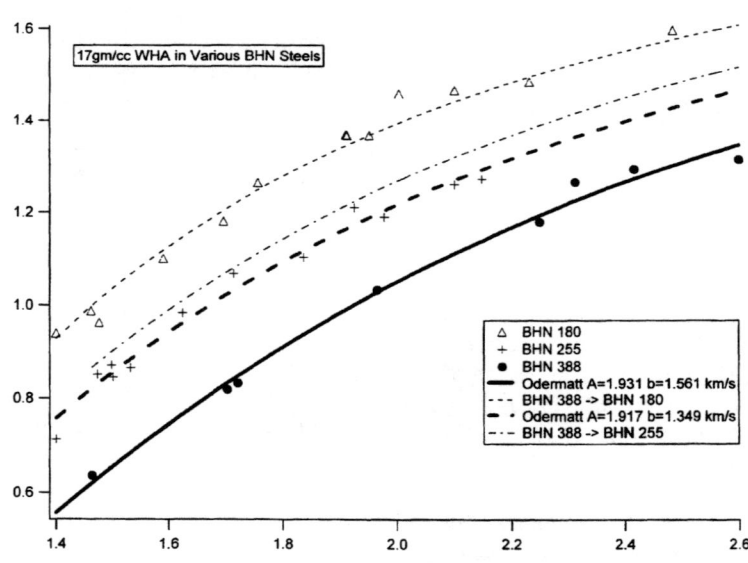

FIGURE 1. Points represent data. Heavy line is Odermatt fit to BHN 388 data. Two light dashed lines are "predictions" using equation (1) for BHN 180 and BHN 255 data. Heavy dashed line is Odermatt fit to 255 data.

Based on the Odermatt fit for BHN 388, using equation (8), "predictions" were made for BHN 255 and 180. Equation (8) predictions fit the BHN 180 data as well as a direct empirical fit. Equation (8) predicts the BHN 255 data to be about 5% higher than a direct Odermatt fit, but the data scatters by almost this much about the fit. We also found that errors were a little larger extrapolating from higher hardness to lower hardness, apparently because of the asymmetry in the second order Taylor terms.

TABLE 1. Odermatt Fits

BHN	a	b (km/s)
388	1.931	1.561
255	1.917	1.349

CONCLUSION

We conclude that using equation (8) with empirical data is a relatively safe procedure to adjust for small changes in target hardness with rod-like projectiles. Naturally, however, one should be alert to threshold conditions when small changes in hardness can result in a change in penetration mode.

ACKNOWLEDGMENTS

This work was supported by the U.S. Army Research Laboratory (ARL) under contract DAAA21-93-C-0101.

REFERENCES

1. Rapacki, Jr., E. J., Frank, K., Leavy, R. Brian, Keele, M. J., and Prifti, J. J., "Armor Steel Hardness Influence on Kinetic Energy Penetration," in *15th Int'l. Symp. Ballistics, Vol. 1*, 1995, 323-330.
2. Littlefield, D. L., Anderson, C. E., Partom, Y., and Bless, S. J., "The Penetration of Steel Targets Finite in Radial Extent," *Int. J. Impact Energy* **19**, 49-62 (1997).
3. Anderson, C. L., Morris, B. L., and Littlefield, D. L., *A Penetration Mechanic Database*, SWRI Report 3593/001, Southwest Research Institute, San Antonio, Texas, 1992.

BALLISTIC TESTING AND HIGH-STRAIN-RATE PROPERTIES OF HOT ISOSTATICALLY PRESSED Ti-6Al-4V

YaBei Gu, Vitali F. Nesterenko and Sastry S. Indrakanti

Department of MAE, University of California, San Diego, CA 92093-0411

Hot isostatically pressed (HIPed) Ti-6Al-4V powder based targets (including composites) have a good ballistic performance against long rod, conical and flat projectiles impact (velocity range ~ 0.4 - 1km/s). Compared to baseline material (MIL-T-9047G), new features such as different shape of craters in long rod penetration tests were observed. The results of compression Hopkinson bar tests, cut from tested targets (final strain controlled tests and hat-shaped specimen tests) are presented with a goal to establish relations between ballistic performance and high strain rate properties of HIPed materials.

INTRODUCTION

Extensive experiment data were obtained about high-strain-rate properties and ballistic performance for Ti-6Al-4V alloy with various microstructures (e.g. equiaxed, acicular and Widmanstatten structures produced by conventional processing methods) [1-4].

The high-strain-rate properties of Ti-6Al-4V are most crucial in the design of components subjected to impact or shock loading. Follansbee and Gray [5] investigated the deformation behavior of Ti-6Al-4V at temperatures 76 and 495 K, strain rates between 0.001 and 3000 s^{-1}, and compressive strains up to 3.0. A deformation model based on the kinetics of dislocation/obstacle interaction and structure evolution was successfully applied. Lee and Lin [6] investigated plastic deformation and fracture behavior of Ti-6Al-4V alloy under high strain rate at various temperatures. The results show that the flow stress of Ti-6Al-4V is sensitive to both temperature and strain rate. It was found that adiabatic shear bands are the major fracture mode at large plastic deformations under high temperature and high strain rate. The rate-dependent deformation and localization of fully dense and porous (7.6% porosity) Ti-6Al-4V was also investigated using a combination of standard testing machines and compression and torsional Kolsky bars [7].

There is no data on the relation between ballistic performance and high strain rate properties of Ti-6Al-4V made from powder. The main goal of this research is to investigate the ballistic performance of Ti-6Al-4V based homogeneous materials obtained by hot isostatic pressing against flat ended, conical and long rod penetration impact.

COMPARISON OF BALLISTIC PERFORMANCE OF HIPed AND BASELINE MATERIAL

Three types of ballistic experiments including flat-ended projectile, conical projectile and long rod projectile tests were conducted to evaluate the ballistic properties of HIPed Ti-6Al-4V material. Flat-ended and conical projectile penetration tests were performed at U.C. Berkeley [8]. Long rod penetration tests were conducted at the UDRI [9,10]. The details about the preparation of targets and testing can be found in papers [8-11].

Relatively large spread of data in plug velocities

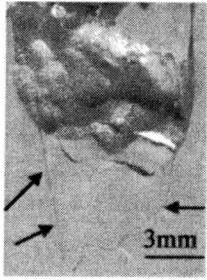 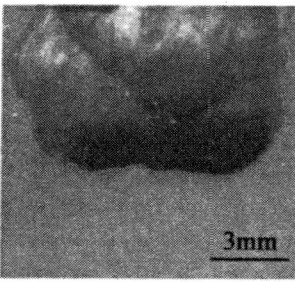

Fig. 1 Different shape of craters after long rod projectile penetration tests. (On the left is baseline material and on the right is HIPed PREP nonmilled material). Arrows show localized shear.

were found after flat-ended projectile penetration tests. For example, one test from PREP nonmilled HIPed Ti-6Al-4V shows only 80 m/s plug velocity while another two similar tests result in 176.8 m/s and 187.2 m/s velocities. This phenomenon also happens in PREP milled HIPed and in PREP ELI-nonmilled HIPed material. Practically in all cases plug velocities in baseline material were higher.

The bottoms of craters in HIPed and baseline materials after long rod projectile penetration tests at 1 km/s are presented in Fig. 1 [9,10]. The crater shapes are quite different. In baseline material, the penetration depth is about 18 mm and additionally shear bands propagated forward to a distance of 6 ~ 8 mm. In the case of HIPed samples, the penetration depth is about 14 ~ 15 mm which is smaller than in baseline material. The bottom part of the craters is relatively flat for HIPed samples and has a few dents left by the projectile (diameter 4.75mm).

Shear bands patterns were found to be quite different in the baseline and HIPed targets after ballistic tests with flat projectile. The shear bands in baseline material exhibit systematic pattern where a dominant shear band runs along the penetration direction and pattern of secondary shear bands develops at 45^0 to penetration path. The secondary shear bands have a length of 40 ~ 100 μm and width of 2 ~ 5 μm.

Shear bands pattern in HIPed material does not show "systematic" behavior. The dominant shear band also runs along the penetration direction, but the secondary shear bands propagated into the target with angles $30^0 \sim 90^0$ and interact with other shear bands developed from the crater or within the grains. The bifurcation and interaction found frequently in the tested HIPed targets are an example of complex shear bands patterns. Twinning was also observed in tested HIPed target inside grains close to the crater. Many of the secondary shear bands is over 100 μm long and over 10 μm thick [11].

Dimple structures on surface of plugs collected after flat-ended penetration tests were evaluated under SEM. Dimples with similar shape and size distribution were found both in baseline and in HIPed material even though the overall shape of the plug appears significantly different [11].

HIGH-STRAIN-RATE DATA

High-strain-rate constitutive relations were obtained using split Hopkinson bar tests for standard baseline material taken from the 50-mm rod (MIL-T-9047G) and from the tested targets made from HIPed material with different velocities of plug. The sample is 4-mm by 4-mm cylinder and the top and bottom surfaces were polished before testing. Samples from baseline material were prepared with axis parallel and normal to the axis of 50-mm rod to examine the texture effect. The results of the stress-strain curves are shown in Fig. 2.

Ductility of PREP nonmilled HIPed material is comparable with the ductility of baseline material in perpendicular direction. PREP nonmilled HIPed material has a comparable strength as baseline material in the parallel direction, but lower than that of the baseline material in the perpendicular direction. The baseline material in the perpendicular direction has low ductility compared to the ductility in the parallel direction. Baseline material has a larger spread of data for ductility than HIPed samples. Phase analysis shows that there is more β-phase in HIPed material (45%) compared with the standard baseline material (39%). This may contribute to the observed lower ductility for HIPed material.

The strain controlled specimen tests also validate that the ductility of PREP nonmilled HIPed material is comparable with the ductility in baseline material in perpendicular direction.

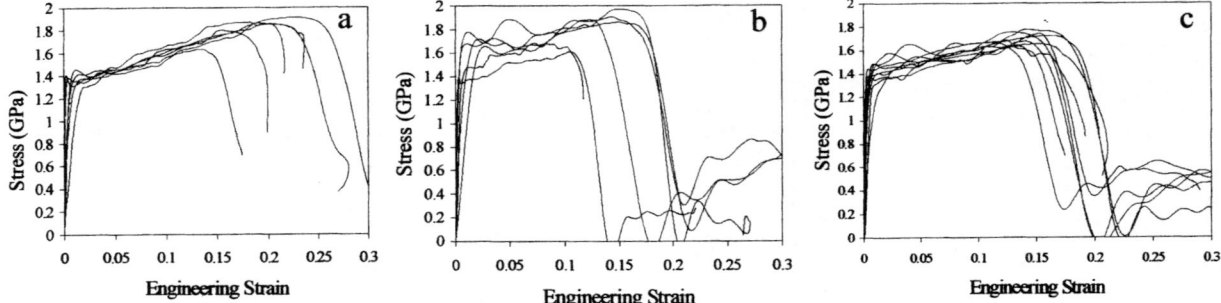

Fig. 2 Constitutive relations in different tests for baseline material at parallel direction (a), baseline material at perpendicular direction (b) and HIPed PREP nonmilled material (c), (samples 6n, 7n and 8n; see Table 2 in [11]).

They both have a failure strain 0.12 ~ 0.17. Failure strain is ~ 0.24 for baseline material in parallel direction. The spontaneous shear bands were initiated inside particular grains and represent the dominant failure mechanism for plastic deformation in HIPed material (Fig. 3).

Hat-shaped specimen tests were used to evaluate shear deformation and failure of the material by forced shear localization in the postcritical region. The main advantage of this test is that the flow stress vs. displacement can be measured and subsequent microstructures can be investigated.

The results presented in Fig. 4 show that PREP nonmilled HIPed material has a relatively lower flow stress compare to baseline material. Baseline material in perpendicular direction has a relatively higher flow stress and similar energy dissipation compare to HIPed material. The area of forced shear localization has the similar thickness in all three cases.

It is also found that larger amount of material is involved into the shear flow in baseline material sheared at perpendicular direction to forging. Only material inside the shear band is heavily deformed at shear direction parallel to forging at the same level of strain (Fig. 5).

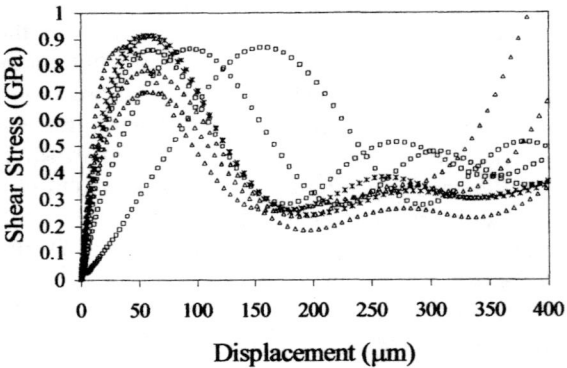

Fig. 4 Shear stress vs. displacement for baseline material at parallel (squares) and perpendicular direction to forging (stars) and HIPed PREP nonmilled material (triangles).

CONCLUSION

HIPed material has a better ballistic performance in comparison with baseline material for three types of ballistic tests. Samples made from ELI powder did not demonstrated better performance in comparison with PREP powders.

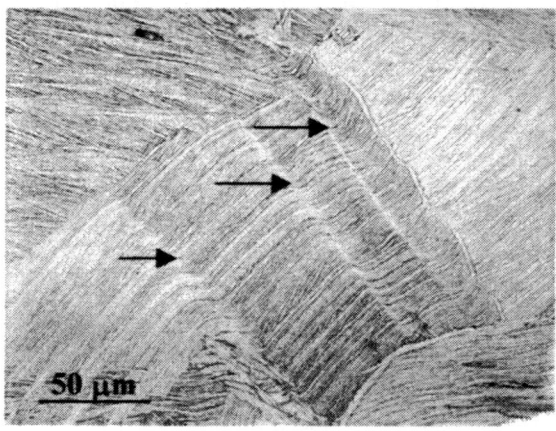

Fig. 3 Nuclei of Spontaneous shear band in HIPed PREP nonmilled material.

 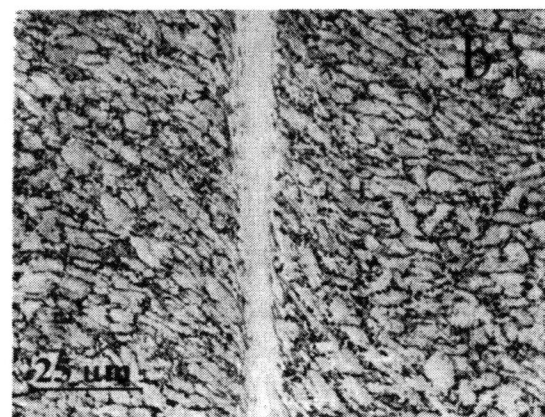

Fig. 5 Forced shear bands developed in baseline material at parallel direction (a) and at perpendicular direction (b) at strain rate ~ 10^3.

PREP nonmilled HIPed material and PREP milled HIPed materials with different microstructures demonstrate the similar ballistic performance. Probably the existence of texture in baseline material and isotropic structure of HIPed material are responsible for different crater shape in long rod penetration test. No correlation was found between shear bands pattern, dimple structure, Hopkinson bar test results in the samples taken from the targets with different plug velocities in flat projectile tests. Lower ductility of HIPed material can be responsible for different shear band pattern in baseline and in HIPed material and the difference in shape of the plugs in flat-ended projectile penetration tests.

ACKNOWLEDGMENTS

The support provided by the ARO, MURI DAAH 04-96-1-0376 (Program manager Dr. David Stepp) is highly appreciated.

REFERENCES

1. Gooch, W.A., Burkins, M.S., and Frank, K., "Ballistic Performance of Titanium Against Laboratory Penetrators." In *Proc. of 1-st Australasian Congress on Applied Mechanics*, Melbourne, Australia, 21-23 February (1996).
2. Meyer, L.W., Krueger, L., Gooch, W. and Burkins, M., *J.Physique IV*, **C3**, pp. 415-422 (1997).
3. Burkins, M., and Love, W., "Effect of Annealing Temperature on the Ballistic Limit Velocity of Ti-6Al-4V ELI", in *Proceedings of 16-th International Symposium on Ballistics*, San Fransisco, CA, 23-27 September (1996).
4. Weerasooriya, T., Magness, L. and Burkins, M., "High Strain-Rate Behavior of Two Ti-6Al-4V Alloys with Different Microstructures" in *Fundamental Issues and Applications of Shock-Wave and High-Strain-Rate Phenomena*, Edited by Staudhammer, K. P., Murr, L. E. and Meyers, M. A., pp. 33-36 (2001).
5. Follansbee, P. S. and Gray, G. T., III, *Metallurgical Transactions A*. **20A**, pp. 863-874, May (1989).
6. Woei-Shyan Lee and Chi-Feng Lin, *Material Science and Engineering*, **A241**, pp. 48-59, (1998).
7. da Silva, M. G. and Ramesh, K. T., *Material Science and Engineering*, **A232**, pp. 11-22 (1997).
8. Nesterenko, V. F., Indrakanti, S. S., Goldsmith, W and Gu, Y., *International conference on Fundamental Issues and Applications of Shock-Wave and High-Strain-Rate Phenomena*, Albuquerque, New Mexico, U.S.A., June (2000).
9. Nesterenko, V. F., Indrakanti, S. S., Brar, S. and Gu, Y., "Long Rod Penetration Test of Hot Isostatically Pressed Ti-based Targets", in *Shock Compression of Condensed Matter*, pp. 419-422, (1999).
10. Nesterenko, V. F., Indrakanti, S. S., Brar, S. and Gu, Y., *Key Engng., Mater.* 177-180, 243-248 (2000).
11. Nesterenko, V. F., Indrakanti, S. S., Goldsmith, W and Gu, Y., in preparation.

DEFORMATION AND DAMAGE OF TWO ALUMINUM ALLOYS FROM BALLISTIC IMPACT

Charles E. Anderson, Jr. and Kathryn A. Dannemann

Southwest Research Institute, Engineering Dynamics Department
P.O. Drawer 28510, San Antonio, TX 78228-0510

A series of impact experiments were conducted on 4.76-mm-thick aluminum plates to investigate the deformation and damage behavior of two aluminum alloys, 6061-T6 and 7075-T6. The Sierra 165 lead-filled bullet was used to load the plates. Impact velocities were varied from approximately 260 m/s to 370 m/s. The flow stress for 7075-T6 aluminum is approximately twice that for 6061-T6 aluminum; however, the ballistic limit velocities differ by only 10%. The 7075-T6 aluminum plates exhibit less deformation than the 6061-T6 plates at the same impact velocity, but at some critical velocity, a through-thickness crack appears in the 7075-T6 plate, ultimately leading to plate perforation. In contrast, the 6061-T6 plates continue to deform and fail by ductile tearing. These differences in damage/failure result in the two alloys having much closer ballistic limit velocities than expected based on differences in strength.

INTRODUCTION

The aluminum alloy 7075-T6 is approximately 85% stronger than aluminum 6061-T6. The initial yield stress is 505 MPa versus 275 MPa for the Al-7075-T6 and Al-6061-T6 alloys, respectively. Ultimate tensile stress values are 570 MPa and 310 MPa, respectively. However, the two alloy plates have almost the same ballistic limit velocity. (For the purposes of this paper, no distinction is made between the ballistic limit velocity, and the velocity at which 50% of the projectiles will perforate the target, i.e., V_{50}). We determined, for 4.76-mm-thick plates, that the ballistic limit velocity for the Al-7075-T6 is 366 m/s; that for the Al-6061-T6 alloy is 330 m/s. An experimental study was initiated to investigate the response of these two alloys to ballistic impact, using a range of impact velocities up to approximately the ballistic limit velocity. The objective was to understand the differences in response of these two alloys, and obtain high-fidelity experimental data that might be used to assess the accuracy of numerical simulations.

EXPERIMENTS

The projectile used for the experimental study was the Sierra 165 ball round. This lead-filled, 0.30-cal bullet is 29.65-mm long and has a total mass of 10.7 grams (165 grains). A thin copper-alloy jacket surrounds the bullet, except for the nose area (spire point).

Approximately 10 targets each were fabricated from 6061-T6 and 7075-T6 aluminum plate. The plates were 20 cm x 20 cm, and were 4.76-mm thick. The plates were held in a target frame and clamped at the edges. A grid pattern was photo-etched on the backside (opposite the impact side) of each plate to facilitate deformation observations measurements. The impact velocity was varied between approximately 270 m/s and 370 m/s.

RESULTS: General Observations

The bullet, because it is quite soft, deforms considerably, flattening into a pancake-like shape. A photograph of the copper jacket and lead filler

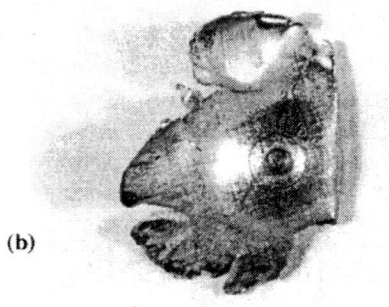

FIGURE 1. Post-test photograph of (a) copper-alloy jacket; (b) lead filler.

are shown in Figs. 1(a) and 1(b), respectively.

The deformation of the aluminum plates increases as the impact velocity increases. In general, the degree of deformation, at the same impact velocity, is greater for the 6061-T6 plates compared to the 7075-T6 plates. At some critical velocity, a through-thickness crack develops. The impact velocity at which the through-thickness crack develops is approximately the same for the two alloys (~321 m/s). However, the maximum deflection of the Al-6061-T6 is approximately three times that of the Al-7075-T6 plate when the through-thickness crack develops.

RESULTS: Quantitative Analysis

The extent of deformation for each test plate was measured with a dial indicator gage. The plates were situated on a flat surface with the bulge extending upwards. Dial indicator readings were recorded every 12.7 mm over a distance of 50.8 mm on either side of the impact location. The deformation measurements were plotted versus location for each impact velocity.

The remaining thickness of the plates at the impact location (there is thinning of the plate beneath the impact point) was also measured using a dial indicator gage. Each plate was situated on a reference surface with the bulge downwards (i.e., adjacent to the flat surface; the opposite orientation from the deformation measurements). A dial indicator reading was obtained for the reference surface. Another reading was taken on the impacted surface. The remaining thickness was determined by taking the difference between the reference and impacted surface readings.

The plate deformations for the 10 experiments conducted on the Al-6061-T6 plates are plotted in Fig. 2. The extent of deformation increases with impact velocity. At an impact velocity of approximately 317 m/s, a through-thickness crack appears in the Al-6061-T6 plate. The horizontal dashed line in the figure denotes the onset of cracking. Figure 3 shows the deformation and through-thickness crack for the 321-m/s experiment.

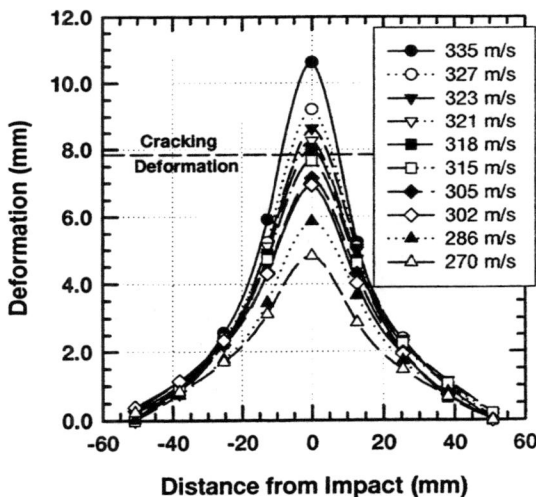

FIGURE 2. Deformation contours for Al-6061-T6 plates.

Similarly, the results of 8 experiments conducted on Al-7075-T6 plates are plotted in Fig. 4. The deformations are considerably less at the lower impact velocities. A through-thickness crack appears in the Al-7075-T6 plates at an impact velocity of approximately 326 m/s, after which the deformation increases rapidly. Figure 5 shows the deformation and through-thickness crack for the 340-m/s experiment.

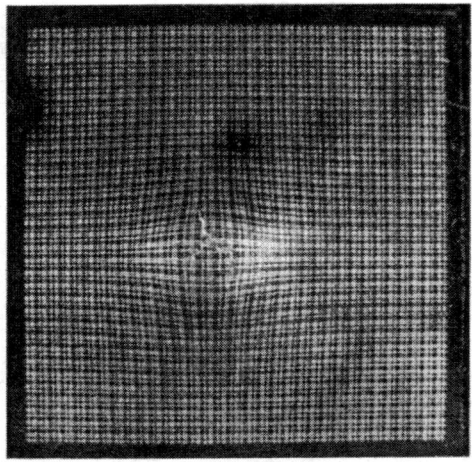

FIGURE 3. Post-test photograph of the backside of an Al-6061-T6 plate; the impact velocity was 321 m/s.

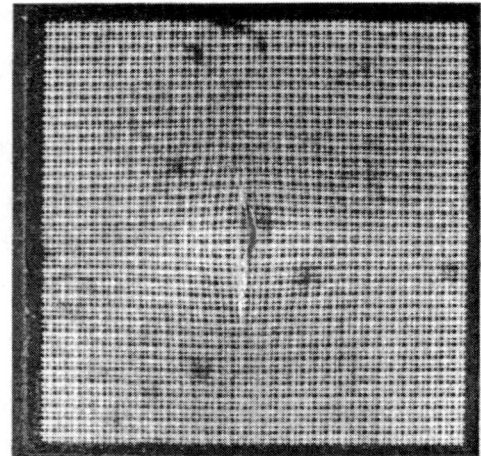

FIGURE 5. Post-test photograph of the backside of an Al-7075-T6 plate; the impact velocity was 340 m/s.

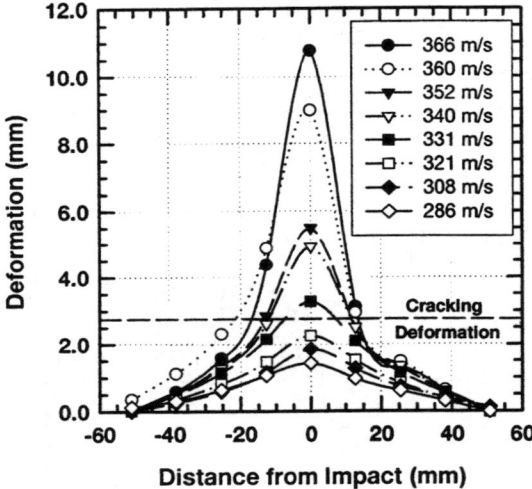

FIGURE 4. Deformation contours for Al-7075-T6 plates.

FIGURE 6. Post-test photograph the backside of an Al-6061-T6 plate; the impact velocity was 335 m/s.

The nature of the deformation, after through-thickness cracking, is different for the two alloys. The differences in deformation magnitudes in Figs. 2 and 4 are approximately 5 mm, except for the plates with excessive cracking. The deformation of the Al-6061-T6 plate continues in a ductile (bulging) mode as the impact velocity increases; e.g., see Fig. 6. However, for the Al-7075-T6 alloy, the stress state is evidently more asymmetric after through-thickness cracking. The stronger 7075-T6 alloy resists bulging deformation initially. But once the through-thickness crack appears, the crack(s) is more easily propagated in the 7075-T6 alloy than in the 6061-T6 alloy. The local deformations around the cracks become quite large as the plate "tears," e.g., contrast Figs. 5 and 7 with Figs. 3 and 6.

The maximum deformations as a function of impact velocity are plotted for the Al-6061-T6 and Al-7075-T6 plates in Fig. 8. The onset of cracking is denoted in the figure, as well as the V_{50} velocity. Note that the maximum deformation (as well as the deformation contours, Figs. 2 and 4) has a more gradual increase for the Al-6061-T6 plates than for Al-7075-T6 plates over the velocity range investigated. However, for both alloys, a change in slope is readily apparent at the impact velocity where the through-thickness crack appears. The change in slope is larger for the 7075-T6 plates. There is another change in slope at approximately the V_{50} velocity for both alloys, although this change is more dramatic for Al-7075-T6.

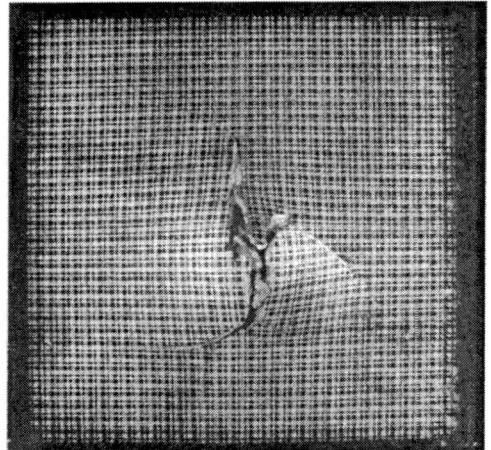

FIGURE 7. Post-test photograph of the backside of an Al-7075-T6 plate; the impact velocity was 360 m/s.

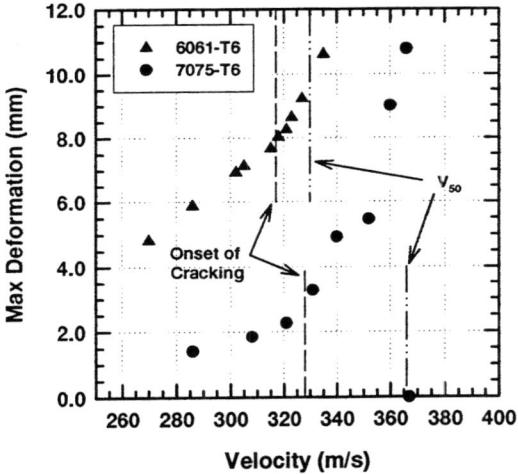

FIGURE 8. Maximum deformation vs. impact velocity.

The plate thickness decreases with penetration. The remaining thickness of the plates is plotted versus impact velocity in Fig. 9; the initial thickness of the plates was 4.76 mm. The data indicate that the Al-6061-T6 plate is penetrated relatively continuously (it is noted, though, that there are slope "discontinuities" at the same velocities as in Fig. 8) until perforation. In contrast, the 7075-T6 alloy re-

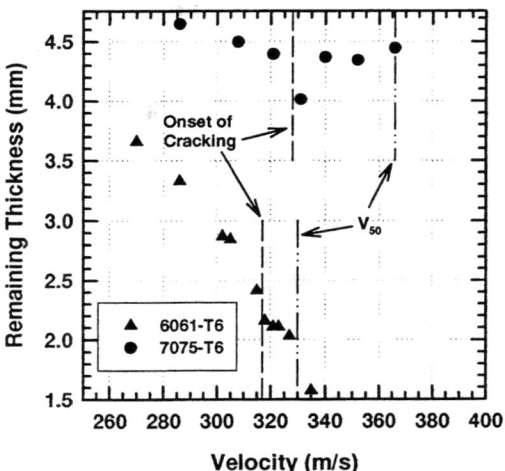

FIGURE 9. Remaining thickness vs. impact velocity.

sists penetration, and instead, perforation results from catastrophic cracking/tearing.

CONCLUSIONS

The deformation behavior resulting from ballistic impact is compared for two aluminum alloys. One alloy (Al-7075-T6) is approximately 85% stronger than the other alloy (Al-6061-T6). The deformation of the plates is consistent with the differences in strength, but the development of through-thickness cracks, at nearly the same impact velocity, and the difference in failure processes, result in relatively similar ballistic limit velocities, 366 m/s (7075-T6) versus 330 m/s (6061-T6).

The stress-strain response as a function of strain rate has been measured for the two alloys, and constitutive constants determined [1]. Numerical simulations have been conducted, and the simulation results compared with the experimental data in Ref. [1].

REFERENCES

1. Dannemann, K. A., Anderson, Jr., C. E., and Johnson, G. R., "Modeling the Ballistic Impact Performance of Two Aluminum Alloys," TMS Fall Meeting, Indianapolis, Indiana, 4-8 Nov. 2001.

RECOVERY OF URANIUM FRAGMENTS

H. R. James, D. H. McElrue and R. E. Winter

AWE, Aldermaston, Reading, Berks, UK

Abstract. We describe a theory for calculating the penetration of fragments into foam. Comparisons with regular projectiles show that the drag term is similar in value to the analogous term in aerodynamics. This, plus the simple model used to describe porosity, enables the theory to be used in predicting the levels of stress present when uranium fragments are arrested in foam catchers. Consequently the theory can be used to assist in the design of catchers which will not distort uranium fragments travelling at 1-3 km/s. The theory is tested against experiments using some current designs.

INTRODUCTION

Explosive accident studies often require experimental determination of the fragmentation characteristics of a "donor" warhead. Information such as fragment size, velocity and trajectory are important when assessing the likely damage that such an event could inflict on its surroundings. A powerful technique for providing such information is to arrest the donor fragments in a medium that is sufficiently soft for the fragments not to undergo significant erosion or break-up, and yet sufficiently robust as to decelerate fragments initially travelling at velocities in the region of 1-3km/s. The design of such a catcher is difficult, and such difficulties are increased when dealing with high-density, pyrophoric materials such as uranium.

Current development of such a catcher at AWE has concentrated on a design which consists of blocks of different low density foams configured in a manner intended to provide increasing resistance to the fragment as its velocity decreases. Theoretical work has been undertaken to describe the stresses induced in the fragment by such a design, and to attempt to identify regimes which will cause severe erosion of the projectile.

Such theory, when verified, can prove a useful aid in further developing the catcher by reducing the number of experiments required to test various foam configurations. Experiments have been carried out to both validate the theory and to provide an initial design for the catcher.

THEORY

The damage inflicted upon a fragment that is abruptly decelerated can be considered as occurring in two separate regimes. In the first the effect of the stresses imparted to the fragment by the initial impact shock have to be evaluated. In the second the deformation can be caused over a longer timescale by the penetration of the fragment into the target. Since in this work the target is porous, a porous model has to be used to generate its equation-of-state (EoS) for use in the shock and penetration calculations. The theory developed in this paper is an extension of that presented in reference 1.

The model of porosity used in this paper has been referred to by a number of names such as "Snowplough" or "locking solid" by different authors in the past. It has the characteristics that the pores are assumed to be massless, and so do not contribute heat during pore collapse. The thermodynamic state of the porous material lies off the EoS surface of the fully dense material, but joins that surface once the pores have been completely closed. A shock in the porous material forms a smooth discontinuity and so the material obeys the Rankine-Hugoniot conservation equations. Any shock strength is sufficient to collapse all the pores, and so the material jumps from its initial porous state onto the fully dense EoS surface.

In this model the foam EoS can be approximated by the use of an incompressible hugoniot. Such a hugoniot will exist at a specific density in the Snowplough model, and numerical tests show that it is a reasonable approximation for most highly porous materials regardless of their actual density. The only compression that then takes place in the shock is between v_o and v_{os} (the reciprocals of the densities of initial and fully dense material).

The projectile will continue to penetrate after the initial shock has attenuated. The penetration will always be supersonic for this model of porosity since any particle velocity generates a shock. A steady state supersonic penetration has a shock velocity that equals the interface velocity. For a material that has strength, the stress behind the shock, σ_s, equals $p_s+2Y/3$ for material undergoing uniaxial stress and having an elastic-perfectly plastic strength model. Y is the dynamic yield strength for the foam and $p_s=\rho_o u_z^2/\mu$ where p_s is the shock pressure, u_z is the interface velocity and $\mu=\rho_{os}/(\rho_{os}-\rho_0)$.

The isentropic, incompressible axisymmetric momentum flow equation can be reduced to

$$\partial \sigma_z = \rho_{os} u_z \partial u_z \qquad (1)$$

in the direction of impact (the z direction) by assuming a steady state and a negligible shear gradient over the area of interest. Integrating in the target between the top of the shock and the interface, and in the projectile between the plastic boundary and the interface gives

$$\sigma_z = \frac{1}{2}\rho_P(u_I - u_z)^2 + \sigma_{HEL} = \frac{1}{2}\rho_{os}u_z^2\left(1-\frac{1}{\mu}\right)^2 + \frac{\rho_o u_z^2}{\mu} + \frac{2Y}{3} \qquad (2)$$

where σ_{HEL} is the Hugoniot Elastic Limit for the projectile and u_I is the initial projectile velocity. Where the right side of (2) tries to drop below σ_{HEL} the projectile ceases to deform and the left side becomes

$$\sigma_z = \rho_P c_L (u_I - u_z) \qquad (3)$$

where c_L is the longitudinal sound velocity.

Although (2) is a steady state equation it is assumed it can be applied to a non-steady penetration provided the projectile deceleration is slow compared to the wave speeds needed to adjust the interface, shock and plastic boundaries. This quasi-steady approach is used in conjunction with

$$m\frac{du_z}{dt} = -\sigma_z C_D A \qquad (4)$$

where m is the projectile mass and A the projected fragment area subjected to the stress σ_z. C_D is analogous to the drag coefficient used in aerodynamics, and will be shown to have very similar values to those obtained in that field. After some manipulation and integration, the above equations give

$$Z_{pen} = -\frac{k_1}{k_2}\ln(\cos k_3) \qquad (5)$$

where Z_{pen} is the total penetration and

$$k = \frac{1}{2}\rho_{os}\left(1-\frac{1}{\mu}\right)^2 + \frac{\rho_o}{\mu}; \qquad k_1 = \sqrt{\frac{2Y}{3k}};$$

$$k_2 = \frac{C_D A\sqrt{2Yk/3}}{m}; \text{ and } k_3 = \tan^{-1}\left\{\sqrt{\frac{3k}{2Y}}u_I\right\}.$$

The velocity above which deformation begins (u_{ID}) is

$$u_{ID} = \frac{\sigma_{HEL}}{\rho_P c_L} + \left(\frac{3\sigma_{HEL} - 2Y}{3\left[0.5\rho_{os}(1-1/\mu)^2 + \rho_o/\mu\right]} \right)^{1/2}$$
(6).

The chief unknowns in the above analysis are C_D and Y. Comparisons with experimental data[1] show that some tuning can take place which produces values for both terms that minimizes the differences between theory and experiment. The following table shows predicted penetration depths compared to radiographic data in which copper spheres penetrate foam for which ρ_o=0.176 g/cc, and ρ_{os}=1.265 g/cc. C_D and Y are tuned to give the experimental penetration distances at 2.21km/s. Thereafter Y is held constant but C_D is tuned to the final penetration distance. As the table shows, intermediate penetrations are then well modelled by the theory provided the sphere does not fragment. Y=0.05GPa in the following. ΔZ is the percentage difference in distance between theory and experiment.

TABLE 1. Comparison of Theory with Experimental Penetration Data.

U_I km/s	C_D	T_1 μs	ΔZ_1 %	T_2 μs	ΔZ_2 %	ΔZ_{pen} %
2.21	0.40	27	-2.43	154.1	-0.46	-**
2.9	0.95	18.5	-11.04	105.4	0	-0.47
3.13	0.81	-	-	107.6	1.43	0
3.27	0.90	-	-	97.3	1.83	0
4.28	1.24*	-	-	100.0	27.32	-1.01

*Sphere fragmented.   ** Complete penetration in experiment.

The peak penetration stress level is comparable with σ_{HEL} at the 2.21km/s impact, but exceeds it for higher velocities indicating that projectile deformation (and probably erosion) has occurred. Radiographs from [1] confirm this conclusion and show the projectile presenting an increasingly flat surface to the foam by 2.9km/s. In aerodynamics[2] the value of C_D for a sphere is 0.47 (over a wide range of Reynolds number), while that for a disc is 1.17. It is interesting to note that as the shape changes from a sphere to disc-like in the above, C_D goes from 0.40 to near unity (discounting the fragmented projectile).

Similar conclusions are reached on the lower value of C_D from experiments where a variety of metals were impacted into strawboard[3]. Strawboard has the consistency of strong cardboard and is used in the catcher designs described below. As it consists of cellulose and glue, ρ_{os} is about 1.5g/cc, while ρ_o is 0.64-0.70g/cc. The velocity range of these experiments is such that σ_{HEL} should not be exceeded for any of the metals used. Consequently the projectile geometry should be largely unaltered during penetration. Y was quickly established as 0.25GPa and held constant for the fits to the complete range of experiments. It was found that C_D varied from 0.44 to 0.60 for spheres, and from 0.70 to 0.86 for cubes. This comparison also established that both theory and experiment scale as $Z_{pen}/m^{1/3}$.

EXPERIMENT

Two variations of foam catcher have so far been tested against uranium tiles. Each tile measured 10mmx10mmx2.37mm, and had a mass of about 4.5gms. An array of either 25 or 21 tiles was driven by an explosive plane wave lens system.

The first catcher design (catcher 1) consisted of 1m of aqueous foam (ρ_o=0.02g/cc), followed by 480mm of a flexible polyurethane-based sponge foam (Intumescent dry foam, ρ_o=0.1g/cc), and backed with 308mm of strawboards. The impact velocity was about 2.3mm/μs, and those tiles that were recovered were mainly fractured and eroded. Four mainly intact tiles were recovered from the strawboard layer but all showed signs of severe erosion. Large scale tumble was observed.

Figure 1 shows an improved catcher design (catcher 2). "Termanto" foam is a rigid PVC-based foam. The Intumescent foam has both dry and wet layers. The latter is intended to provide quenching for the hot uranium fragments. The calculated impact velocity of the tiles was 1.5mm/μs and seven intact tiles were recovered from the penultimate layer. Two of these tiles had hit, but not penetrated, the strawboard. The remaining tiles had appeared to exit through the sides of the target.

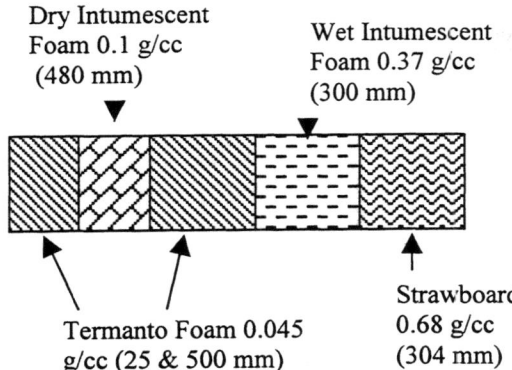

FIGURE 1. Configuration of Catcher 2.

USE OF THEORY IN CATCHER DESIGN

The mechanical data for DU fragments were taken from [4]. From this data σ_{HEL}=0.56 GPa, and so impacts producing penetration stresses below this level are assumed not to cause major projectile erosion. Of course the simple theoretical analysis does not predict pressure gradients across the projectile, or the effect of fragment tumble, both of which will set up bending moments creating deformation. However, it is assumed that *normal* non-tumbling impacts below the σ_{HEL} criterion should not undergo radical deformation. The minimum critical velocities corresponding to this criterion for the target components are seen in the wet intumescent foam (1.3km/s) and in the strawboard (0.86km/s). The initial impact shock will exceed this criterion, but the duration of this shock should not cause major deformation in the fragment and so the analysis below will concentrate on the longer-term penetration phenomena.

TABLE 2: Theoretical Estimate of Catcher Performance.

Table 2 shows theoretical estimates of velocities in km/s seen at the end of each layer in the catcher for different initial fragment velocities. This analysis assumes an impact in either a flat-faced or edge-on mode. The tumble shown in the radiographs indicate that the impact mode of the fragments will lie between these limits. Since the edge-on is the more effective penetration mode, the table gives the upper and lower limits of expected penetration as well as areas where significant erosion can be expected.

DISCUSSION AND CONCLUSIONS

The supersonic penetration through porous media, coupled with the simple EoS approximation, allows the stress equation to be expressed in a readily solvable form. Values of drag are obtained from experiments where the projectile form and trajectory are closely controlled. Such values are comparable to those found from other fields. In more complex situations where fragments tumble, the theory gives an envelope of performance that matches experiment. Catcher 1 is expected to give large scale erosion for 2.3 km/s impacts, while catcher 2 should not for 1.5 km/s fragments – both in line with observation.

REFERENCES

1. Trucano T.G., and Grady D.E., *Int.J.Impact Engng* **17**, 861-872 (1995).
2. Hughes W.F., and Brighton J.A., *Theory and Problems of Fluid Dynamics*, Schaum Publishing Co., New York,1967.
3. McMahon E.G., RARDE Fort Halstead, Private Communication (1969).
4. Tonks D.L., Vorthman J.E., Hixon R., Kelly A. and Zurek A.K.,"Spallation Studies on Shock Loaded U-6 wt% Nb", in *Shock Compression of Condensed Matter –1999*, ed. M.D.Furnish et.al., AIP Conference Proceedings 505, New York, 2000, pp 329-332.

C British Crown Copyright 2001/MoD

Catcher	Vel. Km/s	C_D	Orient-ation	Aq. Foam	Term. Foam	Dry Int.	Term. Foam	Wet Int.	Straw board
1	2.3	0.81	Flat	1.55	-	0.58	-	-	stopped
1	2.3	0.63	Edge	2.13	-	*1.76*	-	-	through
2	1.5	0.81	Flat	-	1.47	0.55	0.31	stopped	
2	1.5	0.63	Edge	-	1.49	1.25	1.14	0.79	stopped
2	2.3	0.81	Flat	-	2.24	0.85	0.51	stopped	
2	2.3	0.63	Edge	-	2.28	1.93	*1.77*	*1.11*	through

1.77 – erosion is expected.

MODELING OF URANIUM ALLOY RESPONSE IN PLANE IMPACT AND REVERSE BALLISTIC EXPERIMENTS

B. Herrmann[1], A. Landau[1], D. Shvarts[1,2], V. Favorsky[2] E. Zaretsky[2]

[1] Nuclear Research Center - Negev, P.O.Box 9001, Beer-Sheva 84106, Israel.
[2] Mechanical Engineering Dept., Ben Gurion University, P.O. Box 653, Beer-Sheva 84105, Israel.

Abstract. The dynamic behavior of a solution heat-treated, water-quenched and aged U-0.75wt%Ti alloy was studied in planar (disk-on-disk) and reverse ballistic (disk-on-rod) impact experiments performed with a 25 mm light-gas gun. The impact velocity ranged from 100 to 500 m/sec. The impacted samples were softly recovered for further metallographic examination. The VISAR records of the sample free surface velocity, obtained in planar impact experiments, were simulated with 1-D hydrocode for calibrating the parameters of modified Steinberg-Cochran-Guinan (SCG) constitutive equation of the alloy. The same SCG equation was employed in 2-D AUTODYN simulation of the alloy response in the reverse ballistic experiments, with VISAR monitoring of the lateral sample surface velocity. Varying the parameters of the strain-dependent failure model allows relating the features of the recorded velocity profiles with the results of the examination of the damaged samples.

INTRODUCTION

Uranium alloys are known to be susceptible to extensive localization of deformation when subjected to high rate loading. The phenomenon is usually attributed to local material softening produced by the strain-induced heat generation.

Lately, special attention was paid to the role of strain-hardening and strain-rate-hardening in the localization of the shear strain [1]. It has been shown that intensification of the hardening results in earlier loss of shear stability and, perhaps, in the shear localization. The correlation between the intensification of the strain hardening and material susceptibility to strain localization due to varying the strain hardening parameters of the modeled material was observed by Zaretsky et al. [2] in numerical simulations of the impact of a tungsten heavy alloy disk on a rod made of the same material ("reverse ballistic impact"). The simulation was performed using the Steinberg-Cochran-Guinan (SCG) [3] constitutive model with a simple dependence of the material yield strength Y on the equivalent plastic strain ε_i: $Y = Y_0(1 + \beta\varepsilon_i)^n$. In the case of the numerical simulation of the reverse ballistic impacts with the impactors and samples made of uranium-titanium alloy, by using the same model with the strain-hardening parameters determined in planar impact experiments, a satisfactory description of the experimental results could not be achieved.

Both the dislocations and the dislocation-related structures [4], such as stacking faults and twins, participate in plastic flow of uranium-based alloys. In contrast to the dislocations-governed plastic flow, resulting in strengthening the metal with the increasing of ε_i, the deformation twinning can effectively either strengthen or weaken the alloy [5].

The purpose of the present work is to find parameters of the constitutive model, additional to those determined in the 1-D experiments, enabling the description of 2-D strain-hardening and failure events in a uranium-titanium alloy

EXPERIMENTAL

In order to establish parameters of SCG constitutive equation the solution heat-treated, water-quenched and aged U-0.75wt%Ti alloy (U-Ti) was previously characterized in planar impact experiments utilizing a 25-mm pneumatic gun. Both the impactors and samples were disks of about 24-mm diameter made of U-Ti. The impactors and samples thickness was 2 and 6-mm, respectively. The reverse ballistic (RB) experiments were performed with the same gun. The U-Ti impactors were 5-mm thick and the U-Ti samples employed were cylinders of 20-mm length and 8-mm diameter. They were impacted at the face of the cylinder. The impact velocity and the impactor-sample misalignment (tilt) were controlled by charged pins. In all impacts, the tilt did not exceed 0.4 mrad. In the planar experiments, the velocity of the free surface of the U-Ti samples was monitored by VISAR [6]. In the RB experiments, the VISAR beam was focused at a point located on the sample cylindrical surface 3.6 mm from the impacted cylinder face. In four RB experiments, the impact velocity was 112, 272, 412 and 472-m/sec. The sample recovered after 112-m/sec impact was found plastically deformed. After the higher impacts the samples were found damaged along the conical surface based on the impacted sample face. At 412 and 472-m/sec the impactors failed by plugging.

The results of the planar impact experiments are shown in Fig. 1a. The stress deviators $s = \sigma - p$, Fig. 1b, were obtained from the experimental velocity profiles by using the known Hugoniot of the alloy to estimate the pressure p and calculating the stress σ and strain ε from momentum and mass conservation laws, considering the compressive part of the profile as a simple centered wave.

SIMULATIONS

The planar impact experiments with U-Ti alloy were simulated numerically, using 1-D finite difference Lagrangian code and material strength $Y = Y_0(1 + \beta\varepsilon_i)^n$ suggested in [3]. Since all the velocity profiles show smooth elastic-plastic transition, some 20% spatial randomization of the Y_0 values was added, in order to achieve better agreement between the modeling and the experiment. It was found, however, that the coincidence may be improved by substituting the above expression by another one, $Y = Y_0 + \beta\varepsilon_i^n$, with no need for any randomization.

The commercial 2-D Lagrangian code AUTODYN-2D was used for numerical simulations of the RB experiments. Since the point of intersection of the VISAR beam with the sample surface moves in Lagrangian coordinate system, the Lagrangian coordinates of this point was determined for each calculation cycle and the normal to the sample axis component of its velocity was saved. The mesh size was 100×100 μm. The 2-D simulations were started with both the equation of state and the SCG-constitutive equation, including the strength of $Y = Y_0(1 + \beta\varepsilon_i)^n$-type, and material

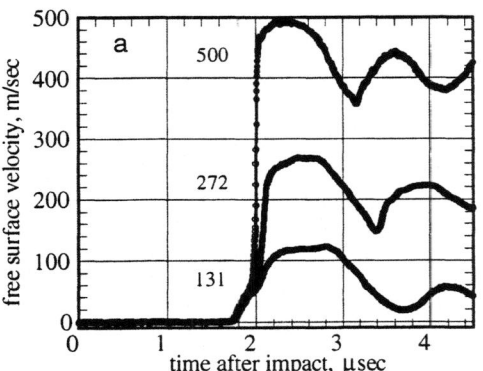
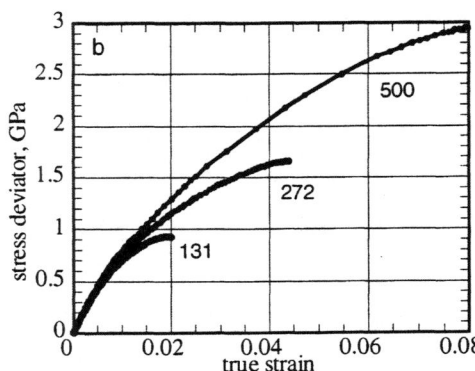

FIGURE 1. Free surface velocity profiles of U-Ti samples, obtained in the planar impact experiments (a) and the stress deviators calculated from the corresponding profiles (b). Impact velocities are shown in m/sec.

parameters suggested for U-Ti alloy by the AUTODYN library. The increase in the strength was limited by Y_{max} whose values for the impact with velocities 112, 272, 412 and 472-m/sec was found equal to 1, 1, 1.2 and 2.5-GPa, respectively. The strength of the $Y = Y_0 + \beta\varepsilon_i^n$ form was also tested. Surprisingly, the 2-D simulations were found insensitive to such substitution. With any form of the strength, the AUTODYN simulations reproduce the general form of the recorded VISAR signal. The coincidence was found reasonable only for 112-m/sec impact. For stronger impacts, the discrepancy between the experimental and the simulated profiles increases with increasing the impact velocity. The VISAR records obtained after RB shots are shown in Fig. 2 together with the simulation results.

PARAMETRIC STUDY

It is apparent from Fig. 2 that the employed constitutive equation fits the data well up to the first maximum of the velocity. The divergence starts with the velocity decrease after the first maximum was achieved. The post-maximum velocity values are supported by the stress signal arriving from the central part of the impactor-sample interface. We assumed that some softening of the material hinders the complete transfer of this signal. A softening factor, acting presumably when the equivalent plastic strain is large, was added to the SCG model. The plastic stain, unlike in the planar impact experiments, may be large during such loading. Simulations, with yield strength of the form $Y = Y_0(1 + \beta\varepsilon_i)^n \exp(1 - B\varepsilon_i^m)$ were performed (Fig. 1a). The exponential form of the factor was chosen in order to preserve positive value of the strength Y for any value of the strain ε_i. It was found that varying the softening parameters in a very wide range has almost no influence on the results of the simulations. Note, that varying the parameters is limited by the interruption of the AUTODYNE calculation due to the mesh distortion.

Failure processes occurring in the internal regions of the sample may also disturb the complete momentum transfer. The simplest (bulk) failure model available in AUTODYN library was introduced into the calculations. Assuming that the failure has to occur at the presence of a sufficient shearing stress and should also be accompanied by large plastic deformation, the failure criterion $F = A\varepsilon_i |\sigma_1 - \sigma_2|$ was chosen. The indexes 1 and 2 signify the local principal stress values. An example of failure-included simulations, with the failure parameter $F = 0.875$ GPa, are shown in Fig. 2b. Two conclusions can be drawn: (i) accounting the failure in the calculations may improve the results

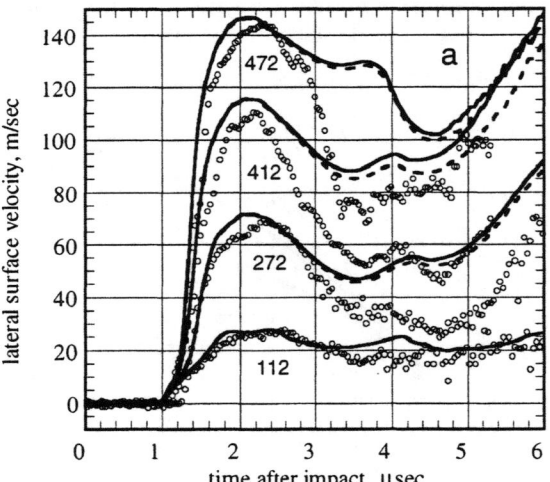 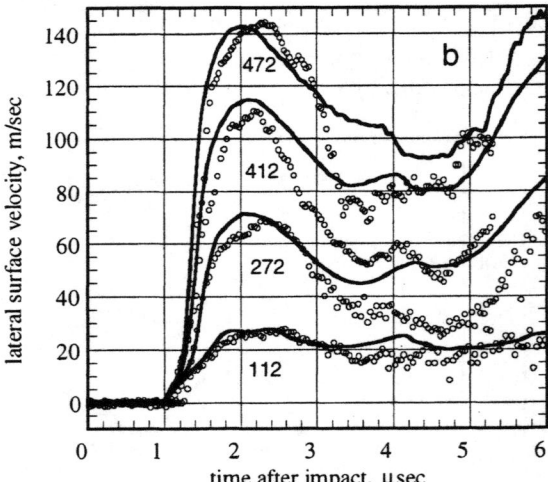

FIGURE 2. Free surface velocity profiles and simulations of U-Ti samples, obtained in the reverse ballistic experiments, with hardening (line) and hardening + softening (dashed line) (a) and with hardening + softening + failure (b). Impact velocities are shown in m/sec.

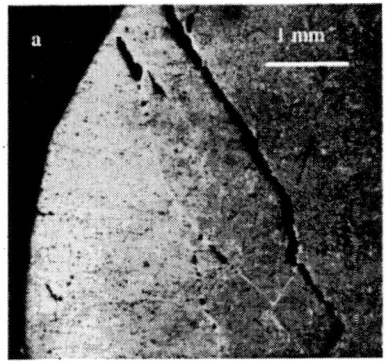 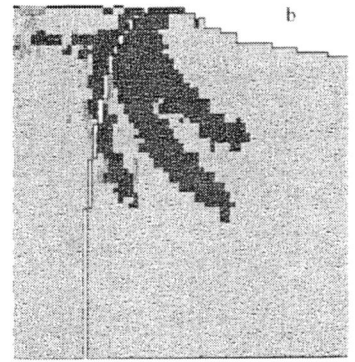

FIGURE 3. A cross-section metallography (a) and a damage map (b) of a U-Ti sample recovered after RB impact at 412 m/sec.

of the simulations and (ii) the failure criterion in these runs is too high to have influence in the three weaker shots. Decreasing the failure criterion results in interrupting the calculations.

Although the shortcomings of the used failure description are clear, the simulation reproduces the main features of the sample damage. The metallograhic cross-section of the U-Ti sample, recovered after 412-m/sec impact, is shown in Fig. 3b, together with the corresponding state of the failure, taking place at the time instant 5.1-μsec after the impact. Several narrow parallel damage paths, inclined to the sample axis at the angle about 60, are present, both in the real sample cross-section and on the failure map. The simulation results also show initiation of the plugging observed in the impactor. Estimation of the velocity of the propagation of the damage regions, both in the impactor and in the sample, yields the value of about 700-750-m/sec.

CONCLUSIONS

Introducing the strength parameters, calibrated on the base of the planar impact experiments, into the 2-D numerical simulation of the velocity profiles obtained in reverse ballistic experiments allows one to reveal the damage processes induced by the ballistic impact.

More accurate modeling of the damage processes is impossible at present because the AUTODYN version that was used exhibited some limitations when rezoning was required due to excessive deformation.

A finer mesh size, than the 100×100 μm that was employed, should be used in the simulations, in order to reproduce fine features of the damage as the width of the widest adiabatic shear band is about 20 μm.

Although the failure threshold criterion used in the present study gives some failure localization, a more comprehensive constitutive description of the failure process is required.

REFERENCES

1. Estrin Y., Molinari A. and Mercier S., *Journ. Eng. Mater. and Technology*, **119** 322-331 (1997).
2. Zaretsky E. et. al,"Lateral Sample Motion in the Plate-Rod Impact" in *SCCM-1999*, edited by M. D. Furnish. et. al., AIP 505-2000, pp. 593-596.
3. Steinberg D.J., Cochran S.J. and Guinan M.W., *Journ. Appl. Phys.*, **51** (1980) 1498.
4. Armstrong P.E., Follansbee P. S. and Zocco T., "A Constitutive Description of the Deformation of α Uranium Based on the Use of MTS as a State Variable" in *IPCS No 102*, edited by J. Harding, Int. Conf. PMHRS, Oxford, 1989, pp. 237-244.
5. Yoo M. H., *Metall. Trans.* **12A**, 409-418 (1981).
6. Barker L.M. and Hollenbach R.E., *Journ. Appl. Phys.*, **43**, 4669 (1972).

ON THE ENTRANCE PHASE IN LONG ROD PENETRATION

Z. Rosenberg and E. Dekel

RAFAEL, P.O. Box 2250, Haifa, Israel

Abstract. The penetration of long rods into semi-infinite targets is a three-stage process, in which the first (entrance) and last are very short, and transient, while the second phase (primary penetration) is a long and quasi-steady process. The present paper summarizes our recent work on the entrance phase using 2D numerical simulations of strengthless steel rods (L/D=5-20) impacting aluminum targets at 1-4 km/s. We look for the significance of this phase as impact velocity, target strength, and penetrator's length are increased. We also show that the target free surface (impact face) is not the cause for the entrance phase. Rather, it is the passage from a cylindrically-shaped to a mushroom-shaped penetrator nose which is responsible for this phase.

INTRODUCTION

The penetration of long rods into semi-infinite targets, is considered as a three-stage process (see Orphal [1], for example). The transient first and third stages are very short compared with the relatively long middle stage which has steady-state characteristics. The one-dimensional model of Tate [2] and Alekseevskii [3] accounts for the main features of the steady state, using a modified Bernoulli equation and considering the deceleration of the rod during penetration. Orphal summarizes much of the work performed on the third stage considering the extra penetration in the target right after the completion of rod consumption. This stage has also been investigated by us recently [4] in a numerical study of the secondary penetration process. The first stage in the penetration process (termed the entrance phase) has received relatively less attention, because it is considered as very short, and thus, less important. Ravid et al. [5] analyzed the transient passage from a 1-D planar shock-wave model, which results at the impact of a cylindrical rod on a flat target, to the 2D penetration mode several microseconds later. Partom [6] has argued recently that, the characteristics of the entrance phase are not due to the free surface of the target. He performed 2D numerical simulations in which he supported the area around the impact area with a rigid wall, and monitored the penetration depths of an L/D=10 tungsten alloy rod impacting a steel target.

The purpose of the work presented here is to highlight some of the typical aspects of the entrance phase by using 2D simulations for long steel rods impacting semi-infinite aluminum targets.

NUMERICAL SIMULATIONS

The 2D simulations were performed with the Eulerian processor of the PISCES 2DELK code. As always, our meshing has 11 cells on the penetrator radius and a similar meshing for the central region of the target (about 5 penetrators diameters). Farther, zones have coarser meshing by a factor of 1.05. The target side and back surfaces are supported by FLOW conditions, thus, guaranteeing its semi-infinite nature.

Our simulations follow the time histories of the penetration velocity, penetration depth, rod length, and the pressure at the penetrator-target interface. From all these, we chose the penetration velocity for our examinations of the entrance phase since it is much more sensitive to subtle changes in the relevant parameters. The penetration process is a good example, since penetration depth histories are much less sensitive than either interface pressures or penetration velocities. We also found that it is much better to simulate penetrations of zero strength rods since these attain a true steady-state penetration right after the entrance phase, which simplifies the distinction between the two phases. Thus, in all simulations presented the steel rods are strengthless, while the aluminum targets have a strength in the 0.4-1.6 GPa range. We used a simple von-Mises strength criterion with no hardening or failure mechanisms. Since we are mainly interested in the first stage of penetration, while interface pressures are very high, fine details in constitutive relations, such as strain to failure have negligible influence.

RESULTS AND DISCUSSION

The first simulations with different L/D rods demonstrate the fact that the entrance phase is not dependent on rod length and is over in a very short time. Figure 1 shows the time histories of the penetration velocities for L/D=5, 10 and 20 steel rods impacting a semi-infinite aluminum target at 3 km/s.

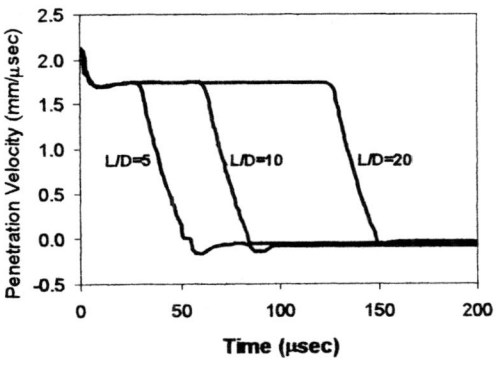

FIGURE 1. Penetration velocity for three L/D rods.

As expected, the relative importance of the entrance phase diminishes as L/D increases. This is the reason for the choice of L/D=5 rods for our next set of simulations with different impact velocities. Figure 2 shows the results of this set for velocities in the range of 1-2.5 km/s. It is clear that the relative importance of the entrance phase decreases with increased velocity, as expected.

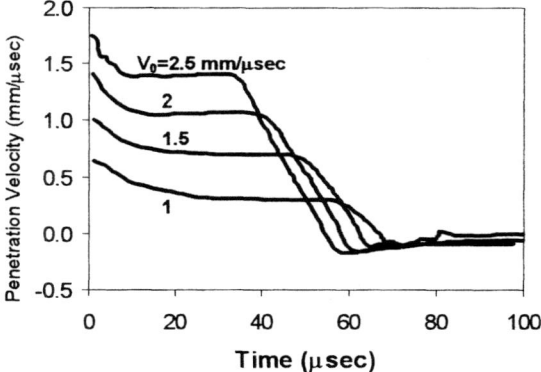

FIGURE 2. Increasing impact velocities.

Thus, as far as the entrance phase is concerned, its contribution diminishes with higher velocities and longer rods. This is the reason for the success of the 1D penetration models for fast long rods and shaped charge jets, for which the penetration is mostly a steady-state process.

We should note here that the time scales for the entrance phase are 10-25 microseconds (see Figure 2). These are much longer than the time scales that Ravid et al. [5] relate to this phase, assuming it lasts as long as rarefactions from penetrator sides reach its center (a few microseconds).

In order to examine the influence of target strength on the entrance phase, we performed several simulations with target strengths in the range of 0.4-1.6 GPa. Figure 3 shows results of these simulations at an impact velocity of 2 km/s. It is evident that with increasing target strength the relative importance of the entrance phase increases. It is interesting to note the effect of target strength on the steady-state penetration velocity which decrease from 1 km/s (for a 0.4 GPa target) to 0.7 km/s, for the strongest target (1.6 GPa).

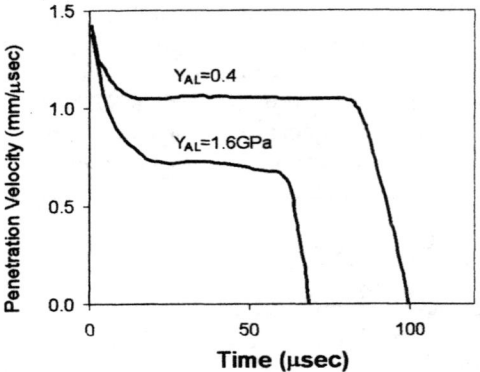

FIGURE 3. The effect of target strength.

The issue we explored next concerns the question that was dealt with by Partom [6]. It is known that penetration is easier at the first stage, where the penetrator is close to the target free surface. The question Partom raised is whether this fact is the cause of the entrance phase. In order to answer this question, we performed a simulation with a target having a deep circular hole in its middle. This way the penetrator hits the target in an area, which is deep inside (33 mm) and the effect of the free surface is diminished. The hole diameter is just slightly larger than that of the rod (9 mm vs. 8.0 mm). Figure 4 shows the results of this simulation, together with the reference simulation for a regular target. In both cases, the strength of the target is 1.2 GPa and the impact velocity is 2 km/s.

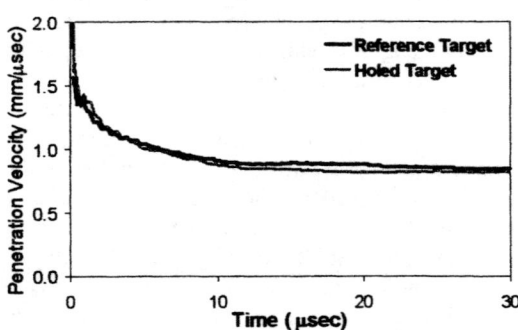

FIGURE 4. Comparison between a regular target and one with a deep circular hole.

Since both simulations resulted in very similar time histories for the penetration velocity, one can conclude that the effect of the free surface of the target is indeed negligible. This result strongly supports the conclusion of Partom [6] who followed the penetration depths of tungsten alloy rods into steel targets, with similar simulations.

The main conclusion from this set of simulations, is that the entrance phase is due to the non-steady nature of the penetration at its first stage. This is the result of the process by which the rod nose is changing from a cylindrical shape to the well-known mushroom shape during the steady-state stage. In order to demonstrate this issue, we performed an extra simulation in which the rod is penetrating first a thin target, 40mm thick. After the rod nose becomes mushroom-shaped, as it leaves the thin target, it then impact a semi-infinite target positioned 40mm away. Figure 5 shows the result of this simulation where zero time for the split target coincides with the mushroom-shaped projectile impacting the semi-infinite target.

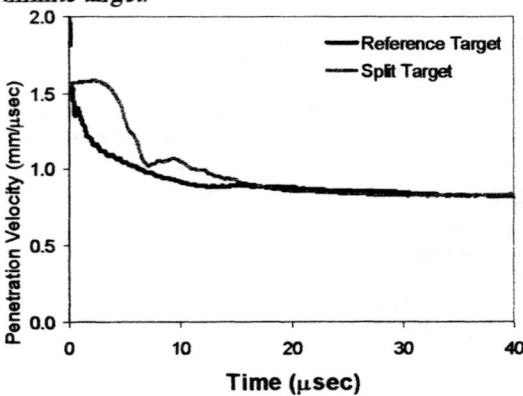

FIGURE 5. A mushroom-shaped rod compared with a cylindrical one.

We can see from the figure that the entrance phase is greatly influenced by the nose shape of the projectile. The mushroom-shaped rod, in the split target, penetrates with a higher velocity during the entrance phase. Thus, it is more effective than the cylindrical rod. Moreover, the penetration velocity history of this rod looks like the ideal two-step history, in which the first (higher) step corresponds to the impact shock, and the second (lower) step to steady-state value.

CONCLUSIONS

The 2D numerical simulations presented here, highlight some of the relevant aspects of the entrance phase in the penetration process of long rods into semi-infinite targets. In particular, it is shown that with increasing impact velocity and rod length, the significance of this phase diminishes, while the opposite is true for increasing target strength. Moreover, we strengthen the claim that the origin of this phase is in the passage from a cylindrical-shaped rod to a mushroom shape.

REFERENCES

1. D. L. Orphal, *Int. J. Impact Eng.* **20**, 601 (1997).
2. A. Tate, *J. Mech. Phys. Solids*, **15**, 387 (1967).
3. V.P. Alekseevskii, *Comb. Explos. and Shock Waves*, **2**, 63 (1966).
4. Z. Rosenberg and E. Dekel, *Int. J. Impact Eng.*, to be published.
5. M. Ravid, S.R. Bodner and I. Holcman, *Int. J. Eng. Sci.*, **25**, 473 (1987).
6. Y. Partom, *Proc. APS Conf. On Shock Waves in Condensed Matter*, eds. S.C. Schmidt and W.C. Tao (New York, 1995), p. 1123.

IMPACT INTERACTION OF PROJECTILE WITH CONDUCTING WALL AT THE PRESENCE OF ELECTRIC CURRENT

Volodymyr T. Chemerys[1], Aleksandr I. Raychenko[2], and Boris S. Karpinos[3]

[1]*Institute of Electrodynamics, 3 Peremoga Ave., Kyiv-57, 03057, Ukraine*
[2]*Institute of Problems for Material Science, 3 Krzhyzhanivsky Str., Kyiv, 03142, Ukraine*
[3]*Institute for Problems of Strength, 2 Timiryazivska Str., Kyiv, 01014, Ukraine*
Ukrainian National Academy of Sciences

Abstract. The paper introduces with schemes of possible electromagnetic armor augmentation. The interaction of projectile with a main wall of target after penetration across the pre-defense layer is of interest here. The same problem is of interest for the current-carrying elements of electric guns. The theoretical analysis is done in the paper for the impact when the kinetic energy of projectile is enough to create the liquid layer in the crater of the wall's metal. Spherical head of projectile and right angle of inclination have been taken for consideration. The solution of problem for the liquid layer of metal around the projectile head has resulted a reduction of the resistant properties of wall material under current influence, in view of electromagnetic pressure appearance, what is directed towards the wall likely the projectile velocity vector.

INTRODUCTION

The physics of penetration at the presence of electric current has not only peculiarities caused by an additional Joule heating, a current is able also to change the energy distribution onto summands. Without touch of the micro aspects of this problem, authors have considered the question about macroscopic effect of electrodynamic forces stipulated by the geometrical peculiarities of current passage along the surface of target, on the process of projectile penetration.

THE OBJECT UNDER CONSIDERATION

This paper consideration can be of interest for two applications of the pulsed power, as rail gun and electromagnetic armor augmentation. There is supposed that in both cases the electric current in the wall of target can be excited by the corresponding pulsed power supply.

Schemes of electrical augmentation of armor

At least two hypothetical schemes of electrical augmentation of armor are known. The first one is based on the "railgun effect" and has a goal to increase a square of impact by some displacement of leading part of projectile in a transversal direction due to appearance of electromagnetic forces *F* as a projectile *2* will close the electric voltage of pulsed power supply *C* applied between main wall *3* and additional pre-defense layer *1* (Fig.1). In the second scheme it is supposed to use the compression of magnetic field confined in the local volume between main wall *3* and additional defense layer *1* after the instant of projectile *2* coming (Fig.2). In this scheme that is necessary to provide the commutation of pulsed power supply *C* just before impact. In the both schemes vector *F* shows the electromagnetic force direction. Let evaluate the second scheme on its principal ability to reduce a destructive action of

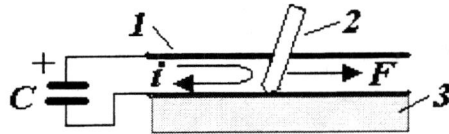

Figure 1. "Railgun effect" for the armor augmentation.

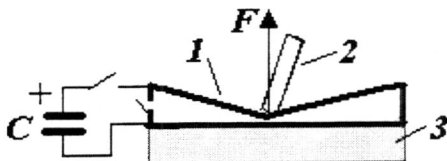

Figure 2. "Field compression effect" for the armor augmentation.

projectile. If the current is switched at instant of projectile touching of pre-defender *1* (Fig.2), the further deformation of layer *1* is going with a closed contour of current in the internal surfaces of layer *1* and wall of target *3*. Let layer *1* is ideally flexible conductor with very low inertia and will suppose the flat (one-dimensional) band of layer. Up to a time of wall touch by layer a cross section of free cavity between them reduces twice. Initial magnetic field B_0 increases there twice in result of the flux compression, if the flux losses are neglected. Magnetic energy in the cavity increases in a second degree on the field, but in view of reduction of the field volume in two time the final magnetic energy is only $W_m = 2 p_m V_0$, where $p_m = B_0^2 / 2\mu_0$ is an initial magnetic pressure in the cavity, V_0 is an initial volume of field. Note, that at two-dimensional deformation of defender a volume of the field would be reduced only in 30%, however the field increase would be less then twice. Thus, in the most optimistic estimation the increment of magnetic energy in the cavity at the field compression by projectile cannot exceed its initial value. Taking out of consideration all additional kinds of projectile work, increment of kinetic energy of projectile is $\Delta W_k = k_s W_m$, where k_s is a coefficient of magnetic energy losses. The average braking electromagnetic force along a displacement of projectile is $F = \Delta W_k / h = k_s \cdot 2 p_m S_i$, here h is displacement of projectile in a normal direction up to closing of cavity, S_i is a square of current-carrying surface of defender layer as acceptor of magnetic pressure. Braking force is more than total force of an initial magnetic pressure in view of the field increase at the compression. The known expression for an inductance of two parallel plates with opposite currents [1] can be rewrite as $L \approx 0.5 \mu_0 S_c / b$, where S_c is a square of cross section of cavity confined by the sides of plates, b is a width of current strip (or, the same, each plate). It gives a possibility to estimate a current needed for opposition to projectile with kinetic energy 10 MJ, with its reduction on 10% due to electromagnetic defense. It would be in general view

$$i = [\Delta W_k \cdot 2b / k_s \mu_0 S_c]^{1/2},$$

and numerically, at $b = 0.05$m, $S_c = 0.07$m x 0.2m, $\Delta W_k = 1$ MJ, $k_s = 0.7$, the needed current is $i = 2.84$ MA. Technically that is not simple to provide this magnitude of current. Nevertheless the considered scheme is not looking more difficult for realization, than scheme in the Fig.1.

Electromagnetic energy influence on the mechanical properties of conductor

Electromagnetic energy is able to change the strength ability of metals not only by addition of source heat losses and electromagnetic forces but also due to influence on the thermodynamic parameters of matter. Correlation between parameters that correspond to different kinds of energy is looking relatively simple at the elastic non-isothermal deformation of material. In a problem of definition for limit state of material the linear concept of accumulation of each kind energy extreme values is not true. That is more correct to use thermodynamic criterion of strength in view of correlation $(Y_{2(\lim)} / Y_{1(\lim) t}) \Delta V = Const$, where $Y_{2(\lim)}$ is summarized mechanical work on deformation up to destruction, $Y_{1(\lim) t}$ is total energy expend on the heating of metal up to melting point in result of different actions, ΔV is relative change of metal volume in result of thermal action, including Joule's losses. That yields to set of interconnection correlations in the form

$$\Delta \sigma_{0,2} / E = f(\alpha T, \xi_e \Psi), \; \Delta \sigma_b / E = f(\alpha T, \xi_e \Psi),$$
$$\Delta \varphi = f(\alpha T, \xi_e \Psi),$$

where $\Delta\sigma_{0,2}$ is deviation of yield point at the action of electric current, $\Delta\sigma_b$ is deviation of ultimate strength, $\Delta\varphi$ is deviation of deformation limits under action of electric current, E is Young's modulus, α is a coefficient of thermal expansion, ξ_e is electrostriction coefficient, T is temperature and Ψ is electric potential. Interpretation of these functional dependencies for specific range of loading (current density, temperature, initial mechanic loads) is the important problem in the evaluation of conductors destruction limits. In application to problem of projectile penetration into metal target the specific factors must be taken into attention in addition to above told ones. Using the theory of similarity, the task about depth of penetration L can be formulated in the form of criterion dependence (1).

$$\frac{L}{l_0} = \zeta \begin{pmatrix} \frac{\sigma_{0,2}}{E}, \frac{\sigma_b}{E}, \vartheta, \alpha T, \gamma T, \frac{P}{S\int_0^T c_p dT}, \\ \frac{Q_{pl}}{T_{pl}}, \frac{\mu UI}{ErR}, \frac{\Delta m E}{\rho_{01}^2 \sigma_e^3}\sqrt{\frac{E}{\rho_{01}}}, \frac{l_0}{d_0}, \\ V_0\sqrt{\frac{\rho_{01}}{E}}, \frac{\rho_{01}V_0^2}{H_1}, \frac{H_2}{H_1}\cdot\chi, \frac{tE}{a_T\rho_{01}}, \\ \frac{j^2 l_0^2 \rho_{01}}{\sigma_e t E^2}, \frac{c_p}{\alpha E}, \frac{jBt}{\rho_{01}V_0}, \mu\sigma_e a_T \end{pmatrix} \quad (1)$$

Arguments in brackets include the criteria of static loading as $\sigma_{0,2}/E$, σ_b/E, ϑ, where ϑ is a transversal contraction at the break, also the correlation between mechanical and thermal processes, as well as between heat energy and energy of phase transition. The second group of criteria gives the relative time characteristic of impact, connection between impact velocity V_0 and elastic waves speed, relative loss of mass Δm in target at the impact, ratio of coefficients for thermal and electromagnetic diffusion in a target. Other values has been used in (1): P is a force of resistance to impact, S is square of impact, c_p is a specific thermal capacity, Q_{pl} is heat of melting, T_{pl} is a melting point, μ is a magnetic permeability of target, U and I are electric field and current, r is a characteristic dimension of target, a_T is a coefficient of temperature conduction, σ_e is an electrical conductivity of target, γ is temperature coefficient of electric resistance, ρ_{01} is a target material density, j and B are a current density and field in a contact zone of target, respectively, t is a time duration of impact, H_1 and H_2 are the static hardness for target and projectile head respectively, χ is a form-factor for the projectile head, l_0 is length of projectile, d_0 is a diameter of projectile. In general, at the temperature rise under action of electric current material has lowering of strength, the increase of strength can be recognized at very high speed of deformation and at the presence of effective covering. With using of theory of similarity experimental data about material durability can be generalized in the view

$$\sigma = \left[1{,}822 + Ei^*(-0{,}1 - k_\varepsilon \varepsilon)\right]^{1+n_T+n_j+n_\varepsilon} \times$$
$$\times E\left(1 - \frac{T}{T_{nn}}\right)^{m_T}\left(1 - \frac{j}{j_{max}}\right)^{m_j}\left(1 + \frac{\dot{\varepsilon}}{\dot{\varepsilon}_{max}}\right)^{m_\varepsilon},$$

here Ei^* (arg) is the integral exponential function, k_ε, n_T, n_j, n_ε, m_T, m_j, m_ε are empirical coefficients for the corresponding factors of influence, j_{max} and $\dot{\varepsilon}_{max}$ are the maximal testing values of electric current density and speed of deformation. Coefficients can be defined from curves of deformation. For orientation, in the conducting materials of electrical engineering there are the magnitudes: $k_\varepsilon \approx 60...110$, $n_T \approx 0...-0{,}15$, $n_j \approx 0...-0{,}1$, $n_\varepsilon \approx 0...0{,}3$, $m_T \approx 0{,}55...0{,}85$, $m_j \approx 1...1{,}15$, $m_\varepsilon \approx 1...2$. The accuracy of approximation for curves of deformation using above expression is going to 5%.

ANALYSIS OF PROJECTILE PENETRATION

The impact velocity of projectile can be enough to create the liquid crater (bath) around a top of head under condition $\rho_c V^2/Y_d \to 10^3$, here ρ_c is a density of projectile material, Y_d is a dynamic yield point. Energy expand on the melting of metal is $W_q \cong (mV^2/2)(1-e^2)$, where e is a coefficient of impact rebound. Using significance of e from [2], it is possible to get the radius of melted zone around top of projectile: $r_e \cong \left(r_i^3 + (3W_d/2\pi L\rho)\right)^{1/3}$,

r_i is radius of projectile head. In assumption of a presence of the liquid metal layer around the top of head when has a form of hemi-sphere the electromagnetic pressure can be calculated for the target with passage one-directional electric current along its surface. The acting forces have been considered in the equation of motion neglecting the viscosity force. Current density distribution in the neighborhood of a projectile top in the cylindrical co-ordinates (r, ϑ) is obtained by the solution of electrodynamic task:

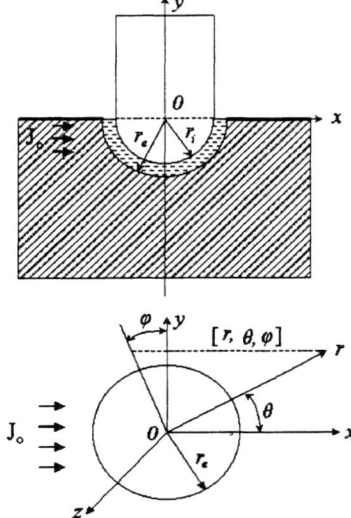

Figure 3. Schematic view of a calculation model.

$$J_e = J_0\left(1 - \frac{\Lambda^3 r_i^3}{r^3}\right)\cos\vartheta\, i_r - J_0\left(1 + \frac{\Lambda^3 r_i^3}{2r^3}\right)\sin\vartheta\, i_\vartheta$$

with a parameter $\Lambda = \left(2\dfrac{\sigma_e - \sigma_i}{2\sigma_e + \sigma_i}\right)^{1/3}$, that includes the electrical conductivity of target σ_e and projectile σ_i. J_0 is undisturbed value of current density on some distance of impact zone. With a proper approximation for the field, a magnetic pressure on the projectile top surface can be found by integration of equation of the forces equilibrium and is given in expression (2), where p_0 is a basic value of pressure:

$$p(r_i, \vartheta)\big|_{r \to r_i} = p_0 - \frac{\mu_e J_0^2 r_i^2}{4}\left(1 - \Lambda^3\right)^2 \sin^2\vartheta, \quad (2)$$

Total force on the projectile head (electromagnetic in sum with hydrostatic one) is respectively

$$F = F_{st} - 6\mu_0 J_0^2 r_i^2 \cdot f(\sigma_R),$$

where $\sigma_R = \sigma_i / \sigma_e$, $f(\sigma_R) = \sigma_R^2 / (2 + \sigma_R)^2$. The function $f(\sigma_R)$ is shown in the Fig.4. The additional electromagnetic force in a liquid metal is directed towards the target, reducing the resulting mechanical resistivity of metal with respect to impact of projectile. In the Fig.4 it is seen that an

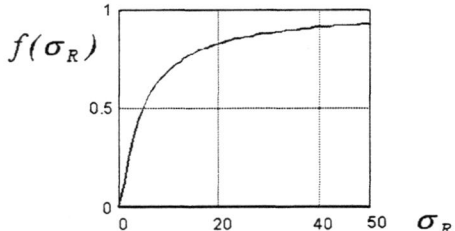

Figure 4. Character of electromagnetic force dependence on the ratio σ_R.

electromagnetic force is going to a stable level at a growth of ratio σ_R up to ∞. If head of projectile doesn't conduct a current ($\sigma_R \to 0$), a total force of resistance is defined by a hydrostatic pressure only: $F_{st} = \pi r_i^2 p_0$.

CONCLUSION

In a macroscopic consideration the analysis of influence for a current flowing along the surface of target has resulted a decrease of mechanical strength of metal wall of target in view of electromagnetic force appearance directed toward a target. Some useful criteria for the impact penetration can be obtained by using the theory of similarity.

REFERENCES

1. Knoepfel, H., *Pulsed High Magnetic Fields*, North-Holland Publ.Comp., Amsterdam – London, 1970.
2. Johnson, K. *Contact Mechanics*, Cambridge University Press (Russian translation: Mir Publishers, Moscow, 1989, 509p.).

THE USE OF THE TAYLOR TEST IN EXPLORING AND VALIDATING THE LARGE-STRAIN, HIGH-STRAIN-RATE CONSTITUTIVE RESPONSE OF MATERIALS

Joseph C. Foster, Jr.[1]
Martin Gilmore[2]
L.L. Wilson[3]

[1]*Air Force Research laboratory/ Munitions Directorate (AFRL/MN)*
Eglin AFB, FL
[2]*Defense Science and Technology Laboratory (DSTL-UK)*
Exchange Scientist @ AFRL/MN
[3]*Science Application Incorporated (SAIC)*
Eglin AFB, FL

Abstract: The characterization of the mechanical response of materials to high rate loading is an experimental challenging task. As the load rate becomes high, the engineering analysis of the results of the experiment places more and more emphasis on understanding the influence of the method of test on the data recovered from the experiment. At very high rates, the inertia of the test specimen dominants the load.[1] Impact testing techniques combined with judicious specimen design provides access to a unique range of strain, strain-rate, and load conditions that have a broad range of engineering applications. This more general interpretation of the classical Taylor test [2] provides opportunities to characterize a variety of materials in a unique range of conditions.

INTRODUCTION

Taylor characterized the difficulty of designing material testing machines as a function of load rates in a James Forrest Lecture in 1945.[1] He concluded that at very high rates of loading "...the inertia of the specimen itself gives rise to changes of stress along its length, which must be taken into account when seeking to interpret experimental results." In essence, the inertia of the specimen is controlling the load cycle and the stress-state is non-equilibrium. This characterization remains valid. It is interesting to note that he focused on the difficulties associated with applying the load quickly as opposed to the high rate deformation response of the materials being tested. The measurement of the mechanical response of materials to high-rate loads drives the test design naturally to impact test techniques. Controlled impact tests to measure the high rate mechanical response of materials have been used in the community for a number of years. The tests provide data in the 10^{4-5} /sec strain rate regime directly above that accessible via Hopkinson bar techniques and below shock experimentation. We have expanded upon the definition of a Taylor test by adopting the spirit of Taylor's 1945 James Forrest Lecture. Experiments designed to study the mechanical response of materials subjected to high-rates of loading where the load cycle is controlled by the inertia of the specimen are herein termed generalized Taylor tests. This more general interpretation of the classical Taylor test [2] provides opportunities to characterize materials in a variety of test conditions that are difficult to obtain via other techniques.

THE EXPERIMENT

Historical

One standard approach used is the impact of a right circular cylinder on either a massive anvil or a receptor rod of identical material, an experiment commonly called a Taylor test after G.I. Taylor. Figure 1 depicts the basic set of experimental parameters recovered from the original experiment.

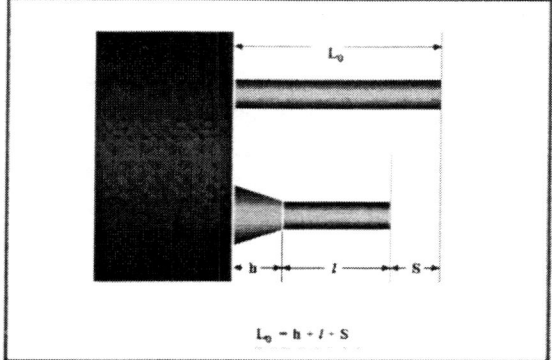

Figure 1. Geometry of the rod-on-anvil experiment together with basic parameters measured from the post-test specimen.

The goal of the experiment was to "extract as much information as possible from measurements of the recovered projectile (*specimen*)." The experiment is used to measure a property associated with a material's plastic response characterized as the dynamic yield stress. The load cycle in the experiment is controlled by the equation of motion (EOM) of the un-deformed section of the specimen. The original EOM was arrived at via an analysis of the elastic ring down of the specimen from the initial impact velocity yielding

$$\frac{dv}{dt} = -\frac{\sigma_0}{\rho l}$$

where : σ_0 = dynamic yield stress

The EOM combined with an assumption that the material passing through the plastic wave front comes to zero velocity in a short distance forms the basis to assess the dynamic yield stress based on the geometry of the deformed specimen. The stress is determined from the experimental measurement of the specimen's final overall length after impact (L_f) compared to its original length (L_0) and the length of the un-deformed section (l_f) using the measured impact velocity (v) and specimen density (ρ) according to

$$\sigma_0' = \rho v^2 \left[\frac{L_0 - l_f}{2(L_0 - L_f)} \frac{1}{\ln\left(L_0 / l_f\right)} \right]$$

assuming uniform deceleration of the un-deformed specimen length.

A first order correction is provided by

$$\sigma_0 = 2\sigma_0' \left[\frac{1}{1 - \frac{l_f}{L_0}} - \frac{1}{\ln\left(L_0 / l_f\right)} \right]$$

that addresses the non-uniform deceleration of the un-deformed specimen length

The strength of this original formulation of the experiment is the relative ease of conducting the experiment and reducing the experimental measurements to mechanical properties data.

Modernization-The Experiment

The Taylor test has survived for the last fifty-five years because it provides data in a unique region of mechanical response that is unobtainable by any other simple test technique. Additions to the fundamental basis of the experiment fall into a.) advanced equations of motion [4,5,6,7,8], b.) alternate approaches to interpreting the results of the experiment[9,10,11,12] and c.) improved test techniques. Many researchers have expanded on the way in which the data is interpreted in terms of a more complete description of the mechanical response of the specimen material. [13,14]

There remains significant benefit in the analysis of the post-test specimen. Improved measurements equipment can be used to accurately map the entire profile of the post-test specimen geometry. Figure 2 is a set of precision profile data taken on the 3.4 hard copper {R_f=103} being used as the driver material in our 'sleeved' Taylor tests.[15]

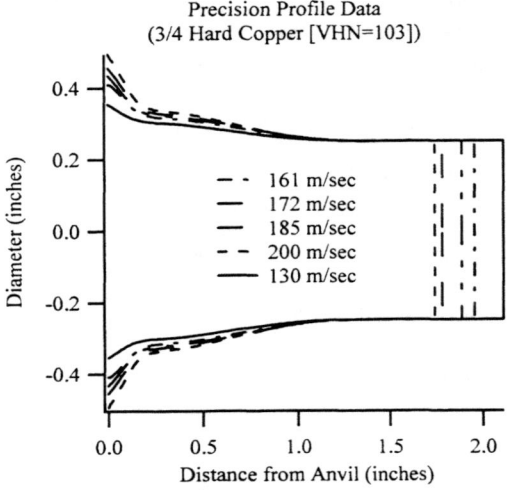

Figure 2. Precision profile data taken from post-test ¾ hard copper specimens at a variety of impact velocities.

This type of data is recovered from the post-test specimen using precision optical scanning techniques. The measurements data can be used to reconstruct the original specimen geometry from the deformed geometry using an approximation that plane sections remain plane.

$$\pi(Z_i - Z_{i-1})\frac{D_0^2}{4} = \frac{\pi}{12}(z_i - z_{i-1})\left[D_i^2 + D_{i-1}^2 + D_i D_{i-1}\right]$$

The (D_i, z_i) data are the data from the optical scanner and Z_i is the axial coordinate in the original specimen which is the referential or Lagrangian coordinates for the experiment.[16] Then, the algorithm for constructing the Lagrangian coordinates from the precision measurements data is

$$Z_i = Z_{i-1} + (z_i - z_{i-1})\left[\frac{1}{3}\left\{\left(\frac{D_i}{D_0}\right)^2 + \left(\frac{D_{i-1}}{D_0}\right)^2 + \left(\frac{D_i}{D_0}\right)\left(\frac{D_{i-1}}{D_0}\right)\right\}\right]$$

Knowledge of the Lagrangian position and the final position permit the construction of the axial displacement

$$U_i(Z) = z_i - Z_i$$

and the strain

$$\varepsilon_{ZZ} \equiv \frac{\partial U_Z}{\partial Z}$$
$$\cong \frac{U_i - U_{i-1}}{Z_i - Z_{i-1}} = \frac{z_i - z_{i-1}}{Z_i - Z_{i-1}} - 1$$

The post-test profile data can now be presented as displacement and incremental strain in the Lagrangian coordinates of the specimen. (Reference figure 3 and figure 4)

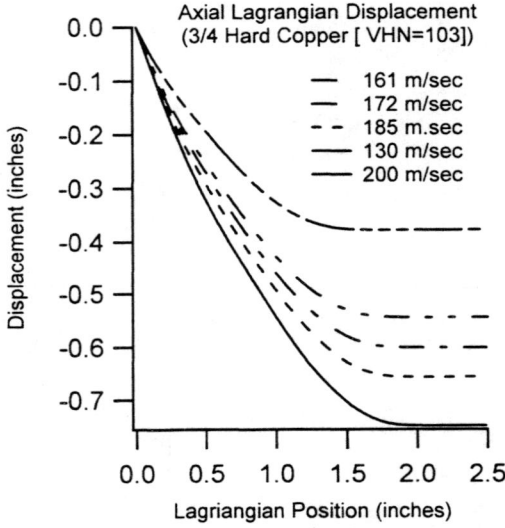

Figure 3. Experimental Lagrangian displacement calculated from the precision profile measurements data on the post-test specimen configuration.

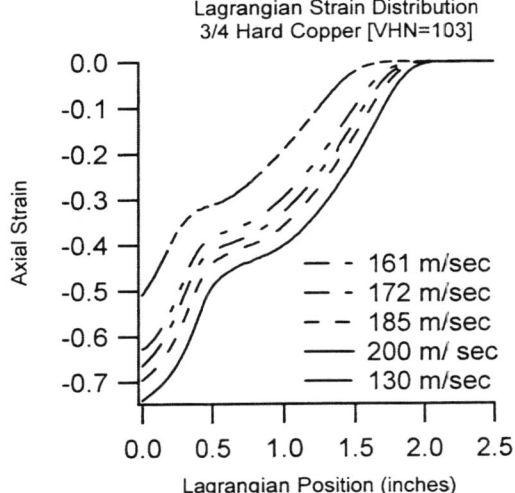

Figure 4. Lagrangian strain distribution calculated from precision profile measurements data on post-test annealed copper specimens at various impact velocities.

Many modelers find this data useful in combination with continuum mechanics codes to develop the constants associated with more complicated constitutive relations describing the yield behavior of the material. The simple geometry of the experiment yields to analysis using continuum codes without burdening the developer with lengthy cycle times in the machine.

Beyond the accumulation of measurements data from the post-test specimen, ultra-high speed photography yields time-resolved deformation states.[17] (Reference figure 5) The time-resolved experimental techniques provide the capability to capture highly transient deformation states such as those in visco-elastic materials that cannot be obtained from post-test measurements. Figure 5 illustrates photographic data taken with a Cordin 330 rotating mirror camera that provides 84 frames of continuous access. The data from the camera can be subject to dimensional analysis similar to that conducted on the post-test specimens albeit with less resolution.

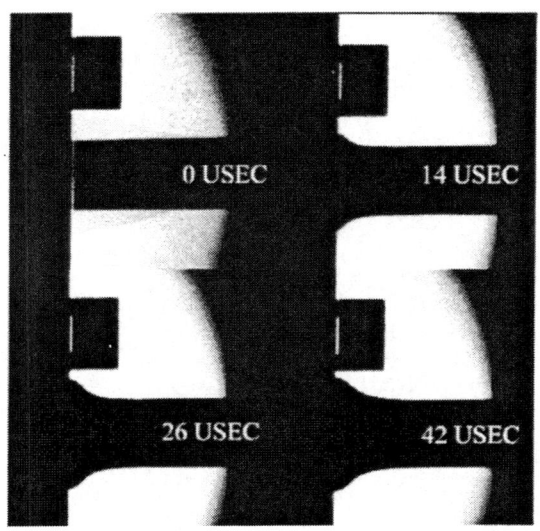

Figure 5. Back-lite high-speed camera data on deforming urethane specimen

The data base provided by the high-speed photography can be differentiated across intermediate phases of the deformation process that represent quasi-steady deformation.[18]

TEST PROTOCOL AND GENERALIZATION

In addition to the physical measurements data recovered from the experiment, if appropriate test protocol is followed, significant data can be can be obtained on the micro-structural changes in the material. Johnson[19] has prescribed a standard materials test protocol for measuring structural response of materials to loads where the instrumentation is unable to time resolve the information. These include:

1.) Pre-load micro-structure characterization
2.) Real-time continuum observation
3.) Post-load micro-structure characterization

This data generally includes one or more of the following: optical microscopy, scanning electron microscopy, and x-ray diffraction studies

When testing energetic materials we have modified Johnson's protocol to read:

1.) Pre-load micro-structure characterization
2.) Real-time continuum observation near threshold
3.) Post-load micro-structure characterization

Figure 6 Pre-test and post-test scanning electron micrographs of PBXN-109 impacted at 69 m/sec using a rod-on-rod test configuration

This same protocol has been used doing dynamic fracture studies with sleeved Taylor specimens where the threshold of interest is now the fracture threshold.[15]

The sleeved Taylor test is basic on our interpretation of Taylor's James Forrest Lecture in that the inertia properties of a well-characterized material (the core) are being exploited to measure the mechanical response of a material of interest (the sleeve). This generalization of test methodology for characterization of the mechanical response of materials to high load rates suggests a variety of geometric configurations that might be used depending on the required data. The combination of test design, test protocol, and formulation of EOM for the design which can be used to describe the load cycle are the fundamental elements of what we have termed herein as 'generalized Taylor tests.'

ACKNOWLEDGMENTS

The work presented in this paper has been funded by a variety of in-house exploratory research projects at the Munitions Directorate of the Air Force Research Laboratory. The researchers are indebted to numerous technicians who provide outstanding support of the research objectives at the Advanced Warheads Experimentation Facility.

REFERENCE

1. Taylor, G.I. *J. Inst. of Civil Engng.* **26**, 486
2. Taylor, G.I. *Proc. R. Soc. London* **Ser. A 194**, 289
4. G. I. Barenblatt and A. I. Islinskii, *Prikladnaya Matematika I Mekhanika* **26**, 497 (1962)
5. T.C.E. Ting, . *J. Appl. Mech. Trans.* ASME **33**, 505 (1966)
6. J.B. Hawkyard, D. Eaton and W. Johnson, *Int. J. Mech. Sci.* **10**, 929 (1968)
7. J.B. Hawkyard. *Int. J. Mech. Sci.* **11**, 313 (1969)
8. Jones, S.E., Gillis, P.P. and Foster Jr., J.C., *J. Appl. Phys.* **61**, 499-502 (1987)
9. J.C. Foster, Jr,, P. J. Maudlin, and S.E. Jones, "On the Taylor test: A continuum analysis of plastic wave propagation" in *Shock Compression of Condensed Matter-1995 by* S. C. Schmidt and W.C. Tao, AIP Press 1996 pp 291
10. S.E. Jones, P.J. Maudlin and J.C. Foster, Jr, *Int. Jour. Imp. Engng.* **19**, no. 2, p. 95 (1997)
11. P. J. Maudlin, S. E. Jones, and J. C. Foster, Jr., *International Journal of Impact Engineering*, 19, no. 3, pp. 231-256, 1997.
12. S.E. Jones, Jeffery A. Drinkard, W.K. Rule and L.L. Wilson, *Int, J. Impact Engng.*, Vol 21, Nos.1 pp 1-13 (1998)
13. Johnson Gordon R. and Holmquist 1988, *J. Appl. Phys.* **64** (8), 15 October
14. William K. Rule and S.E. Jones, , *Int. J. Impact Engng.* Vol. 21, No. 8 pp. 609-624 (1998)
15. Martin Gilmore, Joseph C. Foster, L.L. Wilson, Ian Cullis, "Dynamic Fracture Studies using Sleeved Taylor Specimens" in 2001 Proceeding of the APS Shock Compression of Condensed Matter Topical Subgroup 24-29 June, 2001
16. Malvern, Lawrence, *Introduction to the Mechanics of a Continuous Medium*, Prentice-Hall Inc. 1969, pg.138-141
17. Wilson, L. L., House, J. W. and Nixon. M. E.1989 ,*Time Resolved Deformation from the Cylinder Impact Test*, Air Force Armament Laboratory Report AFATL-TR-89-76
18. J.W. House, B. Aref, J.C. Foster, Jr., and P.P. Gillis, *J. of Strain Analysis*, Vol. 34, No. 5, pp. 337-345
19. Johnson, J.N., "Micromechanical Considerations in Shock Compression of Solids," in *High-Pressure Shock Compression of Solids,* edited by James R Asay and Mohsen Shahinpoor, Springer-Verlag, New York,

DYNAMIC CHARACTERIZATION OF COMPLIANT/BRITTLE MATERIALS USING SPLIT HOPKINSON BAR

N. S. Brar* and Vasant S. Joshi**

University of Dayton Research Institute, Dayton, OH 45469-0182
**Naval Surface Warfare Center, 101 Strauss Avenue, Indian Head, Maryland 20640*

Abstract. High strain rate compression response of thermoplasticolefins (TPO) and explosives (Cast TNT) is determined using the split Hopkinson pressure bar (SHPB). The conventional SHPB technique has been routinely used for measuring high strain rate properties of high strength materials. Attempts are underway in a number of research laboratories to use this technique to determine the high strain rate behavior of compliant materials, such as plastics, rubbers, and foams. A split Hopkinson bar consisting of 15.9-mm 7075 aluminum incident, transmitter, and striker bars was used to determine the compressive response of TNT and other explosives. Initial tests were performed on aluminum 1100-O to compare the stress-strain data with the published data. Stress-strain data on two types of TPOs and TNT at strain rates in the regime 200-2000/s are presented.

INTRODUCTION

Mechanical property determination of extremely brittle as well as soft polymeric material-based explosives has been a topic of considerable interest for investigators in industrial and government research laboratories. Stress-strain data at quasi-static and high strain rates of theses materials are very crucial to generate improved material models required to determine the vulnerability and reactivity of munitions. These two classes of materials are in sharp contrast to the normal metallic materials in their deformation characteristics. During dynamic testing, deformation in soft polymeric materials may be non-linear and strain may be more than 100% [1-2]. In case of brittle materials, deformation is almost linear and the failure strain is relatively small (<5%). Mechanical characterization of both types of materials is difficult due to the problems associated with data acquisition and resolution issues. In the case of polymeric materials, transmitted signals through specimen are extremely small. Brittle materials often fail even before the stress equilibrium is reached at the interfaces between the specimen and the incident and transmitter bars. In both cases, useful data can be obtained only by adapting the basic SHPB system for these specific applications.

The objective of this paper is to present compression stress-strain data on TPOs and TNT at strain rates in the range of 200-2000/s.

EXPERIMENTAL CONFIGURATION

The SHPB system used in this work was originally configured for 10-mm diameter steel pressure bars to test metallic materials. Steel bars were replaced with 15.8-mm diameter magnesium bars to test soft polymeric materials. Mechanical impedance of magnesium matches more closely with that of polymeric materials compared to that of steel or titanium. Although magnesium bar has a better impedance match with polymeric materials, its low yield strength makes it unsuitable at higher strain rates. This led to the choice of high strength aluminum alloy (7075) as pressure bar material. A SHPB system using 15.8-mm diameter pressure bars was used in the final configuration, as shown schematically in Figure 1. The lengths of the incident, trans-mitter, and striker bars were 1.22-m, 1.22-m, and 0.30-m, respectively. The bars are aligned in specially configured bar supports to achieve impact planarity. Pairs of 350-Ω strain gages were mounted on the incident and transmitter pressure bars 0.61-m away from the specimen/bar interfaces. The striker bar is launched at velocities of 5 to 15 m/s using a compressed helium gas gun.

Striker bar impact on the incident bar produces a compressive stress pulse that propagates along the length of the incident bar. The magnitude of the stress pulse σ is given by

$$\sigma = \rho\, C_0\, V_s$$

where ρ (2800 kg/m^3) is the density, C_o (4995 m/s) is the sound speed in bar material (7075 aluminum), and V_s (m/s) is the striker bar velocity. The stress pulse width for a 0.51-m long aluminum striker bar is 200 µs. This compressive pulse in the incident bar subjects the specimen to a compressive load. A portion of the incident compressive pulse, ε_i, is transmitted through the specimen ε_t and the remainder is reflected back in the incident bar ε_r. The amplitude of the incident, reflected, and transmitted pulses are recorded by the strain gages mounted on the pressure bars. Using the recorded strains, the stress (σ), strain (ε) and strain rate ($\dot{\varepsilon}$) in the specimen are determined using the following equations:

$$\sigma(t) = E\frac{A_b}{A_s}\varepsilon_t(t) \quad (1)$$

$$\dot{\varepsilon}(t) = \frac{2\cdot C_o}{L}\varepsilon_r(t) \quad (2)$$

$$\varepsilon(t) = \frac{2\cdot C_o}{L}\int_0^t \varepsilon_r(t)dt \quad (3)$$

where A_b and A_s are the cross-sectional area of the pressure bar and the specimen in the gauge section, respectively. L represents the gauge length of the specimen. Stress, strain, and strain rate are average values, and are determined by assuming a uniform uniaxial stress-state condition [3-4].

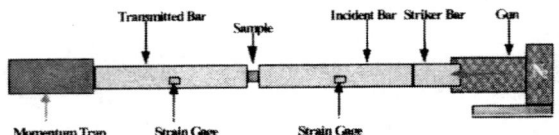

FIGURE 1. Schematic of the SHPB configuration.

RESULTS AND DISCUSSION

Compression tests were performed on three materials: (1) Relatively soft ductile alloy, aluminum 1100-0, (2) Thermoplasticolefin (TPO), and, (3) cast TNT. Aluminum 1100-0 was chosen because of its relatively low flow stress of about 100 MPa. Measured flow stress on 6.35-mm thick and 6.35-mm diameter Al-1100-O specimens of, as shown in Figure 2, at strains of 5% and 10% agree within 2% of the values reported earlier [2]. Such good agreement convinced us that our SHB configuration and stress-strain data analysis are reliable.

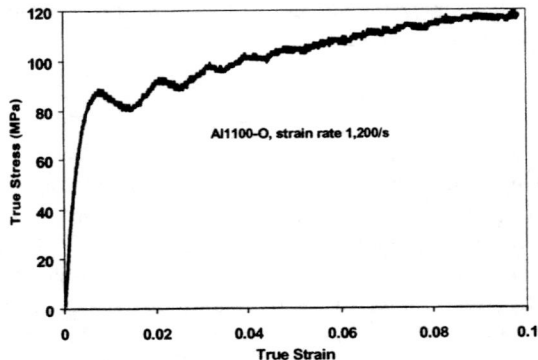

FIGURE 2. Stress-strain curve for Al-1100-0

Test specimens of both types of TPOs (R-2 and R-3) measured 12.7 -mm in diameter and 3.8 -mm in thickness. Experiments were performed at three strain rates ranging from 800/s to 1,500/s. Stress-strain data on R-2 and R-3 TPOs at a strain rate of 1,400/s are shown in Figure 3 and 4 respectively.

Magnitudes of compressive stresses at strains of 10% and 20% compare well to those reported by Brar and Simha [5].

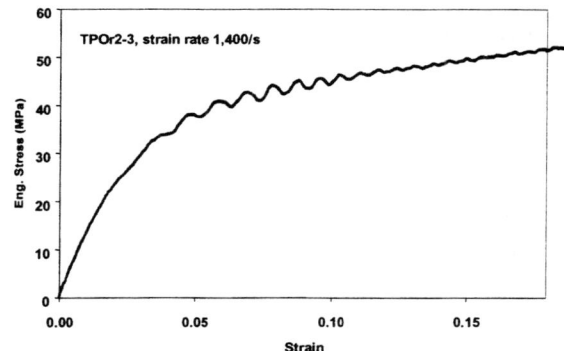

FIGURE 3(a). Stress-strain curves for R-2.

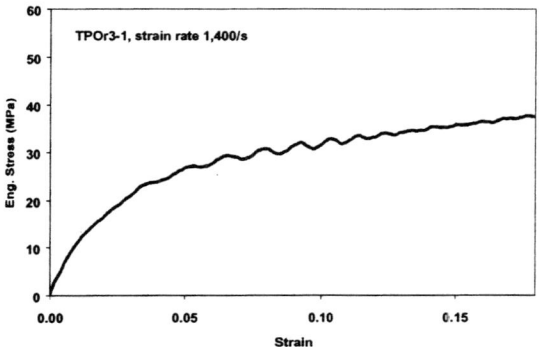

FIGURE 3(b). Stress-strain curves for R-3.

Stress-strain curves for TPOs (Figure 3a and 3b) in the present work are obtained using 15.8-mm diameter aluminum bars and, therefore, are relatively smooth (with higher resolution and strain rates) compared to those determined using 25.4-mm diameter aluminum bars [5].

TNT specimens were fabricated from cast samples. Specimens measured 9.5 -mm in diameter and were 4 -to 9 -mm thick. A Lexan test chamber (300-mm x 250-mm) enclosed the specimen/bar interface portions of the SHB. In the preliminary tests PMMA/plastic disks of appropriate thickness (total of 2 -mm) were placed at the striker bar/incident bar interface to produce long rise time of ~30-40 μs input stress pulses in the incident bar (pulse shaping). Measured stress-strain curves for TNT specimens are shown in Figure 4. The data clearly suggests that specimens fail during the first 30-40 μs of the entire stress pulse of 140 μs corresponding to extremely low failure strain value of (1-1.5%).

A high-speed analog camera (Imacon) was used to observe the failure of TNT specimens during the loading phase of the specimen. It was observed that specimens fail prematurely before reaching the dynamic stress equilibrium even when the loading pulse has a long rise time of ~30-40 μs[6]. As the testing of TNT progressed, steps were taken to improve the pulse shaping process in order to avoid premature failure of the specimens.

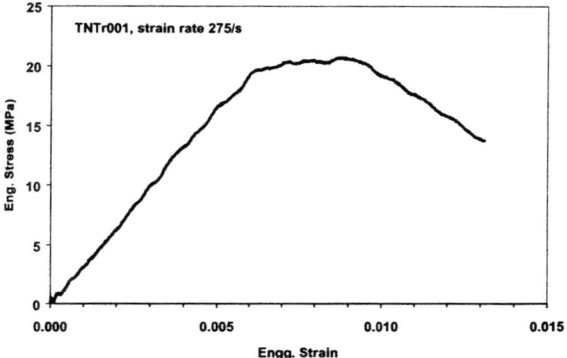

FIGURE 4(a). Stress-strain curves for TNT at a strain rate of 275/s without pulse shaping

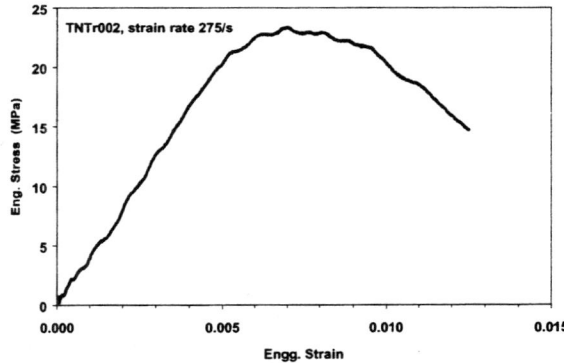

FIGURE 4(b). Stress-strain curves for TNT at a strain rate of 275/s with pulse shaping

Frew [7], based on his experience in characterization of brittle materials (concrete and ceramics), recommended using a linear ramp stress pulse of shorter duration. A simple wave analysis model developed by him suggested placing a 3-mm diameter annealed copper disk of appropriate thickness at the incident/striker bar interface to produce the linear ramp-loading wave. This approach of optimizing the shape of loading pulse is currently being followed to perform further tests on TNT specimens.

CONCLUSIONS

Compression stress-strain curves at strain rates of 200-2000/s for two types of TPOs are generated using SHPB technique. These curves are smoother than those reported earlier [3]. Preliminary high strain rate data on TNT are presented. A proper loading stress pulse shaping procedure needed to avoid premature failure of the TNT specimens during the initial compressive loading phase is currently being pursued.

ACKNOWLEDGEMENTS

This research was supported by NSWC (Indian Head, MD) under the ASEE senior faculty summer fellowship awarded to one of the authors (NSB) during summer 2000. The research was conducted under grant from Navy EOD Tech Center, Indian Head, MD.

REFERENCES

1. Sawas. O.," High Strain Rate Characterization of Low-Density Low-Strength Materials," Ph. D. Thesis, The University of Dayton, Dayton, Ohio (1997).
2. Sawas, O., Brockman, R. A., and Brar N. S., "Dynamic characterization of compliant materials using an All-Polymeric Split Hopkinson Bar," Exp. Mech., **38**, 204-10 (1998).
3. Nicholas, T., "Tensile testing of materials at high rates of strain," Exp. Mech. **21**, 117-185 (1980).
4. Lindholm, U. S., " Some experiments with split Hopkinson pressure bar", J. of Mech. and Phys. of Solids, **12**, 317-22 (1964).
5. Brar, N. S. and H. Simha, "Strain-rate sensitivity of TPO," Proceedings of 6th International Conference on TPO's in Automotive'99, Novi, MI, 120-126, October 1999.
6. Lee, R. J., and Joshi, V. S., "Use of high-speed photography to augment split Hopkinson pressure bar measurements of energetic materials," in this Proceedings.
7. Frew, D., Private communication, 2000.

YIELD AND STRENGTH PROPERTIES OF THE Ti-6-22-22S ALLOY OVER A WIDE STRAIN RATE AND TEMPERATURE RANGE

L. Krüger[1], G. I. Kanel[2], S. V. Razorenov[2], L. Meyer[3], and G. S. Bezrouchko[2]

[1]Nordmetall, Research and Consulting, 09235 Burkhardtsdorf, Germany,
[2]Institute of Problems of Chemical Physics, Chernogolovka, 142432 Russia,
[3]Technical University Chemnitz, Materials and Impact Engineering, 09107 Chemnitz, Germany.

Abstract. A mechanical behavior of the Ti-6-22-22S alloy was studied under uniaxial strain conditions at shock-wave loading and under uniaxial compressive stress conditions over a strain rate range of 10^{-4} s^{-1} to 10^{3} s^{-1}. The test temperature was varied from -175 °C to 620 °C. The strain-rate and the temperature dependencies of the yield stress obtained from the uniaxial stress tests and from the shock-wave experiments are in a good agreement and demonstrate a significant decrease in the yield strength as the temperature increases. This indicates the thermal activation mechanism of plastic deformation of the alloy is maintained at strain rates up to 10^{6} s^{-1}. Variation of sample thickness from 2.24 to 10 mm results in relatively small variations in the dynamic yield strength and the spall strength over the whole temperature range.

INTRODUCTION

On the basis of experiments with titanium alloys up to a strain rate of 300 s^{-1} [1-3] it was concluded the thermally activated mechanism is responsible for the strain rate sensitivity of the flow stress. The examination of the deformed structures revealed that plasticity is the result of interaction of dislocation motion through the lattice and twinning. The extent of twinning increases with increasing strain rate or decreasing temperature. Since titanium alloys are important engineering and armor materials, it would be important to expand their strain-rate and temperature dependencies to ultimately high strain rates of impact loading.

A general trend of the mechanical behavior of shock-wave loaded metals [4-6] is that, whereas the yield strength of pure metals under these conditions does not depend on the temperature or even abnormally increases with the temperature, alloys may exhibit normal behavior with decreasing yield strength as the temperature grows. The resistance to spall fracture of metals usually does not vary much with the temperature up to 85-90% of the melting temperature, T_m, and drops precipitously with the following increase in temperature up to T_m. Unlike in pure metals, experiments with stainless steel demonstrated a decrease in both the dynamic yield strength and the spall strength at the temperature of 980 K to half of their values measured at room temperature [6].

THE MATERIAL

An α-β titanium alloy Ti-6-22-22S with the chemical composition (in wt. %): Al (5.75), Sn (1.6), Zr (1.99), Mo (2.15), Cr (2.10), Si (0.13), Fe (0.04), O (0.082), N (0.006), C (0.009) was prepared in a vacuum arc furnace. The ingot was diffusion annealed at 1100°C for 20 hours. After that, a swaging process at 900°C in the α + β region was performed. Finally, the material was solution annealed and aged. The β-transition temperature of the alloy is 960°C. The final bimodal microstructure contains globular α-phase between the lamella arrangement of α + β phase. The material density is 4.53 g/cm^3, the longitudinal sound speed is $c_l = 6.01 \pm 0.04$ km/s, the bulk sound speed is $c_b = 4.87$ km/s, and the Poisson's ratio is $\nu = 0.327$. All samples were cut out of one massive block of the alloy.

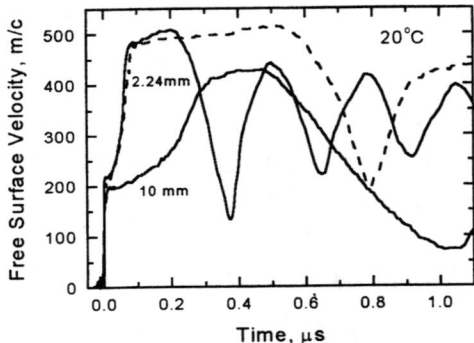

FIGURE 1. Free surface velocity histories of the titanium alloy samples at room temperature.

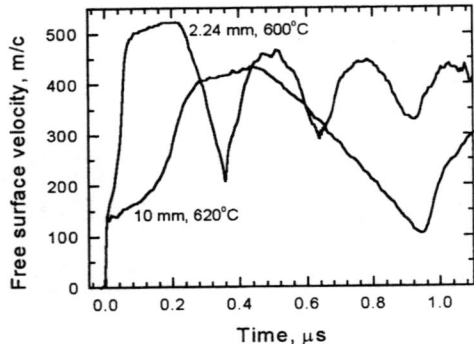

FIGURE 2. Free surface velocity histories of the titanium alloy samples at temperatures of 600°C and 620°C.

EXPERIMENTAL PROCEDURES

Two series of shock-wave experiments with the samples of 10 mm and 2.24 mm in thickness have been carried out with aluminum flyer plates of 2.0 mm and 0.85 mm in thickness at the impact velocity of 630 ± 20 m/s and 680 ± 20 m/s, respectively. Samples were heated with resistive heaters placed on the back surface [5] or cooled by the liquid nitrogen. The power of the resistive heater was 1 kW, which was sufficient to heat the samples to 600°C within 10 to 15 minutes. The temperature of the sample surface was controlled locally at a distance of 7 to 8 mm from the central axis of the sample using a thermocouple of 40 μm in thickness. The free-surface velocity profiles were recorded with the VISAR laser Doppler velocimeter.

The study of material behavior over a range of strain rates of 10^{-4} s^{-1} to 10^{3} s^{-1} was performed using a combination of servohydraulic testing machine, drop-weight tower [7], compression Hopkinson bar and rotating wheel. Tests at high temperatures were conducted by using an induction heating device.

RESULTS OF MEASUREMENTS

Figures 1 and 2 present examples of the free surface velocity histories measured at room temperature and at maximum tests temperatures of 600°C to 620°C. In all cases there are disperse transitions between elastic precursor waves and plastic shock waves that is considered as an evidence of strain hardening behavior of the material. The time interval between the elastic and plastic waves is smaller at elevated temperature as a result of the decrease in the longitudinal sound speed. The shape of velocity histories after spall fracture indicates relatively rapid development of the fracture process. Variation of sample thickness from 2.24 mm to 10 mm results in relatively small variations in the dynamic yield strength and the spall strength over the whole temperature range.

A treatment of high-temperature data has been done accounting for temperature derivatives of the shear modulus according to Ref. [8] where the measured temperature derivative of shear modulus $\frac{\partial G}{\partial T} = -27 \, \text{MPa/K}$ is presented as well as a pre-estimated value of –23 MPa/K. The temperature derivative for bulk modulus was estimated using the relationship

$$\frac{\partial K}{\partial T} \approx -K\alpha\left(\frac{\partial K}{\partial p} - \Gamma\right),$$

where $\frac{\partial K}{\partial p}$ =4.37, the Grueneisen coefficient $\Gamma = 1.23$, and the bulk coefficient of thermal expansion $\alpha = (2.9 \pm 0.4) \times 10^{-5}$ 1/K.

The elastic-plastic waveforms contain information on the incident yield strength and the following strain hardening. In order to get this information the stress-strain diagrams have been recovered from the compressive parts of free surface velocity histories. The estimations have been done within a simple wave approach. The compression wave was considered as a simple centered wave, which is described by a fan of characteristics immediately behind the front of the elastic precursor wave. For simple waves, the longitudinal stress increments $d\sigma$, the strain increments $d\varepsilon_x = -dV/V_0$, the resolved shear stress τ, and the pressure p are related by the equations

$$d\sigma = \rho_0 a_\sigma^2 d\varepsilon_x, \quad \tau = \frac{3}{4}(\sigma - p) \quad (1)$$

where a_σ is the propagation velocity of part of the compression wave at the longitudinal stress σ in Lagrangian coordinates [9]; the stresses are assumed to be positive under compression. For a centered simple wave the propagation velocity a_σ is

$$a_\sigma = \frac{h}{h/c_l + t(\sigma)}, \quad (2)$$

where h is the distance at which the stress history $\sigma(t)$ of a compression wave is analyzed, and t is the time interval after the elastic precursor front. When a free surface velocity history $u_{fs}(t)$ is analyzed instead of a stress history $\sigma(t)$, an approach of

$$u_{fs}(t) = 2u_p(t) \quad \text{and} \quad d\sigma(t) = \rho a_\sigma \cdot du_p(t)$$

may be used. A more detailed analysis accounting for an interaction between the incident compression wave and the reflected unloading wave gives

$$c_\sigma = c_l \frac{2h - c_l t(\sigma)}{2h + c_l t(\sigma)}, \quad (3)$$

Figure 3 shows examples of recovered stress and strain histories at 20°C. The plastic strain γ was calculated according to the relationship

$$d\gamma = d\varepsilon_x - d\tau/G$$

The initial dynamic yield strength values and the flow stresses at 0.2% plastic strain evaluated from the measured free surface velocity histories are presented in Fig. 4.

Figure 5 summarizes the room-temperature yield strength data evaluated from the shock-wave tests under uniaxial strain conditions and from the uniaxial stress tests at lower strain rates. The yield stresses at 0.2% plastic deformations are plotted because it is difficult to determine precisely the incident yield strength values from the Hopkinson bar tests. The strain rates at shock compression have been estimated as average compression rates in middle sections of the samples.

The tensile stress value just before spalling, σ^*, is determined from an analysis invoking the method of characteristics. From the acoustic approach, the following linear approximation [10]

$$\sigma^* = \frac{1}{2}\rho_o c_o \Delta u_{fs}, \quad (4)$$

is used, where c_0 is the sound velocity, ρ_0 is the initial density, and Δu_{fs} is the velocity pullback. However, in the elastic-plastic material the spall pulse front should propagate at the longitudinal elastic wave velocity, c_l, whereas the incident rarefaction plastic wave ahead of it propagates at bulk sound velocity c_b. It was concluded by Stepanov [11], that the relationship to calculate the fracture stress σ^* is

$$\sigma^* = \rho_0 c_b \Delta u_{fs} \frac{1}{1 + c_b/c_l}. \quad (5)$$

A more detailed analysis [12] confirmed validity of the relationship (5) for the case of a

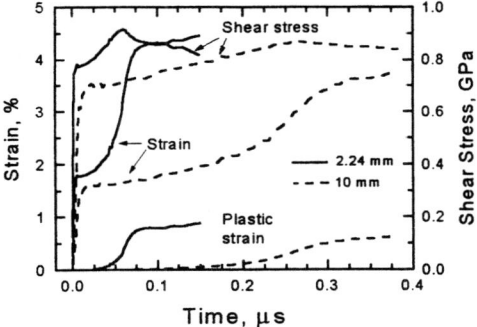

FIGURE 3. Examples of recovered stress and strain histories at 20°C.

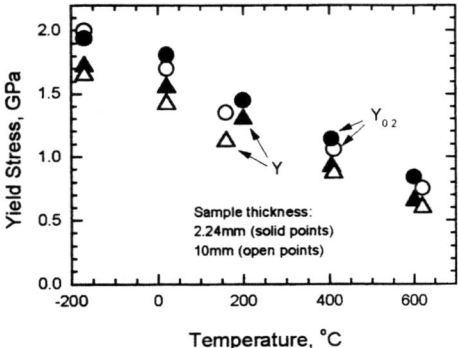

FIGURE 4. Dynamic yield data as a function of temperature for different impact conditions.

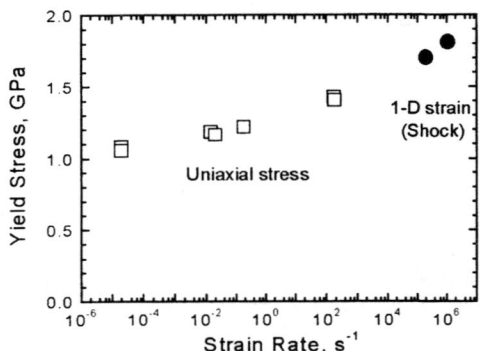

FIGURE 5. Yield strength at 0.2% plastic strain of Ti-6-22-22S alloy as a function of the strain rate.

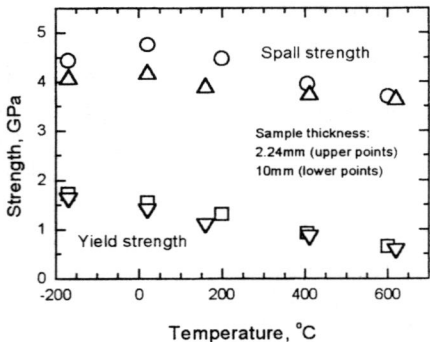

FIGURE 6. Spall strength of Ti-6-22-22S alloy as a function of temperature at two load durations in comparison with the yield strength data.

triangular shape of the incident compression pulse. In the case of trapezoidal shape of the waveform the spall strength may be calculated using relationships

$$\sigma^* = \frac{1}{2}\rho c_b (\Delta u_{fs} + \delta), \quad \delta = \left(\frac{h_{sp}}{c_b} - \frac{h_{sp}}{c_F}\right)\cdot |\dot{u}_1|,$$

$$c_F = c_b c_l \sqrt{\frac{\dot{\sigma}_{x+} - \dot{\sigma}_{x-}}{\dot{\sigma}_{x+}c_l^2 - \dot{\sigma}_{x-}c_b^2}}. \quad (6)$$

Here $\dot{u}_1$ is the free surface velocity derivative ahead of the spall pulse, $c_F < c_l$ is the propagation velocity of the spall pulse front, $\dot{\sigma}_{x+}$ and $\dot{\sigma}_{x-}$ are the stress time derivatives just ahead of the spall pulse front and behind it, respectively, and h_{sp} is the spall plate thickness. The stress derivatives are $\dot{\sigma}_{x+} \approx \rho c_b \dot{u}_1/2$, $\dot{\sigma}_{x-} = \rho c_l \dot{u}_2/2$ near the spall plane and $\dot{\sigma}_{x+} \approx 0$ near the rear free surface of the sample plate. Here $\dot{u}_2$ is the free surface velocity gradient in the spall pulse front.

Figure 6 summarizes the spall strength data over the temperature range from -170°C to 620°C and the yield strength data. Whereas the yield strength varies monotonously with the temperature, the spall data exhibit a maximum tensile strength near the room temperature.

DISCUSSION

The strain-rate and the temperature dependencies of the yield stress obtained from the uniaxial stress tests and from the shock-wave experiments are in a good agreement and demonstrate, in general, a logarithmic dependence over the strain rate range of 10^{-4} s^{-1} to 10^6 s^{-1}. This indicates the thermal activation mechanism of plastic deformation of the alloy is maintained. Note that for pure metals a transition from a logarithmic to linear dependence is often observed in a vicinity of 10^4 s^{-1} strain rate [13]. This explains, why the adiabatic shear bands are relative easily formed in titanium alloys and are not formed in pure metals.

REFERENCES

1. Meyer, L.W. and Chiem, C.Y. In: *Titanium, Science and Technology*, edited by G. Lütjering et al., 1985, pp. 1907-1914.
2. Meyer, L.W. Ibid, pp. 1851-1859.
3. Chichili, D.R., Ramesh, K.T., and Hemker, K.J. *Acta mater.*, pp. 1025-1043 (1998).
4. Rohde, R.W. *Acta Metallurgica*, **17**, 353-363 (1969).
5. Kanel, G.I., Razorenov, S.V. Bogatch, AA., et al. *J. Appl.Phys.*, **79**(11), 8310-8317 (1996).
6. Zhuowei Gu and Xiaogang Jin, In: *Shock compression of condensed matter – 1997*, ed. by S.C. Schmidt et al., AIPCP 429, New York, 1998, 467-470
7. Meyer, L.W and Krüger, L. In: *Mechanical Testing and Evaluation*, ASM Handbook, **8**, 452-454 (2000).
8. Guinan, M. W. and Steinberg, D. J. *J. Phys. Chem. Solids*, **35**, 1501-1512 (1974).
9. Fowles R. and Williams R.F. *J. Appl. Phys.*, **41**(1), 360-363 (1970).
10. Novikov, S.A., Divnov, I.I., and Ivanov, A.G. *Phys. of Metals and Metallography*, **21**(4), 608-615 (1966).
11. Stepanov, GV. *Problems of Strength (USSR)*, No 8, 66-70 (1976).
12. Kanel, G.I. *Journ. of Applied Mech.s and Technical Physics*, **42**(2) 358-362 (2001)
13. Sakino, K. *J. Phys. IV France*, Colloque Pr9, 10, 57-62 (2000).

CHAPTER XXI

LASER–DRIVEN SHOCKS

SUB-PICOSECOND LASER-DRIVEN SHOCKS IN METALS AND ENERGETIC MATERIALS

D.S. Moore[1], David J. Funk[1], K.T. Gahagan[2], J.H. Reho[1],
G.L. Fisher[1], S.D. McGrane[1], and R.L. Rabie[1]

[1]*Dynamic Experimentation Division, Los Alamos National Laboratory, Los Alamos, NM 87545 USA*
[2]*Corning Inc., NYSP-FR-18, Corning, NY 14831 USA*

Abstract. A high-energy sub-picosecond laser was used both to drive a shock into thin film targets and to spectroscopically interrogate the shocked material. Targets were thin films of molecular materials coated or grown upon thin vapor-plated metal films on thin glass substrates, or neat metal films on thin glass substrates. The non-linear optical interaction of the shock-driving laser with the thin glass substrate produced surprisingly flat shock waves. Sub-picosecond time-resolved frequency- and spatial-domain interferometries were used to characterize the shock wave as it transited from the thin metal film into the thin molecular material layer. Overviews of the effect of the pressure-dependent complex index of refraction of the shocked thin film metal layer, ultrafast interferometric interrogation of shocked molecular materials (examples: glycidyl azide polymer and nitrocellulose thin films), and progress in preparation of, as well as the need for, uniform, well oriented, thin energetic material layers appropriate to such highly time-resolved methods are presented.

INTRODUCTION

HERCULES (high explosive reaction chemistry via ultrafast laser excited spectroscopies) is a project at Los Alamos National Laboratory begun in 1998 to utilize recent advances in ultrafast laser technology and thin film sample preparation to try to unravel the details of chemical reactions induced in energetic materials by strong shock waves. Using a portion of the same ultrafast high energy laser pulse to drive the shock wave, diagnose the shock state, and spectroscopically probe the shocked sample, drastically reduces the "time jitter" problem associated with traditional shock wave techniques to sub-picosecond levels, sufficient to time resolve many of the ultrafast processes that occur.

The questions HERCULES hopes to answer are: 1) Is the initiating shock impulsive or more gradual (i.e., what is the rise time of the shock wave stress)? 2) Is a process like multi-phonon up-pumping necessary to excite reaction? 3) Are the observed kinetics in accord with standard transition state theory? 4) Are the reactants and/or products ever in thermodynamic equilibrium on time scales important to the usual HE applications? and 5) Is non-equilibrium temperature (vibrationally hot products) essential to sustaining a detonation wave?

This paper will briefly outline the experimental methods used to produce and characterize the laser induced shocks, both in thin metal films, and in samples composed of energetic materials coated on top of thin metal films. The methods and results are discussed in more detail in other conference papers, to which the reader will be pointed. The implications of the results to date will then be discussed.

EXPERIMENTAL

The entire experimental apparatus is shown schematically in Fig. 1. The laser pulses used in the system are typically 20 mJ at 10 Hz repetition rate, 800 nm wavelength, and 130 fs pulse length, produced by a chirped pulse amplified Ti:sapphire laser system (Spectra Physics). Beam splitters and neutral density filters (absorbing) are used to control the pulse energies in the shock drive, interferometry, and spectroscopic interrogation beams. The sequence of pulses within an individual shock experiment is shown in Fig. 2. After the shock drive pulse strikes the substrate/metal film interface, laser ablation processes launch a shock, which travels through the metal film and arrives at either the free metal surface or the metal/energetic material interface. An interferometry pulse arrives at some

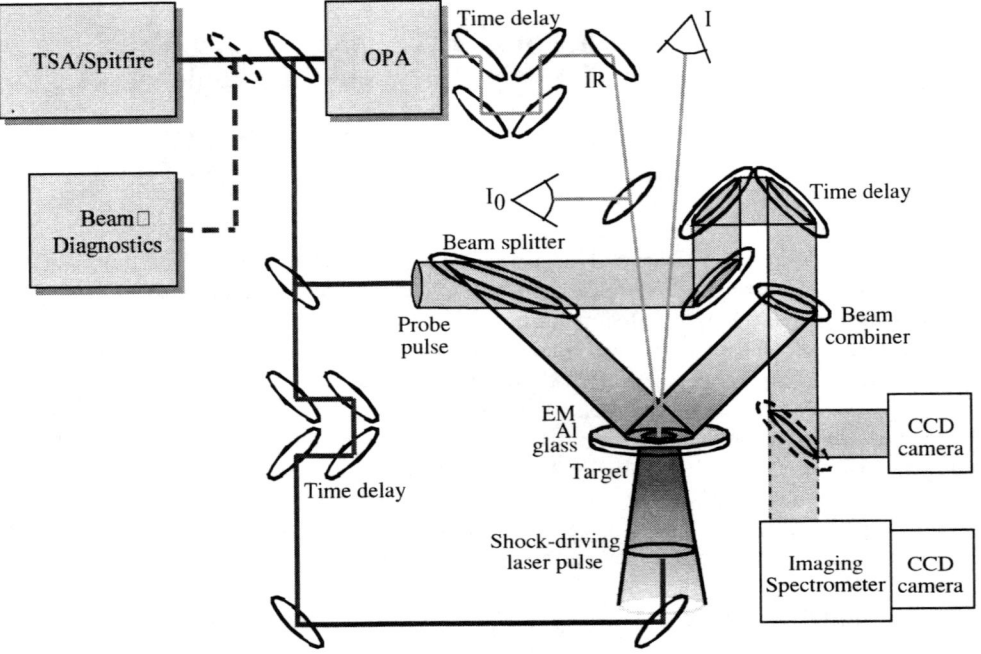

FIGURE 1. Schematic diagram of the ultrafast laser-driven shock experimental apparatus. See text for details.

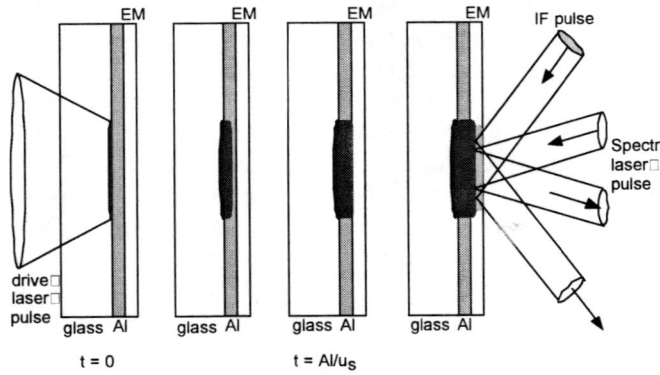

FIGURE 2. Sequence of laser pulses within an individual shock experiment. There are four time snapshots represented: t=0 when the drive laser pulse strikes the glass/Al interface, t < (thickness of Al)/Us, t= (thickness of Al)/Us, and t = when the shock is mostly through the EM layer.

time near when the shock arrives. An optical time delay is used to adjust the relative time of shock breakout and interferometry pulse arrival, in order to build up a time history of shock propagation in step-wise fashion. The target is rastered to a fresh unused spatial location between individual shock experiments. When used, the spectroscopy pulse arrives at the sample at some later time, after the shock has transited an appropriate thickness of overlayer material.

The optical time delays are under computer control to allow several different kinds of time profiles to be recorded. Delaying the shock drive pulse relative to the interferometry pulse results in a record of metal surface motion, and therefore shock state information. Delaying the spectroscopic pulse relative to the shock drive pulse can allow recording of time dependent species concentrations, and therefore kinetics.

Two types of femtosecond (fs) time-resolved interferometry are used. Both utilize a Mach-Zehnder interferometer with the sample in one arm and a time delay in the reference arm. For spectral interferometry (spectral IF, also known as frequency domain interferometry - FDI), the reference arm is slightly longer than the sample arm and the beams are recombined collinearly and focused into a spectrometer. The two probe pulses are stretched in time by the spectrograph, so that they overlap in time and space at the CCD and produce an interference pattern along the wavelength axis. The resulting CCD image for each experiment is analyzed using Fourier transform methods [1-3] to extract the difference in phase between the two probe pulses caused by changes in the film surface position or the optical properties of the metal film. In order to extract the relative contributions of motion and optical properties, data are recorded at two different incidence angles, one near normal incidence and one near the quasi-polarizing angle. Further details and results for several metals can be found in Funk, et al., (Reference 4 and also in this proceedings volume [5]).

For fs time-resolved 2-D spatial interferometry (spatial IF), the reference and sample arms of the interferometer are made the same length, and the beams are combined non-collinearly, so that an interference pattern is produced at the CCD (no spectrograph is used). Two dimensional spatial information about material motion and/or sample optical properties is obtained from the interference patterns using 2-D Fourier transform methods. Again, two different angles of incidence of the sample beam are used to extract the relative contributions of the two effects.

This new method for measuring the shock state of materials with ca. 100 fs time resolution, provides an extraordinary amount of information, and is discussed in detail by Gahagan, et al., in this proceedings volume [6].

Our early ultrafast laser driven shock experiments in thin metal films showed some surprising features. First, the spectral IF data indicated that the shock wave exiting the free metal surface, under a range of drive laser energies, was flat instead of Gaussian in profile, even though the drive laser spatial profile was Gaussian. Serendipitously, the drive laser was focused through the substrate (borosilicate glass microscope slide cover slips of ca. 150 µm thickness) before interacting with the glass-metal interface. Turning the sample around so that the drive laser struck the metal film first and observing the shock as it passed from the metal into the substrate resulted in a Gaussian spatial profile. Apparently, interactions of the femtosecond laser pulse with the substrate material flattened the laser intensity spatial profile, which was confirmed by a series of experiments in uncoated substrates of a variety of materials and thicknesses [see Reference 7]. The planarity of the resulting shock waves is astonishing, measured at < 1 nm RMS over a 75 µm diameter, and ensures subsequent material motion, optical properties, and chemical reaction data are obtained under 1-D strain conditions.

RESULTS

Metal optical properties

The spectral IF (800 nm wavelength and p-polarization) data obtained from a sample of vapor plated aluminium on a borosilicate glass cover slip

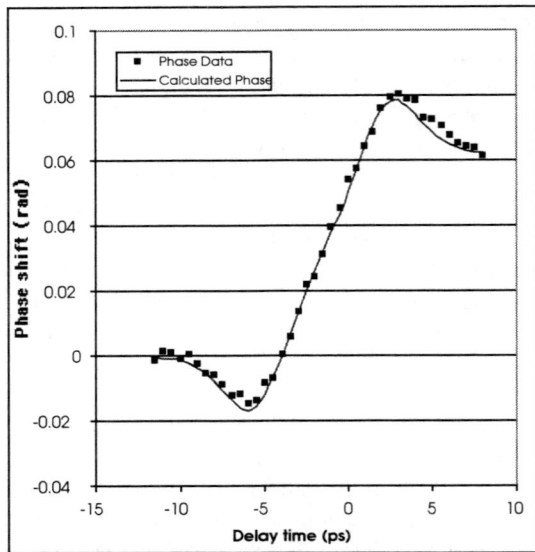

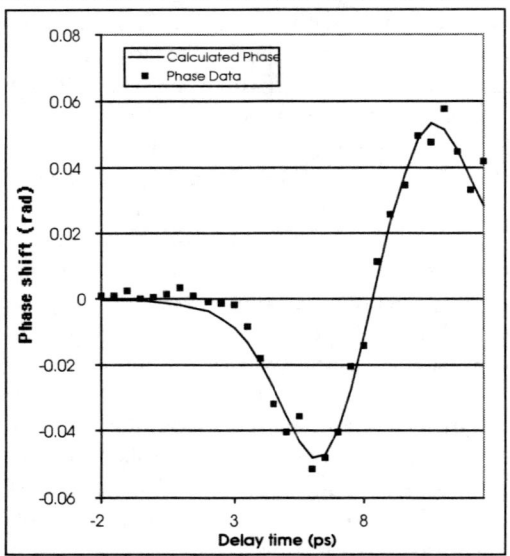

FIGURE 3. Phase shift measured by spectral IF as the shock wave exits the free surface of a 750 nm thick Al layer on borosilicate glass substrate. Probe was p-polarization, 800 nm, and the angle of incidence was top) 32.6°, and bottom) 82.5°. The large negative phase shift is associated with the pressure dependence of the 1.5 eV interband transition in Al [4,5].

substrate as the ultrafast laser driven shock exited the free surface at two different angles of incidence are shown in Fig. 3. These data show an unexpected phase shift during the shock breakout, opposite in sign to that expected (and observed at later times) from material motion. Further spectral IF measurements using 400 nm probe wavelength confirm that this phase shift can be associated with the pressure dependence of the 1.5 eV interband transition in aluminium. For details see References 4 and 5.

This effect has been seen in all metal samples studied to date, although more usually the phase difference caused by the optical properties is of the same sign as the motion. However, making measurements at two or three angles of incidence allows separation of the contributions and yields an effective complex index of refraction for the shocked material. Use of three angles of incidence is more generally useful, as the separation of the contributions does not require assuming a form for the pressure versus time curve.

Energetic polymers

Samples for the energetic polymer (EM) studies were produced by spin coating thin polymer films on top of 250 nm thick Al layers, which had been vapor plated onto borosilicate glass microscope cover slips. These sandwiches were then subjected to ultrafast laser driven shocks, as above. The drive pulse energy was approximately 0.5 mJ into a focal diameter of ca. 75 µm, in order to achieve planar shock loading [7] as the shock wave transited from the aluminium into the polymer.

Spatial IF measurements were made of the aluminium/polymer interface, which was assumed to be the reflecting surface before, during, and after the shock transited from the aluminium into, and progressed through, the polymer. This assumption was checked by obtaining data at two probe angles of incidence, where changes in optical properties of the polymer layer behind the shock should be evident.

Figure 4 shows the interface position (obtained from the phase change data using the ambient literature index of refraction) versus delay time obtained from spatial IF measurements in a nitrocellulose layer spin coated on the aluminium, as well as that obtained in a free aluminium surface (no polymer coating) under the same drive laser conditions. Similar data are also shown for glycidyl azide polymer films. Details are given in Reho, et al., in this proceedings volume [8]. The nitrocellulose layer was measured by thin film UV/vis reflection methods (Filmetrics) to be ca. 550 nm thick, so that the shock wave has not reached the free nitrocellulose surface during the time delays plotted in Fig. 4.

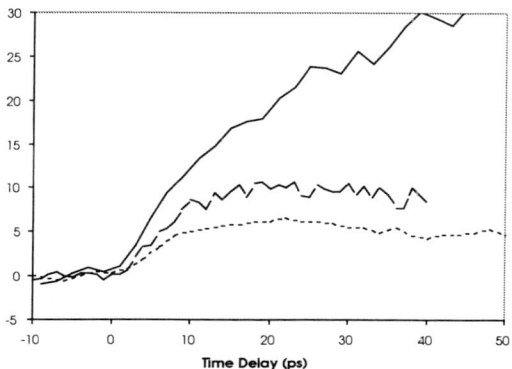

FIGURE 4. Spatial-IF measured Al surface position versus time delay (zero defined to be when the shock arrives at the Al rear surface) for three samples: solid) 250 nm Al layer (no EM layer), dashed) 250 nm Al layer with ca. 1000 nm GAP layer, and dotted) 250 nm Al layer with ca. 550 nm nitrocellulose layer. See text for details.

The main feature to be gleaned from the data given in Fig. 4 is the dramatic difference in the motion of the Al surface when coated with nitrocellulose or GAP. Hydrodynamic simulations of the surface motion were performed in 1-D using an Eulerian-Lagrangian hydrodynamics code. Literature values for the ballistic electron penetration depth, hot electron diffusion distance, and thermal diffusion depth for aluminium [8] were used to estimate an initial effective velocity/temperature profile in the aluminium near the glass-aluminium interface for the hydrodynamic simulations. The Al free surface motion measurements were used as a benchmark, i.e., the peak initial temperature in the ablation region was adjusted so that the simulation most closely matched the experimental data. The resulting curve closely agrees with the free surface Al data in Fig. 4.

That same initial condition was then used in simulations of the aluminium/polymer interface motion. The results are not yet satisfactory, as the simulation predicts larger surface displacements than measured for the nitrocellulose data. However, the Al/NC interface deceleration, recession, and then acceleration at later times may be due to catch up of the rarefaction, and the details of that part of the simulation are still in progress. Data taken at higher incidence angles (75° in air, which is approximately 38° at the Al/NC interface) show a similar displacement trend with time. The reader is directed to Reho et al. in this proceedings volume [9] for further details.

DISCUSSION

Progress has been made in measuring shock dynamics in thin film samples using fs laser techniques to shock load the sample and interrogate the shock state. Some surprises have been discovered. These include the production of extremely flat shock waves by the non-linear optical interaction of 130 fs duration shock drive laser pulses with the borosilicate glass substrate material in our samples and the observation of pressure induced changes in the optical properties of metals as a laser-driven shock exits the metal surface.

These laser-driven shocks are being used to dynamically load layers of energetic polymers. The fs time-resolved interferometric methods measure the time dependent position of the Al-EM interface, to try to understand the loading dynamics. Time resolved infrared absorption spectroscopy of the loaded EM layer is presently being investigated using 130 fs tunable infrared pulses. The time dependence of the disappearance of reactant IR bands, with hopefully the appearance of product IR bands, will be used to extract chemical kinetics. However, the loading history must be understood first, as laser-driven shocks using the methods discussed above are not steady waves.

We have found that detailed knowledge of the microscopic structure of our samples is critical to understand the loading history as measured using 2D spatial interferometry. Spin coating of polymers results in optically transparent layers of precisely measurable thickness, but with unknown density or variation of density with distance through the layer. The position versus time data given in Fig. 4 for GAP and NC films could possibly be explained by density variability through the depth of the coating layer or catch up of the rarefaction in the aluminium driver layer, but it could also indicate reaction. Infrared absorption spectroscopic experiments as well as wave profile measurements versus layer thickness, using a thin vapor plated Al layer on top of the EM layer (both experiments in progress) will help elucidate the actual situation. In addition, we are obtaining spatial interferometric data at three angles of incidence to extract the time-dependent optical properties from the interface motion, and using standard ellipsometry to measure the ambient thin film optical properties and their variation with depth.

Ideally, the thin film samples should be of oriented single crystal material. We are presently working to grow such materials using lattice matching methods and vapor or chemical deposition. Standard surface analytical methods will be used to fully characterize the materials before the shock experiments. Such samples should help alleviate most of the uncertainties such as those seen in our NC and GAP experiments to date.

ACKNOWLEDGEMENT

This research was performed under the auspices of the US Department of Energy.

REFERENCES

1. Gahagan, K.T., Moore, D.S., Funk, D.J., Rabie, R.L., Buelow, S.J., and Nicholson, J.W., Phys. Rev. Lett. **85**, 3205 (2000).
2. Tokunaga E., Terasaki A., and Kobayashi T., Opt. Lett. **17**, 1131 (1992).
3. Geindre J.P., Audebert P., Rousse A., Falliès F., Gauthier J.C., Mysyrowicz A., Dos Santos A., Hamoniaux G., and Antonetti A., Opt. Lett. **19**, 1997 (1994).
4. Funk, D.J., Moore, D.S., Gahagan, K.T., Buelow, S.J., Reho, J.H., Fisher, G.L. and Rabie, R.L., Phys Rev B (in press 2001).
5. Funk, D.J., Moore, D.S., Reho, J.H., Gahagan, K.T., McGrane, S.D., and Rabie, R.L., in this volume.
6. Gahagan, K.T., Reho, J.H., Moore, D.S., Funk, D.J., and Rabie R.L., in this volume.
7. Moore, D.S., Gahagan, K.T., Reho, J.H., Funk, D.J., Buelow, S.J., Rabie, R.L., and Lippert, T., Appl. Phys. Lett. **78**, 40 (2001).
8. Hohlfeld, J., Wellershoff, S.-S., Güdde, J., Conrad, U., Jähnke, V., and Matthias, E., Chem. Phys. 251, 237 (2000).
9. Reho, J.H., Moore, D.S., Funk, D.J., Fisher, G.L., McGrane, S.D., and Rabie, R.L., in this volume.

TIME-RESOLVED MEASUREMENT OF THE LAUNCH OF LASER-DRIVEN FOIL PLATE

Hongliang He, Takamichi Kobayashi, and Toshimori Sekine

Advanced Materials Laboratory, National Institute for Materials Science (NIMS),
1-1 Namiki, Tsukuba, Ibaraki 305-0044, Japan

Abstract. Foil plates (or mini-flyers) have been launched to high velocity by a tabletop laser system, and their velocity histories have been measured by a push-pull type VISAR coupling with an electronic streak camera for recording the quadrature outputs. Typical results are presented for Al (10 μm-thick), Pt (10 μm), Au (5 μm) and Cu (5 μm) foil plates at laser intensities ranging from 20 to 400 GW/cm^2. For the 10 μm-thick Al, velocities over 13 km/s within about 30 ns have been detected with a time resolution of ~300 ps and ~2% error for the peak velocity.

INTRODUCTION

Dynamic high pressures up to tera-pascal (TPa) have been generated with the direct irradiation of strong laser beam on condensed matter in the past decade. This progress offers a new opportunity to study material property under the extreme condition in laboratory. To improve the shock compressing condition of the target sample, an indirect irradiation idea, that we call laser gun, is now being proposed. Analogous to a propellant gun, a foil plate (or mini-flyer) is launched to high velocity driven by pulsed laser beam, and then impacts with a static target. This flyer-impact system may provide a well-controlled, stable shock wave compression for the sample, and also a significant preheating of the sample may be avoided, which is inevitable in the direct irradiation experiments. In order to well characterize the performance of laser gun, a time-resolved measurement on the launch of foil plate is required. A challenge of this measurement, however, comes from: (1) the foil acceleration process being extremely fast that the velocity rises up to a few km/s or even more within a time duration on the order of nanosecond, and (2) the small size of mini-flyer, which is normally hundreds of μm or 1~2 mm in diameter. Previously, the time-resolved velocity profiles have been measured for the foil plates tamped with a substrate at low laser irradiance (<5 GW/cm^2).[1,2] Probing the velocity at higher laser intensity is not satisfactory yet. In this report, we present a measurement on Al, Pt, Au and Cu foil plates irradiated at laser intensities up to 400 GW/cm^2. An optical system that couples with an electronic streak camera (ESC) for recording the quadrature outputs of push-pull type VISAR has been established, which provides high time resolution (300~500 ps) and high accuracy in the velocity history measurement.

EXPERIMENTAL TECHNIQUE

To drive the foil plate, a tabletop-type compact laser system (Q-switched Nd:YAG laser) was used.

The pulsed laser beam with wavelength of 1064 nm, pulse duration (FWHM) of 10 ns and a Gaussian spatial profile was focused onto a foil plate located inside a vacuum target chamber. The focused beam diameter of the driving laser was ~2 mm. Irradiation intensity was from 20 to 400 GW/cm^2 with the changes of laser energy from 8 to 100 J.

To measure the foil plate's velocity history, a push-pull type VISAR was employed. Compared with other velocity interferometers, such as FP[3] (Fabry-Perot interferometer) and ORVIS[4] (optically recording velocity interferometer system), the push-pull type VISAR has great advantages that it provides almost a continuous and an excellent accuracy in velocity measurement because of the usage of four quadrature outputs with special phase difference.[5,6] In the conventional use of push-pull type VISAR, a PMT system, that consists of four photomultiplier tubes (PMTs) and one or two oscilloscopes, is used to record the quadrature outputs.[5,6] It provides a time resolution normally about 2~3 ns. To achieve higher time resolution, in this work the quadrature signals have been recorded by an optical system, that we call ESC system.[7,8] The VISAR's output signals have been transmitted through an optical fiber directly to an ESC. A fast rise-time ESC (C5680 streak camera, HAMAMATSU PHOTONICS K. K, Japan) is used, which provides a time resolution less than 50 ps by using a sweep unit M5677. Considering also the time dispersion in optical fibers and the VISAR's intrinsic rise time, a time resolution of about 300~500 ps is obtained for the present recording system.

RESULTS AND DISCUSSION

Figure 1a shows a typical interferential fringe pattern recorded by the ESC system from a 10 μm-thick Al foil plate irradiated at 340 GW/cm^2 (shot No. 00616s5). The time resolution is about 300 ps given the Velocity Per Fringe (VPF) constant of 10.05 km/s/fringe. Time axis increases from the top to the bottom and the total record is 100 ns. The four outputs of push-pull type VISAR,[6] 1A, 1B, 2A, and 2B, are recorded simultaneously, and fringe patterns are evident by a variation of light intensity with time. The vertical dark lines in outputs 1A, 2A

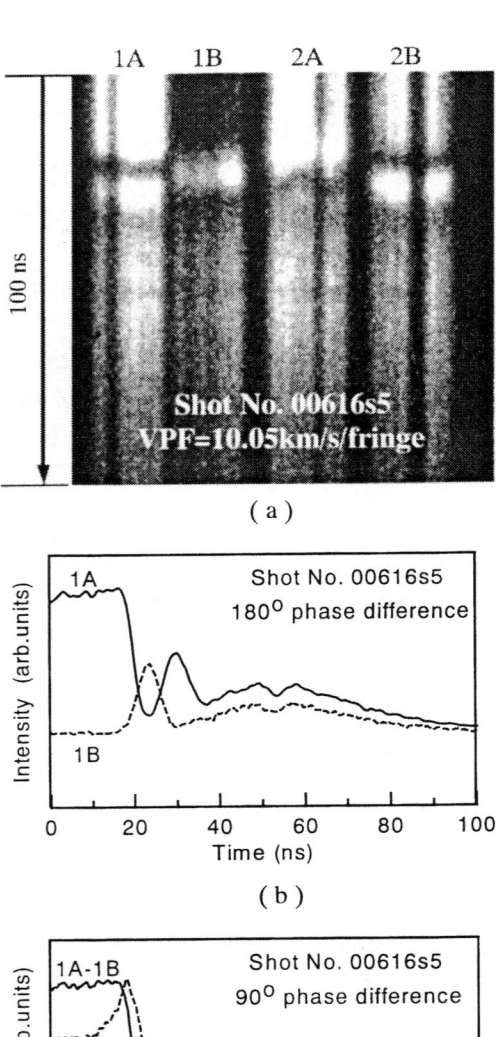

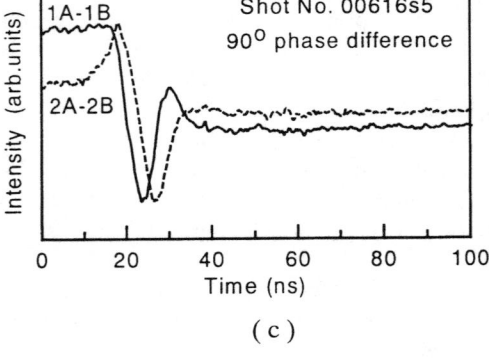

FIGURE 1. VISAR record and analysis in shot 00616s5. (a) Fringe pattern recorded by the ESC system from 10μm-thick Al foil plate driven at 340 GW/cm^2; (b) Digitized data of outputs 1A and 1B; (c) Subtractions of (1A-1B) and (2A-2B).

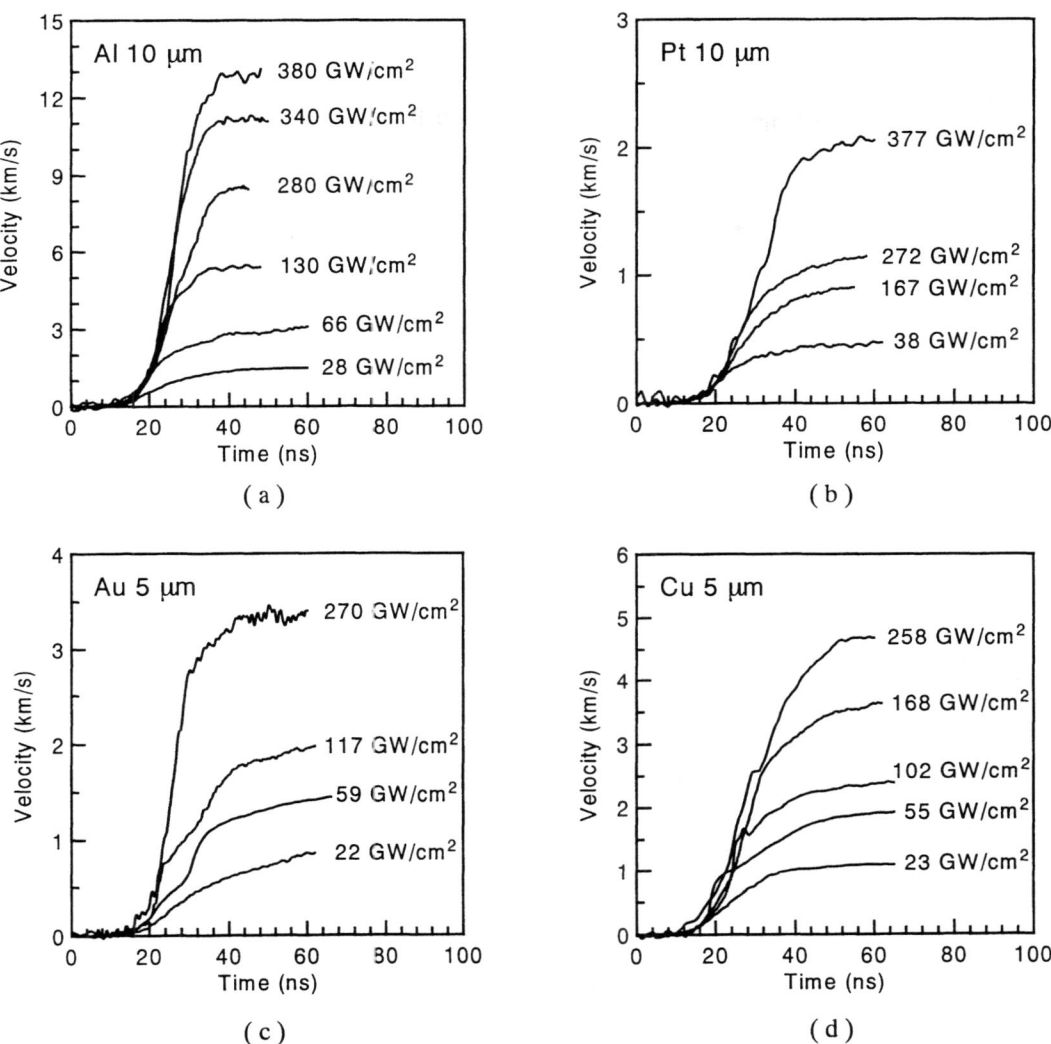

FIGURE 2. Velocity profiles of Al 10μm-thick, Pt 10μm-thick, Au 5μm-thick and Cu 5 μm-thick foil plates.

and 2B are due to the break of fibers or separation at the time of fabrication. The output records are analyzed with an image readout system that yields the data by digitizing the image with 1024 intervals. The time relationship between each interval has been calibrated by the manufacture and no interpolation has been made during our data reduction. Figure 1b is the digitized outputs of 1A and 1B, showing a 180° phase difference. Figure 1c indicates the subtractions of outputs (1A-1B) and (2A-2B), and they exhibit 90° phase relationship.

These results demonstrate the excellent push-pull phase relationship among these quadrature outputs.

The velocity history, v(t), has been computed from the VISAR fringe count, F(t), with the equation,[5,6]

$$v(t-\frac{1}{2}\tau) = VPF \times F(t), \quad (1)$$

where τ is the delay time of VISAR's etalon, and F(t) can be known from,

$$F(t) = \theta(t)/2\pi, \quad (2)$$

and $\theta(t) = \tan^{-1}[D_2(t)/D_1(t)]$. (3)

$D_1(t)$ and $D_2(t)$ are the amplitudes of the two fringe data sets shown in Fig.1c. $D_1(t)$ is the subtraction of (1A-1B), and $D_2(t)$ is (2A-2B). They possess the same amplitude and 90° phase difference. This calculation has been done with a computer program[9] and the result is presented in Fig.2a as indicated by 340 GW/cm^2. Given that about 1.1 fringes were recorded (Fig.1c), the accuracy of peak velocity is about 2% in this shot.[10]

Typical velocity profiles are presented in Fig. 2 for the foil plates Al (10 μm-thick), Pt (10 μm), Au (5 μm) and Cu (5 μm), respectively. Repeated experiments have been conducted at the fixed laser conditions to check the reproducibility. Good coincidence in a series of shots is evident. The difference of acceleration property among the four kinds of foil plate is significant. 10 μm-thick Al foil plate appears to be the most favorite one to generate high velocity under the same laser intensity. A terminal velocity increase from 1.5 to 13 km/s has been observed as the laser intensity increases from 28 to 380 GW/cm^2. 5 μm-thick Cu foil plate is the second candidate to generate high velocity, and then Au (5 μm) and Pt (10 μm), respectively (Fig.2).

In terms of shock impedance, Pt flyer is the highest, then Au, Cu, and Al is the last. However, since Al flyer has significantly higher acceleration velocity than the rest of three, it can generate much stronger dynamic pressure under the same driving laser intensity. With the present tabletop laser system, a shock pressure of ~500 GPa can be generated in a Pt target sample by using an impact of 10 μm-thick Al foil plate at ~13 km/s. Higher shock pressures may be produced by reducing the laser beam diameter down to less than 2 mm or using thinner foil plate. However, the two-dimensional effect in pressure generation will be more significant.

ACKNOWLEDGEMENTS

The authors gratefully acknowledge L. M. Barker at Valyn International for the technical support. This research was supported by a grant from the COE project in Advanced Materials Laboratory, National Institute for Materials Science (NIMS), Japan.

REFERENCES

1. Paisley, D. L., Montoya, N. I., Stahl, D. B., and Garcia, I. A., "Interferometry, streak photography, and stereo photography of laser-driven miniature flying plates", in *19th International Congress on High Speed Photography and Photonics*, SPIE Proceedings Vol. 1358, Cambridge, UK, 1990, pp.760-765
2. Trott, W. M., and Meeks, K. D. J., *Appl. Phys.* **67**, 3297-3301 (1990).
3. Durand, M., and Laharrage, P., *Rev. Sci. Instrum.* **48**, 275-278 (1977)..
4. Bloomqust, D. D., and Sheffield, S. A., *J. Appl. Phys.* **54**, 1717-1722 (1983).
5. Barker, L. M., and Hollenbach, R. E., *J. Appl. Phys.* **43**, 4669-4675 (1972).
6. Hemsing, W. F., *Rev. Sci. Instrum.* **50**, 73-78 (1979).
7. He, H. L., Kobayashi, T. and Sekine, T., *Rev. Sci. Instrum.* **72**, 2032-2035 (2001).
8. He, H. L., Kobayashi, T. and Sekine, T., *Appl. Optics*. (submitted).
9. Barker, L. M., Sandia National Laboratory Report, SAND88-2788 (1988).
10. Barker, L. M., "The Accuracy of VISAR Instrumenttation," in *Shock Compression of Condensed Matter-- 1997*, edited by S. C. Schmidt et al., AIP Conference Proceedings 429, New York, 1998, pp.833-836.

LASER-LAUNCHED FLYER PLATES AND DIRECT LASER SHOCKS FOR DYNAMIC MATERIAL PROPERTY MEASUREMENTS

D. L. Paisley, D. C. Swift, R. P. Johnson, R. A. Kopp and G. A. Kyrala

Los Alamos National Laboratory, NM 87545, USA*

The Trident laser at Los Alamos was used to impart known and controlled shocks in various materials by launching flyer plates or by irradiating the sample directly. Materials investigated include copper, gold, NiTi, SS316, and other metals and alloys. Tensile spall strength, elastic-plastic transition, phase boundaries, and equation of state can be determined with small samples. Using thin samples (0.1 - 1.0 mm thick) as targets, high pressure gradients can be generated with relatively low pressures, resulting in high tensile strain rates (10^5 to $10^8 \, s^{-1}$). Free surface and interface velocities are recorded with point- and line-imaging VISARs. The flexible spatial and temporal pulse profiles of Trident, coupled with the use of laser-launched flyer plates, provides capabilities which complement experiments conducted using gas guns and tensile bars.

INTRODUCTION

A hierarchy of models is used to simulate dynamic loading on shock wave time scales. Coarse features of the material response are modeled with the equation of state, with a constitutive model added as a correction.

Although empirical data have played a dominant role in calibration, theoretical techniques are now being used to devise models and deduce parameters based on a sound physical understanding of material behavior on different length and time scales. An important component of model development is comparison with experimental data which test the components of the model. In the case of crystalline solids – a class which includes most structural materials – it is necessary to predict and measure the response of single crystals to dynamic loading, and to develop and test models of the collective response of a polycrystal ensemble.

The Trident laser at Los Alamos is a useful facility for experiments of this type. Trident is capable of generating fairly well-characterised pressure profiles of $\sim$0.1 to 500 GPa lasting $\sim$0.2 to 200 ns. The laser has three beams for drive or probing, there is a large degree of flexibility in the beam characteristics, and a variety of diagnostics is available. It is also relatively straightforward to recover the sample after shocking, since there is less scope for subsequent damage or confusion from sabots or detonation products as occurs in gun and explosive experiments.

Because of the range of sample sizes which can be used, experiments can probe material response over several orders of magnitude in time, which is important in the study of time-dependent phenomena such as strength and polymorphism. In spall experiments, tensile strain rates from 10^5 to 10^8/s can be produced.

We have performed a range of experiments to characterize dynamic material behavior, mainly in support of the Inertially Confined Fusion (ICF) program. Shock waves are induced in the sample either by using the laser to accelerate a flyer plate which then impacts a stationary target, or by irradiating the sample directly.

DIAGNOSTICS

The main diagnostic used for materials work is velocimetry of the surface of the sample. We measure the velocity history using VISAR systems, in particular a line-imaging VISAR.(1) The line VISAR is used in our flyer experiments to

---

*This work was performed under the auspices of the U.S. Department of Energy under contract # W-7405-ENG-36

measure both the flyer speed and the sample response; the shock speed can also be deduced from its arrival time.

We have also obtained transient X-ray diffraction (TXD) data from direct drive experiments on single crystals. A second beam from Trident is used to generate a hot plasma in a foil target; the plasma radiates predominantly in a single X-ray line which is used to observe the crystal lattice by Bragg scattering. The X-ray wavelength depends on the foil material and the laser energy; results have been obtained with manganese and titanium foils, giving wavelengths of 2.11 and 2.755 Å respectively. We have used two X-ray streak cameras, together with time-integrating films, to obtain records of the scattering angle simultaneously in different directions. Line VISAR records are obtained in the same experiment. (Fig. 1.)

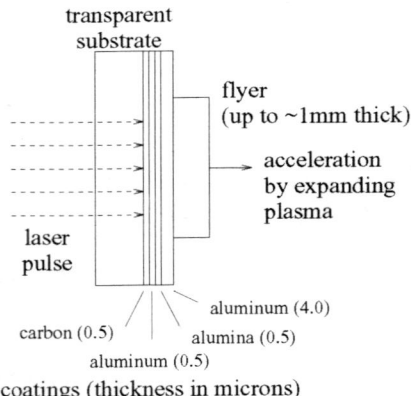

FIGURE 2. Laser-launched flyer plate.

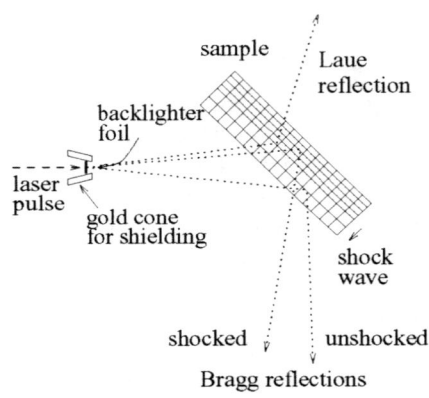

FIGURE 1. Transient X-ray diffraction.

LASER-LAUNCHED FLYER PLATES

Laser energy can be used to launch a flyer plate efficiently by irradiating a working medium through a transparent substrate. The working medium is vaporized, and accelerates the flyer plate as it expands. This scheme can be refined by incorporating extra layers to insulate the flyer from the hot, expanding fluid and to prevent the fluid from escaping sideways. (Fig. 2.)

Trident can deliver up to ~1 kJ of energy, making it possible to accelerate relatively thick flyers and hence span the range of length scales between single crystals through small aggregates of polycrystals to samples fully representative of most materials used in engineering. However, for most of the ICF and plasma applications for which Trident was designed, the pulse length required was only ~1 ns. If a short pulse is used to accelerate a thick plate, the plate will ring up to speed rather than accelerating smoothly, and it may even spall. A system was developed to stretch out the Trident drive by applying a varying waveform to the acousto-optical element used to trigger the laser. This element initially blocks the beam, so no amplification takes place. In normal operation, it is switched to allow photons to pass and thus trigger the laser. In order to generate a long pulse, the element is switched almost off again immediately after triggering to reduce the rate at which the intensity builds up in the laser cavity, and then gradually switched back on to generate the desired pulse shape. In this way, pulses of an approximately Gaussian shape have been generated up to ~2 μs long. We have accelerated copper flyers up to 100 μm thick using a pulse 200 ns wide.

In previous trials with drive pulses up to 60 ns long, sapphire substrates were found to perform well. With 200 ns pulses, the drive efficiency was extremely low using a sapphire substrate. The explanation is not clear, but it is possible that the longer pulse and more massive flyer produced a pressure which was higher or more sustained, and the sapphire changed phase to a structure which was opaque. Flyers were launched successfully using glass and PMMA substrates.

A potential source of concern is the degree to which the flyer may be heated during acceleration. Recovered flyers show no evidence of heating. In experiments performed with copper flyers impacting copper or PMMA targets, the velocity history matched simulations of an initially unheated flyer.

The flyers can be used to impact a variety of types of target. If the flyer is made of the sample material, its equation of state can be measured by impacting it into a transparent window of known properties and observing the velocity history at the impact surface. The equation of state, strength, phase diagram and spall strength can be investigated by impacting the flyer into a target made of the sample material, and measuring the velocity history at the surface of the sample. This may be a free surface or be confined by a transparent window of known properties. (Fig. 3.) We have recently performed experiments on cerium, CeSc, and gold alloys; some samples were recovered (Fig. 4). Analysis is in progress.

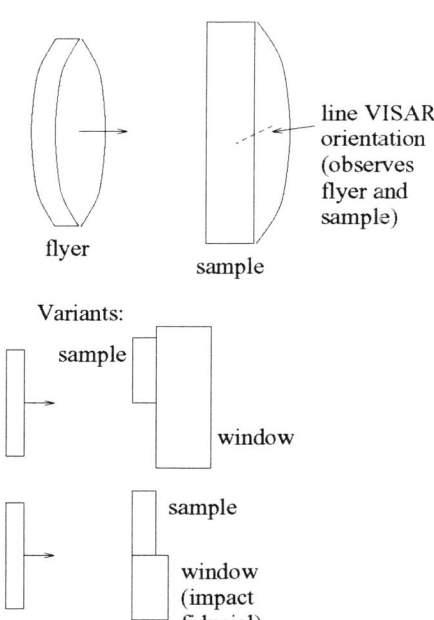

FIGURE 3. Flyer targets for material property measurements.

DIRECT DRIVE

When a laser pulse of sufficient intensity illuminates condensed matter, the surface blows off as plasma and a compression or shock wave is driven into the material. In general a pulse of constant intensity does not drive a shock wave of constant pressure, so the pulse shape is adjusted to produce the desired loading history. The design and interpretation of direct drive experiments require radiation hydrodynamics simulations.

The laser intensity required to generate a shock wave by direct illumination is on the order of 10^{14} W/m^2, which Trident can deliver only as a short pulse. The pulse duration limits the thickness over which a shock of constant pressure can be generated, before the rarefaction from the drive surface makes it decay. In practical terms this thickness depends on the shock speed and sound speed in the shocked state, both of which vary with the material and shock pressure. For most materials the pulse length of ~2 to 4 ns constrains the sample thickness to a few tens of μm.

The Trident beams contain spatial inhomogeneities which may be significant in directly-driven shocks. For material property experiments the drive beam is smoothed spatially using a random-phase plate or Fresnel zone plate. These components operate correctly for a single photon wavelength and spot size on the target, so the spot size (and thus intensity scale) is constrained by the components available: spot diameters of 4, 0.6 and 0.15 mm spot diameter using green light, and 5 mm in the infra-red. A 4 mm spot is sufficient to generate shock waves up to several tens of GPa. A 0.6 mm spot allows shock waves of ~1 TPa to be produced. These diameters are much greater than the sample thickness, so the loading history is 1D over almost the whole diameter, in typical materials experiments.

In the direct drive experiments, the shock propagates from the illuminated surface to the opposite side of the sample. This face may be free or confined by a window. The velocity history is measured at the sample surface with VISARs. We have used wedged or stepped samples to provide better data on the shock speed and evolution of the wave structure. (Fig. 5.)

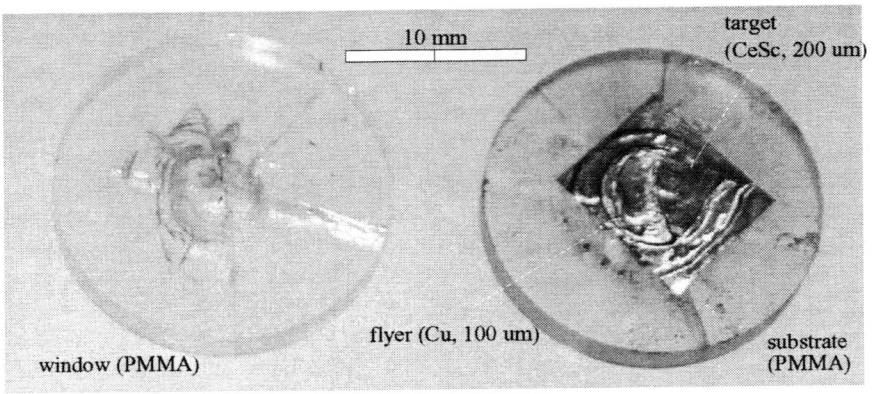

FIGURE 4. Example of parts recovered post-shot.

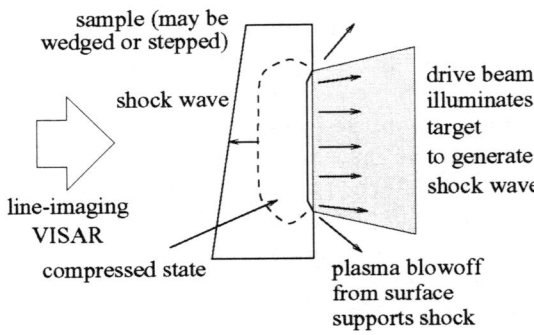

FIGURE 5. Direct drive experiments.

There are several potential pitfalls in direct drive experiments. If the material irradiated by the laser does not have a low atomic number then a significant level of X-rays may be produced. These can pre-heat the material. We have performed experiments driving materials with atomic number up to 29 (copper) with no evidence of preheat. When direct illumination of a material could cause problems, it is possible to mount the material on another of lower atomic number or to coat the sample with a layer of an appropriate material. Another problem is that plasma or hot electrons from the irradiated surface may flow very quickly around the sample – particularly if it is not light-tight – and obscure any optical diagnostics. We had considerable difficulty with experiments on aluminum and silicon driven between several tens and several hundreds of GPa; one palliative was to fit a conductive mask around the driven side of the sample to increase the distance which any plasma or hot electrons had to travel before they could interfere with the diagnostics.

We have performed direct drive experiments on silicon, aluminum, iron, beryllium and Ni-Ti.(2–4)

CONCLUSIONS

We have found the Trident laser to be a useful and versatile facility for measuring the response of materials to dynamic loading. Laser-launched flyer plates allow pressures of up to ~100 GPa to be generated in samples up to ~1 mm thick. Direct laser illumination provides pressures of up to ~1 TPa. Shock wave diagnostics include VISAR velocimetry (point and line), and transient X-ray diffraction.

We have seen no evidence of preheat in flyer plates or directly-driven samples.

REFERENCES

1. Forsman, A.C. and Kyrala, G.A., Phys. Rev. E vol. 63, 056402 (2001).
2. Swift, D.C. and Ackland, G.J., *First principles equations of state for simulations of shock waves*, submitted to Phys. Rev. B (2001).
3. Swift, D.C., Paisley, D.L., Kyrala, G.A. and Hauer, A., this conference.
4. Forsman, A.C., et al, *Simultaneous VISAR and TXD measurements on shocks in silicon crystals*, in preparation (2001).

DEVELOPMENT OF LASER-DRIVEN FLYER TECHNIQUES FOR EQUATION-OF-STATE STUDIES OF MICROSCALE MATERIALS[*]

Wayne M. Trott, Robert E. Setchell, and Archie V. Farnsworth, Jr.

Sandia National Laboratories, Albuquerque, NM 87185-0834

Abstract. Experimental methods utilizing laser-acceleration of flyer plates and high-speed velocity interferometry are being developed for investigation of the dynamic material properties of microscale materials (e.g., polycrystalline silicon). Hugoniot measurements are performed using reverse impact of the laser-driven flyer onto a known witness plate material (e.g., fused silica). A line-imaging optically recording velocity interferometer system (ORVIS) can be used to record both pre-impact flyer velocity and post-impact interface velocity as well as to determine the geometry (along a line segment) of the flyer at impact, providing a viable approach for direct state measurements. Results for 6061-T6 aluminum flyers are consistent with the well-established Hugoniot properties of this material. Methods for fabricating flyers containing thin layers of polycrystalline silicon and for measuring the short duration shock state in this material at impact are discussed.

INTRODUCTION

Methods for pulsed-laser-acceleration of miniature flyer plates provide a promising approach to quantitative studies of material response under shock wave loading (1,2). These methods are particularly suited to investigation of materials that cannot be obtained with sufficient dimensions for use in conventional gas gun tests. An important example is the basic structural material used in surface micromachines, polycrystalline silicon (3). This material is generally prepared in characteristic layer thicknesses of ~2-6μm. Coupled with high-speed, optically recording velocity interferometry, the low temporal jitter available with laser-driven flyer techniques is suited to well-controlled production and measurement of short-duration (few ns) compression states in such microscale materials.

In this paper, we discuss an experimental approach for equation-of-state studies of materials such as polycrystalline silicon. The experimental design utilizes reverse impact of laser-driven flyers onto a well-characterized witness plate material (e.g., fused silica). This configuration can provide a direct state measurement for a sample impacting a window with a well-established Hugoniot, as a result of the boundary condition that axial stress and particle velocity must be continuous across the interface. Important aspects of our efforts include [1] tailoring of the drive laser spatial profile to optimize flyer/impactor planarity, [2] development of methods for consistent, well-characterized flyer fabrication, and [3] execution and interpretation of high-speed witness plate experiments using velocity interferometry. Significant insight into these issues can be obtained from complementary 1-D and 2-D hydrocode simulations of flyer acceleration (4). Results of "baseline" Hugoniot measurements of 6061-T6 aluminum flyers are presented. For this application, we demonstrate a flyer/window and interferometer configuration that allows direct measurement of the flyer impact velocity and post-impact interface velocity on a single record. Initial efforts to obtain Hugoniot measurements from a 3.5-μm layer of polycrystalline silicon coupled to a 30-μm-thick molybdenum flyer are also discussed.

---

[*] Sandia is a multiprogram laboratory operated by Sandia Corporation, a Lockheed Martin Company, for the United States Department of Energy under Contract DE-AC04-94AL85000.

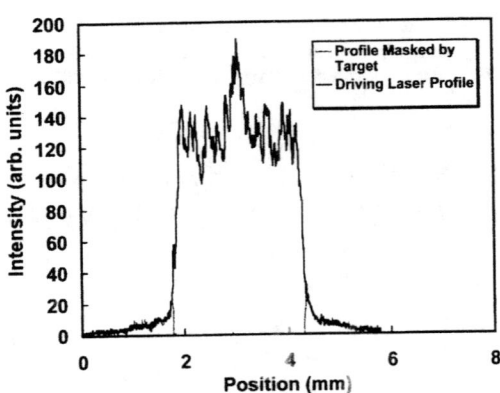

FIGURE 1. Typical spatial profile of drive laser beam used in flyer experiments.

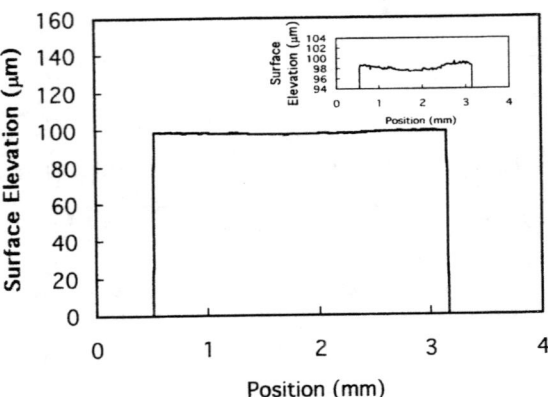

FIGURE 2. Typical surface profile (along one axis) of 6061-T6 Al flyer target prepared by diffusion bonding. Inset: Profile at expanded vertical scale.

LASER-DRIVEN FLYER METHODS

The driving laser used in this work was a Q-switched Nd:Glass oscillator (Lasermetrics Model 9380). The laser output ($\lambda = 1.054 \mu m$) was horizontally polarized and multimode. The spatial intensity distribution of the laser beam at the flyer plane was routinely monitored using a beam profiling system. In an attempt to optimize flyer planarity, we evaluated several optical designs (e.g., microlens arrays, diffractive elements) for beam conditioning in order to achieve a uniform intensity distribution over a flyer diameter of 2.5mm at incident fluences up to 50 J-cm^{-2}. None of these designs provided a true "top hat" distribution for the multimode beam. A typical drive profile is illustrated in Fig. 1. The intensity distribution exhibited an on-axis region of higher intensity but it was free of severe "hot spots." We tailored the beam area to overfill slightly the flyer diameter (Fig. 1). This drive profile provided satisfactory flyer performance. Results of 2-D numerical simulations have shown that the flyer planarity is in fact surprisingly insensitive to the intensity profile and is more dependent on the hydrodynamics associated with expansion of the driving plasma (4).

FLYER TARGET FABRICATION

For initial evaluation of our test methods, flyers of 6061-T6 aluminum were prepared for use in reverse impact studies. For this purpose, 2.5-mm-diameter discs were carefully punched out of a 0.1-mm-thick sheet of this material. We tried several flyer assembly processes utilizing a variety of thin adhesives to bond the discs to fused silica substrates (transparent substrates are used to provide confinement for the driving plasma, thereby increasing the flyer acceleration efficiency). These processes did not provide consistent results in terms of bond strength or target geometry. Our best results were achieved with an adhesive-less process that included the following steps: [1] careful hand-polishing of the 6061-T6 Al discs to a tolerance of a few μm, [2] preparation of slightly oversized 5-μm-thick pads of aluminum on fused silica via physical vapor deposition (PVD), and [3] diffusion bonding of the discs and pads at a temperature of 480°C. This process produced very uniform flyer targets, as illustrated by the typical optical profilometer record shown in Fig. 2.

PVD has also been used to fabricate multiple 30-μm-thick molybdenum flyers intended to carry thin layers (~3.5μm) of polycrystalline silicon to impact onto a witness plate. The polycrystalline silicon layers were prepared via deposition with the substrate temperature held at 300°C. Molybdenum provides a useful base layer for the polycrystalline silicon as a result of its chemical compatibility under high-temperature processing. The thermal expansion characteristics of fused silica preclude its use as a substrate in this process. Satisfactory results were obtained with sapphire and BK-7 glass, however.

DUAL TIME BASE LINE-IMAGING VELOCITY INTERFEROMETRY

Witness plate measurements have been made using a line-imaging optically recording velocity interferometer system (ORVIS). The essential elements of the optical design for this system and methods for image data reduction are described elsewhere (5). Successful measurement of shock states produced by microscale materials in a reverse impact flyer experiment requires an accurate recording of both the flyer impact velocity and the short-duration interface velocity induced by the thin material on impact. Figure 3 displays results of a 1-D hydrocode calculation (4) for acceleration of a composite flyer (consisting of 3-μm-thick silicon coupled to 30-μm-thick molybdenum) and the velocity states produced by this flyer on impact with fused silica. Figure 3b, in particular, illustrates the demanding nature of this application. The time duration of the state induced by the thin silicon is ~1 ns or less. The time required for the flyer acceleration process, on the other hand, is typically much longer (~100 ns).

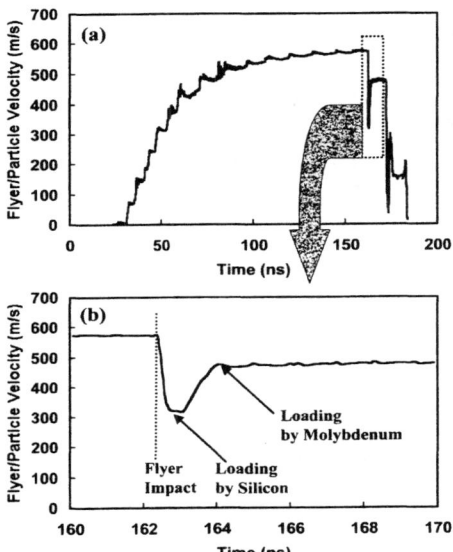

FIGURE 3. 1-D hydrocode calculation of composite flyer (30-μm-thick Mo/3-μm-thick Si) impacting fused silica: (a) curve showing flyer acceleration and impact; (b) expanded view showing short duration state induced by thin Si layer.

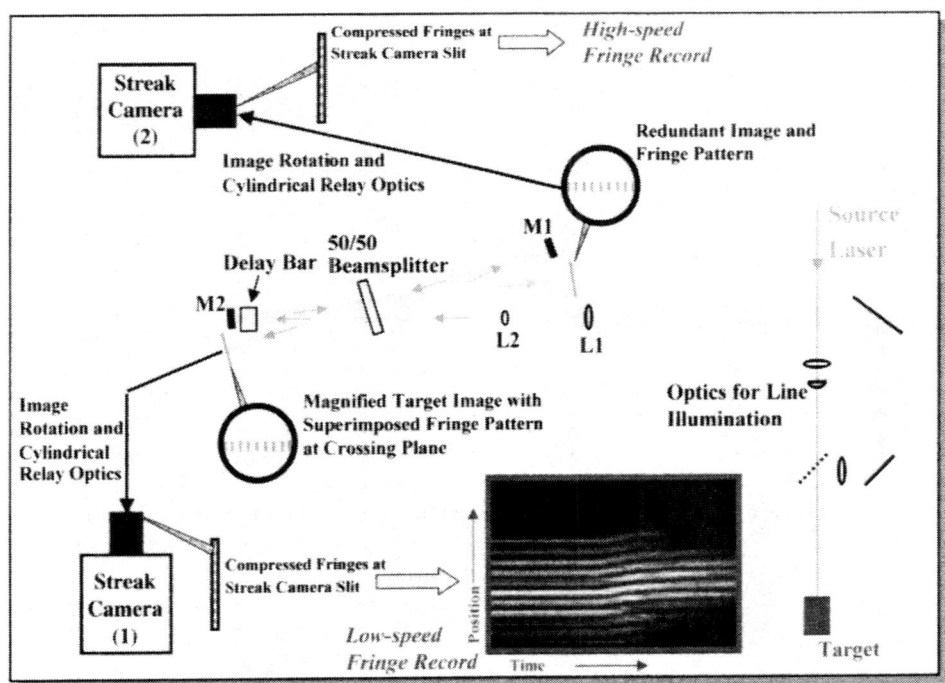

FIGURE 4. Schematic diagram of line-imaging ORVIS with two streak cameras for dual time base recording.

To capture these events, we have employed the system shown in Fig. 4. Fortunately, the ORVIS set-up provides redundant beam recombination legs that carry both magnified image and fringe displacement information. With careful masking of specular reflections from the witness plate window, it is possible to record the flyer acceleration and the post-impact interface velocity on the same image (see below). The streak camera viewing the fringe record on one leg is swept at a relatively slow rate to obtain the full flyer acceleration history, providing an accurate measurement of the impact velocity. The second streak camera is swept at a fast rate to view the impact event at high temporal resolution in order to capture the short-duration particle velocity measurement.

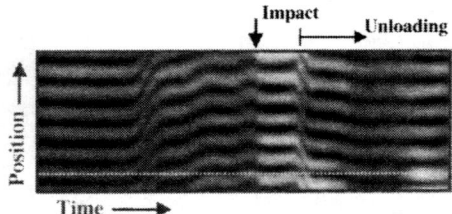

FIGURE 5. ORVIS image record of reverse impact experiment: 6061-T6 Al flyer impacting fused silica.

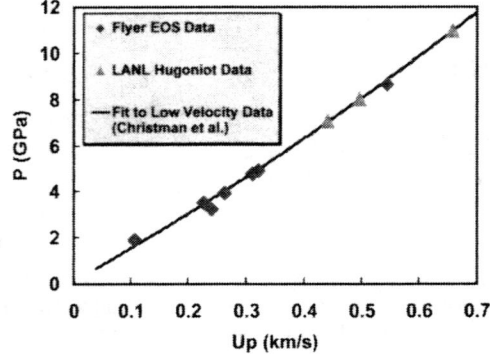

FIGURE 6. Comparison of 6061-T6 Al Hugoniot data obtained in flyer tests with previously reported data.

HUGONIOT MEASUREMENTS

The diagnostic system described above was successfully utilized in the 6061-T6 Al flyer reverse impact experiments. An image record showing the full flyer acceleration history as well as impact and unloading events is shown in Fig. 5. The usual image reduction methods (5) were used to evaluate the flyer velocity at impact as well as the interface velocity (with appropriate refractive index corrections for the interferometer window). The line-imaging ORVIS is very useful for revealing flyer tilt and other 2-D effects. We utilized only "flat" impact regions in determining points for the 6061-T6 Al Hugoniot. The flyer reverse impact results are in excellent agreement with published Hugoniot data (6,7), as shown in Fig. 6. Hydrocode computations indicate that 6061-T6 Al material in the vicinity of the impact surface is only modestly affected by the acceleration process; e.g., this region experiences negligible heating (~8 °C).

Our initial efforts at measuring the short duration shock states for 3.5-μm-thick layers of polycrystalline silicon coupled to 30-μm-thick molybdenum have produced only marginal results thus far; i.e., the transient state was poorly resolved. The hydrocode computations point to one important shortcoming in our target design in that substantial heating of the polysilicon layer likely occurs during acceleration. In future tests, we intend to improve our design by incorporating a thermally insulating layer such as amorphous alumina in the target.

REFERENCES

1. Robbins, D. L., Gehr, R. J., Harper, R. W., Rupp, T. D., Sheffield, S. A., and Stahl, D. B., "Laser-Driven Miniflyer Induced Gold Spall," in *Shock Compression of Condensed Matter--1999*, edited by M. D. Furnish, et al., AIP Conference Proceedings 505, Melville, NY, 2000, pp. 1199-1202.
2. Trott, W. M., Setchell, R. E., and Farnsworth, Jr., A. V., "Investigation of the Effects of Target Material Strength on the Efficiency of Acceleration of Thick Laser-Driven Flyers," in *Shock Compression of Condensed Matter--1999*, edited by M. D. Furnish, et al., AIP Conference Proceedings 505, Melville, NY, 2000, pp. 1203-1206.
3. Garcia, E. J., and Sniegowski, J. J., "Surface Micromachined Microengine," *Sensors and Actuators A* **48**, 203-214 (1995).
4. Farnsworth, Jr., A. V., Setchell, R. E., and Trott, W. M., "A Computational Study of Laser Driven Flyer Plates" (this volume).
5. Trott, W. M., et al., "Dispersive Velocity Measurements in Heterogeneous Materials," Sandia National Laboratories Report, SAND2000-3082, December 2000.
6. Christman, D. R., Isbell, W. M., Babcock, S. G., McMillan, A. R., and Green, S. J., "Final Report Measurements of Dynamic Properties of Materials--Volume III: 6061-T6 Aluminum," General Motors Corporation, MSL-70-23, Vol. III, November 1971.
7. Marsh, S. P., ed., *LASL Shock Hugoniot Data*, U. Calif. Press, Berkeley, CA, 1980, pp. 182-183.

Ultrafast time-resolved 2D spatial interferometry for shock wave characterization in metal films

K.T. Gahagan[1], J. H. Reho[2], D.S. Moore[2], D. J. Funk[2], R.L. Rabie[2]

[1]Corning, Incorporated, Corning, NY 14831
[2]Dynamic Experimentation Division, Los Alamos National Laboratory, Los Alamos, NM 87545

Abstract. We discuss the application of ultrafast time-resolved two-dimensional interferometric microscopy to the measurement of shock wave breakout from thin metal films. This technique allows the construction of a two-dimensional breakout profile for laser generated impulsive shocks with temporal resolution of < 300 fs and out-of-plane spatial resolution of 1.5 nm using 130 fs, 800 nm probe pulses. Constraints placed on the spatial extent of the probe region and on the spatial resolution of the technique by the short duration of the probe pulses will be discussed. In combination with other techniques, such as spectral interferometry, this technique provides a powerful means of investigating shock dynamics in a variety of materials.

INTRODUCTION

The advent of tabletop amplified ultrafast sources has opened the door to new methods of probing the dynamics of surfaces under extreme conditions. In particular, new insights are emerging into the behavior of materials under shock loading conditions on picosecond and femtosecond timescales (1-5).

A new spectroscopic method, frequency-domain interferometry (2-5), has been shown to provide detailed information about the dynamic optical properties and pressure profile of laser-generated pico-shocks propagating in metal thin films. This technique involves imaging the surface onto the entrance slit of an imaging spectrograph and analyzing the resulting spectral content. As such, a line profile through the shock breakout region is obtained. In a typical measurement a time profile of the shock breakout event is built up from many repeated measurements, while precisely adjusting the time delay between the shock generating pulse and the probe pulses. Pointing instability and beam walkoff during the experiment can cause the line profile to intersect the shock profile at different points during the course of the experiment. In addition, the line profile does not provide detailed information about parameters such as shock tilt, or hot spots that a two dimensional image might provide.

To this end, we have developed a complementary technique to provide a two-dimensional image of shock breakout utilizing spatial interferometry with ultrafast pulses. We will describe the technique and its application to the study of shocked aluminium films. A trade-off between spatial resolution, noise and the temporal resolution of the measurement is discussed.

EXPERIMENTAL

Thin aluminium films (250 µm nominal thickness) were vapor plated onto microscope slide cover slips as target samples. A single 800 nm, 130 fs (FWHM), 0.7 mJ laser pulse generated by a seeded, chirped pulse amplified Ti:sapphire laser system (Spectra Physics) was used for both shock-generation and probing. The shock generating pulse (0.2 – 0.5 mJ) was focused onto the front side of the target assembly to a spot size of 75 μm. A small portion of the main pulse (~0.04 mJ) reflected from a beam splitter was passed through an interferometer with the sample in one arm and a variable delay to control temporal overlap in the other. The probe pulse was focused onto the backside of the target at an angle of ~32.6° to a spot size of ~200 μm to circumscribe the region of shock break out. The probe pulse was s-polarized relative to the plane of incidence. An imaging lens and a duplicate in the reference arm were used to image the surface and reference beam onto a CCD camera (Photometrics). The interferogram was stored and processed in real time. The experimental arrangement is shown schematically in Fig. 1.

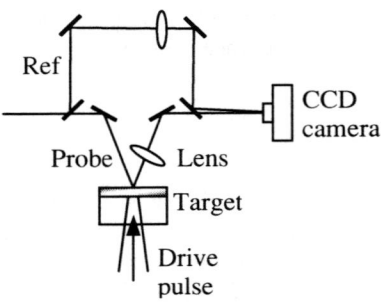

Figure 1. Schematic diagram of the femtosecond laser-driven shock/spatial interferometry experiment on thin film aluminium samples.

The thin film targets were mounted on a computer-controlled x-y translation stage. The target was rastered at 300 μm intervals between experiments so that each laser-driven shock would propagate into undisturbed material.

The 1024 x 1536 camera (9 μm pixel pitch) was arranged with the 1024 height dimension perpendicular to the interference fringes. The interference angle was adjusted to give approximately 3.5 pixels per fringe, corresponding to an angle of ~ 25 mrad. Under these conditions, we found that it was not possible to achieve uniform full depth fringes over the entire 1 cm high array. This results from the fact that the temporal confinement of the pulse reduces the effective overlap of the fields near the edges of the images. This was confirmed by adjusting the time delay between the reference and probe leg. We observe that the region of best contrast would shift from one side of the interferogram to the other as the temporal overlap of the pulses shifted. In order to improve this condition, we narrowed the bandwidth of the seed laser to achieve a ~170 fs (FWHM) pulsewidth. Note that simply adjusting the pulse compressor to vary the pulsewidth did not achieve the desired end. This suggests a fundamental tradeoff between the pulse bandwidth (or minimum pulse duration) and the spatial carrier frequency of the pattern for a given image height.

Using this apparatus a series of interferograms was acquired before and during shock wave breakout. Two-dimensional Fourier analysis was used to separate the carrier frequency contribution to construct a phase and amplitude of the probe pulse. A region surrounding the breakout region was used to perform a two-dimensional background subtraction to remove instrumental artifacts. And finally, to obtain a phase shift associated with surface motion and refractive index dynamics, the central portion of the phase in the shock breakout region was averaged over a 40 μm diameter central area. This result can then be directly compared with phase shifts vs. time measured with spectral interferometry.

RESULTS

A series of phase images corresponding to different time delays is shown in Figure 2. Before breakout, a slight positive phase shift is observed. This phase shift is constant over a long time period (at least 20 ps) prior to breakout suggesting a fast process (electron or x-ray preheating) as an origin. We are currently investigating this effect further. As the shock wave nears the surface, we observe a negative phase shift, first as a ring pattern near the periphery of the shock and then throughout. This negative shift is consistent with spectral interferometry profiles and is believed to be associated with refractive index dynamics. Later in time, the surface motion dominates the phase and a positive shift is observed. It is clear from the later

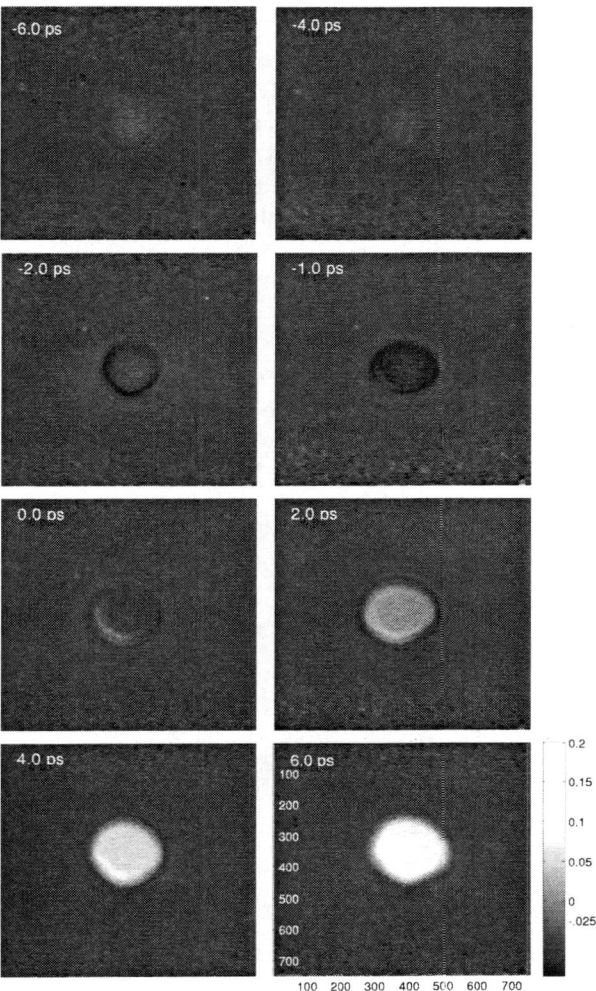

Figure 2. Images of the interferometric phase of a 250 nm Al film during various stages of shock breakout

time images that the shock wave is slightly tilted from top to bottom. An analysis of this tilt reveals that it is on the order of 3-4 nm over 75 μm or 40 μrad.

Figure 3 plots the phase, averaged over the central region of the shock as a function of time. Previous analysis from spectral interferometry has shown that the refractive index change is very nearly linearly proportional to the acceleration of the surface. Using these assumptions, we have plotted a phase due to optical dynamics and a phase due to surface motion on the same graph. Note that the long time behavior, where the surface appears to decelerate gradually, was not used in the fitting of these curves. The noise level of the raw data shows a phase sensitivity of ~3 mrad, or 5 Angstroms of surface displacement. The 10%-90% rise time of the pressure profile is estimated at 3.7 ps with a final free surface velocity of 0.6 nm/ps.

For direct comparison, both the spatial interferometry technique described here and a spectral interferometry measurement were performed on the same sample (not shown). The resulting phase was observed to be identical within the noise limits of the measurements, verifying the consistency of the techniques.

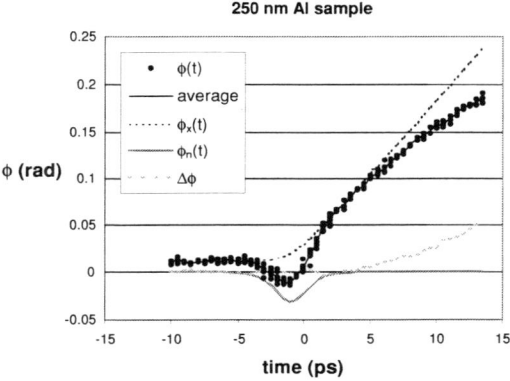

Figure 3. Phase vs. time of shock breakout from a thin Al film. Dots represent raw shot-to-shot data. $\phi_x(t)$ and $\phi_n(t)$ are the contributions to the phase of the surface displacement and refractive index respectively

DISCUSSION

An intriguing feature of the phase profile during breakout is the ring structure of the negative phase shift as the shock wave emerges from the surface. In addition, some profiles show a set of concentric rings. These rings may be related to surface electronic waves created by the shock disturbance at the boundary between shocked and unshocked material. We do not see a consistent growth or motion of these waves with time.

The technique clearly has application to complement spectral interferometry and other techniques for probing picosecond shock waves for the timescales presented here. Probing at shorter timescales, however, will require limiting the spatial extent of the image area or sacrificing spatial resolution.

ACKNOWLEDGEMENT

This research was performed under the auspices of the US Department of Energy.

REFERENCES

1. Tas G., Franken J., Hambir S.A., Dlott D.D., Phys. Rev. Lett. 78, 4585 (1997)
2. Tokunaga E., Terasaki A., and Kobayashi T., Opt. Lett. 17, 1131 (1992)
3. Geindre J.P., Audebert P., Rousse A., Falliès F., Gauthier J.C., Mysyrowicz A., Dos Santos A., Hamoniaux G., and Antonetti A., Opt. Lett. 19, 1997 (1994)
4. Evans R., Badger A.D., Falliès F., Mahdieh M., Hall T.A., Audebert P., Geindre J.-P., Gauthier J.-C., Mysyrowicz A., Grillon G., and Antonetti A., Phys. Rev. Lett. 77, 3359 (1996).
5. Gahagan, K.T., Moore, D.S., Funk, D.J., Rabie, R.L., Buelow, S.J., Nicholson, J.W., Phys. Rev. Lett. 85, 3205 (2000).

A COMPUTATIONAL STUDY OF LASER DRIVEN FLYER PLATES

Archie V. Farnsworth, Jr., Wayne M. Trott, and Robert E. Setchell

Sandia National Laboratories; Albuquerque, NM 87185

Abstract— The existence of both short pulse, high power lasers that can be used drive material flyers to high velocities, and excellent laser diagnostics for determining flyer velocities as a function of time has made the use of such tools an attractive alternative to conventional means for measuring the high pressure properties of materials under shock loading. It provides an especially attractive means for the study of thin materials such as the poly-silicon layers used in the surface micromachine fabrication process. In the current study, our previously developed laser deposition physics addition to the Sandia CTH hydrocode is being used to explore the parameter space of material layering and laser pulse characteristics to develop promising experimental configurations for material property studies. For ordinary materials, thicker foils for acceleration are wanted to provide longer dwell time of the shocked state in a target material. Thermal insulation of the body of the foil from the laser heated accelerating plasma is another desired characteristic for these experiments.

We have performed 1-D studies of material layering with varying thicknesses of the first heated material, Aluminum, and insulating material, Alumina, to find desirable combinations. Other candidate materials exist, of course, and some are currently being considered. Two dimensional studies of flyer acceleration have been performed to guide the selection of laser spacial profiles that may be helpful in providing uniform acceleration across the foil, aiming to provide the maximum attainable flat surface following acceleration.

INTRODUCTION

It has been possible for some time to drive thin flyer plates to km/s velocities using short pulse, high intensity lasers. Likewise, laser velocimeters of various types have been developed, that allow the precision measurement of flyer velocity. These developments invite the investigation of small samples of various materials to determine the high pressure properties of such materials under shock loading, by driving the flyers into target materials of known shock induced behavior. The technique is especially attractive for materials that are inherently thin, that is, materials that are purposely created in very thin layers, for particular applications. The materials used in the development of the micro-machines currently being constructed[1] at Sandia National Laboratories fit in this category. It is known that many materials display different properties when formed in thin layers than are displayed in bulk quantities, so investigation of thin layers for these materials is required.

We are using capabilities previously developed at Sandia, to develop and improve techniques for equation of state (EOS) investigations using laser driven flyers. Coordinated experimental and computational studies are underway, with computations guiding some of the setup parameters for experiments, and experiments both validating the computations and providing areas of increased interest for greater study. The CTH hydrocode as modified by the first author[2] is being used for the computations, and the ORVIS[3] device for experiments. We present here some findings of the computational studies to date.

DISCUSSION

The launching of a laser driven foil is typically done by bringing the laser light through a transparent substrate material onto an absorbing material. The flyer may be of the same material as the first absorbing material, or otherwise. In the case of EOS studies, it is frequently desired to launch a foil that is near room temperature, unheated by the high temperature plasma associated with the launch process. In many of our investigations, we use a thermally insulating material between the launch material and the foil of interest, as is shown in Figure 1. The launch layer material (frequently aluminum in these studies) is followed by a layer of Alumina (or other material) for insulating purposes, and then the foil of interest.

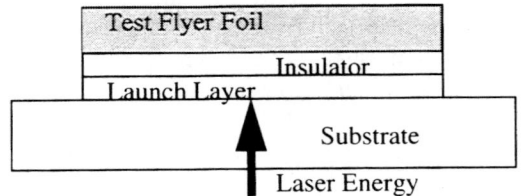

FIGURE 1. Typical layers for EOS experiment.

As we have discussed in a previous presentation[2], the driving plasma is mostly composed of the substrate material, after the initial heating of the launch layer. It is therefore not necessary for the launch layer to be very substantial. Our first optimization study of this arrangement showed one quarter micron of aluminum quite sufficient for the purpose.

The second, insulating layer, however, if it is to prevent heating of the test material flyer, must be considerably thicker. A layer of about 10 microns of alumina seems to be required for the laser pulses studied here. Protecting the foil material is particularly important when studies of very thin materials (a few microns thick, say) are to be performed. The computations show about 3-5 microns of alumina is vaporized by the higher energy pulses studied, while 10 microns prevents significant heating of the flyer. The choice of a thin layer of aluminum backed by an insulator like alumina is also beneficial for maximizing flyer velocity[2], as is seen in Figure 2.

For EOS studies, the observed data is the velocity of the foil surface before and during impact with the transparent target material. This is represented in the

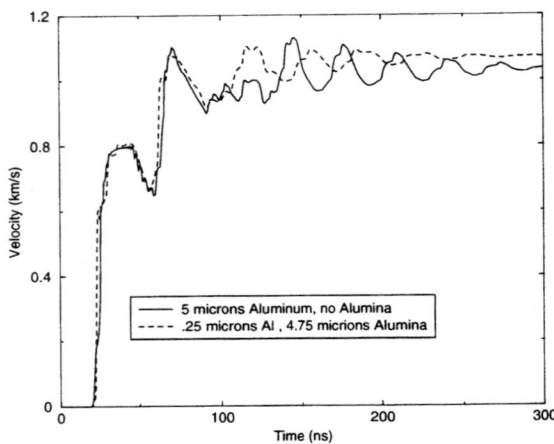

FIGURE 2. Velocity vs. time for acceleration of 100 micron aluminum flyer with and without an insulating layer of alumina.

computations by placing a Lagrangian (material following) tracer point very near the foil surface. When impact occurs, the velocity of the combined materials is a function of the shock properties of the interacting materials. Direct comparison of the computed velocity and the experimental velocity can be made. The experiments are arranged so that the multi-dimensional effects are minimized, so comparison with 1-D computational results is reasonable, but not perfect, since it is not possible to eliminate all multi-dimensional effects in the experiments.

Efforts to compare detailed experimental results with computation are underway. In a recent test for which extensive experimental details were provided, a 1-D computational was performed and the velocity vs. time result is shown and compared with the experimental result in Figure 3. For this experiment, the substrate and target material is fused silica. The accelerated foil material is 92 microns of 6061 T-6 Aluminum, and the launch layer consisted of 5 microns of pure aluminum, vapor deposited on the substrate material. There was no insulating layer in this case, and the foil was heated most of the way through the foil thickness. The driving laser pulse consisted of a 1.6J, 20 ns full width at half maximum pulse of 1.06 micron light.

As illustrated in Figure 3, the magnitudes of the peaks are reasonably well simulated, as is the pulse width. Exact comparisons with timing were not possible since the experiment here lacked detailed correlation between driving laser pulse and the observed signal. The deep valleys between peaks in the computational result are compared with the shallow ones in the experiment. This difference may be the result of two dimensional effects in the experiment, where the spot size observed is about 100 microns in diameter.

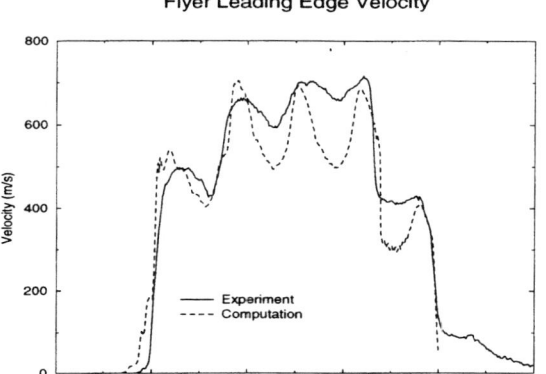

FIGURE 3. Comparison of computed vs. experimentally determined velocity for typical flyer plate experiment. At about 165 ns, impact of the flyer with the target occurs.

We are considering the issues associated with studying the thin layers of silicon frequently used for micro-machine fabrication. One configuration tested computationally consisted of 0.25 microns of Aluminum followed by 0.25 microns of Alumina and a 30 micron layer of molybdenum. The molybdenum has been found experiexperimentallymentally to provide a good support layer for silicon, as deposited in the normal process, consequently it was used here for a supporting material, even though it does not qualify as a good thermal insulator for protecting the silicon foil.

The results of an initial computation is seen in Figure 4

The 30 microns of Moly fails to thermally insulate the silicon in this computation, which is that of a first cut experimental arrangement. However, we infer from these results that a discernible silicon signal

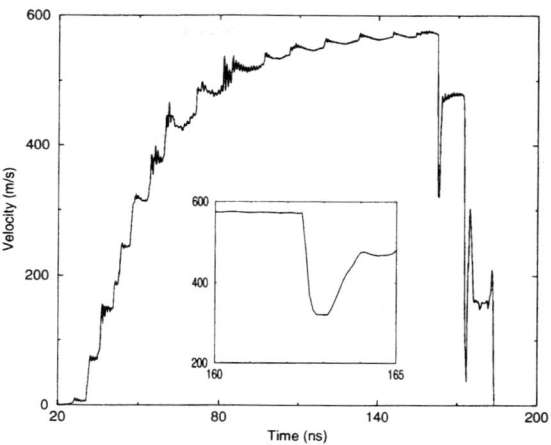

FIGURE 4. Computed velocity vs. time for a silicon flyer on Molybdenum plate. The inset plot shows a magnified time interval containing the silicon impact signal.

may be obtained from the brief dwell time of the silicon impact on a fused silica target. That very brief time interval before the properties of the molybdenum begin to dominate the signal is sufficiently small however, (less than one-half nanosecond) that a significant experimental challenge is indicated in making the measurement. The silicon EOS used for this computation is that of the bulk material, so it is to be expected that the experimental result, when obtained, may differ. Finding that difference is a critical component of this work.

We have performed some 2-D computations, to assess the effects of non-uniformity of the laser profile. It is known that in experiments of this kind, the center of the foil moves slightly ahead of the outer edge. This is in part a result of pressure relief from the foil edge, which then propagates toward the center. We had thought that an extra push at the outer edge, by purposely arranging higher laser intensity there, might be desirable, and is not difficult to obtain from our laser. Two 2-D computations were performed, one with a flat topped profile with a Gaussian decrease of laser intensity near the target edge, and the other with a laser intensity that was volcano like in appearance, flat in the center, but significantly higher at the outer edge (See Figure 5.) The same energy was delivered by each pulse. The comparison is not an ideal one, since more of the energy fell past

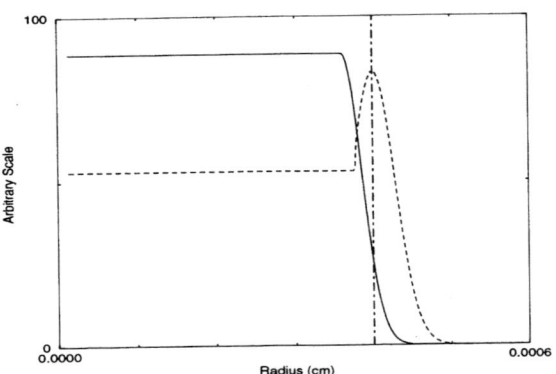

FIGURE 5. Intensity vs. radius for two computations.

the edge of the flyer in the case of the curve peaked at the edge. The results of these computations, with respect to foil flatness, is seen in Figure 6. Surprisingly little effect is seen from the added intensity at the edge, but the effect of the greater intensity at the foil edge is clear in the size of the plasma plume seen there.

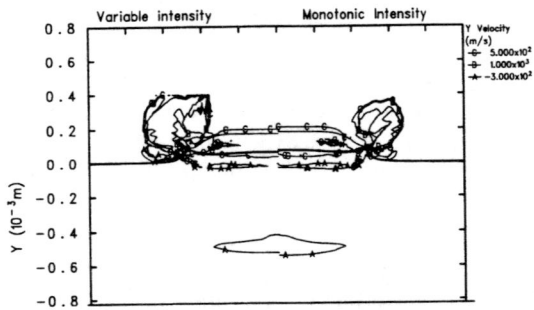

FIGURE 6. Contours of vertical velocity for two computations. The variable intensity on the left (see dashed line of Figure 5), and the monotonic intensity on right side.

FUTURE DIRECTIONS

The tools of this study are being found profitable for exploring the intended parameters. We expect to continue utilizing them to further define materials and layering that will produce favorable experimental results. In particular, we will continue to seek arrangements that will allow good experimental results for very thin materials, and their associated brief signal times.

As more detailed experimental and computational studies are performed, we expect to continue to gain insights into the fundamental processes associated with laser acceleration of thin flyer plates, and therefore be able to better define parameters profitable for equation of state studies.

SUMMARY

The computational and experimental exploration of the parameter space of thin flyers launched by powerful lasers is underway at Sandia National Laboratories. We find that a very thin launch layer is sufficient, since the accelerating plasma consists mostly of substrate material in any case. The presence of a thermally insulating layer seems necessary if thin foils are to be kept near their original state throughout the acceleration process. This layer may be significantly larger than the material being tested. Inherently thin layers may be tested in this way, provided that experimental resolution can be pushed into the sub nanosecond regime, as we expect it can. We also find a rather low sensitivity of the flyer shape at impact time to the details of the laser profile in space, at least near the foil edge.

ACKNOWLEDGMENTS

Sandia is a multiprogram laboratory operated by Sandia Corporation, a Lockheed Martin Company, for the United States Department of Energy under Contract DE-ACO4-94AL85000.

REFERENCES

1. E. J. Garcia and J. J. Sniegowski, "Surface Micromachined Micro Engine", *J. Sensors and Actuators, A*, vol. 48, 203-214, 1995.
2. A. V. Farnsworth, Jr., "Laser Acceleration of Thin Foil", in Shock *Compression of Condensed Matter*, 1991, pp. 1225-1228.
3. D. D. Bloomquist and S. A. Sheffield, *J. Applied Phys.*, Vol 54, 1717-1722, 1983.

MODELLING OF LASER SPALL EXPERIMENTS ON ALUMINIUM

C. M. Robinson

AWE, Aldermaston, Reading, Berks. RG7 4PR. U.K.

Recently a series of shots have been fired on the AWE HELEN 2TW high power laser in order to study the spall of aluminium at high strain rates (1). In the first shot a radiograph was taken which showed a spall layer had formed. Further shots were fired and the free surface velocity of the aluminium was obtained using interferometry. Several of these shots showed that spall had occurred. This paper attempts to model these shots using the extended Johnson spall model (2,3). Previously determined spall parameters (3), which model low strain rate plate impact experiments, are found not to model the spall well, so new spall parameters are determined that match the laser results.

INTRODUCTION

Recently a series of shots have been fired on the AWE HELEN 2TW high power laser in order to study the spall of aluminium at high strain rates ($\sim 10^6$ s^{-1}) (1). The laser irradiates the aluminium, producing a triangular shaped attenuating pressure pulse which travels through the aluminium and is reflected as a tensile wave when it reaches the free surface. If the tension is sufficiently large the aluminium will fail causing the formation of a spall layer. This may be observed in the free surface velocity which will show a sudden jump when the shock wave arrives followed by a gradual fall as the rarefaction wave travels back into the aluminium. If spall occurs the free surface velocity will reach a minimum and then oscillate around a constant value. The oscillation, or ringing, is caused by pressure waves in the spall layer.

This paper considers whether it is possible to model these high strain rate experiments with the extended Johnson spall model (2,3) with spall parameters (3) which model low strain rate ($\sim 10^4$ s^{-1}) plate impact experiments, such as those of Kanel et al (4). The model is found not to model the spall observed in the laser experiments, so a new set of spall parameters are determined which fit these experiments.

EXPERIMENTAL DETAILS

The aluminium sample, 500μm thick, was directly irradiated using a nominal 200ps Gaussian pulse and the laser energy was adjusted for each experiment in order to apply different pressures. In the first experiment an X-ray backlighting technique was used to obtain a radiograph which showed that spall had occurred (3). For the remaining experiments a Michelson interferometer was used to obtain the free surface velocity as a function of time. The lower energy shots showed no signs of spall, spall was observed for the higher energy shots.

CALCULATED PRESSURE PROFILES

A one dimensional radiation-hydrodynamics code including a laser light absorption model was used to calculate the laser induced pressure. Since the nominal laser energy was subject to a large error the laser energy used in the calculation was chosen to match the experimental shock arrival time at the free surface. Absolute timings were not available for all of the shots. However from the shots where absolute timings were available it was observed that the applied pressure was proportional to the maximum free surface velocity. Hence the applied pressure profile for shots where no absolute timings were available were obtained by scaling. An example of a calculated pressure profile 10 μm from the irradiated surface is shown in fig. 1.

THE SPALL MODEL

The spall was accounted for by using the Johnson void growth model (2), with the addition of the failure criterion described by Giles & Maw (3). The model is described briefly here.

The voids in the material are described by a single dependent variable, the distension α, defined as

$$\alpha = \frac{\overline{v}}{v},$$

where v is the specific volume of the material excluding voids and $\overline{v}$ is the mean specific volume of the material including voids. The equation of state of the material is given by

$$\overline{p} = \frac{1}{\alpha} p\left(\frac{\overline{v}}{\alpha}, E\right),$$

where $\overline{p}$ is the mean pressure in the material, E is the internal energy and the function $p(v,E)$ is the equation of state of the material without voids. The distension is described by the following differential equation

$$\dot{\alpha} = -\frac{1}{\eta}(\alpha_0 - 1)^{\frac{2}{3}} \alpha (\alpha - 1)^{\frac{1}{3}} \left(\overline{p} + \frac{a_S}{\alpha} \log\left(\frac{\alpha}{\alpha - 1} \right) \right)$$

when $\overline{p} < -\frac{a_S}{\alpha} \log\left(\frac{\alpha}{\alpha-1} \right)$, $\alpha < \alpha_c$

$\dot{\alpha} = 0$ when $\overline{p} > -\frac{a_S}{\alpha} \log\left(\frac{\alpha}{\alpha-1} \right)$, $\alpha < \alpha_c$

$p\left(\frac{\overline{v}}{\alpha}, E \right) = 0$ when $\alpha > \alpha_c$,

where α_0, a_s, η and α_c are all parameters. The initial distension is α_0. a_s is the spall strength parameter, which determines how large a tension is required before the voids begin to grow. Johnson (2) noted that theoretically a_s should

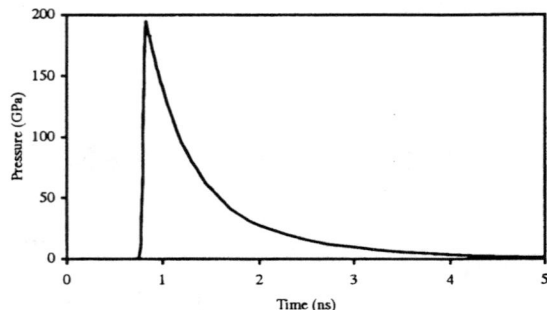

FIGURE 1. Calculated Pressure Profile 10 μm From the Irradiated Surface for a Shot With a Maximum Free Surface Velocity of 1.29 km/s.

equal two thirds of the yield strength, but suggested that it be treated as a free parameter in order to account for localised temperature and strain hardening effects. This approach is taken here. The spall viscosity η determines the rate at which voids grow. The model is extended as suggested by Giles & Maw (3) to include a failure distension α_c. Once the distension reaches this value the voids are assumed to coalesce and the material fails, so for distensions above α_c the material can support no stresses and the pressure is zero.

The material strength is reduced due to the effect of voids as follows

$$\overline{Y} = \begin{cases} \frac{Y}{\alpha}, & \alpha < \alpha_c \\ 0, & \alpha > \alpha_c \end{cases}, \quad \overline{G} = \begin{cases} \frac{G}{\alpha}, & \alpha < \alpha_c \\ 0, & \alpha > \alpha_c \end{cases},$$

where Y and G are respectively the yield strength and shear modulus of the material with no voids and $\overline{Y}$ and $\overline{G}$ are respectively the mean yield strength and mean shear modulus of the material with voids.

SPALL PARAMETERS

Spall parameters for the extended Johnson spall model have been determined for aluminium by Giles & Maw (3), see table 1. These spall parameters predict the low strain rate plate impact spall experiments of Kanel et al (4).

RESULTS

The laser spall experiments have been modelled with a one dimensional Lagrangian hydrocode. The predicted free surface velocity profiles are compared with the interferometer results for two typical shots in figs. 2 and 3. In fig. 2 no spall is observed and the model predicts no void growth and matches the observed pullback. In fig. 3 spall is observed, the calculation predicts the pull back but does not predict the subsequent ringing. At late times the free surface velocity is greater than the calculation made with no spall indicating that void growth has occurred, however the growth is not large enough to cause ringing. This is the case for all shots fired, the model correctly predicts the free surface velocity when there is no spall (the model predicts no void growth), but fails to predict spall and ringing when it occurs (the model predicts void growth that is insufficient to reach the failure distension).

In order to improve the simulation the spall parameters were adjusted to give a better match to the timing and velocity of the first minimum in the free surface velocity profile. The initial distension and spall strength parameter were not changed. Then the tension required for void growth to occur remains unchanged and this ensures that the simulation will still correctly predict the shots where no spall occurred. The spall viscosity and failure distension were adjusted, which allows the voids to grow at a faster rate and allows failure to occur at lower distensions. The new parameters are shown in table 1. The simulation shown in fig. 2 remains unchanged. The simulation with the new spall parameters is shown in fig. 3. It can be seen that the new parameters give an improved spall signature, the first minimum in the velocity is correctly predicted and the period of the ringing is roughly correct. The amplitude of oscillation is too large however.

The rarefaction strain rate, spall strength and

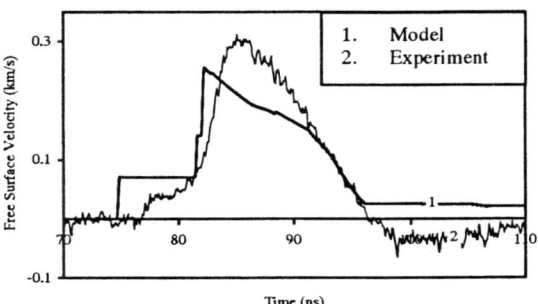

FIGURE 2. Model and Experimental Results. Time Axis of the Experimental Results Shifted to Match the Shock Arrival Times. No Spall is Observed.

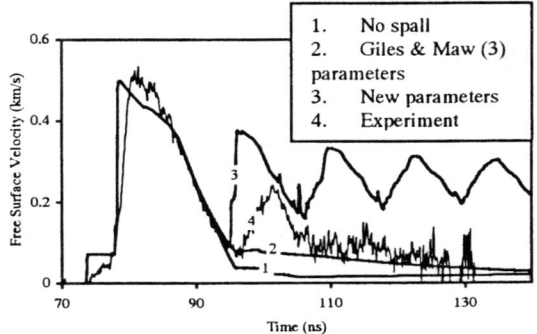

FIGURE 3. Model and Experimental Results. Time Axis of the Experimental Results Shifted to Match the Shock Arrival Times. Spall is Observed.

spall layer thickness are respectively defined as

$$\text{rarefaction strain rate} = \frac{1}{2c_0}\frac{\Delta u}{\Delta t}$$

$$\text{spall strength} = \frac{1}{2}\rho_0 c_0 \Delta u$$

$$\text{spall layer thickness} = \frac{1}{2}c_l T$$

where c_0 is the bulk sound speed (5.41 km/s), ρ_0 is the density (2.70 g/cc) and c_l is the longitudinal sound speed (6.55 km/s). Δu is the difference between the maximum free surface velocity and the free surface velocity of the first minimum, Δt is the difference between the time of the first minimum in the free surface velocity profile and the shock arrival time and T is the

TABLE 1. Spall Parameters for Aluminium.

Parameter	Value determined by Giles & Maw (3)	Value determined in this paper
α_0	1.0003	1.0003
a_s	0.18 GPa	0.18 GPa
η	5×10^{-4} GPa μs	5.4×10^{-4} GPa μs
α_c	1.3	1.0045

period of oscillation of the ringing. The observed strain rates, spall strengths and spall layer thicknesses are compared with the calculated values in table 2. In most cases the calculated rarefaction strain rates are within 10% of the observed values and the calculated spall strengths are within 5% of the observed values. The calculated spall layer thickness are about 20% too small compared to the observed values.

DISCUSSION

The spall parameters of Giles & Maw (3) and the spall parameters derived here have identical initial distensions and spall strength parameters and similar spall viscosities. The only significant difference between the two sets of parameters is the failure distension. The spall parameters of Giles & Maw (3) predict that in the low strain rate experiments there is significant growth of voids which relieves the tension in the aluminium before the failure distension is reached. For the new set of spall parameters the failure distension is very small. Therefore almost no void growth occurs before the material is assumed to have failed, and this failure reduces the tension in the material. This suggests that at high strain rates the spall of the material is not caused by the growth of voids. Possibly the ultimate strength of the material has been reached, as observed by Moshe et al (5)

ACKNOWLEDGMENTS

I would like to thank Peter Graham of AWE for producing the pressure profile calculations.

REFERENCES

1. Evans, A. M., Rothman, S. D., Graham, P., Robinson, C., "Laser Driven Spallation Experiments." in *Shock Compression of Condensed Matter-2001.* submitted.
2. Johnson, J. N., *J. Appl. Phys.* **52 (4)**, 2812-2825 (1981).
3. Giles, A. R., and Maw, J. R., "Modelling the Temperature and Strain Rate Dependence of Spallation in Metals," in *Shock Compression of Condensed Matter-1997.* edited by S. C. Schmidt et al., AIP Conference Proceedings 429, New York, 1998, pp. 243-246.
4. Kanel, G. I., Razorenov, S. V., Utkin, A. V., and Baumung, K., *Shock Wave Profile Data,* Publisher, Scientific Association IVTAN of Russian Academy of Siences, 1996, pp. 68-71.
5. Moshe, E., Eliezer, S., Henis, Z., Werdiger, M., Dekel, E., Horovitz, Y., Maman, S., Goldberg, I. B., Eliezer, D., *Appl. Phys. Letters.* **76 (12)**, 1555-1557 (2000)

TABLE 2. Comparison of Experiment and Simulations for Shots Where Spall was Observed.

Observed Max. Free Surface Velocity (km/s)[a]	Calculated Max. Pressure 10 μm From Irradiated Surface (GPa)	Rarefaction Strain Rate (1/μs)		Spall Strength (GPa)		Spall Thickness (μm)	
		Observed	Calculated	Observed	Calculated	Observed	Calculated
1.50	225.1	4.60	3.71	4.4	3.6	26	27
1.22	183.1	3.28	3.32	3.4	3.4	44	29
1.25	187.6	3.29	3.36	3.4	3.4		28
0.669	100.4	2.70	2.55	3.1	3.1		32
0.541	81.2	2.34	2.31	3.2	3.0		34
0.500[b]	75.0	2.17	2.20	3.1	3.0	52	37
0.602	90.3	2.68	2.43	3.0	3.1	44	35
0.658	98.7	2.85	2.54	3.3	3.1		35
0.654	98.1	2.84	2.52	3.2	3.1	47	34
0.612	91.8	2.22	2.45	2.9	3.1	41	33
0.727	109.1	2.82	2.65	3.1	3.2	42	33

[a] A further six shots with maximum free surface velocities in the range 0.104 km/s-0.299 km/s did not spall.
[b] Observed and calculated free surface velocity shown in fig. 3.

© British Crown Copyright 2001/MOD
Published with the permission of the Controller of Her Britannic Majesty's Stationery Office.

TAKING THIN DIAMONDS TO THEIR LIMIT: COUPLING STATIC-COMPRESSION AND LASER-SHOCK TECHNIQUES TO GENERATE DENSE WATER

Kanani K. M. Lee [1a], L. Robin Benedetti [1b], Andrew Mackinnon [2], Damien Hicks [2], Stephen J. Moon [2], Paul Loubeyre [3], Florent Occelli [3], Agnes Dewaele [3], Gilbert W. Collins [2], and Raymond Jeanloz [1a]

[1a] Department of Earth and Planetary Science and [1b] Department of Physics, University of California, Berkeley, CA 94720-4767
[2] Lawrence Livermore National Laboratory, Livermore, CA 94550
[3] Commissariat à l'énergie atomique, Bruyeres-le-Chatel, France

Abstract. Laser-driven Hugoniot experiments on precompressed samples access thermodynamic conditions unreachable by either static or single-shock compression techniques alone. Recent experiments using Rutherford Appleton Laboratory's Vulcan Laser achieved final pressures up to ~200 GPa and temperatures up to 10,000 K in water samples precompressed to ~1 GPa, thereby validating this new technique. Diamond anvils, used for sample precompression, must be thin in order to avoid rarefaction catch up; but thin diamonds fail under pressure. Anvils no more than 200 μm thick were encased in diamond cells modified to accommodate the laser-beam geometry.

INTRODUCTION

Dynamic experiments generally employ a sample initially at normal density and standard pressure, therefore providing data on the principal Hugoniot. By varying the initial density of the sample through precompression, it is possible to obtain data off the principal Hugoniot, accessing conditions unreachable by either static or single-shock compression. That is, precompression allows an investigation of the entire Pressure-Volume-Temperature (P-V-T) space between the isotherm and Hugoniot.

The method is proven here through experiments on H_2O, which has previously been studied both dynamically and statically to pressures in the multi-Mbar (~10^2 GPa) range. The precompression technique is shown below.

CONSTRAINTS

In laser-driven shock wave experiments, a short, high intensity pulse yields a strong shock. The shock must ideally be steady in order to obtain accurate Hugoniot data. Rutherford Appleton Laboratory's Vulcan Laser generates a $10^{13} - 10^{14}$ W/cm^2, 4 ns long, 400 μm diameter pulse by way of seven beams, six of which come from off-axis. This geometry results in two significant design constraints for precompressed samples. First, the intensity and short pulse length require that the anvil through which the laser shock travels be extremely thin—no more than 200 μm. In order to have a planar shock wave through the sample, the shock wave must propagate through the anvil thickness while the laser is on, thus supporting the wave propagation, else the end-of-pulse rarefaction may catch up. Side rarefaction is also an issue if the laser spot is not large enough to yield a planar front. Second, the laser spot size and beam path require a large aperture radius (hence unsupported anvil radius), r, and aperture angle, θ, in the high-pressure cell itself (Fig. 1).

In order to satisfy the needs for a thin anvil having a large unsupported aperture, we chose the strongest material known, diamond. Use of a diamond anvil cell has numerous advantages, not the least of which are that the basic technology is well established and that the sample can be probed before and during the experiment due to the optical transparency of the anvil. Hence, the thickness, pressure and other characteristics of the pre-compressed sample can be determined at the outset of the Hugoniot experiment.

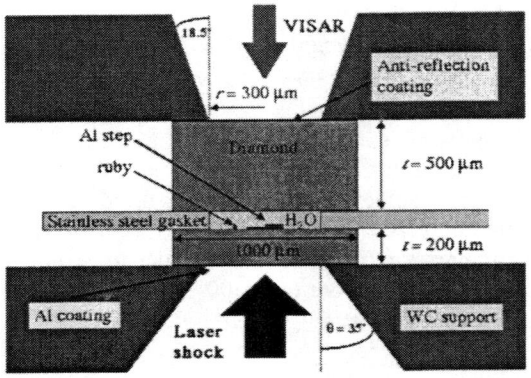

FIGURE 1. Schematic cross-section of diamond-cell configuration used for laser-driven shock experiments on precompressed samples. Wide openings (300 μm radius holes) in tungsten carbide (WC) supports allow ample shock laser entry ($\theta = 35°$ opening) and VISAR access (18.5° opening). Thin diamonds are pushed together to apply pressure on a small sample of water (~30 nL) held in a hole within a stainless steel gasket 100 μm thick. An Al step is glued on the thinnest diamond and used to measure the breakout times (velocities, and ultimately shock pressure) with VISAR (1). A few ruby grains are placed in the sample chamber for precompression pressure measurements via ruby fluorescence (2). There is a 1000 Å Al flash coating on the back side of the thinnest diamond to lower the critical depth of shock ablation, and an anti-reflection coating on the thicker diamond for the VISAR measurement.

A small-radius aperture in the backing for the diamond anvil is problematic because the plasma generated by ablation of the backing plate is a source of high energy x-rays that preheat the sample (3). Also, the blow-off plasma absorbs the laser beam far from the target, thus reducing the shock intensity significantly. In order to avoid preheat, plasma blow-off and laser damage to the DAC, a wide cone angle ($\theta = 35°$) and diamond support hole (300 μm radius) were incorporated into the design of the support plate for the diamond as well as the surrounding diamond cell. This geometry, although accommodating to the laser shock, provides less than optimal support for achieving very high static pressures with thin diamond flats.

Characteristics of Thin Diamond Anvils

The diamond flat can be modeled as a uniformly-loaded circular plate, such that a simple relation exists between the maximum pressure load w and the anvil thickness t

$$w = \frac{k_1 S_m r^2}{t^2} \qquad (1)$$

where S_m is the maximum stress achieved in the diamond (here, the tensile strength of diamond), r is the unsupported radius and k_1 is a constant equal to 0.833 (simply supported disc) or 1.333 (disc with fixed edges) (4). For $r = 300$ μm and $t = 200$ μm and using the value of tensile strength of diamond, 2.8 GPa (5), we arrive at a maximum load of between 1.0 (simply supported) and 1.7 GPa (fixed edges) (Fig. 2).

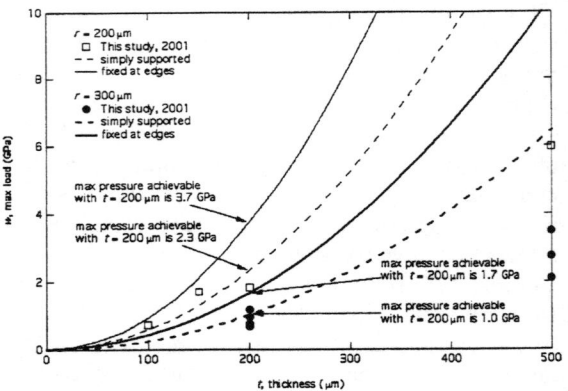

FIGURE 2. Maximum pressure load w achievable assuming a diamond disc of thickness t that is simply supported (dashed line) or fixed at edges (solid line) with an unsupported radius $r = 200$ μm (thin) or 300 μm (thick). Observations of maximum pressures achieved with $r = 200$ μm (open squares) and with $r = 300$ μm (closed circles) are consistent with these expectations.

The thin diamond flats are fragile and flex under pressure (Figs. 3, 4), with the maximum amount of deflection y_m described by,

$$y_m = \frac{k_2 \, w r^4}{E t^3} \quad (2)$$

where E is the elastic modulus (1050 GPa for diamond) and $k_2 = 0.780$ (simply supported) or $k_2 = 0.180$ (fixed edges) (4). Fig. 5 shows good agreement of this simple model with our measurements.

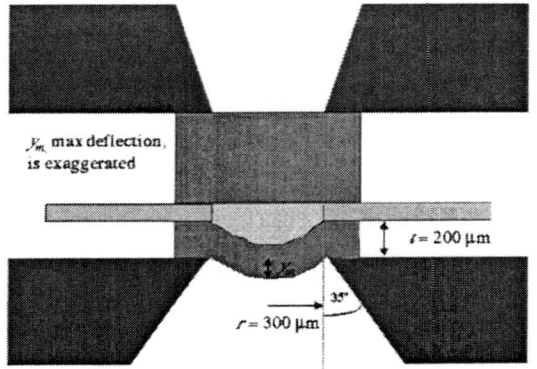

FIGURE 3. Schematic cross-section illustrating flexure of thin diamond flat (highly exaggerated).

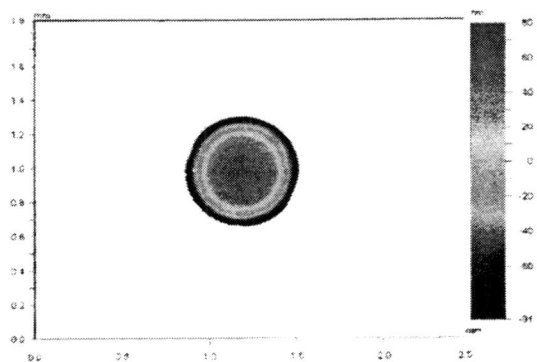

FIGURE 4. Typical white-light interferometric measurement showing the symmetric deflection of a diamond flat under compression: total deflection is 0.170 μm for $t = 200$ μm, $r = 300$ μm at a pressure of 0.4 GPa.

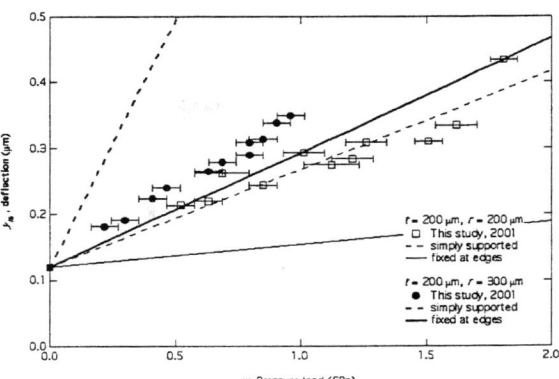

FIGURE 5. Pressure load vs. maximum deflection for $t = 200$ μm, $r = 200$ μm (open squares, thin lines) and $t = 200$ μm, $r = 300$ μm (closed circles, thick lines). Maximum deflection assuming simply-supported (dashed line) and fixed-at-edges (solid line). For the smaller unsupported radius, the measurements agree with the simply-supported theory, whereas for the larger unsupported radius the measurements are nearer to the fixed-at-edges condition.

SAMPLE CHARACTERIZATION

Determining the initial pressure–density–internal energy conditions (P_o, ρ_o, E_o) in the precompressed sample is key for Hugoniot measurements. The precompression pressure P_o was measured in the water via ruby fluorescence (2). Using the equation of state of water by Saul and Wagner (6), the initial density ρ_o and energy E_o were determined.

White-light interferometry was used to determine the product of the index of refraction and distance nd (7). The height of the Al step (known *a priori*) placed into the sample to aid in the VISAR measurements was used as d; determining the index of refraction for the compressed water sample was therefore straightforward (Fig. 6). As the index of refraction of water increases with pressure, it is necessary to obtain an accurate starting value for VISAR measurements, which look through the precompressed water.

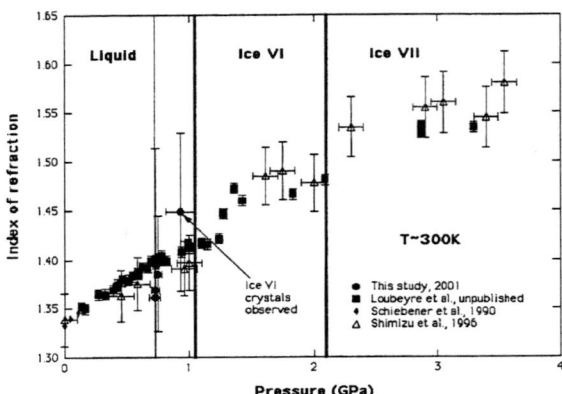

FIGURE 6. Index of refraction as a function of pressure for H_2O water, ice VI and ice VII at room temperature. The values obtained in the present study (closed circles) compare well with previous results of Loubeyre et al., (closed squares) (8); Schiebener et al., (closed diamonds) (9); and Shimizu et al., (open triangles) (10). In one run, ice VI crystals were observed at a pressure less than the nominal freezing pressure, indicating a possibly metastable condition (11).

CONCLUSION

Despite the fragility of the diamond flats, it was possible to precompress water to ~1 GPa (ρ_o ~1.2 g/cm^3), as predicted by the above models. These precompressed samples were laser-shocked up to pressures of ~200 GPa and temperatures to ~10,000 K (see 12, 13). In order to achieve higher initial pressures (and therefore densities), the laser intensity and duration must be increased to allow for thicker diamonds flats. For instance, the maximum load that a 500 μm thick diamond flat could withstand is predicted to lie between 6.5 (simply-supported) and 12.5 GPa (fixed-at-edges).

These results also pave the way for laser-Hugoniot measurements on mixtures, including H_2 and He, at high densities and variable temperatures.

REFERENCES

1. Barker, L. M. and Hollenbach, R. E., *J. Appl. Phys.* **43**, 4669-4675 (1972).
2. Mao, H. K. et al., *J. Appl. Phys.* **49**, 3276-3283 (1978).
3. Moon, S. J. et al., "Computational Design for Laser Produced Shocks in Diamond Anvil Cells," in Shock Compression in Condensed Matter-2001, edited by M. D. Furnish, AIP Conference Proceedings 431, Atlanta, 2002.
4. Timoshenko, S. , *Strength of Materials , Part II: Advanced Theory and Problems*, D. Van Nostrand Company, Princeton, New Jersey, 1956, pp. 76-144.
5. Field, J. E., *The Properties of Diamond*, London Academic, London, 1992.
6. Saul, A. and Wagner, W., *J. Phys. Chem. Ref. Data* **18**, 1537-1564 (1989).
7. Le Toullec, R. et al., *Phys. Rev. B* **40**, 2368-2378 (1989).
8. Loubeyre, P. et al., unpublished (2000).
9. Schiebener, P. et al., *J. Phys. Chem. Ref. Data* **19**, 677-717 (1990).
10. Shimizu, H. et al., *Phys. Rev. B* **53**, 6107-6110 (1996).
11. Tkachev, S. et al., *J. Chem. Phys.* **105**, 3722-3725 (1996).
12. Hicks, D. et al., "Laser Driven Shock Waves in Diamond Anvil Cells," *in preparation*.
13. Henry, E. et al., "Temperature Measurements of Water Under Laser-driven Shock Wave Compression," in Shock Compression in Condensed Matter-2001, edited by M. D. Furnish, AIP Conference Proceedings 431, Atlanta, 2002.

RADIATIVE SHOCK EXPERIMENT USING HIGH POWER LASER

M. Koenig[1], A. Benuzzi-Mounaix[1], N. Grandjouan[1], V. Malka[1], S. Bouquet[2], X. Fleury[2], B. Marchet[2], Ch. Stehlé[3], S. Leygnac[3], C. Michaut[3], J.P. Chièze[4], D. Batani[5], E. Henry[5], T. Hall[6]

[1] *Laboratoire pour l'Utilisation des Lasers Intenses, UMR7605, CNRS – CEA - Université Paris VI - Ecole Polytechnique,, 91128 Palaiseau Cedex, FRANCE*
[2] *CEA DRIF, BP 12, 91680 Bruyères-le-Châtel, FRANCE*
[3] *Département d'Astrophysique Stellaire et Galactique, Observatoire de Paris, 92 Meudon Cedex, FRANCE*
[4] *CEA, Saclay, DSM/DAPNIA/Sap, 91191 Gif-sur-Yvette cedex, France*
[5] *Dipartimento di Fisica 'G. Occhialini', Università di Milano-Bicocca and INFM, Via Emanueli 15, 20126 Milano, Italy*
[6] *University of Essex, Colchester CO4 3SQ, United Kingdom*

Abstract. High power lasers are nowadays tools that can be used to simulate some astrophysical phenomena. In this experiment, the physical parameters have been chosen in order to reproduce radiative shock conditions. The targets were made of a small cell (1 mm^3) filled with Xenon at low pressure (< 1 atm). On the laser side, we had a three layers pusher optimized for reaching conditions where the launched shock in Xe is radiative. Rear side and transverse diagnostics allowed to determine shock and precursor velocities, electronic density along the shock propagation. Comparison with numerical simulations and models are presented. These experiments were performed with the LULI laser.

INTRODUCTION

Radiative shocks are present in advanced star envelopes and contributes to their mass losses. They also exist in pulsed star atmospheres (Miras, RR Lyrae, ...). Laboratory measurement of the evolution and properties of such shocks will provide a better understanding of these stellar objects.

Radiative shocks are present either in dense and hot plasmas where matter is at the local thermal equilibrium (LTE) or in dilute atmospheres where the plasma is at non-LTE. In both cases, a radiative precursor appears before the shock front.

In this experiment, we intend to observe such a feature using the nanosecond laser pulse of the LULI laboratory. Target design is based on semi-analytical model [1] predictions and full radiative hydrodynamic code simulations. We used a Xenon gas cell to observe this radiation phenomenon [2].

RADIATION PHYSICS AND TARGET DESIGN

In a radiative shock, radiation emitted by the hot compressed part of the fluid has a non-negligible effect up or downstream in the flow. This happens only for sufficiently high Mach numbers. Compression ratios are then higher than with classical shocks. When radiative processes are dominant, they modify the structure of the waves. At the shock front, temperature and density discontinuities are smoothed and may disappear under radiation effects.

Recent analytical work [1], adding the effects of radiation energy and pressure to conservation laws for mass, momentum and matter energy derived "generalized" Rankine-Hugoniot relations. In this model, matter is a perfect gas radiating like a black body at local thermal equilibrium (LTE). The radiative properties of the shocks appear when radiation and matter have identical pressures. To reach this condition the shock velocity should exceed a critical velocity defined by:

$$D_{crit} = \left(n\, 7^7 k^4 \left(\frac{3\,c}{4\,\sigma}\right) \left(\frac{1}{6\,\gamma\,\mu}\right)^3 \right)^{1/6} \quad (1)$$

γ is the usual ratio of specific heat for the perfect gas, μ is the atomic mass, k is the Boltzmann constant, σ is the Stefan–Boltzmann constant, c is the speed of light and n is the particle density. For instance, a shock propagating in hydrogen gas at 35 km/s would not have radiative properties if matter density is higher than 10^{14} cm^{-3}. Experimental diagnostics should have access to the structure of shocks in order to demonstrate their radiative properties. This implies to that we choose to drive the shocks in a light and transparent medium (a gas) with an atomic mass μ as high as possible (Xenon).

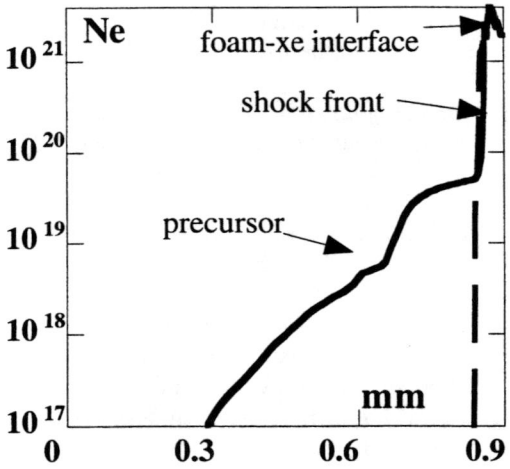

FIGURE 1. Effect of radiation in shocked Xenon given by simulations. Dash and plain lines corresponds to the non radiative and radiative case respectively.

A more quantitative design of the whole experiment has been done with the radiation hydrocode MULTI [3]. A radiative shock wave is produced with less than a hundred joules of pulsed laser light. An optimized three layers pusher drives the shock in Xenon gas initially at rest in a small quartz tube. A detailed comparison with FCI1 (CEA internal code) was performed to validate MULTI as an every day tool to understand and reduce experimental data. The pusher is composed of a polyethylene ablator (2 μm), a titanium X-ray screen (3 μm) and a polyethylene foam accelerator (25 μm). The laser pulse, sent on the ablator side of the pusher, drives a shock in xenon which is at 1/5 of the atmospheric pressure.

The foam-xenon interface that acts as a piston travels with a velocity of 70 km/s. When radiation is included, we observe a long precursor as shown in figure1.

EXPERIMENTAL SET-UP

Experiments have been performed using three of the six available beams of the LULI's Nd-glass laser. The beams were converted at $\lambda = 0.53$ μm, with a maximum total energy $E_{2\omega} \approx 100$ J focused on a same focal spot. The laser pulse was a square pulse with a rise time of 120 ps giving a full width at half maximum (FWHM) of 720 ps. Each beam was focused with a 500 mm lens. We used phase zone plates (PZP) [4], in order to eliminate large scale spatial modulations of intensity and obtained a flat intensity profile in the focal spot [5]. Characteristics of our optical system (lens+PZP) were such that our focal spot had a 500 μm FWHM, with a $\approx$ 250 μm diameter flat region at the center. Spatially averaged intensities between 4-6 10^{13} W/cm^2 were obtained, depending on the laser energy.

The diagnostics used in this experiment are given in Figure2. The self emission diagnostic consisted of a streak camera recording the emitted light from the rear surface of the target at shock breakout.

We had two line VISAR [6] with different sensitivities to infer the shock velocity in the foam and in the Xenon. Finally we implemented a Mach-Zehnder interferometer to determine the electronic density along the shock propagation. Two streak camera were used, one looking at the fringes (LONG), the other one for a transverse image at a given position in the gas (TRANS).

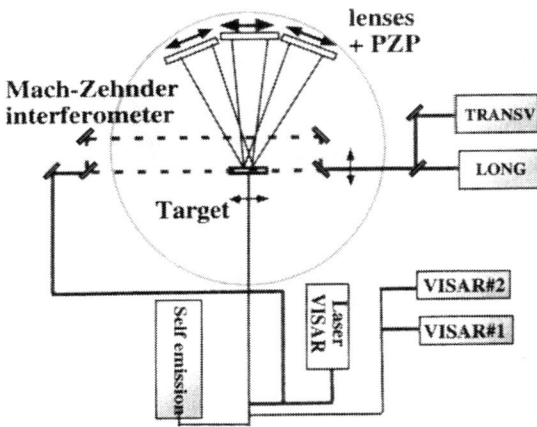

FIGURE 2. Experimental set-up

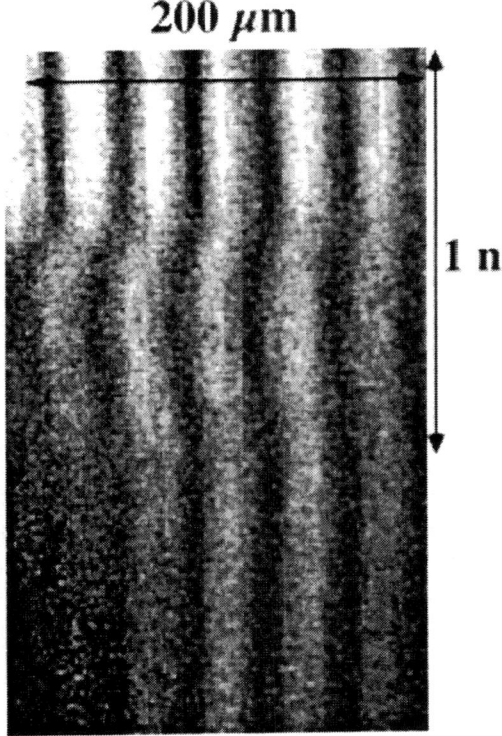

FIGURE 3. VISAR image for a Pa/5 filled gas cell.

With the VISAR, we could determine the piston velocity which drives the shock in the xenon. As we can see in figure 3, the fringes shift to the left side (velocity jump associated with a small deceleration), then to the right (small acceleration) and finally back to the left.

This feature is quite similar to the hydro code prediction of the piston velocity (figure 4.). These three stages correspond to the first shock breaking through at the foam-xenon interface, then a second shock is coming due to reflection on the pusher interface, finally the piston slows down .The mean measured velocity is approximately 67 km/s.

Code computed velocities are in good agreement with these experimental values and suggest (see Fig. 1) that the shock in xenon is strong enough to have a precursor. According to the expected shock temperatures (10-15 eV), the shock front in xenon is subcritical ($N_e \approx 10^{21}$ cm^{-3}) when the precursor zone is much less than critical ($N_e \leq 10^{20}$ cm^{-3}).

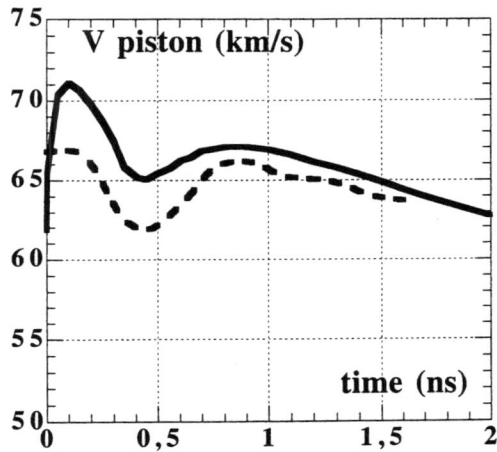

FIGURE 4. Foam-Xenon interface velocity. Plain and dashed lines are 1D simulations and experimental data respectively

Indeed, according to equation (1), the radiative process becomes important when the shock velocity is high enough for a given initial pressure. With our experimental parameters, $D_{crit} \approx 25$ km/s. In the experiment, the measured shock velocity is twice this value so we are confident to be in a regime where radiation effects in the shock are important.

To the precursor, whose position is defined by a change in density (fringe shift), we can associate a much higher velocity than the shock velocity. A fit of the trajectory, represented by the plain curve in figure 5, gives an initial velocity ≈ 140 km/s. It is again very close to the hydro code prediction with a time decrease due to piston velocity slowing down and possible 2D effects. Those effects can be assessed by our transverse diagnostic. Here we are looking at one longitudinal position in the cell ($\approx$ 100-200 mm from the foam interface) with spatial information in the transverse direction.

The fringe shift in figure 5 is directly related to the electronic density N_e per plasma length; in our case one fringe corresponds to $\approx 4.5 \; 10^{21}$ e-/cm^3/µm. Assuming a 200 µm plasma created by the precursor, one can deduce the variation of N_e versus time (figure 6.).

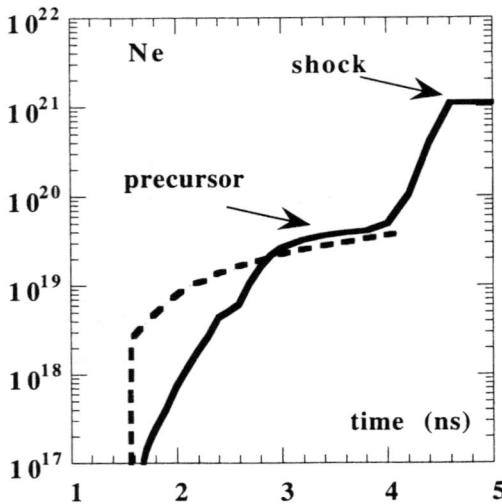

FIGURE 6, Electronic density vs time at a given position inside Xenon (200 µm from the foam interface). Dashed and plain lines are experiment and simulations results respectively.

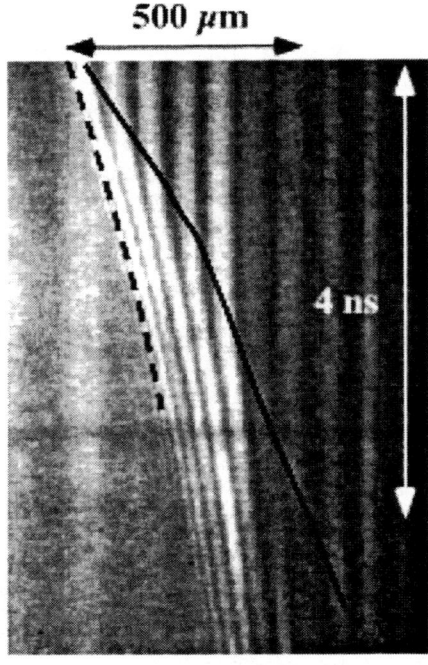

FIGURE 5. Interferometry in Xenon along the laser axis. Dashed line and solid lines correspond to shock front and precursor trajectories respectively

We observe a reasonable agreement between hydrodynamic simulations and the experimental results. However at late time, one can expect 2D effects to become important.

CONCLUSIONS

In this experiment, we observed a radiative precursor preceding a strong shock wave in the xenon gas cell. The results are in general agreement with numerical simulations or semi-analytical models. However the low number of shots cannot give us a detailed picture of the radiation effects in our experimental conditions.

ACKNOWLEDGEMENTS

The authors would like to thank F. Gex (OPM/GASGAL), L. Poles (CEA/VALDUC), B. Cathala (CEA/CESTA) for their fundamental contribution to the target fabrication. Also Ph. Moreau (LULI) has to be mentioned for his contribution to the success of the experiment.

REFERENCES

1 Bouquet, S., et al., *Astroph. J. Supp.* **127**, 245 (2000).
2 Bozier, J. C., et al., *Phys. Rev. Lett.* **57**, 1304 (1986).
3 Ramis, R., et al., *Comp. Phys. Comm.* **49**, 475 (1988).
4 Bett, T. H., et al., *Appl. Opt.* **34**, 4025 (1995).
5 Koenig, M., et al., *Phys. Rev. E* **50**, R3314 (1994).
6 Celliers, P. M., et al., *Applied Phys. Lett.* **73**, 1320 (1998).

1 – 10 MBAR LASER-DRIVEN SHOCKS USING THE JANUS LASER FACILITY

James Dunn[1], Dwight F. Price[1], Stephen J. Moon[1], Robert C. Cauble[1], Paul T. Springer[1], and Andrew Ng[2]

[1] *University of California, Lawrence Livermore National Laboratory, Livermore, CA 94551*
[2] *Dept. Of Physics and Astronomy, University of British Columbia, Vancouver, Canada*

Abstract. We report preliminary results using the Lawrence Livermore National Laboratory (LLNL) Janus laser facility to generate high pressure laser-driven shocks in the 1 – 10 Mbar regime. These experiments address various issues, including shock steadiness, planarity, uniformity and low target pre-heat, important for making precision EOS measurements on a small (E < 250 J) laser facility. A brief description of the experimental techniques, target design and measurements will be given.

INTRODUCTION

The use of high-power laser pulses focused to high intensities is a well-established technique and one way to shock compress materials at the Mbar (100 GPa) level [1–3]. Many direct-drive laser experiments have been performed to characterize the principal Hugoniots of Al [4], Cu [5] and liquid D_2 [6] to compare with equation-of-state (EOS). Indirect-drive experiments using laser beams focused in hohlraum cavities have also been conducted to improve the shock uniformity [5, 7, 8].

Laboratory gas gun experiments have been conducted to benchmark Hugoniot measurements for various materials including Al, Cu, and Ta undergoing shock compression in the 0 – 5 Mbar regime [9]. The gas gun type experiments have a number of important properties that allow precise determination of the shock pressure and EOS properties: (1) the initial conditions before the shock arrives are well characterized; (2) the launched shocks are planar, uniform, and steady; (3) the particle and shock velocities are accurately measured, the former determined by radiography of the impactor in flight. The number of gas gun shots, in common with large laser facilities, tends to be limited as a result of cost and setup time. Further review of gas gun experiments and additional references can be found in ref. [10].

Smaller laser facilities have the advantage of allowing a larger number of shots on target than the larger national facilities e.g. OMEGA and previously on NOVA. These laser shots are at lower shock pressures but more time can be dedicated to studying important details of the experiment, in particular, shock steadiness, planarity, uniformity and low target pre-heat. In this work, we address some of these issues for laser-driven shocks using the LLNL Janus laser facility in an initial set of experiments.

EXPERIMENTAL DESCRIPTION

The shock experiments were conducted on the LLNL Janus laser. This is a two-beam facility with pulse durations available from 100 ps to 7 ns (FWHM). For this work, a single beam of a 5 ns square pulse shape with a ~200 ps risetime was generated by utilizing a fast Pockels Cell switchout from a 7 ns (FWHM) Gaussian laser pulse. The 5 ns square pulse was determined from LASNEX hydrodynamic code simulations to give the desired combination of steady shocks and optimum pressures for the experiment. Output energy up to

220 J, 1064 nm wavelength, was available with 82.5% delivered on target. Laser energy losses were mainly associated with the uncoated debris shield and the Phase Zone Plate (PZP) optic. The laser repetition rate was 1 shot/30 minutes with 8 shots/day being regularly achieved. This shot rate together with the total number of shots available is substantially higher than larger laser facilities and gas gun experiments.

The shock pressure is determined by the maximum intensity of the focused laser beam hence high pressures of 1 – 10 Mbars or higher require intensities of $10^{13} - 10^{14}$ W cm^{-2}. A large focal spot, width w, is essential to launch a planar shock without shock erosion at the edges through a target of thickness z. Generally, $w > 5z$ is desired for good planarity in the center part of the laser focus driving the shock. For this experiment with target thickness z of ~ 100 µm the highest pressure possible was less important than having large, smooth, planar shocks. A ~500µm (FWHM) smoothed focal spot with a Super Gaussian (n=3.4) profile was achieved by using a PZP optic in combination with an f=34.3 cm focal length, f#3 aspherical lens [11], shown in Fig. 1. The PZP optic contains an array of Fresnel lenses which by sampling the laser spatial beam profile at different positions maps any low frequency beam non-uniformities into a uniform high-frequency speckle pattern at the focus [12, 13]. Plasma ablation and shock propagation through the initial few microns of target smoothes the high frequency micron scale speckle leaving a planar uniform shock.

LASNEX simulations were performed to design a target to minimize target pre-heat at the sample before the arrival of the shock. This is shown in Fig. 2. The generic target design consists of a 10 µm C_8H_8 (parylene-N, ρ=1.11g/cm^3) ablator layer coated onto a 25 µm Al (1100 alloy or high purity) pusher

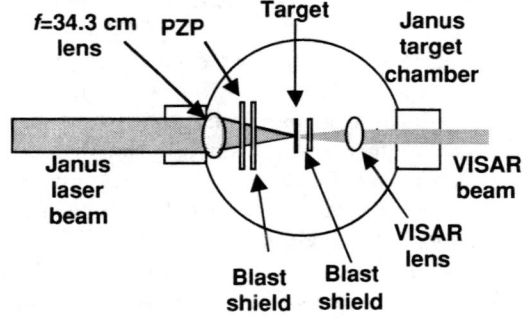

FIGURE 1. Setup showing Janus drive and VISAR probe.

layer and finally the sample to be studied. The low Z CH ablator gives good absorption of the laser energy into launching a shock but also minimizes the coupling of the laser energy into keV x-rays which together with hot electrons from the laser-produced plasma can be a source of sample pre-heat. The simulations indicated that a 25 µm thick Al pusher was sufficient to filter the keV x-rays and keep pre-heat at the pusher/sample interface to a minimum.

The main shock diagnostic was a Velocity Interferometer for Any Reflector (VISAR) probe beam that monitored the back surface of the target [14, 15]. The velocity of the interface between the sample or the free surface can be measured and thus the particle velocity determined. A caveat is that the surface has to remain reflective during the motion. A powerful application of velocity interferometry is the ability to measure shock propagation inside a transparent medium if the shock pressure is high enough to metallize the material [6, 15]. The VISAR instrumentation used here was similar to that of Celliers et al [15]. A 14 ns (FWHM), 532 nm

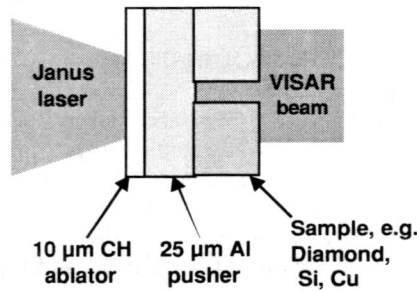

FIGURE 2. Target design showing CH ablator, Al pusher and sample to be shocked.

wavelength beam from a frequency-doubled Q-switched Nd:yttrium-aluminum-garnet (YAG) laser was synchronized to the main Janus laser drive to probe the target back surface during and after the launch of the shock wave. A variety of etalon lengths were used from 0.125 – 8 cm giving a velocity sensitivity of 38 – 0.6 µm/ns/fringe, respectively. The back surface was imaged with 10× magnification using an f=15 cm achromat and imaged onto the entrance slit of an S-20 photocathode Kentech or Hamamatsu streak camera. A time-resolved 1-dimensional line image of the shock breakout at the back of the target was obtained.

A custom target wheel consisting of a 5 cm diameter disk with 12 target positions was designed.

Each target sat in a 1 cm diameter disk with a 0.3 cm aperture on center to allow the laser drive beam onto the ablator side of the target. Eleven targets were mounted radially on the target wheel in machined recesses with a cross-wire alignment fiducial on the 12th position. The target alignment and positioning could be performed offline. After each laser shot, an encoded stepper-motor rotation stage allowed the next target to be aligned giving a fast time between shots. An anti-reflection (AR) coated fused silica blast shield placed behind the target prevented target debris from coating the VISAR imaging lens.

RESULTS AND DISCUSSION

A simple target using 10 µm CH ablator vapor-deposited on various Al pusher foil thicknesses of 25 – 100 µm was constructed to determine shock planarity. The critical element in achieving smooth, planar shocks was the focusing of the main optic (f#3 asphere) and the Phase Zone Plate combination on the front of the target. The VISAR imaging achromat was used to focus the probe beam on the back surface of the target using a CCD imaging system built into the VISAR design. Then the same optic was used to focus the main Janus drive laser. Optimum (smooth) beam quality is achieved away from best focus [12]. Two sets of experiments were performed where the drive laser was focused in either the converging or diverging beam. The second diverging beam experiments gave more consistent, repeatable results for planar shocks.

A planar shock launched through a thin ~25 µm Al foil was fairly straightforward as a result of the fast shock transit time (~2 ns) and large w/z ratio (~14). A thicker Al foil is more demanding to maintain shock planarity as a result of progressive erosion at the edges of the shock as it propagates through the material. Figure 3 shows the shock breakout for a 10 µm CH/100 µm Al foil driven at a laser intensity of 1.2×10^{13} W cm^{-2} generating a shock pressure in the 1 – 1.5 Mbar range. At the left side, the start of the Janus 5 ns square laser drive is indicated at 0 ns and the shock breakout of the Al foil is 8.7 ns later. At shock breakout the VISAR signal disappears simultaneously in a uniform and planar manner across about 400 µm of the foil. The VISAR probe is still incident on the foil, as indicated by the intensity marker on the right. The loss of the VISAR signal may be due to a drop in the Al

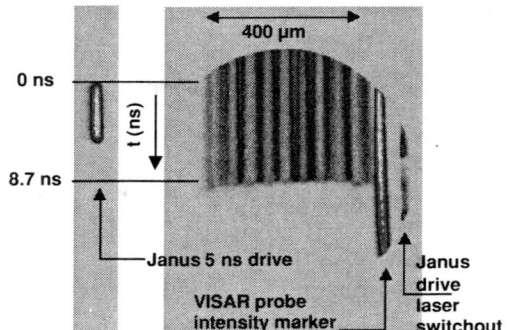

FIGURE 3. Streaked image showing planar shock breakout of 10 µm CH/100 µm Al foil at t = 8.7 ns.

reflectivity from shock melting or absorption in a dense vapor layer close to the foil surface. Melting on the shock Hugoniot is predicted to begin at 1.2 Mbars and end at about 1.55 Mbars [16].

Al foil targets of different thickness, with and without CH ablator, were irradiated at 1.2×10^{13} W cm^{-2} to determine the effectiveness of a low Z ablator for reducing pre-heat. A 5.5 cm etalon was used in the VISAR to give a velocity sensitivity of 0.92 µm/ns/fringe. Motion of the back surface of the foil before shock breakout would indicate target pre-heat. Figure 4(a) shows that with the 10 µm CH ablator on a 25 µm foil no pre-heat could be detected before shock breakout at 4.05 ns. The laser pulse starts at 2 ns. This gives an upper limit on the pre-heat of $\Delta T \sim 0.01$ eV. Figure 4(b) for a thinner 12.5 µm Al foil with no CH ablator clearly shows fringe motion starting 0.48 ns before breakout and reaching

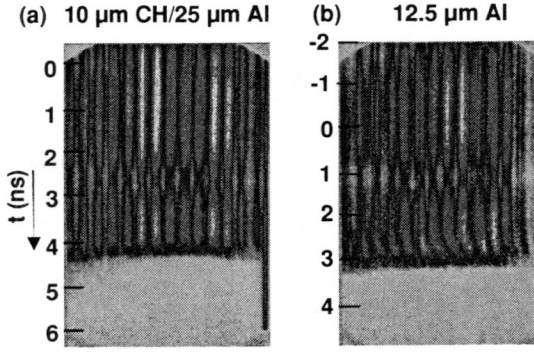

FIGURE 4. (a) Streaked image of 10 µm CH /25 µm Al shows no back surface motion before breakout at 4.05 ns. (b) 12.5 µm Al with no ablator showing back surface movement before breakout at 3.0 ns. In each case the laser pulse starts at 2 ns. Note that the change in the fringe reflectivity at 2.5 ns on Fig. 4(a) and at 1.0 ns on Fig. 4(b) is due to an artifact of the streak camera.

a maximum velocity of 0.73 µm/ns at 3.0 ns. Estimates of the pre-heat level are 0.05 – 0.1 eV. Further analysis and modeling of these results are under way but the effect of the ablator in reducing pre-heat is demonstrated.

Shock velocity and steadiness were determined by measuring the time of shock breakout from a stepped Al target of maximum step thickness 125 µm with a 10 µm CH ablator. Figure 5 shows that for 10^{13} W cm^{-2}, u_s = 10.65 µm/ns and the velocity was constant to within 2.5%. This corresponds to a shock pressure of 1.12 Mbar (based on the u_s vs u_p relations from ref. [9]). A maximum pressure of 1.6 Mbar was achieved in Al in this experiment though higher pressure ~4 Mbar was achieved when higher density samples e.g. diamond were placed behind the Al pusher layer.

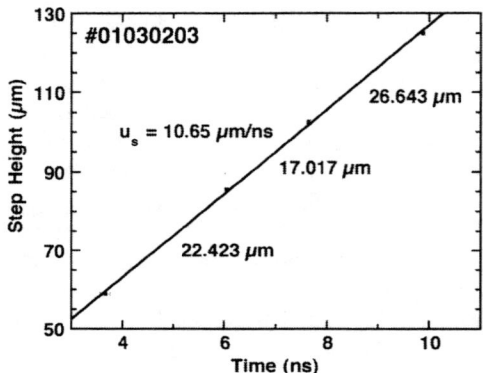

FIGURE 5. Shock breakout for Al step height versus breakout time.

CONCLUSIONS

We have conducted a few Megabar pressure experiments on the LLNL Janus facility at intensities ~1 – 2 × 10^{13} W cm^{-2}. The target design has been optimized under these conditions to give steady, planar shocks with very low pre-heat. More detailed analysis and comparisons with simulations will be reported at a later date.

ACKNOWLEDGMENTS

The authors would like to thank Jim Emig, Derek Decker, and Al Ellis for their technical contributions to instrumentation and target fabrication in this research. This work was performed under the auspices of the US Department of Energy by the University of California Lawrence Livermore National Laboratory under Contract No. W-7405-Eng-48.

REFERENCES

1. Veeser, L., and Solem, J., *Phys. Rev. Lett.* **40**, 1391 – 1394 (1978).
2. Trainor, R. J., Shaner, J. W., Auerbach, J. M., and Holmes, N. C., *Phys. Rev. Lett.* **42**, 1154 – 1157 (1979).
3. Cottet, F., Romain, J. P., Fabbro, R., and Faral, B., *Phys. Rev. Lett.* **52**, 1884 – 1887 (1984).
4. Ng, A., Parfeniuk, D., and Da Silva, L., *Phys. Rev. Lett.* **54**, 2604 – 2607 (1985).
5. Benuzzi, A., Löwer, T., Koenig, M., Faral, B., Batani, D., Beretta, D., Danson, C., and Pepler, D., *Phys. Rev. E.* **54**, 2162 – 2165 (1996).
6. Da Silva, L. B., Celliers, P., Collins, G. W., Budil, K. S., Holmes, N. C., Barbee Jr., T.W., Hammel, B. A., Kilkenny, J. D., Wallace, R. J., Ross, M., Cauble, R., Ng, A., and Chiu, G., *Phys. Rev. Lett.* **78**, 484 – 487 (1997).
7. Löwer, T., Sigel, R., Eidmann, K., Földes, I. B., Hüller, S., Massen, J., Tsakiris, G. D., Witkowski, S., Preuss, W., Nishimura, H., Shigara, H., Kato, Y., Nakai, S., and Endo, T., *Phys. Rev. Lett.* **72**, 3186 – 3189 (1994).
8. Evans, A. M., Freeman, N. J., Graham, P., Horsfield, C. J., Rothman, S. D., Thomas, B. R., and Tyrrell, A.J., *Laser and Particle Beams*, **14**, 113 – 123 (1996).
9. Mitchell, A. C., and Nellis, W. J., *J. Appl. Phys.* **52**, 3363 – 3374 (1981).
10. Nellis, W. J., *Encyclopedia. Appl. Phys.* **18**, 541 – 554 (1997).
11. Dunn, J., unpublished data (2001).
12. Stevenson, R. M., Norman, M. J., Bett, T. H., Pepler, D. A., Danson, C. N., Ross, I. N., *Opt. Lett.* **19**, 363 (1994).
13. Koenig, M., Faral, B., Boudenne, J.-M., Batani, D., Benuzzi, A., and Bossi, S., *Phys. Rev. E.* **50**, R3314 – 3317 (1994).
14. Barker, L.M., and Hollenbach, R.E., *J. Appl. Phys.* **43**, 4669 (1972).
15. Celliers, P. M., Collins, G. W., Da Silva, L. B., Gold, D. M., and Cauble, R., *Appl. Phys. Lett.* **73**, 1320 – 1322 (1998).
16. Moriarty, J. A., Young, D.A., and Ross, M., *Phys. Rev. B* **30**, 578 – 588 (1984).

TRANSITION FROM EXPANSION TO SHOCK COMPRESSION IN LASER IRRADIATED Si BY MULTIPLE SHOTS

Akio Yazaki, Hiroaki Kishimura, Yoichiro Hironaka, Fumikazu Saito, Kazutaka G. Nakamura, and Ken-ichi Kondo

Materials and Structures Laboratory, Tokyo Institute of Technology, Nagatsuta, Midori, Yokohama, 226-8503, Japan

Abstract. Picosecond time-resolved X-ray diffraction measurements have been performed on laser irradiated Si single crystal. Two types of experiments were performed in air. In one set of experiments, single shot irradiation on the silicon (111) crystal was performed at an irradiance range of 1 – 10 GW/cm^2. The results of X-ray diffraction measurement showed both thermal expansion and transient elastic shock compression. It is deduced that the ablation threshold falls in the range in irradiance of 1 – 10 GW/cm^2. The second set of experiments was performed with multiple shot irradiation in the range in irradiance of GW/cm^2. Lattice compression due to the multiple laser irradiation was observed. This result indicates that multiple irradiation causes reduction of the ablation threshold.

INTRODUCTION

In recent years, advances in the development of picosecond sources of hard X-rays have opened the possibility of the direct observation of the changes in material structure in the nanosecond to picosecond time scale. Time resolved X-ray diffraction studies have been reported in recent years [1-3]. Laser heating was studied and lattice expansion was reported by single laser irradiation on a Si (111) at a few GW/cm^2.[4-6] On the other hand, Hironaka et al.[7] studied multiple laser shots on Si(111) at the same power density (4 GW/cm^2) and observed lattce compression up to −1.05%. In the present paper, we investigated transformation from expansion to compression of Si(111) by laser irradiation using picosecond time-resolved X-ray diffraction. Various numbers of multiple laser irradiations and irradiation powers were examined. The mechanism of shock generation was also discussed.

EXPERIMENTAL

Time resolved X-ray diffraction with ultra-short X-ray pulses (1.93 Å, 10 ps, 10 Hz) is used to probe structural dynamics in Si pumped by laser pulses (780 nm, 300ps). Details of the experimental setup are shown elsewhere [8]. The 300 ps pulsed beam is divided into two beams by a beam splitter. One is used for sample excitation after passing through an optical delay line, and another is compressed to about 50 fs and irradiated on an iron target at 10^{17} W/cm^2 for X-ray generation. The pulse width of the X-rays was measured to be about 10 ps. The Fe Kα line was used for the diffraction experiment. A 300 ps laser pulse was focused to a spot size of 1 $\times$4, 2$\times$5 and 1$\times$3 mm on a target in air with an energy of 60 - 100 mJ/pulse. The axis of irradiation was normal to the surface. Two types of the experiments were performed under varying conditions of laser irradiations. In one set of experiments, single shot irradiations, varying power density, were performed. The other set of experiments was performed with multiple laser shots. The samples used were n-type and non-doped Si (111) wafers with the thickness of 860 and 525 μm, respectively. The X-ray pulse was aimed at the center of the laser focal spot on the Si (111) and the diffracted X-rays were detected with an LN$_2$-cooled CCD area image sensor (512$\times$512 channels of 25 μm $\times$ 25 μm size). Data are accumulated for 600 shots. We define the time zero at the arrival of the peak of pump pulse and X-ray producing pulse at the position of sample.

RESULTS AND DISCUSSION

Single shot experiments

Results of single laser shot irradiation experiments are shown in Fig 1. Figure 1. (a) shows scattering spectrum on the condition for laser irradiation with power density of 5 GW/cm^2 at time delay 250ps. The X-ray diffraction peak was shifted to lower Bragg angles relative to the spectrum from the unirradiated crystal. Figure 1. (b) shows the result of X-ray diffraction under irradiation with power density of 10 GW/cm^2. The new signal appeared at a higher diffracted angle. The strain profile inside the crystal can be estimated from the X-ray diffraction data. (a) In the case of irradiance of 5 GW/cm^2, apparent diffraction signal at a lower angle corresponds to lattice expansion, (b) while new signals at higher angles observed in irradiance of 10 GW/cm^2 provides evidence that lattice compression had occurred. X-ray diffraction was analyzed by a code based on dynamical diffraction theory and obtained strain distributions are shown in

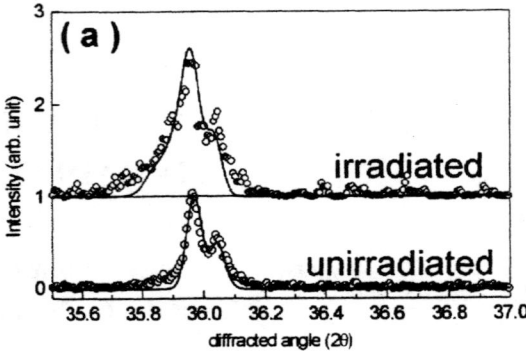

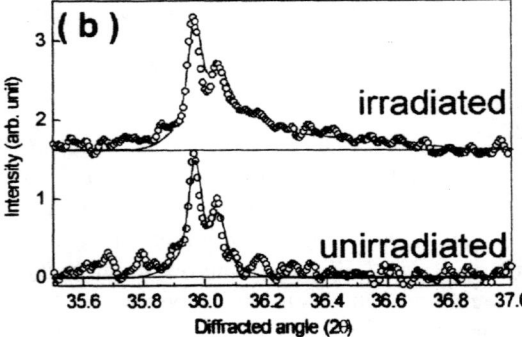

FIGURE 1. Scattering spectra for the laser heated Silicon under single laser shot irradiation. The conditions are (a) power density of 5 GW/cm^2 and the time delay of 250ps. (b) 10 GW/cm^2, 350 ps. The circles are the experimental results, and the solid lines are calculated X-ray diffraction profiles.

Fig 2. Figure 2. (a) indicates that thermal expansion due to laser heating occurred at this range in irradiance and delay time, and the maximum change in the strain and the surface temperature were estimated to be ~0.5% and 1500 K, respectively. It was also suggested that the absorption coefficient of silicon is increasing to 1.2 $\times 10^4$ cm^{-1}, ten times as large as that to 780 nm light at 300 K. Similar laser-induced lattice expansion has been observed in Si. Larson et al.[4] studied time resolved X-ray diffraction of 15 ns laser irradiated Si at 0.1 GW/cm^2 to find the lattice temperature in Si and observed thermal expansion due to the heating to reach the melting point.

On the other hand, at a power density of 10 GW/cm^2, the X-ray diffraction shows lattice

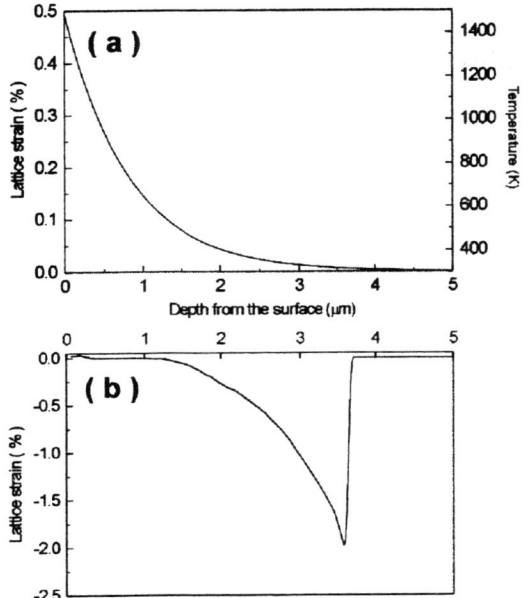

FIGURE 2. Strain Distributions under the surface of single shot irradiated Silicon. The conditions are (a) with power density of 5 GW/cm^2 and the time delay 250ps. (b) 10 GW/cm^2, 350 ps.

compression. The maximum compression of 2.1% was estimated from the calculation. To account for the phenomena observed in the present experiments, the absorption coefficient was estimated using the values of laser energy density deposited in the sample, enthalpy changes, and latent heat of silicon. The observed shock compression in this experiment is due to laser ablation. The absorption depth of 780 nm light was estimated to be several hundreds of nanometers (~ 0.4 µm). The ablation occurred in the thin surface layer (several hundreds of nanometer) at the power density of 10 GW/cm^2 (2 J/cm^2 for 300 ps pulsed laser). Pronko et al.[9] reported that the pulsed laser (780nm, 300ps) induced dielectric breakdown threshold of silicon was ~3 J/cm^2, which provides further evidence of laser ablation occurring in this range of laser irradiation.

Multiple shots experiments

Figure 3. shows the X-ray diffracted spectrum from the silicon (111) surface under multiple laser irradiations (300 shots) with a power density of 4 GW/cm^2. Multiple shot experiments were performed for several sets of shots (20 - 300 shots) with the same power density at delay time 350 ps. In all experiments, the shifts of the diffracted spectrum to higher Bragg angle due to laser shocked compression were observed and strains were estimated to be ~ −1.4 %. The generation of shock compression was observed even at 20-shots as reduction of ablation threshold was induced by the surface damage in laser irradiated area.

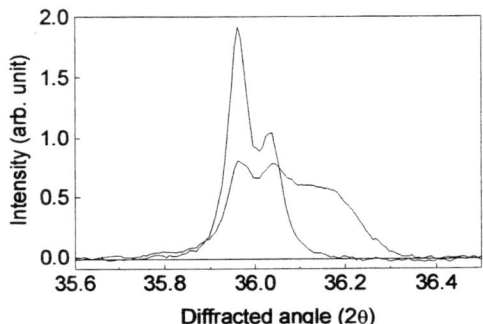

FIGURE 3. Typical X-ray diffraction spectrum on the Si (111) under multiple laser irradiation (300 shots) with power density of 4 GW/cm^2. The time delay of the X-ray probe was 350ps.

ACKNOWLEDGEMENT

We thank Y. Okano, H. Kawano, and M. Hasegawa for their help on experiments and valuable discussions. This work has been supported by CREST (Core Research for Evolutional Science and Technology) program organized by Japan Science and Technology Corporation (JST).

REFERENCES

1) Chin, A. H., Schoenlein, R. W., Glover, T. E., Balling, P., Leemans, W. P., and Shank, C. V., *Phys. Rev. Lett.* **83**, 336-339 (1999).
2) Siders, C. W., Cavalleri, A., Sokorowski-Tinten, K., Toth, Cs., Guo, T., Kammler, M., Horn von Hoegen,

M., Willson, K.R., von der Linde, D., and Barty, C. P. J., *Science* **286**, 1340-1342 (1999).

3) Chen, P., Tomov, I. V., and Rentzepis, P. M., *J. Phys. Chem.* **103**, 2359-2363 (1999).

4) Larson, B. C., White, C. W., Noggle, T. S., Barhorst, J. F., and Mills, D. M., *Appl. Phys. Lett* **42**, 282-284 (1983).

5) Wark, J. S., Whitlock, R. R., Hauer, A. A., Swain, J. E., and Solone, P. J., *Phys. Rev. B.* **40**, 5705-5714 (1989).

6) Lunney, J. G., Dobson, P. J., Hares, J. D., Tabatabaei S.D., and Eason, R. W., *Opt. Comm.* **58**, 269-272 (1986).

7) Hironaka, Y., Yazaki, A., Saito, F., Nakamura, K. G., Kondo, K., Takenaka, H., and Yoshida, M., *Appl. Phys. Lett.* **77**, 1967-1969 (2000).

8) Hironaka, Y., Yazaki, A., Saito, F., Nakamura, K. G., and Kondo, K., *Jpn .J. Appl. Phys.* **39**, 984-986 (2000).

9) Pronko, P. P., VanRompay, P. A., Horvath, C., Loesei, F., Juhasz, T., Liu, X., and Mourou, G., *Phys. Rev. B.* **58**, 2387-2390 (1998).

CHAPTER XXII

EQUATION OF STATE AND GEOPHYSICS

EVIDENCE FOR KINETIC EFFECTS ON SHOCK WAVE PROPAGATION IN TECTOSILICATES

Paul S. DeCarli [1,2], Emma Bowden [2], Thomas G. Sharp [3], Adrian P. Jones [2], and G. David Price [2]

[1] SRI International, Menlo Park, CA 94025, USA, [2] Dept of Geological Sciences, University College London, Gower Street, London WC1E 6BT, [3] Department of Geology, Arizona State University, Tempe, AZ 85287, USA

Abstract: The question of whether phase transition kinetics can affect shock wave propagation has been around for about 50 years. Some workers have speculated that shock compression is fundamentally different from static compression; others cite evidence that static and dynamic transitions follow the same rules. Metastable high-pressure phases that are found in large (long-duration shock) impact craters constrain the post-shock temperature histories of the neighboring rock. The post-shock temperature is a function of the area enclosed by loading and unloading paths. We use petrographic evidence to constrain the unloading paths. The presence of metastable phases then serves to constrain possible loading paths. The limited range of possible loading paths indicates that there must be a large kinetic effect on shock wave propagation in tectosilicates.

INTRODUCTION

It has long been accepted that Hugoniot discontinuities can be interpreted in terms of dynamic phase transitions. To account for the rapidity of these transitions, some workers have speculated that shock wave compression was fundamentally different from static compression. (1) Thus, one could accept the possibility that the reconstructive phase transitions of silicates, sluggish under static high-pressure conditions, could occur rapidly under shock compression.

Jeanloz presented evidence in support of a contrary view, that the Hugoniot data on olivines, pyroxenes and quartz did not represent evidence for extraordinarily rapid reconstructive phase transitions. (2) Based on shock recovery studies, he inferred that a mineral could deform to a volume nearly equal to that of a high-pressure phase without undergoing a reconstructive phase transition.

As a working hypothesis, we assume that phase transitions under shock loading are governed by the same thermodynamic and kinetic factors as phase transitions under static loading. We test this hypothesis by looking for experimental evidence of kinetic effects. We have found indirect evidence of kinetic effects on Hugoniot measurements on rocks and minerals. Kinetic effects are also inferred from the results of shock recovery experiments and from studies of natural impact craters.

LABORATORY STUDIES

We note that in laboratory shock experiments, practical considerations limit the duration of peak shock pressure to a few microseconds. However, it is possible to study kinetic effects via variation of the initial temperature of a sample. Two independent sets of shock recovery experiments on preheated quartz showed that the conversion to diaplectic glass (formed by an inferred solid-solid transformation) increased with increasing preheat temperature. (3,4) Subsequent work confirmed the quartz results and showed a similar initial temperature effect on the conversion of feldspar to diaplectic glass.

Note that the conventional wisdom, among those who have compared Hugoniot and release adiabat measurements with the results of shock recovery experiments, is that diaplectic glass represents material that transformed at pressure to a dense phase and expanded to glass on release of pressure. In the case of quartz, this dense phase is presumed to be stishovite-like, with 6:3 O:Si coordination, but not necessarily having long-range crystalline order. Observed steep release adiabats indicate that the dense material does not begin to revert to lower coordination until pressure falls below about 7 GPa. (5,6)

One may also infer evidence of kinetic effects from a comparison of Hugoniot data on an initially porous material with data on the solid. The initially porous material is much hotter, at a given shock pressure, than the solid. One would expect the initially porous material to have a higher volume (because of thermal effects), at a given pressure, than the initially solid material. We compare data on Coconino sandstone (porous polycrystalline quartz, 2.0 g/cm^3) with data on Arkansas novaculite (non-porous polycrystalline quartz, 2.65 g/cm^3). (7) The predicted relationship, initially solid denser than initially porous, holds up to about 10 GPa. At higher pressures, the sandstone becomes the denser material. At 10 GPa, the internal energy content of the sandstone is about 700 J/g higher than the novaculite. One may infer that the sandstone begins to transform to 6:3 coordination at a lower pressure than the novaculite.

LONG-DURATION SHOCKS IN NATURE

Although long-duration laboratory shock experiments may be impractical, we have another source of information in the form of large natural impact craters. (8) These craters are generally identified by the presence of impact-metamorphosed rocks and minerals. Impact metamorphosed rocks have been extensively studied during the past 40 years. The question we address in this paper is whether the study of impact-metamorphosed material found in large natural craters reveals anomalies that might be ascribed to the operation of kinetic effects during a relatively long-duration pressure pulse. We begin by reviewing the relationship between crater dimensions and shock pressure duration.

Various scaling rules, based on calculations of large cratering events and on extrapolations of small-scale experiments, relate the crater size to the kinetic energy of the impactor. (9) One can thus bracket likely ranges of density and impact velocity for asteroidal (meteorite-like) objects to infer the approximate size of the impacting object. The effective duration (in the region of maximum pressure duration) of the high pressure shock produced by the impact is of the order of the shock wave transit time through the impactor.

For a ca. 1 km diameter crater, such as the Arizona Meteor Crater, the impactor will have a diameter of a few tens of meters. The shock wave (or peak release wave) velocity will be about 10 km/s, and the maximum pressure duration in the material directly below the impact point will be in the millisecond regime. For a 100 km diameter crater, such as the Popigai crater, the impactor will be of the order of 10 km diameter, and the maximum pressure duration will be of the order of a second.

SHOCK METAMORPHISM IN LARGE IMPACT CRATERS

Numerous studies of shock metamorphism reveal differences between field observations and laboratory calibration studies. (10, 11) It is customary to ascribe these differences to the longer duration of the natural events. However, quantitative interpretations of these differences have hitherto not been attempted.

We have been impressed by the relative abundance of "fragile" metastable high-pressure phases in large impact craters. These phases include diamond, stishovite, coesite, and the alpha-PbO_2 structure polymorph of TiO_2. Simple estimates of post-shock temperature, based on laboratory calibrations of shock metamorphic effects in surrounding rock, imply that the most fragile phases, stishovite and the alpha-PbO_2 structured TiO_2, should definitely have inverted to low-pressure forms. Diamond and coesite can withstand much higher post-shock temperatures but are nevertheless found in environments in which their survival appears problematic. The possibility of quenching is ruled out in most cases by simple heat flow calculations. In the absence of other possible explanations, one must postulate that the post-shock temperatures of the rock were actually much lower than the laboratory calibrations would imply.

A BOOTSTRAP ESTIMATE OF KINETIC EFFECTS ON SHOCK WAVE PROPAGATION

Here we present the hypothesis: Kinetic effects on shock wave propagation can account for the survival of a fragile high-pressure mineral in a shock-metamorphosed rock. To test this hypothesis we examine a sample of shock-metamorphosed Coconino sandstone from Meteor Crater, Arizona.

Optical and X-ray diffraction analyses of the sample indicate that it consists of approximately 66 % diaplectic glass, 14 % crystalline quartz, 19 % coesite, and 1 % stishovite, after correcting for 2 % calcite that was deposited post-impact. We make the customary assumption that the coesite formed on release, as the pressure decayed into the coesite stability field. Thus, at pressure, 86 % of the quartz is presumed to have transformed to the 6:3 Si:O coordination as noted above. Using available Hugoniot release data (5) together with static compression data (12), we can approximate an appropriate release adiabat for the observed post-impact mineralogy. The peak pressure in the sample is then determined to be 31 GPa, the intersection of the release adiabat with the Hugoniot. The waste heat, the net internal energy increase after compression and release, is the area between loading and release paths, as shown in Figure 1.

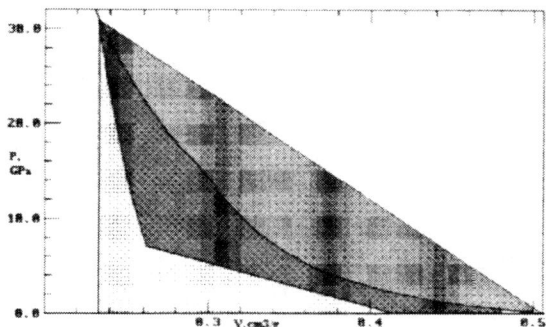

Figure 1. Internal energy increase after release for dry Coconino sandstone shocked to 31 GPa. The darker hatched regions represent the waste heat, the area between the Rayleigh line and the release adiabat.

The waste heat, 3.3 kJ/g, corresponds to a post-shock temperature of ~2800 K, corresponding to superheated liquid and obviously much too high to account for the observed mineralogy. However, we assumed dry Coconino sandstone. Kieffer has inferred that the sandstone was water-saturated at the time of impact. (13)

In the absence of experimental data, we approximate the Hugoniot and release of water-saturated Coconino by the well-known and generally successful technique of volume addition of the components. We assumed the release behavior of the water to be the same as the compression behavior, neglecting the large expansion as the water flashes into steam at low pressure on release. The peak pressure of the sample is determined to be 32 GPa; the waste heat estimate is shown in Figure 2.

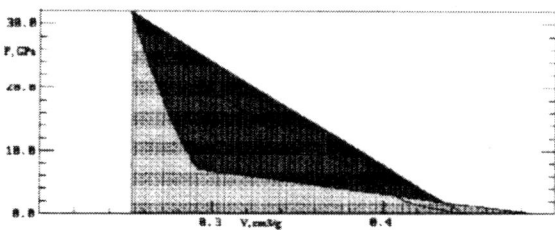

Figure 2. Internal energy increase after release for water-saturated Coconino sandstone shocked to 32 GPa. The darker hatched regions correspond to the area between the Rayleigh line and the release adiabat.

Water saturation of the initially 25 % porous Coconino sandstone reduces the waste heat by about a factor of 2, to ~ 1.7 kJ/g. The corresponding post-shock temperature, ~1400 K, is still too hot for survival of either stishovite or coesite. Coesite survival might be explained as indicative of about a 200 K uncertainty in our post-shock temperature estimate combined with relatively rapid post-shock cooling of the sample. However, the upper limit of post-shock temperature for stishovite survival is only about 900 K. The discrepancy seems much too large to attribute to uncertainties in our approximations of the Hugoniot and release adiabat of water-saturated Coconino sandstone.

We began this section with the hypothesis that kinetic effects could account for the survival of fragile high-pressure phases like stishovite. In the absence of an *a priori* basis for inferring the details of the kinetic effects, we take the bootstrap approach. We will assume an arbitrary but reasonable kinetic effect and then calculate the post-shock temperature. If we can find a combination of a reasonable kinetic effect with a sufficiently low post-shock temperature, we have effectively confirmed the possibility that kinetic effects could account for the survival of fragile high-pressure phases. As we noted earlier, the observed mineralogy serves as a basis for constraining the release path. One also requires that the loading path intersect the release path. The initial part of the loading path would consist of a Rayleigh line connecting the initial state to a peak stress state on the Hugoniot. This state would be identical to that obtained in microsecond-duration loading. Over the millisecond duration of pressure in the Meteor Crater impact, the stress would relax as transformation to 6:3 O:Si coordination progressed. Figure 3 represents our guess at the effects of kinetics at a peak stress of 15 GPa.

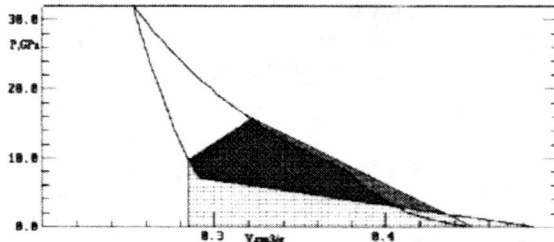

Figure 3. Effects of kinetics on the internal energy increase of water-saturated Coconino sandstone after pressure release. The darker cross-hatched regions represent the waste heat, the area enclosed by the Rayleigh line, the stress relaxation line, and the release adiabat.

The assumption of kinetic effects reduces the waste heat to ~1.14 kJ/g. If one assumes that the water and quartz achieve thermal equilibrium, the post-shock temperature is ~900 K. This is about the upper temperature limit for stishovite survival. At this point, we conclude that we have adequately tested our initial hypothesis: that the survival of fragile high-pressure phases in the impact-metamorphosed rocks of large impact could indeed be accounted for by the assumption of kinetic effects on dynamic phase transitions.

DISCUSSION

We picked the example of Meteor Crater impact-metamorphosed Coconino sandstone for several reasons. First, we have adequate Hugoniot and release adiabat data on Coconino sandstone and other quartz rocks. Second, the Meteor Crater impact is the smallest natural impact for which we have inferred possible evidence of kinetic effects. The impact that produced Meteor Crater was approximately equivalent in energy to about a 2 MT thermonuclear explosion.

Further examination of possible kinetic effects on wave propagation in tectosilicates (open-structured minerals like quartz and feldspars) appears warranted by the need for confidence in the validity of calculations of ground motion produced by large events. The kinetic effect that we have inferred is so large that it may well be detectable in appropriately designed large-scale high explosive experiments.

REFERENCES

1. Alder, B. J., "Physics Experiments With Strong Pressure Pulses", in *Solids Under Pressure*, edited by Paul, W., and Warshauer, D. W., McGraw-Hill, New York, 1963, pp. 385-420
2. Jeanloz, R., *J. Geophys. Res.* **85**, 3163-3176 (1980)
3. Langenhorst, F., et al, *Nature* **356**, 507-509 (1992)
4. Gratz, A. J., et al, *Physics and Chemistry of Minerals* **19**, 267-288 (1992)
5. Grady, D. E., et al, *J. Geophys. Res.* **79**, 322-338, (1974)
6. Grady, D. E., and Murri, W. J., *Geophys. Res. Lett.* **3**, 472-474, (1976)
7. Ahrens, T. J., and Gregson, V. G., *J. Geophys. Res.* **69**, 4839-4874 (1964)
8. Grieve, R. A. F., and Shoemaker, E. M., "The Record of Past Impacts on Earth", in *Hazards Due to Asteroids and Comets*, edited by Gehrels, T., Univ. of Arizona Press, Tucson, 1994, pp. 417-462
9. Melosh, H. J., *Impact Cratering: A Geologic Process*, Oxford University Press, New York 1989, Chapter VII
10. Stöffler, D., and Langenhorst, F., *Meteoritics* **29**, 155-181 (1994)
11. Huffman, A. R., and Reimold, W. U., *Tectonophysics* **256**, 165-217 (1996)
12. Knittle, E., "Static Compression Measurements of Equations of State", in *Mineral Physics and Crystallography AGU Reference Shelf 2*, edited by Ahrens, T. J., American Geophysical Union, Washington, 1995, pp 98-142
13. Kieffer, S. W., et al, *Contributions to Mineralogy and Petrology* **59**, 41-93 (1976)

THE PRINCIPAL HUGONIOT AND DYNAMIC STRENGTH OF DOLERITE UNDER SHOCK COMPRESSION

K. Tsembelis, W. G. Proud and J.E. Field

PCS, Cavendish Laboratory, Madingley Road, Cambridge, CB3 0HE. UK.

Abstract. A series of plate impact experiments was performed on Dolerite (diabase) igneous rock. Longitudinal stresses were measured using embedded manganin stress gauges up to *ca.* 11 GPa. In addition, lateral stresses were also measured up to *ca.* 7 GPa. In combination with the longitudinal stresses, these results have been used to obtain the material shear stress under shock compression. Results indicate that the longitudinal behaviour is elastic for the stress range involved although shear stresses indicate deviation from elastic loading for longitudinal stresses higher than *ca.* 4.3 GPa. The results are then compared and contrasted to data for other geologic materials.

INTRODUCTION

The shock properties of geological materials have long been a source of interest. Traditionally, the main driving forces have been planetary impact and geological research. Recently, there has been a growing interest in the shock properties of concrete, where geological materials are added as aggregates [1-2]. In addition, most available information consists of Equation of State (EoS) data. There are few data on the dynamic strength of such brittle materials because of the difficulty in obtaining such results. However, such results are needed to help develop constitutive models for these materials. In the last fifteen years a technique has been developed [3] using manganin gauges to measure the *lateral* stresses in materials under shock loading. Combining both Hugoniot and lateral data, shear stress information can be obtained. In this paper, results are presented on Hugoniot and lateral experiments performed on Dolerite. Dolerite (also known as Diabase) is a fine-to medium-grained, dark grey to black intrusive igneous rock [4]. Chemically and mineralogically, it closely resembles the volcanic rock basalt, but it is somewhat coarser and contains glass. With increase in grain size it resembles gabbro.

EXPERIMENTAL PROCEDURE

All the impact experiments were carried out in the plate impact gun facility at the University of Cambridge [6], which consists of a single stage 50 mm bore light gas gun. The gun is capable of achieving velocities up to 1200 m s^{-1}. The impactor materials consisted of copper and tungsten. Impact velocities were measured to an accuracy of 0.5% using a sequential pin-shorting method and tilt was arranged to be less than 1 mrad by means of an adjustable specimen mount. To measure the Hugoniot of Dolerite, manganin stress gauges

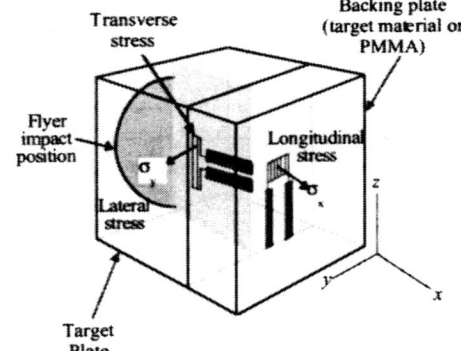

Figure 1. Target configuration

(MicroMeasurements type LM-SS-210FD-050) were embedded between tiles 8 and 17 mm thick. One sample was prepared with the stress gauge supported on the rear surface with a block of Polymethylmethacrylate (PMMA). In that configuration the gauge had a faster rise time due to the near impedance match of the PMMA, epoxy adhesive and gauge package. Material specimens for lateral gauge experiments were sectioned in two, and commercial stress gauges (J2M-SS-580SF-025) were introduced 3 and 8.2 mm from the impact surface of each sample. Samples were assembled, for both configurations, using a low viscosity epoxy with a curing time of approximately 24 hours. Lateral gauge data were reduced using the analysis of Rosenberg and Partom [3]. The shear stress (τ) of the material can thus be calculated through knowledge of the longitudinal (σ_x) and lateral stresses (σ_y) through the relation,

$$2\tau = \sigma_x - \sigma_y \quad (1)$$

Our method of determining the shear stress has the advantage over previous calculations of being direct since no computation of the hydrostat is required.

MATERIAL DATA

Dolerite tested in this study, was supplied by Concrete Structures Section (CSS), Department of Civil & Environmental Engineering, Imperial College, London, UK as a part of a large block weighting over 20kg. It was then cut into smaller specimens with dimensions 8-20 mm thick by 50 mm x 50 mm. Density and ultrasonic measurements were performed after grinding the samples. Several samples were used. The density was 2894 ± 27 kg m^{-3}, while the longitudinal and shear elastic wave speeds, determined using ultrasonic transducers, were 5.89 ± 0.07 and 3.34 ± 0.11 mm µs^{-1}, respectively.

RESULTS AND DISCUSSION

Table 1 summarises the impact conditions and Hugoniot stresses for the longitudinal experiments, while Table 2 summarises the impact conditions, lateral stresses and shear stresses obtained using equation 1 and the gauge data. Figures 2 and 3 illustrate some typical longitudinal and lateral stress wave profiles for experiments 1Hdol/1Tdol and 4Hdol/4Tdol, respectively (for impact conditions, see Tables 1 and 2). The solid trace corresponds to the longitudinal stress while the dotted traces correspond to the lateral stresses at two different positions. It can be seen that the longitudinal stresses have higher values than the lateral ones and their difference leads to the shear stress inside the material according to equation 1.

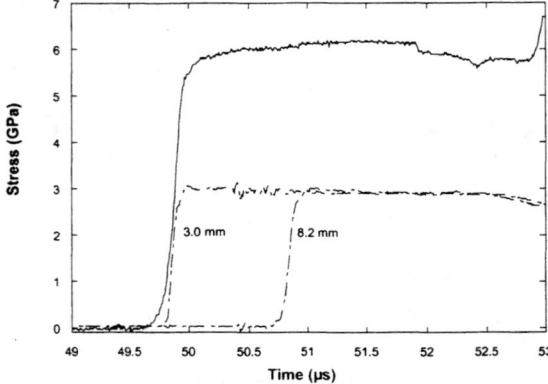

Figure 2. Stress Wave Profiles for experiments 1Hdol/1Tdol (see Tables 1 and 2 for impact conditions).

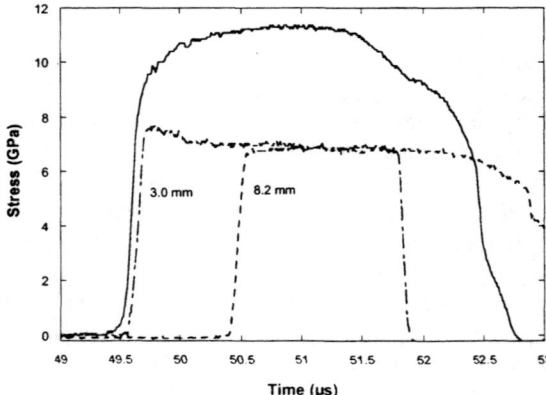

Figure 3. Stress Wave Profiles for experiments 4Hdol/4Tdol (see Tables 1 and 2 for impact conditions).

Figure 4 illustrates the Dolerite Hugoniot curve together with the Hugoniot data for Gabbro and Diabase [6-8]. It can be seen that all data are tightly grouped together. In addition, the Dolerite data have been fitted with the elastic impedance of

Table 1. Experimental parameters and results for longitudinal data and Hugoniot points

Shot no.	Impactor material	Target Front Sample	Target Backing Sample	Impact Velocity (m s^{-1})	Hugoniot Stress (GPa) ± 3%	Particle Velocity (mm µs^{-1}) ± 3%
1Hdol	10 mm Cu	8.26 mm Dol	17.16 mm Dol	519	6.08	0.35
2 Hdol	10 mm Cu	8.20 mm Dol	17.22 mm Dol	702	8.39	0.48
3 Hdol	10 mm Cu	7.69 mm Dol	17.25 mm Dol	833	10.17	0.57
4 Hdol	6 mm W	7.35 mm Dol	8.40 mm Dol	815	11.34	0.67
1Bdol	10 mm Cu	6.8 mm Dol	12 mm PMMA	451	5.16	0.31

Table 2. Experimental parameters and results for lateral data and shear stresses

Shot no.	Impactor material	Impact Velocity (m s^{-1})	Lateral Stress (GPa) ± 4%	2*Shear Stress (GPa) ± 6%
1Tdol	10 mm Cu	521	2.90	3.18
2 Tdol	10 mm Cu	703	4.49	3.90
3 Tdol	10 mm Cu	835	5.68	4.49
4 Tdol	6 mm W	814	6.86	4.48
5Tdol	10 mm Cu	265	1.01	2.12

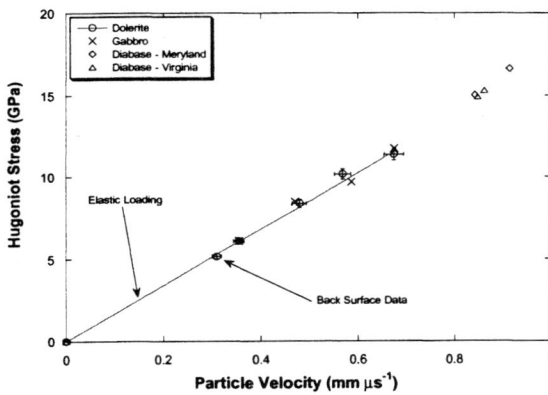

Figure 4. Dolerite Hugoniot

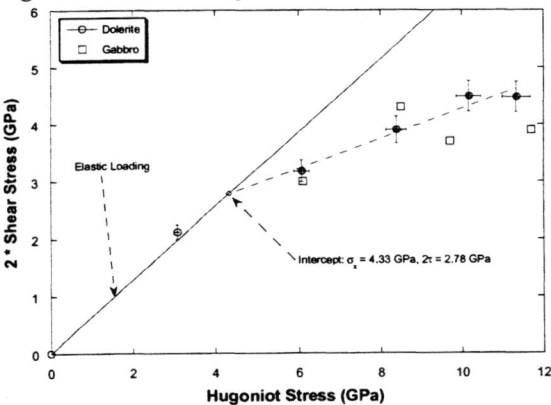

Figure 5. Dolerite Shear Stress

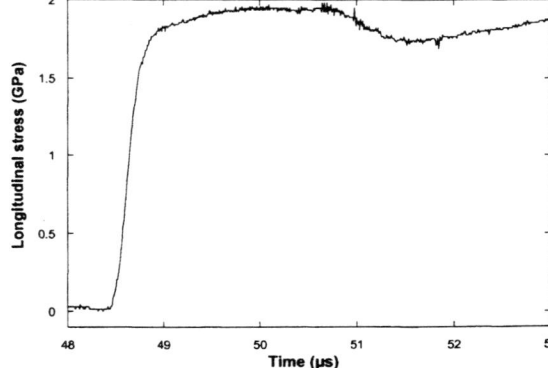

Figure 6. Stress Wave Profile for experiment 1Bdol (see Table 1 for impact conditions)

the material such as $\sigma_x = \rho_0 U_p C_L$, where ρ_0 is the initial density of the material, U_p is the particle velocity and C_L is the longitudinal wave speed. The agreement with this fit is excellent suggesting elastic loading. Although no Hugoniot Elastic Limit (HEL) data are available for Dolerite, HELs for other igneous geological materials have been quoted; for instance, basalt [9] has an HEL in the vicinity of 5 GPa, while jadeite has a quoted HEL [10] in the range 5.8-7.2 GPa. It is possible that the elastic and shock impedances of Dolerite are similar and thus make it difficult to resolve the change in slope at the HEL in the trace.

Figure 5 illustrates the Dolerite and Gabbro [7] shear stress vs. Hugoniot Stress. The Dolerite data have been fitted to the elastic loading line using

$$2\tau = \frac{1-2\nu}{1-\nu}\sigma_x \quad (2).$$

It can be seen that above a Hugoniot stress of *ca.* 4.3 GPa the material behaviour deviates from the purely elastic loading suggesting a process such as fracture or damage in the shock front, which results in reducing the dynamic strength of Dolerite. Note that when lateral stress measurements are taken below 4.3 GPa, the shear stress lies on the elastic loading line. This result can lead to the conclusion that a Hugoniot stress of 4.3 GPa is a possible HEL value.

For that reason, an extra longitudinal shot was performed where the Dolerite was backed by PMMA. Because of similar impedance between the gauge package and PMMA the gauge rise time was *ca.* 10ns, compared to 200 ns for a fully embedded gauge. The measured stress in the PMMA (σ_p) was converted to stress in the Dolerite (σ_D) through the well-known relation

$$\sigma_D = \frac{Z_D + Z_P}{2Z_P}\sigma_P, \quad (3)$$

where Z_D and Z_P are the elastic and shock impedances of the Dolerite and PMMA, respectively. The trace is illustrated in Figure 6. The stress induced in the Dolerite was 5.11 GPa, higher than the presumed HEL. However, no two-wave structure was seen, reinforcing the assumption that the elastic and shock impedances are similar. Experiments with VISAR are under way to resolve this discrepancy.

CONCLUSIONS

Plate impact experiments have been presented to assess the longitudinal and deviatoric behaviour of the Dolerite. Results indicate that the Hugoniot curve is elastic up to 11 GPa. However, shear stress data show a deviation from elastic loading at a stress of 4.3 GPa. It can thus be concluded that the HEL is around that value and the gauges were unable to resolve the two-wave structure because of similar elastic and shock impedances.

ACKNOWLEDGEMENTS

The Defence and Evaluation Agency, UK has sponsored this work, under contract WSS/U3257. Dr. A. Pullen from Imperial College provided the cement samples. Dr J. Sheridan, C. O'Carroll, I.G. Cullis and P.D. Church are thanked for their interest. Finally, we thank D.L.A. Cross and R. Flaxman for technical support.

REFERENCES

1. Tsembelis, K., Millett, J.C.F., Proud, W.G. and Field, J.E., *Shock Compression of Condensed Matter-1999*, (eds. M.D. Furnish, L.C. Chhabildas and R.S. Hixson), 1267.
2. Field, JE, Tsembelis, K., Proud, W.G., *Proc. Of SHOCK 2001 – APS 12th Topical Conference on Shock Compression of Condensed Matter*, Atlanta, 24-29 June, 2001.
3. Z. Rosenberg, and Y. Partom., J. Appl. Phys., **58**, 3072 (1985).
4. Encyclopaedia Britannica, Inc, 2000.
5. Bourne, N. K., Rosenberg, Z., Johnson, D. J., Field, J. E., Timbs, A. E., and Flaxman, R. P., *Meas. Sci. Technol.* **6**, 1462 (1995).
6. Millett, J.C.F., Tsembelis, K., Bourne, N.K., and Field, J.E., *Shock Compression of Condensed Matter-1999*, (eds. M.D. Furnish, L.C. Chhabildas and R.S. Hixson), 1247.
7. Millett, J.C.F., Tsembelis, K. and Bourne, N.K., *Journal Appl. Phys.*, **87**, 3678 (2000).
8. March, S.P., *LASL Shock Hugoniot Data* (University of California Press, LA, 1980).
9. Nakasawa, N., Watanabe, S., Kato., M., Iijima, Y., Kobayashi, T. and Sekine, T., *Planet. Space. Sci.*, **45**, 1489 (1997).
10. Takasawa, E., Sekine, T., Kobayashi, T. and Zhu, Y., *J. Geophys. Res. B*, **103**, 12261 (1998).

EXPLOSION IN THE GRANITE FIELD: HARDENING AND SOFTENING BEHAVIOR IN ROCKS.

Ilya N. Lomov, Tarabay H. Antoun, Lewis A. Glenn

Lawrence Livermore National Laboratory, Livermore, California 94550, USA

Abstract. Properties of rock materials under quasistatic conditions are well characterized in laboratory experiments. Unfortunately, quasistatic data alone are not sufficient to calibrate models for use to describe inelastic wave propagation associated with conventional and nuclear explosions, or with impact. First, rock properties are size-dependent. Properties measured using laboratory samples on the order of a few centimeters in size need to be modified to adequately describe wave propagation in a problem on the order of a few hundred meters in size. Second, there is lack of data about the damage (softening) behavior of rock because most laboratory tests focus on the pre-peak hardening region with very little emphasis on the post-peak softening region. This paper presents a model for granite that accounts for both the hardening and softening of geologic materials, and also provides a simple description of rubblized rock. The model is shown to reproduce results of quasistatic triaxial experiments as well as peak velocity and peak displacement attenuation from a compendium of dynamic wave propagation experiments that includes US and French nuclear tests in granite.

INTRODUCTION

Modeling the dynamic response of rock materials is a challenging area of research. Since most strength measurements in rock materials are performed for intact samples under static conditions, the models based on these data should account for possible scale and rate effects when being applied to simulation of the dynamic response of large-scale rock masses. Unlike intact rock samples, rock masses may contain discontinuities that may reduce the strength.

Several attempts were made in the past to simulate wave propagation in rock media [1,2]. The standard approach was to optimize the model to reproduce wave profiles at several different ranges away from the source. The ability to reproduce specific characteristics of the profile (peak values, width, rebound phase, damping) was considered important for understanding the physics of the problem. However, experimental data show significant scatter even in wave profile measurements made at the same range, but at different azimuths for a single event. On the other hand, ensemble data of peak velocity and peak displacement attenuation for hard rock fall into a relatively small band. Hence, this data was chosen to be a reference for the model calibration in the present paper.

THE CONSTITUTIVE MODEL

The constitutive equations developed here are non-linear, thermodynamically consistent, and properly invariant under superposed rigid body motions. The equations are valid for large deformations and they are hyperelastic in the sense that the stress tensor is related to a derivative of the Helmholtz free energy. We assume that the material is isotropic and apply the mathematical structure of plasticity theory to capture the basic features of the mechanical response of geological materials. The mathematical framework used to develop the model and a detailed description of the model equations can be found in [3].

The deviatoric behavior of the rock is described using an elastic-viscoplastic model, coupled with a damage model, and the volumetric behavior is described using an equation of state, coupled with a porous compaction and bulking model. Initial yielding is followed with a plastic strain hardening phase that persists until the loading path intersects the failure surface. Thereafter, damage accumulation causes gradual strength reduction from the failure surface down to a residual value equivalent to a small fraction of the undamaged strength.

The equation of state, which describes the solid rock behavior, is supplemented with an analytic porous compaction model that describes the relationship between pressure and porosity. Also included in the volumetric behavior description is a dilatancy model that relates bulking to plastic distortion in such a way as to ensure thermomechanical consistency with the second law of thermodynamics.

Strength of material

As described in [3], the physical phenomena that influence the yield strength Y are taken into account using a multiplicative form with Y given by:

$$Y = Y_0 F_0(\xi,p) F_1(\xi,p) F_2(\varepsilon_p) F_3(\rho,\varepsilon), \quad (1)$$

where Y_0 is the initial yield strength of the rock at zero pressure, the functions $F_0(\xi,p)$, $F_1(\xi,p)$, $F_2(\varepsilon_p)$, and $F_3(\rho,\varepsilon)$ respectively account for the effects of scaling, hardening, damage and melting on the strength and failure of the rock

In Equation 1, $F_0(\varepsilon_p,p)$ is a scaling function equal to unity for intact rock samples and decreases monotonically as a function of pressure and inelastic deformation. This function was incorporated into the model because our simulation results showed that the yield and failure stresses measured statically using relatively small, defect-free samples had to be reduced to satisfactorily reproduce the dynamic data. This is in line with experimental observations that show the strength of granite and other geologic materials to be size-dependent, decreasing with increasing specimen dimensions.

F_1 is a hardening function expressed in terms of the hardening parameter ξ as follows:

$$F_1 = 1 + (k_1(p) - 1)\xi$$

where k_1 is a function that expresses the relationship between the initial yield and the failure strength of the rock, both of which are varying functions of pressure as shown in Figure 1. The hardening parameter ξ varies between 0 and 1; it is determined by an evolution equation of the form

$$\dot{\xi} = k_2 (1-\xi) \dot{\varepsilon}_p / f_1(p)$$

where $\dot{\varepsilon}_p$ is the plastic strain rate and k_2 is a model parameter. The pressure hardening function $f_1(p)$, which expresses the dependence of the yield strength on applied pressure, is determined from laboratory measurements on small rock samples [4].

The damage function F_2 controls the strength degradation of the rock after damage begins to accumulate. It is expressed in terms of the plastic strain, ε_p, using the simple relation

$$F_2 = \max \begin{cases} 1 - k_5 \varepsilon_p & (2a) \\ \alpha p / Y(p) & (2b) \end{cases}$$

where k_5 and α are model parameters, and $Y(p)$ is the pressure-dependent, but undamaged value of the yield strength. The function in Equation (2a) takes on values ranging between 1 for the intact material, and 0 for the fully damaged material. The function in Equation (2b) represents the residual strength of the damaged material. It is a linear function of pressure and it represents only a small fraction of the strength of the undamaged rock. The parameter α in this equation can be viewed as the friction coefficient of the damaged material. A value of $\alpha = 0.1$ was used in the simulations discussed here.

F_3 models the effect of material melting (i.e., thermal softening) near the source region.

Porous compaction and bulking

The total gas porosity is separated into two parts, ϕ_1 and ϕ_2:

$$\varphi = \varphi_1 + \varphi_2.$$

ϕ_1 describes changes in porosity associated with the compaction of existing pores using the modified $p-\alpha$ model [3], whereas ϕ_2 describes changes in porosity associated with bulking (increasing porosity under positive pressure) and it is proportional to the rate of dissipation due to plastic deformation.

The rate at which bulking may proceed is constrained by the second law of thermodynamics, which governs the entropy production of dissipative thermodynamic processes. The evolution of the porosity component ϕ_2, which models the porosity due to bulking, is described using the following relation:

$$\dot{\phi}_2 = (1-\phi) H(\phi_2) \times$$
$$\times \left[\frac{m_d \sigma_{eff} \dot{\varepsilon}_p \langle \phi^* - \phi_2 \rangle}{\max(p^*, p)} - m_c (1-F_2) \langle -D \cdot I \rangle \right]$$

In this equation, σ_{eff} is von Mises stress and $\dot{\varepsilon}_p$ is the plastic strain rate. ϕ^* is a model parameter that specifies the maximum bulking porosity that can be achieved, m_d determines the rate of bulking and m_c is used to control the rate of recompaction of the bulking porosity, a process that takes place only when the material is damaged and accelerates as damage increases and $F_2 \rightarrow 0$.

SIMULATION RESULTS

Peak velocity and peak displacement measurements from deeply buried nuclear explosions [5] are shown in Figures 1 and 2. These data are assembled from events with 2 orders of magnitude yield range and do not exhibit significant scale dependence. The data shown in Figs. 1 and 2 are well represented by a linear fit in log-log space, with all the points falling within a factor of ±2 from the line. An automatic optimization procedure, based on this linear fit to the velocity and displacement data, was used to calibrate characteristic model parameters. The solid lines in Figs. 1 and 2 depict the simulated peak velocity and displacement attenuation as a function of slant range. This model is in good agreement with the ensemble of data shown in the figures, and is therefore believed to be a good representative of the overall behavior of granite. Hence this model can be used for predicting attenuation of shock waves in hard rock when specific information about the geology and material properties of the location of

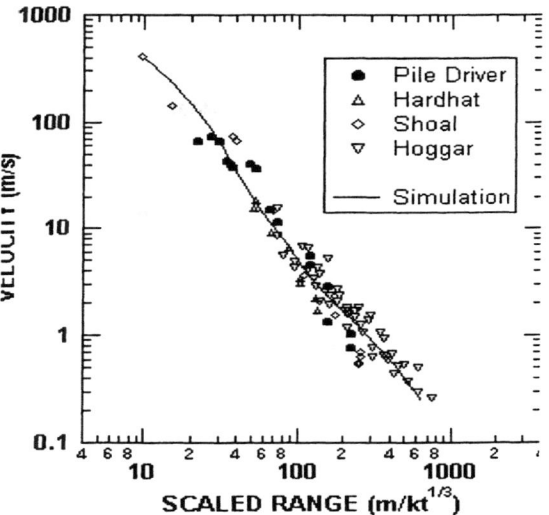

FIGURE 1. Comparison of simulated peak velocity attenuation with measurements from several spherical wave experiments in granite.

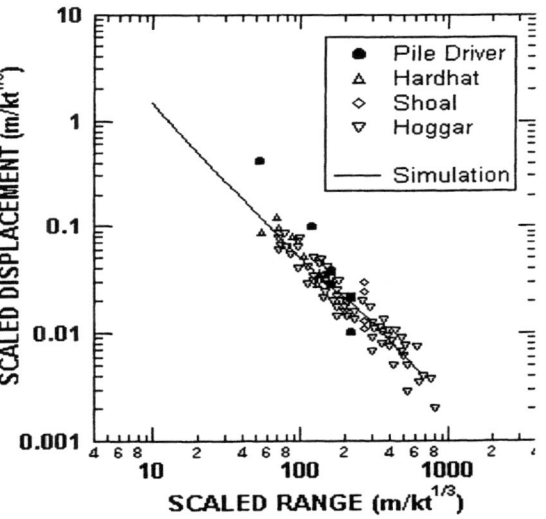

FIGURE 2. Comparison of simulated peak displacement attenuation with measurements from several spherical wave experiments in granite.

interest are unknown. Other possible applications of the model include studying several other effects like complex geologies, depth of burial and scaling.

Figures 3 and 4 show the results of a parameter sensitivity study that further illustrates some of the effects discussed in the preceding paragraphs. Specifically, these two figures compare the peak velocity and peak displacement attenuations simulated using the calibrated model with those simulated using a version of the model in which a specific feature is disabled. The effects of porous compaction, bulking, and scaling are investigated in this manner and as shown in the figures, each of these model features has a pronounced effect on the simulated response.

Attempts to calibrate the model in the absence of any of these features were not successful because it was not possible to obtain good agreement with both peak velocity and peak displacement attenuation simultaneously. This emphasizes the fact that all of the seemingly complex model features are necessary for achieving a consistent description of the dynamic behavior of hard rocks.

Although the model has not been yet thoroughly validated for geologic materials other than granite, we believe that the model is sufficiently flexible that it can be used to represent the behavior of a wide range of geologic materials. This stems from the fact that the model incorporates many of the phenomenological features normally associated with the behavior of geologic materials.

ACKNOWLEDGMENTS

This work was performed under the auspices of the U.S. Department of Energy by the University of California, Lawrence Livermore National Laboratory under contract No. W-7405-Eng-48.

REFERENCES

1. Stevens, J. L., Rimer, N., and Day, S. M., "Constraints on Modeling of Underground Explosions in Granite," Report SSS-R-87-8312, S-Cubed division, Maxwell Labs, 1986.
2. Vorobiev, O.Yu, Antoun, T. H., Lomov I.N, Glenn, L.A, "A Strength and damage model for rock under dynamic loading", in *Shock Compression in Condensed Matter-1999*, edited by S. C. Furnish et al., AIP Conference Proceedings 505, New York: American Institute of Physics, 2000, pp. 317-320.
3. Rubin, M.B, Vorobiev, O.Yu., Glenn, L.A., 2000, "Mechanical and numerical modeling of a porous elastic-viscoplastic material with tensile failure," *Int. J. Solids and Structure*, Vol.37, pp.1841-1871.
4. Schock, R. N., Heard, H. C., and Stephens, D. R., 1973, "Stress-Strain Behavior of a Granodiorite and Two Craywackes on Compression to 20 Kilobars," *J. Geophys. Res.*, Vol.78(36), pp. 5922-5941.

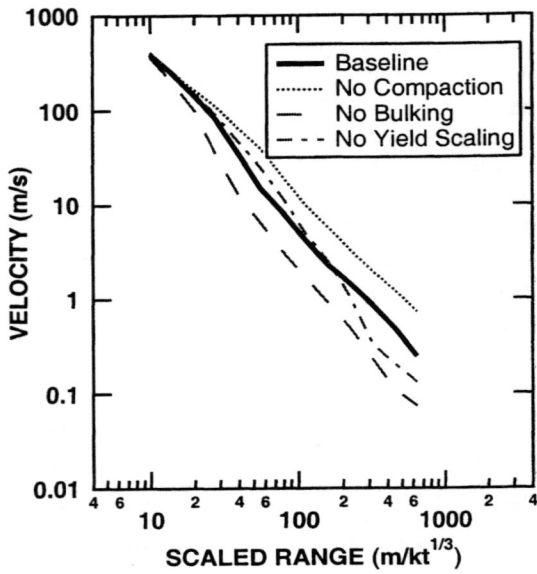

FIGURE 3. Effect of compaction bulking, and yield strength scaling on the simulated peak velocity attenuation as a function of scaled range.

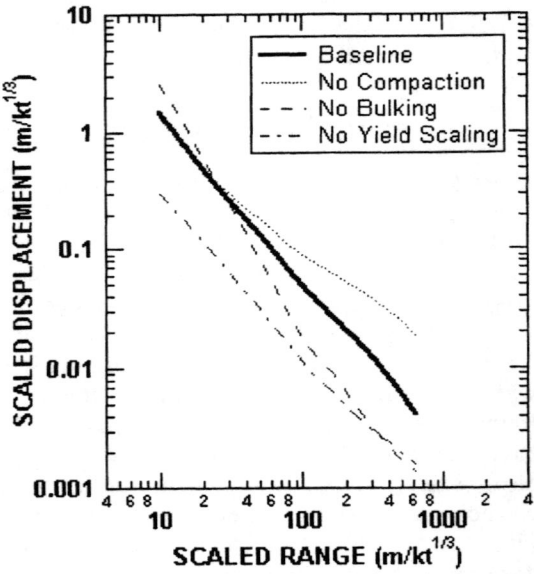

FIGURE 4. Effect of compaction bulking, and yield strength scaling on the simulated peak displacement attenuation as a function of scaled range.

5. Heuze, F. E. Review Of Geomechanics Data From French Nuclear Explosions In The Hoggar Granite, With Some Comparisons To Tests In United States Granite. Report Ucid-19812, LLNL, 1983.

DEPTH OF CRACKING BENEATH IMPACT CRATERS: NEW CONSTRAINT FOR IMPACT VELOCITY

Thomas J. Ahrens*, Kaiwen Xia*, and Demirkan Coker†

*Seismological Laboratory 252-21, California Institute of Technology, Pasadena, CA 91125
†Guggenheim Aeronautics Laboratory 105-50, California Institute of Technology, Pasadena, CA 91125

Abstract. Both small-scale impact craters in the laboratory and less than 5 km in diameter bowl-shaped craters on the Earth are strength (of rock) controlled. In the strength regime, crater volumes are nearly proportional to impactor kinetic energy. The depth of the cracked rock zone beneath such craters depends on both impactor energy and velocity. Thus determination of the maximum zone of cracking constrains impact velocity. We show this dependency for small-scale laboratory craters where the cracked zone is delineated via ultrasonic methods. The 1 km-deep cracked zone beneath Meteor Crater is found to be consistent with the crater scaling of Schmidt (1) and previous shock attenuation calculations.

INTRODUCTION

Impact-induced fractures in rock have long been recognized beneath terrestrial impact craters. Simmons et al. (2) pointed out that the multiply impacted near-surface rocks on the Moon were fractured to depths of tens of kilometers and the fracture density controls the near-surface seismic velocity structure. Dvorak and Phillips (3) discovered cracks beneath young fresh craters that were unfilled on the Moon, or filled with air or water on the Earth gave rise to significant (negative) gravity anomalies. Ahrens and Rubin (4) demonstrated that the seismic velocity of rocks beneath small laboratory, strength-limited, craters increased with depth. This results from the near-surface zone of cracking beneath craters. The intensity of cracking decreases with increasing depth reflecting the decreasing dynamic tensile stresses with depth in the rock. The volume of strength-controlled impact craters has been demonstrated to be nearly proportional to impactor energy for soft and hard rocks (5).

In the present report we show that the depth of cracking is related to the spatial attenuation rate of the impact-induced shock wave. This is, in turn, related to the impact velocity. Thus given the dimensions of the crater, and the depth of rock cracking, in principle, these data can be combined to provide constraints on impactor velocity. Since on the Earth, the asteroidal impact velocity range is from ~18 to 25 km/sec and cometary impactors encounter the Earth from 30 to 75 km/sec, it may now be possible to differentiate between the different craters produced by these objects. Traces of siderophilic elements from the impactor sometimes are preserved in the impact-induced melt. For some craters, samples of the impactor are discovered within the ejecta blanket. Fragments of Meteor Crater's impactor (Canyon Diablo meteorite) were strewn around the crater rim, when it was discovered in ~1880.

On a larger scale, on the Earth, bowl-shaped impact craters with diameters up to several kilometers appear to be strength controlled. Beneath these craters the maximum depth of crack damage is approximately proportional to crater diameter. For example, the 1.2 km diameter Meteor Crater in Arizona displays a zone ~1 km deep of severely cracked rock. This feature was outlined by Ackermann et al. (6) using seismic refraction methods. The maximum depth of damage for Meteor Crater and other terrestrial impact structures (Table 1) versus crater diameter are plotted in Fig. 1. At diameters greater than ~10 km, the initial bowl-shape of impact craters are usually collapsed and the crater floors are flattened. Larger diameter craters also display central peaks and concentric rings as a result of the gravitation stresses that dominate the final crater shapes.

TABLE 1. Depth of Damage (L_D) for Terrestrial Impact Craters

Crater Name	Crater Diameter (km)	Maximum Depth of Damage L_D (km)
Meteor, USA	1.2	1.0
Brent, CAN	3.0	0.3
West Hawk Lake, CAN	3.7	1.25
Gow Lake, CAN	4.0	0.9
Upheaval Dome, USA	5.0	2.0
Soderfjarden, FI	5.5	0.5
Gosses Bluff, AU	22.0	5.0
Ries, DE	23.0	6.0
Charlevoix, CAN	46.0	8.0
Manicouagan, CAN	65.0	9.0
Chicxulub, MEX	165.0	11.0
Vredefort, SA	300.0	14.0

Shock-induced cracking beneath small-scale craters excavated in rock samples appears to be controlled by the same physics as the cracking observed beneath strength-controlled natural craters up to several kilometers in diameter on the Earth. We studied the extent of impact-induced cracking beneath craters so as to employ these data to specify the impact conditions required to induce a given crater. We are especially interested in applying the seismic data on terrestrial craters. In the future we expect additional seismic data for crater structures on the Earth and on other planets to become available.

Here we summarize two methods used to measure the shock damage beneath laboratory induced craters in rock targets. We relate the resulting seismic velocity profiles (reflecting shock damage versus depth) to a simple model of shock attenuation and propose a model to relate shock attenuation rate and impactor velocity to the intensity and distribution of shock damage.

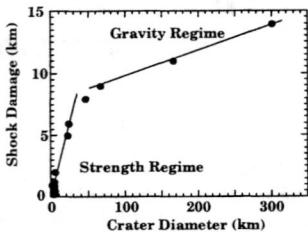

FIGURE 1. Depth of shock damage, versus, crater diameter for terrestrial impact craters (Table 1). Strength regime corresponds to bowl-shaped strength-controlled simple craters. Gravity regime corresponds to complex (central peaks and multi-ringed) craters that have suffered major failure from gravitational stresses.

EXPERIMENTS

San Marcos gabbro (7) targets (~15 cm on side) were impacted with 0.8 g aluminum projectiles at velocities of 0.8 to 1.2 km/sec. The resulting impact-induced bowl-shaped craters, ~ 5 cm^3, in volume displayed zones of cracked rock beneath the craters (Fig. 2). Budiansky and O'Connell (8) first showed that dry circular cracks in solids (rocks) cause the bulk and shear moduli, $\bar{K}$ and $\bar{G}$, to be decreased relative to that of the uncracked rock moduli (K and G). Hence, the elastic wave velocities are strongly reduced by cracks in rock. Cracking is defined in terms of the damage parameter $\varepsilon = N \langle a^3 \rangle$, where N is the number of (circular) cracks per unit volume of average radius, a. Thus

$$\bar{K} = K / \left[1 + \frac{16}{9} \frac{(1-\bar{v})^2}{(1-2\bar{v})} \varepsilon \right] \quad (1)$$

and

$$\bar{G} = G \cdot \left[1 - \left(\frac{32}{45}\right) \cdot \frac{(1-\bar{v})(5-\bar{v})}{(2-\bar{v})} \varepsilon \right] \quad (2)$$

where v and $\bar{v}$ are the uncracked and cracked Poisson's ratio and

$$\bar{v} = v \left[1 - \left(\frac{16}{9}\right) \varepsilon \right] \quad (3)$$

We employed two methods to measure the distribution of compressional wave velocity $C_p = \sqrt{\left(\overline{K} + \frac{4}{3}\overline{G}\right)/\rho}$, where ρ is the density of the rock. These are the tomographic and dicing methods outlined below.

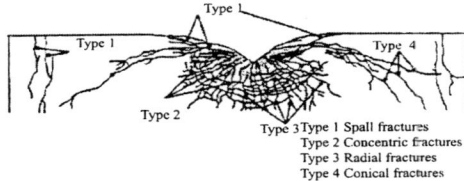

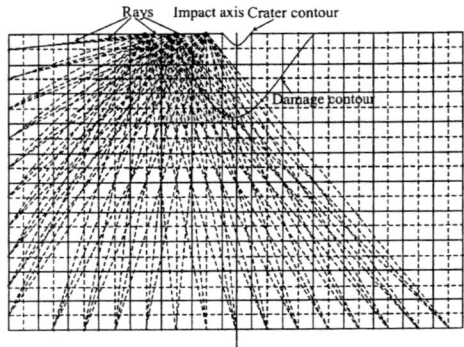

FIGURE 3. Tomographic ray diagram. Three stations on upper surface site of pulsed excitation of compressional waves. Wave detection stations (24) along left side and bottom of target block interrogated target along diagonal dashed lines (rays). One-centimeter inversion grid, crater and damage contours are also shown.

FIGURE 2. Cross-section of laboratory impact crater in San Marcos gabbro, demonstrating the four types of crack failure discussed by Ahrens and Rubin (4) (after Polanskey and Ahrens (9)).

To map the region of velocity deficit below the crater, we conducted a small-scale seismic survey. We measured the compressional wave velocity by deploying a pulsed 480 erg (~0.2 MHz) ultrasonic source (10) at three stations on the impact surface to the left of the crater in Fig. 3. The pulsed excitation signal was recorded at the 24 stations (Fig. 3) on the left side and bottom of the target block. We detected compressional waves that propagated along the assumed 72 straight line paths also is shown in Fig. 3. This data set (measured to a precision of ±0.05 μs) was inverted to obtain a 2-dimensional compressional wave velocity map over the 15 x 15 cm biaxially symmetric (about the axis of impact) grid of Fig. 3, using the least squares damping method of Menke (11). The velocity obtained for the heavily damaged regions centered at the 1, 2, and 3 cm cells, below the crater (for one sample) are plotted in Fig. 4.

Subsequently, for one of the experiments, the impacted rock sample was sawn open and ~ 1 cm cubes of rock were cut from a series of 6 positions directly beneath the axis of the crater. Compressional-wave travel times and hence propagation velocities were measured (± 0.1 km/s) and these are also plotted in Fig. 4.

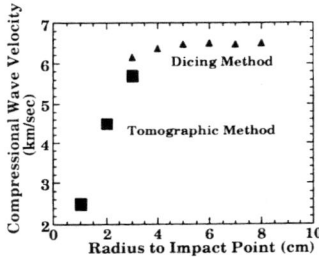

FIGURE 4. Compressional wave velocity below strength-controlled impact craters in rock. Tomographic inversion results are seen to delineate heavily damaged region beneath crater whereas dicing method provides complimentary data for deeper less damaged rock.

APPLICATION TO METEOR CRATER

The decay of an impact-induced shock wave was described by Ahrens and O'Keefe (12) using the relation

$$P(r) = P_0 (r/r_0)^{-n} \qquad (4)$$

where r_0 is the radius of a spherical impactor (3.5 mm, here) and P_0 is the contact impact pressure calculated from the shock-wave impedance-match solution. We define P_c as the critical tensile stress required for cracking and assume that when the compressional pulse decays such that $|P_c| = P(r)$, the depth is the maximum depth of cracking. This occurs at $r = L_D$, for the deepest extending cracks (radial) (see Fig. 2).

Our experiments yielded maximum depths of cracking over the range, 3.5 to 4.5 cm, from impacts at 0.8 to 1.0 km/sec. Upon fitting to Equation 4, we found n = 1.6 and P_c = 0.17 GPa.

Using the value of L_D from Table 1 for Meteor Crater, we assume P_c (for the Coconino sandstone unit) ranges from 0.10 to 0.15 GPa. Fixing the impact velocity value at ~20 km/s, (we know Canyon Diablo meteorite was initially a metallic asteroid), the value of the attenuation index concordant with Shoemaker's (13) value of the initial radius of the impactor of 11 m yields attenuation indices in the 1.75 to 1.85 range (Fig. 5). A more recent estimate of r_0 = 31 m based on using an improved theory of crater scaling yields n = 2.4 to 2.5 The latter are in good accordance with Ahrens and O'Keefe's (12) results for impact of an iron meteorite on a silicate half-space between 15 and 45 km/s.

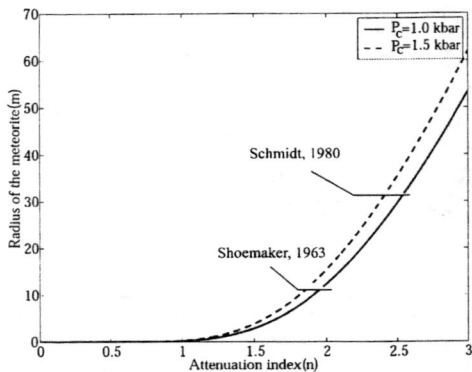

FIGURE 5. Inferred 20 km/sec iron impactor radius versus shock attenuation index (Eq. 4). Peak shock pressure at depth of maximum extent of damage for Meteor Crater (assumed solid, 1.0 kbar, and dashed, 1.5 kbar strength). Impactor radius of 11 and 31 m, from scaling of Shoemaker (13) and Schmidt (1) indicated.

CONCLUSIONS

1. In the strength controlled cratering regime (crater diameter, <5 km, which also extends also to laboratory samples) the maximum depth of shock-induced cracking corresponds closely to the radius at which the amplitude of the decaying shock wave is approximately equal to the rock dynamic tensile strength.

2. Since the shock attenuation parameter depends on relative shock impedance of impactor and target rock, as well as impactor velocity, if the impactor type can be constrained from either recovery of impactor fragments or the geochemical signature in the impact melt, the depth of crack damage (taken with impactor energy inferred from crater volume) constrains impactor velocity.

3. In the case of Meteor Crater, where the maximum depth of crack damage is 1 km, using the crater scaling of Schmidt (1) to infer a 31 m radius iron-nickel impactor predicts a shock attenuation index of 2.4 to 2.5 for tensile strengths of 1.0 to 1.5 kbar. This agrees with calculations of the attenuation indices of 2.4 and 2.6 obtained for iron impacting silicate at 15 and 45 km/sec, respectively.

ACKNOWLEDGMENTS

Research supported by NASA. Contribution number 8816, Division of Geological and Planetary Sciences.

REFERENCES

1. Schmidt, R.M., *Proc. Lunar Planet. Sci. Conf., 11th*, 2099-2128 (1980).
2. Simmons, G., Todd, T. and Wang, H., *Science*, **182**, 158-161 (1973).
3. Dvorak, J. and Phillips, R.J., *Geophys. Res. Lett.*, **4**, 380-382 (1977).
4. Ahrens, T.J. and Rubin, A.M., *J. Geophys. Res.*, **98**, 1185-1203 (1993).
5. Holsapple, K.A., *Ann. Rev. Earth Planet. Sci.*, **21**, 333-373 (1993).
6. Ackermann, H.D., Godson, R.H. and Watkins, J.S., *J. Geophys. Res.*, **80**, 765-775 (1975).
7. Lange, M.A., Ahrens, T.J. and Boslough, M.B., *Icarus*, **58**, 383-395 (1984).
8. Budiansky, B. and O'Connell, R.J., *Int. Journal Solids and Structures*, **12**, 81-97 (1976).
9. Polanskey, C. and Ahrens, T.J., *Icarus*, **87**, 140-155 (1990).
10. Xia, K. and Ahrens, T.J., *Geophys. Res. Lett.*, accepted for publication (2001).
11. Menke, W., *Geophysical Data Analysis: Discrete Inverse Theory*, San Diego: Academic Press, 1989.
12. Ahrens, T.J. and O'Keefe, J.D., in *Impact and Explosion Cratering*, Roddy, D. J., R. O. Pepin and R. B. Merrill (eds.), Pergamon, New York, 1977, pp. 639-656.
13. Shoemaker, E.M., in *The Solar System: Moon, Meteorites, and Comets*, Middlehurst, B. M. and G. P. Kuiper (eds.), University of Chicago Press, Chicago, IL, 1963, pp. 301-336.

SHOCK FLATTENING OF SPHERES IN POROUS MEDIA : IMPLICATIONS FOR FLATTENED CHONDRULES

Toshimori Sekine[1], Naru Hirata[2], Akira Yamaguchi[3], Takamichi Kobayashi[1], Hongliang He[1], and Zhi-ping Tang[4]

[1] *Advanced Materials Laboratory, National Institute for Materials Science, Namiki 1-1, Tsukuba 305-0044, Japan*
[2] *National Space Development of Japan, Sengen 2-1-1, Tsukuba 305-8505, Japan*
[3] *National Institute for Polar Research, Kaga 1-9-10, Itabashi-ku, Tokyo 173-8515, Japan*
[4] *Department of Modern Mechanics, University of Science and Technology of China, Hefei, Anhui 230026, P.R. China*

Abstract. Shock deformation of spherical particles has been investigated in the model systems of fused silica beads in porous metal powders by experimental observations and by numerical simulation for application to chondrules flattening in some primitive meteorites. Peak shock pressure and porosity in initial sample play an important role to deform spherical particles. A comparison of the results between the experimental observation and the numerical simulations indicates that shock deformation is plastic and quenchable.

INTRODUCTION

Primitive meteorites are full of tiny igneous spherules called chondrules, consisting mainly of olivine and pyroxene. Chondrules were made by some pervasive process in the early solar system that formed melted silicate droplets [1]. Some chondules displays considerable flattening indicating deformation of spherules. These features are closely related to the degree of shock metamorphism observed in the host meteorite [2, 3]. Experimental observations of shock-induced flattening of chondrules in meteorites have been carried out [4, 5] and revealed a linear relationship between shock peak pressure and the degree of flattening of chondrules. The experimental systems are heterogeneous in terms of grain sizes and materials and seems to be difficult to carry out a numerical simulation of modeling for shock-flattening. It is not clear whether the shock-induced deformation can be quenched without any significant alternation.

Here we investigate a model system experimentally and at the same time simulate shock-induced deformation of spheres by the help of a computer code. In this paper we present the experimental and simulated results and compare them with the observations in meteorites.

EXPERIMENTAL

We employed a mixture of copper powder and fused silica beads as the starting material. The average grain size of the copper powders is about 10 μm. The diameter of fused silica beads ranges briefly between 100 and 200 μm with a mean aspect ratio (short axis / long axis) of 0.91±0.08. The mixture was pressed into a steel container at pressures of 0.1 ~ 0.5 GPa to control the initial porosity. The amount of beads was so small that beads do not contact each other. The density of thus obtained samples (~ 12mm in diameter, ~ 2 mm thick) in the containers were 4.4 ~ 6.1 g/cm^3 (8-34% porosity).

Shock recovery experiments were performed using a 30-mm bore single-stage propellant gun. The flyer plates were steels (SUS304) and 4 mm thick. The peak pressure was estimated by measured impact velocity of flyer and the impedance match solution. The impact velocity ranged between 0.7 and 2.0 km/sec and the pressure ranges between 14 and 49 GPa. Recovered samples were cut parallel to the shock compression axis and polished in order to observe the shapes of beads by SEM.

EXPERIMENTAL RESULTS

Figures 1 (a) to (c) show the textures of polished sections of shocked samples with initial porosity of about 13%. In the shocked samples copper powders are well compacted and no pores are present. The beads displayed flattened deformation and preferred orientations. We measured and averaged the aspect ratios of beads. Figure 2 illustrates a deformed bead quenched from 22 GPa, indicating asymmetrical flattening. The ratio is summarized in Figure 6, and a significant change can be seen at pressures of between 22 and 31 GPa. It was not successful to recover samples from 49 GPa.

To evaluate the effects of initial porosity of sample, samples with higher (34 and 27 %) and lower (8%) porosity were investigated at 22 GPa and 32 GPa, respectively. Figures 3 and 4 illustrate some deformed beads as a function of porosity at a shock pressure of 31-32 GPa. It was not successful to recover the samples with 34% porosity from higher shock pressures. The measured aspect ratios are shown in Figure 6.

To compare the effect of samples with different shock impedance, we have carried out several shots

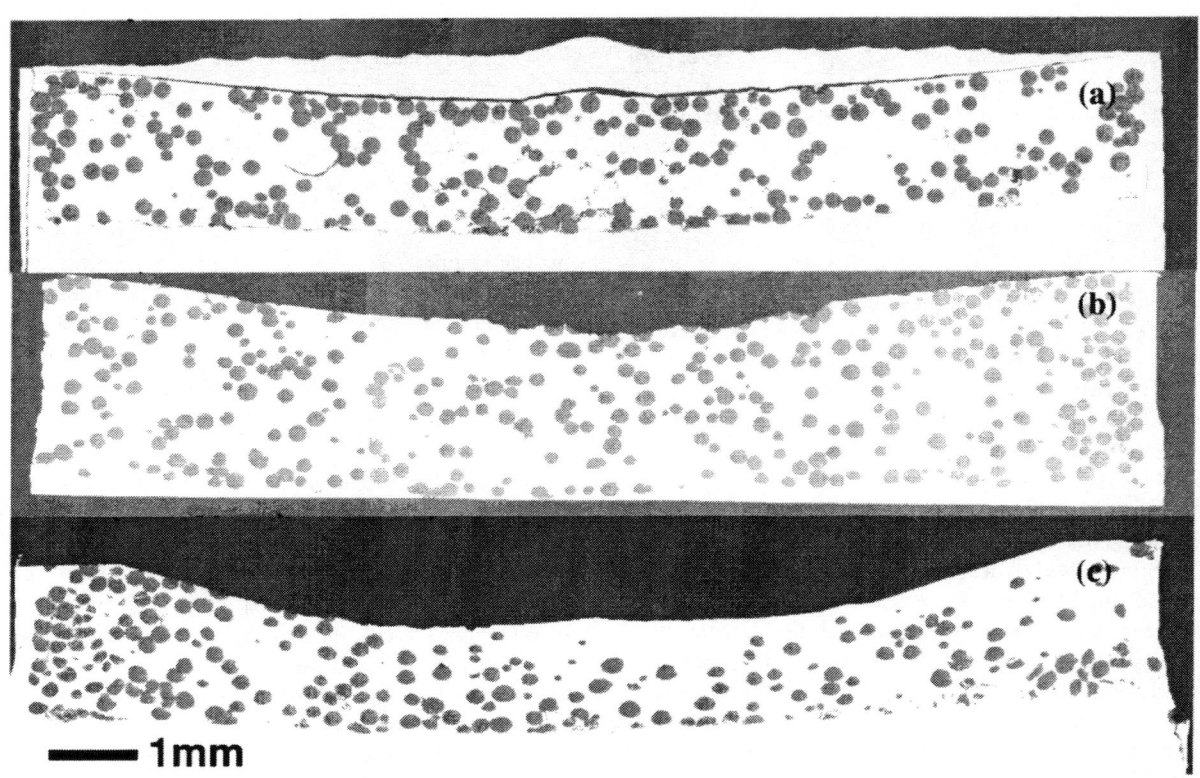

FIGURE1. Cross sections of post-shock samples at 14 GPa (a), 22 GPa (b), and 31 GPa (c). Fused silica beads (dark dots) are in copper powders (light area) with an initial porosity of about 13%. Most bead diameters range between 100 and 200 μm.

using aluminum containers with mixtures of silica beads and aluminum powders, but it was quite difficult to recover samples from pressures as low as 18 GPa.

NUMERICAL SIMULATIONS

Discrete Meso-Dynamic Method (DM^2) has been applied to model the deformation of the beads embedded in porous powders such as copper and aluminum powders.

The details of the method has been published elsewhere [4]. We have investigated the deformation of the beads as a function of time, and impact velocity. To make a simple configuration, the assembly of a flyer of SUS304 (0.5mm thick), a mixture of metal powder and beads (0.4 mm thick), and container (0.9 mm thick) are used at impact velocities of 0.7, 1.0 and 1.5 km/sec.

In the first series of calculations for fused silica beads in a porous copper powders (18% porosity), the aspect ratios decrease with increasing impact velocity (peak shock pressure). They do not change at times of 0.2 and 0.4 μsec after impacted, as shown in Figure 5.

However, beyond about 0.5μm later after impacted, the ratio returns to nearly one, suggesting that the deformed beads rebound. The time for beads to start to rebound corresponds to the arrival time of the refraction wave originated from the back of a flyer plate. A second series of calculations for the beads in porous aluminum powders (18% porosity) indicate that the obtained aspect ratio is quite similar to the results for the copper powder. The distribution of porosity around a bead seemed to be very important to the initial stage of deformation.

The simulated changes of aspect ratio before rarefaction wave arrives are indicated in Figure 6, to compare with the experimental observations. The results are consistent with the experimental observations although the simulation indicates rebounding deformation after subjected to pressure release. Additional simulation on the effect of different containers was carried out at a pressure of 14 GPa. The mean aspect ratio of beads at 0.2 μsec after impacted is 0.86 in a steel container and 0.83 in aluminum container, and beads in latter container indicate to rebound faster and greater.

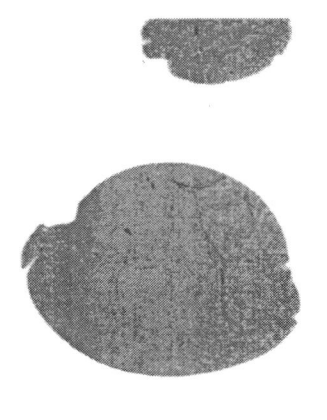

FIGURE 2. Asymmetrically deformed bead in the post-sample from 2 GPa (Figure 1.b). The bead width is about 200 μm.

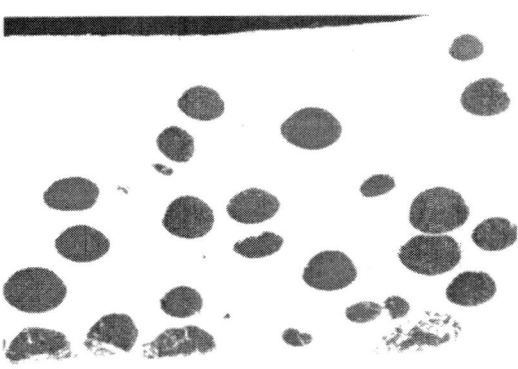

FIGURE 3. Some deformed beads in the post-shock sample with initial porosity of 13%, recovered from 31 GPa (same as in Fig. 1c).

FIGURE 4. Some deformed beads in the post-shock sample with initial porosity of 34%, recovered from 32 GPa.

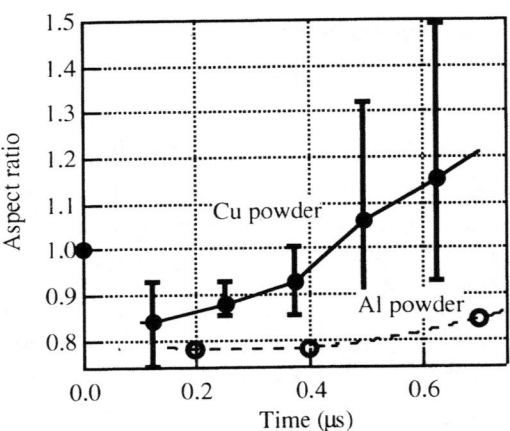

FIGURE 5. Simulated changes of the aspect ratio of beads as a function of time. Solid curve is for fused silica beads in porous copper powders, and broken curve for the beads in porous aluminum powders 0.5 mm thick steel flyer impacts on a 0.4 mm thick sample backed on steel plate at 1.5 km/sec.

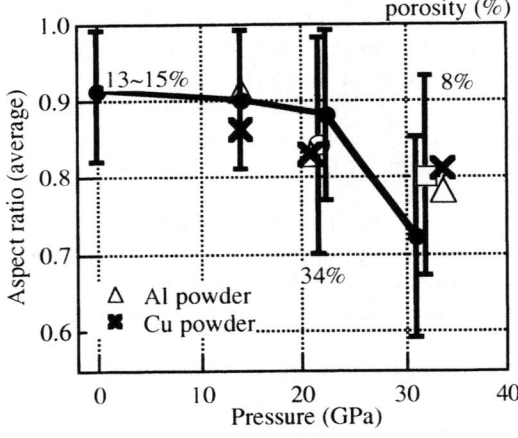

FIGURE 6. Summary of the change of aspect ratio of beads as a function of shock pressure. Solid circles are beads in copper powders with initial porosity of 13~15%. Open circle is for beads in 34% porosity of copper powders and open square for beads in 8% porosity of copper powders. Crosses and open triangles are for simulated results on beads in 15% porosity copper powders and aluminum powders in steel containers, respectively.

DISCUSSION

Our present experimental and simulation results indicate that beads in porous metal powders change the aspect ratio by shock compression. The change of the aspect ratio is a plastic deformation and can be quenchable. The distribution of pores contacting on each bead may play an important role for the initial deformation. The shock pressure is also a key factor to the change of the aspect ratio.

Nakamura et al. [5] and Tomeoka et al. [6] have carried out experiments on Allende and Murshison, respectively. Their data indicate that the mean aspect ratio of chondrules increases linearly with increasing shock pressure up to about 30 GPa. The change of the aspect ratio is much greater in natural samples, and it can be recognized at a lower onset pressure of about 10 GPa. This may be due to a higher porosity of the natural meteorites. The porosity of Allend and Murchison meteorites are ~23% and ~26%, respectively. The distribution of pores also may affect. If it is heterogeneous, then the stress distribution is also heterogeneous and more shear stress is expected during shock compression. If the beads are subjected to heterogeneous pressure, the flattening should not be observed. Shock loading is uniaxial compression, and provides a chance for beads to flatten. The degree depends on factors such as peak pressure, porosity, and so on, as indicated by both the present experimental observations and simulations.

REFERENCES

1. Hewins, R.H., Jones, R.H., Scott, E.R. (eds), *Chondrules and the Protoplanetary Disk*, Cambridge University Press, Cambridge, 1996.
2. Martin, P.M. and Mills, A.A., *Earth Planet. Sci. Lett.* **51**, 18-25 (1980).
3. Sneyd, D.S., McSween, H.Y., Sugiura, N., Strangway, S.W., *Meteoritics* **23**, 139-149 (1988).
4. Tang, Z.P., Horie, Y., and Psakhie, S.G., "Discrete Meso-Element Modeling of Shock Processes in Powders," in *High-Pressure Shock Compression of Solids IV*, edited by L. Davison, Y. Horie, and M. Shahinpoor, New York 1997, pp. 143-175.
5. Nakamura, T., Tomeoka, K., Sekine, T., and Takeda, H., *Meteoritics* **30**, 334-347 (1995).
6. Tomeoka, K., Yamahana, Y., and Sekine, T., *Geochim. Cosmochim. Acta* **63**, 3683-3703 (1999).

THE POSSIBLE COMPOSITION AND THERMAL STRUCTURE OF THE EARTH'S LOWER MANTLE AND CORE

Zizheng Gong[1], Xijun Li[2], Fuqian Jing[3,1]

[1] *Laboratory of High Pressure Physics, Southwest Jiaotong University, Chengdu 610031, China*
[2] *Institute of Electronic Engineering, China Academy of Engineering Physics, Maianyang, Sichuan 621900,*
[3] *Laboratory for Shock Wave and Detonation Physics Research, Southwest Institute of Fluid physics, Maianyang, Sichuan 621900, China*

Abstract. In order to constraint on the possible mineral composition, thermal structure of the Earth's interior, equation of state, sound velocity and shock temperature and melting measurements of perovskite-enstatite (Mg, Fe)SiO_3 and porous iron, were performed through shock wave experiments. A fully new mineralogical model of the lower mantle is proposed The temperature of the core-mantle boundary (core side) is 3750($\pm$150) K, the inner-outer core boundary is 5350($\pm$150) K, and the center of the core is 5500($\pm$150) K.

INTRODUCTION

For more than 30 years shock wave has played an important and effective role in the study of the Earth's science. Up to now, two aspects of the Earth's deep interior are still unknown: (1) Mineralogical composition and (2) Temperature profiles. Comparing the in situ determination of physical properties (density, bulk and shear moduli etc.) of candidate mineral assemblages with that of PREM (Preliminary reference Earth Model [1]), is the main way to determine the possible composition of Earth's deep interior.

(Mg, Fe) SiO_3 with perovskite (pv) structure has been considered to be main phase of the lower mantle. Equation of state (EOS) and elastic properties are increasingly well constrained [2, 3, 4]. Few studies, however, have been able to achieve sound velocities on (Mg, Fe) SiO_3-pv Under the actual pressure and temperature condition of the lower mantle. In this paper, shock wave study of enstatite (Mg$_{0.92}$, Fe$_{0.08}$)SiO_3 pv at pressure from 40 GPa to 140GPa were reported and its geophysical implications are discussed.

The melting point of iron at the pressure of the outer (liquid) core-inner (solid) core (330GPa) was suggested to provide a constraint on the temperature estimation [5]. However, there was a great gap of it between former shock wave compression (SWC) data (around 7500K)[6,7,8] and diamond anvil cell (DAC) results (around 4800K)[9]. This deviation was testified and discussed through shock wave experimental results of porous iron. The temperature profiles of the Earth's core was suggested in term of corrected melting curve of iron.

NEW MINERALOGY MODEL OF THE LOWER MANTLE

Shock Wave Experiments of Enstatite-pv.

The enstatite samples studied in this experiment were made of the natural enstatite minerals, which were collected from Mine, Changjiakou, Hebei Province, China. After purification, grinding and chemical treatment, the tiny powder of enstatite (<0.05mm in diameter) was hot pressed into disk-like sample, at 56MPa and 1700K using graphite cylindrical mould at vacuum condition. The average density of initial samples is 3.08g/cm^3(its crystal density is 3.27g/cm^3).The chemical composition are SiO_2 (54.72%), MgO (31.09%), FeO (5.00%), Al_2O_3 (4.02%), Fe_2O_3 (2.44%), CaO (1.70%), and d values of x-ray Diffraction are 3.167(100), 2.876(64), 2.499(6) and 2.940(5). It can be represented typically as $(Mg_{0.92}, Fe_{0.08})SiO_3$.

(1) Hugoniot Equation of State of Enstatite -pv.

The Hugoniot EOS experiments were carried out with 37-mm two-stage light gas gun using metal flyer plate bearing projectiles to impact samples at speeds of up to 6.5km/s. In all experiments, shock wave velocity, D, in the sample was measured using the electrical probe technique. The particle velocity behind the shock front, u, and pressure-density states were calculated through the impedance-match method [10]. 8 shots of impact experiments was conducted, and together with the experimental data of McQueen [4] and Watt [2], a linear relation between D and u (both in units of km/s) of En-pv

$$D = 5.13 + 1.2u \qquad (1)$$

is drawn in Fig.1. This shows that no phase change happens for enstatite-pv in the temperature and pressure ranges in the lower mantle (40-140GPa).

(2) Sound Velocity of Enstatite-pv

The compressional sound velocity v_P for the enstitate specimens was measured by the optical analyzer techniques under shock loading [11]. Three samples with the same diameter (15mm) and about 2, 3, 4 mm thick (stepped-sample) are arranged on the same plane. The initially transparent bromoform (BF) is used as the optical analyzer medium. Three thin optical fibers are used to monitor the light radiation from BF. Each fiber records the radiation history for only one step of the stepped-sample.

The dependence of compressional sound velocity v_P of enstatite-pv on Hugoniot pressure fitted for five shots can be described by [12]:

$$\ln v_P \text{ (km/s)} = 2.030 + 0.104 \ln P \text{(GPa)} \qquad (2)$$

When compared it with the calculated bulk velocity, no evidence of shock-induced melting is found. This means what we measured is really the compressional wave of sample. At the experimental pressure range, the compressional wave velocity of enstatite varies linearly with density (seeing Fig.2), satisfied Birch's Law [13]: $v_P = 6.213 + 1.212\rho$, where ρ is the corresponding density, which means enstatite (pv) is stable in the lower mantle conditions.

(3) Shock Temperature and Melting of En.-pv

The shock temperature T_H of enstatite-pv were measured using a six-channel high sensitivity tran-

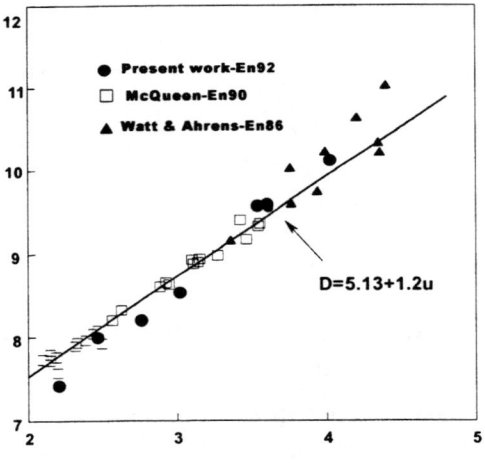

FIGURE 1 All present shock velocity-particle velocity data and linear fits for enstitate-pv.

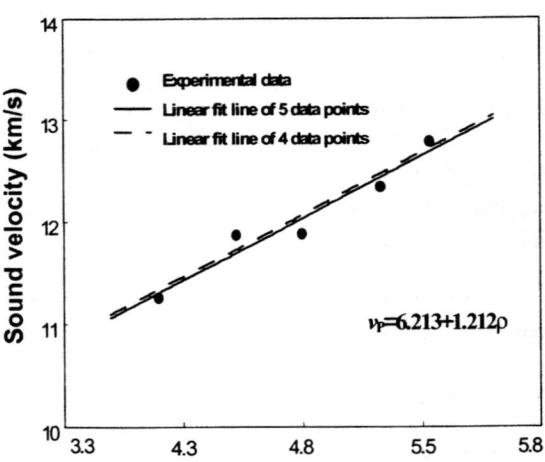

FIGURE 2 Relationship between compression sound velocity and density for enstatite-pv

sient optical pyrometer. Three samples with the same diameter (15mm) and about 2, 3, 4 mm of thickness are arranged on the same plane, and LiF window with the same diameter and about 8, 7, 6 mm of thickness were backed on them, respectively.

The parallelism is about 5μm and 1μm, and the plainness is less than 2μm and 0.5μm, for samples and windows, respectively. Therefore, the width of gap between sample/window none-ideal interface is about 1μm. T_H of samples are calculated according to the three layer heat conduction model for sample-gap-window [14]. Measured T_H increases continuously with increasing pressure from 3319±160K at 46 GPa to 8272±420K at 140 GPa. The P-T Hugoniot lies under the melting line measured by static high pressure, inferring that no melting or decomposing happened in our experimental pressure range. According to the shock temperature results, melting point of perovskite $(Mg_{0.92}, Fe_{0.08})SiO_3$ at core-mantle boundary (CMB) pressure (136 GPa) should be higher than 7600 K at least, this is in very good agreement with the extrapolation of 8000±500K of Zerr & Boehler's experiments [15], and the extrapolation of 4500±100K of Knittle & Jeanloz' [16] was negated.

Geophysical Implications

(1) A Fully New Mineralogical Model of the Lower Mantle [17]

(1) Both the corrected Hugoniot density and compressional wave velocity are compared with PREM. Both the profiles of density and P wave velocity of $(Mg_{0.92}, Fe_{0.08})SiO_3$-pv are parallel with those of PREM. The density is by 1.6% on average higher and P wave velocity is by 1.81% on average higher than those of PREM, respectively. Based on density and compressional velocity constraints, Reuss-Voigt-Hill average conclude that the lower mantle is mainly composed of 89.69%-88.65% (wt.% or mol.%) $(Mg_{0.92}, Fe_{0.08})SiO_3$-Pv and 10.31%-11.35% $(Mg_{0.92}, Fe_{0.08})O$, and the lower mantle is chondrite-rich. This is consistent with most of the previous results.

(2) As shown in Fig. 3, the profile of bulk modulus (K_S) was obviously divided into two parts by that of PREM at depth of about 1771±100 km:it

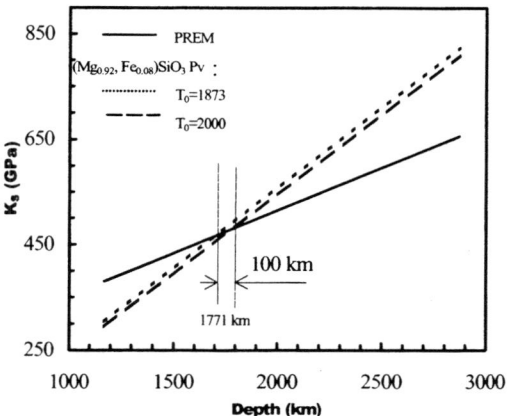

FIGURE 3 $K_S(P)$ of enstatite-pv. compared with PREM

was lower than that of PREM at depth shallower than 1771 km and the deviation decreases with depth, the largest deviation is -9.71% and the average the -4.13%. At 1771±100km depth it is equal to that of PREM. Below 1771 km depth, it is higher than that of PREM and the deviation increase with depth, the largest deviation is 8.22% and the average 5.46%.

(3) Based on the above mentioned, a fully new mineralogy model of the lower mantle is proposed (seeing Fig.4): ① $(Mg_{0.92}, Fe_{0.08})SiO_3$-Pv is the main mineral phase (not less than 85%). ② A chemical boundary exists at the depth of about

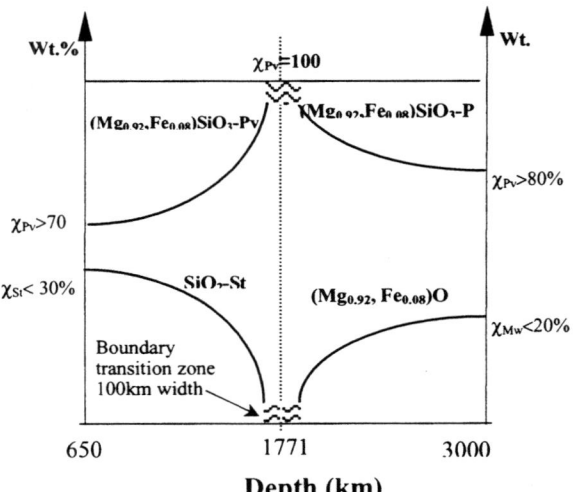

FIGURE 4 A fully new mineralogy model of the lower mantle

1771±100km in the middle of lower mantle. ③ Above the boundary stishovite SiO_2 is the second main mineral phase. SiO_2 (st) has a largest amount (<30%) at the top of lower mantle and its amount decreases with depth to the smallest (the absence is not excluded) at the boundary. ④ Below the boundary $(Mg_{0.92}, Fe_{0.08})O$ (Mw) is the second main mineral phase. Mw has a smallest amount (the absence is not excluded) at the boundary and its amount increases with depth to the largest amount (<20%) at the bottom of lower mantle. This new model is supported by the latest seismological detection data.

(2) Lateral Thermal Heterogeneity in the Lower Mantle [18]

In terms of the definition of sound velocity and thermodynamics, the temperature coefficients of sound velocity of perovskite-enstatite under high pressure were obtained [18] $|(\partial v_P/\partial T)_P|<0.25$ m $/K \bullet S^{-1}$ when pressure is high than 100GPa. On the basis of our data, we conclude that the compressional wave velocity anomaly of 0.1-0.2% in the deep lower mantle [19] and 2% in the D" region [20] would imply lateral thermal heterogeneity with an amplitude of 53-106K and 1066K in these regions respectively.

POSSIBLE COMPOSITION AND TEMPERATURE PROFILE OF THE CORE

Possible Composition of the Core

It was generally accepted that about 90%(wt.) of the core is iron, a small amount of light element is present in it. According to the last point of view, (Fe-S-O) or (Fe-S-Si) three element system were suggested to be the possible composition of the core. But we don't know what mineral styles they exist. As shock Hugoniot of Nandan Iron meteorite (Fe 92.5%, Ni6.8%, Co0.47%, wt.) from 60 to 208GPa showed that its pressure-density fits with that of PREM [21]. So the iron-nickel alloy (Fe93%, Ni7%) core is possible.

Thermal Structure of the Core

(1) Melting Point of Iron at 330GPa

As it was mentioned above, there is a large difference in experimental results between extrapolation of SWC and DCA. Is this a systematic deviation between the two methods? Can this be corrected?

For a kind of porous iron with average initial density ρ_0=6.904g /cm^3, in the pressure range P_H<122GPa, the measured sound speed (using optical analysis method with Al_2O_3 windows) is the longitudinal wave speed in nature, which can be empirically fitted as v_P= 5.951+ 1.224$\ln P_H$− 0.0349$(\ln P_H)^2$, where P_H is in unite of GPa, v_P of km/s; in the pressure range of P_H >157GPa, the measured sound speed is its bulk sound speed C_B in nature [22]. This means the shocked porous iron is in solid states when P_H <122GPa and transforms into liquid state in the pressure range 122-157GPa. Following the interface temperature model under the case of shock-induced melting [23], two melting points of iron were measured that are (171.4GPa, 5550-5730K) and (98GPa, 4470-4610K). The two melting points are within the scope of the SWC's data extrapolation. Both the high pressure sound speed and melting temperature experiments support the confidence of the SWC's data extrapolation, or, in other words, the systematic deviation between the SWC's data and the DAC's melting data is indeed a phenomena of objective reality. According to overheating homogeneously nucleation and catastrophe model, overhot melting point T_M^{Over} in SWC is 1.212 time of real melting point T_M for iron [24], and according to surface melting model, the observed results T_M^{Exp} in DAC should be 0.9

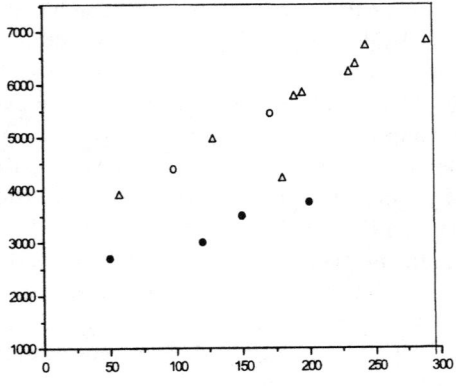

FIGURE 5 systematic deviation of melting data of iron between the SWC method and DAC method.

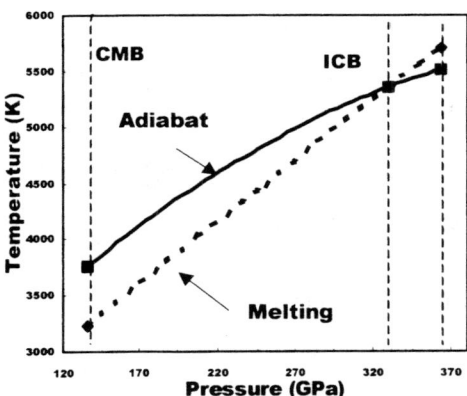

FIGURE 6 Temperature gradient in the Earth's core

time of the real melting point T_M [25]. It is surprising that these two corrected data are almost in coincidence with each other. The melting curve of iron can be fitted form corrected data as: $T_M = 2523(\pm153) + 11.679 (\pm2.204) P - 0.00157 (\pm0.00609) P^2$. From this equation, T_M of pure iron at 330GPa is 6200K. This value agrees well with that many others have found.

(2) Temperature Profile of the Core

It was reported that depression of melting point by impurities is about 700-1000K [26], So temperature at the inner-outer core boundary is constrained to be 5350(±150) K. Based on the assumption of adiabatic compression, the temperature at the boundary of the core-mantle (core side), and the center of the core are around 3750(±150) K, 5500(±150) K, respectively, with $\gamma=1.3$ and 1.23 [26], respectively. Temperature gradient of the Earth's core is shown in Fig.6.

ACKNOWLEDGMENTS

This research was supported by the National Natural Science Foundation of China under Grant No. 10032040.

REFERENCES

[1] Dziewonski, A. M. and Anderson, D. L., *Phys. Earth Planet. Interiors*, **25**, 295 (1981).

[2] Fiquet, G., Dewaele, A., and Andrault, D., et al., *Geophys. Res. Lett.*, **27**, 21 (2000).

[3] Watt, J. P. and Atrens, T. J., *J. Geophys. Res.*, **91**, 7495 (1986).

[4] McQueen, R. G., Marsh, S. P., et al.: *J. Geophys. Res.*, **72**, 4999 (1967).

[5] Brich, F., *J. Geophys. Res.*, **57**, 227-286 (1952).

[6] Williams, Q., Jeanloz, R., and Bass, J. D., et al., *Science*, **236**, 181 (1987).

[7] Bass, J. D., Svendsen, B., and Ahrens, T.J., in *High Pressure Research in Mineral Physics*, Manghnani, M. H. And Y. Syono (eds.), Terra Secentific, Tokyo, 1987, pp.393-402.

[8] Yoo, C.S., Holmes, N. C., and Ross, M., et al., *Phys. Rev. Lett.*, **70**, 3931-3934 (1993).

[9] Boehler, R., *Nature*, **363**, 534-536 (1993)

[10] Jing, F. Q., *Introduction to Experimental Equation of State*, second editor, Academic Press, Beijing, 1999, Chap 4

[11] McQueen, R. G, Hopson, J. W. et al., *Rev. Sci. Instrum*, **53**, 245 (1982).

[12] Gong, Z.Z., Jing, F.Q.et al., *Chin. Phys. Lett.*, **16** (9), 695~697 (1999).

[13] Campell, A. J.and Heinz, D.L., *Science*, **257** 66 (1992).

[14] Tan, H. and Dai, C.D., *Chinese J. High pressure Phys.*, **14**, 81 (2000), (in Chinese).

[15] Zerr, A. and Boehler, R., *Science*, **262**, 553 (1993).

[16] Knittle, E. and Jeanloz, R., *Geophys. Res. Lett.*, **16**, 421 (1989).

[17] Gong, Z.Z., *Equation of State, Sound velocity and melting of Enstatite at High Pressure: Constraints on the Possible Composition and thermal structure of the earth's lower Mantle*, Postdoctoral Research Report, Institute of Geochemistry, Chinese Academy of Sciences. 1999.

[18] Gong, Z. Z., Jing, F. Q., et al., *Chin. Phys. Lett.*, **17**(3) 218-220 (2000).

[19] Dziewonski, A. M. and J. H. Wooshouse, *Science*, **236**, 37 (1987).

[20] .Lay, T., *Rev. Geophys., Suppl.*, 325 (1995).

[21] Fu, S. Q., Jin, X.G.and Wang D. D., *Chinese J. Geophys* **36** (2), 158-163 (1993).

[22] Li, X.J., Fu, Q.J., et al., *Chin. Phys. Lett.*, **18**(9) (2001).will be published.

[23] Tan, H. And Ahrens, T.J., *High Press. Res.* **2**, 159-182 (1990).

[24] Lu Ke and Li Yi, *Phys. Rev. Lett.*, **80**, 4470 (1998).

[25] Williams, Q. and Knittle, E., *J. Geophys. Res.*, **96**, 2171-2184 (1990).

[26] Anderson, O.L., *Phys. Earth Planet.Interiors*, **109**,179-197 (1998).

MOLECULAR DYNAMICS MODELING OF IMPACT-INDUCED SHOCK WAVES IN HYDROCARBONS

Mark L. Elert[1], Sergey Zybin[2], and C. T. White[3]

[1]Chemistry Department, U. S. Naval Academy, Annapolis, MD 21402
[2]Department of Chemistry, The George Washington University, Washington, D.C. 20052
[3]Code 6189, Naval Research Laboratory, Washington, D.C. 20375

Abstract. We use nonequilibrium molecular dynamics (MD) simulations to study the behavior of hydrocarbons under shock compression and spallation processes in shock and rarefaction waves generated by the high-velocity impact of a flyer plate into a target material. The interatomic forces were introduced using a recently modified reactive empirical bond order (REBO) potential with intermolecular interactions, termed the adaptive intermolecular REBO potential (AIREBO). This potential allows us to simulate as many as ten thousand molecules on a single processor, providing a relatively large cross-section at the shock front in hydrocarbon solids. We performed plane-wave impact experiments with different flyer velocities and observed the chemical dissociation of methane and acetylene molecules in the shock layer, followed by polymerization into carbon chains for certain flyer velocities. The hydrocarbon oligomers survive into the rarefaction region, indicating that stable molecular products have been formed. These results may be significant for the understanding of shock-induced chemical reactions resulting from meteorite impact in planetary atmospheres and methane ice surfaces.

INTRODUCTION

The chemistry of hydrocarbons subjected to shock impact may be relevant to our understanding of the composition and evolution of planetary atmospheres and to the formation of life. Comets and carbonaceous asteroids are known to be rich in organic compounds, but these materials may have undergone pyrolysis, or alternatively may have formed more complex compounds via shock synthesis in the atmosphere, during impact onto the surface of the early earth [1,2]. In addition to its possible cosmological interest, shock wave impact on hydrocabons may provide a unique environment for the synthesis of organic compounds under novel kinetically-controlled conditions.

To effectively study the shock-induced chemistry of condensed-phase hydrocarbons using molecular dynamics, it is necessary to simulate the behavior of several thousand atoms for at least several picoseconds. A large cross section is necessary to allow the formation of oligomers without spurious effects from periodic boundary conditions, and a sufficiently long time scale is required to determine whether reaction products survive into the rarefaction region behind the advancing shock front. Although workers at Los Alamos National Laboratory have recently demonstrated the feasibility of reactive shock simulations employing up to 576 benzene molecules [3] and 1728 methane molecules [4] for times on the order of one picosecond using a semiempirical tight-binding method, larger simulations must rely on empirical potentials. In the present study we use the reactive empirical bond order (REBO) potential for hydrocarbons developed by Brenner [5] and modified by Stuart, Tutein, and Harrison [6] to include intermolecular and torsional effects. This potential can accurately model bond-breaking and bond-forming processes in hydrocarbons, including changes in the hybridization of carbon atoms. The potential has been successfully applied to such diverse problems as the study of phase transitions in carbon [7,8], the use of carbon nanotubes as STM tips [9], and the compression and friction of anchored hydrocarbon chains on diamond surfaces [10]. Preliminary two-dimensional simulations of shock-induced chemistry in acetylene using this potential have been reported previously [11].

CALCULATIONS

MD simulations of shock impact on solid methane and acetylene were carried out by impacting crystals of the subject hydrocarbon with a flyer plate of the same material. Periodic boundary conditions were employed in the two directions perpendicular to that of shock wave propagation. Flyer plates were between four and eight unit cells in thickness, and the cross section of the periodic supercell was eight unit cells in each of the transverse dimensions. The time step for integration of the equations of motion was between 0.05 and 0.10 fs, depending on flyer plate impact speed.

For acetylene, shock wave chemistry was investigated as a function of flyer plate impact speed and thickness. Some polymerization was observed for impact speeds as low as ten km/s and flyer plates four unit cells in width. At higher impact speeds, the fraction of carbon atoms incorporated into chains at least three atoms in length increased significantly, as shown in Fig. 1.

The fraction of reacting carbon atoms depended also on the width of the impacting flyer plate, which determines the total energy delivered to the target crystal. Figure 2 shows the fraction of carbon atoms in oligomers of three or more carbons at comparable times in simulations with different flyer plate widths.

Most of the hydrocarbon oligomers formed in the acetylene simulations were quite short, and were of course hydrogen-deficient as a result of the 1:1 carbon-hydrogen stoichiometry of the reactant material. Typically about ten percent of the carbons participating in chains of three or more atoms were found in chains of eight or more, and no oligomers longer than about fifteen carbons were found. Most of the reaction products were straight chain polymers, although some ring structures were also formed. Four-carbon chains were the dominant product at an impact speed of 12 km/s, but at higher impact speeds the distribution favored three-carbon chains, as shown in Fig. 3. In fact, the data suggest that product hydrocarbons of maximum complexity

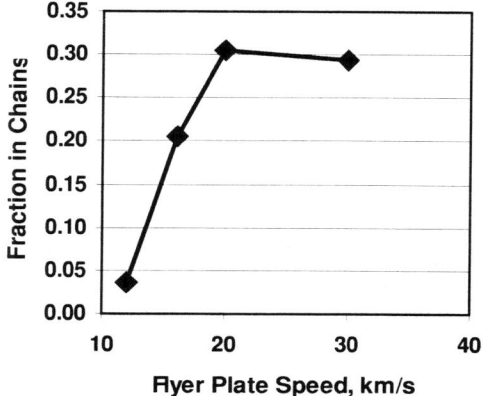

FIGURE 1. The fraction of carbon atoms in the target acetylene crystal which are involved in chains of at least three carbons in length, as a function of flyer plate speed. All measurements were made at a point approximately 1.5 ps after the start of the simulation, when most of the energy of the flyer plate had been dissipated in the target crystal. Flyer plates were five unit cells thick.

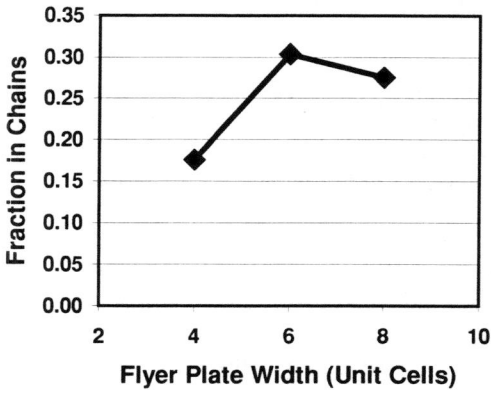

FIGURE 2. The fraction of carbon atoms in the acetylene crystal participating in chains of three or more carbons after passage of the shock wave, as a function of flyer plate thickness, for an impact speed of 20 km/s.

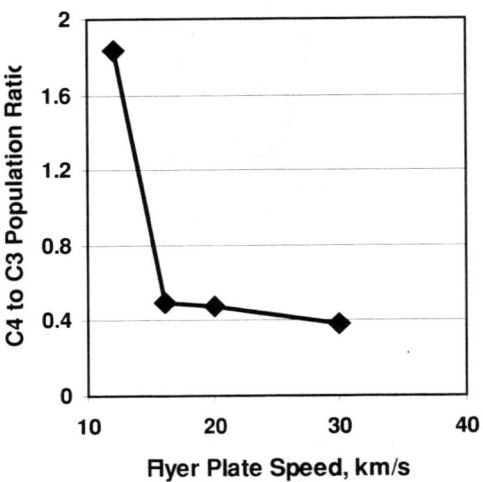

FIGURE 3. The ratio of the number of carbon atoms found in four-carbon chains to the number in three-carbon chains, as a function of flyer plate impact speed, for the same four simulation snapshots described in Fig. 1.

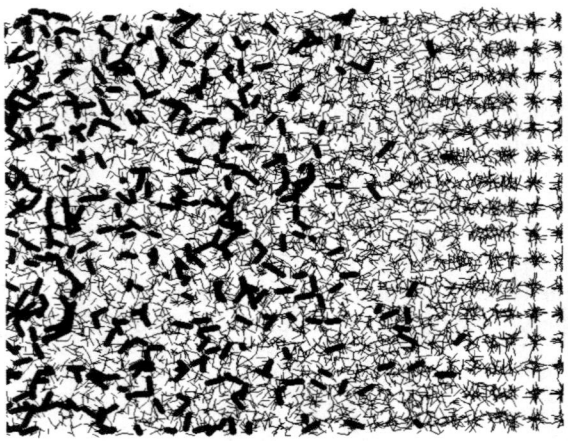

FIGURE 4. Carbon-carbon bonds resulting from shock impact of methane at a flyer plate impact speed of 30 km/s are shown as heavy lines. Carbon-hydrogen bonds are shown as lighter lines.

are formed preferentially when the incident energy density is only moderately greater than the threshold value needed to induce polymerization.

The results for simulations of shock impact in methane are qualitatively different from those in acetylene. With no pi bonding and a high H:C ratio, methane is much less likely than unsaturated hydrocarbons to undergo polymerization reactions. Simulations with flyer plate impact speeds of twenty km/s produced essentially no carbon-carbon bonding, and even impacts of 25 km/s produced only C_2 products. More significant chemistry was found with flyer impacts of 30 km/s. A sample frame from such a simulation is shown in Fig. 4. In this figure, the shock wave is proceeding from left to right through the methane crystal. A few unit cells of the unshocked crystal are visible at right. (In this view, looking through a thickness of eight unit cells, the free rotation of methane molecules in their lattice positions is apparent.) Carbon-carbon bonds, shown as thick lines in the figure, have formed behind the shock front, and a few examples of C_3 and C_4 chains are evident.

The products resulting from the 30 km/s impact simulations are strikingly similar to those predicted by Kress et al. [4] in their double-shock simulations of methane using a semiempirical tight-binding method. This lends further support to the conclusion of those authors that the initial products of shock-induced chemistry in methane are molecular hydrogen and hydrocarbon oligomers, although experimental studies [12, 13] indicate that solid amorphous carbon is formed at much longer time scales.

CONCLUSION

MD simulations of shock-induced chemistry in hydrocarbons using the AIREBO potential [6] can produce information relevant to the study of organic material processing in meteor and comet impacts with planetary atmospheres. Acetylene is found to readily form hydrocarbon oligomers at impact speeds which are modest on a planetary scale, whereas methane is unreactive up to much higher flyer plate velocities. Although acetylene polymerization is thought to lead eventually to the formation of polycyclic aromatic hydrocarbons (PAHs) in interplanetary dust particles and large solar system objects [14], the present study indicates that PAHs are not the initial products of shock impact on acetylene.

ACKNOWLEDGMENTS

This work was supported by the Office of Naval Research. MLE received additional support from the Naval Academy Research Council.

REFERENCES

1. Chyba, C., and Sagan, C., *Nature* **355**, 125-132 (1992).
2. Chyba, C., Thomas, P. J., Brookshaw, L., and Sagan, C., *Science* **249**, 366-373 (1990).
3. Bickham, S. R., Kress, J. D., and Collins, L. A., *J. Chem. Phys.* **112**, 9695-9698 (2000).
4. Kress, J. D., Bickham, S. R., Collins, L. A., and Holian, B. L., *Phys. Rev. Lett.* **83**, 3896-3899 (1999).
5. Brenner, D. W., *Phys. Rev. B* **42**, 9458-9471 (1990).
6. Stuart, S. J., Tutein, A. B., and Harrison, J. A., *J. Chem. Phys.* **112**, 6472-6486 (2000).
7. Glosli, J. N. and Ree, F. H., *Phys. Rev. Lett.* **82**, 4659-4662 (1999).
8. Glosli, J. N. and Ree, F. H., *J. Chem. Phys.* **110**, 441-446 (1999).
9. Harrison, J. A., Stuart, S. J., Robertson, D. H., and White, C. T., *J. Phys. Chem. B* **101**, 9682-9685 (1997).
10. Tutein, A. B., Stuart, S. J., and Harrison, J. A., *J. Phys. Chem. B* **103**, 11357-11365 (1999); Tutein, A. B., Stuart, S. J., and Harrison, J. A., *Langmuir* **16**, 291-296 (2000).
11. Elert, M. L., Swanson, D. R., and White, C. T., "Molecular Dynamics Simulation of Shock-Induced Chemistry in Acetylene," in Shock Compression of Condensed Matter – 1999, edited by M. D. Furnish, L. C. Chhabildas, and R. S. Hixson, AIP Conference Proceedings 505, New York, 2000, pp. 283-286.
12. Nellis, W. J., Ree, F. H., van Thiel, M., and Mitchell, A. C., *J. Chem. Phys.* **75**, 3055-3063 (1981).
13. Nellis, W. J., Hamilton, D. C., and Mitchell, A. C., *J. Chem. Phys.* **115**, 1015-1019 (2001).
14. Sagan, C., Khare, B. N., Thompson, W. R., McDonald, G. D., Wing, M. R., Bada, J. L., Vo-Dinh, T., and Arakawa, E. T., *Astrophys. J.* **414**, 399-405 (1993).

HIGH INTENSITY X-RAY COUPLING TO METEORITE TARGETS

J. L. Remo[1] and M. D. Furnish[2]

[1] Harvard Smithsonian Center for Astrophysics, Planetary Science Division, Mail Stop 18, 60 Garden Street, Cambridge Massachusetts 02138.
[2] MS 1168, Sandia National Laboratories, P.O. Box 5800, Albuquerque NM 87185-1168

Abstract. Experimental results of shock wave effects from high intensity (70 -215 GW) soft X-ray irradiation on several meteorite targets are presented. From inhomogeneous materials, useful data on particle velocity and in-situ velocity were obtained and permitted the computation of the yield stress, shock wave velocity, compression, as well as the momentum and energy coupling coefficients.

INTRODUCTION

High intensity (≥ 200 GW/cm^2) X-ray pulses generated from an exploding wire/hohlraum configuration at the Sandia Z-machine have been used to generate shock wave driven high pressures (multi megabar range) on various test samples and structures in order to determine the equation of state (EOS) and constitutive properties of materials at these high pressures[1]. Following this lead, we report on the utilization of the Sandia Z-machine to irradiate several meteorite specimens with soft X-rays (Plankian and line emission) in order to study the meteorite targets' response to high rates of dynamic loading provided by the ablation driven shock waves and the ensuing high pressure generation throughout the target sample. Previous work on these same meteorite targets used pulsed lasers to generate pressures from 0.7 to 11 GPa[2], the results of which provided significant insights into the response of different meteorite material categories[3] to high strain rate dynamic loading.

The rationale for these Z-pinch experiments is an outgrowth of a suggestion in 1995[4] that soft X-ray hohlraums be used to provide experimental approaches to understanding how the microstructures of near-earth objects (NEOs) respond to high pressure and loading conditions, using meteorites as asteroid analogs. This current series of experiments is anticipated to lead to an empirical understanding of the high pressure thermodynamics and material properties, such as material strength and isentropic compression and decompression of several different meteorite materials.

Of particular importance for this research is the determination of the momentum coupling coefficient, C_M, for the NEO material categories when subjected to intense (soft) X-ray irradiation. Knowledge of C_M is absolutely necessary to calculate orbital adjustments of potentially hazardous NEOs. Another objective is to gain an understanding of the EOS and constitutive properties of the different meteorite categories, which will help in modeling the dynamic response of asteroids to a high energy density interaction. These objectives will be somewhat difficult to achieve due to the inhomogeneous and irregular nature of these naturally occurring materials. Nonetheless, the results of the initial experiments appear to be encouraging. It is noted that this experimental approach provides significant advances towards understanding high energy density X-ray coupling to heterogeneous materials in general as well as for momentum transfer, heating, phase changes, and radiative scattering interactions with materials encountered in space.

Other applications include the interpretation of momentum coupling and related interactions from strong X-radiation with primordial solar nebula material and the interstellar medium.

DESCRIPTION OF THE SANDIA Z-PINCH EXPERIMENT

The experimental objective is to demonstrate the feasibility of obtaining reliable measurements of shock Hugoniots for meteorite materials experiencing ablative loading in order to determine their EOS and momentum coupling coefficients. The Sandia Z machine is a 4.5 MV accelerator using Marx generators to store capacitive energies of about 11MJ which can produce currents of about 20 MA within the thin conductive wires between the anode and cathode (see Fig. 1) over a time scale of about 100 ns. Usually, a few hundred wires are used to generate the Z pinch source within the primary hohlraum whose typical diameters are 2 - 5 cm with 1- 2 cm heights and contains the radiation produced by the imploded pinch. After implosion of a (tungsten) wire array, a Z pinch Planckian-like radiation source with a 2 mm diameter is formed on axis within the primary hohlraum with temperatures of about 150 eV.

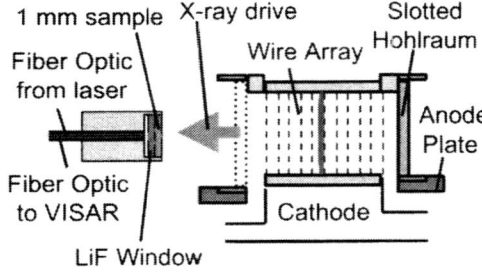

FIGURE 1. Configuration of Z-pinch experiment. Components on the right side of the figure are cylindrically symmetric about the heavy line at the center of the array (location of the pinch).

X-ray radiation is delivered to the sample through slots in the primary hohlraum. For the present experiments, no radiation filtration was performed. Instrumentation used to diagnose the sample response was comprised of a VISAR interferometer measuring the velocity of a spot at the back of the sample. For details of Z pinch instrumentation available, one is referred to many sources such as Konrad[5].

EXPERIMENT RESULTS

Velocity Measurements

Observed velocity profiles are shown in Figure 2, and appear to correspond to attenuating waves. In most cases dual-delay VISAR (velocity interferometry) instrumentation was used to measure velocity histories.

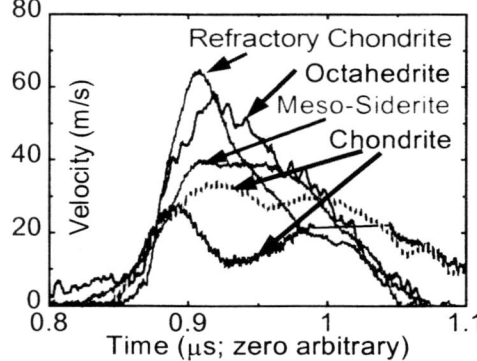

FIGURE 2. Observed velocity histories.

Shocked States

Basic experimental results for high intensity soft X-ray coupling to meteorites targets are summarized in Table 1, which lists estimated stress, P, in gigapascals (GPa), the observed (VISAR) particle velocity, V_p, and the inferred (in situ) velocity, V.

TABLE 1. Shocked states achieved in present tests.

Sample	CV3 Allende		Refr. Chond (LL6) Tuxtuac	Meso siderite Vaca- muerta	FeNi (Og) Odessa
Test #	Z675	Z676	Z676	Z675	Z636
Sample	1/2	1/1	2/1	3/1	8/1
P GPa	0.5	0.5	1.0	0.53	1.20
V_P m/s	33	33	64	39	55
V m/s	30	30	58	39	40
D km/s	5.73	5.73	5.71	3.78	4.16
ρ_0 g/cm^3	2.91	2.91	3.02	3.60	7.21
ρ/ρ_0	1.006	1.006	1.012	1.013	1.013

The shock wave velocity D may be obtained from the momentum conservation relation:

$$P = \rho_0 D V, \qquad (1)$$

where ρ_0 is the pre-shock density. The post shock density, ρ, can be obtained from the mass conservation equation:

$$\rho = \rho_0 D/(D - V) \qquad (2)$$

All of the particle velocities listed in Table 1 appear to be attenuating (transient) waves, with the overtaking release wave corresponding to the decay of the Z-pinch emission. These particle velocities are corrected for the mechanical effect of the windows (which reduces the particle velocity at the interface observed by VISAR by 30 – 50%. The windows were necessary to preserve the reflecting surface, allowing the velocity measurements.

The inferred in-situ velocity can be interpreted as representing a lower bound for the ultimate velocity a thin slice of sample material would have reached under an extended X-ray pulse. Although peak stress is relatively low for natural (meteorite) materials, as compared to pure materials, they nonetheless possess high shock velocities because their particle velocity is very low as compared to pure materials.

Momentum and Energy Coupling Coefficients

Momentum and energy coupling coefficients may be calculated by two methods. The first (SS) treats the x-ray drive as a steady input stress; the second (I), as an impulse. The best value for application to problems is probably intermediate between the values thereby computed. Inputs and results from both methods are shown in Table 2.

In the steady input stress method, the input intensity I is taken as a constant. The momentum coupling coefficient, representing the momentum uptake of the target, is the ratio of the pressure P to the radiation intensity I:

$$C_{M,SS} = P/I = \rho_0 D V/I \qquad (3)$$

Here, D, V, I and P are (respectively) the shock velocity, the in-situ material velocity, the radiation intensity, and the resultant pressure.

The energy coupling coefficient, C_E, is the fraction of original input energy coupled to the target such that,

$$C_E = E_{Sam}/F = \tfrac{1}{2} \rho_0 (d) V^2 /E_{In}, \qquad (4)$$

where F is the (time-integrated) X-ray energy incident on the target (fluence), E_{Sam} is the kinetic energy imparted to the target sample, and d is the sample thickness.

The Z-pinch is not a point source at these ranges R from the pinch, but is intermediate between point and cylindrical, so the intensity varies approximately as $R^{-3/2}$.

In the impulse method of calculating coupling coefficients, the x-ray pulse is taken as providing a brief impulse which accelerates the plate to a limiting velocity taken as the in-situ material velocity just below the monitored surface. Thus,

$$C_{M,I} = \rho_0 (d) D V/F. \qquad (5)$$

TABLE 2. Computation of the momentum, C_M, and energy, C_E, coupling coefficients

Sample	Chondrite (CV3) Allende	Chondrite (LL6) Tuxtuac	Refr. Chond	Mesosiderite Vacamuerta	FeNi (Og) Odessa
Test #	Z675	Z676	Z676	Z675	Z636
Sample	1/2	1/1	2/1	3/1	8/1
d mm*	1.012	1.014	1.01	1.013	1.508
E kJ	867	1116	1116	867	1187
R cm	14	14	14	14	10
Fluence J/cm²	352	453	453	352	945
τ (ns)	5.03	5.03	5.03	5.03	4.40
I GW/cm²	70	90	90	70	215
$C_{M,SS}$ s/m×10⁻⁵	0.050	0.031	0.078	0.080	0.054
$C_{M,I}$ s/m×10⁻⁵	2.51	1.56	3.91	4.0	4.6
C_E ×10⁻³	0.38	0.19	1.13	0.79	0.92

*Radii of all targets was about 3 mm, yielding an area of about 0.283 cm². The volume of each target was 0.029 cm³ except for sample 8 which was 0.038 cm³

Comparison with Laser Results

In analogous experiments conducted with a laser[2], the coupling coefficients may be worked out in a parallel manner. Samples were subjected to a 22 J/cm^2 fluence of 1054 nm Nd-glass laser light (approx. 20 ns). The configuration is shown in Fig. 3. Note that the presence of the quartz window substantially increases the coupling coefficients.

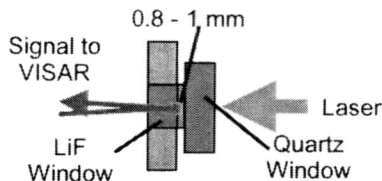

FIGURE 3. Configuration of laser experiment.

Experiment parameters and inferred coupling coefficients are shown in Table 3. The method for calculating the coupling coefficients follows Eqs. 4 and 5 (i.e. impulse assumption). It is worth noting that the momentum coupling coefficients are roughly 50× those obtained for the soft X-rays (impulse calculation), and the energy coupling coefficients are 20× those from the soft X-rays.

TABLE 3. Computation of the momentum, C_M, and energy, C_E, coupling coefficients from the laser experiments

Sample	Odessa (Fe/Ni)		Gibeon (Fe/Ni)	Meso-siderite	LL6 Stony
Test #	N20		A1A	A1B	A2A
	C	D			
Density g/cm^3	7.2	7.2	7.2	3.6	3.02
d mm*	1.0	1.0	0.8	1.0	1.0
V_P m/s	64	72	42	50	42
V m/s	48	54	31	45	40
$C_{M,I}$ s/m ×10^{-5}	157	177	81	74	55
C_E ×10^{-3}	38	48	13	17	11

CONCLUSIONS

Five parasitic experiments were conducted to begin to assess X-ray coupling for representative meteorite compositions. This study used a combination of thermal and line emissions from the Z-pinch. Both momentum and energy coupling coefficients were smaller than for 1 micron laser light (by approximately an order of magnitude). It will be necessary to establish coupling behavior with only line emission, or only blackbody emission, to better understand the physics of this system. Future experiments of value would include those using coupons of pure metal and homogeneous silicates at comparable fluences (there exist data for much higher X-ray fluences as well as for comparable laser fluences).

ACKNOWLEDGEMENTS

This work was supported in part by Sandia National Laboratories, a Lockheed Martin Company, for the United States Department of Energy under contract DE-AC04-94-AL85000.

REFERENCES

1. (e.g.) M. D. Furnish, R. J. Lawrence, C. A. Hall, J. R. Asay, D. L. Barker, G. A. Mize, E. A. Marsh and M. A. Bernard, Radiation-driven shock and debris propagation down a partitioned pipe, *Int. J. Impact Engrg.*, 26, 189-200, 2001.
2. Remo, J. L., High-Power-Pulsed 1054 nm Laser Induced Shock Pressure and Momentum and Energy Coupling to Iron and Stony Meteorites, *Lasers and Particle Beams*, **17** 25-44, 1999.
3. Remo, J. L., Classifying and Modeling NEO Materials Properties and Interactions in " Hazards Due to Comets and Asteroids," T. Gehrels, ed., 551-596, Univ. Arizona Press, Tucson 1994.
4. Hammerling, P. and Remo, J. L., Laboratory Planetary Physics, in "Near Earth Objects; The United Nations International Conference," J. L. Remo ed. 585 - 602, NY. Acad. Sci, NY 1997.
5. Konrad, C. H., J. R. Asay, C. A. Hall, W. M. Trott, B. F. Clark, G. A. Chandler, K. G. Holland, K. J. Fleming, J. S. Lash, L. C. Chhabildas and T. G. Trucano, Use of Z-pinch sources for high-pressure shock wave studies, Sandia National Laboratories report, SAND98-0047, 1998.

THE DYNAMIC STRENGTH OF CEMENT PASTE UNDER SHOCK COMPRESSION

K. Tsembelis, W. G. Proud and J.E. Field

PCS, Cavendish Laboratory, Madingley Road, Cambridge, CB3 0HE. UK.

Abstract. A series of plate impact experiments on cement paste (grout) has been performed to assess the dynamic strength of this material. Lateral stresses have been directly measured by means of embedded manganin stress gauges. In combination with longitudinal stresses, measured previously [1], these results have been used to obtain shear strength under shock loading. Results indicate that the material is behaving in an inelastic manner with the shear strength increasing with increasing pressure.

INTRODUCTION

Considerable interest in characterising the dynamic loading of concrete under impact conditions exists because of its extensive use as a structural material [2-5]. Concrete is a heterogeneous material containing aggregates in a cement matrix. Therefore, characterisation under dynamic conditions is complicated compared to homogeneous materials. For instance, impedance differences inside the concrete emanating from its different constituents lead to variations in the particle velocities, longitudinal and lateral stresses. One way to study this material is to average these variations using a plate reverberation technique, where a disc-shaped concrete specimen is mounted on the projectile and undergoes planar impact on a stationary target (PMMA, copper, tantalum) which is fitted with diagnostics [2-5]. However, only the Hugoniot curve (longitudinal behaviour) can be found using this technique. For this reason, the material understanding has been gradually built up starting from studies of the matrix (cement paste) and individual aggregates. In this paper, results are presented on the lateral behaviour of the matrix material, which are combined with published longitudinal data under the same impact conditions to determine the shear strength of the cement paste.

EXPERIMENTAL PROCEDURE

All the impact experiments were carried out in the plate impact gun facility at the University of Cambridge [6], which consists of a single stage 50-mm bore, light gas gun. The gun is capable of achieving velocities up to 1200 ms^{-1}. The impactor materials consisted of Polymethylmethacrylate (PMMA), aluminium, or copper. Impact velocities were measured to an accuracy of 0.5% using a sequential pin-shorting method and tilt was fixed to be less than 1 mrad by means of an adjustable specimen mount. The cement paste specimen were 20 mm thick by 50 mm x 50 mm.

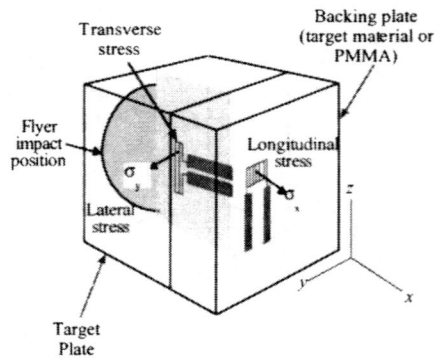

Figure 1. Target configuration

Each sample was sectioned in two, and commercial stress gauges (MicroMeasurements type J2M-SS-580SF-025) were introduced at fixed distances between 2 mm and 7 mm from the impact surface of the sample. Samples were then assembled using a low viscosity epoxy with a curing time of approximately 24 hours. Lateral gauge data was reduced using the analysis of Rosenberg and Partom [7]. The gauge mounting positions and sample configurations are shown in figure 1. The shear strength ( ) of a material under one-dimensional shock loading can be calculated from knowledge of the longitudinal (σ_x) and lateral stresses (σ_y) through the relation,

$$2\tau = \sigma_x - \sigma_y \qquad (1)$$

Our method of determining the shear stress has the advantage over previous calculations of being direct since no computation of the hydrostat is required.

MATERIAL DATA

Cement paste tested in this study was prepared and supplied by Concrete Structures Section (CSS), Department of Civil & Environmental Engineering, Imperial College, London, UK. The paste had a water-to-cement ratio of 0.35 by weight. Specimens were cured for 21 days in a water-bath at 20 ^{0}C. Density and ultrasonic measurements were performed after grinding samples from different batches. The density was 2.0 ± 0.2 g cm^{-3}, while the longitudinal and shear elastic wave velocities, were 3.7 ± 0.2 and 2.2 ± 0.2 km s^{-1}, respectively. Density variations resulted from different initial porosity. The density variations were in agreement with independent measurements performed by CSS.

RESULTS AND DISCUSSION

Table 1 summarises the impact conditions and lateral stresses as obtained from the gauges. In addition, the corresponding longitudinal stresses (Hugoniot curve) taken from [1] are also given and illustrated in Figure 2. Figures 3 and 4 illustrate some typical stress wave profiles for experiments 9TCu and 10TCu respectively (for impact conditions, see Table 1). The solid trace corresponds to the longitudinal stress while the dotted trace corresponds to the lateral stress as obtained in [1]. It can be seen that the longitudinal stresses have higher values than the lateral ones and their difference leads to the shear stress inside the material according to equation 1.

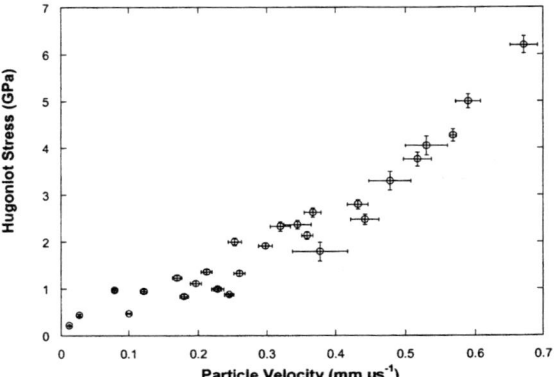

Figure 2. Cement Paste Hugoniot

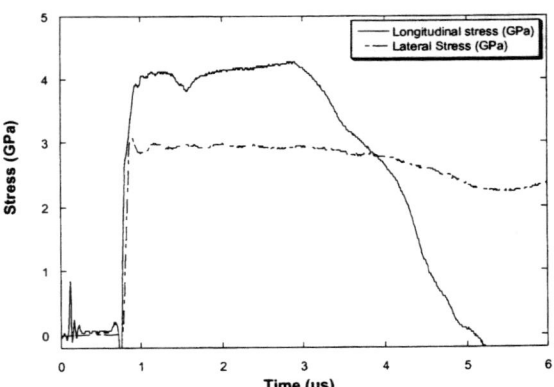

Figure 3. Stress Wave Profiles for experiment 9TCu.

The results are presented in Figure 5. The shear stress data calculated from equation 1 are plotted against longitudinal stress. As a comparison, the data are also fitted to the assumed elastic behaviour given by equation,

$$\sigma_y = \frac{\nu}{1-\nu}\sigma_x, \qquad (2a)$$

Table 1. Experimental Parameters, Lateral Stresses and corresponding Hugoniot Data

Shot no.	Impactor material and thickness (mm)	Impact Velocity (m s^{-1})	Lateral Stress (GPa)	Corresponding Longitudinal Stress (GPa)
1TAl	10 mm Al	352	1.00 ± 0.03	1.34 ± 0.06
2TAl	10 mm Al	512	2.00 ± 0.06	2.25 ± 0.12
3TAl	10 mm Al	261	0.52 ± 0.02	0.84 ± 0.04
4TAl	10 mm Al	438	1.39 ± 0.04	1.92 ± 0.06
5TAl	10 mm Al	440	1.35 ± 0.03	1.92 ± 0.06
6TCu	10 mm Cu	499	2.28 ± 0.05	2.64 ± 0.16
7TCu	10 mm Cu	569	2.66 ± 0.08	3.30 ± 0.20
8TAl	10 mm Al	514	1.64 ± 0.14	2.25 ± 0.12
9TCu	10 mm Cu	635	2.94 ± 0.15	4.05 ± 0.20
10TCu	6 mm Cu	724	3.75 ± 0.15	5.00 ± 0.15
11TPMMA	10 mm PMMA	229	0.09 ± 0.01	0.47 ± 0.01
12TCu	10 mm Cu	846	4.69 ± 0.40	6.20 ± 0.19

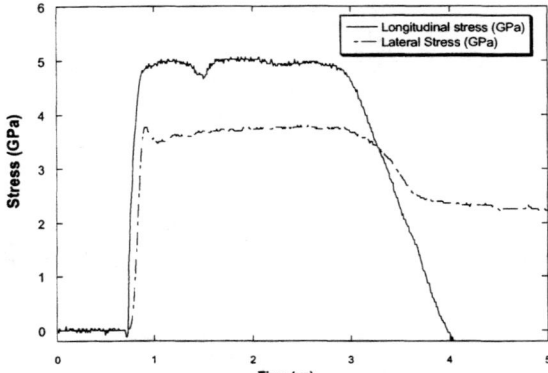

Figure 4. Stress Wave Profiles for experiment 10TCu.

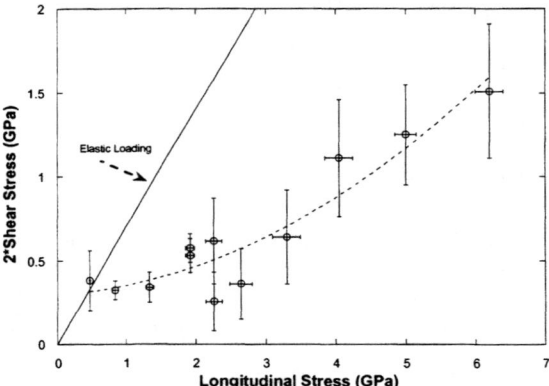

Figure 5. Cement Paste Shear Response

$$2\tau = \frac{1-2\nu}{1-\nu}\sigma_x \quad (2b)$$

where is the Poisson's ratio of the material (0.23). It can be seen that, within the data scatter due to the nature of the cement paste, the material behaviour deviates from the purely elastic behaviour around the HEL, which has previously [1] been measured to be 0.30 ± 0.05 GPa. In addition, the shear stress increases with increasing impact stress. Therefore, there must be some process such as fracture, or pore collapse, which reduces the strength, while the shear stress exhibits pressure dependence, which is indicative of the brittle nature of the material. Similar behaviour has also been observed in certain filled glasses [8] where the shear stress of the damaged material deviates from the elastic loading line and also increases with pressure.

CONCLUSIONS

Plate impact experiments have been presented to assess the shear behaviour of the cement paste. It is observed that above its HEL the material behaves inelastically indicating a process such as fracture or pore collapse reducing the strength. In addition, the shear stress increases with increasing pressure.

ACKNOWLEDGEMENTS

The Defence and Evaluation Research Agency, UK has sponsored this work, under contract WSS/U3257. Dr. A. Pullen from Imperial College provided the cement samples. Dr J. Sheridan, C. O'Carroll, I.G. Cullis and P.D. Church are thanked for their interest. Finally, we thank D.L.A. Cross and R. Flaxman for technical support.

REFERENCES

1. Tsembelis, K, Millett, J.C.F., Proud, W.G. and Field, J.E.. "The Shock Hugoniot Properties of Cement Paste up to 5 GPa",. in *Shock Compression of Condensed Matter – 1999*, (Edited by M.D. Furnish, L.C. Chhabildas and R.S. Hixson), pp. 1267-1270.
2. Grady, D. E., "Impact Compression Properties of Concrete", in *Proceedings of the Sixth International Symposium on Interaction of Nonnuclear Munitions with Structures, Panama City Beach, Florida,* pp. 172-175, May 3-7 (1993)
3. Hall, C. A., Chhabildas, L. C., and Reinhart, W. D., "Shock Hugoniot and Release States in Concrete Mixtures with Different Aggregate Sizes from 3 to 23 GPa," in *Shock Compression in Condensed Matter-1997*, edited by S. C. Schmidt et al., AIP Conference Proceedings 429, New York, 1998, pp. 119-122.
4. Grady, D. E., "Shock Equation of State Properties of Concrete," in *Structures under Shock and Impact IV*, edited by N. Jones et al., Computational Mechanics Publications, Southampton, 1996, pp. 405-414.
5. Kipp, M. E., Chhabildas, L. C., and Reinhart, W. D., "Elastic Shock Response and Spall Strength of Concrete," in *Shock Compression in Condensed Matter-1997*, edited by S. C. Schmidt et al., AIP Conference Proceedings 429, New York, 1998, pp. 557-560.
6. Bourne, N. K., Rosenberg, Z., Johnson, D. J., Field, J. E., Timbs, A. E., and Flaxman, R. P., *Meas. Sci. Technol.* **6**, 1462-1470 (1995).
7. Z. Rosenberg, and Y. Partom., J. Appl. Phys., **58**, 3072-3076 (1985).
8. Radford, D.D., Proud, W.G. and Field, J.E., To appear in: *Proc. Of SHOCK 2001 – APS 12th Topical Conference on Shock Compression of Condensed Matter*, Atlanta, 24-29 June, 2001.

PARTICIPANT LIST

Robert Abernathy, EMRTC, New Mexico Tech, Socorro, NM 87801, USA, 505-835-5728, 505-835-5630, robert@emrtc.nmt.edu

Tae-Young Ahn, , Taejon, 305-600, SO KOREA, 82 42 821 4192, 82 42 821 2390, leejupark@hapmail.net

Thomas J. Ahrens, Cal Tech, 1200 E. California Blvd., Pasadena, CA 91125, USA

Joseph A. Akins, CalTech, MC252-21, Pasadena, CA 91125, USA, 626-395-3825, 626-585-1917, sstewart@caltech.edu

Aleksey Y. Aleinikov, Sarov Open Computer Ctr, 53 Zernov St., Nizhnii Novgorod Region, Sarov 607190, RUSSIA, 7-83-304-3834, 7-831-304-1849, sel@socc.ru

David J. Alexander, Los Alamos National Laboratory, MS-G770, MST-6, Los Alamos, NM 87545, USA, 505-667-8828, 505-667-5268, djalexander@lanl.gov

Bill Anderson, Los Alamos National Laboratory, MS-P952, DX-1, Los Alamos, NM 87545, USA, 505-667-5460, 505-667-6372, wvanderson@lanl.gov

Mark U. Anderson, Sandia National Laboratories, Box 5800 MS, Albuquerque, NM 87185, USA, 505-844-5726, 505-845-3528, muander@sandia.gov

Charles E. Anderson Jr., Southwest Research Inst., Drawer 28510, San Antonio, TX 78228-0510, USA, 210-522-2313, 210-522-6290, canderson@swri.edu

Bill Andrzejewski, Sandia National Laboratories, Box 5800 MS, Albuquerque, NM 87185, USA, 505-844-8839, 505-844-5924, wjandrz@sandia.gov

Tommy Ao, Univ. British Columbia, 6224 Agricultural Rd, Vancouver, BC U6T1Z1, CANADA, 604-822-2945, 604-822-5324, tao@physics.ubc.ca

Chantel Aracne-Ruddle, Lawrence Livermore National Laboratory, Box 808, MS L-364 Livermore, CA 94551, USA, 925-422-3446, 925-423-1057, aracneruddle1@llnl.gov

Benjamin Arad, Soreq NRC, Plasma Physics, Yavne, 81800, ISRAEL, 972-8-943-4774, 972-8-943-4775, arad@ndc.soreq.gov.il

Marco Arienti, CalTech, 1200 E. California Blvd., Pasadena, CA 91125, USA, 626-395-4788, 626 449 2677, arienti@galcit@caltech.edu

Ron Armstrong, AFRL/MNME, 2306 Perimeter Rd., Eglin AFB, FL 32542, USA, 850-882-4212 x 201, 850 -882-3540, ronald.armstrong2@eglin.af.mil

Werner Arnold, EADS-TDW, Box 1340, Schrobenhausen, 86523, GERMANY, 49 8252 99 6267, 49 8252 99 6696, werner.arnold@tdw.lfk.eads.net

Blaine Asay, Los Alamos National Laboratory, MS-C920, DX-2, Los Alamos, NM 87545, USA, 505-667-3266, 505-667-0500, bwa@lanl.gov

James R. Asay, Sandia National Laboratories, Box 5800 MS 1181, Albuquerque, NM 87185-1181, USA, 505-844-1506, 505-845-7003, jrasay@sandia.gov

Gilles Avrillaud, SCA ITHPP, Le hameau de Drele, Thegra, 46500, FRANCE, 33 5 65 334330, 335 65334340, avrillaud@ithpp.fr

Joseph E. Backofen, Brigs Co., 2668 Petersborough St., Herndon, VA 20171, USA, 703-476-6448, 76616.3175@compuserve.com

Mel Baer, Sandia National Laboratories, Box 5800 MS 0836, Albuquerque, NM 87185, USA, 505-844-5223, 505-844-8251, mrbaer@sandia.gov

James E. Bailey, Sandia National Laboratories, Box 5800 MS 1196, Albuquerque, NM 87185-1196, USA, 505-845-7230, 505-845-7820, jebaile@sandia.gov

Amit Bandyopadhyay, Washington State University, Inst. for Shock Physics, Box 642816, Pullman, WA 99164-2816, USA, 509-335-7217, 509-335-6115, amitband@mme.wsu.edu

Eli Bar-on, Rafael Ballistics Ctr, Atzman, D. A. Misgau, 20170, ISRAEL, 972-4990-9701, 972-4878-213, ebaron@rafael.gov.il

Nicolas R. Barnes, AWE, PLC, Aldermaston, Reading RG7 4PR, UK, 44 118 982 4144, 44 118 982 5206, nich.barnes@awe.co.uk

Peter Bartkowski, US Army Research Lab, Aberdeen Proving Ground, MD 21005, USA, 410-278-6216, 410-278-6061, pbartko@arl.mil

Marina Bastea, Lawrence Livermore National Laboratory, L-415, Box 808, Livermore, CA 94551, USA, 925-424-2802, 925-422-6594, bastea1@llnl.gov

Francois Bauer, Inst. de Allemand de Recherches, 5 rue du General Cassagnou, St. Louis, 68301, FRANCE, 33 389 695174, 33 389 695002,

Dennis W. Baum, Lawrence Livermore National Laboratory, L-099, Box 808, Livermore, CA 94551, USA, 925-423-2236, 925-423-1845, baum1@llnl.gov

Steve Beissel, AHPCRC/Network Computing Serv., 1200 Washington Ave. So., Minneapolis, MN 55415, USA, 612-337-3411, 612-337-3400, sbeissel@networkcs.com

Jim Belak, Lawrence Livermore National Laboratory, L-45, Box 808, Livermore, CA 94551, USA, 925-422-6061, 925-422-2851, belak@llnl.gov

Robert Belmas, CEA/Le Ripault, BP 16, Monts, 37260, FRANCE, 33 2 47 34 47 01, 33 2 47 34 51 04, belmasr@ripault.cea.fr

Langdon Bennett, Los Alamos National Laboratory, MS-F664, Los Alamos, NM 87545, USA, 505-665-3475, 505-667-4420, langdonb@lanl.gov

David Benson, UCSD, 9500 Gilman, Dept. of Mech & Aero Eng, LaJolla, CA 92093-0411, USA, 858-534-5928, 858-534-7078, dbenson@ucsd.edu

Alessandra Benuzzi-Mounaix, Laboratoire LULI, Ecole Polytechnique, Palaiseau, 91128, FRANCE, 33 1 69333560, 33 1 69333009, alessandra.benuzzi-maunaix@polytechnique.fr

Thomas Bergstresser, Sandia National Laboratories, Box 5800 MS 1168, Albuquerque, NM 87185, USA, 505-845-7563, 505-845-7685, tbergst@sandia.gov

Nitin Bhate, General Electric R&D, Bldg K-1, Rm 3424, One Re4search Circle, Niskayuna, NY 12309, USA, 518-387-4702, 518-387-5459, bhateni@crd.ge.com

James P. Billingsley, US Army AMCOM, AMSAM-RD-SS-AA, Redstone Arsenal, AL 35898, USA, 256-876-5210, 256-876-5210,

Jennifer G. Blank, Lawrence Livermore National Laboratory, L-415, Box 808, Livermore, CA 94551, USA, 925-423-8566, 925-422-6594, blank4@llnl.gov

Stephan Bless, U of Texas--Austin, 3725 W. Braker Lane, Austin, TX 78759, USA, 512-471-9060, 512-471-9096, bless@iat.utexas.edu

Bill Blumenthal, Los Alamos National Laboratory, MS-G755, Los Alamos, NM 87545, USA, 505-667-0986, 505-667-8021, blumenthal@lanl.gov

Thomas Boehly, Laboratory for Laser Energetics, 250 E. River Rd., Rochester, NY 14623, USA, 716 275-3421, 716-275-5960, trb@lle.rochester.edu

David Boness, Seattle University, Dept. of Physics, eattle, WA 98122, USA, 206-296-5924, 206-296-2129, dboness@seattleu.edu

J. Michael Boteler, Army Research Laboratory, Box 374, Darlington, MD 21034, USA, 410-457-4277, jboteler@arl.mil

Neil K. Bourne, RMCS Cranfield Univ, Shrivenham, Swindon, Oxon SN6 8LA, UK, 44 1793 785350, 44 1793 785772,

Viviane Bouyer, LEEE--Universite Paris 10, 1, Chemin Desvallieies, Ville d'Avray, 92100, FRANCE, 33 1 47097013, 33 147091645, viviane.bouyer@aia.u-paris10.fr

David K. Bradley, Lawrence Livermore National Laboratory, L-472, Box 808, Livermore, CA 94551, USA, 925-424-5837, 925-423-62132, bradley9@llnl.gov

Keith Bradley, Lawrence Livermore National Laboratory, L-15, Box 808, Livermore, CA 94551, USA, 925422-3581, 925-422-5102, llnl.gov

Martin Braithwaite, RMCS Cranfield Univ, Shrivenham, Swindon, Oxon SN6 8LA, UK, 44 1793 785220, 44 1793 785772, m-braithwaite@rmcs.cranfield.ac.uk

Rebecca M. Brannon, Sandia National Laboratories, Box 5800 MS 0820, Albuquerque, NM 87185, USA, 505-844-5095, 505-844-5095, rmbrann@sandia.gov

N. Singh Brar, Univ. Dayton Research Inst, MS SPC 1911, 300 College Park Ave., Dayton, OH 45469-0182, USA, 937-229-3554, 937-229-3869, brarns@udri.udayton.edu

Richard Browning, Los Alamos National Laboratory, MS-P946, Los Alamos, NM 87545, USA, 505-667-4618, 505-665-2137, rvb@lanl.gov

Mike Burkett, Los Alamos National Laboratory, MS-F664, X-4, Los Alamos, NM 87545, USA, 505-665-6336, 505-667-4420, mwb@lanl.gov

Rick Burmeister, Cooke Corp., 1099 Centre Rd. Ste 100, Auburn Hills, MI 48326-2670, USA, 248-276-8020, 248-276-8825, info@cookecorp.com

Francois Buy, CEA Valduc, BP 14, Is sur Tille, 21120, FRANCE, 33 380235039, 33 380235278, buy@valduc.cpa.fr

Carl Cady, Los Alamos National Laboratory, MS-G755, Los Alamos, NM 87545, USA, 505-667-6369, 505-667-8021, cady@lanl.gov

Daniel Carroll, Sandia National Laboratories, Box 5800 MS 0819, Albuquerque, NM 87185-0819, USA, 505, 505, decarro@sandia.gov

Erik P. Carton, TNO Prins Mauritz Lab, Box 45, Rijswijk, 2280AA, NETHERLANDS, 31 15 284 3355, 31 15 2843997, carton@pml.tno.nl

Richard A. Catanach, Los Alamos National Laboratory, MS-P952, Los Alamos, NM 87545, USA, 505-667-9263, 505-667-6372, riochcat@lanl.gov

Robert Cauble, Lawrence Livermore National Laboratory, L-,41 Box 808, Livermore, CA 94551, USA, 925-422-4724, 925-423-2260, cauble1@llnl.gov

James Cazamias, Lawrence Livermore National Laboratory, L-414, Box 808, Livermore, CA 94551, USA, 925-423-2079, 925-422-5940, cazamias1llnl.gov

Peter Celliers, Lawrence Livermore National Laboratory, MS L399, Box 808, Livermore, CA 94551, USA, 925-424-4531, 925-424-2778, celliers1@llnl.gov

George Paul Chambers, NSWC--IH, 101 Strauss, Indian Head, MD 20640, USA, 301-744-2379, 301-744-6399, chambersgp@ih.navy.mil

S. K. Chan, Orica Canada, 301 Hotel de Ville, Brownsburg, Quebec J0V 1AD, CANADA, 450-533-1338, 450-533-5951, jim.chan@orica.com

Elaine A. Chandler, Lawrence Livermore National Laboratory, L-097, Box 808, Livermore, CA 94551, USA, 925-422-2482, 925-422-0779, eachandler@llnl.gov

Pierre Charrue, CEA/Le Ripault, BP 16, Monts, 37260, FRANCE, 33 247 344 603, 33 247 345 104, pierre.charrue@cea.fr

Ricky Chau, Lawrence Livermore National Laboratory, L-045 , Box 808, Livermore, CA 94551, USA, 925-423-4388, 925-422-2851, chau2@llnl.gov

Fran Chavez, Sandia National Laboratories, Box 5800 MS 1203, Albuquerque, NM 87185, USA, 505-844-27987, 505-844-8119, fcchave@sandia.gov

Volodymyr T. Chemerys, Inst. of Electrodynamics, UNAS, 56 Peremoga Ave., Kyiv, 02090, UKRAINE, 380 44 552 4039, 380 44 552 4039, vchemer@sabbo.net

Lalit C. Chhabildas, Sandia National Laboratories, Box 5800 MS-1181, Albuquerque, NM 87185-1181, USA, 505-844-4147, lcchhab@sandia.gov

Steven K. Chidester, Lawrence Livermore National Laboratory, L-125 Box 808, Livermore, CA 94551, USA, 925-422-0865, 925-423-4097, chidester1@llnl.gov

Philip Church, DERA, CDT, Q14, Ft. Halstead, Sevenoaks, Kent TN14 7BP, UK, 44 1959 515138, 44 1959 516050, pdchurch@dera.gov.uk

Richard A. Clegg, Century Dynamics Ltd, Horst Rd, Horsham, W.Sussex RH12 2DT, UK, 44 1403 270066, 44 1408 270099, all@centdyn.demon.co.uk

Brad Clements, Los Alamos National Laboratory, MS-B221, T-1, Los Alamos, NM 87545, USA, 505-667-8836, 505-665-5757, bclements@lanl.gov

Carl Cline, Lawrence Livermore National Laboratory, 728 Liquid Amber Pl., Danville, CA 94506-4528, USA, 925, 925, cline1@llnl.gov

Steve Coffey, NSWC--IH, 101 Strauss Code 920C2 , Indian Head, MD 20640-5035, USA, 301-744-6779, 301-744-4451, coffeycs@ih.navy.mil

Gilbert Collins, Lawrence Livermore National Lab., L-481, Box 808, Livermore, CA 94551, USA, 925-423-2204, 925-423 -6319, collins7@llnl.gov

Malcolm Cook, DERA, Ft. Halstead, Bldg X3, Room 11, Sevenoaks, Kent TN14 7BP, UK, 44 1959 515100, 44 1959 616041, mdcook@dera.gov.uk

Sara Cordova, Sandia National Laboratories, Box 5800 MS 1181, Albuquerque, NM 87185-1181, USA, 505-284-5898, 505-845-7003, smcordo@sandia.gov

John Corley, AFRL/MNME-Ernest Mach Inst., MNME, Eglin AFB, FL 32542, USA, 850-882-5969, 850-882-3540,

Jonathan Crowhurst, University of Washington, Box 351700, Seattle, WA 98195, USA, 206-543-6673, jcc4908@u.washington.edu

Andrew Crowson, US Army Research Office, Box 12211, Research Trianngle Park, NC 27709, USA, 919-549-4261, 919-549-4384, crowson@aro-emhi.army.mil

Jay Dallman, Los Alamos National Laboratory, MSP915, DX-DO, Los Alamos, NM 87545, USA, 505-667-4831, 505-665-3407, dallman@lanl.gov

Dattatraya P. Dandekar, US Army Research Lab, AMSRL-WM-TD, Aberdeen Proving Ground, MD 21005-5066, USA, 410-278-3803, 410-278-6952, ddandek@arl.army.mil

Kathryn Dannemann, Southwest Research Inst., 6220 Calebra Rd., San Antonio, TX 78228, USA, 210-522-2523, 210-522-6290, kdannemann@swri.org

Jean-Paul Davis, Sandia National Laboratories, Box 5800 MS 1181, Albuquerque, NM 87185-1181, USA, 505-284-3892, 505-845-7003, jpdavis@sandia.gov

Jeffery Davis, NAWC-Weapons Div, 1 Administration Circle, China Lake, CA 93555-6100, USA, 760-939-3357, 760-939-2597, davisjj@navair.navy.mil

Lloyd Davis, Los Alamos National Laboratory, MS-P952, DX-1, Los Alamos, NM 87545, USA, 505-665-3907, 505-667-6372, lldavis@lanl.gov

William C. Davis, Los Alamos National Laboratory, 693 46th St., Los Alamos, NM 87544, USA, 505-662-0542, daviswc@lanl.gov

Lee Davison, 39 Canoncito Vista Rd., Tijeras, NM 87059, USA, 505-286-4313, leedavison@aol.com

Robert H. Day, Los Alamos National Laboratory, MS-P915, DX-DO, Los Alamos, NM 87545, USA, 505-667-5653, 505-665-3407, day_r@lanl.gov

William E. Deal, Los Alamos National Laboratory, MS-P915, DX-DO, Los Alamos, NM 87545, USA, 505-665-6277, 505-665-3407, william.deal@gte.net

Paul S. DeCarli, SRI Int./Univ. College London, 333 Ravenswood Ave., Menlo Park, CA 94025, USA, 650-859-3171, 650-859-3200, pdecarli@unix.sri.com

Paul N. Demmie, Sandia National Laboratories, Box 5800 MS 0820, Albuquerque, NM 87185, USA, 505-844-7400, 505-844-0918, pademmi@sandia.gov

Christophe Denoual, CEA-DAN/DPTA, BP 12, Bruyeres le Chatel, F 91680, FRANCE, 33-1-69 26 7355, 33-1-6926 70 77, christophe.denoual@cea.fr

Jerry Dick, Los Alamos National Laboratory, MS-P952, DX-1, Los Alamos, NM 87545, USA, 505-667-18641, 505-667-6372, jjd@lanl.gov

Richard D. Dick, 9737 Academy St. NW, Albuquerque, NM 87114, USA, 505-897-9443, 505-898-0905, dikdik@nmcg.net

Peter Dickson, Los Alamos National Laboratory, MS-J585, DX-2, Los Alamos, NM 87545, USA, 505-665-7830, 505-665-4817, pmd@lanl.gov

Jaap Dijkhuis, University of Utrecht, Box 80000, Utrecht, 3508TA, NETHERLANDS, 313 2532319, 313 2537468, j.i.dijkhuis@phys.ww.no

Jow-Lian Ding, Washington State University, Inst. for Shock Physics, Box 642816, Pullman, WA 99164-2816, USA, 509-335-7217, 509-335-6115, ding@mme.wsu.edu

Dan Dolan, Washington State University, Inst. for Shock Physics, Box 642816, Pullman, WA 99164-2816, USA, 509-335-7217, 509-335-6115, ddolan@mail.wsu.edu

A. Yu Dolgoborodov, Inst. Chemical Physics, Kosygin St. 4, Moscow, 117334, RUSSIA, 7 095 939 72 52, 7 095 938 2156, aol@cwenter.chph.ras.ru

Zbigniew Dreger, Washington State University, Inst. for Shock Physics, Box 642816, Pullman, WA 99164-2816, USA, 509-335-7217, 509-335-6115, dreger@wsu.edu

Vladimir V. Dremov, Inst. Tech Physics, RFNC, 13, Vasilieva St., Snezhinsk, Chelyabinsk Region 456770, RUSSIA, 7 351 72 329 19, v.v.dzyomov@vniitf.ru

Oleg B. Drennov, VNIIEF, RFNC, Mira pr. 3, Nizhni Novgorod Region, Sarov, 607190, RUSSIA, 7-831 30 45009, root@gdd.vniief.ru

James Dunn, Lawrence Livermore National Laboratory, L-251, Box 808, Livermore, CA 94551, USA, 925423-1557, 925-422-2253, dunn6@llnl.gov

Sunil Dwivedi, Washington State University, Inst. for Shock Physics, Box 642816, Pullman, WA 99164-2816, USA, 509-335-7217, 509-335-6115, dwivedi@mail.wsu.edu

Mike Edwards, Cranfield University, RMCS-Shrivenham, Swindon, Wilts SN6 8LA, UK, 44 1793 785331, 44 1793 785772

Mark Elert, US Naval Academy, Chemistry Dept. 572 Holloway Rd., Annapolis, MD 21402, USA, 410-293-6616, 410-293-2218, elert@usna.edu

Henry Emeric, Laboratoire LULI, Ecole Poytechnique, Palaiseau, 91128, FRANCE, +33169334618, +33169333009, emeric.henry@polytechnique.fr

Anthony Esposito, Lawrence Livermore National Laboratory, L-268, Box 808, Livermore, CA 94551, USA, 925-424-3497, 925-422-6810, esposito4@llnl.gov

Andrew Evans, AWE, Aldermaston, Bldg. N75, Reading, Berkshire RG7 4PR, UK, 44-1189 825514, 44-1189-82-4843, andrew.m.evans@awe.co.uk

Eric P. Fahrenthold, University of Texas, Dept. of ME, Austin, TX 78712, USA, 512-471-3064, 512-471-8727, epfahren@mail.utexas.edu

Dan Farber, Lawrence Livermore National Lab., L-201, Box 808, Livermore, CA 94551, USA, 925-424-2256, 925-423-1057, farber2@llnl.gov

Archie Farnsworth, Sandia National Laboratories, Box 5800 MS 0820, Albuquerque, NM 87185, USA, 505-865-6160, 505-865-6160, avfarns@juno.com

J. Paul Farrell, Brookhaven Technology Group Inc., 120 Lake Ave. So, Nesconset, NY 11767, USA, 631-979-9405, pfarrell@btg.cc

Oleg Fat'yanov, Washington State University, Inst. for Shock Physics, Box 642816, Pullman, WA 99164 -2816, USA, 509-335-7217, 509-335-6115, shef@mail.wsu.edu

Alexey Fedorov, RFNC-VNIIEF, Mira pr. 32 Nizhni Novgorod, Sarov, 607190, RUSSIA

Justin Fellows, DERA, Ft. Halstead, Sevenoaks, Kent TN14 7BP, UK, 44 1959 515100, 44 1959 616041, mdcook@dera.gov.uk

Eric Ferm, Los Alamos National Laboratory, MS-P240, DX-3, Los Alamos, NM 87545, USA, 505-667-3343, 505-667-5539, enf@lanl.gov

Louis Ferranti, Ga Tech, School of MSE, 771 Ferst Dr., Atlanta, GA 30332-0245, USA, 404-894-1475, 404-894-9140, gte162s@prism.gatech.edu

Dawn Flicker, Los Alamos National Laboratory, MS-F664, Los Alamos, NM 87545, USA, 505-667-4206, 505-667-4420, dgf@lanl.gov

Jerry W. Forbes, Lawrence Livermore National Laboratory, L-282, Box 808, Livermore, CA 94551, USA, 925-423-4906, 925-424-3281, forbes1@llnl.gov

Vladimir E. Fortov, Inst. Chemical Physics, Institutskii Pr. 14, Chernogolovka, Moscow 142432, RUSSIA, 7 095 913 23 22, 7 095 913 2322, fortov@ficp.ac.ru

Joe Foster Jr., AFRL/MN, 101 Eglin Blvd., Eglin AFB, FL 32542, USA, 850-882-9643, 850-883-1381

Robert E. Franz, 16 Fox Hill Ct., Perry Hall, MD 21128, USA, 410-592-6614, 410-592-6614, elainefranz@aol.com

Laurence Fried, Lawrence Livermore National Laboratory, L- 282, Box 808, Livermore, CA 94551, USA, 925-422-7796, 925-424-3281, lfried@llnl.gov

David Frost, McGill University, 817 Sherbrooke St. W., Montreal, Quebec H3A 2K6, CANADA, 514-398-6279, 514-398-7365, david@mecheng.mcgill.ca

David J. Funk, Los Alamos National Laboratory, MS-C920, DX-2, Los Alamos, NM 87545, USA, 505-667-9659, 505-667-0500, djf@lanl.gov

Michael D. Furnish, Sandia National Laboratories, Box 5800 MS 1168, Albuquerque, NM 87185-1168, USA, 505-844-2877, 505-845-7685, mdfurni@sandia.gov

Kevin Gahagan, Corning Inc., SP-AR-01-1, Corning, NY 14831, USA, 607 -248-1165, 607-974-3385, gahagankt@corning.com

Raymond Gamache, NSWC--Dahlgren, Code G-22., Bldg. 221, 17320 Dahlgren Rd., Dahlgren, VA 22448, USA, 540-653-2959, 540-653-4662, gamacherm@nswc.navy.mil

Frank Garcia, Lawrence Livermore National Laboratory, L-283, Box 808, Livermore, CA 94551, USA, 925-423-1821, 925-424-3281, garcia13llnl.gov

Russell J. Gehr, Honeywell FM&T/NM, 3500 Trinity, Ste. C3, Los Alamos, NM 87544, USA, 505-663-0321, 505-661-6955, rgehr@kcp.com

Vladimir Georgevich, Lawrence Livermore National Laboratory, L-099, Box 808, Livermore, CA 94551, USA, 925-423-2916, 925-422-3389, georgevich1@llnl.gov

Bence Gerber, Century Dynamics, 2333 San Ramon Blvd, San Ramon, CA 94583, USA, 925-552-1600, 925-552-1609, big@autodyn.com

Jean Gerin-Roze, CEA-DIF-DCSA, BP 12, Bruyeres le Chatel, 91680, FRANCE, 33 1 69267090

Timothy C. Germann, Los Alamos National Laboratory, MS-D413, X-7, Los Alamos, NM 87545, USA, 505-665-9772, 505-667-3726, tcg@lanl.gov

M. J. Gifford, University of Cambridge, Madingley Rd., Cambridge, CB3 OHE, UK, 44-1223-337, 44-1223-350

John J. Gilman, UCLA, 6532 Boelter Hall, Los Angeles, CA 90085, USA, 310-825-9608, 310-206-7353, gilman@seas.ucla.edu

Martin Gilmore, DSTL/AFRL, 101 W. Eglin Blvd., Eglin AFB, FL 32542, USA, 850-882-9443x241, 850-883-1381, martin.gilmore@eglin.AF.mil

Albert Ginzburg, Rafael, Box 2250, Haifa, 31021, ISRAEL, 972 4 8792566, 972 4 8792842

Michael F. Gogulya, Inst. Chemical Physics, Kosygin St. 4, Moscow, 117334, RUSSIA, 7 095 939 72 89, 7 095 938 2156, aol@cwenter.chph.ras.ru

Jose Gois, University of Coimbra, ME Dept., Pinhal de Marrocos, Polo II, Coimbra, 3030, PORTUGAL, 351 239 941234, 351 239 944031, jose.gois@dem.uc.pt

H. T. Goldrein, University of Cambridge, Madingley Rd., Cambridge, CB3 OHE, UK, 44-1223-337336, 44-1223-350266, kes24@phy.cam.ac.uk

Zizheng Gong, Southwest Jiaotong University, Laboratory of High Pressure Physics, Chengdu, Sichuan 610031, PRC, 86 28 7603401, 86 28 7603401, gongzz@263.net

Alexander Gonor, University of Toronto, 624-523 Finch Ave. W, Toronto, ON M2R 1N4, CANADA, 416-630-3515, 416-667-7799, agonor@attcanada.net

Dennis Grady, Applied Research Assoc., 4300 San Mateo Blvd NE, Ste A-220, Albuquerque, NM 87110, USA, 505-883-3636, 505-872-0794, dgrady@ara.com

Stephen Grantham, University of Cambridge, Madingley Rd., Cambridge, CB3 OHE, UK, 44-1223-337366, 44-1223-350266, kes24@phy.cam.ac.uk

Ian N. Gray, AWE, PLC, Aldermaston, Reading RG7 4PR, UK, 44 118 982 7599, 44 118 982 4836, ian.gray@awe.co.uk

G. T. Rusty Gray III, Los Alamos National Laboratory, MS-G755, Los Alamos, NM 87545, USA, 505-667-5452, 505-667-8021, rusty@lanl.gov

Carl Greeff, Los Alamos National Laboratory, MS-D413, X-7, Los Alamos, NM 87545, USA, 505-665-1849, 505-667-3726, greeff@lanl.gov

M. Greenaway, University of Cambridge, Madingley Rd., Cambridge, CB3 OHE, UK, 44-1223-337, 44-1223-350, mwg24@phy.cam.ac.uk

Daniel Greenwood, Lawrence Livermore National Laboratory, L-283, Box 808, Livermore, CA 94551, USA, 925, 925, llnl.gov

Ella Grier, Los Alamos National Laboratory, 450 Noble Blvd., Carlisle, PA 17101, USA, 717-245-2804, 717-243-3845, ellagrier@lanl.gov

Yuri Gruzdkov, Washington State University, Inst. for Shock Physics, Box 642816, Pullman, WA 99164-2816, USA, 509-335-7217, 509-335-6115, gruzdkov@mail.wsu.edu

YaBei Gu, U of CA--San Diego, 9226 E. Regents Rd., LaJolla, CA 92037, USA, 858 822-2019, ygu@ucsd.edu

Mustafa Guden, University of Delaware, Ctr for Composite Material, Newark, DE 19716, USA, 302-831-8805, 302-831-8525, guden@ccm.udel.edu

Yogendra M. Gupta, Washington State University, Inst. for Shock Physics, Box 642816, Pullman, WA 99164-2816, USA, 509-335-3140, 509-335-6115, ymgupta@wsu.edu

Rick Gustavsen, Los Alamos National Laboratory, MS-P952, Los Alamos, NM 87545, USA, 505-667-2086, 505-667-6372, rgus@lanl.gov

Clint Hall, Sandia National Laboratories, Box 5800 MS 1181, Albuquerque, NM 87185-1181, USA, 505-845-3300, 505-845-7003, chall@sandia.gov

James Hammerberg, Los Alamos National Laboratory, MS-D413, Los Alamos, NM 87545, USA, 505-667-0676, jeh@lanl.gov

Sathya Hanagud, University of Florida, Grad. Eng & Res Ctr, 1350 N. Poquito Rd, Shalimar, FL 32579, USA, 850-833-9350

Donald W. Hansen, Lawrence Livermore National Laboratory, L-282, Box 808, Livermore, CA 94551, USA, 925-423-8023, 925-422-5922, hansen6@llnl.gov

David L. Hanson, Sandia National Laboratories, Box 5800 MS 1193, Albuquerque, NM 87185-1193, USA, 505-845-7547, 505-845-8467, dlhanso@sandia.gov

Robert J. Hardy, University of Nebraska, Lincoln, NE 68588-0111, USA, 402-472-5614, 402-472-2879, rhardy1@unl@edu

David E. Hare, Lawrence Livermore National Laboratory, L415, Box 808, Livermore, CA 94551, USA, 925-423-6828, 925-422-6594, hare2@llnl.gov

E. J. Harris, AWE, Aldermaston, Reading, Berkshire RG7 4PR, UK, 44 118982 5596, 44 1189824536, ernst.j.harris@awe.co.uk

Peter J. Haskins, DERA, Ftr. Halstead, Bldg X51, Sevenoaks, Kent TN14 7BP, UK, 44 1959 515199, 44 1959 516041, pjhaskins@dera.gov.uk

Dennis Hayes, Sandia Natl/Los Alamos Natl Lab, Box 591, Tijeras, NM 87059, USA, 505-281-9282, 505-286-3164, dennis@nmia.com

Kevin Hayes, Cordin Camera, Inc., 2230 South 3270 West, Salt Lake City, UT 84119-8270, USA, 801-972-5272, 801-972-8270, cordin@cordin.com

David Hebert, CEA./CESTA, BP 2, LeBarp, 33114, FRANCE, 33 557046981, 33 57045447, davidhebert@libertysurf.fr

Heidi Hegewald, Washington State University, Inst. for Shock Physics, Box 642816, Pullman, WA 99164-2816, USA, 509-335-7217, 509-335-6115, i9636289@wsunix.wsu.edu

Naoki Hemmi, Washington State University, Inst. for Shock Physics, Box 642816, Pullman, WA 99164-2816, USA, 509-335-7217, 509-335-6115, nhemmi@wsu.edu

Rudy Henninger, Los Alamos National Laboratory, MS-D413, CCS-2, Los Alamos, NM 87545, USA, 505-665-1562, 505-667-3726, rjh@lanl.gov

Bryan F. Henson, Los Alamos National Laboratory, MS-J567, Los Alamos, NM 87545, USA, 505-665-4837, 505-667-0440, henson@lanl.gov

Pierre Louis Hereil, CEG de Gramat, Gramat, 46500, FRANCE, 33 5 65105386, 33 5 65 105313, hereilpl@cegramat.fr

Sarah Heretz, Centra Technology, Inc., 83 Cambridge St. Ste 1A, Burlington, MA 01803, USA, 781-272-7887, 781-272-7883, heretzs@centrama.com

Benny Herrmann, Box 3377, Beer-Sheva, 84133, ISRAEL, 972-8-6235344, bpher@barak-online.net

Olivier Heuze, CEA/DAM, BP 12, Bruyeres le Chatel, 91680, FRANCE, , heuze@bruyeres.cea.fr

Damien Hicks, Lawrence Livermore National Laboratory, L-447, Box 808, Livermore, CA 94551, USA, 925-424-5220, 925-423-8451, hicks13@llnl.gov

Andrew J. Higgins, McGill University, 817 Sherbrooke St. W
Montreal, Montreal, Quebec H3A 2K6, CANADA, 514-398-6297, 514-398-6297, higgins@mecheng.mcgill.ca

Larry Hill, Los Alamos National Laboratory, MS-P952, Los Alamos, NM 87545, USA, 505-665-1086, 505-667-6372, lgh@lanl.gov

Yoichiro Hironaka, Tokyo Inst. of Technology, Materials Structure Lab, 4259 Nagatsuta, Yokohama, 226-8503, JAPAN, 81 45 924 5382, 81 45 924 5339, hironak1@rlem.titech.ac.jp

Tom Hitchcock, Office of Secretary of Defense, 3090 Defense Pentagon Room 3B1060, Washington, DC 20301-3090, USA, 703-695-1407, 703-614-3496, tom.hitchcock@osd.mil

Dieter H. H. Hoffmann, Technische University Darmstadt, Darmstadt, , GERMANY

Brad Lee Holian, Los Alamos National Laboratory, MS-B268, T-12, Los Alamos, NM 87545, USA, 505-667-9237, 505-665-3909, blh@lanl.gov

Knut B. Holm, Norwegian Defence Res. Estab., Box 25, Kjeller, N-2027, NORWAY, 47 63807622, 47 63807509, kbh@ffi.no

William Holt, NSWC--DD, 906 Carol Lane, Fredericksburg, VA 22405-1618, USA, 540-653-8687, 540-653-4662, wholt@nswc.navy.mil

Tomotaka Homae, Tokyo Inst. of Technology, Materials Structure Lab, 4259 Nagatsuta, Yokohama, 226-8503, JAPAN, 81 45 924 5382, 81 45 924 5360, homae1@rlem.titech.ac.jp

Irene Hooton, Defense Res. Est. Suffield, Box 4000, Stn Main, Medicine Hat, Alberta T1A 8K6, CANADA, 405-544-4728, 403-544-4821, irene.hooton@dres.dnd.ca

Yuki Horie, Los Alamos National Laboratory, MS-D413, X-7, Los Alamos, NM 87545, USA, 505-667-6422, 505-667-3726, horie@lanl.gov

Mike Howard, Lawrence Livermore National Laboratory, L-282, Box 808, Livermore, CA 94551, USA, 925-422-4138, 925-424-3281, howard11@llnl.gov

George M. Hrbek, Los Alamos National Laboratory, MS-F664, Los Alamos, NM 87545, USA, 505-667-6898, 505-667-4420, hrbek@lanl.gov

Roger Hsiao, Boeing Co., Box 3999, Seattle, WA 98124, USA, 253-773-8409, 253-773-8419, yu-yuan.r.hsiao@boeing.com

Greg Hutchens, 331 Old Ticket Pl, Aiken, SC 29803, USA, 803-208-8934, 803-208-8091, greg.hutchens@srs.gov

Deanne J. Idar, Los Alamos National Laboratory, MS-P915, DX-DO, Los Alamos, NM 87545, USA, 505-667-7695, 505-665-3407, djidar@lanl.gov

Edward J. Idar, Los Alamos National Laboratory, MS-F664, Los Alamos, NM 87545, USA, 505-667-1721, 505-667-4420, esi@lanl.gov

Hugh James, AWE, Aldermaston, Bldg. E1, Reading, Berkshire RG7 4PR, UK, 44-118-9826475, 44-118-9824820, hugh.james@awe.co.uk

Brian Jensen, Washington State University, Inst. for Shock Physics, Box 642816, Pullman, WA 99164 -2816, USA, 509-335-7217, 509-335-6115, bjjensen@mail.wsu.edu

Francois-Xavier Jette, McGill University, 817 Sherbrooke St. W, Montreal, Quebec H3A 2K6, CANADA, 514 -398-2003, 514-398-6297, fjette@po-box.mcgill.ca

Henry John, Remington Arms Co. Inc., 2592 Arkansas Hwy. Box 400, Lonoke, AR 72086-0400, USA, 501--676-4209, 270-765-4420

Scott Jones, Washington State University, Inst. for Shock Physics, Box 642816, Pullman, WA 99164 -2816, USA, 509-335-7217, 509-335-6115, scjones@wsu.edu

Jennifer Jordan, Ga Tech, School of MSE, 771 Ferst Dr., Atlanta, GA 30332-0245, USA, 404-894-1475, 404-894-9140, gto595c@prism.gatech.edu

Vasant Joshi, NSWC-IH, Bldg. 600, 101 Strauss, Indian Head, MD 20640, USA, 301-744-6769, 301-744-4203, joshivs@ih.navy.mil

Nimesh Juthani, Cooke Corp., 1099 Centre Rd. Ste 100, Auburn Hills, MI 48326-2670, USA, 248-276-8020, 248-276-8825, info@cookecorp.com

Kai Kadau, Los Alamos National Laboratory, MS-B262, T-11, Los Alamos, NM 87545, USA, 505-665-0354, 505-665-4063, kai@lviking.anl.gov

Daniel Kalantar, Lawrence Livermore National Laboratory, L-472, Box 808, Livermore, CA 94551, USA, 925-422-6147, 925-423-6212, kalantar1@llnl.gov

Gennadi I. Kanel, Inst. High Energy Densities, 13/19 Izhorskaya, Moscow, 127412, RUSSIA, 7 095 4832295, kanel@icp.ac.ru

Randall J. Kanzleiter, Los Alamos National Laboratory, MS-B220, X-2, Los Alamos, NM 87545, USA, 505-665-7700, 505-665-2227, kanzlr@lanl.gov

Daniela Kartoon, NRCN, Box 9001, Beer-Sheva, 84190, ISRAEL, 972-8-6568845, 972-8-6567878, danyk@bgumail.bgu.ac.il

Yukio Kato, NOF Corp., 61-1 Kitakomatsudani, Taketoyo-cho, Chite-gun, Aichi 470-2398, JAPAN, 81 569 72 0917, 81 569 73 7376, yukio_kato@nof.co.jp

John M. Kelley, Naval Surface Warfare Center, 101 Strauss Ave., Indian Head, MD 20640, USA, 301-744-1366, 301-744-4717, kelleyjm@ih.navy.mil

Gregory B. Kennedy, GA Tech, 771 Ferst Dr. NW, Atlanta, GA 30332, USA, 404-894-1475, 404-894-9140, gte290r@prism.gatech.edu

James E. Kennedy, Los Alamos National Laboratory, MS-P950, DX-1, Los Alamos, NM 87545, USA, 505-667-1468, 505-667-6301, jkennedy@lanl.gov

Tatyana A. Khantouleva, St. Petersburg State Univ., Inst. Math & Mechanics, Bibliotechnaja pl. 2, St. Petersburg, 198904, RUSSIA, 7 812 4284245, khan@math.spbu.ru

Konstantin V. Khishchenko, Inst. for High Energy Densities, 13/19 Izhorskaya, Moscow, 127412, RUSSIA, 7 095 4858563, 7 095 4857990, konst@hedric.msk.su

Jong Han Kim, U of CA, San Diego, Mech & Aero Eng., La Jolla, CA 92093-0418, USA, 858-534-6091, 858-534-5698, hanbellkim@hanmail.net

Vladimir Y. Klimenko, Inst. Chemical Physics, High Pressure Ctr, 4 Kosygin St., Moscow, 117334, RUSSIA, 7 488 40 93, klimenko@center.chph.ras.ru

Kinnan Kline, Graduate Engr & Rsch Cntr Univ of Florida, 1350 N Poquito Rd, Shalimar, FL 32579, kline@gerc.eng.ufl.edu

Marcus Knudson, Sandia National Laboratories, Box 5800 MS 1181, Albuquerque, NM 87185-1181, USA, 505-845-7796, 505-845-7003, mdknuds@sandia.gov

Takamichi Kobayaski, Natl. Inst. Materials Science, 1-1 Namiki, Tsukuba, Ibaraki 305-00444, JAPAN, 81-298-51-3351, 81-298-51-2768, kobayashi.takamichi@nims.go.jp

Edward M. Kober, Los Alamos National Laboratory, MS-B214, Los Alamos, NM 87545, USA, 505-667-5140, 505-667-1483, emk@lanl.gov

Michel Koenig, Laboratoire LULI, Ecole Polytechnique, Palaiseau, 91128, FRANCE, 33 1 69334799, 33 1 69333009, michel.koenig@polytechnique.fr

Frank Kosel, DRS Hadland Inc., 20480 Pacifica Dr., Cupertino, CA 95014, USA, 408-996-0901, 408-996-0889, drshadland@compuserve.com

Joel D. Kress, Los Alamos National Laboratory, MS-B268, Los Alamos, NM 87545, USA, 505-667-8906, 505-665-3909, jdk@lanl.gov

Shiro Kubota, Kyushu University, Dept. of Earth Resources Eng., 6-10-1 Hakozaki, Higashikyu 812-8581, JAPAN, 81-92-642-3626, 81-92-642-3614, kubota@mine.kyushu-u.ac.jp

Kurtis Kuhrts, BWXT-Pantex, Box 30020, Amarillo, TX 79120, USA, 806-477-3520, 806-477-5447, kkuhrts@pantex.com

Maija M. Kukla, Michigan Tech University, 1400 Townsend Dr., Houghton, MI 49931, USA, 906-487-2494, 906-487-2949, mmkukla@mtu.edu

Oyeon Kum, Washington State University, Inst. for Shock Physics, Box 642816, Pullman, WA 99164-2816, USA, 509-335-7217, 509-335-6115, okum@wsu.edu

Andrei L. Kutepov, Inst. Tech Physics, RFNC, 13, Vasilieva St., Snezhinsk, Chelyabinsk Region 456770, RUSSIA, 7 351 72 329 19, a.l.kutepov@vniitf.ru

Pierre L'Eplattenier, Centre d'Etudes de Gramat, Gramat, 46500, FRANCE, 33 5 65105370, 33 5 65105409, leplattp@cegramat.fr

Leo Laine, Anker-Zemer Engineering AS, Grindbakken 1, Oslo, 0764, NORWAY, 47 2213 9580, 47 22 13 9595, l.laine@anker-zemr.no

Y. Lanzerotti, US Army ARDEC, Box 3022, Picatinny Arsenal, NJ 07806-5000, USA, 973-724-4625, 973-724-4308, ylanzero@pica.army.mil

R. Jeffery Lawrence, Sandia National Laboratories, Box 5800 MS 1186, Albuquerque, NM 87185-1186, USA, 505-844-0127, 505-845-7820, rjlawre@sandia.gov

Maxim Lebedev, Natl. Inst. Adv. Industrial Science & Tech, Namiki 1-2, Tsukuba, Ibaraki 305-8564, JAPAN, 81-298 61 7852, 81 298 61 7129, maxim-lebedev@aist.go.jp

John H. Lee, McGill University, Dept. of ME, 817 Sherbrooke St. W, Montreal, Quebec H3A 2K6, CANADA, 514-398-6298, 514-398-7365, jhslee@mecheng.mcgill.ca

Kanani K. M. Lee, Univ. of CA Berkeley, 307 McCone, Berkeley, CA 94720-4767, USA, 510-643-8325, 510-643-9980, leeka@uclink.berkeley.edu

Larry M. Lee, Ktech Corp., 2201 Buena Vista SE, Albuquerque, NM 87106, USA, 505-998-5830, 505-998-5848, ktech@ktech.com

Peter Lee, NIMIC, NATO QT, Bruxelles, B-1110, BELGIUM, 32-2-707-5495, 32-2-707-5363, p.lee@hq.nato.int

Richard Lee, NSWC--IH, 101 Strauss, Indian Head, MD 20640, USA, 301-744-2380, 301-744-6399, leerj@ih.navy.mil

Ray Lemke, Sandia National Laboratories, Box 5800 MS-1186, Albuquerque, NM 87185, USA, 505-845-7423, 505-845-7820, rwlemke@sandia.gov

Pavel Levashov, Inst. for High Energy Densities, 13/19 Izhorskaya, Moscow, 127412, RUSSIA, 7 095 4858563, 7 095 4857990, pasha@hedric.msk.su

David Levi-Hevroni, NRCN, Box 9001, Beer-Sheva, 84190, ISRAEL, 972-8-6567418, 972-8-6567878, hevronid@bezeqint.net

David Levron, Nuclear Research Ctr-Negev, Box 9001, Beer-Sheva, 84190, ISRAEL

Fabrice Llorca, CEA Valduc, BP 16, Is sur Tille, 21120, FRANCE, 33 3 802 34938, 33 3 802 35216

Peter Lomdahl, Los Alamos National Laboratory, MS-B262, T-11, Los Alamos, NM 87545, USA, 505-665-0461, 505-665-4063, pxl@lanl.gov

Igor V. Lomonosov, Inst. Chemical Physics, 14 Institutskii Pr., Chernogolovka, Moscow 142432, RUSSIA, 7 095 913 2322, ivl@ficp.ac.ru

Ilya N. Lomov, Lawrence Livermore National Laboratory, L-206, Box 808, Livermore, CA 94551, USA, 925-423-7856, 925-422-3118, lomov1@llnl.gov

Thomas Lorenz, Lawrence Livermore National Laboratory, L-472, Box 808, Livermore, CA 94551, USA, 925424-4200, 925-423-6212, lorenz3@llnl.gov

Gabi Luttwak, Rafael, Box 2250, Haifa, 31021, ISRAEL, 972 487 92   460, 972 48702113, gabilo@rafael.co.il

Steve Lutz, Bechtel Nevada, 5520 Ekwill St., Ste. B, Santa Barbara, CA 93111, USA, 805-681-2244, 805-681-2241

Catherine Mabire, CEA de Gramat, Gramat, 46500, FRANCE, 33 5 65105318, 33 5 65 105343, mabirec@cegramat.fr

Jonathan Lee Mace, Los Alamos National Laboratory, MS-C920, DX-2, Los Alamos, NM 87545, USA, 505-667-2030, 505-667-0500, jonathan@lanl.gov

Andrew MacKinnon, Lawrence Livermore National Laboratory, L-481, Box 808, Livermore, CA 94551, USA, 025-424-2711, 925-423-6319, mackinnon2@llnl.gov

Dolores Maes, Sandia National Laboratories, Box 5800 MS 0145, Albuquerque, NM 87185, USA, 505-845-0491, 505-284-6778, mdmaes@sandia.gov

Jean-Bernard Maillet, CEA-DIF, BP 12, Bruyeres le Chatel, 91680, FRANCE, 33 169267336, 33 169267090, jean-bernard.maillet@cea.fr

Joseph T. Mang, Los Alamos National Laboratory, MS-C920, DX-2, Los Alamos, NM 87545, USA, 505-665-6856, 505-667-0500, jtmang@lanl.gov

Eugene Martinez, CEA/DIF, Bruyeres-le-Chatel, BP 12 91680, FRANCE, 33 1 69 26 50 92, 33 1 69 26 70 95, martinez@bruyeres.cea.fr

Eric M. Mas, Los Alamos National Laboratory, MS-B221, T-1, Los Alamos, NM 87545, USA, 505-665-5018, 505-665-5757, mas@lanl.gov

Erik Matheson, Lockheed Martin Space Sys. Co., 1111 Lockhead Martin Way, Sunnyvale, CA 94089, USA, 408-756-0896, 408-756-2224, erik.matheson@lmco.com

John Maw, AWE, Aldermaston, Reading, Berkshire RG7 4PR, UK, 44 1189 8277655, 44 1189 827820, john.maw@awe.co.uk

Stephane Mazevet, Los Alamos National Laboratory, MS-B268, Los Alamos, NM 87545, USA, 505-667-0956, 505-665-3501, smuzevet@lanl.gov

Matthew McCluskey, Washington State University, Inst. for Shock Physics, Box 642816, Pullman, WA 99164 -2816, USA, 509-335-7217, 509-335-6115, mattmcc@wsu.edu

Shawn McGrane, Los Alamos National Laboratory, MS-C920, Los Alamos, NM 87545, USA, 505-665-6086, mcgrane@lanl.gov

Dan Meiron, Caltech, MS 08-31, Pasadena, CA 91125, USA, 616-395-8157, 616-568-9102, dcm@!its.caltech.edu

Ralph Menikoff, Los Alamos National Laboratory, MS-B214, Los Alamos, NM 87545, USA, 505-667-7761, 505-667-1483, rtm@lanl.gov

Yuri Mescheryakov, Inst. of Mechanical Eng, Problems RAS, V.O. Bolshoi 61, Saint Petersburg, 199178, RUSSIA, 812-321-4765, 812-321-4771, ymesch@impact.ipme.ru

Marc A. Meyers, UCSD, Dept of MAE, LaJolla, CA 92053, USA, 858-534-4719, 858-534-5698

Anatoly Mikhailov, RFNC-VNIIEF, Mira pr. 33 Nizhni Novgorod, Sarov, 607190, RUSSIA, 7 831 304 5009, 7 831 304 5958, mikha@gdd.vniief.ru

Pietro Paolo Milella, ANPA, via C. Pavese 305, Rome, 00144, ITALY, 39 650072628, 39 6 50072649, milella@anpa.it

Phil Miller, NSWC-IH, Code 90, 101 Strauss Ave., Indian Head, MD 20640, USA, 301-744-4323, 301-744-4445

Jeremy Millett, RMCS Cranfield Univ, Shrivenham, Swindon, Oxon SN6 8LA, UK, 44 1793 785350, 44 1793 785772

Alec Milne, FGE Ltd., 83 Market St., St. Andrews, Fife Ky169NX, UK, 44 1334 460800, 44 1334 460813, alec@fges.demon.co.uk

Karel Minnaar, GA Tech, School of ME, 801 Ferst Dr. NW, Atlanta, GA 30309, USA, 404-894-3647, 404-894-0186, gt1825b@prism.gatech.edu

Stephen T. Montgomery, Sandia National Laboratories, Box 5800 MS 0521, Org. 2561, Albuquerque, NM 87185-0521, USA, 505-844-9062, 505-844-3894, stmontg@sandia.gov

Stephen Moon, Lawrence Livermore National Laboratory, L-041, Box 808, Livermore, CA 94551, USA, 925-424-4856, 925, moon1@llnl.gov

David S. Moore, Los Alamos National Laboratory, MS-C920, Los Alamos, NM 87545, USA, 505-665-6089, 505-667-0500, moored@lanl.gov

Eric Morano, CalTech, 1200 E. California Blvd., Pasadena, CA 91125, USA, 626-395-4788, 626 449 2677, eric@galcit@caltech.edu

Yosuhito Mori, Kyushu University, Dept. of Aeronautics & Astronautics, 6-10-1 Hakozaki, Higashiku, Fukuoka 812-8581, JAPAN, 81-92-642-3808, 81-92-642-4143, mori@aero.kyushu-u.ac.jp

Tatsumi Moritoh, Tokyo Inst. of Technology, Materials Structure Lab, 4259 Nagatsuta, Yokohama, 226-8503, JAPAN, 81 45 924 5382, 81 45 924 5339, moritoh@kkhp8.rlem.titech.ac.jp

Joseph P. Morris, Lawrence Livermore National Lab., L-206, Box 808, Livermore, CA 94551, USA, 925-424-4581, 925-422-3118, morris50@llnl.gov

Robi Mulford, Los Alamos National Laboratory, MS-E530, Los Alamos, NM 87545, USA, 505-667-7909, 505-665-4459, mulford@lanl.gov

Kunihito Nagayama, Kyushu University, Dept. of Aeronautics & Astronautics, Hakozaki, Higashiku 812-8581, JAPAN, 81-92-642-3804, 81-92-642-4143, nagayama@aero.kyushu-u.ac.jp

Kazutaka Nakamura, Tokyo Inst. of Technology, Materials Structure Lab, 4259 Nagatsuta, Yokohama, 226-8503, JAPAN, 81 45 924 5397, 81 45 924 5360, nakamur2@rlem.titech.ac.jp

Bill Nellis, Lawrence Livermore National Laboratory, L-45, Box 808, Livermore, CA 94551, USA, 925-422-7200, 925-422-2851, nellis1@llnl.gov

Vitali F. Nesterenko, UCSD, 9500 Gilman Dr., LaJolla, CA 92093-0411, USA, 858-822-0289, 858-534-5698, vnestere@ucsd.edu

Andrew Ng, Univ. of British Columbia, Dept. of Physics, 6224 Agricultural Rd., Vancouver, BC V6T 1Z1, CANADA, 604--822-3191, 604--822-5324, nga@physics.ubc.ca

Jeffrey Nguyen, Lawrence Livermore National Laboratory, L-045, Box 808, Livermore, CA 94551, USA, 925, 925, llnl.gov

Dmitry Nikolaev, Inst. Chemical Physics, Chernogolovka, Moscow 142432, RUSSIA, 7 095 913 23 22, 7 095 913 2322, nik@ficp.ac.ru

Angela M. Niles, Lawrence Livermore National Laboratory, L- 283 , Box 808, Livermore, CA 94551, USA, 925-422-7383, 925-424-3281, niles4@llnl.gov

Brendan O'Toole, UNLV, 4505 Maryland Pkwy., Las Vegas, NV 89154-9027, USA, 702-895-3885, 702-895-3936, bj@me.unlv.edu

Andy Obst, Los Alamos National Laboratory, MS-H803, P-23, Los Alamos, NM 87545, USA, 505-667-1330, 505-665-1421, obst@lanl.gov

Toshiyuki Ogura, Tokyo Inst. of Technology, Materials Structure Lab, 4259 Nagatsuta, Yokohama, 226-8503, JAPAN, 81 45 924 5382, 81 45 924 5339

Vladimir Okhitin, Bauman Moscow State Tech Univ, c/o Centra Tech. C83 Cambridge St. Ste 1A, Burlington, MA 01803, RUSSIA, 781-272-7887, 781-272-7883, okhitinv@centrama.com

Dennis L. Orphal, International Research Assocs., 4450 Black Ave., Pleasanton, CA 94566, USA, 925-485-0130, 925-485-0133, dorphal@aol.com

Michael Ortiz, Cal Tech, 1200 E. California Blvd MS 105-50, Pasadena, CA 91125-0050, USA, 626-395-4530, 626-449-6359, ortiz@aero.caltech.edu

Troy Oxby, Michigan Tech Univ, 1400 Townsend Dr., Houghton, MI 49931, USA, 906-482-3346, 906-487-2949, tjoxby@mtu.edu

Gerry I. Pangilinan, NSWC--IH, 101 Strauss, Code 920N, Indian Head, MD 20640, USA, 301-744-4413, 301-744-4445, pangilinangi@ih.navy.mil

Lee-Ju Park, , Taejon, 305-600, SO KOREA, 82 42 821 4192, 82 42 821 2390, leejupark@hapmail.net

Yehuda Partom, Rafael, Box 2250, Haifa, 31021, ISRAEL, 972-4-879-2672, 972-4-879-2113, yehudap@rafael.co.il

Su M. Peiris, NSWC--IH, Code 920P, Bldg. 600, Indian Head, MD 20640, USA, 301-744-4252, 301-744-4445, peirissm@ih.navy.mil

Robert A. Pelak, Los Alamos National Laboratory, MS-P940, DX-3, Los Alamos, NM 87545, USA, 505-665-1984, 505-665-3359, pelak@lanl.gov

Hongying Peng, Washington State University, Inst. for Shock Physics, Box 642816, Pullman, WA 99164 -2816, USA, 509-335-7217, 509-335-6115, appeng@wsu.edu

Warren Perger, Michigan Tech Univ, 1400 Townsend Dr., Houghton, MI 49931, USA, 906-482-3346, 906-487-2949, wfp@mtu.edu

Paul D. Peterson, Los Alamos National Laboratory, MS-C920, Los Alamos, NM 87545, USA, 505-667-4411, 505-667-0500, pdp@lanl.gov

Alexandre V. Petrovtsev, Inst. Tech Physics, RFNC, 13, Vasilieva St., Snezhinsk, Chelyabinsk Region 456770, RUSSIA, 7 351 72 329 30, a.v.petrovtsev@vniitf.ru

Alan Picklesimer, Los Alamos National Laboratory, MS-F613, NW-DP, Los Alamos, NM 87545, USA, 505-665-0577, 505-665-5916, apickle@lanl.gov

Leta Picklesimer, Los Alamos National Laboratory, MS-P950 DX-1, Los Alamos, NM 87545, USA, 505-667-0102, 505-667-6301, lpickle@lanl.gov

Ian Pickup, DERA, Chobham Lane, Chertsey, Surrey KT160EE, UK, 01 344 75 6696, 01 344 63 3335, impickup@dera.gov.uk

Igor Plaksin, University of Coimbra, ME Dept., Pinhal DG Marrolos Polo II, Coimbra, 3030, PORTUGAL, 351 010815871, igor.plaksin@mail.dem.uc.pt

Viktoz I. Postnov, Inst. Chemical Physics, Chernogolovka, Moscow 142432, RUSSIA, 7 095 913 23 22, 7 095 913 2322, postnov@iep.ac.ru

W. G. Proud, University of Cambridge, Madingley Rd., Cambridge, CB3 OHE, UK, 44-1223-337336, 44-1223-350266, kes24@phy.cam.ac.uk

Andrew Radchenko, Tomsk Scientific Centre, Dept. for Structural Macrokinetics, 10/3 Academichesky Ave., Tomsk, 634055, RUSSIA, 7 3822 257276, 7 3822 259838, andrew@dsm.tsc.ru

Darren Radford, University of Cambridge, Madingley Rd., Cambridge, CB3 OHE, UK, 44-1223-337336, 44-1223-350266, kes24@phy.cam.ac.uk

Philip Rae, University of Cambridge, Madingley Rd., Cambridge, CB3 OHE, UK, 44-1223-337, 44-1223-350

A. Rajendran, Army Research Office, Box 12211, Durham, NC 27709-2211, USA, 919-549-4346, 919-54999-4248, raj@arl.aro.army.mil

Oyvind Ranestad, Hagglunds Moelv, Box 244, Moelv, N-2391, NORWAY, 47 62354641, 47 62354601, oyvind.ranestad@haggmo.no

G. Ravichandran, CalTech, 105-50 Galcit, Pasadena, CA 91125, USA, 626-395-4525, 626-449-6359, ravi@caltech.edu

Sergey V. Razorenov, Inst. Chemical Physics, Chernogolovka, Moscow 142432, RUSSIA, 7 095 913 23 22, 7 095 913 2322, razsv@ficp.ac.ru

Evan Reed, MIT, 12-111, 77 Mssachusetts Ave., Cambridge, MA 02139, USA, 617-253-5482, 617-253-2562, ean@mit.ed

James H. Reho, Los Alamos National Laboratory, MS-C920. DX-2, Los Alamos, NM 87545, USA, 505-665-6102, 505-667-6372, jreho@lanl.gov

Bill Reinhart, Sandia National Laboratories, 1515 Eubank SE, MS 1181, Albuquerque, NM 87123-1181, USA, 505-284-3185, 505-845-3412, wdreinh@sandia.gov

David Reisman, Lawrence Livermore National Lab., L-041, Box 808, Livermore, CA 94551, USA

Michelle Repp, Washington State University, Inst. for Shock Physics, Box 642816, Pullman, WA 99164 -2816, USA, 509-335-7217, 509-335-6115, mrepp@wsu.edu

Jose Ribeiro, University of Coimbra, ME Dept., Pinhal DG Marrolos Polo II, Coimbra, 3030, PORTUGAL, 351 239 790724, 351 239 790701, jose.baranda@mail.dem.uc.pt

Paulo A. Rigg, Los Alamos National Laboratory, MS-P952, Los Alamos, NM 87545, USA, 505-665-3940, 505-667-6372, prigg@lanl.gov

Alita M. Roach, Los Alamos National Laboratory, MS-P915, DX-DO, Los Alamos, NM 87545, USA, 505-665-6277, 505-665-3407, alita@lanl.gov

David L. Robbins, Los Alamos National Laboratory, MS-P952, DX-1, Los Alamos, NM 87545, USA, 505-665-5431, 505-667-6372, robbins@lanl.gov

Joshua Robbins, Sandia National Laboratories, Box 5800 MS 0819, Albuquerque, NM 87185, USA, 505-284-5653, 505-844-0918, sandia.gov

Chris Robinson, AWE, Aldermaston, Reading, Berkshire RG7 4PR, UK, 44 118982-7813, 44 118982-4820, christopher.robinson@awe.co.uk

Dirk Robinson, Sandia National Laboratories, Box 5800 MS 1181, Albuquerque, NM 87185, USA, 505-845-7721, 505-845-7685, djrobin@sandia.gov

Keith M. Roessig, AFRL/Munitions Directorate, 101 Eglin Blvd., Eglin AFB, FL 32542-6810, USA, 850-882-9643ex266, 850-883-1381, roessig@eglin.af.mil

Seth Root, University of Nebraska, Dept of Physics, Lincoln, NE 68588-0111, USA, 402-472-5614, 402-472-2879, root1@bigred.unl.edu

Zvi Rosenberg, Rafael, Box 2250, Haifa, 31021, ISRAEL, 972 8374743, 972 48795 289, zvir@rafael.co.il

Johannes Roth, ITAP, Pfaffenvaldring 57, Stuttgart, 70550, GERMANY, 49 711 6855258, 49 711 685 5271, johannes@itap.physik.uni-stuttart.de

Stephen Rothman, AWE, Aldermaston, Reading, Berkshire RG7 4PR, UK, 44 118 982 7199, 44 118 982 4844, steve.rothman@awe.co.uk

Ted D. Rupp, Honeywell FM&T/NM, 3500 Trinity, Ste C3, Los Alamos, NM 87544, USA, 505-661207, 505-661-6955, trupp@kcp.com

Rod Russell, Institute for Advanced Technology, Suite 400 3925 W. Braker, Auatin, TX 78759, USA, 512-232-4406, rod_russell@iat.utexas.edu

Thomas P. Russell, NSWC--IH, Code 90, 101 Strauss, Indian Head, MD 20640, USA, 301-744-4323, 301-744-4445, russelltp@ih.navy.mil

Oren Sadot, NRCN, Box 9001, Beer-Sheva, 84190, ISRAEL, 972-8-6567278, 972-8-6567878, sorens@bgumail.bgu.ac.il

Darren Salisbury, AWE, Aldermaston, Reading, Berkshire RG7 4PR, UK, 44 118982 6999, 44 1189824836, darren.salisbury@awe.co.uk

Harold Sandusky, NSWC--IH, 101 Strauss, Indian Head, MD 20640-5035, USA, 301-744-2378, 301-744-6300, sanduskyhw@ih.navy.mil

Andreas Sandvik, Anker-Zemer Engineering AS, Grindbakken 1, Oslo, 0764, NORWAY, 47 2213 9582, 47 22 13 9595, a.sandvik@anker-zemr.no

Mike Scheidler, US Army Reseach Lab, APG, MD 21005, USA, 410-306-0794, 410-278-6952, mjs@arl.army.mily

Matthew Schneider, U of CA, San Diego, Mat.Sci.& Eng. UCSD-0418, La Jolla, CA 92093, USA, 858-534-6091, 858-534-5698, m1schnei@ucsd.edu

Bill Schonberg, U of Missouri--Rolla, CE Dept., Rolla, MO 65409, USA, 573-341-4787, 573-341-4729, wschon@umr.edu

Adam J. Schwartz, Lawrence Livermore National Laboratory, L- 355, Box 808, Livermore, CA 94551, USA, 925-423-3454, 925-424-4737, ajschwartz@llnl.gov

Lynn Seaman, SRI Int, 333 Ravenswood, Menlo Park, CA 94025, USA, 415-859-3587, 415-859-2260, lynn.seaman@sri.com

Toshimori Sekine, Natl. Inst. Materials Science, Namiki 1-1, Tsukuba, 305-0044, JAPAN, 81-298-51-3354, 81-298-51-2768, sekine.toshimori@nims.go.jp

Aleksandr A. Selezenev, Sarov Open Computer Ctr, 53 Zernov St., Nizhnii Novgorod Region, Sarov 607190, RUSSIA, 7-83-304-3834, 7-831-304-1849, sel@socc.ru

Victor Selivanov, Centra Technology, Inc., 83 Cambridge St. Ste 1A, Burlington, MA 01803, USA, 781-272-7887, 781-272-7883,

Robert E. Setchell, Sandia National Laboratories, Box 5800 MS 1421, Albuquerque, NM 87185, USA, 505-844-3847, 505-844-4045, rsetch@sandia.gov

Thomas D. Sewell, Los Alamos National Laboratory, MS-B214, Los Alamos, NM 87545, USA, 505-667-8205, 505-667-1483, sewell@lanl.gov

Moshe Shapira, Soreq, Propulsion Phy. Div, Yavne, 81800, ISRAEL, 972-8-9434285, 972-8-9434227, shapira2@012.met.il

J. Sharma, NSWC--Carderock, MacArthur Blvd., West Bethesda, MD 20817, USA, 301-227-4452, 301-227-4732, sharmj@nswccd.navy.mil

Milton S. Shaw, Los Alamos National Laboratory, MS-B214, Los Alamos, NM 87545, USA, 505-667-5093, 505-667-1483, mss@lanl.gov

Steve Sheffield, Los Alamos National Laboratory, MS-P952, Los Alamos, NM 87545, USA, 505-665-0350, 505-667-6372, ssheffield@lanl.gov

Hari Simha, Washington State University, Inst. for Shock Physics, Box 642816, Pullman, WA 99164 -2816, USA, 509-335-7217, 509-335-6115, simha@wsu.edu

Philippe Simonetti, CEA/Le Ripault, BP 16, Monts, 37260, FRANCE, 33 2 47 36 47 37, 323 2 4734 5142, simonett@ripault.cea.fr

Vladimir Skripnyak, Tomsk State University, 36 Lenin, Tomsk, 634050, RUSSIA, 7 382 2 420 680, 7 382 2 410 129, skrp@ftf.tsu.ru

Laura Smilowitz, Los Alamos National Laboratory, MS-J585, Los Alamos, NM 87545, USA, 505-667-5207, 505-665-4817, smilo@lanl.gov

Steve Son, Los Alamos National Laboratory, MS-C920, DX-2, Los Alamos, NM 87545, USA, 505-665-0380, 505-667-0500, son@lanl.gov

Laurent Soulard, CEA DAN, BP 12, Bruyeres-le-Chatel, 91680, FRANCE, 33-1-6926-338, 33-1-6926-70-71, laurent.soulard@cea.fr

Chad Sparks, Bell Helicoper Textron, Inc., MS 1401, Box 482, Ft. Worth, TX 76101, USA, 817-280-2408, 817-278-2408, csparks@bellhelicoper.textron.com

David W. Spicer, Centra Technology, Inc., 83 Cambridge St. Ste 1A, Burlington, MA 01803, USA, 781-272-7887, 781-272-7883, spicerd@centrama.com

John Steindel, Centra Technology, Inc., 83 Cambridge St. Ste 1A, Burlington, MA 01803, USA, 781-272-7887, 781-272-7883, steindelj@centrama.com

Chris Stennett, DERA, Ft. Halstead, Sevenoaks, Kent TN14 7BP, UK, 44 1959 515100, 44 1959 616041, cstennett@dera.gov.uk

David M. Stepp, US Army Research Office, Box 12211, Research Triangle Park, NC 22709-2211, USA, 919-549-4329, 919-549-4399, steppd@arl.aro.army.mil

Sarah T. Stewart, CalTech, MC150-21, Pasadena, CA 91125, USA, 626-395-3992, 626-585-1917, sstewart@caltech.edu

Youri V. Sud'enkov, St. Petersburg State Univ., Inst. Math & Mechanics, Bibliotechnaja pl. 2, St. Petersburg, 198904, RUSSIA, 7 812 4284245, spm@unicorn.math.spbu.ru

Gerrit Sutherland, NSWC--IH, Code 9230E, 101 Strauss, Indian Head, MD 20640, USA, 301-744-2382, 301-744-6399, sutherlandgt@ih.navy.mil

David Swanson, Univ of Nebraska--Lincoln, 1331 N. 38, Lincoln, NE 68503, USA, 402-472-5006, 402-472-1718, dswanson4@unl.edu

Damian Swift, Los Alamos National Laboratory, MS-E526, P-24, Los Alamos, NM 87545, USA, 505-667-1279, 505-665-3552, dswift@lanl.gov

Lori L. Switzer, Lawrence Livermore National Laboratory, L-851, Box 808, Livermore, CA 94551, USA, 925-423-5348, 925-423-8852, switzer2@llnl.gov

Izhar H. Syed, Univ of Dayton Research Inst, M/S SPC, 300 College Park Ave., Dayton, OH 45409-0182, USA, 937-229-3554, 937-229-3869, syedih@udri.udayton.edu

Katsumi Tanaka, Natl. Inst. Adv. Science & Texh, 1-1-1Higashi Tsukuba Central 5, Tsukuba, Ibaraki 305-8565, JAPAN, 81 298 61 4697, 81 298 61 4697, tanaka-katsumi@aist.go.up

Zhiping Tang, Univ Science & Tech, Dept. of Modern Mech USTC, Hefei, Anhui 230026, PRC, 86 551 3601289, 86 551 3606459, zptang@ustc.edu.cn

Dome Tanguy, CECAM, ENS Lyon, 46 alle d'Italie, Lyon, 69364, FRANCE, 33 4 7272 86 35, 33 4 7272 86 36, tamgiu@cecam.fr

Craig M. Tarver, Lawrence Livermore National Laboratory, L-282, Box 808, Livermore, CA 94551, USA, 925-423-3259, 925-424-3281, tarver1@llnl.gov

Douglas G. Tasker, Los Alamos National Laboratory, MS J566, DX-3, Los Alamos, NM 87545, USA, 505-665-2859, 505-665-3050, tasker@lanl.gov

Peter Taylor, AWE, Aldermaston, Reading, Berkshire RG7 4PR, UK, 44 118982 5395, 44 1189824836, peter.taylor@awe.co.uk

Vladimir Tchijov, National Univ. of Mexico, Campus Cuautitlan, Av. 1 de Mayo s/n, Cuautitlan, Izca Pli 54700, MEXICO, 52 5623 2037, 52 5623 2037, tchijov@scrvldor.unum.mx

Francesco Teodori, University of Bologna, via dei Colli 16, Bologna, 40136, ITALY, 39 51 6441711, 39 51 6441747, francesco.teodori@mail.ing.unibo.it

V. Ternovoi, Inst. Chemical Physics, Chernogolovka, Moscow 142432, RUSSIA, 7 095 913 23 22, 7 095 913 2322, ternovoi@ficp.ac.ru

Naresh N. Thadhani, Ga Tech, 771 Ferst Dr., Atlanta, GA 30332-0245, USA, 404-894-2651, 404-894-9140, naresh.thadhani@mse.gatech.edu

Wm. Richards Thissell, Los Alamos National Laboratory, MS-G755 MST-8, Los Alamos, NM 87545, USA, 505-667-9767, 505-667-8021, thissell@lanl.gov

Keith Thomas, Los Alamos National Laboratory, MS-P950, Los Alamos, NM 87545, USA, 505-665-5248, 505-667-6301, thomask@lanl.gov

Tom F. Thornhill, Sandia National Laboratories, Box 5800 MS-1181, Albuquerque, NM 87185-1181, USA, 505-845-3354, 505-845-3429, tfthorn@sandia.gov

Bob Tokheim, SRI Int., 333 Ravenswood Ave., Menlo Park, CA 94025, USA, 650-859-3239, 650-859-2343, robert.tokheim@sri.com

Davis Tonks, Los Alamos National Laboratory, MS-D413, Los Alamos, NM 87545, USA, 505-665-8481, 505-667-3736, tonks@lanl.gov

Wayne M. Trott, Sandia National Laboratories, Box 5800 MS 0834, Albuquerque, NM 87185, USA, 505-844-9556, 505-844-8251, wmtrott@sandia.gov

K. Tsembelis, University of Cambridge, Madingley Rd., Cambridge, CB3 OHE, UK, 44-1223-337366, 44-1223-350266, kes24@phy.cam.ac.uk

Paul A. Urtiew, Lawrence Livermore National Laboratory, L- 282 , Box 808, Livermore, CA 94551, USA, 925-423-0333, 925-424-3281, urtiew1llnl.gov

Alexandr V. Utkin, Inst. Chemical Physics, Chernogolovka, Moscow 142432, RUSSIA, 7 095 913 23 22, 7 095 913 2322, utkin@ficp.ac.ru

Ranji Vaidyanathan, Advanced Ceramics Research Inc., 3292 E. Hemisphere Loop, Tucson, AZ 85706, USA, 520-434-6350, 520-434-6355, r.vanidyanathan@acrtucson.com

Steve Valone, Los Alamos National Laboratory, MS-G755, Los Alamos, NM 87545, USA, 505-667-2067, 505-667-8621, smv@lanl.gov

Kevin S. Vandersall, Lawrence Livermore National Laboratory, L-282 , Box 808, Livermore, CA 94551, USA, 925-422-3337, 925-424-3281, vandersall1@llnl.gov

William Von Holle, Defense Nuclear Fac. Safety Bd, 625 Indiana Ave. NW, Washington, DC 20004, USA, 202-694-7146, 202-208-6518, williamv@dnfsb.com

Boris L. Voronin, Sarov Open Computer Ctr, 53 Zernov St., Nizhnii Novgorod Region, Sarov 607190, RUSSIA, 7-83-304-3834, 7-831-304-1849, sel@socc.ru

Kunihico Wakabayashi, Tokyo Inst. of Tech, Materials Structure Lab, 4259 Nagatusta, Yokohama, 226-8503, JAPAN, 81 45 924 5382, 81 45 924 5339, k-wakabayashi@aist.go.jp

James D. Walker, Southwest Research Inst., 6220 Culebra Rd., San Antonio, TX 78228, USA, 210-522-2051, 210-522-6290, jwalker@swri.edu

Alan Wan, Lawrence Livermore National Laboratory, L-15 , Box 808, Livermore, CA 94551, USA, 925-423-3342, 925-422-5102, wan1@llnl.gov

Yi Wang, Uppsala University, Box 530, Dept. of Physics, Condensed Matter Theory, Uppsala, S-751 21, SWEDEN, 46 18 47 3567, 46 18 4713524, wangy@fysik.uu.se

Dongqing Wei, CERCA, 5160 Blv Decarie, Ste. 400, Montreal, Quebec H32 2H9, CANADA, 514-484-2399, 514-484-6647, dongqing@cerca.umontreal.ca

Chris Weickert, Defense Res. Est. Suffield, Box 4000, Stn Main, Medicine Hat, Alberta T1A 8K6, CANADA, 405-544-5331, 403-544-4821, dres.dnd.ca

Meir Werdiger, Soreq NRC, Plasma Dept., Soreq, Yavne 81900, ISRAEL, 972-8-343-4753, 372-8-343-4775, meir@soreq.gov.ol

Christopher Werner, Los Alamos National Laboratory, MS-D413, Los Alamos, NM 87545, USA, 505-667-6892, 505-667-3726, cwerner@lanl.gov

Carter T. White, Naval Research Laboratory, Code 6189, Washington, DC 20375-5320, USA, 202-767-3270, 202-767-1716, carter.white@nrl.navy.mil

Nicholas Whitworth, AWE, Aldermaston, Bldg. E3, Reading, Berkshire RG7 4PR, UK, 44 1189 826404, 44 1189 824820, nick.whitworth@awe.co.uk

Donald A. Wiegand, Picatinny Arsenal, Bldg. 3022, Picatinny Arsenal, NJ 07806-5000, USA, 973-724-3336, 973-724-5869, dwiegand@pica.army.mil

Mark Wilke, Los Alamos National Laboratory, MS-H803, P-23, Los Alamos, NM 87545, USA, 505-667-1509, 505-665-4121, wilke@lanl.gov

Pharis E. Williams, EMRTC, New Mexico Tech, Socorro, NM 87801, USA, 505-835-5774, 505-835-5630, pharis@emrtc.nmt.edu

William H. Wilson, AFRL--Munitions Directorate, Eglin AFB, FL 32542, USA, 850-882-4212,ex 246, 850-882-3540, wilsonwh@eglin.af.mil

Michael Winey, Washington State University, Inst. for Shock Physics, Box 642816, Pullman, WA 99164 -2816, USA, 509-335-7217, 509-335-6115, mwiney@wsu.edu

Nancy Winfree, Dominca, 9813 Admiral Dewey Ave. NE , Albuquerque, NM 87111, USA, 505-822-0005, 505-822-0462, nancy@dominca.com

Ron Winter, AWE, Aldermaston, Reading, Berkshire RG7 4PR, UK, 44 118982 5493, 44 118982 4836, rwinter@awe.co.uk

Tom Woo, University of W. Ontario, Dept. of Chemistry, London, Ontario N6A5B7, CANADA, 519-661-2111ex86310, 519-661-3023, twoo@uwo.ca

Diana Woody, NAWC-Weapons Div, 1 Administration Circle, China Lake, CA 93555-6100, USA, 760-939-7856, 760-939-2597, woodyd1@navair.navy.mil

Steve Wortley, AWE, Bldg B8C, Aldermaston, Reading RG7 4PR, UK, 44 1189 24287, 44 1189 82600, steve.wortley@awe.co.uk

Daniel Wright, University of Cambridge, Madingley Rd., Cambridge, CB3 OHE, UK, 44-1223-337336, 44-1223-350266, kes24@phy.cam.ac.uk

Xiao Xu, GA Tech, 771 Ferst Dr. NW, Atlanta, GA 30332, USA, 404-894-1475, 404-894-9140, gt0389C@prism.gatech.edu

Qing Xue, U of CA San Diego, Mats Sci & Eng. 9500 Gilman Dr., LaJolla, CA 92093-0418, USA, 858-534-6091, 858-534-5698, qxue@ucsd.edu

Wenbo Yang, Schlumberger, 14910 Airline Rd., Rosharon, TX 77584-1590, USA, 281-285-5253, 281-285-5453, yang@rosharon.oilfield.slb.com

Kazushige Yano, Los Alamos National Laboratory, MS-D413, Los Alamos, NM 87545, USA, 505-665-7875, 505-665-3726, kyano@lanl.gov

Jin Yao, General Dynamics--OTS, 4565 Commercial Dr., Niceville, FL 32578, USA, 850-897-6243, 850-897-6299

Jerome D. Yatteau, Applied Research Assoc., 5941 S. Middlefield Rd. Ste. 100, Littleton, CO 80123-2877, USA, 303-795-8106, 303-795-8159, jyatteau@ara.com

Akio Yazaki, Tokyo Inst. of Technology, Materials Structure Lab, 4259 Nagatsuta, Yokohama, 226-8503, JAPAN, 81 45 924 5382, 81 45 924 5360, yazaki@oldkkhp8.rlem.titech.ac.jp

Akio Yoshinaka, McGill University, 817 Sherbrooke St. W, Montreal, Quebec H3A 2K6, CANADA, 514-398-2003, 5 14-398-6297, superdet@hotmail.com

Eugene Zaretsky, Ben-Gurion University, Box 653, Beer-Sheva, 84105, ISRAEL, 972 8 6477102, 972 8 647 7100, zheka@bgumail.bgu.ae.il

Joseph M. Zaug, Lawrence Livermore National Laboratory, L-282, Box 808, Livermore, CA 94551, USA, 925-423-4428, 925-424-3281, zaug1@llnl.gov

Frank J. Zerilli, NSWC--IH, 101 Strauss Code 9220C, Indian Head, MD 20640-5035, USA, 301-744-6762, 301-744-4717, zerillifj@ih.navy.mil

Fan Zhang, Defense Res. Est. Suffield, Box 4000, Stn Main, Medicine Hat, Alberta T1A 8K6, CANADA, 405-544-4887, 403-544-4821, fzhang@dres.dnd.ca

Mikhail Zhernokletov, RFNC-VNIIEF, Mira pr. 32 Nizhni Novgorod, Sarov, 607190, RUSSIA, 7 831 304 5009, 7 831 304 5958, mikha@gdd.vniief.ru

Min Zhou, GA Tech, School of ME, Atlanta, GA 30033, USA, 404-894-3294, 404-894-0186, min-zhou@me.gatech.edu

Shiming Zhuang, CalTech, GALCIT 205-45, Pasadena, CA 91125, USA, 6266-395-4768, zugsm@aero.caltech.edu

Alexie Zibarov, Tula State Tech. Univ., c/o Centra Tech. 83 Cambridge St. Ste 1A, Burlington, MA 01803, RUSSIA, 781-272-7887, 781-272-7883, zibarova@centrama.com

Kurt Zimmerman, Washington State University, Inst. for Shock Physics, Box 642816, Pullman, WA 99164-2816, USA, 509-335-7217, 509-335-6115, kurtz@mail.wsu.edu

Marvin A. Zocher, Los Alamos National Laboratory, MS-D413, X-7, Los Alamos, NM 87545, USA, 505-665-3472, 505-667-3726, zocher@lanl.gov

Sergey Zybin, US Naval Academy, Chemistry Dept. 572 Holloway Rd., Annapolis, MD 21402-5026, USA, 410-293-6614, 410-293-2218, zybin@usna.edu

AUTHOR INDEX

A

Ahrens, T. J., 1393
Ahuja, R., 67
Aidun, J. B., 197
Akedo, J., 1101
Albers, R. C., 225
Alcon, R. R., 999, 1019, 1051
Aleinikov, A. Y., 374
Alexander, D. J., 499, 630
Allan, N. L., 185
Allen, A. M., 615
Alrick, K. R., 1247
Aminov, Y. A., 875
Anderson, M. U., 669
Anderson, W. W., 1247
Anderson Jr., C. E., 275, 1279, 1298
Andrew, M. I., 841
Andrews, T. D., 487, 511
Antoun, T. H., 287, 1389
Ao, T., 1243
Aracne, C. M., 1015
Arad, B., 583
Arienti, M., 251
Arkhipov, V. I., 962
Armstrong, R. W., 657, 837
Arnold, W., 527
Asay, B. W., 1065, 1069, 1077
Asay, J. R., 26, 79, 221, 849, 1141, 1163
Atchison, W., 143
Atisivan, R., 697
Averin, A. N., 439
Avrillaud, G., 1173, 1188

B

Backofen, J. E., 954, 958
Baer, M. R., 713, 845, 1051
Bailey, J. E., 1141
Bandyopadhyay, A., 697
Barabanov, R. A., 374
Barlow, A. J., 507
Barnes, N., 135, 649
Bar-On, E., 739
Bartkowski, P. T., 779
Batani, D., 83, 1367

Batchelor, K., 1185
Baudin, G., 169, 1223
Bauer, F., 1149
Baumung, K., 503, 603
Becker, S., 1169
Bedrov, D., 399, 403
Beissel, S. R., 1287
Belcher, I., 649
Belmas, R., 439
Benedetti, L. R., 1363
Beno, T., 811
Benson, D. J., 1087
Benuzzi-Mounaix, A., 83, 1367
Bergstresser, T., 1169
Bergues, D., 1173
Bernard, S., 367
Bezrouchko, G. S., 1327
Bhate, N., 339
Billingsley, J. P., 735
Bless, S. J., 767, 787, 811, 1291
Blottiau, P., 99
Bluhm, H. J., 503, 603
Blumenthal, W. R., 539, 661, 665, 821
Boboridis, K., 1247
Boehly, T., 619
Boluijt, A., 1127
Bonitz, M., 119
Bouquet, S., 1367
Bourne, N. K., 131, 135, 213, 479, 511, 523, 575, 579, 634, 649, 653, 717, 743, 771, 775, 914, 1267
Boustie, M., 83
Bouyer, V., 1223
Bowden, E., 1381
Bowers, R., 143
Braithwaite, M., 185
Brannon, R. M., 197, 201
Brar, N. S., 693, 1323
Brazhnikov, M. A., 962
Briggs, R. I., 890, 1047
Brooks, R., 475
Browning, R. V., 987
Budge, K. G., 291
Burkett, M. W., 279
Butnev, O. I., 374
Buttler, W. T., 1247

Buy, F., 319, 323, 327, 638
Bychenkov, V. A., 591

C

Cady, C. M., 539, 661, 665, 725, 821
Campos, J., 721, 898, 918, 922
Carley, D. J., 507
Carroll, D. E., 75, 307
Carton, E. P., 1127
Casas-Cordero, M., 946
Castañeda, J. N., 845
Catanach, R. A., 906
Cauble, R. C., 849, 1371
Cazamias, J. U., 491, 587, 767, 787, 1291
Chakravarty, A., 1007
Chambers, G. P., 894, 1073
Charest, J. A., 1153
Cheese, P. J., 1047
Chemerys, V. T., 1314
Chhabildas, L. C., 75, 205, 307, 483, 515, 787, 791, 845
Chidester, S. K., 886
Chièze, J., 1367
Church, P. D., 487, 511
Clancy, S. P., 279
Clegg, R. A., 685
Clements, B. E., 427, 535, 539, 661
Clifton, R. J., 339
Cochrane, K. R., 291
Coffey, C. S., 563, 837, 1003
Cogar, J. R., 483
Coker, D., 1393
Collins, G. W., 1363
Collins, L. A., 91, 99
Cook, M. D., 391, 890, 1047
Cooper, G., 131
Corley, J., 705
Cox, D. E., 669
Cruz León, G., 241
Cunningham, B., 1137
Curran, D. R., 607

D

Dandekar, D. P., 779, 783, 791
Dannemann, K. A., 275, 729, 1298
Dariel, M. P., 1119
Davis, J. J., 942, 950
Davis, J.-P., 79, 221, 1163
Davis, L. L., 165, 1051
Davison, L., 20
DeCarli, P. S., 1381
Deeney, C., 1163
Dekel, E., 583, 1310
Demmie, P. N., 311
Denoual, C., 495
De Rességuier, T., 83
Dewaele, A., 1363
Diani, J. M., 495
Dick, J. J., 411, 817
Dickson, P. M., 1065, 1069, 1077
Dmitireva, M. A., 1093
Do, I., 1087
Dolgoborodov, A. Y., 868, 962
Dremov, V. V., 87
Drennov, O. B., 595
Dudnikov, V., 1185
Dunn, J., 1371

E

Edwards, M. R., 523
Elban, W. L., 837
Elert, M. L., 355, 1406
Eliezer, S., 583
Engelke, R., 1051
Erickson, L., 1157
Esposito, A. P., 1015
Evans, A. M., 79
Evans, D. J., 1011
Everett, R., 475

F

Faral, B., 83
Farber, D. L., 856, 1015
Farnsworth Jr., A. V., 1347, 1355
Farre, J., 638
Farrell, J. P., 1185
Favorsky, V., 1306
Fedorov, A. V., 910
Fellows, J., 391, 1047
Ferm, E. N., 966
Ferranti, L., 755
Field, J. E., 747, 807, 878, 1007, 1035, 1105, 1385, 1414
Filimonov, A. S., 95, 107
Filin, V. P., 439
Filinov, V. S., 119
Fisher, G. L., 1219, 1333
Fiske, P. S., 491, 767
Fleury, X., 1367
Flores, P. A., 221
Forbes, J. W., 11, 153, 849, 882, 886, 902, 1019, 1039, 1043, 1137, 1157
Fortov, V. E., 71, 107, 111, 119, 233, 237, 759, 763, 938
Foster Jr., J. C., 519, 829, 1031, 1318
Frage, N., 1119
Fried, L. E., 161, 177, 343, 385
Frost, D. L., 946
Funk, D. J., 1219, 1227, 1333, 1351
Furnish, M. D., 205, 849, 1410

G

Gahagan, K. T., 1227, 1333, 1351
Gamache, R. M., 1196
Garcia, F., 153, 849, 886, 902, 1019, 1039, 1043, 1137, 1157
Garcia, F. G., 882
Garcia, I. A., 1031
Garmasheva, N. V., 439
Garza, R. G., 886
Gazonas, G., 689
Gehr, R. J., 499, 1145
Geltmacher, A., 475
Germann, T. C., 333, 351, 359
Gifford, M. J., 878, 1007, 1035
Giles, A. R., 303
Gilman, J. J., 36
Gilmore, M. R., 519, 1318
Glenn, L. A., 287, 1389
Gogulya, M. F., 962
Góis, J., 898, 922
Goldrein, H. T., 825, 1105
Goldthorpe, B., 487
Golubev, V. K., 374
Gong, P., 259
Gong, Z., 1401
Gonor, A. L., 63, 423, 934
Gorshkov, M. M., 875
Goutelle, J. C., 169
Goveas, S. G., 1035
Grady, D. E., 515, 709, 783, 799
Graham, P., 79
Grandjouan, N., 83, 1367
Grantham, S. G., 803, 1105
Gray III, G. T., 131, 315, 479, 535, 539, 575, 579, 634, 653, 661, 665, 725, 775, 821
Greeff, C. W., 225
Green, L., 902, 1157
Greenaway, M. W., 1007, 1035
Greening, D., 983
Greenwood, D. W., 882, 886, 902, 1137, 1157
Gregori, F., 615, 619
Grove, D. J., 779
Gu, Y., 1294
Gudarenko, L. F., 127
Guirguis, R. H., 864, 926
Gunger, M. E., 435
Gupta, Y. M., 3, 697
Gustavsen, R. L., 999, 1019
Guyot, F., 83
Guzik, J., 143

H

Haill, T. A., 291
Hall, C. A., 79, 849, 1141, 1163, 1177
Hall, T., 83, 1367
Hallouin, M., 83
Hammerberg, J. E., 359, 1212
Hanrahan, R. J., 499
Hansen, D. W., 177
Hanson, D. L., 1141
Hardy, R. J., 363
Hare, D. E., 1231

Harris, E. J., 419
Harrison, J. A., 355
Haskins, P. J., 391, 890, 1047
Hauer, A., 1192
Hayes, D. B., 221, 1163, 1177
Hayhurst, C. J., 685
He, H., 1255, 1339, 1397
Henis, Z., 583
Henninger, R. J., 299
Henry, E., 83, 1367
Henson, B. F., 1065, 1069, 1077
Héreil, P.-L., 229, 1173, 1235
Herrmann, B., 623, 1306
Hervé, P., 1223
Heuzé, O., 169, 450
Hickman, R. J., 1141
Hicks, D., 1363
Hiermaier, S., 705
Higgins, A. J., 946, 1023
Hill, L. G., 149, 165, 906
Hirabayashi, K., 1131
Hirata, N., 1397
Hironaka, Y., 1181, 1375
Hirosaki, Y., 930
Hjelm, R. P., 833
Hodowany, J., 557
Hogan, G., 966
Holian, B. L., 351, 359
Holian, K. S., 279
Holmes, N. C., 1231, 1251
Holmquist, T. J., 1287
Holt, W. H., 1196
Homae, T., 1131
Hooton, I., 423
Hoover, S. M., 837
Horie, Y., 259, 371, 411, 553, 983
Horovitz, Y., 583
Howard, W. M., 161, 177
Hrbek, G. M., 115, 139
Huser, G., 83
Hutchens, G. J., 442

I

Idar, D. J., 665, 821
Indrakanti, S. S., 1294
Inou, K., 995
Itoh, S., 930

J

Jacquez, B. L., 725, 821
Jalinaud, T., 79
James, H. R., 1302
Jeanloz, R., 1363
Jetté, F. X., 1023
Jing, F., 1401
Joannopoulos, J. D., 343, 385
Johansson, B., 67
Johnson, G. R., 1287
Johnson, J. D., 99
Johnson, R. P., 1343
Johnston, R. R., 1141
Jones, A. P., 1381
Jones, H. D., 103
Jones, I. P., 634
Jordan, J. L., 1097
Joshi, V. S., 701, 860, 864, 1323

K

Kad, B. K., 615, 619
Kagan, K. L., 237
Kai, K., 351
Kalantar, D. H., 615, 619
Kanel, G. I., 503, 603, 1327
Kanzleiter, R., 143
Karpinos, B. S., 1314
Kato, Y., 930
Kawai, N., 1204
Keller, A. R., 795
Kelley, J. M., 864
Kelly, A. M., 499
Kennedy, G., 755
Kennedy, J. E., 1031
Kerley, G. I., 75, 307
Khantouleva, T. A., 263
Khishchenko, K. V., 71, 111, 759
Kimmel, G., 623
King, N., 966
Kishimura, H., 1375
Klein, B., 1119
Kline, K., 411
Knudson, M. D., 79, 1141, 1163
Kobayashi, T., 1255, 1339, 1397
Kobenko, S. V., 543

Koenig, M., 83, 1367
Kolesnikov, S. A., 938
Kondo, K.-I., 1131, 1181, 1204, 1215, 1259, 1263, 1375
Konrad, C. H., 787
Kopp, R. A., 1343
Kovalenko, G. V., 591, 875
Kress, J. D., 91, 99
Krivosheina, M. N., 543
Krüger, L., 1327
Kubota, S., 468
Kudelkin, V. G., 127
Kuklja, M. M., 454, 599
Kumar, M., 615
Kuryanchik, A. N., 237
Kutepov, A. L., 87, 245
Kutepova, S., 245
Kyrala, G. A., 1192, 1343

L

Laine, L., 431
Lamoreaux, S. K., 1247
Landau, A., 623, 1306
Landsberg, A. M., 926
Lankford Jr., J., 729
Lanzerotti, Y., 853
Lassalle, F., 1173
Lawrence, R. J., 205, 291
Lebedev, M., 1101
Lebedeva, T. S., 763
Lee, J. H. S., 1023
Lee, K. K. M., 1363
Lee, R. J., 701, 860, 894
Lee, S-H., 1231
Le Gallic, C., 1223
Leitsin, V. N., 1093
Lemke, R. L., 1163
L'Eplattenier, P., 1173, 1188
Levashov, P. R., 71, 111, 119
Leygnac, S., 1367
Li, X., 1401
Link, R., 934
Liu, C., 661, 725
Llorca, F., 319, 323, 327, 638
Loboiko, B. G., 439
Lomdahl, P. S., 351
Lomonosov, I. V., 71, 111, 237, 759
Lomov, I. N., 287, 1389

Lopez, M. F., 479, 665
Lorenz, K. T., 615
Loubeyre, P., 1363
Loveridge, A., 615
Luttwak, G., 255, 283
Lutz, S. S., 1239

M

Mabire, C., 229, 1235
Macdougall, D. A. S., 475
Mace, J. L., 1153
MacFarlane, J. J., 291
Mackinnon, A., 1363
Maillet, J.-B., 367
Makhov, M. N., 962
Malka, V., 1367
Malone, R. M., 531
Mang, J. T., 833
Mangeant, C., 1173
Mann, G. A., 483
Marchet, B., 83, 1367
Markland, L. S., 507, 841
Marshakov, V. N., 868
Marshall, B. R., 1247
Martin, E. S., 1031
Martinez, A. R., 817
Martinez, E., 1200
Mas, E. M., 427, 535, 539, 661
Matheson, E. R., 464
Mathieu, D., 439
Matsui, K., 468
Matuska, D. A., 435
Maudlin, P. J., 279
Maw, J. R., 1027
Mazevet, S., 91, 99
McCahan, S., 946
McDonald, J., 1185
McElrue, D. H., 1302
McGrane, S. D., 1227, 1333
McNelley, T. R., 571
Medvedev, A. B., 763
Mehlhorn, T. A., 291
Mendes, R., 721, 918, 922
Menikoff, R., 399, 979
Mescheryakov, Y. I., 267
Meyer, L., 1327
Meyers, M. A., 567, 571, 615, 619, 1087
Michaut, C., 1367
Mikhaylov, A. L., 547, 595

Milella, P. P., 642
Miller, D., 475
Miller, P. J., 950
Millett, J. C. F., 131, 135, 213, 479, 511, 523, 575, 579, 634, 649, 653, 717
Milne, A. M., 914, 1011
Minich, R. W., 491, 531
Minnaar, K., 1208
Mintsev, V. B., 107
Mochalov, M. A., 763
Mock Jr., W., 1196
Molinari, V., 271
Montgomery, S. L., 1247
Montgomery, S. T., 197, 201, 205
Moon, S. J., 1363, 1371
Moore, D. S., 1219, 1227, 1333, 1351
Morano, E. O., 446
Mori, Y., 673
Moritoh, T., 1204
Morris, C. L., 966
Morris, J. P., 287
Mort, P., 653
Moshe, E., 583
Mukai, T., 725
Mulford, R. N., 415
Murata, K., 930
Murray, N. H., 747
Murray, S. B., 946

N

Nagayama, K., 468, 673, 995
Nahme, H., 685
Nakahara, M., 995
Nakamura, K. G., 1131, 1181, 1204, 1215, 1259, 1263, 1375
Namkung, J., 1003
Narayan, S., 423
Nesterenko, V. F., 567, 1294
Ng, A., 53, 1243, 1371
Nguyen, J. H., 1251
Nicholls, A. E., 729
Nikitenko, Y. R., 875
Nikolaev, D. N., 59
Niles, A. M., 886, 1137
Niwase, K., 1131
Nizovtsev, P. N., 595

O

Obst, A. W., 1247
Occelli, F., 1363
Ogura, T., 1215
Okamoto, A., 1131
Ornthanalai, C., 946
Orphal, D. L., 1279
Oxby, T. J., 894

P

Paisley, D. L., 1192, 1343
Palmer, J., 79
Palmer, S. J. P., 825
Pangilinan, G. I., 181
Parker, K. W., 79
Partom, Y., 460, 739, 1283
Pavlovskii, M. N., 759
Payton, J. R., 1247
Pazuchanic, P., 966
Pedroso, L., 922
Peiris, S. M., 181
Pelak, R. A., 1212
Perez-Prado, M. T., 571
Perger, W. F., 894
Pershin, S. V., 938
Peterson, P. D., 821
Petrovtsev, A. V., 87, 591
Phillips, R., 339
Pickup, I. M., 771
Plaksin, I., 721, 898, 918, 922
Pollaine, S., 615
Portugal, A., 922
Postnov, V. I., 233, 237
Pragnell, H., 507
Price, D. F., 1371
Price, G. D., 1381
Primas, L. E., 1239
Proud, W. G., 487, 747, 803, 807, 825, 878, 1007, 1035, 1081, 1105, 1385, 1414
Pyalling, A. A., 59, 95, 107

Q

Quintana, J. P., 966

R

Rabie, R. L., 1219, 1227, 1333, 1351
Radchenko, A. V., 543
Radford, D. D., 807
Rae, P. J., 825
Raevskii, V. A., 595
Ranestad, Ø., 431
Ravichandran, G., 557, 619, 709
Raychenko, A. I., 1314
Razorenov, S. V., 503, 603, 1327
Reaugh, J. E., 1015
Reed, E. J., 343, 385
Reho, J. H., 1219, 1227, 1333, 1351
Reinhart, W. D., 75, 307, 483, 515, 787, 791
Reisman, D. B., 221, 849
Remington, B. A., 615, 619
Remo, J. L., 1410
Renlund, A. M., 1051
Resnyansky, A. D., 315, 717, 743
Riad Manaa, M., 385
Ribeiro, J., 721, 918, 922
Riedel, W., 705
Rieker, T. P., 833
Rightley, P. M., 1212, 1239
Robbins, D. L., 499, 630, 1145
Robbins, J., 201, 205
Robinson, A. C., 197
Robinson, C. M., 1359
Robinson, J. M., 1077
Rodríguez Romo, S., 241
Roemer, E. L., 821
Roessig, K. M., 829, 973
Romero, J. L., 531
Root, S., 363
Rosakis, A. J., 557
Rosakis, P., 557
Rosenberg, J. T., 464
Rosenberg, Z., 575, 1310
Roth, J., 378
Rothman, S. D., 79
Rubin, M. B., 739
Rupp, T. D., 499, 1145
Russell, R., 755, 811
Russell, T. P., 181
Rykovanov, G. N., 875

S

Saito, F., 1181, 1375
Salisbury, D. A., 303, 419, 841, 999
Sandusky, H. W., 1073
Sandvik, A., 431
Santiago, F., 1196
Sapozhnikov, A. T., 87
Saw, C. K., 856
Scammon, R. J., 987
Scheidler, M., 689
Schneider, M. S., 619
Schwartz, A. J., 491
Seaman, L., 607
Sekine, T., 1113, 1255, 1339, 1397
Selezenev, A. A., 374
Servas, J.-M., 1200
Setchell, R. E., 191, 201, 209, 669, 1347, 1355
Sewell, T. D., 399, 403
Sharma, J., 563, 837, 853
Sharp, T. G., 1381
Shaw, M. S., 157
Sheffield, S. A., 499, 999, 1019, 1051, 1145
Shepherd, J. E., 251, 446
Shimada, H., 468
Shuykin, A. N., 763
Shvarts, D., 623, 1306
Simakov, G. V., 759
Simões, P., 922
Simonetti, P., 439
Sims, C. E., 185
Skidmore, C. B., 833
Skripnyak, E. G., 751
Skripnyak, V. A., 751, 1093
Skryl, Y., 599
Slanik, M., 946
Smedley, J., 1185
Smilowitz, L. B., 1065, 1069, 1077
Smirnov, G. S., 547
Smith, G. D., 399, 403
Snekkevik, A., 431
Snow, R. C., 499
Softley, I., 1011
Son, S. F., 833, 1059
Sorenson, D. S., 531
Souers, P. C., 161, 902
Soulard, L., 173, 347

Springer, P. T., 1371
Srinivasan-Rao, T., 1185
Stacy, H., 966
Stahl, D. B., 499, 1145
Stehlé, C., 1367
Stennett, C., 1047
Stevens, G. S., 579
Struve, K., 849
Stryk, R. A., 1287
Stuivinga, M., 1127
Sud'enkov, Y., 627
Sutherland, G. T., 1055
Swanson, D. R., 363, 395
Swift, D. C., 415, 1192, 1343
Swizter, L. L., 886
Syed, I. H., 693

T

Takenaka, H., 1215
Tang, Z.-P., 259, 371, 679, 1397
Tarver, C. M., 42, 153, 849, 882, 886, 1019, 1039, 1043
Taylor, P., 419, 507, 841, 999
Tchijov, V., 241
Tellier, L., 1065
Teodori, F., 271
Ternovoi, V. Y., 59, 95, 107
Thadhani, N. N., 755, 1097, 1109, 1123
Thibault, P. A., 934
Thissell, W. R., 475, 611
Thoma, K., 705
Thomas, K. A., 1031
Thompson, D. G., 821
Thornhill, T. F., 515
Tomasini, M., 83
Tonks, D. L., 475, 611
Townsend, D., 1267
Trinkle, D. R., 225
Trott, W. M., 205, 483, 713, 791, 845, 1347, 1355
Trucano, T. G., 291
Trujillo, C. P., 725
Tsembelis, K., 1385, 1414
Tunnell, T. W., 531
Turley, W. D., 1239
Tuttle, B. A., 209

U

Urtiew, P. A., 153, 882, 902, 1039, 1043, 1157
Utkin, A. V., 938

V

Valone, S. M., 295
Vandersall, K. S., 153, 882, 902, 1109, 1137, 1157
Vanpoperynghe, J., 1188
Vaughan, B. A. M., 747
Vecchio, K. S., 479
Venkert, A., 623
Venturini, E. L., 209
Vitello, P. A., 161
Voigt, J. A., 209
Vollrath, I., 1243
Voronin, B. L., 374

W

Wakabayashi, K., 1259, 1263
Walker, J. D., 275, 1273, 1279
Wang, W. W., 259, 371, 411, 679
Wang, Y., 67
Wark, J. S., 615, 619
Watts, P. W., 221
Webb, D. J., 1231
Wei, D., 407
Weickert, C. A., 954, 958
Weidemaier, P., 705
Werdiger, M., 583
White, C. T., 355, 395, 1406
Whitworth, N. J., 991
Wilke, M. D., 1247
Wilson, L. L., 519, 1318
Wilson, L. T., 515
Wilson, W. H., 1137
Winfree, N. A., 75, 307
Winter, R. E., 303, 419, 507, 841, 999, 1302
Woo, T. K., 407
Wood, W. W., 99
Woody, D., 942
Wright, W. J., 821

X

Xia, K., 1393
Xu, X., 1123
Xu, Y., 571
Xue, Q., 567, 571

Y

Yakushev, V. V., 233, 237
Yakusheva, T. I., 233, 237
Yamaguchi, A., 1397
Yankelevsky, D. Z., 739
Yano, K., 553, 983
Yao, J., 435
Yazaki, A., 1181, 1375
Yoshida, M., 1131
Yoshinaka, A. C., 1023

Z

Zaikin, V. T., 875
Zaretsky, E., 217, 623, 1119, 1306
Zaug, J. M., 177, 856, 1015
Zerilli, F. J., 181, 657
Zeuch, D. H., 201
Zhang, F., 407, 934, 946, 1023
Zhernokletov, M. V., 759, 763
Zhiembetov, A. K., 547
Zhou, M., 755, 795, 1208
Zhuang, S., 709
Zhukova, T. V., 751
Zumbro, J. D., 966
Zurek, A. K., 475, 611
Zybin, S. V., 355, 1406

SUBJECT INDEX

A

ab initio calculations
    electronic structure, 3, 454
    equation of state, 87
    molecular collisions, 407
    structural stability, 245
    uniaxial Hugoniostat method, 367
ablation
    aluminum, 468
    copper, 619
    PETN explosive, 995
    silicon, 1375
absorption spectroscopy, 763
acceleration
    explosives, 853
    magnetic, 1169
    ultra-fine particles, 1101
acetylene, 1406
acoustic impedance, 689
adhesion, 1101
adiabatic release experiment, 79
adiabatic shear bands, 567, 571
aerogels, 763
aerosol deposition method, 1101
aerospace materials, 131
alumina
    /aluminum composite, 697
    bar impact test, 787
    flyer material, 1355
    grain size, 751
    Hugoniot elastic limit, 735
    spall strength, 747
    /titanium boride composite, 755, 795
aluminum
    ablation, 468
    additives, 890, 962
    /alumina composite, 697
    coating, 995
    combustion, 1011
    dislocations, 339
    failure, 275
    foam, 725, 729
    foil plate, 1339, 1355
    grain size, 531
    Hugoniot, 53, 1163, 1347
    Hugoniot elastic limit, 26, 503, 603
    laser-driven shock, 1333
    lattice dynamics, 339
    melting, 374
    nanocrystalline, 942, 950
    /nickel powder, 1093
    penetration experiment, 1310
    plasma, 53
    /polycarbonate composite, 709
    proton beam impact, 271
    /RDX mixture, 423
    recovery, 615
    shock breakout, 1351
    spall strength, 603, 1359
    strain rate, 507
    strength, 583
    vaporization, 63
    velocity transmission factor, 934
    voids, 607
aluminum alloys
    ballistic impact, 1298
    grain size, 634
    hardening, 627
    oblique impact, 595
    quasicrystals, 378
    spall strength, 523
    strain rate, 557
aluminum nitride, failure, 771
aluminum oxynitride, 767
American Physical Society, 11
ammonium nitrate
    emulsion explosives, 930
    /fuel oil, 165, 906
    hot spots, 1081
ANFO explosive
    cylinder test, 165
    detonation, 906
approximate blast theory, 442
aquarium test, 837
armor
    cassette, 1283
    electromagnetic, 1314
    materials, 689, 701, 747
astrophysical phenomena, 1367
atomic force microscopy
    explosives, 837
    molecular crystals, 563
    polymers, 1219
augmented plane wave method, 245

austenitic transformations, 351
automobile materials, 131
awards, 3, 16

B

backward integration technique, 1177
ballistic impact chamber test, 942, 1003
baseline material, 1087, 1294
benzene, 1263
beryllium
    failure, 275
    velocity history, 1192
binders
    GAP, 918
    HTPB, 1047
    KEL-F, 653, 942
    modeling, 661
    /PMMA interface, 829, 973
    stiffness, 415
    viscoelasticity, 427
    Viton, 942
birefringence, 1251
bismuth
    equation of state, 1145
    phase diagrams, 111
bistrinitroethylnitramine explosive, 962
blast, 950
boiling point, 63
Boltzmann superposition principle, 661
boron carbide
    failure, 775
    Hugoniot elastic limit, 735
    penetration experiment, 1287
    spallation, 779
borosilicate glass, 803
boundary value problem, 1089
bremsstrahlung, 1185
brittle materials
    failure, 775, 811
    fracture, 743
    plastics, 739
    short pulse propagation, 751
BTATZ explosive, 1059
bubbles, 914

C

cadmium sulphide, 3
capacitor bank, Shiva Star, 143
carbon
    equation of state, 759
    phase transition, 1131
    resistor gauge, 882, 886, 1137
cavitation, 495, 547
cement, 1414
ceramics
    alumina, 751, 755, 787
    aluminum nitride, 771
    aluminum oxynitride, 767
    boron carbide, 735, 775, 779
    Hugoniot elastic limit, 735, 739
    PBZT, 311
    penetration experiment, 1287
    PSZ, 735
    PZT, 191, 197, 201, 205, 209, 1153
    silicon carbide, 751
    spall strength, 747
    synthesis, 1097, 1101
    titanium boride, 755
    transparent, 767
    tungsten carbide, 783
    two-phase, 755
    zirconium oxide, 751
Chapman–Jouguet state, 20, 42, 177, 395, 450, 910, 962, 1031
chondrules, 1397
Cindy test, 1059
cinematography, 323
Clausius–Clapeyron equation, 63
clay, 431
clusters, void, 611
coatings
    aluminum, 995
    polymer, 1219
combustion
    aluminum, 1011
    explosives, 1059
    PBX 9501, 833
    solid propellant, 868
compaction
    ceramics, 739
    sand, 431
    titanium carbide, 1127
composite materials
    alumina/aluminum, 697
    aluminum foam, 725, 729
    armors, 689
    ceramic/metal, 1119
    explosives, 169, 431, 685, 701, 705, 853, 1055
    fiber-reinforced, 275, 1208

graphite/polymer, 693
interface scattering, 709
modeling, 679, 685
PBX 9501/binder, 427
titanium boride/alumina, 795
Composition A3 explosive, 853
Composition B3 explosive, 169, 431
compressibility
  aerogel, 763
  aluminum foam, 725, 729
  glycerin, 864
  graphite, 759
  HMX explosive, 399
  iron, 26
  Monte Carlo method, 185
  PMMA, 665
  powders, 1087, 1093, 1109
computer aided design, 697
computer codes
  ABAQUS, 946
  ADIFOR, 299
  ALE, 161, 255, 303, 882, 1087
  ALEGRA, 197, 201, 205, 291, 311
  ASCI, 311
  AUTODYN, 255, 283, 287, 431, 685, 705, 1283, 1306
  BRIGS, 954, 958
  CALE, 291
  CAST, 519
  CAVEAT, 255
  CEA, 319
  CHEETA, 311
  CHEETAH, 161, 165, 177, 890, 1011
  CORVUS, 303, 507
  CRYSTAL95/98, 454
  CTH, 275, 307, 464, 713, 1031, 1239, 1355
  CW2, 450
  0D, 1173, 1188
  DDEM2, 411
  DM2, 26, 259, 371, 553, 679
  DYNA, 487, 511, 689, 709, 721, 743
  EMMA, 311
  EPIC, 535, 539
  HESIONE, 638
  HULL, 527
  hydrocode, 143, 161, 275, 279, 299, 319, 415, 450, 487, 511, 527, 591, 841, 914, 991, 1188, 1279, 1359, 1367
  KARASH, 439
  KHT, 930
  LASNEX, 1371
  MESA, 299, 966
  MMALE, 283
  MSC/DYTRAN, 255
  MULTI, 83
  NEMD, 359
  ODE, 295
  PETRA, 291, 419
  PIMM, 439
  PISCES, 460, 1310
  RAGE, 143, 291
  RAVEN MHD, 143
  RRCK, 1173
  SESAME, 53, 83, 91, 143
  SHAKE, 399, 403
  SPECT3D, 291
  S-TAMP, 173
  TAPP, 1173
  TEPLA, 535
  VASP, 99
  VOLNA, 875
  WONDY, 221, 1177
condensed matter, response to shock, 3, 11
conservation laws, 139, 435
constitutive modeling
  ballistic impact, 1306
  composite materials, 679, 685
  dynamic fracture, 519
  elastic-plastic properties, 979
  equation of solid state, 642
  expanding shell test, 327
  grain boundaries, 553
  high explosives, 539
  KS-32 explosive, 705
  mesoscale, 415
  nonlinear viscoelasticity, 371
  PBX 9501 explosive, 411
  plastic deformation, 279, 319
  polymers, 657
  porosity, 717, 739
  rock explosion, 1389
  shock initiation, 415
  viscoelasticity, 495
continuum mechanics, 3
  classical, 20
  models, 26
  molecular dynamics, 363
  steady flow, 395
convection, 926, 1093
cook-off experiment, 882, 1059, 1073

copper
   dynamic fracture, 475
   elastoplasticity, 323
   fluid–solid interactions, 251
   foil plate, 1339
   fracture, 519
   hardening, 627
   Hugoniot, 67, 169
   Hugoniot elastic limit, 503
   mechanical properties, 319
   melting, 374
   plastic deformation, 619
   recovery, 615
   shear strength, 575
   spall, 487, 491
   strain rate, 327
   strength, 583
   Taylor test, 1318
corner turning experiment, 966
coupled damage and reaction model, 464
cryogenic systems, 1141
crystal structure
   diaminodinitroethylene, 181
   PZT, 1101
   uranium, 245
cutting techniques, 287
cylinder test
   aluminum, 1011
   ANFO explosive, 165
   detonation, 954, 958
   fragmentation, 515, 527
   modeling, 299
cylindrical configuration, 799, 902, 1065

D

deflagration-to-detonation transition, 391, 875, 926, 1059
density functional method, 91, 99, 343, 385
detonation
   Chapman–Jouguet, 20
   convective, 926
   deflagration, 878
   delayed, 464
   electromagnetic radiation, 894
   emulsion explosives, 930
   equation of state, 450
   failure, 898
   front curvature, 906, 926
   heterogeneous, 423
   initiation, 1023, 1069
   isentropic, 875
   mesoscale, 918, 922
   methanol, 177
   microsamples, 918
   modeling, 161, 385, 431
   molecular dynamics, 333
   momentum transfer, 934
   Monte Carlo method, 157
   particle momentum flux, 946
   powder, 995
   pressure wave, 902
   propulsion model, 954, 958
   proton radiography, 966
   simulation, 415
   steady state, 395, 423, 1223
   theory, 42
   transient, 1031
   velocity, 890
   wave front, 910
deuterium, liquid, 91, 1141
diaminodinitroethylene, 181
diamond
   amorphous, 1131
   compressibility, 759
   Hugoniot elastic limit, 735
   phase transition, 233
   shock wave structure, 355
diamond anvil cell, 1015, 1363
dielectric polarization, 347
differential thermal analysis, 1097, 1113
digital speckle x-ray flash photography, 303, 1105
diopside, melting point, 185
discrete element method, 259, 371, 411, 553, 679
discrete meso-dynamic method, 1397
dislocations
   aluminum, 339
   copper, 619
   RDX explosive, 454
   screw, 36
   titanium, 623
dispersive waves, 845
displacement measurement, 803, 1105, 1208, 1219
dissociation
   fluids, 91, 99
   hydrocarbons, 1406
dolerite, 1385
drop weight apparatus, 1137
Dyneema UD-HB25 composite, 685

E

Earth's interior, 1401
EDC-32 explosive, 419
EDC-37 explosive
   initiation, 415, 999, 1027
   undetonated, 841
effective medium theory, 427, 539
ejecta
   dynamic properties, 1247
   particle size, 531
elastic displacements, 36
elastic-plastic models, 26, 36
elastomers, 131
electrical conductivity
   graphite, 233
   helium, 107
   lithium, 237
   phase transition, 233, 237
   plasma, 53
   PMMA, 1267
   PZT, 191, 197, 201, 209
   RDX explosive, 894
electromagnetic armor, 1314
electromagnetic radiation, emisson, 894
electromechanical devices, 311
electron backscattering diffraction, 571
electron beams, pulsed, 1185
electronic structure
   ab initio calculations, 3
   HMX explosive, 385
   molecular liquids, 91
   RDX explosive, 454
electron temperature, 1243
ellipsometry, 1247, 1251
embedded-atom method, 339, 351
emission, electromagnetic, 894
emission, optical
   lead, 229
   preshocked explosive, 1023
   RDX explosive, 894
   sapphire, 1231
   sodium chloride, 1215
   surface, 1247
   syntactic foams, 721
emulsion explosives, 930
energy dispersive spectroscopy, 795
enstatite, 1401
enthalpy, explosives, 439
epoxies
   deviatoric response, 649
   encapsulants, 205, 669
   /graphite composite, 693
   Hugoniot, 135, 649
   /Kevlar flyer, 685
   syntactic foams, 721
   temperature effects, 669
equation of state (EOS)
   ANFO explosive, 165
   bismuth, 1145
   carbon, 759
   Clausius–Clapeyron, 59
   compaction, 431
   composite materials, 685
   copper, 319
   detonation products, 450
   explosives, 157
   FOX-7 explosive, 181
   iron, 83, 87, 127
   Jones–Wilkins–Lee, 165, 1039
   $K(\Box,P)$ type, 115, 139
   lead, 79
   liquid helium-4, 1141
   liquid metals, 71
   liquid nitrogen, 91, 99
   LX-17 explosive, 153
   measurements, 1141, 1145
   mechanical, 642
   metals, 111
   methanol, 177
   microscale materials, 1347
   molecular liquids, 91
   multiphase, 87, 111, 591, 759
   nickel, 59
   PBX 9501 explosive, 149
   plastic, 36
   porous metals, 67
   QEOS, 53
   rock, 1401
   silicon oxide, 763
   sugar, 713
   Tait, 115, 139
   temperature-dependent, 169
   thermochemical, 161
   tin, 143
   titanium, 225
   titanium alloys, 75
   tungsten carbide, 783
   urea nitrate, 161
   vanadium, 71
   Vinet, 185

water, 103
wide-range, 127
equations of motion, 1177
estane, 653, 661, 821
Eulerian representation, 20, 435
eutectics, steel, 479
expanding shell test, 232, 327
expansion-to-shock transition, 1375
explosions, rock, 1389
explosive
    burning rate, 1011
    charge, 42, 303, 507, 721
    detonation, 323, 817, 938
    fragmentation, 547
    /inert particle mixture, 423
    molecule, 42
    propulsion, 954, 958
    switch, 1173
explosive materials
    ANFO, 165, 906
    bistrinitroethylnitramine, 962
    BTATZ, 1059
    combustion, 1059
    Composition A3, 853
    Composition B3, 169, 431
    crystalline, 1003
    diaminodinitroethylene, 181
    EDC-32, 419
    EDC-37, 415, 841, 999, 1027
    emulsion, 930
    glycerin, 864
    heterogeneous, 42
    HMX, 42, 157, 169, 385, 399, 403, 415, 817, 841, 845, 918, 922, 938, 979, 1015, 1047, 1069
    HNS, 1007, 1035
    homogeneous, 42
    IRX-4, 1055
    isopropyl nitrate, 1051
    KS-32, 705
    LX-04, 42, 849, 886
    LX-14, 853
    LX-17, 42, 153, 902, 942, 1019
    nitroguanidine, 962
    nitromethane, 173, 347, 385, 407, 817, 890, 898, 914, 946, 1015, 1023, 1223, 1255
    NM/PMMA-GBM mixture, 817, 898
    PAX-2, 853
    PAX-3, 853
    PBX, 679, 918, 922
    PBX 9404, 415, 991
    PBX 9501, 149, 157, 411, 415, 427, 450, 539, 547, 661, 821, 825, 882, 886, 1043, 1059, 1065, 1077
    PBX 9502, 157, 460, 966
    PBXN-5, 1031
    PBXN-109, 950, 973, 1073
    PBXN-110, 701
    PBXW-115, 169
    PBXW-128, 701, 860
    PETN, 42, 468, 878, 910, 926, 995, 1007
    plastic bonded, 701, 821, 829, 841, 853, 973, 1031, 1055
    predetonating, 894
    radially-graded, 926
    RDX, 423, 454, 837, 878, 894, 910, 934, 938, 950, 983, 1007, 1047
    TATB, 42, 157, 849, 875, 942, 1019, 1039
    TNAZ, 853
    TNETB, 938
    TNT, 860, 1047, 1323
    ZOX, 938

F

fabrics, 1273, 1279
Fabry–Perot interferometry, 910
Fermi statistics, 119
ferroelectric materials
    PBZT, 311
    PZT, 191, 197, 201, 205, 209, 1101, 1153
fiber-reinforced composites
    delamination, 1208
    failure, 275
finite-element methods, 197, 535, 539, 543, 608, 689, 709, 946, 1087, 1208, 1287
fluctuation-dissipation theorem, 36
fluids
    compressible, 20
    interfaces, 295
    modeling, 946
    shock propagation, 26
fluid–solid coupling, 251
fluorescent x-ray source, 1185
flyers
    high-speed, 1163, 1169
    Kevlar/epoxy, 685
    magnetically driven, 1141
    miniature, 630, 1339
    multiflyer experiments, 1145

foams
  aluminum, 725, 729
  multiple shock, 419
  penetration experiment, 1302
  syntactic, 721
foils
  aluminum, 507, 583
  bismuth, 1145
  copper, 583
  gauges, 886
  plate, 1333, 1355
  uranium, 499
Fokker–Planck equation, 271
FOX-7 explosive, 181
fractography, 697, 795
fragmentation
  controlled, 527
  cylinder test, 515
  expanding cylinders, 799
  liquid metals, 547
  nitromethane, 407
  uranium, 1302
frequency domain interferometry, 1227
friction
  inter-granular, 987
  internal, 36
  sliding, 507
fullerenes, 1131

G

gas
  analysis, 1196
  cavity collapse, 914
  driver, 1204
  modeling, 259, 679
  propulsion, 954, 958
gas gun, 153, 191, 209, 267, 483, 491, 515, 523, 653, 673, 747, 755, 771, 779, 845, 914, 999, 1043, 1051, 1149, 1196
  compact, 1204
  three-stage, 75, 307, 1169
gas phase element modeling, 679
gauges
  cantilever, 946
  carbon foil, 886
  carbon resistor, 882, 886, 1137
  electromagnetic, 999, 1019
  magnetic, 411, 1051
  manganin, 131, 135, 479, 511, 523, 649, 653, 709, 717, 747, 771, 787, 807, 902, 1019, 1039, 1043, 1055, 1157, 1385
  particle streak, 946
  piezoresistive, 153, 864
  PVDF, 303, 673, 841, 868, 930, 1109, 1123, 1149
  strain, 878
  transverse, 767
Gauss' mechanics, 173
generalized gradient approximation, 91, 99, 385
generalized Maxwell model, 427, 661, 679
generalized particle algorithm, 1287
geological materials
  chondrules, 1397
  dolerite, 1385
  enstatite, 1401
  granite, 1389
  iron, 1401
  meteorites, 1410
  rocks, 1393
  tectosilicates, 1381
glass
  additives, 890
  borosilicate, 803
  microballoons, 817, 898
  /polycarbonate composite, 709
  Pyrex, 743
  silica-based, 807
glycerin, 864
glycidyl azide, 1219, 1333
Godunov technique, 255
gold
  foil plate, 1339
  optical properties, 1227
graded elastic solids, 689
grain/particle size
  aluminum, 531, 942, 950, 962
  aluminum alloys, 627
  austenite, 351
  ceramics, 351, 755, 1101
  copper, 475, 491, 627
  ejecta, 531
  HNS explosive, 1035
  iron, 333, 553, 627
  modeling, 427
  nickel, 627
  PZT, 197

    scale effect, 26
    steel, 571
    tin, 531
    titanium aluminide, 634
granite, 1389
granular materials
    ammonium nitrate, 1081
    HMX explosive, 979
    inter-granular friction, 987
    internal deformation, 1105
    sugar, 26
graphite
    compressibility, 759
    phase transition, 233
    /polymer laminate, 693
group theory, 115, 139
Grüneisen
    equation of state, 20
    parameter, 20, 99, 169, 442
Gurney model, 954, 958

H

hardness
    aluminum alloy, 627
    copper, 627
    iron, 627
    nickel, 627
    polymers, 657
    PZT, 1101
    rocks, 1389
    targets, 1291
Hartree–Fock method, 454
hazard assessment, 464
hazardous materials, 1145
heat transfer
    convective, 1093
    explosives, 833
HELEN facility, 79, 1359
helium
    driver gas, 1204
    liquid, 1141
    phase diagrams, 107
    shock velocity, 59
Hellmann–Feynmann forces, 91, 99
highly compressed matter, 83, 91, 1163
highly confined gap experiments, 1059
historical discussions
    APS, 11
    U.S. and Soviet Union, 3

HMX explosive
    chemical decomposition, 1069
    detonation, 157, 918, 938
    dispersive waves, 845
    elasticity, 399
    electronic structure, 385
    fragment impact, 1047
    hot spots, 42
    Hugoniot, 169
    mesoscale detonation, 922
    phase transitions, 856
    plastic bonded, 841
    plastic deformation, 1003
    /PMMA interface, 817
    polymorphism, 403
    reaction propagation rate, 1015
    shock initiation, 415
    wave profiles, 979
HNS explosive
    shock duration, 1007
    shock sensitivity, 1035
hohlraum, 291, 1410
holography, 531
Homalite bar, 811
Hopkinson bar
    aluminum foam, 725, 729
    binder, 661
    brittle materials, 1323
    composite materials, 701, 795, 1208
    copper, 638
    direct tension split, 693
    explosives, 701, 860
    gauge calibration, 1137
    polymers, 665
    simulation, 315, 539
    titanium aluminide, 638
hot spots
    ammonium nitrate, 1081
    detonation, 910
    grain distribution, 979
    HMX explosive, 42
    initiation, 991
    modeling, 439, 446, 983
    thermodynamics, 599
    unknown-to-detonation transition, 464
Hugoniostat method, 367
Hugoniot
    aluminum, 53, 1163, 1347
    Composition-B3 explosive, 169
    copper, 67, 169

curve, 20
deuterium, 91
diamond, 355
epoxy resin, 135, 649
estane, 653
glass, 807
HMX explosive, 169
iron, 83, 87, 127, 442
IRX-4 explosive, 1055
isentropic compression experiments, *versus*, 1163
isopropyl nitrate, 1051
Kel-F-800, 653
melting, 374
molecular dynamics, 343
molybdenum, 67
nickel, 67
nickel alloys, 579
nitrogen, 91, 99
nitromethane, 173
oxygen, 91
PBXW-115 explosive, 169
piston simulation, 395
polychloroprene, 131
porous materials, 717
porous metals, 67, 79
potassium chloride, 213
principal, 79, 83, 91, 99, 1385
PZT, 191, 1153
rock, 1381, 1385
silicon nitride, 1113
tantalum alloys, 169
Teflon, 169
tin, 367
titanium, 225
titanium alloys, 75, 579
water, 103, 1363
Hugoniot elastic limit, 26
  alumina, 735, 739
  aluminum, 503, 523, 603
  aluminum nitride, 771
  aluminum oxynitride, 767
  boron carbide, 735, 739
  ceramics, 735, 739, 751, 755
  composite materials, 1119
  copper, 503
  diamond, 735
  minerals, 735
  PSZ, 735
  quartz, 735
  sapphire, 735
  silicon carbide, 735
  titanium aluminide, 634
  tungsten carbide, 783
  upper limit, 735
hydrocarbons, 1406
hydrodynamic shock, 115, 139
hydrogen
  driver gas, 1204
  plasma, 119
hypervelocity impact, 75, 307, 685, 1113

I
ICE. *See* isentropic compression
ice
  phase transition, 241
  refraction index, 1363
impact
  ballistic, 743, 942, 1003, 1279, 1294, 1298, 1306
  bar, 787, 811
  cylinder, 205, 515, 531, 1097
  electrostatic particle, 1101
  fragment, 1047
  graded targets, 689
  high-speed, 1163, 1169, 1204, 1406
  hypervelocity, 75, 307, 685, 1113
  inverse, 705
  jet, 914
  meteorite, 1381, 1393, 1397, 1406
  multi-dimensional, 255, 886
  non-symmetric, 1149, 1200
  oblique, 595
  one-dimensional, 689
  piston, 395, 864, 1204
  planar, 623, 630, 673, 1119, 1306
  plate, 3, 131, 135, 229, 319, 475, 487, 511, 523, 638, 697, 755, 767, 807, 1043, 1235
  projectile, 307, 886, 1047, 1204, 1273, 1283, 1287, 1294, 1314
  reverse, 669, 1149, 1306
  rod, 464, 587, 743, 787, 803, 1105
  sleeved, 519
  Steven, 42, 886
  symmetric, 411, 483, 511, 673, 791, 1149
  Taylor, 279, 299, 464, 519, 587, 1318
impedance matching method, 875
impulsive stimulated light scattering, 177
indium, liquid, 547

inertially confined fusion, 1343
infrared images, 1239
infrared spectroscopy, 1247, 1263
initiation
    chemical decomposition, 1069
    crystalline explosives, 1003
    detonation, 1023
    double shock, 999
    EDC-37 explosive, 1027
    hot spots, 991, 1081
    isopropyl nitrate, 1051
    laser, 468, 995
    modeling, 415, 983, 987, 1019
    PBXN-109 explosive, 973
    RDX explosive, 454
    self-ignition, 1065
    TATB explosive, 1039
instrumented experiments, 1109, 1123
interatomic potential, 177, 333
interfaces
    binder/PMMA, 829
    fluid, 295
    fluid–solid, 251
    HMX/PMMA, 817
    Kapton/PMMA, 713
    modeling, 283
    oblique impact, 595
    sliding, 507
    tin/lithium fluoride, 1235
    wave scattering, 709
interferometry
    frequency domain, 1227, 1333
    microwave, 1065
    moiré, 825
    two-dimensional, 1227, 1351
intermolecular potential, 1406
internal friction, 36
iron
    compressibility, 11, 26
    Earth's mantle, 1401
    equation of state, 83, 87, 127
    grain size, 333, 553, 627
    hardening, 627
    Hugoniot, 83, 87, 442
    phase diagrams, 87
    phase transitions, 287, 591
    shear strength, 575
    spallation, 487, 591
IRX-4 explosive, stresses, 1055
isentropic compression, 20
    copper, 575
    detonation, 875
    experiments (ICE), 1163
    iron, 575
    liquid nitrogen, 1141
    lithium, 237
    LX-04 explosive, 849
    nickel, 59
    optimization, 1188
    power generators, 1173
    steel, 575
    TATB explosive, 849
    tin, 221
    Z accelerator, 79, 1141, 1163
isobaric expansion method, 71
isopropyl nitrate explosive, 1051

J

Janus laser facility, 1371
jets, 291, 914
Jones–Wilkins–Lee EOS, 165, 1039

K

Kapton/PMMA interface, 713
Kel-F binder, 653, 942, 979, 999
Kolsky bar, 557
KS-32 explosive, 705

L

Lagrangian
    analysis of shock initiation, 1027
    meshless methods, 1287
    representation, 20, 435
    shock speed relation, 20
    sound speed, 1163
laminates, 693, 1208
laser Doppler velocimetry, 26, 323, 1200
laser-driven flyer plates, 1007, 1035
laser-driven initiation, 468, 995
laser-driven shock, 53, 79, 83, 499, 615, 619, 630, 1219, 1227, 1333, 1339, 1343, 1347, 1355, 1359, 1363, 1371
laser heating, 1227
laser-induced shock, 583, 1259, 1263, 1351
laser irradiation, 1181, 1192, 1343, 1367, 1375
laser polarimetry, 1247
laser triggered x-ray system, 1185
laser vibrometer, 1208
lattice compression
    molecular crystals, 563

silicon, 1375
lattice disintegration phenomena, 735
lattice dynamics, 339
lattice parameters
   beryllium, 1192
   HMX explosive, 403
Laves crystals, 378
lead
   equation of state, 79
   liquid, 547
   melting, 229
   phase diagrams, 111
Lennard-Jones systems, 367
Lie group analysis, 139
liner, cylindrical, 143, 149
Line VISAR/ORVIS. See VISAR, ORVIS or VISAR/ORVIS
liquid metals
   equation of state, 71
   fragmentation, 547
   tin, 221
liquids
   molecular, 91
   polar, 347
lithium, phase transition, 237
lithium fluoride
   birefringence, 1251
   window, 1235
LX-04 explosive
   isentropic compression, 849
   pressure histories, 42
   Steven test, 886
LX-14 explosive, 853
LX-17 explosive
   additives, 942
   detonation, 902
   initiation, 1019
   interface velocity, 42
   re-shock states, 153

M

Mach–Zehnder interferometer, 1333, 1367
magic angle spinning NMR spectroscopy, 1113
magnesium
   Hugoniot elastic limit, 603
   phase diagrams, 111
   spall strength, 603
   velocity transmission factor, 934
magnetic flux compression, 1188
magneto-hydrodynamic effects, 1163, 1173

manganese sulphide, 479, 535
martensitic transformations, 351, 571
Maxwell medium model, 263, 411
mean-field potential approach, 67
mechanical properties, 36
   crack growth, 563, 833, 1059, 1393
   deformation, 26, 36, 279, 315, 557, 619, 630, 725, 860, 1003, 1208, 1281, 1298
   elasticity, 399.755, 543, 669
   elastic-plastic properties, 26, 303, 323, 557, 627, 638, 979
   failure, 275, 464, 563, 771, 775, 795, 811, 817, 825, 1208, 1298, 1306
   fracture, 475, 479, 515, 519, 535, 543, 611, 630, 743, 853
   fragmentation, 515, 527, 547, 799
   hardness, 627, 657, 1101, 1291
   Hugoniot elastic limit, 26, 503, 523, 603, 735, 751, 755, 767, 771, 783, 1119
   plasticity, 279, 319, 333, 355, 495, 519, 563, 26837
   Poisson's ratio, 579
   shear banding, 567, 571, 587, 1294
   shear strength, 36, 131, 135, 213, 575, 634, 649, 653, 771, 1414
   shear stress, 339, 385, 411, 807, 973
   spall, 267, 475, 479, 483, 487, 491, 503, 511, 523, 591, 603, 630, 697, 747, 779, 1119, 1327
   strain rate, 427, 495, 503, 507, 515, 523, 557, 575, 583, 623, 642, 665, 679, 693, 725, 1294, 1318, 1323
   strength, 543, 775, 787, 795, 821, 1389
   stresses, 197, 209, 213, 217, 579, 673, 689, 717, 829, 1055, 1109, 1157, 1181, 1385, 1410
   stress-strain, 315, 661, 665, 693, 701, 729, 1323
   tensile strength, 511, 697, 755, 783, 821
   viscoelasticity, 371, 411, 427, 495, 591, 649, 657, 661, 679, 705, 1149
   yield strength, 603, 642, 767, 1327
melting
   aluminum, 603
   carbon, 759
   diopside, 185
   lead, 225, 229
   magnesium, 603
   metals, 111
   molecular dynamics, 374

rock, 1401
silicon, 1109
zinc, 603
mesoscale, 26
    compressed powder, 1093
    constitutive modeling, 415
    detonation, 918, 922
    dispersive waves, 845
    energy exchange, 263, 267
    heterogeneous materials, 713
    PZT, 197
    simulation, 679, 979
metallography, 630, 1306
metastable phases
    meteor craters, 1381
    silicon, 343
    titanium, 225
meteorite impact, 1381, 1393, 1397, 1406
meteorite targets, 1410
methane, 1406
method of cells, 427, 535, 539
microchannels, 1059
microdetonics, 1031
micro-gap test, 918, 922
micromechanics, 427, 539, 739, 751, 987
microstructure
    aluminum alloys, 627
    aluminum foam, 729
    ceramics, 197, 209, 755
    composite materials, 697, 795
    copper, 475, 491, 619, 627
    defects, 333, 454, 483, 615
    inclusions, 479, 535
    iron, 553, 627
    nickel, 627
    PBX 9501 explosive, 539, 825, 833
    porosity, 197, 209
    RDX explosive, 454
    stacking faults, 571, 619
    steel, 267, 479, 571, 630
    tantalum, 483
    texture, 475
    titanium, 623
    titanium aluminide, 634
    twinning, 571, 619, 623, 627
    uranium, 499
    vacancies, 599
    voids, 483, 495, 607, 611, 615, 833
microwave interferometry, 1065

Mie–Grüneisen equation, 20, 169, 251, 275, 442, 450, 519, 979
mine clearing vehicles, 431
minerals
    Earth's interior, 1401
    Hugoniot elastic limit, 735
    phase transitions, 1381
mixtures
    explosive, 817, 898
    explosives/metal particles, 934
    modeling, 127
    molybdenum/silicon, 1109
    nickel/titanium powder, 1123
    nitromethane/additives, 890
    RDX/aluminum, 423
    reactive, 763
    titanium/carbon powder, 1127
mixture theory, 1087
modeling
    ALOX test, 205
    brittle materials, 743
    burn rate, 1015
    ceramics, 751
    chemical decomposition, 1069
    cook-off, 1073
    delamination, 1208
    detonation, 177, 431, 446, 875, 890, 930, 966
    discrete element method, 259, 371, 411, 553, 679
    elastoplasticity, 323
    electromechanical devices, 311
    equation of state, 169, 185, 759
    excited states, 385
    explosive propulsion, 954, 958
    fabrics, 1279
    failure, 275
    flier plate launch, 307
    foam, 1302
    fracture, 535, 543
    fragmentation, 527
    fused deposition, 697
    gas, 259
    gas cavity collapse, 914
    gas–particle flow, 946
    high explosives, 427
    Hopkinson bar loading, 315
    hot spots, 983
    impact sensitivity, 439
    initiation, 987, 991, 1019, 1039, 1043
    interfaces, 283, 709

isentropic compression, 849
laser-driven shock, 1355, 1371
laser initiation, 468
laser spall, 1359
lattice defects, 454
liquid–solid transition, 221
magnetic flux compression, 1188
magnetohydrodynamic, 1173
mesoscale, 713
meteorite impact, 1397, 1406
microstructure, 638
mineralogical, 1401
mixtures, 934
multi-dimensional impact, 255
oblique impact, 595
penetration experiment, 1273, 1287, 1310
phase transition, 241
powder compression, 1087, 1093
pressure history, 1177
projectile, 1273
PZT depoling, 197, 201, 205
radiation-driven jetting, 291
radiative shock, 1367
reaction history, 841
reaction rate, 460
refraction shock wave, 287
re-shock states, 153
rod-plate interaction, 1283
Saturn plane wave lens, 419
shape charge, 435
shear banding, 567
shock trajectories, 143
sliding friction, 507
spallation, 464, 487, 511, 779
steady-state detonation, 423
Steven test, 886
surface geometry, 1239
syntactic foams, 721
Taylor test, 299
thermal explosion, 882, 1069
thin-wall tubes, 303
transient detonation, 1031
two-phase flow, 1011
vacancy diffusion, 599
VISAR records, 1177
viscoelasticity, 591
viscosity, 295
void coalescence, 611
void growth, 607
wave profiles, 979

moiré interferometry, 825
molecular crystals, 333, 563
molecular dynamics
    bond distortion, 385
    deflagration-to-detonation transition, 391
    detonation, 333, 395
    dislocations, 339
    dusty plasmas, 359
    elasticity, 399
    hydrocarbons, 1406
    iron, 83
    Laves crystals, 378
    liquid nitrogen, 99
    melting, 374
    metastable phases, 343
    molecular collisions, 407
    molecular liquids, 91
    phase transitions, 333, 351
    plasticity, 333
    polymorphism, 403
    quasicrystals, 378
    shock front, 363
    shock polarization, 173, 347
    shock wave structure, 355
    uniaxial Hugoniostat method, 367
molybdenum
    Hugoniot, 67
    phase diagrams, 111
    /silicon powder, 1109
Monte Carlo method
    compressibility, 185
    equation of state, 157
    hydrogen plasma, 119

N
nanocrystalline materials
    aluminum, 942, 950
    ceramics, 751
    silicon nitride, 1113
    steel, 571
nano-indentation, 837
nanoscale thermites, 1059
nickel
    /aluminum powder, 1093
    critical point, 59
    hardening, 627
    Hugoniot, 67
    optical properties, 1227
    /titanium powder, 1123
nickel alloys, phase transitions, 579

nitrobenzene, 1255
nitrocellulose, 1219, 1333
nitrogen, liquid, 91, 95, 99, 1141
nitroguanidine explosive, 962
nitromethane
    detonation, 890, 946
    electronic excitations, 385
    emission spectroscopy, 1223
    gas cavity collapse, 914
    Hugoniot, 173
    molecular collisions, 407
    /PMMA-GMB mixture, 817, 898
    polarization, 347
    reaction propagation rate, 1015
    reflected shock, 1023
    vibrational spectra, 1255
noise generation, 1157
nonhomogeneous deformation, 26
nonlinear wave propagation, 20
non-Newtonian fluids, 295
Nova facility, 291, 615
nuclear test, 1389
nylon, 673

O

oil-gas platform removal, 287
OMEGA laser, 615
optical extinction, 1231
optical fibers, 75, 918, 922, 1139, 1141
optical films, laser irradiation, 259, 679
optical microscopy
    aluminum foam, 725
    ceramics, 755
    FTIR, 1219
    PBX 9501 explosive, 825, 833
    titanium, 623
optical probes, 1243
optical properties
    aluminum, 1333
    cadmium sulphide, 3
    gold, 1227
    lead, 229
    liquid nitrogen, 95
    lithium fluoride, 1251
    nickel, 1227
    ruby, 3
    sapphire, 1231
    surface, 1239, 1247
optical radiometry, 1215
optical spectroscopy, 1141, 1219, 1223, 1255

ORVIS. *See also* VISAR
    aluminum, 583
    copper, 583
    line-imaging, 791, 845, 1347
    porous materials, 713
    tantalum, 483
oxygen, liquid, 91

P

Padé approximation, 169
partially stabilized zirconia, 735
particulates, 705, 829, 973
PAX-2 explosive, 853
PAX-3 explosive, 853
PBX explosive
    mesoscale detonation, 922
    microsamples, 918
    strain rate, 679
PBX 9404 explosive
    hot spots, 991
    initiation, 415
PBX 9501 explosive
    binder, 661
    detonation, 157
    equation of state, 450
    fragmentation, 547
    impact response, 411
    mechanical properties, 821
    modeling, 427, 539
    moiré interferometry, 825
    phase transition, 1077
    porosity, 833
    pressure waves, 882
    reaction violence, 1059, 1065
    reactive flow modeling, 1043
    Sandwich test, 149
    shock initiation, 415
    Steven test, 886
PBX 9502 explosive
    corner turning experiment, 966
    detonation, 157
    initiation, 1019
    plastic deformation, 1003
    reaction rate, 460
PBXN-5 explosive, 1031
PBXN-109 explosive
    blast, 950, 1073
    cook-off, 1073
    mesoscale structure, 973
    plastic deformation, 1003

PBXN-110 explosive, 701
PBXW-115 explosive, 169
PBXW-128 explosive
   Hopkinson bar, 860
   plastic bonded, 701
PBZT, electromechanical effect, 311
peek /graphite composite, 693
Pegasus facility, 531
penetration experiments
   electric current, 1314
   entrance phase, 1310
   fabrics, 1279
   impact velocity, 1291
   modeling, 1273, 1287
   rod-plate interaction, 1283
   theory, 1302
percolation theory, 531, 611
perturbation theory, 103, 454
PETN explosive
   ablation, 995
   detonation, 878, 910, 926
   laser initiation, 468
   reaction rate, 42
   shock duration, 1007
petrography, 1381
phase diagrams
   bismuth, 111
   helium, 107
   iron, 87
   lead, 111, 229
   molybdenum, 111
   tungsten, 111
   uranium, 111
phase transitions, 63
   alpha–omega, 225
   bismuth, 1145
   cadmium sulphide, 3
   carbon, 1131
   diamond, 233
   fluid–solid coupling, 251
   fullerenes, 1131
   graphite, 233, 759
   HMX explosive, 856
   ice, 241
   iron, 287
   lead, 229
   liquid–solid, 221
   lithium, 237
   martensitic, 351, 571
   minerals, 1381
   modeling, 343
   molecular dynamics, 333
   PBX 9501 explosive, 1077
   polymorphous, 399, 403, 591, 759
   potassium chloride, 213, 217
   PZT, 191, 197, 201, 205, 209
   silicon nitride, 1113
   TATB explosive, 856
   tin, 221
   titanium, 225
   uranium, 245
photo-chronograms, 922
photoelastic technique, 829, 975
photography
   digital speckle x-ray, 803, 1105
   high-speed, 743, 811, 829, 860, 894, 914, 973, 1035, 1081
   streak, 165, 205, 721, 791, 898, 918, 922, 995
piezoelectric materials, 1149, 1153
piezopin, 1153
piston test, 395, 864, 1204
planetary atmosphere, 1406
plasma
   brightness temperature, 83
   dusty, 359
   electron temperature, 1243
   hydrogen, 119
   strongly coupled, 53
plastics
   failure, 739
   sensitizers, 930, 973
platinum, foil plate, 1339
PMMA
   /aluminum, 419
   /binder interface, 829
   /ceramic interface, 713
   compressibility, 665
   cover plate, 575
   disk geometry, 973
   electrical conductivity, 1267
   /HMX explosive interface, 817
   /Kapton interface, 713
   /metal interface, 511
   rough surface, 995
   stresses, 523
pneumatic gun, 623
polarimetry, 1247
polycarbonate
   compressibility, 665
   /glass composite, 709

/steel composite, 709
stress relaxation, 673
polychloroprene, 131
polyethylene, 673
polymers
   binder, 661, 918, 942
   chemical reactions, 1219
   compressibility, 665
   constitutive modeling, 657
   decomposition, 1196
   epoxy resins, 649, 669
   /graphite laminate, 693
   molecular weight, 821
   Raman spectra, 1259
   shear-rate-dependent viscosity, 295
   shear strength, 653
   strain rate, 1323
   stress relaxation, 673
polymorphs, 399, 403, 591, 759
polytetrafluoroethylene
   constitutive modeling, 657
   decomposition, 1196
   Raman spectra, 1259
porous materials
   alumina/aluminum composite, 697
   decomposition, 1196
   explosives, 833
   heterogeneous, 713
   Hugoniot, 67, 79, 717
   Hugoniot elastic limit, 739
   iron, 127
   nickel, 59
   PZT, 191, 197, 205, 209
   silica beads, 1397
potassium chloride, 213, 217
powder gun, 229, 1200, 1235
powders
   chemical reactions, 1123
   compression, 1087, 1093
   densification, 1097
   laser ablation, 995
   synthesis, 1113
power generators, 1173
probability distribution functions, 26, 713
Prony series kernel, 987
propellant
   combustion, 868
   damage, 464
   gun, 1113, 1255
propulsion test, 954, 958

proton beams, 271
proton radiography, 966
pyrometry, 229, 763, 1169, 1231, 1235
pyrotechnic devices, 323
PZT
   depoling, 197, 201, 205
   disks, 1153
   phase transformation, 191, 209
   synthesis, 1101

Q
quantum chemistry, 403
quartz, 735
quasicrystals, 378
quasi-static tests, 693

R
Raman spectra
   benzene, 1263
   FOX-7 explosive, 181
   fullerene, 1131
   nitrobenzene, 1255
   nitromethane, 1255
   polymers, 1259
Rankine–Hugoniot equation, 20, 91, 99, 169, 395
Rayleigh line, 20, 395, 411
RDX explosive
   blast, 950
   crystal defects, 454
   detonation, 423, 878, 910, 934, 938
   electromagnetic properties, 894
   fracture, 837
   hot spots, 973
   melt-cast, 1047
   shock duration, 1007
reaction
   burning, 942
   under compression, 36
   decomposition, 385, 1069
   exothermic, 391
   fragmentation, 407
   heterogeneity, 446
   history, 1027
   low level threshold, 886
   rate, 415
      1015
   shock-induced, 1087, 1109, 1123
   synthesis, 1097
   temperature dependent, 460

undetonated explosive, 841
violence, 1059, 1065, 1073
zone structure, 926, 938
reactive burn model, 419, 841, 1019
reactive empirical bond order potential, 355, 391, 1406
reactive flow model, 415, 886, 1039
real-time measurements, 26, 1185, 1255
recovery experiments
    aluminum, 615
    composite materials, 697
    copper, 319, 615, 619, 638
    fullerene, 1131
    silica beads, 1397
    silicon, 615
    silicon nitride, 1113
    steel, 515
    titanium alloys, 587
    uranium, 499
reflectivity measurement, 83
refraction index, 1333, 1363
resonance theory, 263
rezoning, 435
R-lines, 3
rocks
    cracking, 1393
    hardening, 1389
    Hugoniot, 1385
ruby, 3
R-value problem, 279

S

sand, 431, 717, 1105
Sandwich test, 149
sapphire
    Hugoniot elastic limit, 735
    multi-demensional effects, 791
    optical extinction, 1231
Saturn plane wave lens, 419
scaling, 26
    Z-pinch, 291
scanning electron microscopy
    composite materials, 795
    PBXN-109 explosive, 973
    titanium, 623
second harmonic generation, 856, 1077
self-organization, 721, 918
self-sustained high-temperature synthesis, 1127
sensitizers, void size, 930
servohydraulic test, 705

shaped charge, modeling, 435
shape memory alloys, 579
Shiva Star capacitor bank, 143
shock, 3
    aerogel, 763
    aluminum, 26, 53, 374, 503, 507, 531, 615
    aluminum nitride, 771
    beryllium, 1192
    boron carbide, 775
    cadmium sulphide, 3
    ceramics, 191, 197, 201, 205, 209
    composite materials, 685, 709, 713
    copper, 319, 374, 503, 615, 638
    deuterium, 91
    dusty plasmas, 359
    EDC-37 explosive, 999
    epoxies, 669
    epoxy resin, 135, 649
    foams, 419, 721
    front, 26, 36, 363
    glass, 807
    graphite, 233
    helium, 107
    HMX explosive, 385, 841
    ice, 241
    iron, 83, 287, 553
    Laves crystals, 378
    lead, 229
    lithium, 237
    LX 17 explosive, 153
    multiple particles, 934
    nickel, 59
    nitrogen, 91, 95, 99
    nitromethane, 173, 347, 385, 914
    oxygen, 91
    polars, 251
    polymers, 295, 673
    porous materials, 713
    potassium chloride, 213, 217
    powders, 995, 1087
    quasicrystals, 378
    radiative, 1367
    RDX explosive, 894
    rocks, 1381, 1385
    ruby, 3
    sapphire, 1231
    sensitivity, 1007
    silicon, 53, 615
    sodium chloride, 115, 139
    solid propellant, 868

    steel, 303
    tin, 531, 1235, 1239
    titanium, 225, 623
    titanium aluminide, 634
    tube test, 950
    tungsten carbide, 783
    uranium, 499
    weak, 42
shock-to-detonation transition, 464, 922, 1035, 1047, 1223
shock wave paradigms, 26
silicon
    flyers, 1347
    laser irradiation, 1375
    metastable phases, 343
    /molybdenum powder, 1109
    plasma, 53
    recovery, 615
    reflectivity, 1243
    x-ray diffraction, 1181
silicon carbide
    grain size, 751
    Hugoniot elastic limit, 735
silicon nitride, 1113
silicon oxide, aerogel, 763
sliding friction, 507
small-angle scattering, 833
smooth compression waves, 20
smooth particle hydrodynamics method, 435
sodium chloride
    equation of state, 115, 139
    shock temperature, 1215
sound speed
    aerogel, 763
    enstatite, 1401
    HMX explosive, 403
    isopropyl nitrate, 1051
    Lagrangian, 1163
    methanol, 177
    PBX 9501 explosive, 450
    tin, 1200
spall
    aluminum, 503, 1359
    aluminum alloys, 523
    boron carbide, 779
    ceramics, 747, 755
    composite materials, 1119
    copper, 475, 487, 491, 503
    iron, 487
    modeling, 464
    pressure effects, 491
    steel, 479, 630
    tantalum, 483
    titanium alloys, 511, 1327
    uranium, 499
specific heat, 245
statistical mechanics
    expanding cylinders, 799
    non-equilibrium, 263
steel
    fracture, 519, 535, 543, 630
    fragmentation, 527
    grain size, 351
    mechanical strength, 303
    microstructure, 571
    /polycarbonate composite, 709
    powder, 1087
    shear bands, 567, 571
    shear strength, 575
    spall, 267, 479
    strain rate, 515, 642
    /titanium carbide composite, 1119
Steven test, 42, 886
streak photography
    ALOX test, 205
    detonation, 898, 918, 922, 991
    lateral unloading, 791
    syntactic foams, 721
    wall expansion, 165
sublimation, 403
sugar
    dispersive waves, 845
    equation of state, 713
    /HTPB composite, 705
    simulation, 26
superheated solid states, 603
superseismic loading, 251
supersonic mode, 595
surface burn reaction model, 460
synergetic effect, 918
syntactic foams, 721
synthesis
    ceramics, 1097, 1101
    self-sustained, 1127
    silicon nitride, 1113
SYRINX project, 1173

T

tantalum
    elastoplasticity, 323

    equation of state, 71
    free surface velocities, 42
    spall, 483
    strain rate, 327
    Taylor cylinder test, 279
tantalum alloys, 169
TATB explosive
    additives, 942
    detonation, 157, 875
    hot spots, 42
    initiation, 1019, 1039
    isentropic compression, 849
    phase transitions, 856
Taylor impact test, 279, 299, 464, 519, 587, 1318
Taylor series expansion, 1291
tectosilicates, 1381
Teflon
    acceptor, 882
    Hugoniot, 169
    insulation, 886
temperature
    control, 1141, 1145
    Earth's core, 1401
    electron, 1243
    epoxies, 669
    flyer plate, 1169
    fullerene, 1131
    gas cavity collapse, 914
    lead, 229
    liquid nitrogen, 95
    nitromethane, 1223
    PBX 9501 explosive, 1065
    PBXN-109 explosive, 1073
    polymers, 665, 821
    probability distribution functions, 713
    sodium chloride, 1215
    surface, 1247
    tin, 1235
    void growth, 607
TEXT experiments, 882
thermal activation model, 657
thermal cameras, 1239
thermal conductivity, sapphire, 1231
thermal diffusion analysis, 1131
thermal expansion
    diopside, 185
    HMX explosive, 403
    iron, 87
    PBXN-109 explosive, 1073
    silicon, 1375

thermal explosion, 882, 1069
thermo-chemical modeling, 987
thermocouples, 878, 882, 1066, 1077, 1145
thermodynamic properties
    aerogel, 763
    fluids, 20
    helium, 107
    hydrogen plasma, 119
    iron, 127
    metals, 111
    nitromethane, 173
    uranium, 245
    water, 103
thermography, infrared, 557
thermogravimetry, 1113
thermo-mechanical properties, 427
thermoplasticolefin, 1323
thermoplastics, 665, 1267
thermosets, 811
thin films
    aluminum, 468, 1333, 1351
    gold, 1227
    nickel, 1227
    polymers, 1219, 1333
    sapphire, 1231
tight-binding method, 343
time-resolved measurements, 3, 53, 237, 713,
    1109, 1163, 1181, 1208, 1219, 1223,
    1259, 1263, 1333, 1339, 1351
tin
    equation of state, 143
    grain size, 531
    Hugoniot, 367
    phase transitions, 221
    sound speed, 1200
    surface geometry, 1239
    temperature, 1235
titanium
    equation of state, 225
    /nickel powder, 1123
    phase transition, 225
    shear bands, 567
    strain rate, 557
    twinning, 623
titanium alloys
    ballistic test, 1294
    elastoplasticity, 323
    equation of state, 75
    fragmentation, 527
    Hugoniot, 579

phase transitions, 579
plates, 307
shear bands, 567, 587
spallation, 511, 1327
strain rate, 327
titanium aluminide, 634
titanium aluminum nitride, 1097
titanium boride
/alumina, 755
/alumina composite, 795
titanium carbide
/steel composite, 1119
synthesis, 1127
titanium silicon carbide, 1097
TNAZ explosive, 853
TNETB explosive, 938
TNT explosive
Hopkinson bar, 860
melt-cast, 1047
plastic deformation, 1003
strain rate, 1323
tomographic methods, 1393
transmission electron microscopy
copper, 615, 619
PZT, 1101
steel, 571
titanium, 623
Trident laser, 1192, 1343
tungsten
equation of state, 71
penetration experiment, 1287
phase diagrams, 111
tungsten carbide, 783

U

ultrafast spatial interferometry, 1227
unknown-to-detonation transition (XDT), 464
uranium
fragmentation, 1302
phase diagrams, 111
spall, 499
structural stability, 245
uranium alloys, 1306
urea nitrate, 161

V

vanadium, equation of state, 71
vaporization, high-pressure, 63
vibrational spectra
HMX explosive, 385

nitro compounds, 1255
nitromethane, 385
VISAR. *See also* ORVIS
aluminum oxynitride, 767
back-surface response, 630
bar impact test, 787
beryllium, 1192
boron carbide, 779
composite materials, 709
detonation, 938
expansion rate, 515
fiber-optic probe, 79
flier plate velocity, 307
highly compressed iron, 83
isentropic compression, 849
laser-driven shock, 615, 1371
line-imaging, 26, 1192, 1343, 1367
liquid–solid transition, 221
melting, 229
meteorite targets, 1410
modeling, 1177
PBX 9501 explosive, 149
potassium chloride, 217
push-pull, 1339
PZT, 191, 201, 205, 209
spallation, 475, 487, 491, 499, 755, 1119
temperature effects, 669
uranium alloy, 1306
Z accelerator, 1141
VISAR/ORVIS, Line, 26, 205, 483
viscous fluid, 295
Viton binder, 942
von Neumann spike pressure, 161, 423, 910, 938

W

wall expansion, 165
Warm Dense Matter, 53
water
dense, 1363
equation of state, 103
waveforms
optimized, 1188
steady, 20
transmitted, 669
wave profile
copper, 491
double shock, 999
HMX explosive, 979
iron, 591
laser-driven shock, 615

LX-17 explosive, 902, 1019
PETN explosive, 910
potassium chloride, 217
PZT, 205, 209
RDX explosive, 910
titanium, 225
uranium, 499

X
xenon, 1367
x-ray diffraction
   beryllium, 1192
   FOX-7 explosive, 181
   HMX explosive, 856
   laser-driven shock, 615
   PZT, 1101
   silicon, 1181
   silicon nitride, 1113
   TATB explosive, 856
   time-resolved, 1181, 1375

titanium, 623
titanium carbide, 1127
transient, 1192, 1344
x-ray flash photography, 803, 1105
x-ray imaging, 547
x-ray irradiation, 1410
x-ray radiography, 291, 307, 507, 515, 1359
x-ray sources, 1185

Z
Z accelerator, 849, 1141, 1163, 1177, 1188, 1410
zinc
   Hugoniot elastic limit, 603
   spall strength, 603
zirconium
   failure, 275
   fragmentation, 527
zirconium oxide, 751
ZOX explosive, 938
Z-pinch experiment, 291, 1410